Instrumentation Reference Book

Instrumentation Reference Book

Fourth Edition

Edited by
Walt Boyes

AMSTERDAM • BOSTON • HEIDELBERG • LONDON • NEW YORK • OXFORD • PARIS
SAN DIEGO • SAN FRANCISCO • SINGAPORE • SYDNEY • TOKYO

Butterworth-Heinemann is an imprint of Elsevier

Butterworth-Heinemann is an imprint of Elsevier
30 Corporate Drive, Suite 400, Burlington, MA 01803, USA
Linacre House, Jordan Hill, Oxford OX2 8DP, UK

Copyright © 2010 Elsevier Inc. All rights reserved.

No part of this publication may be reproduced or transmitted in any form or by any means, electronic or mechanical, including photocopying, recording, or any information storage and retrieval system, without permission in writing from the publisher. Details on how to seek permission, further information about the Publisher's permissions policies and our arrangements with organizations such as the Copyright Clearance Center and the Copyright Licensing Agency, can be found at our website: www.elsevier.com/permissions.

This book and the individual contributions contained in it are protected under copyright by the Publisher (other than as may be noted herein).

Notices
Knowledge and best practice in this field are constantly changing. As new research and experience broaden our understanding, changes in research methods, professional practices, or medical treatment may become necessary. Practitioners and researchers must always rely on their own experience and knowledge in evaluating and using any information, methods, compounds, or experiments described herein. In using such information or methods they should be mindful of their own safety and the safety of others, including parties for whom they have a professional responsibility.

To the fullest extent of the law, neither the Publisher nor the authors, contributors, or editors, assume any liability for any injury and/or damage to persons or property as a matter of products liability, negligence or otherwise, or from any use or operation of any methods, products, instructions, or ideas contained in the material herein.

Library of Congress Cataloging-in-Publication Data
Instrumentation reference book / [edited by] Walt Boyes. —4th ed.
   p. cm.
   Includes bibliographical references and index.
   ISBN 978-0-7506-8308-1
   1. Physical instruments—Handbooks, manuals, etc. 2. Engineering instruments—Handbooks, manuals, etc. I. Boyes, Walt.
II. Title.
   QC53.I574 2010
   530'.7—dc22

2009029513

British Library Cataloguing-in-Publication Data
A catalogue record for this book is available from the British Library.

ISBN: 978-0-7506-8308-1

For information on all Butterworth–Heinemann publications
visit our Web site at *www.elsevierdirect.com*

Printed in the United States of America
09 10 11 12 13   10 9 8 7 6 5 4 3 2 1

Typeset by: diacriTech, India

Working together to grow
libraries in developing countries

www.elsevier.com | www.bookaid.org | www.sabre.org

ELSEVIER   BOOK AID International   Sabre Foundation

Contents

Preface xvii
Contributors xix
Introduction xxi

Part I
The Automation Knowledge Base

1. The Automation Practicum
W. Boyes

1.1 Introduction 3
1.2 Job Descriptions 4
1.3 Careers and Career Paths 4
    1.3.1 ISA Certified Automation Professional (CAP) Classification System 5
1.4 Where Automation Fits in the Extended Enterprise 13
1.5 Manufacturing Execution Systems and Manufacturing Operations Management 14
    1.5.1 Introduction 14
    1.5.2 Manufacturing Execution Systems (MES) and Manufacturing Operations Management (MOM) 15
    1.5.3 The Connected Enterprise 15
Suggested Reading 18

2. Basic Principles of Industrial Automation
W. Boyes

2.1 Introduction 19
2.2 Standards 19
2.3 Sensor and System Design, Installation, and Commissioning 20
    2.3.1 The Basics 20
    2.3.2 Identification of the Application 20
    2.3.3 Selection of the Appropriate Sensor/Transmitter 20
    2.3.4 Selection of the Final Control Element 20
    2.3.5 Selection of the Controller and Control Methodology 20
    2.3.6 Design of the Installation 20
    2.3.7 Installing, Commissioning, and Calibrating the System 21
2.4 Maintenance and Operation 21
    2.4.1 Introduction 21
    2.4.2 Life-cycle Optimization 21
    2.4.3 Reliability Engineering 21
    2.4.4 Asset Management, Asset Optimization, and Plant Optimization 21
Suggested Reading 21

3. Measurement Methods and Control Strategies
W. Boyes

3.1 Introduction 23
3.2 Measurement and Field Calibration Methodology 23
3.3 Process Control Strategies 23
3.4 Advanced Control Strategies 24
Suggested Reading 24

4. Simulation and Design Software
M. Berutti

4.1 Introduction 25
4.2 Simulation 25
4.3 Best Practices for Simulation Systems in Automation 25
4.4 Ground-up Testing and Training 26
4.5 Simulation System Selection 26
4.6 Simulation for Automation in the Validated Industries 26
4.7 Conclusion 26

5. Security for Industrial Automation
W. Boyes and J. Weiss

5.1 The Security Problem 27
5.2 An Analysis of the Security Needs of Industrial Automation 28
5.3 Some Recommendations for Industrial Automation Security 28

Part II
Mechanical Measurements

6. Measurement of Flow
G. Fowles and W. H. Boyes

6.1	Introduction	31
6.2	Basic Principles of Flow Measurement	31
	6.2.1 Streamlined and Turbulent Flow	31
	6.2.2 Viscosity	32
	6.2.3 Bernoulli's Theorem	33
	6.2.4 Practical Realization of Equations	34
	6.2.5 Modification of Flow Equations to Apply to Gases	35
6.3	Fluid Flow in Closed Pipes	36
	6.3.1 Differential-Pressure Devices	36
	6.3.2 Rotating Mechanical Meters for Liquids	43
	6.3.3 Rotating Mechanical Meters for Gases	48
	6.3.4 Electronic Flowmeters	51
	6.3.5 Mass Flowmeters	58
6.4	Flow in Open Channels	60
	6.4.1 Head/Area Method	60
	6.4.2 Velocity/Area Methods	63
	6.4.3 Dilution Gauging	64
6.5	Point Velocity Measurement	64
	6.5.1 Laser Doppler Anemometer	64
	6.5.2 Hotwire Anemometer	64
	6.5.3 Pitot Tube	64
	6.5.4 Electromagnetic Velocity Probe	65
	6.5.5 Insertion Turbine	65
	6.5.6 Propeller-Type Current Meter	66
	6.5.7 Insertion Vortex	66
	6.5.8 Ultrasonic Doppler Velocity Probe	66
6.6	Flowmeter Calibration Methods	66
	6.6.1 Flowmeter Calibration Methods for Liquids	66
	6.6.2 Flowmeter Calibration Methods for Gases	67
	References	68
	Further Reading	68

7. Measurement of Viscosity
K. Walters and W. M. Jones

7.1	Introduction	69
7.2	Newtonian and Non-Newtonian Behavior	69
7.3	Measurement of the Shear Viscosity	71
	7.3.1 Capillary Viscometer	71
	7.3.2 Couette Viscometer	72
	7.3.3 Cone-and-plate Viscometer	72
	7.3.4 Parallel-plate Viscometer	73
7.4	Shop-Floor Viscometers	73
7.5	Measurement of the Extensional Viscosity	74
7.6	Measurement of Viscosity Under Extremes of Temperature and Pressure	74
7.7	Online Measurements	74
7.8	Accuracy and Range	74
	References	75
	Further Reading	75

8. Measurement of Length
P. H. Sydenham

8.1	Introduction	77
8.2	The Nature of Length	78
8.3	Derived Measurements	79
	8.3.1 Derived from Length Measurement Alone	79
8.4	Standards and Calibration of Length	80
8.5	Practice of Length Measurement for Industrial Use	81
	8.5.1 General Remarks	81
	8.5.2 Mechanical Length-Measuring Equipment	81
	8.5.3 Electronic Length Measurement	82
	8.5.4 Use of Electromagnetic and Acoustic Radiation	87
	8.5.5 Miscellaneous Methods	90
8.6	Automatic Gauging Systems	91
	References	92
	Further Reading	92

9. Measurement of Strain
B. E. Noltingk

9.1	Strain	93
9.2	Bonded Resistance Strain Gauges	93
	9.2.1 Wire Gauges	94
	9.2.2 Foil Gauges	94
	9.2.3 Semiconductor Gauges	94
	9.2.4 Rosettes	95
	9.2.5 Residual Stress Measurement	95
9.3	Gauge Characteristics	95
	9.3.1 Range	95
	9.3.2 Cross-sensitivity	96
	9.3.3 Temperature Sensitivity	96
	9.3.4 Response Times	96
9.4	Installation	96
9.5	Circuits for Strain Gauges	98
9.6	Vibrating Wire Strain Gauge	98
9.7	Capacitive Strain Gauges	99
9.8	Surveys of Whole Surfaces	99
	9.8.1 Brittle Lacquer	99
	9.8.2 Patterns on Surfaces	99
9.9	Photoelasticity	100
	References	101

10. Measurement of Level and Volume
P. H. Sydenham and W. Boyes

10.1	Introduction	103
10.2	Practice of Level Measurement	103

10.2.1	Installation	103
10.2.2	Sources of Error	104

10.3 Calibration of Level-Measuring Systems — 106
10.4 Methods Providing Full-Range Level Measurement — 107
- 10.4.1 Sight Gauges — 107
- 10.4.2 Float-driven Instruments — 107
- 10.4.3 Capacitance Probes — 108
- 10.4.4 Upthrust Buoyancy — 109
- 10.4.5 Pressure Sensing — 109
- 10.4.6 Microwave and Ultrasonic, Time-Transit Methods — 109
- 10.4.7 Force or Position Balance — 110

10.5 Methods Providing Short-Range Detection — 110
- 10.5.1 Magnetic — 110
- 10.5.2 Electrical Conductivity — 110
- 10.5.3 Infrared — 111
- 10.5.4 Radio Frequency — 111
- 10.5.5 Miscellaneous Methods — 112

References — 112

11. Vibration

P. H. Sydenham

11.1 Introduction — 113
- 11.1.1 Physical Considerations — 113
- 11.1.2 Practical Problems of Installation — 116
- 11.1.3 Areas of Application — 116

11.2 Amplitude Calibration — 117
- 11.2.1 Accelerometer Calibration — 117
- 11.2.2 Shock Calibration — 117
- 11.2.3 Force Calibration — 117

11.3 Sensor Practice — 118
- 11.3.1 Mass-Spring Seismic Sensors — 118
- 11.3.2 Displacement Measurement — 120
- 11.3.3 Velocity Measurement — 120
- 11.3.4 Acceleration Measurement — 121
- 11.3.5 Measurement of Shock — 124

11.4 Literature — 124
References — 125
Further Reading — 125

12. Measurement of Force

C. S. Bahra and J. Paros

12.1 Basic Concepts — 127
12.2 Force Measurement Methods — 127
12.3 Lever-Balance Methods — 127
- 12.3.1 Equal-lever Balance — 127
- 12.3.2 Unequal-lever Balance — 128
- 12.3.3 Compound lever Balance — 128

12.4 Force-Balance Methods — 128
12.5 Hydraulic Pressure Measurement — 129
12.6 Acceleration Measurement — 129
12.7 Elastic Elements — 129
- 12.7.1 Spring Balances — 129
- 12.7.2 Proving Rings — 129
- 12.7.3 Piezoelectric Transducers — 130
- 12.7.4 Strain-gauge Load Cells — 130

12.8 Further Developments — 133
References — 133

13. Measurement of Density

E. H. Higham and W. Boyes

13.1 General — 135
13.2 Measurement of Density Using Weight — 135
13.3 Measurement of Density Using Buoyancy — 136
13.4 Measurement of Density Using a Hydrostatic Head — 137
- 13.4.1 General Differential Pressure Transmitter Methods — 137
- 13.4.2 DP Transmitter with Overflow Tank — 138
- 13.4.3 DP Transmitter with a Wet Leg — 138
- 13.4.4 DP Transmitter with a Pressure Repeater — 139
- 13.4.5 DP Transmitter with Flanged or Extended Diaphragm — 139
- 13.4.6 DP Transmitter with Pressure Seals — 139
- 13.4.7 DP Transmitter with Bubble Tubes — 139
- 13.4.8 Other Process Considerations — 140

13.5 Measurement of Density Using Radiation — 140
13.6 Measurement of Density Using Resonant Elements — 140
- 13.6.1 Liquid Density Measurement — 140
- 13.6.2 Gas Density Measurements — 141
- 13.6.3 Relative Density of Gases — 143

Further Reading — 143

14. Measurement of Pressure

E. H. Higham and J. M. Paros

14.1 What is Pressure? — 145
14.2 Pressure Measurement — 145
- 14.2.1 Pressure Measurements by Balancing a Column of Liquid of Known Density — 145
- 14.2.2 Pressure Measurements by Allowing the Unknown Pressure to Act on a Known Area and Measuring the Resultant Force — 147
- 14.2.3 Pressure Measurement by Allowing the Unknown Pressure to Act on a Flexible Member and Measuring the Resultant Motion — 149
- 14.2.4 Pressure Measurement by Allowing the Unknown Pressure to Act on an Elastic Member and Measuring the Resultant Stress or Strain — 155

14.3 Pressure transmitters — 158
- 14.3.1 Pneumatic Motion-Balance Pressure Transmitters — 159

14.3.2 Pneumatic Force-Balance Pressure Transmitters — 159
14.3.3 Force-Measuring Pressure Transmitters — 160
14.3.4 Digital Pressure Transducers — 162
References — 163
Further Reading — 163

15. Measurement of Vacuum

D. J. Pacey

15.1 Introduction — 165
15.1.1 Systems of Measurement — 165
15.1.2 Methods of Measurement — 165
15.1.3 Choice of Nonabsolute Gauges — 166
15.1.4 Accuracy of Measurement — 166
15.2 Absolute Gauges — 166
15.2.1 Mechanical Gauges — 166
15.2.2 Liquid Manometers — 167
15.2.3 The McLeod Gauge (1878) — 167
15.3 Nonabsolute Gauges — 169
15.3.1 Thermal Conductivity Gauges — 169
15.3.2 Ionization Gauges — 170
References — 173

16. Particle Sizing

W. L. Snowsill

16.1 Introduction — 175
16.2 Characterization of Particles — 175
16.2.1 Statistical Mean Diameters — 176
16.3 Terminal Velocity — 176
16.4 Optical Effects Caused by Particles — 177
16.5 Particle Shape — 177
16.6 Methods for Characterizing a Group of Particles — 178
16.6.1 Gaussian or Normal Distributions — 178
16.6.2 Log-Normal Distributions — 179
16.6.3 Rosin–Rammler Distributions — 180
16.7 Analysis Methods that Measure Size Directly — 180
16.7.1 Sieving — 180
16.7.2 Microscope Counting — 181
16.7.3 Direct Optical Methods — 183
16.8 Analysis Methods that Measure Terminal Velocity — 183
16.8.1 Sedimentation — 183
16.8.2 Elutriation — 187
16.8.3 Impaction — 188
16.9 Analysis Methods that Infer Size from Some Other Property — 188
16.9.1 Coulter Counter — 188
16.9.2 Hiac Automatic Particle Sizer — 188
16.9.3 Climet — 189
16.9.4 Adsorption Methods — 189
References — 189
Further Reading — 189

17. Fiber Optics in Sensor Instrumentation

B. T. Meggitt

17.1 Introduction — 191
17.2 Principles of Optical Fiber Sensing — 192
17.2.1 Sensor Classification — 192
17.2.2 Modulation Parameters — 192
17.2.3 Performance Criteria — 193
17.3 Interferometric Sensing Approach — 193
17.3.1 Heterodyne Interferometry — 194
17.3.2 Pseudoheterodyne Interferometry — 194
17.3.3 White-Light Interferometry — 195
17.3.4 Central Fringe Identification — 201
17.4 Doppler Anemometry — 202
17.4.1 Introduction — 202
17.4.2 Particle Size — 203
17.4.3 Fluid Flow — 204
17.4.4 Vibration Monitoring — 206
17.5 In-Fiber Sensing Structures — 210
17.5.1 Introduction — 210
17.5.2 Fiber Fabry–Perot Sensing Element — 210
17.5.3 Fiber Bragg Grating Sensing Element — 212
References — 215

18. Nanotechnology for Sensors

W. Boyes

18.1 Introduction — 217
18.2 What is Nanotechnology? — 217
18.3 Nanotechnology for Pressure Transmitters — 217
18.4 Microelectromechanical Systems (MEMS) — 217
18.5 MEMS Sensors Today — 218

19. Microprocessor-Based and Intelligent Transmitters

E. H. Higham and J. Berge

19.1 Introduction — 219
19.2 Terminology — 220
19.3 Background Information — 221
19.4 Attributes and Features of Microprocessor-Based and Intelligent Transmitters — 222
19.4.1 Microprocessor-Based Features — 222
19.4.2 Intelligent Features — 223
19.5 Microprocessor-Based and Intelligent Temperature Transmitters — 224
19.6 Microprocessor-Based and Intelligent Pressure and Differential Transmitters — 226
19.7 Microprocessor-Based and Intelligent Flowmeters — 229
19.7.1 Coriolis Mass Flowmeters — 229
19.7.2 Electromagnetic Flowmeters — 233
19.7.3 Vortex Flowmeters — 234

19.8 Other Microprocessor-Based and Intelligent Transmitters **236**
19.8.1 Density Transmitters 236
19.8.2 Microprocessor-Based and Intelligent Liquid Level Measurement Systems 239
19.9 Other Microprocessor-Based and Intelligent Measurement Systems **240**
19.10 Fieldbus **241**
19.10.1 Background 241
19.10.2 Introduction to the Concept of a Fieldbus 241
19.10.3 Current Digital Multiplexing Technology 241
19.10.4 The HART Protocol 243
19.11 User Experience with Microprocessor-Based and Intelligent Transmitters **246**
19.12 Fieldbus Function and Benefits **247**
19.12.1 Foundation Fieldbus and Profibus-PA 247
19.12.2 Field-Mounted Control 248
19.12.3 Future of Analog Instruments 249
19.12.4 Sensor Validation 249
19.12.5 Plant Diagnostics 249
19.12.6 Handheld Interfaces (Handheld Terminals or Handheld Communicators) 249
19.12.7 Measuring Directives 250
19.12.8 Further Developments of Intelligent Transmitters 250
19.12.9 Integration of Intelligent Transmitters into Instrument Management Systems 250
References 251

20. Industrial Wireless Technology and Planning

D. R. Kaufman

20.1 Introduction **253**
20.2 The History of Wireless **253**
20.3 The Basics **254**
20.3.1 Radio Frequency Signals 254
20.3.2 Radio Bands 254
20.3.3 Radio Noise 255
20.3.4 Radio Signal-to-Noise Ratio (SNR) 255
20.3.5 Wireless Reliability 256
20.3.6 Fixed Frequencies 256
20.3.7 Spread Spectrum 256
20.3.8 Security 257
20.3.9 Antennas 258
20.3.10 Antenna Connection 260
20.3.11 Commissioning 261
20.3.12 Mesh Technologies 262
20.3.13 System Management 262
20.3.14 System Interfaces 262
20.3.15 Standards and Specifications 263
20.4 Planning for Wireless **263**
20.4.1 Imagine the Possibilities 264
20.4.2 Getting Ready for Wireless 264
References 265

Part III
Measurement of Temperature and Chemical Composition

21. Temperature Measurement

C. Hagart-Alexander

21.1 Temperature and Heat **269**
21.1.1 Application Considerations 269
21.1.2 Definitions 269
21.1.3 Radiation 271
21.2 Temperature Scales **272**
21.2.1 Celsius Temperature Scale 272
21.2.2 Kelvin, Absolute, or Thermodynamic Temperature Scale 272
21.2.3 International Practical Temperature Scale of 1968 (IPTS-68) 273
21.2.4 Fahrenheit and Rankine Scales 273
21.2.5 Realization of Temperature Measurement 274
21.3 Measurement Techniques: Direct Effects **274**
21.3.1 Liquid-in-Glass Thermometers 274
21.3.2 Liquid-Filled Dial Thermometers 278
21.3.3 Gas-Filled Instruments 281
21.3.4 Vapor Pressure Thermometers 282
21.3.5 Solid Expansion 285
21.4 Measurement Techniques: Electrical **286**
21.4.1 Resistance Thermometers 286
21.4.2 Thermistors 290
21.4.3 Semiconductor Temperature Measurement 291
21.5 Measurement Techniques: Thermocouples **293**
21.5.1 Thermoelectric Effects 293
21.5.2 Thermocouple Materials 299
21.5.3 Thermocouple Construction 301
21.6 Measurement Techniques: Radiation Thermometers **306**
21.6.1 Introduction 306
21.6.2 Radiation Thermometer Types 307
21.7 Temperature Measurement Considerations **319**
21.7.1 Readout 319
21.7.2 Sensor Location Considerations 320
21.7.3 Miscellaneous Measurement Techniques 324
References 326
Further Reading 326

22. Chemical Analysis: Introduction

W. G. Cummings; edited by I. Verhappen

22.1	Introduction to Chemical Analysis	327
22.2	Chromatography	328
	22.2.1 General Chromatography	328
	22.2.2 Paper Chromatography and Thin-Layer Chromatography	328
22.3	Polarography and Anodic Stripping Voltammetry	331
	22.3.1 Polarography	331
	22.3.2 Anodic Stripping Voltammetry	334
22.4	Thermal Analysis	335
	Further Reading	339

23. Chemical Analysis: Spectroscopy

A. C. Smith; edited by I. Verhappen

23.1	Introduction	341
23.2	Absorption and Reflection Techniques	341
	23.2.1 Infrared	341
	23.2.2 Absorption in UV, Visible, and IR	346
	23.2.3 Absorption in the Visible and Ultraviolet	348
	23.2.4 Measurements Based on Reflected Radiation	348
	23.2.5 Chemiluminescence	349
23.3	Atomic Techniques: Emission, Absorption, and Fluorescence	349
	23.3.1 Atomic Emission Spectroscopy	349
	23.3.2 Atomic Absorption Spectroscopy	351
	23.3.3 Atomic Fluorescence Spectroscopy	352
23.4	X-Ray Spectroscopy	353
	23.4.1 X-ray Fluorescence Spectroscopy	353
	23.4.2 X-ray Diffraction	355
23.5	Photo-Acoustic Spectroscopy	355
23.6	Microwave Spectroscopy	355
	23.6.1 Electron Paramagnetic Resonance (EPR)	356
	23.6.2 Nuclear Magnetic Resonance Spectroscopy	357
23.7	Neutron Activation	357
23.8	Mass Spectrometers	357
	23.8.1 Principle of the Classical Instrument	358
	23.8.2 Inlet Systems	359
	23.8.3 Ion Sources	359
	23.8.4 Separation of the Ions	359
	23.8.5 Other Methods of Separation of Ions	361
	References	362
	Further Reading	362

24. Chemical Analysis: Electrochemical Techniques

W. G. Cummings and K. Torrance; edited by I. Verhappen

24.1	Acids and Alkalis	363
24.2	Ionization of Water	364
24.3	Electrical Conductivity	364
	24.3.1 Electrical Conduction in Liquids	364
	24.3.2 Conductivity of Solutions	365
	24.3.3 Practical Measurement of Electrical Conductivity	365
	24.3.4 Applications of Conductivity Measurement	372
24.4	The Concept of pH	375
	24.4.1 General Theory	375
	24.4.2 Practical Specification of a pH Scale	376
	24.4.3 pH Standards	376
	24.4.4 Neutralization	376
	24.4.5 Hydrolysis	376
	24.4.6 Common Ion Effect	378
	24.4.7 Buffer Solutions	378
24.5	Electrode Potentials	378
	24.5.1 General Theory	378
	24.5.2 Variation of Electrode Potential with Ion Activity (The Nernst Equation)	380
24.6	Ion-Selective Electrodes	380
	24.6.1 Glass Electrodes	381
	24.6.2 Solid-State Electrodes	381
	24.6.3 Heterogeneous Membrane Electrodes	381
	24.6.4 Liquid Ion Exchange Electrodes	381
	24.6.5 Gas-Sensing Membrane Electrodes	381
	24.6.6 Redox Electrodes	382
24.7	Potentiometry and Specific Ion Measurement	382
	24.7.1 Reference Electrodes	382
	24.7.2 Measurement of pH	384
	24.7.3 Measurement of Redox Potential	390
	24.7.4 Determination of Ions by Ion-Selective Electrodes	390
24.8	Common Electrochemical Analyzers	393
	24.8.1 Residual Chlorine Analyzer	393
	24.8.2 Polarographic Process Oxygen Analyzer	395
	24.8.3 High-temperature Ceramic Sensor Oxygen Probes	396
	24.8.4 Fuel Cell Oxygen-measuring Instruments	397
	24.8.5 Hersch Cell for Oxygen Measurement	397
	24.8.6 Sensor for Oxygen Dissolved in Water	397
	24.8.7 Coulometric Measurement of Moisture in Gases and Liquids	399
	Further Reading	399

25. Chemical Analysis: Gas Analysis

C. K. Laird; edited by I. Verhappen

25.1	Introduction	401
25.2	Separation of Gaseous Mixtures	402
	25.2.1 Gas Chromatography	402
25.3	Detectors	404
	25.3.1 Thermal Conductivity Detector (TCD)	404
	25.3.2 Flame Ionization Detector (FID)	406
	25.3.3 Photo-Ionization Detector (PID)	407

	25.3.4	Helium Ionization Detector	408
	25.3.5	Electron Capture Detector	409
	25.3.6	Flame Photometric Detector (FPD)	409
	25.3.7	Ultrasonic Detector	410
	25.3.8	Catalytic Detector (Pellistor)	411
	25.3.9	Semiconductor Detector	411
	25.3.10	Properties and Applications of Gas Detectors	412
25.4	Process Chromatography	412
	25.4.1	Sampling System	414
	25.4.2	Carrier Gas	416
	25.4.3	Chromatographic Column	417
	25.4.4	Controlled Temperature Enclosures	417
	25.4.5	Detectors	417
	25.4.6	Programmers	418
	25.4.7	Data-Processing Systems	418
	25.4.8	Operation of a Typical Process Chromatograph	419
25.5	Special Gas Analyzers	421
	25.5.1	Paramagnetic Oxygen Analyzers	421
	25.5.2	Ozone Analyzer	424
	25.5.3	Oxides of Nitrogen Analyzer	425
	25.5.4	Summary of Special Gas Analyzers	426
25.6	Calibration of Gas Analyzers	426
	25.6.1	Static Methods	427
	25.6.2	Dynamic Methods	427
	Further Reading	428

26. Chemical Analysis: Moisture Measurement

D. B. Meadowcroft; edited by I. Verhappen

26.1	Introduction	429
26.2	Definitions	429
	26.2.1 Gases	429
	26.2.2 Liquids and Solids	430
26.3	Measurement Techniques	431
	26.3.1 Gases	431
	26.3.2 Liquids	433
	26.3.3 Solids	434
26.4	Calibration	435
	26.4.1 Gases	435
	26.4.2 Liquids	436
	26.4.3 Solids	436
	References	436

Part IV
Electrical and Radiation Measurements

27. Electrical Measurements

M. L. Sanderson

27.1	Units and Standards of Electrical Measurement	439
	27.1.1 SI Electrical Units	439
	27.1.2 Realization of the SI Base Unit	439
	27.1.3 National Primary Standards	440
27.2	Measurement of DC and AC current and Voltage Using Indicating Instruments	444
	27.2.1 Permanent Magnet-Moving Coil Instruments	445
	27.2.2 Moving-Iron Instruments	448
	27.2.3 AC Range Extension Using Current and Voltage Transformers	452
	27.2.4 Dynamometer Instruments	454
	27.2.5 Thermocouple Instruments	454
	27.2.6 Electrostatic Instruments	455
27.3	Digital Voltmeters and Digital Multimeters	455
	27.3.1 Analog-to-Digital Conversion Techniques	456
	27.3.2 Elements in DVMs and DMMs	460
	27.3.3 DVM and DMM Specifications	464
27.4	Power Measurement	465
	27.4.1 The Three-Voltmeter Method of Power Measurement	465
	27.4.2 Direct-Indicating Analog Wattmeters	465
	27.4.3 Connection of Wattmeters	467
	27.4.4 Three-Phase Power Measurement	468
	27.4.5 Electronic Wattmeters	469
	27.4.6 High-Frequency Power Measurement	470
27.5	Measurement of Electrical Energy	472
27.6	Power-Factor Measurement	473
27.7	The Measurement of Resistance, Capacitance, and Inductance	474
	27.7.1 DC Bridge Measurements	474
	27.7.2 AC Equivalent Circuits of Resistors, Capacitors, and Inductors	477
	27.7.3 Four-Arm AC Bridge Measurements	478
	27.7.4 Transformer Ratio Bridges	482
	27.7.5 High-Frequency Impedance Measurement	487
27.8	Digital Frequency and Period/Time-Interval Measurement	489
	27.8.1 Frequency Counters and Universal Timer/Counters	489
	27.8.2 Time-Interval Averaging	492
	27.8.3 Microwave-Frequency Measurement	495
27.9	Frequency and Phase Measurement Using an Oscilloscope	497
	References	497
	Further Reading	498

28. Optical Measurements

A. W. S. Tarrant

28.1	Introduction	499
28.2	Light Sources	499
	28.2.1 Incandescent Lamps	500
	28.2.2 Discharge Lamps	500
	28.2.3 Electronic Sources: Light-emitting Diodes	501
	28.2.4 Lasers	501

28.3 Detectors 502
    28.3.1 Photomultipliers 502
    28.3.2 Photovoltaic and Photoconductive Detectors (Photodiodes) 503
    28.3.3 Pyroelectric Detectors 504
    28.3.4 Array Detectors 505
28.4 Detector Techniques 506
    28.4.1 Detector Circuit Time Constants 506
    28.4.2 Detector Cooling 506
    28.4.3 Beam Chopping and Phase-Sensitive Detection 507
    28.4.4 The Boxcar Detector 507
    28.4.5 Photon Counting 508
28.5 Intensity Measurement 508
    28.5.1 Photometers 509
    28.5.2 Ultraviolet Intensity Measurements 509
    28.5.3 Color-Temperature Meters 510
28.6 Wavelength and Color 510
    28.6.1 Spectrophotometers 510
    28.6.2 Spectroradiometers 512
    28.6.3 The Measurement of Color 512
28.7 Measurement of Optical Properties 514
    28.7.1 Refractometers 514
    28.7.2 Polarimeters 516
28.8 Thermal Imaging Techniques 518
References 519

29. Nuclear Instrumentation Technology

D. Aliaga Kelly and W. Boyes

29.1 Introduction 521
    29.1.1 Statistics of Counting 521
    29.1.2 Classification of Detectors 524
    29.1.3 Health and Safety 525
29.2 Detectors 526
    29.2.1 Gas Detectors 526
    29.2.2 Scintillation Detectors 528
    29.2.3 Solid-state Detectors 532
    29.2.4 Detector Applications 533
29.3 Electronics 541
    29.3.1 Electronics Assemblies 541
    29.3.2 Power Supplies 542
    29.3.3 Amplifiers 543
    29.3.4 Sealers 543
    29.3.5 Pulse-Height Analyzers 543
    29.3.6 Special Electronic Units 544
References 547
Further Reading 547

30. Measurements Employing Nuclear Techniques

D. Aliaga Kelly and W. Boyes

30.1 Introduction 549
    30.1.1 Radioactive Measurement Relations 550
    30.1.2 Optimum Time of Measurement 551
    30.1.3 Accuracy/Precision of Measurements 551
    30.1.4 Measurements on Fluids in Containers 551
30.2 Materials Analysis 552
    30.2.1 Activation Analysis 552
    30.2.2 X-ray Fluorescence Analysis 553
    30.2.3 Moisture Measurement: By Neutrons 555
    30.2.4 Measurement of Sulfur Contents of Liquid Hydrocarbons 557
    30.2.5 The Radioisotope Calcium Monitor 558
    30.2.6 Wear and Abrasion 559
    30.2.7 Leak Detection 559
30.3 Mechanical Measurements 559
    30.3.1 Level Measurement 559
    30.3.2 Measurement of Flow 560
    30.3.3 Mass and Thickness 561
30.4 Miscellaneous Measurements 563
    30.4.1 Field-survey Instruments 563
    30.4.2 Dating of Archaeological or Geological Specimens 563
    30.4.3 Static Elimination 565
References 565

31. Non-Destructive Testing

Scottish School of Non-Destructive Testing

31.1 Introduction 567
31.2 Visual Examination 568
31.3 Surface-Inspection Methods 568
    31.3.1 Visual Techniques 568
    31.3.2 Magnetic Flux Methods 569
    31.3.3 Potential Drop Techniques 570
    31.3.4 Eddy-Current Testing 570
31.4 Ultrasonics 571
    31.4.1 General Principles of Ultrasonics 571
    31.4.2 The Ultrasonic Test Equipment Controls and Visual Presentation 574
    31.4.3 Probe Construction 576
    31.4.4 Ultrasonic Spectroscopy Techniques 577
    31.4.5 Applications of Ultrasonic Spectroscopy 578
    31.4.6 Other Ways of Presenting Information from Ultrasonics 579
    31.4.7 Automated Ultrasonic Testing 580
    31.4.8 Acoustic Emission 580
31.5 Radiography 580
    31.5.1 Gamma Rays 581
    31.5.2 X-rays 582
    31.5.3 Sensitivity and IQI 582
    31.5.4 Xerography 585
    31.5.5 Fluoroscopic and Image-Intensification Methods 585
31.6 Underwater Non-Destructive Testing 586
    31.6.1 Diver Operations and Communication 587
    31.6.2 Visual Examination 587
    31.6.3 Photography 587
    31.6.4 Magnetic Particle Inspection (MPI) 588

31.6.5	Ultrasonics	588
31.6.6	Corrosion Protection	588
31.6.7	Other Non-Destructive Testing Techniques	589
31.7	Developments	590
31.8	Certification of Personnel	590
	References	591
	Further Reading	592

32. Noise Measurement

J. Kuehn

32.1	Sound and Sound Fields	593
32.1.1	The Nature of Sound	593
32.1.2	Quantities Characterizing a Sound Source or Sound Field	594
32.1.3	Velocity of Propagation of Sound Waves	594
32.1.4	Selecting the Quantities of Interest	595
32.2	Instrumentation for the Measurement of Sound-Pressure Level	596
32.2.1	Microphones	596
	Appendix 32.1	597
32.2.2	Frequency Weighting Networks and Filters	600
32.2.3	Sound-Level Meters	601
32.2.4	Noise-Exposure Meters/Noise-Dose Meters	603
32.2.5	Acoustic Calibrators	603
32.3	Frequency Analyzers	604
32.3.1	Octave Band Analyzers	604
32.3.2	Third-Octave Analyzers	605
32.3.3	Narrow-Band Analyzers	606
32.3.4	Fast Fourier Transform Analyzers	607
32.4	Recorders	607
32.4.1	Level Recorders	607
32.4.2	XY Plotters	607
32.4.3	Digital Transient Recorders	607
32.4.4	Tape Recorders	608
32.5	Sound-Intensity Analyzers	608
32.6	Calibration of Measuring Instruments	609
32.6.1	Formal Calibration	609
32.6.2	Field Calibration	609
32.6.3	System Calibration	609
32.6.4	Field-System Calibration	609
32.7	The Measurement of Sound-Pressure Level and Sound Level	609
32.7.1	Time Averaging	610
32.7.2	Long Time Averaging	611
32.7.3	Statistical Distribution and Percentiles	611
32.7.4	Space Averaging	611
32.7.5	Determination of Sound Power	611
32.7.6	Measurement of Sound Power by Means of Sound Intensity	612
32.8	Effect of Environmental Conditions on Measurements	613
32.8.1	Temperature	613
32.8.2	Humidity and Rain	613
32.8.3	Wind	613
32.8.4	Other Noises	613
	References	614
	Further Reading	614

Part V
Controllers, Actuators, and Final Control Elements

33. Field Controllers, Hardware and Software

W. Boyes

33.1	Introduction	617
33.2	Field Controllers, Hardware, and Software	617

34. Advanced Control for the Plant Floor

Dr. James R. Ford, P. E.

34.1	Introduction	619
34.2	Early Developments	619
34.3	The Need for Process Control	619
34.4	Unmeasured Disturbances	620
34.5	Automatic Control Valves	620
34.6	Types of Feedback Control	621
34.7	Measured Disturbances	621
34.8	The Need for Models	623
34.9	The Emergence of MPC	623
34.10	MPC vs. ARC	623
34.11	Hierarchy	624
34.12	Other Problems with MPC	625
34.13	Where We Are Today	626
34.14	Recommendations for Using MPC	626
34.15	What's in Store for the Next 40 Years?	627

35. Batch Process Control

W. H. Boyes

35.1	Introduction	629
	Further Reading	630

36. Applying Control Valves

B. G. Liptak; edited by W. H. Boyes

36.1	Introduction	631
36.2	Valve Types and Characteristics	631
36.3	Distortion of Valve Characteristics	633
36.4	Rangeability	634
36.5	Loop Tuning	634
36.6	Positioning Positioners	635
36.7	Smarter Smart Valves	635
36.8	Valves Serve as Flowmeters	635
	Further Reading	636

Part VI
Automation and Control Systems

37. Design and Construction of Instruments

C. I. Daykin and W. H. Boyes

37.1	Introduction	639
37.2	Instrument Design	639
	37.2.1 The Designer's Viewpoint	639
	37.2.2 Marketing	640
	37.2.3 Special Instruments	640
37.3	Elements of Construction	640
	37.3.1 Electronic Components and Printed Circuits	640
	37.3.2 Surface-Mounted Assemblies	642
	37.3.3 Interconnections	642
	37.3.4 Materials	643
	37.3.5 Mechanical Manufacturing Processes	644
	37.3.6 Functional Components	646
37.4	Construction of Electronic Instruments	647
	37.4.1 Site Mounting	647
	37.4.2 Panel Mounting	647
	37.4.3 Bench-Mounting Instruments	647
	37.4.4 Rack-Mounting Instruments	649
	37.4.5 Portable Instruments	649
	37.4.6 Encapsulation	650
37.5	Mechanical Instruments	650
	37.5.1 Kinematic Design	650
	37.5.2 Proximity Transducer	651
	37.5.3 Load Cell	651
	37.5.4 Combined Actuator Transducer	652
	References	653

38. Instrument Installation and Commissioning

A. Danielsson

38.1	Introduction	655
38.2	General Requirements	655
38.3	Storage and Protection	655
38.4	Mounting and Accessibility	655
38.5	Piping Systems	656
	38.5.1 Air Supplies	656
	38.5.2 Pneumatic Signals	656
	38.5.3 Impulse Lines	656
38.6	Cabling	657
	38.6.1 General Requirements	657
	38.6.2 Cable Types	658
	38.6.3 Cable Segregation	658
38.7	Grounding	658
	38.7.1 General Requirements	658
38.8	Testing and Pre-Commissioning	658
	38.8.1 General	658
	38.8.2 Pre-Installation Testing	658
	38.8.3 Piping and Cable Testing	659
	38.8.4 Loop Testing	659
38.9	Plant Commissioning	660
	References	660

39. Sampling

J. G. Giles

39.1	Introduction	661
	39.1.1 Importance of Sampling	661
	39.1.2 Representative Sample	661
	39.1.3 Parts of Analysis Equipment	662
	39.1.4 Time Lags	662
	39.1.5 Construction Materials	663
39.2	Sample System Components	664
	39.2.1 Probes	664
	39.2.2 Filters	665
	39.2.3 Coalescers	666
	39.2.4 Coolers	666
	39.2.5 Pumps, Gas	666
	39.2.6 Pumps, Liquid	668
	39.2.7 Flow Measurement and Indication	669
	39.2.8 Pressure Reduction and Vaporization	670
	39.2.9 Sample Lines, Tube and Pipe Fitting	670
39.3	Typical Sample Systems	672
	39.3.1 Gases	672
	39.3.2 Liquids	674
	References	676

40. Telemetry

M. L. Sanderson

40.1	Introduction	677
40.2	Communication Channels	679
	40.2.1 Transmission Lines	679
	40.2.2 Radio Frequency Transmission	681
	40.2.3 Fiber-Optic Communication	681
40.3	Signal Multiplexing	684
40.4	Pulse Encoding	685
40.5	Carrier Wave Modulation	687
40.6	Error Detection and Correction Codes	688
40.7	Direct Analog Signal Transmission	689
40.8	Frequency Transmission	690
40.9	Digital Signal Transmission	690
	40.9.1 Modems	692
	40.9.2 Data Transmission and Interfacing Standards	693
	References	697
	Further Reading	697

41. Display and Recording

M. L. Sanderson

41.1	Introduction	699
41.2	Indicating Devices	699
41.3	Light-Emitting Diodes (LCDs)	700
41.4	Liquid Crystal Displays (LCDs)	702

41.5	Plasma Displays	703
41.6	Cathode Ray Tubes (CRTs)	704
	41.6.1 Color Displays	705
	41.6.2 Oscilloscopes	706
	41.6.3 Storage Oscilloscopes	707
	41.6.4 Sampling Oscilloscopes	708
	41.6.5 Digitizing Oscilloscopes	708
	41.6.6 Visual Display Units (VDUs)	709
	41.6.7 Graphical Displays	709
41.7	Graphical Recorders	709
	41.7.1 Strip Chart Recorders	710
	41.7.2 Circular Chart Recorders	711
	41.7.3 Galvanometer Recorders	711
	41.7.4 X–Y Recorders	712
41.8	Magnetic Recording	712
41.9	Transient/Waveform Recorders	713
41.10	Data Loggers	713
	References	714

42. Pneumatic Instrumentation

E. H. Higham; edited by
W. L. Mostia Jr., PE

42.1	Basic Characteristics	715
42.2	Pneumatic Measurement and Control Systems	716
42.3	Principal Measurements	717
	42.3.1 Introduction	717
	42.3.2 Temperature	717
	42.3.3 Pressure Measurement	718
	42.3.4 Level Measurements	721
	42.3.5 Buoyancy Measurements	721
	42.3.6 Target Flow Transmitter	721
	42.3.7 Speed	722
42.4	Pneumatic Transmission	722
42.5	Pneumatic Controllers	723
	42.5.1 Motion-Balance Controllers	723
	42.5.2 Force-Balance Controllers	725
42.6	Signal Conditioning	729
	42.6.1 Integrators	729
	42.6.2 Analog Square Root Extractor	729
	42.6.3 Pneumatic Summing Unit and Dynamic Compensator	729
	42.6.4 Pneumatic-to-Current Converters	730
42.7	Electropneumatic Interface	732
	42.7.1 Diaphragm Motor Actuators	732
	42.7.2 Pneumatic Valve Positioner	733
	42.7.3 Electropneumatic Converters	734
	42.7.4 Electropneumatic Positioners	735
	References	735

43. Reliability in Instrumentation and Control

J. Cluley

43.1	Reliability Principles and Terminology	737
	43.1.1 Definition of Reliability	737
	43.1.2 Reliability and MTBF	737
	43.1.3 The Exponential Failure Law	738
	43.1.4 Availability	739
	43.1.5 Choosing Optimum Reliability	739
	43.1.6 Compound Systems	740
43.2	Reliability Assessment	742
	43.2.1 Component Failure Rates	742
	43.2.2 Variation of Failure Rate with Time	742
	43.2.3 Failure Modes	743
	43.2.4 The Effect of Temperature on Failure Rates	743
	43.2.5 Estimating Component Temperature	744
	43.2.6 The Effect of Operating Voltage on Failure Rates	745
	43.2.7 Accelerated Life Tests	745
	43.2.8 Component Screening	746
	43.2.9 Confidence Limits and Confidence Level	746
	43.2.10 Assembly Screening	746
	43.2.11 Dealing with the Wear-out Phase	747
	43.2.12 Estimating System Failure Rate	747
	43.2.13 Parallel Systems	748
	43.2.14 Environmental Testing	748
43.3	System Design	749
	43.3.1 Signal Coding	749
	43.3.2 Digitally Coded Systems	750
	43.3.3 Performance Margins in System Design	750
	43.3.4 Coping with Tolerance	751
	43.3.5 Component Tolerances	751
	43.3.6 Temperature Effects	752
	43.3.7 Design Automation	753
	43.3.8 Built-in Test Equipment	754
	43.3.9 Sneak Circuits	754
43.4	Building High-Reliability Systems	755
	43.4.1 Reliability Budgets	755
	43.4.2 Component Selection	755
	43.4.3 The Use of Redundancy	756
	43.4.4 Redundancy with Majority Voting	757
	43.4.5 The Level of Redundancy	758
	43.4.6 Analog Redundancy	758
	43.4.7 Common Mode Faults	759
43.5	The Human Operator in Control and Instrumentation	760
	43.5.1 The Scope for Automation	760
	43.5.2 Features of the Human Operator	760
	43.5.3 User-Friendly Design	762
	43.5.4 Visual Displays	764
	43.5.5 Safety Procedures	764
43.6	Safety Monitoring	765
	43.6.1 Types of Failure	765
	43.6.2 Designing Fail-Safe Systems	765
	43.6.3 Relay Tripping Circuits	766
	43.6.4 Mechanical Fail-Safe Devices	766
	43.6.5 Control System Faults	767
	43.6.6 Circuit Fault Analysis	767
43.7	Software Reliability	768
	43.7.1 Comparison with Hardware Reliability	768

43.7.2	The Distinction between Faults and Failures	769
43.7.3	Typical Failure Intensities	769
43.7.4	High-Reliability Software	769
43.7.5	Estimating the Number of Faults	769
43.7.6	Structured Programming	770
43.7.7	Failure-Tolerant Systems	771

43.8 Electronic and Avionic Systems — 771
- 43.8.1 Radio Transmitters — 771
- 43.8.2 Satellite Links — 772
- 43.8.3 Aircraft Control Systems — 772
- 43.8.4 Railway Signaling and Control — 774
- 43.8.5 Robotic Systems — 775

43.9 Nuclear Reactor Control Systems — 776
- 43.9.1 Requirements for Reactor Control — 776
- 43.9.2 Principles of Reactor Control — 776
- 43.9.3 Types of Failure — 779
- 43.9.4 Common Mode Faults — 779
- 43.9.5 Reactor Protection Logic — 781

43.10 Process and Plant Control — 782
- 43.10.1 Additional Hazards in Chemical Plants — 782
- 43.10.2 Hazardous Areas — 782
- 43.10.3 Risks to Life — 783
- 43.10.4 The Oil Industry — 783
- 43.10.5 Reliability of Oil Supply — 784
- 43.10.6 Electrostatic Hazards — 785
- 43.10.7 The Use of Redundancy — 786

References — 786
British Standards — 787
British Standard Codes of Practice — 787
European and Harmonized Standards — 787

44. Safety

L. C. Towle

44.1 Introduction — 789
44.2 Electrocution Risk — 790
- 44.2.1 Earthing (Grounding) and Bonding — 791

44.3 Flammable Atmospheres — 791
44.4 Other Safety Aspects — 795
44.5 Conclusion — 796

References — 796
Further Reading — 796

45. EMC

T. Williams

45.1 Introduction — 797
- 45.1.1 Compatibility between Systems — 797
- 45.1.2 The Scope of EMC — 798

45.2 Interference Coupling Mechanisms — 801
- 45.2.1 Source and Victim — 801
- 45.2.2 Emissions — 805
- 45.2.3 Susceptibility — 808

45.3 Circuits, Layout, and Grounding — 814
- 45.3.1 Layout and Grounding — 815
- 45.3.2 Digital and Analog Circuit Design — 824

45.4 Interfaces, Filtering, and Shielding — 841
- 45.4.1 Cables and Connectors — 841
- 45.4.2 Filtering — 848
- 45.4.3 Shielding — 858

45.5 The Regulatory Framework — 865
- 45.5.1 Customer Requirements — 865
- 45.5.2 The EMC Directive — 865
- 45.5.3 Standards Relating to the EMC Directive — 869

References — 870
Further Reading — 871

Appendices

A. General Instrumentation Books — 873
B. Professional Societies and Associations — 879
C. The Institute of Measurement and Control — 883
- Role and Objectives — 883
- History — 883
- Qualifications — 884
- Chartered Status for Individuals — 884
- Incorporated Engineers and Engineering Technicians — 884
- Membership — 884
- Corporate Members — 884
- Honorary Fellow — 884
- Fellows — 884
- Members — 884
- Noncorporate Members — 885
- Companions — 885
- Graduates — 885
- Licentiates — 885
- Associates — 885
- Students — 885
- Affiliates — 885
- Subscribers — 885
- Application for Membership — 885
- National and International Technical Events — 885
- Local Sections — 885
- Publications — 885
- Advice and Information — 886
- Awards and Prizes — 886
- Government and Administration — 886

D. International Society of Automation, Formerly Instrument Society of America — 887
- Training — 887
- Standards and Practices — 887
- Publications — 888

Index — 889

Preface

PREFACE TO THE FOURTH EDITION

In this fourth edition of the *Instrumentation Reference Book* we have attempted to maintain the one-volume scheme with which we began while expanding the work to match the current view of the automation practitioner in the process industries.

In the process industries, practitioners are now required to have knowledge and skills far outside the "instrumentation and control" area. Typically, automation practitioners have been required to be familiar with enterprise organization and integration, so the instruments and control systems under their purview can easily transfer and receive needed information and instructions from anywhere throughout the extended enterprise. They have needed substantially more experience in programming and use of computers and, since the first edition of this work was published, an entirely new subdiscipline of automation has been created: industrial networking.

In fact, the very name of the profession has changed. In 2008, the venerable Instrumentation Society of America changed its official name to the International Society of Automation in recognition of this fact.

The authors and the editor hope that this volume and the guidance it provides will be of benefit to all practitioners of automation in the process industries.

The editor wishes to thank Elsevier, and his understanding and long-suffering publisher, Matthew Hart, and the authors who contributed to this volume.

—*W. H. Boyes, ISA Fellow*
Editor

PREFACE TO THE THIRD EDITION

This edition is not completely new. The second edition built on the first, and so does this edition. This work has been almost entirely one of "internationalizing" a work mainly written for the United Kingdom. New matter has been added, especially in the areas of analyzers, level and flowmeters, and fieldbus. References to standards are various, and British Standards are often referenced. International standards are in flux, and most standards bodies are striving to have equivalent standards throughout the world. The reader is encouraged to refer to IEC, ANSI, or other standards when only a British Standard is shown. The ubiquity of the World Wide Web has made it possible for any standard anywhere to be located and purchase or, in some cases, read online free, so it has not been necessary to cross-reference standards liberally in this work.

The editor wants to thank all the new contributors, attributed and not, for their advice, suggestions, and corrections. He fondly wishes that he has caught all the typographical errors, but knows that is unlikely. Last, the Editor wants to thank his several editors at Butterworth-Heinemann for their patience, as well as Michael Forster, the publisher.

—*W. H. Boyes*
Maple Valley, Washington
2002

PREFACE TO THE SECOND EDITION

E. B. Jones's writings on instrument technology go back at least to 1953. He was something of a pioneer in producing high-level material that could guide those studying his subjects. He had both practical experience of his subject and had taught it at college, and this enabled him to lay down a foundation that could be built on for more than 40 years. I must express my thanks that the first edition of the *Instrumentation Reference Book*, which E. B. Jones's work was molded into, has sold well from 1988 to 1994.

This book has been accepted as one of the Butterworth-Heinemann series of reference books—a goodly number of volumes covering much of technology. Such books need updating to keep abreast of developments, and this first updating calls for celebration!

There were several aspects that needed enlarging and several completely new chapters were needed. It might be remarked that a number of new books, relevant to the whole field of instrumentation, have appeared recently, and these have been added to the list. Does this signify a growing recognition of the place of instrumentation?

Many people should be thanked for their work that has brought together this new edition. Collaboration with the Institute of Measurement and Control has been established, and this means that the book is now produced under their sponsorship. Of course, those who have written, or revised what they had written before, deserve my gratitude for their response. I would also like to say thank you to the Butterworth-Heinemann staff for their cooperation.

—*B. E. N. Dorking*

PREFACE TO THE FIRST EDITION

Instrumentation is not a clearly defined subject, having what might be called a "fuzzy frontier" with many other subjects. Look for books about it, and in most libraries you are liable to find them widely separated along the shelves, classified under several different headings. Instrumentation is barely recognized as a science or technology in its own right. That raises some difficulties for writers in the field and indeed for would-be readers. We hope that what we are offering here will prove to have helped with clarification.

A reference book should of course be there for people to refer to for the information they need. The spectrum is wide: students, instrument engineers, instrument users, and potential users who just want to explore possibilities. And the information needed in real life is a mixture of technical and commercial matters. So while the major part of the *Instrumentation Reference Book* is a technical introduction to many facets of the subject, there is also a commercial part where manufacturers and so on are listed. Instrumentation is evolving, perhaps even faster than most technologies, emphasizing the importance of relevant research; we have tried to recognize that by facilitating contact with universities and other places spearheading development.

One need for information is to ascertain where more information can be gained. We have catered for this with references at the ends of chapters to more specialized books.

Many agents have come together to produce the *Instrumentation Reference Book* and to whom thanks are due: those who have written, those who have drawn, and those who have painstakingly checked facts. I should especially thank Caroline Mallinder and Elizabeth Alderton who produced order out of chaos in the compilation of long lists of names and addresses. Thanks should also go elsewhere in the Butterworth hierarchy for the original germ of the idea that this could be a good addition to their family of reference books. In a familiar tradition, I thank my wife for her tolerance and patience about time-consuming activities such as telephoning, typing, and traveling–or at the least for limiting her natural intolerance and impatience of my excessive indulgence in them!

—*B. E. N. Dorking*

Contributors

C. S. Bahra, BSc, MSc, CEng, MIMechE, was formerly Development Manager at Transducer Systems Ltd.

J. Barron, BA, MA (Cantab), is a Lecturer at the University of Cambridge.

Jonas Berge, Senior Engineer, Emerson Process Management.

Martin Berutti, Director of Marketing, Mynah Technologies Inc., Chesterfield, MO, is an expert on medium and high resolution process simulation systems.

Walt Boyes, Principal, Spitzer and Boyes LLC, Aurora, Ill., is an ISA Fellow and Editor in Chief of *Control* magazine and www.controlglobal.com and is a recognized industry analyst and consultant. He has over 30 years experience in sales, marketing, technical support, new product development, and management in the instrumentation industries.

G. Burns, BSc, PhD, AMIEE, Glasgow College of Technology.

J. C. Cluley, MSc, CEng, MIEE, FBCS, was formerly a Senior Lecturer in the Department of Electronic and Electrical Engineering, University of Birmingham.

R. Cumming, BSc, FIQA, Scottish School of Nondestructive Testing.

W. G. Cummings, BSc, CChem, FRSC, MInstE, MinstMC, former Head of the Analytical Chemistry Section at Central Electricity Research Laboratories.

A. Danielsson, CEng, FIMechE, FInstMC, is with Wimpey Engineering Ltd. He was a member of the BS working party developing the Code of Practice for Instrumentation in Process Control Systems: Installation-Design.

C. I. Daykin, MA, is Director of Research and Development at Automatic Systems Laboratories Ltd.

Dr. Stanley Dolin, Scientist, Omega Engineering, Stamford, Conn., is an expert on the measurement of temperature.

James R. Ford, PhD, Director of Strategic Initiatives, Maverick Technologies Inc., Columbia, Ill., is a professional engineer and an expert on Advanced Process Control techniques and practices.

T. Fountain, BEng, AMIEE, is the Technical Manager for National Instruments UK Corp., where he has worked since 1989. Before that he was a design engineer for Control Universal, interfacing computers to real-world applications.

G. Fowles was formerly a Senior Development Engineer with the Severn-Trent Water Authority after some time as a Section Leader in the Instrumentation Group of the Water Research Centre.

Charlie Gifford, 21st Century Manufacturing Technologies, Hailey, Id., is a leading expert on Manufacturing Operations Management and the chief editor of Hitchhiking through Manufacturing, ISA Press, 2008.

J. G. Giles, TEng, has been with Ludlam Sysco Ltd. for a number of years.

Sir Claud Hagart-Alexander, Bt, BA, MInstMC, DL, formerly worked in instrumentation with ICI Ltd. He was then a director of Instrumentation Systems Ltd. He is now retired.

D. R. Heath, BSc, PhD, is with Rank Xerox Ltd.

E. H. Higham, MA, CEng, FIEE, MIMechE, MInstMC, is a Senior Research Fellow in the School of Engineering at the University of Sussex, after a long career with Foxboro Great Britain Ltd.

W. M. Jones, BSc, DPhil, FInstP, is a Reader in the Physics Department at the University College of Wales.

David Kaufman, Director of New Business Development, Honeywell Process Solutions, Phoenix, Az., is an officer of the Wireless Compliance Institute, a member of the leadership of ISA100, the industrial wireless standard, and an expert on industrial wireless networking.

D. Aliaga Kelly, BSc, CPhys, MInstP, MAmPhys-Soc, MSRP, FSAS, is now retired after working for many years as Chief Physicist with Nuclear Enterprises Ltd.

C. Kindell is with AMP of Great Britain Ltd.

E. G. Kingham, CEng, FIEE, was formerly at the Central Electricity Research Laboratories.

T. Kingham, is with AMP of Great Britain Ltd.

J. Kuehn, FInst Accoust, is Managing Director of Bruel & Kjaer (UK) Ltd.

C. K. Laird, BSc, PhD, CChem, MRSC, works in the Chemistry Branch at Central Electricity Research Laboratories.

F. F. Mazda, DFH, MPhil, CEng, MIEE, MBIM, is with Rank Xerox Ltd.

W. McEwan, BSc, CEng, MIMechE, FweldInst, Director Scottish School of Non-destructive Testing.

A. McNab, BSc, PhD, University of Strathclyde.

D. B. Meadowcroft, BSc, PhD, CPhys, FInstP, FICorrST works in the Chemistry Branch at Central Electricity Research Laboratories.

B. T. Meggitt, BSc, MSc, PhD, is Development Manager of LM Technology Ltd. and Visiting Professor in the Department of Electronic and Electrical Engineering, City University, London.

Alan Montgomery, Sales Manager, Lumberg Canada Ltd., is a long-time sales and marketing expert in the instrumentation field, and is an expert on modern industrial connectors.

William L. Mostia, Principal, WLM Engineering, Kemah, Tex. Mr. Mostia is an independent consulting engineer and an expert on pneumatic instrumentation, among other specialties.

G. Muir, BSc, MSc, MIM, MInstNDT, MWeldInst, CEng, FIQA, Scottish School of Non-destructive Testing.

B. E. Noltingk, BSc, PhD, CEng, FIEE, FInstP, is now a Consultant after some time as Head of the Instrumentation Section at the Central Electricity Research Laboratories.

Eoin O'Riain, Publisher, *Readout Magazine*.

D. J. Pacey, BSc, FInst P, was, until recently, a Senior Lecturer in the Physics Department at Brunel University.

Dr. Jerry Paros, President, Paroscientific Corp., Redmond, Wash., is founder of Paroscientific, a leading-edge pressure sensor manufacturer, and one of the leading experts on pressure measurement.

J. Riley is with AMP of Great Britain Ltd.

M. L. Sanderson, BSc, PhD, is Director of the Centre for Fluid Instrumentation at Cranfield Institute of Technology.

M. G. Say, MSc, PhD, CEng, ACGI, DIC, FIEE, FRSE, is Professor Emeritus of Electrical Engineering at Heriot-Watt University.

R. Service, MSc, FInstNDT, MWeldInst, MIM, MICP, CEng, FIQA, Scottish School of Non-destructive Testing.

A. C. Smith, BSc, CChem, FRSC, MInstP, former Head of the Analytical Chemistry Section at Central Electricity Research Laboratories.

W. L. Snowsill, BSc, was formerly a Research Officer in the Control and Instrumentation Branch of the Central Electricity Research Laboratories.

K. R. Sturley, BSc, PhD, FIEE, FIEEE, is a Telecommunications Consultant.

P. H. Sydenham, ME, PhD, FInstMC, FIIC, AMIAust, is Head of and Professor at the School of Electronic Engineering in the South Australian Institute of Technology.

A. W. S. Tarrant, BSc, PhD, CPhys, FInstP, FCIBSE, is Director of the Engineering Optics Research Group at the University of Surrey.

M. Tooley, BA, is Dean of the Technology Department at Brooklands College and the author of numerous electronics and computing books.

K. Torrance, BSc, PhD, is in the Materials Branch at Central Electricity Research Laboratories.

L. C. Towle, BSc, CEng, MIMechE, MIEE, MInstMC, is a Director of the MTL Instruments Group Ltd.

L. W. Turner, CEng, FIEE, FRTS, is a Consultant Engineer.

Ian Verhappen, ICE-Pros Ltd., Edmonton, AB Canada, is the former Chair of the Fieldbus Foundation User Group, and an industrial networking consultant. Verhappen is an ISA Fellow, and is an expert on all manner of process analyzers.

K. Walters, MSc, PhD, is a Professor in the Department of Mathematics at the University College of Wales.

Joseph Weiss, principal, Applied Control Solutions LLC, is an industry expert on control systems and electronic security of control systems, with more than 35 years of experience in the energy industry. He is a member of the Standards and Practices Board of ISA, the International Society of Automation.

T. Williams, BSc, CEng, MIEE, formerly with Rosemount, is a consultant in electromagnetic compatability design and training with Elmac Services, Chichester.

Shari L. S. Worthington, President, Telesian Technology Inc., is an expert on electronic enablement of manufacturing and marketing in the high technology industries.

Introduction

1. TECHNIQUES AND APPLICATIONS

We can look at instrumentation work in two ways: by *techniques* or by *applications*. When we consider instrumentation by technique, we survey one scientific field, such as radioactivity or ultrasonics, and look at all the ways in which it can be used to make useful measurements. When we study instrumentation by application, we cover the various techniques to measure a particular quantity. Under flowmetering, for instance, we look at many methods, including tracers, ultrasonics, or pressure measurement. This book is mainly applications oriented, but in a few cases, notably pneumatics and the employment of nuclear technology, the technique has been the primary unifying theme.

2. ACCURACY

The most important question in instrumentation is the accuracy with which a measurement is made. It is such a universal issue that we will talk about it now as well as in the individual chapters to follow. Instrument engineers should be skeptical of accuracy claims, and they should hesitate to accept their own reasoning about the systems they have assembled. They should demand evidence—and preferably proof. Above all, they should be clear in their own minds about the level of accuracy needed to perform a job. Too much accuracy will unnecessarily increase costs; too little may cause performance errors that make the project unworkable.

Accuracy is important but complex. We must first distinguish between *systematic* and *random* errors in an instrument. Systematic error is the error inherent in the operation of the instrument, and calibrating can eliminate it. We discuss calibration in several later chapters. Calibration is the comparison of the reading of the instrument in question to a known standard and the maintenance of the evidentiary chain from that standard. We call this *traceability*.

The phrase *random errors* implies the action of probability. Some variations in readings, though clearly observed, are difficult to explain, but most random errors can be treated statistically without knowing their cause. In most cases it is assumed that the probability of error is such that errors in individual measurements have a normal distribution about the mean, which zero if there is no systematic error.

This implies that we should quote errors based on a certain probability of the whereabouts of the true value. The probability grows steadily wider as the range where it might be also grows wider.

When we consider a measurement chain with several links, the two approaches give increasingly different figures. For if we think of possibilities/impossibilities, we must allow that the errors in each link can be extreme and in the same direction, calling for a simple addition when calculating the possible total error. On the other hand, this is *improbable*, so the "chain error" that corresponds to a given probability, e_c, is appreciably smaller. In fact, statistically,

$$e_c = \sqrt{e_1^2 + e_2^2 + \cdots}$$

where e_1, e_2, and so on are the errors in the different links, each corresponding to the same probability as e_c.

We can think of *influence quantities* as the causes of random errors. Most devices that measure a physical quantity are *influenced* by other quantities. Even in the simple case of a tape measure, the tape itself is *influenced* by temperature. Thus, a tape measure will give a false reading unless the influence is allowed for. Instruments should be as insensitive as possible to influence quantities, and users should be aware of them. The effects of these influence quantities can often be reduced by calibrating under conditions as close as possible to the live measurement application. Influence quantities can often be quite complicated. It might not only be the temperature than can affect the instrument, but the *change* in temperature. Even the *rate* of change of the temperature can be the critical component of this influence quantity. To make it even more complex, we must also consider the differential between the temperatures of the various instruments that make up the system.

One particular factor that could be thought of as an influence quantity is the direction in which the quantity to be measured is changing. Many instruments give slightly different readings according to whether, as it changes, the particular value of interest is approached from above or below. This phenomenon is called *hysteresis*.

If we assume that the instrument output is exactly proportional to a quantity, and we find discrepancies, this is called *nonlinearity error*. Nonlinearity error is the maximum departure of the true input/output curve from the idealized straight line approximating it.

It may be noted that this does not cover changes in *incremental gain*, the term used for the local slope of the input/output curve. Special cases of the accuracy of conversion from digital to analog signals, and vice versa, are discussed in Sections 29.3.1 and 29.4.5 of Part 4. Calibration at sufficient intermediate points in the range of an instrument can cover systematic nonlinearity.

Microprocessor-based instrumentation has reduced the problem of systematic nonlinearity to a simple issue. Most modern instruments have the internal processing capability to do at least a multipoint breakpoint linearization. Many can even host and process complex linearization equations of third order or higher.

Special terms used in the preceding discussion are defined in BS 5233, several ANSI standards, and in the ISA *Dictionary of Instrumentation*, along with numerous others.

The general approach to errors that we have outlined follows a statistical approach to a static situation.

Communications theory emphasizes working frequencies and time available, and this approach to error is gaining importance in instrumentation technology as instruments become more intelligent. Sensors connected to digital electronics have little or no error from electronic noise, but most accurate results can still be expected from longer measurement times.

Instrument engineers must be very wary of measuring the wrong thing! Even a highly accurate measurement of the wrong quantity may cause serious process upsets. Significantly for instruments used for control, Heisenberg's law applies on the macro level as well as on the subatomic. The operation of measurement can often disturb the quantity measured.

This can happen in most fields: A flowmeter can obstruct flow and reduce the velocity to be measured, an over-large temperature sensor can cool the material studied, or a low-impedance voltmeter can reduce the potential it is monitoring. Part of the instrument engineer's task is to foresee and avoid errors resulting from the effect instrument has on the system it is being used to study.

3. ENVIRONMENT

Instrument engineers must select their devices based on the environment in which they will be installed. In plants there will be extremes of temperature, vibration, dust, chemicals, and abuse. Instruments for use in plants are very different from those that are designed for laboratory use.

Two kinds of ill effects arise from badly selected instruments: false readings from exceptional values of influence quantities and the irreversible failure of the instrument itself.

Sometimes manufacturers specify limits to working conditions. Sometimes instrument engineers must make their own judgments. When working close to the limits of the working conditions of the equipment, a wise engineer derates the performance of the system or designs environmental mitigation.

Because instrumentation engineering is a practical discipline, a key feature of any system design must be the reliability of the equipment. Reliability is the likelihood of the instrument, or the system, continuing to work satisfactorily over long periods. We discuss reliability deeply in Part 4. It must always be taken into account in selecting instruments and designing systems for any application.

4. UNITS

The introductory chapters to some books have discussed the theme of what systems of units are used therein. Fortunately the question is becoming obsolete because SI units are adopted nearly everywhere, and certainly in this book. In the United States and a few other areas, where other units still have some usage, we have listed the relationships for the benefit of those who are still more at home with the older expressions.

REFERENCES

British Standards Institution, *Glossary of terms used in Metrology*, BS 5233 (1975).

Dietrich, D. F., *Uncertainty, Calibration and Probability: the Statistics of Scientific and Industrial Measurement*, Adam Hilger, London (1973).

Instrumentation, Systems and Automation Society (ISA), *The ISA Comprehensive Dictionary of Measurement and Control*, 3rd ed.; online edition, www.isa.org.

Topping, J., *Errors of Observation and their Treatment*, Chapman and Hall, London (1972).

The Automation Knowledge Base

Part I

Chapter 1

The Automation Practicum

W. Boyes

1.1 INTRODUCTION

In the years since this book was first published, there have been incredible changes in technology, in sociology, and in the way we work, based on those changes. Who in the early 1970s would have imagined that automation professionals would be looking at the outputs of sensors on handheld devices the size of the "communicators" on the science fiction TV show *Star Trek*? Yet by late 2007, automation professionals could do just that (see Figure 1.1).

There is now no way to be competitive in manufacturing, or no way to do science or medicine, without sensors, instruments, transmitters, and automation. The broad practice of automation, which includes instrumentation, control, measurement, and integration of plant floor and manufacturing operations management data, has grown up entirely since the first edition of this book was published.

So, what exactly is automation, and why do we do it? According to the dictionary,[1] *automation* has three definitions: 1: the technique of making an apparatus, a process, or a system operate automatically; 2: the state of being operated automatically; 3: automatically controlled operation of an apparatus, process, or system by mechanical or electronic devices that take the place of human labor.

How do we do this? We substitute sensors for the calibrated eyeball of human beings. We connect those sensors to input/output devices that are, in turn, connected to controllers. The controllers are programmed to make sense of the sensor readings and convert them into actions to be taken by final control elements. Those actions, in turn, are measured by the sensors, and the process repeats.

Although it is true that automation has replaced much human labor, it is not a replacement for human beings. Rather, the human ability to visualize, interpret, rationalize, and codify has been moved up the value chain from actually pushing buttons and pulling levers on the factory floor to designing and operating sensors, controllers, computers, and final control elements that can do those things. Meanwhile, automation has become ubiquitous and essential.

Yet as of this writing there is a serious, worldwide shortage of automation professionals who have training, experience, and interest in working with sensors, instrumentation, field controllers, control systems, and manufacturing automation in general.

There are a number of reasons for this shortage, including the generally accepted misunderstanding that the profession of automation is not necessarily recognized as a profession at all. Electrical engineers practice automation.

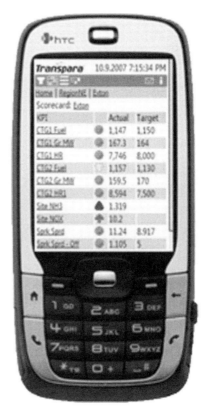

FIGURE 1.1 Courtesy of Transpara Corporation.

---

1. Merriam-Webster Online Dictionary, 2008.

Some mechanical engineers do, too. Many nonengineers also practice automation, as technicians. Many people who are not engineers but have some other technical training have found careers in automation.

Automation is really a multidisciplinary profession, pulling its knowledge base from many different disciplines, including mechanical engineering, electrical engineering, systems engineering, safety engineering, chemical engineering, and many more areas.

What we hope to present in this book is a single-volume overview of the automation profession—specifically, the manufacturing automation profession and the tools and techniques of the automation professional.

It must be said that there is a closely allied discipline that is shared by both manufacturing and theater. It is the discipline of *mechatronics*. Gerhard Schweitzer, Emeritus Professor of Mechanics and Professor of Robotics at the ETH Zurich, defines mechatronics this way:

> Mechatronics is an interdisciplinary area of engineering that combines mechanical and electrical engineering and computer science. A typical mechatronic system picks up signals from the environment, processes them to generate output signals, transforming them for example into forces, motions and actions.
>
> It is the extension and the completion of mechanical systems with sensors and microcomputers which is the most important aspect. The fact that such a system picks up changes in its environment by sensors, and reacts to their signals using the appropriate information processing, makes it different from conventional machines.
>
> Examples of mechatronic systems are robots, digitally controlled combustion engines, machine tools with self-adaptive tools, contact-free magnetic bearings, automated guided vehicles, etc. Typical for such a product is the high amount of system knowledge and software that is necessary for its design. Furthermore, and this is most essential, software has become an integral part of the product itself, necessary for its function and operation. It is fully justified to say software has become an actual "machine element."[2]

This interdisciplinary area, which so obviously shares so many of the techniques and components of manufacturing automation, has also shared in the reluctance of many engineering schools to teach the subject as a separate discipline. Fewer than six institutions of higher learning in North America, for example, teach automation or mechatronics as separate disciplines. One university, the University of California at Santa Cruz, actually offers a graduate degree in mechatronics—from the Theater Arts Department.

In this text we also try to provide insight into ways to enter into a career in manufacturing automation other than "falling into it," as so many practitioners have.

1.2 JOB DESCRIPTIONS

Because industrial automation, instrumentation, and controls are truly multidisciplinary, there are many potential job descriptions that plant the job holder squarely in the role of automation professional. There are electricians, electrical engineers, chemical engineers, biochemical engineers, control system technicians, maintenance technicians, operators, reliability engineers, asset and management engineers, biologists, chemists, statisticians, manufacturing, industrial, and civil and mechanical engineers who have become involved in automation and consider themselves automation professionals. System engineers, system analysts, system integrators—all are automation professionals working in industrial automation.

1.3 CAREERS AND CAREER PATHS

A common thread that runs through surveys of how practitioners entered the automation profession is that they were doing something else, got tapped to do an automation project, found they were good at it, and fell into doing more automation projects. There are very few schools that offer careers in automation. Some technical schools and trade schools do, but few universities do.

Many automation professionals enter nonengineering-level automation careers via the military. Training in electronics, maintenance, building automation, and automation in most of the Western militaries is excellent and can easily transfer to a career in industrial automation. For example, the building automation controls on a large military base are very similar to those found in the office and laboratory space in an industrial plant. Reactor control technicians from the nuclear navies of the world are already experienced process control technicians, and their skills transfer to industrial automation in the process environment.

Engineering professionals usually also enter the automation profession by studying something else. Many schools, such as Visvesaraya Technological University in India, offer courses in industrial automation (usually focusing on robotics and mechatronics) as part of another degree course. Southern Alberta Institute of Technology (SAIT) in Canada, the University of Greenwich in the United Kingdom, and several others offer degrees and advanced degrees in control or automation. Mostly, control is covered in electrical engineering and in chemical engineering curricula, if it is covered at all.

The International Society of Automation (or ISA, formerly the Instrumentation, Systems and Automation Society and before that the Instrument Society of America) has addressed the multidisciplinary nature of the automation professional by establishing two global certification programs.

The Certified Control Systems Technician, or CCST, program serves to benchmark skills in the process industries for technicians and operator-level personnel: "ISA's Certified Control Systems Technician Program (CCST) offers third-party recognition of technicians' knowledge and skills

---

2. www.mcgs.ch/mechatronics_definition.html.

in automation and control."[3] The certification is divided into seven functional domains of expertise: calibration, loop checking, troubleshooting, startup, maintenance/repair, project organization, and administration. The CCST is achieving global recognition as an employment certification.

Although ISA is the Accreditation Board for Engineering and Technology (ABET) curriculum designer for the Control System Engineering examination, ISA recognized in the early 2000s the fact that the U.S. engineering licensure program was not global and didn't easily transfer to the global engineering environment. Even in the United States, only 44 of 50 states offer licensing to control system engineers. ISA set out to create a nonlicensure-based certification program for automation professionals. The Certified Automation Professional (CAP) program was designed to offer an accreditation in automation on a global basis: "ISA certification as a Certified Automation Professional (CAP) will provide an unbiased, third-party, objective assessment and confirmation of your skills as an automation professional. Automation professionals are responsible for the direction, definition, design, development/application, deployment, documentation, and support of systems, software, and equipment used in control systems, manufacturing information systems, systems integration, and operational consulting."[4]

The following guidelines list the experience and expertise required to achieve CAP certification. They are offered here by permission from ISA as a guide to the knowledge required to become an automation professional.

1.3.1 ISA Certified Automation Professional (CAP) Classification System[5]

Domain I: Feasibility Study. Identify, scope, and justify the automation project.

Task 1. Define the preliminary scope through currently established work practices in order to meet the business need.

Knowledge of:

1. Established work practices
2. Basic process and/or equipment
3. Project management methodology
4. Automation opportunity identification techniques (e.g., dynamic performance measures)
5. Control and information technologies (MES) and equipment

Skill in:

1. Automating process and/or equipment
2. Developing value analyses

Task 2. Determine the degree of automation required through cost/benefit analysis in order to meet the business need.

Knowledge of:

1. Various degrees of automation
2. Various cost/benefit tools
3. Control and information technologies (MES) and equipment
4. Information technology and equipment

Skill in:

1. Analyzing cost versus benefit (e.g., life cycle analysis)
2. Choosing the degree of automation
3. Estimating the cost of control equipment and software

Task 3. Develop a preliminary automation strategy that matches the degree of automation required by considering an array of options and selecting the most reasonable option in order to prepare feasibility estimates.

Knowledge of:

1. Control strategies
2. Principles of measurement
3. Electrical components
4. Control components
5. Various degrees of automation

Skill in:

1. Evaluating different control strategies
2. Selecting appropriate measurements
3. Selecting appropriate components
4. Articulating concepts

Task 4. Conduct technical studies for the preliminary automation strategy by gathering data and conducting an appropriate analysis relative to requirements in order to define development needs and risks.

Knowledge of:

1. Process control theories
2. Machine control theories and mechatronics
3. Risk assessment techniques

Skill in:

1. Conducting technical studies
2. Conducting risk analyses
3. Defining primary control strategies

Task 5. Perform a justification analysis by generating a feasibility cost estimate and using an accepted financial model to determine project viability.

Knowledge of:

1. Financial models (e.g., ROI, NPV)
2. Business drivers
3. Costs of control equipment
4. Estimating techniques

---

3. From www.isa.org.
4. From www.isa.org.
5. ISA Certified Automation Professional (CAP) Classification System, from www.isa.org/~/CAPClassificationSystemWEB.pdf.

Skill in:

1. Estimating the cost of the system
2. Running the financial model
3. Evaluating the results of the financial analysis for the automation portion of the project

Task 6. Create a conceptual summary document by reporting preliminary decisions and assumptions in order to facilitate "go/no go" decision making.

Knowledge of:

1. Conceptual summary outlines

Skill in:

1. Writing in a technical and effective manner
2. Compiling and summarizing information efficiently
3. Presenting information

Domain II

Task 1. Determine operational strategies through discussion with key stakeholders and using appropriate documentation in order to create and communicate design requirements.

Knowledge of:

1. Interviewing techniques
2. Different operating strategies
3. Team leadership and alignment

Skill in:

1. Leading an individual or group discussion
2. Communicating effectively
3. Writing in a technical and effective manner
4. Building consensus
5. Interpreting the data from interviews

Task 2. Analyze alternative technical solutions by conducting detailed studies in order to define the final automation strategy.

Knowledge of:

1. Automation techniques
2. Control theories
3. Modeling and simulation techniques
4. Basic control elements (e.g., sensors, instruments, actuators, control systems, drive systems, HMI, batch control, machine control)
5. Marketplace products available
6. Process and/or equipment operations

Skill in:

1. Applying and evaluating automation solutions
2. Making intelligent decisions
3. Using the different modeling tools
4. Determining when modeling is needed

Task 3. Establish detailed requirements and data including network architecture, communication concepts, safety concepts, standards, vendor preferences, instrument and equipment data sheets, reporting and information needs, and security architecture through established practices in order to form the basis of the design.

Knowledge of:

1. Network architecture
2. Communication protocols, including field level
3. Safety concepts
4. Industry standards and codes
5. Security requirements
6. Safety standards (e.g., ISAM, ANSI, NFPA)
7. Control systems security practices

Skill in:

1. Conducting safety analyses
2. Determining which data is important to capture
3. Selecting applicable standards and codes
4. Identifying new guidelines that need to be developed
5. Defining information needed for reports
6. Completing instrument and equipment data sheets

Task 4. Generate a project cost estimate by gathering cost information in order to determine continued project viability.

Knowledge of:

1. Control system costs
2. Estimating techniques
3. Available templates and tools

Skill in:

1. Creating cost estimates
2. Evaluating project viability

Task 5. Summarize project requirements by creating a basis-of-design document and a user-requirements document in order to launch the design phase.

Knowledge of:

1. Basis of design outlines
2. User-requirements document outlines

Skill in:

1. Writing in a technical and effective manner
2. Compiling and summarizing information
3. Making effective presentations

Domain III

Task 1. Perform safety and/or hazard analyses, security analyses, and regulatory compliance assessments by identifying key issues and risks in order to comply with applicable standards, policies, and regulations.

Knowledge of:

1. Applicable standards (e.g., ISA S84, IEC 61508, 21 CFR Part 11, NFPA)

2. Environmental standards (EPA)
3. Electrical, electrical equipment, enclosure, and electrical classification standards (e.g., UL/FM, NEC, NEMA)

Skill in:

1. Participating in a Hazard Operability Review
2. Analyzing safety integrity levels
3. Analyzing hazards
4. Assessing security requirements or relevant security issues
5. Applying regulations to design

Task 2. Establish standards, templates, and guidelines as applied to the automation system using the information gathered in the definition stage and considering human-factor effects in order to satisfy customer design criteria and preferences.

Knowledge of:

1. Process Industry Practices (PIP) (Construction Industry Institute)
2. IEC 61131 programming languages
3. Customer standards
4. Vendor standards
5. Template development methodology
6. Field devices
7. Control valves
8. Electrical standards (NEC)
9. Instrument selection and sizing tools
10. ISA standards (e.g., S88)

Skill in:

1. Developing programming standards
2. Selecting and sizing instrument equipment
3. Designing low-voltage electrical systems
4. Preparing drawings using AutoCAD software

Task 3. Create detailed equipment specifications and instrument data sheets based on vendor selection criteria, characteristics and conditions of the physical environment, regulations, and performance requirements in order to purchase equipment and support system design and development.

Knowledge of:

1. Field devices
2. Control valves
3. Electrical standards (NEC)
4. Instrument selection and sizing tools
5. Vendors' offerings
6. Motor and drive selection sizing tools

Skill in:

1. Selecting and sizing motors and drives
2. Selecting and sizing instrument equipment
3. Designing low-voltage electrical systems
4. Selecting and sizing computers
5. Selecting and sizing control equipment

6. Evaluating vendor alternatives
7. Selecting or sizing of input/output signal devices and/or conditioners

Task 4. Define the data structure layout and data flow model considering the volume and type of data involved in order to provide specifications for hardware selection and software development.

Knowledge of:

1. Data requirements of system to be automated
2. Data structures of control systems
3. Data flow of control systems
4. Productivity tools and software (e.g., InTools, AutoCAD)
5. Entity relationship diagrams

Skill in:

1. Modeling data
2. Tuning and normalizing databases

Task 5. Select the physical communication media, network architecture, and protocols based on data requirements in order to complete system design and support system development.

Knowledge of:

1. Vendor protocols
2. Ethernet and other open networks (e.g., DeviceNet)
3. Physical requirements for networks/media
4. Physical topology rules/limitations
5. Network design
6. Security requirements
7. Backup practices
8. Grounding and bonding practices

Skill in:

1. Designing networks based on chosen protocols

Task 6. Develop a functional description of the automation solution (e.g., control scheme, alarms, HMI, reports) using rules established in the definition stage in order to guide development and programming.

Knowledge of:

1. Control theory
2. Visualization, alarming, database/reporting techniques
3. Documentation standards
4. Vendors' capabilities for their hardware and software products
5. General control strategies used within the industry
6. Process/equipment to be automated
7. Operating philosophy

Skill in:

1. Writing functional descriptions
2. Interpreting design specifications and user requirements
3. Communicating the functional description to stakeholders

Task 7. Design the test plan using chosen methodologies in order to execute appropriate testing relative to functional requirements.

Knowledge of:

1. Relevant test standards
2. Simulation tools
3. Process Industry Practices (PIP) (Construction Industry Institute)
4. General software testing procedures
5. Functional description of the system/equipment to be automated

Skill in:

1. Writing test plans
2. Developing tests that validate that the system works as specified

Task 8. Perform the detailed design for the project by converting the engineering and system design into purchase requisitions, drawings, panel designs, and installation details consistent with the specification and functional descriptions in order to provide detailed information for development and deployment.

Knowledge of:

1. Field devices, control devices, visualization devices, computers, and networks
2. Installation standards and recommended practices
3. Electrical and wiring practices
4. Specific customer preferences
5. Functional requirements of the system/equipment to be automated
6. Applicable construction codes
7. Documentation standards

Skill in:

1. Performing detailed design work
2. Documenting the design

Task 9. Prepare comprehensive construction work packages by organizing the detailed design information and documents in order to release project for construction.

Knowledge of:

1. Applicable construction practices
2. Documentation standards

Skill in:

1. Assembling construction work packages

Domain IV: Development. Software development and coding.

Task 1. Develop Human Machine Interface (HMI) in accordance with the design documents in order to meet the functional requirements.

Knowledge of:

1. Specific HMI software products
2. Tag definition schemes
3. Programming structure techniques
4. Network communications
5. Alarming schemes
6. Report configurations
7. Presentation techniques
8. Database fundamentals
9. Computer operating systems
10. Human factors
11. HMI supplier options

Skill in:

1. Presenting data in a logical and aesthetic fashion
2. Creating intuitive navigation menus
3. Implementing connections to remote devices
4. Documenting configuration and programming
5. Programming configurations

Task 2. Develop database and reporting functions in accordance with the design documents in order to meet the functional requirements.

Knowledge of:

1. Relational database theory
2. Specific database software products
3. Specific reporting products
4. Programming/scripting structure techniques
5. Network communications
6. Structured query language
7. Report configurations
8. Entity diagram techniques
9. Computer operating systems
10. Data mapping

Skill in:

1. Presenting data in a logical and aesthetic fashion
2. Administrating databases
3. Implementing connections to remote applications
4. Writing queries
5. Creating reports and formatting/printing specifications for report output
6. Documenting database configuration
7. Designing databases
8. Interpreting functional description

Task 3. Develop control configuration or programming in accordance with the design documents in order to meet the functional requirements.

Knowledge of:

1. Specific control software products
2. Tag definition schemes

3. Programming structure techniques
4. Network communications
5. Alarming schemes
6. I/O structure
7. Memory addressing schemes
8. Hardware configuration
9. Computer operating systems
10. Processor capabilities
11. Standard nomenclature (e.g., ISA)
12. Process/equipment to be automated

Skill in:

1. Interpreting functional description
2. Interpreting control strategies and logic drawings
3. Programming and/or configuration capabilities
4. Implementing connections to remote devices
5. Documenting configuration and programs
6. Interpreting P & IDs
7. Interfacing systems

Task 4. Implement data transfer methodology that maximizes throughput and ensures data integrity using communication protocols and specifications in order to assure efficiency and reliability.

Knowledge of:

1. Specific networking software products (e.g., I/O servers)
2. Network topology
3. Network protocols
4. Physical media specifications (e.g., copper, fiber, RF, IR)
5. Computer operating systems
6. Interfacing and gateways
7. Data mapping

Skill in:

1. Analyzing throughput
2. Ensuring data integrity
3. Troubleshooting
4. Documenting configuration
5. Configuring network products
6. Interfacing systems
7. Manipulating data

Task 5. Implement security methodology in accordance with stakeholder requirements in order to mitigate loss and risk.

Knowledge of:

1. Basic system/network security techniques
2. Customer security procedures
3. Control user-level access privileges
4. Regulatory expectations (e.g., 29 CFR Part 11)
5. Industry standards (e.g., ISA)

Skill in:

1. Documenting security configuration
2. Configuring/programming of security system
3. Implementing security features

Task 6. Review configuration and programming using defined practices in order to establish compliance with functional requirements.

Knowledge of:

1. Specific control software products
2. Specific HMI software products
3. Specific database software products
4. Specific reporting products
5. Programming structure techniques
6. Network communication
7. Alarming schemes
8. I/O structure
9. Memory addressing schemes
10. Hardware configurations
11. Computer operating systems
12. Defined practices
13. Functional requirements of system/equipment to be automated

Skill in:

1. Programming and/or configuration capabilities
2. Documenting configuration and programs
3. Reviewing programming/configuration for compliance with design requirements

Task 7. Test the automation system using the test plan in order to determine compliance with functional requirements.

Knowledge of:

1. Testing techniques
2. Specific control software products
3. Specific HMI software products
4. Specific database software products
5. Specific reporting products
6. Network communications
7. Alarming schemes
8. I/O structure
9. Memory addressing schemes
10. Hardware configurations
11. Computer operating systems
12. Functional requirements of system/equipment to be automated

Skill in:

1. Writing test plans
2. Executing test plans
3. Documenting test results

4. Programming and/or configuration capabilities
5. Implementing connections to remote devices
6. Interpreting functional requirements of system/equipment to be automated
7. Interpreting P & IDs

Task 8. Assemble all required documentation and user manuals created during the development process in order to transfer essential knowledge to customers and end users.

Knowledge of:

1. General understanding of automation systems
2. Computer operating systems
3. Documentation practices
4. Operations procedures
5. Functional requirements of system/equipment to be automated

Skill in:

1. Documenting technical information for non-technical audience
2. Using documentation tools
3. Organizing material for readability

Domain V

Task 1. Perform receipt verification of all field devices by comparing vendor records against design specifications in order to ensure that devices are as specified.

Knowledge of:

1. Field devices (e.g., transmitters, final control valves, controllers, variable speed drives, servo motors)
2. Design specifications

Skill in:

1. Interpreting specifications and vendor documents
2. Resolving differences

Task 2. Perform physical inspection of installed equipment against construction drawings in order to ensure installation in accordance with design drawings and specifications.

Knowledge of:

1. Construction documentation
2. Installation practices (e.g., field devices, computer hardware, cabling)
3. Applicable codes and regulations

Skill in:

1. Interpreting construction drawings
2. Comparing physical implementation to drawings
3. Interpreting codes and regulations (e.g., NEC, building codes, OSHA)
4. Interpreting installation guidelines

Task 3. Install configuration and programs by loading them into the target devices in order to prepare for testing.

Knowledge of:

1. Control system (e.g., PLC, DCS, PC)
2. System administration

Skill in:

1. Installing software
2. Verifying software installation
3. Versioning techniques and revision control
4. Troubleshooting (i.e., resolving issues and retesting)

Task 4. Solve unforeseen problems identified during installation using troubleshooting skills in order to correct deficiencies.

Knowledge of:

1. Troubleshooting techniques
2. Problem-solving strategies
3. Critical thinking
4. Processes, equipment, configurations, and programming
5. Debugging techniques

Skill in:

1. Solving problems
2. Determining root causes
3. Ferreting out information
4. Communicating with facility personnel
5. Implementing problem solutions
6. Documenting problems and solutions

Task 5. Test configuration and programming in accordance with the design documents by executing the test plan in order to verify that the system operates as specified.

Knowledge of:

1. Programming and configuration
2. Test methodology (e.g., factory acceptance test, site acceptance test, unit-level testing, system-level testing)
3. Test plan for the system/equipment to be automated
4. System to be tested
5. Applicable regulatory requirements relative to testing

Skill in:

1. Executing test plans
2. Documenting test results
3. Troubleshooting (e.g., resolving issues and retesting)
4. Writing test plans

Task 6. Test communication systems and field devices in accordance with design specifications in order to ensure proper operation.

Knowledge of:

1. Test methodology

2. Communication networks and protocols
3. Field devices and their performance requirements
4. Regulatory requirements relative to testing

Skill in:

1. Verifying network integrity and data flow integrity
2. Conducting field device tests
3. Comparing test results to design specifications
4. Documenting test results
5. Troubleshooting (i.e., resolving issues and retesting)
6. Writing test plans

Task 7. Test all safety elements and systems by executing test plans in order to ensure that safety functions operate as designed.

Knowledge of:

1. Applicable safety
2. Safety system design
3. Safety elements
4. Test methodology
5. Facility safety procedures
6. Regulatory requirements relative to testing

Skill in:

1. Executing test plans
2. Documenting test results
3. Testing safety systems
4. Troubleshooting (i.e., resolving issues and retesting)
5. Writing test plans

Task 8. Test all security features by executing test plans in order to ensure that security functions operate as designed.

Knowledge of:

1. Applicable security standards
2. Security system design
3. Test methodology
4. Vulnerability assessments
5. Regulatory requirements relative to testing

Skill in:

1. Executing test plans
2. Documenting test results
3. Testing security features
4. Troubleshooting (i.e., resolving issues and retesting)
5. Writing test plans

Task 9. Provide initial training for facility personnel in system operation and maintenance through classroom and hands-on training in order to ensure proper use of the system.

Knowledge of:

1. Instructional techniques
2. Automation systems
3. Networking and data communications
4. Automation maintenance techniques

5. System/equipment to be automated
6. Operating and maintenance procedures

Skill in:

1. Communicating with trainees
2. Organizing instructional materials
3. Instructing

Task 10. Execute system-level tests in accordance with the test plan in order to ensure the entire system functions as designed.

Knowledge of:

1. Test methodology
2. Field devices
3. System/equipment to be automated
4. Networking and data communications
5. Safety systems
6. Security systems
7. Regulatory requirements relative to testing

Skill in:

1. Executing test plans
2. Documenting test results
3. Testing of entire systems
4. Communicating final results to facility personnel
5. Troubleshooting (i.e., resolving issues and retesting)
6. Writing test plans

Task 11. Troubleshoot problems identified during testing using a structured methodology in order to correct system deficiencies.

Knowledge of:

1. Troubleshooting techniques
2. Processes, equipment, configurations, and programming

Skill in:

1. Solving problems
2. Determining root causes
3. Communicating with facility personnel
4. Implementing problem solutions
5. Documenting test results

Task 12. Make necessary adjustments using applicable tools and techniques in order to demonstrate system performance and turn the automated system over to operations.

Knowledge of:

1. Loop tuning methods/control theory
2. Control system hardware
3. Computer system performance tuning
4. User requirements

5. System/equipment to be automated

Skill in:

1. Tuning control loops
2. Adjusting final control elements
3. Optimizing software performance
4. Communicating final system performance results

Domain VI: Operation and Maintenance. Long-term support of the system.

Task 1. Verify system performance and records periodically using established procedures in order to ensure compliance with standards, regulations, and best practices.

Knowledge of:

1. Applicable standards
2. Performance metrics and acceptable limits
3. Records and record locations
4. Established procedures and purposes of procedures

Skill in:

1. Communicating orally and written
2. Auditing the system/equipment
3. Analyzing data and drawing conclusions

Task 2. Provide technical support for facility personnel by applying system expertise in order to maximize system availability.

Knowledge of:

1. All system components
2. Processes and equipment
3. Automation system functionality
4. Other support resources
5. Control systems theories and applications
6. Analytical troubleshooting and root-cause analyses

Skill in:

1. Troubleshooting (i.e., resolving issues and retesting)
2. Investigating and listening
3. Programming and configuring automation system components

Task 3. Perform training needs analysis periodically for facility personnel using skill assessments in order to establish objectives for the training program.

Knowledge of:

1. Personnel training requirements
2. Automation system technology
3. Assessment frequency
4. Assessment methodologies

Skill in:

1. Interviewing
2. Assessing level of skills

Task 4. Provide training for facility personnel by addressing identified objectives in order to ensure the skill level of personnel is adequate for the technology and products used in the system.

Knowledge of:

1. Training resources
2. Subject matter and training objectives
3. Teaching methodology

Skill in:

1. Writing training objectives
2. Creating the training
3. Organizing training classes (e.g., securing demos, preparing materials, securing space)
4. Delivering training effectively
5. Answering questions effectively

Task 5. Monitor performance using software and hardware diagnostic tools in order to support early detection of potential problems.

Knowledge of:

1. Automation systems
2. Performance metrics
3. Software and hardware diagnostic tools
4. Potential problem indicators
5. Baseline/normal system performance
6. Acceptable performance limits

Skill in:

1. Using the software and hardware diagnostic tools
2. Analyzing data
3. Troubleshooting (i.e., resolving issues and retesting)

Task 6. Perform periodic inspections and tests in accordance with written standards and procedures in order to verify system or component performance against requirements.

Knowledge of:

1. Performance requirements
2. Inspection and test methodologies
3. Acceptable standards

Skill in:

1. Testing and inspecting
2. Analyzing test results
3. Communicating effectively with others in written or oral form

Task 7. Perform continuous improvement by working with facility personnel in order to increase capacity, reliability, and/or efficiency.

Knowledge of:

1. Performance metrics
2. Control theories

3. System/equipment operations
4. Business needs
5. Optimization tools and methods

Skill in:

1. Analyzing data
2. Programming and configuring
3. Communicating effectively with others
4. Implementing continuous improvement procedures

Task 8. Document lessons learned by reviewing the project with all stakeholders in order to improve future projects.

Knowledge of:

1. Project review methodology
2. Project history
3. Project methodology and work processes
4. Project metrics

Skill in:

1. Communicating effectively with others
2. Configuring and programming
3. Documenting lessons learned
4. Writing and summarizing

Task 9. Maintain licenses, updates, and service contracts for software and equipment by reviewing both internal and external options in order to meet expectations for capability and availability.

Knowledge of:

1. Installed base of system equipment and software
2. Support agreements
3. Internal and external support resources
4. Life-cycle state and support level (including vendor product plans and future changes)

Skill in:

1. Organizing and scheduling
2. Programming and configuring
3. Applying software updates (i.e., keys, patches)

Task 10. Determine the need for spare parts based on an assessment of installed base and probability of failure in order to maximize system availability and minimize cost.

Knowledge of:

1. Critical system components
2. Installed base of system equipment and software
3. Component availability
4. Reliability analysis
5. Sourcing of spare parts

Skill in:

1. Acquiring and organizing information
2. Analyzing data

Task 11. Provide a system management plan by performing preventive maintenance, implementing backups, and designing recovery plans in order to avoid and recover from system failures.

Knowledge of:

1. Automation systems
2. Acceptable system downtime
3. Preventive and maintenance procedures
4. Backup practices (e.g., frequency, storage media, storage location)

Skill in:

1. Acquiring and organizing
2. Leading
3. Managing crises
4. Performing backups and restores
5. Using system tools

Task 12. Follow a process for authorization and implementation of changes in accordance with established standards or practices in order to safeguard system and documentation integrity.

Knowledge of:

1. Management of change procedures
2. Automation systems and documentation
3. Configuration management practices

Skill in:

1. Programming and configuring
2. Updating documentation

The Certified Automation Professional program offers certification training, reviews, and certification examinations regularly scheduled on a global basis. Interested people should contact ISA for information.

1.4 WHERE AUTOMATION FITS IN THE EXTENDED ENTERPRISE

In the late 1980s, work began at Purdue University, under the direction of Dr. Theodore J. Williams, on modeling computer-integrated manufacturing enterprises.[6] The resultant model describes five basic layers making up the enterprise (see Figure 1.2):

Level 0: The Process
Level 1: Direct Control
Level 2: Process Supervision
Level 3: Production Supervision

---

6. This effort was documented in *A Reference Model for Computer Integrated Manufacturing (CIM): A Description from the Viewpoint of Industrial Automation*, Theodore J. Williams, Ph.D., editor, ISA, 1989.

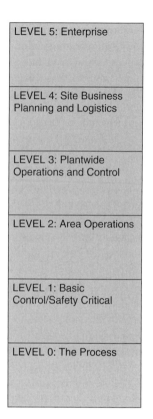

FIGURE 1.2 The Purdue Model.

Level 4: Plant Management and Scheduling

Level 0 consists of the plant infrastructure as it is used for manufacturing control. This level includes all the machinery, sensors, controls, valves, indicators, motors, drives, and so forth.

Level 1 is the system that directly controls the manufacturing process, including input/output devices, plant networks, single loop controllers, programmable controllers, and process automation systems.

Level 2 is supervisory control. This is the realm of the Distributed Control System (DCS) with process visualization (human/machine interface, or HMI) and advanced process control functions.

These three levels have traditionally been considered the realm of industrial automation. The two levels above them have often been considered differently, since they are often under the control of different departments in the enterprise. However, the two decades since the Purdue Research Foundation CIM Model was developed have brought significant changes, and the line between Level 2 and below and Level 3 and above is considerably blurrier than it was in 1989.

In 1989, basic control was performed by local control loops, with supervision from SCADA (in some industries, otherwise known as Supervisory Control and Data Acquisition) or DCSes that were proprietary, used proprietary networks and processors, and were designed as standalone systems and devices.

In 2009, the standard for control systems is Microsoft Windows-based operating systems, running proprietary Windows-based software on commercial, off-the-shelf (COTS) computers, all connected via Ethernet networks to local controllers that are either proprietary or that run some version of Windows themselves. What this has made possible is increasingly integrated operations, from the plant floor to the enterprise resource planning (ERP) system, because the entire enterprise is typically using Ethernet networks and Windows-based software systems running on Windows operating systems on COTS computers.

To deal with this additional complexity, organizations like MESA International (formerly known as the Manufacturing Execution Systems Association) have created new models of the manufacturing enterprise (see Figure 1.3). These models are based on information flows and information use within the enterprise.

The MESA model squeezes all industrial automation into something called Manufacturing/Production and adds two layers above it. Immediately above the Manufacturing/Production layer is Manufacturing/Production Operations, which includes parts of Level 2 and Level 3 of the Purdue Reference Model. Above that is the Business Operations layer. MESA postulates another layer, which its model shows off to one side, that modulates the activities of the other three: the Strategic Initiatives layer.

The MESA model shows "events" being the information transmitted to the next layer from Manufacturing/Operations in a unidirectional format. Bidirectional information flow is shown from the Business Operations layer to the Manufacturing/Production Operations layer. Unidirectional information flow is shown from the Strategic Initiatives layer to the rest of the model.

Since 1989, significant work has been done by many organizations in this area, including WBF (formerly the World Batch Forum); ISA, whose standards, ISA88 and ISA95, are the basis for operating manufacturing languages; the Machinery Information Management Open Systems Alliance (MIMOSA); the Organization for Machine Automation and Control (OMAC); and others.

1.5 MANUFACTURING EXECUTION SYSTEMS AND MANUFACTURING OPERATIONS MANAGEMENT

C. Gifford, S. L. S. Worthington, and W. H. Boyes

1.5.1 Introduction

For more than 30 years, companies have found extremely valuable productivity gains in automating their plant floor

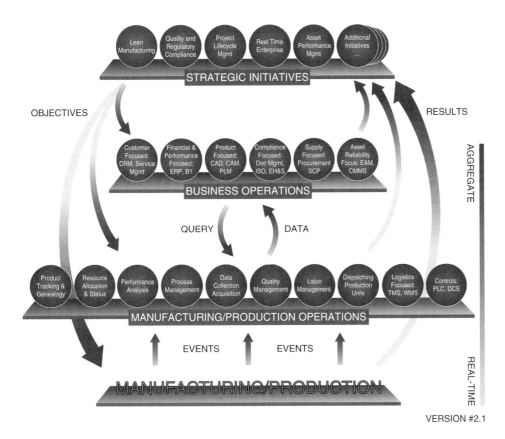

FIGURE 1.3 The MESA Model.

processes. This has been true whether the plant floor processes were continuous, batch, hybrid, or discrete manufacturing. No manufacturing company anywhere in the world would consider operating its plant floor processes manually, except under extreme emergency conditions, and even then a complete shutdown of all systems would be preferable to many plant managers.

From the enterprise level in the Purdue Model (see Figure 1.2) down to the plant floor and from the plant floor up to the enterprise level, there are significant areas in which connectivity and real-time information transfer improve performance and productivity.

1.5.2 Manufacturing Execution Systems (MES) and Manufacturing Operations Management (MOM)

At one time, the theoretical discussion was limited to connecting the manufacturing floor to the production scheduling systems and then to the enterprise accounting systems. This was referred to as *manufacturing execution systems*, or MES. MES implementations were difficult and often returned less than expected return on investment (ROI). Recent thought has centered on a new acronym, MOM, which stands for *manufacturing operations management*. But even MOM does not completely describe a fully connected enterprise.

The question that is often asked is, "What are the benefits of integrating the manufacturing enterprise?"

In "Mastering the MOM Model," a chapter written for the next edition of *The Automation Book of Knowledge*, to be published in 2009, Charlie Gifford writes, "Effectiveness in manufacturing companies is only partially based on equipment control capability. In an environment that executes as little as 20% make-to-order orders (80% make-to-stock), resource optimization becomes critical to effectiveness. Manufacturing companies must be efficient at coordinating and controlling personnel, materials and equipment across different operations and control systems in order to reach their maximum potential."[7]

1.5.3 The Connected Enterprise[8]

To transform a company into a true multidomain B2B extended enterprise is a complex exercise. Here we have provided a 20-point checklist that will help transform a traditional manufacturing organization into a fully extended

---

7. Gifford, Charlie, "Mastering the MOM Model," from *The Automation Book of Knowledge*, 3rd ed., to be published by ISA Press in 2009.
8. This section contains material that originally appeared in substantially different form in *e-Business in Manufacturing*, Shari L. S. Worthington and Walt Boyes, ISA Press, 2001.

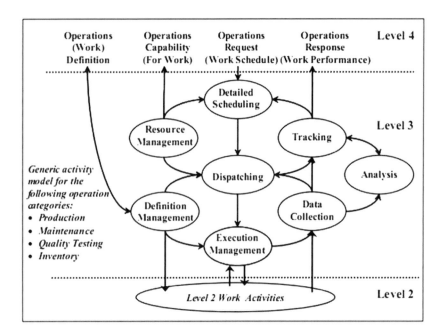

FIGURE 1.4 ISA-95 generic detailed work activity model (part 3) for MOM ANSI/ISA95. Used with permission of ISA.

enterprise with a completely integrated supply chain and full B2B data interchange capability.

To begin, a company must develop, integrate, and maintain metrics vertically through the single domain of the enterprise and then horizontally across the extended domains of the company's supplier and customer base. At each step of the process, the company must achieve buy-in, develop detailed requirements and business cases, and apply the metrics to determine how much progress has been made toward an extended enterprise. The following details the key points that a company must include in any such checklist for measuring its progress toward becoming a true e-business.

- *Achieve upper management buy-in.* Before anything else, upper management must be educated in the theory of manufacturing enterprise integration. The entire c-level (CEO, COO, CFO, CIO, etc.) must be clear that integration of the enterprise is a process, not a project, that has great possibility of reward but also a possibility of failure.

  The board and the executive committee must be committed to, and clearly understand, the process of building an infrastructure for cultural change in the corporation. They must also understand the process that must be followed to become a high-response extended enterprise.

- *Perform a technology maturity and gap analysis.* The team responsible for enterprise transformation has to perform two assessments before it can proceed to implementation:

A. Feasibility Assessment:

1. Identify the main corporate strategy for the extended enterprise.
2. Identify the gap between the technologies and infrastructure that the company currently uses and those that are necessary for an extended enterprise.
3. Identify the current and future role of the internal supply chain.
4. Identify the gap between the technologies and infrastructure that are currently available and those that are required to fully integrate the enterprise supply chain.
5. Identify all the corporate views of the current and future state of the integrated supply chain and clearly define their business roles in the extended enterprise model.
6. Determine the costs/benefits of switching to an integrated supply chain model for the operation of the extended enterprise.

B. Requirements Assessment:

1. Develop a plan to scale the IT systems, enterprisewide, and produce a reliability analysis.
2. Determine the probable failure points and data storage and bandwidth load requirements.
3. Perform an assessment of the skill sets of the in-house team and determine how to remedy any gaps identified.

Once the preliminary work is done and upper management has been fully educated about and has bought into the project, the enterprise team can begin to look at specific tactics.

- *Identify the business drivers and quantify the return on investment (ROI) for the enterprise's integrated supply chain strategy.*

1. Perform a supply chain optimization study to determine the business drivers and choke points for the supply chain, both vertically in the enterprise itself and horizontally across the supplier and customer base of the enterprise.

2. Perform a plant-level feasibility assessment to identify the requirements for supply chain functionality at the plant level in the real world.
3. Create a preliminary data interchange model and requirements.
4. Do a real-world check by determining the ROI of each function in the proposed supply chain integration.
5. Make sure that the manufacturing arm of the enterprise is fully aligned with the supply chain business strategy and correct any outstanding issues.

- *Develop an extended enterprise and supply chain strategy and model that is designed to be "low maintenance."*

  1. Write the integrated supply chain strategy and implementation plan, complete with phases, goals, and objectives, and the metrics for each.
  2. Write the manufacturing execution infrastructure (MEI) plan. This process includes creating a project steering team, writing an operations manual for the steering team, developing communications criteria, creating a change management system, and establishing a version control system for both software and manuals. This plan should also include complete details about the document management system, including the documentation standards needed to fulfill the project goals.
  3. Last, the MEI plan should include complete details about the management of the project, including chain of command, responsibilities, authority, and reporting.

Once the plan is written, the enterprise team can move to the physical pilot stage. At this point, every enterprise will need to do some experimenting to determine the most effective way to integrate the supply chain and become an extended enterprise.

- *Develop an MEI plan.* In the next phase, the team must develop a manufacturing execution infrastructure (MEI) plan during the pilot phase that will respond appropriately to the dynamic change/knowledge demands of a multidomain environment. In the pilot phase, the earlier enterprisewide planning cycle takes concrete form in a single plant environment. So, we will see some of the same assessments and reports, but this time they will be focused on a single plant entity in the enterprise.

  1. Perform a plant-level feasibility and readiness assessment.
  2. Perform a plant-level functional requirements assessment that includes detailed requirements for the entire plant's operations and the business case and ROI for each requirement.
  3. Determine the pilot project's functionality and the phases, goals, and objectives that are necessary to achieve the desired functionality.
  4. Draft a plant manufacturing execution system (MES) design specification.
  5. Determine the software selection criteria for integrating the supply chain.
  6. Apply the criteria and produce a software selection process report.
  7. Design and deploy the pilot manufacturing execution system.
  8. Benchmark the deployment and produce a performance assessment.
  9. Perform a real-world analysis of what was learned and earned by the pilot project, and determine what changes are required to the pilot MOM model.

Once the enterprise team has worked through several iterations of pilot projects, it can begin to design an enterprisewide support system.

- *Design a global support system (GSS) for manufacturing.* In an integrated manufacturing enterprise, individual plants and departments cannot be allowed to do things differently than the rest of the enterprise. Global standards for work rules, training, operations, maintenance, and accounting must be designed, trained on, and adhered to, with audits and accountability, or the enterprise integration project will surely fail.

  1. Draft a GSS design specification.
  2. Draft a complete set of infrastructure requirements that will function globally throughout the enterprise.
  3. Draft an implementation plan and deploy the support system within the enterprise.

- *Finalize the development of data interchange requirements* between the MES, ERP, SCM, CRM, and logistics and fulfillment applications by using the data acquired during the pilot testing of plantwide systems.

  1. Develop an extended strategy for an enterprise application interface (EAI) layer that may incorporate the company's suppliers and customers into a scalable XML data set and schema. Finalize the specifications for the integrated supply chain (ISC) data interchange design.
  2. Figure out all the ways the project can fail, with a failure effort and mode analysis (FEMA). Now, in the period before the enterprisewide implementation, the enterprise team must go back to the pilot process. This time, however, they are focusing on a multiplant extended enterprise pilot project.
  3. Extend the MEI, global support system (GSS), EAI layer, and the failure effort and mode analysis (FEMA) into a multiplant, multidomain pilot project.

- Perform a multiplant plant-level feasibility and readiness assessment.

- Identify detailed requirements for all the plants in the project, including business case and ROI for each requirement.
- Draft a multiplant design specification.
- Draft an integrated supply chain specification for all the plants in the project.
- Draft and deploy the multiplant MES.
- Draft and deploy the multiplant integrated supply chain (ISC).
- Benchmark and measure the performance of the system "as designed."
- Finally, it becomes possible for the enterprise team to develop the corporate deployment plan and to fully implement the systems they have piloted throughout the enterprise. To do this successfully, they will need to add some new items to the "checklist":

  1. Develop rapid application deployment (RAD) teams to respond as quickly as possible to the need for enterprisewide deployments of new required applications.
  2. Develop and publish the corporate deployment plan. The company's education and training requirements are critical to the success of an enterprisewide deployment of new integrated systems. A complete enterprise training plan must be created that includes end-user training, superuser training, MEI and GSS training, and the relevant documentation for each training program.
  3. The multiplant pilot MEI and GSS must be scaled to corporatewide levels. This project phase must include determining enterprisewide data storage and load requirements, a complete scaling and reliability plan, and benchmark criteria for the scale-up.
  4. Actually deploy the MEI and GSS into the enterprise and begin widespread use.

- Next, after it has transformed the enterprise's infrastructure and technology, the enterprise team can look outward to the supplier and customer base of the company.

  1. Select preferred suppliers and customers that can be integrated into the enterprise system's EAI/XML schema. Work with each supplier and customer to implement the data-interchange process and train personnel in using the system.
  2. Negotiate the key process indicators (KPI) with supply chain partners by creating a stakeholders' steering team that consists of members of the internal and external supply chain partners. Publish the extended ISC data interchange model and requirements through this team. Have this team identify detailed requirements on a plant-by-plant basis, including a business case and ROI for each requirement.
  3. This steering team should develop the specifications for the final pilot test: a pilot extended enterprise (EE), including ISC pilot implementation specifications and a performance assessment of the system "as designed."
  4. Develop and track B2B performance metrics using the pilot EE as a model, and create benchmarks and tracking criteria for a full implementation of the extended enterprise.
  5. Develop a relationship management department to administer and expand the B2B extended enterprise.

SUGGESTED READING

Gifford, Charles, *The Hitchhiker's Guide to Manufacturing Operations Management,* ISA Press, Research Triangle Park, NC, 2007.

Chapter 2

Basic Principles of Industrial Automation

W. Boyes

2.1 INTRODUCTION

There are many types of automation, broadly defined. Industrial automation, and to some extent its related discipline of building automation, carries some specific principles. The most important skills and principles are discussed in Chapter 1 of this book.

It is critical to recognize that industrial automation differs from other automation strategies, especially in the enterprise or office automation disciplines. Industrial automation generally deals with the automation of complex processes, in costly infrastructure programs, and with design life cycles in excess of 30 years. Automation systems installed at automotive assembly plants in the late 1970s were still being used in 2008. Similarly, automation systems installed in continuous and batch process plants in the 1970s and 1980s continued to be used in 2008. Essentially, this means that it is not possible to easily perform rip-and-replace upgrades to automation systems in industrial controls, whereas it is simpler in many cases to do such rip-and-replace in enterprise automation systems, such as sales force automation or even enterprise requirements planning (ERP) systems when a new generation of computers is released or when Microsoft releases a new version of Windows. In the industrial automation environment, these kinds of upgrades are simply not practical.

2.2 STANDARDS

Over the past three decades, there has been a strong movement toward standards-based design, both of field instruments and controls themselves and the systems to which they belong. The use of and the insistence on recognized standards for sensor design, control system operation, and system design and integration have reduced costs, improved reliability, and enhanced productivity in industrial automation.

There are several standards-making bodies that create standards for industrial automation. They include the International Electrotechnical Commission (IEC) and the International Standards Organization (ISO). Other standards bodies include CENELEC, EEMUA, and the various national standards bodies, such as NIST, ANSI, the HART Communication Foundation, and NEC in the United States; BSI in the United Kingdom; CSA in Canada; DIN, VDE, and DKE in Germany; JIS in Japan; and several standards organizations belonging to the governments of China and India, among others.

For process automation, one of the significant standards organizations is ISA, the International Society of Automation. ISA's standards are in use globally for a variety of automation operations in the process industries, from process and instrumentation diagram symbology (ISA5 and ISA20) to alarm management (ISA18) to control valve design (ISA75), fieldbus (ISA50), industrial wireless (ISA100), and cyber security for industrial automation (ISA99).

Three of the most important global standards developed by ISA are the ISA84 standard on safety instrumented systems, the ISA88 standard on batch manufacturing, and the ISA95 standard on manufacturing operations language.

Other organizations that are similar to standards bodies but do not make actual standards include NAMUR, OMAC, WBF (formerly World Batch Forum), WIB (the Instrument Users' Association), and others.

In addition, with the interpenetration of COTS computing devices in industrial automation, IEEE standards, as well as standards for the design and manufacture of personal computers, have become of interest and importance to the automation professional.

There are also *de facto* standards such as the Microsoft Windows operating system, OPC (originally a Microsoft "standard" called Object Linking and Embedding for Process Control, or OLE for Process Control, and now called simply OPC), and OPC UA (Universal Architecture).

It is important for the process automation professional to keep current with standards that impinge on the automation system purview.

2.3 SENSOR AND SYSTEM DESIGN, INSTALLATION, AND COMMISSIONING

It is not generally in the purview of the automation professional to actually design sensors. This is most commonly done by automation and instrumentation vendors. What are in the purview of the automation professional are system design, installation, and commissioning. Failure to correctly install a sensor or final control element can lead to serious consequences, including damage to the sensor, the control element, or the process and infrastructure themselves.

2.3.1 The Basics

The basics of sensor and system design are:

- Identification of the application
- Selection of the appropriate sensor/transmitter
- Selection of the final control element
- Selection of the controller and control methodology
- Design of the installation
- Installing, commissioning, and calibrating the system

2.3.2 Identification of the Application

Most maintenance problems in automation appear to result from improper identification of the application parameters. This leads to incorrect selection of sensors and controls and improper design of the installation. For example, it is impossible to produce an operational flow control loop if the flowmeter is being made both inaccurate and nonlinear by having been installed in a location immediately downstream of a major flow perturbation producer such as a butterfly valve or two 90-degree elbows in series. The most common mistake automation professionals make is to start with their favorite sensors and controls and try to make them fit the application.

2.3.3 Selection of the Appropriate Sensor/Transmitter

The selection of the most appropriate sensor and transmitter combination is another common error point. Once the application parameters are known, it is important to select the most correct sensor and transmitter for those parameters. There are 11 basic types of flow measurement devices and a similar number of level measurement principles being used in modern automation systems. This is because it is often necessary to use a "niche" instrument in a particular application. There are very few locations where a gamma nuclear-level gauge is the most correct device to measure level, but there are a number where the gamma nuclear principle is the only practical way to achieve the measurement. Part of the automation professional's skill set is the applications knowledge and expertise to be able to make the proper selection of sensors and transmitters.

2.3.4 Selection of the Final Control Element

Selection of the final control element is just as important as selection of the transmitter and sensor and is equally based on the application parameters. The final control element can be a control valve, an on/off valve, a temperature control device such as a heater, or a pump in a process automation application. It can be a relay, a PLC ladder circuit, or a stepper motor or other motion control device in a discrete automation application. Whatever the application, the selection of the final control element is critical to the success of the installation.

Sometimes, too, the selection of the final control element is predicated on factors outside the strict control loop. For example, the use of a modulating control valve versus the use of a variable-speed drive-controlled pump can make the difference between high energy usage in that control loop and less energy usage. Sometimes this difference can represent a significant cost saving.

2.3.5 Selection of the Controller and Control Methodology

Many automation professionals forget that the selection of the control methodology is as important as the selection of the rest of the control loop.

Using an advanced process control system over the top of a PID control loop when a simple on/off deadband control will work is an example of the need to evaluate the controller and the control methodology based on the application parameters.

2.3.6 Design of the Installation

As important as any other factor, properly designing the installation is critical to the implementation of a successful control loop. Proper design includes physical design within the process.

Not locating the sensor at an appropriate point in the process is a common error point. Locating a pH sensor on the opposite side of a 1,000-gallon tank from the chemical injection point is an example. The pH sensor will have to wait until the chemical injection has changed the pH in the entire vessel as well as the inlet and outlet piping before it sees the change. This could take hours. A loop lag time that long will cause the loop to be dysfunctional.

Another example of improper location is to locate the transmitter or final control element in a place where it is difficult or impossible for operations and maintenance personnel to reach it after startup. Installations must be designed with an eye to ease of maintenance and calibration. A sensor mounted 40 feet off the ground that requires a cherry-picker crane to reach isn't a good installation.

Another example of improper installation is to place a device, such as a flowmeter, where the process flow must be stopped to remove the flowmeter for repair. Bypass lines should be installed around most sensors and final control elements.

2.3.7 Installing, Commissioning, and Calibrating the System

Installation of the system needs to be done in accordance with both the manufacturers' instructions and good trade craft practices, and any codes that are applicable. In hazardous areas, applicable national electrical codes as well as any plant specific codes must be followed. Calibration should be done during commissioning and at regularly scheduled intervals over the lifecycle of the installation.

2.4 MAINTENANCE AND OPERATION

2.4.1 Introduction

Automation professionals in the 21st century may find themselves working in maintenance or operations rather than in engineering, design, or instrumentation and controls. It is important for automation professionals to understand the issues and principles of maintenance of automation systems, in both continuous and batch process and discrete factory automation. These principles are similar to equipment maintenance principles and have changed from maintenance practices of 20 years ago. Then maintenance was done on a reactive basis—that is, if it broke down, it was fixed. In some cases, a proactive maintenance scheme was used. In this practice, critical automation assets would be replaced at specific intervals, regardless of whether they were working or not. This led to additional expense as systems and components were pulled out when they were still operational. Recent practice has become that of predictive maintenance. Predictive maintenance uses the recorded trends of physical measurements compared to defined engineering limits to determine how to analyze and correct a problem before failure occurs. This practice, where asset management software is used along with sensor and system diagnostics to determine the point at which the automation asset must be replaced, is called *life-cycle maintenance* or *life-cycle optimization*.

2.4.2 Life-cycle Optimization

In any automation system, there is a recognized pattern to the life cycles of all the components. This pattern forms the well-known "bathtub curve." There are significant numbers of "infant mortality" failures at the start of the curve; then, as each component ages, there are relatively few failures. Close to the end of the product's life span, the curve rises, forming the other side of the "bathtub." Using predictive maintenance techniques, it is possible to improve the operational efficiency and availability of the entire system by monitoring physical parameters of selected components. For example, it is clear that the mean time between failures (MTBF) of most electronics is significantly longer than the design life of the automation system as a whole, after infant mortality. This means that it is possible to essentially eliminate the controller as a failure-prone point in the system and concentrate on what components have much shorter MTBF ratings, such as rotating machinery, control valves, and the like.

2.4.3 Reliability Engineering

For the automation professional, *reliability* is defined as the probability that an automation device will perform its intended function over a specified time period under conditions that are clearly understood. *Reliability engineering* is the branch of engineering that designs to meet a specified probability of performance, with an expressed statistical confidence level. Reliability engineering is central to the maintenance of automation systems.

2.4.4 Asset Management, Asset Optimization, and Plant Optimization

Asset management systems have grown into detailed, layered software systems that are fully integrated into the sensor networks, measure parameters such as vibration and software diagnostics from sensors and final control elements, and are even integrated into the maintenance systems of plants. A modern asset management system can start with a reading on a flow sensor that is out of range, be traced to a faulty control valve, and initiate a work order to have the control valve repaired, all without human intervention.

This has made it possible to perform workable asset optimization on systems as large and complex as the automation and control system for a major refinery or chemical process plant. Using the techniques of reliability engineering and predictive maintenance, it is possible to maximize the amount of time that the automation system is working properly—the uptime of the system.

Asset optimization is conjoined to another subdiscipline of the automation professional: plant optimization. Using the control system and the asset management system, it is possible to operate the entire plant at its maximum practical level of performance.

SUGGESTED READING

Mather, Darryl, *Lean Strategies for Asset Reliability: Asset Resource Planning,* Industrial Press Inc., 2009.

EAM Resource Center, *The Business Impact of Enterprise Asset Management,* EAM, 2008.

Chapter 3

Measurement Methods and Control Strategies

W. Boyes

3.1 INTRODUCTION

Measurement methods for automation are somewhat different from those designed for use in laboratories and test centers.

Specific to automation, measurement methods that work well in a laboratory might not work at all in a process or factory-floor environment. It might not be possible to know all the variables acting on a measurement to determine the degree of error (uncertainty) of that measurement. For example, a flowmeter may be calibrated at the factory with an accuracy of $\pm 0.05\%$ of actual flow. Yet in the field, once installed, an automation professional might be lucky to be able to calibrate the flowmeter to $\pm 10\%$ of actual flow because of the conditions of installation. Because control strategies are often continuous, it is often impossible to remove the sensor from the line for calibration.

3.2 MEASUREMENT AND FIELD CALIBRATION METHODOLOGY

In many cases, then, field calibration methods are expedients designed to determine not the absolute accuracy of the measurement but the repeatability of the measurement. Especially in process applications, repeatability is far more critical to the control scheme than absolute accuracy. It is often not possible to do more than one or two calibration runs *in situ* in a process application. It often means that the calibration and statistical repeatability of the transmitter is what is checked in the field, rather than the accuracy of the entire sensor element.

3.3 PROCESS CONTROL STRATEGIES

Basic process control strategies include on/off control; deadband control; proportional, integral, derivative (PID) control; and its derivatives.

On/off control is simple and effective but may not be able to respond to rapid changes in the measured variable (known as PV, or *process variable*). The next iteration is a type of on/off control called *deadband*, or *hysteresis control*. In this method, either the "on" or the "off" action is delayed until a prescribed limit set point is reached, either ascending or descending. Often multiple limit set points are defined, such as a level application with "high" level, "high-high" level, and "high-overflow" level set points. Each of the set points is defined as requiring a specific action.

Feedback control is used with a desired set point from which deviation is not desired. When the measured variable deviates from the set point, the controller output drives the measured variable back toward the set point. Most of the feedback control algorithms in use are some form of PID algorithm, of which there are three basic types: the *standard*, or *ideal*, form, sometimes called the *ISA form*; the *interactive* form, which was the predominant form for analog controllers; and the *parallel* form, which is rarely found in industrial process control.

In the PID algorithm, the proportional term provides most of the control while the integral function and the derivative function provide additional correction. In practice, the proportional and integral terms do most of the control; the derivative term is often set to 0.

PID loops contain one measured variable, one controller, and one final control element. This is the basic "control loop" in automation. PID loops need to be "tuned"; there are several tuning algorithms, such as Ziegler-Nichols and others, that allow the loop to be tuned. Many vendors today provide automatic loop-tuning products in their control software offerings.

PID feedback controllers work well when there are few process disturbances. When the process is upset or is regularly discontinuous, it is necessary to look at other types of controllers. Some of these include ratio, feed forward, and cascade control. In *ratio control*, which is most often found in blending of two process streams, the basic process stream provides the pacing for the process while the flow rates for the other streams are modulated to make sure that they are in a specific ratio to the basic process stream. *Feed forward control*, or *open loop control*, uses the rate of fall-off from the set point (a disturbance in the process) to manipulate the controlled variable. An example is the use of a flowmeter to control the injection of a chemical additive downstream of the flowmeter. There must be some model of the process so that the effect of the flow change can be used to induce the correct effect on the process downstream.

Combining feed forward and feedback control in one integrated control loop is called *cascade control*. In this scheme, the major correction is done by feed forward control, and the minor correction (sometimes called *trim*) is done by the feedback loop. An example is the use of flow to control the feed of a chemical additive while using an analyzer downstream of the addition point to modulate the set point of the flow controller.

3.4 ADVANCED CONTROL STRATEGIES

Since the 1960s, advances in modeling the behavior of processes have permitted a wholly new class of control strategies, called *advanced process control*, or APC. These control strategies are almost always layered over the basic PID algorithm and the standard control loop. These APC strategies include fuzzy logic, adaptive control, and model predictive control.

Conceived in 1964 by University of California at Berkeley scientist Lotfi Zadeh, *fuzzy logic* is based on the concept of fuzzy sets, where membership in the set is based on probabilities or degrees of truth rather than "yes" or "no."[1] Because multiple fuzzy logic sets appear to be able to learn, they are often regarded as a crude form of artificial intelligence. In process automation, only four rules are required for a fuzzy logic controller:[2]

Rule 1: If the error is negative and the change in error is negative, the change in output is positive.
Rule 2: If the error is negative and the change in error is positive, the change in output is zero.
Rule 3: If the error is positive and the change in error is negative, the change in output is zero.
Rule 4: If the error is positive and the change in error is positive, the change in output is negative.

Adaptive control is somewhat loosely defined as any algorithm in which the controller's tuning has been altered. Another term for adaptive controllers is *self-tuning controllers*.

Model predictive control uses historicized incremental models of the process to be controlled where the change in a variable can be predicted. When the MPC controller is initialized, the model parameters are set to match the actual performance of the plant. According to Gregory McMillan, "MPC sees future trajectory based on past moves of manipulated variables and present changes in disturbance variables as inputs to a linear model. It provides an integral-only type of control."[3]

These advanced control strategies can often improve loop performance but, beyond that, they are also useful in optimizing performance of whole groups of loops, entire processes, and even entire plants themselves.

SUGGESTED READING

Dieck, Ronald H., *Measurement Uncertainty, Methods and Applications*, 4th ed., ISA Press, Research Triangle Park, NC, 2007.

Trevathan, Vernon L., editor, *A Guide to the Automation Body of Knowledge*, 2nd ed., ISA Press, Research Triangle Park, NC, 2006.

Blevins, Terry, and McMillan, Gregory, et al., *Advanced Control Unleashed: Plant Performance Management for Optimum Benefit*, ISA Press, Research Triangle Park, NC, 2003.

---

1. *Britannica Concise Encyclopedia*, quoted in www.answers.com

2. McMillan, Gregory K., "Advanced Process Control," *in A Guide to the Automation Body of Knowledge*, 2nd ed., ISA Press, Research Triangle Park, NC, 2006

3. *Ibid*

Chapter 4

Simulation and Design Software[1]

M. Berutti

4.1 INTRODUCTION

Chemical engineers are accustomed to software for designing processes and simulation. Simulation systems such as Matlab and Aspen Plus are commonly referenced in chemical engineering curricula as required courseware and study tools. Automation professionals are also becoming used to applying simulation to operator training, system testing, and commissioning of plant process control systems. Plant design simulation programs are substantially different from systems used for training and commissioning. Many of the most common plant design simulation programs are steady-state, low-resolution simulations that are not usable for automation or plant life-cycle management.

4.2 SIMULATION

Simulation is usually integrated into the plant life cycle at the front-end engineering and design stage and used to test application software using a simulated I/O system and process models in an offline environment. The same simulation system can then be used to train operations staff on the automation and control systems and the application software that will be running on the hardware platform. In the most advanced cases, integration with manufacturing execution systems (MES) and electronic batch records (EBR) systems can be tested while the operations staff is trained.

Once installed, the simulation system can be used to test and validate upgrades to the control system before they are installed. The simulator then becomes an effective tool for testing control system modifications in a controlled, offline environment. In addition, plant operations staff and new operators can become qualified on new enhancements and certified on the existing operating system. The simulation system can be used as a test bed to try new control strategies, build new product recipes, and design new interlock strategies prior to proposing those changes as projects to production management. The simulator can also be an effective risk management tool by providing the ability to conduct failure testing in an offline environment rather than on the operating process.

Simulation's ROI has been proven to be effectual and substantial across all process industries, from batch to continuous processes. The savings come from identifying and correcting automation system errors prior to startup and commissioning and by identifying so-called "sleeping" errors, or inadequacies in the system's application software. Additional savings come from accelerating operators' learning curves and the ability to train operators on upset or emergency conditions they normally do not encounter in day-to-day operations. This reduces operator errors in responding to abnormal situations.

4.3 BEST PRACTICES FOR SIMULATION SYSTEMS IN AUTOMATION

Nonintrusive simulation interfaces allow the user to test the control system configuration without making any modifications to the configuration database. As far as the operator is concerned, the system is "live"; as far as the database is concerned, there are no changes. By nature, a nonintrusive simulation interface will provide a virtual I/O interface to the process controller application code that supports I/O and process simulation and modeling. It allows the application software to run in a normal mode without any modification so that the testing application software is identical to the production application software. In addition, the nonintrusive interface will not produce "dead code" that formerly was necessary to spoof the control system during testing.

This type of simulation interface supports complete and thorough testing of the application software and provides

---

1. This material appeared in somewhat different form in a white paper, "Optimizing Results in Automation Projects with Simulation," by Martin Berutti, Mynah Technologies, 2006, and published on www.controlglobal.com

a benchmark of process controller performance, including CPU and memory loading, application software order execution, and timing.

4.4 GROUND-UP TESTING AND TRAINING

Ground-up testing and training is an incremental approach, integrated with the automation project life cycle. This approach to application software testing and training has several benefits. Ground-up testing allows identification and correction of issues early in the project, when they can be corrected before being propagated throughout the system.

Additionally, ground-up testing allows the end-user and operations staff to gain acceptance and familiarity with the automation system throughout the entire project instead of at one final acceptance test.

Best practices dictate that training and testing are inextricably linked throughout the automation project. Here are some general guidelines for following this best practice:

- *Control modules* are the base-level database elements of the process automation system. These include motors, discrete valves, analog loops, and monitoring points. Testing of these elements can be effectively accomplished with simple tieback simulations and automated test scripts. Operator training on these elements can also bolster buy-in of the automation system and provide a review of usability features.
- *Equipment modules* are the next level of an automation system and generally refer to simple continuous unit operations such as charging paths, valve manifolds, and package equipment. Testing these elements can be effectively accomplished with tieback simulations and limited process dynamics. Operator training on these elements is also valuable.
- *Sequence, batch controls, and continuous advanced controls* are the next layer of the automation system. Effective testing of these controls generally requires a mass balance simulation with effective temperature and pressure dynamics models. Operator training at this level is necessary due to the complexity of the controls and the user interface.
- *MES applications and business system integration* is the final layer of most automation systems. Mass and heat balance models are usually required for effective testing of this layer. Training at this level may be extended beyond the operations staff to include quality assurance, information technology, and other affected departments.
- *Display elements* for each layer of the automation should be tested with the database elements listed here. In other words, control module faceplates are tested with the control modules, and batch help screens are tested with the batch controls.

4.5 SIMULATION SYSTEM SELECTION

The proven best practice is to use actual automation system controllers (or equivalent soft controllers) and application software with a simulation "companion" system. Simulation systems that use a "rehosted" automation system should be eliminated for system testing and avoided for operator training. Use of the actual automation system components with a simulation system allows effective testing and training on HMI use, display access familiarity, process and emergency procedures, response to process upsets, and control system dynamics. This approach builds automation system confidence in operations staff, resulting in more effective use of the automation system and greater benefits.

4.6 SIMULATION FOR AUTOMATION IN THE VALIDATED INDUSTRIES

Automation system users and integrators for the validated industries need to be concerned about the GAMP4 Guidelines when they are applying simulation systems to their automation projects.

The GAMP4 Guidelines clearly state that simulation systems are allowable tools for automation system testing. The guidelines also make two requirements for the treatment of the automation system application software. First, they require that the application software be "frozen" prior to software integration and system acceptance testing. Second, they require the removal of "dead code" prior to testing. These two requirements dictate the use of nonintrusive simulation interfaces.

The GAMP4 Guidelines also state several requirements for the supplier of simulation systems for testing of automation projects. The supplier must have a documented quality and software development program in line with industry best practices. The product used should be designed specifically for process control system testing and operator training. Finally, the product should be a commercially available, off-the-shelf (COTS) tool, delivered in validated, tested object code.

Additionally, operator training management modules allow comprehensive development of structured operator training sessions with scripted scenarios and process events. Large or small automation projects can both use a simulation system with a scalable client/server architecture and a stable, repeatable simulation engine.

4.7 CONCLUSION

The use of simulation systems for testing and training process automation projects has been proven to reduce time to market and increase business results. The same systems can be utilized in automation life-cycle management to reduce operational costs and improve product quality.

Chapter 5

Security for Industrial Automation[1]

W. Boyes and J. Weiss

One of the very largest problems facing the automation professional is that the control systems in plants and the SCADA systems that tie together decentralized facilities such as power, oil, and gas pipelines and water distribution and wastewater collection systems were designed to be open, robust, and easily operated and repaired, but not necessarily secure.

5.1 THE SECURITY PROBLEM

For example, in August 2008, Dr. Nate Kube and Bryan Singer of Wurldtech demonstrated at the ACS Cyber Security Conference that a properly designed safety instrumented system that had received TÜV certification could very easily be hacked. The unidentified system failed *unsafely* in less than 26 seconds after the attack commenced. Note that "properly designed" meant that the controller was designed to be robust and safe and to operate properly. Operating cyber-securely was not one of the design elements.

For quite some time, Schweitzer Engineering Laboratories has had a utility on its website, www.selinc.com, that allowed SEL Internet-enabled relays to be programmed via a Telnet client by any authorized user. Recently, several security researchers found and acted on exploits against Telnet; SEL has now taken the utility down to protect the users.

It isn't the power industry alone that faces these issues, although the critical infrastructure in the power industry is certainly one of the largest targets. These cyber incidents have happened in many process industry verticals, whether they've been admitted to or not.

History shows that it is much more likely to be an internal accident or error that produces the problem.

In 1999, an operator for the Olympic Pipeline Company in Bellingham, Washington, was installing a patch on his pipeline SCADA system. Unknown to him, the scan rate of the SCADA system slowed to the point where a leak alarm failed to reach the SCADA HMI until after the ignition of the leak and the deaths of three people as well as numerous injuries. This is a classic cyber accident.

On January 26, 2000, the Hatch Nuclear Power Station experienced a cyber event. A Wonderware HMI workstation running on the plant local area network (LAN) was patched, experienced instability because of the patch, and rebooted. It was connected via a firewall directly into the OSI PI database that Hatch used as the plant historian. So was an Allen-Bradley PLC, which was caused to reboot. When it rebooted, it reinitialized all the valve positioners, and with all the valves closed the main feedwater pumps shut down, exactly as they were supposed to do, scramming the reactor.

At Brown's Ferry Nuclear Station in August 2006, a broadcast storm apparently caused by the plant's IT department "pinging" the network in a standard network quality control procedure caused a similar PLC to fail, shutting off the feedwater pumps and … you guessed it, scramming the reactor. It is troubling that Hatch and Brown's Ferry had similar incidents six years apart.

Lest one conclude that this is all about the power industry and the oil and gas industry, there is the case of Maroochy Shire in Western Australia. From the official MITRE report of the incident, coauthored by Joe Weiss, here is what happened: Vitek Boden worked for Hunter Watertech, an Australian firm that installed a SCADA system for the Maroochy Shire Council in Queensland, Australia. Boden applied for a job with the Maroochy Shire Council. The Council decided not to hire him. Consequently, Boden decided to get even with both the Council and his former employer. He packed his car with stolen radio equipment attached to a (possibly stolen) computer. He drove around the area on at least 46 occasions from February 28 to April 23, 2000, issuing radio commands to the sewage equipment he (probably) helped install. Boden caused 800,000 liters of raw sewage to

---

1. Some of this material originally appeared in somewhat different form in the November 2008 issue of *Control* magazine in an article coauthored by Walt Boyes and Joe Weiss and from the text of a speech given by Walt Boyes at the 2008 TÜV Safety Symposium.

spill out into local parks, rivers, and even the grounds of a Hyatt Regency hotel. Boden coincidentally got caught when a policeman pulled him over for a traffic violation after one of his attacks. A judge sentenced him to two years in jail and ordered him to reimburse the Council for cleanup.

There is evidence of more than 100 cyber incidents, whether intentional, malicious, or accidental, in the real-time ACS database maintained by Joe Weiss. These include the Northeast power outage and the Florida power outage in 2008. It is worth noting that neither event has been described as a cyber event by the owners of the power companies and transmission companies involved.

5.2 AN ANALYSIS OF THE SECURITY NEEDS OF INDUSTRIAL AUTOMATION

Industrial automation systems (or *industrial control systems*, abbreviated ICS) are an integral part of the industrial infrastructure supporting the nation's livelihood and economy. They aren't going away, and starting over from scratch isn't an option. UCSs are "systems of systems" and need to be operated in a safe, efficient, and secure manner. The sometimes competing goals of reliability and security are not just a North American issue, they are truly a global issue. A number of North American control system suppliers have development activities in countries with dubious credentials; for example, a major North American control system supplier has a major code-writing office in China, and a European RTU manufacturer uses code written in Iran.

Though sharing basic constructs with enterprise IT business systems, ICSs are technically, administratively, and functionally different systems. Vulnerability disclosure philosophies are different and can have devastating consequences to critical infrastructure. A major concern is the dearth of an educated workforce; there are very few control system cyber security experts (probably fewer than 100) and currently no university curricula or ICS cyber security personnel certifications. Efforts to secure these critical systems are too diffuse and do not specifically target the unique ICS aspects. The lack of ICS security expertise extends into the government arena, which has focused on repackaging IT solutions.

The successful convergence of IT and ICS systems and organizations is expected to enable the promised secure productivity benefits with technologies such as the *smart grid*. However, the convergence of mainstream IT and ICS systems requires both mainstream and control system expertise, acknowledging the operating differences and accepting the similarities. One can view current ICS cyber security as being where mainstream IT security was 15 years ago; it is in the formative stage and needs support to leapfrog the previous IT learning curve. Regulatory incentives and industry self-interest are necessary to create an atmosphere for adequately securing critical infrastructures. However, regulation will also be required.

5.3 SOME RECOMMENDATIONS FOR INDUSTRIAL AUTOMATION SECURITY

The following recommendations, taken from a report[2] to the bipartisan commission producing position papers for the Obama administration, can provide steps to improve the security and reliability of these very critical systems, and most of them are adoptable by any process industry business unit:

- Develop a clear understanding of ICS cyber security.
- Develop a clear understanding of the associated impacts on system reliability and safety on the part of industry, government, and private citizens.
- Define cyber threats in the broadest possible terms, including intentional, unintentional, natural, and other electronic threats, such as electromagnetic pulse (EMP) and electronic warfare against wireless devices.
- Develop security technologies and best practices for the field devices based on actual and expected ICS cyber incidents.
- Develop academic curricula in ICS cyber security.
- Leverage appropriate IT technologies and best practices for securing workstations using commercial off-the-shelf (COTS) operating systems.
- Establish standard certification metrics for ICS processes, systems, personnel, and cyber security.
- Promote/mandate adoption of the NIST Risk Management Framework for all critical infrastructures, or at least the industrial infrastructure subset.
- Establish a global, nongovernmental Cyber Incident Response Team (CIRT) for control systems, staffed with control system expertise for vulnerability disclosure and information sharing.
- Establish a means for vetting ICS experts rather than using traditional security clearances.
- Provide regulation and incentives for cyber security of critical infrastructure industries.
- Establish, promote, and support an open demonstration facility dedicated to best practices for ICS systems.
- Include subject matter experts with control system experience at high-level cyber security planning sessions.
- Change the culture of manufacturing in critical industries so that security is considered as important as performance and safety.
- Develop guidelines similar to that of the Sarbanes-Oxley Act for adequately securing ICS environments.

Like process safety, process security is itself a process and must become part of a culture of inherent safety and security.

---

2. This report was authored by Joe Weiss, with assistance from several other cyber security experts. The full text of the report is available at www.controlglobal.com

Part II

Mechanical Measurements

Chapter 6

Measurement of Flow

G. Fowles and W. H. Boyes

6.1 INTRODUCTION

Flow measurement is a technique used in any process requiring the transport of a material from one point to another (for example, bulk supply of oil from a road tanker to a garage holding tank). It can be used for quantifying a charge for material supplied or maintaining and controlling a specific rate of flow. In many processes, plant efficiency depends on being able to measure and control flow accurately.

Properly designed flow measurement systems are compatible with the process or material they are measuring. They must also be capable of producing the accuracy and repeatability that are most appropriate for the application.

It is often said that the "ideal flowmeter should be nonintrusive, inexpensive, have absolute accuracy, infinite repeatability, and run forever without maintenance." Unfortunately, such a device does not yet exist, although some manufacturers might claim that it does. Over recent years, however, many improvements have been made to established systems, and new products utilizing novel techniques are continually being introduced onto the market. The "ideal" flowmeter might not in fact be so far away, and now more than ever, potential users must be fully aware of the systems at their disposal.

6.2 BASIC PRINCIPLES OF FLOW MEASUREMENT

We need to spend a short time with the basics of flow measurement theory before looking at the operation of the various types of available measurement systems. *Flow* can be measured as either a volumetric quantity or an instantaneous velocity (this is normally translated into a *flow rate*). You can see the interdependence of these measurements in Figure 6.1.

$$\text{flow rate} = \text{velocity} \times \text{area} = \frac{(m)}{(s)} \cdot m^2 = \frac{(m^3)}{(s)}$$

$$\text{quantity} = \text{flow rate} \times \text{time} = \frac{(m^3)}{(s)} \cdot s = m^3$$

If, as shown here, flow rate is recorded for a period of time, the quantity is equal to the area under the curve (shaded area in the figure). This can be established automatically by many instruments, and the process is called *integration*. The integrator of an instrument may carry it out either electrically or mechanically.

6.2.1 Streamlined and Turbulent Flow

Streamlined flow in a liquid is a phenomenon best described by example. Reynolds did a considerable amount of work on this subject, and Figure 6.2 illustrates the principle of streamlined flow (also called *laminar flow*).

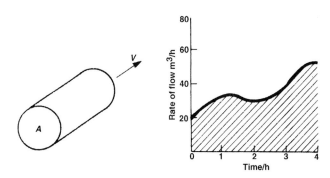

FIGURE 6.1 Flow-time graph.

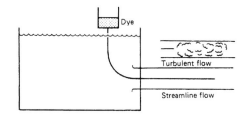

FIGURE 6.2 Reynolds's experiment.

A thin filament of colored liquid is introduced into a quantity of water flowing through a smooth glass tube. The paths of all fluid particles will be parallel to the tube walls, and therefore the colored liquid travels in a straight line, almost as if it were a tube within a tube. However, this state is velocity- and viscosity-dependent, and as velocity is increased, a point is reached (critical velocity) when the colored liquid will appear to disperse and mix with the carrier liquid. At this point the motion of the particles of fluid is not all parallel to the tube walls but also has a transverse velocity. This form of flow pattern is called *turbulent flow*.

Summarizing, therefore, for velocities below the critical velocity, flow is said to be streamlined or laminar, and for velocities above the critical value, flow is said to be turbulent, a situation that is most common in practice.

Reynolds formulated his data in a dimensionless form:

$$Re = \frac{(D \cdot v \cdot \rho)}{(\mu)} \qquad (6.1)$$

where Re is the Reynolds number, D is the diameter of the throat of the installation, v is velocity, ρ is density of fluid, and μ is absolute viscosity. Flow of fluid in pipes is expected to be laminar if the Reynolds number is less than 2000 and turbulent if it is greater than 4000. Between these values is the critical zone. If systems have the same Reynolds number and are geometrically similar, they are said to have *dynamic similarity*.

6.2.1.1 Flow Profile

The velocity across the diameter of a pipe varies due to many influence quantities. The distribution is termed the *velocity profile* of the system. For laminar flow the profile is parabolic in nature. The velocity at the center of the pipe is approximately twice the mean velocity. For turbulent flow, after a sufficient straight pipe run, the flow profile becomes fully developed.

The concept of "fully developed flow" is critical to good flow measurement system design. In a fully developed flow, the velocity at the center of the pipe is only about 1.2 times the mean velocity. This is the preferred flow measurement situation. It permits the most accurate, most repeatable, and most linear measurement of flow.

6.2.1.2 Energy of a Fluid in Motion

Let's look at the forms in which energy is represented in a fluid in motion. This will help us understand the use of the Reynolds number in universal flow formulas. The basic types of energy associated with a moving fluid are:

- Potential energy or potential head
- Kinetic energy
- Pressure energy
- Heat energy

6.2.1.3 Potential Energy

The fluid has potential energy by virtue of its position or height above some fixed level. For example, 1 m³ of liquid of density ρ_1 kg/m³ will have a mass of ρ_1 kg and would require a force of 9.81 ρ_1 N to support it at a point where the gravitational constant g is 9.81 m/s. Therefore, if it is at a height of z meters above a reference plane, it would have 9.81 $\rho_1 z$ joules of energy by virtue of its height.

6.2.1.4 Kinetic Energy

A fluid has kinetic energy by virtue of its motion. Therefore, 1 m³ of fluid of density ρ_1 kg/m³ with a velocity V_1 m/s would have a kinetic energy of $(1)/(2)\rho_1 V_1^2$ joules.

6.2.1.5 Pressure Energy

A fluid has pressure energy by virtue of its pressure. For example, a fluid having a volume v_1 m³ and a pressure of ρ_1 N/m² would have a pressure energy of $\rho_1 v_1$ joules.

6.2.1.6 Internal Energy

The fluid will also have energy by virtue of its temperature (i.e., heat energy). If there is resistance to flow in the form of friction, other forms of internal energy will be converted into heat energy.

6.2.1.7 Total Energy

The total energy E of a fluid is given by the equation:

total energy (E) = potential energy
+ kinetic energy
+ pressure energy
+ internal energy

$$E = \text{P.E.} + \text{K.E.} + \text{PR.E.} + \text{I.E.} \qquad (6.2)$$

6.2.2 Viscosity

Viscosity is the frictional resistance that exists in a flowing fluid. It is discussed in more detail in the next chapter. Briefly, the particles of fluid actually in contact with the walls of the channel are at rest while those at the center of the channel move at maximum velocity. Thus the layers of fluid near the center, which are moving at maximum velocity, will be slowed by the slower-moving layers, and the slower-moving layers will be sped up by the faster-moving layers.

Dynamic viscosity of a fluid is expressed in units of Ns/m². Thus a fluid has a dynamic viscosity of 1 Ns/m² if a force a 1 N is required to move a plane of 1 m² in area at a speed of 1 m/s parallel to a fixed plane, the moving plane

CHAPTER | 6 Measurement of Flow

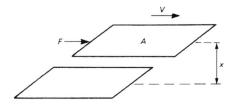

FIGURE 6.3 Determination of dynamic viscosity.

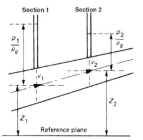

FIGURE 6.4 Hydraulic conditions for pipe flow.

being 1 m away from the fixed plane and the space between the planes being completely filled with the fluid. This is illustrated diagrammatically in Figure 6.3.

Thus for parallel flow lines:

$$\text{dynamic viscosity } \mu = \frac{\text{force}(F)}{\text{area}(A) \times \text{velocity}(v)} \quad (6.3)$$

or, if a velocity gradient exists,

$$\mu = \frac{F}{A \, dv/dx} \quad (6.4)$$

Kinematic viscosity is the ratio of the dynamic viscosity of a fluid to its density at the same temperature.

$$\text{kinematic viscosity at } T°C$$

$$= \frac{\text{dynamic viscosity at } T°C}{\text{density at } T°C} \quad (6.5)$$

For liquids, the viscosity decreases with increase of temperature at constant pressure, whereas for gases, viscosity will increase with increasing temperature at a constant pressure.

It is viscosity that is responsible for the damping out or suppression of flow disturbances caused by bends and valves in a pipe; the energy that existed in the swirling liquid is changed into heat energy. This is the reason manufacturers of flow instruments require stated distances ahead and behind the installation point of a flowmeter. What they are trying to achieve is to allow fluid viscosity to have time to work to suppress flow disturbances and permit accurate and repeatable readings.

6.2.3 Bernoulli's Theorem

All fluid flow formulas in a closed pipe are based on *Bernoulli's theorem*. This states that in a steady flow, without friction, the sum of potential energy, kinetic energy, and pressure energy is a constant along any streamline. If we have a closed pipe or channel (Figure 6.4) in which there are two sections due to the placement of a restriction, orifice, or hydraulic gradient, there is a pressure or head loss in the transition from the first section to the second. If 1 kg of fluid enters the pipe at the first section, 1 kg of fluid must leave at the second.

The energy of the fluid at Section 1

= potential energy + kinetic energy

+ pressure energy + internal energy

$$= 1 \cdot Z_1 \cdot g + \frac{1}{2} \cdot 1 \cdot V_1^2 + p_1 \cdot v_1 + I_1 \quad (6.6)$$

The energy of the fluid at Section 2

$$= 1 \cdot Z_2 \cdot g + \frac{1}{2} \cdot 1 \cdot V_2^2 + p_2 \cdot v_2 + I_2 \quad (6.7)$$

and since energy cannot leave the channel nor be created or destroyed,

Total energy at Section 1 = Total energy at Section 2

$$Z_1 \cdot g + \frac{V_1^2}{1} + p_1 \cdot v_1 + I_1$$
$$= Z_2 \cdot g + \frac{V_2^2}{2} + p_2 \cdot v_2 + I_2 \quad (6.8)$$

Now, if the temperature of the fluid remains the same, the internal energy remains the same and

$$I_1 = I_2 \quad (6.9)$$

and Equation (6.8) reduces to

$$Z_1 \cdot g + \frac{V_1^2}{1} + p_1 \cdot v_1 = Z_2 \cdot g + \frac{V_2^2}{2} + p_2 \cdot v_2 \quad (6.10)$$

This equation applies to liquids and ideal gases.

Now consider liquids only. These can be regarded as being incompressible and their density and specific volume will remain constant along the channel and

$$v_1 = v_2 = \frac{1}{\rho_1} = \frac{1}{\rho_2} + \frac{1}{\rho} \quad (6.11)$$

and Equation (6.10) may be rewritten as

$$Z_1 \cdot g + \frac{V_1^2}{1} + \frac{p_1}{\rho} = Z_2 \cdot g + \frac{V_2^2}{2} + \frac{p_2}{\rho} \quad (6.12)$$

Dividing by g, this becomes

$$Z_1 + \frac{V_1^2}{2g} + \frac{p_1}{\rho \cdot g} = Z_2 + \frac{V_2^2}{2g} + \frac{p_2}{\rho \cdot g} \quad (6.13)$$

Referring back to Figure 6.4, it is obvious that there is a height differential between the upstream and downstream vertical connections representing Sections 1 and 2 of the fluid. Considering first the conditions at the upstream tapping, the fluid will rise in the tube to a height $p_1/\rho \cdot g$ above the tapping or $p_1/\rho \cdot g + Z_1$ above the horizontal level taken as the reference plane. Similarly, the fluid will rise to a height $p_2/\rho \cdot g$ or $p_2/\rho \cdot g + Z_2$ in the vertical tube at the downstream tapping.

The differential head will be given by

$$h = \left(\frac{p_1}{\rho \cdot g} + Z_1\right) - \left(\frac{p_2}{\rho \cdot g} + Z_2\right) \quad (6.14)$$

but from Equation (6.13) we have

$$\left(\frac{p_1}{\rho \cdot g} + Z_1\right) + \frac{V_1^2}{2g} = \left(\frac{p_2}{\rho \cdot g} + Z_2\right) + \frac{V_2^2}{2g}$$

or

$$\left(\frac{p_1}{\rho \cdot g} + Z_1\right) - \left(\frac{p_2}{\rho \cdot g} + Z_2\right) = \frac{V_2^2}{2g} - \frac{V_1^2}{2g}$$

Therefore

$$h = \frac{V_2^2}{2g} - \frac{V_1^2}{2g} \quad (6.15)$$

and

$$V_2^2 - V_1^2 = 2gh \quad (6.16)$$

Now the volume of liquid flowing along the channel per second will be given by Q m³ where

$$Q = A_1 \cdot V_1 = A_2 \cdot V_2$$

or

$$V_1 = \frac{A_2 \cdot V_2}{A_1}$$

Now, substituting this value in Equation (6.16):

$$V_2^2 - V_2^2 \frac{A_2^2}{A_1^2} = 2gh$$

or

$$V_2^2 \left(1 - \frac{A_2^2}{A_1^2}\right) = 2gh \quad (6.17)$$

and dividing by $\left(1 - A_2^2/A_1^2\right)$, Equation (6.17) becomes

$$V_2^2 = \frac{2gh}{1 - A_2^2/A_1^2} \quad (6.18)$$

and taking the square root of both sides

$$V_2 = \frac{\sqrt{2gh}}{\sqrt{\left(1 - A_2^2/A_1^2\right)}} \quad (6.19)$$

Now A_2/A_1 is the ratio (area of Section 2)/(area of Section 1) and is often represented by the symbol m. Therefore

$$\left(1 - \frac{A_2^2}{A_1^2}\right) = 1 - m^2$$

and

$$\frac{1}{\sqrt{\left[1 - \left(A_2^2/A_1^2\right)\right]}} \text{ may be written as } \frac{1}{\sqrt{(1 - m^2)}}$$

This is termed the *velocity of approach factor*, often represented by E. Equation (6.19) may be written

$$V_2 = E\sqrt{2gh} \quad (6.20)$$

and

$$Q = A_2 \cdot V_2 = A_2 \cdot E\sqrt{2gh} \text{ m}^3/\text{s} \quad (6.21)$$

Mass of liquid flowing per second $= W = \rho \cdot Q$

$= A_2 \cdot \rho \cdot E\sqrt{2gh}$ kg also since $\Delta p = h\rho$,

$$Q = A_2 \cdot E\sqrt{\frac{2g\Delta p}{\rho}} \text{ m}^3/\text{s} \quad (6.22)$$

$$W = A_2 \cdot E\sqrt{2g\rho \cdot \Delta p} \text{ kg/s} \quad (6.23)$$

6.2.4 Practical Realization of Equations

The foregoing equations apply only to streamlined (or laminar) flow. To determine actual flow, it is necessary to take into account various other parameters. In practice, flow is rarely streamlined but is turbulent. However, the velocities of particles across the stream will be entirely random and will not affect the rate of flow very much.

In developing the equations, effects of viscosity have also been neglected. In an actual fluid the loss of head between sections will be greater than that which would take place in a fluid free from viscosity.

To correct for these and other effects, another factor is introduced into the equations for flow. This factor is the discharge coefficient C and is given by this equation:

Discharge coefficient:

$$C = \frac{\text{actual mass rate of flow}}{\text{theoretical mass rate of flow}}$$

CHAPTER | 6 Measurement of Flow

or if the conditions of temperature, density, and the like are the same at both sections, it may be written in terms of volume:

$$C = \frac{\text{actual volume flowing}}{\text{theoretical volume flowing}}$$

It is possible to determine C experimentally by actual tests. It is a function of pipe size, type of pressure tappings, and the Reynolds number.

Equation (6.22) is modified and becomes

$$Q = C \cdot A_2 \cdot E \sqrt{\frac{2g \cdot \Delta p}{\rho}} \quad (6.24)$$

This is true for flow systems where the Reynolds number is above a certain value (20,000 or above for orifice plates). For lower Reynolds numbers and for very small or rough pipes, the basic coefficient is multiplied by a correction factor Z whose value depends on the area ratio, the Reynolds number, and the size and roughness of the pipe. Values for both C and Z are listed with other relevant data in BS 1042 Part 1 1964.

We can use differential pressure to measure flow. Here's a practical example:

Internal diameter of upstream pipe: D mm
Orifice or throat diameter: d mm
Pressure differential produced: h mm water gauge
Density of fluid at upstream tapping: ρ kg/m^2
Absolute pressure at upstream tapping: p bar

Then, introducing the discharge coefficient C, the correction factor and the numerical constant, the equation for quantity rate of flow Q m^3/h becomes

$$Q = 0.01252 C \cdot Z \cdot E \cdot d^2 \sqrt{\frac{h}{\rho}} \quad (6.25)$$

and the weight or mass rate of the flow W kg/h is given by

$$W = 0.01252 C \cdot Z \cdot E \cdot d^2 \sqrt{h\rho} \quad (6.26)$$

6.2.5 Modification of Flow Equations to Apply to Gases

Gases are compressible; liquids, mostly, are not. If the gas under consideration can be regarded as an ideal gas (most gases are ideal when well away from their critical temperatures and pressures), the gas obeys several very important gas laws. These laws will now be stated.

6.2.5.1 Dry Gases

(a) *Boyle's law* This law states that the volume of any given mass of gas will be inversely proportional to its absolute pressure, provided temperature remains constant. Thus if a certain mass of gas occupies a volume v_0 at an absolute pressure p_0 and a volume v_1 at an absolute pressure p, then

$$p_0 \cdot v_0 = p \cdot v_1 \quad \text{or} \quad v_1 = v_0 \cdot p_0 p \quad (6.27)$$

(b) *Charles's law* This law states that if the volume of a given mass of gas occupies a volume v_1 at a temperature T_0 Kelvin, its volume v at T Kelvin is given by

$$\frac{v_1}{T_0} = \frac{v}{T} \quad \text{or} \quad v = v_1 \cdot \frac{T}{T_0} \quad (6.28)$$

(c) *The ideal gas law* In the general case, p, v, and T change. Suppose a mass of gas at pressure p_0 and temperature T_0 Kelvin has a volume v_0, and the mass of gas at pressure p and temperature T has a volume v, and that the change from the first set of conditions to the second set of conditions takes place in two stages.

(1) Change the pressure from p_0 to p at a constant temperature. Let the new volume be v_1. From Boyle's law:

$$p_0 \cdot v_0 = p \cdot v_1 \quad \text{or} \quad v_1 = v_0 \cdot \frac{p_0}{p}$$

(2) Change the temperature from T_0 to T at constant pressure. From Charles's law:

$$\frac{v_1}{T_0} = \frac{v}{T}$$

Hence, equating the two values of v_1

$$v_0 \cdot \frac{p_0}{p} = v \cdot \frac{T_0}{T}$$

$$p_0 \cdot \frac{v_0}{T_0} = \frac{pv}{T} = \text{constant} \quad (6.29)$$

If the quantity of gas considered is 1 mole, i.e., the quantity of gas that contains as many molecules as there are atoms in 0.012 kg of carbon-12, this constant is represented by R, the gas constant, and Equation (6.29) becomes:

$$pv = R_0 \cdot T$$

where $R_0 = 8.314$ J/Mol K and p is in N/m^2 and v is in m^3.

(d) *Adiabatic expansion* When a gas is flowing through a primary element, the change in pressure takes place too rapidly for the gas to absorb heat from its surroundings. When it expands owing to the reduction in pressure, it does work, so that if it does not receive energy it must use its own heat energy, and its temperature will fall. Thus the expansion that takes place owing to the fall in pressure does not obey Boyle's law, which applies only to an expansion at constant temperature. Instead it obeys the law for adiabatic expansion of a gas:

$$p_1 \cdot v_1^\gamma = p_2 \cdot v_2^\gamma \quad \text{or} \quad p \cdot v^\gamma = \text{constant} \quad (6.30)$$

where γ is the ratio of the specific heats of the gas:

$$\gamma = \frac{\text{specific heat of a gas at constant pressure}}{\text{specific heat of a gas at constant volume}}$$

and has a value of 1.40 for dry air and other diatomic gases, 1.66 for monatomic gases such as helium, and about 1.33 for triatomic gases such as carbon dioxide.

If a metered fluid is not incompressible, another factor is introduced into the flow equations. This factor is necessary to correct for the change in volume due to the expansion of the fluid while passing through the restriction. This factor is called the *expansibility factor* ε and has a value of unity (1) for incompressible fluids. For ideal compressible fluids expanding without any change of state, the value can be calculated from the equation

$$\varepsilon = \sqrt{\left(\frac{\gamma r^{2/\gamma}}{\gamma-1}\frac{1-m^2}{1-m^2 r^{2/\gamma}}\frac{1-r^{(\gamma-1)/\gamma}}{1-r}\right)}$$

where r is the ratio of the absolute pressures at the upstream and downstream tappings (i.e., $r = p_1/p_2$) and γ is the ratio of the specific heat of the fluid at constant pressure to that at constant volume. This is detailed in BS 1042 Part 1 1964.

To apply working fluid flow equations to both liquids and gases, the factor ε is introduced and the equations become:

$$Q = 0.0125\, CZ\varepsilon E d^2 \sqrt{h/\rho} \text{ m}^3/\text{h} \quad (6.32)$$

$$W = 0.01252\, CZ\varepsilon E d^2 \sqrt{h\rho} \text{ kg/h} \quad (6.33)$$

$$\varepsilon = 1 \text{ for liquids}$$

6.2.5.2 Critical Flow of Compressible Fluids

For flow through a convergent tube such as a nozzle, the value of r at the throat cannot be less than a critical value r_c. When the pressure at the throat is equal to this critical fraction of the upstream pressure, the rate of flow is a maximum and cannot be further increased except by raising the upstream pressure. The critical pressure ratio is given by the equation

$$2r_c^{(1-\gamma)/\gamma} + (\gamma-1)m^2 \cdot r_c^{2/\gamma} = \gamma - 1 \quad (6.34)$$

The value of r is about 0.5, but it increases slightly with increase of m and with decrease of specific heat ratio. Values of r are tabulated in BS 1042 Part 1 1964.

The basic equation for critical flow is obtained by substituting $(1 - r_c)\rho$ for Δp in Equation (6.23), substituting r_c for r in Equation (6.31), and the equation becomes

$$W = 1.252 U \cdot d^2 \sqrt{\rho p} \text{ kg/h} \quad (6.35)$$

where

$$U = C\sqrt{(\gamma/2) r_c \cdot (\gamma-1)/\gamma} \quad (6.36)$$

The volume rate of flow (in m³/h) is obtained by dividing the weight ratio of flow by the density (in kg/m³) of the fluid at the reference conditions.

6.2.5.3 Departure from Gas Laws

At room temperature and at absolute pressures less than 10 bar, most common gases except carbon dioxide behave sufficiently like an ideal gas that the error in flow calculations brought about by departure from the ideal gas laws is less than 1 percent. To correct for departure from the ideal gas laws, a deviation coefficient K (given in BS 1042 Part 1 1964) is used in the calculation of densities of gases where the departure is significant. For ideal gases, $K = 1$.

6.2.5.4 Wet Gases

The preceding modification applies to dry gases. In practice, many gases are wet, since they are a mixture of gas and water vapor. Partial pressure due to saturated water vapor does not obey Boyle's law.

Gas humidity is discussed in Chapter 11. If the temperature and absolute pressure at the upstream tapping and the state of humidity of the gas are known, a correction factor can be worked out and applied to obtain the actual mass of gas flowing.

Gas density is given by the equation

$$\rho = 6.196\left[\delta\frac{(p-p_v)}{k \cdot T} + \frac{0.622 p_v}{T}\right] \text{ kg/m}^3 \quad (6.37)$$

where δ is specific gravity of dry gas relative to air, T is temperature in Kelvin, p is pressure in mbar at the upstream tapping, p_v is partial pressure in mbar of the water vapor, k is the gas law deviation at temperature T, and p is gas density. For dry gas pv is zero and the equation becomes

$$\rho = 6.196 \frac{\delta p}{kT} \text{ kg/m}^3 \quad (6.38)$$

6.3 FLUID FLOW IN CLOSED PIPES

6.3.1 Differential-Pressure Devices

Differential pressure devices using a constriction in the pipeline have been the most common technique for measuring fluid flow. Recently, other devices have made substantial inroads in the basic measurement of fluids. Differential pressure is still a widely used technique, with even some new devices that have been introduced in the recent past. A recent estimate puts the use of differential pressure devices to measure flow in the petrochemical industry at over 70 percent of all flow devices.

As already shown in the derivation of Bernoulli's equation in the previous section, a constriction will cause an increase in fluid velocity in the area of that constriction, which in turn will result in a corresponding pressure drop across the constriction. This differential pressure (DP) is a function of the flow velocity and density of the fluid and is shown to be a square root relationship; see Equation (6.24).

A flowmeter in this category would normally comprise a primary element to develop a differential pressure and a secondary element to measure it. The secondary element is effectively a pressure transducer, and operational techniques are discussed in Chapter 14, so no further coverage will be given here. However, there are various types of primary elements, and these deserve further consideration. The main types of interest are orifice plate, venturi, nozzle, Dall, rotameter, gate meter, Gilflo element, target meter, and V-Cone.

6.3.1.1 Orifice Plate

An orifice plate in its simplest form is a thin steel plate with a circular orifice of known dimensions located centrally in the plate. This is termed a *concentric orifice plate*; see Figure 6.5(a). The plate would normally be clamped between adjacent flange fittings in a pipeline, a vent hole and drain hole being provided to prevent solids building up and gas pockets developing in the system; see Figure 6.5(b).

The differential pressure is measured by suitably located pressure tappings on the pipeline on either side of the orifice plate. These may be located in various positions, depending on the application (e.g., corner, D and $D/2$, or flange tappings), and reference should be made to BS 1042 Part 1 1964 for correct application. Flow rate is determined from Equation (6.24).

This type of orifice plate is inadequate to cope with difficult conditions experienced in metering dirty or viscous fluids and gives a poor disposal rate of condensate in flowing steam and vapors. Several design modifications can overcome these problems, in the form of segmental or eccentric orifice plates, as shown in Figure 6.5(a).

The segmental orifice provides a method for measuring the flow of liquids with solids in suspension. It takes the form of a plate that covers the upper cross-section of the pipe, leaving the lower portion open for the passage of solids to prevent their buildup.

The eccentric orifice is used on installations where condensed liquids are present in gas-flow measurement or where undissolved gases are present in the measurement of liquid flow. It is also useful where pipeline drainage is required.

To sum up the orifice plate:

Advantages:

- Inherently simple in operation
- No moving parts
- Long-term reliability
- Inexpensive

Disadvantages:

- Square-root relationship
- Poor turndown ratio
- Critical installation requirements
- High irrecoverable pressure loss

6.3.1.2 Venturi Tube

The classical venturi tube is shown in Figure 6.6. It comprises a cylindrical inlet section followed by a convergent entrance into a cylindrical throat and a divergent outlet section. A complete specification may be found by reference to BS 1042 Part 1 1964; relevant details are repeated here:

A. Diameter of throat. The diameter d of the throat shall be not less than $0.224D$ and not greater than $0.742D$, where D is the entrance diameter.

B. Length of throat. The throat shall have a length of $1.0d$.

C. Cylindrical entrance section. This section shall have an internal diameter D and a length of not less than $1.0d$.

D. Conical section. This shall have a taper of $10½°$. Its length is therefore $2.70(D - d)$ within $\pm 0.24(D - d)$.

E. Divergent outlet section. The outlet section shall have an inclined angle of not less than $5°$ and not greater than $15°$. Its length shall be such that the exit diameter is not less than $1.5d$.

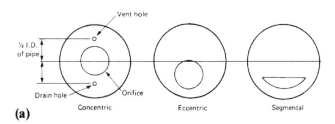

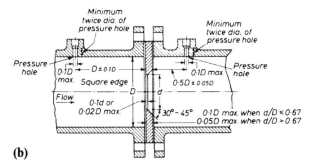

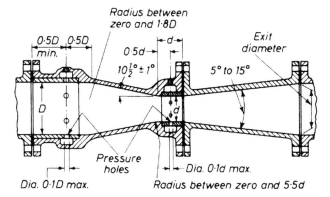

FIGURE 6.5 (a) Orifice plate types. (b) Concentric orifice plate with D and $D/2$ tappings mounted between flange plates. Courtesy of British Standards Institution.

FIGURE 6.6 Venturi tube. Courtesy of British Standards Institution.

In operation the fluid passes through the convergent entrance, increasing velocity as it does so, resulting in a differential pressure between the inlet and throat. This differential pressure is monitored in the same way as for the orifice plate, the relationship between flow rate and differential being as defined in Equation (6.24).

Location of Pressure Tappings The upstream pressure tapping is located in the cylindrical entrance section of the tube $0.5D$ upstream of the convergent section and the downstream pressure tapping is located in the throat at a distance $0.5D$ downstream of the convergent section. Pressure tappings should be sized so as to avoid accidental blockage.

Generally the tappings are not in the form of a single hole but several equally spaced holes connected together in the form of an annular ring, sometimes called a *piezometer ring*. This has the advantage of giving a true mean value of pressure at the measuring section.

Application The venturi is used for applications in which there is a high solids content or high pressure recovery is desirable. The venturi is inherently a low head-loss device and can result in an appreciable saving of energy.

To sum up the venturi tube:

Advantages:

- Simple in operation
- Low head loss
- Tolerance of high solids content
- Long-term reliability
- No moving parts

Disadvantages:

- Expensive
- Square-root pressure-velocity relationship
- Poor turndown ratio
- Critical installation requirements

6.3.1.3 Nozzles

The other most common use of the venturi effect is the venturi nozzle.

Venturi Nozzle This is in effect a shortened venturi tube. The entrance cone is much shorter and has a curved profile. The inlet pressure tap is located at the mouth of the inlet cone and the low-pressure tap in the plane of minimum section, as shown in Figure 6.7. This reduction in size is taken a stage further in the flow nozzle.

Flow Nozzle Overall length is again greatly reduced. The entrance cone is bell-shaped and there is no exit cone. This is illustrated in Figure 6.8. The flow nozzle is not suitable for viscous liquids, but for other applications it is considerably cheaper than the standard venturi tube. Also, due to the smooth entrance cone, there is less resistance to fluid flow

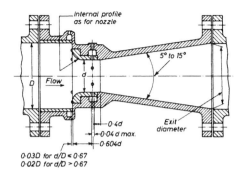

FIGURE 6.7 Venturi nozzle. Courtesy of British Standards Institution.

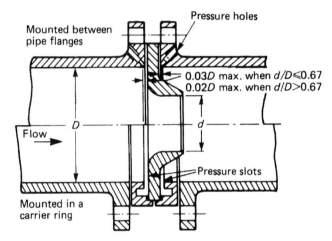

FIGURE 6.8 Flow nozzle. Courtesy of British Standards Institution.

through the nozzle, and a lower value of m may be used for a given rate of flow. Its main area of use therefore is in high-velocity mains where it will produce a substantially smaller pressure drop than an orifice plate of similar m number.

6.3.1.4 Dall Tube

A Dall tube is another variation of the venturi tube and gives a higher differential pressure but a lower head loss than the conventional venturi tube. Figure 6.9 shows a cross-section of a typical Dall flow tube. It consists of a short, straight inlet section, a convergent entrance section, a narrow throat annulus, and a short divergent recovery cone. The whole device is about 2 pipe-diameters long.

A shortened version of the Dall tube, the Dall orifice or insert, is also available; it is only 0.3 pipe-diameter long. All the essential Dall tube features are retained in a truncated format, as shown in Figure 6.10. Venturi tubes, venturi nozzles, Dall tubes, and other modifications of the venturi effect are rarely used outside the municipal wastewater and mining industries. There is even a version of a venturi tube combined with a venturi flume called a DataGator that is useful for any pipe, full or not. In this device, the inlet fills up simultaneously with the throat, permitting measurement in subcritical flow as though the device were a venturi flume and above critical flow as though the device were a venturi

CHAPTER | 6 Measurement of Flow

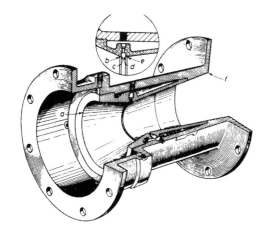

FIGURE 6.9 Dall tube. Courtesy of ABB.

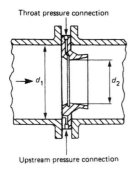

FIGURE 6.10 Dall insert. Courtesy of British Standards Institution.

tube. In the "transition zone" between sub- and supercritical flow, the design of the unit permits a reasonably accurate measurement. This design won an R&D100 Award in 1993 as one of the 100 most important engineering innovations of the year.

Pressure Loss All the differential pressure devices discussed so far cause an irrecoverable pressure loss of varying degrees. In operation it is advantageous to keep this loss as low as possible, which will often be a major factor in the selection criteria of a primary element. The pressure loss curves for nozzles, orifices, and venturi tubes are given in Figure 6.11.

Installation Requirements As already indicated, installation requirements for differential-pressure devices are quite critical. It is advisable to install primary elements as far downstream as possible from flow disturbances, such as bends, valves, and reducers. These requirements are tabulated in considerable detail in BS 1042 Part 1 1964 and are reproduced in part in Appendix 6.1. It is critical for the instrument engineer to be aware that these requirements are rules of thumb, and even slavish adherence to them may not produce measurement free from hydraulics-induced error. From a practical point of view, the best measurement is the one with the longest upstream straight run and the longest downstream straight run.

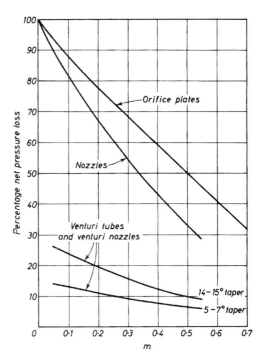

FIGURE 6.11 Net pressure loss as a percentage of pressure difference. Courtesy of British Standards Institution.

6.3.1.5 Variable-Orifice Meters

So far the devices discussed have relied on a constriction in the flowstream causing a differential pressure varying with flow rate. Another category of differential-pressure device relies on maintaining a nominally constant differential pressure by allowing effective area to increase with flow. The principal devices to be considered are the rotameter, gate meter, and Gilflo.

Rotameter This device is shown schematically in Figure 6.12(a). In a tapered tube the upward stream of fluid supports the float where the force on its mass due to gravity is balanced against the flow force determined by the annular area between the float and the tube and the velocity of the stream. The float's position in the tube is measured by a graduated scale, and its position is taken as an indication of flow rate.

Many refinements are possible, including the use of magnetic coupling between the float and external devices to translate vertical movement into horizontal and develop either electrical transmission or alarm actuation. Tube materials can be either metal or glass, depending on application. Figure 6.12(b) shows an exploded view of a typical rotameter.

Gate Meter In this type of meter the area of the orifice may be varied by lowering a gate either manually or by an automatically controlled electric motor. The gate is moved so as to maintain a constant pressure drop across the orifice. The pressure drop is measured by pressure tappings located upstream and downstream of the gate, as shown

Appendix 6.1 Minimum lengths of straight pipeline upstream of device*

	Minimum number of pipe diameters for Cases A to F listed below								(b) Minimum length between first upstream fitting and next upstream fitting
	(a) Minimum length of straight pipe immediately upstream of device								
Diameter ratio d/D less than:	0.22	0.32	0.45	0.55	0.63	0.7	0.77	0.84	
Area ratio m less than:	0.05†	0.1	0.2	0.3	0.4	0.5	0.6	0.7	
Fittings producing symmetrical disturbances									
Case A. Reducer (reducing not more than $0.5D$ over a length of $3D$); enlarger (enlarging not more than $2D$ over a length of $1.5D$)	16	16	18	20	23	26	29	33	13
Any pressure difference device having an area ratio m not less than 0.3									
Case B. Gate valve fully open (for ¾ closed, see Case H)	12	12	12	13	16	20	27	38	10
Case C. Globe valve fully open (for ¾ closed, see Case J)	18	18	20	23	27	32	40	49	16
Case D. Reducer (any reduction, including from a large space)	25	25	25	25	25	26	29	33	13
Fittings producing asymmetrical disturbances in one plane									
Case E. Single bend up to 90°, elbow, Y-junction, T-junction (flow in either but not both branches)	10	10	13	16	22	29	41	56	15
Case F. Two or more bends in the same plane, single bend of more than 90°, swan	14	15	18	22	28	36	46	57	18
Fittings producing asymmetrical disturbances and swirling motion									
Case G‡. Two or more bends, elbows, loops, or Y-junctions in different planes, T-junction with flow in both branches	34	35	38	44	52	63	76	89	32
Case H‡. Gate valve up to ¾ closed§ (for fully open, see Case B)	40	40	40	41	46	52	60	70	26
Case J‡. Globe valve up to ¾ closed§ (for fully open, see Case C)	12	14	19	26	36	60	80	100	30
Other fittings									
Case K. All other findings (provided there is no swirling motion)	100	100	100	100	100	100	100	100	50

*See Subclauses 47b and 47c.
†For area ratio less than 0.015 or diameter ratios less than 0.125, see Subclause 47b.
‡If swirling motion is eliminated by a flow straightener (Appendix F) installed downstream of these fittings, they may be treated as Class F, B, and C, respectively.
§The valve is regarded as three quarters closed when the area of the opening is one quarter of that when fully open.
nb: Extracts from British Standards are reproduced by permission of the British Standards Institution, 2 Park Street, London, W1A 2BS, from which complete copies can be obtained.

CHAPTER | 6 Measurement of Flow

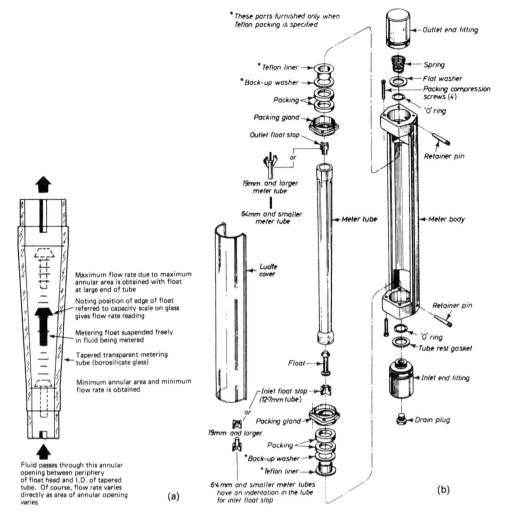

FIGURE 6.12 (a) Rotameter principle of operation. Courtesy of ABB Instrument Group. (b) Rotameter-exploded view. Courtesy of ABB Instrument Group.

in Figure 6.13(a). The position of the gate is indicated by a scale. As the rate of flow through the orifice increases, the area of the orifice is increased. If all other factors in Equation (6.21) except area A_2 are kept constant, the flow through the orifice will depend on the product $A_2 \cdot E$, or $A_2/\sqrt{[1 - (A_2/A_1)^2]}$. As A_2 increases, $(A_2/A_1)^2$ increases and $[1 - (A_2/A_1)^2]$ decreases; therefore $1_2/\sqrt{[1 - (A_2/A_1)^2]}$ increases.

The relationship between A_2 and flow is not linear. If the vertical movement of the gate is to be directly proportional to the rate of flow, the width of the opening A_2 must decrease toward the top, as shown in Figure 6.13(a).

The flow through the meter can be made to depend directly on the area of the orifice A_2 if, instead of the normal static pressure being measured at the upstream tapping, the impact pressure is measured. To do this the upstream tap is made in the form of a tube with its open end facing directly into the flow, as shown in Figure 6.13(b). It is in effect a pitot tube (see the section on point-velocity measurement).

The differential pressure is given by Equation (6.15), where h is the amount the pressure at the upstream tap is greater than that at the downstream tap:

$$h = \frac{V_2^2}{2g} - \frac{V_1^2}{2g} \quad (6.39)$$

Now, at the impact port, $V_2 = 0$; therefore

$$h_1 = \frac{V_1}{2g}$$

where h_1 is the amount the impact pressure is greater than the normal upstream static pressure. Thus the difference between impact pressure and the pressure measured at the downstream tap will be h_2 where

$$h_2 = h + h_1$$

$$= \frac{V_2^2}{2g} - \frac{V_1^2}{2g} + \frac{V_1^2}{2g} = \frac{V_2^2}{2g} \quad (6.40)$$

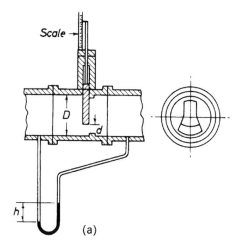

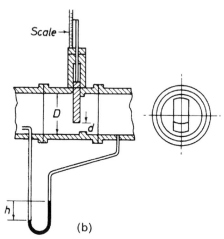

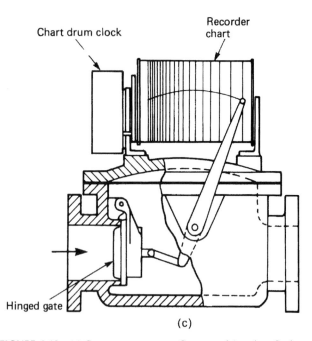

FIGURE 6.13 (a) Gate-type area meter. Courtesy of American Society of Mechanical Engineers. (b) Gate-type area meter corrected for velocity of approach. Courtesy of American Society of Mechanical Engineers. (c) Weight-controlled hinged-gate meter.

Therefore, the velocity V_2 through the section A_2 is given by $V_2 = \sqrt{(2g \cdot h_2)}$. The normal flow equations for the type of installation shown in Figure 6.13(b) will be the same for other orifices, but the velocity of approach factor is 1 and flow is directly proportional to A_2. The opening of the gate may therefore be made rectangular, and the vertical movement will be directly proportional to flow.

The hinged gate meter is another version of this type of device. Here a weighted gate is placed in the flowstream, its deflection being proportional to flow. A mechanical linkage between the gate and a recorder head provides flow indication. It is primarily used for applications in water mains where the user is interested in step changes rather than absolute flow accuracy. The essential features of this device are shown in Figure 6.13(c).

The "Gilflo" Primary Sensor The Gilflo metering principle was developed to overcome the limitations of the square law fixed orifice plate in the mid-1960s. Its construction is in two forms: A and B. The Gilflo A, Figure 6.14(a), sizes 10 to 40 mm, has an orifice mounted to a strong linear bellows fixed at one end and with a shaped cone positioned concentrically in it. Under flow conditions, the orifice moves axially along the cone creating a variable annulus across which the differential pressure varies. Such is the relationship of change that the differential pressure is directly proportional to flow rate, enabling a rangeability of up to 100:1.

The Gilflo B, Figure 6.14(b), sizes 40 to 300 mm standard, has a fixed orifice with a shaped cone moving axially against the resistance of a spring, again producing a linear differential pressure and a range of up to 100:1.

The Gilflo A has a water equivalent range of 0–5 to 0–350 liters/minute; the Gilflo B range is 0–100 to 0–17500 liters/minute.

The main application for Gilflo-based systems is on saturated and superheated steam, with pressures up to 200 bar and temperatures up to ±500°C.

6.3.1.6 Target Flowmeter

Although not strictly a differential-pressure device, this is generally categorized under that general heading. The primary and secondary elements form an integral unit, and differential pressure tappings are not required. It is particularly suited for measuring the flow of high-viscosity liquids: hot asphalt, tars, oils, and slurries at pressures up to 100 bar and Reynolds numbers as low as 2000. Figure 6.15 shows the meter and working principles.

The liquid impinging on the target will be brought to rest so that pressure increases by $V^2/2g$ in terms of head of liquid so that the force F on the target will be

$$F = \frac{K\gamma' V_1^2 A_t}{2N} \qquad (6.41)$$

CHAPTER | 6 Measurement of Flow

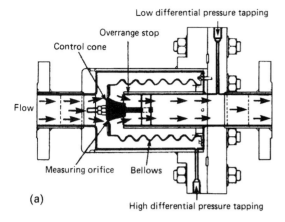

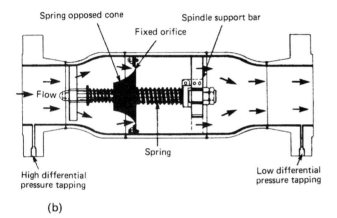

FIGURE 6.14 (a) The essentials of Gilflo A. As flow increases the measuring orifice moves along the control cone against the spring bellows. Courtesy of Gervase Instruments Ltd. (b) Gilflo B extends the principle to higher flow. Now the orifice is fixed and the control cone moves against the spring. Courtesy of Gervase Instruments Ltd.

FIGURE 6.15 A Target flowmeter with an electronic transmitter. Courtesy of the Venture Measurement Division of Alliant Inc.

where γ is the mass per unit volume in kg/m³. The area of the target is A_t measured in m³, K is a constant, and V_1 is the velocity in m/s of the liquid through the annular ring between target and pipe.

If the pipe diameter is D m and the target diameter d m, then area A of the annular space equals $\pi(D^2 - d^2)/4$ m².

Therefore, volume flow rate is:

$$Q = A \cdot V_1 = \frac{(D^2 - d^2)}{4}\sqrt{\frac{8F}{K\gamma'\pi d^2}}$$

$$= \frac{C(D^2 - d^2)}{d^2}\sqrt{\frac{F}{\gamma'}} \text{ m}^3/\text{s} \quad (6.42)$$

where C is a new constant including the numerical factors. Mass flow rate is:

$$W = Q\gamma' = \frac{C(D^2 - d^2)}{d}\sqrt{F\gamma'} \text{ kg/s} \quad (6.43)$$

The force F is balanced through the force bar and measured by a balanced strain gauge bridge whose output signal is proportional to the square root of flow.

Available flow ranges vary from 0–52.7 to 0–123 liters/minute for the 19 mm size at temperatures up to 400°C to a range of 0–682 to 0–2273 liters/minute for the 100 mm size at temperatures up to 260°F. Meters are also available for gas flow.

The overall accuracy of the meter is ±0.5 percent, with repeatability of ±0.1 percent.

Target flowmeters are in use in applications as diverse as supersaturated two-phase steam and municipal water distribution. Wet chlorine gas and liquefied chlorine gas are also applications for this type of device. The shape of the target, which produces the repeatability of the device, is empirical and highly proprietary among manufacturers.

6.3.2 Rotating Mechanical Meters for Liquids

Rotating mechanical flowmeters derive a signal from a moving rotor that is rotated at a speed proportional to the fluid flow velocity. Most of these meters are velocity-measuring devices except for positive displacement meters, which are quantity or volumetric in operation. The principal types are positive displacement, rotating vane, angled propeller, bypass, helix, and turbine meters.

6.3.2.1 Positive Displacement Meters

Positive displacement meters are widely used on applications where high accuracy and good repeatability are required. Accuracy is not affected by pulsating flow, and accurate measurement is possible at higher liquid viscosities than with many other flowmeters. Positive displacement meters are frequently used in oil and water undertakings for accounting purposes.

The principle of the measurement is that as the liquid flows through the meter, it moves a measuring element that seals off the measuring chamber into a series of measuring compartments, which are successively filled and emptied. Thus for each complete cycle of the measuring element a fixed quantity of liquid is permitted to pass from the inlet to the outlet of the meter. The seal between the measuring element and the measuring chamber is provided by a film of the measured liquid. The number of cycles of the measuring element is indicated by several possible means, including a pointer moving over a dial driven from the measuring element by suitable gearing and a magnetically coupled sensor connected to an electronic indicator or "flow computer."

The *extent of error*, defined as the difference between the indicated quantity and the true quantity and expressed as a percentage of the true quantity, is dependent on many factors, among them being:

A. The amount of clearance between the rotor and the measuring chamber through which liquid can pass unmetered.
B. The amount of torque required to drive the register. The greater the torque, the greater the pressure drop across the measuring element, which in turn determines the leakage rate past the rotor. This is one reason that electronic readout devices have become much more common in recent years, since they eliminate this error factor.
C. The viscosity of the liquid to be measured. Increase in viscosity will also result in increased pressure drop across the measuring element, but this is compensated for by the reduction in flow through the rotor clearances for a given pressure drop.

The accuracy of measurement attained with a positive displacement meter varies very considerably from one design to another, with the nature and condition of the liquid measured, and with the rate of flow. Great care should be taken to choose the correct meter for an application.

The most common forms of positive displacement meters are rotary piston, reciprocating piston, nutating disc, fluted spiral rotor, sliding vane, rotating vane, and oval gear.

Rotary Piston The rotary-piston flowmeter is most common in the water industry, where it is used for metering domestic supplies. It consists of a cylindrical working chamber that houses a hollow cylindrical piston of equal length. The central hub of the piston is guided in a circular motion by two short inner cylinders. The piston and cylinder are alternately filled and emptied by the fluid passing through the meter. A slot in the sidewall of the piston is removed so that a partition extending inward from the bore of the working chamber can be inserted. This has the effect of restricting the movement of the piston to a sliding motion along the partition. The rotary movement of the piston is transmitted via a permanent-magnet coupling from the drive shaft to a

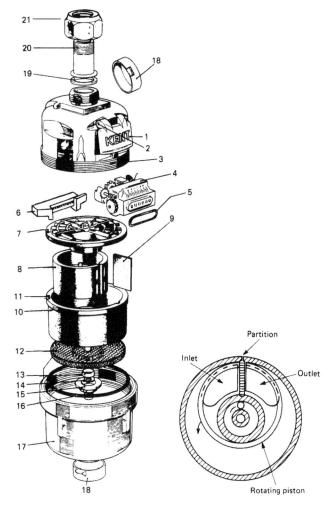

FIGURE 6.16 Rotary-piston positive displacement meter. Courtesy of ABB Instrument Group. 1. Lid. 2. Hinge pin. 3. Counter housing complete with lid and hinge pin. 4. Counter with worm reduction gear and washer. 5. Counter washer. 6. Ramp assembly. 7. Top plate assembly comprising top plate only; driving spindle; driving dog; dog retaining clip. 8. Piston. 9. Shutter. 10. Working chamber only. 11. Locating pin. 12. Strainer-plastic. Strainer-copper. 13. Strainer cap. 14. Circlip. 15. Nonreturn valve. 16. O ring. 17. Chamber housing. 18. Protective caps for end threads.

mechanical register or electronic readout device. The basic design and principle of operation of this meter is shown diagrammatically in Figure 6.16.

Reciprocating Piston A reciprocating meter can be either of single- or multi-piston type, this being dependent on the application. This type of meter exhibits a wide turndown ratio (e.g., 300:1), with extreme accuracy of ±0.1 percent, and can be used for a wide range of liquids. Figure 6.17 illustrates the operating principle of this type of meter.

Suppose the piston is at the bottom of its stroke. The valve is so arranged that inlet liquid is admitted below the piston, causing it to travel upward and the liquid above the piston to be discharged to the outlet pipe. When the piston has reached the limit of its travel, the top of the cylinder is cut off from

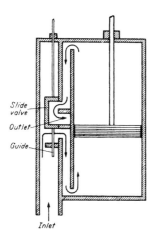

FIGURE 6.17 Reciprocating-piston meter.

the outlet side and opened to the inlet liquid supply. At the same time the bottom of the cylinder is opened to the outlet side but cut off from the inlet liquid. The pressure of the incoming liquid will therefore drive the piston downward, discharging the liquid from below the piston to the outlet pipe. The process repeats.

As the piston reciprocates, a ratchet attached to the piston rod provides an actuating force for an incremental counter, each count representing a predetermined quantity of liquid. Newer devices use magnetically coupled sensors—Hall-effect or Wiegand-effect types being quite common—or optical encoders to produce the count rate.

Nutating-Disc Type This type of meter is similar in principle to the rotary-piston type. In this case, however, the gear train is driven not by a rotating piston but by a movable disc mounted on a concentric sphere. The basic construction is shown in Figure 6.18.

The liquid enters the left side of the meter, alternately above and below the disc, forcing it to rock (*nutate*) in a circular path without rotating about its own axis. The disc is contained in a spherical working chamber and is restricted from rotating about its own axis by a radial partition that extends vertically across the chamber. The disc is slotted to fit over this partition. The spindle protruding from the sphere traces a circular path and is used to drive a geared register.

This type of meter can be used for a wide variety of liquids—disc and body materials being chosen to suit.

Fluted-Spiral-Rotor Type (Rotating-Impeller Type) The principle of this type of meter is shown in Figure 6.19. The meter consists of two fluted rotors supported in sleeve-type bearings and mounted so as to rotate rather like gears in a liquid-tight case. The clearance between the rotors and measuring chambers is kept to a minimum. The shape of the rotors is designed so that a uniform uninterrupted rotation is produced by the liquid. The impellers in turn rotate the index of a counter, which shows the total measured quantity.

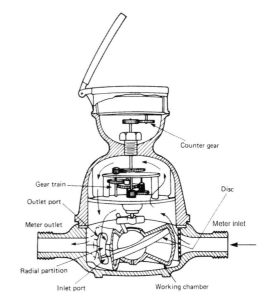

FIGURE 6.18 Nutating-disc meter.

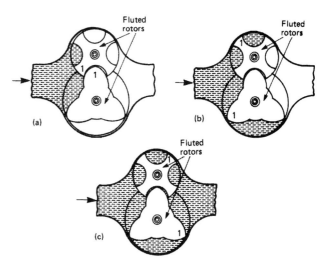

FIGURE 6.19 Fluted-spiral-rotor type of meter.

This type of meter is used mainly for measuring crude and refined petroleum products covering a range of flows up to 3000 m³/h at pressures up to 80 bar.

Sliding-Vane Type The principle of this type is illustrated in Figure 6.20. It consists of an accurately machined body containing a rotor revolving on ball bearings. The rotor has four evenly spaced slots, forming guides for four vanes. The vanes are in contact with a fixed cam. The four cam-followers follow the contour of the cam, causing the vanes to move radially. This ensures that during transition through the measuring chamber the vanes are in contact with the chamber wall.

The liquid impact on the blades causes the rotor to revolve, allowing a quantity of liquid to be discharged. The number of revolutions of the rotor is a measure of the volume of liquid passed through the meter.

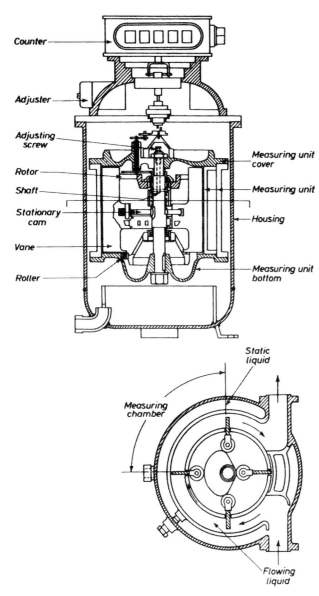

FIGURE 6.20 Sliding-vane type meter. Courtesy of Wayne Tank & Pump Co.

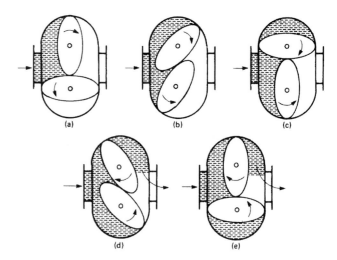

FIGURE 6.21 Oval-gear meter.

Rotating-Vane Type This meter is similar in principle to the sliding-vane meter, but the measuring chambers are formed by four half-moon-shaped vanes spaced equidistant on the rotor circumference. As the rotor is revolved, the vanes turn to form sealed chambers between the rotor and the meter body. Accuracy of ±0.1 percent is possible down to 20 percent of the rated capacity of the meter.

Oval-Gear Type This type of meter consists of two intermeshing oval gearwheels that are rotated by the fluid passing through it. This means that for each revolution of the pair of wheels, a specific quantity of liquid is carried through the meter. This is shown diagrammatically in Figure 6.21. The number of revolutions is a precise measurement of the quantity of liquid passed. A spindle extended from one of the gears can be used to determine the number of revolutions and convert them to engineering units by suitable gearing.

Oval-gear meters are available in a wide range of materials, in sizes from 10 to 400 mm and suitable for pressures up to 60 bar and flows up to 1200 m^3/h. Accuracy of ±0.25 percent of rate of flow can be achieved.

6.3.2.2 Rotating Vane Meters

The *rotating vane* type of meter operates on the principle that the incoming liquid is directed to impinge tangentially on the periphery of a free-spinning rotor. The rotation is monitored by magnetic or photoelectric pickup, the frequency of the output being proportional to flow rate, or alternatively by a mechanical register connected through gearing to the rotor assembly, as shown in Figure 6.22.

Accuracy is dependent on calibration, and turndown ratios up to 20:1 can be achieved. This device is particularly suited to low flow rates.

6.3.2.3 Angled-Propeller Meters

The *propeller* flowmeter comprises a Y-type body, with all components apart from the propeller being out of the liquid stream. The construction of this type of meter is shown in Figure 6.23. The propeller has three blades and is designed to give maximum clearance in the measuring chamber, thereby allowing maximum tolerance of suspended particles. The propeller body is angled at 45° to the main flowstream, and liquid passing through the meter rotates it at a speed proportional to flow rate. As the propeller goes through each revolution, encapsulated magnets generate pulses through a pickup device, with the number of pulses proportional to flow rate.

6.3.2.4 Bypass Meters

In a *bypass meter* (also known as a *shunt meter*), a proportion of the liquid is diverted from the main flowstream by an orifice plate into a bypass configuration. The liquid is concentrated through nozzles to impinge on the rotors of

CHAPTER | 6 Measurement of Flow

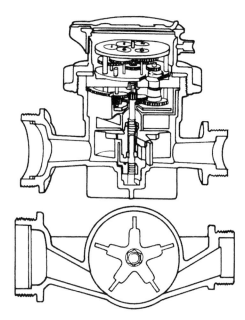

FIGURE 6.22 Rotating-vane type meter.

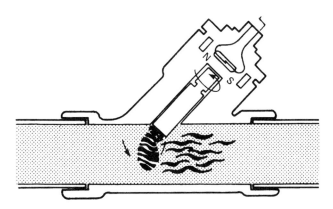

FIGURE 6.23 Angled-propeller meter.

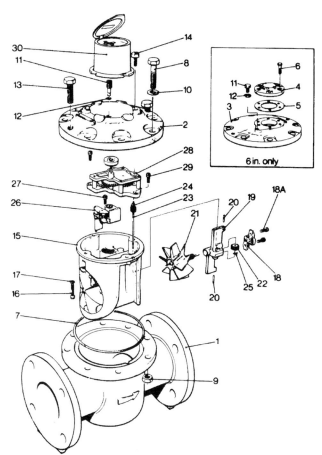

FIGURE 6.24 Helix meter, exploded view. 1. Body. 2. Top cover with regulator plug and regulator sealing ring. 3. Top cover plate. 4. Joint plate. 5. Joint plate gasket. 6. Joint plate screws. 7. Top cover sealing ring. 8. Body bolt. 9. Body bolt unit. 10. Body bolt washer. 11. Regulator plug. 12. Regulator plug sealing ring. 13. Joint breaking screw. 14. Counter box screw. 15. Measuring element. 16. Element securing screw. 17. Element securing screw washer. 18. Back bearing cap assembly. 19. Back vane support. 20. Tubular dowel pin. 21. Vane. 22. Worm wheel. 23. Vertical worm shaft. 24. First pinion. 25. Drive clip. 26. Regulator assembly. 27. Regulator assembly screw. 28. Undergear. 29. Undergear securing screw. 30. Register.

a small turbine located in the bypass, the rotation of the turbine being proportional to flow rate.

This type of device can give moderate accuracy over a 5:1 turndown ratio and is suitable for liquids, gases, and steam. Bypass meters have been used with other shunt-meter devices, including Coanda-effect oscillatory flowmeters, rotameters, ultrasonic meters, and positive displacement meters and multijets.

6.3.2.5 Helix Meters

In a *helix meter*, the measuring element takes the form of a helical vane mounted centrally in the measuring chamber with its axis along the direction of flow, as shown in Figure 6.24. The vane consists of a hollow cylinder with accurately formed wings. Owing to the effect of the buoyancy of the liquid on the cylinder, friction between its spindle and the sleeve bearings is small. The water is directed evenly onto the vanes by means of guides.

Transmission of the rotation from the under-gear to the meter register is by means of ceramic magnetic coupling.

The body of the meter is cast iron, and the mechanism and body cover are of thermoplastic injection molding. The meter causes only small head loss in operation and is suited for use in water-distribution mains. It is available in sizes from 40 mm up to 300 mm, respective maximum flow rates being 24 m^3/h and 1540 m^3/h, with accuracy of ±2 percent over a 20:1 turndown ratio.

6.3.2.6 Turbine Meters

A *turbine meter* consists of a practically friction-free rotor pivoted along the axis of the meter tube and designed in such a way that the rate of rotation of the rotor is proportional to the rate of flow of fluid through the meter. This rotational speed is sensed by means of an electric pick-off

In many similar product designs, the rotor is designed so that the pressure distribution of the process liquid helps suspend the rotor in an "axial" floating position, thereby eliminating end-thrust and wear, improving repeatability, and extending the linear flow range. This is illustrated in Figure 6.25(b).

As the liquid flows through the meter, there is a small, gradual pressure loss up to Point A caused by the rotor hangers and housing. At this point the area through which flow can take place reduces and velocity increases, resulting in a pressure minimum at Point B. By the time the liquid reaches the downstream edge of the rotor (C), the flow pattern has reestablished itself and a small pressure recovery occurs, causing the rotor to move hard upstream in opposition to the downstream forces. To counteract this upstream force, the rotor hub is designed to be slightly larger in diameter than the outside diameter of the deflector cone to provide an additional downstream force. A hydraulic balance point is reached, with the rotor floating completely clear of any end stops.

The turbine meter is available in a range of sizes up to 500 mm, with linearity better than ±0.25 percent and repeatability better than ±0.02 percent, and can be bidirectional in operation. To ensure optimum operation of the meter, it is necessary to provide a straight pipe section of 10 pipe-diameters upstream and 5 pipe-diameters downstream of the meter. The addition of flow is sometimes necessary.

6.3.3 Rotating Mechanical Meters for Gases

The principal types to be discussed are positive displacement, deflecting vane, rotating vane, and turbine.

6.3.3.1 Positive Displacement Meters

Three main types of meter come under the heading of positive displacement. They are diaphragm meters, wet gas meters (liquid sealed drum), and rotary displacement meters.

Diaphragm meters (*bellows type*) This type of meter has remained fundamentally the same for over 100 years and is probably the most common kind of meter in existence. It is used in the United Kingdom for metering the supply of gas to domestic and commercial users.

The meter comprises a metal case with an upper and a lower section. The lower section consists of four chambers, two of which are enclosed by flexible diaphragms that expand and contract as they are charged and discharged with the gas being metered. Figure 6.26 illustrates the meter at four stages of its operating cycle.

Mechanical readout is obtained by linking the diaphragms to suitable gearing, since each cycle of the diaphragms discharges a known quantity of gas. This type of meter is of necessity highly accurate and trouble-free,

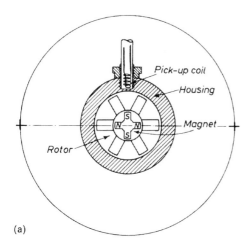

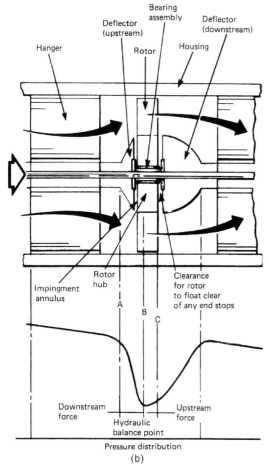

FIGURE 6.25 (a) Principle of operation of turbine meter. (b) Pressure distribution through turbine meter.

coil fitted to the outside of the meter housing, as shown in Figure 6.25(a).

The only moving component in the meter is the rotor, and the only component subject to wear is the rotor bearing assembly. However, with careful choice of materials (e.g., tungsten carbide for bearings) the meter should be capable of operating for up to five years without failure.

CHAPTER | 6 Measurement of Flow

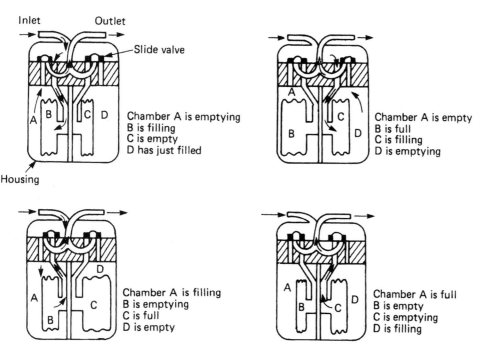

FIGURE 6.26 Diaphragm meter-stages of operation.

and the performance is governed by the regulations of the Department of Trade and Industry.

Liquid Sealed Drum This type of meter differs from the bellows type of meter in that the sealing medium for the measuring chambers is not solid but is water or some other suitable liquid.

The instrument is shown in section in Figure 6.27. It consists of an outer chamber of tinned brass plate or Staybrite steel sheeting containing a rotary portion. This rotating part consists of shaped partitions forming four measuring chambers made of light-gauge tinplate or Staybrite steel, balanced about a center spindle so that it can rotate freely. Gas enters by the gas inlet near the center and leaves by the outlet pipe at the top of the outer casing. The measuring chambers are sealed off by water or other suitable liquid, which fills the outer chamber to just above the center line. The level of the water is so arranged that when one chamber becomes unsealed to the outlet side, the partition between it and the next chamber seals it off from the inlet side. Thus each measuring chamber will, during the course of a rotation, deliver a definite volume of gas from the inlet side to the outlet side of the instrument. The actual volume delivered will depend on the size of the chamber and the level of the water in the instrument. The level of the water is therefore critical and is maintained at the correct value by means of a hook type of level indicator in a side chamber, which is connected to the main chamber of the instrument. If the level becomes very low, the measuring chambers will become unsealed and gas can pass freely through the instrument without being measured; if the level is too high, the volume delivered at each rotation will be too small, and water may pass back down

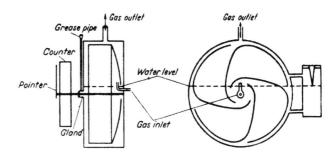

FIGURE 6.27 Liquid sealed drum type gas meter.

the inlet pipe. The correct calibration is obtained by adjusting the water level.

When a partition reaches a position where a small sealed chamber is formed connected to the inlet side, there is a greater pressure on the inlet side than on the outlet side. There will therefore be a force that moves the partition in an anticlockwise direction, thus increasing the volume of the chamber. This movement continues until the chamber is sealed off from the inlet pipe but opened up to the outlet side; at the same time the chamber has become open to the inlet gas but sealed off from the outlet side. This produces continuous rotation. The rotation operates a counter that indicates complete rotations and fractions of rotation and can be calibrated in actual volume units. The spindle between the rotor and the counter is usually made of brass and passes through a grease-packed gland. The friction of this gland, together with the friction in the counter gearing, will determine the pressure drop across the meter, which is found to be almost independent of the speed of rotation. This friction must be kept as low as possible, for if there is a large pressure difference between inlet and outlet sides of the meter, the level of the water in the measuring chambers will be forced down, causing errors in the volume delivered; and at low rates of flow the meter will rotate in a jerky manner.

It is very difficult to produce partitions of such a shape that the meter delivers accurate amounts for fractions of a rotation; consequently the meter is only approximately correct when fractions of a rotation are involved.

The mass of gas delivered will depend on the temperature and pressure of the gas passing through the meter. The volume of gas is measured at the inlet pressure of the meter, so if the temperature and the density of the gas at STP are known, it is not difficult to calculate the mass of gas measured. The gas will, of course, be saturated with water vapor, and this must be taken into account in finding the partial pressure of the gas.

Rotating-Impeller Type This type of meter is similar in principle to the rotating-impeller type meter for liquids and could be described as a two-toothed gear pump. It is shown schematically in Figure 6.28. Although the meter is usually manufactured almost entirely from cast iron, other materials can be used. The meter basically consists of two impellers housed in a casing and supported on rolling element bearings. A clearance of a few thousandths of an inch between the impellers and the casing prevents wear, with the result that the calibration of the meter remains constant throughout its life. The leakage rate is only a small fraction of 1 percent, and this is compensated for in the gearing counter ratio. Each lobe of the impellers has a scraper tip machined onto its periphery to prevent deposits forming in the measuring chamber. The impellers are timed relative to each other by gears fitted to one or both ends of the impeller shafts.

The impellers are caused to rotate by the decrease in pressure, which is created at the meter outlet following a consumer's use of gas. Each time an impeller passes through the vertical position, a pocket of gas is momentarily trapped between the impeller and the casing. Four pockets of gas are therefore trapped and expelled during each complete revolution of the index shaft. The rotation of the impellers is transmitted to the meter counter by suitable gearing so that the counter reads directly in cubic feet. As the meter records the quantity of gas passing through it at the conditions prevailing at the inlet, it is necessary to correct the volume indicated by the meter index for various factors. These are normally pressure, temperature, and compressibility. Corrections can be carried out manually if the conditions within the meter are constant. Alternatively, the correction can be made continuously and automatically by small mechanical or electronic computers if conditions within the meter vary continuously and by relatively large amounts. Meters can also drive, through external gearing, various types of pressure- or temperature-recording devices as required.

Meters of this type are usually available in pressures up to 60 bar and will measure flow rates from approximately $12\,m^3/h$ up to $10,000\,m^3/h$. Within these flow rates the meters will have a guaranteed accuracy of ± 1.0 percent, over a range of from 5 to 100 percent of maximum capacity. The pressure drop across the meter at maximum capacity is always less than 50 mm wg. These capacities and the pressure loss information are for meters operating at low pressure; the values would be subject to the effects of gas density at high pressure.

6.3.3.2 Deflecting-Vane Type: Velometers

The principle of this type of instrument is similar to that of the same instrument for liquids. The construction, however, has to be different because the density of a gas is usually considerably less than that of a liquid. As the force per unit area acting on the vane depends on the rate of change of momentum and momentum is mass multiplied by velocity, the force will depend on the density and on the velocity of the impinging gas. The velocity of gas flow in a main is usually very much greater (6 to 10 times) than that of liquid flow, but this is not sufficient to compensate for the greatly reduced density. (Density of dry air at 0°C and 760 mm is 0.0013 g/ml; density of water is 1 g/ml.)

The vane must therefore be considerably larger when used for gases or be considerably reduced in weight. The restoring force must also be made small if an appreciable deflection is to be obtained.

The simple velometer consists of a light vane that travels in a shaped channel. Gas flowing through the channel deflects the vane according to the velocity and density of the gas, the shape of the channel, and the restoring torque of the hairspring attached to the pivot of the vane.

The velometer is usually attached to a "duct jet," which consists of two tubes placed so that the open end of one faces upstream while the open end of the other points downstream. The velometer then measures the rate of flow through the pair of tubes, and because this depends on the

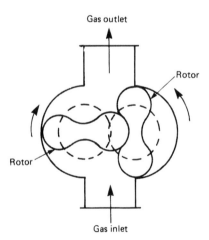

FIGURE 6.28 Rotary displacement meter.

lengths and sizes of connecting pipes and the resistance and location of the pressure holes, each assembly needs individual calibration.

The main disadvantages of this simple velometer are the effects of hot or corrosive gases on the vane and channel. This disadvantage may be overcome by measuring the flow of air through the velometer produced by a differential air pressure equal to that produced by the "duct jet." In this way the hot gases do not pass through the instrument, and so it is not damaged.

6.3.3.3 Rotating-Vane Type

Anemometers As in the case of the deflecting-vane type, the force available from gases to produce the rotation of a vane is considerably less than that available in the measurement of liquids. The vanes must therefore be made light or have a large surface area. The rotor as a whole must be accurately balanced, and the bearings must be as friction-free as possible and may be in the form of a multicap or multiple-fan blade design, the speed of rotation being proportional to air speed.

Rotary Gas Meter The rotary meter is a development of the air meter type of anemometer and is shown in Figure 6.29. It consists of three main assemblies: the body, the measuring element, and the multipoint index driven through the intergearing. The lower casing (1) has integral inline flanges (2) and is completed by the bonnet (3) with index glass (4) and bezel (5).

The measuring element is made up of an internal tubular body (6), which directs the flow of gas through a series of circular ports (7) onto a vaned anemometer (8). The anemometer is carried by a pivot (9) that runs in a sapphire–agate bearing assembly (10), the upper end being steadied by a bronze bush (11).

The multipointer index (12) is driven by an intergear (13) supported between index plates (14). The index assembly is positioned by pillars (15), which are secured to the top flange of the internal tubular body.

The meter casing is made of cast iron; the anemometer is made from aluminum. The larger sizes have a separate internal tubular body made from cast iron, with a brass or mild steel skirt that forms part of the overall measuring element.

Its area of application is in the measurement of gas flow in industrial and commercial installations at pressures up to 1.5 bar and flows up to 200 m^3/h, giving accuracy of ±2 percent over a flow range of 10:1.

6.3.3.4 Turbine Meters

The gas turbine meter operates on the same principle as the liquid turbine meter, although the design is somewhat different since the densities of gases are much lower than those of liquids; high gas velocities are required to turn the rotor blades.

6.3.4 Electronic Flowmeters

Either the principle of operation of flowmeters in this category is electronically based or the primary sensing is by means of an electronic device. Most of the flowmeters discussed in this section have undergone considerable development in the last five years, and the techniques outlined are a growth area in flowmetering applications. They include electromagnetic flowmeters, ultrasonic flowmeters, oscillatory flowmeters, and cross-correlation techniques. It is important to note, however, that there has been very limited development of new techniques in flowmetering since the early 1980s, due in part to concentration of effort on the design of other sensors and control systems.

6.3.4.1 Electromagnetic Flowmeters

The principle of operation of this type of flowmeter is based on Faraday's law of electromagnetic induction, which states that if an electric conductor moves in a magnetic field, an electromotive force (EMF) is induced, the amplitude of which is dependent on the force of the magnetic field, the velocity of the movement, and the length of the conductor, such that

$$E \propto BlV \qquad (6.44)$$

where E is EMF, B is magnetic field density, l is length of conductor, and V is the rate at which the conductor is cutting the magnetic field. The direction of the EMF with respect to the movement and the magnetic field is given by Fleming's right-hand generator rule.

If the conductor now takes the form of a conductive liquid, an EMF is generated in accordance with Faraday's law. It is useful at this time to refer to BS 5792 1980, which states: "If the magnetic field is perpendicular to an electrically

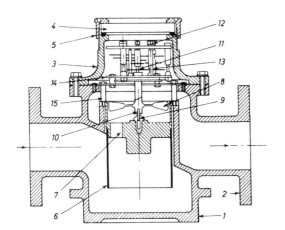

FIGURE 6.29 Diagrammatic section of a rotary gas meter. Courtesy of Parkinson & Cowan Computers.

insulating tube through which a conductive liquid is flowing, a maximum potential difference may be measured between two electrodes positioned on the wall of the tube such that the diameter joining the electrodes is orthogonal to the magnetic field. The potential difference is proportional to the magnetic field strength, the axial velocity, and the distance between the electrodes." Hence the axial velocity and rate of flow can be determined. This principle is illustrated in Figure 6.30(a).

Figure 6.30(b) shows the basic construction of an electromagnetic flowmeter. It consists of a primary device, which contains the pipe through which the liquid passes, the measurement electrodes, and the magnetic field coils and a secondary device, which provides the field-coil excitation and amplifies the output of the primary device and converts it to a form suitable for display, transmission, and totalization.

The flow tube, which is effectively a pipe section, is lined with some suitable insulating material (dependent on liquid type) to prevent short-circuiting of the electrodes, which are normally button type and mounted flush with the liner. The field coils wound around the outside of the flow tube are usually epoxy resin encapsulated to prevent damage by damp or liquid submersion.

Field-Coil excitation To develop a suitable magnetic field across the pipeline, it is necessary to drive the field coil with some form of electrical excitation. It is not possible to use pure DC excitation due to the resulting polarization effect on electrodes and subsequent electrochemical action, so some form of AC excitation is employed. The most common techniques are sinusoidal and nonsinusoidal (square wave, pulsed DC, or trapezoidal).

Sinusoidal AC excitation Most early electromagnetic flowmeters used standard 50 Hz mains voltage as an excitation source for the field coils, and in fact most systems in use

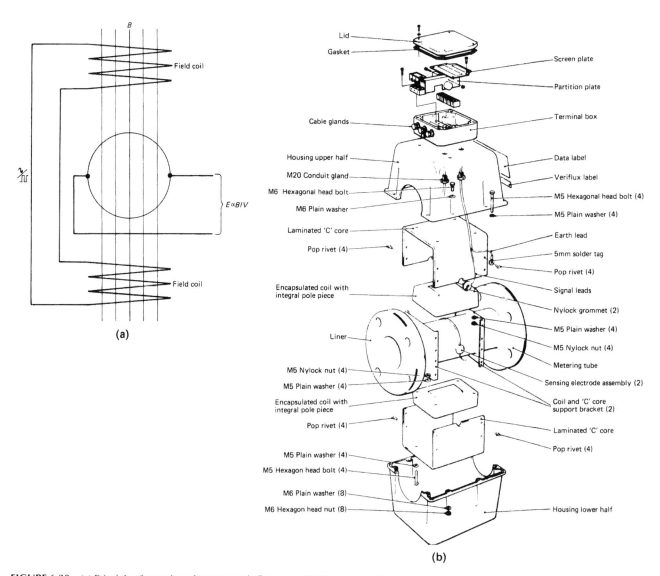

FIGURE 6.30 (a) Principle of operation: electromagnetic flowmeter. (b) Electromagnetic flowmeter detector head: exploded view.

today operate on this principle. The signal voltage will also be AC and is normally capacitively coupled to the secondary electronics to avoid any DC interfering potentials. This type of system has several disadvantages. Due to AC excitation, the transformer effect produces interfering voltages. These are caused by stray pickup by the signal cables from the varying magnetic field. It has a high power consumption and suffers from zero drift caused by the previously mentioned interfering voltages and electrode contamination. This necessitates manual zero control adjustment.

These problems have now been largely overcome by the use of nonsinusoidal excitation.

Nonsinusoidal Excitation Here it is possible to arrange that rate of change of flux density $dB/dt = 0$ for part of the excitation cycle; therefore, there is no transformer action during this period. The flow signal is sampled during these periods and is effectively free from induced error voltages.

Square-wave, pulsed, and trapezoidal excitations have all been employed initially at frequencies around 50 Hz, but most manufacturers have now opted for low-frequency systems (2–7 Hz) offering the benefits of minimum power consumption (i.e., only 20 percent of the power used by a comparative 50 Hz system), automatic compensation for interfering voltages, automatic zero adjustment, and tolerance of light buildup of material on electrode surfaces.

An example of this type of technique is illustrated in Figure 6.31, where square-wave excitation is used. The DC supply to the coils is switched on and off at approximately 2.6 Hz, with polarity reversal every cycle. Figure 6.31(a) shows the ideal current waveform for pulsed DC excitation, but, because of the inductance of the coils, this waveform cannot be entirely achieved. The solution as shown in Figure 6.31(b) is to power the field coils from a constant-current source giving a near square-wave excitation. The signal produced at the measuring electrodes is shown in Figure 6.31(c). The signal is sampled at five points during each measurement cycle, as shown, and microprocessor techniques are utilized to evaluate and separate the true flow signal from the combined flow and zero signals, as shown in the equation in Figure 6.31(c).

Area of Application Electromagnetic flowmeters are suitable for measuring a wide variety of liquids such as dirty liquids, pastes, acids, slurries, and alkalis; accuracy is largely unaffected by changes in temperature, pressure, viscosity, density, or conductivity. However, in the case of the latter, conductivities must be greater than 1 micromho/cm.

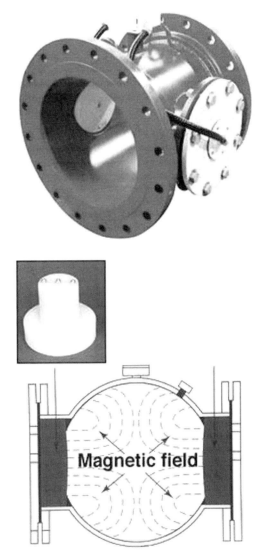

FIGURE 6.32 Encapsulated coil magmeter. Courtesy of ISCO Inc.

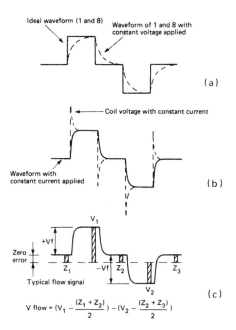

FIGURE 6.31 Electromagnetic flowmeter: pulsed DC excitation. Courtesy of Flowmetering Instruments Ltd.

Installation The primary element can be mounted in any attitude in the pipework, although care should be taken to ensure that when the flowmeter is mounted horizontally, the axis of the electrodes are in the horizontal plane.

Where buildup of deposits on the electrodes is a recurring problem, there exist three alternatives for consideration:

A. Ultrasonic cleaning of electrodes.
B. Utilize capacitive electrodes that do not come into contact with the flowstream, and therefore insulating coatings have no effect.
C. Removable electrodes, inserted through a hottap valve assembly, enabling the electrodes to be withdrawn from the primary and physically examined and cleaned, then reinserted under pressure and without stopping the flow.

It should be noted that on insulated pipelines, earthing rings will normally be required to ensure that the flowmeter body is at the same potential as that of the flouring liquid to prevent circulating current and interfering voltages. Recently, a magnetic flowmeter design was introduced that relies on a self-contained coil-and-electrode package, mounted at 180° to a similar package, across the centerline of the flow tube. This design does not require a fully lined flow tube and appears to have some advantages in terms of cost in medium- and larger-sized applications.

The accuracy of the flowmeter can be affected by flow profile, and the user should allow at least 10 straight-pipe diameters upstream and 5 straight-pipe diameters downstream of the primary element to ensure optimum conditions. In addition, to ensure system accuracy, it is essential that the primary element should remain filled with the liquid being metered at all times. Entrained gases will cause similar inaccuracy.

For further information on installation requirements, the reader is referred to the relevant sections of BS 5792 1980.

Flowmeters are available in sizes from 32 mm to 1200 mm nominal bore to handle flow velocities from 0–0.5 m/s to 0–10 m/s with accuracy of ±1 percent over a 10:1 turndown ratio.

6.3.4.2 Ultrasonic Flowmeters

Ultrasonic flowmeters measure the velocity of a flowing medium by monitoring interaction between the flowstream and an ultrasonic sound wave transmitted into or through it. Many techniques exist; the two most commonly applied are Doppler and transmissive (time of flight). These will now be dealt with separately.

Doppler Flowmeters These devices use the well-known Doppler effect, which states that the frequency of sound changes if its source or reflector moves relative to the listener or monitor. The magnitude of the frequency change is an indication of the speed of the sound source or sound reflector.

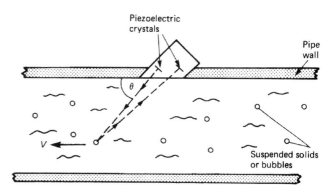

FIGURE 6.33 Principle of operation: Doppler meter.

In practice the Doppler flowmeter comprises a housing in which two piezoelectric crystals are potted, one a transmitter and the other a receiver, with the whole assembly located on the pipe wall, as shown in Figure 6.33. The transmitter transmits ultrasonic waves of frequency F_1 at an angle θ to the flowstream. If the flowstream contains particles, entrained gas, or other discontinuities, some of the transmitted energy will be reflected back to the receiver. If the fluid is travelling at velocity V, the frequency of the reflected sound as monitored by the receiver can be shown to be F_2 such that

$$F_2 = F_1 \pm 2V \cdot \cos\theta \cdot \frac{F_1}{C}$$

where C is the velocity of sound in the fluid. Rearranging:

$$V = \frac{C(F_2 - F_1)}{2 \cdot F_1 \cdot \cos\theta}$$

which shows that velocity is proportional to the frequency change.

The Doppler meter is normally used as an inexpensive clamp-on flowmeter, the only operational constraints being that the flowstream must contain discontinuities of some kind (the device will not monitor clear liquids), and the pipeline must be acoustically transmissive.

Accuracy and repeatability of the Doppler meter are somewhat suspect and difficult to quantify, because its operation is dependent on flow profile, particle size, and suspended solids concentration. However, under ideal conditions and given the facility to calibrate *in situ*, accuracies of ±5 percent should be attainable. This type of flowmeter is most suitable for use as a flow switch or for flow indication where absolute accuracy is not required.

Transmissive Flowmeters Transmissive devices differ from Doppler flowmeters in that they rely on transmission of an ultrasonic pulse through the flowstream and therefore do not depend on discontinuities or entrained particles in the flowstream for operation.

The principle of operation is based on the transmission of an ultrasonic sound wave between two points, first in the direction of flow and then of opposing flow. In each case the time of flight of the sound wave between the two points will have been modified by the velocity of the flowing medium, and the difference between the flight times can be shown to be directly proportional to flow velocity.

In practice, the sound waves are not generated in the direction of flow but at an angle across it, as shown in Figure 6.34. Pulse transit times downstream T_1 and upstream T_2 along a path length D can be expressed as $T_1 = D/(C + V)$ and $T_1 = D/(C - V)$, where C is the velocity of sound in the fluid and V is the fluid velocity. Now:

$$T = T_1 - T_2 = \left(\frac{2DV}{C^2 - V^2}\right) \quad (6.45)$$

Since V_2 is very small compared to C^2, it can be ignored. It is convenient to develop the expression in relation to frequency and remove the dependency on the velocity of sound (C). Since $F_1 = 1/T_1$ and $F_2 = 1/T_2$ and average fluid velocity $\overline{V} = V/\cos \theta$, Equation (6.44) is developed to:

$$F_1 - F_2 = \left(\frac{2\overline{V} \cos \theta}{D}\right)$$

The frequency difference is calculated by an electronic converter, which gives an analog output proportional to average fluid velocity. A practical realization of this technique operates in the following manner.

A voltage-controlled oscillator generates electronic pulses from which two consecutive pulses are selected. The first of these is used to operate a piezoelectric ceramic crystal transducer, which projects an ultrasonic beam across the liquid flowing in a pipe. This ultrasonic pulse is then received on the other side of the pipe, where it is converted back to an electronic pulse. The latter is then received by the "first-arrival" electronics, comparing its arrival time with the second pulse received directly. If the two pulses are received at the same time, the period of time between them equates to the time taken for the first pulse to travel to its transducer and be converted to ultrasound, to travel across the flowstream, to be reconverted back to an electronic pulse, and to travel back to the first-arrival position.

Should the second pulse arrive before the first one, the time between pulses is too short. Then the first-arrival electronics will step down the voltage to the voltage-controlled oscillator (VCO), reducing the resulting frequency. The electronics will continue to reduce voltage to the VCO in steps, until the first and second pulses are received at the first-arrival electronics at the same time. At this point, the periodic time of the frequency will be the same as the ultrasonic flight time plus the electronic delay time.

If a similar electronic circuit is now used to project an ultrasonic pulse in the opposite direction to that shown, another frequency will be obtained that, when subtracted from the first, will give a direct measure of the velocity of the fluid in the pipe, since the electronic delays will cancel out.

In practice, the piezoelectric ceramic transducers used act as both transmitters and receivers of the ultrasonic signals and thus only one is required on each side of the pipe.

Typically the flowmeter will consist of a flowtube containing a pair of externally mounted, ultrasonic transducers and a separate electronic converter/transmitter, as shown in Figure 6.35(a). Transducers may be wetted or nonwetted and consist of a piezoelectric crystal sized to give the desired frequency (typically 1–5 MHz for liquids and 0.2–0.5 MHz for gases). Figure 6.35(b) shows a typical transducer assembly.

(a)

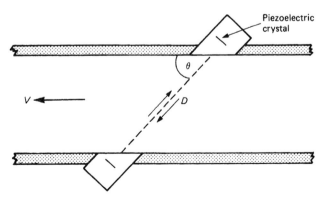

FIGURE 6.34 Principle of operation: time-of-flight ultrasonic flowmeter.

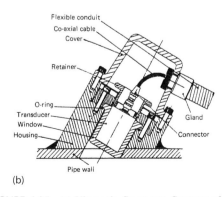

(b)

FIGURE 6.35 (a) Ultrasonic flowmeter. Courtesy of Sparling Inc. (b) Transducer assembly.

Due to the fact that the flowmeter measures velocity across the center of the pipe, it is susceptible to flow profile effects, and care should be taken to ensure sufficient length of straight pipe upstream and downstream of the flowtube, to minimize such effects. To overcome this problem, some manufacturers use multiple-beam techniques in which several chordal velocities are measured and the average computed. However, it is still good practice to allow for approximately 10 upstream and 5 downstream diameters of straight pipe. Furthermore, since this type of flowmeter relies on transmission through the flowing medium, fluids with a high solids or gas-bubble content cannot be metered.

This type of flowmeter can be obtained for use on liquids or gases for pipe sizes from 75 mm nominal bore up to 1500 mm or more for special applications, and it is bidirectional in operation. Accuracy of better than ±1 percent of flow rate can be achieved over a flow range of 0.2 to 12 meters per second.

This technique has also been successfully applied to open channel and river flow and is also now readily available as a clamp-on flowmeter for closed pipes, but accuracy is dependent on knowledge of each installation, and *in situ* calibration is desirable.

6.3.4.3 Oscillatory "Fluidic" Flowmeters

The operating principle of flowmeters in this category is based on the fact that if an obstruction of known geometry is placed in the flowstream, the fluid will start to oscillate in a predictable manner. The degree of oscillation is related to fluid flow rate. The three main types of flowmeter in this category are vortex-shedding flowmeters, swirl flowmeters, and the several Coanda effect meters that are now available.

The Vortex Flowmeter This type of flowmeter operates on the principle that if a bluff (i.e., nonstreamlined) body is placed in a flowstream, vortices will be detached or shed from the body. The principle is illustrated in Figure 6.36.

The vortices are shed alternately to each side of the bluff body, the rate of shedding being directly proportional to flow velocity. If this body is fitted centrally into a pipeline, the vortex-shedding frequency is a measure of the flow rate.

Any bluff body can be used to generate vortices in a flowstream, but for these vortices to be regular and well defined requires careful design. Essentially, the body must be nonstreamlined, symmetrical, and capable of generating vortices for a wide Reynolds number range. The most commonly adopted bluff body designs are shown in Figure 6.37.

These designs all attempt to enhance the vortex-shedding effect to ensure regularity or simplify the detection technique. If the design (d) is considered, it will be noted that a second nonstreamlined body is placed just downstream of the vortex-shedding body. Its effect is to reinforce and stabilize the shedding. The width of the bluff body is determined by pipe size, and a rule-of-thumb guide is that the ratio of body width to pipe diameter should not be less than 0.2.

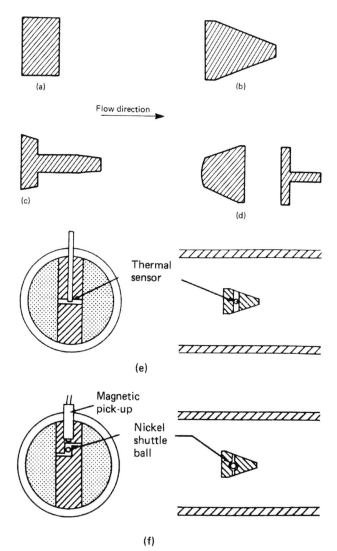

FIGURE 6.37 (a)–(d) Bluff body shapes. (e) Thermal sensor. Courtesy of Actaris Neptune Ltd. (f) Shuttle ball sensor. Courtesy of Actaris Neptune Ltd.

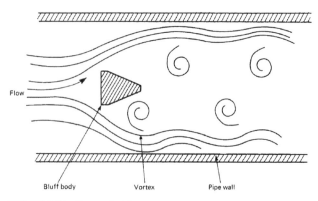

FIGURE 6.36 Vortex shedding.

Sensing Methods Once the bluff-body type has been selected we must adopt a technique to detect the vortices. Various methods exist, the more popular techniques being as follows:

A. *Ultrasonic.* Where the vortices pass through an ultrasonic beam and cause refraction of this beam, resulting in modulation of the beam amplitude.
B. *Thermal* (Figure 6.37(e)). Where a thermistor-type sensor is located in a through passage across the bluff body and behind its face. The heated thermistor will sense alternating vortices due to the cooling effect caused by their passage, and an electrical pulse output is obtained.
C. *Oscillating disc.* Sensing ports on both sides of the flow element cause a small disc to oscillate. A variable-reluctance pickup detects the disc's oscillation. This type is particularly suited to steam or wet-gas flow.
D. *Capacitance.* Metal diaphragms are welded on opposite sides of the bluff body, the small gaps between the diaphragms and the body being filled with oil. Interconnecting ports allow transfer of oil between the two sides. An electrode is placed close to each plate and the oil used as a dielectric. The vortices alternately deform the diaphragm plates, causing a capacitance change between the diaphragm and the electrode. The frequency of changes in capacitance is equal to the shedding frequency.
E. *Strain.* Here the bluff body is designed such that the alternating pressures associated with vortex shedding are applied to a cantilevered section to the rear of the body. The alternating vortices create a cyclic strain on the rear of the body, which is monitored by an internal strain gauge.
F. *Shuttle ball* (Figure 6.37(f)). The shuttle technique uses the alternating pressures caused by vortex shedding to drive a magnetic shuttle up and down the axis of a flow element. The motion of the shuttle is detected by a magnetic pickup.

The output derived from the primary sensor is a low-frequency signal dependent on flow; this is then applied to conditioning electronics to provide either analog or digital output for display and transmission. The calibration factor (pulses per m^3) for the vortex meter is determined by the dimensions and geometry of the bluff body and will not change.

Installation parameters for vortex flowmeters are quite critical. Pipe flange gaskets upstream and at the transmitter should not protrude into the flow, and to ensure a uniform velocity profile there should be 20 diameters of straight pipe upstream and 5 diameters downstream. Flow straighteners can be used to reduce this requirement if necessary.

The vortex flowmeter has wide-ranging applications in both gas and liquid measurement, providing the Reynolds number lies between 2×10^3 and 1×10^5 for gases and

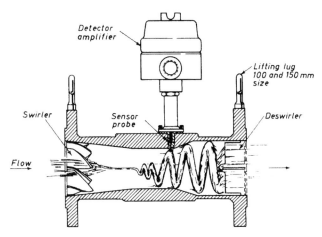

FIGURE 6.38 Cutaway view of the swirlmeter. Courtesy of ABB Instrument Group.

4×10^3 and 1.4×10^5 for liquids. The output of the meter is independent of the density, temperature, and pressure of the flowing fluid and represents the flow rate to better than ±1 percent of full scale, giving turndown ratios in excess of 20:1.

The Swirlmeter Another meter that depends on the oscillatory nature of fluids is the swirlmeter, shown in Figure 6.38. A swirl is imparted to the body of flowing fluid by the curved inlet blades, which give a tangential component to the fluid flow. Initially the axis of the fluid rotation is the center line of the meter, but a change in the direction of the rotational axis (precession) takes place when the rotating liquid enters the enlargement, causing the region of highest velocity to rotate about the meter axis. This produces an oscillation or precession, the frequency of which is proportional to the volumetric flow rate. The sensor, which is a bead thermistor heated by a constant-current source, converts the instantaneous velocity changes into a proportional electrical pulse output. The number of pulses generated is directly proportional to the volumetric flow.

The operating range of the swirlmeter depends on the specific application, but typical for liquids are 3.5 to 4.0 liters per minute for the 25 mm size, ranging to 1700 to 13,000 liters per minute for the 300 mm size. Typical gas flow ranges are 3 to 35 m^3/h for the 25 mm size, ranging to 300 to 9000 m^3/h for the 300 mm size. Accuracy of ±1 percent of rate is possible, with repeatability of ±0.25 percent of rate.

The Coanda Effect Meters The Coanda effect produces a fluidic oscillator for which the frequency is linear with the volumetric flow rate of fluid. The Coanda effect is a hydraulic feedback circuit. A chamber is designed with a left-hand and a right-hand feedback channel. A jet of water flows through the chamber, and because of the feedback channels, some of the water will impact the jet from the side.

FIGURE 6.39 Coanda Effect Fluidic Meter. Courtesy of Fluidic Flowmeters LLC.

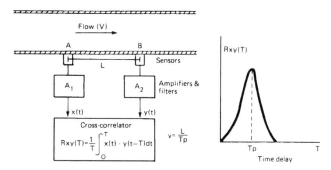

FIGURE 6.40 Cross-correlation meter.

This causes a pressure differential between one side of the jet and the other, and the jet "flips" back and forth in the chamber. The frequency of this flipping is proportional to the flow through the chamber. Several means exist to measure this oscillation, including electromagnetic sensors and piezo-resistive pressure transducers. This Coanda effect is extremely linear and accurate across at least a 300:1 range. It is reasonably viscosity independent, too, and can be made simply and inexpensively.

Typically, small fluidic meters can be made so inexpensively, in fact, that fluidic flowmeters are being promoted as a replacement for the inexpensive positive displacement meters currently used as domestic water meters. Several companies have developed fluidic flowmeters as extremely inexpensive replacements for AGA-approved diaphragm-type gas meters for household metering.

Coanda effect meters are insensitive to temperature change, too. A fluidic flowmeter is being marketed as an inexpensive BTU (heat) meter for district heating applications. Coanda effect meters become more expensive as their physical size increases. Above 50 mm diameter, they are more expensive in general than positive displacement meters. Currently, the only designs available above 50 mm are "bypass designs" that use a small diameter coanda effect meter as a bypass around a flow restriction in a larger pipeline. Meters up to 250 mm diameter have been designed in this fashion. These meters exhibit rangeability of over 100:1, with accuracies (when corrected electronically for linearity shift) of 0.5% of indicated flow rate. See Figure 6.39.

6.3.4.4 Cross-Correlation

In most flowing fluids there exist naturally occurring random fluctuations such as density, turbulence, and temperature, which can be detected by suitably located transducers. If two such transducers are installed in a pipeline separated by a distance L, as shown in Figure 6.40, the upstream transducer will pick up a random fluctuation t seconds before the downstream transducer, and the distance between the transducers divided by the transit time t will yield flow velocity. In practice the random fluctuations will not be stable and are compared in a cross-correlator that has a peak response at transit time T_p and correlation velocity $V = L/T_p$ meters per second.

This is effectively a nonintrusive measurement and could in principle be developed to measure flow of most fluids. Very few commercial cross-correlation systems are in use for flow measurement because of the slow response time of such systems. However, with the use of microprocessor techniques, processing speed has been increased significantly, and several manufacturers are now producing commercial systems for industrial use. Techniques for effecting the cross-correlation operation are discussed in Part 4 of this book.

6.3.5 Mass Flowmeters

The measurement of mass flow rate can have certain advantages over volume flow rate, i.e., pressure, temperature, and specific gravity do not have to be considered. The main interfering parameter to be avoided is that of two-phase flow, in which gas/liquid, gas/solid, or liquid/solid mixtures are flowing together in the same pipe. The two phases may be travelling at different velocities and even in different directions. This problem is beyond the scope of this book, but the user should be aware of the problem and ensure where possible that the flow is as near homogeneous as possible (by pipe-sizing or meter-positioning) or that the two phases are separately metered.

Methods of measurement can be categorized under two main headings: true mass-flow measurement, in which the measured parameter is directly related to mass flow rate, and inferential mass-flow measurement, in which volume flow rate and fluid density are measured and combined to give mass flow rate. Since volume flow rate and density measurement are discussed elsewhere, only true mass-flow measurement is dealt with here.

6.3.5.1 True Mass-Flow Measurement Methods

Fluid-momentum Methods An *angular momentum* type of device consists of two turbines on separate axial shafts in the meter body. The upstream turbine is rotated at constant speed and imparts a swirling motion to the fluid passing through it. On reaching the downstream turbine, the swirling fluid attempts to impart motion onto it; however, this turbine is constrained from rotating by a calibrated spring. The meter is designed such that on leaving the downstream turbine, all angular velocity will have been removed from the fluid, and the torque produced on it is proportional to mass flow.

This type of device can be used for both gases and liquids with accuracies of ±1 percent.

Mass flowmeters in the category of *gyroscopic/Coriolis mass flowmeters* use the measurement of torque developed when subjecting the fluid stream to a Coriolis acceleration,* as a measure of mass flow rate.

An early application of this technique is illustrated in Figure 6.41.

The fluid enters a T-shaped tube, with flow equally divided down each side of the T, and then recombines into a main flowstream at the outlet from the meter. The whole assembly is rotated at constant speed, causing an angular displacement of the T-tube that is attached to the meter casing through a torque tube. The torque produced is proportional to mass flow rate.

This design suffered from various problems, mainly due to poor sealing of rotating joints or inadequate speed control. However, recent developments have overcome these problems, as shown in Figure 6.42.

The mass flowmeter consists of a U-tube and a T-shaped leaf spring as opposite legs of a tuning fork. An electromagnet is used to excite the tuning fork, thereby subjecting each particle within the pipe to a Coriolis-type acceleration. The resulting forces cause an angular deflection in the U-tube inversely proportional to the stiffness of the pipe and proportional to the mass flow rate. This movement is picked up by optical transducers mounted on opposite sides of the U-tube, the output being a pulse that is width-modulated proportional to mass flow rate. An oscillator/counter digitizes the pulse width and provides an output suitable for display purposes.

This system can be used to measure the flow of liquids or gases, and accuracies better than ±0.5 percent of full scale are possible. Even more recent developments include "straight-through" designs (see Figure 6.43) that have produced similar performance to the U-tube designs. Several manufacturers now offer these designs.

In addition, with better signal processing technologies, Coriolis mass meters have now begun to be used to measure gas flows, with apparently excellent results.

In liquid flow measurement, even in slurries, Coriolis mass flowmeters have nearly completely replaced other types of mass flow measurements such as dual-turbine or volumetric/density combinations.

Pressure Differential Methods In its classical form, this meter consists of four matched orifice plates installed in a Wheatstone bridge arrangement. A pump is used to transfer fluid at a known rate from one branch of the bridge into another to create a reference flow. The resultant differential pressure measured across the bridge is proportional to mass flow rate.

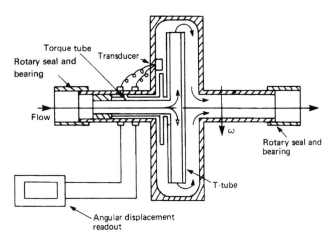

FIGURE 6.41 Early form of Coriolis mass flowmeter.

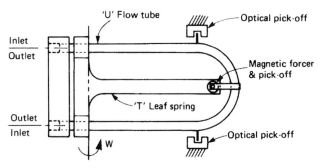

FIGURE 6.42 Gyroscopic/Coriolis mass flowmeter.

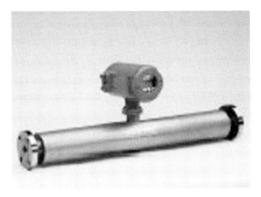

FIGURE 6.43 Straight Tube Coriolis Mass Flowmeter. Courtesy of Krohne America Inc.

---

*On a rotating surface there is an inertial force acting on a body at right angles to its direction of motion, in addition to the ordinary effects of motion of the body. This force is known as a *Coriolis force*.

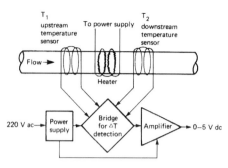

FIGURE 6.44 Thermal mass flowmeter. Courtesy of Emerson Process Management.

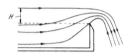

FIGURE 6.45 Rectangular notch, showing top and bottom of contraction.

Thermal Mass Flowmeters This version of a mass flowmeter consists of a flowtube, an upstream and downstream temperature sensor, and a heat source, as illustrated in Figure 6.44. The temperature sensors are effectively active arms of a Wheatstone bridge. They are mounted equidistant from the constant-temperature heat source such that for no-flow conditions, heat received by each sensor is the same and the bridge remains in balance. However, with increasing flow, the downstream sensor receives progressively more heat than the upstream sensor, causing an imbalance to occur in the bridge circuit. The temperature difference is proportional to mass flow rate, and an electrical output representing this is developed by the bridge circuit.

This type of mass flowmeter is most commonly applied to the measurement of gas flows within the ranges 2.5×10^{-10} to 5×10^{-3} kg/s, and accuracy of ± 1 percent of full scale is attainable. Some thermal flowmeters are also used for liquid flow measurements, including very low flow rates.

6.4 FLOW IN OPEN CHANNELS

Flow measurement in open channels is a requirement normally associated with the water and wastewater industry. Flow in rivers, sewers (part-filled pipes), and regular-shaped channels may be measured by the following methods:

A. *Head/area method.* Where a structure is built into the flowstream to develop a unique head/flow relationship, as in:

1. The weir, which is merely a dam over which liquid is allowed to flow, the depth of liquid over the sill of the weir being a measure of the rate of flow.
2. The hydraulic flume, an example being the venturi flume, in which the channel is given the same form in the horizontal plane as a section of a venturi tube while the bottom of the channel is given a gentle slope up the throat.

B. *Velocity/area method.* Where measurement of both variables, that is, head and velocity, is combined with the known geometry of a structure to determine flow.

C. *Dilution gauging.*

6.4.1 Head/Area Method

6.4.1.1 Weirs

Weirs may have a variety of forms and are classified according to the shape of the notch or opening. The simplest is the *rectangular notch* or in certain cases the *square notch*.

The *V* or *triangular notch* is a V-shaped notch with the apex downward. It is used to measure rates of flow that may become very small. Owing to the shape of the notch, the head is greater at small rates of flow with this type than it would be for the rectangular notch.

Notches of other forms, which may be trapezoidal or parabolic, are designed so that they have a constant discharge coefficient or a head that is directly proportional to the rate of flow.

The velocity of the liquid increases as it passes over the weir because the center of gravity of the liquid falls. Liquid that was originally at the level of the surface above the weir can be regarded as having fallen to the level of the center of pressure of the issuing stream. The head of liquid producing the flow is therefore equal to the vertical distance from the center of pressure of the issuing stream to the level of the surface of the liquid upstream.

If the height of the center of pressure above the sill can be regarded as being a constant fraction of the height of the surface of the liquid above the sill of the weir, then the height of the surface above the sill will give a measure of the differential pressure producing the flow. If single particles are considered, some will have fallen a distance greater than the average, but this is compensated for by the fact that others have fallen a smaller distance.

The term *head of a weir* is usually taken to mean the same as the depth of the weir and is measured by the height of the liquid above the level of the sill of the weir, just upstream of where it begins to curve over the weir, and is denoted by *H* and usually expressed in units of length such as meters.

Rectangular Notch Consider the flow over the weir in exactly the same way as the flow through other primary differential-pressure elements. If the cross-section of the stream approaching the weir is large in comparison with the area of the stream over the weir, the velocity V_1 at Section 1 upstream can be neglected in comparison with the velocity V_2 over the weir, and in Equation (6.17) $V_1 = 0$ and the equation becomes:

$$V_2^2 = 2gh \quad \text{or} \quad V_2 = \sqrt{(2gh)}$$

The quantity of liquid flowing over the weir will be given by:

$$Q = A_2 V_2$$

But the area of the stream is BH, where H is the depth over the weir and B the breadth of the weir, and h is a definite fraction of H.

By calculus it can be shown that for a rectangular notch

$$Q = \frac{2}{3} BH \sqrt{(2gH)} \qquad (6.46)$$

$$= \frac{2}{3} B \sqrt{(2gH^3)} \, \text{m}^3/\text{s} \qquad (6.47)$$

The actual flow over the weir is less than that given by Equation (6.45), for the following reasons:

A. The area of the stream is not BH but something less since the stream contracts at both the top and bottom as it flows over the weir, as shown in Figure 6.46, making the effective depth at the weir less than H.

B. Owing to friction between the liquid and the sides of the channel, the velocity at the sides of the channel will be less than that at the middle. This effect may be reduced by making the notch narrower than the width of the stream, as shown in Figure 6.47. This, however, produces side-contraction of the stream. Therefore, $B_1 = B$ should be at least equal to $4H$ when the side contraction is equal to $0.1H$ on both sides, so that the effective width becomes $B-0.2H$.

When required to suppress side contraction and make the measurement more reliable, plates may be fitted as shown in Figure 6.47 so as to make the stream move parallel to the plates as it approaches the weir.

To allow for the difference between the actual rate of flow and the theoretical rate of flow, the discharge coefficient C, defined as before, is introduced, and Equation (6.46) becomes:

$$Q = \frac{2}{3} CB \sqrt{(2gH^3)} \, \text{m}^3/\text{s} \qquad (6.48)$$

The value of C will vary with H and will be influenced by the following factors, which must remain constant in any installation if its accuracy is to be maintained: (a) the relative sharpness of the upstream edge of the weir crest, and (b) the width of the weir sill. Both of these factors influence the bottom contraction and influence C, so the weir sill should be inspected from time to time to see that it is free from damage.

In developing the preceding equations, we assumed that the velocity of the liquid upstream of the weir could be neglected. As the rate of flow increases, this is no longer possible, and a velocity of approach factor must be introduced. This will influence the value of C, and as the velocity of approach increases it will cause the observed head to become less than the true or total head so that a correcting factor must be introduced.

Triangular Notch If the angle of the triangular notch is θ, as shown in Figure 6.48, $B = 2H \tan(\theta/2)$. The position of the center of pressure of the issuing stream will now be at a different height above the bottom of the notch from what it was for the rectangular notch. It can be shown by calculus that the numerical factor involved in the equation is now $(4)/(15)$. Substituting this factor and the new value of A_2 in Equation (6.47):

$$Q = \frac{4}{15} CB \sqrt{(2gH^3)} \, \text{m}^3/\text{s}$$

$$= \frac{4}{15} C 2H \tan \frac{\theta}{2} \sqrt{(2gH^3)}$$

$$= \frac{8}{15} C \tan \frac{\theta}{2} \sqrt{(2gH^5)} \qquad (6.49)$$

Experiments have shown that θ should have a value between 35° and 120° for satisfactory operation of this type of installation.

Although the cross-section of the stream from a triangular weir remains geometrically similar for all values of H, the value of C is influenced by H. The variation of C is from 0.57 to 0.64 and takes into account the contraction of the stream.

If the velocity of approach is not negligible, the value of H must be suitably corrected, as in the case of the rectangular weir.

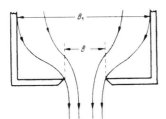

FIGURE 6.46 Rectangular notch, showing side-contraction.

FIGURE 6.47 Rectangular notch, showing side plates.

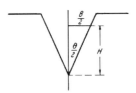

FIGURE 6.48 Triangular notch (V-notch).

Installation and operation of weirs The following points are relevant to installing and operating weirs:

A. Upstream of a weir there should be a wide, deep, and straight channel of uniform cross-section, long enough to ensure that the velocity distribution in the stream is uniform. This approach channel may be made shorter if baffle plates are placed across it at the inlet end to break up currents in the stream.
B. Where debris is likely to be brought down by the stream, a screen should be placed across the approach channel to prevent the debris reaching the weir. This screen should be cleaned as often as necessary.
C. The upstream edge of the notch should be maintained square or sharp-edged, according to the type of installation.
D. The weir crest should be level from end to end.
E. The channel end wall on which the notch plate is mounted should be cut away so that the stream may fall freely and not adhere to the wall. To ensure that this happens, a vent may be arranged in the side wall of the channel so that the space under the falling water is open to the atmosphere.
F. Neither the bed nor the sides of the channel downstream from the weir should be nearer the weir than 150 mm, and the water level downstream should be at least 75 mm below the weir sill.
G. The head H may be measured by measuring the height of the level of the stream above the level of the weir sill, sufficiently far back from the weir to ensure that the surface is unaffected by the flow. This measurement is usually made at a distance of at least $6H$ upstream of the weir. It may be made by any appropriate method for liquids, as described in the section on level measurement: for example, the hook gauge, float-operated mechanisms, air purge systems ("bubblers"), or ultrasonic techniques. It is often more convenient to measure the level of the liquid in a "stilling well" alongside the channel at the appropriate distance above the notch. This well is connected to the weir chamber by a small pipe or opening near the bottom. Liquid will rise in the well to the same height as in the weir chamber and will be practically undisturbed by currents in the stream.

6.4.1.2 Hydraulic Flumes

Where the rate of fall of a stream is so slight that there is very little head available for operating a measuring device or where the stream carries a large quantity of silt or debris, a *flume* is often much more satisfactory than a weir. Several flumes have been designed, but the only one we shall consider here is the venturi flume. This may have more than one form, but where it is flat-bottomed and of the form shown in Figure 6.49 the volume rate of flow is given by the equation:

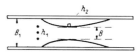

FIGURE 6.49 Hydraulic flume (venturi type).

$$Q = CBh_2 \sqrt{\frac{2g(h_1 - h_2)}{1 - (Bh_2/B_1 h_1)^2}} \text{ m}^3/\text{s} \quad (6.50)$$

where B_1 is width of channel, B is width of the throat, h_1 is depth of water measured immediately upstream of the entrance to the converging section, and h_2 is minimum depth of water in the throat. C is the discharge coefficient for which the value will depend on the particular outline of the channel and the pattern of the flow. Tests on a model of the flume may be used to determine the coefficient, provided that the flow in the model and in the full-sized flume are dynamically similar.

The depths of water h_1 and h_2 are measured as in the case of the weir by measuring the level in wells at the side of the main channel. These wells are connected to the channel by small pipes opening into the channel near or at the bottom.

As in the case of the closed venturi tube, a certain minimum uninterrupted length of channel is required before the venturi is reached, in order that the stream may be free from waves and vortices.

By carefully designing the flume, it is possible to simplify the actual instrument required to indicate the flow. If the channel is designed in such a manner that the depth in the exit channel at all rates of flow is less than a certain percentage of the depth in the entrance channel, the flume will function as a free-discharge outlet. Under these conditions, the upstream depth is independent of the downstream conditions, and the depth of water in the throat will maintain itself at a certain critical value, at which the energy of the water is at the minimum, whatever the rate of flow. When this is so, the quantity of water flowing through the channel is a function of the upstream depth h_1 only and may be expressed by the equation:

$$Q = kh_1^{3/2}$$

where k is a constant for a particular installation and can be determined.

It is now necessary to measure h_1 only, which can be done by means of a float in a well, connected to the upstream portion of the channel. This float operates an indicated recording and integrating instrument.

Other means of sensing the height in a flume or weir include up-looking ultrasonic sensors mounted in the bottom of the channel. More often used are down-looking ultrasonic sensors mounted above the flume. Direct pressure transducers mounted at the bottom of the channel or in a standpipe can also be used. Other methods, such as RF admittance or capacitance slides, are used as well.

The channel is usually constructed of concrete and the surface on the inside of the channel made smooth to reduce the friction between water and channel. Flumes of this kind are used largely for measuring flow of water or sewerage and may be made in a very large variety of sizes to measure anything from the flow of a small stream to that of a large river.

6.4.1.3 The DataGator Flowmeter

In the early 1990s experimentation showed that a combination venturi flume and venturi tube could be constructed such that the signal from three pressure transducers could be used to measure the flow through the tube in any flow regime: subcritical flow, supercritical flow, and surcharge. By making the flow tube symmetrical, it was shown to be possible to measure flow in either direction with the same accuracy. This patented device, called a *DataGator flowmeter* (see Figure 6.50), can be used to monitor flow in manholes. It has the advantage over any other portable sewer flow-monitoring device of being traceable to the U.S. National Institute of Standards and Testing (NIST) since it is a primary device like a flume or flow tube.

6.4.2 Velocity/Area Methods

In velocity/area methods, volume flow rate is determined by measurement of the two variables concerned (mean velocity and head), since the rate of flow is given by the equation

$$Q = V \cdot A \text{ m}^3$$

where area A is proportional to head or level.

The head/level measurement can be made by many of the conventional level devices described in Chapter 10 and therefore are not dealt with here. Three general techniques are used for velocity measurement: turbine current meter, electromagnetic, and ultrasonic. The techniques have already been discussed in the section on closed pipe flow, so we describe application only here.

6.4.2.1 Turbine Current Meter

In a current-meter gauging, the meter is used to give point velocity. The meter is sited in a predetermined cross-section in the flowstream and the velocity obtained. Since the meter only measures point velocity, it is necessary to sample throughout the cross-section to obtain mean velocity.

The velocities that can be measured in this way range from 0.03 to 3.0 m/s for a turbine meter with a propeller of 50 mm diameter. The disadvantage of a current-meter gauging is that it is a point and not a continuous measurement of discharge.

6.4.2.2 Electromagnetic Method

In this technique, Faraday's law of electromagnetic induction is utilized in the same way as for closed-pipe flow measurement (Section 6.3.4.1). That is, $E \propto BlV$, where E is EMF generated, B is magnetic field strength, l is width of river or channel in meters, and V is average velocity of the flowstream.

This equation applies only if the bed of the channel is insulated, similar to the requirement for pipe flowmeters. In practice it is costly to insulate a riverbed, and where this cannot be done, riverbed conductivity has to be measured to compensate for the resultant signal attenuation.

In an operational system, a large coil buried under the channel is used to produce a vertical magnetic field. The flow of water throughout the magnetic field causes an EMF to be set up between the banks of the river. This potential is sensed by a pickup electrode at each bank. This concept is shown diagrammatically in Figure 6.51.

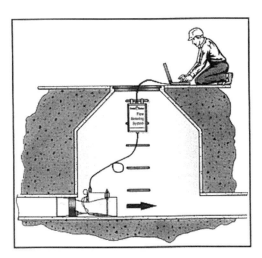

FIGURE 6.50 DataGator FlowTube. Courtesy of Renaissance Instruments.

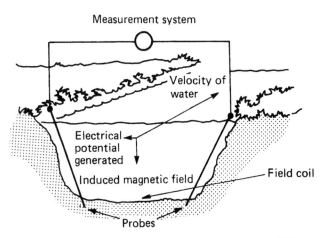

FIGURE 6.51 Principle of electromagnetic gauge. Courtesy of Plessey Electronic Systems Ltd.

6.4.2.3 Ultrasonic Method

As for closed-pipe flow, two techniques are available: single-path and multipath, both relying on time-of-flight techniques, as described in Section 6.3.4.2. Transducers capable of transmitting and receiving acoustic pulses are staggered along either bank of the river or channel. In practice the acoustic path is approximately 60° to the direction of flow, but angles between 30° and 60° could be utilized. The smaller the angle, the longer the acoustic path. Path lengths up to 400 meters can be achieved. New spool piece designs have included corner targets and other devices to improve the accuracy of the signal. Recently, clamp-on transit-time flow sensors have been adapted to work directly on the high-purity tubing used in the semiconductor manufacturing and in pharmaceutical industries. Correlation flowmeters have also been constructed using these new techniques.

6.4.3 Dilution Gauging

This technique is covered in detail in section 6.6 on flow calibration, but basically the principle involves injecting a tracer element such as brine, salt, or radioactive solution and estimating the degree of dilution caused by the flowing liquid.

6.5 POINT VELOCITY MEASUREMENT

In flow studies and survey work, it is often desirable to be able to measure the velocity of liquids at points within the flow pattern inside both pipes and open channels to determine either mean velocity or flow profile. The following techniques are most common: laser Doppler anemometer, hotwire anemometer, pitot tube, insertion electromagnetic, insertion turbine, propeller-type current meter, insertion vortex, and Doppler velocity probe.

6.5.1 Laser Doppler Anemometer

This device uses the Doppler shift of light scattered by moving particles in the flowstream to determine particle velocity and hence fluid flow velocity. It can be used for both gas and liquid flow studies and is used in both research and industrial applications.

Laser Doppler is a noncontact technique and is particularly suited to velocity studies in systems that would not allow the installation of a more conventional system—for example, around propellers and in turbines.

6.5.2 Hotwire Anemometer

The hotwire anemometer is widely used for flow studies in both gas and liquid systems. Its principle of operation is that a small, electrically heated element is placed within the flowstream; the wire sensor is typically 5μm diameter and approximately 5 mm long. As flow velocity increases, it tends to cool the heated element. This change in temperature causes a change in resistance of the element proportional to flow velocity.

6.5.3 Pitot Tube

The pitot tube is a device for measuring the total pressure in a flowstream (i.e., impact/velocity pressure and static pressure); the principle of its operation is as follows.

If a tube is placed with its open end facing into the flowstream (Figure 6.52), the fluid impinging on the open end will be brought to rest and its kinetic energy converted into pressure energy. The pressure buildup in the tube will be greater than that in the free stream by an amount termed the *impact pressure*. If the static pressure is also measured, the differential pressure between that measured by the pitot tube and the static pressure will be a measure of the impact pressure and therefore the velocity of the stream. In Equation (6.15), h, the pressure differential or impact pressure developed, is given by $h = (V_2^2/2g) - (V_1^2/2g)$, where $V_2 = 0$. Therefore, $V_1^2/2g$, that is, the pressure increases by $V_1^2/2g$. The negative sign indicates that it is an increase in pressure and not a decrease.

Increase in head:

$$h = V_1^2/2g \quad \text{or} \quad V_1^2 = 2gh \quad \text{i.e.} \quad V_1 = \sqrt{(2gh)} \quad (6.51)$$

However, since this is an intrusive device, not all of the flowstream will be brought to rest on the impact post; some will be deflected round it. A coefficient C is introduced to compensate for this, and Equation (6.50) becomes:

$$V_1 = C\sqrt{(2gh)} \quad (6.52)$$

If the pitot tube is to be used as a permanent device for measuring the flow in a pipeline, the relationship between the velocity at the point of its location to the mean velocity must be determined. This is achieved by traversing the pipe and sampling velocity at several points in the pipe, thereby determining flow profile and mean velocity.

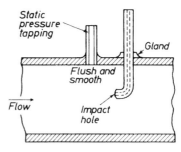

FIGURE 6.52 Single-hole pitot tube.

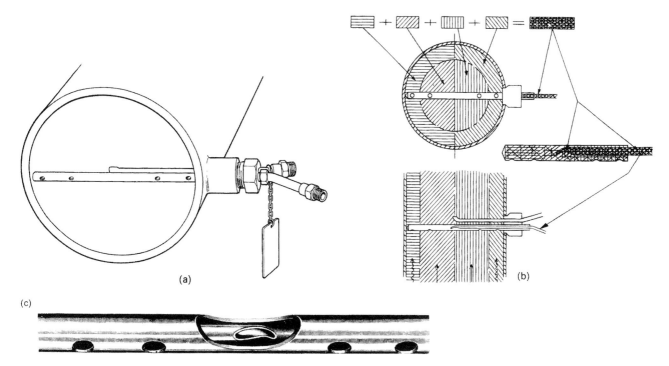

FIGURE 6.53 The Annubar. Courtesy of Emerson Process Management.

For more permanent types of pitot-tube installation, a multiport pitot tube (such as an Annubar) may be used as shown in Figure 6.53. The pressure holes are located in such a way that they measure the representative dynamic pressure of equal annuli. The dynamic pressure obtained at the four holes facing into the stream is then averaged by means of the "interpolating" inner tube (Figure 6.53(b)), which is connected to the high-pressure side of the manometer.

The low-pressure side of the manometer is connected to the downstream element, which measures the static pressure less the suction pressure. In this way a differential pressure representing the mean velocity along the tube is obtained, enabling the flow to be obtained with an accuracy of ±1 percent of actual flow.

6.5.4 Electromagnetic Velocity Probe

This type of device is basically an inside-out version of the electromagnetic pipeline flowmeter discussed earlier, the operating principle being the same. The velocity probe consists of either a cylindrical or an ellipsoidal sensor shape that houses the field coil and two diametrically opposed pickup electrodes.

The field coil develops an electromagnetic field in the region of the sensor, and the electrodes pick up a generated voltage that is proportional to the point velocity. The probe system can be used for either open-channel or closed-pipe flow of conducting liquids. It should be noted, however, that the accuracy of a point-velocity magnetic flowmeter

FIGURE 6.54 Multiple Sensor Averaging Insertion Magmeter. Courtesy of Marsh-McBirney Inc.

is approximately similar to that of a paddlewheel or other point-velocity meter. Although it shares a measurement technology with a highly accurate flowmeter, it is not one. Recently, a combination of the multiple-port concept of an Annubar-type meter with the point velocity magnetic flowmeter has been released, with excellent results. See Figure 6.54.

6.5.5 Insertion Turbine

The operating principle for this device is the same as for a full-bore pipeline flowmeter. It is used normally for pipe-flow velocity measurement in liquids and consists of a small turbine housed in a protective rotor cage, as shown

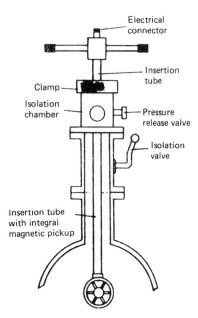

FIGURE 6.55 Insertion turbine flowmeter.

FIGURE 6.56 Propeller-type current meter. Courtesy of Nixon Instrumentation Ltd.

in Figure 6.55. In normal application the turbine meter is inserted through a gate valve assembly on the pipeline; hence it can be installed under pressure and can be precisely located for carrying out a flow traverse. Also, given suitable conditions, it can be used as a permanent flowmetering device in the same way as the pitot tube. The velocity of the turbine is proportional to liquid velocity, but a correction factor is introduced to compensate for errors caused by blockage in the flowstream caused by the turbine assembly.

6.5.6 Propeller-Type Current Meter

Similar to the turbine in operation, this type of velocity probe typically consists of a five-bladed PVC rotor (Figure 6.56) mounted in a shrouded frame. This device is most commonly used for river or stream gauging and has the ability to measure flow velocities as low as 2.5 cm/s. Propeller meters are often used as mainline meters in water distribution systems and in irrigation and canal systems as inexpensive alternatives to turbine and magnetic flowmeters.

6.5.7 Insertion Vortex

Operating on the same principle as the full-bore vortex meter, the insertion vortex meter consists of a short length of stainless-steel tube surrounding a centrally situated bluff body. Fluid flow through the tube causes vortex shedding. The device is normally inserted into a main pipeline via a flanged T-piece and is suitable for pipelines of 200 mm bore and above. It is capable of measuring flow velocities from 0.1 m/s up to 20 m/s for liquids and from 1 m/s to 40 m/s for gases.

6.5.8 Ultrasonic Doppler Velocity Probe

This device is more commonly used for open-channel velocity measurement and consists of a streamlined housing for the Doppler meter.

6.6 FLOWMETER CALIBRATION METHODS

There are various methods available for the calibration of flowmeters, and the requirement can be split into two distinct categories: *in situ* and laboratory. Calibration of liquid flowmeters is generally somewhat more straightforward than that of gas flowmeters since liquids can be stored in open vessels and water can often be utilized as the calibrating liquid.

6.6.1 Flowmeter Calibration Methods for Liquids

The main principles used for liquid flowmeter calibration are *in situ*: insertion-point velocity and dilution gauging/tracer method; laboratory: master meter, volumetric, gravimetric, and pipe prover.

6.6.1.1 In situ *Calibration Methods*

Insertion-Point Velocity One of the simpler methods of *in situ* flowmeter calibration utilizes point-velocity measuring devices (see Section 1.5) where the calibration device chosen is positioned in the flowstream adjacent to the flowmeter being calibrated and such that mean flow velocity can be measured. In difficult situations a flow traverse can be carried out to determine flow profile and mean flow velocity.

Dilution Gauging/Tracer Method This technique can be applied to closed-pipe and open-channel flowmeter calibration. A suitable tracer (chemical or radioactive) is injected at an accurately measured constant rate, and samples are taken from the flowstream at a point downstream of the injection point, where complete mixing of the injected tracer will have taken place. By measuring the tracer concentration in the samples, the tracer dilution can be established, and from this dilution and the injection rate the volumetric flow can be calculated. This principle is illustrated in Figure 6.57. Alternatively, a pulse of tracer material may be added to the flowstream, and the time taken for the tracer to travel a known distance and reach a maximum concentration is a measure of the flow velocity.

6.6.1.2 Laboratory Calibration Methods

Master Meter For this technique a meter of known accuracy is used as a calibration standard. The meter to be calibrated and the master meter are connected in series and are therefore subject to the same flow regime. It must be borne in mind that to ensure consistent accurate calibration, the master meter itself must be subject to periodic recalibration.

Volumetric Method In this technique, flow of liquid through the meter being calibrated is diverted into a tank of known volume. When full, this known volume can be compared with the integrated quantity registered by the flowmeter being calibrated.

Gravimetric Method Where the flow of liquid through the meter being calibrated is diverted into a vessel that can be weighed either continuously or after a predetermined time, the weight of the liquid is compared with the registered reading of the flowmeter being calibrated (see Figure 6.58).

Pipe Prover This device, sometimes known as a *meter prover*, consists of a U-shaped length of pipe and a piston or elastic sphere. The flowmeter to be calibrated is installed on the inlet to the prover, and the sphere is forced to travel the length of the pipe by the flowing liquid. Switches are inserted near both ends of the pipe and operate when the sphere passes them. The swept volume of the pipe between the two switches is determined by initial calibration, and this known volume is compared with that registered by the flowmeter during calibration. A typical pipe-prover loop is shown in Figure 6.59.

6.6.2 Flowmeter Calibration Methods for Gases

Methods suitable for gas flowmeter calibration are *in situ*: as for liquids; and laboratory: soap-film burette, water-displacement method, and gravimetric.

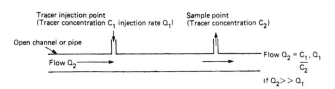

FIGURE 6.57 Dilution gauging by tracer injection.

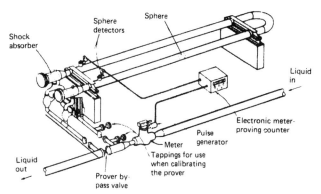

FIGURE 6.59 Pipe prover.

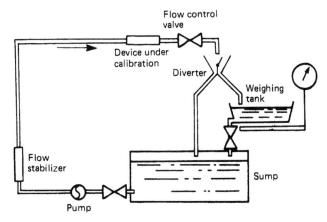

FIGURE 6.58 Flowmeter calibration by weighing. Courtesy of British Standards Institution.

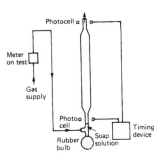

FIGURE 6.60 Gas flowmeter calibration: soap-film burette.

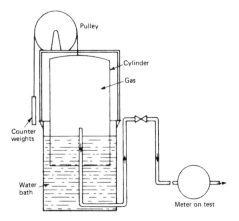

FIGURE 6.61 Water displacement method (bell prover).

6.6.2.1 Laboratory Calibration Methods

Soap-Film Burette This method is used to calibrate measurement systems with gas flows in the range of 10^{-7} to 10^{-4} m³/s. Gas flow from the meter on test is passed through a burette mounted in the vertical plane. As the gas enters the burette, a soap film is formed across the tube and travels up it at the same velocity as the gas. By measuring the time of transit of the soap film between graduations of the burette, it is possible to determine flow rate. A typical calibration system is illustrated in Figure 6.60.

Water-Displacement Method In this method a cylinder closed at one end is inverted over a water bath, as shown in Figure 6.61. As the cylinder is lowered into the bath, a trapped volume of gas is developed. This gas can escape via a pipe connected to the cylinder out through the flowmeter being calibrated. The time of the fall of the cylinder combined with the knowledge of the volume/length relationship leads to a determination of the amount of gas displaced, which can be compared with that measured by the flowmeter under calibration.

Gravimetric Method Here gas is diverted via the meter under test into a gas-collecting vessel over a measured period of time. When the collecting vessel is weighed before diversion and again after diversion, the difference will be due to the enclosed gas, and flow can be determined. This flow can then be compared with that measured by the flowmeter.

It should be noted that the cost of developing laboratory flow calibration systems as outlined can be quite prohibitive; it may be somewhat more cost-effective to have systems calibrated by the various national standards laboratories (such as NIST, NEL, and SIRA) or by manufacturers rather than committing capital to what may be an infrequently used system.

REFERENCES

BS 1042, *Methods for the Measurement of Fluid Flow in Pipes*, Part 1: Orifice Plates, Nozzles & Venturi Tubes, Part 2a: Pitot Tubes (1964).
BS 3680, *Methods of Measurement of Liquid Flow in Open Channels* (1969–1983).
BS 5781, *Specification for Measurement & Calibration Systems* (1979).
BS 5792, *Specification for Electromagnetic Flowmeters* (1980).
BS 6199, *Measurement of Liquid Flow in Closed Conduits Using Weighting and Volumetric Methods* (1981).
Cheremisinoff, N. P., *Applied Fluid Flow Measurement*, Dekker (1979).
Durrani, T. S. and Greated, C. A., *Laser Systems in Flow Measurement*, Plenum (1977).
Haywood, A. T. J., *Flowmeter: A Basic Guide and Sourcebook for Users*, Macmillan (1979).
Henderson, F. M., *Open Channel Flow*, Macmillan (1966).
Holland, F. A., *Fluid Flow for Chemical Engineers*, Arnold (1973).
International Organization for Standardization, ISO 3354 (1975), Measurement of Clean Water Flow in Closed Conduits (Velocity Area Method Using Current Meters).
Linford, A., *Flow Measurement and Meters*, E. & F. N. Spon.
Miller, R. W., *Flow Measurement Engineering Handbook*, McGraw-Hill (1982).
Shercliff, J. A., *The Theory of Electromagnetic Flow Measurement*, Cambridge University Press (1962).
Watrasiewisy, B. M. and Rudd, M. J., *Laser Doppler Measurements*, Butterworth (1975).

FURTHER READING

Akers, P. et al., *Weirs and Flumes for Flow Measurement*, Wiley (1978).
Baker, R. C., *Introductory Guide to Flow Measurement*, Mechanical Engineering Publications (1989).
Fowles, G., *Flow, Level and Pressure Measurement in the Water Industry*, Butterworth-Heinemann (1993).
Furness, R. A., *Fluid Flow Measurement*, Longman (1989).
Spitzer, D., *Flow Measurement*, Instrument Society of America (1991).
Spitzer, D., *Industrial Flow Measurement*, Instrument Society of America (1990).

Chapter 7

Measurement of Viscosity

K. Walters and W. M. Jones

7.1 INTRODUCTION

In his *Principia*, published in 1687, Sir Isaac Newton postulated that "the resistance which arises from the lack of slipperiness of the parts of the liquid, other things being equal, is proportional to the velocity with which parts of the liquid are separated from one another" (see Figure 7.1). This "lack of slipperiness" is what we now call *viscosity*. The motion in Figure 7.1 is referred to as *steady simple shear flow*, and if τ is the relevant shear stress producing the motion and γ is the velocity gradient ($\gamma = U/d$), we have

$$\tau = \eta\gamma \tag{7.1}$$

η is sometimes called the *coefficient of viscosity*, but it is now more commonly referred to simply as the viscosity. An instrument designed to measure viscosity is called a *viscometer*. A viscometer is a special type of *rheometer* (an instrument for measuring rheological properties), which is limited to the measurement of viscosity.

The SI units of viscosity are the pascal second = 1 Nsm^{-2} (= 1 kgm^{-1}s^{-1} and Nsm^{-2}). The CGS unit is the poise (= 0.1 kgm^{-1}s^{-1}) or the poiseuille (= 1 Nsm^{-2}). The units of kinematic viscosity υ (= η/ρ, where ρ is the density) are m^{2}s^{-1}. The CGS unit is the stokes (St) and 1 cSt = 10^{-6} m^{2}s^{-1}.

For simple liquids such as water, the viscosity can depend on the pressure and temperature, but not on the velocity gradient (i.e., shear rate). If such materials satisfy certain further formal requirements (e.g., that they are inelastic), they are referred to as *Newtonian* viscous fluids. Most viscometers were originally designed to study these simple Newtonian fluids. It is now common knowledge, however, that most fluid-like materials have a much more complex behavior, and this is characterized by the adjective *non-Newtonian*. The most common expression of non-Newtonian behavior is that the viscosity is now dependent on the shear rate γ, and it is usual to refer to the *apparent viscosity* $\eta(\gamma)$ of such fluids, where, for the motion of Figure 7.1,

$$\tau = \eta(\gamma)\gamma \tag{7.2}$$

In the next section, we argue that the concept of viscosity is intimately related to the flow field under investigation (e.g., whether it is steady simple shear flow or not), and in many cases it is more appropriate and convenient to define an *extensional viscosity* η_ε corresponding to a steady uniaxial extensional flow. Now, although there is a simple relation between the (extensional) viscosity η_ε and the (shear) viscosity η in the case of Newtonian liquids (in fact, $\eta_\varepsilon = 3\eta$ for Newtonian liquids), such is not the case in general for non-Newtonian liquids, and this has been one of the motivations behind the emergence of a number of *extensional viscometers* in recent years (see Section 7.5).

Most fluids of industrial importance can be classified as non-Newtonian: liquid detergents, multigrade oils, paints, printing inks, and molten plastics are obvious examples (see, for example, Walters, 1980), and no chapter on the measurement of viscosity would be complete without a full discussion of the application of viscometry to these complex fluids. This will necessitate an initial discussion of such important concepts as yield stress and thixotropy (which are intimately related to the concept of viscosity), as undertaken in the next section.

7.2 NEWTONIAN AND NON-NEWTONIAN BEHAVIOR

For Newtonian liquids, there is a linear relation between shear stress γ and shear rate γ. For most non-Newtonian materials, the *shear-thinning* behavior shown schematically in Figure 7.2 pertains. Such behavior can be represented by

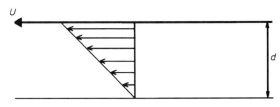

FIGURE 7.1 Newton's postulate.

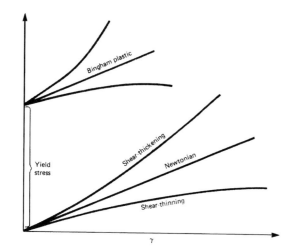

FIGURE 7.2 Representative (τ, γ) rheograms.

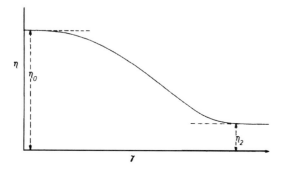

FIGURE 7.3 Schematic diagram of typical shear-thinning behavior.

the viscosity/shear-rate rheogram of Figure 7.3, where we see that the viscosity falls from a "zero-shear" value η_0 to a lower (second-Newtonian) value η_2. The term *pseudo-plasticity* was once used extensively to describe such behavior, but this terminology is now less popular. In the lubrication literature, shear thinning is often referred to as *temporary viscosity loss*.

Some non-Newtonian fluids—corn-flour suspensions, for example—show the opposite type of behavior in which the viscosity increases with shear rate (Figure 7.2). This is called *shear thickening*. In old-fashioned texts, the term *dilatancy* was often used to describe this behavior.

For many materials over a limited shear-rate range, a logarithmic plot of τ against γ is linear, so that

$$\tau = K\gamma^n \quad (7.3)$$

When $n > 1$, these so-called "power-law fluids" are shear-thinning, and when $n < 1$, they are shear-thinning.

An important class of materials will not flow until a critical stress, called the *yield stress*, is exceeded. These "plastic" materials can exhibit various kinds of behavior above the yield stress, as shown in Figure 7.2. If the rheogram above the yield stress is a straight line, we have what is commonly referred to as a *Bingham plastic material*.

In addition to the various possibilities shown in Figure 7.2, there are also important "time-dependent" effects exhibited

 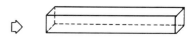

FIGURE 7.4 Uniaxial extensional deformation.

by some materials; these can be grouped under the headings *thixotropy* and *antithixotropy*. The shearing of some materials at a *constant* rate can result in a substantial lowering of the viscosity with time, with a gradual return to the initial viscosity when the shearing is stopped. This is called *thixotropy*. Paints are the most obvious examples of thixotropic materials. As the name suggests, *antithixotropy* involves an *increase* in viscosity with time at a constant rate of shear.

Clearly, the measurement of the shear viscosity within an industrial context is important and requires an understanding of material behavior. Is the material Newtonian or non-Newtonian? Is thixotropy important? Other questions come to mind.

Many industrial processes involve more extensional deformation than shear flow, and this has been the motivation behind the search for extensional viscometers, which are constructed to estimate a material's resistance to a stretching motion of the sort shown schematically in Figure 7.4. In this case, it is again necessary to define an appropriate stress T and rate of strain k and to define the extensional viscosity η_ε by

$$T = \eta_\varepsilon k \quad (7.4)$$

For a Newtonian liquid, η_ε is a constant ($\equiv 3\eta$). The extensional viscosity of some non-Newtonian liquids can take very high values, and it is this exceptional resistance to stretching in some materials, together with the practical importance of extensional flow, that makes the study of extensional viscosity so important. The reader is referred to the book *Elongational Flows* by Petrie (1979) for a detailed treatise on the subject. *Rheometers for Molten Plastics*, by Dealy (1982), on polymer-melt rheometry is also recommended in this context.

A detailed assessment of the importance of non-Newtonian effects is given in the text *Rheometry: Industrial Applications* (Walters, 1980), which contains a general discussion of basic principles in addition to an in-depth study of various industrial applications.

The popular book on viscometry by Van Wazer et al. (1963) and that of Wilkinson (1960) on non-Newtonian flow are now out of date in some limited respects, but they have stood the test of time remarkably well and are recommended to readers, provided the dates of publication of the books are appreciated. More modern treatments, developed from different but complementary viewpoints, are given in the books by Lodge (1974), Walters (1975), and Whorlow (1980). Again, the text by Dealy (1982) to is limited to polymer-melt rheometry, but much of the book is of general interest to those concerned with the measurement of viscosity.

7.3 MEASUREMENT OF THE SHEAR VISCOSITY

It is clearly impracticable to construct viscometers with the infinite planar geometry associated with Newton's postulate (Figure 7.1), especially in the case of mobile liquid systems, and this has led to the search for convenient geometries and flows that have the same basic steady simple shear flow structure. This problem has now been resolved, and a number of the so-called "viscometric flows" have been used as the basis for viscometer design. (The basic mathematics is nontrivial and may be found in the texts by Coleman et al., 1966; Lodge, 1974; and Walters, 1975.) Most popular have been (1) capillary (or Poiseuille) flow, (2) circular Couette flow, and (3) cone-and-plate flow. For convenience, we briefly describe each of these flows and give the simple operating formulae for Newtonian liquids, referring the reader to detailed texts for the extensions to non-Newtonian liquids. We also include, in Section 7.3.4, a discussion of the parallel-plate rheometer, which approximates closely the flow associated with Newton's postulate.

7.3.1 Capillary Viscometer

Consider a long capillary with a circular crosssection of radius a. Fluid is forced through the capillary by the application of an axial pressure drop. This pressure drop P is measured over a length L of the capillary, far enough away from both entrance and exit for the flow to be regarded as "fully developed" steady simple shear flow. The volume rate of flow Q through the capillary is measured for each pressure gradient P/L and the viscosity η for a Newtonian liquid can then be determined from the so-called Hagen-Poiseuille law:

$$Q = \frac{\pi P a^4}{8 \eta L} \quad (7.5)$$

The nontrivial extensions to (2.5) when the fluid is non-Newtonian may be found in Walters (1975), Whorlow (1980), and Coleman et al. (1966). For example, in the case of the power-law fluid (2.3), the formula is given by

$$Q = \frac{\pi n a^3}{(3n+1)} \left(\frac{ap}{2KL} \right)^{1/n} \quad (7.6)$$

One of the major advantages of the capillary viscometer is that relatively high shear rates can be attained.

Often, it is not possible to determine the pressure gradient over a restricted section of the capillary; it is then necessary, especially in the case of non-Newtonian liquids, to carefully study the pressure losses in the entry and exit regions before the results can be interpreted correctly (see, for example, Dealy, 1982, and Whorlow, 1980). Other possible sources of error include viscous heating and flow instabilities. These and other potential problems are discussed in detail by Dealy (1982), Walters (1975), and Whorlow (1980).

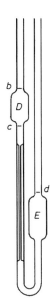

FIGURE 7.5 Schematic diagram of an Ostwald viscometer.

The so-called "kinetic-energy correction" is important when it is not possible to limit the pressure drop measurement to the steady simple shear flow region and when this is taken over the complete length L of the capillary. For a *Newtonian* fluid, the kinetic energy correction is given (approximately) by

$$P = P_0 - \frac{1.1 \rho Q^2}{\pi^2 a^4} \quad (7.7)$$

where P is the pressure drop required in (7.5), P_0 is the measured pressure drop, and ρ is the density of the fluid.

Since a gas is highly compressible, it is more convenient to measure the *mass* rate of flow, m. Equation (7.5) has then to be replaced by (see, for example, Massey, 1968)

$$\eta = \frac{\pi a^4 \bar{p} M P}{8 \dot{m} R T L} \quad (7.8)$$

where $\bar{p}$ is the mean pressure in the pipe, M is the molecular weight of the gas, R is the gas constant per mole, and T is the Kelvin temperature. The kinetic-energy correction (7.7) is still valid and must be borne in mind, but in the case of a gas, this correction is usually very small. A "slip correction" is also potentially important in the case of gases, but only at low pressures.

In commercial capillary viscometers for nongaseous materials, the liquids usually flow through the capillaries under gravity. A good example is the Ostwald viscometer (Figure 7.5). In this, b, c, and d are fixed marks, and there are reservoirs at D and E. The amount of liquid must be such that at equilibrium one meniscus is at d. To operate, the liquid is sucked or blown so that the other meniscus is now a few millimeters above b. The time t for the level to fall from b to c is measured. The operating formula is of the form

$$\upsilon = At - B/t \quad (7.9)$$

where υ is the kinematic viscosity ($\equiv \eta/\rho$). The second term on the right-hand side of Equation (7.9) is a correction factor for end effects. For any particular viscometer, A and B are given as calibration constants. Viscometers with pipes of different radii are supplied according to British Standards specifications; a "recommended procedure" is also given in B.S. Publication 188:1957.

Relying on gravity flow alone limits the range of measurable stress to between 1 and 15 Nm^{-2}. The upper limit can be increased to 50 Nm^{-2} by applying a known steady pressure of inert gas over the left-hand side of the U-tube during operation.

7.3.2 Couette Viscometer

The most popular rotational viscometer is the Couette concentric-cylinder viscometer. Fluid is placed in the annulus between two concentric cylinders (regarded as infinite in the interpretation of data), which are in relative rotation about their common axis. It is usual for the outer cylinder to rotate and for the torque required to keep the inner cylinder stationary to be measured, but there are variants, as in the Brookfield viscometer, for example, where a cylindrical bob (or sometimes a disc) is rotated in an expanse of test liquid and the torque on this same bob is recorded; see Section 7.4.

If the outer cylinder of radius r_0 rotates with angular velocity Ω_0 and the inner cylinder of radius r_1 is stationary, the torque C per unit length of cylinder on the inner cylinder for a Newtonian liquid is given by

$$C = \frac{4\pi \Omega_0 r_1^2 r_0^2 \eta}{\left(r_0^2 - r_1^2\right)} \quad (7.10)$$

so that measurement of C at each rotational speed Ω_0 can be used to determine the viscosity η. The extensions to (7.10) when the fluid is non-Newtonian are again nontrivial (unless the annular gap is very small), but the relevant analysis is contained in many texts (see, for example, Walters, 1975, and Whorlow, 1980). With reference to possible sources of error, end effects are obvious candidates, as are flow instabilities, misalignment of axes, and viscous heating. Detailed discussions of possible sources of error are to be found in Dealy (1982), Walters (1975), and Whorlow (1980).

7.3.3 Cone-and-plate Viscometer*

Consider the cone-and-plate arrangement shown schematically in Figure 7.6. The cone rotates with angular velocity

---

*The torsional-flow rheometer in which the test fluid is contained between parallel plates is similar in operation to the cone-and-plate rheometer, but the data interpretation is less straightforward, except in the Newtonian case (see, for example, Walters, 1975)

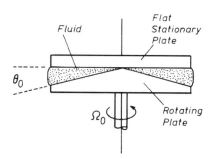

FIGURE 7.6 Basic cone-and-plate geometry.

Ω_0, and the torque C required to keep the plate stationary is measured. The gap angle θ_0 is usually very small ($<4°$), and, in the interpretation of results, edge effects are neglected. It is then easy to show that for a Newtonian liquid, the operating formula is

$$C = \frac{2\pi a^3 \Omega_0}{3\theta_0}\eta \quad (7.11)$$

where a is the radius of the cone.

In contrast to the capillary-flow and Couette flow situations, the operating formula for non-Newtonian fluids is very similar to (7.11) and is in fact given by

$$C = \frac{2}{3}a^3 \gamma \eta(\gamma) \quad (7.12)$$

where the shear rate γ is given by

$$\gamma = \Omega_0/\theta_0 \quad (7.13)$$

which is (approximately) constant throughout the test fluid, provided θ_0 is small ($<4°$, say). This is an important factor in explaining the popularity of cone-and-plate flow in non-Newtonian viscometry. Indeed, it is apparent from (7.12) and (7.13) that measurements of the torque C as a function of rotational speed Ω_0 immediately yield apparent viscosity/shear-rate data.

Sources of error in the cone-and-plate viscometer have been discussed in detail by Walters (1975) and Whorlow (1980). Measurements on all fluids are limited to modest shear rates (<100 s^{-1}), and this upper bound is significantly lower for some fluids, such as polymer melts.

Time-dependent effects such as thixotropy are notoriously difficult to study in a systematic way, and the constant shear rate in the gap of a cone-and-plate viscometer at least removes one of the complicating factors. The cone-and-plate geometry is therefore recommended for the study of time-dependent effects.

For the rotational viscometer designs discussed thus far, the shear rate is fixed and the corresponding stress is measured. For plastic materials with a yield stress, this might not be the most convenient procedure, and the last decade has seen the emergence of constant-stress devices, in which the shear *stress* is controlled and the resulting motion (i.e., shear rate) recorded. The Deer rheometer is the best known of the constant-stress devices; at least three versions of such an instrument are now commercially available. The cone-and-plate geometry is basic in current instruments.

7.3.4 Parallel-plate Viscometer

In the parallel-plate rheometer, the test fluid is contained between two parallel plates mounted vertically; one plate is free to move in the vertical direction so that the flow is of the plane-Couette type and approximates that associated with Newton's postulate.

A mass M is attached to the moving plate (of area A), and this produces a displacement x of the plate in a time t. If the plates are separated by a distance h, the relevant shear stress τ is given by

$$\tau = Mg/A \tag{7.14}$$

and the shear rate γ by

$$\gamma = x/th \tag{7.15}$$

so that the viscosity η is determined from

$$\eta = Mg\,th/xA \tag{7.16}$$

Clearly, this technique is only applicable to "stiff" systems, that is, liquids of very high viscosity.

7.4 SHOP-FLOOR VISCOMETERS

A number of ad hoc industrial viscometers are very popular at the shop-floor level of industrial practice; these usually provide a very simple and convenient method of determining the viscosity of *Newtonian* liquids. The emphasis on Newtonian is important because their application to non-Newtonian systems is far less straightforward (see, for example, Walters and Barnes, 1980). Three broad types of industrial viscometer can be identified (see Figure 7.7). The first type comprises simple rotational devices such as the Brookfield viscometer, which can be adapted in favorable circumstances to provide the apparent viscosity of non-Newtonian systems (see, for example, Williams, 1979). The instrument is shown schematically in Figure 7.8. The pointer and the dial rotate together. When the disc is immersed, the test fluid exerts a torque on the disc. This twists the spring, and the pointer is displaced relative to the dial. For Newtonian liquids (and for non-Newtonian liquids in favorable circumstances), the pointer displacement can be directly related to the viscosity of the test sample.

The second type of industrial viscometer involves what we might loosely call *flow-through constrictions* and is typified by the Ford-cup arrangement. The idea of measuring

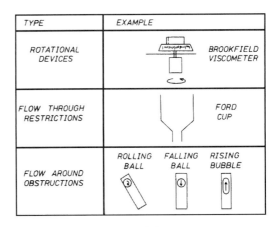

FIGURE 7.7 Classes of industrial viscometers.

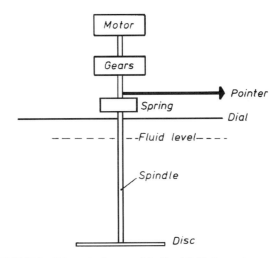

FIGURE 7.8 Schematic diagram of the Brookfield viscometer.

the viscosity of a liquid by timing its efflux through a hole at the bottom of a cup is very attractive. It is simple to operate and inexpensive, and the apparatus can be made very robust. Historically, the cup device was probably one of the first forms of viscometer ever used; today there are more than 50 versions of the so-called flow cups.

Often, results from flow cups are simply expressed as "time in seconds" (e.g., "Redwood seconds"), but for Newtonian liquids these can be converted to kinematic viscosity v through an (approximate) formula of the form

$$v = At - B/t \tag{7.17}$$

where A and B are constants that depend on the cup geometry (see, for example, Walters and Barnes, 1980).

The second term on the right-hand side of (7.17) is essentially a kinetic-energy correction. For Newtonian liquids, A and B can be determined by carrying out standard experiments on liquids of known kinematic viscosity.

A major disadvantage of the standard Ford cup so far as *non*-Newtonian liquids are concerned is that only one time can be taken, i.e., the time taken for the cup to empty. Such a measurement leads to a single (averaged) viscosity for a complicated deformation regime, and this is difficult to interpret consistently for rheologically complex fluids. Indeed, liquids with different rheologies as regard shear viscosity, extensional viscosity, and elasticity may behave in an identical fashion in a Ford-cup experiment (Walters and Barnes, 1980), so shop-floor instruments of this sort should be used with extreme caution when dealing with *non*-Newtonian liquids. The same applies to the "flow-around-obstacle" viscometers of Figure 7.7. Typical examples of this type of viscometer are the Glen Creston falling-ball viscometer and the Hoeppler rolling-ball instrument (see, for example, Van Wazer et al., 1963, and Cheng, 1979). Rising-bubble techniques may also be included in this category.

7.5 MEASUREMENT OF THE EXTENSIONAL VISCOSITY

Many industrial processes, especially in the polymer industries, involve a high extension-flow component and there is an acknowledged need for *extensional viscometers*. The construction of such devices is, however, fraught with difficulties. For example, it is difficult to generate an extensional deformation over a sufficient deformation rate range. Indeed, many of the most popular and sophisticated devices for work on polymer melts (such as those constructed at BASF in Germany) cannot reach the steady state required to determine the extensional viscosity η_ε defined in (7.4). Therefore, they are, as yet, of unproven utility. A full discussion of the subject of extensional viscometry within the context of polymer-melt rheology is provided by Dealy (1982).

In the case of more mobile liquid systems, it is difficult to generate flows which approximate to steady uniaxial extension; the most that can reasonably be hoped for is that instruments will provide an *estimate* of a fluid's resistance to stretching flows (see, for example, Chapter 1 of Walters, 1980). With this important proviso, the Ferguson Spin-Line Rheometer is a commercially available instrument that can be used on mobile liquids to provide extensional viscosity information.

7.6 MEASUREMENT OF VISCOSITY UNDER EXTREMES OF TEMPERATURE AND PRESSURE

Any of the techniques discussed to this point in the chapter can be adapted to study the effect of temperature and pressure on viscosity, provided that the apparatus can accommodate the prevailing extremes.

It is important to emphasize that viscosity is very sensitive to temperature. For example, the viscosity of water changes by 3 percent per Kelvin. It is therefore essential to control the temperature and to measure it accurately.

Pressure is also an important variable in some studies. In the case of lubricating oils, for example, high pressures are experienced during use, and it is necessary to know the pressure dependence of viscosity for these fluids.

At temperatures near absolute zero, measurements have been concerned with the viscosity of liquid helium. Recently, special techniques have been developed. Webeler and Allen (1972) measured the attenuation of torsional vibrations initiated by a cylindrical quartz crystal. Vibrating-wire viscometers have also been used. The resonant frequency and damping of the oscillations of a wire vibrating transversely in a fluid depend on the viscosity and density of the fluid. References to this and other work are given in Bennemann and Ketterson (1976).

To study the effect of high pressure, Abbot et al. (1981) and Dandridge and Jackson (1981) have observed the rate of fall of a sphere in lubricants exposed to high pressures (~3 GPa). Galvin, Hutton, and Jones (1981) used a capillary viscometer at high pressure to study liquids over a range of temperatures (0 to 150°C) and shear rates (0 to 3×10^5 sec^{-1}) with pressures up to 0.2 GPa. Kamal and Nyun (1980) have also adapted a capillary viscometer for high-pressure work.

7.7 ONLINE MEASUREMENTS

It is frequently necessary to monitor the viscosity of a fluid "online" in a number of applications, particularly when the constitution or temperature of the fluid is likely to change. Of the viscometers described in this chapter, the capillary viscometer and the concentric-cylinder viscometer are those most conveniently adapted for such a purpose. For the former, for example, the capillary can be installed directly in series with the flow and the pressure difference recorded using suitably placed transducers and recorders. The corresponding flow rate can be obtained from a metering pump.

Care must be taken with the online concentric cylinder apparatus because the interpretation of data from the resulting *helical* flow is not easy.

Other online methods involve obstacles in the flow channel; for example, a float in a conical tube will arrive at an equilibrium position in the tube depending on the rate of flow and the kinematic viscosity of the fluid. The parallel-plate viscometer has also been adapted for online measurement. These and other online techniques are considered in detail in *The Instrument Manual* (1975).

7.8 ACCURACY AND RANGE

The ultimate absolute accuracy obtained in any one instrument cannot be categorically stated in a general way. For example, using the Ostwald viscometer, reproducible measurements of time can be made to 0.3 percent. But to achieve this absolutely, the viscometer and the fluid must be scrupulously clean and the precise amount of fluid must be used. The temperature within the viscometer must also be uniform and

TABLE 7.1 Manufacturer's listed viscosity values by instrument type

Viscometer type	Lowest viscosity (poise)	Highest viscosity (poise)	Shear-rate range (S^{-1})
Capillary	2×10^{-3}	10^3	1 to 1.5×10^4
Couette	5×10^{-3}	4×10^7	10^{-2} to 10^4
Cone-and-plate	10^{-3}	10^{10}	10^{-4} to 10^3
Brookfield type	10^{-2}	5×10^5	10^{-3} to 10^6
Falling-ball, rolling-ball	10^{-4}	10^4	Indeterminate

be known to within 0.1 K. Obviously, this can be achieved, but an operator might well settle for 1 to 2 percent accuracy and find this satisfactory for his purpose with less restriction on temperature measurement and thermostating. Similar arguments apply to the Couette viscometer, but here, even with precise research instruments, an accuracy of 1 percent requires very careful experimentation.

The range of viscosities and rates of shear attainable in any type of viscometer depend on the dimensions of the viscometer—for example, the radius of the capillary in the capillary viscometer and the gap in a Couette viscometer.

By way of illustration, we conclude with a table of values (see Table 7.1) claimed by manufacturers of instruments within each type, but we emphasize that no one instrument will achieve the entire range quoted.

REFERENCES

Abbott, L. H., Newhall, D. H., Zibberstein, V. A., and Dill, J. F., *A.S.L.E. Trans.*, **24**, 125 (1981).
Barnes, H. A., Hutton, J. F., and Walters, K., *Introduction to Rheology*, Elsevier, New York (1989).
Bennemann, K. H., and Ketterson J. B., (eds), *The Physics of Liquid and Solid Helium*, Wiley, New York (Part 1, 1976; Part 2, 1978).
Cheng, D. C.-H., "A comparison of 14 commercial viscometers and a homemade instrument," Warren Spring Laboratory LR 282 (MH) (1979).
Coleman, B. D., Markovitz, H., and Noll, W., *Viscometric Flows of Non-Newtonian Fluids*, Springer-Verlag, Berlin (1966).
Dandridge, A., and Jackson, D. A., *J. Phys. D*, **14**, 829 (1981).
Dealy, J. M., *Rheometers for Molten Plastics*, Van Nostrand, New York (1982).
Galvin, G. D., Hutton, J. F., and Jones, B. J., *Non-Newtonian Fluid Mechanics*, **8**, 11 (1981).
The Instrument Manual, United Trade Press, p. 62 (5th ed., 1975).
Kamal, M. R., and Nyun, H., *Polymer Eng. and Science*, **20**, 109 (1980).
Lodge, A. S., *Body Tensor Fields in Continuum Mechanics*, Academic Press, New York (1974).
Massey, R. S., *Mechanics of Fluids*, Van Nostrand, New York (1968).
Petrie, C. J. S., *Elongational Flows*, Pitman, London (1979).
Van Wazer, J. R., Lyon, J. W., Kim, K. Y., and Colwell, R. E., *Viscosity and Flow Measurement*, Wiley-Interscience, New York (1963).
Walters, K., *Rheometry*, Chapman & Hall, London (1975).
Walters, K. (ed.), *Rheometry: Industrial Applications*, Wiley, Chichester, U.K. (1980).
Walters, K., and Barnes, H. A., *Proc. 8th Int. Cong. on Rheology, Naples, Italy*, p. 45, Plenum Press, New York (1980).
Webeler, R. W. H., and Allen, G., *Phys. Rev.*, **A5**, 1820 (1972).
Whorlow, R. W., *Rheological Techniques*, Wiley, New York (1980).
Wilkinson, W. L., *Non-Newtonian Fluids*, Pergamon Press, Oxford (1960).
Williams, R. W., *Rheol. Acta*, **18**, 345 (1979).

FURTHER READING

Bourne, M. C., *Food Texture and Viscosity: Concept and Measurement*, Academic Press, New York (1983).
Malkin, A. Y., *Experimental Methods of Polymer Physics: Measurement of Mechanical Properties, Viscosity and Diffusion*, Prentice-Hall, Englewood Cliffs, N.J. (1984).

Chapter 8

Measurement of Length

P. H. Sydenham

8.1 INTRODUCTION

Length is probably the most measured physical parameter. This parameter is known under many alternative names—displacement, movement, motion.

Length is often the intermediate stage of systems used to measure other parameters. For example, a common method of measuring fluid pressure is to use the force of the pressure to elongate a metal element, a length sensor then being used to give an electrical output related to pressure.

Older methods were largely mechanical, giving readout suited to an observer's eyes. The possibility of using electrical and radiation techniques to give electronic outputs is now much wider. Pneumatic techniques are also quite widely used, and these are discussed in Part 4.

Length can now be measured through over 30 decadic orders. Figure 8.1 is a chart of some common methods and their ranges of use. In most cases only two to three decades can be covered with a specific geometrical scaling of a sensor's configuration.

This chapter introduces the reader to the common methods that are used in the micrometer-to-subkilometer range.

For further reading, it may be noted that most instrumentation books contain one chapter or more on length measurement of the modern forms, examples being Mansfield (1973), Norton (1969), Oliver (1971), and Sydenham (1983, 1984). Mechanical methodology is more generally reported in the earlier literature on the subjects of mechanical measurements, toolroom gauging, and optical tooling. Some such books are Batson and Hyde (1931), Hume (1970), Kissam (1962), Rolt (1929), and Sharp (1970). In this aspect of length measurement the value of the older books should not be overlooked, since they provide basic understanding of a technique that is still relevant today in proper application of modern electronic methods.

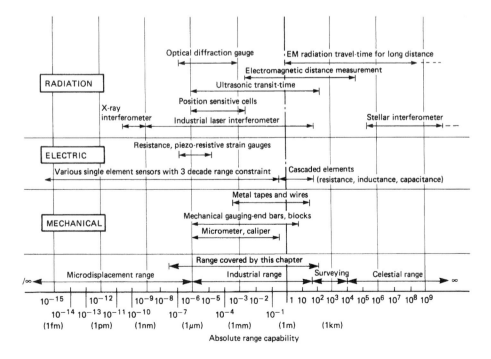

FIGURE 8.1 Ranges and methods of measuring length.

©2010 Elsevier Inc. All rights reserved.
DOI: 10.1016/B978-0-7506-8308-1.00008-5

For the microdisplacement range, see Garratt (1979), Sydenham (1969; 1972), and Woltring (1975); for larger ranges, see Sydenham (1968) and Sydenham (1971). Neubert (1975) has written an excellent analysis of transducers, including those used for length measurement.

8.2 THE NATURE OF LENGTH

Efficient and faithful measurement requires an understanding of the nature of the parameter and of the pitfalls that can arise for that particular physical system domain.

Length, as a measured parameter, is generally so self-evident that very little is ever written about it at the philosophical level. Measurement of length is apparently simple to conceptualize, and it appears easy to devise methods for converting the measured value into an appropriate signal.

Space can be described in terms of three length parameters. Three coordinate numbers describe the position of a point in space, regardless of the kind of coordinate framework used to define that point's coordinates. The number of coordinates can be reduced if the measurement required is in two dimensions. Measuring position along a defined straight line only requires one length-sensing system channel; to plot position in a defined plane requires two sensors.

Length measurements fall into two kinds: those requiring determination of the absolute value in terms of the defined international standard and those that determine a change in length of a gauge length interval (relative length). For relative length there is no need to determine the gauge interval length to high accuracy. Measuring the length of a structure in absolute terms is a different kind of problem from measuring strains induced in the structure.

Descriptive terminology is needed to simplify general description of the measuring range of a length sensor. Classification into microdisplacement, industrial, surveying, navigation, and celestial categories is included in Figure 8.1.

The actual range of a length sensor is not necessarily that of the size of the task. For example, to measure strain over a long test interval may use a long-range, fixed-length, standard structure that is compared with the object of interest using a short-range sensor to detect the small differences that occur (see Figure 8.2(b)). Absolute whole length measurement, Figure 8.2(a), requires a sensor of longer range.

It is often possible to measure a large length by adding together successive intervals—for example, by using a single ruler to span a length greater than itself. The same concept is often applied using multiple electronic sensors placed in line or by stepping a fixed interval along the whole distance, counting the number of coarse intervals, and subdividing the last partial interval by some other sensor that has finer sensing detail.

When a noncontacting sensor is used (see Figure 8.3), the length measurement is made by a method that does

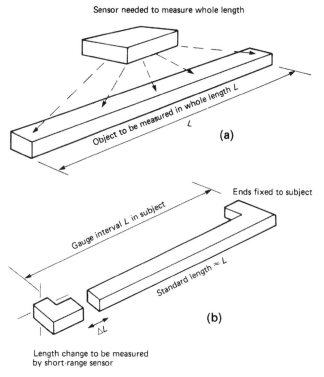

FIGURE 8.2 Absolute (a) and relative (b) length-measurement situations.

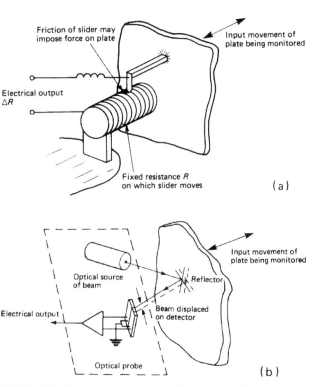

FIGURE 8.3 Examples of contacting and noncontacting length measurements. (a) Contacting, using variable resistance. (b) Noncontacting, using an optical probe.

not mechanically contact the subject. An example is the use of an optical interferometer to monitor position of a machine slide. It does not impose significant force on the slide and, as such, does not alter the measured value by its presence.

Contacting methods must be used with some caution lest they alter the measurement value due to the mechanical forces imposed by their presence.

8.3 DERIVED MEASUREMENTS

8.3.1 Derived from Length Measurement Alone

Length (m) comes into other measurement parameters, including relative length change (m/m), area (m^2), volume (m^3), angle (m/m), velocity (m^{-1}), and acceleration (m^{-2}). To measure position, several coordinate systems can be adopted. Figure 8.4 shows those commonly used. In each instance the general position of a point P will need three measurement numbers, each being measured by separate sensing channels.

The Cartesian (or rectangular) system shown in Figure 8.4(a) is that most adopted for ranges less than a few tens of meters. Beyond that absolute size it becomes very difficult to establish an adequately stable and calibratable framework. Errors can arise from lack of right angles between axes, from errors of length sensing along an axis, and from the imperfection of projection out from an axis to the point.

The polar system of Figure 8.4(b) avoids the need for an all-encompassing framework, replacing that problem with the practical need for a reference base from which two angles and a length are determined. Errors arise here in definition of the two angles and in the length measurement, which, now, is not restricted to a slide-way. Practical angle measurement reaches practical and cost barriers at around one arc-second of discrimination. This method is well suited to such applications as radar tracking of aircraft or plotting of location under the sea.

The preceding two systems of coordinate framework are those mostly adopted. A third alternative, which is less used, has, in principle, the fewest error sources. This is the triangular system shown as Figure 8.4(c). In this method three lengths are measured from a triangle formed of three fixed lengths. Errors arise only in the three length measurements with respect to the base triangle and in their definition in space. Where two or more points in space are to be monitored, their relative position can be obtained accurately, even if the base triangle moves in space. The major practical problem in adopting this method is that the three length measurements each require tracking arrangements to keep them following the point. The accuracy of pointing, however, is only subject to easily tolerated cosine forms of error,

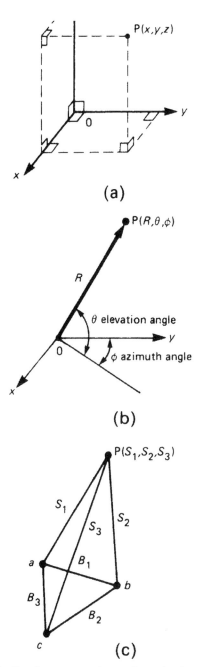

FIGURE 8.4 Coordinate systems that can be used to locate position in space. (a) Cartesian, rectangular, frame for three lengths. (b) Two polar directions and a length. (c) Triangulated lengths from a base triangle.

which allow relatively poor following ability to give quite reasonable values.

The three alternatives can also be combined to provide other arrangements, but in each case there will always be the need to measure three variables (as combinations of at least one length with length and/or angle) to define point position in a general manner. Where a translational freedom is constrained, the need to measure reduces to a simpler arrangement needing fewer sensing channels.

8.4 STANDARDS AND CALIBRATION OF LENGTH

With very little exception, length measurements are now standardized according to SI measurement unit definitions, length being one of the seven base units. It is defined in terms of the unit called the *meter*.

Until early 1982, the meter was defined in terms of a given number of wavelengths of krypton-86 radiation. Over the 1970s, however, it became clear that there were improved methods available that would enable definition with reduced uncertainty.

The first was to use, in the manner already adopted, the radiation available from a suitable wavelength-stabilized laser source, since this is easier to produce and is more reproducible than krypton radiation. At first sight this might be an obvious choice to adopt, but a quite different approach was also available, that which was recommended in 1982.

The numerical value of the speed of light c ($c = 299\ 792\ 458\ m^{-1}$) is the result of numerical standards chosen for the standards of time and of length. Thus the speed of light, as a numerical value, is not a fundamental constant.

Time standards (parts in 10^{14} uncertainty) are more reproducible in terms of uncertainty than length (parts in 10^8 uncertainty), so if the speed of light is defined as a fixed number, then in principle the time standard will serve as the length standard provided suitable apparatus exists to make the conversion from time to length via the constant c.

Suitable equipment and experimental procedures have now been proven as workable. By choosing a convenient value for c that suited measurement needs (that given earlier), in 1982 the signatories of the committee responsible for standardization of the meter agreed that the new definition should be, "The meter is the length of the path traveled by light in vacuum during the fraction (1/299, 792, 458) of a second."

This new definition takes reproducibility of length definition from parts in 10^8 to parts in 10^{10}. In common industrial practice, few people require the full capability of the standard, but adequate margin is needed to provide for loss of uncertainty each time the standards are transferred to a more suitable apparatus.

To establish the base standard takes many months of careful experimental work using very expensive apparatus. This method is not applicable to the industrial workplace due to reasons of cost, time, and apparatus complexity. The next level of uncertainty down is, however, relatively easily provided in the form of the industrial laser interferometer that has been on the market for several years. Figure 8.5 is such an equipment in an industrial application. The nature of the laser system shown is such that it has been given approval by the National Institute for Standards and Testing (NIST) for use without traceable calibration, since, provided the lengths concerned are not too small, it can give an uncertainty of around 1 part in 10^8, which is adequate for most industrial measurements.

FIGURE 8.5 Laser interferometer being used to measure length in an industrial application. Courtesy of Agilent, Inc.

Optical interferometer systems operate with wavelengths on the order of 600 nm. Subdivision of the wavelength becomes difficult at around 1/1000 of the wavelength, making calibration of submicrometer displacements very much less certain than parts in 10^8. In practice, lengths of a micrometer cannot be calibrated to much better than 1 percent of the range.

Laser interferometers are easy to use and very precise, but they, too, are often not applicable due to high cost and unsuitability of equipment. A less expensive calibration method that sits below the interferometer in the traceable chain uses the mechanical slip and gauge block family. These are specially treated and finished pieces of steel whose lengths between defined surfaces are measured using a more certain method, such as an interferometer. The length values are recorded on calibration certificates. In turn the blocks are used as standards to calibrate micrometers, go/no-go gauges, and electronic transducers. Mechanical gauges can provide of the order of 1 in 10^6 uncertainties.

For lengths over a few meters, solid mechanical bars are less suitable as standard lengths due to handling reasons. Flexible tapes are used; they are calibrated against the laser interferometer in standards facilities such as that shown in Figure 8.6. Tapes are relatively cheap and easy to use in the field compared with the laser interferometer. They can be calibrated to the order of a part in 10^6.

For industrial use, little difficulty will be experienced in obtaining calibration of a length-measuring device. Probably the most serious problem to be faced is that good calibration requires considerable time: The standard under calibration must be observed for a time to ensure that it does have the long-term stability needed to hold the calibration.

8.5 PRACTICE OF LENGTH MEASUREMENT FOR INDUSTRIAL USE

8.5.1 General Remarks

A large proportion of industrial range measurements can be performed quite adequately using simple mechanical gauging and measuring instruments. If, however, the requirement is for automatic measurement such as is needed in automatic inspection or in closed-loop control, the manual methods must be replaced by transducer forms of length sensor.

In many applications the needed speed of response is far greater than the traditional mechanical methods can yield. Numerically controlled mills, for instance, could not function without the use of electronic sensors that transduce the various axial dimensions into control signals.

Initially—that is, in the 1950s—the cost of electronic sensors greatly exceeded that of the traditional mechanical measuring tools, and their servicing required a new breed of technician. Most of these earlier shortcomings are now removed, and today the use of electronic sensing can be more productive than the use of manually read micrometers and scales because of the reduced cost of the electronic part of the sensing system and the need for more automatic data processing. There can be little doubt that solely mechanical instruments will gradually become less attractive in many uses.

8.5.2 Mechanical Length-Measuring Equipment

Measurement of length from a micrometer to fractional meters can be performed inexpensively using an appropriate

FIGURE 8.6 Tape-calibration facility at the National Measurement Laboratory, New South Wales, Australia. Courtesy of CSIRO.

mechanical instrument. These group into the familiar range of internal and external micrometers, sliding-jaw calipers, and dial gauges. Accuracy obtained with these depends a great deal on the quality of manufacture.

Modern improvements, which have been incorporated into the more expensive units, include addition of electronic transduction to give direct digital readout of the value, making them easier to read and suitable for automatic recording.

Another improvement, for the larger-throat micrometers, has been the use of carbon fiber for throat construction. This has enabled the frame to be lighter for a given size, allowing larger units and increased precision.

For the very best accuracy and precision work, use is made of measuring machines incorporating manually or automatically read optical scales. Figure 8.7 shows the modern form of such measuring machines. This is capable of a guaranteed accuracy of around 1 μm in length measurements up to its full range capability of 300 mm. Larger machines are also made that cover the range of around 4 m. These machines have also been adapted to provide electronic readout; see Section 8.6.

Where measurement of complex shapes is important, the use of measuring machines can be quite tedious; more speedy, direct methods can be used when the accuracy needed is not of the highest limits. In this aspect of toolroom measurement, the optical profile projector may be applicable.

In these (see Figure 8.8(a)), the outline, Figure 8.8(b), of the article to be measured is projected onto a large screen. This can then be compared with a profile template placed around the image. It is also possible to project an image of the surface of an article, Figure 8.8(c), onto the viewing screen. The two forms of lighting can also be used in combination.

8.5.3 Electronic Length Measurement

Any physical principle that relates a length to a physical variable has the potential to be used for converting length to another equivalent signal. The most used output signal is the electronic form. Thus, most length sensors use a transduction relationship that converts length into an electrical entity, either by a direct path or via one or more indirect conversion stages.

Most methods for smaller ranges make use of electromechanical structures, in which electrical resistance, inductance, or capacitance vary, or make use of time and spatial properties of radiation. Basic cells of such units are often combined to form larger-range devices that have similar discrimination and dynamic range to those given by the best mechanical measuring machines.

For best results, differential methods are utilized where practicable because they reduce the inherent errors (no transducers are perfect!) of the various systems by providing an in-built mechanism that compensates for some deficiencies of the adopted transducer principle. For example, to measure displacement it is possible to use two electrical plates forming a capacitor. As their separation varies, the capacitance alters to give a corresponding change in electrical signal. To use only one plate pair makes the system directly susceptible to variations in the dielectric constant of the material between the plates; small changes in the moisture content in an air gap can give rise to considerable error. By placing two plate pairs in a differential

FIGURE 8.7 Automatically-read length-measuring machine incorporating ruled scales. Courtesy of SIP Society Genevoise d'Instruments de Physique.

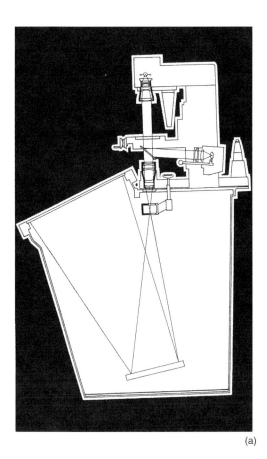

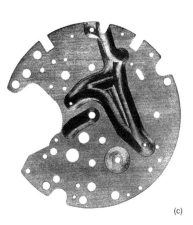

FIGURE 8.8 Measurement of geometry and lengths using the optical projector. (a) Optical system schematic. (b) Projected and enlarged image of profile of a component. (c) Oblique episcopic lighting to show surface detail. Courtesy of Henri Hauser Ltd.

connection, the effect of the air moisture can largely be cancelled out.

8.5.3.1 Electrical Resistance

In essence, some mechanical arrangement is made in which the electrical resistance between two ends of an interval is made to vary as the interval changes length.

Methods divide into two groups: those in which the resistance of the whole sensor structure remains constant, length being taken off as the changing position of a contact point, and those in which the bulk properties of the structure are made to change as the whole structure changes length.

In the first category is the slide-wire, in which a single wire is used. A coiled system can provide a larger resistance gradient, which is generally more suited to signal levels and impedance of practical electronic circuitry.

Figure 8.9(a) is a general schematic of sliding-contact length sensors. A standard voltage V_s is applied across the whole length of the resistance unit. The output voltage V_{out} will be related to the length l being measured as follows:

$$V_{out} = (V_s / L) \cdot l$$

Given that V_s and L are constant, V_{out} gives a direct measure of length l. The resistance unit can be supplied, V_s, with either DC or AC voltage. Errors can arise in the transduction process due to non-uniform heating effects causing resistance and length L change to the unit, but probably the most important point is that the readout circuit must not load the resistance; in that case the output-to-length relationship does not hold in a linear manner as shown previously.

Sliding-contact sensors are generally inexpensive but can suffer from granularity as the contact moves from wire to wire, in wound forms, and from noise caused by the mechanical contact of the wiper moving on the surface of the wire. Wear can also be a problem, as can the finite force imposed by the need to maintain the wiper in adequate contact. These practical reasons often rule them out as serious contenders for an application. Their use is, however, very simple to understand and apply. The gradient of the resistance with position along the unit can be made to vary according to logarithmic, sine, cosine, and other progressions. The concept can be formed in either a linear or a rotary form. Discrimination clearly depends on the granularity of the wire diameter in the wound types; one manufacturer has reduced this by sliding a contact along (rather than across) the wound wire as a continuous motion.

Resistance units can cover the range from around a millimeter to a meter, with discrimination on the order of up to 1/1000 of the length. Nonlinearity errors are on the same order.

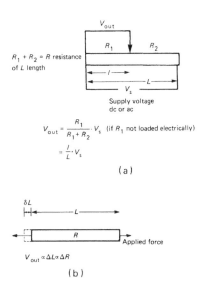

FIGURE 8.9 Electrical-resistance length sensors. (a) Sliding contact along fixed resistance unit. (b) Resistance change due to change of bulk properties of a resistance element induced by strain of the element.

The frequency response of such units depends more on the mechanical mass to be moved during dynamic changes, because the electrical part can be made to have low inductance and capacitance, these being the two electrical elements that decide the storage of electrical energy and hence slowness of electrical response.

Signal-to-noise performance can be quite reasonable but not as good as can be obtained with alternative inductive and capacitive methods; the impedance of a resistance unit set to give fine discrimination generally is required to be high with subsequent inherent resistance noise generation. These units are variously described in most general instrumentation texts. The alternative method, Figure 8.9(b), uses strain of the bulk properties of the resistance, the most used method being the resistance strain gauge. Because strain gauges are the subject of Chapter 4, they are not discussed further at this stage.

An alternative bulk resistance method that sometimes has application is to use a material such as carbon in disc form, using the input length change to alter the force of surface contact between the discs. This alters the pile resistance. The method requires considerable force from the input and, therefore, has restricted application. It does, however, have high electric current-carrying capability and can often be used to directly drive quite powerful control circuits without the need for electronic amplification. The bulk properties method can only transduce small relative length changes of an interval. Practical reasons generally restrict its use to gauge intervals of a few millimeters and to strains of that interval of around 1 percent.

8.5.3.2 Electrical Magnetic Inductive Processes

In general, the two main groups that use electrical inductive processes are those that vary the inductance value by geometry change and those that generate a signal by the law of electromagnetic induction.

An electrical inductance circuit component is formed by current-carrying wire(s) producing a magnetic field that tends to impede current change in dynamic current situations. The use of a magnetic circuit enhances the effect; it is not absolutely essential but is generally found in inductive sensors. Change of the magnetic field distribution of an inductor changes its inductance. Length sensors use this principle by using a length change of the mechanical structure of an inductance to vary the inductance. This can be achieved by varying the turns, changing the magnetic circuit reluctance, or inducing effects by mutual inductance. Various forms of electric circuit are then applied to convert the inductance change to an electronic output signal.

Figure 8.10 shows these three options in their primitive form. In some applications, such simple arrangements may suffice, but the addition of balanced, differential arrangements and use of phase-sensitive detection, where applicable, are often very cost-effective because the performance and stability are greatly improved. Now described, in more detail, are examples of the mainly used forms of Figures 8.10(b) and 8.10(c).

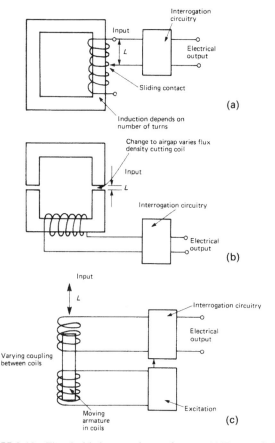

FIGURE 8.10 Electrical-inductance forms of sensor. (a) Turns variation with length change. (b) Reluctance variation with length change. (c) Mutual inductance change with length change.

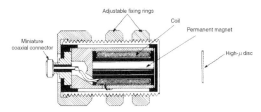

FIGURE 8.11 Magnetic-reluctance proximity sensor. Courtesy of Bruel & Kjaer Ltd.

Figure 8.11 shows a single-coil proximity detector that is placed close to a suitable, high magnetic permeability plate attached to or part of the subject. The sensor would be mounted around 2 mm from the plate. As the plate moves relative to the unit, the reluctance of the iron circuit, formed by the unit, the plate, and the air-gap, varies as the air-gap changes. When the unit has a permanent magnet included in the magnetic circuit, movement will generate a voltage without need for separate electronic excitation. It will not, however, produce a distance measurement when the system is stationary unless excited by a continuous AC carrier signal.

Where possible, two similar variable-reluctance units are preferred, mounted on each side of the moving object and connected into a bridge configuration giving common-mode rejection of unwanted induced noise pickup. These arrangements are but two of many possible forms that have been applied. Variable-reluctance methods are characterized by their relatively short range, poor linearity over longer ranges, and the possible need to move a considerable mass in making the measurement with consequent restricted dynamic performance.

Mutual-inductance methods also exist in very many forms, including the equivalent of Figure 8.11. Probably the most used is the linear variable-differential transformer (LVDT). Figure 8.12 shows a cross-section through a typical unit mounted for monitoring length change of a tensile test specimen. A magnetic material core, normally a ferrite rod, moves inside three coils placed end to end. The center coil is fed from an AC excitation supply, thus inducing voltages into the outer two coils. (It can also be wound over the other two outer coils.) The two generated voltages will be equal when the core is positioned symmetrically. The voltage rises in one coil relative to the other when the core is off-center. The difference between the two voltages is, therefore, related to the position of the core, and the relation can be made linear. Without circuitry to detect in which direction the core has moved from the null position, the output will be an AC signal of the excitation frequency, which changes in amplitude with position and has direction in the signal as its phase.

Practical use generally requires a DC output signal (actually a signal having frequency components in it that are present in the measured value's movement) with direction information as signal polarity. This is easily achieved, at marginal additional expense, by the use of *phase-

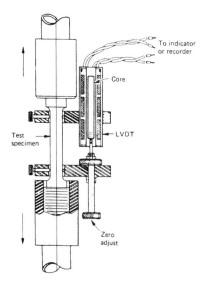

FIGURE 8.12 Cross-section of an LVDT inductive length sensor used to measure length change of a tensile test specimen. Courtesy of Schaevitz Engineering.

sensitive detection* (also known as *lock-in detection* or *carrier demodulation*). Figure 8.13(a) shows a block diagram of the subsystem elements that form a complete LVDT length-measuring system. Figure 8.13(b) shows the output relationship with position of the core. Modern units now often supply the phase-sensitive detection circuits inside the case of the sensor; these are known as DC LVDT units. Considerable detail of the operation of these and variations on the theme are available in Herceg (1976). Detail of phase-sensitive detection is included in Part 4 and Sydenham (1982b), where further references can be found in a chapter on signal detection by D. Munroe.

A simpler nontransformer form of the LVDT arrangement can be used in which the need for a separate central excitation coil is avoided. With reference to the LVDT unit shown in Figure 8.12, the two outer coils only would be used, their inner ends being joined to form a center-tapped linearly wound inductor. This split inductor is then placed in an AC bridge to which phase-sensitive detection is applied in a similar manner to recover a polarized DC output signal.

Inductive sensors of the mutual-inductance form are manufactured in a vast range of sizes, providing for length detection from as small as atomic diameters (subnanometer) to a maximum range of around ± 250 mm. They are extremely robust, quite linear, very reliable, and extremely sensitive if well designed. By mounting the core on the measured value object and the body on the reference frame, it is also possible to arrange for noninteracting, noncontact measurement. Frequency response depends on the carrier frequency used to modulate the coils; general practice sets that at around 50 kHz as the upper limit, but this is not a strong constraint on design. Attention may also need to be paid to mechanical resonances of the transducer.

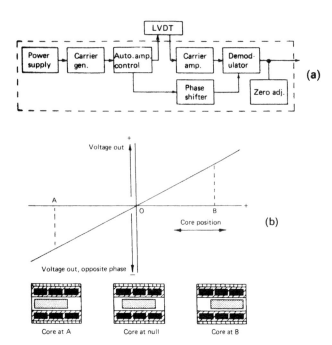

FIGURE 8.13 Phase-sensitive detection system used with practical LVDTs. (a) Block diagram of electronic system. (b) Input/output characteristics after phase-sensitive detection. Courtesy of Schaevitz Engineering.

The second class of magnetically inductive length sensors, the magnetoelectric kind, are those in which part of the electromagnetic circuit moves to generate a voltage caused by flux cutting. For these the relevant basic expression is $e = -(N\, d\phi/dt)$, where e is the voltage generated as N turns are cut by flux ϕ in time t. These are strictly velocity, not displacement, sensors, since the output is proportional to velocity. Integration of the signal (provided it has sufficient magnitude to be detected, since at low speeds the signal induced may be very small) will yield displacement. This form of sensor is covered in more detail in Chapter 6.

Magnetoelectric sensors are prone to stray magnetic field pickup that can introduce flux cutting that appears to be a signal. Good shielding is essential to reduce effects of directly detected fields, as well as for the more subtle induced eddy current fields that are produced in adjacent nonmagnetic materials if they are electrically conducting. Magnetic shielding is a highly developed engineering process that requires special materials and heat treatments to obtain good shielding. Manufacturers are able to provide units incorporating shielding and to advise users in this aspect. It is not simply a matter of placing the unit in a thick, magnetic material case! Herceg (1976) is a starting point on effective use of LVDT units and other inductive sensors.

When larger range is required than can be conveniently accommodated by the single unit (desired discrimination will restrict range to a given length of basic cell), it is possible to add inductive sensor cells, end to end, using digital logic to count the cells along the interval, with analog interrogation of the cell at the end of the measured value interval. This hybrid approach was extensively developed to provide for the meter distances required in numerically controlled machine tools. A form of long inductive sensor that has stood the test of time is the flat, "printed" winding that is traversed by a sensing short, flat coil. Each cell of this continuous grid comprises a coil pair overlapped by the sense coil that forms a flat profile LVDT form of sensor.

Angles can be measured electromagnetically using devices called *synchros*. These inherently include means for transmitting the information to a distance and are therefore described under Telemetry, in Part 4.

8.5.3.3 Electrical Capacitance Sensors

Electrical capacitance stores electrical energy in the electric field form; *electrical inductors* store energy in the magnetic field form. Electromagnetic theory relates the two forms of field; thus most concepts applied to magnetic-inductance sensing are applicable to electrical capacitance structures.

It is, therefore, also possible to sense length change by using the input length to alter the structure of a capacitance assembly or to cause shape change to a solid material, thereby directly generating charge. The electrical capacitance of a structure formed by two electrically conducting plates is given by

$$C = \frac{\varepsilon A}{l}$$

where C is the capacitance and ε the dielectric constant of the material between the plates in the area A where they overlap at a distance of separation l.

Thus a length sensor can be formed by varying the value of l, which gives an inverse relationship between length input and capacitance change or by varying one length of the two plates that combine to provide the overlap area A; this latter alternative can give a direct relationship. It is also sometimes practical to vary ε by inserting a moving piece of different dielectric into the capacitance structure.

Simpler forms use a single capacitance structure, but for these, variation in ε can introduce error due to humidity and pressure changes of air, the most commonly used dielectric. Differential systems are more satisfactory. Figure 8.14 shows the basic configuration of differential capacitance sensors. They can be formed from flat plates or from cylindrical sections, whichever is convenient. The cylindrical form of the second design has been used in a highly accurate alternative to the LVDT. The guard rings shown are not needed if some nonlinearity can be accepted.

Capacitance systems are characterized by high output impedance, need for relatively high excitation voltages, accuracy of plate manufacture, and small plate clearance dimensions.

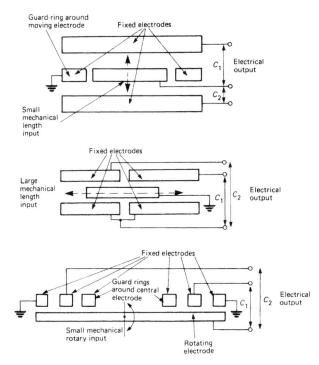

FIGURE 8.14 Some differential capacitance length-sensing structures.

Potential sensitivity of the alternatives (inductive and capacitive) is virtually the same in practice, each having its own particular signal-to-noise problems. For low-sensitivity use, capacitance devices are more easily manufactured than the inductive alternatives.

Noise occurs in capacitance systems from charge pickup produced by stray voltage potentials in the high impedance of the capacitance assembly. It is reduced by appropriate shielding and use of earthed guard plates that surround the active plates to collect and dump the unwanted charge. Capacitance structures lend themselves more to original equipment design than as ready-made sensor units applied after the basic plant is designed. This is because the layout of the working plant to be sensed can often directly provide one or more plates of the sensor as a noncontacting arrangement. For example, to monitor position of a pendulum in a tilt sensor, it is straightforward to use the pendulum bob as the central plate that moves inside two plates, added one to each side.

In general, therefore, it will be found that commercial sensor makers offer more inductive systems than capacitive alternatives, whereas in scientific and research work the tendency is to use capacitance systems because they are easier to implement at the prototype stage. Commercial suppliers also want to offer a product that is self-contained and ready to apply and, therefore, a unit that can be verified before delivery.

At extreme limits of discrimination (subnanometers), capacitance sensing, if properly designed, can be shown to be superior to inductive arrangements. As with inductive systems, they also need AC excitation to obtain a length change response for slowly moving events.

Forces exerted by the magnetic and electric fields of the two alternatives can be designed to be virtually nonexistent. Use in sensitive force balance systems allows the same plates to be used to apply a DC voltage to the central plate of a differential system. The plate can thus be forced back to the central null position at the same time as a higher-frequency excitation signal is applied for position detection.

Although the plate-separation capacitance method fundamentally provides an inverse relationship to length change, the signal can be made directly proportional by placing the sensing capacitance in the feedback path of an operational amplifier.

8.5.4 Use of Electromagnetic and Acoustic Radiation

Radiation ranging from the relatively long radio wavelengths to the short-wavelength X-rays in the electromagnetic (EM) radiation spectrum and from audio to megahertz frequencies in the acoustic spectrum has been used in various ways to measure length.

In the industrial range, the main methods adopted have been those based on optical and near-optical radiation, microwave EM, and acoustic radiation. These are now discussed in turn.

8.5.4.1 Position-sensitive Photocells

An optical beam, here to be interpreted as ranging in wavelength from infrared ($\approx 10\,\mu$m) to the short visible ($\approx 4\,\mu$m), can be used in two basically different ways to measure length. The beam can be used to sense movements occurring transverse to it or longitudinally with it.

Various position-sensitive optical detectors have been devised to sense transverse motion. Their range is relatively small, on the order of only millimeters. They are simple to devise and apply and can yield discrimination and stability on the order of a micrometer.

Figure 8.15 outlines the features of the structure of the three basic kinds that have been designed. Consider that of Figure 8.15(a). A beam with uniform radiation intensity across its cross-section falls on two equal characteristic photocells. When the beam straddles the two cells, thereby providing equal illumination of each, the differentially connected output will be zero. At that point any common-mode noise signals will be largely cancelled out by the system. Such systems have good null stability. As the beam moves to one side of the null, the differential output rises proportionally until all the beam's illumination falls on one cell alone. Direction of movement is established by the polarity of the output signal. Once the beam has become fully placed on one cell, the output is saturated and remains at its maximum. These cells can be manufactured from one silicon slice by sawing or diffusing a nonconducting barrier in the top

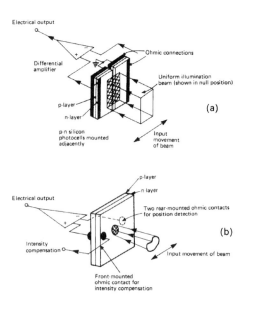

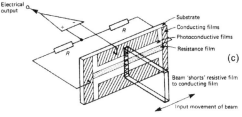

FIGURE 8.15 Optical position-sensitive detectors. (a) Split cell. (b) Lateral effect cell. (c) Photopotentiometer.

junction layer or can be made from separate cells placed side by side. Four cells, placed in two perpendicular directions in a plane, can be used to sense two axes of motion.

Linearity of the output of these cells depends on their terminating conditions. Working range can be seen to be equal to twice the beam width, which should not exceed the width of the half-detector size. Sensitivity depends on the level of beam illumination, so it is important to have constant beam intensity to obtain good results.

In some applications the light beam cross-section may vary with distance to the cell. In such cases the so-called Wallmark or lateral effect cell, Figure 8.15(b), may be more appropriate. In this case the two contacts are not rectifying as in the Figure 8.15(a) case but are, instead, ohmic.

It has been shown that the voltage produced between the two ohmic contacts is related to the position of the centroid of the beam's energy and to the intensity of the whole beam. Addition of the rectifying contact on the other side of the cell enables correction to be made for intensity changes, making this form of cell able to track the movements of a spot of radiation that changes in both intensity and size. Here also the beam must be smaller than the full working region of the cell surface. Detection limits are similar to those of the split cell form.

This cell has enjoyed a resurgence of interest in its design, for the original logarithmic voltage-to-position characteristic can quite easily be arranged to be effectively linear by driving the cell into the appropriate impedance amplifier. It, too, is able to sense the motion of a beam in two axes simultaneously by the use of two additional contacts placed at right angles to those shown.

Optical position-sensitive photocells such as these have found extensive use in conjunction with laser sources of radiation to align floors, ceilings, and pipes and to make precise mechanical measurement of geometry deviations in mechanical structures.

The third form of optical position detector is the photopotentiometer, shown in Figure 8.15(c). This form, although invented several years before microelectronic methods (it uses thick-film methods of manufacture), has also found new interest due to its printable form. The input beam of light falls across the junction of the conducting and resistive films, causing, in effect, a wiper contact action in a Wheatstone bridge circuit. The contact is frictionless and virtually stepless. The range of these units is larger than for the position-sensitive photocells, but they are rather specialized; few are offered on the market. Their response is somewhat slow (10 ms) compared with the cells detailed earlier (which have micro-second full scale times) due to the time response of the photoconductive materials used. The light beam can be arranged to move by the use of a moving linear shutter, a moving light source, or a rotating mirror.

Moiré fringe position-sensing methods use mechanical shuttering produced by ruled lines on a scale. These produce varying intensity signals at a reference position location and are in a fixed phase relationship; see Figure 8.16. These signals are interrogated to give coarse, whole-line cycle counts, with cycle division ranging from simple four-times digital division up to around 1 part in 100 of a cycle by the use of analog subdivision methods. Moiré fringe scales are able to provide large range (by butting sections) and a discrimination level at the subdivision chosen, which can be as small as 1 μm. Accuracy is limited by the placement of the lines on the scale, by the mounting of the scale, and by the temperature effects that alter its length. The method is suitable for both linear and rotary forms of sensing.

Although there was much developmental activity in Moiré methods in the 1950–1965 period, their design has stabilized now that the practicalities have been realized. Moiré methods have also found use for strain investigation, a subject discussed in Chapter 4.

The easily procured coherent properties of continuous wave laser radiation have provided a simple yet highly effective method for monitoring the diameter of small objects, such as wire, as they are being drawn. The radiation, being coherent, diffracts at the intersection with the wire, producing a number of diffraction beams at points past the wire. Figure 8.17 shows the method. As the wire size reduces, the

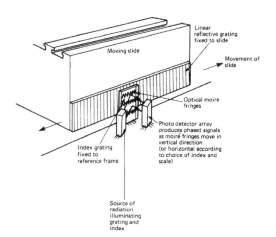

FIGURE 8.16 Moiré fringes used to measure length.

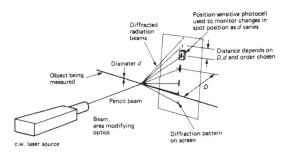

FIGURE 8.17 Use of diffraction to gauge fine diameters.

spots diverge, giving this method improved precision as the object size becomes smaller.

The position of the spots can be sensed with a position-sensitive photocell, as described earlier. Where digital output of beam movement is needed, it is also possible to use a linear array of photodetectors, each operating its own level detector or being scanned, sequentially, into a single output channel. Linear arrays with over 200 elements are now commonplace.

Optical lenses and mirrors can be used to alter the beam movement geometry to scale the subject's movement amplitude to suit that of an optical position detector.

8.5.4.2 Interferometry and Transit-Time Methods

Where long length (meters and above) measurements are needed, it is possible to use a suitable beam of radiation sensed in its longitudinal direction. Several different methods are available.

If the beam has time-coherent properties (either superimposed or inherently in the carrier), it will then be possible to use interference methods to detect phase differences between a reference part of the beam and that sent to the subject to be reflected back. Laser, microwave sources, and coherently generated acoustic radiation each can be used in this interference mode. The shorter the wavelength of the radiation, the smaller the potential discrimination that can be obtained. Thus with optical interferometers it is practical to detect length differences as small as a nanometer but only around a few millimeters when microwave radiation is used.

Figure 8.18 shows the basic layout of the optical elements needed to form a laser-based interferometer. That shown incorporates frequency stabilization and Zeeman splitting of the radiation features that give the system a highly stable and accurate measurement capability and a frequency, rather than amplitude form, of output representing length. A commercial unit is shown in Figure 8.5. Corner cubes are used instead of the flat mirrors originally used in the classical Michelson interferometer. They make adjustment very straightforward, for the angle of the cube to the incoming radiation is not critical. Usually one corner cube is held fixed to provide a reference arm. The other corner cube must be translated from a datum position to give length readings, for the method is inherently incremental. Allowing both corner cubes to move enables the system to be configured to also measure small angles to extreme precision.

The laser interferometer is, without doubt, the superior length-measuring instrument for general-purpose industrial work, but its cost, line-of-sight beam movement restriction, incremental nature, and need for path condition control (for the best work) does eliminate it from universal use. Its dynamic measuring range covers from micro-meters to tens of meters, a range not provided by any other electronic length sensor. Recent improvements in laser system design have resulted in practical measurement systems, even for industrial processes such as level in tanks and open channels.

Interferometry requires a reflector that gives adequate energy return without wavefront distortion. At optical wavelengths the reflecting surface must be of optical quality. Where very fine, micrometer discrimination is not required, the use of microwave radiation allows interferometry systems that can operate directly on to the normal machined surface of components being machined. Acoustic methods can yield satisfactory results when the required accuracy is only on the order of 1 part in 1,000.

Radiation methods can also use the time of flight of the radiation. For light this is around 300 mm in a nanosecond and for acoustic vibration from 300 mm to 6 m in a millisecond, depending on the medium. In "time-of-flight" methods the radiation, which can here be incoherent, is modulated at a convenient frequency or simply pulsed. The radiation returning from the surface of interest is detected and the elapsed time to go and return is used to calculate the distance between the source and the target. These methods do not have the same discrimination potential that is offered by interferometry but can provide, in certain applications (for EDM systems used over kilometer ranges), uncertainty on the order of a few parts in 10^6. Using more than one modulation frequency, it is possible to provide absolute ranging by this method, a feature that is clearly required for long-distance measurements. The need for a controlled movement path over which the reflector must traverse the whole distance, as is required in incremental interferometers, is unworkable in surveying operations.

The interference concept shown in Figure 8.18 for one-dimensional movement can be extended to three-dimensional measurement application in the holograph method. Holography provides a photographic image, in a plate known as a *hologram*, that captures the three-dimensional geometric detail of an object. The object of interest is flooded with coherent radiation, of which some reflects from the surfaces, to be optically combined with reference radiation taken directly from the source. Because the two are coherent radiations, their wavefronts combine to form a flat, two-dimensional interference pattern. The hologram bears little pictorial resemblance to the original object and has a most unexpected property. When the hologram is illuminated by coherent light, the object can be seen, by looking through the illuminated hologram, as an apparent three-dimensional object.

This basic procedure can be used in several forms to provide highly discriminating measurements of the shape of objects.

A first method places a similar object to that for which a hologram has been made in the image space of that of the hologram. This, in effect, superimposes the standard object over the real object. Differences between the two can then be decided by eye. This is not a very precise method, but it does suit some inspection needs.

In another method for using holography, a second hologram is formed on the same plate on which the first was exposed. The combined pair is developed as a single plate. When viewed, as explained earlier, this will reproduce an apparent object on which are superimposed fringes that represent shape differences between the two units. Each fringe width, as a guide, represents detailed differences on the order of the wavelength of the radiation. This form of holography is, therefore, a very powerful method for detecting small differences. It has been used, for example, to detect imperfections in car tires (by slightly altering the internal pressure) and to investigate shape changes in gas cylinders. It is very suitable for nondestructive testing but is expensive and somewhat slow in its use.

Fast-moving objects can also be gauged using optical holography in the so-called time-lapse pulse holography method. Two holograms are exposed on top of each other on an undeveloped plate, as mentioned earlier, but in this situation they are formed by the same object, which presents itself periodically at known times—for example, a turbine blade rotating inside an aircraft engine. The laser source is pulsed as the object passes using synchronized electronic circuitry.

Holography is suitable for use with any form of coherent radiation; optical, microwave, and acoustic systems have been reported. It is also possible to mix the radiations used at various stages to produce and view the hologram with different absolute size scales. For example, a seafloor sand-profile mapping system (see Figure 8.19) uses an acoustic interference hologram, which is then viewed by optical radiation for reasons of convenience.

The most serious disadvantages of holography are the cost of the apparatus, slowness to produce an output, and difficulties in obtaining numerical measurements from the recorded information.

Optical ways of measuring length are also discussed in Chapter 17.

8.5.5 Miscellaneous Methods

The preceding descriptions have shown that even for a few restricted classes of sensor, there are many principles that

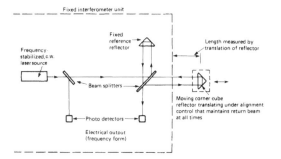

FIGURE 8.18 Basic layout of the frequency output form of laser length-measuring interferometer.

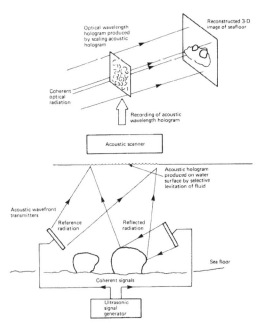

FIGURE 8.19 Holography applied to seafloor mapping. Two radiations are used to convert the image size for reasons of convenience.

can be used to produce transduced length signals. A comprehensive coverage would require several volumes on this parameter alone. This short subsection is included to emphasize the availability of many more methods that may be appropriate in given circumstances. Many of the unusual methods are less likely to be marketed for potential sales and would not justify quantity manufacture. In applications of the aerospace industry, in original equipment needs of science, and in industrial testing, development, and isolated applications, they may be the most viable methods to adopt. It should not be construed that lack of commercial interest implies that a method is necessarily unworkable. Here are a few of these less commonly used methods.

Magnetoresistive sensing elements are those in which their electrical resistance varies with the level of ambient magnetic field. These can be used as linear or as proximity sensors by moving the sensor relative to a field that is usually provided by a permanent magnet.

Thickness of a layer being deposited in a deposition chamber can be measured by several means. One way is to monitor the mass buildup on a test sample placed in the chamber alongside that being coated. Another method is to directly monitor the change in optical transmission during deposition.

Statistical calculation on the signal strength of an ionizing radiation source can be used to determine distance from the source.

Pressure formed or liquid displaced can be used to drive a pressure- or volume-sensitive device to measure movement of the driving element. This method has been used in the measurement of volumetric earth strain, since it can provide very sensitive detection due to the cube-law relationship existing between volume change input and length output.

The following chapter deals specifically with strain measurements. Chapter 11, on vibration, and Chapter 10, on level measurement, each include descriptions of length sensors.

8.6 AUTOMATIC GAUGING SYSTEMS

Toolroom and factory gauging has its roots in the use of manually read measuring machines and tools such as those shown in Figures 8.7 and 8.8. These required, in their non-electronic forms, high levels of skill in their use and are very time-consuming. In general, however, they can yield the best results, given that the best machines are used.

The advent of electronic sensing of length and angle has gradually introduced a transformation in the measurement practices in the toolroom and on the factory floor. The cost of providing, using manual procedures, the very many inspection measurements needed in many modern production processes has often proven to be uneconomical and far too slow to suit automatic production plants. For example, piston manufacture for a car engine can require over 20 length parameters to be measured for each piston produced

on the final grinding machine. The time available to make the measurements is on the order of fractions of minutes. It has, in such instances, become cost-effective to design and install automatic measuring machines that generate the necessary extensive data.

Automatic measuring systems are characterized by their ability to deliver electronic signals representing one or many length dimensions. In simple applications the operator places the component in a preset, multiprobe system. In a totally automated plant, use is often made of pick-and-place robots to load the inspection machines.

Automatic inspection systems began their development in the 1950s, when they were required to complement the then emerging numerically controlled metalworking machine tools. They used similar measuring sensors as the tools but differed from the metalworking machine in several ways.

Where inspection machines are hand operated, the operator can work best when the system effectively presents no significant inertial forces to the input probe as it is moved. This can be achieved by a design that minimizes the moving masses or by the use of closed-loop sensor control that effectively reduces the sluggish feel due to the inertia of the moving mass present. For small systems (those around a meter in capacity), multipoint inspection needs can be met economically by the use of short-range length sensors. These come into contact with the surfaces to be measured as the component is placed in the test setup. Values are recorded, stored, and analyzed. The component may need to be rotated to give total coverage of the surfaces of interest. Figure 8.20 shows such an apparatus being used to automatically inspect several length dimensions of a gearbox shaft.

When the size of the object to be inspected is large, the use of multiple probes can be too expensive, and a single

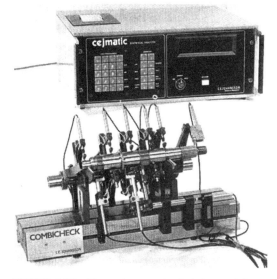

FIGURE 8.20 Electronic gauge heads being used in a versatile test apparatus set up to inspect several length parameters of a gearbox shaft. Courtesy of C. E. Johansson.

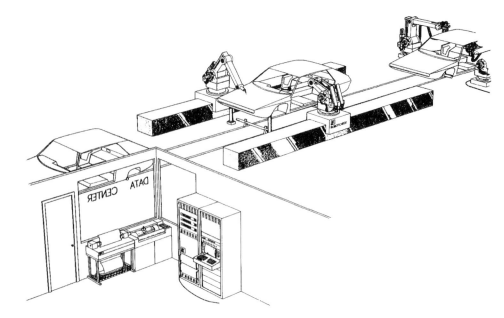

FIGURE 8.21 Robot, pick-and-place arm adapted to a production-line inspection measurement task. Courtesy of LK Tool Company and ASEA.

probe may be used to check given locations as a serial operation. Manual methods of point-to-point movement have, in some applications, given way to automatic, surface contour-following probes and to the use of robot arms that are preprogrammed to move as required (see Figure 8.21).

The reliability of electronics, its cost-effectiveness, and its capability to be rapidly structured into circuits that can suit new data-processing, recording, and display situations have made transducer measurement of length parameters a strong competitor for the traditional manually operated measuring machines. In most cases, however, the quality of length measurements made with automatic transducer methods still largely rests on the mechanical structures of the sensing systems and on the user's appreciation of the effects of temperature, operation of transducer, and presentation to the subject, all of which can generate sources of error.

REFERENCES

Batson, R. G., and Hyde, J. H., *Mechanical Testing, Vol. 1: Testing of Materials of Construction*, Chapman & Hall, London (1931).

Garratt, J. D., "Survey of displacement transducers below 50 mm," *J. Phys. E: Sci. Instrum.*, **12**, 563–574 (1979).

Herceg, E. E., *Handbook of Measurement and Control*, Schaevitz Engineering, Pennsauken (1976).

Hume, K. J., *Engineering Metrology*, Macdonald, London (1970).

Kissam, P., *Optical Tooling for Precise Manufacture and Alignment*, McGraw-Hill, New York (1962).

Mansfield, P. H., *Electrical Transducers for Industrial Measurement*, Butterworth, London (1973).

Neubert, K. K. P., *Instrument Transducers*, Clarendon Press, Oxford, 2nd ed. (1975).

Norton, H. N., *Handbook of Transducers for Electronic Measuring Systems*, Prentice-Hall, Englewood Cliffs, N.J. (1969).

Oliver, F. J., *Practical Instrumentation Transducers*, Pitman, London (1971).

Rolt, R. H., *Gauges and Fine Measurements*, 2 vols, Macmillan, London (1929).

Sharp, K. W. B., *Practical Engineering Metrology*, Pitman, London (1970).

Sydenham, P. H., "Linear and angular transducers for positional control in the decameter range," *Proc. IEE*, **115**, 7, 1056–1066 (1968).

Sydenham, P. H., "Position sensitive photo cells and their application to static and dynamic dimensional metrology," *Optica Acta*, **16**, 3, 377–389 (1969).

Sydenham, P. H., "Review of geophysical strain measurement," Bull, N. Z., *Soc. Earthquake Engng.*, **4**, 1, 2–14 (1971).

Sydenham, P. H., "Microdisplacement transducers," *J. Phys., E. Sci., Instrum.*, **5**, 721–733 (1972).

Sydenham, P. H., *Transducers in Measurement and Control*, Adam Hilger, Bristol; or ISA, Research Triangle (1984).

Sydenham, P. H., "The literature of instrument science and technology," *J. Phys. E.: Sci. Instrum.*, **15**, 487–491 (1982a).

Sydenham, P. H., *Handbook of Fundamentals of Measurement Systems Vol. 1 Theoretical Fundamentals*, Wiley, Chichester (1982b).

Sydenham, P. H., *Handbook of Measurement Science, Vol. 2: Fundamentals of Practice*, Wiley, Chichester, U.K. (1983).

Woltring, H. J., "Single and dual-axis lateral photo-detectors of rectangular shape," *IEEE Trans Ed.* 581–590 (1975).

FURTHER READING

Kafari, O., and Glatt, I., *Physics of Moiré Metrology*, Wiley, Chichester (1990).

Rueger, J. M., *Electronic Distance Measurement*, Springer-Verlag, New York (1990).

Small, G. D., *Survey of Precision Linear Displacement Measurement*, ERA Technology (1988).

Walcher, H., *Position Sensing: Angle and Distance Measurement for Engineers*, Butterworth-Heinemann, Oxford (1994).

Whitehouse, D. J., *Handbook of Surface Metrology*, Institute of Physics Publishing, London (1994).

Williams, D. C., *Optical Methods in Engineering Metrology*, Chapman & Hall, London (1992).

Chapter 9

Measurement of Strain

B. E. Noltingk

9.1 STRAIN

A particular case of length measurement is the determination of strains, that is, the small changes in the dimensions of solid bodies as they are subjected to forces. The emphasis on such measurements comes from the importance of knowing whether a structure is strong enough for its purpose or whether it could fail in use.

The interrelation between stress (the force per unit area) and strain (the fractional change in dimension) is a complex one, involving in general three dimensions, particularly if the material concerned is not isotropic, that is, does not have the same properties in all directions. A simple stress/strain concept is of a uniform bar stretched lengthwise, for which Young's modulus of elasticity is defined as the ratio stress: strain, that is, the force per unit of cross-sectional area divided by the fractional change in length:

$$E = \frac{F}{A} \div \frac{\Delta l}{l}$$

The longitudinal extension is accompanied by a transverse contraction. The ratio of the two fractions (transverse contraction)/(longitudinal extension) is called *Poisson's ratio*, denoted by μ, and is commonly about 0.3. We have talked of increases in length, called *positive strain;* similar behavior occurs in compression, which is accompanied by a transverse expansion.

Another concept is that of *shear*. Consider the block PQRS largely constrained in a holder (see Figure 9.1). If this is subjected to a force F as shown, it will distort, PQ moving to P'Q'. The *shear strain* is the ratio PP'/PT, that is, the angle PTP' or $\Delta \theta$ (which equals angle QUQ'), and the *modulus of rigidity* is defined as (shear stress)/(shear strain), or

$$C = \frac{F}{A} \div \Delta \theta$$

when A is the area PQ × depth of block. In practical situations, shear strain is often accompanied by bending, the magnitude of which is governed by Young's modulus.

There is some general concern with stress and strain at all points in a solid body, but there is a particular interest in measuring strains on surfaces. It is only there that conventional strain gauges can readily be used and wide experience has been built up in interpreting results gained with them. At a surface, the strain normal to it can be calculated because the stress is zero (apart from the exceptional case of a high fluid pressure being applied), but we still do not usually know the direction or magnitude of strains in the plane of the surface, so that for a complete analysis three strains must be measured in different directions.

The measurement of strain is also discussed in Chapter 17.

9.2 BONDED RESISTANCE STRAIN GAUGES

In the early 1940s, the bonded resistance strain gauge was introduced; it has dominated the field of strain measurement ever since. Its principle can be seen in Figure 9.2. A resistor R is bonded to an insulator I, which in turn is fixed to the substrate S whose strain is to be measured. (The word *substrate* is not used universally with this meaning; it is adopted here for its convenience and brevity.) When S is

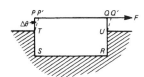

FIGURE 9.1 Shear strain.

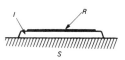

FIGURE 9.2 Principle of resistance strain gauge.

strained, the change in length is communicated to R if the bonding is adequate; it can be shown that the strain will be transmitted accurately, even through a mechanically compliant insulator provided there is sufficient overlap, that is, if I is larger than R by several times the thickness of either of them. Strains of interest are commonly very small; for elastic behavior, where concern is usually concentrated, strains do not exceed about 10^{-3}. Many metals break if they are stretched by a few percent, and changes in length of a few parts in a million are sometimes of interest, but when these are used to produce even small changes in the resistance of R, we can take advantage of the precision with which resistance can be measured to get a precise figure for strain.

The resistance of a conductor of length l, cross-sectional area A, and resistivity ρ is

$$R = \frac{\rho l}{A}$$

When a strain $\Delta l/l$ is imparted, it causes a fractional change of resistance

$$\frac{\Delta R}{R} = \left(1 + 2\mu + \frac{1}{\rho}\frac{\Delta \rho}{\Delta l}\right)\Delta l$$

since there will be a Poisson contraction in A and there may also be a change in resistivity. The ratio $(\Delta R/R)/(\Delta l/l)$ is called the *gauge factor* of a strain gauge. If there were no change in resistivity, it would be $1 + 2\mu$ about 1.6 for most metals, whereas it is found to be 2 or more, demonstrating that the sensitivity of strain gauges is increased by a change in ρ.

Nickel alloys are commonly used as the strain-sensitive conductor, notably Nichrome (nickel-chromium) and Constantan (copper-nickel), paper, epoxy resins and polyamide films are used for the backing insulator.

Strain gauges are available commercially as precision tools; units supplied in one batch have closely similar characteristics, notably a quoted gauge factor.

9.2.1 Wire Gauges

It is easier to measure resistances accurately when their values are not too low; this will also help avoid complications in allowing for the effect of lead resistances in series with the active gauge element. On the other hand, to measure strain effectively "at a point," gauges should not be too big; this calls for dimensions on the order of a centimeter, or perhaps only a few millimeters where very localized strains are to be studied. Both considerations point to the need for very fine wire, and diameters of 15–30 micrometers are used. The effective length is increased by having several elements side by side, as shown in Figure 9.3. Larger tags are attached at

FIGURE 9.3 Layout of wire gauge.

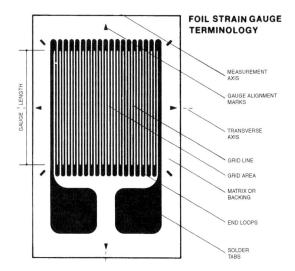

FIGURE 9.4 Shape of foil gauge. Courtesy of Micro-Measurements Division, Measurements Group Inc.

the ends of the strain-sensitive wire, to which leads can be connected.

9.2.2 Foil Gauges

An alternative to using wire is to produce the conductor from a foil—typically 4 micrometers thick—by etching. Figure 9.4 illustrates a typical shape. Foil gauges have the advantage that their flatness makes adhesion easier and improves heat dissipation (see the following discussion) as well as allowing a wider choice of shape and having the tags for the leads integral with the strain-sensitive conductor; they are in fact more widely used now than wire gauges.

9.2.3 Semiconductor Gauges

Another version of strain gauge employs semiconductor material, commonly silicon. Because the resistivity is higher, the sensitive element can be shorter, wider, and simpler (see Figure 9.5). The great advantage of semiconductor strain gauges is that their resistivity can be very sensitive to strain, allowing them to have gauge factors many times (typically 50 times) greater than those of simple metals, but they tend to have higher temperature sensitivity and are less linear. They

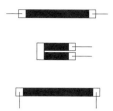

FIGURE 9.5 Examples of semiconductor gauges. Courtesy of Kulite Semiconductor Products Inc.

can be made integral with structural components and are used in this way for pressure measurement (see Chapter 14).

9.2.4 Rosettes

We pointed out earlier that a full analysis of strain involves measurements in more than one direction. In fact, three measurements are required on a surface because strain can be represented as an ellipse, for which the magnitudes and directions of the axes must be established. The directions chosen for strain measurements are commonly either at 120° or at 45° and 90° to each other.

If we are dealing with large structures, it might be expected that strain will vary only gradually across a surface, and three closely spaced individual gauges can be thought of as referring to the same point. When there is little room to spare, it is desirable to have the three gauges constructed integrally, which simplifies installation anyway. Such a unit is called a *rosette*. The three units may be either close together in one plane or actually stacked on top of each other (see Figure 9.6).

9.2.5 Residual Stress Measurement

The state of the surface at the time when a strain gauge is bonded to it has, of course, to be taken as the strain zero relative to which subsequent changes are measured. The gauge essentially measures increments of strain with increments of load. For many purposes of calculating stresses and predicting life, this is the most important thing to do.

However, during fabrication and before a gauge can be attached, some stresses can be locked up in certain parts, and it might be desirable to know these. This cannot be done nondestructively with any accuracy, but if we deliberately remove some material, the observed strain changes in neighboring material can tell us what forces were previously applied through the now absent material. One technique is to strain-gauge a small area of interest, noting the changes in the gauge readings as that area is freed by trepanning. An alternative procedure is to drill a simple hole inside an array of strain gauges that remain attached to the main surface; changes in the strain they show can again indicate what the residual stress was. An array for this purpose is shown in Figure 9.7.

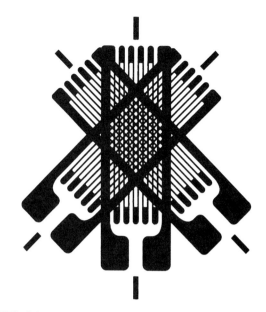

FIGURE 9.6 Rosette of gauges. Courtesy of Micro-Measurements Division, Measurements Group Inc.

FIGURE 9.7 Array of gauges for measuring residual stress. Courtesy of Micro-Measurements Division, Measurements Group Inc.

9.3 GAUGE CHARACTERISTICS

We have discussed the *gauge factor* at some length; that is what enables resistance to be used at all to measure strain. Other features of strain gauges are important for successful instrumentation. Information about the characteristics of particular gauges is available from manufacturers.

9.3.1 Range

The materials that strain gauges are made from cannot be expected to stretch by more than a few percent at most and still retain their properties in linear relationships; generally, nonlinearity is introduced before permanent damage occurs. Metals vary in the strain range over which they can be used; semiconductors have an appreciably shorter range. Although their limited range is an obvious theoretical restriction on the use of strain gauges, they can in fact cover most of the common field of interest for metals and other hard structural materials. Strain gauges are not generally suitable for use on rubber.

9.3.2 Cross-sensitivity

We have so far described the action of a strain gauge in terms of strain in the direction of the length of its conductor; this is the strain it is intended to measure. But, as explained, some strain is generally also present in the substrate in a direction at right angles to this, and gauges are liable to respond in some degree. For one thing, part of the conducting path may be in that direction; for another, the variation of resistivity with strain is a complex phenomenon. The cross-sensitivity of a gauge is seldom more than a few percent of its direct sensitivity and for foil gauges can be very small, but it should be taken into account for the most accurate work.

9.3.3 Temperature Sensitivity

The resistance of a strain gauge, as of most things, varies with temperature. The magnitude of the effect may be comparable with the variations from the strain to be measured, and a great deal of strain-gauge technology has been devoted to ensuring that results are not falsified in this way.

Several effects must be taken account of. Not only does the resistance of an unstrained conductor vary with temperature, but the expansion coefficients of the gauge material and of the substrate it is bonded to mean that temperature changes cause dimensional changes apart from those, resulting from stress, that it is desired to measure.

It is possible to eliminate these errors by compensation. Gauge resistance is commonly measured in a bridge circuit (see the following discussion), and if one of the adjacent bridge arms consists of a similar strain gauge (called a *dummy*) mounted on similar but unstressed material whose temperature follows that of the surface being strained, then thermal but not strain effects will cancel and be eliminated from the output.

Self-temperature compensated gauges are made in which the conductor material is heat treated to make its resistivity change with temperature in such a way as to balance out the resistance change from thermal expansion. Because the expansion coefficient of the substrate has an important effect, these gauges are specified for use on a particular material. The commonly matched materials are ferritic steel (coefficient $11 \times 10^{-6} K^{-1}$), austenitic steel ($16 \times 10^{-6} K^{-1}$), and aluminum ($23 \times 10^{-6} K^{-1}$).

9.3.4 Response Times

In practice, there are few fields of study in which strain gauges do not respond quickly enough to follow the strain that has been imposed. An ultimate limit to usefulness is set by the finite time taken for stress waves to travel through the substrate, which means that different parts of a strain gauge could be measuring different phases of a high-frequency stress cycle. But with stress-wave velocities (in metals) on the order of 5000 m/s, a 10 mm gauge can be thought of as giving a point measurement at frequencies up to 10–20 kHz. Of course, it is necessary that the measuring circuits used should be able to handle high-frequency signals.

It must be noted that strain gauges essentially measure the change in strain from the moment when they are fixed on. They do not give absolute readings.

Very slowly varying strains present particular measurement problems. If a strain gauge is to be used over periods of months or years without an opportunity to check back to its zero reading, errors will be introduced if the zero has drifted. Several factors can contribute to this: creep in the cement or the conductor, corrosion, or other structural changes. Drift performance depends on the quality of the installation; provided that it has been carried out to high standards, gauges used at room temperature should have their zero constant to a strain of about 10^{-6} over months. At high temperatures it is a different matter; gauges using ceramic bonding can be used with difficulty up to 500/600°C, but high-temperature operation is a specialized matter.

9.4 INSTALLATION

Sometimes strain gauges are incorporated into some measuring device from the design stage. More often they are used for a stress survey of a pre-existing structure. In either case it is most important to pay very close attention to correct mounting of the gauges and other details of installation. The whole operation depends on a small unit adhering securely to a surface, generally of metal. Very small changes in electrical resistance have then to be measured, necessitating close control of any possible resistances that may be either in series or parallel to that of interest—that is, from leads or leakage.

Of course, it is important to ensure that a gauge is mounted at the correct site—often best identified by tape. It may be noted that gauges can be mounted on cylindrical surfaces with quite small radii, but any double curvature makes fixing very difficult. We have already referred to the use of another gauge for temperature compensation. The introduction of any such "dummies" must be thought out; it is possible that an active gauge can be used for compensation, thus doubling the signal, if a place can be identified where the strain will be equal but opposite to that at the primary site—for example, the opposite side of a bending beam.

The surface where the gauge is to be fixed must be thoroughly cleaned—probably best by abrasion followed by chemical degreasing. Commonly used cements are cellulose nitrate up to 100°C, epoxy up to 200°C, and ceramic above that where special techniques must be used. Gauge manufacturers may specify a particular cement for use with their product.

After the gauge is fixed down, its leads should be fastened in position and connected (by soldering or spot-welding) to the gauge. It is most important for leads to be mounted securely to withstand the vibration to which they might be subject; in practice there are more failures of leads than in strain gauges themselves.

Unless the installation is in a friendly environment, it must then be protected by covering with wax, rubber, or some such material. The chief purpose of this covering is to exclude moisture. Moisture could cause corrosion and, also serious, an electrical leakage conductance. It must be remembered that 10^8 ohms introduced in parallel with a 350-ohm gauge appears as a three-in-a-million reduction in the latter; such a paralleling can be caused between leads or by an earth leakage, depending on the circuit configuration, and gives a false indication of strain of 3×10^{-6}.

The various stages of installation are illustrated in Figure 9.8.

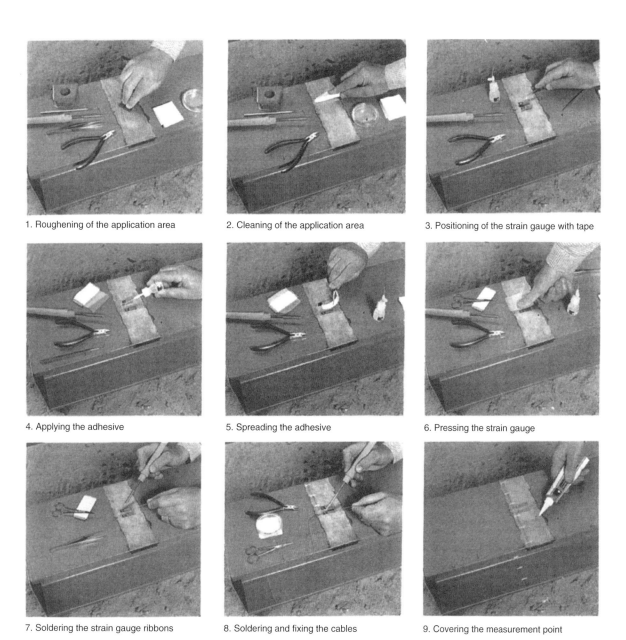

FIGURE 9.8 Stages and installing gauges. Courtesy of HBM.

9.5 CIRCUITS FOR STRAIN GAUGES

For measurement of its small resistance changes, a strain gauge is generally connected in a Wheatstone bridge. This may be energized with DC, but AC—at frequencies on the order of kilohertz—is more common; AC has the advantage of avoiding errors from the thermocouple potentials that can arise in the leads when the junctions of dissimilar metals are at different temperatures.

Gauges are often mounted some distance from their associated measuring equipment, and care must be taken that the long leads involved do not introduce errors. In a simple gauge configuration (see Figure 9.9), the two leads will be directly in series with the live gauge, and any changes in their resistance, for instance from temperature, will be indistinguishable from strain. This can be overcome by having three leads to the gauge (see Figure 9.10); one lead is now in series with each of two adjacent arms, thus giving compensation (for equal-ratio arms) provided that changes in one lead are reproduced in the other. The third lead, going to the power source, is not critical. These are called *quarter-bridge* arrangements.

A *half-bridge* setup is sometimes used (see Figure 9.11). This is when two strain gauges are both used in the measurement, as explained earlier.

The third possibility is a *full-bridge* setup (see Figure 9.12), when all four arms consist of gauges at the measurement site, the four long leads being those connecting the power source and out-of-balance detector. Their resistances are not critical,

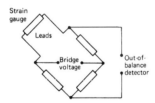

FIGURE 9.9 Simple bridge circuit (quarter-bridge).

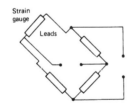

FIGURE 9.10 Quarter-bridge circuit with three long leads.

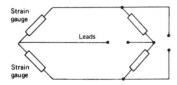

FIGURE 9.11 Half-bridge circuit.

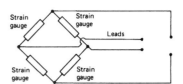

FIGURE 9.12 Full-bridge circuit.

so it is not necessary even to ensure that changes are equal. As with most bridge circuits, the power source and the detector can be interchanged. We have called the power source *bridge voltage*, implying that the supply is at constant potential; it can alternatively come as a constant current, which has some advantages for linearity.

To take up component tolerance, bridges can be balanced by fine adjustment of series or parallel elements in the arms. Instead, the zero can be set within the amplifier that commonly forms part of the detector. Changing a high resistance across a strain gauge can be used to simulate a known strain and so calibrate all the circuit side of the measuring system. It is possible to have measurements made in terms of the adjustment needed to rebalance a bridge after a strain has occurred, but more often the magnitude of the out-of-balance signal is used as an indication.

The larger the voltage or current applied to a strain-gauge bridge, the higher its sensitivity. The practical limit is set by self-heating in the gauge. If a gauge is appreciably hotter than its substrate, temperature errors are introduced. Compensation from a similar effect in another gauge cannot be relied on, because the cooling is unlikely to be identical in the two cases. Self-heating varies a great deal with the details of an installation, but, with metal substrates, it can generally be ignored below 1 milliwatt per square millimeter of gauge area.

We have described the basic circuitry as it concerns a single strain gauge. Tests are often made involving large numbers of gauges. For these, there is available elaborate equipment that allows a multiplicity of gauges to be scanned in quick succession or simultaneous recordings to be made of a number of high-speed dynamic strain measurements.

9.6 VIBRATING WIRE STRAIN GAUGE

Although bonded-resistance strain gauges are the type that has much the widest application, one or two other principles are made use of in certain situations. One of these is that of the vibrating-wire strain gauge.

If a wire is put under tension, its natural frequency of vibration (in its first mode) is

$$f = \frac{1}{2l}\sqrt{\frac{T}{m}}$$

where l is its length, T its tension, and m its mass per unit length.

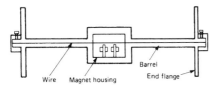

FIGURE 9.13 Vibrating-wire strain gauge. Courtesy of Strainstall Ltd.

FIGURE 9.14 Capacitive strain gauge. Courtesy of G.V. Planar Ltd.

The fixing points of the wire can be bonded to material of which the strain changes are to be measured. As the latter stretches, l changes, and, more significantly, T changes following Hooke's law. Strain can then be determined by monitoring natural frequency, easily done if the wire is magnetic and a solenoid is placed nearby to attract it. Wire diameter is typically 0.25 mm, operating frequency 1 kHz. The commonest circuit arrangement is to excite the wire with a single current pulse, measuring the frequency of the damped vibrations that ensue. A sketch of a typical device is given in Figure 9.13.

With the number of items that go to make up a vibrating-wire gauge, it is considerably larger (typically 100 mm long) than a bonded-resistance gauge. Because of the large force needed to stretch it, thought must be given to the gauge's mounting.

In fact, the largest application of such gauges has been to the measurement of *internal* strains in concrete, where both these factors are attended to. By embedding the gauge in the concrete when the concrete is cast, with electrical leads coming out to some accessible point, good bonding to the end-points is ensured. A large gauge length is desirable to average the properties of the material, which is very inhomogeneous on a scale up to a few centimeters. By choosing appropriate dimensions for the components of a vibrating-wire strain gauge, it is possible to make its effective elastic modulus the same as that of the concrete it is embedded in; the stress distribution in the bulk material will not then be changed by the presence of the gauge. A strain range up to 0.5 percent can be covered.

It has been found that vibrating-wire strain gauges can be very stable; the yielding that might be expected in the vibrating wire can be eliminated by pre-straining before assembly. With careful installation and use at room temperature, the drift over months can correspond to strains of less than 10^{-6}.

9.7 CAPACITIVE STRAIN GAUGES

It is possible to design a device to be fixed to a substrate so that when the latter is strained, the electrical *capacitance* (rather than the *resistance*) of the former is changed. Figure 9.14 is a diagram showing the principles of such a gauge. When the feet are moved nearer together, the arched strips change curvature, and the gap between the capacitor plates P changes. The greater complexity makes these devices several times more expensive than simple bonded-resistance strain gauges, and they are seldom used except when their unique characteristic of low drift at high temperature (up to 600°C) is important. They are most commonly fixed to metal structures by spot welding. Although the capacitance is only about a picofarad, it can be measured accurately using appropriate bridge circuits; because both plates are live to earth, the effects of cable capacitance can be largely eliminated.

9.8 SURVEYS OF WHOLE SURFACES

A strain gauge only gives information about what is happening in a small region. Sometimes it is desirable to take an overview of a large area.

9.8.1 Brittle Lacquer

One way of surveying a large area is to use brittle lacquer, though the technique is much less accurate than using strain gauges and does not work at all unless the strains are large. It has particular value for deciding where to put strain gauges for more accurate measurements.

A layer of special lacquer is carefully and consistently applied to the surface to be studied. The lacquer is dried under controlled conditions, whereupon it becomes brittle and therefore cracks if and when the part is strained above a certain threshold. Moreover, the higher the surface strain, the closer together are the cracks. It is best to coat a calibration bar in the same way and at the same time as the test part. Bending the bar then gives a range of known strains along it, and the crack pattern observed at different points on the live surface can be compared with the pattern on the calibration bar.

In this way, the critical points at which there are the highest stresses on the surface of a structure can be quickly recognized and the strain levels identified within about ±25 percent (of those levels) over a range of strains from 0.05 percent to 0.2 percent. Because the cracks form perpendicularly to the maximum principal stress, the technique has the considerable additional advantage of showing the directions of principal stresses all over a structure. Figure 9.15 shows the sort of crack pattern that can be observed.

9.8.2 Patterns on Surfaces

Large strains can be determined simply by inscribing a pattern on a surface and noting how it changes. Figure 9.16 shows the changes of shape observed in one such investigation.

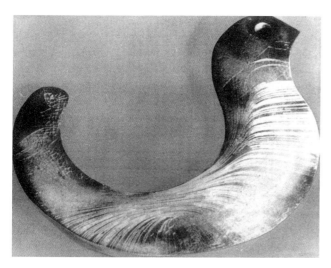

FIGURE 9.15 Brittle lacquer used on a specimen. Courtesy, Photolastic Division, Measurements Group Inc.

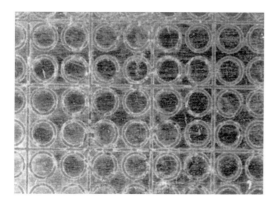

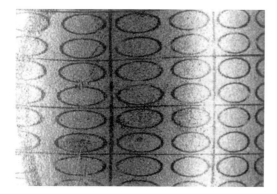

FIGURE 9.16 Distortion of a pattern under large strain. Courtesy of South Australian Institute of Technology.

Using Moiré fringes can increase the sensitivity to smaller strains. When a fine grating is seen through another grating that is comparable but not identical, dark and brighter regions—called *fringes*—will alternate. The dark regions correspond to where spaces in one grating block light that has come through gaps in the other grating; bright regions arise when the gaps are superposed. The small difference between the gratings can be one of orientation or of separation of their elements. For instance, if one grating has 1,000 lines per centimeter and its neighbor has 1,001 lines, there will be a dark fringe every centimeter, with bright ones in between them. Now suppose a 1,000-line grating is etched on a surface and another placed just above it; a strain of 10^{-3} in the surface will change the former into a 1,001-line grating and mean that fringes (seen in this case in light reflected off the surface) appear every centimeter. The larger the strain, the more closely spaced the fringes.

Fringes appearing parallel to the original grating lines will be a measure of the direct strain of one grating relative to the other. Fringes appearing perpendicular to the original grating lines will indicate rotation of one grating relative to the other. In general the fringes will be at an angle to the grating lines; therefore one pair of gratings will give a direct strain and a shear strain component. Two pairs, that is, a grid or mesh, will give two direct strains and the associated shear strain and will thus permit a complete surface strain determination to be made, that is, principal strains and principal strain directions.

9.9 PHOTOELASTICITY

For many years, the phenomenon of *photoelasticity* has been employed in experimental studies of stress distribution in bodies of various shapes. The effect used is the birefringence that occurs in some materials and its dependence on stress.

Light is in general made up of components polarized in different directions; if in a material the velocities of these components are different, the material is said to be *birefringent*. The birefringence is increased (often from zero in stress-free material) by stress, commonly being proportional to the difference in stress in the two directions of polarization. The effect was originally discovered in glass; synthetic materials, notably epoxy resins, are much more commonly used now.

In practice, a model of the structure to be examined is made out of photoelastic material. This is placed in a rig, or *polariscope*, such as the one shown in Figure 9.17, and loaded in a way that corresponds to the load imposed on the original. By recombining components of the light ray with polarizations parallel to the two principal stresses, fringes are produced, and their position and number give information about strain in the model. The first-order fringe occurs when there is a phase difference of 360° between the components; the *n*th order occurs when there is a 360°*n* difference.

As discussed for strain gauges, both the direction and the magnitude of the principal stresses come into stress analysis, and this complicates the situation. To find directions, a polariscope system as indicated in Figure 9.18 can be used; if the axes of polarizer and analyzer are at right angles and parallel to the principal stresses, there will be no

FIGURE 9.17 Polariscope in use. Courtesy of Sharples Stress Engineers Ltd.

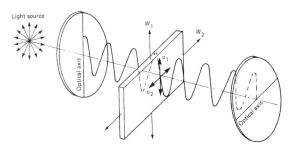

FIGURE 9.18 Use of a polariscope to determine principal stress directions. Courtesy of Sharples Stress Engineers Ltd.

interference, just a black spot, whatever the load may be. So the *isoclinics*—loci of points having the same direction of principal stresses—can be established.

For example, if a two-dimensional model of a loaded notched beam were examined in such a crossed polariscope, two types of fringes would be observed: a black fringe, the isoclinic, joining all points where the principal stress directions were parallel to the axes of polarization, and colored fringes, the *isochromatics*, contours of equal principal stress difference. The first-order isochromatic would pass through all points of the model where the stress had a particular value of $P-Q$, where P and Q are the two principal stresses. Similarly, the nth-order isochromatic would pass through all points where the stress had n times that value.

Using simple tensile calibration strips it is possible to determine the value that corresponds to each fringe order. Since the stress normal to the unloaded boundaries of the model is zero, that is, $Q = 0$, it is a relatively simple matter to determine the stress all along the unloaded boundaries. Determination of the stresses in the interior of the model is also possible but requires complex stress separation.

Normally a monochromatic light source is used, but white light has the advantage that the first-order fringe can be distinguished from higher orders.

Birefringence all along the ray path through the model will contribute to the phase difference between the two optical components. The effect measured is therefore an integral of stress along that path. This means that fringe patterns in photoelasticity are most easily interpreted with thin—or effectively two-dimensional—models. There is a technique for studies in three dimensions; this is to load the model in an oven at a temperature high enough to "anneal" out the birefringence. Subsequent slow cooling means that when the load is removed at room temperature, birefringence is locked into the unloaded model, which can then be carefully sliced; examination of each slice in a polariscope will show the original stresses in that particular region.

REFERENCES

Holister, G. S., *Experimental Stress Analysis, Principles and Methods*, Cambridge University Press, Cambridge (1967).

Kuske, A., and G. Robertson, *Photoelastic Stress Analysis*, Wiley, New York (1974).

Theocaris, P. S., *Moiré Fringes in Strain Analysis*, Pergamon Press, Oxford (1969).

Window, A. L., and G. S. Holister (eds.), *Strain Gauge Technology*, Allied Science Publishers, London (1982).

Chapter 10

Measurement of Level and Volume

P. H. Sydenham and W. Boyes

10.1 INTRODUCTION

Many industrial and scientific processes require knowledge of the quantity of content of tanks and other containers. In many instances it is not possible or practical to directly view the interior. Parameters of interest are generally level of the contents, volume, or simply the presence of substances. Sensors may be needed (see Figure 10.1) to detect high- or low-level states for alarm or control use, to provide proportional readout of the level with respect to a chosen datum, for automatic control purposes, or for manually read records.

Simple installations may be able to adopt inexpensive manual methods such as a dipstick. Automatic control applications, however, will require control signals for operation of a process actuator or alarm. For these cases, several options of output signal are available; they include indirect electric contacts and electronic proportional outputs as well as direct flow and pneumatic valving.

As well as the obvious liquid substances such as water, oil, petroleum, and milk, level sensors have been successfully applied to measurement of solids such as flour, mineral ores, food grain, and even potatoes and coal. Two-phase systems are also often measured—for example, liquid and froth levels in beer making and for mineral slurries.

Due to the extensive need for this basic process parameter, many types of level instruments are available; their installation details are important. A useful tutorial introduction to level measurement is available in Lazenby (1980) and Norton (1969). O'Higgins (1966) and Miller (1975) also make valuable contributions.

10.2 PRACTICE OF LEVEL MEASUREMENT

10.2.1 Installation

Suppliers of level-sensing systems generally provide good design support for installation, enabling the prospective user to appreciate the practical problems that arise and offering wide variation in the sensor packaging and systems

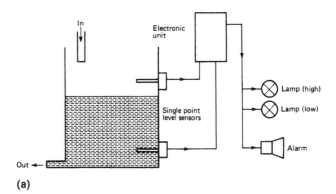

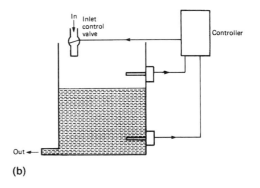

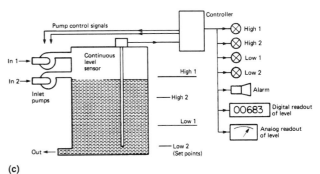

FIGURE 10.1 Schematic diagrams of level instrument installations. (a) High–low level detectors driving lamps and an alarm. (b) High–low detectors operating on inlet valve. (c) Full control of two inlet pumps using a continuous detector readout over full range of capacity, one or more settable high–low trigger positions, and appropriate lamps and alarms.

arrangements. However, it is often difficult to select the appropriate sensor for a particular system, even with support from a supplier. If the system is other than a simple one, great care should be taken in product selection. Then, even greater care should be taken in the application of the sensor itself.

Put the sensor where it will not be affected by the turbulence caused by the product flowing in and out of the vessel. Positioning to control errors is also important. For example, when a stilling tube is placed outside the container, its contents may be at a different temperature from those in the container. The complete sensor may need to be fully removable without imposing the need to empty the container.

It is often necessary to incorporate followers, such as shown in Figure 10.2, in the sensing system to constrain the unwanted degrees of freedom of such components as floats.

Corrosion effects on the components of the sensing arrangements, caused by the contents, must also be carefully considered. High temperatures, corrosive materials, and abrasion in granular-material measurement can progressively alter the characteristics of the system by producing undue friction, changing the mass of floats and simply reducing the system to an unworkable state.

In recent years this has caused more reliance on noncontacting or noninvasive means of measuring level. When these devices are properly applied, they have the potential for indefinitely long operational life.

The level sensor should preferably have built-in backlash, which provides a toggle action (hysteresis) so that the sensing contacts for simple high- or low-level alarms or switch controls do not dither at the control point. Some systems incorporate this feature in their mechanical design (an example is given later, in Figure 10.5), others in their electronics.

Nuclear level gauges must be installed in accordance with the specific regulatory guidelines of each country. In the United States, those guidelines are set by the Nuclear Regulatory Commission (NRC). These gauges use very small-strength sources and pose an absolutely minimal hazard in themselves. They are described in Part 3.

If done without due consideration, the installation itself may introduce a hazard due to the nature of the mechanical components of the system causing blockages in it.

10.2.2 Sources of Error

A first group of errors are those associated with problems of defining the distributed contents by use of a single measurement parameter made at one point in the extended surface of the whole. As a general guide, definition of the surface, which is used to decide the contents, can be made to around 0.5 mm in industrial circumstances. Methods that sense the surface can be in error due to surface tension effects that introduce hysteresis. Where the quantity of a particular substance is of concern, any buildup of sediment and other unwanted residues introduces error.

Granular materials will not generally flow to form a flat surface as do liquids. The angle of repose of the material and the form of input and output porting will give rise to errors in volume calculations if the actual geometry is not allowed for in use of a single point sensor.

Turbulence occurring at the sensor, caused by material flow in the container or by vibrations acting on the container, may also be a source of error. It is common practice to mount a mechanical level sensor in some form of integrating chamber that smoothes out transient dynamic variations. A common method is the use of a stilling pipe or well that is allowed to fill to the same level as the contents via small holes (see Figures 10.3 and 10.4). The time rate responses of

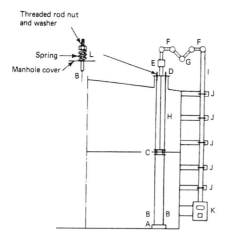

FIGURE 10.2 Installation of float-type level indicator with guide wires on a fixed-roof tank. A. Guide wire anchor. Wires may be anchored to a heavy weight or the bottom of the tank. B. Float guide wires. C. Float having sliding guides which move freely on the guide wires. D. Manhole sufficiently large for float and anchor weight to pass through. E. Flexible joint. F. Pulley housing. G. Vapor seal (if required). H. Float tape. I. Tape conduit. J. Sliding guides. K. Gauge head. L. Guide wire tension adjustment.

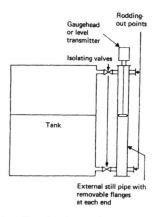

FIGURE 10.3 Integrating chamber used to average transient variation. Internal still pipe.

CHAPTER | 10 Measurement of Level and Volume

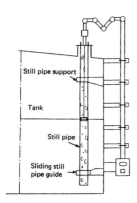

FIGURE 10.4 Integrating chamber used to average transient motion. External still pipe.

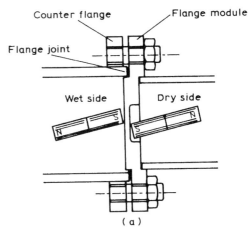

such still tubes, however, become important in fast-moving systems because they can introduce phase-shift and amplitude attenuation.

Changes of the mass of floats due to sediment buildup or corrosion will alter the depth of immersion of float sensors. A systematic error also exists due to the actual depth to which the float sinks to provide its necessary buoyancy force. This varies with material density, which often varies with temperature.

A second class of errors arises due to temperature and, to a lesser extent, pressure changes to the contents. Where the required measurement is a determination of the volume or mass of the contents, use is made of level as an indirect step toward that need. All materials change volume with changing temperature. It may therefore be necessary to provide temperature measurements so that level outputs can be corrected.

For some forms of level sensor, external still tubes should be situated to retain the same temperature as that of the tank because localized heating or cooling can cause the contents of the still tube to have a different density from that existing in the tank. Methods that are based on use of buoyancy chambers, which produce a measurement force rather than following the surface, will produce force outputs that vary with temperature due to altered buoyancy upthrust as the density of the fluid changes.

Floats are generally made from waterproofed cork, stainless steel, copper, and plastic materials. The material used may need to be corrosion-resistant. Where the contents are particularly corrosive or otherwise inhospitable to the components of the sensing systems, it is preferable to reduce, to the absolute minimum, the number of subsystem parts that are actually immersed in the contents.

Considerable use is made of magnetic coupling between the guided float and the follower. Figure 10.5 shows one such arrangement.

Nuclear level gauging offers the distinct advantage (see Figure 10.6) that no part of the level-detecting system need be inside the container. It is discussed further in Part 3.

FIGURE 10.5 Repulsion of like magnetic poles providing nonimmersed sensing and toggle action to a float sensor. One commercial form. Courtesy of BESTA Ltd.

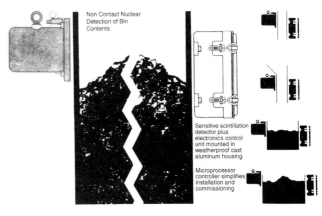

FIGURE 10.6 Nuclear gauging system needs mounting only on the outside of existing pipes or containers. Courtesy of Mineral Control Instrumentation.

Finally, on general choice of level-sensing system, Table 10.1, from Lazenby (1980), provides guidelines for selecting appropriate methods.

TABLE 10.1 Some guidelines for selecting a suitable level sensor. Adapted from Lazenby (1980)

Is remote control or indication desirable?	A yes excludes mechanical float gauges, sight glasses, dipsticks, and other devices designed specifically for local control.
With level indicators, is time spent in taking a reading important?	A yes excludes dipsticks, sight glasses, mechanical float gauges, and certain balance systems.
Can the sensor contact the material being measured?	A no eliminates all but ultrasonics, radiation, radar, and optical and load cells.
Must weight be measured rather than height?	A yes means the choice is limited to load cells, but in uniform-sided tanks, other devices such as capacitance meters can be calibrated for weight, particularly if a liquid is being measured.
Are there objections to mechanical moving parts?	A yes gives one a choice of sight glasses and capacitance, ultrasonic, radiation, conductivity, radar, load-cell, optical, thermistor, and bubbler devices.
Is the application in a liquid?	A yes eliminates vibrators and certain paddle types.
Is the application for a powdered or granular material?	A yes eliminates dipsticks, sight glasses, floats, themistors, conductivity devices, pressure (except pressure switches) instruments, bubblers, displacers.
Do level indicators have to be accurate to about 2 percent?	A yes eliminates thermistors, vibrators, paddles, optical devices, suspended tilting switches, and conductivity instruments, as those types only provide control. All other types can be considered, but the poorest accuracies probably come from float gauges and radiation instruments.
Do level indicators need to have an accuracy that is a lot better than 1 percent?	A yes reduces the list to dipsticks, some of the displacers, and balance devices.

10.3 CALIBRATION OF LEVEL-MEASURING SYSTEMS

Contents that are traded for money, such as petrochemicals, foods, milk, and alcohol, must be measured to standards set by the relevant weights-and-measures authority. Official approval of the measuring system and its procedures of use and calibration is required. In such cases the intrinsic value of the materials will decide the accuracy of such measurements, which often means that the system and calibrations must comply to very strict codes and be of the highest accuracy possible.

The use of the indirect process of determining volumetric or mass contents, based on a level measurement, means that a conversion coefficient or chart of coefficients must be prepared so that the level measurements can be converted into the required measurement form.

Calibration tables for a large fabricated tank are most easily prepared using the original engineering construction drawings. This, however, is not an accurate or reliable method. This is especially true because of the difficulty in obtaining accurate "as-built" details on many vessels other than custom-built process reactor vessels.

A more accurate and traceable method (known as *strapping*) is to actually survey the container's dimensions after it is built. This can be a time-consuming task. The values can be used to calculate the volume corresponding to different levels. From this is compiled a conversion chart for manual use or a "lookup table" for computer use.

By far the most accurate method, however, is a direct volumetric calibration of the container by which a suitable fluid (usually water) is pumped in and out to provide two readings, the tank passing it through an accurate flow-metering station. While this is in process, level data are recorded, enabling the conversion factors to be provided for each level measurement value. These results will often require correction for temperature, as has already been discussed. Another volumetric calibration method is to pump the liquid into a tanker vessel, usually a truck, which can then itself be weighed and the tare of the empty vessel subtracted to produce an accurate measure of volume.

Highly accurate level measurement requires continuous monitoring of the various error sources described earlier so that ongoing corrections can be made. A continuous maintenance program is needed to clean floats and electrodes and to remove unwanted sediment.

In many instances the use of hand dipping is seen as the ongoing calibration check of the level measurement. For this, rods or tapes are used to observe the point at which the contents wet the surface along the mechanical member. Obviously this cannot be used for dry substances; for those, the rod or tape is lowered until the end rests on the surface.

In each case it is essential to initially establish a permanent measurement datum, either as the bottom of the tank where the rod strikes or as a fiducial mark at the top. This mark needs to be related to the transducer system's readout and to the original calibration.

10.4 METHODS PROVIDING FULL-RANGE LEVEL MEASUREMENT

Methods used to measure or control level in a container can be divided into those that measure a continuous range of level and those that measure a small change or point level. These full-range or continuous level methods have found wide acceptance.

10.4.1 Sight Gauges

A simple, externally mounted sight glass can be used for reading the level of contents within a closed container, such as a steam boiler. This generally consists of a tube of toughened (usually borosilicate) glass connected through unions and valves into the tank wall. The diameter of the tube must be large enough not to cause "climb" of the contents due to capillary action. The level will follow that of the contents. Figure 10.7(a) gives the basic structure of such a device. Where the contents are under pressure, it will be necessary to use safety devices to control pressure release in case of tube breakage. In addition, when the contents are corrosive, methods other than simple glass sight gauges should be considered. Valves are usually incorporated to allow the whole fitting to be removed without having to depressurize the container. Figure 10.7(b) shows a configuration incorporating these features.

A modern development of this sight-gauge concept is the magnetic level indicator shown in Figure 10.8. As the float, containing a bar magnet, follows the liquid surface in the still tube, the individual, magnetically actuated flaps rotate to expose a differently colored surface that is coated with luminous paint. The float is chosen to suit the specific gravity of the fluid.

The magnetic action also operates individual switches for control and alarm purposes. Discrimination is to around 5 mm. Magnets must not be operated beyond their Curie point, at which temperature they lose their desired properties. As a guide these systems can measure liquids under pressures up to 300 bars and at temperatures up to 400°C.

In some circumstances it may be possible to view the surface of the liquid from above but not from the side. In this case a hook gauge (see Figure 10.9) can be used to enable the observer to detect when the end of the dipstick rod just breaks the surface.

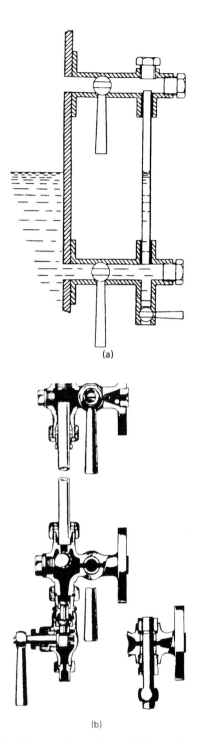

FIGURE 10.7 Sight-glass level indicator. (a) Basic schematic. (b) Sight-glass with automatic cut-off. Courtesy of Hopkinsons Ltd.

10.4.2 Float-driven Instruments

The magnetic indicator described previously is one of a class of level indicators that use a float to follow the liquid surface. Where a float is used to drive a mechanical linkage that operates a remotely located readout device of the linkage motion, there is need to ensure that the linkage geometry does not alter the force loading imposed on the float, because this will alter its immersion depth and introduce error. Frictional forces exerted by the linkage can also introduce error.

Compensation for changes in linkage weight as a float moves is achieved by using such mechanisms as counterbalance masses and springs. Figure 10.10 shows the

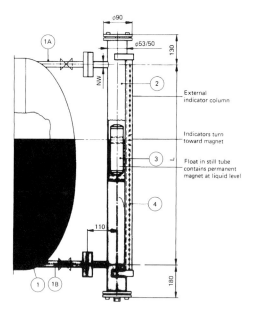

FIGURE 10.8 Schematic of magnetic level indicator installation. Courtesy of Weka-Besta Ltd.

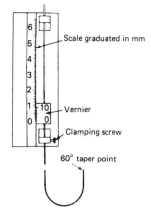

FIGURE 10.9 Hook-type level indicator.

construction of a sophisticated form that uses a pre-wound Neg'ator (also called a Tensator) spring torque motor that has its torque characteristic tailored to vary as more tape is to be supported during windout.

The production costs of precision mechanical systems can make them less attractive than electronic equivalents, but such systems do have the advantage that no electrical power supply is needed. Previously, another advantage of mechanical level systems was the fact that a wider range of plant operators easily understood them. This is no longer necessarily the case. It is commonly now recommended that noncontacting or noninvasive level measurements be the measurements of first choice where possible.

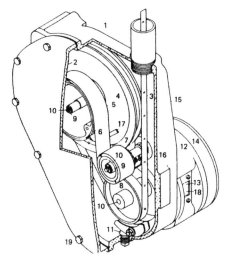

FIGURE 10.10 Spring torque motor compensated float-type transfer system. 1. Precision cast main housing. 2. Side cover. 3. Perforated steel tape type 316 stainless. 4. Molded thermosetting phenolic tape drum. 5. Broad Neg'ator Motor, stainless steel. 6. Power drum. 7. Storage drum. 8. Precision made sprocket. 9. P.T.F.E. bearings. 10. Type 316 stainless steel shafts. 11. Drain plug. 12. Digital counter housing. 13. Reading window. 14. Stainless steel band covers adjustment slots. 15. Operation checker handle (out of view). 16. Operation checker internal assembly. 17. Neg'ator motor guide, stainless steel. 18. Counter assembly in the chamber beyond tank pressure and vapors. 19. Cap screws drilled for sealing. Courtesy of Whessoe Ltd.

10.4.3 Capacitance Probes

The electrical capacitance C between two adjacent electrically conducting surfaces of area A, separated by distance d, is given by

$$C = \varepsilon \frac{A}{d}$$

The constant of proportionality ε is the dielectric constant of the material between the plates. An electrode is suspended in the container, electrically insulated from it. The presence of liquid or granular material around the electrode alters the capacitance between the electrode and the walls. The capacitance is sensed by electronic circuitry. Figure 10.11 is a cutaway view of one form.

The electrode is tailored to the situation; forms include rigid metal rods, flexible cables, and shielded tubes. Capacitance sensors rely on uniform contact being maintained between the contents and a long, thin electrode. Where they are used for level sensing of granular materials such as wheat, the material has a tendency to pile nonuniformly around the electrode, producing what is known as *ratholing*. Placing the electrode at an angle to the vertical helps reduce this effect because it alters the angle of repose of the material, helping it to follow the stem more consistently. Because the method provides continuous readout of level over its full electrode

CHAPTER | 10 Measurement of Level and Volume

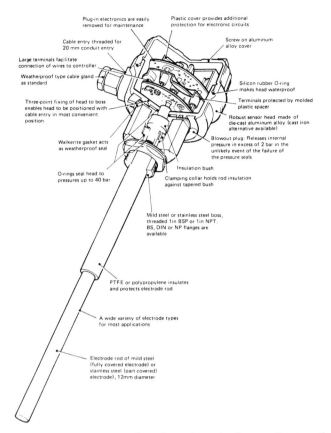

FIGURE 10.11 Cutaway view of capacitance level sensor. Courtesy of Kent Industrial Measurements Ltd.

length, circuitry can also be used to provide multiple on/off setpoints for alarms and control functions. The same principle is used for single point sensing, in which case a simpler electrode and circuitry can be used. Electrical potential and power are usually low enough to eliminate hazards.

10.4.3.1 Weighing of the Contents

The volume of a container's contents can, of course, be inferred from weight measurements; these are discussed in Chapter 13.

10.4.4 Upthrust Buoyancy

A long, vertical tubular float will exert an upward force proportional to the depth of immersion in the fluid. These are also sometimes referred to as *displacers*. The float does not rise to follow the surface but is used instead to exert a force that is usually converted into a torque by a radius arm with a counteracting torque shaft. Force balance can also be used to determine the upthrust force. Figure 10.12 is an assembly view of one design.

Upthrust depends on the specific gravity of the fluid, so instruments employing it must be calibrated for a stated

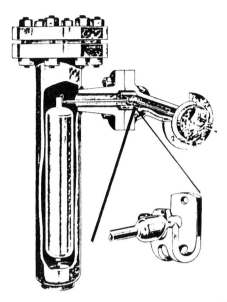

FIGURE 10.12 Assembly view of Fisher torque tube unit. Courtesy of GEC Elliot Control Valves Ltd.

density. Density varies with temperature. For the best accuracy, correction is needed; some reduction in the actual error magnitude, however, occurs due to the float becoming a little larger in volume as its temperature increases.

10.4.5 Pressure Sensing

Providing that the contents behave as a liquid that flows to equalize pressures at a given depth (some granular materials may not fulfill this requirement), pressure acting on a given area at the bottom of a tank is proportional only to the density of the fluid and the head of pressure. In most cases density can be assumed to be uniform, thereby allowing a pressure sensor, placed on the bottom, to be used as a measure of tank level. Pressure gauges are described in Chapter 14.

Lying in the same class are purge methods. The pressure needed to discharge gas or liquid from a nozzle placed at the bottom of the tank depends on the head of liquid above and its density. *Bubblers*, as these are called, are simple to arrange and give readout as a pressure-gauge reading that can be read directly or transduced into a more suitable form of signal.

Obviously bubblers do not work in granular materials. The addition of small quantities of liquid or gas must not significantly affect the contents of the tank.

10.4.6 Microwave and Ultrasonic, Time-Transit Methods

A source of coherent radiation, such as ultrasound or microwaves, can be used to measure the distance from the surface to the top of the tank or the depth of the contents. In essence, a pulse of radiation is transmitted down to the surface, where some proportion is bounced back by the reflecting interface formed at the surface. The same concept can be used with the waves being sent upward through the material, to be reflected downward from the surface. With relatively sophisticated electronic circuitry it is possible to measure the flight time and, given that the velocity of the waves is known, the distance may then be calculated.

Many variations exist on this basic theme. The choice of radiation, use from above or from below, and of frequency depend much on the material and the accuracy sought.

Although pulses are sent, the repetition rate is fast enough for the output to appear continuous. The method can be made suitable for use in hazardous regions.

10.4.7 Force or Position Balance

In these methods a short-range sensor, such as a float resting in the surface or a surface sensor of an electronic nature, is used to provide automatic control signals that take in or let out cable or wire so that the sensor is held at the same position relative to the surface. Figure 10.13 gives the arrangement of one such system that uses a radio-frequency surface sensor to detect the surface.

Self-balancing level sensors offer extreme ranges, and variable forces exerted by changing mechanical linkage geometry are made negligible. Very high accuracies can be provided, the method being virtually an automated tape measure.

10.5 METHODS PROVIDING SHORT-RANGE DETECTION

In many applications there is only a need to sense the presence or absence of contents to operate on/off switches or alarms. In some continuous-reading systems, a short proportional range is needed to control the driven measuring member. This section addresses short-range detectors of level.

10.5.1 Magnetic

Movement of a permanent magnet floating in the surface of the liquid can be sensed by using the magnet to operate a switch contact set. Figure 10.5 shows a system actuated by a rising radius arm and incorporating a toggle snap action. An alternative arrangement uses the rising magnet to close a magnetic reed switch contact, as shown in Figure 10.14, either as a coaxial arrangement or as a proximity sensor.

10.5.2 Electrical Conductivity

Liquids such as sewage, sea water, and town supply water, which contain dissolved salts, have conductivities higher than pure water. The conductivity of most liquids is much

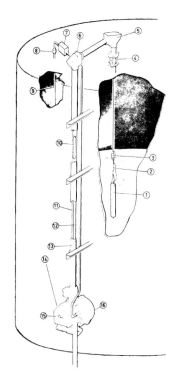

FIGURE 10.13 Schematic of self-balancing level gauge using RF surface sensing. Courtesy of GEC-Elliot Process Instruments Ltd. 1. Sensing element. 2. Tape insulator. 3. Tape counter-weight. 4. Flexible coupling used on fixed roof only. 5. Pulley box over tank. 6. Pulley box over tank-side unit. 7. Temperature cable junction box. 8. Temperature bulb mounting kit. 9. Averaging resistance thermometer bulb. 10. Cable counterweight. 11. Stainless steel perforated measuring tape. 12. Radio frequency cable. 13. 65 mm dia. standpipe. 14. Servo-electronic box. 15. Level indication. 16. Tape retrieval housing.

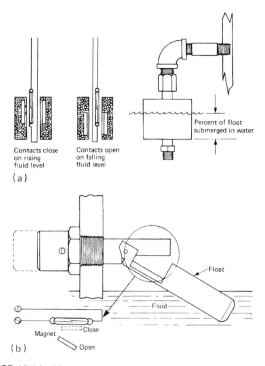

FIGURE 10.14 Magnetic level switch using magnetic reed contact set. (a) Coaxial. (b) Proximity.

higher than that of air, so an electrical circuit, depending on current flow, can discriminate between air and liquid and so detect the interface. Figure 10.15 is a multiple-probe system used to distinguish the various layers in a pulp-froth-air system, as is found in mineral processing.

Conductivity probes are used for digital monitoring of the level of boiler water. Conductivity can also be used to provide continuous range measurement, because as the liquid rises up an electrode, the resistivity between the electrode and a reference surface changes in a proportional manner.

10.5.3 Infrared

When fluid wets an optical surface, the reflectance of that surface changes considerably, enabling detection of liquid when it rises to cover the optical component.

The optical arrangement that is commonly used is a prism, arranged as shown in Figure 10.16. Infrared radiation, easily produced with light-emitting diodes and readily detected, is used. When the prism outer surface is wetted, the majority of the radiation passes through the quartz-glass prism into the liquid, dropping the signal level given out by the photocell.

This method does not require the installation of electrical connections into the tank and, therefore, lends itself to use where intrinsically safe operation is needed. Typical discrimination is around 1 mm.

10.5.4 Radio Frequency

This form of surface sensor is used in the system shown in Figure 10.13. The tank gauge unit contains a radio frequency (RF) oscillator tuned to around 160 MHz. Its signal, modulated at 50 Hz, is transmitted to the sensing probe located on the end of the cable line. The probe is a tuned antenna set to be resonant at the carrier frequency. When the tip is brought close to the liquid, its resonant frequency is altered. Demodulation at the probe produces a 50 Hz signal that is fed back along the cable as a voltage level, depending on the relationship between oscillation frequency and the resonance of the antenna. This is compared with a reference voltage

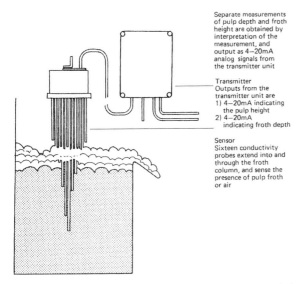

FIGURE 10.15 Multiple-conductivity probe sensing pulp and froth layers over set increments. Courtesy of Mineral Control Instrumentation.

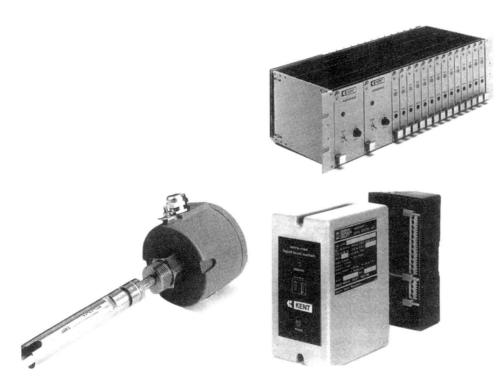

FIGURE 10.16 Infrared discrete-position level sensor. Principle and physical equipment. Courtesy of ABB Instrument Group.

to produce an error signal that is used to drive the cable to place the probe at the present null position with respect to the surface.

10.5.5 Miscellaneous Methods

The following are some of the other principles that have been used to sense the presence of a liquid. The turning moment exerted to rotate a turning paddle will vary when material begins to cover the paddle. The resonant frequency of a vibrating tuning fork will change as it becomes immersed. The electrical resistance of a heated thermistor will vary depending on its surroundings.

REFERENCES

Boyes, W. H., "The state of the art in level measurement," *Flow Control* (February 1999).

Lazenby, B., "Level monitoring and control," *Chemical Engineering*, **87**, 1, 88–96 (1980).

Miller, J. T. (ed.), *The Instrument Manual*, United Trade Press (5th ed., 1975).

Norton, H. N., *Handbook of Transducers for Electronic Measuring Systems*, Prentice-Hall, Englewood Cliffs, N.J. (1969).

O'Higgins, P. J., *Basic Instrumentation-Industrial Measurement*, McGraw-Hill, New York (1966).

Chapter 11

Vibration

P. H. Sydenham

11.1 INTRODUCTION

11.1.1 Physical Considerations

Vibration is the oscillatory motion of objects. Several different measurable parameters may be of interest: relative position, velocity, acceleration, jerk (the derivative of acceleration), and dynamic force are those most generally desired.

For each parameter it may be the instantaneous value, the average value, or some other descriptor that is needed. Accuracy on the order of 1 part in 100 is generally all that is called for.

Vibration, in the general sense, occurs as periodic oscillation, as random motion, or as transient motion, the latter more normally being referred to as shock when the transient is large in amplitude and brief in duration.

Vibration can occur in linear or rotational forms of motion, the two being termed, respectively, translational or torsional vibrations. In many ways the basic understanding of each is similar because a rotational system also concerns displacements. Translational forms are outlined in the following description. There will usually exist an equivalent rotational system for all arrangements described.

In vibration measurement it is important to decide whether or not a physically attached mechanical sensor, corresponding to a contacting or noncontacting technique, can be used.

Adequate measurement of vibration can be a most complex problem. The requirement is to determine features of motion of a point or an extended object in space relative to a reference framework (see Figure 11.1).

A point in space has three degrees of freedom. It can translate in one or more of three directions when referred to the Cartesian coordinate system. Rotation of a point has no meaning in this case. Thus to monitor free motion of a point object requires three measurement-sensing channels.

If the object of interest has significant physical size, it must be treated as an extended object in which the rotations about each of the three axes, described previously, provide a further three degrees of freedom. Thus to monitor the free motion of a realistic object may need up to six sensors, one for each degree of freedom.

In practice some degrees of freedom may be nominally constrained (but are they really?), possibly eliminating the need for some of the six sensors. Practical installation should always contain a test that evaluates the degree of actual constraint because sensors will often produce some level of output for the directions of vibration they are not primarily measuring. This is called their *cross-axis coupling factor, transverse response,* or some such terminology.

In many installations the resultant of the motion vector may lie in a constant fixed direction with time. In such cases, in principle, only one sensor will be required, provided it can be mounted to sense in that direction. If not, as is often the

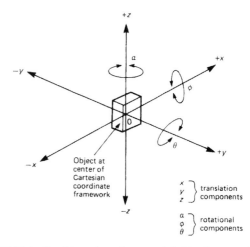

FIGURE 11.1 Possible motions of an extended object in space relative to a Cartesian framework.

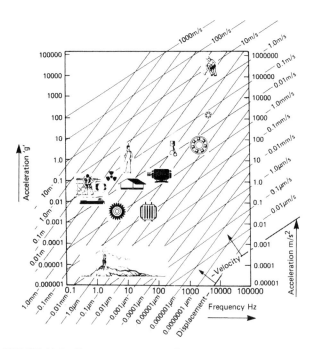

FIGURE 11.2 Frequency spectrum and magnitude of vibration parameters. Courtesy of Bruel & Kjaer.

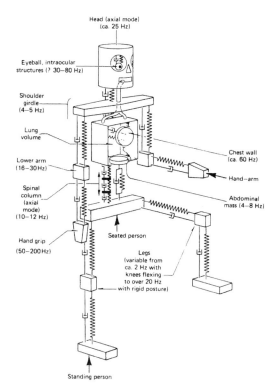

FIGURE 11.3 Mechanical systems can be modeled in terms of springs, masses, and dampers. This is a model of a human body being vibrated upward from the lower foot. Courtesy of Bruel & Kjaer.

case, more than one unit will be required, the collective signals then being combined to produce the single resultant.

The potential frequency spectrum of vibration parameters extends, as shown in Figure 11.2, from very slow motions through frequencies experienced in machine tools and similar mechanical structures to the supersonic megahertz frequencies of ultrasound. It is not possible to cover this range with general-purpose sensors. Each application will need careful consideration of its many parameters to decide which kind of sensor should be applied to make the best measurement.

A complicating factor in vibration measurement can be the distributed nature of mechanical systems. This leads to complex patterns of vibration, demanding care in the positioning of sensors.

Mechanical systems, including human forms, given as an example in Figure 11.3, comprise mass, spring compliance (or stiffness), and damping components. In the simplest case, where only one degree of freedom exists, linear behavior of this combination can be well described using linear mathematical theory to model the time behavior as the result of force excitation or some initial position displacement.

Vibration can be measured by direct comparison of instantaneous dimensional parameters relative to some adequately fixed datum point in space. The fixed point can be on an "independent" measurement framework (fixed reference method) or can be a part that remains stationary because of its high inertia (seismic system).

In general, a second-order linear system output response q_o is related to an input function q_i by the differential equation

$$\frac{a_2 d^2 q_o}{dt^2} + \frac{a_1 d q_o}{dt} + a_o q_o = q_i$$

(spring-mass-damper system) (input driving function)

For the specific mechanical system of interest here, given in Figure 11.4, this becomes

$$\frac{m d^2 x_0}{dt^2} + \frac{c d x_0}{dt} = k_s x_0 = q_i$$

where m is the effective mass (which may need to include part of the mass of the spring element or be composed entirely of it), c is the viscous damping factor, and k_s the spring compliance (expressed here as length change per unit of force applied).

Where the damping effect is negligible, the system will have a frequency at which it will naturally vibrate if excited by a pulse input. This natural frequency ω_n is given by

$$\omega_n = \sqrt{\frac{k_s}{m}}$$

CHAPTER | 11 Vibration

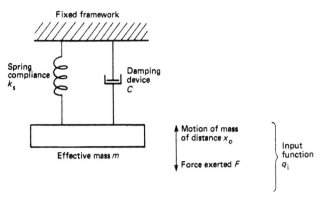

FIGURE 11.4 One-degree-of-freedom, spring, mass, and damper system model.

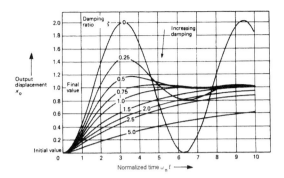

FIGURE 11.5 Displacement responses of second-order system to input step of force.

The presence of damping will alter this value, but as the damping rises the system is less able to provide continuous oscillation.

The static sensitivity is given by the spring constant, either as k_s the spring compliance or as its reciprocal, that is, expressed as force per unit extension.

The influence of damping is easily described by a dimensionless number, called the *damping ratio*, which is given by

$$\xi = \frac{c}{2\sqrt{k_s \cdot m}}$$

It is usually quoted in a form that relates its magnitude with respect to that at $\xi = 1$.

These three important parameters are features of the spring-mass-damper system and are independent of the input driving function.

Such systems have been extensively analyzed when excited by the commonly met input forcing functions (step, impulse, ramp, sinusoid). A more general theory for handling any input function other than these is also available. In practice the step, impulse, and continuous sinusoidal responses are used in analyses because they are reasonably easy to apply in theory and in practical use.

As the damping factor ξ increases, the response to a transient step force input (applied to the mass) can vary from sinusoidal oscillation at one extreme (underdamped) to a very sluggish climb to the final value (overdamped). These responses are plotted in Figure 11.5. In the case of continuous sinusoidal force input, the system frequency response varies as shown in Figure 11.6. Note the resonance buildup at ω_n, which is limited by the degree of damping that exists. Thus the damping of the system to be measured or of the sensor, if it is of the seismic kind, can be of importance as a modifier of likely system responses. As damping increases, the system response takes on the form of the lower first-order, exponential response system, and it cannot oscillate.

Useful introductions to this aspect of vibrations are to be found in Oliver (1971), the dynamic behavior of systems

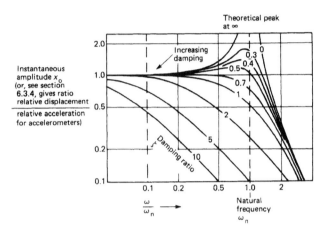

FIGURE 11.6 Displacement responses of second-order system to continuous sinusoidal force input. The same curves relative displacements of a seismic mass to the acceleration of the mass. (See Section 11.3.4.)

being expounded in more depth in Crandall (1959), Harris and Crede (1961), Sydenham (1983), Trampe-Broch (1980), and Wallace (1970).

The preceding discussion in respect to vibration of the measurand is also the basis of understanding the operation of seismic vibration sensors, as we will see later.

It is a property of second-order systems, therefore, to have a natural frequency of vibration. This is the frequency at which they vibrate when given impulse energy that is not overridden by continuous forced vibrations. Thus a sensing system that is second order and not damped will (due to noise energy inputs) produce outputs at its natural frequency that are not correlated with frequencies occurring in the system of interest. Use of seismic vibration sensors must, therefore, recognize these limitations.

In practice it is also often more convenient to sense vibration by an indirect means and obtain the desired unit by mathematical processing. For example, accelerometers are conveniently used to obtain forces (from force = mass × acceleration) and hence stresses and strains. Acceleration signals can be twice integrated with respect to time to yield displacement. Sensors that operate as velocity transducers can yield displacement by single integration.

Integration is generally preferred to differentiation because the former averages random noise to a smaller value compared to the signal, whereas the latter, in reverse, can deteriorate the signal-to-noise ratio. Mathematical signal manipulation is common practice in vibration measurement as a means to derive other related variables.

11.1.2 Practical Problems of Installation

With vibration measurement it is all too easy to produce incorrect data. This section addresses several important installation conditions that should be carefully studied for each new application.

11.1.2.1 Cross-Coupling

Transducers may exhibit cross-axis coupling. Wise practice, where possible, includes a test that vibrates the sensor in a direction perpendicular to the direction of normal use. Rotational sensitivity may also be important. These tests can be avoided each time they are used if the sensors are precalibrated for this source of error and, of course, are still within calibration. Sensors that have no such quoted parameter should be regarded as potential sources of error until proven otherwise.

11.1.2.2 Coupling Compliance

The compliance of the bond made between the sensor and the surface it is mounted on must be adequately stiff. If it is not, the surface and the sensor form a system that can vibrate in unpredictable ways. For example, an insufficiently stiff mounting can give results that produce much lower-frequency components than truly exist. In extreme cases the sensor can be shaken free as it builds up the unexpectedly low-resonance frequency of the joint to dangerous amplitude levels. As a guide the joint should be at least 10 times stiffer than the sensor so that the resonant frequency of the joint is well above that of the sensor.

11.1.2.3 Cables and Preamplifiers

Certain types of sensors, notably the piezoelectric kind, are sensitive to spurious variation in capacitance and charge. Sources of such charges are the triboelectric effect of vibrating cables (special kinds are used, the design of which allows for movement of the cable), varying relative humidity that alters electric field leakage (this becomes important in designing long-term installations), and preamplifier input condition variations.

11.1.2.4 Influence Errors

Ideally the sensor should operate in a perfect environment wherein sources of external error, called *influence parameters*, do not occur. In vibration sensing, possible influence error sources include temperature variation of the sensor, possible magnetic field fluctuations (especially at radio frequency), and existing background acoustic noise vibrations. Each of these might induce erroneous signals.

A good test for influence parameters is to fully connect the system, observing the output when the measurand of interest is known to be at zero level. Where practical, the important error inputs can be systematically varied to see the sensor response. Many a vibration measurement has finally been seen to be worthless because some form of influence error turned out to be larger than the true signal from the measurand. Vibrations apparently occurring at electric mains frequency (50 or 60 Hz) and harmonics thereof are most suspect. Measurement of mechanical vibration at these frequencies is particularly difficult because of the need to separate true signal from influence error noise.

11.1.2.5 Subject Loading by the Sensor

Vibration sensors contain mass. Because this mass is made smaller, the sensitivity usually falls. Addition of mass to a vibrating system can load the mass of that system, causing shifts in frequency. For this reason manufacturers offer a wide range of attached-type sensors. Provided the mass added is, say, 5 percent or less of the mass of interest, the results will be reasonable. Cables can also reduce mechanical compliance, reducing the system amplitude. Where a system is particularly sensitive to loading, the use of non-contact, fixed-reference methods may be the only way to make a satisfactory measurement.

11.1.2.6 Time to Reach Equilibrium

When damping of a structure is small, the time taken for a resonance to build up to its peak value is large. When using forced vibration to seek such a resonance, it is therefore important not to sweep the excitation input frequency too rapidly.

11.1.3 Areas of Application

In searching for information about a measurement technique, it is usually helpful to have an appreciation of the allied fields that use the same equipment. Vibration, of course, will be of interest in very many applications, but a small number can be singled out as the main areas to which commercial marketing forces have been directed.

11.1.3.1 Machine Health Monitoring

A significant field of interest is that of *machine health* or *condition* monitoring; failures can often be avoided by "listening" to the sounds and vibrations made by the system. An example is shown in Figure 11.7. Vibration and other forms

CHAPTER | 11 Vibration

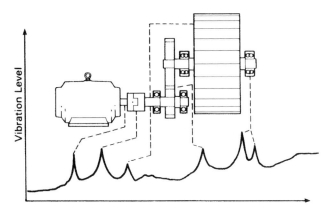

FIGURE 11.7 In machine health monitoring the normal vibration levels of parts of the installation are recorded to provide a normal signature. Variations of this indicate changes in mechanical conditioning. Courtesy of Bruel & Kjaer.

of sensor are applied to the operating system, first while running in early life and then at periodic intervals during life. If the frequency/amplitude data (the so-called *signature*) has changed, this can provide diagnostic information suggesting which component is beginning to fail. The component can then be conveniently replaced before a major, untimely breakdown occurs. Introduction to this aspect is to be found in Bently Nevada (1982) and Wells (1981).

11.2 AMPLITUDE CALIBRATION

Static amplitude (displacement) is easily calibrated using a standardized micrometer, displacement sensor, or optical interferometry. Dynamic calibrations may be made either by comparison, using a technique of known accuracy and frequency response, or by using a calibrated vibration generator.

11.2.1 Accelerometer Calibration

Figure 11.8 shows outlines of three methods for the calibration of accelerometers and other vibration-measuring sensors. Calibration is normally performed at 500 rad s^{-1}.

Other methods that can be used are to subject the accelerometer to accelerations produced by the earth's force. Simple pseudostatic rotation of an accelerometer in the vertical plane will produce accelerations in the 0 to ±1 g range (g is used here for the earth's acceleration). Larger values can be obtained by whirling the accelerometer on the extremity of a rotating arm of a calibrating centrifuge, or it can be mounted on the end of a hanging pendulum.

11.2.2 Shock Calibration

Short-duration acceleration, as produced by impact, requires different approaches to calibration. Accelerations can exceed 10,000 g and can last for only a few milliseconds.

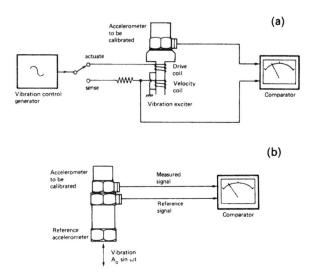

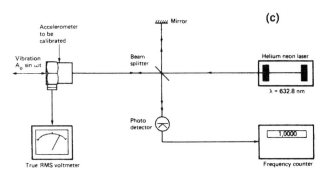

FIGURE 11.8 Two alternatives for calibrating accelerometers. (a) Calibrated vibration exciter shaking accelerometer at calibrated levels—the reciprocity method. (b) Back-to-back calibration of a calibrated accelerometer against one to be calibrated–comparison method. (c) Absolute measurement using optical interferometry.

A commonly used method is to produce a calibrated shock by allowing a steel ball to free-fall on to an anvil on which is mounted the sensor. This method provides an absolute calibration, but, as with all of the previously described methods, it has uncertainties associated with the practical method. In this case one source of error is caused by the difficulty of releasing a ball to begin its downward path without imparting some velocity at time zero.

11.2.3 Force Calibration

Static forces can be calibrated by applying "dead weights" to the force sensor, the "weights" being calibrated masses. (See Chapter 12.)

Dynamic forces arising in vibration can more easily be determined using the relationship force = mass × acceleration. A shaking table is used to produce known accelerations on a known mass. In this way the forces exerted on the accelerometer can be determined along with the corresponding output voltage or current needed to produce transducer sensitivity constants.

Space does not permit greater explanation here, but several detailed accounts of vibration sensor calibration are available in the literature: Endevco (1980), Harris and Crede (1961), Herceg (1972), Norton (1969), Oliver (1971), and Trampe-Broch (1980). National and international standards are extensively listed in Bruel and Kjaer (1981).

11.3 SENSOR PRACTICE

11.3.1 Mass-Spring Seismic Sensors

Whereas the fixed reference methods do have some relevance in the practical measurement of vibration, the need for a convenient datum is very often not able to be met. In the majority of vibration measurements, use is made of the mass-spring, seismic sensor system.

Given the correct spring-mass-damping combination, a seismic system attached to a vibrating surface can yield displacement, velocity, or acceleration data. Unfortunately, the conflicting needs of the three do not enable one single design to be used for all three cases. However, it is often possible to derive one variable from another by mathematical operations on the data.

Two forms of seismic sensor exist. The first, called *open loop*, makes use of the unmodified response of the mass moving relative to the case to operate either a displacement or a velocity-sensing transducer. The second form closes the loop (and is, therefore, referred to as a *closed-loop* or *servo seismic sensor*) using the output signal to produce an internal force that retains the mass in the same relative position with respect to the case, the magnitude of the force being a measure of the vibration parameter.

11.3.1.1 Open-Loop Sensors

The fundamental arrangement of the open-loop seismic sensor form is as given in Figure 11.9. Actual construction can vary widely, depending on how the spring force and damping are provided and on the form of the sensor used.

The spring element can be produced as a distinct mechanical element. Figure 11.10 is an example made with flexure strips; alternatively, perforated membranes, helical coils, torsional strips, and the like can be used. Otherwise, the compliance of the mass itself may be the spring—for example, in the piezoelectric crystal, which also acts as the sensing element.

Important design parameters of a spring are the compliance, amplitude range, fatigue life, constancy of rate with time, temperature, and other influence effects and the suitability to be packaged to produce a suitable sensor unit. Except for the highest natural frequency sensors, the masses used can be regarded as completely rigid compared with the spring element.

Rotary forms of the linear arrangement, shown in Figure 11.9, are also available.

Sensing methods that have been used include the electrical-resistance sliding potentiometer, variable inductance (see, for instance, Figure 11.10), variable reluctance, variable capacitance, electrical metallic strain gauges (bonded and unbonded, as in Figure 11.11) and semiconductor strain gauges, piezoelectric crystal and magnetostrictive elements, position-sensitive optical detectors, and electro-magneto principles (which provide direct velocity sensing).

Sensors are often encapsulated. The encapsulation takes many forms, ranging from miniature units of total weight of around 1 g through to 0.5 kg units, where the sensing mass must be physically large. Because simultaneous measurements in two or three directions are often required, seismic sensors are also made that consist of two or three units, as shown in Figure 11.12, mounted in different directions.

Compensation for temperature is needed in many designs. This is either performed in the electronic circuitry or by incorporating some form of thermomechanical device into the spring-mass-sensor layout.

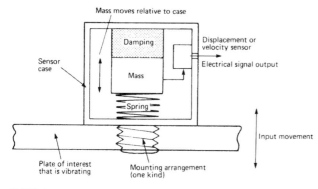

FIGURE 11.9 Schematic layout of open-loop, seismic-form, vibration sensor.

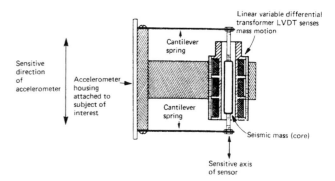

FIGURE 11.10 Diagrammatic view of a spring-mass seismic sensor that uses parallel flexure-strip spring suspension and inductive sensor of mass displacement. Courtesy of Schaevitz Engineering.

11.3.1.2 Servo Accelerometers

The performance of open-loop seismic sensors can be improved with respect to their sensitivity, accuracy, and output signal amplitude by forming the design into a closed-loop form.

Figure 11.13 gives the schematic diagram of one kind of closed-loop system that is based on a moving-coil actuator and capacitance displacement sensor. The mass on which the acceleration is to be exerted is able to rotate on the end of a freely supported arm. This is attached to the electrical coil placed in the permanent magnetic field supplied by the magnet assembly. Acceleration applied to the mass attempts to rotate the arm, causing displacement. This unbalances the capacitive displacement sensor monitoring the relative position of the mass. Displacement signals produce an input to the difference-sensing amplifier. The amplifier drives a corresponding electric current into the coil, causing the arm to rotate back to the null displacement position. Provided that the loop response is rapid enough, the mass will be retained in a nearly constant place relative to the displacement sensor. Acceleration variations are thereby converted to variations in coil current. In this way the displacement sensor is used in the preferred null-balance mode in which error of linearity and temperature shift are largely avoided. Only the more easily achieved proportionality between coil current and force is important. Servo instruments are further described in Jones (1982), Herceg (1972), and Norton (1969).

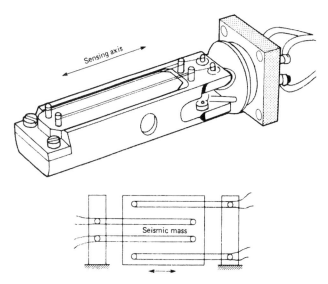

FIGURE 11.11 Displacement sensing of mass motion in accelerometers can be achieved by many methods. This unit uses unbonded strain gauges. Courtesy of Statham Instruments Inc.

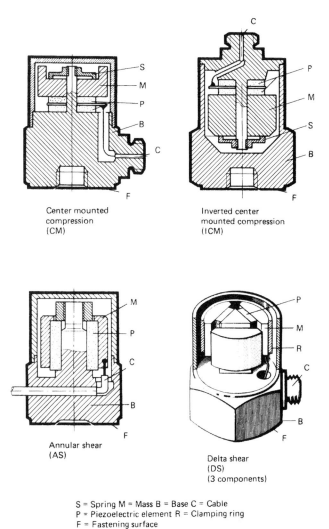

S = Spring M = Mass B = Base C = Cable
P = Piezoelectric element R = Clamping ring
F = Fastening surface

FIGURE 11.12 Examples of single-and three-axis accelerometers based on the piezoelectric sensor. Courtesy of Bruel & Kjaer.

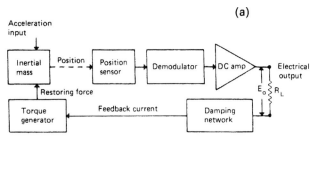
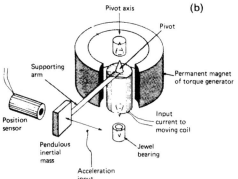

FIGURE 11.13 Basic component layout of one form of closed-loop accelerometer. Courtesy of Schaevitz Engineering.

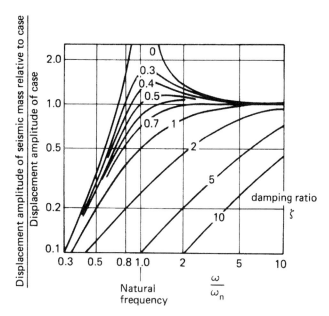

FIGURE 11.14 Responses relevant to displacement and velocity sensing with seismic sensors.

11.3.2 Displacement Measurement

Where a fixed reference method is applicable, it is possible to employ any suitable displacement sensor. Spurious phase shift in the system may be a concern, but in some applications phase is unimportant. The reader is directed to Chapter 8 for an introduction to displacement devices.

It is sometimes more convenient to integrate the signal from a velocity transducer mounted to make use of a fixed reference.

Where a fixed reference method is inconvenient, one of several forms of seismic sensor system can be employed as follows.

The second-order equations of motion, given in Section 11.1.1 for a mass moving relative to a fixed reference frame (the mode for studying the movement of vibrating objects), can be reworked to provide response curves relating the displacement amplitude of the seismic mass to the amplitude of its case. This is the seismic sensor output. Figure 11.14 is the family of response curves showing the effects of operating frequency and degree of damping. Given that the case is moving in sympathy with the surface of interest, it can be seen, from the curves, that for input vibration frequencies well above the natural frequency of the seismic sensor, the measured output displacements will be a true indication (within a percent or so) of the movements of the surface. This form of seismic displacement sensor is also often called a *vibrometer*. It is possible to lower the frequency of operation by using a damping factor with a nominal value of 0.5. This, however, does introduce more rapidly changing phase shift errors with frequency, which may be important. The lowest frequency of use above which the response remains virtually flat is seen to be where the various damping factor curves approach the horizontal line equal to unity ratio, to within the allowable signal tolerance.

Given that the chosen damping remains constant and that a system does follow the second-order response, it is also possible to provide electronic frequency compensation that can further lower the useful frequency of operation by a small amount.

Thus to directly measure displacement amplitudes with a seismic sensor, it must have a natural frequency set to below the lowest frequency of interest in the subject's vibration spectrum. In this mode the seismic mass virtually remains stationary in space, acting as a fixed reference point. It is also clear that the seismic sensor cannot measure very low frequencies, since it is not possible to construct an economical system having a low enough resonant frequency.

The curves are theoretical perfections and would apparently indicate that the seismic sensor, in this case, will have flat response out to infinite frequency. This is not the case in practice because as the frequency of vibration rises, the seismic sensor structure begins to resonate at other frequencies, caused by such mechanisms as the spring vibrating in modes other than the fundamental of the system.

Given that the low-frequency range of accelerometers can extend down to less than 1 Hz (see Section 11.3.4), it may often be more practical to twice integrate an accelerometer signal to derive displacement amplitude rather than use a direct-reading displacement design of seismic sensor.

Vibration measurement is also discussed in Chapter 17.

11.3.3 Velocity Measurement

The prime method used to generate a direct velocity signal uses the law of electromagnetic induction. This gives the electrical voltage e generated as N turns of an electric coil cut magnetic flux ϕ over time t as

$$e = -N\frac{d\phi}{dt}$$

Velocity sensors are self-generating. They produce a voltage output that is proportional to the velocity at which a set of turns moves through a constant and uniform magnetic field.

There are many forms of this kind of sensor. The commonly used moving-coil arrangement comprises a cylindrical coil vibrating inside a magnetic field that is produced by a permanent electromagnet. A commonly seen arrangement, shown in Figure 11.15(a), is typified by the reversible-role moving-coil, loudspeaker movement. For this form the output voltage V_{out} is given by

$$V_{out} = -Blv \cdot 10^{-9}$$

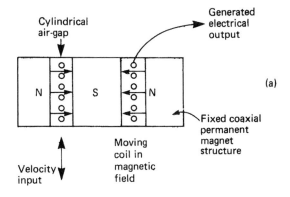

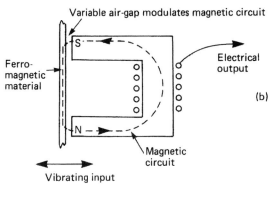

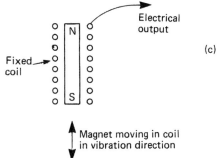

FIGURE 11.15 Forms of velocity sensor. (a) Moving coil, (b) Variable reluctance, (c) Moving permanent magnet.

where B is the flux density in Tesla, l the effective length of conductor contributing in total flux cutting, and v the instantaneous velocity, expressed in ms^{-1}, of the coil relative to the magnet. Given that the design ensures that the field is uniform in the path of the fixed conductor length, the sensor can provide very linear output.

Velocity sensors were adopted in early seismology studies because of their inherently high output at relatively low velocities. The coil impedance will generally be low, enabling signals with good signal-to-noise ratios to be generated along with reduction of error caused by variations in lead length and type. They are, however, large, with resultant mass and rigidity problems. They tend to have relatively low resonant frequencies (tens of Hertz), which restricts use to the lower frequencies. Output tends to be small at the higher frequencies. It will be apparent that these sensors cannot produce signals at zero velocity because no relative movement occurs to generate flux cutting.

A second variation of the self-generating velocity sensor, the variable-reluctance method, uses a series magnetic circuit containing a permanent magnet to provide permanent magnetic bias. A part of this circuit, the armature, is made so that the effective air gap is varied by the motion to be monitored. Around the magnetic circuit is placed a pickup coil. As the armature moves, the resulting flux variation cuts the coil, generating a signal that can be tailored by appropriate design to be linear with vibration amplitude. This form of design has the advantage that the armature can readily be made as part of the structure to be monitored, as shown in Figure 11.15(b). This version is not particularly sensitive, for the air gap must be at least as large as the vibration amplitude. For example, a unit of around 12 mm diameter, when used with a high magnetic permeability moving disc set at 2 mm distance, will produce an output of around 150 mV/m s^{-1}.

A third method uses a permanent magnet as the mass supported on springs. One example is shown in Figure 11.15(c). Vibration causes the magnet to move relative to the fixed coil, thereby generating a velocity signal. This form can produce high outputs, one make having a sensitivity of around 5 V/m s^{-1}.

Where a fixed reference cannot be used this form of sensor, instead of a displacement sensor, can be built into the seismic sensor arrangement. In such cases the vibrating seismic sensor will then directly produce velocity signals. These will follow the general responses given in Figure 11.14. From those curves it can be seen that there is a reasonably flat response above the natural frequency, which is inherently quite low.

11.3.4 Acceleration Measurement

The fixed-reference method of measuring acceleration is rarely used, most determinations being made with the seismic form of sensor. For the seismic sensor system, the mass and the spring compliance are fixed. Consideration of the $F = m \cdot a$ law and spring compliance shows that displacement of the mass relative to the sensor case is proportional to the acceleration of the case. This means that the curves, plotted in Figure 11.6 (for sinusoidal input of force to a second-order system), are also applicable as output response curves of accelerometers using displacement sensing. In this use the vertical axis is interpreted as the relative displacement of the mass from the case for a given acceleration of the case.

The curves show that a seismic sensor will provide a constant sensitivity output representing sensor acceleration from very low frequencies to near the natural frequency of

FIGURE 11.16 A range of accelerometers is required to cover the full needs of vibration measurement. Courtesy of Inspek Supplies, New South Wales.

the spring-mass arrangement used. Again, the damping ratio can be optimized at around 0.5–0.6 and electronic compensation added (if needed) to raise the upper limit a little further than the resonance point.

At first sight it might therefore appear that a single, general-purpose design that has a very high resonant frequency could be made. This, however, is not the case, because the deflection of the spring (which is a major factor deciding the system output sensitivity) is proportional to $1/w_n^2$. In practice this means that as the upper useful frequency limit is extended, the sensor sensitivity falls off. Electronic amplification allows low signal output to be used but with additional cost to the total measuring system.

At the low-frequency end of the accelerometer response, the transducers become ineffective because the accelerations produce too small a displacement to be observed against the background noise level.

11.3.4.1 Typical Sensors

As a guide to the ranges of capability that are available, one major manufacturer's catalog offers accelerometers with sensitivities ranging from a small 30 μV/ms^{-2} through 1 V/ms^{-2}, with corresponding sensor weights of 3 g and 500 g and useful frequency ranges of 1–60 000 Hz and 0.2–1000 Hz. Sensors have been constructed for even higher frequencies, but these must be regarded as special designs. A selection is shown in Figure 11.16.

The many constraints placed on the various performance parameters of a particular seismic sensor can be shown on a single chart such as Figure 11.17, from Harris and Crede (1961).

The accelerometer spring is often required to be stiff compared with that of the seismic displacement sensor, so it will not always need to use coiling, a device for decreasing the inherent spring constant of a material. Accelerometer springs may occur as stamped, rigid plates; as flat, cusped spring washers; or as a sufficiently compliant clamping bolt. In the case of piezosensitive material, use is often made of the compliance of the material.

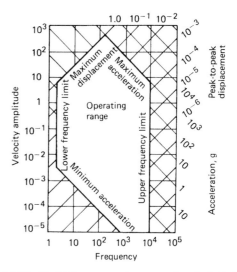

FIGURE 11.17 Useful linear operating range of an individual seismic vibration sensor can be characterized with this form of chart. Courtesy of McGraw-Hill.

11.3.4.2 Response to Complex Waveforms

The given response curves relate to seismic sensors excited by sinusoidal signals. To predict the behavior of a certain sensor, such as an accelerometer used to measure other continuous or discrete waveforms, it is first necessary to break down the waveform into its Fourier components. The response, in terms of amplitude and phase, to each of these components is then added to arrive at the resultant response. It has been stated that damping can be added to extend the useful bandwidth of a seismic sensor. However, where this is done it generally increases the phase shift variation with frequency. A signal comprising many frequencies will, therefore, produce an output that depends largely on the damping and natural frequency values of the sensor. A number of responses are plotted, such as that in Figure 11.18, in Harris and Crede (1961), to which the reader is referred. Generally the damping value for best all-round results is that near the critical value.

11.3.4.3 The Piezoelectric Sensor

Numerous sensing methods have been devised to measure the motion of the mass in a seismic sensor. We discuss here the most commonly used method: others are described in Endevco (1980), Harris and Crede (1961), Herceg (1972), Norton (1969), and Oliver (1971).

Force applied to certain crystalline substances such as quartz produces between two surfaces of a suitably shaped crystal an electric charge that is proportional to the force. This charge is contained in the internal electrical capacitance formed by the high-dielectric material and two deposited conducting surfaces. The descriptive mathematical relation for this effect is

$$q = a \cdot F \cdot K_s$$

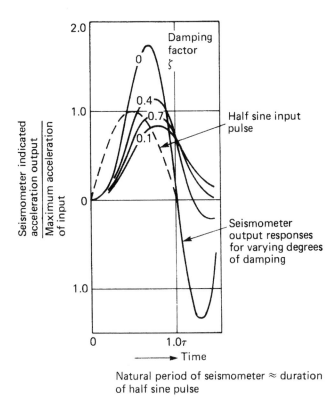

FIGURE 11.18 Example of response, at various damping factor levels, of a seismic accelerometer to a complex forcing input-a half-sine wave of similar period to that of the natural resonance period of the sensor.

equivalent is used. The nature of the system provides no true dc response.

In practice the PZT sensors used to measure acceleration can be operated down to around 0.1 Hz, dependent on the amplitude to be measured. With natural resonant frequency that can be made relatively very high (up to 100,000 Hz in some designs), PZT sensors provide a useful frequency range that can cover most vibration needs. The system response, however, relies not only on the sensor but on the cables and the preamplifier used with the PZT unit.

To produce the charge, the PZT material can be used in pure compression, shear, or bending. Figure 11.12 gives some examples of commercially available PZT accelerometers. The sensor design is amenable to the combination of three units, giving the three translation components of vibration.

PZT material itself contributes only on the order of 0.03 of critical damping. If no additional damping is added, PZT transducers must not be used too close to their resonant frequency. Mounting arrangements within the case will also add some damping. Some designs use an additional spring element. Some use an additional spring to precompress the PZT element so that it remains biased in compression under all working amplitudes; this makes for more linear operation.

Typical sensor sensitivities range from 0.003 pC/m s^{-2} up to 1000 pC/m s^{-2}, implying that the following preamplifier units will also need to vary considerably.

11.3.4.4 Amplifiers for Piezoelectric Sensors

An amplifier for reading out the state of the PZT sensor is one that has very high input impedance, an adequate frequency response, and low output impedance. Adjustment of gain and filtering action and integration to yield velocity and displacement are usually also needed to provide easy use for a variety of applications. Figure 11.19 is a typical system incorporating most features that might be needed.

The amplifier could be designed to see the sensor either as a voltage source or as a charge source. The latter is preferred for using modern electronic-feedback operational amplifier techniques; the effect of cable, sensor, and amplifier capacitances can be made negligible (which, in the voltage-reading method, is not the case). Cable length is, therefore, of no consequence. This is justified as follows.

Figure 11.20 is the relevant equivalent circuit for a PZT accelerometer that is connected to an operational amplifier (the preamplifier) via a cable. It includes the dominant capacitances that occur.

It can be shown (see Trampe-Broch, 1980, for example) that the use of feedback in this way and a very high amplifier gain A gives

$$e_0 = \frac{S_q a}{C_f}$$

where q is the electrical charge generated by force F (in Newtons) applied across the faces of a piezoelectric device with a mechanical compliance of spring rate K_s (mN^{-1}) and a more complex material constant a (of dimensions C m^{-1}).

The constant a depends on many factors, including the geometry of the crystal, position of electrodes, and material used. Typical materials now used (natural quartz is less sensitive and, therefore, less applicable) include barium titanate with controlled impurities, lead zirconate, lead niobate, and many that are trade secrets. The material is made from loose powder that, after shaping, is fired at very high temperature. While cooling, the blocks are subjected to an electric field that polarizes the substance.

The sensitivity of these so-called *PZT materials* is temperature-dependent through the charge sensitivity and the capacitance value, both of which alter with temperature. These changes do not follow simple linear laws. Such materials have a critical temperature, called the *Curie point*. They must never be taken above it. The Curie point varies, from 120°C for the simpler barium titanate forms ranging up to values close to 600°C. For the interested reader more explanation can be found in Bruel and Kjaer (1976), Endevco (1980), Harris and Crede (1961), Klaasen (1978), and Trampe-Broch (1980) and in the detailed information provided by the makers of PZT materials.

To read the charge of a PZT sensor, an electronic amplifier that converts charge magnitude to a voltage

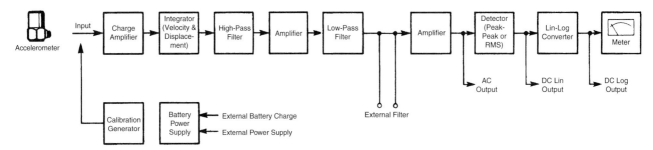

FIGURE 11.19 Block diagram of vibration measuring system showing functions that may be required. Courtesy of Bruel & Kjaer.

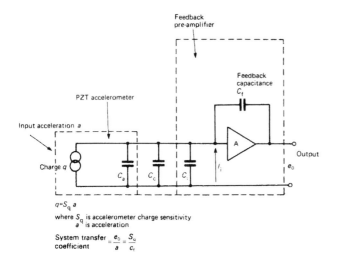

FIGURE 11.20 Equivalent circuit for piezoelectric sensor when interrogated, as a charge-generating device, by an operational amplifier technique.

This shows that the user need only define the sensor-charge sensitivity S_q and the feedback capacitance C_f to be able to relate output voltage from the preamplifier to the acceleration of the sensor.

11.3.5 Measurement of Shock

Shock is a sudden short impulse of applied force that generates very large acceleration (100,000 g can arise) and is not recurrent. It can be regarded as a once-only occurrence of a vibration waveform, although sometimes it is used to describe a short burst of oscillation.

Understanding the behavior of a given vibration sensor requires Fourier analysis of its response to a truncated wave shape. The mathematics becomes more complex. Theoretical study does lead to the generalization that as the waveform becomes more like a single pulse of high magnitude and very short duration, the frequency band of the sensor must be widened if the delivered output is to be a satisfactory replica of the actual input vibration parameter. Fidelity increases as the period of the natural frequency of the sensor becomes shorter than the pulse length. An idea of the variation of responses with natural frequency and damping is available in graphs given in Harris and Crede (1961). An example is shown in Figure 11.18.

The very large forces exerted on the transducer require a design that recognizes the need to withstand large transient forces without altering mechanical strains in the sensor system.

Well-designed shock sensors can accurately measure single half-sine-wave pulses as short as 5 μs. Some amount of ringing in the output is usually tolerated to provide measurement of very short duration shocks.

11.4 LITERATURE

There are many general books on the kinds of transducers that are in use. An IMEKO bibliography, Sydenham (1983), is a useful entry point to the literature of measurement technology.

Of the many general instrument texts that are available, very few actually address the subject of vibration as a distinct chapter. Where included, relevant material will be found under such headings as velocity and acceleration measurement, accelerometers, position sensing, and piezoelectric systems. Texts containing a chapter-length introductory discussion include Herceg (1972), Norton (1969), and Oliver (1971).

There are, as would be expected, some (but only a few) works entirely devoted to vibration and related measurands. The following will be of value to readers who require more than the restricted introduction that a chapter such as this provides: Bruel and Kjaer (1975, 1982), Endevco (1980), Harris and Crede (1961), Trampe-Broch (1980), and Wallace (1970).

The various trade houses that manufacture vibration-measuring and -testing equipment also often provide extensive literature and other forms of training aids to assist the uncertain user.

REFERENCES

Bently Nevada, *Bently Book One* (application notes on vibration and machines), Bently Nevada, Minden, Nev. (1982).

Bruel & Kjaer, *Vibration Testing Systems*, Bruel & Kjaer, Naerum, Denmark (1975).

Bruel & Kjaer, *Piezoelectric Accelerometer and Vibration Preamplifier Handbook*, Bruel & Kjaer, Naerum, Denmark (1976).

Bruel & Kjaer, *Acoustics, Vibration & Shock, Luminance and Contrast, National and International Standards and Recommendations*, Bruel & Kjaer, Naerum, Denmark (1981).

Bruel & Kjaer, *Measuring Vibration-an Elementary Introduction*, Bruel & Kjaer, Naerum, Denmark (1982).

Crandall, S. H., *Random Vibration*, Wiley, New York (1959).

Endevco, *Shock and Vibration Measurement Technology*, Endevco Dynamic Instrument Division, San Juan Capistrano, Calif. (1980).

Harris, C. M. and Crede, C. E., *Shock and Vibration Handbook Vol. 1, Basic Theory and Measurements*, McGraw-Hill, New York (1961, reprinted in 1976).

Herceg, E. E., *Handbook of Measurement and Control*, HB-72, Schaevitz Engineering, Pennsauken, N.J. (1972, revised 1976).

Jones, B. E. (ed.), "Feedback in instruments and its applications" in *Instrument Science and Technology*, Adam Hilger, Bristol, U.K. (1982).

Klaasen, K. B., "Piezoelectric accelerometers" in *Modern Electronic Measuring Systems*, Regtien, P. P. L. (ed.), Delft University Press, Delft (1978).

Norton, H. N., *Handbook of Transducers for Electronic Measuring Systems*, Prentice-Hall, Englewood Cliffs, N.J. (1969).

Oliver, F. J., *Practical Instrumentation Transducers*, Pitman, London (1971).

Sydenham, P. H., *Handbook of Measurement Science-Vol. 2. Fundamentals of Practice*, Wiley, Chichester, U.K. (1983).

Trampe-Broch, J., *Mechanical Vibration and Shock Measurements*, Bruel & Kjaer, Naerum, Denmark (1980).

Wallace, R. H., *Understanding and Measuring Vibrations*, Wykeham Publications, London (1970).

Wells, P., "Machine condition monitoring," Proceedings of structured course, Chisholm Institute of Technology, Victoria, Australia (1981).

FURTHER READING

Smith, J. D., *Vibration Measurement and Analysis*, Butterworth-Heinemann, Oxford (1989).

Wowk, V., *Machinery Vibration*, McGraw-Hill, New York (1991).

Chapter 12

Measurement of Force

C. S. Bahra and J. Paros

12.1 BASIC CONCEPTS

If a body is released, it will start to fall with an acceleration due to gravity or acceleration of free fall of its location. We denote by g the resultant acceleration due to attraction of the earth on the body and the component of acceleration due to rotation of the earth about its axis. The value of g varies with location and height, and this variation is about 0.5 percent between the equator and the poles. The approximate value of g is 9.81 m/s^2. A knowledge of the precise value of g is necessary to determine gravitational forces acting on known masses at rest, relative to the surface of the earth, in order to establish practical standards of force. Practical standards of dead-weight calibration of force-measuring systems or devices are based on this observation.

It is necessary to make a clear distinction between the units of weight-measuring (mass-measuring) and force-measuring systems. The weight-measuring systems are calibrated in kilograms; the force-measuring systems are in Newtons. Mass, force, and weight are defined as follows:

Mass. The mass of a body is defined as the quantity of matter in that body, and it remains unchanged when taken to any location. The unit of mass is the kilogram (kg).
Force. Force is that which produces or tends to produce a change of velocity in a body at rest or in motion. Force has magnitude, direction, and a point of application. It is related to the mass of a body through Newton's second law of motion, which gives: force = mass × acceleration.
Unit of Force. In the International System of units, the unit of force is the Newton (N), and it is that force which, when applied to a mass of one kilogram, gives it an acceleration of one meter per second per second (m/s^2).
Weight. Weight F of a body of mass m at rest relative to the surface of the earth is defined as the force exerted on it by gravity: $F = mg$, where g is the acceleration due to gravity.

The main purpose of this chapter is to review the most commonly used force measurement methods and to discuss briefly the principles employed in their design, limitations, and use. It is not intended to give a detailed description of mathematical and physical concepts but provides enough information to allow an interested reader to read further.

12.2 FORCE MEASUREMENT METHODS

Force measurement methods may be divided into two categories: direct comparison and indirect comparison. In a direct comparison method, an unknown force is directly compared with a gravitational force acting on a known mass. A simple analytical balance is an example of this method. An indirect comparison method involves the use of calibrated masses or transducers. A summary of indirect comparison methods is given below:

- Lever-balance methods.
- Force-balance method.
- Hydraulic pressure measurement.
- Acceleration measurement.
- Elastic elements.

Note that the lever-balance methods include examples of both direct and indirect comparisons, but to maintain continuity of information, they are described under one heading.

12.3 LEVER-BALANCE METHODS

12.3.1 Equal-lever Balance

A simple analytical balance is an example of an *equal-lever balance*, which consists of a "rigid" beam pivoted on a knife edge, as shown in Figure 12.1. An unknown force F_1 is compared directly with a known force F_2. When the beam is in equilibrium, the sum of moments about the pivot is zero.

$$F_1 a - F_2 a = 0$$
$$\therefore F_1 = F_2$$

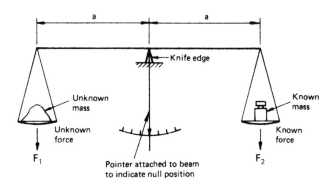

FIGURE 12.1 Equal-lever balance.

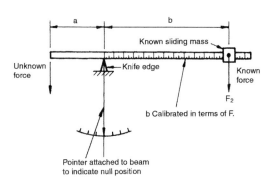

FIGURE 12.2 Unequal-lever balance.

This type of balance is mainly used for weighing chemicals. It gives direct reading and can weigh up to about 1,000 kg with high accuracy if required.

12.3.2 Unequal-lever Balance

Figure 12.2 shows a typical arrangement of an *unequal-lever balance*, which can be used for measuring large masses or forces. The balance is obtained by sliding a known mass along the lever. At equilibrium, we have

$$F_1 a = F_2 b$$
$$F_1 = F_2 b/a$$
$$\therefore F_1 \mu b$$

The right-hand side of the beam can therefore be used as a measure of force.

This type of balance is extensively used in materials-testing machines and weight measurement. The balance is normally bulky and heavy but can be made very accurate.

12.3.3 Compound Lever Balance

Figure 12.3 shows a compound lever balance that is used for measuring very large masses or forces. Using a number of ratio levers, the applied force is reduced to a level that is just sufficient to actuate a spring within the indicator dial head. The balance is calibrated in units of mass.

12.4 FORCE-BALANCE METHODS

Figure 12.4 shows an electronic force-balance type of force-measuring system. The displacement caused by the applied force is sensed by a displacement transducer. The displacement transducer output is fed to a servo amplifier to give an output current that flows through a restoring coil and exerts a force on the permanent magnet. This force always adjusts itself to balance the applied force. In equilibrium, the current

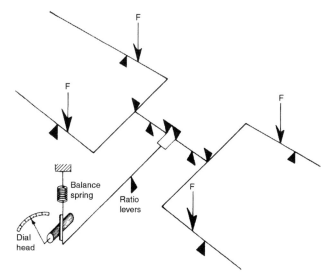

FIGURE 12.3 Compound-lever balance.

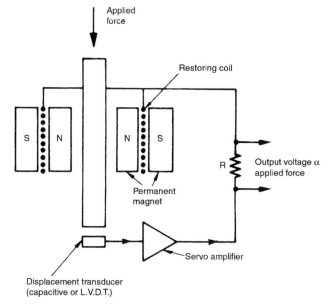

FIGURE 12.4 Force-balance system.

flowing through the force coil is directly proportional to the applied force. The same current flows through the resistor R, and the voltage drop across this resistor is a measure of the applied force. Being a feedback system, such a device must be thoroughly checked for stability.

The force-balance system gives high stability, high accuracy, and negligible displacement and is suitable for both static and dynamic force measurement. The range of this type of instrument is from 0.1 N to about 1 kN. It is normally bulky and heavy and tends to be expensive.

12.5 HYDRAULIC PRESSURE MEASUREMENT

The change in pressure due to the applied force may be used for force measurement. Figure 12.5 shows a general arrangement of a hydraulic load cell. An oil-filled chamber is connected to a pressure gauge and is sealed by a diaphragm. The applied force produces a pressure increase in the confined oil and is indicated on the pressure gauge, which is calibrated to give direct reading of force. If an electrical output is required, an electrical pressure transducer may be used in place of the pressure gauge.

Hydraulic load cells are stiff, with virtually no operational movement, and they can give local or remote indication. They are available in force ranges up to 5 MN with system accuracy on the order of 0.25 to 1.0 percent.

12.6 ACCELERATION MEASUREMENT

As mentioned earlier, force is a product of mass and acceleration. If the acceleration $\ddot{x}$ of a body of known mass m is known, the force Fx causing this acceleration can be found from the relationship

$$Fx = m\ddot{x}$$

The acceleration is measured by using a calibrated accelerometer, as shown in Figure 12.6. In practice, this method may be used for measuring dynamic forces associated with vibrating masses and is discussed further in Chapter 11.

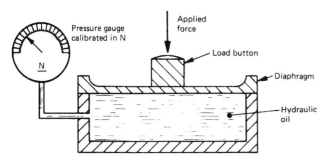

FIGURE 12.5 Force measurement using hydraulic load cell.

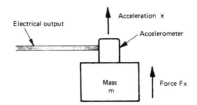

FIGURE 12.6 Force measurement using accelerometer.

12.7 ELASTIC ELEMENTS

A measuring system basically consists of three elements: a transducer, a signal conditioner, and a display or recorder. In this section, we discuss various types of transducers based on small displacements of elastic elements. In general, a *transducer* is defined as a device that changes information from one form to another. For the purpose of this discussion, a force transducer is defined as a device in which the magnitude of the applied force is converted into an electrical output proportional to the applied force. Transducers are divided into two classes: active and passive. A passive transducer requires an external excitation voltage, whereas an active transducer does not require an electrical input.

In general, a transducer consists of two parts: a primary elastic element that converts the applied force into a displacement and a secondary sensing element that converts the displacement into an electrical output. The elastic behavior of the elastic element is governed by Hooke's law, which states that the relationship between the applied force and the displacement is linear, provided the elastic limit of the material is not exceeded. The displacement may be sensed by various transducing techniques; some of them are examined in this section.

12.7.1 Spring Balances

The extension of a spring may be used as a measure of the applied force. This technique is employed in the design of a spring balance, as shown in Figure 12.7. This type of balance is a relatively low-cost, low-accuracy device and can be used for static force measurement.

12.7.2 Proving Rings

A *proving ring* is a high-grade, steel ring-shaped element with integral loading bosses, as shown in Figure 12.8. Under the action of a diametral force, the ring tends to distort. The amount of distortion is directly proportional to the applied force. For low-accuracy requirements, the distortion is measured using a dial gauge or a micrometer; for high-accuracy applications, a displacement transducer such as a linear variable differential transformer may be used. See Chapter 8.

Proving rings are high-precision devices that are extensively used to calibrate materials-testing machines. They

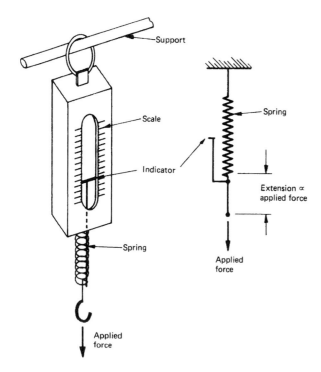

FIGURE 12.7 Spring balance.

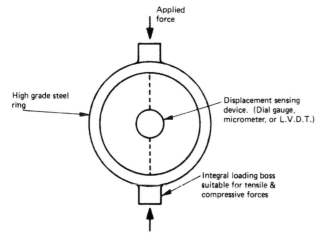

FIGURE 12.8 Proving ring fitted with displacement-sensing device.

may be used both in tension and compression, with a compressive force range on the order of 2 kN to 2000 kN and accuracy from 0.2 to 0.5 percent.

12.7.3 Piezoelectric Transducers

A typical arrangement of a piezoelectric transducer is shown in Figure 12.9. When the transducer is subjected to the applied force, a proportional electrical charge appears on the faces of the piezoelectric element. The charge is also a function of force direction. The piezoelectric transducer differs from a conventional (passive) transducer in two respects.

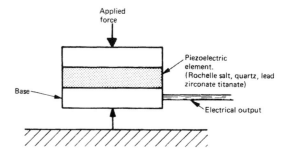

FIGURE 12.9 Piezoelectric force transducer.

First, it is an active system; second, the deflection at rated load is no more than a few thousandths of a millimeter, whereas the corresponding deflection for the conventional system may amount to several tenths of a millimeter.

This type of transducer has high stiffness, resulting in a very high resonant frequency. Because charge can leak away through imperfect insulation, it is unsuitable for measuring steady forces. It is mainly used in vibration studies and is discussed further in Chapter 11. It is small, rugged in construction, sensitive to temperature changes, and capable of measuring compressive forces from a few kilonewtons to about 1 meganewton, with accuracy from 0.5 to 1.5 percent.

12.7.4 Strain-gauge Load Cells

12.7.4.1 Design

Bonded strain-gauge load cells are devices producing an electrical output which changes in magnitude when a force or weight is applied, and which may be displayed on a readout instrument or used in a control device. The heart of the load cell is the bonded-foil strain gauge which is an extremely sensitive device, whose electrical resistance changes in direct proportion to the applied force. See Chapter 9.

A load cell comprises an elastic element, normally machined from a single billet of high-tensile steel alloy, precipitation-hardening stainless steel beryllium copper, or other suitable material, heat-treated to optimize thermal and mechanical properties. The element may take many forms, such as hollow or solid column, cantilever, diaphragm, shear member, or ring. The design of the element is dependent on the load range, type of loading, and operational requirements. The gauges are bonded on to the element to measure the strains generated and are usually connected into a four-arm Wheatstone bridge configuration. On larger elements, to get a true average of the strains, often 8, 16, or even 32 gauges are used. To illustrate the working principle, a cantilever load cell is shown in Figure 12.10. Figure 12.11 shows a bridge circuit diagram that includes compensation resistors for zero balance and changes of zero and sensitivity with temperature. To achieve high performance and stability and to minimize glue line thickness, the gauges are often installed on flat-sided elements.

CHAPTER | 12 Measurement of Force

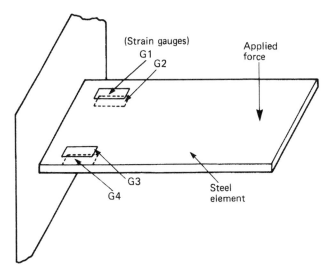

FIGURE 12.10 Cantilever load cell.

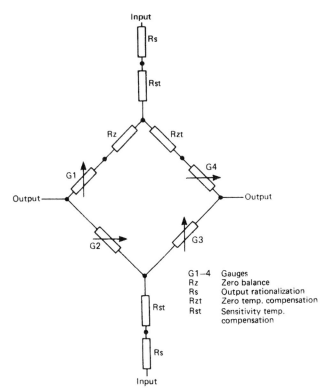

FIGURE 12.11 Load cell bridge circuit with compensation resistors.

The complete assembly is housed within a protective case that has sealing sufficient to exclude the external environment but capable of allowing the deformation of the element to occur when the force is applied. In some cases, restraining diaphragms minimize the effect of side loading.

After assembly, the elements are subjected to a long series of thermal and load cycling to ensure that remaining "locked-up" stresses in the element and bonding are relieved so that the units will give excellent long-term zero stability.

FIGURE 12.12 Compression load cells.

Figure 12.12 shows some commercially available compression load cells that have been successfully used for monitoring the tension in the mooring legs of a North Sea platform.

12.7.4.2 Selection and Installation

There are five basic types of cell on the market: compression, tension, universal (both compression and tension), bending, and shear. The main factors influencing the selection of cell type are:

- The ease and convenience (and hence the cost) of incorporating a cell into the weigher structure.
- Whether the required rated load and accuracy can be obtained in the type of cell.

Other considerations include low profile, overload capacity, resistance to side loads, environmental protection, and a wide operating temperature range.

To retain its performance, a cell should be correctly installed into the weigher structure. This means the structure of the weigher, such as vessel, bin, hopper, or platform, is the governing factor in the arrangement of the load cells. The supporting structure is also to be considered, since it will carry the full weight of the vessel and contents. Difficulties caused by misapplication leading to poor performance and unreliability fall into three main headings:

- A nonaxial load is applied.
- Side loads are affecting the weight reading.
- Free-axial movement of the load is restricted.

Figure 12.13 shows how normal, nonaxial, and side loading affects a column stress member. Under normal loading

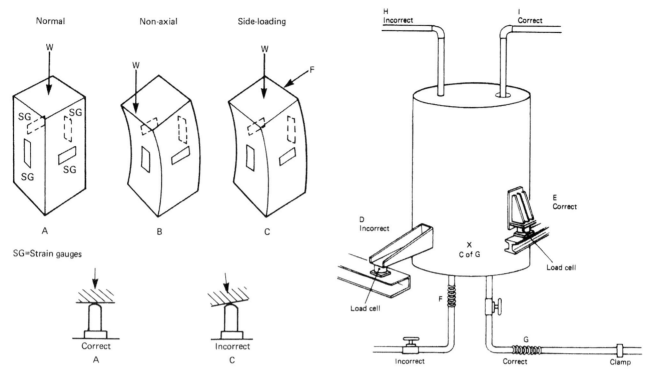

FIGURE 12.13 Effects of normal, nonaxial, and side loading.

FIGURE 12.14 Examples of correct and incorrect fitments.

conditions (A) the active strain gauges go into equal compression; however, under nonaxial (B) or side-loading (C) conditions, asymmetrical compression results, causing readout errors.

Examples of correct and incorrect fitments are shown in Figure 12.14. The support bracket D is cantilevered out too far and is liable to bend under load. The bracket is applying a load to the side of the vessel, which itself exaggerates this effect because the vessel is not strong enough to support it. The beam also deflects under load, rotating the load cell away from the vertical. The correct example, E, shows how the errors can be overcome.

In weighing installations it is important that there is unimpeded vertical movement of the weigh vessel. Obviously this is not possible where there are pipe fittings or stay rods on the vessel, but the vertical stiffness must be kept within allowable limits. One of the most satisfactory ways of reducing the spring rates is to fit flexible couplings in the pipework, preferably in a horizontal mode, and after (for example) the discharge valve so that they are not subject to varying stiffness due to varying pressure (see F and G in Figure 12.14). Where possible, entry pipes should be free of contact with the vessel (refer to H and I).

12.7.4.3 Applications

Load cells have many applications, including weight and force measurement, weigh platforms, process control systems, monorail weighing, belt weighers, aircraft, freight and baggage weighing, and conversion of a mechanical scale to an electronic scale. Over the past few years, the industrial weighing field has been dominated by load cells because electrical outputs are ideal for local and remote indication and to interface with microprocessors and minicomputers.

Key features of load cells are:

- Load range 5 N to 40 MN.
- Accuracy 0.01 to 1.0 percent.
- Rugged and compact construction.
- No moving parts and negligible deflection under load.
- Hermetically sealed and thermal compensation.
- High resistance to side loads; withstands overloads.

12.7.4.4 Calibration

Calibration is a process that involves obtaining and recording the load cell output while a direct known input is applied in a well-defined environment. The load cell output is directly compared against a primary or secondary standard of force. A primary standard of force includes dead-weight machines with force range up to about 500 kN; higher forces are achieved with machines having hydraulic or mechanical amplification.

A secondary standard of force involves the use of high-precision load cells and proving rings with a

TABLE 12.1 Summary of main parameters of force-measuring methods

Method	Type of loading	Force range, N (approx.)	Accuracy % (approx.)	Size
Lever balance	Static	0.001 to 150 k	Very high	Bulky and heavy
Force balance	Static/dynamic	0.1 to 1 k	Very high	Bulky and heavy
Hydraulic load cell	Static/dynamic	5 k to 5 M	0.25 to 1.0	Compact and stiff
Spring balance	Static	0.1 to 10 k	Low	Large and heavy
Proving ring	Static	2 k to 2 M	0.2 to 0.5	Compact
Piezoelectric transducer	Dynamic	5 k to 1 M	0.5 to 1.5	Small
Strain-gauge load cell	Static/dynamic	5 to 40 M	0.01 to 1.0	Compact and stiff

calibration standard directly traceable to the National Institute for Standards and Testing in Gaithersburg, Maryland, or the equivalent standards in other countries. The choice of the standards to be used for a particular calibration depends on the range and the location of the device to be calibrated.

The foregoing has indicated some force-measurement methods. Others are many and varied, and no attempt has been made to cover all types here. To simplify the selection of a method for a particular application, in Table 12.1 we summarize the main parameters of the methods discussed.

12.8 FURTHER DEVELOPMENTS

Advancing technology, improvements in manufacturing techniques, and new materials have permitted increased accuracy and improved design of bonded strain-gauge load cells since their introduction about 30 years ago. Microprocessor-enabled and controlled load cells have become ubiquitous.

New transducing techniques are being constantly researched; a number of them have been well studied or are being considered, including gyroscopic force transducers, fiber optics, microwave cavity resonators, and thin-film transducing techniques. The thin-film techniques are well documented and therefore are briefly discussed.

Pressure transducers based on vacuum-deposited thin-film gauges are commercially available, and attempts are being made to apply these techniques to load cells. The advantages of these techniques are as follows:

- Very small gauge and high bridge resistance.
- Intimate contact between the element and gauge. No hysteresis or creep of a glue line.
- Wide temperature range ($-200°C$ to $+200°C$).
- Excellent long-term stability of the bridge.
- Suitability for mass production.

The techniques are capital-intensive and are generally suitable for low force ranges.

REFERENCES

Adams, L. F., *Engineering Measurements and Instrumentation*, The English Universities Press, London (1975).

Cerni, R. H., and Foster, L. E., *Instrumentation for Engineering Measurement*, John Wiley, London (1962).

Mansfield, P. H., *Electrical Transducers for Industrial Measurement*, Butterworth, London (1973).

Neubert, H. K. P., *Instrument Transducers*, Clarendon Press, Oxford (2nd ed., 1975).

WEIGHTECH 79, *Proceedings of the Conference on Weighing and Force Measurement: Hotel Metropole*, Brighton, England, 24–26 September 1979.

Chapter 13

Measurement of Density

E. H. Higham and W. Boyes

13.1 GENERAL

The measurement and control of liquid density are critical to a large number of industrial processes. But although density in itself can be of interest, it is usually more important as a way of inferring composition or concentration of chemicals in solution, or of solids in suspension. In addition, density is often used as a component in the measurement of mass flow. Because the density of gases is very small, the instruments for that measurement have to be very sensitive to small changes. They are dealt with separately at the end of the chapter.

In considering the measurement and control of density or relative density* of liquids, the units used in the two factors should be borne in mind. Density is defined as the mass per unit volume of a liquid and is expressed in such units as kg/m^3, g/l or g/ml.

Relative density, on the other hand, is the ratio of the mass of a volume of liquid to the mass of an equal volume of water at 4°C (or some other specified temperature), the relative density of water being taken as 1.0. Both density and relative density are temperature-dependent and, for high precision, the temperature at which a measurement is made will have to be known so that any necessary compensation can be introduced.

The majority of industrial liquid-density instruments are based on the measurement of weight, buoyancy, or hydrostatic head, but measuring systems based on resonant elements or radiation techniques are also used. In recent years, the increasing popularity of Coriolis force-based mass flowmeters has led to their being one of the most common ways to measure density, derived, of course, from the mass flow value.

---

* The term *specific gravity* is often used for relative density. However, it is not included in the SI System of Units, and BS350 points out that it is commonly used when the reference substance is water. *Specific gravity* is far more commonly used in the United States, where American Standard Units prevail.

13.2 MEASUREMENT OF DENSITY USING WEIGHT

The actual weighing of a sample of known volume is perhaps the simplest practical application of this principle. Various methods for continuous weighing have been devised, but the most successful involves the use of a horizontal U-shaped tube with flexible couplings at a pivot point.

One example of this type of instrument is the Fisher-Rosemount Mark V gravitrol density meter[†] shown in Figure 13.1. In it, the process fluid passes via flexible connectors into the tube loop, which is supported toward the curved end on a link associated with the force-balance measuring system. In the pneumatic version of the instrument, the link is attached toward one end of the weighbeam, which itself is supported on cross-flexure pivots and carries an adjustable counterbalance weight on the opposite side. Also attached to the weighbeam is a dashpot to reduce the effect of vibration induced by the flow of the process fluid or by the environment in which the instrument is located.

In operation, the counterbalance weight is positioned to achieve balance with the tube loop filled with fluid, having a density within the desired working range, and the span adjustment is set to its midposition. Balance is achieved when the force applied by the feedback bellows via the pivot and span-adjustment mechanism to the weighbeam causes it to take up a position in which the feedback loop, comprising the flapper nozzle and pneumatic relay, generates a pressure that is both applied to the feedback bellows and used as the output signal. A subsequent increase in the density of the process fluid causes a minute clockwise rotation of the weighbeam, with the result that the flapper is brought closer to the nozzle and so increases the back pressure. This change is amplified by the relay and applied to the feedback bellows, which in turn applies an increased force via the span-adjustment system until balance is restored.

[†] No longer available

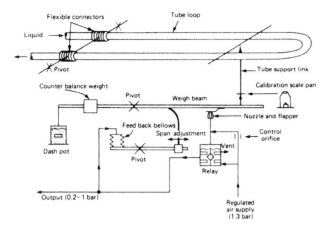

FIGURE 13.1 Gravitrol density meter. Courtesy of Fisher-Rosemount Inc.

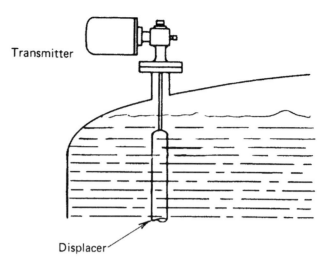

FIGURE 13.2 Buoyancy transducer and transmitter with tank.

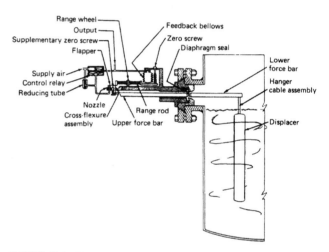

FIGURE 13.3 Buoyancy transducer and transmitter installation. Courtesy of Invensys Inc.

An electronic force-balance system is also available that serves the same function as the pneumatic force-balance system just described. The basic calibration constants for each instrument are determined at the factory in terms of the weight equivalent to a density of 1.0 kg/dm^3. To adjust the instrument for any particular application, the tube loop is first emptied. Then weights corresponding to the lower-range value are added to the calibration scale pan, and the counterbalance weight is adjusted to achieve balance.

Further weights are then added, representing the required span, and the setting of the span adjustment is varied until balance is restored. The two procedures are repeated until the required precision is achieved. The pneumatic output, typically 20–100 kPa, then measures the change in density of the flowing fluid. It can be adjusted to operate for spans between 0.02 and 0.5 kg dm^3 and for fluids having densities up to 1.6 kg/dm^3. The instrument is of course suitable for measurement on "clean" liquids as well as slurries or fluids with entrained solid matter. In the former case a minimum flow velocity of 1.1 m/s and in the latter case at least 2.2 m/s is recommended, to avoid deposition of the entrained solids.

13.3 MEASUREMENT OF DENSITY USING BUOYANCY

Buoyancy transmitters operate on the basis of Archimedes' principle: that a body immersed in a liquid is buoyed upward by a force equal to the weight of the liquid displaced. The cross-sectional area of a buoyancy transmitter displacer is constant over its working length so that the buoyant force is proportional to the liquid density; see Figure 13.2.

With the arrangement of the force-balance mechanism shown in Figure 13.3, the force on the transmitter force bar must always be in the downward direction. Thus the displacer element must always be heavier than the liquid it displaces. Displacers are available in a wide selection of lengths and diameters to satisfy a variety of process requirements.

Buoyancy transmitters are available for mounting either on the side of a vessel or for top entry and can be installed on vessels with special linings such as glass—vessels in which a lower connection is not possible. They are also suitable for density measurements in enclosed vessels where either the pressure or level may fluctuate, and they avoid the need for equalizing legs or connections for secondary compensating instrumentation such as repeaters. These transmitters are also suitable for applications involving high temperatures.

Turbulence is sometimes a problem for buoyancy transmitters. When turbulence occurs, the most simple (and often the least expensive) solution is to install a stilling well or guide rings. Another alternative is to use a cage-mounted buoyancy transmitter, as shown in Figure 13.4. With this configuration, the measurement is outside the vessel and therefore isolated from the turbulence.

CHAPTER | 13 Measurement of Density

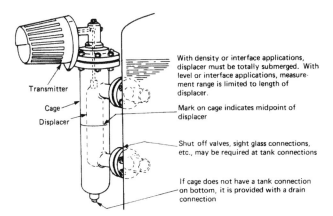

FIGURE 13.4 Buoyancy transducer and transmitter with external cage. Courtesy of Invensys Inc.

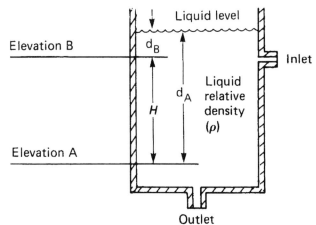

FIGURE 13.5 Density measurement–hydrostatic head.

13.4 MEASUREMENT OF DENSITY USING A HYDROSTATIC HEAD

The hydrostatic-head method, which continuously measures the pressure variations by a fixed height of liquid, has proved to be suitable for many industrial processes and continues to be a most commonly used method. Briefly, the principle, illustrated in Figure 13.5, is as follows:

The difference in pressure between any two elevations (A and B) below the surface is equal to the difference in liquid head pressure between these elevations. This is true regardless of variation in liquid level above elevation B. This difference in elevation is represented by dimension H. Dimension H must be multiplied by the relative density of the liquid to obtain the difference in head. This is usually measured in terms of millimeters of water.

To measure the change in head resulting from a change in relative density from ρ_1 to ρ_2 it is necessary only to multiply H by the difference between ρ_1 and ρ_2. Thus

$$P = H(\rho_2 - \rho_1)$$

and if both H and P are measured in millimeters, the change in density is

$$(\rho_2 - \rho_1) = P/H$$

It is common practice to measure only the span of actual density changes. Therefore, the instrument "zero" is "suppressed" to the minimum head pressure to be encountered; this allows the entire instrument measurement span to be devoted to the differential caused by density changes. For example, if ρ_1 is 0.6 and H is 3 meters, the zero suppression value should be 1.8 meters of water. The two principal relationships that must be considered in selecting a measuring device are:

$$\text{span} = H(\rho_2 - \rho_1)$$
$$\text{zero suppression value} = H \cdot \rho_1$$

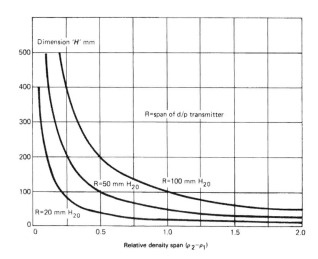

FIGURE 13.6 Relative density span versus H for various spans.

For a given instrument span, a low-density span requires a greater H dimension (a deeper tank). For a given density span, a low-span measuring device permits a shallower tank. Figure 13.6 shows values of H plotted against gravity spans.

13.4.1 General Differential Pressure Transmitter Methods

There is a variety of system arrangements for determining density from measurements of hydrostatic heads using differential pressure (DP) transmitters. Although flange-mounted DP transmitters are often preferred, pipe-connected transmitters can be used on liquids where crystallization or precipitation in stagnant pockets will not occur.

These DP transmitter methods are usually superior to those in which the DP transmitter operates in conjunction with bubble tubes and can be used in either pressure or vacuum systems. However, all require the dimensions of the process vessel to provide a sufficient change of head to satisfy the minimum span of the transmitter.

13.4.2 DP Transmitter with Overflow Tank

Constant-level overflow tanks permit the simplest instrumentation, as shown in Figure 13.7. Only one DP transmitter is required. With H as the height of liquid above the transmitter, the equations are still:

$$\text{span} = H(\rho_2 - \rho_1)$$
$$\text{zero suppression value} = H \cdot \rho_1$$

13.4.3 DP Transmitter with a Wet Leg

Applications with level or static pressure variations require compensation. There are three basic arrangements for density measurement under these conditions. First, when a seal fluid can be chosen that is always denser than the process fluid and will not mix with it, the method shown in Figure 13.8 is adequate. This method is used extensively on hydrocarbons with water in the wet leg. For a wet-leg fluid of specific gravity ρ_s, an elevated zero transmitter must be used. The equations become

$$\text{span} = H(\rho_2 - \rho_1)$$
$$\text{zero elevation value} = H(\rho_s - \rho_1)$$

When there is no suitable wet-leg seal fluid but the process liquid will tolerate a liquid purge, the method shown in Figure 13.9 can be used. To ensure that the process liquid does not enter the purged wet leg, piping to the process vessel should include an appropriate barrier, either gooseneck or trap, as shown in Figure 13.10. Elevation or suppression of the transmitter will depend on the difference in specific gravity of the seal and process liquids. Here, the equations are

$$\text{span} = H(\rho_2 - \rho_1)$$
$$\text{zero suppression value} = H(\rho_1 - \rho_s), \text{ when } \rho_1 > \rho_s$$
$$\text{zero elevation value} = H(\rho_s - \rho_1), \text{ when } \rho_s > \rho_1$$

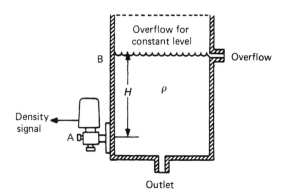

FIGURE 13.7 Density measurement with constant head.

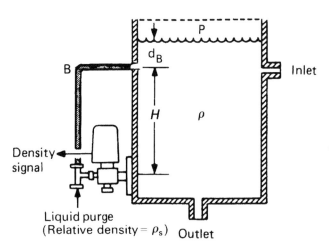

FIGURE 13.9 Density measurement with purge liquid.

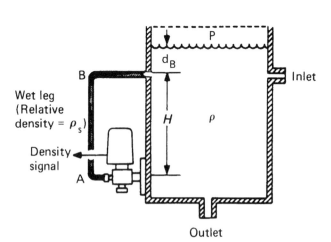

FIGURE 13.8 Density measurement with wet leg.

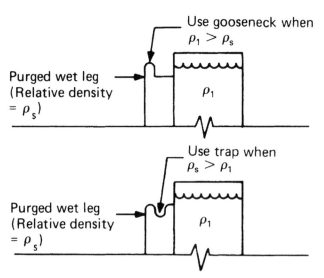

FIGURE 13.10 Purge system with gooseneck and trap.

Ideally, the purge liquid has a specific gravity equal to ρ_1, which eliminates the need for either suppression or elevation.

13.4.4 DP Transmitter with a Pressure Repeater

Figure 13.11 shows the use of a pressure repeater for the upper connection. In one form, this instrument reproduces any pressure existing at the B connection, from full vacuum to about 250 Pa positive pressure. In another form this instrument will reproduce any pressure from 7 kPa to 700 kPa. The repeater transmits the total pressure at elevation B to the low-pressure side of the DP transmitter. In this way, the pressure at elevation B is subtracted from the pressure at elevation A.

The lower transmitter, therefore, measures density (or $H \cdot \rho$, where ρ is the specific gravity of the liquid). The equations for the lower transmitter are:

$$\text{span} = H(\rho_2 - \rho_1)$$

$$\text{zero suppression value} = H \cdot \rho_1$$

The equation for the upper repeater is:

$$\text{output (maximum)} = (d_B \max)/(\rho_2) + P\max$$

where d_B is the distance from elevation B to the liquid surface, and P is the static pressure on the tank, if any.

Special consideration must be given when the repeater method is used for vacuum applications, where the total pressure on the repeater is less than atmospheric. In some instances density measurement is still possible. Vacuum application necessitates biasing of the repeater signal or providing a vacuum source for the repeater relay. In this case, there are restrictions on allowable gravity spans and tank depths.

13.4.5 DP Transmitter with Flanged or Extended Diaphragm

Standard flanged and extended diaphragm transmitter applications are illustrated in Figures 13.12(a) and (b), respectively. An extended diaphragm transmitter may be desirable to place the capsule flush with or inside the inner wall of the tank. With this instrument, pockets in front of the capsule where buildup may occur are eliminated.

13.4.6 DP Transmitter with Pressure Seals

If the process conditions are such that the process fluid must not be carried from the process vessel to the DP transmitter, a transmitter fitted with pressure seals can be used, as shown in Figure 13.13. Apart from the additional cost, the pressure seals reduce the sensitivity of the measurement, and any mismatch in the two capillary systems can cause further errors. However, the system can be used for either open or closed vessels.

13.4.7 DP Transmitter with Bubble Tubes

This very simple system, illustrated in Figure 13.14, involves two open-ended tubes, terminated with V notches. These are immersed in the liquid with the V notches separated by a known fixed vertical distance H and purged with a low but steady flow of air (or inert gas) at a suitable pressure.

A DP transmitter connected between these tubes, with the higher-pressure side associated with the lower V notch, measures the difference Δp in hydrostatic pressure at the two points. This is equal to the density × the vertical distance between the two V notches:

$$\text{density} = \Delta p/H$$

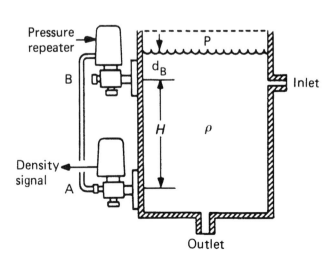

FIGURE 13.11 Density measurement with pressure repeater.

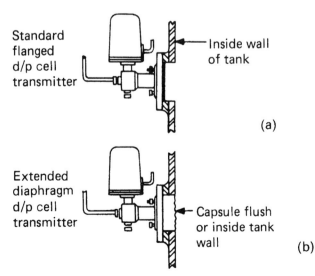

FIGURE 13.12 DP cell with flanged or extended diaphragm.

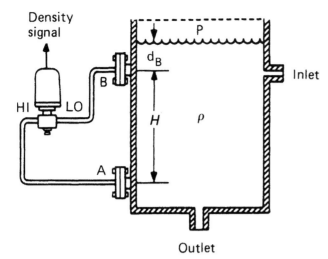

FIGURE 13.13 DP cell with pressure seals.

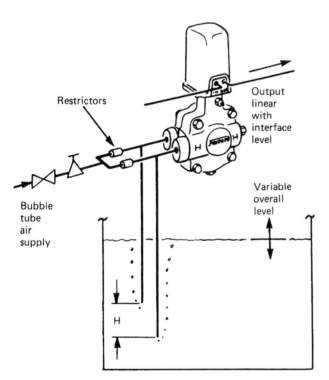

FIGURE 13.14 DP cell with bubble tubes.

Although this method is very simple and effective, it is unsuitable for closed vessels or for liquids that may crystallize or involve precipitation, which might block the bubble tubes and so give rise to erroneous results.

13.4.8 Other Process Considerations

Agitation in a process tank where density measurement is made must be sufficient to ensure uniformity of the liquid. But the velocity of fluid at the points where head pressure is measured must be sufficiently low to avoid a significant measurement error. Locations of side-mounted transmitters should be sufficiently high above the bottom of the tank to avoid errors due to them becoming submerged in the sediment that tends to collect there.

13.5 MEASUREMENT OF DENSITY USING RADIATION

Density measurements by this method are based on the principle that absorption of gamma radiation increases with increasing specific gravity of the material measured. These measurements are discussed in Part 3.

The principal instrumentation includes a constant gamma source, a detector, and an indicating or recording instrument. Variations in radiation passing through a fixed volume of flowing process liquid are converted into a proportional electrical signal by the detector.

13.6 MEASUREMENT OF DENSITY USING RESONANT ELEMENTS

Several density-measuring instruments are based on the measurement of the resonant frequency of an oscillating system such as a tube filled with the fluid under test or a cylinder completely immersed in the medium. Examples of each are described in the following sections.

13.6.1 Liquid Density Measurement

The Solartron 7835 liquid density transducer is shown in Figure 13.15. The sensing element comprises a single smooth-bore tube through which flows the fluid to be measured. The tube is fixed at each end into heavy nodal masses, which are isolated from the outer case by bellows and ligaments. Located along the tube are the electromagnetic drive and pickup coil assemblies. In operation, the amplifier maintains the tube oscillating at its natural frequency.

Since the natural frequency of oscillation of the tube is a function of the mass per unit length, it must also be a function of the density of the flowing fluid. It also follows that the tube should be fabricated from material having a low and stable coefficient of expansion. If for reasons of corrosion or wear this is not possible, it is important that the temperature is measured and a suitable correction applied to the density value determined from the resonant frequency.

Typically, the tube vibrates at about 1.3 kHz (when filled with water) and with an amplitude of about 0.025 mm. Densities up to 3000 kg/m^3 can be measured with an accuracy of 0.2 kg/m^3 and a repeatability of 0.02 kg/m^3. This contrasts with

CHAPTER | 13 Measurement of Density

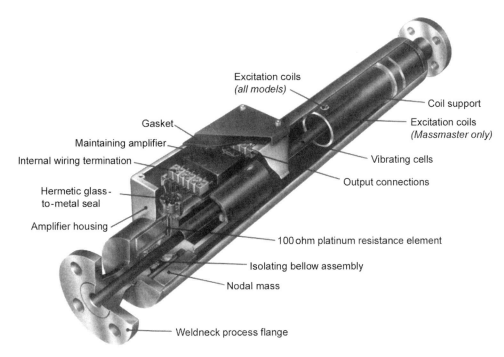

FIGURE 13.15 Solartron 7835 liquid density transducer. Courtesy of Solartron Transducers.

accuracies of only about 1 percent of span that can be achieved with other methods, unless extreme care is taken.

The response is continuous throughout its operating range with no adjustments of span or zero. Recalibration is effected by adjustment of the constants in the associated readout or signal conditioning circuits. The density-frequency relation is given by

$$\rho = K_0 \left(\frac{T^2}{T_0^2} - 1 \right)$$

where ρ is the density of the measured fluid; K_0 is constant for the transducer. T_0 is the time period of oscillation under vacuum conditions, and T is the time period of oscillation under operating conditions.

It is noteworthy that, although the relation between density and the period of the oscillation strictly obeys a square law, it is linear within 2 percent for a change in density of 20 percent. For narrower spans the error is proportionally smaller.

13.6.2 Gas Density Measurements

The relationship between temperature, pressure, and volume of a gas is given by

$$PV = nZR_0T$$

where P is the absolute pressure, V is the volume, and n is the number of moles. Z is the compressibility factor, R_0 is the universal gas constant, and T is the absolute temperature.

Use of the mole in this equation eliminates the need for determining individual gas constants, and the relationship among the mass of a gas m, its molecular weight Mw, and number of moles is given by

$$n = m/Mw$$

When the compressibility factor Z is 1.0, the gas is called *ideal* or *perfect*. When the specific heat is assumed to be only temperature dependent, the gas is referred to as *ideal*. If the *ideal* relative density RD of a gas is defined as the ratio of molecular weight of the gas to that of air, then

$$RD = \frac{Mw_{gas}}{Mw_{air}}$$

whereas the *real* relative density is defined as the ratio of the density of the gas to that of air, which is

$$RD = \frac{\rho_{gas}}{\rho_{air}}$$

for a particular temperature and pressure.

The preceding equation can be rearranged as a density equation, thus

$$\rho = \frac{m}{V} = \frac{SGMw_{air}P}{ZR_0T}$$

Most relative density measuring instruments operate at pressures and temperatures close to ambient conditions and hence measure *real* relative density rather than the *ideal* relative density, which is based on molecular weights and does not take into account the small effects of compressibility. Hence,

$$RD_{(real)} = \left(\frac{\rho_{gas}}{\rho_{air}}\right)_{TP}$$

where T and P are close to ambient conditions. Substituting the equation leads to P:

$$RD_{(ideal)} = \left(\frac{ZT}{P_{gas}}\right) \times \left(\frac{P}{ZT_{air}}\right)_{TP} \times RD_{(real)}$$

For most practical applications, this leads to

$$RD = \left(\frac{Z_{gas}}{Z_{air}}\right)_{TP} RD_{(ideal)}$$

Thus, the signal from the density transducer provides an indication of the molecular weight or specific gravity of the sampled gas.

The measurement can be applied to almost any gas provided that it is clean, dry, and noncorrosive. The accuracy is typically 0.1 percent of reading and the repeatability is 0.02 percent.

Measuring the lower densities of gases requires a more sensitive sensing element than that described for measurements on liquids. The Solartron 7812 gas density transducer shown in Figure 13.16 achieves this task by using a thin-walled cylinder resonated in the hoop or radial mode. The maximum amplitude of vibration occurs at the middle of the cylinder with nodes at each end, and it is therefore clamped at one end with a free node-forming ring at the other end.

The cylinder is immersed in the gas whose density is to be measured, and it is thus not stressed due to the pressure of the gas. Gas in contact with the cylinder is brought into oscillation and effectively increases the mass of the vibrating system, thereby reducing its resonant frequency.

Oscillation is maintained electromagnetically by positioning drive and pickup coils inside the cylinder and connecting them to a maintaining amplifier. The coils are mounted at right angles to each other to minimize stray coupling and are phased so that the induced signal is proportional to the velocity, thereby reducing the effect of viscous damping.

A low-temperature coefficient is obtained by constructing the cylinder from material having a low-temperature coefficient of expansion. The cylinder wall thickness varies from 0.05 to 0.15 mm according to the required density range, the corresponding density ranges varying from 0 to 60 kg/m³ and 40 to 400 kg/m³.

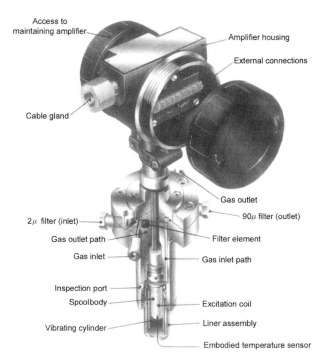

FIGURE 13.16 Solartron 7812 gas density transducer. Courtesy of Solartron Transducers.

The relation between the time period of oscillation of the transducer and gas density ρ is given by

$$\rho = 2d_0 \frac{(\tau - \tau_0)}{(\tau_0)} \left[1 + \frac{K}{2}\left(\frac{\tau - \tau_0}{\tau_0}\right)\right]$$

where τ is the measured time period of oscillation, τ_0 is the time period of oscillation under vacuum conditions, and d_0 and K are the calibration constants for each transducer.

An alternative method for measuring gas density involves a cylindrical test cell in which a hollow spinner is rotated at constant speed. This develops a differential pressure between the center and ends of the spinner that is directly proportional to the density of the gas and can be measured by any standard differential pressure-measuring device. A calibration constant converts the differential pressure to density for the actual conditions of temperature and pressure in the cell.

A sample flow of gas through the cell is induced by connecting it across a small restriction inserted in the main line to create an adequate pressure drop. The restriction could be determined from the square root of the differential pressure across the orifice plate multiplied by the differential pressure developed in the density cell. However, it is important to ensure that the flow of the gas through the density cell is not a significant proportion of the total flow. It is also important to apply a correction if there is any difference between the temperature and pressure of the gas in the density transducer and that in the main stream.

13.6.3 Relative Density of Gases

The Solartron 3096 specific gravity transducer, shown in Figure 13.17, utilizes the density sensor described in the previous section to measure relative density of gases. In it, the sample of gas and the reference gas are stabilized at the same temperature by coils within thermal insulation. The reference chamber is a constant volume containing a fixed quantity of gas; any variation in temperature is compensated by a change in pressure, which is transmitted to the sample gas by a flexible diaphragm. Having achieved pressure and temperature equalization by using a reference gas, a direct relationship between density and relative density can be realized.

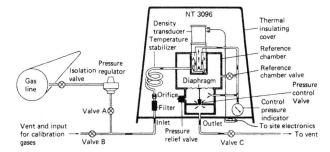

FIGURE 13.17 Solartron 3096 specific gravity transducer. Courtesy of Solartron Transducers.

FURTHER READING

Petroleum Measurement Manual, Part 7: Density, Wiley, Chichester, U.K., Section 2 (1984); Section 1 (1985).

Chapter 14

Measurement of Pressure

E. H. Higham and J. M. Paros

14.1 WHAT IS PRESSURE?

When a fluid is in contact with a boundary, it produces a force at right angles to that boundary. The force per unit area is called the *pressure*. In the past, the distinction between mass and force has been blurred because we live in an environment in which every object is subjected to gravity and is accelerated toward the center of the earth unless restrained. As explained in Chapter 8, the confusion is avoided in the SI system of units (Systeme International d'Unités), where the unit of force is the Newton and the unit of area is a square meter, so that pressure, being force per unit area, is measured in Newtons per square meter and the unit, known as the Pascal, is independent of the acceleration due to gravity.

The relation between the Pascal and other units used for pressure measurements is shown in Table 14.1.

There are three categories of pressure measurement: absolute pressure, gauge pressure, and differential pressure. The *absolute pressure* is the difference between the pressure at a particular point in a fluid and the absolute zero of pressure, that is, a complete vacuum. A barometer is one example of an absolute pressure gauge because the height of the column of mercury measures the difference between the atmospheric pressure and the "zero" pressure of the Torricellian vacuum that exists above the mercury column.

When the pressure-measuring device measures the difference between the unknown pressure and local atmospheric pressure, the measurement is known as *gauge pressure*.

When the pressure-measuring device measures the difference between two unknown pressures, neither of which is atmospheric pressure, the measurement is known as the *differential pressure*.

A mercury manometer is used in Figure 14.1 to illustrate these three forms of measurement.

14.2 PRESSURE MEASUREMENT

There are three basic methods for measuring pressure. The simplest method involves balancing the unknown pressure against the pressure produced by a column of liquid of known density. The second method involves allowing the unknown pressure to act on a known area and measuring the resultant force either directly or indirectly. The third method involves allowing the unknown pressure to act on an elastic member (of known area) and measuring the resultant stress or strain. Examples of these methods are described in the following sections.

14.2.1 Pressure Measurements by Balancing a Column of Liquid of Known Density

The simplest form of instrument for this type of measurement is the U-tube. Consider a simple U-tube containing a liquid of density ρ, as shown in Figure 14.2. Points A and B

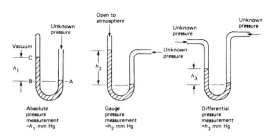

FIGURE 14.1 Comparison of types of pressure measurements.

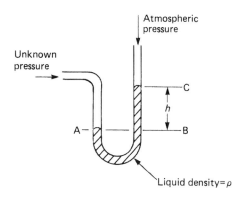

FIGURE 14.2 Simple U-tube manometer.

TABLE 14.1 Pressure measurements

	Pascal (Pa)	Bar (bar)	Millibar (mbar)	Standard atmosphere (atm)	Kilogram force per square cm (kgf/cm²)	Pound force per square inch (lbf/in²)	Torr	Millimeter of water (mmH$_2$O)	Millimeter of mercury (mmHg)	Inch of water (inH$_2$O)	Inch of mercury (inHg)
Pa	1	10^{-5}	10^{-2}	9.86923×10	1.01972×10^{-5}	1.45038×10^{-4}	7.50062×10^{-3}	1.01972×10^{-1}	7.50062×10^{-3}	4.01463×10^{-3}	2.95300×10^{-4}
bar	10^5	1	10^3	9.86923×10^{-1}	1.01972	14.5038	7.50062×10^{-2}	1.01972×10^{-2}	7.50062×10^{-2}	4.0163×10^{-8}	29.5300
mbar	10^2	10^{-3}	1	9.86923×10^{-4}	1.01972×10^{-3}	1.45038×10^{-2}	7.50062×10^{-1}	1.01972×10	7.50062×10^{-5}	4.01462×10^{-1}	2.95300×10^{-2}
atm	1.01325×10^5	1.01325	1.01325×10^3	1	1.03323	1.46959×10	7.60000×10^2	1.03323×10^5	760	4.06783×10^2	29.9213
kgf/cm²	98066.5	0.980665	980665	0.967841	1	14.2233	735.559×10^2	10^4	7.35559×10^2	3.93700×10^2	28.9590
lbf/in²	6894.76	0.0689476	68.9476	6.80460×10^{-2}	$7.03070 10^{-2}$	1	51.7149×10^2	7.03069	51.7149	27.6798	2.03602
torr	133.322	1.33322×10^{-3}	1.33322	1.31579×10^{-3}	1.35951×10^{-3}	1.93368×10^{-2}	1	13.5951	1	53.5240	3.93701×10^{-2}
mmH$_2$O	9.80665	9.80665×10^{-5}	9.80665×10^2	9.67841×10^{-5}	$10^{-4} \times 10^{-3}$	1.42233×10^{-2}	7.35559	1×10^{-2}	7.35559×10^{-2}	3.93701×0^{-3}	2.89590
mmHg	133.322	1.33322×10^{-3}	1.33322×10^{-3}	1.31579×10^{-3}	1.35951×10^{-2}	1.93368	1	13.5951	1×10^{-2}	53.5240	3.93701
inH$_2$O	249.089	2.49089×10^{-3}	2.49089×10^{-3}	2.45831×10^{-3}	2.54×10^{-2}	3.61272	1.86832	25.4	1.86832	1×10^{-2}	7.35559
inHg	33.8639	3.38639×10^{-2}	33.8639	3.34211×10^{-2}	3.45316×10^{-2}	0.491154	25.4000	3.45316×10^{-2}	25.4000	13.5951	1

n.b.: Extracts from British Standards are reproduced by permission of the British Standards Institution, 2 Park Street, London W1A 2BS, from which complete copies can be obtained.

are at the same horizontal level, and the liquid at C stands at a height h mm above B.

Then the pressure at A = the pressure at B
= atmospheric pressure
    + pressure due to column of liquid BC
= atmospheric pressure + $h\rho$

If the liquid is water the unit of measure is mmH$_2$O, and if the liquid is mercury, the unit of measure is mmHg. The corresponding SI unit is the Pascal and

$$1\,mmH_2O = 9.806\,65\,Pa$$
$$1\,mmHg = 133.332\,Pa$$

For a system such as this, it must be assumed that the density of the fluid in the left-hand leg of the manometer (Figure 14.2) is negligible compared with the manometer liquid. If this is not so, a correction must be applied to allow for the pressure due to the fluid in the gauge and connecting pipes. Referring to Figure 14.3, we have

Pressure at A = pressure at B

P(gauge pressure) = $\rho_1 h_1$ + atmospheric pressure
$= \rho_2 h$ + atmospheric pressure

or

$$P = \rho_2 h - \rho_1 h_1$$

(gauge pressure, because the atmospheric pressure is superimposed on each manometer leg measurement).

If the manometer limbs have different diameters, as in the case for a well-type manometer, shown in Figure 14.4, the rise in one leg does not equal the fall in the other. If the well has a cross-sectional area A and the tube an area a, the loss of liquid in one unit must equal the gain of liquid in the other. Hence $h_m A = h_2 a$ so that $h_2 = h_m A/a$.

For a simple U-tube measurement, the applied pressure $P = (h_2 + h_m)\rho$. If the left-hand leg of the manometer becomes a wet leg with fluid density, then

$$P + (h_1 + h_2)\rho_2 = (h_2 + h_m)\rho_1$$

so that

$$P = (h_2 + h_m)\rho_1 - (h_1 + h_2)\rho_2$$

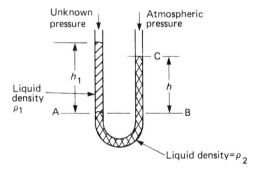

FIGURE 14.3 Manometer with wet leg connection.

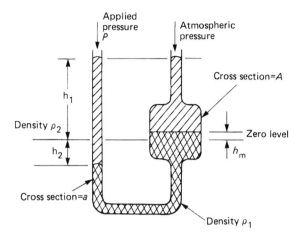

FIGURE 14.4 Manometer with limbs of different diameters.

If both manometer legs are wet, then

$$P + (h_1 + h_2)\rho_2 = (h_2 + h_m)\rho_1 + (h_1 - h_m)\rho_2$$
$$P = (h_2 + h_m)\rho_1 + (h_1 - h_m)\rho_2 - (h_1 + h_2)\rho_2$$
$$= h_2\rho_1 + h_m\rho_1 + h_1\rho_2 - h_m\rho_2 - h_1\rho_2 - h_2\rho_2$$
$$= \rho_1(h_2 + h_m) - \rho_2(h_m + h_2)$$
$$= (h_2 + h_m)(\rho_1 - \rho_2)$$
$$= h_m(A/a + 1)(\rho_1 - \rho_2)$$

Effect of Temperature

The effect of variations in temperature has been neglected so far, but for accurate work the effect of temperature on the densities of the fluids in the manometer must be taken into account, and the effect of temperature on the scale should not be overlooked. For most applications, it is sufficient to consider the effect of temperature only on the manometer liquid, in which case the density ρ at any temperature T can be taken to be:

$$\rho = \frac{\rho_0}{1 + \beta T - T_0}$$

where ρ_0 is the density at base conditions, β is the coefficient of cubic expansion, T_0 is the base temperature, and T is the actual temperature.

14.2.2 Pressure Measurements by Allowing the Unknown Pressure to Act on a Known Area and Measuring the Resultant Force

14.2.2.1 Deadweight Testers

The simplest technique for determining a pressure by measuring the force that is generated when it acts on a known

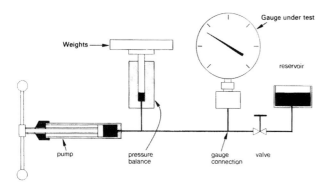

FIGURE 14.5 Basic system of a deadweight tester. Courtesy of Budenberg Gauge Co. Ltd.

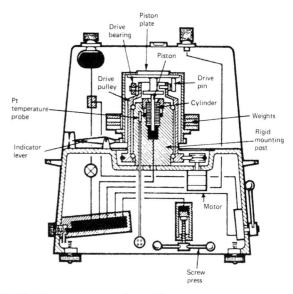

FIGURE 14.6 Arrangement of a precision deadweight tester. Courtesy of Desgranges and Huot.

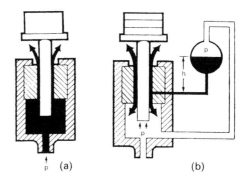

FIGURE 14.7 Lubrication of the piston. (a) Oil-operated system. (b) Gas-operated system. Courtesy of Desgranges and Huot.

area is illustrated by the *deadweight tester*, but this system is used for calibrating instruments rather than measuring unknown pressures.

The basic system is shown diagrammatically in Figure 14.5. It comprises a priming pump and reservoir, an isolating valve, the piston carrying the weight, a screw press, and the gauge under test. In operation, the screw press is set to its zero position, weights representing the desired pressure are applied to the piston, and the priming pump is operated to pressurize the system. The valve is then shut and the screw press is adjusted until the pressure in the system is sufficient to raise the piston off its stops. If the frictional forces on the piston are neglected, the pressure acting on it is p Newtons per square meter, and if its area is a square meters, the resultant force is pa N. This will support a weight $W = pa$ N.

The accuracy depends on the precision with which the piston and its associated cylinder are manufactured and on eliminating the effect of friction by rotating the piston while the reading is taken.

The Desgranges and Huot range of primary pressure standards is a very refined version of the deadweight testers. Figure 14.6 shows a sectional drawing of an oil-operated standard. For this degree of precision, it is important to ensure that the piston area and gravitational forces are constant so that the basic relation between the mass applied to the piston and the measured pressure is maintained. The instrument therefore includes leveling screws and bubble indicators.

Side stresses on the piston are avoided by loading the principal weights on a bell so that their center of gravity is well below that of the piston. Only the fractional weights are placed directly on the piston plate, and the larger of these are designed to stack precisely on the center line.

The mobility of the piston in the cylinder assembly determines the sensitivity of the instrument; this requires an annulus that is lubricated by liquid, even when gas pressures are being measured. Figures 14.7(a) and (b) show how this is achieved.

The system for liquids is conventional, but for gases, lubricant in the reservoir passes into the annulus between the piston and cylinder. The gas pressure is applied to both the piston and the reservoir so that there is always a small hydraulic head to cause lubricant to flow into the assembly.

Rotation of the piston in the cylinder sets up radial forces in the lubricating fluid that tend to keep the piston centered, but the speed of rotation should be constant and the drive itself should not impart vibration or spurious forces. This is achieved by arranging the motor to drive the cylinder pulley via an oval drive pulley, which is therefore alternatively accelerating and decelerating. The final drive is via the bearing on to the pin secured to the piston plate. In this way, once the piston is in motion, it rotates freely until it has lost sufficient momentum for the drive bearing to impart a small impulse, which accelerates the piston. This ensures that it is rotating freely for at least 90 percent of the time.

The piston and cylinder are machined from tungsten carbide to a tolerance of 0.1 micrometer so that the typical clearance between them is 0.5 micrometer. A balance indicator that tracks a soft iron band set in the bell shows the position of the piston and allows fluid head corrections for the most precise measurements.

The principal weights are fabricated in stainless steel and supplied in sets up to 50 kg, according to the model chosen. The mass of the bell (typically 0.8 kg) and the piston plate assembly (typically 0.2 kg) must be added to the applied mass.

A complete set of piston and cylinder assemblies allows measurements to be made in the ranges from 0.1 to 50 bar to 2.0 to 1000 bar, the uncertainty of measurement being $\pm 5 \times 10^{-4}$ or less for the "N" class instruments and $\pm 1 \times 10^{-4}$ or less for the "S" class instruments.

14.2.3 Pressure Measurement by Allowing the Unknown Pressure to Act on a Flexible Member and Measuring the Resultant Motion

The great majority of pressure gauges utilize a Bourdon tube, stacked diaphragms, or a bellows to sense the pressure. The applied pressure causes a change in the shape of the sensor that is used to move a pointer with respect to a scale.

14.2.3.1 Bourdon Tubes

The simplest form of Bourdon tube comprises a tube of oval cross-section bent into a circle. One end is sealed and attached via an adjustable connecting link to the lower end of a pivoted quadrant. The upper part of the quadrant is the toothed segment that engages in the teeth of the central pinion, which carries the pointer that moves with respect to a fixed scale. Backlash between the quadrant and pinion is minimized by a delicate hairspring. The other end of the tube is open so that the pressure to be measured can be applied via the block to which it is fixed and which also carries the pressure connection and provides the datum for measurement of the deflection.

If the internal pressure exceeds the external pressure, the shape of the tube changes from oval toward circular, with the result that it becomes straighter. The movement of the free end drives the pointer mechanism so that the pointer moves with respect to the scale. If the internal pressure is less than the external pressure, the free end of the tube moves toward the block, causing the pointer to move in the opposite direction.

The material from which the tube is formed must have stable elastic properties and be selected to suit the fluid for which the pressure is to be measured. Phosphor bronze, beryllium copper, and stainless steel are used most widely, but for applications involving particularly corrosive fluids, alloys such as K-Monel are used. The thickness of the tube and the material from which it is to be fabricated are selected according to the pressure range, but the actual dimensions of the tube determine the force available to drive the pointer mechanism. The construction of a typical gauge is shown in Figure 14.8.

The performance of pressure gauges of this type varies widely, not only as a result of their basic design and materials of construction but also because of the conditions under which they are used. The principal sources of error are hysteresis in the Bourdon tube, changes in its sensitivity due to changes of temperature, frictional effects, and backlash in the pointer mechanism. A typical accuracy is ±2 percent of span.

Much higher precision can be achieved by attention to detail; one example is illustrated in Figure 14.9, which shows a gauge for measuring absolute pressure. It includes two Bourdon tubes, one being completely evacuated and sealed to

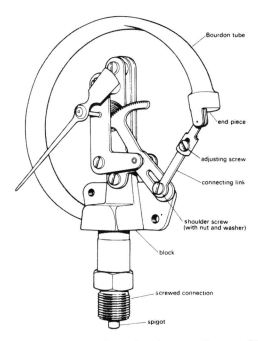

FIGURE 14.8 Mechanism of Bourdon tube gauge. Courtesy of Budenberg Gauge Co. Ltd.

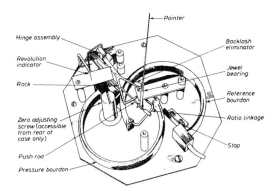

FIGURE 14.9 Precision absolute pressure gauge. Courtesy of U.S. Filter Corp.

provide the reference while the unknown pressure is applied to the other Bourdon tube. The free ends of the Bourdon tubes are connected by a ratio linkage that, through a push rod, transmits the difference in the movement of the free ends to a rack assembly, which in turn rotates the pinion and pointer. Jewel bearings are used to minimize friction, and backlash is eliminated by maintaining a uniform tension for all positions of the rack and pinion through the use of a nylon thread to connect a spring on the rack with a grooved pulley on the pinion shaft.

The Bourdon tubes are made of Ni-Span C, which has a very low thermoelastic coefficient (change in modulus of elasticity with temperature) and good resistance to corrosion. As both Bourdon tubes are subjected to the same atmospheric pressure, the instrument maintains its accuracy for barometric pressure changes of ±130 mmHg. The dial diameter is 216 mm, and the full range of the instrument is covered by two revolutions of the pointer, giving an effective scale length of 1.36 m. The sensitivity is 0.0125 percent and the accuracy 0.1 percent of full scale. The ambient temperature effect is less than 0.01 percent of full scale per Kelvin.

14.2.3.2 Spiral and Helical Bourdon Tubes

The amount of movement of the free end of a Bourdon tube varies inversely with the wall thickness and is dependent on the cross-sectional shape. It also varies directly with the angle subtended by the arc through which the tube is formed. By using a helix or spiral to increase the effective angular length of the tube, the movement of the free end is similarly increased and the need for further magnification is reduced. Examples of these constructions are shown in Figures 14.10 and 14.11. They avoid the necessity for the toothed quadrant, with the consequent reduction of backlash and frictional errors. In general, the spiral configuration is used for low pressures and the helical form for high pressures.

14.2.3.3 Diaphragm Pressure Elements

There are two basic categories of diaphragm elements, namely stiff metallic diaphragms and slack diaphragms associated with drive plates.

FIGURE 14.10 Helical Bourdon tube. Courtesy of Invensys Inc.

FIGURE 14.11 Spiral Bourdon tube. Courtesy of Invensys Inc.

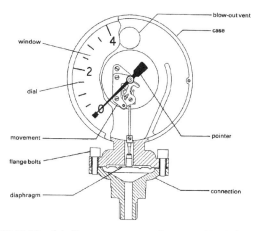

FIGURE 14.12 Schaffer pressure gauge. Courtesy of Budenberg Gauge Co. Ltd.

The simplest form of diaphragm gauge is the Schaffer gauge, shown in Figure 14.12. It consists of a heat-treated stainless-steel corrugated diaphragm about 65 mm in diameter and held between two flanges. The unknown pressure is applied to the underside of the diaphragm, and the resultant movement of the center of the diaphragm is transmitted through a linkage to drive the pointer, as in the Bourdon gauge. The upper flange is shaped to provide protection against the application of over-range pressures.

In the Schaffer gauge it is the elastic properties of the metallic diaphragm that govern the range and accuracy of the measurement. An aneroid barometer (Figure 14.13) also uses a corrugated diaphragm, but it is supplemented by a spring. The element consists of a flat, circular capsule with a corrugated lid and base and is evacuated before being sealed. It is prevented from collapse by a spring that is anchored to a bridge and attached to the top center of the capsule. Also attached at this point is a lever that acts through a bell crank-and-lever mechanism to rotate the pointer. When the

CHAPTER | 14 Measurement of Pressure

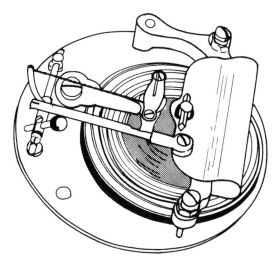

FIGURE 14.13 Aneroid barometer.

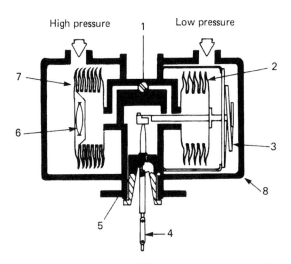

FIGURE 14.15 Components of a differential pressure transmitter. Courtesy of Invensys Inc.

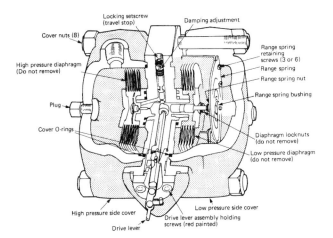

FIGURE 14.14 Diaphragm-type differential pressure transmitter. Courtesy of Invensys Inc.

atmospheric pressure increases, the capsule contracts, causing the pointer to rotate in one direction. Conversely, when the atmospheric pressure falls, the capsule expands and the pointer is driven in the opposite direction.

A further example of an instrument employing stiff diaphragms augmented by springs is shown in Figures 14.14 and 14.15. This instrument has largely superseded the bell-type mercury pressure manometer that was previously widely used for measuring differential pressures associated with orifice-plate flowmeters, partly because of the increased cost but more particularly because of the health hazards associated with the use of mercury.

The diaphragm elements (7) and (2) are made up from pairs of corrugated diaphragms with a spacing ring stitch-welded at the central hole. These assemblies are then stitch-welded at their circumference to form a stack. This configuration ensures that when excess pressure is applied to the stack, the individual corrugations nest together while the stack spacing rings come together to form a metal-to-metal stop.

The diaphragm stacks (7) and (2) are mounted on the central body, together with the range spring (3) and drive unit (4). Pressure-tight covers (8) form the high- and low-pressure chambers. The diaphragm stacks (2) and (7) are interconnected via the damping valve (1) and fitted internally with a liquid that remains fluid under normal ambient conditions.

An increase in pressure in the high-pressure chamber compresses the diaphragm stack (7), and in so doing displaces fluid via the damping valve (1) into the stack (2), causing it to expand until the force exerted by the range spring balances the initial change in pressure. The deflection of the range spring is transmitted to the inner end of the drive unit, which, being pivoted at a sealed flexure (5), transfers the motion to the outer end of the drive shaft (4), where it can be used to operate a pen arm.

A bimetallic temperature compensator (6) is mounted inside the stack (7) and adjusts the volume of that stack to compensate for the change in volume of the fill liquid resulting from a change of temperature. The instrument is suitable for operating at pressures up to 140 bar, and spans between 50 and 500 mbar can be provided by selecting suitable combinations of the range springs, which are fabricated from Ni-Span C to make them substantially insensitive to changes of temperature.

Bellows Elements

With the development of the hydraulic method for forming bellows, many of the pressure-sensing capsules previously fabricated from corrugated diaphragms have been replaced by bellows, which are available in a variety of materials. The spring rate or modulus of compression of a bellows varies directly as the modulus of elasticity of the material from which it is formed and proportionally to the third power of the wall thickness. It is also inversely proportional to the number of convolutions and to the square of the outside diameter of the bellows.

The combined effect of variations in the elastic properties of the materials of construction and manufacturing tolerance results in appreciable variations in the bellows spring rate, not only from one batch to another but also within a batch. For some applications this may not be particularly significant, but when it is, the effect can be reduced by incorporating a powerful spring into the assembly.

Figure 14.16 shows a pneumatic receiver, that is, a unit specifically designed for measurements in the range 20 to 100 kPa, which is one of the standard ranges for transmission in pneumatic systems.

Figure 14.17 shows a bellows assembly for the measurement of absolute pressure. It comprises two carefully matched stainless-steel bellows, one of which is evacuated to a pressure of less than 0.05 mmHg and sealed. The unknown pressure is applied to the other bellows. The two bellows are mounted within a frame and connected via a yoke, which transmits the bellows motion via a link to a pointer or the pen arm of a recorder.

14.2.3.4 Low Pressure Range Elements

The Model FCO52 Transmitter from Furness Controls, shown in Figure 14.18(a), is available with spans between 10 Pa and 20 kPa (1 mmH$_2$O and 2000 mmH$_2$O). The transducer assembly, shown in Figure 14.18(b), incorporates a metal diaphragm that is stretched and electron-beam-welded into a carrier. Capacitor electrodes mounted on the transducer shells are positioned on either side of the diaphragm

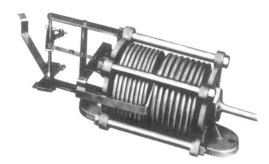

FIGURE 14.17 Bellows assembly for an absolute pressure gauge. Courtesy of Invensys Inc.

and adjusted so that the requisite change in capacitance is obtained for the specified pressure range. The electrodes are connected in a Wheatstone bridge network, which is operated at 1.7 MHz and is balanced when the input differential pressure is zero. The subsequent application of a differential pressure changes the capacitances of the electrodes, which unbalances the bridge network. The out-of-balance signal is amplified and converted into either a 0–10 Vdc or a 4–20 mA output signal.

The Ashcroft Model XLdp Low Pressure Transmitter, shown in Figure 14.19(a), uses Si-Glass technology to form a capacitive silicon sensor. The sensor is a thin micro-machined silicon diaphragm that is electrostatically bonded between two glass plates, as shown in Figure 14.19(b). The glass plates are each sputtered with aluminum, and with the silicon diaphragm positioned centrally, each forms a parallel plate capacitor. Application of a differential pressure causes the diaphragm to move closer to one electrode and further away from the other, thereby changing the respective capacitances. The movement, which is extremely small (only a few micrometers), is detected by electronic circuits that are built into an application-specific integrated circuit (ASIC), which generates an output signal in the range 4–20 mA, 1–5 Vdc or 1–6 Vdc. The measurement spans of these transmitters are from 25 Pa to 12.5 kPa (2.5 mmH$_2$O to 1250 mmH$_2$O).

In the rather unlikely event that a pneumatic signal proportional to the measured differential pressure is required, the 4–20 mA output signal from one of these transmitters would be applied to an electropneumatic converter of the type described in Chapter 42.

14.2.3.5 Capacitance Manometers

The application of electronic techniques to measure the deflection of a diaphragm and hence to infer pressure has resulted in major improvements in both sensitivity and resolution as well as providing means for compensating for nonlinear effects. One of the devices in which these techniques have been applied is the capacitance manometer shown diagrammatically in Figure 14.20.

For such a sensor it is important that the diaphragm and sensor body are capable of withstanding a wide range of

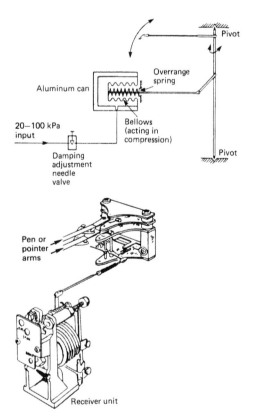

FIGURE 14.16 Pneumatic receiver using a bellows. Courtesy of Invensys Inc.

CHAPTER | 14 Measurement of Pressure

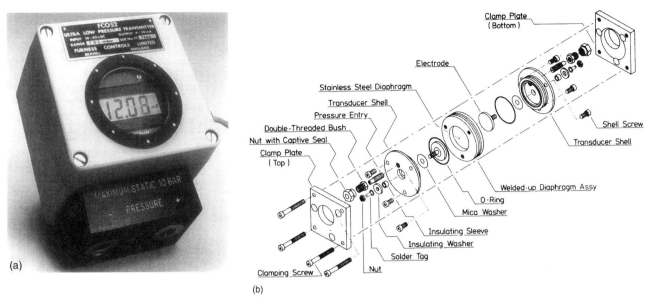

FIGURE 14.18 (a) FC052 ultra-low-pressure transmitter. (b) Exploded view of the sensor. Courtesy of Furness Controls Ltd.

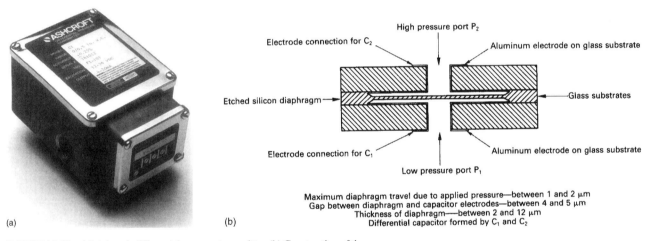

FIGURE 14.19 (a) Ashcroft differential pressure transmitter. (b) Construction of the sensor.

process fluids, including those that are highly corrosive. It is also important for them to have thermal coefficients that closely match those of the electrode assembly and screening material. Inconel is a suitable material for the body and diaphragm; Fosterite with either nickel or palladium is used for the electrode assembly. With these materials, pressures as low as 10^{-3} Pa can be measured reliably.

The tensioned metal diaphragm is welded into the sensor body, and the electrode assembly is located in the body at the correct position with respect to the diaphragm. If the sensor is to be used for absolute pressure measurements, the sensor-body assembly is completed by welding in place a cover that carries the two electrode connections and the getter assembly. If on the other hand the sensor is to be used for differential pressure measurements, provision is made for connecting the reference pressure.

The hysteresis error for such a sensor is normally less than 0.01 percent of the reading; for sensors having spans greater than 100 Pa, the error is almost immeasurable. The nonlinearity is the largest source of error in the system apart from temperature effects and is usually on the order of 0.05 percent of reading, minimized by selective adjustments in the associated electronic circuits.

Errors due to ambient temperature changes affect both the zero and span. Selection of the optimum materials of construction results in a zero error of approximately 0.02 percent of span per Kelvin and a span error of approximately 0.06 percent of span per Kelvin. The span error can be reduced to 0.0005 percent by including a temperature sensor in the body of the pressure sensor and developing a corresponding correction in the measuring circuits. The zero error can be reduced to 0.002 percent by including a nulling circuit.

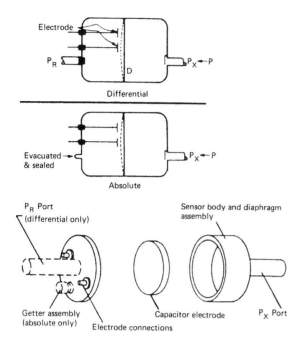

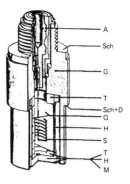

FIGURE 14.21 Pressure transducer using transverse piezoelectric effect of quartz. Courtesy of Kistler Instruments Ltd.

FIGURE 14.20 Capacitance manometer sensor. Courtesy of MKS Instruments Inc.

14.2.3.6 Quartz Electrostatic Pressure Sensors

Dynamic pressure measurements can be made with quartz sensors in which the applied force causes an electrostatic charge to be developed across the crystal, which is then measured by a charge amplifier and the resultant signal used to provide an indication of the applied force.

The Kistler type 601 and 707 series shown in Figure 14.21 are examples of quartz electrostatic sensors. The assemblies utilize the transverse piezoelectric effect illustrated in Figure 14.22. The application of a force F in the direction of one of the neutral axes Y sets up an electrostatic charge on the surfaces of the polar axis x at right angles to it. The magnitude of this charge depends on the dimensions of the quartz crystal, and by selecting a suitable shape it is possible to secure a high charge yield combined with good linearity and low temperature sensitivity. Similarly, the principle of the longitudinal piezoelectric effect is illustrated in Figure 14.23.

A typical transducer is assembled from three quartz stacks Q (see Figure 14.21) joined rigidly to the holder G by means of a preload sleeve H and temperature compensator T. The pressure to be measured acts on the diaphragm M, where it is converted into the force that is applied to the three quartz stacks. The contact faces of the quartz are coated with silver, and a central noble metal coil S conducts charge to the connector A.

The outer faces of the quartz are connected to the housing. With this configuration linearities of between 0.2 and 0.3 percent are achieved for spans up to 25 MPa, and the sensors have a uniform response up to about 30 kHz, with

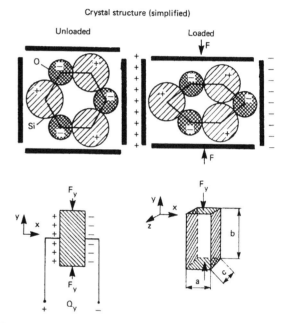

FIGURE 14.22 Principle of transverse piezoelectric effect. Courtesy of Kistler Instruments Ltd.

a peak of about 100 kHz. Because there must be a finite leakage resistance across the sensor, such devices cannot be used for static measurements. The low-frequency limit is on the order of 1 Hz, depending on the sensitivity. The type of charge amplifier associated with these sensors is shown in Figure 14.24. It comprises a high-gain operational amplifier with MOSFET input stage to ensure that the input impedance is very high as well as capacitor feedback to ensure that the charge generated on the quartz transducer is virtually completely compensated. It can be shown that the output voltage from the amplifier is $-Q/C_g$, where Q is the charge generated by the quartz sensor and C_g is the feedback capacitance. Thus the system is essentially insensitive to the influence of the input cable impedance.

Sensors such as these are characterized by their high stability, wide dynamic range, good temperature stability, good linearity, and low hysteresis. They are available in a very wide variety of configurations for dynamic pressure ranges from 200 kPa to 100 MPa.

CHAPTER | 14 Measurement of Pressure

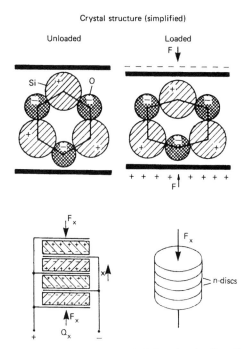

FIGURE 14.23 Principle of longitudinal piezoelectric effect. Courtesy of Kistler Instruments Ltd.

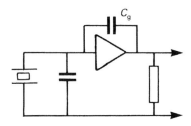

FIGURE 14.24 Charge amplifier associated with piezoelectric effect sensor.

FIGURE 14.25 Piezoresistive pressure transducer. Courtesy of Kistler Instruments Ltd.

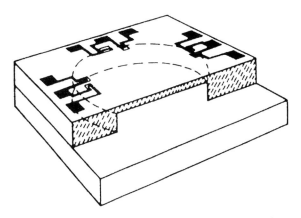

FIGURE 14.26 Schematic drawing of a pressure-sensing element. Courtesy of Kistler Instruments Ltd.

14.2.4 Pressure Measurement by Allowing the Unknown Pressure to Act on an Elastic Member and Measuring the Resultant Stress or Strain

14.2.4.1 Piezoresistive Pressure Sensors

For many metals and some other solid materials, the resistivity changes substantially when subjected to mechanical stress. Strain gauges, as described in Chapter 9, involve this phenomenon, but the particular characteristics of silicon allow construction of a thin diaphragm that can be deflected by an applied pressure and can have resistors diffused into it to provide a means for sensing the deflection. An example of this is the Kistler 4000 series, shown in Figure 14.25, for which the pressure-sensing element is shown in Figure 14.26.

Because the stress varies across the diaphragm, four pairs of resistors are diffused into a wafer of n-type silicon, each pair having one resistor with its principal component radial and one with its principal component circumferential. As described later, this provides means for compensating the temperature sensitivity of the silicon. Mechanically they form part of the diaphragm, but they are isolated electrically by the p–n junction so that they function as strain gauges. The diaphragm is formed by cutting a cylindrical recess on the rear surface of the wafer using initially ultrasonic or high-speed diamond machining and, finally, chemical etching. This unit is then bonded to a similar unprocessed chip so that a homogeneous element is produced. If it is desired to measure absolute pressures, the bonding is effected under a vacuum. Otherwise the cavity behind the diaphragm is connected via a hole in the base chip and a bonded tube to the atmospheric or reference pressure. The schematic arrangements of two such transducers are shown in Figure 14.27.

The mechanical strength of silicon depends largely on the state of the surface; in general this imposes an upper limit of about 100 MPa on the pressures that can be measured safely by the sensors. The lower limit is about 20 kPa and is determined by the minimum thickness to which the diaphragm can be reliably manufactured.

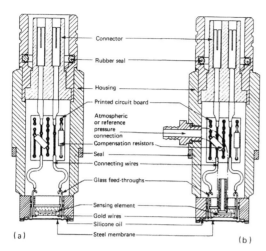

FIGURE 14.27 Cross-section of piezoresistive pressure transducer, (a) For absolute pressure. (b) For gauge pressure. Courtesy of Kistler Instruments Ltd.

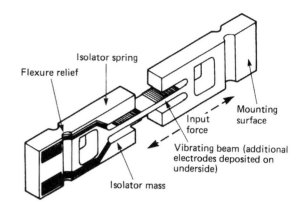

FIGURE 14.28 Pressure transducer utilizing strain-gauge sensor. Courtesy of TransInstruments.

Both the gauge factor G and resistance R of the diffused resistors are sensitive to changes of temperature, and the sensors need to be associated with some form of compensating circuits. In some instances this is provided by discrete resistors closely associated with the gauge itself. Others utilize hybrid circuits, part of which may be on the chip itself. Individual compensation is always required for the zero offset, the measurement span, and temperature stabilization of the zero. Further improvement in the performance can be achieved by compensating for the nonlinearity and the effect of temperature on the span.

14.2.4.2 Strain-Gauge Pressure Sensors

Another group of pressure sensors is based on strain-gauge technology (see Chapter 9), in which the resistance-type strain sensors are connected in a Wheatstone bridge network. To achieve the required long-term stability and freedom from hysteresis, the strain sensors must have a molecular bond to the deflecting member, which, in addition, must provide the necessary electrical isolation over the operating temperature range of the transducer.

This can be achieved by first sputtering the electrical isolation layer on the stainless-steel sensor beam or diaphragm and then sputtering the thin-film strain-gauge sensors on top of that. An example of this type of sensor is the TransInstruments 4000 series shown in Figure 14.28.

The pressure inlet adaptor is fabricated from precipitation-hardened stainless steel and has a deep recess between the mounting thread and diaphragm chamber to isolate the force-summing diaphragm from the mounting and other environmental stresses.

The transducer is modular in construction to allow the use of alternative diaphragm configurations and materials. For most applications the diaphragm is stainless steel and the thickness is selected according to the required measurement range. For some applications, enhanced corrosion resistance is required, in which case Inconel 625 or other similar alloys may be used as the diaphragm material, but to retain the same margin of safety a thicker member is usually required, which in turn reduces the sensitivity.

The sensor is a sputtered thin-film strain gauge in which the strain-gauge pattern is bonded into the structure of the sensor assembly on a molecular basis and the sensor assembly itself is welded into the remaining structure of the transducer. The stainless-steel header, which contains the electrical feed-through to the temperature-compensation compartment, is also welded into the structure of the transducer.

This welding, in conjunction with the ceramic firing technique used for the electrical feed-through connections, provides secondary containment security of 50 MPa for absolute gauges and those with a sealed reference chamber.

Sensors of this type are available with ranges from 0 to 100 kPa up to 0 to 60 MPa, with maximum nonlinearity and hysteresis of 0.25 to 0.15 percent, respectively, and a repeatability of 0.05 percent of span.

The maximum temperature effect is 0.15 percent of span per Kelvin.

14.2.4.3 High-Accuracy Digital Quartz Crystal Pressure Sensors

Performance and utilization factors may be used to differentiate between digital and analog instrumentation requirements. Performance considerations include resolution, accuracy, and susceptibility to environmental errors. Utilization factors include ease of measurement, signal transmission, equipment interfaces, and physical characteristics.

Some advantages of digital sensors relate to the precision with which measurements can be made in the time domain. Frequencies can be routinely generated and measured to a part in 10 billion, whereas analog voltages and resistances are commonly measured to a part per 1 million. Thus digital-type transducers have a huge inherent advantage in resolution and accuracy compared to analog sensors.

Frequency signals are less susceptible to interference, easier to transmit over long distances, and easily interfaced to counter-timers, telemetry, and digital computer systems.

The use of digital pressure transducers has grown dramatically with the trend toward digital data acquisition and control systems. Inherently digital sensors such as frequency output devices have been combined with microprocessor-based systems to provide unparalleled accuracy and performance, even under extreme environmental conditions. The design and performance requirements of these advanced transducers include:

- Digital-type output.
- Accuracy comparable to primary standards.
- Highly reliable and simple design.
- Insensitivity to environmental factors.
- Minimum size, weight, and power consumption.
- Ease and utility of readout.

Over the last three decades, Paroscientific Inc. has developed and produced digital quartz crystal pressure sensors featuring resolution better than one part per million and accuracy better than 0.01 percent. This remarkable performance is achieved through the use of a precision quartz crystal resonator for which the frequency of oscillation varies with pressure-induced stress. Quartz crystals were chosen for the sensing elements because of their remarkable repeatability, low hysteresis, and excellent stability. The resonant frequency outputs are maintained and detected with oscillator electronics similar to those used in precision clocks and counters.

Several single- or dual-beam load-sensitive resonators have been developed. The single-beam resonator is shown diagrammatically in Figure 14.29. It depends for its operation on a fixed beam oscillating in its first flexural mode with an integral isolation system that effectively decouples it from the structures to which it is attached. The entire sensor is fabricated from a single piece of quartz to minimize energy loss to the mounting surfaces. The beam is driven piezoelectrically to achieve and maintain beam oscillations. Figure 14.29 shows the placement of electrodes on the beam; Figure 14.30 illustrates the response to the imposed electric field from the oscillator electronics. The double-ended tuning fork (DETF) shown in Figure 14.31 consists of two identical beams driven piezoelectrically in 180° phase opposition such that very little energy is transmitted to the

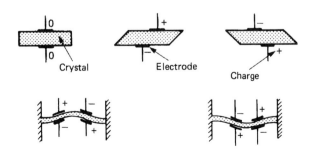

FIGURE 14.29 Resonant piezoelectric force sensor. Courtesy of Paro-scientific Inc.

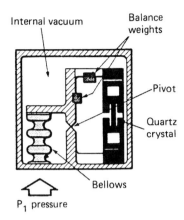

FIGURE 14.30 Oscillator mode for piezoelectric force sensor. Courtesy of Paroscientific Inc.

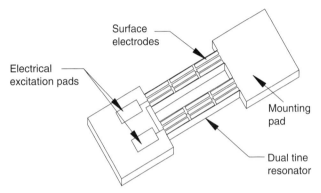

FIGURE 14.31 Double-ended tuning fork force sensor. Courtesy of Paroscientific Inc.

mounting pads. The high Q resonant frequency, like that of a violin string, is a function of the applied load that increases with tension and decreases with compressive forces.

Although the resonator will operate at normal ambient pressure, its performance improves significantly in a vacuum. This is because the elimination of air loading and damping effects permit operation at Q values of over 20,000. In addition, the vacuum improves stability by precluding the absorption or evaporation of molecules from the surface. In such small resonators, a layer of contaminant only one molecule deep has a discernible effect.

The crystallographic orientation of the quartz is chosen to minimize temperature-induced errors. To achieve maximum performance over a wide range of environmental conditions, the small residual thermal effects of the quartz load-sensitive resonator may be compensated for using a torsional-mode quartz resonator that provides a frequency output related only to temperature. Thus the two frequency outputs from the transducer represent applied load (with some temperature effect) and temperature (with no load effect). The two signals contain all the information necessary to eliminate temperature errors.

As shown in Figure 14.32, pressure transducer mechanisms employ bellows or Bourdon tubes as the pressure-to-load generators. Pressure acts on the effective area of the

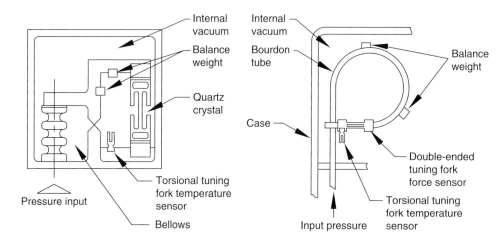

FIGURE 14.32 High-accuracy quartz crystal pressure sensor mechanisms. Courtesy of Paroscientific Inc.

bellows to generate a force and torque about the pivot and compressively stress the resonator. The change in frequency of the quartz crystal oscillator is a measure of the applied pressure. Similarly, pressure applied to the Bourdon tube generates an uncoiling force that applies tension to the quartz crystal to increase its resonant frequency. Temperature-sensitive crystals are used for thermal compensation. The mechanisms are acceleration compensated with balance weights to reduce the effects of shock and vibration. The transducers are hermetically sealed and evacuated to eliminate air damping and maximize the Q of the resonators. The internal vacuum also serves as an excellent reference for the absolute pressure transducer configurations.

Gauge pressure transducers can be made by placing an opposing bellows on the other side of the lever arm that is exposed to ambient atmospheric pressure. With one bellows counteracting the other, only the difference in the two bellows' pressures is transmitted to the crystal resonator. Differential sensors operating at high line pressure employ isolating mechanisms to allow the quartz crystal to measure the differential force while operating in an evacuated enclosure.

Each transducer produces two frequency outputs: one for pressure and one for temperature. The pressure resonator has a nominal frequency of 38 kilohertz, which changes by about 10 percent when full-scale pressure is applied. The temperature signal has a nominal frequency of 172 kilohertz and changes by about 50 ppm per degree Centigrade. As described in Section 14.3.4, microprocessor-based intelligent electronics are available with countertimer circuitry to measure transducer frequency or period outputs, storage of the conformance and thermal compensation algorithm, calibration coefficients, and command and control software to process the outputs in a variety of digital formats.

Overall performance is comparable to the primary standards. Because the quartz resonator in the pressure transducer provides high constraint to the pressure-generating element (bellows or Bourdon tube), there is very little mechanical motion under load (several microns). This increases repeatability and reduces hysteresis. The crystals themselves produce the kind of stability characteristic of any high-quality quartz resonator. In addition, the high Q values eliminate extraneous noise on the output signals, resulting in high resolution. The use of a frequency output quartz temperature sensor for temperature compensation allows accuracies of 0.01 percent of full scale to be achieved over the entire operating range.

Absolute, gauge, and differential transducers are available with full-scale pressure ranges from a fraction of an atmosphere (15 kPa) to thousands of atmospheres (276 MPa).

14.3 PRESSURE TRANSMITTERS

In the process industries, it is often necessary to transmit the measurement signal from a sensor over a substantial distance so that it can be used to implement a control function or can be combined with other measurement signals in a more complex scheme.

The initial development of such transmission systems was required for the petroleum and petrochemical industries, where pneumatic control schemes were used most widely because they could be installed in plants where explosive or hazardous conditions could arise and the diaphragm actuator provided a powerful and fast-acting device for driving the final operator. It followed that the first transmission systems to be evolved were pneumatic and were based on the standardized signal range 20 to 100 kPa (3 to 15 psig).

Early transmitters utilized a motion-balance system; that is, one in which the primary element produces a movement proportional to the measured quantity, such as a Bourdon tube, in which movement of the free end is proportional to the applied pressure. However, these transmitters were rather sensitive to vibration and have, in general, been superseded by force-balance systems. But because of the time delay and response lag that occur, pneumatic transmission itself is unsuitable when the distance involved exceeds a few hundred meters.

Consequently, an equivalent electronic system has been evolved. In this, a current in the range of 4 to 20 mAdc and proportional to the span of the measured quantity is generated by the sensor and transmitted over a two-wire system. The advantage of this system is that there is virtually no delay or response lag and the transmitted signal is not affected by changes in the characteristic of the transmission line. In addition, there is sufficient power below the live zero (i.e., 4 mA) to operate the sensing device. Such systems have the additional advantage that in complex control schemes, they are more easily configured than the corresponding pneumatic transmitters.

The growth in digital computers and control systems has generated a need for intelligent, digital output pressure transmitters. Since 1994, many pressure transmitters have been installed that use for their primary means of communication some form of digital fieldbus such as the Profibus or Foundation Fieldbus. It is expected that these intelligent transmitters will eventually supersede the 4–20 mAdc standard (ISA S50) and the remaining pneumatic transmitters in use. Telemetry and pneumatic systems are discussed further in Chapters 40 and 42.

14.3.1 Pneumatic Motion-Balance Pressure Transmitters

Figure 14.33 shows the arrangement of a typical pneumatic motion-balance transmitter in which the sensor is a spiral Bourdon tube. Changes in the measured variable, which could be pressure or, in the case of a filled thermal system, temperature, cause the free end of the Bourdon tube to move. This movement is transmitted via a linkage to the lever that pivots about the axis A. The free end of this lever bears on a second lever that is pivoted at its center so that the movement is thus transmitted to a third lever that is free to pivot about the axis C. The initial movement is transferred to the flapper of the flapper/nozzle system. If, as a result, the gap between the flapper and nozzle is increased, the nozzle back-pressure falls, which in turn causes the output pressure from the control relay to fall. As this pressure is applied to the bellows, the change causes the lever pivoted about axis B to retract so that the lever pivoted around axis C moves the flapper toward the nozzle. This causes the nozzle back-pressure to rise until equilibrium is established. For each value of the measurement there is a definite flapper/nozzle relationship and therefore a definite output signal.

14.3.2 Pneumatic Force-Balance Pressure Transmitters

There are many designs of pneumatic force-balance transmitters, but in the Invensys Inc. design the same force-balance mechanism is used in all the pressure and differential pressure transmitters. This design is shown in Figure 14.34. Its basic function is to convert a force applied to its input point into a proportional pneumatic signal for transmission, such as 20 to 100 kPa (3 to 15 psig).

The force to be measured may be generated by a Bourdon tube, a bellows, or a diaphragm assembly and applied to the free end of the force bar. This is pivoted at the diaphragm seal, which in some instruments also provides the interface between process fluid and the force-balance mechanism so that an initial displacement arising from the applied force appears amplified at the top of the force bar, where it is transmitted via the flexure connector to the top of the range rod. If the applied force causes movement to the right, the flapper uncovers the nozzle, with the result that the nozzle

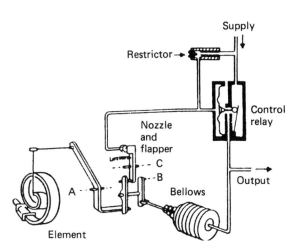

FIGURE 14.33 Arrangement of pneumatic motion-balance transmitter. Courtesy of Invensys Inc.

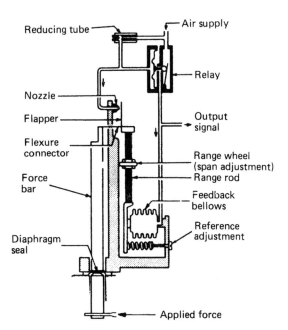

FIGURE 14.34 Arrangement of pneumatic force-balance transmitter. Courtesy of Invensys Inc.

back-pressure falls. This change is magnified by the "relay," the output of which is applied to the feedback bellows, thereby producing a force that balances the force applied initially. The output signal is taken from the "relay," and by varying the setting of the range wheel, the sensitivity or span can be adjusted through a range of about 10 to 1. By varying the primary element pressures from about 1.3 kPa to 85 Mpa, differential pressures from 1 kPa to 14 MPa may be measured.

Figures 14.35–14.38 show some of the alternative primary elements that can be used in conjunction with this force-balance mechanism to measure gauge, differential, and absolute (high and low) pressures.

14.3.3 Force-Measuring Pressure Transmitters

In addition to the previously described force-balance pressure transmitters, there are now transmitters that measure pressure by measuring the deflection of an elastic member resulting from the applied pressure. One of these is the Invensys Foxboro 820 series of transmitters, in which the force is applied to a pre-stressed wire located in the field of a permanent magnet. The wire is an integral part of an oscillator circuit that causes the wire to oscillate at its resonant (or natural) frequency. For

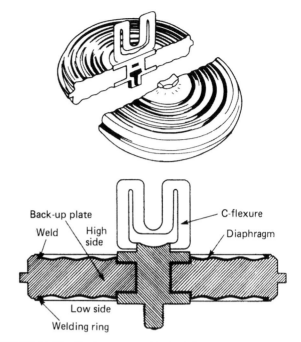

FIGURE 14.37 Diaphragm assembly for differential pressure measurements. Courtesy of Invensys Inc.

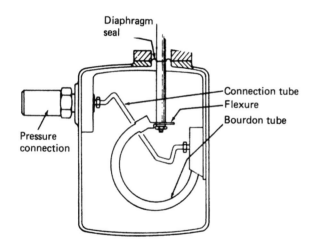

FIGURE 14.35 Bourdon tube primary element arranged for operation in conjunction with a force-balance mechanism. Courtesy of Invensys Inc.

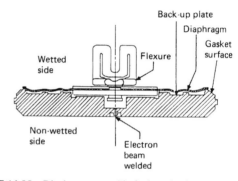

FIGURE 14.38 Diaphragm assembly for low absolute pressure measurements. Courtesy of Invensys Inc.

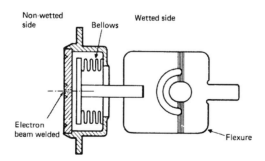

FIGURE 14.36 Bellows type primary element for absolute pressure measurements. Courtesy of Invensys Inc.

an ideal system, the resonant frequency is a function of the length, the square root of tension, and the mass of the wire.

The associated electronic circuits include the oscillator as well as the components to convert oscillator frequency into a standard transmission signal such as 4 to 20 mAdc. As shown in Figure 14.39, the oscillator signal passes via a pulse shaper to two frequency converters arranged in cascade, each of which produces an output proportional to the product of the applied frequency and its input voltage so that the output of the second converter is proportional to the square of the frequency and therefore to the tension in the wire. The voltage is therefore directly proportional to the force produced by the primary element, which in turn is proportional to the measured pressure.

The configurations of the resonant-wire system for primary elements such as a helical Bourdon tube to measure gauge pressure, differential pressure, and absolute pressure are shown in Figures 14.40–14.43. Vibrating wires are also used as strain gauges, as discussed in Chapter 9.

A second category of pressure transmitters involves the measurement of the deflection of a sensing diaphragm that is arranged as the movable electrode between the fixed plates of a differential capacitor. An example of this setup is the Siemens Teleperm K Transmitter. The arrangement of the measuring cell for differential pressures and absolute pressures is shown in Figure 14.44.

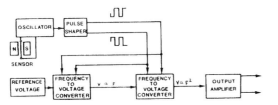

FIGURE 14.39 Functional diagram of electronic circuit for resonant wire pressure transmitter. Courtesy of Invensys Inc.

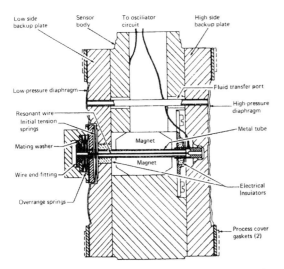

FIGURE 14.40 Arrangement of diaphragm assembly for differential pressure measurements. Courtesy of Invensys Inc.

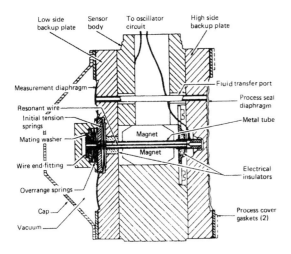

FIGURE 14.41 Arrangement of diaphragm assembly for absolute pressure measurements. Courtesy of Invensys Inc.

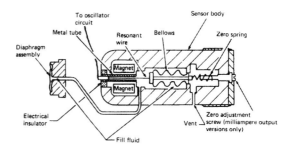

FIGURE 14.42 Arrangement of gauge pressure element for resonant wire sensor. Courtesy of Invensys Inc.

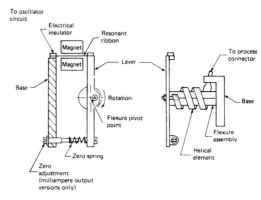

FIGURE 14.43 Arrangement of helical Bourdon tube for resonant wire sensor. Courtesy of Invensys Inc.

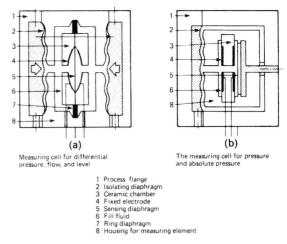

FIGURE 14.44 Arrangement of capacitance type differential and absolute pressure sensors. Courtesy of Siemens Ltd.

It is a flat cylindrical unit sealed at both ends by flexible corrugated diaphragms that provide the interface between the process fluid and the sensor. Under overload conditions, these seat on matching corrugations machined in the housing. The sensor comprises a hollow ceramic chamber divided into two by the sensing diaphragm. The interiors of the chambers are filled with liquid and sealed by a ring diaphragm. The interior walls of both chambers are metalized and form the fixed plates of the differential capacitor;

the sensing diaphragm forms the movable plates of the capacitor.

When the measuring cell is subjected to a differential pressure, the sensing diaphragm is displaced slightly, causing a change in capacitance that is converted by the associated measuring circuit into a standard transmission signal such as 4 to 20 mAdc.

For measuring absolute pressures, one side of the measuring cell is evacuated to provide the reference pressure; for measuring gauge pressures, one side is vented to atmosphere.

Under these conditions the stiffness of the isolating diaphragm determines the range. For high pressures the diaphragm is augmented by a spring.

Figure 14.45 shows the basic circuit for absolute pressure and gauge pressure transmitters. The sensing diaphragm acts as the moving electrode in the capacitor detector. The effective values of the capacitors are as follows:

$$C_1 = \frac{A\varepsilon}{d_0} - C_s$$

$$C_2 = \frac{A\varepsilon}{d_0 + \Delta d} + C_s$$

where A is effective electrode area, ε is permittivity of the dielectric fluid, d_0 is effective distance between fixed electrodes and sensing electrode, Δd is displacement of sensing electrode, and C_s is stray capacitance. From this it follows that

$$\frac{C_1 - C_2}{C_2 - C_s} = \frac{\Delta d}{d_0}$$

which is the same as the deflection constant for the sensing diaphragm so that Δd is proportional to the differential pressure, which can therefore be measured by an all-bridge network. To compensate for the effect of stray capacitances, a capacitor C is included in the bridge circuit but supplied with a voltage in antiphase to that applied to C_1 and C_2.

If the impedances of the capacitors C_1, C_2, and C_3 are high compared with their associated resistors, the currents flowing through them are proportional to the capacitances,

so the output from amplifier U_1 is proportional to $(i_1 - i_2)$ and from amplifier $U_2 (i_2 - i_c)$. When these two signals are applied to a dividing stage, the resultant signal is proportional to the displacement of the sensing electrode and hence to the applied pressure. For the differential pressure transmitter, both C_1 and C_2 are variable, but it can be shown that

$$\frac{\Delta d}{d_0} = \frac{C_1 - C_2}{C_1 + C_2 - 2C_s}$$

Applying the same conditions as before, this leads to

$$\frac{\Delta d}{d_0} = \frac{i_1 - i_2}{i_1 + i_2 - iC}$$

Because $(i_1 - i_2)$ and $(i_1 + i_2 - i_c)$ are proportional to the input signals of amplifier U_1 and U_2, it follows that the output from the dividing stage is proportional to the applied differential pressure.

Most differential pressure transmitters are used in conjunction with orifice plates to measure flow. If, therefore, the output from the dividing stage is applied to a square root extracting circuit, its output becomes proportional to flow rate.

14.3.4 Digital Pressure Transducers

Until recently, digital data acquisition and control systems were designed with their measurement and control intelligence localized in a central computer. Transducers were required to do little more than sense and report to the computer, their analog outputs converted to digital format for interface compatibility. Today's "smart" transmitters offer users a new order of power and flexibility. Placing the intelligence within the transmitter allows the user to easily configure a distributed measurement and control system from the host computer via a two-way digital bus.

Inherently digital pressure transducers, such as the quartz crystal frequency-output sensors described in Section 14.2.4.3, have been developed that offer significant benefits over the older-style analog devices. These high-accuracy digital pressure transducers have been combined with a built-in microprocessor capability to yield intelligent pressure transmitters. Operating under field conditions, these transmitters can offer performance comparable to the primary standards.

The electronics architecture of an intelligent digital pressure transmitter is shown in Figure 14.46. The digital interface board contains a precision clock, counter, microprocessor, RS-232 serial port, RS-485 serial port, and EPROM and EEPROM memory for storing the operating program and calibration coefficients.

The digital interface board uses the two continuous frequency output signals provided by the pressure transducer (corresponding to pressure and the sensor's internal temperature) to calculate fully corrected pressure. The microprocessor

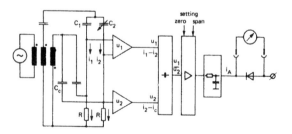

FIGURE 14.45 Functional diagram of electronic circuit for gauge and absolute pressure capacitance sensor. Courtesy of Siemens Ltd.

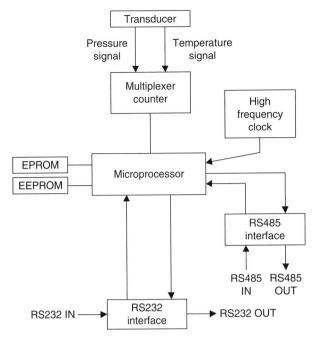

FIGURE 14.46 Digital interface board. Courtesy of Paroscientific Inc.

monitors incoming commands from the computer. When a sampling command is received, the microprocessor selects the appropriate frequency signal source and makes a period measurement using a high-frequency time-base counter and a user-specified integration time. When the period measurements are completed, the microprocessor makes the appropriate calculations and transmits the data on the RS-232/485 bus. The RS-232/RS485 interfaces allow complete remote configuration and control of all transmitter operations, including resolution, sample rate, integration time, baud rate, pressure adders, and pressure multipliers. Resolution is programmable from 0.05 to 100 parts per million, depending on system requirements. Baud rates up to 115.2K baud can be selected. Pressure data are available in eight different or user-defined engineering units. The command set includes single sample requests for pressure and temperature, continuous samples, and sample and hold. Up to 100 samples per second can be obtained with special burst sampling commands. The internal serial bus communicates with the outside world by means of a digital fieldbus connection, such as Profibus or Foundation Fieldbus.

REFERENCES

An Introduction to Process Control, Pub. 105B, The Foxboro Company (1986).
Busse, D. W., "Quartz transducers for precision under pressure," *Mechanical Engineering*, **109**, No. 5 (May 1987).
Gillum, D. R., *Industrial Pressure Measurement*, Instrument Society of America (1982).
Hewson, J. E., *Process Instrumentation Manifolds: their Selection and Use*, Instrument Society of America (1981).
Lyons, J. L., *The Designer's Handbook of Pressure-Sensing Devices*, Van Nostrand Reinhold, New York (1980).
Neubert, H. K. P., *Instrument Transducers, an Introduction to their Performance and Design*, Clarendon Press, Oxford (1975).

FURTHER READING

British Geotechnical Society, *Pressuremeters*, American Society of Civil Engineers, New York (1990).

Chapter 15

Measurement of Vacuum

D. J. Pacey

15.1 INTRODUCTION

15.1.1 Systems of Measurement

The measurement of vacuum is the measurement of the range of pressures below atmospheric. *Pressure* is defined as force divided by area. The American National Standard unit is the pound/inch^2, or psi; the SI unit is the Newton/meter^2 (Nm^{-2}) or Pascal (Pa). Pressure may also be stated in terms of the height of a column of a suitable liquid, such as mercury or water, that the pressure will support. The relation between pressure units currently in use is shown in Table 15.1.

In engineering, it has long been customary to take atmospheric pressure as the reference and to express pressures below this as "pounds per square inch of vacuum," or "inches of vacuum" when using a specified liquid. The continual changes in atmospheric pressure, however, will lead to inaccuracy unless they are allowed for. It is preferable to use zero pressure as the reference and to measure pressure above this level. Pressures expressed in this way are called *absolute pressures*.

15.1.2 Methods of Measurement

Since pressure is defined to be force/area, its measurement directly or indirectly involves the measurement of the force exerted on a known area. A gauge that does this is called an *absolute gauge* and allows the pressure to be obtained from a *reading* and known physical quantities associated with the gauge, such as areas, lengths, sometimes temperatures, elastic constants, and so on. The pressure, when obtained, is independent of the composition of the gas or vapor that is present.

Many technological applications of vacuums use the long free paths or low molecular incidence rates that vacuums make available. These require pressures that are only a very small fraction of atmospheric, where the force exerted by the gas is too small to be measured, making absolute gauges unusable. In such cases nonabsolute gauges are used that measure pressure indirectly by measuring a pressure-dependent physical property of the gas, such as thermal conductivity, ionizability, or viscosity. These gauges always require calibration against an absolute gauge for each gas that is to be measured. Commercial gauges are usually calibrated by the manufacturer using dry air and will give true readings only when dry air is present. In practice it is difficult to be certain of the composition of the gases in vacuum apparatus, thereby causing errors. This problem is overcome in the following way: When a gauge using variation of thermal conductivity indicates a pressure of 10^{-1} Pa, this is recorded as an *equivalent dry air pressure* of 10^{-1} Pa. This means that the thermal conductivity of the unknown gases present in the vacuum apparatus has the same value as that of air at 10^{-1} Pa, not that the pressure is 10^{-1} Pa.

TABLE 15.1 Relations between pressure units

	N/m^2(Pa)	torr	mb	atm
N/m^2 (Pa)	1	7.50×10^{-3}	10^{-2}	9.87×10^{-6}
torr	133.3	1	1.333	1.316×10^{-3}
mb	100	0.750	1	9.87×10^{-4}
atm	1.013×10^5	760	1.013×10^3	1

15.1.3 Choice of Nonabsolute Gauges

Since the gauge just referred to measures thermal conductivity, it is particularly useful for use on vacuum apparatus used for making vacuum flasks or in which low-temperature experiments are carried out and in which thermal conductivity plays an important part. Similarly, an ionization gauge would be suitable in the case of apparatus used for making radio valves and cathode ray tubes in which the ionizability of the gases is important. In general, it is desirable to match, as far as possible, the physical processes in the gauge with those in the vacuum apparatus.

15.1.4 Accuracy of Measurement

Having chosen a suitable gauge, it is necessary to ensure that the pressure in the gauge head is the same as that in the vacuum apparatus. First, the gauge head is connected at a point as close as possible to the point at which the pressure is to be measured and by the shortest and widest tube available. Second, sufficient time must be allowed for pressure equilibrium to be obtained. This is particularly important when the pressure is below 10^{-1} Pa and when ionization gauges, which interact strongly with the vacuum apparatus, are used. Times of several minutes are often required. When nonabsolute gauges are used, even under ideal conditions, the accuracy is rarely better than ±20 percent, and in a carelessly operated ionization gauge they're worse than ±50 percent. Representative values for the midrange accuracy of various gauges are given in Table 15.2, along with other useful information.

15.2 ABSOLUTE GAUGES

15.2.1 Mechanical Gauges

These gauges measure the pressure of gases and vapors by using the mechanical deformation of tubes or diaphragms

TABLE 15.2 Properties of gauges

Gauge	Pressure range (Pa)	Accuracy ±%	Cost*	Principal advantages	Principal limitations
Bourdon tube	10^5–10^2	10	A	Simple; robust.	Poor accuracy below 100 Pa.
Quartz spiral	10^5–10	10	B	Reads differential pressures.	Rather fragile.
Diaphragm	10^5–10	5	B	Good general-purpose gauge.	Zero setting varies.
Spinning rotor	10–10^4	<5	F	Sensitive and accurate.	Long response time.
Liquid manometers	10^5–10^2	5–10	A	Simple; direct reading.	Vapor may contaminate vacuum.
McLeod	10^5–5×10^{-4}	5–10	C	Wide pressure range; used for calibration.	Intermittent; measures gas pressures only.
Thermocouple	10^3–10^{-1}	20	B	Simple; robust; inexpensive.	Response not instantaneous. Reading depends on gas species.
Pirani	10^3–10^{-2}	10	C	Robust.	Zero variation due to filament contamination. Reading depends on gas species.
Thermistor	10^3–10^{-2}	10	C	Fast response; low current consumption.	Reading depends on gas species.
Discharge tube	10^3–10^1	20	B	Very simple; robust.	Limited pressure range. Reading depends on gas species.
Penning	1–10^{-5}	10–20	C	Sensitive; simple; robust.	Large pumping effect; hysteresis. Reading depends on gas species.
Hot-cathode ion	10^2–10^{-6}	10–30	E	Sensitive linear scale; instantaneous response.	Filament easily damaged; needs skillful use; reading depends on gas species.
Bayard–Albert	1–10^{-8}	10–30	D	As above, with better lowpressure performance.	As above.
Capacitance nanometers	10^5–10^{-3}	<5	E	Sensitive and accurate.	Reading depends on gas species.

*Scale of Costs (£): A 0, 50; B 50, 200; C 200, 400; D 400, 600; E 1,000, 4,000; F 7,000, 10,000.

when exposed to a pressure difference. If one side of the sensitive element is exposed to a good vacuum, the gauge is absolute.

15.2.1.1 The Bourdon Tube Gauge

A conventional gauge of the Bourdon tube type, if carefully made, can be used to measure pressure down to 100 Pa. Its construction is described in Chapter 14 on pressure measurement.

15.2.1.2 The Quartz Spiral Gauge

This gauge measures differential pressures over a range of 100 Pa from any chosen reference pressure. It is suitable for use with corrosive gases or vapors.

Construction The sensitive element is a helix of 0.5 mm diameter quartz tubing, usually 20 mm in diameter and 30 mm long, to which the vacuum is applied internally. The helix coils and uncoils in response to pressure changes, the motion being measured by observing the movement of a light spot reflected from a small mirror attached to its lower end. The whole assembly is mounted in a clear glass or quartz enclosure, which can be brought to any desired reference pressure. If this is zero, the gauge indicates absolute pressure.

15.2.1.3 Diaphragm Gauge

This gauge measures pressures of gases and vapors down to 10 Pa. Its construction is described in Chapter 14.

15.2.1.4 Capacitance Manometers

Capacitance manometers are among the most accurate and sensitive gauges available for the measurement of medium vacuums (see Chapter 14 for a description of their operation). Gauge heads are available with various full-scale ranges and usually operate over three to four decades. Heads with differing accuracies, reflected in the price, are also often available.

15.2.2 Liquid Manometers

Liquid manometer gauges measure the pressure of gases and vapors from atmospheric to about 1 Pa by balancing the force exerted by the gas or vapor against the weight of a column of liquid, usually mercury, water, or oil. These devices provide the simplest possible means of pressure measurement.

Construction The construction of various forms of liquid manometer is described in Chapter 14.

Operation For measuring relative pressures, the open manometer shown in Figure 15.1(a) may be used. In this case the difference h in levels may be taken to express the *vacuum* directly in inches of water, or by use of the formula $p = h\rho g$, where ρ is the density of the liquid and g is the acceleration due to gravity, the vacuum may be expressed in SI units. The measurement of absolute pressures requires a vacuum reference that may be obtained in several ways, as shown in Figures 15.1(b)–(d). A barometer tube is used in diagram (b), immersed in the same liquid pool, in this case usually mercury, as the manometer tube. A more compact form is provided by the closed manometer, shown in (c). This again uses mercury, and the space in the closed limb is evacuated. A useful version of the closed manometer that can be used with oil is shown in (d), where the tap may be opened when the apparatus is at zero pressure and closed immediately before taking measurements. When oil is used in vacuum measurement, difficulty will be experienced with the liberation of dissolved gases, which must be removed slowly by gradual reduction of the pressure.

15.2.3 The McLeod Gauge (1878)

This gauge measures the pressure of gases only, from 5×10^{-4} Pa to atmospheric, by measuring the force exerted by a sample of the gas of known volume after a known degree of compression.

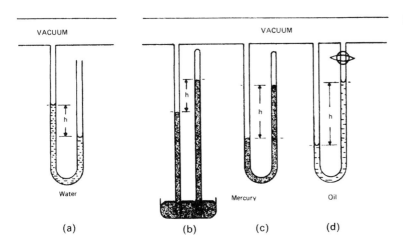

FIGURE 15.1 Liquid manometers.

Construction Construction is of glass, as shown in Figure 15.2, and uses mercury as the manometric liquid. The measuring capillary E and the reference capillary F are cut from the same length of tube, to equalize capillary effects. A trap R refrigerated with liquid nitrogen or solid carbon dioxide excludes vapors and prevents the escape of mercury vapor from the gauge. The tap T maintains the vacuum in the gauge when it is transferred to another system.

Operation The mercury normally stands at A, allowing the bulb and measuring capillary to attain the pressure p in the vacuum apparatus; at pressures below 10^{-2} Pa several minutes are required for this process. To take a reading, raise the mercury by slowly admitting air to the mercury reservoir M. When the mercury passes B, a sample of gas of volume V is isolated and the pressure indicated will be that in the gauge at this instant. The mercury in the reference capillary is then brought to O, the level of the top of the measuring capillary, and the length h of the enclosed gas column of area a is measured. The mercury is then returned to A by reducing the pressure in the reservoir M, thus preparing the gauge for a further measurement.

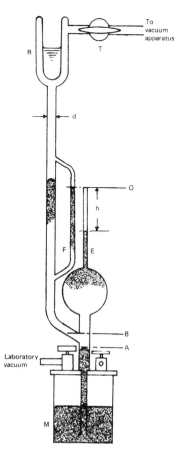

FIGURE 15.2 The McLeod gauge.

15.2.3.1 Calculation of the Pressure

Applying Boyle's law to the isolated sample of gas, we have

original pressure × original volume
= final pressure × final volume

or

$$pV = (h\rho g + p)ah$$

where ρ is the density of mercury and g is the acceleration due to gravity.

Thus

$$p = \frac{ah^2}{V - ah}\rho g$$

When measuring low pressures, the final volume ah is very much less than the original volume V.

Hence,

$$P = \frac{a}{V} \cdot \rho g \cdot h^2$$

showing that $p \propto h^2$ and giving a square-law scale.

The value of a is found by weighing a pellet of mercury of known length inserted in the measuring capillary, and V is determined by weighing the quantity of distilled water that fills the bulb and measuring capillary from the level A to its closed end. Both of these measurements are carried out by the manufacturer, who then calculates values of h corresponding to a series of known pressures that can then be read directly on the scale.

15.2.3.2 The Ishii Effect

The use of a refrigerated trap with this gauge, which is necessary in almost every instance, leads to a serious underestimation of the pressure, which is greater for gases with large molecules. First noted by Gaede in 1915 and thoroughly investigated by Ishii and Nakayama in 1962, the effect arises from the movement of mercury vapor from the pool at A, which passes up the connecting tube and then condenses in the refrigerated trap. Gas molecules encountered by the mercury vapor stream are carried along with it and are removed from the bulb, producing a lowering of the pressure there. The effect is greater for large molecules, since they provide large targets for the mercury vapor stream. The error may be reduced by reducing the mercury vapor flow, by cooling the mercury pool at A artificially, or by reducing the diameter d of the connecting tube. In the latter case, the response time of the gauge to pressure changes will be lengthened.

Approximate errors for a gauge in which $d = 1$ cm and the mercury temperature is 300°K are -4 percent for helium, -25 percent for nitrogen, and -40 percent for xenon.

However, the McLeod gauge, although popular in the past, is less used nowadays due to the difficulty of its operation, its delicate structure and size, and its hazardous nature (it contains mercury).

15.3 NONABSOLUTE GAUGES

15.3.1 Thermal Conductivity Gauges

These gauges measure the pressure of gases and vapors from 1,000 Pa to 10^{-1} Pa by using the changes in *thermal conductivity* that take place over this range. Separate calibration against an absolute gauge is required for each gas. Since the sensitive element used is an electrically heated wire, these gauges are known as *hotwire gauges*.

15.3.1.1 Thermocouple Gauge (Voege, 1906)

Construction An electrically heated wire operating at a temperature of about 320°K is mounted inside a glass or metal envelope connected to the vacuum apparatus. A thermocouple attached to the center of the wire enables its variation in temperature due to pressure changes to be observed. For simplicity the construction shown in Figure 15.3 may be used. Four parallel lead-through wires pass through one end of the envelope, and two noble metal thermocouple wires are fixed across diagonal pairs. The wires are welded at their intersection so that one dissimilar pair forms the hotwire and the other the thermocouple.

Operation A stabilized electrical supply S is connected to the hotwire and adjusted to bring it to the required temperature. The resistance R connected in series with the millivoltmeter M is used to set the zero of the instrument. A rise in the pressure increases the heat loss from the wire, causing a fall of temperature. This results in a reduction of thermocouple output registered by M, which is scaled to read the pressure of dry air.

15.3.1.2 The Pirani Gauge (Pirani, 1906)

Construction An electrically heated platinum or tungsten wire, operating at a temperature of 320°K, is mounted along the axis of a glass or metal tube, which is connected to the vacuum apparatus. Changes in the pressure cause temperature changes in the wire, which are followed by using the corresponding changes in its electrical resistance. The wire temperature may also be affected by variations of room temperature; these are compensated by use of an identical dummy gauge head sealed off at a low pressure. The gauge and dummy heads are shown in Figures 15.4 and 15.5. For reasons of economy, the dummy head is often replaced by a bobbin of wire that has the same resistance and temperature coefficient of resistance, mounted close to the gauge head.

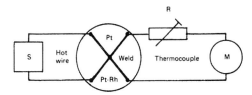

FIGURE 15.3 The Thermocouple gauge.

Operation The gauge and dummy heads form adjacent arms of a Wheatstone bridge circuit, as shown in Figure 15.6. This arrangement also compensates for the effects of temperature changes on the resistance of the leads connecting the gauge head to the control unit, thereby allowing remote indications of pressure. Resistances Y and Z form the other arms of the bridge, Z being variable for use in zero setting. Power is obtained from a stabilized supply S, and the meter M measures the out-of-balance current in the bridge.

FIGURE 15.4 Pirani gauge head.

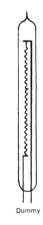

FIGURE 15.5 Pirani dummy head.

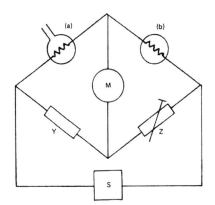

FIGURE 15.6 Pirani gauge circuit.

The wire is brought to its operating temperature, and the bridge is balanced, giving zero current through M when the pressure is low. A rise in pressure causes an increase of heat loss and a fall in temperature of the wire. If the power input is kept constant, the wire temperature falls, causing a fall of resistance that produces a current through M, which is calibrated to read pressure of dry air. Alternatively, the fall in wire temperature may be opposed by increasing the input voltage so that the wire temperature remains constant. The input voltage then depends on the pressure, and the meter measuring the voltage can be scaled to read pressure. The constant balance of the bridge is maintained by a simple electronic circuit. This arrangement is effective in extending the high-pressure sensitivity of the gauge to 10^4 Pa or higher, since the wire temperature is maintained at this end of the pressure range.

15.3.1.3 The Thermistor Gauge

This gauge closely resembles the Pirani gauge in its construction except that a small bead of semiconducting material takes the place of the metal wire as the sensitive element. The thermistor bead is made of a mixture of metallic oxides, about 0.2 mm in diameter. It is mounted on two platinum wires, 0.02 mm in diameter, so that a current may be passed through it. Since the semiconductor has a much greater temperature coefficient of resistance than a metal, a greater sensitivity is obtained. Furthermore, on account of its small size, it requires less power, allowing the use of batteries, and it responds more rapidly to sudden changes of pressure.

15.3.2 Ionization Gauges

These gauges measure the pressure of gases and vapors over the range 10^3 Pa to 10^{-8} Pa by using the current carried by ions formed in the gas by the impact of electrons. In cold-cathode gauges the electrons are released from the cathode by the impact of ions; in hot-cathode gauges the electrons are emitted by a heated filament.

15.3.2.1 The Discharge-Tube Gauge

Construction This is the simplest of the cold-cathode ionization gauges and operates over the range from 10^3 Pa to 10^{-8} Pa. The gauge head shown in Figure 15.7 consists of a glass tube, about 15 cm long and 1 cm in diameter, connected to the vacuum apparatus. A flat or cylindrical metal electrode attached to a glass/metal seal is mounted at each end. Aluminum is preferable because it does not readily disintegrate to form metal films on the gauge walls during use. A stable power supply with an output of 2.0 kV at 2.0 mA is connected across the electrodes, in series with a resistor R of about 2 MΩ to limit the current and a 1 mA meter M scaled to read the pressure.

Operation When the gauge is operating, several distinct luminous glows appear in the tube, the colors of which depend on the gases present. These glows, called the positive column P and negative glow N, result from ionization of the gas. The process is illustrated in Figure 15.8, where a positive ion striking the cathode C releases an electron. The electron is accelerated toward the anode and after traveling some distance encounters a gas molecule and produces the ionization that forms the negative glow. Ions from the negative glow are attracted to the cathode, where further electrons are emitted. This process, though continuous when established, requires some initial ions to start it. These may be formed by traces of radioactivity in the environment, though some delay may be experienced after switching on. The operating pressure range is determined by the electron path lengths. Above 10^3 Pa the motion of the electrons is impeded by the large number of gas molecules; below about 10^{-1} Pa the electrons travel the whole length of the tube without meeting a gas molecule, and ionization ceases.

15.3.2.2 The Penning Ionization Gauge (Penning, 1937)

This cold-cathode gauge is sensitive, simple, and robust and therefore finds wide industrial application. It measures the pressure of gases and vapors over the range from 1 Pa to 10^{-5} Pa, as shown in Figure 15.9.

Construction A glass or metal envelope connected to the vacuum apparatus houses two parallel cathode plates C of nonmagnetic material separated by about 2 cm and connected electrically. Midway between these and parallel to them is a wire anode ring A. Attached to the outside of the envelope is a permanent-magnet assembly that produces a transverse magnetic flux density B of about 0.03 T. This greatly increases the electron path length and enables a glow discharge to be maintained at low pressures. An alternative construction, of greater

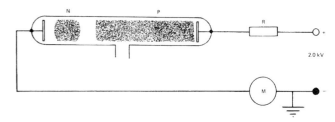

FIGURE 15.7 The Discharge-tube gauge.

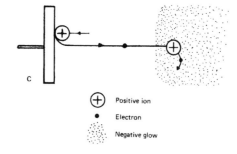

FIGURE 15.8 The production of carries in the glow discharge.

sensitivity due to Klemperer (1947), is shown in Figure 15.10. This uses a nonmagnetic cylindrical cathode C about 30 mm in diameter and 50 mm long, which may form the gauge envelope, along the axis of which is a stiff wire anode A, 1 mm in diameter. An axial magnetic flux density B of about 0.03 T is provided by a cylindrical permanent magnet or solenoid.

Operation A stable 2.0 kV power supply capable of supplying 2 mA is connected in series with a 1 mA meter M scaled to read pressure. The 2 Ω ballast resistor R limits the current to 1 mA at the upper end of the pressure range. Electron paths in the gauge are shown in Figure 15.11. The electrode assembly, shown in section, is divided into two regions for purposes of explanation. On the right-hand side, the magnetic field is imagined to be absent, and an electron from the cathode oscillates through the plane of the anode ring several times before collection, thereby increasing the electron path length. On the left-hand side, the presence of the magnetic flux causes a helical motion around the oscillatory path, causing a still greater increase. The combined effect of these two processes is to bring about electron paths many meters in length, confined within a small volume. All gas discharges are subject to abrupt changes in form when the pressure varies. These mode changes lead to a sudden change in gauge current of 5 to 10 percent as the pressure rises, which is not reversed until the pressure is reduced below the level at which it occurred. The effect, shown in Figure 15.12, is known as *hysteresis* and causes ambiguity, since a given pressure p is associated with two slightly different currents i_1 and i_2. By careful design, the effect may be minimized and made to appear outside the operating range of the gauge.

15.3.2.3 The Hot-Cathode Ionization Gauge (Buckley, 1916)

Construction The gauge head shown in Figure 15.13 is a special triode valve, usually with a hard glass envelope. Stainless steel or a nude form of gauge in which the gauge electrodes are inserted directly into the vacuum vessel may also be used. The filament F, of heavy-gauge tungsten, operates at about 2000°K, and may be readily damaged by accidental inrushes of air. A filament of iridium coated with thorium oxide operating at a

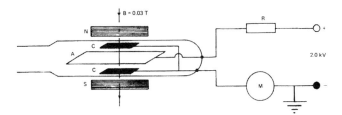

FIGURE 15.9 The Penning gauge.

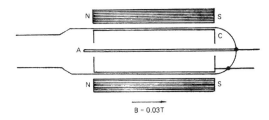

FIGURE 15.10 Cylindrical form of Penning gauge.

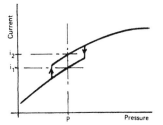

FIGURE 15.12 Hysteresis in the Penning gauge.

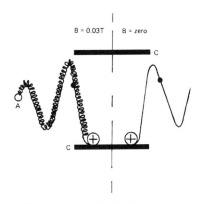

FIGURE 15.11 Electron paths in the Penning gauge.

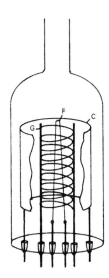

FIGURE 15.13 The Hot-cathode ionization gauge.

lower temperature is almost indestructible. Around the filament is a molybdenum or tungsten grid G, also heavily constructed, and outside the grid is a cylindrical ion collector C of nickel. Since the ion current received by this electrode is very small, special care is taken with its insulation.

Operation The gauge head is furnished with stable electrical supplies, as shown in Figure 15.14. The filament is heated to produce electrons, which are attracted to the grid, where a fraction of them are immediately collected. The remainder oscillate several times through the grid wires before collection, forming ions by collision with the gas molecules. The electron current i is measured by M_1 and is usually between 0.1 mA and 5.0 mA. Ions formed between the grid and the ion collector constitute an ion current i_+, shown by M_2. Electrons are prevented from reaching the ion collector by the application of a negative bias of 20 V. Ions formed between the filament and the grid are attracted by the filament, where their impact etches its surface and shortens its life. This is particularly the case when the gauge is operated at pressures above 1 Pa and if active gases such as oxygen are present.

Outgassing Since the gauge is highly sensitive, the gas molecules covering the electrodes and envelope must be removed by heating them to the highest safe temperature. For the filament, the required temperature can be obtained by increasing the filament current, while the grid and ion-collector may be heated by electron bombardment, using the filament as the source. Commercial gauge control units make provision for this treatment. The envelope is heated in an oven or by means of a hot-air gun.

Pumping During operation, the gauge removes gas molecules from the vacuum apparatus and thus behaves as a pump. Two processes are involved: At the filament, one process takes place in which molecules of active gases combine to form stable solid compounds; in the other, at the ion collector, the positive ions embed themselves beneath its surface. The speed of pumping can be reduced by lowering the filament temperature and by reducing the rate of collection of ions.

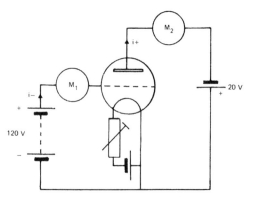

FIGURE 15.14 Hot-cathode ionization gauge.

Relationship Between Ion Current and Pressure If the pressure is p, the ion current i_+, and the electron current i_-, it is found for a given gas—say, nitrogen—that

$$i_+ \propto p i_- \qquad (15.1)$$

Therefore

$$i_+ = K p i_- \qquad (15.2)$$

where K is a constant called the *gauge factor* for nitrogen. The SI unit is Pa^{-1}, and its value for an average gauge is $0.1\ Pa^{-1}$. The value for a particular gauge is given by the maker. For other gases,

$$i_+ = CK p i_- \qquad (15.3)$$

where C is the relative sensitivity, with respect to nitrogen. Its approximate value for various gases is as follows:

Gas	He	H_2	N_2	Air	Ar	Xe	Organic Vapors
C	0.16	0.25	1.00	1.02	1.10	3.50	>4.0

Equation (15.3) shows that for a given gas and value of i_-

$$i_+ \propto p$$

This valuable property of the gauge is obtained by stabilizing i_- by means of an electronic servo system that controls the filament temperature and can switch the filament off if the gauge pressure becomes excessive.

15.3.2.4 The Bayard-Alpert Ionization Gauge (1950)

The Soft X-ray Effect The conventional ionization gauge described in Section 15.3.2.3 is not able to measure pressures below 10^{-6} Pa, due to the presence of a spurious current in the ion collector circuit produced by processes occurring in the gauge and independent of the presence of gas. This current is produced by soft X-rays generated when electrons strike the grid. The phenomenon is called *the soft X-ray effect*. The wavelength λ of this radiation is given by

$$\lambda = 1200/V \text{ nm} \qquad (15.4)$$

where V is the grid potential. If this is 120 volts, $\lambda = 10$ nm. Radiation of this wavelength cannot escape through the gauge envelope, but when it is absorbed by the ion collector it causes the emission of photoelectrons, which are collected by the grid. In the collector circuit, the loss of an electron cannot be distinguished from the gain of a positive ion, so the process results in a steady spurious ion current superimposed on the true ion current. The spurious current is about 10^{-10} A and is of the same order of magnitude as true ion current at 10^{-6} Pa. It is therefore difficult to measure pressures below this value with this type of gauge. A modified design due to Bayard and Alpert, shown in Figure 15.15, enables the area of the ion collector, and hence the spurious current, to be reduced by a factor of 10^3, thereby extending the range to 10^{-9} Pa.

Construction The filament F is mounted outside a cylindrical grid G, having a fine wire collector C of 0.1 mm diameter tungsten-mounted along its axis. The electrodes are mounted in a glass envelope that has a transparent conducting coating W on its inner wall to stabilize the surface potential.

Operation The gauge is operated in the same manner as a conventional ionization gauge. Electrons from the filament oscillate through the grid volume and form ions there. These ions are pushed by the positive potential of the grid toward the ion collector, which forms their only means of escape. Thus despite its small area, the electrode is an extremely good collector of ions.

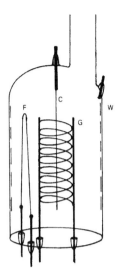

FIGURE 15.15 The Bayard-Alpert ionization gauge.

REFERENCES

Carpenter, L. G., *Vacuum Technology*, Hilger, Bristol (1970).

Leck, J. H., *Total and Partial Pressure Measurement in Vacuum Systems*, Blackie, Glasgow (1989).

Pirani, M. and Yarwood, J., *Principles of Vacuum Engineering*, Chapman & Hall, London (1961).

Ward, L. and Bunn, J. P., *Introduction to the Theory and Practice of High Vacuum Technology*, Butterworth, London (1967).

Chapter 16

Particle Sizing

W. L. Snowsill

16.1 INTRODUCTION

The size of particles is an extremely important factor in their behavior. To name but a few examples, size affects particles' chemical reactivity, their optical properties, their performance in a gas stream, and the electrical charge they can acquire. The methods used for assessing size are often based on one or more of these effects.

Particulate technology is a complex subject, and the major factor in this complexity is the variety of the physical and chemical properties of the particles. What appears to the naked eye as a simple gray powder can be a fascinating variety of shapes, colors, and sizes when viewed under a microscope. Particles can be solid or hollow, or filled with gas. The surface structure, porosity, specific gravity, and the like can have a profound effect on particles' behavior. Their ability to absorb moisture or to react with other chemicals in the environment or with each other can make handling very difficult as well as actually affecting the size of the particles. The size analyst has to combat the problem of particles adhering to each other because of chemical reactions, mechanical bonding, or electrostatic charging, and the problem increases as the size decreases. At the same time he must be aware that with friable particles, the forces applied to keep particles separate may be enough to break them.

Sampling is a crucial factor when measurements are made on particles. The essential points are:

- To be of any value at all, the sample must be representative of the source.
- Steps must be taken to avoid the sample changing its character before or during analysis.
- Particulate material, when poured, vibrated, or moved in any way, tends to segregate itself. The coarser particles tend to flow down the outside of heaps, rise to the top of any vibrating regime, and be thrown to the outside when leaving a belt feeder. These factors need to be given careful consideration, especially in attempting to subdivide samples.

16.2 CHARACTERIZATION OF PARTICLES

Most particles are not regularly shaped, so it is not possible to describe the size uniquely. To overcome this problem, the standard procedure is to use the diameter of equivalent spheres. However, an irregularly shaped particle can have an almost limitless number of different equivalent spheres, depending on the particular parameter chosen for equivalence.

For example, the diameter of a sphere with an equivalent volume would be different from that with an equivalent surface area.

Consider a cubic particle with edge of length x. The diameter of an equivalent volume sphere would be $x(6/\pi)^{1/3}$, that is, $1.24x$. The diameter of an equivalent surface area sphere would be $x(6/\pi)^{1/2}$, that is, $1.38x$. The chosen equivalent is usually related to the method of analysis. It is sensible to select the method of analysis to suit the purpose of the measurement, but in some cases this is complicated by practical and economic considerations.

Sometimes the equivalent diameter is not particularly relevant to the process, whereas the actual measurement made is relevant. In such cases, the size is sometimes quoted in terms of the parameter measured. A good example of this idea is terminal velocity (see Section 16.3). If, for example, information is required to assess the aerodynamic effect of a gas stream on particles, terminal velocity is more relevant than particle size. Even if the particles are spherical, conversion can be complicated by the possible variations in particle density. The term *vel* is sometimes used to denote particle size. A 1-vel particle has a freefalling speed of 10 mm s^{-1} in still, dry air at STP.

When equivalent diameters are quoted, it is important that the basis (equivalent mass, volume, surface area, projected area, and so on) is clearly stated.

16.2.1 Statistical Mean Diameters

Microscopic examination of an irregularly shaped particle suggests other methods of assessing the mean diameter. Consider a large number of identical particles, "truly" randomly oriented on a microscope slide. The mean of a given measurement made on each of the particles but in the same direction (relative to the microscope) would yield a statistical mean diameter. The following is a series of statistical mean diameters that have been proposed (see Figure 16.1):

- *Ferêt diameter*. The mean of the overall width of a particle measured in all directions.
- *Martin's diameter*. The mean of the length of a chord bisecting the projected area of the particle, measured in all directions.
- *Projected area diameter*. The mean of the diameters of circles having the same area as the particle viewed in all directions.
- *Image shear diameter*. The mean of the distances that the image of a particle needs to be moved so that it does not overlap the original outline of the particle, measured in all directions.

In microscopy, because particles tend to lie in a stable position on the slide, measurements of a group of particles, as described previously would not be "truly" randomly oriented. In these circumstances, the preceding diameters are *two-dimensional statistical mean diameters*.

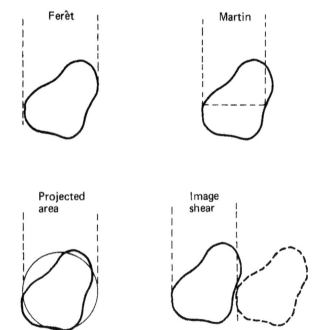

FIGURE 16.1 Statistical diameters.

16.3 TERMINAL VELOCITY

The terminal velocity of a particle is that velocity resulting from the action of accelerating and drag forces. Most commonly it is the freefalling speed of a particle in still air under the action of gravity. The relationship between the terminal velocity of a particle and its diameter depends on whether the flow local to the particle is laminar or turbulent. In laminar flow, the particle falls more quickly than in turbulent flow, where particles tend to align themselves for maximum drag and the drag is increased by eddies in the wake of the particles.

The general equation for the drag force F on a particle is:

$$F = Kd^n V^n \eta^{2-n} \rho_0^{n-1}$$

where K is the drag coefficient, depending on shape, surface, and so on; d is the particle dimension (diameter of a sphere); V is relative velocity; η is fluid viscosity; ρ_0 is fluid density; and n varies from 1 for laminar flow to 2 for turbulent flow.

For some regularly shaped particles, K can be calculated. For example, for a sphere in laminar flow, $K = 3\pi$.

Hence from the preceding, we find that for laminar flow spheres,

$$F = 3\pi d V \eta$$

which is known as Stokes's law.

By equating drag force and gravitational force, we can show

$$3\pi d V_T \eta = \frac{\pi d^3}{6(\rho - \rho_0)g}$$

where ρ is particle density, and V_T is terminal velocity. Thus,

$$V_T = (\rho - \rho_0)\frac{gd^2}{18\eta}$$

If the terminal velocity of irregularly shaped particles is measured together with ρ and η, the value obtained for d is the Stokes diameter. Sometimes the term *aerodynamic diameter* is used, denoting an equivalent Stokes sphere with unit density. Stokes diameters measured with spheres are found to be accurate (errors < 2 percent) if the Reynolds number

$$Re = \frac{\rho_0 V d}{\eta}$$

is less than 0.2. At higher values of Re, the calculated diameters are too small. As Re increases, n increases progressively to 2 for $Re > 1{,}000$ when the motion is fully turbulent, and according to Newton the value of K for spheres reduces to $\pi/16$.

For very small particles where the size approaches the mean free path of the fluid molecules (~0.1 μm for dry air at STP), the drag force is less than that predicted by Stokes. Cunningham devised a correction for Stokes's equation:

$$F = 3\pi dV\eta \frac{1}{1 + (b\lambda/d)}$$

where λ is the mean free path and b depends on the fluid (e.g., for air at STP dry, $b \simeq 1.7$).

16.4 OPTICAL EFFECTS CAUSED BY PARTICLES

When light passes through a suspension of particles, some is absorbed, some scattered, and a proportion is unaffected, the relative proportions depending on the particle size, the wavelength of the light, and the refractive indices of the media. The molecules of the fluid also scatter light.

Some optical size-analysis methods infer size from measurements of the transmitted, that is, unaffected light; others measure the scattered light. Some operate on suspensions of particles, others on individual particles.

The theory of light-scattering by particles is complicated. Rayleigh's treatment, which applies only to particles for which the diameter $d \ll \lambda$ (the wavelength), shows that the intensity of scattered light is proportional to d^6/λ^4. It also shows that the scattering intensity varies with the observation angle, and this also depends on d. As size d approaches λ, however, the more rigorous treatment of the Mie Solution indicates that the scattering intensity becomes proportional to d^2, that is, particle cross-sectional area, but that the effective area is different from the geometrical area by a factor K, known as the *scattering coefficient* or *efficiency factor*, which incorporates d, λ, and the refractive index. Where $d \ll \lambda$ the two theories are similar. In the region around $d = \lambda$, however, K oscillates (typically between about 1.5 and 5), tending toward a mean of 2. Beyond about $d = 5\lambda$, the value of K becomes virtually 2, that is, the effective cross-sectional area of a particle is twice the geometrical area. As d/λ increases, the preferred scattering angle reduces and becomes more distinct and forward scattering predominates (diffraction).

If the light is not monochromatic, the oscillation of K is smoothed to a mean of about 2.

The ratio of the intensity of the transmitted light I_T to the incident light I_0 is given by the Lambert-Beer law:

$$\frac{I_T}{I_0} = \exp\left(-K\frac{a}{A}\right)$$

where a is the total projected area of the particles in the light beam, A is the area of the beam, and again K is the scattering coefficient. This is often simplified to

$$\text{optical density } D = \log_{10} I_0/I_T$$
$$= 0.4343K(a/A)$$

The scattering coefficient is sometimes called the *particle extinction coefficient*. This should not be confused with the extinction coefficient ξ. If the transmission intensity of a beam of light changes from I_0 to I_t in a path length L,

$$I_T/I_0 = \xi L$$

where K is contained within ξ.

Extinction in I_0/I_t is the Napierian equivalent of optical density.

Although the value of K has been shown to be virtually 2, the scattering angle for larger particles (~30 μm) is small and about half the light is forward-scattered. It follows that depending on the observation distance and the size of the sensor, much of the forward-scattered light could be received and the effective value of K in the above expression could be as low as 1. It will be apparent that the effect of a distribution of particles on light transmission is not a simple function of the projected area.

Bearing in mind the above limitations on K, it is possible to estimate the transmitted light intensity through a distribution of particles by summing the area concentrations within size bands. In each band of mean diameter d, the effective area a/A is $1.5\ KcL/\rho d$, where c is the mass concentration, L is the optical path length, and ρ is the particle density.

16.5 PARTICLE SHAPE

Although we can attribute to a particle an equivalent diameter—for example, a Ferêt diameter d_F—this does not uniquely define the particle. Two particles with the same Ferêt diameter can have very different shapes. We can say that the volume of an equivalent sphere is

$$\frac{\pi}{6}(d_F)^3$$

but we must recognize that the actual volume V is probably very different. Heywood has proposed the use of shape coefficients. We can assign a coefficient $\alpha_{V,F}$ to a particle such that

$$V = \alpha_{V,F}(d_F)^3$$

Thus, if we use another method of size analysis that in fact measures particle volume V, knowing d_F we can calculate $\alpha_{V,F}$. Similarly, by measuring particle surface area S, we can assign a coefficient $\alpha_{S,F}$ so that

$$S = \alpha_{S,F}(d_F)^2$$

$\alpha_{V,F}$ is called the *volume shape coefficient* (based on Ferêt diameter), and $\alpha_{S,F}$ is called the *surface shape coefficient* (based on Ferêt diameter).

Clearly, there are other shape coefficients, and they can be associated with other diameters.

The ratio $\alpha_S/\alpha_V = \alpha_{S,V}$ is called the *surface volume shape coefficient*.

The subject is covered by BS 4359 (1970) Pt III, which includes definitions and tables of the various coefficients for a number of regular shapes—cubes, ellipsoids, tetrahedra, and so on—and a number of commonly occurring particles. The coefficients α_S, α_V and $\alpha_{S,V}$ together provide a very good indication of particle shape in a quantified form.

16.6 METHODS FOR CHARACTERIZING A GROUP OF PARTICLES

We have already established a number of alternative "diameters" to be used to characterize particles. There are also several ways of characterizing groups of particles. They are all assessments of the quantities of particles within "diameter" bands, but the quantities can be numbers of particles, mass of particles, volume, surface area, and so on. As with particle equivalent diameters, it is important that the basis of the analysis is made clear.

There are also several methods for expressing the results of a size analysis. Perhaps the most obvious is tabulation; a contrived example is given in Table 16.1, which shows the masses of particles contained within 5 μm size fractions from 0 to 40 μm. The main disadvantage is that it requires considerable experience to recognize what could be important differences between samples. Such differences are much more readily apparent if the results are plotted graphically. One method is to plot the quantity obtained, whether mass, volume, surface area, or number of particles in each size fraction against size, both on a linear scale. This is called a *relative frequency plot*, and Table 16.1 has been transferred in this way to Figure 16.2.

Students with a little understanding of statistics will be tempted to compare this with a normal or Gaussian distribution (also shown). In practice, Gaussian distributions are not very common with powder samples, but this simple example is useful to illustrate a principle.

16.6.1 Gaussian or Normal Distributions

The equation for a Gaussian distribution curve is

$$y = \frac{1}{\sigma\sqrt{2\pi}} \exp -\left[\frac{(x-\bar{x})^2}{(2\sigma^2)}\right]$$

where $\int y\,dx$, the area under the curve, represents the total quantity of sample (again, number, mass, volume, etc.) and is made equal to 1. The symbol $\bar{x}$ represents the arithmetic mean of the distribution, and σ, the standard deviation of the distribution, is a measure of the spread. These two

TABLE 16.1 Alternative methods of tabulating the same size analysis

(a)		(b)	
Size band (μm)	% mass in band	Stated size	% less than stated size
0–5	0.1	5	0.1
5–10	2.4	10	2.5
10–15	7.5	15	10.0
15–20	50.0	20	60.0
20–25	27.0	25	87.0
25–30	12.5	30	99.5
>30	0.5		

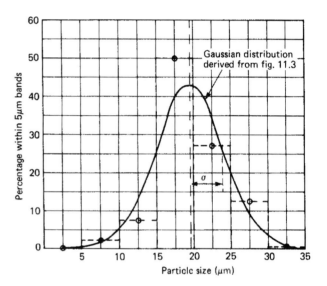

FIGURE 16.2 Relative percentage mass-frequency plot.

parameters uniquely define a Gaussian distribution. It can be shown that 68.26 percent of the total area under the curve is contained between the boundaries $x = \bar{x} \pm \sigma$. In this case we have plotted the values of $y\delta x$ for equal 5 μm increments. We could just as easily have drawn a histogram. At this point it should be stated that if any one of the distributions should turn out to be "normal" or Gaussian, none of the other plots, that is, number, volume, or surface area distributions, will be Gaussian.

An advantage of this presentation is that small differences between samples would be readily apparent. However, it would be useful to be able to easily measure the values of x and σ, and this is not the case with the preceding. Two alternatives are possible. One is to plot a cumulative percentage frequency diagram, again on linear axes, as in Figure 16.3. In this case one plots the percentage less (or greater) than given sizes. Alternatively, one can plot the same information on linear-probability paper where one axis, the percentage frequency axis, is designed so that a Gaussian distribution will give a straight line, as in Figure 16.4. In an inexact science such as size analysis, the latter has distinct advantages, but in either case the arithmetic mean $\bar{x}$ is the value of x at the 50 percent point, and the value of σ can be deduced as follows. Since 68.26 percent of a normal distribution is contained between the values $x = \bar{x} + \sigma$ and $x = \bar{x} - \sigma$, it follows that

$$\sigma = x_{84\%} - \bar{x} = \bar{x} - x_{16\%}$$
$$= \frac{1}{2}(x_{84\%} - x_{16\%})$$

because $x_{84\%} - x_{16\%}$ covers the range of 68 percent of the total quantity.

The closeness of fit to a Gaussian distribution is much more obvious in Figure 16.4 than in Figures 16.2 and 16.3. With probability paper, small differences or errors at either extreme produce an exaggerated effect on the shape of the line. This paper can still be used when the distribution is not "normal," but in this case the line will not be straight and standard deviation is no longer meaningful. If the distribution is not "normal," the 50 percent size is not the arithmetic mean but is termed the median size. The arithmetic mean needs to be calculated from

$$\bar{x} = \sum \frac{(\text{percentage in size fraction} \times \text{mean of size fraction})}{100}$$

and the basis on which it is calculated (mass, surface, area, volume, or particle number) has to be stated. Each will give a different mean and median value.

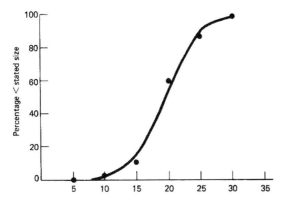

FIGURE 16.3 Cumulative percentage mass-frequency plot using linear scales.

16.6.2 Log-Normal Distributions

It is unusual for powders to occur as Gaussian distributions. A plot as in Figure 16.2 would typically be skewed toward the smaller particle sizes. Experience has shown, however, that powder distributions often tend to be log-normal. Thus a percentage frequency plot with a logarithmic axis for the particle size reproduces a close approximation to a symmetrical curve, and a cumulative percentage plot on log-probability paper often approximates to a straight line (see Figure 16.5).

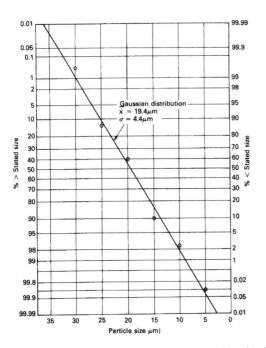

FIGURE 16.4 Cumulative percentage mass-frequency plot using linear x percentage scales.

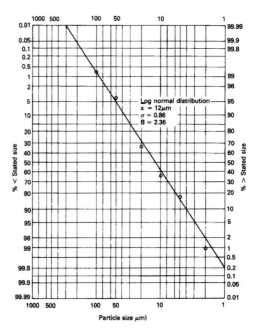

FIGURE 16.5 Cumulative percentage mass–frequency plot using base 10 log x percentage scales.

In a true log-normal distribution, the equation becomes

$$y = \frac{1}{\sigma\sqrt{2\pi}} \exp\left[\frac{(\ln x - \overline{\ln x})^2}{2\sigma^2}\right]$$

where now it is $\int y d(\ln x)$, the area under the curve using a log axis, which represents the total quantity, and σ now refers to the log distribution and is not the same as before. The expression $\overline{\ln x}$ is the arithmetic mean of the logarithms of the size so that $\bar{x}$ is now the geometric mean of the distribution. On a cumulative percentage diagram, $\bar{x}$, the geometric mean particle size is the 50 percent size and σ is found from

$$\sigma = \ln x_{84} - \ln \bar{x} - \ln \bar{x} - \ln x_{16}$$
$$= \ln \frac{x_{84}}{\bar{x}} = \ln \frac{\bar{x}}{x_{16}}$$
$$= \frac{1}{2} \ln \frac{x_{84}}{x_{16}}$$

If x is plotted on base-10 logarithm × probability paper,

$$\sigma = \frac{1}{2} \ln 10 \log_{10} \frac{x_{84}}{x_{16}}$$
$$= 1.15 \log_{10} \frac{x_{84}}{x_{16}}$$

Again $\bar{x}$ and σ define the distribution.

Sometimes σ is replaced by $\ln B$ to show that it is the standard deviation of a log-normal distribution:

$$B = \sqrt{(x_{84}/x_{16})}$$

Again, if the cumulative percentage plot is not truly linear, the derivation of the standard deviation is not truly meaningful and the 50 percent size is then the median size. However, in practice such curves are commonly used for comparing size analyzes, and it is sometimes useful for mathematical treatment to draw an approximate straight line.

A feature of a log-normal distribution is that if one method of treatment—for example, a mass particle size analysis—demonstrates log-normal properties, all the other methods will also be log-normal. Clearly the values of $\bar{x}$ will be different. Log-probability diagrams are particularly useful when the range of particle sizes is large.

16.6.3 Rosin–Rammler Distributions

Some distributions are extremely skewed—for example, ground coal. Rosin and Rammler have developed a further method for obtaining a straight-line cumulative percentage frequency graph. If the percentage over size x is R, it has been found that

$$\log \log (100/R) = K + n \log x$$

where K is a constant and n a characteristic for the material.

The Rosin–Rammler distribution is included for completeness, but its use is not generally recommended.

Sometimes when a distribution covers a wide range of sizes, more than one analysis method has to be used. It is not unusual for a discontinuity to occur in the graphs at the change-over point, and this can be due to shape or density effects (see the discussion of shape factor, Section 16.5).

16.7 ANALYSIS METHODS THAT MEASURE SIZE DIRECTLY

16.7.1 Sieving

Sieving is the cheapest, most popular, and probably the most easily understood method of size analysis. It also covers a very wide range of sizes, it being possible to buy sieves (screens) ranging in mesh size from 5 μm up to several centimeters. However, sieving of fine materials requires special techniques, and the British Standard 410:(1962) indicates a lower limit of μm. Sieves are made in a variety of materials, from nonmetallic (e.g., polyester) to stainless steel. The common method of construction is woven wire or fabric, but the smallest mesh sizes are electroformed holes in plates. The British Standard gives minimum tolerances on mesh size, wire spacing, and so on. American, German, and ISO standards are also applicable and carry very similar criteria. The British Standard nomenclature is based on the number of wires in the mesh per inch. Thus BS sieve number 200 has 200 wires per inch and with a specified nominal wire diameter of 52 μm has a nominal aperture size of 75 μm square. In principle, all particles less than 75 μm diameter in a sample of spherical particles will pass through a BS number 200 sieve. The sample is placed in the uppermost of a stack of sieves covering the range of diameters of interest, arranged in ascending order of size from the bottom. The powder is totally enclosed by means of a sealed base and lid. The stack is agitated until the particles have found their appropriate level and the mass in each sieve noted. The tolerance on mesh size introduces a measure of uncertainty on the bandwidths, and clearly irregularly shaped particles with one or more dimensions larger than the nominal size could still pass through. It is customary therefore to quote particle size when sieving in terms of "sieve" diameters.

Sieving is by no means as straightforward as it might first appear. For example, it is relatively easy for large particles to block the apertures to the passage of small particles (blinding), and there is a statistical uncertainty whether a small particle will find its way to a vacant hole. Both of these are very dependent on the quantity of material placed on the sieve, and as a general rule this quantity should be kept small. It is not possible to give an arbitrary figure for the quantity, since it will depend on the material and its size distribution, particle shape and surface structure, its adhesive qualities, and, to some extent, the sieve itself. The same comments apply to sieving time. Optimum times and quantities can only be found by experiment, assessing the variation in size grading produced by both factors. Generally a reduction in the quantity is more advantageous than an increase in sieving time, and it is normally possible to obtain repeatable results with less than about 10 minutes' sieving. The analyst is cautioned that some friable materials break up under the action of sieving.

A number of manufacturers produce sieves and sieving systems using various methods of mechanical agitation, some of which include a rotary action. The objective, apart from reducing the tedium, is to increase the probability of particles finding vacant holes by causing them to jump off the mesh and to return in a different position. Figure 16.6 is an example of one that uses vibration. The vibration can be adjusted in amplitude, and it can be pulsed.

A feature of any vibrating mechanism is that parts of it can resonate and this is particularly relevant to sieving, where it is possible for the sieve surface to contain systems of nodes and anti-nodes so that parts of the surface are virtually stationary. This can be controlled to some extent by adjustment of the amplitude, but although the top sieve surface may be visible through a transparent lid, the lower surfaces are not and they will have different nodal patterns. One solution to this dilemma is to introduce into each sieve a small number (5 to 10) of 10 mm diameter agate spheres. These are light enough not to damage the sieves or, except in very friable materials, the particles, but they break up the nodal patterns and therefore increase the effective area of the sieve.

A useful feature of dry sieving is that it can be used to obtain closely sized samples for experimental purposes in reasonable quantities.

FIGURE 16.6 Fritsch Analysette sieve shaker.

Although most sieving is performed in the dry state, some difficult materials and certainly much finer sieves can be used in conjunction with a liquid, usually water, in which the particles are not soluble. The lid and base of the sieve stack are replaced by fitments adapted for the introduction and drainage of the liquid with a pump if necessary.

Sieving systems are now commercially available that introduce either air or liquid movement alternately up and down through the sieves to prevent blinding and to assist the particles through the sieves.

Results are usually quoted in terms of percentage of total mass in each size range, the material being carefully removed from the sieve with the aid of a fine soft brush. In wet sieving, the material is washed out, filtered, dried at 105°C, and weighed. Many powders are hygroscopic, so the precaution of drying and keeping them in a dessicator until cool and therefore ready for weighing is a good general principle. Convection currents from a hot sample will cause significant errors.

16.7.2 Microscope Counting

16.7.2.1 Basic Methods

With modern microscopes, the analyst can enjoy the benefits of a wide range of magnifications, a large optical field, and stereoscopic vision and zoom facilities, together with back and top illumination. These and calibrated field stop graticules have considerably eased the strain of microscope counting as a method of size analysis, but it is still one of the most tedious. However, it has the advantage that, as well as being able to size particles, the microscope offers the possibility of minute examination of their shape, surface structure, color, and so on. It is often possible to identify probable sources of particles.

The optical microscope can be used to examine particles down to sizes approaching the wavelength of light. At these very small sizes, however, interference and diffraction effects cause significant errors, and below the wavelength of light the particles are not resolvable. Microscope counting is covered by British Standard 3406.

Smaller particles, down to 0.001 μm diameter, can be examined using the electron microscope.

The two major disadvantages of microscopy are (1) the restricted depth of focus, which means that examination of a sample with a wide size range involves continual refocusing, and (2) the real possibility of missing "out-of-focus" particles during a scan; it depends more than most other methods on good representativeness.

Several techniques are available for the preparation of slides, the all-important factor being that the sample is fully representative. The particles also need to be well separated from each other. A common method is to place a small fraction of the sample onto the slide with one drop of an organic fluid such as methanol or propanol, which disperses the particles. Subsequent evaporation leaves the particles suitably positioned. The fluid obviously must not react with the particles, but it must have the ability to "wet" them. Agitation with a soft brush can help. If the particles have a sticky coating on them, it may be necessary to remove this first by washing. One technique is to agitate, perhaps in water, to allow ample time for all the particles to settle, and then to pour off the fluid carefully, repeating as necessary. Obviously the particles must not be soluble in the fluid. Sometimes the material is first agitated in a fluid and then one drop of the particle-laden fluid transferred to the slide. Representativeness can only be tested by the repeatability of results from a number of samples. Techniques have been devised for transferring samples from suspension onto films within the suspension. It is sometimes possible to collect samples directly onto sticky slides coated with grease, gelatin, or even rubber solution.

Earlier methods of microscope counting involved the use of an optical micrometer by which a cross-hair could be aligned with each side of each particle in turn and the difference measured. As can be imagined, this process was slow and tedious. Now the most commonly used methods involve calibrated field graticules. The graticules are engraved with a scale (see Figure 16.7), nominally divided, for example, into 20 μm, 100 μm, and 1 mm steps. Calibration is dependent on the magnification, which is usually finely adjustable by slightly moving the objective relative to the eyepiece or by adjusting the zoom, if available. Calibration is effected by comparing the field graticule with a stage graticule, similarly and accurately engraved. When set, the stage graticule is replaced by the sample slide.

The slide is scanned in lateral strips, each strip an order or so wider than the largest particles, the objective being to cover the whole of the slide area containing particles. Typically one edge of a chosen reference particle will be aligned with a major graticule line using the longitudinal stage adjustment. The slide will then be traversed laterally along that line, and all particles to the right of that line will be counted and measured using the eyepiece scale. The slide will then be traversed longitudinally to the right until the original particle is in the same relative position but, for example, five major graticule lines further over; the counting process is repeated for particles within the strip formed by the two lines. The process involves selecting new reference particles as necessary. To avoid duplication, if a particle lies on one of the strip edgelines, it is counted as though it were in the strip to the right. Particles are allocated to size bands suitably chosen to give, say, 10 points on the distribution curve. The tedium is relieved if operators work in pairs, one observing, one recording, alternately.

Some graticules have been designed containing systems of opaque and open circles, the sizes arranged in various orders of progression. This can assist the classification of particles by comparison into size bands, each bounded by one of the circles.

In sizing irregularly shaped particles, microscope counting introduces a bias because the particles tend to lie in their most stable orientation. By measuring a distribution of randomly oriented particles on a slide along a fixed direction, one obtains a two-dimensional statistical mean diameter.

This method of using a line graticule measures the mean two-dimensional Ferêt diameter, the direction being fixed parallel with the long edge of the slide. It is equally possible to measure the mean two-dimensional Martin's diameter, projected area, or image shear diameter.

To obtain three-dimensional mean diameters, it is necessary to take special steps to collect and fix the particles on the slide in a truly random fashion.

16.7.2.2 Semiautomatic Microscope Size Analyzers

Semiautomatic methods are those in which the actual counting is still done by the analyst, but the task, especially the recording, is simplified or speeded up.

The Watson image-shearing eyepiece is a device that replaces the normal eyepiece of a microscope and produces a double image, the separation of which can be adjusted using a calibrated knob along a fixed direction, which again can be preset. The image spacing can be calibrated using a stage graticule. In the Watson eyepiece the images are colored red and green to distinguish them. The technique in this case is to divide the slide into a number of equal areas, to set the shear spacing at one of a range of values, and to count the number of particles in each area with image-shear diameters less than or equal to each of the range.

Some methods have been developed for use with photomicrographs, particularly with the electron microscope. Usually an enlargement of the print or a projection of it onto a

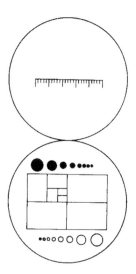

FIGURE 16.7 Examples of eyepiece graticules.

screen is analyzed using comparison aids. The Zeiss–Endter analyzer uses a calibrated iris to produce a variable-diameter light spot, which can be adjusted to suit each particle. The adjustment is coupled electrically via a multiple switch to eight counters so that pressing a foot switch automatically allocates the particle to one of eight preset ranges of size. A hole is punched in the area of each counted particle to avoid duplication.

16.7.2.3 Automatic Microscope Size Analyzers

Several systems have been developed for the automatic scanning of either the microscope field or of photomicrographs. In one type, the system is similar to a television scanning system; indeed, the field appears on a television screen. Changes in intensity during the scan can be converted to particle size and the information analyzed in a computer to yield details of size in terms of a number of the statistical diameters. It can also calculate shape factors. One system, the Quantimet, now uses digital techniques, dividing the field into 650,000 units.

16.7.3 Direct Optical Methods

Malvern Instruments Ltd. uses a laser and forwardscattering to analyze particles in suspension. The parallel light from the laser passes through a lens, producing an intense spot at the focus. Light falling on any particle is diffracted and is brought to a focus in a system of Fraunhofer diffraction rings centered on the axis and in the focal plane of the lens. The diameters correspond to the preferred scattering angles, which are a function of the diameters of the particles. The intensity of each ring is proportional to the total cross-sectional area of the particles of that size. The variation of intensity therefore reflects the size distribution. Irregularly shaped particles produce blurred rings. A multiringed sensor located in the focal plane of the lens passes information to a computer, which calculates the volume distribution. It will also assess the type of distribution, whether normal, log-normal, or Rosin–Rammler.

The position of the particles relative to the axis of the lens does not affect the diffraction pattern, so movement at any velocity is of no consequence. The method therefore works "online" and has been used to analyze oil fuel from a spray nozzle. The claimed range is from 1 to more than 500 μm.

There are distinct advantages in conducting a size analysis "online." Apart from obtaining the results generally more quickly, particles as they occur in a process are often agglomerated, that is, mechanically bound together to form much larger groups that exhibit markedly different behavioral patterns. This is important, for example, in pollution studies. Most laboratory techniques have to disperse agglomerates produced in sampling and storage. This can be avoided with an online process.

16.8 ANALYSIS METHODS THAT MEASURE TERMINAL VELOCITY

As already discussed, the terminal velocity of a particle is related to its size and represents a useful method of analysis, particularly if the area of interest is aerodynamic. Methods can be characterized broadly into sedimentation, elutriation, and impaction.

16.8.1 Sedimentation

A group of particles, settling, for example, under the influence of gravity, segregates according to the terminal velocities of the particles. This phenomenon can be used in three ways to grade the particles:

1. The particles and the settling medium are first thoroughly mixed, and changes in characteristics of the settling medium with time and depth are then measured.
2. The particles and settling medium are mixed as in (1), and measurements are then made on the sediment collecting at the base of the fluid column.
3. The particles are introduced at the top of the fluid column and their arrival at the base of the column is monitored.

Group (1) is sometimes termed *incremental*, that is, increments of the sedimenting fluid are analyzed.

Group (2) is sometimes termed *cumulative,* referring to the cumulative effect on the bottom of the column.

Group (3) is also cumulative, but it is sometimes distinguished by the term *two-layer,* that is, at the initiation of the experiment there are two separate fluids, the upper one thin compared with the lower and containing all the particles, the lower one clear.

16.8.1.1 Incremental Methods

Consider at time $t = 0$ a homogeneous distribution of particles containing some special ones with terminal velocity V. Ignoring the minute acceleration period, after a time t_1 all the special particles will have fallen a distance $h = Vt_1$. The concentration of those special ones below depth h will have remained unchanged except on the bottom. Above depth h, however, all the special particles will have gone. The same argument applies to any sized particle except that the values of h and V are obviously different.

It follows that a measurement of the concentration of all the particles at depth h and time t is a measurement of the concentration of those particles with a terminal velocity less than V; therefore we have a method of measuring the cumulative distribution.

The following methods use this general principle.

Andreasen's Pipette This consists of a relatively large (~550 ml) glass container with a pipette fused into its ground-glass stopper. A 1 percent concentration of the sample, suitably dispersed in a chosen liquid, is poured into the container to a set level exactly 200 mm above the lower tip of the pipette. Means are provided to facilitate the withdrawal of precise 10 ml aliquots from the container via the pipette.

After repeated inversions of the container to give thorough mixing, the particles are allowed to sediment. At preselected times after this, such as 1 minute, 2 minutes, 4 minutes, and so on, 10 ml samples are withdrawn and their solids content weighed. Corrections are applied for the change in depth as samples are removed. Samples are removed slowly over a 20-second period, centered around the selected times to minimize errors caused by the disturbance at the pipette tip. The results yield a cumulative mass/terminal velocity distribution that can be converted to mass size and so on, as already discussed. With suitable choice of liquid, the method can be used for particles ranging in size from about 1–100 μm. The conditions should be controlled to be Stokesian, that is, laminar flow, and of course the terminal velocity is appropriate to the fluid conditions; in other words, it depends on the liquid density and viscosity, which are also dependent on temperature.

Density-measuring Methods Several techniques involve the measurement of the density of the fluid/particle mixture at different times or at different depths at the same time. Initially, after mixing, the density is uniform and depends on the fluid, the particle densities, and the mass concentration of particles. After a period of sedimentation, the density varies with depth.

One method of measuring the density uses a hydrometer. This is complicated by allowance having to be made for the change in the overall height of the fluid caused by the immersion of the hydrometer tube. In addition, any intruding object causes a discontinuity in the settling medium, some particles settling on the upper surface of the hydrometer bulb, none settling into the volume immediately below the bulb. Motion around the bulb is not vertical. These problems tend to outweigh the basic simplicity of the method.

A neat method of hydrometry that overcomes the overall height problem uses a number of individual hydrometers, commonly called *divers*. Each diver consists of a small body that is totally immersed in the fluid but with its density individually adjusted so that after the preselected time, the divers will be distributed vertically within the sedimenting column, each at the depth appropriate to its density. Accurate measurement of the positions yields the required information. The main problem in this case is being able to see the divers.

A specific-gravity balance can be used. The change with time in the buoyancy of a ball suspended from one arm of a balance at a fixed depth again gives the information required.

Photosedimentation In a photosedimentometer, the sedimentation is observed by a lamp and photocell system (see Section 16.4). The observation distance is small, and for particles greater in size than about 15 μm, the value of K, the scattering coefficient, progressively reduces from 2 to 1. We know that

$$\text{optical density } D = 04343\, Ka/A$$

where a/A is the area concentration. With no particles present in the liquid, let the values of D and K be D_0 and K_0. With all particles present thoroughly mixed, let the corresponding values be D_1 and K_1. At time t and depth h, $h = Vt$, where V is the upper limit of the terminal velocity of the particles present. If the corresponding values of D and K are D_v and K_v, the fractional surface area of particles with terminal velocity less than V is given by

$$\left(\frac{D_V}{K_V} - \frac{D_0}{K_0}\right) \bigg/ \left(\frac{D_1}{K_1} - \frac{D_0}{K_0}\right)$$

We thus have a method of measuring cumulative surface area terminal velocity distribution.

Proprietary sedimentometers are available to measure D at a fixed height or scan the whole settlement zone. It is usual to assume $K_V = K_1 = K_0$ and to compensate the result appropriately from supplied tables or graphs. Most photosedimentometers use narrow-beam optics in an attempt to restrict the light to maintain the value of K as 2. The wide-angle scanning photosedimentometer (WASP) has the photocell close to the fluid so that most of the diffracted light is also received and the value of K is nearer 1. The 200 mm settling column is scanned automatically at a fixed rate and the optical density continuously recorded, giving a graph that can then be evaluated as a cumulative mass size or cumulative surface area size distribution.

X-ray sedimentation is similar to photosedimentation except that X-rays replace light and the intensity of transmission is dependent on the mass of the particles rather than the surface area. Again,

$$I_T = I_0 \exp(-Kc)$$

where c is the mass concentration of the particles and K is a constant. The X-ray density is

$$D = \log_{10} \frac{I_0}{I_T}$$

16.8.1.2 Cumulative Methods

Sedimentation Balance Consider at time $t = 0$ a homogeneous suspension of particles contained in a column that includes in its base a balance pan.

Let W_1 be the mass of particles with terminal velocity greater than V_1. If h is the height of the column, at time $t_1 = h/V_1$ all those particles will have arrived on the balance pan. However, the mass M_1 on the pan will also include a fraction of smaller particles that started partway down. It can be shown that

$$M_1 = W_1 + t\frac{dM}{dt}$$

and measurements of M and t can be used to evaluate W. British Standard 3406 Part 2 suggests that values of M should be observed at times t following a geometrical progression—for example, 1, 2, 4, 8, etc., seconds. Then t/dt is constant, in this case 2. It follows that, comparing the nth and the $(n-1)$th terms in the time progression,

$$W_n = M_n - 2(M_n - M_{n-1})$$

The final value of M is assumed to be equal to the initial mass introduced.

An alternative method, useful if M is continuously recorded, is to construct tangents, as in Figure 16.8. Then W is the intercept on the M axis. Unfortunately, because of the inaccuracy of drawing tangents, the method is not very precise, especially if the overall time is protracted, with a wide size distribution. The method can be improved by replotting M against $\ln t$ instead of t.

Since

$$\frac{dM}{dt} = \frac{1}{t}\frac{dM}{d(\ln t)}$$

the preceding expression can be rewritten

$$M = W + \frac{dM}{d(\ln t)}$$

A plot of M against t on logarithmic paper (see Figure 16.9) enables tangents to be drawn with greater precision, making it possible to compute $dM/d(\ln t)$, the gradient at time t. From a further plot of $dM/d(\ln t)$ against $\ln t$ on the same graph, W can be derived by difference. The method relies on none of the initial material being lost to the sides of the column or around the edges of the pan, and the initial quantity beneath the pan is insignificant. These factors do lead to errors.

Several commercial liquid sedimentation balances are available, notably Sartorious, Shimadzu, and Bostock. They have means for introducing the homogenous suspension into the column above the balance pan. In some, the fluid beneath the pan initially contains no dust. The pan is counterbalanced incrementally by small amounts to minimize pan movement, although some is bound to occur and this causes pumping, that is, some of the finer particles transfer from above to below the pan. All the columns are contained in thermostatic jackets, constancy of temperature being important to the constancy of fluid viscosity and density.

Sedimentation Columns A sedimentation column works on the same principle as a sedimentation balance, but instead of weighing the sediment continuously the sediment

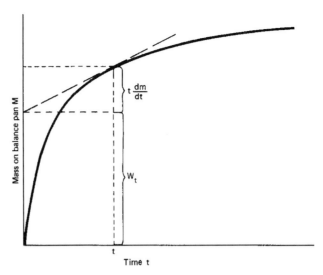

FIGURE 16.8 Sedimentation balance: plot of mass against time.

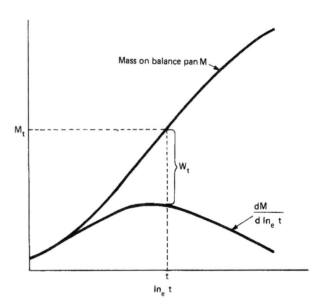

FIGURE 16.9 Sedimentation balance: plots of M and $dM/d \ln t$ against $\ln t$.

is removed at preset times and weighed externally, enabling a higher-quality balance to be used.

The ICI sedimentation column (see Figure 16.10) tapers at the bottom to a narrow tube a few centimeters long. It is fitted with taps top and bottom. A side branch in the narrow section connects to a clear fluid reservoir. The dust sample is introduced into the main column and mixed using a rising current of compressed air through Tap A, with Tap B also open and Tap C closed. At time $t = 0$, Taps A and B are closed and C is opened. Particles sediment into the narrow tube below the side branch. At preset times, Tap A is opened, allowing clear fluid to wash the sediment into a centrifuge tube or beaker. Negligible sedimenting fluid is lost. The sediment is filtered, dried, and weighed.

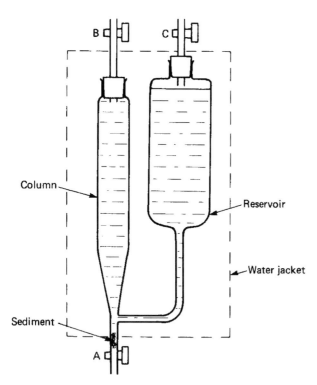

FIGURE 16.10 ICI sedimentation column.

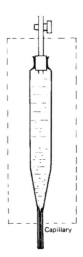

FIGURE 16.11 BCURA sedimentation column.

The BCURA sedimentation column also uses an air supply for mixing (see Figure 16.11). In this case, however, the lower tap and narrow tube are replaced by a length of 1 mm capillary tubing. With the top tap closed, surface tension in the capillary supports the column of fluid. At the prescribed times, a container of clear fluid is brought up around the open capillary, breaking the surface tension, and the sediment passes out into the container. The container is then removed. In principle no sedimenting fluid is lost, but usually a small initial loss occurs until a partial vacuum forms at the top of the column.

The preceding systems have the advantage of cheapness, but they are subject to errors because the tapered walls affect sedimentation and particles adhere to the tapered walls.

Manometric methods have been used to assess the sedimentation. A manometer fused into the side of the column near the base and filled with the clear fluid will register the change in mean pressure of the column with time; this of course depends on the mass of material still in suspension. Pressure differences are very small.

Beta particle backscattering has been used to measure the mass of material at the base of a column. The intensity is proportional to the atomic number and the thickness, that is, the weight of sediment, provided that the thickness does not build up to saturation level.

Decanting If a homogenous fluid/sample mixture is allowed to settle for time t_1 seconds and the fluid down to a depth h is then removed using a pipette, particles removed would all have a terminal velocity less than h/t_1. In the decanting method, this process is repeated several times, replacing the removed fluid with clear fluid and remixing until the supernatant fluid is clear. The process is then repeated but with a shorter time, t_2. The removed fluids are analyzed for dust content, each containing the fraction of the total mass of material of terminal velocity between h/t_n and h/t_{n-1}. The accuracy depends on the precision of h, t, the rate of removal of fluid, and the number of repeated decantations at each value of t, so it is not high.

Two-layer Methods If the upper layer is significantly thinner than the lower layer at time $t = 0$, then after a time $t = t_1$, the only material to have reached the base of the column will be those particles with terminal velocity greater than h/t_1, where h is the height of the column, and a measurement of the weight of those particles gives the cumulative distribution directly. Liquid two-layer methods are not common because of the difficulties of arranging the starting condition. The Granulometer uses a shutter system to maintain separation until $t = 0$ and a manometer to measure the pressure changes as particles sediment.

The Sharples Micromerograph uses an air column approximately 2 m tall. The sample is injected using a puff of compressed nitrogen between two concentric cones to disperse the particles and break up the agglomerates. An electronic balance coupled to a pen recorder monitors the arrival of particles at the base of the column. The column is jacketed to maintain thermal stability. Errors are experienced due to the loss of some fine material to the walls of the column. These can be reduced to some extent by antistatic treatment. The time scale of the recorder is adjustable—fast for the initial phase, slow for the later phase of the sedimentation. With friable particles care has to be exercised at the injection point.

Centrifugal Methods For small particles, gravitational systems are very slowacting. There is also a lower limit to the size of particle that can be measured because of the effects of Brownian motion and convection. Although it is possible to use gravitational systems for particles as small as 1 μm in water and about 0.5 μm in air, in practice the lower limit is usually taken to be about 5 μm. These problems can be reduced by the use of centrifugal enhancement of the settling velocity.

The theory for centrifugal settling from a homogeneous fluid mixture is complicated in particular because the terminal velocity varies with distance from the center of rotation. Approximations have to be made, limiting the usefulness of the techniques.

The theory for two-layer systems is exact; therefore these are more attractive. Unfortunately, a phenomenon known as *streaming* can occur, in which particles tend to agglomerate and accelerate in a bunch instead of individually, behaving as a large particle; this renders the results useless. Streaming has been prevented by using extremely low concentrations of particles in the starting layer. A technique using a third, intermediate layer of immiscible fluid has also been used successfully.

Theories always depend on the applicability of Stokes's law to the particle motion, which imposes a restriction on the maximum particle size, depending on the speed of rotation. Further problems exist for larger particles with respect to the acceleration and deceleration of the centrifuge.

In spite of the many problems, the techniques have advantages for very small particles, as small as 0.01 μm, and systems are available that paralleling many of the gravitational techniques. The most promising appear to be the three-layer methods and the use of optical detection devices.

16.8.2 Elutriation

A group of particles suspended in a fluid that is moving upward at a velocity V will undergo separation—those particles with a terminal velocity less than V traveling upward, the others settling downward. This process is called *elutriation*. The fluid is usually water or air, depending on the particle sizes. Strictly, an elutriator is a classifier rather than a particle sizer. It divides the group of particles into those above and those below a given cut size. In the perfect system, the fluid would clarify except for the few particles with terminal velocity exactly V, and the settled particles could be removed and would contain no undersized particles. The system, however, is not perfect. Proper Stokesian conditions do not exist because practical considerations make the tubes too short compared with their diameters. Furthermore, the velocity at the cylinder walls is considerably less than at the center, causing a circulation of some particles up the center and down the walls. The cut point therefore is not sharp. A multiple elutriator consists of a number of elutriators connected in series but with increasing diameters. Each section the refore collects progressively smaller particles.

Both water and air elutriators are commercially available. In use, the sample is introduced into the smallest section and the fluid velocity set to a preset value. When the fluid has cleared in all the sections, the system is switched off and the sediment in each weighed. A filter usually terminates the last section. In some designs, the separating section is short and is followed by a conical reducing zone to hasten removal of the undersize particles to the next section.

The elutriators described so far are counter-flow elutriators. Acceleration (i.e., gravity) and drag are in opposite directions. Elutriators have been designed for transverse flow. If the sample of articles is introduced slowly into the upper level of a horizontal laminar stream of fluid, two-layer sedimentation takes place, but the sediment is separated horizontally, different-sized particles following different trajectories. The particles are collected on judiciously placed preweighed sticky plates.

16.8.2.1 Centrifugal Elutriation

Elutriation is also used in the centrifugal field, usually with air as the fluid. In principle, air with entrained particles travels inward against the centrifugal force. Particles with sufficiently low terminal velocities also pass inward; the rest pass outward.

The natural flow of air in a spinning system is radially outward. Therefore, to obtain a counterflow, the air must be driven. In some systems a separate pump is used. In others, the air is introduced at the center, passes radially outward under centrifugal force, turns 180° to travel radially inward for a short distance, on the order of half a radius, and finally passes through another 180° to pass to the circumference. Elutriation takes place in the inward-flowing section. In this case, no pump is necessary, because the net airflow is outward.

Adjustment of either rotation speed or air velocity will affect the cut size. Air velocity is usually set by a throttling mechanism.

A variety of centrifuge systems is available, their design and size making them particularly suitable for different size ranges. Some, for example, are very sensitive in the range 2–12 μm; others are better at larger sizes. Some are capable of a continuous and large throughput that makes them especially suitable for producing quantities of closelysized particles. The devices are then used for their original purpose, which is classification rather than size grading.

A cyclone is a centrifugal device normally used for extracting dust from carrier gases. It consists of a conically shaped vessel. The dusty gas is drawn tangentially into the base of the cone, takes a helical route toward the apex, where the gas turns sharply back along the axis, and is withdrawn axially through the base. The device is a classifier in which only dust with terminal velocity less than a given value can pass through the formed vortex and out with the gas. The particle cut-off diameter is calculable for given conditions. Systems have been designed using a set of cyclones in series with increasing gas velocities for size analysis.

The Cyclosizer Analyzer uses liquid instead of gas and has five equal-sized cyclones with different inlet and outlet diameters to obtain the velocity variation. The cyclones are arranged apex uppermost. Thus the coarse particles retained in a given section fall back many times for reclassification, thereby obtaining good separation. In this case the range is 8–50 μm.

16.8.3 Impaction

When a fluid containing a suspension of particles is made to turn a corner, particles with terminal velocity in excess of a value determined by the fluid velocity and the geometry of the bend are deposited or impacted. A cascade impactor consists of a series of orifices, each accurately positioned above a collector plate. The orifices can be round holes or slots. The holes in successive stages are reduced in size to increase the impaction velocity. The particles pass through the holes and are either deposited on the adjacent plate or passed on to the next stage. There are typically between 6 and 10 stages covering an aerodynamic size range from about 0.4 μm to 15 μm.

The Andersen cascade impactor is designed to work "online," incorporating a nozzle for isokinetic sampling from a gas stream. A precyclone removes particles >15 μm and a filter catches those <0.4 μm. The Sierra, designed for room or atmospheric air measurement, covers a range 0.05 μm to 10 μm. The collected particles are removed and weighed. California Measurements Inc. markets an instrument with piezoelectric crystal mass monitors at each of 10 stages giving immediate automatic readout.

Impaction surfaces frequently require the aid of an adhesive to prevent re-entrainment from one stage to the next.

16.9 ANALYSIS METHODS THAT INFER SIZE FROM SOME OTHER PROPERTY

The methods discussed so far either measure size directly or measure a fluid-dynamic effect dependent on the particle terminal velocity, which, although dependent on size, is also affected by density, shape, and surface structure. The following methods do not measure size directly nor are they dependent on terminal velocity.

16.9.1 Coulter Counter

The Coulter counter uses the principle that the electrical resistance of a conducting liquid is increased by the addition of an insulating material. Particles are assessed individually. To obtain adequate sensitivity, the volume of liquid measured must be similar to the volume of the particle.

These criteria are achieved by containing the electrolyte in two chambers separated by a narrow channel containing an orifice, the dimensions of which are accurately known. An electric current from a constant-current source passes through the orifice from one chamber to the other. The voltage across the orifice is therefore directly proportional to the resistance of the orifice.

The sample, suitably dispersed, is placed in one of the chambers. An accurately controlled volume of the well-agitated electrolyte is then passed through the orifice. The concentration of the sample (on the order of 0.1 percent) is such that particles pass through individually. Each particle causes a voltage pulse, and a pulse-height analyzer increments one of a set of counters, each representing a size maximum.

The theory of the Coulter counter is complicated, particularly for randomly shaped particles, but it has been shown that to a first approximation, the pulse height is directly proportional to particle volume, errors being less than 6 percent when the particle size is less than 40 percent of the orifice diameter. This size limitation also represents a reasonable practical limitation to avoid blockage of the orifice. Although the resistivity of the particles should affect the result, in practice surface-film effects make this insignificant. The method also works with conducting particles.

The lower limit on particle size is set by the electronic noise in the circuit and in practical terms is usually taken to be about 4 percent of the orifice diameter. Orifices are available ranging in size from 10 μm up to 1 mm, giving a particle-size range (using different orifices) from 0.4 μm to 400 μm. Samples containing a wide range of sizes need to be wet-sieved to remove those larger than 40 percent orifice size. With small sizes, as always, care has to be exercised to avoid contamination. Bubbles can cause false signals. Aqueous or organic electrolytes can be used. It is usual to calibrate the instrument using a standard sample of latex polymer particles. The technique permits several runs on each sample/electrolyte mix, and it is easy to test the effect of changes of concentration, which should be negligible if dilution is adequate.

Models are available with 16-channel resolution and the output can be in tabular form, giving particle frequency volume or particle cumulative volume, or can be in the form of an automatic plot.

16.9.2 Hiac Automatic Particle Sizer

The Pacific Scientific Hiac-Royco analyzer, now manufactured by a Danaher division, can be considered the optical equivalent of the Coulter. In this device, the cylindrical orifice is replaced by a two-dimensional funnel-shaped orifice that guides the particle stream through a very narrow light beam located at its throat. A sensor measures the obscuration caused by each particle as it passes through. The responses are proportional to the particle cross-sectional areas and are sorted by a pulse-height analyzer. A range of orifices is available to suit particle sizes from 2 μm up to 9 mm, each covering a size range in the ratio 30:1. Although the measurement on a given particle is along one axis, an irregularly shaped particle will not tend to be oriented in any particular way, so statistically the area measured will be a mean cross-sectional area. The optical method has an advantage over the conductivity method in that it can operate in any liquid or gas provided it is translucent over the very short optical path length (typically 2–3 mm of fluid). Scattering-coefficient problems are reduced by calibration using standard samples. The instrument has been used "online" to measure the contamination of hydraulic fluid. The number of particles in a fixed volume is found by counting while timing at a constant flow rate.

16.9.3 Climet

The Climet method involves measuring the light scattered from individual particles that are directed accurately through one focus of an elliptical mirror. Light is focused onto the particles, and it is claimed that about 90 percent of the scattered light is detected by a photomultiplier at the other focus of the ellipse. Direct light is masked. The response is pulse-height analyzed.

16.9.4 Adsorption Methods

In some processes, a knowledge of the surface area of particles is more important than the actual size. Optical techniques may not be appropriate, since these tend to give the "smoothed" surface area rather than the true surface area—including roughness and pores. Some gases *adsorb* onto the surface of substances, that is, they form a layer that may be only one molecule thick on the surface. This is not to be confused with *absorption*, in which the substance is porous and the gases penetrate the substance. The subject is too large for a complete treatment here. The principle is that the particles are first "cleaned" of all previous adsorbates and then treated with a specific adsorbate such as nitrogen, which is then measured by difference from the adsorbate source. If the layer is monomolecular, the surface area is readily calculated; corrections have to be applied because this is generally not the case. Results are given in terms of surface area/volume or surface area/mass.

REFERENCES

Allen, T., *Particle Size Measurement*, Chapman & Hall, London (1981).
British Standards, *Methods for the Determination of Particle Size Distribution*, BS 3406 Parts 1–7.
Hinds, W. C., *Aerosol Technology: Properties, Behavior and Measurement of Airborne Particles*, Wiley, Chichester, U.K. (1982).
Silverman, L., Billings, C. E., and M. W. First, *Particle Size Analysis in Industrial Hygiene*, Academic Press, New York (1972).
Willeke, K., and P. Baron (eds.), *Aerosol Measurement: Principles, Techniques and Applications*, Van Nostrand, New York.

FURTHER READING

Barth, H. G., *Modern Methods of Particle Size Analysis*, Wiley, Chichester, U.K. (1984).
Berhardt, I. C., *Particle Size Analysis: Classification and Sedimentation Techniques*, Chapman & Hall, London (1994).
Hireleman, E. D., W. D. Bachalo, and P. G. Felton (eds.), *Liquid Particle Size Measurement Techniques, Vol. 2*, American Society for Testing and Materials, Philadelphia, PA (1990).
Syvitski, J. P. M. (ed.), *Principles, Methods and Applications of Particle Size Analysis*, Cambridge University Press, Cambridge (1991).

Chapter 17

Fiber Optics in Sensor Instrumentation

B. T. Meggitt

17.1 INTRODUCTION

Fiber optic instrumentation has progressively evolved since early 1980 following the successful development of optical fiber networks for telecommunications applications. To a large degree, optical fiber sensor instrumentation has taken advantage of the maturing component market for telecommunications use, and its progress has been dictated by it. Initially a significant level of research effort was put into developing sensor systems based on the use of multimoded optical fibers, but as the bias in long-haul telecommunications moved toward the use of single-moded optical fibers for improved performance, so optical fiber sensor systems began to accept the use of single-mode techniques as well as multimode fiber methods. In addition, the development of compact disc (CD) systems has made available low-cost, high-reliability laser diode devices (both single-mode and multimode) that have added further to the advancement of optical fiber sensor systems. The initial research work carried out on optical fiber sensor instrumentation included a wide range of ideas and methods, many taken over from more traditional open path optical-sensing methods.

By the end of the first decade of research into this new sensing method, the ideas and techniques followed a degree of rationalization as the advantages of certain methods became favored for specific applications (see, for example, Grattan and Meggitt, 1994). The early optimism over the introduction of this new measurement method did not initially live up to the earlier expectations. The subject was then largely technology driven, with factors that make for a successful commercial development being largely overlooked. The conservative nature of users of the existing electronic and electrical sensor systems meant that optical fiber sensor systems, despite their advantages over the electronic rivals, had to compete on a level playing field; they had to compete on price maintainability and reliability as well as performance criteria. The products therefore had to display credible advantages over competing systems or had to be able to perform certain measurement feats that were not possible with the techniques already available.

Despite the conservative attitude shown to the early work on optical fiber sensors, however, some commercial instruments did emerge for specialist applications, and this has led, or is leading, to internationally accepted measurement techniques. Some of these methods, as will be seen, are specific for a single measurement parameter (e.g., vibration, acoustic sensing, etc.), whereas others offer the credibility of being developed into a generic measurement technique that can be used to monitor a wide range of parameters (e.g., temperature, pressure, or strain).

When considering design options to adopt for a particular sensor system, several performance-and market-related factors need to be taken into account. These can be summarized as:

- Resolution requirement.
- Dynamic range capability.
- Response time requirements.
- Operating environment conditions.
- Reliability.
- Long-term stability.
- Production capability.
- Maintenance requirements.
- Target costs.
- Market acceptability.

In reality, the chosen method will lead to a compromise on these criteria, depending on the most dominant requirement factors. One should not underestimate the final two items (cost and market acceptance), since these have been two of the most overriding factors that have slowed the general acceptance of optical fiber sensor systems. However impressive the operating performance and technology design behind a sensor system, the end users will have their own unique set of selection criteria, and unavoidable cost of ownership (both capital and subsequent maintenance costs) will play a leading role in this process. Many large end users have remained conservative in their approach to new sensor technology. They often will not be prepared to take unnecessary risks on (to them) unproven techniques, bearing in mind that systems deployed today may be expected to last for the next 20 years with only a minimum degree of maintenance. Despite this conservatism, optical fiber sensors are progressively finding new application areas due to their unique performance advantages.

17.2 PRINCIPLES OF OPTICAL FIBER SENSING

17.2.1 Sensor Classification

Basically, optical fiber sensors consist of an optical source coupled to an optical fiber transmission line that directs the radiation to a sensor head, as shown in Figure 17.1. The light is then returned after being modified in some way by the sensor interaction via the optical fiber, in either a reflective or transmission mode, to a photodetector and subsequent electronic processing system. The signal processing system then detects, demodulates, and analyzes the changes introduced in the optical signal by the sensor head and then relates this to a change in the measurand field of interest.

The sensor head can be either a *point sensor*, shown in Figure 17.2(a), which makes the measurement in a localized region in space, or it may be a *distributed sensor*, shown in Figure 17.2(b), which has the ability to make measurements along a length of the optical fiber. The sensor head may also be either an *extrinsic sensor* to the fiber and consist of bulk optical components configured into a sensing mechanism, or it may be an *intrinsic sensor* to the fiber, where the measurement process takes place within the optical fiber medium. In the former case the optical fiber forms the function of transmitting the light to and from the sensor head; in the latter it actively takes part in the sensing process where the inherent optical properties of the fiber are utilized.

In addition, the optical fiber can be either single mode or multimode, the choice depending largely on the application and the measurement method being employed. In general, single-mode fiber is used for intrinsic optical fiber sensors such as interferometric methods, whereas multimode fibers tend to be used in extrinsic sensor systems and transmit the light to and from the sensor head. However, there are some notable exceptions to this generalization—microbend sensor systems, for example.

17.2.2 Modulation Parameters

Fiber optic sensors operate by the modulation of an optical carrier signal by some optical mechanism present in the sensing region that is itself responsive to the external parametric measurand field. Subsequent signal processing of the modulated carrier signal then relates these changes to variations in the measurand field of interest. There are a limited number of such possible optical properties that can be modulated in an optical sensor system. These can be identified as follows:

- Intensity modulation.
- Wavelength/frequency modulation.
- Temporal modulation.
- Phase modulation.
- Polarization modulation.

All these optical modulation parameters are well known in optical metrology. The purpose in optical fiber sensors is to adopt and extend such methods for use with the optical fiber medium.

The first method in the list, intensity modulation, is perhaps the simplest technique to consider for optical fiber use (see, e.g., Senior and Murtaza, 1989). However, since there are many processes in an optical fiber network, such

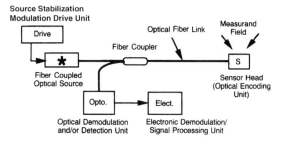

FIGURE 17.1 Schematic diagram of general optical fiber sensor system operating in the reflective mode.

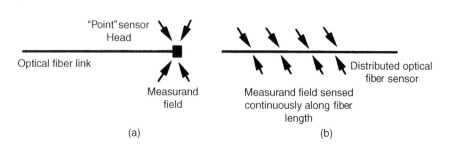

FIGURE 17.2 Illustration of (a) a "point" optical fiber sensor and (b) a distributive optical fiber sensor system.

as coupler loss and bend loss, that can also modulate the intensity of the transmitted radiation, it cannot be used directly without providing an additional processing technique that can unambiguously identify the intensity changes induced solely by the measure and interaction of interest. The most common method for providing this facility in the intensity-modulated sensor is to use a two-wavelength referencing technique (e.g., for gas sensing, see Bianco and Baldini, 1993). Here, two different wavelength sources are used (see Figure 17.3) whereby one, the signal channel, has its wavelength variably absorbed by the sensor interaction and the second channel acts as a reference wavelength that is unaffected by the sensing interaction and against which the sensing signal can be normalized for system losses. It is, however, necessary to ensure that the optical fiber link to and from the sensor head has the same spectral response to both beam wavelengths. This requires the two beams to have closely spaced wavelengths; these are usually produced from the spectral division of a single source element.

In principle, the other listed optical modulation techniques are immune to intensity modulation changes in the fiber link, provided that some form of temporal modulation is applied to the radiation and that the link does not introduce addition modulation effects.

17.2.3 Performance Criteria

The processing of signal and noise components of the modulated signals returned from an optical fiber sensor are similar to those commonly found in electronic sensor systems utilizing a reference measurement channel. These components are referred to as:

- Common mode modulation.
- Differential mode modulation.

The *common mode* signal is one that has the same noise spectrum for both output channels, the signal and reference channels, and can be eliminated by ratioing of (dividing) the two signals once they have experienced the same system gain. The *differential mode* output, on the other hand, will reflect the changes that have been induced by the sensor head on the signal output only, with the reference beam ideally being unaffected by the sensor changes. Therefore, the division process that removes the common mode signal effects will leave unaltered the differential mode effects. The residual noise output from this processing stage will reflect the degree to which the noise spectrum is common to both signals and the degree to which the reference signal is unperturbed by the sensing mechanism.

As mentioned earlier, sensor systems each have their own related operating performance factors. These can be identified as those factors that help define a measurement point; they include:

- Resolution, responsivity, sensitivity.
- Accuracy.
- Reproducibility.

These are distinct performance-related factors and not merely different ways of expressing the same quantity. Two other important performance criteria are:

- Dynamic range.
- Low cross-sensitivity to other parameters.

Sensor cross-sensitivity relates to the fact that sensors are usually required to measure one specific variable parameter in the presence of other changing conditions such as pressure in the presence of temperature and vibration and so on; minimizing such undesirable effects is important in any sensor design exercise.

17.3 INTERFEROMETRIC SENSING APPROACH

The use of interferometric techniques in optical fiber sensing has become a well-established and one of the most widely

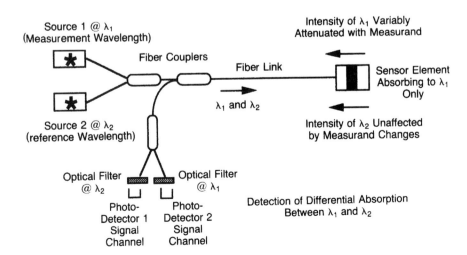

FIGURE 17.3 Typical configuration for a two-wavelength intensity modulated optical fiber sensor system.

used methods, since they can be applied to a large range of measurement parameters providing both a high resolution and large dynamic range capability. Initially, single-mode fiber interferometric systems were demonstrated (e.g., Jackson et al., 1980; Giallorenzi et al., 1982), but more recently multimode methods have also shown themselves to be equally viable (e.g., Boheim, 1985) under the right operating conditions. There are three basic modulation methods suitable for interferometric sensor systems that can produce a carrier signal, the phase of which is then used to monitor for optical path length changes in the associated sensing interferometer. These methods are *heterodyne techniques,* which are useful with gas laser sources; *pseudoheterodyne techniques* for use with injection-current modulated single-mode laser diode devices; and *"white-light" interferometric techniques* for use with low-coherence broadband sources such as light-emitting diode (LED) and multimode laser diode devices.

17.3.1 Heterodyne Interferometry

In this technique two (or more) laser sources are used such that interferometric mixing occurs to produce a heterodyned output carrier signal. The laser sources can either be separate devices having narrow and stable spectral lines, or they can be the same source beam (see Figure 17.4) that has been amplitude divided with one beam then frequency shifted, for example, by an acousto-optic Bragg cell modulator (see, e.g., Tsubokawa et al., 1988; Meggitt et al., 1991). For source wavelengths of λ_1 and λ_2, with a wavelength separation of $\Delta\lambda$, the associated beat frequency Δf_{Het} between the two optical frequencies is then given by:

$$\Delta f_{Het} = (f_1 - f_2) = \frac{\Delta\lambda}{\lambda_1 \lambda_2} c \quad (17.1)$$

and where c is the velocity of light. In the case where the two wavelengths are closely spaced, such as in the use of a Bragg cell element, the associated heterodyned wavelength (or synthetic wavelength) of the beat frequency is given by:

$$\lambda_{Syn} = \frac{\lambda^2}{\Delta\lambda} \quad (17.2)$$

For the case mentioned previously, where a Bragg cell element introduces a relative frequency shift of, say, 80 MHz between the two beams (typically between 40 MHz and 1 GHz), the associated synthetic wavelength caused by the beating between the beams is 3.75 m in air. It is also possible to down-convert the RF output signal (MHz) to a lower and more manageable intermediate frequency (kHz) by mixing the output signal in a double-balanced mixer with a stable RF oscillator having a slightly differing frequency. Mixing the 80 MHz carrier with a 80.002 MHz local oscillator, for example, will produce a carrier signal at 2 kHz. Using APD photodetectors it is also possible to carry out this down-conversion within the detector device itself by applying the mixing frequency directly onto one terminal of the detector while the device actively detects the RF modulated photo signal. This type of mixing is particularly useful in considering long path length interferometers such as free-space ranging devices and Doppler anemometers, used for vibration analysis, where optical fibers can be conveniently used in aspects of the signal processing.

17.3.2 Pseudoheterodyne Interferometry

Pseudoheterodyne interferometry is an interferometric technique that utilizes an optical source with its emitted radiation frequency modulated. This can either be in the form of a direct frequency modulation of the source output wavelength, by modulation of its drive injection current, or by use of a fixed frequency source but with a Doppler frequency shift introduced by reflection from an oscillating mirror element. The former case has been treated extensively in optical fiber interferometric sensors (e.g., Dandridge and Goldberge, 1982; Kersey et al., 1983). A semiconductor laser diode device has its output frequency modulated by changes in either its drive current or the device temperature and experiences a frequency shift of about 3 GHz/mA for current changes and about 0.25 nm/°C variation for temperature changes. It is usual to stabilize the device temperature by mounting the device on a Peltier unit, with a thermistor-based feedback control circuit giving temperature stabilization down to typically 1/100°C (about 2.5 pm). When a serrodyne current ramp (a sawtooth ramp with fast flyback) is applied to the laser diode, the output optical frequency of the device follows a similar modulation and can be conveniently coupled into a single-mode optical fiber core with typically 10–20 percent launch efficiency.

A typical optical fiber pseudoheterodyne sensing scheme is shown in Figure 17.5, which shows a reflective Michelson, but a Fabry–Perot or Mach–Zehnder interferometer can also

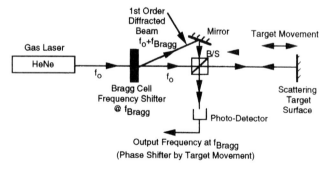

FIGURE 17.4 Illustration of a typical heterodyne processing scheme for generating an intermediate carrier signal in a displacement measurement device.

CHAPTER | 17 Fiber Optics in Sensor Instrumentation

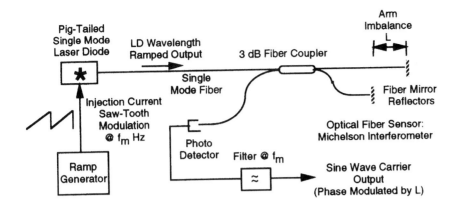

FIGURE 17.5 Optical fiber Michelson interferometer illustrating a pseudo-heterodyne modulation scheme using an unbalanced interferometer (Cork et al., 1983).

be used. The method has been the subject of much early work on optical fiber interferometric sensors. Examples of its use are in the all-fiber Mach–Zehnder hydrophone (e.g., Yurek et al., 1990) and a Michelson fiber interferometer for quasi-static temperature measurement (Cork et al., 1983). In this method it is necessary to provide an arm imbalance L in the fiber interferometer to facilitate the signal processing and to produce the required carrier signal for monitoring the optical phase changes induced in the sensing interferometer. For an optical path imbalance (nL) in the Michelson interferometer, where n is the fiber core refractive index and with a wavelength sawtooth ramp of 10 GHz frequency modulation ($\Delta\lambda = 0.02$ nm in wavelength) representing a 0.003 percent depth of modulation, the corresponding change in output optical phase $\Delta\phi$ is given by:

$$\Delta\phi = \frac{4\pi n L}{\lambda^2}\Delta\lambda \quad (17.3)$$

For a 2π change in output phase per wavelength ramp of the laser diode, the output interference signal from the sensor will transverse one complete fringe. It can be seen from the preceding equality that, for the parameters given, the imbalance length of the interferometer needs to be about 10 mm or greater. Smaller cavity lengths are possible but require larger peak wavelength modulations and hence heavier current modulations, and in any event the limit will be a few millimeters. A drawback in applying the current modulation is that not only does the process induce the required wavelength modulation, it also induces unwanted intensity modulation across the output waveform (at the same frequency) that will need to be compensated for in the processing electronics. Although this method has seen successful application in the sensing of dynamic parameters, notably for acoustic detection in hydrophone developments, its use in the monitoring of low-frequency, quasi-static measurands such as temperature, pressure, and strain fields has seen less progress.

This lack of progress results from a serious limitation of the current modulation method, which requires a high degree of wavelength stabilization of the laser diode source. For example, even without the necessary wavelength modulation, there is a stringent requirement on the degree of wavelength stabilization needed. To provide a δN resolution in the interferometric fringe, the required wavelength stability $\delta\lambda$ of the source wavelength is related to the cavity length L by:

$$\delta\lambda = \frac{\lambda^2}{2nL}\delta N \quad (17.4)$$

For a modest 1/100 of a fringe resolution and a 1 cm arm imbalance, the required stabilization in the central wavelength is therefore about 2×10^{-13} m. Considering that the variation in source wavelength is about 0.02 nm/mA and about 0.3 nm/°C, this implies a control in the drive current of $<10\ \mu A$ and at a temperature of $<1/1{,}000°C$. These represent demanding control conditions on the laser diode source, especially when it is realized that the drive current of the laser diode needs to be linearly modulated over a 2×10^{-2} nm range at the same time as its central wavelength is stabilized to the indicated levels. Attempts have been made to stabilize the central wavelength of the source to atomic absorption lines (see, e.g., Villeneuve and Tetu, 1987) and to a linearly scanned Fabry–Perot reference interferometer (Change and Shay, 1988). However, the stringent control qrequirements needed for this technique have led to the rise in an alternative interferometric method—that based on the "white-light" technique that makes use of a broad spectral bandwidth source. This method is considered in the following sections.

17.3.3 White-Light Interferometry

17.3.3.1 Introduction

"White-light" optical fiber interferometry has established itself as a powerful sensing technique in the development of a wide range of sensing systems. The method was initially confined to the use of single-mode optical fiber components (see Al-Chalabi et al., 1983). More recently this "white-light" or low-coherence interferometric method has attracted a broad interest (e.g., Bosselmann and Ulrich, 1984; Mariller and Lequine, 1987; Valleut et al., 1987) due to its ability to overcome some of the major limitations in the use of single-mode laser diodes. These include a greatly reduced degree of wavelength stabilization of the source and the elimination of feedback problems in the lasing cavity, since the white-light system can operate with multimode laser diodes or LED devices. White-light interferometry is dependent on the relatively short coherence lengths of this type of source and operates by connecting the source, sensing interferometers, and processing system via an optical fiber network to establish a remote sensing device.

17.3.3.2 Temporally Scanned Method

To conveniently measure optical phase changes introduced in the sensing interferometer, a carrier signal is produced by modulation of the path imbalance in a second processing interferometer, as described shortly. In the conventional white-light interferometric method this is carried out by the periodic displacement of one interferometer mirror in a linear ramp fashion by a piezoelectric modulator, therefore producing a sinusoidal fringe output signal. This type of system is classified as the *temporal fringe* method (e.g., Boheim and Fritsch, 1987). A major advantage of the white-light interferometric technique is its relative insensitivity to wavelength fluctuations of the source.

Basically, the white-light technique has a sensing interferometer and a reference or processing interferometer along with a broad spectral bandwidth source such as an LED, a multimode laser diode, or a superradiant source (see, e.g., Ning et al., 1989). The broadband source is launched into the core of the optical fiber and transmitted to the sensor head interferometer, as shown in Figure 17.6(a). The path imbalance of the sensing interferometer, L_1, is made a distance greater than the coherence length, l_c, of the source radiation such that light reflected back from the two cavity mirrors does not interfere but is passed down the connecting fiber leads and via a directional coupler into the processing or reference interferometer cavity. The imbalance, L_2, of this interferometer is made comparable to that of the sensing interferometer and necessarily within the coherence length of the source, $\pm l_c/2$. Here part of the radiation is brought back into temporal coherence by the interaction of the two interferometers, and the resulting interference signal is detected on the output photodetector. Modulation of either the sensor or processing interferometer path imbalance will lead to a modulation in the phase of the output interference signal. A carrier signal is then introduced on the output signal by a serrodyne displacement modulation on one mirror of the reference interferometer. Using a piezoelectric transducer, the mirror is linearly displaced over ½ wavelength range (with fast flyback) such that the output interference pattern is driven over one complete interferometric fringe, as shown in Figure 17.6(b) (Meggitt, 1991).

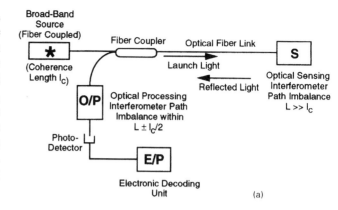

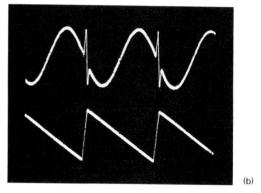

FIGURE 17.6 (a) Basic configuration of a "white-light" interferometric sensor system using two coupled interferometers. (b) Output temporal fringe pattern (lower trace) from a "white-light" interferometric sensor system using PZT sawtooth displacement ramp of $\lambda/2$ (upper trace) (Meggitt, 1991).

The phase of this induced carrier signal is then modulated by optical path-length changes in the sensing interferometer in response to changes in the measurand field. The characteristic equation of the processing system is then:

$$I_{o/p} = I_o \left\{ 1 + \exp\left(-\left[\frac{\sigma(nL_1 - nL_2)}{2}\right]^2\right) \right. \\ \left. \times \cos\left(\omega_c t - \frac{4\pi(nL_1 - nL_2)}{\lambda_o}\right) \right\} \quad (17.5)$$

where σ represents the half-width of the source spectrum at which the optical power falls to $1/e$ of its maximum value at k_o, and ω_c is the induced carrier signal. The exponential term represents the visibility of the fringes; in white-light interferometry this factor is usually between ½ and ⅓ in fringe modulation depth. It is seen that the phase of the cosine fringes is a function of the path difference between the imbalance of the sensing and reference interferometers. Since the coherence length of a low-coherence source is typically between 20 and 50 μm, it is seen that the interferometer imbalance in the sensor can be less than 0.1 mm and, therefore, represents a near-point sensing device. In addition, it is the difference in the path-length imbalance between the two interferometers that is sensitive to fluctuations in the source central wavelength. This quantity will necessarily be less than the coherence length of the source (<100 μm) and typically less than 10 μm. Therefore, it is seen that, compared with the wavelength modulated pseudoheterodyne laser diode technique, the wavelength stability requirements will be in the ratio of their respective cavity imbalances, that is, <1,000 times. It is this factor that has largely been responsible for the intense recent interest in the low-coherence method. It should also be pointed out that unlike the current-modulated pseudoheterodyne method, the low-coherence method can operate, under suitable conditions, with multimode as well as single-mode optical fibers. The requirement here is that the light returned from the optical sensing interferometer satisfies the common mode condition such that both reflected beams from the two interferometer mirrors follow exactly the same modal path when transmitted back through the multimode optical fiber; if not, sufficient mode mixing occurs in the fiber length to evenly distribute the back-reflected light into all the fiber modes (Chen et al., 1992a).

One of the first successful practical demonstrations of the capability of the white-light technique was reported by Boheim (1985). In this system a multimode optical fiber link was used with a Michelson reference interferometer and a low-finesse Fabry–Perot sensor cavity. Further analysis was then given in subsequent publications, and the system was later applied to a high-temperature sensor (Boheim and Fritsch, 1987) and a pressure-sensing device (Boheim et al., 1987). A similar approach using single-mode fiber techniques has been used by Meggitt (1991) in a prototype system in physiological applications for temperature and pressure measurement. The basic configuration used in the high-temperature sensor work (Boheim, 1986) is illustrated in Figure 17.7. Here, an 840 nm wavelength LED source is collimated and then passed through a Michelson reference interferometer before being coupled into an optical fiber directional coupler, the output of which feeds two interferometer cavities: one a reference Fabry–Perot cavity and the other the operating high-temperature Fabry–Perot sensor. The Fabry–Perot cavity sensor element was composed of a thin SiC film etalon (Boheim, 1986). The reference cavity is temperature controlled and its output interference signal is used in a feedback control to stabilize the mean imbalance of the Michelson processing interferometer. The piezoelectric modulator PZT1 drives one mirror of the Michelson in a serrodyne (sawtooth) waveform to scan a complete output interference sinusoidal temporal fringe in the outputs from both the cavity elements. By utilizing the temperature-stabilized reference sensor output and by interacting with PZT2, the feedback control circuit compensates for both thermal and vibration instabilities in the processing interferometer.

The sensor element in this work was separated from the fiber sensor lead by an air path, since the elevated temperature of 1,000°C used was well above the survival temperature of the fiber network. The collimated output from the fiber sensor lead was reflected back from the remotely placed SiC sensor etalon. The SiC etalon was 18 μm thick and deposited on a silicon substrate. The reflected radiation was relaunched back into the optical fiber and returned through a second directional coupler to an output photodetector. Phase movement in the output fringe pattern of the sensor cavity was a direct measure of temperature of the sensor etalon. Measurements were made over the temperature range 20–1,000°C and compared with a thermocouple device cemented to the SiC sensor etalon. With a 1 s measurement time constant, a long-term temperature resolution of 0.5°C was reported.

Meggitt (1991) has reported the development of a temperature sensor system using single-mode fiber techniques and a multimode laser source, a capacitive feedback stabilized low-finesse Fabry–Perot processing interferometer, and a miniature sensing head (1 mm diameter by 10 cm in length) designed for physiological uses, covering the range 20–50°C with a 0.01°C resolution. Use with a small pressure sensor head (3 mm diameter) was also reported for a pressure range of 250 torr, with a 1 torr resolution using the same processing unit.

In a second application, as a pressure sensor device, shown in Figure 17.8 (Boheim et al., 1987), the sensor

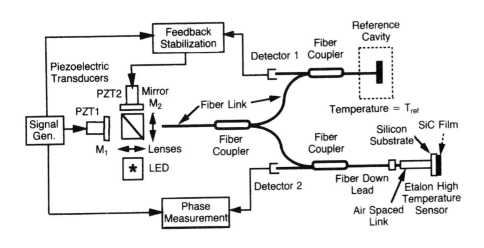

FIGURE 17.7 Low-coherence high temperature optical fiber sensor system (Boheim, 1986) using a SiC thin film etalon sensing element.

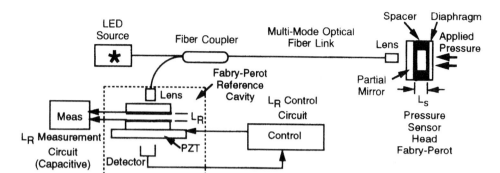

FIGURE 17.8 Optical fiber "white-light" interferometric pressure sensor (Boheim et al., 1987).

cavity used was composed of a low-finesse Fabry–Perot etalon with a first semireflecting mirror (reflectivity 0.06) and a second mirror formed from a deformable, polished steel diaphragm taken from a commercial pressure transducer with a reflectivity of 0.4 and a thickness and diameter of 2.5 mm and 2.5 cm, respectively. The pressure sensitivity of the diaphragm was −3.3 nm kPa; the mirror separation L_s in the sensor was 90 μm. The interferometer output function from the dual Fabry–Perot interferometer cavities is given by (see, e.g., Meggitt, 1991):

$$I_T = 2R\frac{(1-R)}{(1+R)^2} - 2\frac{(1-R)^2}{(1+R)^2} \sum_{m=1}^{\infty} R^{2m} \exp[-(\sigma m \varepsilon)^2] \cos(2k_o m \varepsilon) \quad (17.6)$$

where $\varepsilon = \varepsilon_{ac} + \varepsilon_o$ is the error term between the two cavities, $\varepsilon_o = (l_s - l_r)$ is the low-frequency term, and $\varepsilon_{ac} = \alpha \sin \omega t$ is the high-frequency term; α and ω are the modulation amplitude and frequency, respectively. Since low-finesse cavities are used in white-light interferometry with low R values (about 0.5), the exponential term decreases very rapidly and effectively only the $m = 1$ term is retained, giving a sinusoidal form for the interferometer output (similar to that given by the Michelson processing interferometer), the amplitude of which is proportional to exp $[-(\sigma \varepsilon)^2]$ and is a maximum at $\varepsilon = 0$.

The etalon gaps of the pressure cavity decrease proportionally with increased applied pressure. The reference cavity separation is adjusted by means of the PZT transducer to null the path length difference between them. The cavity length of the reference etalon is then determined by measurement of its effective electrical capacitance through the use of metal semireflective coatings used as the cavity's low-finesse mirrors. The signal processing in this sensor scheme has two components: first, the optical feedback control to keep the reference cavity balanced with the (variable) sensor cavity length, and second, the electrical measurement circuit to monitor the changing capacitance of the reference cavity.

Here a fixed amplitude sinusoidal oscillator at 70 Hz modulates the reference cavity length through the PZT transducer element. In addition, a low-frequency voltage is applied to the PZT element to control the cavity length, to balance it with the sensor cavity. This effective error voltage is produced through a monitoring circuit by processing the photodetector output in a sequence of amplification, high-pass filtering, rectification, peak following, demodulation of the instantaneous fringe waveform amplitude function, and, finally, a servo-control that integrates the demodulated output, thereby producing the PZT bias voltage to null the path difference between the two cavities. Measurement of the resulting reference cavity gap is by a voltage-controlled oscillator (VCO), the output frequency of which is applied across the reference cavity capacitance. Both this voltage and that of the VCO drive voltage are separately converted by RMS-to-DC elements; their difference is used by the servo-control device to maintain the impedance of the capacitor at a constant value by means of changing the VCO output frequency of the feedback circuit. The resulting VCO frequency is then directly proportional (via a scaling factor) to the reference cavity separation l_R.

Experiments on the performance of the system reported a pressure dynamic range of 0–3.8 MPa with a linear dependence of VCO output frequency, with applied pressure having a least-mean-square fit of 26.932 MPa − 1.02082 kPa/Hz × f_{VCO}, a deviation of 0.1 percent of full scale (3.4 kPa), and a short-term noise stability of about 3 kPa peak to peak. Reversible thermal effects were evident in the sensing and reference cavities, and at $P = 0$ these were reported as 1 kPa/°C and 12 kPa/°C, respectively.

17.3.3.3 Electronically Scanned Method

More recently, a second white-light sensing method has been established; it eliminates the use of the mechanical scanned interferometer mirror and is termed the *electronically scanned* system (Koch and Ulrich, 1990; Chen et al., 1990). This type of system therefore has no moving parts and offers a more rugged and stable configuration than the temporal domain approach. By slightly tilting one mirror of the processing interferometer and passing an expanded beam through it, a *spatial fringe* pattern is created in the output beam that is then imaged onto a linear CCD array device. With a suitable system design and with use of a low-coherence source, the imaged fringe pattern displays a Gaussian intensity profile that is localized about a limited region across the CCD array. The spatial fringe pattern envelope then moves across the array pixel structure with optical path-length changes in the sensing interferometer cavity. By tracking the center fringe of the interference pattern envelope, phase movements can be monitored in response to changes in the sensing environment. The system is shown schematically in Figure 17.9.

The processing interferometer can take on many forms by selecting different interferometer cavities for the processing interferometer. The most conventional choice is that of a Michelson processing interferometer configuration, since it is relatively straightforward to set up experimentally. Design of this interferometer will determine the achievable resolution and dynamic range of the sensor system through choice of the arm imbalance and tilt angle of the mirror. Figure 17.10 illustrates the geometry of the system. The optical path difference between the sensing and processing interferometers is given by:

$$\mathrm{OPD} = [(\delta_1 - \delta_2) + 2y\beta] \quad (17.7)$$

where δ_1 is the path imbalance of the sensing interferometer, δ_2 is the mean imbalance of the processing interferometer, β is the mirror tilt angle, and y is the distance along the CCD array. Because δ_1 varies with changes in the sensing interferometer, the matching balance point of the processing interferometer moves along the CCD array. For a broadband source having a Gaussian spectral profile of

$$I(k) = I_0 \exp\left[-\left(\frac{k - k_0}{\sigma/2}\right)^2\right] \quad (17.8)$$

where k_0 is the central wave number and σ represents the half-width of the spectrum $(k - k_0)$ at which the optical intensity falls to $1/e$ of the maximum value at k_0, the spatial interference pattern appearing across the CCD array is then expressed as

$$I_{y,\delta} = I_0\{1 + \gamma\xi v(\delta_1 - \delta_2 + 2y\beta) \\ \times \cos[k(\delta_1 - \delta_2 + 2y\beta)]\} \quad (17.9)$$

where γ is an additional term that corresponds to the *spatial coherence* across the beam width (varying between 0 and 1, decreasing with increasing β) and ξ is dependent on the *sampling factor* (N pixels/fringe) associated with the discrete nature of the CCD pixel array given by

$$\xi\left(\frac{P}{b}\right) = \sin c\left(\frac{p\pi}{b}\right) \text{ and } \xi \simeq 1 \text{ for } N > 10 \text{ pixels/fringe} \quad (17.10)$$

and $v(\delta_1 - \delta_2 + 2y\beta)$ is the visibility function associated with the *temporal coherence* properties of the source and is given by

$$v(\delta_1 - \delta_2 + 2y\beta) = \exp\left\{-\left[\frac{\sigma}{2}(\delta_1 - \delta_2 + 2y\beta)\right]^2\right\} \quad (17.11)$$

The fringe period will therefore have a spatial width b equal to

$$b = \left(\frac{\lambda_0}{2\beta}\right) \quad (17.12)$$

where λ_0 is the central wavelength of the source. For a CCD array that is composed of M pixel elements, each of width p, the number of pixel elements N per fringe (sampling factor) is given by

$$N = \left(\frac{b}{p}\right) = \left(\frac{\lambda_0}{2p\beta}\right) \text{ pixels/fringe} \quad (17.13)$$

A typical fringe pattern produced by this technique is illustrated in Figure 17.11, where the cosine fringes contained under the Gaussian intensity envelope can be seen.

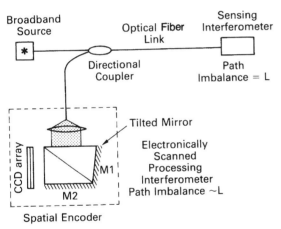

FIGURE 17.9 Schematic of the spatial-domain optical fiber "white-light" sensor system using a CCD array in the electronically scanned configuration (Chen et al., 1990).

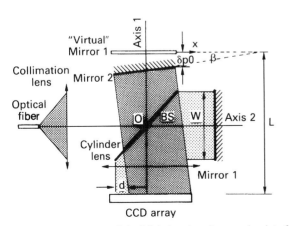

FIGURE 17.10 Geometry of the Michelson-based processing interferometer in the electronically scanned technique.

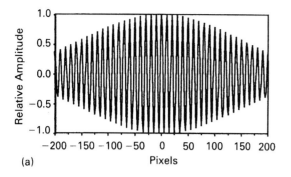

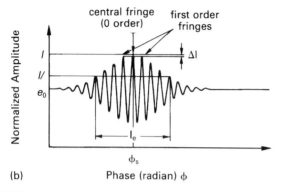

FIGURE 17.11 (a) Typical output interferogram for multimode laser diode or LED source, and (b) structural features of interferogram in "white-light" interferometry.

The dynamic range R of the sensing system is limited to the maximum scan range of the processing interferometer, is dependent on the tilt angle β and the array width (Mp), and is given by:

$$R \leq \left(\frac{2M\beta p}{\lambda_0}\right) \quad (17.14)$$

Clearly the greater the tilt angle β, the larger the dynamic range R. However, the larger β becomes, the smaller the fringe sampling factor N and, therefore, the fewer pixels used to define each interferometric fringe period b. Ordinarily, a sampling factor of >20 along with fringe-smoothing techniques will give a good fringe definition. When too few pixels sample each fringe, good definition would seem to be lost. However, an interesting case arises when N approaches 2 pixels/fringe. Under these conditions, a beating effect is observed between the discrete nature of the array elements and the periodicity of the interference fringe pattern. This effect gives rise to a moiré-type fringe pattern (Chen et al., 1990), as illustrated in Figure 17.13. The phase change observed in the moiré pattern correlates directly to that of the original fringe pattern. In addition, the moiré fringes are sharper than the straight interference pattern and it becomes easier, in principle, to track the phase changes in the pattern and to have knowledge of the fringe number by following the pattern centroid. From the earlier discussion, the

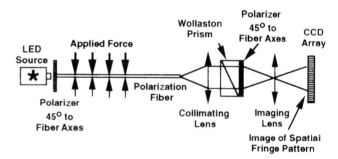

FIGURE 17.12 Noise-resistant, electronically scanned interferometric system for force measurement (Danliker et al., 1992).

Published work on the optical fiber "white-light" electronically scanned technique includes its application to a displacement sensor covering a range of 75 μm with a 0.02 μm resolution (Koch and Ulrich, 1990), a fiber optic pressure sensor (Valluet et al., 1987), a refractometer for measuring the refractive index of air (Trouchet et al., 1992), and the measurement of absolute force (Danliker et al., 1992). The latter work is of particular interest since it utilizes a solid construct for the reference interferometer design based around a Wollaston prism, as illustrated in Figure 17.12. It is inherently resistant to vibrational noise and easy to set up, although the thermal-induced refractive index changes inherent in such dielectric materials will be present, unlike in the Fizeau type interferometer (e.g., Chen et al., 1991), from which, due to the use of an air spaced cavity, such thermal effects are absent. Here (see Figure 17.12) the processing interferometer is formed by the optical path difference between the two orthogonally polarized output beams that occurs progressively across the Wollaston prism aperture. A dichroic polarized element oriented at 45° to the two polarization axes then induces interference between the two orthogonal components. The sensing element is formed in the intervening polarization-maintaining fiber where the applied force will induce a variable optical path-length shift between the two propagating eigenmodes, although in the reported work a Soleil-Babinet compensator was used to simulate this action.

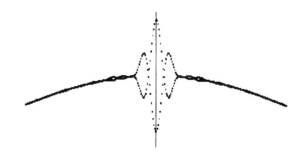

FIGURE 17.13 Moiré pattern from beating between an interferometric fringe pattern and a CCD pixel array structure for a large mirror tilt angle (Chen et al., 1990).

sampling factor affects the fringe pattern intensity through the factor $\xi(p/b)$. Expressing (p/b) as the factor K/N, where K and N are integers:

$$\xi\left(\frac{P}{b}\right) = \sin c\left(\frac{p\pi}{b}\right) = \frac{2}{\pi(2K+1)} \quad (17.15)$$

This leads to the description of the interference fringe pattern intensity function at each pixel number i being given by the expression

$$I(i) = p\left\{1 + \frac{(i)^i}{\pi(K+1/2)} \gamma v_i \cos\left[2\pi\left(\frac{2y}{\gamma_0} + \frac{1}{N}\right)\right]\right\} \quad (17.16)$$

Here N is the number of pixel elements per moiré fringe and y is the pixel distance along the CCD array. If we choose N (moiré) $= N$ (interferometric), as used previously, the dynamic range is increased by a factor of $N(K+1/2)$. Although the same degree of resolution is maintained, the fringe visibility is reduced by a factor of $\pi(K+1/2)$.

17.3.4 Central Fringe Identification

17.3.4.1 Introduction

Interferometry allows changes in the sensor measurand to be monitored through phase changes in the interference pattern, but there is a problem when it comes to making absolute measurements due to the fact that the cosine fringes are generally indistinguishable. Although in principle low-coherent techniques permit the identification of individual fringes through the Gaussian fringe envelope, it is not straightforward to implement this feature. Figure 17.11(a) shows a typical output fringe interferogram from the low-coherence sensor system. The fringe pattern will be similar for both the temporal and spatially scanned systems. It can be seen that using an LED (or multimode laser diode) source, it's not easy to track the central fringe of this profile, especially in the presence of excess noise, since there is a group of central fringes having similar amplitudes. The amplitude difference between the central and first neighboring fringe (Figure 17.11(b)) is given by:

$$\Delta I = \left[1 - \exp\left(\frac{2\lambda}{L_c}\right)^2\right] \quad (17.17)$$

where L_c is the coherence length of the source. When considering fringe identification methods it is necessary to ensure that they can operate with noise levels up to about 10 percent. We have recently demonstrated two methods for identifying the central fringe: the centroid method and the two-wavelength beat method (Chen et al., 1992b, 1993).

17.3.4.2 Centroid Method

The first of these methods (Chen et al., 1992b) identifies the central fringe by finding the *center of gravity* of the Gaussian fringe pattern envelope (see Figure 17.14). To identify the central fringe correctly, it is necessary to locate the centroid to $\pm 1/2$ of a fringe period of the center fringe. The fringe pattern is first "rectified" so that the lower negative-going fringe section is reflected back to produce all positive peaks (Figure 17.14, lower trace). Next each peak height is represented by a vertical line at the peak center. Using this set of discrete data points $(1, 2, ..., i)$, each of amplitude $a[i]$, the centroid or symmetric center (Figure 17.14, vertical line) of the fringe pattern can be computed using the algorithm:

$$C = \frac{\sum(i|a(i)|)}{\sum|a(i)|} \quad (17.18)$$

By computer simulation, it has been shown that this procedure can correctly identify the central fringe in the presence of 10 percent noise (signal/noise ratio -20 dB) with a certainty of 99.2 percent. For an increased signal/noise ratio of <-23 dB, this error rate falls to below 10^{-4}.

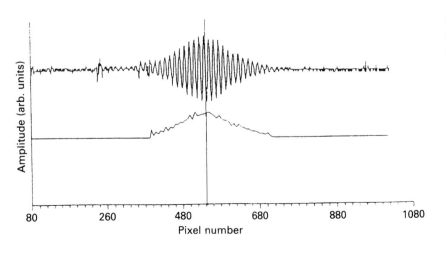

FIGURE 17.14 Central fringe identification method based on the centroid technique, from an electronically scanned system (Chen et al., 1992b).

17.3.4.3 Two-Wavelength Beat Method

The second method (Chen et al., 1993) uses two sources of widely spaced wavelengths, which both pass through the optical fiber sensor network. Each source will produce its own interferogram at the output of the processing interferometer, and each will have a different fringe period corresponding to the source wavelength. Since the two interference patterns are superimposed on each other, a modified pattern will result due to the fringe period beating effect. Although each source produces a fringe pattern of the form shown in Figure 17.11, the superimposed patterns will have the form (Figure 17.15):

$$I(\delta) = \exp\left[-\left(\frac{2\Delta\delta}{L_c}\right)^2\right]\cos\left(\frac{2\pi\Delta\delta}{\lambda_{sy}}\right)\cos\left(\frac{2\pi\Delta\delta}{\lambda}\right)$$

where

$$\lambda_{sy} = \frac{2\lambda_1\lambda_2}{\Delta\lambda}; \quad \lambda = \frac{(\lambda_1+\lambda_2)}{2}; \quad \Delta\lambda = (\lambda_1-\lambda_2) \quad (17.19)$$

In using the amplitude intensity threshold method to detect the dominant central fringe, it should be noted that the first fringe neighbor is not always the second most prominent fringe. It competes with the first synthetic wavelength fringe for this position, and the latter dominates as the wavelength difference $\Delta\lambda$ becomes large. The amplitude difference between the *central* and *first-order* fringe now becomes:

$$\Delta I = 1 - \exp\left[-\left(\frac{2\lambda}{L_c}\right)^2\right]\cos\left(2\pi\frac{\Delta\lambda}{2\lambda}\right) \quad (17.20)$$

for

$$I(\delta \pm \lambda) > I(\delta \pm \lambda_{sy}/2)$$

and when the *first synthetic wave* dominates:

$$\Delta I = 1 - \exp\left[-\left(\frac{2\lambda}{L_c}\frac{\lambda}{\Delta\lambda}\right)^2\right] \quad (17.21)$$

for

$$I(\delta \pm \lambda) < I(\delta \pm \lambda_{sy}/2)$$

In the latter case it can be seen that the central fringe is now enhanced by a factor of approximately $(\lambda/\Delta\lambda)^2$ over the single-wavelength case (e.g., about 40 for $\lambda_1 = 670$ nm and $\lambda_2 = 830$ nm) for large wavelength differences. In both cases a minimum signal-noise ratio of $-20\log(\Delta I/2)$ is required to identify the central fringe directly through its amplitude. The results are shown graphically in Figure 17.16: The upper curve represents the SNR_{min} of a single multimode laser diode and the four lower curves represent the SNR_{min} for the system using dual sources with $\Delta\lambda/\lambda$ ratios of 0.24, 0.19, 0.15, and 0.08, respectively. The shaded bands represent the coherence length ranges of typical types of low-coherence sources, that

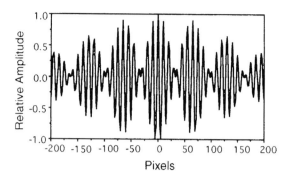

FIGURE 17.15 Modified interferogram produced by the two-wavelength beat-method, showing central fringe enhancement (Chen et al., 1993).

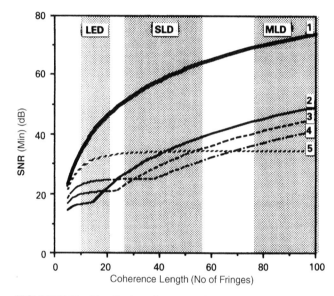

FIGURE 17.16 Signal/noise ratio versus coherence length: enhancement effect of the two-wavelength method for different source separations $\lambda\Delta$ (Chen et al., 1993).

is, LED, SLD, and multimode laser diodes. The discontinuity in the curves is a direct result of the competition between the first neighbor and the first synthetic fringes. The main attraction of this two-wavelength method is that, because the central fringe is the point at which both interferometers are in balance, the variations in the source wavelength will have a minimum effect on the measurement process, and therefore it is not necessary to provide a high degree of wavelength stabilization to use the method successfully.

17.4 DOPPLER ANEMOMETRY

17.4.1 Introduction

The use of optical fibers in anemometry applications provides two main operating advantages. The first is that it allows Doppler-type measurements to be made in remote and inaccessible areas. The second is that it allows measurements to be made in environments that are normally opaque to optical radiation, and the fiber probe can be introduced within the medium to make *in vivo* measurements. This latter point is particularly relevant to liquids for which laminar flow rates, for example, are difficult to measure across the flow area.

Here three specific applications of optical fibers in anemometry are described. The first is their use in the measurement of the *size of particles* in a fluid medium; the second considers the measurement of fluid flow velocities; and the third is the use of optical fibers in three applications to *vibration* monitoring.

17.4.2 Particle Size

This technique (Ross et al., 1978; Dyott, 1978) is a particularly interesting and useful application of optical fiber anemometry and relies on the Doppler shift introduced in radiation reflected from suspended particles moving under the influence of Brownian motion. It has been used in many applications: for sizing suspended particles in emulsions, oils, and bacteria as well as for determining the mobility of spermuzola in cattle breeding use. The fiber is introduced into the liquid medium containing the particles, and laser light is directed down the fiber such that it is both partially reflected from the fiber end face and backscattered from the particles in suspension (see Figure 17.17). The penetration depth of the radiation into the liquid will depend on the concentration of particles and the absorbency of the medium. Due to the numerical aperture of the fiber, the emitted cone of light will have a maximum distance of penetration before the backscattered captured light falls off appreciably. For a clear liquid, this is about 2 mm for a 90 percent reduction in capture efficiency. For higher attenuating solutions, this distance falls to 100 μm or less. The application of optical fiber anemometry to Brownian motion is significant since there is no net flow of the fluid medium. In contrast, in using such techniques in fluid flow applications, there is the additional complication that the fiber end face disturbs the flow velocities about the fiber tip, producing a stagnation region and thus giving erroneous results. When measuring, for example, fluid flow rates such as arterial blood flow, methods of overcoming this problem need to be addressed; these are discussed further later on.

Doppler anemometry relates the velocity v of a moving, scattering surface to the induced Doppler frequency shift f_d by the relationship

$$f_d = \frac{2vn_1}{\lambda_0} \cos\theta \quad (17.22)$$

where n_1 is the refractive index of the liquid medium. Due to the numerical aperture ($NA = \sin\theta_m$) of the optical fiber, there will be a range of angles up to which the light can enter the fiber given by θ_m. Consequently, there will be a spread in the Doppler frequency Δf_d of:

$$\frac{\Delta f_d}{f_d} = 1 - \cos\left[\sin^{-1}\left(\frac{NA}{n_1}\right)\right] \quad (17.23)$$

For an NA of 0.15 and when operating in a water medium ($n = 1.33$), the normalized frequency spread is 0.0065.

In considering the sizing of particles in the region of a 5–2000 nm diameter suspended in a liquid medium, the frequency spectrum produced from the motion of the particles under the action of Brownian motion can be monitored through the Doppler frequency spectrum observations. A typical power spectrum for such particle motion is illustrated in Figure 17.18. The diffusion coefficient y of a spherical particle of radius a can be expressed according to the Stokes-Einstein relation as

$$y = \frac{kT}{6\pi\eta a} \quad (17.24)$$

where k is Boltzman's constant, T is the absolute temperature, and η is the absolute viscosity of the fluid medium. For a particle distribution with a diffusion coefficient of $p(y)$, the autocorrelation function of scattered light amplitude $C(\tau)$ resulting from a suspension of particles undergoing Brownian motion is given by

$$C(t) = \int_0^\infty p(y) e^{-Q^2 yt} dy \quad (17.25)$$

where Q is the Bragg wave number ($Q = 4\pi n/\lambda$).

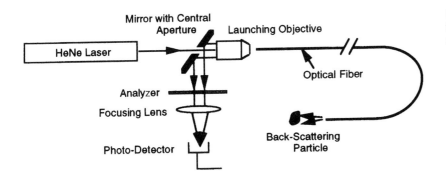

FIGURE 17.17 Illustration of an optical configuration of the fiber optic Doppler anemometer (FODA) used for particle-size analysis (Dyott, 1978).

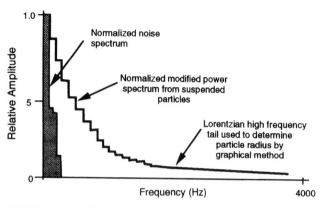

FIGURE 17.18 Illustration of a FODA output signal showing a Lorentzian power spectrum from suspended particles moving under the action of Brownian motion (Ross et al., 1978).

Using the power spectrum of the scattered light with an amplitude of

$$S(\omega) = 2\int_0^\infty C(t) \cos \omega t \, dt \quad (17.26)$$

it can be shown that the expression for the normalized modified Lorentzian power spectrum is:

$$S(\omega) = \frac{\omega^2}{\omega_0^2 + \omega^2}\left\{1 + \left[\frac{\omega^2(\sigma_0/a)^2}{\omega_0^2 + \omega^2}\right] \times \left[\frac{\omega^2 - 3\omega_0^2}{\omega^2 + \omega_0^2}\right]\right\} \quad (17.27)$$

where $\omega_0 = Q^2 kT/6\pi\eta a$ for particles with a mean radius a, with a standard distribution of the radius of for $\sigma_a^2 < a$.

The particle radii and the radii standard deviation were determined by a computer least-squares fit, and good correlation was found for the calculated Lorentzian with the measured power spectrum (Ross et al., 1978). A second and more convenient method of determining a and σ_a from the integrated Lorentzian power spectrum expressed in Equation (17.27) is to plot the inverse Lorentzian as a function of ω^2 from the data "tail" in Figure 17.18 and use the high-frequency linear asymptotic expansion:

$$S^{-1}[\omega(\text{large})] = \frac{1 + 6(\sigma_0/a)^2}{[1 + (\sigma_0/a)^2]^2} + \frac{\omega^2/\omega_0^2}{[1 + (\sigma_0/a)^2]} \quad (17.28)$$

The mean particle radius a and standard deviation σa were then obtained graphically from the slope and intercept, respectively, as:

$$\text{Slope} = \frac{1/\omega_0^2}{1 + (\sigma_a/\bar{a})^2}$$

$$\text{Intercept} = \frac{1 + 6(\sigma_a/\bar{a})^2}{1 + (\sigma_a/\bar{a})^2}\omega_0^2 \quad (17.29)$$

Small differences between using the computer least-squares fit and the graphical method were reported, with a discrepancy of about 3.5 percent in determining the mean radius. The computer curve-fitting method was likely to be the more accurate, although the graphical method is much simpler.

Experiments using syton particles of approximately 46 nm radius over a 1:2048 dilution range from undiluted syton over 12 subsequent dilutions, each diluted with an equal volume of distilled water, have been reported. Good correlation was observed for the mean radius (46.5 nm) and the standard deviation (20.8 nm) over the spread of radii, although particle clamping in the undiluted specimen was believed to have distorted the result for this initial sample.

17.4.3 Fluid Flow

The extension of the FODA-type technique to fluid flow measurement has met with some success, although problems have been found in that the fiber disturbs the flow around its end face. This problem stems from the fact that a stagnation region exists within a few hundred micrometers of the fiber end face. Since the depth of penetration of the radiation from the fiber into the fluid medium ranges from about 100 μm to about 1 mm, a range of velocities will generally be observed in the Doppler-shifted backscattered radiation, depending on the absorbency of the fluid suspension. FODA has been applied to the measurement of blood flow (Kilpatrick et al., 1982), where the penetration depth of HeNe laser radiation at 632.8 nm is on the order of 300 μm from the fiber tip. Here the fiber probe was introduced into a simulated blood flowstream and the backscattered power spectrum recorded with a spectrum analyzer. A frequency spectrum of the form shown in Figure 17.19 was obtained, showing an exponential-type decrease with increased frequency. The free-flow velocity was identified as the maximum observed frequency; it was assumed that the depth of penetration of the radiation extended beyond the disturbed flow region. To identify the maximum frequency more precisely, a second

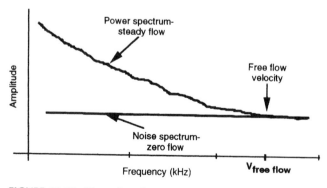

FIGURE 17.19 Illustration of a typical FODA output power spectrum for liquid flow measurement, showing free-flow velocity (Kilpatrick et al., 1982).

power spectrum was recorded, this time with the probe in blood at a zero flow velocity, thus giving the noise associated spectrum of the system (see Figure 17.19). The point at which the signal power spectrum overlapped with the noise spectrum was assigned to be the free-flow Doppler frequency indicated by the point V_{free} flow. Due to the noise of the system, this point contains some degree of uncertainty. Another interesting feature of these measurements was a series of readings of blood-flow velocity taken *in vivo* as the fiber probe was moved progressively across the diameter of a 3 mm glass tube. The readings showed a velocity gradient of the flow across the tube diameter; a schematic illustration of the reported results is shown in Figure 17.20. In addition, with the fiber probe positioned facing into the blood flow, nonlinear blood flow was observed that could be calibrated up to flow velocities of <0.15 m/s, the nonlinear behavior showing that laminar flow was not reestablished in the measurement region about the fiber tip except at very low velocities. With the fiber oriented against the blood flow, the response was linear up to flow rates of at least 1 m/s. Since arterial blood will be pulsating, the fiber probe will always experience flow with and against its direction of orientation.

To overcome the problem associated with the flow disturbance near the fiber probe tip, a measurement geometry has been devised that observes the free-flow velocity beyond this disturbed region (Kajiya et al., 1988) using the configuration shown in Figure 17.21. This was achieved by the use of two optical fiber probes placed side by side in the blood flow. One delivers the radiation to the flow region; the other collects the backscattered light only over a limited part of the illuminated penetration volume, as illustrated in Figure 17.22(a). The cone of radiation launched by the delivery fiber will overlap with the cone of acceptance of the collection fiber at some distance from the fiber end faces. If this overlap region is sufficiently far away from the fiber ends to be in the free-flow region, an unambiguous Doppler signal will be observed, as depicted in Figure 17.22(b). It was found that when two smaller-diameter (62.5 μm diameter; 50 mm core) optical fibers were placed adjacent to and touching each other, a Doppler frequency peak was observed with good correspondence to the free-flow velocity. When larger-diameter (125 μm) fibers were used, no Doppler signals were observed. With the 62.5 μm fibers, Doppler signals were seen when the direction of flow was both aligned with and against the fiber orientation. The calibration was performed with the fiber probe at an angle of 60° in a rotating turntable containing blood, and a good correlation was observed between the Doppler frequencies and the equivalent flow

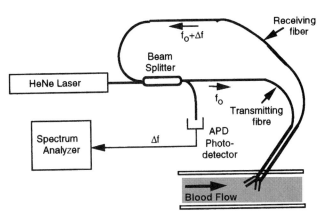

FIGURE 17.21 Schematic illustration of the dual-fiber method for Doppler blood-flow velocity measurement (Kajiya et al., 1988).

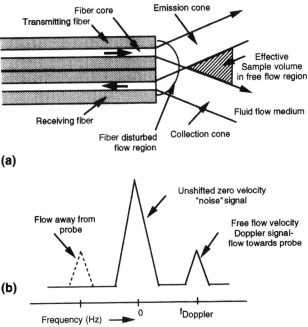

FIGURE 17.22 Schematic of (a) the sampling volume for the dual-fiber method in the free-flow velocity region, and (b) a spectrum analyzer type output signal indicating a free-flow Doppler signal (Kajiyal et al., 1988).

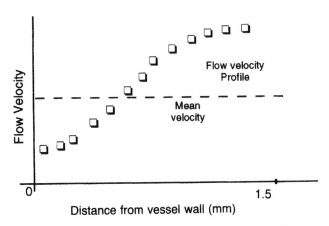

FIGURE 17.20 Schematic illustration of a FODA velocity profile across the radius of flow in a 3 mm diameter tube (Kilpatrick et al., 1982).

velocities, with correlation coefficients of $r = 0.985$ in the direction of flow and $r = 0.998$ when against the direction of flow. This optical fiber measurement system can thus be used to make accurate flow velocity measurements, even in disturbed flow fields.

17.4.4 Vibration Monitoring

There have been several successful examples of the application of optical fiber sensor systems to the noncontact measurement of vibration, and at least one commercial instrument has been marketed. Laser Doppler velocimetry (LDV) is a well-established technique for noncontact surface vibration measurement. Observing the Doppler shift effects in the light frequency modulation, it is necessary to introduce a carrier frequency modulation to determine the direction of the surface displacement. As we will see, this can be achieved by either a heterodyne method using a discrete frequency modulator such as a Bragg cell, or it can be introduced by a pseudoheterodyne approach by modulating the phase of the interferometer path imbalance. Here we will look at three such systems, each of which has its own particular area of application and all of which are based on a form of laser Doppler anemometry.

17.4.4.1 Heterodyne Modulation: Vibration Monitoring

This approach utilizes an acousto-optic Bragg cell frequency modulator device to provide the required high-frequency carrier signal. The phase of the carrier is then modulated by the radiation reflected or scattered from the vibrating target surface. The principles of this approach have been the subject of recent developments (e.g., Jackson and Meggitt, 1986; Fox, 1987). The system (shown schematically in Figure 17.23) essentially consists of a modified Mach–Zehnder bulk optic interferometer. To ensure a wavelength stabilized light source, a HeNe laser is used. Light from the laser is first passed through a quarter-wave plate to produce a circularly polarized output beam and then into a polarizing beam splitter (PBSI). The output of the PBSI is composed of two polarized beams propagating at 90° to each other, one of which is horizontally polarized and the other vertically polarized. One beam is passed through an appropriately oriented acousto-optic Bragg cell in which only the first-order diffracted output beam is allowed to transfer to the interferometer output. It is first reflected from the interferometer mirror and through a second (unpolarizing) beam splitter (BS3), where it is further divided to give to output beams that impinge on the two photodetectors (PD1 and PD2).

The second beam from PBSI is allowed to pass through the polarizing beam splitter (PBS2) oriented along the light polarizing axis and, with the aid of the focusing objective lens, is launched into one polarization axis of the polarization maintaining optical fiber. It is then transmitted through the optical fiber in this polarization mode to the fiber end, where it is focused on the vibrating target surface by a second lens element. The light is then reflected or scattered from this surface and relaunched back into the optical fiber. The presence of the quarter-wave plate at the fiber end ensures that the polarization axis of the light is rotated through 90° such that when it is launched back into the fiber, it propagates back down the fiber in the orthogonal polarization mode.

When this light emerges at the other fiber end, it is collimated by the focusing objective and then reflected through

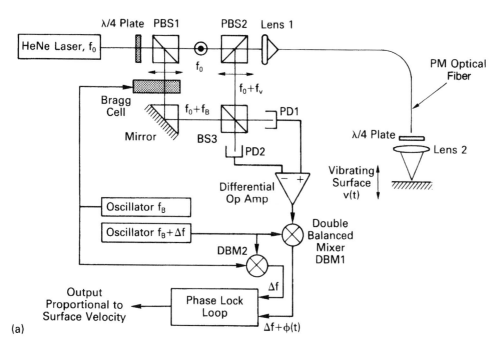

FIGURE 17.23 (a) Schematic of a heterodyne vibration monitor system using a Bragg cell frequency shifter and phase lock loop demodulator. PBS = polarizing beam splitter; PD = photodetector. (b) Spectrum analyzer PD output of a heterodyne vibration monitor (a) for sinusoidal target vibration (central frequency, $f_B = 80$ MHz; Jackson and Meggitt, 1986).

FIGURE 17.23 *continued*

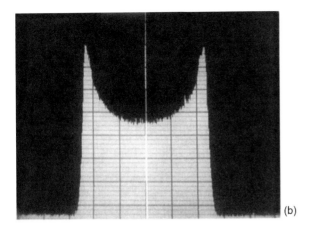

(b)

90° by the polarizing beam splitter (PBS2). The light then passes to the final beam splitter (BS3) and its output is divided between the two photodetectors, along with the radiation that has passed through the Bragg cell arm. The two component beams falling on each photodetector have traversed different paths in the Mach–Zehnder interferometer and therefore have different frequencies and phases but the same polarization states; consequently they interfere coherently on the surface of the detectors. The interference effect thus produces a modulated carrier at the Bragg cell frequency f_B, the phase of which is modulated by the Doppler effect induced in the reflected/scattered light from the vibrating target surface. Therefore, around the Bragg frequency f_B there is a time-varying modulated signal with a frequency shift δ_f given by:

$$\delta f = \frac{2v_s(t)}{(\lambda_o)} \cos\theta \qquad (17.30)$$

where v_s is the target surface velocity, λ_o is the wavelength of the HeNe source, and θ is the angle that the light makes with the reflecting surface. A typical spectrum analyzer output signal is shown in Figure 17.23(b) for a sinusoidally modulated target vibration ($f_B = 80$ MHz; Jackson and Meggitt, 1986). Due to the fiber NA, there will be a small group of such angles around the normal incident beam that will lead to a small spread in the observed Doppler shifts. As the target surface vibrates it will go through a range of surface displacements that will induce in the reflected beam a corresponding time-varying optical phase difference $\phi_s(t)$ in the interferometer output $I_{o/p}$, which can be expressed as:

$$I_{o/p}(t) = I_1 + I_2 \pm 2\beta\sqrt{I_1 I_2}\cos\left[2\pi f_B + \phi_s(t) + \phi_o\right] \qquad (17.31)$$

where I_1 and I_2 are the output photodiode currents, β is the interface mixing efficiency, and ϕ_o is the static interferometer phase imbalance.

The presence of the carrier signal f_B allows the direction of the motion of the vibrating surface to be determined unambiguously. Since the beat frequency is proportional to, $\sqrt{(I_1 I_2)}$, it is possible to improve the detection of the weak Doppler signal I_2 by increasing the intensity of the reference beam I_1. There is a fundamental limit to which the output signal i_s can be increased relative to the noise component i_n, which is related to the quantum detection limit of the photodetector, given by

$$\left(\frac{i_s^2}{i_n^2}\right)_{max} = \frac{(\eta I_2)}{(h\nu\Delta f)} \qquad (17.32)$$

where η is the quantum efficiency of the detector, Δf is the pass band of the detector and preamplifier electronics, $h\nu$ is the light quantum, and I_2 is the weak Doppler signal.

The output signal is demodulated by a phaselock loop (PLL) that produces an error signal proportional to the surface velocity $v_s(t)$. The PPL can respond to, typically, ±20 percent of the carrier signal while still maintaining good linearity along with a wide frequency response. The design of the instrument permits the detection of surface velocities in the range 0.316 to 3.16×10^{-5} m/s over a frequency range of 10^{-1} Hz to 30 kHz. The linear response region of such a vibration sensor is illustrated by the neumonic shown in Figure 17.24, where frequency, displacement amplitude, and surface velocity are related graphically.

For example, for a target surface moving in a periodic sinusoidal motion such that

$$\phi_s(t) = \frac{4\pi A}{\lambda_o}\sin(2\pi f_s t) \qquad (17.33)$$

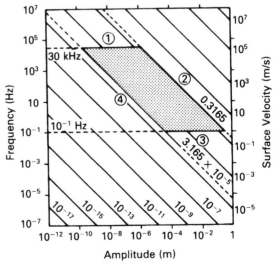

FIGURE 17.24 Neumonic showing the performance regime with regard to frequency, amplitude, and surface velocity related to the linear response region of the PPL shown in Figure 17.23.

the resulting dynamic component of the output signal is given by

$$I_{o/p}(t) \alpha \cos\left\{2\pi f_B t + \frac{4\pi A}{\lambda_o}\left[\sin(2\pi f_s t + \phi_o)\right]\right\} \quad (17.34)$$

The output spectrum is composed of a carrier centered at frequency f_B with sidebands at frequency differences of $f_B \pm f_s, f_B \pm 2f_s, f_B \pm 3f_s$, and so on. It is possible to select the first term $f_B \pm f_s$ by use of a band-pass filter centered at f_B and with a bandwidth of less than $\pm 2f_s$ (max). The output frequency modulated signal can be conveniently demodulated by use of a PLL where the output error signal is proportional to the surface velocity v_s. However, since the Bragg cell frequency shift is on the order of tens of megahertz (typically 40 MHz–100 MHz) and the maximum frequency shift $f_{s\,(max)}$ is <1 MHz, it is difficult to design a PLL to cover this range. Therefore, the Bragg carrier frequency is first down-converted using conventional double-balanced mixers (DBM1 and DBM2) and a second oscillator offset from the Bragg frequency by Δf (i.e., at $f_B + \Delta f$). The mixer output is therefore centered on the intermediate frequency of Δf, and this is conveniently set at a maximum of 5 MHz for which PLLs are available, having a maximum modulation depth of 20 percent as described (<1 MHz).

17.4.4.2 Pseudoheterodyne Modulation: Vibration Monitoring

A different form of optical fiber vibration monitor has been described (Meggitt et al., 1989) for use as a noncontacting reference-grade vibration sensor. It was initially designed for use in calibrating secondary-grade accelerometers such as piezoelectric accelerometers, but it has a general applicability. The advantage in using an optical approach for a reference-grade device is that the system is capable of making displacement measurements, which are referred only to the wavelength of the radiation used, in this case a HeNe gas laser at 623.8 nm. It can also be configured to provide calibrations that are independent of the temperature of the environment, allowing the temperature dependence of the piezoelectric devices to be characterized.

The vibration monitor is based on a form of Michelson optical fiber interferometer, as shown in Figure 17.25. Here the HeNe laser light is launched into one port of a single-mode fiber coupler. Radiation is directed via an output port onto the vibrating surface being investigated, from which it is back-reflected into the optical fiber coupler and transmitted back to a photodetector element. The second output port of the fiber coupler is wound several times around a piezoelectrical cylinder that acts as a fiber stretcher under the influence of a periodic modulating voltage. Some of the HeNe light will travel in this second arm and be reflected back along its path by a silvered reflecting end coating deposited on the fiber end face. This back-reflected light will also be directed to the photodetector, where it will interfere coherently with the light that is Doppler shifted by the vibrating surface. If a high-frequency sinusoidal voltage is applied to the piezoelectric modulator of 100 kHz, the phase $\phi(t)$ of the light will be modulated due to the modulated interferometer arm imbalance as

$$\phi(t) = \frac{4\pi n}{\lambda_o} L_o \sin(\omega_m t) \quad (17.35)$$

where L_o is the peak fiber length extension due to the pzt element. Similarly, light reflected back from the vibrating target will also be phase modulated but at a somewhat lower frequency due to the displacement effect of the vibrating surface modulating the arm imbalance L_s of the interferometer, giving rise to a second time-dependent phase component $\theta(t)$ of the form

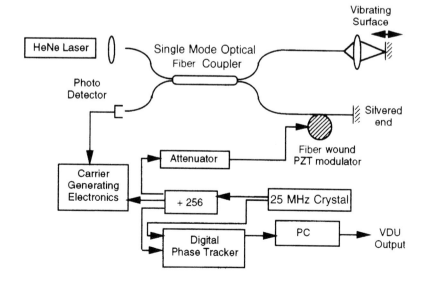

FIGURE 17.25 Schematic of a reference grade fiber optic vibration sensor (Meggitt et al., 1989).

$$\theta(t) = \frac{4\pi L_s(t)}{\lambda_o} \quad (17.36)$$

If the two arms of the interferometer are equal, $L_s(t)$ will represent the time-dependent displacement of the vibrating surface. The output photodiode current I_{pd} will then have the form, from the transfer function of the Michelson interferometer, of

$$I_{pd} = A\left\{\left(1 + K\cos\left[\theta(t) + \phi_0 \sin \omega_m t\right]\right)\right\} \quad (17.37)$$

Using standard trigonometric identities, this can be expressed as

$$I_{pd} = A\left\{1 + K\left[\cos\theta\cos(\phi\sin\omega_m t) - \sin\theta\sin(\phi\sin\omega_m t)\right]\right\} \quad (17.38)$$

and can be expanded in terms of a Bessel function series:

$$I_{pd}(t) \propto \left\{1 + K\left[J_o(\phi_o) + 2\sum_{n=1}^{\infty} J_{2n}(\phi_o)\cos 2n\omega_m t\right]\cos\phi_T \right.$$
$$\left. - K\left[2\sum_{n=0}^{\infty} J_{2n-1}(\phi_o)\sin(2n+1)\omega_m t\right]\phi_T\right\} \quad (17.39)$$

where $J_o(\phi_o), J_1(\phi_o), \ldots, J_{2n}(\phi_o)$ are Bessel functions of the first order of argument ϕ_o.

The output signal I_{pd} is first gated with a 50:50 mark space ratio square wave and then band-pass filtered at a frequency of twice the modulation frequency, that is, at $2\omega_m$, to give

$$S(t) = \left[A(\phi_o)\cos\theta\cos(2\omega_m t) - B(\phi_o)\sin\theta\cos(2\omega_m t)\right] \quad (17.40)$$

By setting $A(\phi_o) = B(\phi_o) = R$ by adjusting the voltage on the PZT modulator to give $\phi_o = 2.82$ radians, the output of the instrument is then given by the reduced form of

$$S(t) = R\cos(2\omega_m t + \theta) \quad (17.41)$$

which represents a carrier of frequency $2\omega_m$ that is phase modulated by the vibrating surface $\theta = 4\pi L_s(t)\lambda_o$. An electronic phase tracker is used to follow the phase movement induced by the vibrating surface. To achieve this objective, the phase modulated carrier (PCM) square-wave output is monitored against the reference wave derived from the PZT master oscillator. The phase of the PMC relative to the reference is measured by use of the master clock running at 25 MHz with a 1/126 measurement increment using the rising edges of the two waveforms as trigger points. By storing two successive phase measurements in buffer memories, the direction of the surface displacement can be determined by comparison of their two most significant bits. The resolution of the system corresponds to about 3° in phase measurement, or 2.47 nm (i.e. $\lambda_o/126$). The noise floor of the processing system was determined to be equivalent to approximately 1 nm in surface displacement. Since the gated output of the detector produces two such PMC signals moving in opposite directions, it is possible to process both to halve the resolution of the system. The performance of the system is limited by the maximum slew rate that can be satisfactorily processed due to the bandwidth limitation of the band-pass filter of about 30 kHz that must accept only the second harmonic $2\omega_m$ (about 200 kHz) of the PZT modulation frequency and reject the adjacent fundamental ω_m and the third harmonic $3\omega_m$. Using a bandwidth of 30 kHz, this corresponded to a displacement amplitude of 1 μm at 5 kHz and down to 0.5 mm displacement amplitude at 10 kHz vibration frequency. Results indicate that an error in surface amplitude of less than 1 percent was present over a 2 kHz bandwidth for a scaled 1 g surface acceleration; this rises to <3 percent over a 5 kHz bandwidth when measured against a reference-grade piezoelectric accelerometer under standard conditions.

17.4.4.3 Frequency Modulated Laser Diode Vibration Monitoring

In frequency modulated laser diode vibration monitoring (Laming et al., 1985), a frequency modulated single-mode laser diode is used as the source element. No separate frequency shifter element is required, and the setup is suitable as a handheld vibration monitor that is insensitive to external vibrations. It is claimed to be robust, lightweight, and inexpensive. The device, illustrated in Figure 17.26, uses a pseudoheterodyne approach whereby the emitted optical frequency of a single-mode laser diode is modulated by changes induced in its injection current. The emitted radiation is launched into a fiber pigtail that is connected to a single-mode fiber directional coupler. Only one output arm of the coupler is used, the other being immersed in an index matching liquid to suppress any back-reflections. The light transmitted by the output port is collimated by a Selfoc graded index lens. The collimated light is directed through free space onto the vibrating target surface, where it is scattered back into the fiber and directed by the coupler onto a photodetector element at the end of the fourth coupler arm. A cavity is set up between the fiber-Selfoc interface and the target surface, ensuring a common fiber path for both reference and signal beams. The laser diode injection current is sinusoidally modulated such that the optical frequency of the output also follows a sine modulation (about 1 GHz/mA). This method avoids the linearity problems associated with linearly ramped phase modulation but requires additional signal processing to produce the phase modulated carrier signal. The output optical frequency $v(t)$ from the laser diode has the form

$$v(t) = v_o + \Delta v \sin \omega_m t \quad (17.42)$$

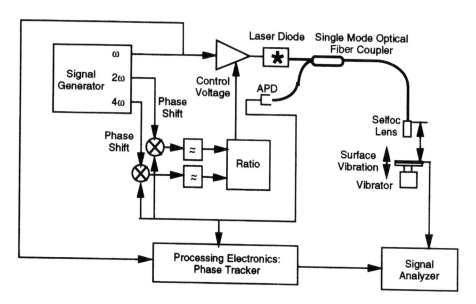

FIGURE 17.26 Schematic diagram of a frequency modulated laser diode fiber optic vibration probe (Laming et al., 1985).

where ν_o is the mean laser frequency and $\Delta\nu$ is the peak frequency shift. The corresponding phase $\phi(t)$ of the interference between the reference and signal reflections from the sensing cavity of length l is given by:

$$\phi(t) = \frac{4\pi l}{c}(\nu_o + \Delta\nu \sin\omega_m t) = \phi_T + \phi_m \sin\omega_m t \quad (17.43)$$

The transfer function of the low-finesse Fabry–Perot cavity can then be expressed as:

$$I(t) \propto [1 + K\cos(\phi_m \sin\omega_m t + \phi_T)] \quad (17.44)$$

where K is the fringe visibility, which depends on the amplitude of the scattered signal; the constant of proportionality depends on the optical power and detector sensitivity. This signal can then be processed in a similar way to that described previously for the reference-grade accelerometer, that is, by first expanding the transfer function in terms of Bessel functions similar to Equation (17.39).

By gating $I(t)$ with a square wave at frequency ω_m and adjusting the laser drive frequency Δ_ν, the phase excursion ω_m can be maintained at 2.82 radians. Finally, the output is band-pass-filtered at twice the laser modulation frequency, that is, $2\omega_m$, giving

$$I(t) \propto KJ_2(\phi_m)\cos(2\omega_m t - \phi_T) \quad (17.45)$$

Hence, the output represents a carrier at frequency $2\omega_m$ that is phase modulated by the surface displacement l through ϕ_T, since

$$\frac{d\phi_T}{dt} = \frac{4\pi\nu_o}{c}\frac{dl}{dt} \quad (17.46)$$

A signal generator produces outputs at ω_m, $2\omega_m$ and $4\omega_m$. To control the laser output frequency to maintain ϕ_m at 2.82 radians, the photodetector output is mixed in two operations with $2\omega_m$ and $4\omega_m$ and their respective outputs are low band-pass-filtered to give the $J_2(\phi_m)\cos\phi_T$ and $J_4(\phi_m)\cos\phi_T$, and their ratio is used to control the AGC amplifier driving the laser diode. In this way, any temperature effects in the laser diode and variations in the sensing cavity are compensated for. A tracker is used that measures $d\phi_T/dt$, which gives an output voltage proportional to surface velocity. The vibration sensor is designed to operate with a sensor cavity of 50–300 mm. Vibration frequencies up to 20 kHz are measured; measurement is limited to the region in which the frequency tracker gives good linearity, with a maximum surface velocity up to 0.2 m/s being reported.

17.5 IN-FIBER SENSING STRUCTURES

17.5.1 Introduction

In-fiber sensing structures, such as the Fabry–Perot and the Bragg grating elements, are proving invaluable devices, especially when the fibers are embedded in or attached to composite or metallic components for monitoring structural integrity. When the sensor is integrated into the fiber structure, the sensor size is comparable to that of the original fiber dimensions and thus will only minimally perturb the host structure. In addition, it is possible to produce many such short-length sensors into a given fiber length to form a quasi-distributed array of sensing elements. It is with this background that we now discuss in some detail the sensing mechanisms and structures of the in-fiber Fabry–Perot and the fiber Bragg grating devices.

17.5.2 Fiber Fabry–Perot Sensing Element

The fiber Fabry–Perot sensor (Kist et al., 1985; Lee and Taylor, 1988) is a useful sensing device because it is an inline sensing element (see Figure 17.27). It is essentially an

CHAPTER | 17 Fiber Optics in Sensor Instrumentation

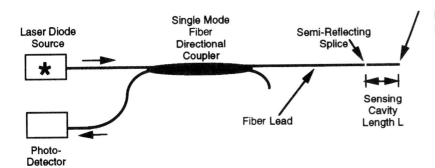

FIGURE 17.27 Schematic diagram of a fiber Fabry–Perot sensing system.

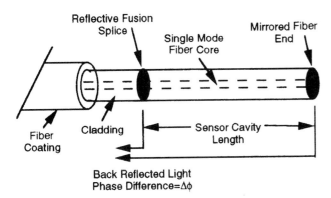

FIGURE 17.28 Illustration of a fiber Fabry–Perot sensing structure (e.g., Lee and Taylor, 1988).

optical cavity that is defined by two semireflecting parallel mirrors (see Figure 17.28). More usually, the Fabry–Perot cavity (at least in bulk optic form) has highly reflecting mirrors of reflectivity R such that the device has a high finesse, and consequently its reflection/transmission is spectrally selective and serves as an interference filter element. The intensity I_T of radiation at wavelength λ that is transmitted by the Fabry–Perot device is given as a function of optical phase change by:

$$I_T = I_0 \frac{(1 - R^2)}{[(1 - R)^2 + 4R \sin(\phi/2)]} \quad (17.47)$$

where ϕ is related to the cavity length L, its refractive index n, and optical wavelength λ by:

$$\phi = \frac{4\pi nL}{\lambda} \quad (17.48)$$

In its use with optical fibers, the cavity is formed in a short length (1–30 mm) of optical fiber that has partially reflecting coated ends (both aluminum and TiO_2 have been used; Lee et al., 1989), which is then fusion-spliced onto the end of the connecting fiber (see Figure 17.28). For example, with 100 nm deposited in TiO_2 layers, after fusing and splicing a 1 mm Fabry–Perot fiber cavity into a fiber length, 2 percent reflectivity was reported for the fused fiber cavity mirrors. This arrangement provides a low-finesse cavity with mirror reflectivities of about 10 percent or less. Under these conditions the Fabry–Perot cavity acts essentially as a two-beam interferometer, since multiple reflections in the cavity are weak and the transfer function for the optical phase difference ϕ of the back-reflected intensity I_r has the cosine fringe form

$$I_T = I_0[1 + \nu \cos \phi] \quad (17.49)$$

where I_0 is the mean return intensity and ν is the fringe visibility (≤ 1). It is possible to detect changes in the optical phase difference of the interferometer output using interferometric techniques such as the single-mode laser diode pseudoheterodyne method or the broadband, white-light interferometric approach. In the former (Kist et al., 1985), the output wavelength of the laser diode is ramped by applying a serrodyne input current ramp that modulates its output frequency Δf by about 3 GHz/mA, where frequency modulations of 10–100 GHz are possible. The back-reflected radiation from the Fabry–Perot cavity has a periodic sine wave form, as described previously. Alternatively, the white-light technique (Lee and Taylor, 1991) can be used as described earlier. Either way, the change in the phase of the output fringe pattern can be monitored and related to the measurand of interest. For temperature measurement, the change in phase has the form

$$\Delta \phi = \frac{4\pi}{\lambda}\left(L\frac{\partial n}{\partial T} + n\frac{\partial L}{\partial T}\right)\Delta T - \frac{4\pi}{\lambda^2}nL\, d\lambda \quad (17.50)$$

where the last term is associated with the effect of wavelength changes in the source radiation and can be ignored if the source wavelength fluctuation is small or negligible. The change in refractive index (dn/dT) dominates the phase change at temperatures above about 20°C, since the thermal expansion (dL/dT) of the fiber length is an order of magnitude lower for a silica optical fiber and where the temperature dependence of the refractive index associated phase change is about 100 rad/°C. However, at lower temperatures (<20°C), the temperature coefficient of the refractive index is low and the fiber contracts with decreasing temperature (negative value of thermal expansion), resulting in a more nonlinear optical path length change with temperature. In high-temperature sensor applications using a novel pulsed modulation technique over a temperature range of 1,225°C

($-200°C$ to $1{,}050°C$), Lee et al. (1988) found a 1 mm fiber cavity length to give a fringe extinction ratio of 20:1.

Pressure effects have also been monitored using the fiber Fabry–Perot sensor where a 30 cm diameter coil was compressed between two metal plates (Yoshimo et al., 1982) and a sensitivity of 0.04 rad/Pa/m was reported.

Similarly, longitudinal strain (the ratio of the fiber extension ΔL to its original length L) can also be monitored through observation of the induced phase changes in the fiber interferometer output (Measures, 1992). In this case it is the strain-optic coefficient ($\varepsilon = \Delta L/L$) that is of interest. Under the condition that zero strain is coupled to the fiber in directions other than along the longitudinal axis, that is, a uniaxial stress loading, the corresponding phase change is given by

$$\Delta \phi_s = \frac{4\pi n L}{\lambda}\left\{1 - \frac{n^2}{2}\left[p_{12} - \nu(p_{11} + p_{12})\right]\right\} \quad (17.51)$$

where p_{11} and p_{12} are the strain-optic coefficients and ν is Poisson's ratio.

In general, the cosine form of the fringe output means that the phase change will be multifunctional, and ambiguities can arise when the phase is greater than 2π. As discussed earlier, dual-wavelength interrogation can be used to remove such processing uncertainties.

17.5.3 Fiber Bragg Grating Sensing Element

A second example of an in-fiber sensing structure is the holographic fiber Bragg grating. This device is essentially a linear grating structure that is formed in the core of an optical fiber through the periodic modulation of its refractive index along a fixed length of the fiber (see Figure 17.29). The periodic modulation of the core refractive index is produced by the interaction of UV laser radiation with defect structures in some doped silica optical fibers through a photorefractive process (see Morey et al., 1989). The absorption of radiation about the 245 nm wavelength, for example, is associated with the germaniumoxygen vacancy defect band structure. Grating structures are usually written in the core of silica fibers by transverse or side illumination. An interference pattern is formed between two intersecting coherent beams of UV laser radiation, as shown in Figure 17.30. The period of the interference pattern λ_{Bragg} is determined by the radiation wavelength λ_{UV}, the core index n_c, and the angle of intersection of the two overlapping beams, θ:

$$\lambda_{Bragg} = \lambda_{UV}\frac{n_c}{\sin\theta} \quad (17.52)$$

It can be seen that by varying the angle of intersection of the two interfering beams, the period of the grating, and thus the wavelength at which the grating structure will reflect incident radiation propagating along the optical fiber, can be altered. The wavelength λ_{Bragg} at which a grating period Λ back-reflects is given by

$$\lambda_{Bragg} = 2n_c\Lambda \quad (17.53)$$

The efficiency R of the grating structure in reflecting back a particular wavelength λ_{Bragg} is determined by the following: the length L of the grating structure, and hence the number of grating elements; the reflectivity of each element, that is, the refractive index depth of variation ($\Delta n/n_c$) produced by the writing process; and the fraction of the integrated mode intensity that is concentrated in the fiber core $\eta(V)$, $\eta(V) \simeq 1 - 1/V^2$, where the normalized fiber frequency $V \geq 2.4$. The expression given by Morey et al. (1989) is

$$R = \tan h^2\left[\frac{\pi L \Delta n \eta(V)}{\lambda_{Bragg}}\right] \quad (17.54)$$

The rate at which a Bragg grating can be written in the fiber core depends on the radiation power and exposure time. For CW radiation at 244 nm, a power above 10 mW allows the grating to form quickly (Meltz et al., 1989). For example, with a power of 23 mW and a 10 s exposure, transmission at the Bragg wavelength was reported to decrease to 0.65. With a 5 min exposure at an average power of 18.5mW in a 6.6 mol percent GeO_2 doped fiber of core 2.6 μm, the reflected and transmitted spectra were found to be complementary and a line width of approximately 5×10^{-11} was reported, giving a fractional bandwidth of about 10^{-4}. Several methods

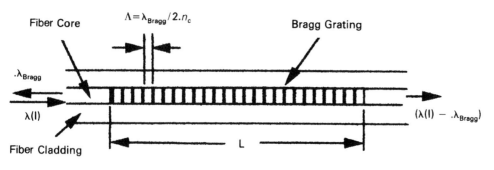

FIGURE 17.29 Structure of a holographic in-fiber Bragg grating.

CHAPTER | 17 Fiber Optics in Sensor Instrumentation

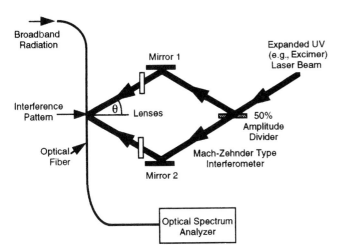

FIGURE 17.30 Optical configuration for writing holographic optical fiber Bragg gratings (Meltz et al., 1989).

have been reported to increase further the reflectivity of the gratings, with index changes up to 10^{-2} being observed with techniques such as boron co-doping and dehydrogenation of the fiber to increase its photosensitivity.

Gratings have also been successfully produced using pulsed laser sources, such as excimer lasers having a pulse duration of approximately 20 ns (Askins et al., 1992; Archambault et al., 1993). Subsequent work (Archambault et al., 1993b) has shown that there are two basic types of possible grating structures: Type I and Type II, formed in the low- and high-pulse energy regimes, respectively. Pulse energies below about 20 mJ were shown to generate Type I gratings in a similar way to the method using continuous wave sources having a narrow spectral bandwidth and moderate modulation depths up to about 75 percent. For pulse energies above 40 mJ, Type II gratings were formed, having very high reflectivities (approaching 100 percent) and with a much broader fractional bandwidth (about 6×10^{-3}). In these Type II gratings a structural transformation occurs at the cladding–core interface on one side of the fiber, with the result that below the Bragg wavelength light is coupled strongly to the cladding, whereas above the Bragg spectral bandwidth light is transmitted in the fiber core.

Two significant features are evident for the use of the pulsed laser grating method. The first is that there is sufficient energy per pulse to allow a single shot formation of the grating structure. This feature permits the production of the gratings in a fiber as it emerges from the fiber drawing tower, thus allowing any number of gratings with chosen separations to be made in a quasi-distributed fashion. This is a significant process since, to write gratings in ready-pulled fiber, the secondary coating must first be removed because it absorbs the UV radiation. This latter fact limits the possibility of creating large numbers of grating arrays in an optical fiber due to the need first to strip off the fiber coating and then to replace it after exposure; this will weaken the mechanical strength of the fibers.

The second feature is that Type II grating structures exhibit the property of surviving without signs of annealing up to temperatures of 800°C and are removed only after being held at temperatures of 1,000°C. In Type I gratings, temperatures above 450°C anneal the photorefractive grating structure within seconds. Therefore, it is possible to consider Type II fiber gratings for use in harsh environments, especially where high temperatures are encountered (e.g., process control, high-temperature piping such as power generation, and aeronautical engine monitoring applications).

The availability of the technology to produce in-fiber grating structures has permitted their use in a number of sensing applications, including temperature sensing (Morey et al., 1989), strain monitoring (Kersey et al., 1992), and high-pressure sensing (Xu et al., 1993). The parameter measured in each case is the spectral change λ_{Bragg} in the central Bragg wavelength λ_{Bragg} of the grating structure. In temperature-sensing applications (Morey et al., 1989), the fractional change in the Bragg wavelength is given by the expression

$$\frac{\Delta \lambda_{\text{Bragg}}}{\lambda_{\text{Bragg}}} = (\alpha + \xi)\Delta T \qquad (17.55)$$

where α is the thermal expansion coefficient and ξ is the thermo-optic coefficient. The latter is the dominant fiber effect, and wavelength shifts of 0.006 nm/°C have been reported at a wavelength of 550 nm.

When considering strain-monitoring applications, as shown in Figure 17.31(a) (Morey et al., 1989; Measures, 1992; Kersey et al., 1992; Measures et al., 1992), the fractional change in Bragg wavelength is given by

$$\frac{\Delta \lambda_{\text{Bragg}}}{\lambda_{\text{Bragg}}} = [1 - p_e]\varepsilon \qquad (17.56)$$

where ε is the applied longitudinal train on the fiber and p_e is the effective photoelastic coefficient having the form

$$p_e = \left(\frac{n^2}{2}\right)[p_{12} - v(p_{11} + p_{12})] \qquad (17.57)$$

where p_{11} and p_{12} are the strain–optic tensor coefficients, v is Poisson's ratio, and n is the core refractive index. Typically p_e has a value of 0.22 and $\lambda_{\text{Bragg}}/\lambda_{\text{Bragg}}$ is related to the applied stress by the relation $\Delta \lambda_{\text{Bragg}}/\lambda_{\text{Bragg}} = 7.5 \times 10^{-8}/\varepsilon$.

The measurement of strain is complicated by the effect of the thermal expansion of the body under measurement, which produces an effect referred to as the *apparent strain*. It is necessary to remove the latter in the measurement to reveal the *true strain* effect. This is conventionally done by providing a second dummy sensor (Measures, 1992), which is placed in the vicinity of the strain monitor but desensitized to the strain effect, thus providing only a measure of the temperature environment. It is then necessary to subtract the thermal "apparent strain" measured by the dummy sensor

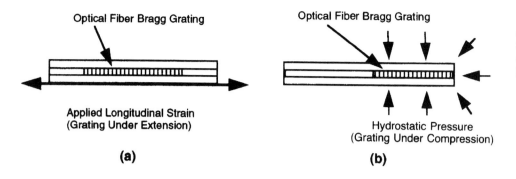

FIGURE 17.31 Illustration of a fiber Bragg grating used for monitoring (a) longitudinal strain (Measures et al., 1992), and (b) hydrostatic pressure (Xu et al., 1993).

from the combined "real and apparent strain" measured by the first strain sensor.

It is also possible to write Bragg gratings in polarization maintaining optical fibers that have two orthogonal polarization propagating modes (Morey et al., 1989; Measures et al., 1992). Since each polarization eigenmode has a slightly different associated refractive index, the optical path grating period will be different for the two axes; the Bragg wavelength differences between the two modes are typically on the order of 0.1 nm.

As already mentioned, fiber gratings can also be used to monitor high-pressure environments (see Figure 17.32(b)) resulting from the structural changes induced in the fiber by the surrounding hydrostatic pressure (Xu et al., 1993). In this case the fiber is compressed under the action of the pressure force and the fractional change in Bragg wavelength is then given by

$$\frac{\Delta \lambda_{Bragg}}{\lambda_{Bragg}} = \left(\frac{1}{\Lambda}\frac{\partial \Lambda}{\partial P} + \frac{1}{n}\frac{\partial n}{\partial P}\right)\Delta P \quad (17.58)$$

where Λ is the spatial period of the grating and n is the core refractive index. It has then been shown that this fractional change can be related to the mechanical properties of the fiber by

$$\frac{\Delta \lambda_{Bragg}}{\lambda_{Bragg}\Delta P} = -\frac{(1-2v)}{E} + \frac{n^2}{2E}(1-2v)(2p_{12}+p_{11}) \quad (17.59)$$

where E is Young's modulus and the other symbols have the assignments given earlier. Measurements of the fractional change when the grating structure was subjected to pressures of up to 70 Mpa showed a linear response, with a figure of -1.98×10^{-6} MPa. By contrast, the temperature effect gave a fractional change of $+6.72 \times 10^{-6}/°C$ (0.01045 nm/°C). Therefore, as with the measurement of strain, to obtain a true pressure value there is a need to eliminate the effect of temperature on the pressure reading.

To measure the change in the central Bragg reflected wavelength it is convenient to use a conventional spectrometer or optical spectrum analyzer. These devices, however,

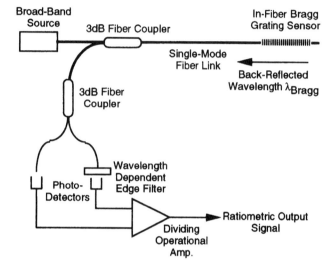

FIGURE 17.32 Radiometric wavelength-dependent detection system for demodulating the reflected Bragg wavelength (Measures et al., 1992).

have a limited response time (up to the order of milliseconds) and are costly items. For a low-cost sensor application, wavelength demodulation has been demonstrated by the use of appropriately selected optical edge filters (see Figure 17.32; Measures, 1992). The spectral properties of the edge filters are chosen such that the filter operates in the middle of the extreme wavelength shift of the Bragg grating element and is placed immediately prior to the photodetector at the return fiber output. Hence, as the Bragg wavelength shifts occur in response to the changing measurand, the amount of light intensity passed by the edge filter also changes. For a linear filter function

$$F(\lambda) = A(\lambda - \lambda_0) \quad (17.60)$$

where A is the filter slope response and λ_0 is the wavelength for which $F(\lambda)$ is zero. The ratio of the intensity passed by the filter I_F to that of the total light intensity I_T returned (reference signal) is given by

$$\frac{I_T}{I_F} = A(\lambda_{Bragg} - \lambda_0 + \Delta\lambda/\pi) \quad (17.61)$$

assuming that the narrow backscattered Bragg peak has a Gaussian profile. When an RG830 infrared high-pass filter was used in the wavelength ratiometric detection system with a Bragg grating of sensitivity of 0.65 pm/μstrain, the linear filtering range corresponded to a dynamic strain measurement range of about 35,000 μstrain. When tested on a Lexan beam it was shown to have a strain resolution of about 1 percent of the linear strain range of the filter element, although it is expected that this figure can be reduced by an order of magnitude through improved noise performance of the processing electronics.

REFERENCES

Al-Chalabi, S. A., B. Culshaw, and D. E. N. Davies, "Partial coherence sources in interferometric sensors," *1st International Conference on Optical Fiber Sensors IEE*, London, 132–136 (1983).

Archambault, I. L., L. Reekie, and P. St J. Russell, "100% Reflectivity Bragg reflectors produced in optical fibers by single excimer laser pulses," *Elect. Lett.*, **29**, 535–455 (1993a).

Archambault, I. L., L. Reekie, and P. St J. Russel, "High reflectivity and narrow bandwidth fiber gratings written by a single excimer pulse," *Elect. Lett.*, **29**, 28–29 (1993b).

Askins, C. G., T. E. Tsai, G. M. Williams, M. A. Putnam, M. Bashkansky, and E. J. Friebele, "Fiber Bragg reflectors prepared by a single excimer pulse," *Opt. Lett.*, **17**, 833–835 (1992).

Bianco, A. D. and F. Baldini, "A portable optical fiber sensor for oxygen detection," *9th Optical Fiber Sensor Conference*, Ferenzi, Italy, 197–200 (1993).

Boheim, G., "Remote displacement measurement using a passive interferometer with a fiber optic link," *Appl. Opt.*, **24**, 2335 (1985).

Boheim, G., "Fiber-optic thermometer using semiconductor etalon sensor," *Elect. Lett.*, **22**, 238–239 (1986).

Boheim, G. and K. Fritsch, "Fiber-optic temperature sensor using a spectrum-modulated semiconductor etalon," *SPIE Fiber Optic and Laser Sensors V*, **838**, 238–246 (1987).

Boheim, G., K. Fritsch, and R. N. Poorman, "Fiber-linked interferometric pressure sensor," *Rev. Sci. Instrum.*, **59**, 1655–1659 (1987).

Bosselmann, T. and R. Ulrich, "High accuracy position-sensing with fiber-coupled white-light interferometers," *2nd International Conference on Optical Fiber Sensors*, Stuttgart, 361–364 (1984).

Change, Y. C. and T. M. Shay, "Frequency stabilisation of a laser diode to a Fabry–Perot interferometer," *Opt. Eng.*, **27**, 424–427 (1988).

Chen, S., B. T. Meggitt, and A. J. Rogers, "Electronically scanned white-light interferometry with enhanced dynamic range," *Elec. Lett.*, **26**, 1663–1664 (1990).

Chen, S., A. W. Palmer, K. T. V. Grattan, and B. T. Meggitt, "Study of electronically-scanned optical-fiber white light Fizeau interferometer," *Elect. Lett.*, **27**, 1032–1304 (1991).

Chen, S., A. W. Palmer, K. T. V. Grattan, and B. T. Meggitt, "An extrinsic optical fiber interferometric sensor that uses multimode optical fiber: system and sensing head design for low noise operation," *Opt. Lett.*, **17**, 701–703 (1992a).

Chen, S., A. J. Palmer, K. T. V. Grattan, and B. T. Meggitt, "Digital processing techniques for electronically scanned optical fiber white-light interferometry," *Appl. Opt.*, **31**, 6003–6010 (1992b).

Chen, S., B. T. Meggitt, K. T. V. Grattan, and A. W. Palmer, "Instantaneous fringe order identification using dual broad-band sources with widely spaced wavelengths," *Elect. ELT.*, **29**, 334–335 (1993).

Cork, M., A. D. Kersey, D. A. Jackson, and J. D. C. Jones, "All-fiber 'Michelson' thermometer," *Elect. Lett.*, **19**, 471–471 (1983).

Dandridge, A. and L. Goldberge, *Elect Lett.*, **18**, 304–344 (1982).

Danliker, R., E. Zimmermann, and G. Frosio, "Noise-resistant signal processing for electronically scanned white-light interferometry," *8th Optical Fiber Sensor Conference (OFS-8) IEEE*, Monterey, 53–56 (1992).

Dyott, R. B., "The fiber optic Doppler anemometer," *Microwave Opt. Acoust.*, **2**, 13–18 (1978).

Fox, N., "Optical Vibration Monitor," MSc dissertation, City University, London (1987).

Giallorenzi, T. G., J. A. Bucaro, A. Dandridge, G. H. Sigel, J. H. Gale, S. C. Rashleigh, and R. G. Priest, "Optical fiber sensor technology," *IEEE J. Quant. Elect.*, **18**, 626 (1982).

Grattan, K. T. V. and B. T. Meggitt (eds), *Optical Fiber Sensor Technology*, Chapman & l Hall, London (1994).

Jackson, D. A. and B. T. Meggitt, *Non-Contacting Optical Velocimeter for Remote Vibration* Analysis, Sira (Internal Report), Sira/Kent University Development Report (1986).

Jackson, D. A., A. Dandridge, and S. N. Sheem, "Measurement of small phase shifts using a single-mode optical fiber interferometer," *Opt. Lett.*, **6**, 139 (1980).

Kajiya, F., O. Hiramatsu, Y. Ogasawara, K. Mito, and K. Tsujinka, "Dual-fiber laser velocimeter and its application to the measurement of coronary blood velocity," *Biorrheology*, Pergamon Press, 227–235 (1988).

Kersey, A. D., T. A. Berkoff, and W. W. Morey, High resolution fiber-grating based strain sensor with interferometric wavelength-shift detection," *Elect. Lett.*, **28**, 236–238 (1992).

Kersey, A. D., J. D. C. Jones, M. Corke, and D. A. Jackson, "Fiber optic sensor using heterodyne signal processing without resource to specific frequency shifting element," *Proceedings of Optical Technology in Process Control (BHRA)*, The Hague, 111–119 (1983).

Kilpatrick, D., J. V. Tyberge, and W. W. Parmley, "Blood velocity measurement by fiber optic laser Doppler anemometer," *IEEE Trans. Biomedical Eng.*, **BEM-29**, 142–145 (1982).

Kist, R., H. Ramakrishnan, and H. Wolfelschneider, "The fiber Fabry–Perot and its applications as a fiberoptic sensor element," *SPIE Fiber Optic Sensors*, **585**, 126–133 (1985).

Koch, A. and R. Ulrich, "Displacement sensor with electronically scanned white-light interferometer," *SPIE Fiber Optic Sensors IV*, **1267**, 126–133 (1990).

Laming, R. I., M. P. Gold, D. A. Payne, and N. A. Haniwell, "Fiber Optic vibration probe," *SPIE, Fiber Optic Sensors*, **586**, 38–44 (1985).

Lee, C. E., and H. F. Taylor, *Elect. Lett.*, **24**, 198 (1988).

Lee, C. E., and H. F. Taylor, "Fiber-optic Fabry–Perot sensor using a low-coherence light source," *J. Light-Wave Technol.*, **9**, 129–134 (1991).

Lee, C. E., R. A. Atkins, and H. F. Taylor, "Performance of a fiber-optic temperature sensor from –200°C to 150°C," *Opt. Lett.*, **18**, 1038–1040 (1988).

Lee, C. E., H. F. Taylor, A. M. Markus, and E. Udd, "Optical-fiber Fabry–Perot embedded sensor," *Opt. Lett.*, **24**, 1225–1227 (1989).

Mariller, C. and W. Lequine, "Fiber-optic 'white-light' birefringent temperature sensor," *SPIE Fiber Optic Sensors 1*, **798**, 121–130 (1987).

Measures, R. M., "Smart composite structures with embedded sensors", *Composite Eng.*, **2**, 597–618 (1992).

Measures, R. M., S. Melle, and K. Lui, "Wavelength demodulated Bragg grating fiber optic sensing systems for addressing smart structure critical issues," *Smart Mater. Struct.*, **1**, 36–44 (1992).

Meggitt, B. T., "Optical fiber sensors for temperature and pressure measurement," *ESETEC Report No. 8043/88/NL/PB, Minimal Invasive Diagnostics*, **2**, 2.6–2.12 (1991).

Meggitt, B. T., A. C. Lewin, and D. A. Jackson, "A fiber optic non-contacting reference grade vibration sensor," *SPIE*, Olympia, **1120**, 307–315 (1989).

Meggitt, B. T., W. J. O. Boyle, K. T. V. Grattan, A. E. Bauch, and A. W. Palmer, "Heterodyne processing scheme for low coherence interferometric sensor systems," *IEE Proc.-J.*, **138**, 393–395 (1991).

Melle, S. M., K. Lui, and R. M. Measures, "A passive wavelength demodulation system for guided-wave Bragg grating sensors," *IEEE Photonics Tech. Lett.*, **4**, 516–518 (1992).

Meltz, G., W. W. Morey, and W. H. Glen, "Formation of Bragg gratings in optical fibers by a transverse holographic method," *Opt. Lett.*, **34**, 823–825 (1989).

Morey, W. W., G. Meltz, and W. H. Glenn, "Fiber optic Bragg grating sensors," *SPIE Fiber Optic and Laser Sensors VI*, **1169**, 98–107 (1989).

Ning, Y., K. T. V. Grattan, B. T. Meggitt, and A. W. Palmer, "Characteristics of laser diodes for interferometric use," *Appl. Opt.*, **28**, 3657–3661 (1989).

Ross, D. A., H. S. Dhadwal, and R. B. Dyott, "The determination of the mean and standard deviation of the size distribution of a colloidal suspension of submicron particles using the fiber optic Doppler anemometer, FODA," *J. Colloid. Interface Sci.*, **64**, 533–542 (1978).

Senior, J. W., and G. Murtaza, "Dual wavelength intensity modulated optical fiber sensor system," *SPIE Fiber Optics '89*, **1120**, 332–337 (1989).

Trouchet, D., F. X. Desforges, P. Graindorge, and H. C. Lefevre, "Remote fiber optic measurement of air index with white-light interferometry," *8th Optical Fiber Sensor Conference (OFS-8) IEEE*, Monterey, 57–60 (1992).

Tsubokawa, M., T. Higashi, and Y. Negishi, "Mode coupling due to external forces distributed along a polarisation-maintaining fiber: an evaluation," *Appl. Opt.*, **27**, 166–173 (1988).

Valluet, M. T., P. Graindorge, and H. J. Ardity, "Fiber optic pressure sensor using white-light interferometry," *SPIE Fiber Optic and Laser Sensors V*, **838**, 78–83 (1987).

Xu, M. G., L. Reekie, Y. T. Chow, and J. P. Dakin, "Optical in-fiber grating high pressure," *Elect. Lett.*, **29**, 389–399 (1993).

Yoshimo, T., K. Kurosawa, K. Itoh, and S. K. Sheen, "Fiber-optic Fabry–Perot interferomet its sensor applications," *IEEE J. Quantum Elec.*, **QE-18**, 1624–1633 (1982).

Yurek, A. M., A. B. Tveten, and A. Dandridge, "High performance fiber optic hydrophones in the arctic environment," *Proc. OFS '90*, 321 (1990).

Chapter 18

Nanotechnology for Sensors

W. Boyes

18.1 INTRODUCTION

Much has changed since the publication of the first edition of this book, but nothing has altered more radically than the field of nanotechnology. Today nanotechnology is beginning to have an impact on sensors, sensor design, and operation.

18.2 WHAT IS NANOTECHNOLOGY?

Nanotechnology, as defined by the *As in Encyclopedia Britannica*, is "Manipulation of atoms, molecules, and materials to form structures on the scale of nanometres (billionths of a metre)."[1] In his 1959 lecture, "There's Plenty of Room at the Bottom,"[2] Prof. Richard Feynman described a future filled with "nanomachines"—devices that could be constructed and operated on the molecular level.

Although Feynman's nanomachines have yet to materialize, nanotechnology itself is appearing more and more frequently in everyday life. Since the early 2000s clothing has been made with antistain coatings created using nanowhiskers, and nanocoatings have been used for sunblock and as antibacterial coatings in bandages.

18.3 NANOTECHNOLOGY FOR PRESSURE TRANSMITTERS

Nanotechnology has also made an appearance in automation. Newly designed ABB pressure transmitters are equipped with Diaflex-coated diaphragms. Diaflex is a nanostructured material with a hardness of 4,000 HV, which is similar to diamond. Diaflex is produced via a patented process that consists of a quaternary coating based on titanium and silicon (SiTiN) of the nitrides class. Diaflex is deposited using a technique called *physical vapor deposition* (PVD) on the

---

1. *nanotechnology*, Encyclopedia Britannica Online, March 15, 2009; www.britannica.com/EBchecked/topic/962484/nanotechnology
2. www.its.caltech.edu/~feynman/plenty.html

FIGURE 18.1 Nanodeposition on pressure transducer diaphragm. Courtesy of ABB Inc.

standard 316L stainless steel diaphragm. The Diaflex coating thickness is only 3 μm to 5 μm thick, so it allows a very good spring rate value, according to ABB. Diaflex's physical characteristics remain stable, between –100°C and +700°C (–148°F to +1,292°F), but the transmitter maximum working process temperatures are, as always, limited by the fill fluid.

18.4 MICROELECTROMECHANICAL SYSTEMS (MEMS)

One step up from nanotechnology is the world of microelectromechanical systems, or MEMS. This is where we currently find Feynman's small machines. In the automation environment, MEMS have already been used to create many kinds of sensors, with many more likely to come. MEMS sensors have the property of not only being very small in comparison with their standard technology predecessors but also of having very low power consumption, which makes them attractive for battery operation. They are also becoming much easier to manufacture at low cost, which makes the vision of Honeywell's David Kaufman of "lick-and-stick sensors" closer to a reality.

18.5 MEMS SENSORS TODAY

In 2009, there are MEMS technology sensors for pressure, temperature, viscosity, and such analytical measurements as pH, ORP, REDOX, and even complex chemical compounds. For example, Figure 18.2 shows a MEMS pressure transmitter from Turck Inc. Notice how small and lightweight it seems compared to more traditional pressure transmitter designs. Turck claims a lower price as well.

In 2006, BiODE, later the SenGenuity division of Vectron International, introduced a MEMS and nanotechnology device for the measurement of viscosity. This solid-state sensor used a technology called *acoustic wave technology* to produce an online viscometer with no moving parts and, according to the company, requiring only minimal maintenance and calibration.

In the future, automation professionals should expect to see many more MEMS- and nanotechnology-based products for sensors and perhaps also for other devices.

Nanotechnology expert and former Digital Equipment Corp. and Compaq Fellow Jeffrey R. Harrow has theorized on "smart" materials that can be bonded to the surface of a vessel and that will provide a digitized "window" into the vessel, complete with metadata on composition, performance algorithms, and control functions.[3] The technology of small machines will have a significant impact on automation in the 21st century.

FIGURE 18.2 MEMS Pressure Sensor. Courtesy of Turck Inc.

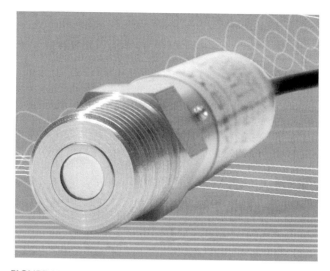

FIGURE 18.3 MEMS Viscosity Sensor. Courtesy of SenGenuity.

---

3. www.controlglobal.com/articles/2007/252.html

Chapter 19

Microprocessor-Based and Intelligent Transmitters

E. H. Higham and J. Berge

19.1 INTRODUCTION

The evolution in the design of transmitters has been influenced, on one hand, by the requirements of users for improved performance coupled with reduced cost of ownership and, on the other, by developments that have taken place in adjacent technologies such as computer-aided design (CAD), microelectronics, materials science, and communication technologies. The most significant advances have resulted from the emergence of low-power microprocessors and analog-to-digital converters that, in conjunction with the basic sensor circuits, can function on the limited power (typically less than 40 mW) available at the transmitter in a conventional 4–20 mA measurement circuit. This has provided two distinct routes for improving the performance of transmitters: (1) by enabling nonlinear sensor characteristics to be corrected, and (2) by enabling a secondary sensor to be included so that secondary effects on the primary sensor can be compensated. To differentiate the conventional transmitters (see Figure 19.1) from those in which corrections are applied to the primary sensor signal, using a microprocessor to process information that is embedded in memory (see Figure 19.2), or those in which a microprocessor is used in conjunction with a secondary sensor to derive corrections for the primary sensor signal (see Figure 19.3), the term *microprocessor-based* has come into use.

The fact that a microprocessor can be incorporated in a transmitter has also provided an opportunity to move from a regime in which only the measurement signal is transferred from the transmitter to a receiver, such as an indicator or

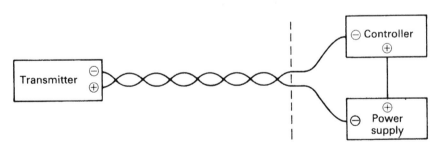

FIGURE 19.1 A conventional 4–20 mA transmitter circuit.

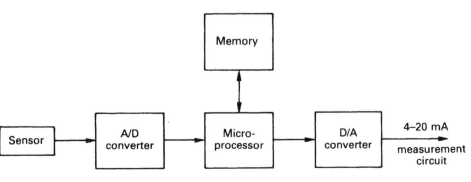

FIGURE 19.2 Components of a microprocessor-based transmitter.

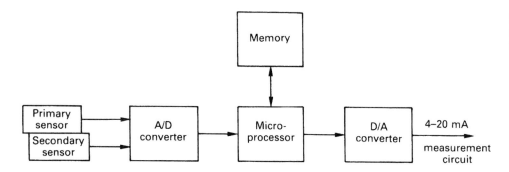

FIGURE 19.3 Components of a microprocessor-based transmitter with a secondary sensor.

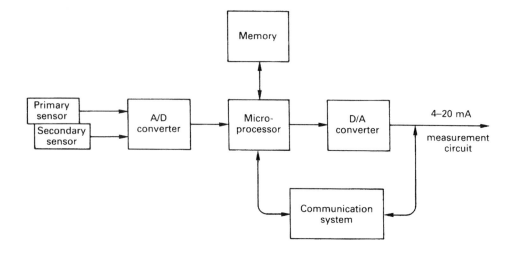

FIGURE 19.4 Components of a "smart" or "intelligent" transmitter with a communication facility.

controller, to one in which the microprocessor not only implements the microprocessor-based functions mentioned previously, it also manages a communication facility. This enables data specific to the transmitter itself, such as its type, serial number, and so on, to be stored at the transmitter and accessed via the measurement loop in which it is installed, as shown in Figure 19.4. Other functions, such as setting or resetting the zero and span, details of the location and application, and running diagnostic routines to give warning of malfunctioning, can also be implemented. The term *intelligent* has come to be used to identify such transmitters.

A further evolution that is now gaining acceptance is to multiplex the transmitter outputs onto a network or *fieldbus* instead of connecting the transmitters via individual circuits to the control room. For the concept to realize its full potential, an international standard is required to ensure that transmitters from different manufacturers are (1) *interchangeable*, that is, a transmitter from one manufacturer can be replaced by a transmitter from another manufacturer without any change to the system; or (2) *interoperable*, that is, a transmitter from one manufacturer can communicate with a device or host from another manufacturer. Work started in 1985 to develop a single international standard, but in the intervening period, proprietary and national standards have been developed and are competing to become the accepted industry standard.

19.2 TERMINOLOGY

There is a lack of consistency in the terminology used to describe the various attributes and features of these new transmitters; therefore, in the context of this chapter the following interpretations are used:

Sensor. A component that converts one physical parameter (e.g., pressure) into another parameter (e.g., electrical resistance).

Primary sensor. The sensor that responds principally to the physical parameter to be measured.

Secondary sensor. The sensor mounted adjacent to the primary sensor to measure the physical parameter that adversely affects the basic characteristic of the primary sensor (e.g., the effect of temperature on a pressure sensor).

Transmitter. A field-mounted device used to sense a physical parameter (e.g., temperature) at the point where it is mounted and to provide a signal in a 4–20 mA circuit that is a function (usually proportional) of the parameter.

Microprocessor-based transmitter. A transmitter in which a microprocessor system is used to correct nonlinearity errors of the primary sensor through interpolation of calibration data held in memory or to compensate for the effect of secondary influences on the primary sensor by incorporating a secondary sensor adjacent to the primary sensor and interpolating stored calibration data for both the primary and secondary sensors.

Intelligent transmitter. A transmitter in which the functions of a microprocessor system are shared among (1) deriving the primary measurement signal; (2) storing information regarding the transmitter itself, its application data, and its location; and (3) managing a communication system that enables two-way communication between the transmitter and either an interface unit connected at any access point in the measurement loop or at the control room.

Smart transmitter. An intelligent transmitter with simultaneous digital communication superimposed over the analog signal, that is, smart is a special case of intelligent.

Fieldbus network. A single communication medium such as a twisted pair of copper conductors, coaxial cable, or the like, which carries information such as address, control data, and process parameter data between a number of transmitters, actuators, controllers, and so on.

19.3 BACKGROUND INFORMATION

The range of microprocessor-based measuring systems now in use covers a wide variety of applications, but those that are smart or intelligent have been developed specifically for use in the process industries; in this chapter consideration is restricted to this particular range. The types and relative numbers of process measurements vary from one industry to another, but Figure 19.5 (based on transmitter sales data) shows that the most widely used measurements are temperature, pressure and differential pressure, flow, level, density, and analysis.

The principal development of smart transmitters has been concentrated on temperature, pressure, and differential pressure transmitters. For temperature transmitters, it is the data regarding the characteristic curves for various thermocouples, platinum resistance, and other temperature detectors that have to be embedded in memory, for interpolation by the microprocessor throughout the selected range of the transmitter. For pressure and differential pressure transmitters, it is the calibration data for both the primary and secondary sensors that have to be embedded in memory for interpolation by the microprocessor throughout the selected range of the transmitter. Differential pressure transmitters are particularly important because about half of the level measurements and about two-thirds of flow measurements are based on this measurement, although this latter proportion is declining.

The second important development arises from the potential benefits to be derived from being able to communicate with a remote instrument over the same network that carries the measurement signal. Hitherto, in the process industries, instruments were regarded as devices that are located in remote positions and connected by a twisted pair of wires to the point where the measurement signal is required. The technology that enables two types of information to be communicated over the same pair of wires, without mutual interference, has long been available but only recently has it been applied and exploited for process measurements.

Having developed transmitters in which microprocessors are used to enhance the performance by linearizing the basic sensor characteristic or compensating for secondary influences, it is a logical progression to add further memory so that other data, specific to the transmitter itself, such as its type, serial number, and so on, are stored at the transmitter. The microprocessor can be used to control a communication facility, so that information stored at the transmitter can be interrogated via the measurement circuit in which it is installed, and other functions, such as setting or resetting the zero and span, accessing data regarding the application, installed position, and service history can be implemented. Again, the term *intelligent* has come into use to identify transmitters with such facilities.

The concept of transmitters that not only provide a measurement signal but are also capable of two-way communication over the same circuit raised two issues, namely: (1) the need for a standard that will enable transmitters from different manufacturers to be interchanged (as is possible with the old 4–20 mA transmission standard); and (2) the need for a standard that enables the enhanced capabilities to be exploited. It also led to the possibility of replacing the individual circuits between transmitters and control rooms with a data highway. The term *fieldbus* has come into use to describe any form of data highway that supports communication between field devices and a control room.

A fieldbus has many advantages over the conventional 4–20 mA system, the most immediate of which are the reduced cost of wiring, commissioning, and maintenance as well as increased versatility and enhanced functionality. Work started in 1985 to develop a single international

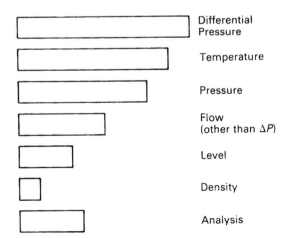

FIGURE 19.5 Relative numbers of process transmitters.

standard, but in the intervening period proprietary and national standards were developed that are now competing to become the accepted industry standard. In spite of this confused situation, the introduction of fieldbuses is now having a dramatic effect on the control industry as advances in adjacent technologies, such as signal processing, neural networks, fuzzy logic, and sensor arrays, improve the power and versatility of measurement and control systems. One of the first field communication protocols to be successfully exploited commercially was the HART Field Communication Protocol, supported by the HART Communication Foundation, an outline description of which is given in Section 19.10.4.

One of the many problems that has existed at large process plants is the variety of transmitters needed to meet the operational requirements and the costly inventory of complete instruments as well as spare components that have to be carried. This is due in part to the fact that, hitherto, most of the transmitters were analog devices and their basic design was such that the range of span adjustment was limited, so that many different primary sensors were required. The introduction of a secondary sensor and a microprocessor system has greatly increased rangeability and accuracy of transmitters so that, for pressure or differential pressure measurements, the operational requirements in most process plants can be covered by two transmitters of each type.

The 4–20 mA transmission standard provided an excellent basis for the development of process instrumentation during the past three decades. Its advantages include the fact that the measurement signal is not affected by changes in the loop resistance, transmitters from one manufacturer can be interchanged with those from another, and, within quite wide limits, individual instruments such as alarms and indicators can be added without affecting the accuracy of measurement. Furthermore, the power needed to energize the sensor and the signal conditioning circuits is provided within the live zero. However, the only information that can be communicated is the measurement signal, which is transmitted from the transmitter to the center or control room.

A dominant influence in the evolution of process measurement systems is the need to reduce the cost of ownership, that is, the cost of the transmitter itself plus the cost of installing and commissioning it in the plant and subsequently maintaining it.

Developments in the basic sensing elements have exploited CAD methods to enhance silicon strain-gauge technologies and the microfabrication of both capacitance and resonant sensing elements. These have progressed in parallel with developments in solid-state electronics, particularly the advances in low-power microprocessors and analog-to-digital converters. A further way of reducing the cost of manufacture is to reduce the weight and size of a device. If the transmitters that were generally available a decade ago are compared with those currently available, the improvements that have been achieved are significant.

19.4 ATTRIBUTES AND FEATURES OF MICROPROCESSOR-BASED AND INTELLIGENT TRANSMITTERS

Attributes and features that are provided in microprocessor-based and intelligent transmitters are described in this section, but the detailed implementations may differ from one manufacturer to another as well as from one type of transmitter to another, and some of the features may not be provided. Most of the intelligent features or attributes can be addressed from an appropriate handheld terminal (sometimes described as a handheld communicator or a handheld interface) or from a suitably equipped control room or computer system, although, in some instances, the communication can be effected via a locally mounted operator interface.

19.4.1 Microprocessor-based Features

19.4.1.1 Linearization, Characterization, and Correction of the Primary Sensor Characteristic

The presence of memory and computing power at the transmitter permits the signal to be conditioned before onward transmission. For example, the primary sensor signal may have a known nonlinear relationship with the measured variable. The most common examples of this are resistance temperature detectors, thermocouples, and the square law inherent in the relation between flow rate and differential pressure developed across an orifice plate or other head-producing primary element. In addition, however, actual calibration information for an individual transmitter may be stored in memory and used to enhance the accuracy of the output signal.

In those cases where the measurement system comprises two discrete units, such as a primary sensor and a separate transmitter that incorporates the computation unit, provision would be made to enter the constants of the primary sensor into the computation unit of the transmitter so that either unit could be changed.

As well as these purely mathematical manipulations of the original data, auxiliary internal measurements may be made of line pressure or temperature to permit the output to be corrected for the effect of these quantities on the transmitter's performance. That is a separate activity from correcting the resulting measurement for changes in the fluid properties with temperature or pressure, although some instruments may permit access to these auxiliary measurements over the communication link so that they could be used externally for that purpose.

The provision of linearized and corrected signals to a control system means that it is not burdened with the need to perform these computations, and the type or make of transmitter fitted in a particular location is of no significance to the control system, since it can assume that the received signal will always be linearly representative of the process parameter.

19.4.1.2 Inclusion of Control Functions and Other Algorithms

The microprocessors used in microprocessor-based or intelligent transmitters are more than capable of carrying out the relatively simple computation involved in on/off, two-term, or three-term control, and some instruments now available provide this feature. Tuning of the loop is then implemented using the communication link.

Until the advent of FOUNDATION Fieldbus, relatively few manufacturers were offering transmitters with this feature, so they were therefore rare in large and complex plants, which required more complex and sophisticated control than could be provided in early transmitter mounted devices. For small self-contained applications, the built-in PID controller in the Smar LD301 pressure transmitter has proved very useful. With FOUNDATION Fieldbus there is a move toward further field-mounted computation and data reduction with gas massflow calculations, for example, being carried out *in situ*. In a control system based on the FOUNDATION Fieldbus protocol, such as Smar's SYSTEM302, the regulatory control is primarily performed by the Fieldbus control valve positioner, such as the Smar FY302.

19.4.1.3 Expression of the Measurement in Engineering Units

There are several ways in which this measurement can be effected.

Adjustment of the 4–20 mA Range Using the span and zero setting functions, it is possible to set the values for the measured quantity at which the output shall be 4 mA and 20 mA, respectively. This is particularly useful in matching existing recorders or indicators.

Digital Communication It is generally possible to set the scaling of an instrument so that the measured variable (e.g., flow rate) shown on the handheld communicator or on the operator display appears in the desired or specified units of measure, with the abbreviation for the unit alongside (e.g., 7.5 kg/s). This can be set quite independently of the scaling of the 4–20 mA signal and avoids calculation errors by the operators converting from percentage flow rates to engineering units.

Pulse Outputs for Totalization Many flowmeters have provision for a pulse output at a frequency proportional to the flow rate. These pulses may be counted externally or in the instrument itself to provide a totalized flow readout. The scaling of this output can be chosen so that the interval between pulses represents a specific volume or mass of fluid. This feature is useful in that it permits simple counters to be used to record the total and permits a suitable pulsating rate to be chosen for electromechanical counters.

Failsafe Features All the intelligent flowmeters provide some internal diagnostic routines and are capable of flagging problems. In some instruments it may be possible to specify what should happen to the output under certain fault conditions—for example, "go to 3.6 mA" or "go to 21 mA."

These sections are so application specific that no general conclusion can be drawn. In some cases it may be important to stop the process; in others this might not be feasible and an assumed value for the process parameter might have to be substituted during failure.

There were initially some fears that these more complex instruments would fail more frequently than their analog counterparts. However, more than a decade of operation has proven them reliable, and when they do fail, much more information regarding the nature of the fault will usually be available.

19.4.2 Intelligent Features

The use of digital communication superimposed on the 4–20 mA measurement signal not only enables full advantage to be taken of the enhanced performance of the sensors, it also permits a wide range of information to be extracted from the transmitter on request and adjustments to be made to the mode of operation. Typical examples of these features are described in the following sections.

19.4.2.1 Adjustment of Span and Zero

This feature permits the full range of the analog output signal (usually 4–20 mA) to be used for a range that is less than the total measurement range of the primary sensor. It is usually employed to change the span without change to the zero, but a suppressed zero may be useful when the process variable has to be closely controlled around a specific value and there is no interest in values of the process variable outside that narrow band.

More commonly, it is used to permit the same instrument to be used for either large or small measurement ranges and has special significance when the transfer characteristic of the sensor is nonlinear. This is particularly true of orifice plate flowmeters in which the flow rate computation involves taking the square root of the signal representing the differential pressure.

Before the advent of intelligent transmitters, an orifice metering system for a flow turndown of 20:1 might have required three differential pressure transmitters with overlapping ranges. One intelligent transmitter can now provide the same analog outputs but with digital communication of the range setting. In general, this feature is only needed if an analog output is required, since a digital representation of the differential pressure is also available from the transmitter and this does not require any span and zero adjustment. An important result of this versatility that applies to all intelligent transmitters is that one model can be configured to deal with a wide range of applications, so fewer types and ranges of transmitters have to be held as spares at a process plant.

19.4.2.2 Adjustment of Damping, Time Constant, or Response Time

Most transmitters provide a selectable damping, time constant, or response time to permit fluctuations of the measured variable or electronic noise to be reduced in the output signal. Most manufacturers use a filter that may be represented by a single lag, but some use an adaptive filter, with different responses to small and large changes.

Selection of a long time constant will produce a smooth measurement signal but may mask the onset of instability in the process variable. When the measurement signal is being used as the input to a flow control loop, a long time constant may make tuning of the loop more difficult. Changing the time constant will almost certainly affect the tuning and result in either poor control or instability.

19.4.2.3 Diagnostic Routines and Status Information

There are two types of information; one concerns the normal status of the instrument and the other provides diagnosis of fault or unusual process conditions. Both are primarily of interest in plant maintenance and are therefore considered together.

Examples of the status information are as follows:

- Transmitter model and serial number.
- Primary variable and units of measure.
- Transmitter range and damping time constant.
- Primary variable corresponding to 4 and 20 mA.
- Plant tag number.
- Materials of construction of wetted parts.
- Software revision number.
- Date of last calibration.

The ability to call up this information assists in keeping an up-to-date plant instrumentation maintenance log that can be checked by actual interrogation to ensure that the type, range, and so on of the plant-mounted equipment is in line with the plant records as well as the operational requirements. Some of the status information, such as the model number, serial number, and materials of construction, where this is critical, is permanent. Other details are entered or modified when the transmitter is first installed or removed from service.

Diagnostic information is concerned with actual operation of the transmitter and changes much more frequently. Some diagnostic routines prevent discrepant configuration information from being entered (e.g., the measured value corresponding to 4 mA being set higher than that for 20 mA). Others warn of unexpected situations, such as reverse flow when the flowmeter is not set up to measure bidirectional flow; others may report internal failures in the electronics, such as failure to write to a memory location.

Many of these routines run continuously and set a flag immediately when a fault is identified. Others may be requested by the operator via a handheld terminal or from the control room, making it possible to check the operation of a transmitter without leaving the control room.

Note: It is important to appreciate that the information given in the following sections has been summarized from the various published descriptions and specifications. It is provided for guidance; therefore not only is it incomplete, it might also become out of date. Readers should refer to current manufacturers' documentation before embarking on any application.

19.5 MICROPROCESSOR-BASED AND INTELLIGENT TEMPERATURE TRANSMITTERS

Several companies manufacture transmitters of this type, but the units manufactured by MTL (see References) have been chosen as examples. There are two models, the MTL 414 and the MTL 418, both of which are extremely accurate two-wire 4–20 mA transmitters. The outline performance characteristics are set out in Table 19.1, and the MTL 418 transmitter itself is shown in Figure 19.6. The transmitters are designed for use in hazardous areas and can be installed in zone 0 for use with sensors also installed in zone 0, as defined by the IEC Standard 79–10. (Details of these requirements are given in Chapter 44.

FIGURE 19.6 The MTL 418 temperature transmitter in the surface mounting enclosure (cover removed) and the stem-mounted enclosure. Courtesy of Measurement Technology Ltd.

CHAPTER | 19 Microprocessor-Based and Intelligent Transmitters

TABLE 19.1 Outline performance specifications for the MTL 414 and MTL 418 transmitters

Accuracy	(MTL 414) ±0.1% of mV or ohm range
	(MTL 418) ±0.05% of mV or ohm range
Repeatability	(MTL 414) ±0.05%
	(MTL 418) ±0.025%
Supply voltage	12–42 Vdc
Analog output signal	4–20 mA
Fault indication	Selectable, output driven to 21 mA or 3.9 mA
Input isolation	500 V
Update time	Less than 0.5 s

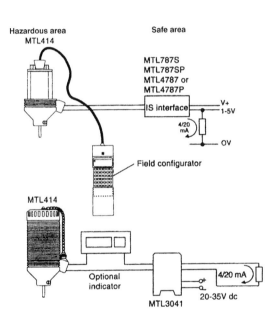

FIGURE 19.7 Installation of the MTL 414 transmitter, with a safety barrier or an isolator. Courtesy of Measurement Technology Ltd.

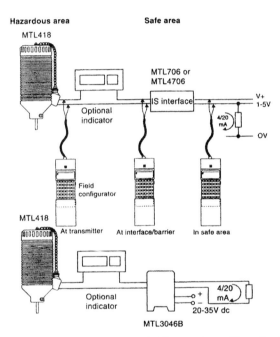

FIGURE 19.8 Installation of the MTL 418 transmitter, with a safety barrier or an isolator. Courtesy of Measurement Technology Ltd.

The MTL 414 is a conventional transmitter, requiring at least 12 Vdc at 20 mA. It can be used with the MTL 787S+ safety barrier or the MTL 3041 isolating interface units, as shown in Figure 19.7. It can also be configured for input type, range, and the like before or after installation. A local intrinsically safe indicator, such as the MTL 634 or MTL 681/2, can be included in the loop, provided that the interface unit or barrier gives sufficient voltage at 20 mA.

The MTL 418 combines the versatility of the MTL 414 with even greater accuracy and the ability to communicate digitally over the current measurement loop using the HART Communication Protocol, further details of which are given in Section 19.10.4. As shown in Figure 19.8, it can be configured from any point in the current loop provided that any intervening safety barriers, such as the MTL 706, or isolators, such as the MTL 3046B, can pass the communication signals. The unit implements all the 'universal commands' specified by the HART protocol as well as selected 'common practice' and 'transmitter-specific' commands, to provide a full range of functions. One special function is 'custom linearization,' by which points are chosen to modify the sensor characteristic. The input types, input ranges, spans, and minimum absolute errors for the two transmitters are set out in Table 19.2.

Configuration is implemented using the MTL 611 Field Configurator (these devices are also known as *handheld terminals* and *handheld interfaces*), shown in Figure 19.9. It is based on the Psion Organizer Model LZ, fitted with a CNF Interface. When used with the MTL 414, it is supplied with

TABLE 19.2 Range specification

Type	Input range		Span		Minimum absolute error value	
	Min	Max	Min	Max	MTL 414	MTL 418
E	−175°C	+1,000°C	50°C	1,175°C	±0.4°C	±0.2°C
J	−185°C	+1,200°C	60°C	1,385°C	±0.4°C	±0.2°C
K	−175°C	+1,372°C	60°C	1,547°C	±0.4°C	±0.2°C
T	−170°C	+400°C	75°C	570°C	±0.4°C	±0.2°C
R	+125°C	+1,768°C	360°C	1,643°C	±1.2°C	±0.6°C
S	+150°C	+1,768°C	370°C	1,618°C	±1.2°C	±0.6°C
B	+700°C	+1,820°C	700°C	1,120°C	±1.6°C	±0.8°C
N	0°C	+1,300°C	180°C	1,300°C	±0.4°C	±0.2°C
Pt 100 IEC	−200°C	+850°C	6°C	1,050°C	±0.2°C	±0.1°C
Pt 100 MINCO 11-100	−200°C	+850°C	6°C	1,050°C	±0.2°C	±0.1°C
Ni 120 MINCO 7-120	−70°C	300°C	3.5°C	370°C	±0.2°C	±0.1°C
mV	−15 mV	+150 mV	3 mV	165 mV	±16 μV	±8 μV
Ω	9 Ω	+500 Ω	2.5 Ω	500 Ω	±0.05 Ω	±0.025 Ω

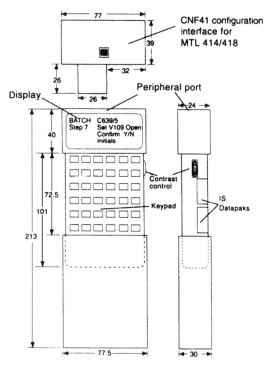

FIGURE 19.9 The MTL 611 field configurator. Courtesy of Measurement Technology Ltd.

an extension connector that plugs directly into the configuration port on the transmitter. For the MTL 418, the extension cable is fitted with two clips that can be connected at any access point in the measurement loop in either hazardous or safe areas.

Configuration information is presented on a four-line dot matrix display with 20 characters on each line, all options being selected from menus and submenus. Testing and monitoring can also be carried out with the configurator. The range and versatility of the options are shown in Figure 19.10. Configuration options include sensor type, choice of input values corresponding to both 4 and 20 mA, selection of the output span and the alarm conditions, and identification of the application, the variables being set directly from the configurator keyboard.

Testing options include facilities for checking the communication link, the loop current, and the transmitter self-diagnostic routines. The monitoring facilities include the continuous display of output current, measured temperature, measured millivolt, or input resistance value. Output data can be logged and stored in the configurator memory for subsequent transfer to a PC for analysis.

19.6 MICROPROCESSOR-BASED AND INTELLIGENT PRESSURE AND DIFFERENTIAL TRANSMITTERS

Conventional pressure and differential pressure transmitters are described in Chapter 9, but now there are numerous microprocessor-based and intelligent versions. The Rosemount Model 3051C (see References) is an excellent example of these and serves to illustrate the high standard of performance that is now achieved. The functional diagram of these transmitters is shown in Figure 19.11.

CHAPTER | 19 Microprocessor-Based and Intelligent Transmitters

The sensor module on which these transmitters are based is shown in Figure 19.12. In it, the process pressure is transmitted through the isolating diaphragm and fill fluid to the sensing diaphragm in the center of the capacitance cell. Electrodes on both sides of the sensing diaphragm detect its position, and the differential capacitance between the sensing diaphragm and the electrodes is directly proportional to the differential pressure.

The capacitance cell is laser welded and isolated mechanically, electrically, and thermally from the process medium and the external environment. Mechanical and thermal isolation is achieved by moving it away from the process flange to a position in the neck of the electronics housing. Glass-sealed pressure transport tubes and insulated cell mountings provide electrical isolation and improve the performance and transient protection. The signal from a temperature sensor incorporated in the cell is used to compensate for thermal effects.

During the characterization process at the factory, all sensors are run through pressure and temperature cycles over the entire operating range. Data from these cycles are used to generate correction coefficients, which are then stored in the sensor module memory to ensure precise signal correction during operation. This sensor module memory also facilitates repair, because all the module characteristics are stored with the module so that the electronics can be replaced without having to recalibrate or substitute different correction PROMs. Also located in the sensor are the electronics that convert the differential capacitance and temperature input signals directly into digital formats for further processing by the electronics module.

The electronics module consists of a single board incorporating ASIC and other surface-mounted components. This module accepts the digital input signals from the sensor module, along with the correction coefficients, and then corrects and linearizes the signal. The output section of the electronics module converts the digital signal into a 4–20 mA output circuit and handles the communication with the handheld terminal, which may be connected at any access point in the 4–20 mA measurement loop, as shown in Figure 19.13.

Configuration data are stored in nonvolatile EEPROM memory in the electronics module. These data are retained when the transmitter power is interrupted, so the transmitter

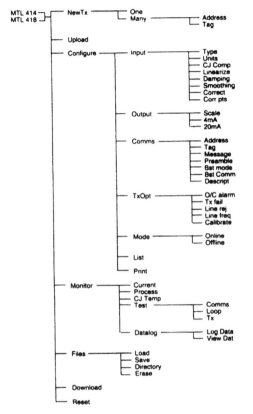

FIGURE 19.10 Configuration options for the MTL 414 and MTL 418 transmitter. Courtesy of Measurement Technology Ltd.

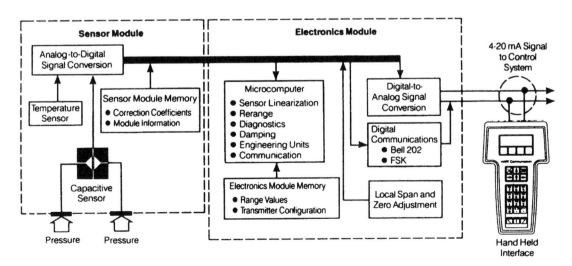

FIGURE 19.11 Functional diagram of the Model 3051C transmitter. Courtesy of Fisher-Rosemount.

is functional immediately on power-up. The process variable is stored as digital data, enabling precise corrections and engineering unit conversions to be made. The corrected data are then converted to a standard 4–20 mA current in the measurement circuit. The sensor signal can be accessed directly as a digital signal by the handheld terminal, bypassing the digital-to-analog conversion and thereby providing higher accuracy.

The Model 3051C transmitters communicate via the HART protocol, which uses the industry-standard Bell 202 frequency shift keying (FSK) technique (see Section 19.10.4). Remote communication is accomplished by superimposing a high-frequency signal on top of the 4–20 mA output signal, as shown in Figure 19.14. The Rosemount implementation of this technique allows simultaneous communications and output without compromising loop integrity.

The HART protocol, further details of which are given in Section 19.10, allows the user easy access to the configuration, test, and format capabilities of the Model 3051C.

The configuration consists of two parts. First the transmitter operational parameters are set; these include:

- 4 and 20 mA points.
- Damping.
- Selection of engineering units.

Second, informational data can be entered into the transmitter to allow identification and physical description of the transmitter. These data include:

- Tag (eight alphanumeric characters).
- Descriptor (15 alphanumeric characters).
- Message (32 alphanumeric characters).
- Date.
- Integral meter installation.
- Flange type.
- Flange material.
- Drain/vent material.
- O-ring material.
- Remote seal information.

In addition to these configurable parameters, the Model 3051C software contains several kinds of information that are not user-changeable, such as transmitter type, sensor limits, minimum span, fill fluid, isolator material, module serial number, and software revision number.

Continuous diagnostic tests are run by the transmitter. In the event of fault, a user-selected analog output warning is activated. The handheld terminal can then be used to interrogate the transmitter to determine the fault. The transmitter outputs specific information to the handheld terminal or control system to facilitate fast and easy correction of the fault. If the operator believes there is a fault in the loop, the transmitter can be directed to give specific outputs for loop testing.

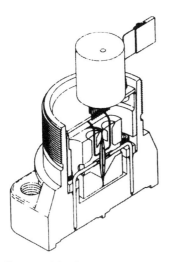

FIGURE 19.12 Sensor module of the Model 3051C transmitter. Courtesy of Fisher-Rosemount.

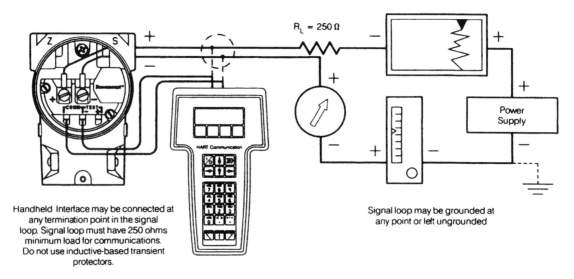

FIGURE 19.13 Connection of the transmitter in a measurement loop. Courtesy of Fisher-Rosemount.

The format function is used during the initial setup of the transmitter and for maintenance of the digital electronics. It allows the sensor and the 4–20 mA output to be trimmed to meet plant pressure standards. The characterize function allows the user to prevent accidental or deliberate adjustment of the 4 and 20 mA set points.

The three versions of the transmitter have many common features, but the sensor modules for the gauge and differential pressure transmitters differ from those incorporated in the absolute pressure transmitter. Details of the sensor ranges and sensor limits are set out in Table 19.3 and the principal specification features in Table 19.4, but this information is for illustration only and must be confirmed for any specific application. An exploded view of the transmitter is shown in Figure 19.15.

19.7 MICROPROCESSOR-BASED AND INTELLIGENT FLOWMETERS

Although the majority of flow measurements in the process industries continue to be based on orifice plate/differential pressure systems, other types of flowmeter, such as electromagnetic, vortex, and Coriolis mass flowmeters, are becoming used more widely. Most types of flowmeter are available from several manufacturers; the following are typical examples.

19.7.1 Coriolis Mass Flowmeters

The first practical Coriolis mass flowmeters were introduced by MicroMotion in 1977 (see References) and were the first

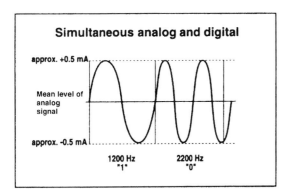

FIGURE 19.14 Frequency shift keying (FSK) signal superimposed on the 4–20 mA analog measurement signal. Courtesy of Fisher-Rosemount.

TABLE 19.3 Range limits for the sensor modules for the differential, gauge, and absolute pressure transmitters

	Model 3501 CD differential pressure transmitter		Model 3501 CG gauge pressure transmitter		Model 3501A absolute pressure transmitter	
Sensor code	Minimum	Maximum	Minimum	Maximum	Minimum	Maximum
0	NA	NA	NA	NA	1.15 kPa	34.6 kPa
1	210 Pa	6,220 Pa	NA	NA	6.89 kPa	207 kPa
2	2.07 kPa	62.2 kPa	2.07 kPa	62.2 kPa	34.5 kPa	1.034 MPa
3	8.28 kPa	248 kPa	8.28 kPa	248 kPa	186 kPa	5.51 MPa
4	69.0 kPa	2,070 kPa	69.0 kPa	2,070 kPa	913 kPa	27.6 MPa
5	460 kPa	13.8 MPa	460 kPa	13.8 MPa	NA	NA

NA = Not available.

TABLE 19.4 Outline performance specifications of the 3501C series of transmitters

Service	Liquid or gas
Output	Two-wire 4–20 mAdc with the digital value of the process variable superimposed
Power supply	10.5–55 Vdc
Temperature limits	Process: −40°C to +110°C Ambient: −40°C to +85°C
Damping	Adjustable from 0 to 16 s
Accuracy	±0.075% of spans for spans from 1:1 to 10:1 of URL
Stability	Range 2 & 3: ±0.1% of URL for 12 months Range 4 & 5: ±0.2% of URL for 12 months

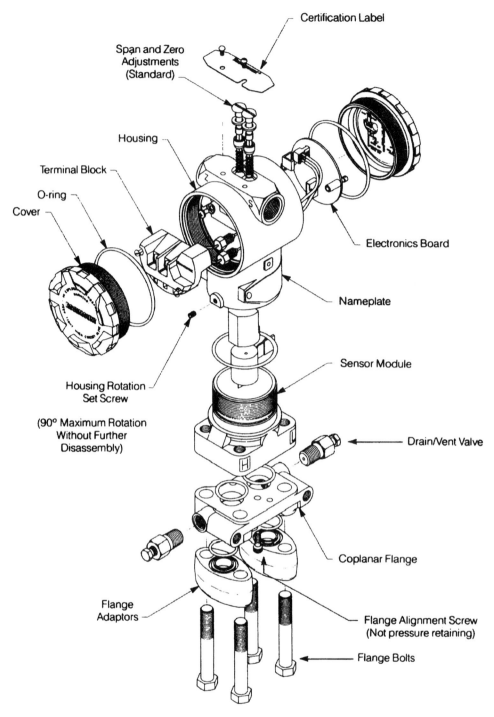

FIGURE 19.15 Exploded view of the model 3051 CD differential pressure transmitter. Courtesy of Fisher-Rosemount.

to provide direct, online measurement of mass flow rate, which is so important in many processes. The flowmetering system comprises a sensor and a signal processing transmitter. In most of the transmitters, the sensor consists of two flow tubes (Figure 19.16) that are enclosed in a sensor housing. The flow tubes vibrate at their natural frequency with an amplitude of less than 2.5 mm, being driven electromagnetically via a coil mounted at the center of the bend in the tube. Krohne was later the first to introduce a Coriolis unit with a single straight-through tube, also suitable for high flow rates, high-viscosity liquids, and pasty products.

The fluid flows into the sensor tube and is forced to take on the vertical momentum of the vibrating tube. When the tube is moving upward during half its vibration cycle, the fluid resists being forced upward by pushing down on the tube. Having the tube's upward momentum as it travels around the bend, the fluid flowing out of the sensor resists having its vertical motion decreased by pushing on the tube, causing it to twist. When the tube is moving downward during the second half of its vibration cycle, it is twisted in the opposite direction, as shown in Figure 19.17. This effect is known as the *Coriolis effect*.

The amount of sensor tube twist is directly proportional to the mass flow rate of the flowing fluid. Electromagnetic sensors on each side of the flow tube pick up the phase shift, which is proportional to the mass flow rate. At no flow, there is no tube twist, with the result that there is no phase difference between the two signals. With flow, twist occurs, resulting in a phase difference that is proportional to the mass flow rate.

Because the sensor tubes are fixed at one end and free at the other, the system can be regarded as a spring and a mass assembly. Once placed in motion, the assembly vibrates as its resonant frequency, which is a function of the tube geometry, material of construction, and mass. The total mass comprises that of the tube plus that of the flowing fluid, and for a particular sensor, the mass of the tube is constant and the dimensions are known, so that the frequency of oscillation can be related to the fluid density. However, the modulus of elasticity of the tube material varies with temperature, and therefore a temperature sensor is included in the assembly so that a correction can be applied to the density measurement for this effect. Thus the complete sensor assembly provides the signals from which the mass flow rate, density, and temperature of the flowing fluid can be determined by the associated transmitter.

There are four types of sensor: the CMF series, the D series, the DL series, and the DT series.

The "ELITE" (CMF) series of mass flowmeters is characterized by improved accuracy (±0.15% ± sensor stability), reduced pressure drop, and wide rangeability (80:1). The sensitivity to field effects such as vibration and temperature has been reduced, and for increased safety, the sensors are enclosed in pressure containment housing (see Table 19.5).

The D series of sensors is available in a variety of line sizes (Table 19.6) that enable measurement of massflow rates from 0.11 b/min to 25 000 lb/min. The rangeability of the individual sensors is 20:1 and the accuracy of measurement is ±0.20% of rate ± sensor stability. There are alternative materials of construction for the wetted parts, and the sensor itself is enclosed in an hermetically sealed case.

The DL series sensors (see Table 19.7) are constructed with a continuous stainless-steel flow tube, and the design meets the 3A Sanitary Standards. The singleflow path facilitates cleaning and can be "pigged." As with the D series, the rangeability of the individual sensors is 20:1 and the accuracy of measurement is ±0.20% of rate ± sensor stability.

The DT series of sensors has been designed to operate at up to 800°F (426°C). It is a dual-tube design using Hastelloy C-22 tubing, 316L stainless-steel manifolds, and flanges. The sensor is enclosed in a hermetically sealed 304 stainless-steel case.

The model RTF9739 transmitter can be used in conjunction with any of these sensors. The model RFT9712 transmitter is designed to work with the D, DL, and DT series of sensors. Both transmitters require an AC power supply (15 W maximum) or a DC supply of 12–30 V (14 W maximum).

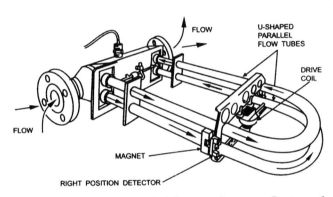

FIGURE 19.16 MicroMotionCoriolis mass flowmeter. Courtesy of Fisher-Rosemount.

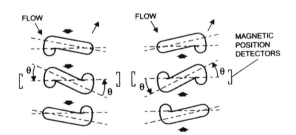

FIGURE 19.17 Coriolis effect. Courtesy of Fisher-Rosemount.

TABLE 19.5 Outline details of the ELITE (CMF) series of mass flow sensors

Model code	Nominal flow range (lb/min)	Usable range (lb/min)	Zero stability (lb/min)	Pressure rating (psi)	Temperature rating (°F)
CMF025	0–40	0–80	0.002	1,450	−400 to +400
CMF050	0–125	0–250	0.006	1,450	−400 to +400
CMF100	0–500	0–1,000	0.025	1,450	−400 to +400
CMF200	0–1,600	0–3,200	0.08	1,450	−400 to +400
CMF300	0–5,000	0–10,000	0.25	1,450	−400 to +400

TABLE 19.6 Outline details of the D series of mass flow sensors

	Maximum flow range (lb/min)	Minimum flow range* (lb/min)	Zero stability (lb/min)	Pressure rating (psi)	Temperature rating (°F)	Wetted parts
Standard pressure						
D6	0–2	0–0.1	0.0002	2,600	−400 to 400	316L
D12	0–11	0–0.55	0.0001	1,700	−400 to 400	316L
D25	0–45	0–2.25	0.005	1,900	−400 to 350	316L
D40	0–90	0–4.5	0.009	1,250	−400 to 350	316L
D65	0–300	0–15	0.03	2,250	−400 to 350	316L
D100	0–1,000	0–50	0.10	2,250	−400 to 400	316L
D150	0–2,800	0–140	0.30	1,500	−400 to 400	316L
D300	0–7,000	0.350	0.70	740	−400 to 400	316L
D600	0–25,000	0–1,250	2.5	625	−400 to 400	316L
Hastelloy C						
D6	0–2	0–0.1	0.0002	2,600	−400 to 400	Hastelloy C
D12	0–13	0–0.65	0.001	1,700	−400 to 400	Hastelloy C
D25	0–45	0–2.25	0.005	1,900	−400 to 350	Hastelloy C
D40	0–120	0–6	0.01	1,250	−400 to 400	Hastelloy C
D100	0–1,000	0–50	0.1	2,250	−400 to 400	Hastelloy C
D150	0–3,000	0–150	0.3	1,500	−400 to 400	Hastelloy C
D300	0–7,200	0–360	0.7	740	−400 to 400	Hastelloy C
Lined						
D150	0–2,800	0–140	0.3	1,000	32 to 250	Tefzel
D300	0–7000	0–350	0.7	740	32 to 250	Tefzel
High pressure						
DH6	0–2	0–0.2	0.0004	5,700	−400 to 400	316L
DH12	0–11	0–1.2	0.002	5,700	−400 to 400	316L
DH25	0–45	0–3	0.0006	4,000	−400 to 350	316L
DH40	0–90	0–10	0.02	2,800	−400 to 400	316L
DH100	0–1,000	0–150	0.3	5,600	−400 to 400	316L
DH150	0–2,800	0–600	1.2	5,600	−400 to 400	316L
DH300	0–7,000	0–2,000	4.0	4,000	−400 to 400	316L

*The lowest flow range over which a flowmeter can be spanned.

TABLE 19.7 Outline details of the DL series of mass flow sensors

	Maximum flow range (lb/min)	Minimum flow range* (lb/min)	Zero stability (lb/min)	Pressure rating (psi)	Temperature rating (°F)	Wetted parts
DL65	0–250	0–12.5	0.025	1,500	−400 to 350	316L
DL100	0–800	0–40	0.08	900	−400 to 350	316L
DL200	0–3,500	0–175	0.35	740	−400 to 400	316L

*The lowest flow range over which a flowmeter can be spanned.

TABLE 19.8 Outline details of the DT series of mass flow sensors

	Maximum flow range (lb/min)	Minimum flow range* (lb/min)	Zero stability (lb/min)	Pressure rating (psi)	Temperature rating (°F)	Wetted parts
DL65	0–300	0–15	0.03	900	32 to 800	Hastelloy C/316L†
DT100	0–800	0–40	0.08	900	32 to 800	Hastelloy C/316L
DT150	0–1,400	0–70	0.14	600	32 to 800	Hastelloy C/316L

*The lowest flow range over which a flowmeter can be spanned.
†The tubing is Hastelloy C and manifold castings and flanges are 316L stainless steel.

The RFT9739 has two independently configured outputs, each of which can represent mass flow rate, density, or temperature. There is also an isolated 0–15 V frequency output that is scalable from 1 to 10,000 Hz and a control output that can represent flow direction or fault alarm. Digital communication can be implemented using the Bell 202 communication standard signal (see Section 19.10.4) superimposed on the 4–20 mA circuit or via an RS 485 port. The RFT9712 has similar features except that it has only one analog output, which can be configured to represent mass flow rate, density, or temperature.

Using the Rosemount Model 268 handheld terminal, the transmitter operational parameters that are set include:

- 4 and 20 mA points for the analog output(s).
- Damping.
- Selection of engineering units.

The following informational data can be entered to allow identification and physical description of the transmitter:

- Tag (eight alphanumeric characters).
- Descriptor (15 alphanumeric characters).
- Message (32 alphanumeric characters).
- Date.

Data that are stored in nonvolatile memory include:

- Transmitter model number.
- Line size.
- Material of construction.
- Sensor limits.
- Minimum span.
- Transmitter software revision level.

19.7.2 Electromagnetic Flowmeters

Since the first applications of electromagnetic flowmeters in the process industries during the mid-1950s, their design has progressed through several evolutionary phases and their performance has now reached a very high standard. As described in Chapter 6, their principal attributes are that the head loss is negligible and the output signal is proportional to flow rate over a very wide range. The sensor is relatively insensitive to velocity profile and hence to changes in viscosity, with the result that the installation requirements are less severe than for most other types of flowmeter. Provided that the fluid conductivity is above a specified minimum value (typically 5 μS/cm), the accuracy of measurement is typically better than 1 percent. However, the system is not suitable for gases.

Electromagnetic flowmeters differ from pressure and many other transmitters in that they require considerably more power than is available in a conventional 4–20 mA measurement circuit. However, some manufacturers have developed transmitters that include intelligent features; the Invensys (formerly Foxboro) 8000 Series (see References) is representative of them.

The system involves two components: the flowtube, which is mounted in the pipeline and senses the flow, and the transmitter, which takes the small signal developed between the electrodes by the flowing fluid and converts it into an output signal such as 4–20 mA in a conventional current loop or a pulse output signal. The two units can be combined, as shown in Figure 19.18, or mounted separately.

The transmitter provides three electrical outputs: (1) an analog signal in the range 4–20 mA current into a 300 Ω load when powered internally or into 1800 Ω load when powered by an external 50 V DC supply; (2) a digital signal that is superimposed on the 4–20 mA measurement loop; and (3) a low- or high-rate pulse output that may be powered internally or externally. There is an LCD display that shows the flow rate either as a percentage of the upper range value or in engineering units.

Other features include a built-in calibrator that enables the units of measure and the upper-range value of the system to be set for a particular application. During power-up and periodically during normal operation, diagnostic routines are run to identify and isolate faults in either the transmitter or the flowtube. Means are also provided to prevent false measurements when the flowtube has run empty and to cut off the output signal when the flow rate falls below 2 percent of the upper range value.

The transmitter communicates bidirectionally to a handheld terminal (HHT), shown in Figure 19.19, or to the Invensys (formerly Foxboro) I/A system, using the frequency shift keying technique with the signals superimposed on the 4–20 mA measurement loop. The handheld terminal can be connected at any point in a 4–20 mA measurement circuit in a general-purpose, ordinary location or a Division 2 area.

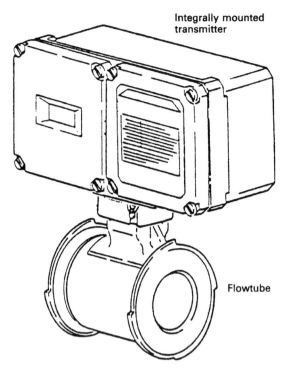

FIGURE 19.18 Electromagnetic flowmeter with transmitter mounted integrally with the flowtube. Courtesy of the Invensys (formerly Foxboro) Company.

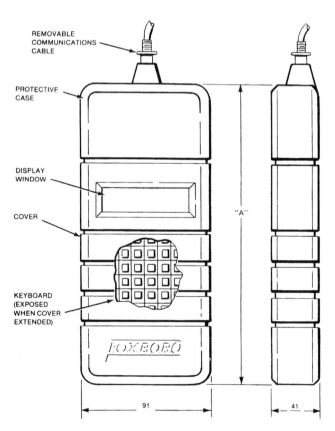

FIGURE 19.19 Handheld terminal. Courtesy of the Invensys (formerly Foxboro) Company.

The information that can be displayed includes:

- Status of the continuous self-diagnostic routine.
- Day, date, and time.

In addition, the following information can be displayed and/or configured:

- Measurement output (in engineering units).
- Pulse output parameters.
- Flowtube lining material.
- Flowtube maximum flow range.
- Calibrated upper range value.
- Selected upper range value.
- Response time (electronic damping).
- Nominal line size.
- Device name.
- Tag name.
- Tag number.
- Location.

Details of the Invensys (formerly Foxboro) 8000 series of flowmeters are given in Tables 19.9 and 19.10.

19.7.3 Vortex Flowmeters

The principal attributes of this type of flowmeter are that it has no moving parts and the output signal is proportional to flow rate over a range that varies between 10:1 and 15:1. It is insensitive to moderate changes of the fluid density and viscosity and is suitable for measurement of both liquid and gas flows. However, in practice it is restricted to line sizes between 15 and 300 mm, although at present this range is not available from any one manufacturer. As with the electromagnetic flowmeter, these flowmeters do not require a secondary sensor to enhance their performance. However, the inclusion of a microprocessor system in the transmitter not only enables the slight nonlinearity of the primary element to be corrected, it also enables a communication facility to be added. An example of such a flowmeter is the Yokogawa YEWFLO (see References), which is illustrated in Figure 19.20.

The measurement is based on the detection of the vortices shed from a bluff body, set diametrically across the flowtube and known as a *Karman street of vortices*. The frequency of vortex shedding is proportional to the flow velocity and inversely proportional to the width of the bluff body. The relationship, known as the *Strouhal number*, is constant over a wide range of Reynolds numbers.

There are two versions of the flowmeter: one has the transmitter mounted integral with the detector, and the other has the transmitter mounted separately from detector, for use when the temperature of the process fluid is high. The YEWFLO is designed to operate in a conventional 4–20 mA measurement circuit and the converter provides both analog

CHAPTER | 19 Microprocessor-Based and Intelligent Transmitters

TABLE 19.9 Line sizes and upper range value limits of the Foxboro 8000 series of electromagnetic flowtubes

Flowtube code	Nominal line size (mm)	Upper range values (l/min) Min	
		Min.	Max.
Ceramic lining			
800H	15	4.0	80
8001	25	14	280
8001H	40	34	680
8002	50	51	1,000
8003	80	125	2,500
8004	100	220	4,400
8006	150	490	8,750
PTFE lining			
8006	150	600	12,000
8008	200	1,030	20,600
8040	250	1,645	32,800
8012	300	2,350	46,900
Polyurethane lining			
8006	150	465	9,250
8008	200	870	17,300
8010	250	1,430	28,600
8012	300	2,100	41,800

TABLE 19.10 Outline performance specifications for Foxboro 8000 series of electromagnetic flowmeters

Accuracy	Pulse output: ±1.0% of flow rate
	Analog: As pulse output, ±0.1% of URV
Repeatability	0.05% of span
Power consumption	Less than 10 W
Response time	1s for output to rise from 10% to 90% for 100% step change of input
Output damping	Adjustable from 1 to 50 s

and pulse outputs, as well as a single line LC display which can be set to show either the totalized flows, or the instantaneous flow rate in engineering units or as a percentage of the span. Three keys are provided for entering the flow meter parameters, and communication is via the Yokogawa BRAIN handheld terminal, which may be connected to the measurement circuit at any intermediate access point, as shown in Figure 19.21.

The information regarding a transmitter that is accessible but not changeable through the communication system includes:

- Model code and line size.
- Type of output.
- Process connections.
- Shedder and body materials.
- Calibration factor.

Information that can be displayed and is specific to an application includes:

- Tag and location.
- Application.
- Process fluid.
- Span.
- Units of measure.
- Flow rate.
- Totalized flow.
- Error messages/status of diagnostic routines.
- Damping time constant.

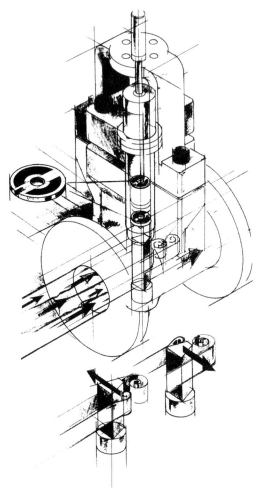

FIGURE 19.20 Illustration of the Yokogawa vortex flowmeter. Courtesy of Yokogawa Corp of America Inc.

The outline performance specifications are given in Table 19.11, and the relationship between flow rate and pressure loss factor for the various line sizes is given in Figure 19.22.

19.8 OTHER MICROPROCESSOR-BASED AND INTELLIGENT TRANSMITTERS

19.8.1 Density Transmitters

There are three principal methods for measuring the density of process fluids. The most widely applied method involves the measurement of the differential pressure between two points with a known vertical separation (H in Figures 19.23 and 19.24), as described in Chapter 13. In such cases, the liquid density is a function of the pressure difference divided by the vertical height (H) between the points of measurement. If a smart or intelligent transmitter of the type described is used, full advantage can be taken of the facility to identify the transmitter details, its location, application, zero and span, units of measure, damping, and so on. Although the transmitter would almost certainly include a secondary temperature sensor, it is unlikely that the temperature measurement signal would be sufficiently close to that of the process fluid for it to be used for that purpose.

A practical difficulty that is likely to arise in using a differential pressure transmitter in this way is that the process fluid in these connections is static and therefore may become blocked due to sedimentation or solidification of the process fluid as a result of changes in ambient conditions. This can be mitigated by using a differential pressure transmitter fitted with pressure seals and capillary connections,

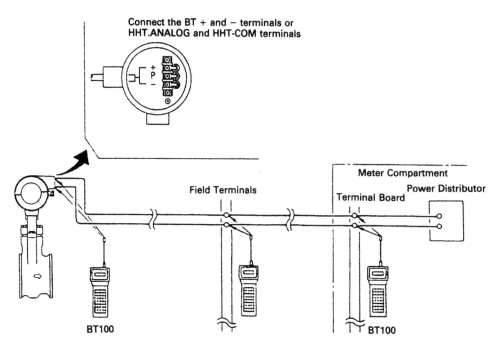

FIGURE 19.21 Arrangement of the measurement circuit for a vortex flowmeter. Courtesy of Yokogawa Corp of America Inc.

CHAPTER | 19 Microprocessor-Based and Intelligent Transmitters

TABLE 19.11 Outline performance specifications of the YEWFLO vortex flowmeter	
Service	Liquid, gas, or steam
Accuracy	Liquids: ±1.0% of reading
	Gas and steam: ±1.0% for velocities up to 35 m/s
	±1.5% for velocities from 35 to 80 m/s
Repeatability	±0.2% of reading
Process temperature limits	−40°C to +400°C
Analog output	4–20 mA DC two-wire system, 20–42 V DC (delay time 0.5 s)
Pulse output	Voltage pulse, three-wire system (delay time 0.5 s, scaled and unscaled)

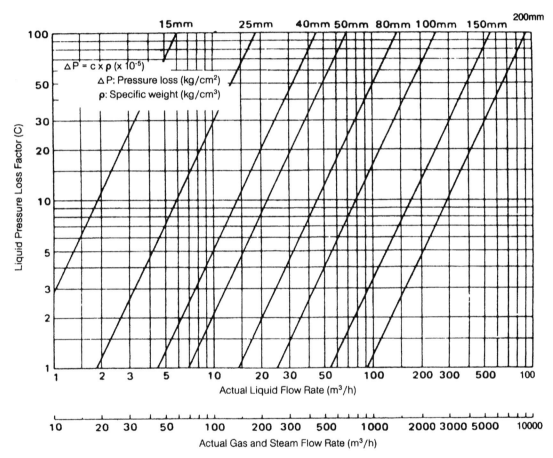

FIGURE 19.22 Pressure loss factor versus flow rate for the range of flowmeter line sizes. Courtesy of Yokogawa Corp of America Inc.

as shown in Figure 19.23. The disadvantages of this arrangement are the higher cost of a transmitter fitted with the pressure seals and the reduced accuracy of the system. An alternative approach is to use two flange-mounted pressure transmitters, mounted a known vertical distance apart, as shown in Figure 19.24. In this case, the pressure difference is determined by subtracting the pressure measured at the upper level from that measured at the lower level. Thanks to modern multivariable technology, a third option has become available only in the past few years. The SMAR DT301 is a single integrated smart device with an insertion probe that has large diaphragms and a true process temperature sensor to overcome the previously mentioned problems and at the same time compute density, referred density, and concentration using the onboard microprocessor. For either method to be successful, it is essential to ensure that the liquid level does not fall below the position of either the upper pressure seal or the connection to the upper pressure transmitter.

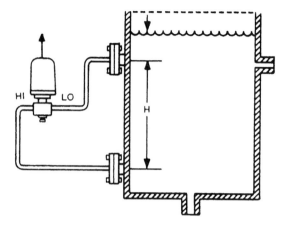

FIGURE 19.23 Measurement of liquid density using a differential pressure transmitter fitted with pressure seals.

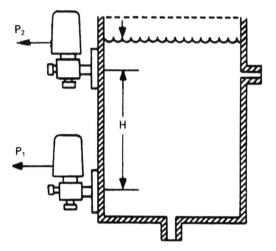

FIGURE 19.24 Measurement of liquid density using two pressure transmitters.

The Solartron 7835 series of liquid density transducers (see References), referred to in Chapter 13, represents an entirely different approach to the measurement of liquid density. It utilizes a straight length of pipe that is maintained in resonance by an electronic feedback system. A change in the density of the fluid in the tube changes the mass of the resonating element; this, in turn, changes the resonant frequency. Refinements in the construction of the resonant tube, the positioning of the excitation coils and the motion sensors, and the inclusion of a secondary temperature sensor have resulted in a transmitter that is very accurate, stable, and repeatable.

However, it differs from the microprocessor-based and intelligent transmitters described previously in that it operates in conjunction with a separate signal converter that provides the power to operate the circuits that maintain the tube in vibration. The temperature is sensed by a four-wire, 100 Ω platinum RTD and transmitted direct to the converter while the resonant frequency signal is superimposed on the power supply lead.

The converter computes the line density, the transducer temperature correction, and the transducer temperature correction, when this is measured by a separate pressure transmitter. It is provided with an RS 232 C or RS 485/422 communication port, but no provision is made to store the range or type of information that is stored in some other intelligent transmitters.

The Solartron Type 7826 insertion liquid density transducers (see References) also employ a resonant sensor based on a tuning fork mounted on a standard flange for insertion into a pipe or tank. The tines are excited into oscillation by a piezoresistive element positioned at the root of the tines. The oscillation is detected by a second piezo element, and the fork is maintained at its resonant frequency by a phase-corrected amplifier. As the fluid density varies it changes the vibrating mass, which in turn changes the resonant frequency. The unit differs from transmitters described previously in that they operate in conjunction with a separate signal converter that provides the power to operate the circuits, which maintain the tube in vibration. The temperature is sensed by a four-wire, 100 Ω platinum RTD and transmitted directly to the converter, while the resonant frequency signal is superimposed on the power supply lead.

For the measurement of gas density, there is a different type of insertion transducer, also described in Chapter 13. The Solartron Type 7812 gas density transducer (see References) is based on a thin-walled metal cylinder, contained in a constant volume chamber and maintained in resonance at its natural frequency. The gas sample is brought to the sensor via a sampling loop so that the gas flows over both surfaces of the cylinder. The mass of gas in contact with the cylinder, which depends on the gas density, changes the resonant frequency of the sensor. The electronic circuits that maintain the cylinder in resonance are mounted on the stem or flange carrying the sensor and are connected to a remote mounted converter, which provides the power for the electronic circuits and also computes the density, taking into account the operating temperature and pressure.

For gas density measurements, it is important to know both the temperature and pressure. The former is measured by a four-wire, 100 Ω platinum RTD located in the wall of the chamber, but the latter has to be measured by a separate pressure transmitter.

γ-Ray absorption can be used to measure the density of moving or flowing liquids, slurries, or solids, of which the Thermo Measuretech (formerly TN and KAYRAY) Model 3660 system (see References) provides an example. In it, the γ-beam emitted from a source is directed through the process pipe toward the detector/electronics. The amount of γ-radiation that passes completely through the pipe and its contents varies inversely with the density of the material within the pipe. The detector/electronics contain a scintillation counter that, when subjected to the γ-beam, produces light photons that are amplified through a scintillation sensor. The number of pulses from the scintillation sensor is directly proportional to the intensity of the received beam.

These pulses are conditioned, counted, and scaled by of the density of the process fluid. The associated microprocessor-based transmitter conditions the signal further and generates a conventional 4–20 mA analog output signal, but the unit also includes features of a smart transmitter so that the density measurement can also be provided digitally using the HART communication protocol, either superimposed on the 4–20 mA signal or via a separate intrinsically safe RS 423 port on the transmitter.

The equipment is mounted around an existing pipe and so avoids the need to modify a plant or to interrupt its operation during installation. Because the system is not intrusive, it is not affected by the process pressure, viscosity, corrosiveness, or abrasiveness.

Using the HART protocol, the transmitter parameters are entered in the configuration phase and include:

- Model number and serial number.
- Date and time.
- Reference material.
- γ-Source details.
- External input (RTD or corresponding 4–20 mA temperature signal).

In the characterization phase, the following details are entered:

- Units of measure.
- Span of 4–20 mA output.
- Temperature compensation characteristic.
- Damping (constants or the adaptive implementation).

The application data that are entered include:

- Tag (an eight-character field).
- Descriptor (a 16-character field).
- Message (a 32-character field).
- Date.
- Pipe description.
- Process description.

The digital trim involves:

- Reference to the process.
- Development of the process density calibration curve.
- Calibration of the temperature input.
- Calibration of the 4–20 mA output.

19.8.2 Microprocessor-based and Intelligent Liquid Level Measurement Systems

A great many level measurements that are made for process control purposes are based on the measurement of hydrostatic pressure and are referred to briefly in Chapter 12. For an open vessel, a gauge pressure transmitter mounted either in the base or close to it, as shown in Figure 19.25, is used to avoid the effects of ambient pressure. The liquid level is proportional to the pressure divided by the liquid density. For a closed vessel a second transmitter should be mounted in the roof, as shown in Figure 19.26, or above the highest level to be reached by the liquid. Alternatively, a differential pressure transmitter can be used, with the high side at the level tap and the low side to the top of the tank.

For these measurements, the value of the liquid density must be assumed, but if the microprocessor-based and intelligent transmitters described previously are used, information regarding the application, tag and location, process liquid, span and units of measure, and so on can be stored, in addition to the data regarding the transmitter model and serial number and the like. However, it is important to remember that changes in the temperature of the stored liquid may lead to changes in density, which, in turn, may result in an unacceptable loss of accuracy. A third transmitter located at a fixed distance above the bottom transmitter can be used to obtain the density and make the necessary correction.

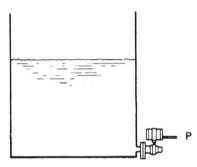

FIGURE 19.25 Measurement of liquid level in an open vessel using a single pressure transmitter.

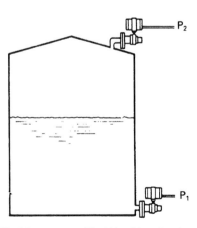

FIGURE 19.26 Measurement of liquid level in a closed vessel using two pressure transmitters.

19.9 OTHER MICROPROCESSOR-BASED AND INTELLIGENT MEASUREMENT SYSTEMS

An interesting application of microprocessor-based and intelligent instrumentation occurs in the metering of natural gas. A typical example is the Spectra-Tek Flow Computer (see References), shown in Figure 19.27.

The preferred method for fiscal metering of gas flow is by the measurement of the differential pressure across an orifice plate. The traditional approach to achieve high accuracy over a wide flow range has been to install several differential pressure transmitters, with overlapping ranges, across a single orifice plate, as shown in Figure 19.28. With the improvements that have been made in the basic sensors and the introduction of smart transmitters, it has become possible to use a single transmitter, the output from which is a conventional 4–20 mA analog signal with the digital measurement signal superimposed on it.

Taking the Honeywell ST3000 Model STD120, the reference accuracy for the digital mode is 0.0625 percent of calibrated span or 0.125 percent of reading for spans between 62 mbar and 1,000 mbar. This means that the transmitter can be calibrated for the differential pressure corresponding to the maximum flow rate and operated over a wide range of flows with optimal accuracy.

The Spectra-Tek Sentinel Flow Computer provides the facilities to communicate digitally with the transmitter. It can either use the transmitter's analog output, switching the range of the transmitter as the flow rate varies, or it can use the digital value received from the transmitter, in which case no range changing is necessary. The unit also provides similar facilities for temperature and pressure transmitters, both of which could be "intelligent," and it also accepts the frequency signal from a gas density transmitter.

The computer calculates the mass flow rate in accordance with ISO 5167, with temperature correction for pipe and orifice bores. Flowing density, standard density, and standard compressibility are calculated by the primary method defined in AGA 8, using measured pressure and composition data for up to 20 components. The calculated standard density is used to derive standard volume. Calorific value is calculated in accordance with ISO 6976 using the measured temperature, pressure, and gas composition data. Calorific value is used to calculate energy flow. Mass, standard volume, and energy are each available as a cumulative total, a daily total, and the instantaneous flow rates. Analog outputs are available for each flow rate, together with an incremental pulse output for mass flow rate.

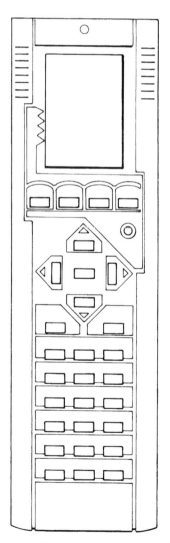

FIGURE 19.27 Spectra-Tek flow computer. Courtesy of Spectra-Tek Ltd.

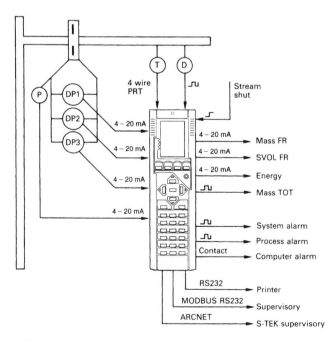

FIGURE 19.28 Natural gas cost-metering installation. Courtesy of Spectra-Tek Ltd.

Three separate alarm outputs are provided to indicate computer alarm (generated by internal diagnostic routines), system alarm (generated by a fault in the internal or external measurement circuits), and process alarms (generated by process conditions moving out of limits). These hardwired alarm outputs operate in conjunction with the flow computer's front panel display and may be connected to an external annunciator system.

19.10 FIELDBUS

19.10.1 Background

At present, the majority of microprocessor-based and intelligent transmitters operate in conventional 4–20 mA measurement circuits, but it is inevitable that the advantages to be gained from multiplexing signals from sensors, actuators, and controllers will be a feature of new process plants as well as plant upgrades. However, to implement these changes, a common communication system, or *fieldbus*, needs to be established.

There are many advantages of a fieldbus over conventional analog systems, the most immediate of which are reduced cost of installation and maintenance, increased flexibility, and enhanced functionality, such as self-diagnostics, calibration, and condition monitoring. The additional benefits available with these microprocessor-based and intelligent devices will also expand the use of sensors and control systems.

Work to develop a single international fieldbus standard started in 1985. The intervening period saw the development of a variety of local national standards and proprietary industrial standards, leading to the present situation, in which several standards are competing to become the accepted industry standard. In spite of this, the fieldbus is now having a dramatic effect on the control industry.

19.10.2 Introduction to the Concept of a Fieldbus

Conventional transmitters, actuators, alarm and limit switches, and controllers are connected directly to a control room via individual cables, as shown in Figure 19.29. The fieldbus is based on technology that allows devices to share a common communication medium, such as a single twisted-pair cable, so that transmitters can multiplex their data with other devices, as shown in Figure 19.30. The reduction in cabling and installation costs is evident.

For the system to find acceptance in the process industries, not only it is important that equipment from a variety of suppliers be capable of operating in the system; equipment from one manufacturer must also work together with equipment from another. This point emphasizes the importance of establishing an international fieldbus standard; however, at present, the cost and functionality benefits associated with an open system are not being fully realized because of the competing open systems that are currently available.

19.10.3 Current Digital Multiplexing Technology

Fieldbus protocols and systems have been developed in line with the International Standards Organization/Open System Interconnection (ISO/OSI) seven-layer model shown in Figure 19.31, similar to the MAP/TOP standards. It divides the features of any communication protocol into seven distinct layers, from the physical to the application layer. A

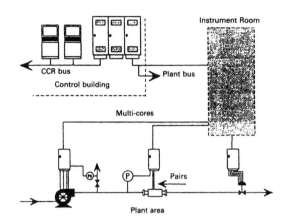

FIGURE 19.29 Traditional method of interconnecting transmitters, actuators, controllers, and so on.

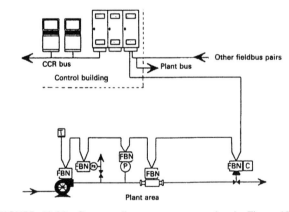

FIGURE 19.30 Corresponding arrangement to that in Figure 19.29, based on the use of a fieldbus.

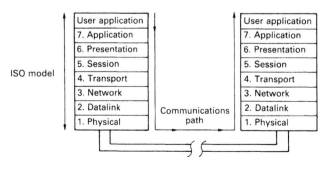

FIGURE 19.31 ISO/OSI seven-layer model.

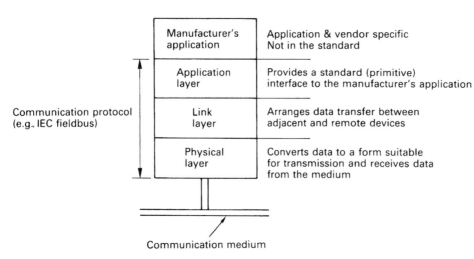

FIGURE 19.32 Three-layer implementation of the fieldbus concept.

number of fieldbus concepts have been chosen to implement only three of these layers shown in Figure 19.32. Layer 1, the physical layer, specifies the connection medium; Layer 7, the application layer, specifies the interface between the protocol and the application running it; and Layer 2, the datalink layer, specifies the interconnection between the two. It has been generally recognized that for a protocol to be useful to end users, an eighth layer needs to be designed to allow user-friendly interaction with the communication system, effectively making it transparent to the user.

The physical layer can take many forms such as twisted pair, coaxial cable, radio link, infrared, or optical fiber. Each medium may then have several different specifications to allow for variations in performance requirements, such as different data transmission speeds.

There are many alternative fieldbus protocols, including FOUNDATION Fieldbus-H1/HSE, World FIP/FIP, Profibus-FMS/DP/PA, Lon-Works-based, P-NET, CAN-based, HART, BIT-BUS, Modbus-RTU/TCP, and Ethernet-based protocols. Note that CAN and Ethernet are only physical layers, not complete protocols. However, these serve as a basis for many protocols, such as Device Net and SDS (CAN), FOUNDATION-HSE, PROFINet, and Modbus/TCP (Ethernet, used at the host level as opposed to the field level). The most obvious reason for the variety of protocols is that they have been designed with different applications in mind and optimized for specific features such as security, low cost, and high number of connected devices. Therefore, each standard may have advantages to suit the priorities of a particular application. At the time of this writing, as many as eight different process fieldbus and three automation bus technologies are set to become part of the IEC 61158 and IEC 62026 standards, respectively. Unless one single standard becomes a clear leader, it might be necessary for suppliers to provide interfaces into whatever protocol their customers use.

One of the key issues is that of interchangeability and interoperability. *Interchangeability* allows one manufacturer's devices to be substituted for those from another manufacturer on the bus, without the need to reconfigure the system. This has considerable benefits for a user who wants to keep the cost of standard components to a minimum and to avoid having to retrain their maintenance staff to any significant extent. There are two keys to achieving this task: The user needs to select powerful substitute instruments with so-called instantiable function blocks that provide the functionality of the previous device, and the user must, from the beginning, select a system with an open host that requires only the standard device description and capabilities files to interact with a device. If the host requires proprietary files, the user's choice will be limited.

Interoperability allows devices from different manufacturers to communicate with one another without the need for special drivers.

Interoperability will allow a manufacturer to compete in the market on the basis of features exclusive to its product and the added value content of the device rather than simply based on the cost of the device. However, users are concerned that this will complicate the situation and that if a special feature is present, it may be used at some stage, possibly as a workaround, and leave them subject to a monopoly of supplier. On the other hand, in a networked system there is always a possibility to perform the missing function in another device. Moreover, communications are only one aspect of interchangeability. If the length of the flowtube or temperature range is different, one device cannot replace the other anyway.

For small manufacturers, membership fees for the various organizations and the cost of development tools and technical staff to operate them may represent a significant investment. This will deter manufacturers from becoming involved until an obvious standard emerges.

The following outline description and information are reproduced, with permission, from the bulletin prepared by the HART Communication Foundation.

CHAPTER | 19 Microprocessor-Based and Intelligent Transmitters

19.10.4 The HART Protocol

19.10.4.1 Method of Operation

The HART protocol operates using the frequency shift keying (FSK) principle, which is based on the Bell 202 (Bell 1976) Communication Standard. The digital signal is made up of two frequencies (1200 Hz and 2200 Hz), representing bits 1 and 0, respectively. Sine waves of these frequencies are superimposed on the dc analog signal cables to give simultaneous analog and digital communications. Because the average value of the FSK signal is always zero, the 4–20 mA signal is not affected (Figure 19.33).

This produces genuine, simultaneous communication with a response time of approximately 500 ms for each field device, without interrupting any analog signal transmission that might be taking place.

Up to two master devices may be connected to each HART loop. The primary one is generally a management system or a PC; the secondary one can be a hand held terminal or laptop computer. A standard handheld terminal (called the HART Communicator) is available to make field operations as uniform as possible. Further networking options are provided by gateways.

Point-to-Point Figure 19.34 shows some examples of point-to-point mode. The conventional 4–20 mA signal continues to be used for analog transmission while measurement, adjustment, and equipment data are transferred digitally. The analog signal remains unaffected and can be used for control in the normal way. HART data give access to maintenance, diagnostic, and other operational data.

Multidrop This mode requires only a single pair of wires and, if applicable, safety barriers and an auxiliary power supply for up to 15 field devices (see Figure 19.35). Multidrop connection is particularly useful for supervising installations that are widely spaced, such as pipelines, feeding stations, and tank farms.

HART instruments can be used in either mode. In point-to-point operations, the field device has address 0, setting the current output to 4–20 mA. In multidrop mode, all device addresses are greater than zero and each device sets its output current to 4 mA. For this mode of operation, controllers and indicators must be equipped with a HART modem.

HART devices can communicate using company-leased telephone lines (Bell, 1973). In this situation only a local power supply is required by the field device and the master can be many kilometers away. However, most European countries do not permit Bell 202 signals to be used with

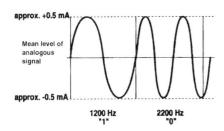

FIGURE 19.33 Because the mean harmonic signal value is zero, digital communication made no difference to any existing analog signal.

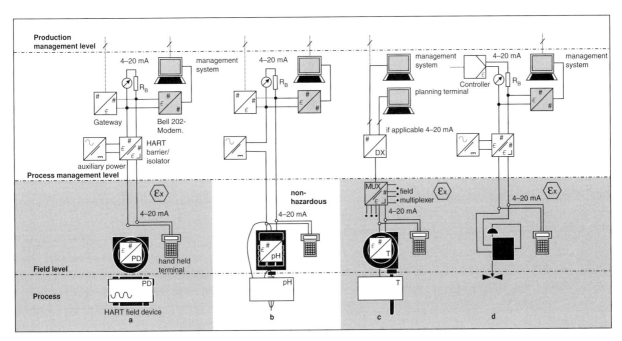

FIGURE 19.34 Point-to-point mode; with provision for one 4–20 mA device and up to two masters, such as one management system and a handheld terminal.

national carrier equipment, so HART products should not be used in this way.

Any number of field devices can be operated on leased lines as long as they are individually supplied with auxiliary power independently of the communication. If only one power supply is used for all the field devices, their number is limited to 15.

19.10.4.2 HART Protocol Structure

HART follows the basic OSI reference model, developed by the ISO (DIN ISO 7498). The OSI model provides the structure and elements of a communication system. The HART protocol uses a reduced OSI model, implementing only Layers 1, 2, and 7 (see Table 19.12).

Layer 1, the physical layer, operates on the FSK principle, based on the Bell 202 Communication Standard:

- Data transfer rate: 1200 bit/s.
- Logic 0 frequency: 2200 Hz.
- Logic 1 frequency: 1200 Hz.

The vast majority of existing wiring is used for this type of digital communication. For short distances, unshielded, 0.2 mm² two-wire lines are suitable. For longer distances

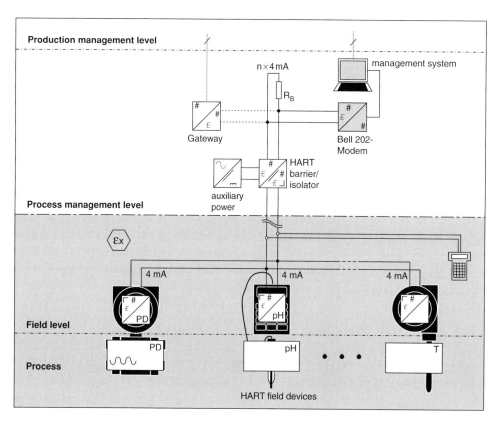

FIGURE 19.35 With multidrop mode, installation costs are considerably reduced. As many as 15 field devices can be operated from one auxiliary power supply. Management systems and handheld terminals can be used.

TABLE 19.12 Layers 1–7 of the OSI reference model; the HART protocol implements Layers 1, 2, and 7

	Layer	Function	HART
7.	Application	Provides formatted data	HART instructions
6.	Presentation	Converts data	
5.	Session	Handles the dialogue	
4.	Transport	Secures the transport connection	
3.	Network	Establishes network connections	
2.	Link	Establishes the data link connection	HART protocol regulations
1.	Physical	Connects the equipment	Bell 202

(up to 1,500 m), single, shielded bundles of 0.2 mm² twisted pairs can be used. Beyond this, distances up to 3,000 m can be covered using single, shielded, twisted 0.5 mm² pairs.

A total resistance of between 230 Ω and 1100 Ω must be available in the communication circuit, as indicated by R_B in Figures 19.34 and 19.35.

Layer 2, the link layer, establishes the format for a HART message. HART is a master/slave protocol. All the communication activities originate from a master, that is, a display terminal. This addresses a field device (slave), which interprets the command message and sends a response.

The structure of these messages can be seen in Figure 19.36. In multidrop mode this can accommodate the addresses for several field devices and terminals.

A specific size of operand is required to enable the field device to carry out the HART instruction. The byte count indicates the number of subsequent status and data bytes.

Layer 2 improves transmission reliability by adding the parity character derived from all the preceding characters; each character also receives a bit for odd parity.

The individual characters are:

- 1 start bit.
- 8 data bits.
- 1 bit for odd parity.
- 1 stop bit.

Layer 7, the application layer, brings the HART instruction set into play. The master sends messages with requests for specified values, actual values, and any other data or parameters available from the device. The field interprets these instructions as defined in the HART protocol. The response message provides the master with status information and data from the slave.

To make interaction between HART-compatible devices as efficient as possible, classes of conformity have been established for masters and classes of commands for slaves. There are six classes of conformity for a master, as shown in Figure 19.37. For slave devices, logical, uniform communication is provided by the following command sets:

- *Universal commands.* Understood by all field devices.
- *Common practice commands.* Provide functions that can be carried out by many, though not all, field devices.

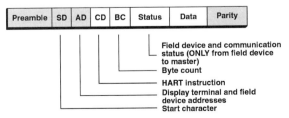

FIGURE 19.36 The HART message structure offers a high degree of data integrity.

Together these commands comprise a library of the most common field device functions.

- *Device-specific commands.* Provide functions that are restricted to an individual device, permitting incorporation of special features that are accessible by all users.

Examples of all three command sets can usually be found in a field device, including all universal commands, some common-practice commands, and any necessary device-specific commands.

19.10.4.3 Operating Conditions

The HART Standard (HART Communication Foundation) requires Level 3 resistance to interference in the lines in accordance with IEC 801-3 and 801-4. This satisfies the general requirement for noise resistance.

Connecting or disconnecting a user, or even a breakdown of communication does not interfere with transmission between the other units.

Intrinsically safe applications deserve special mention. Barriers or isolators must be able to transmit the Bell 202 frequencies in both directions (see Figure 19.34). As shown in Figure 19.35, in multidrop mode it is also possible to interconnect field devices in accordance with DIN VDE 0165.

19.10.4.4 Technical Data

Data Transmission

Type of Data Transmission. Frequency shift keying (FSK) in accordance with Bell 202, relating to the transfer rate and the frequency for bit information 0 and 1.
Transfer Rate. 1200 bit/s.
0 bit Information Frequency. 2200 Hz.
1 bit Information Frequency. 1200 Hz.
Signal Structure. 1 start bit, 8 data bits, 1 bit for odd parity, 1 stop bit.
Transfer Rate for Simple Variables. Approximately 2/s.
Maximum Number of Units in Bus Mode. With a central power supply, 15.
Multiple Variable Specification. Maximum number of variables per field unit (one modem), 256; maximum number of variables per message, 4.
Maximum Number of Master Systems. Two.

Data Integrity

Physical Layer. Error rate destination circuit $1/10^5$ bit.
Link Layer. Recognizes all groups of up to three corrupt bits and practically all longer and multiple groups.
Application Layer. Communication status transmitted in a response message.

Hardware Recommendations

The types of connection and their length limitations are given in Table 19.13.

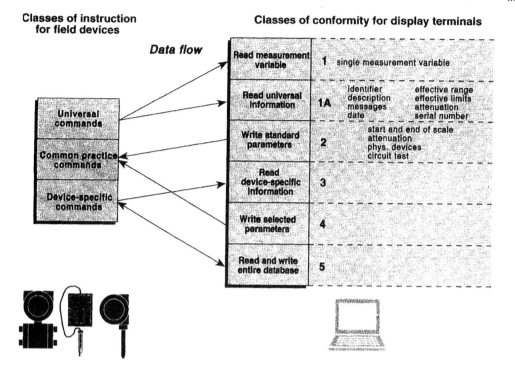

FIGURE 19.37 Classes of instruction and classes of conformity.

TABLE 19.13 Type of connection and length limitations in the HART protocol		
Distance (m)	Line type	Minimum conducting area (AWG/mm²)
≤1.500	Multiple two-wire, twisted, common shielding	24/0.2
>1.50 ≤3.000	Single two-wire, twisted, shielded	20/0.5

The following rule of thumb for determining the maximum line length for a particular application can be taken from the restrictions governing the signal:

$$l = \frac{60 \cdot 10^6}{(RC)} - \frac{(C + 10.000)}{C}$$

where l is the length (m), R is the resistance (Ω) of the load plus internal resistance from the barrier/isolator, C is the line capacity (pF/m), and C_f is the maximum internal capacitance for the smart field units (pF).

Consider the example of a pressure transducer, a control system, and a simple shielded pair with $R = 250\,\Omega$, $C = 150$, and $C_f = 5.000$ pF:

$$l = \frac{65 \times 10^6}{(250 \times 150)} - \frac{(5.000 + 10.000)}{150}$$
$$l = 1633\,\text{m}$$

In intrinsically safe applications, there may be further restrictions.

For an in-depth examination of whether a particular hookup will work, refer to the specification for the physical layer in the HART document (HART Communication Foundation).

19.11 USER EXPERIENCE WITH MICROPROCESSOR-BASED AND INTELLIGENT TRANSMITTERS

Reports of users' experiences are dominated by references to differential pressure measurements used for flow measurements. This arises not only because of the large number in use but also because these were the first microprocessor-based and intelligent transmitters to become available. The ability to set the range of an instrument before or at the time of installation was one of the earliest and is still one of the most valuable benefits of the new technology. From the user's viewpoint, orders for instrumentation can be released earlier because there is no penalty for ordering the wrong range, and the holding of spares is greatly reduced. Perhaps the most dramatic change has been in temperature transmitters, some of which can accommodate virtually any type and range of thermocouple or resistance temperature detector likely to be used in a plant.

There are fewer reports of the benefits derived from the ability to re-range the transmitter during plant operation, except for some batch processes in which a large turndown is involved and one transmitter can be used instead of two. In such instances, the use of a single instrument not only saves the cost of an instrument but also reduces the cost and complication of the installation.

The most significant benefit arising from the use of "intelligent" transmitters is in the commissioning time. In some instances this more than compensates for the additional cost of the transmitters compared with their analog counterparts. The improvement arises from the ability to address the instrument from the control room or the marshalling cabinet and to confirm that the correct type and range have been installed and that the system responds correctly when the transmitter output is forced to 4 or 20 mA. With analog instrumentation, this would involve obtaining access to the transmitter, perhaps with a ladder, and making adjustments while using a radio link with the control room. The difference can be even more marked for transmitters involved in a complex re-ranging sequence, which can now be controlled without anyone being present at the point of measurement.

Although one might expect these more complex instruments to be less reliable than the older, simpler ones, user perceptions indicate the opposite. This may be due to the fact that the built-in diagnostic routines provide an immediate warning of faults and that the status of the transmitter can be checked without having to go to its installed position, take off its cover, and carry out tests—or worse, disconnect it and take it to a workshop for maintenance. Even if one disconnected it and found no fault, the transmitter will have been identified as one that requires attention. More seriously, it is quite likely that a fault will be introduced by the unnecessary handling.

In general, users have recognized a reduced need for repair and maintenance. This may be due in part to the communication facility of the intelligent transmitters, which facilitates the keeping of better maintenance records.

19.12 FIELDBUS FUNCTION AND BENEFITS

Of the completely digital fieldbus technologies we've mentioned, only FOUNDATION Fieldbus and Profibus-PA are used for process instrumentation such as valve positioners and the transmitters discussed in this chapter. Both of these are based on the IEC 61158-2 physical layer standard, giving them the unique capability to provide both power and communication on the same pair of wires, be intrinsically safe, and run as far as 1.9 km using ordinary twisted-pair cable. Other fieldbus technologies are used in discrete factory automation and building automation. The main difference between FOUNDATION Fieldbus and Profibus-PA is that FOUNDATION Fieldbus is not only a communications protocol, it is also a programming language form building control strategies.

19.12.1 FOUNDATION Fieldbus and Profibus-PA

The IEC 61158-2 deals only with the physical layer that is based on the Manchester coding principle of multiplexing the data stream with a clock signal to achieve synchronous data transmission. Data bits of 1 and 0 are represented by falling and rising edges, respectively transmitted at 31.25 kbit/s superimposed on the 9-32 Vdc power. Because the Manchester waveform has a 50 percent duty cycle, the dc power remains sufficiently stable.

Several devices may be multidropped on a network. If the devices are not drawing power from the bus, as many as 32 are possible. If the devices are bus powered, the number should be limited to 12–16; that is also a good number for best performance. When intrinsic safety is required, the power is limited and therefore only four devices can be connected on the hazardous area side segment of a linear barrier such as the Smar SB302, or eight if a trapezoidal FISCO (Fieldbus Intrinsic Safety Concept) barrier such as the Smar DF47 is used. The previously mentioned barriers have an important feature in that they both have repeaters built in on the safe side, allowing them to be multidropped on a safe-side network segment such that a total network of 16 devices can still be achieved per interface port, keeping the interface quantity and system cost down.

Because of the relatively low speed, no special wiring or connectors are required for a fieldbus. Normal shielded twisted-pair instrument cable can be used for installation. Use of 0.8 mm^2 (AWG# 18) gauge wire is recommended to achieve a full 1.9 km overall bus length for a fully populated network. For shorter distances or fewer devices, thinner wires such as those in multicore cables can be used. The latter is ideal for the "homerun" cable from the field junction box to the control room. Distance can be increased using up to four repeaters in series.

The fieldbus network segment must be terminated in each end. In the far fieldend, this is done using a terminator such as the Smar BT302. In the hostend, the terminator should preferably already be built into the network linking device or power supply subsystem. Power for the field instruments is provided from the field end. For nonintrinsically safe installations, the power comes from a regular 24 Vdc power supply through the power supply impedance module, such as the Smar DF53, whose purpose is to prevent the dc power supply from short-circuiting the ac communication signal. Two DF53 units may be connected in parallel to provide redundant power for high availability. There is no limitation for the current that can be drawn by a field device, but selecting devices that consume

12 mA or less is required to achieve long wire runs and to put several devices on a safety barrier. In older systems fitted with a fieldbus, the power supply subsystem is implemented using external components.

The purpose of a linking device is to bridge communication between one or more field-level networks up to the host-level network, e.g., from an H1 to HSE (100 Mbit/s on Ethernet) FOUNDATION Fieldbus or from Profibus PA to DP (up to 12 Mbit/s on RS485). A linking device has the intelligence to buffer messages, allowing them to be transmitted on the two network levels at different speeds. A simpler device called a *coupler* does not provide this buffering, thus forcing the host-level network to operate at a much lower speed close to that of the field-level network. The host-level network in turn connects the linking devices to the workstations.

At first glance there may be some concern regarding the availability of multidropping devices related to several loops on a single pair of wires. However, this is no different from connecting devices related to several loops to the same I/O module in a traditional architecture. A module failure or accidental removal would affect all loops. To increase availability for so-called critical loops, users typically spread the devices related to these on separate segments such that each segment affects only a single critical loop, the rest of the devices on a segment being noncritical control or monitoring. This is particularly common in the petrochemical and chemical industries, where safety barriers are used that conveniently divide the fieldbus network into several isolated segments.

Because FOUNDATION Fieldbus H1 and Profibus-PA have identical physical layers, the networking accessories, such as repeaters, power supply impedances, safety barriers, and terminators, are the same. However, the data link and application layers are different, which gives them different characteristics.

19.12.1.1 FOUNDATION *Fieldbus H1*

FOUNDATION Fieldbus allows multiple link masters to be connected on the network. At any one time only one of them acts as a Link Active Schedule (LAS). The LAS controls the communication, ensuring that messages are sent at the exact right time. This ensures that control variables are transmitted in a precise interval, eliminating jitter. FOUNDATION Fieldbus synchronizes the function block execution with the communication, minimizing delays even further. Unscheduled traffic does not affect scheduled traffic. FOUNDATION Fieldbus provides for three types of communication relationships:

- *Client/server.* One device, called a *client*, requests a value such as a tuning parameter requested by the host from a positioner on the network, called a *server*, which responds with the value.
- *Publisher/subscriber.* Initiated by the LAS schedule, one device, called a *publisher*, broadcasts a message such as a process variable from a transmitter on the network. All devices that need the variable, called *subscribers*, such as a valve positioner with a controller in it, take this message. Thus several devices can receive the same variable in a single communication. Direct peer-to-peer communication eliminates the need for a request by a centralized device and subsequent retransmission to each individual user after the response. This reduces the communication load. The publisher/subscriber functionality is a key enabling control in the field devices.
- *Report distribution.* When an event has occurred, such as a diagnostic alarm, and the device receives the token it transmits its message, e.g., to the host. This eliminates the need for the host to frequently poll a device to catch infrequent events, thereby reducing the communication load.

System management is responsible for automatic device detection and address assignment, making the addition and removal of devices a plug-and-play operation eliminating human errors. System management also takes care of time synchronization between all devices as well as block and publishing schedule to ensure an exact execution.

To ensure interoperability, all device types have their communications conformance tested. In addition, all device types have their function block programming checked and registered.

19.12.1.2 *Profibus-PA*

Profibus allows multiple masters to be connected to the network, but only one can write values. A token is passed from master to master, and the master that holds the token has the right to access the bus. The token rotation time is not fixed but varies at an average of 50 percent. A master/slave communication relationship is always used. The master requests a value such as the process variable requested by the central controller from a transmitter on the network, called the *slave*, which responds with the value. One master can also write, for example, a manipulated variable into a valve positioner.

To ensure interoperability, all device types have their communications conformed and tested.

19.12.2 Field-mounted Control

The FOUNDATION Fieldbus specification is unique among fieldbus technologies in that it is the only one to specify a communications protocol and is also a programming language for building control strategies. This is a graphical programming language using the familiar function block concept: an extensive library of function blocks for inputs, outputs, control, arithmetic, and selection, and so on have been specified, along with guidelines on how nonstandard function blocks shall be designed to be interoperable. The function blocks primarily execute in field devices such as transmitters and positioners, thereby reducing the need for, and offloading, centralized controllers. The FOUNDATION Fieldbus thus creates a new architecture that takes the old concept of DCS a step further, becoming more distributed and more digital. This networked architecture is often referred to as Field Control System, or FCS. Powerful control strategies can be built by linking blocks from several devices. Field control is also the key to the scalability of the FCS. A traditional control system has finite resources, so adding field instruments and loops makes it slower and forces costly expansion of controllers and I/O subsystems. However, in an FCS, each added device provides more computing power and memory, making the system larger, with minor negative impact on performance. Because I/O and controller hardware is minimized, expansion is comparatively economical. Interoperability and level of system integration also benefits because the semantics of the data are defined; for example, for any manufacturer, controller mode is set by the MODE_BLK parameter, and the code for Manual is always the same.

There is some doubt whether, in safety-critical applications, communication among transmitter, controller, and final control element via a shared fieldbus is acceptable. At the time of this writing, no fieldbus technology has been approved for certified critical control systems, nor for field control, device configuration, measurement variables, or actuation. Once more, reliability for multidrop networks is available; this may be considered. Certainly, the increased diagnostics that can be communicated using a fieldbus can increase both safety and availability of systems, and control in the field is more fault-tolerant than the centralized controls of the legacy DCS.

19.12.3 Future of Analog Instruments

Although the number of microprocessor-based and intelligent transmitters being installed is rising steadily, it will be a long time before the transition is complete because of the large installed base of traditional instruments. Smar provides a range of current-to-fieldbus (IF302), fieldbus-to-current (FI302), and fieldbus-to-pneumatic signal converters (FP302) for interfacing between the old analog world and the FOUNDATION or Profibus fieldbus. These are ideal not only for migration of an existing plant but also to interface devices not yet available in a fieldbus version into a new installation. There was initially some concern regarding the use of software-based instruments in safety-critical applications because of the possibility that they may fail in some unexpected manner. However, several years of reliable operation and refined design techniques have resulted and the first few safety-certified HART devices have now emerged.

19.12.4 Sensor Validation

Many of the diagnostic routines applied in early microprocessor-based and intelligent transmitters concerned the electronics rather than the primary sensor.

However, based on years of experience and study of sensor failure modes, some manufacturers now offer transmitters that diagnose not only themselves but also the sensor. An added advantage of the FOUNDATION Fieldbus is that it not only specifies how to present the status to the user—the function block programming language has also been designed to provide automatic shut down action in case of sensor failure. Particularly interesting is the development seen in valve positioners over the past few years. For example, the Smar FY302 pneumatic valve positioner performs diagnostics on itself, the actuator, and the valve, thereby providing the operator with status for the entire valve package.

19.12.5 Plant Diagnostics

A considerable amount of potentially useful information regarding a process, beyond the value of the process parameter being measured, is available at the interface between the process and the primary sensor. This information can be recovered by spectral and statistical analysis of the measurement signal if the frequency response of the transmitter is adequate. In effect, it provides a window through which the status of the associated process can be assessed and changes detected, to provide warning of incipient failures. However, since most of the transmitters in use at present have been designed to operate in a process control loop, and since it is generally accepted that there is no useful information in those components of the measurement signal having frequencies greater than about 5 Hz, the amount of useful information that can be recovered using currently available transmitters is very restricted. However, devices that locally sample and cache information at a higher speed during the test phase and then communicate it at a more leisurely pace are already on the drawing board.

19.12.6 Handheld Interfaces (Handheld Terminals or Handheld Communicators)

Using fieldbus technology, there is an unbroken digital chain from the field instruments to the operator work station. Therefore it is possible to diagnose, configure, and otherwise interrogate a device from the computer without having to venture into the field with a handheld. There is no need to find the right connection point at the junction box. The workstation runs the engineering and maintenance application that allows the user to configure and diagnose the devices and manage maintenance. Because these tools can be loaded with Device Description (DD) files that describe the devices in the field (or GSD in the case of Profibus), a single tool such as the Smar SYSCCON software can configure a plethora of devices from different manufacturers without interoperability problems. An additional advantage can be achieved using the FOUNDATION Fieldbus. Because the control strategy programming language is part of the same specification as the communications for device configuration and maintenance, strategy building and device management, as well as network management, can be performed from the same single software tool. In the case of the Smar SYSCON software, this means that users can drag and drop from the strategy directly into the devices.

19.12.7 Measuring Directives

It is likely that many process measurements that at present are *not* subject to weights and measures surveillance will become so in future. Directives requiring further integration of diagnostic routines into these instruments, data backup, and so on are to be expected.

19.12.8 Further Developments of Intelligent Transmitters

The accuracy and versatility achieved by the present generation of microprocessor-based and intelligent transmitters is such that there is little real advantage to be gained from further improvement in these respects. For temperature and differential pressure transmitters for which accuracies of 0.1 percent of the measured value are claimed, the point has been reached where it is necessary to question whether the transmitted signal has the significance implied by the number of digits in the displayed value.

For most process measurements, the noise component can range from 1 percent to 30 percent of the mean value, and the damping applied may be anything from 1 to 60 s. It is then necessary to consider how realistically the displayed value represents the process variable.

The focus has now shifted toward the valve positioners that until now have been contributing significantly to significant process variability. Until recently positioners, due to their imprecise mechanics and simple analog electronics, have been unable to respond to demands in position change by less than a few percent. Modern positioners such as the Smar FY302 use a noncontact position-sensing method that eliminates mechanical linkages, and they perform all characterization and positioning algorithms in software, thereby being able to modulate with a precision of a fraction of a percent.

19.12.9 Integration of Intelligent Transmitters into Instrument Management Systems

The situation is quite different when we consider the advantages to be gained from integrating information from a number of intelligent transmitters into the management system. The real benefits from the use of intelligent transmitters arise from the fact that so much basic data regarding the transmitter itself, such as the model number, serial number, tag, location, materials of construction, and so on, and its application, such as the zero and span, units of measure, and damping, are stored in the memory of the transmitter electronics and can be accessed via the two-way communication system.

For an individual transmitter operating in a 4–20 mA measurement loop, this means that its identity can be identified and its operation monitored from any junction box or set of terminals in the measurement loop, thereby avoiding the need to gain access to the transmitter itself to check its status or to make an adjustment to its operating range. If, however, the transmitter is connected directly to a fieldbus, together with other transmitters and actuators, as shown in Figure 19.38, the same information can be accessed at any junction on the fieldbus or via an interface unit. This opens the opportunity to collect data from all the sensors and actuators in the control system and analyze that data using a variety of statistical and knowledge-based techniques, to optimize the maintenance and technical management of the process plant.

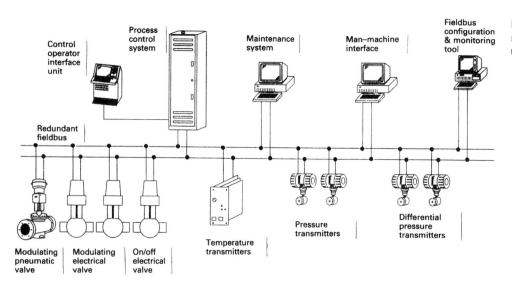

FIGURE 19.38 Typical arrangement of the equipment connections to a fieldbus.

Although regulatory control is primarily done by the instruments in the field, operation is still conveniently centralized. The operators monitor the process and initiate control actions for the field-level and host-level networks. Setpoint, control mode, and tuning parameters set at the workstations are communicated to the function blocks executing in the field. The control strategy in the field devices executes several times per second, whereas the process information is updated on the operator display and for historical trending only once per second.

The fact that so much basic data regarding a transmitter can be stored at the transmitter itself and accessed via a fieldbus, coupled with the instrument management systems that are now available, provides an opportunity for plant managers to enhance the management of their processes by optimizing the control regime and reducing unscheduled shutdowns through the availability of better information regarding the status of the transmitters, actuators, and the process equipment.

REFERENCES

Bell, PUB 41004, "Data communications using voice-band private line channels," *Appendix to Bell System Technical Reference PUB 41004* (October 1973).

Bell, "Data sets 202S and 202T interface specification," *Bell System Technical Reference: PUB 41212* (July 1976).

DIN ISO 7498: *Informations verarbeitung, Kommunikation Offener Systems*, Basis-Referenzmodell, Beuth Verlag, Berlin.

DTI, *Fieldbus; The Executive Guide.*

Fisher-Rosemount Ltd., Heath Place, Bognor Regis, West Sussex PO22 9SH, U.K. (*Product Data Sheets 62067, 62071, 62072, 62077, 62078,* and *11014*).

Fisher-Rosemount Ltd., Horsefield Way, Bredbury Industrial Estate, Stockport SK6 2SU, U.K. (for *Coriolis Mass Flowmeter* product information).

HART Communication Foundation, *HART Smart Communications Protocol Specification.*

Invensys (formerly Foxboro) Great Britain Ltd., Manor Royal, Crawley, West Sussex RH10 2SJ, U.K. (*PSS 1-6F2 A* and *PSS 2A-1Z3A*).

Laboratory of the Government Chemist, *Multiplexing for Control in Industrial Environments.*

MTL, Measurement Technology Limited, Power Court, Luton, Bedfordshire LU1 3JJ, U.K.

Smar International, Houston (*Fieldbus Book 2000* and *HART Book 2000*).

Solartron Transducers, Victoria Road, Farnborough, Hants GU14 7PW, U.K. (*Bulletins 6513, 6516, 6517,* and *6519*).

Yokogawa U.K. Ltd, Stuart Road, Manor Park, Runcorn, Cheshire, WA7 1TR, U.K. (*GS 1F2B4-E-H* and *GS 1 COA10-E*).

Chapter 20

Industrial Wireless Technology and Planning

D. R. Kaufman

20.1 INTRODUCTION

The pervasive use of wireless technology in industrial applications is inevitable. The technology is maturing at an extremely rapid pace due to the commercial market explosion of wireless-based products for personal and home use across the last few decades. It all started with primarily one commercial need that began in the 1960s: opening your garage door remotely from inside your car.

Fast-forward several decades: Wireless technologies have become pervasive in our personal lives. We use wireless to change the channels on our TVs, call friends on our cell phones, or connect to the Internet via WiFi on our laptops. This personal pervasive usage has generated a huge commercial wireless market with a never-ending list of new product offerings from very large corporations. The competitiveness has caused the wireless component technologies to standardize and improve in terms of reliability, security, and power management, which are the basic functional needs for use in the industrial markets.

Now couple that technology improvement with personal wireless product use by plant engineers and operations personnel, and you have a revolution of ideas and thoughts around how to consider using wireless technology for more than just walkie-talkie communications in a plant. Industrial workers now think about how to use wireless to do their jobs much as they do for their personal lives. They are getting more comfortable with the technology with each passing day and, as with any key asset, will use it whenever and wherever to improve their business.

Therefore, the pervasive use of industrial wireless is inevitable. Just like *Star Trek's* Borg assimilation process, by which they integrate beings and cultures into a "collective," so will wireless technology be integrated into normal day-to-day plant operations; "resistance is futile."

This chapter will help you through this industrial wireless evolution by discussing the fundamental basics of wireless technology and its use in an industrial facility. The chapter covers wireless history, technology, terminology, standards, and top planning considerations for industrial deployment.

20.2 THE HISTORY OF WIRELESS

Wireless is not a new technology; it is merely relatively new to industrial applications. Actually, the term *wireless* started as an easy reference for a transceiver that is a dual-purpose device combining both transmitting and receiving.

It all started with the demonstration of the theory of electromagnetic waves by Heinrich Rudolf Hertz in 1888. This demonstration proved that electromagnetic waves could be transmitted through space in straight lines and received electronically. In 1895, Guglielmo Marconi, an Italian inventor, began experimenting with the technology and was granted the world's first wireless telegraphy patent by the British Patent Office. Marconi is generally credited with the development of radio by most scholars and historians. He established the Wireless Telegraph and Signal Company Ltd., the first radio factory in the world.

In 1901 signals were received across the Atlantic, and in 1905 the first wireless distress signal was sent using Morse code. Wireless technology eventually progressed as an invaluable tool used by the U.S. military. The military configured wireless signals to transmit data over a medium that had complex encryption, which makes unauthorized access to network traffic almost impossible. This type of technology was introduced during World War II when the U.S. Army began sending battle plans over enemy lines and when Navy ships instructed their fleets from shore to shore.

Wireless proved so valuable as a secure communications medium that many businesses and schools thought it could expand their computing arena by expanding their wired local area networks (LANs) using wireless LANs. The first wireless LAN came together in 1971, when networking technologies met radio communications at the University of Hawaii as a research project called ALOHAnet. The bidirectional star topology of the system included seven computers deployed over four islands to communicate with the central computer on Oahu Island without using phone lines. Thus wireless technology as we know it began its journey into houses, classrooms, businesses, and now industrial facilities around the world.

20.3 THE BASICS

Wikipedia states that "Wireless communication is the transfer of *information* over a distance without the use of electrical conductors or '**wires**.' The distances involved may be short (a few meters as in television remote control) or very long (thousands or even millions of kilometers for radio communications)."

But how does it actually work? In its simplest form, a source device creates electromagnetic waves that travel through air at close to the speed of light and are received at a destination device. The source device can be any wire or conducting object such as an antenna that conducts alternating current creating electromagnetic radiation or "waves" propagated at the same frequency as the electric current. The wave is characterized by the wavelength and frequency, which are inversely proportional, so the shorter the wavelength, the higher the frequency. Thus the wavelength for a 900 MHz device is longer than that of a 2.4 GHz device. In general, signals with longer wavelengths travel a greater distance and penetrate through and around objects better than signals with shorter wavelengths.

Data is modulated or coded using conventional binary data (1s and 0s) onto a radio frequency carrier. The wireless information is transmitted as a radio message using these 1s and 0s to represent the payload or actual message, plus additional data that controls the message handling and synchronization.

20.3.1 Radio Frequency Signals

Radio frequency (RF) signals have two common measurements: frequency and "strength." Many signals are a mixture of different frequencies and different strengths.

Frequency is measured in Hertz (Hz), meaning 1 cycle per second. The radio spectrum is broken into groups, with names such as HF (high frequency), VHF (very high frequency), and UHF (ultra-high frequency). Graphically, the radio spectrum is illustrated using a logarithmic scale rather than linear. Most industrial wireless products are found in the upper VHF and UHF frequencies.

The most common measurement of RF strength is in mW (milliwatts) of RF power. Again, strength is expressed in a logarithmic scale using decibels (dB). RF signal strength (or RF "signal") is expressed as dBm, with a reference of 1 mW of RF power. That is, 0 dBm = 1 mW RF power. RF signal strength in dBm is the logarithm of the RF power in mW.

$$dBm = 10 \log_{10}[RF\ Power\ in\ mW]$$

Being a log scale, doubling signal strength adds another 3 dB. So, increasing strength 4 times adds 6 dB (2×3), and increasing 8 times adds 9 dB (3×3). Similarly, halving a signal removes 3 dB.

20.3.2 Radio Bands

The use of wireless devices is heavily regulated throughout the world. Each country has a government department responsible for deciding where and how wireless devices can be used and in what parts of the radio spectrum. Most countries have allocated parts of the spectrum for open use, or "license-free" use. Other parts of the spectrum can only be used with permission or "license" for each individual application.

Most wireless products for short-range industrial and commercial applications use the license-free areas of the spectrum to avoid the delay, cost, and hassle of obtaining licenses. The license-free areas of the spectrum are also known as industrial, scientific, and medical (ISM) bands. In many countries several ISM bands are available in different parts of the spectrum.

The radio spectrum is split into frequency bands, each of which is split into frequency channels. The width of each channel is normally regulated. The channel width dictates how fast data can be transmitted; the wider a channel, the higher the data rate.

The common frequency bands for industrial wireless applications are:

- 220 MHz band in China: licensed.
- 433 MHz band in Europe and some other countries: license free.
- 400–500 MHz: various parts are available in most parts of the world as licensed channels.
- 869 MHz band in Europe: license free.
- 915 MHz band in North and South America and some other countries: license free.
- 2.4 GHz band allowed in most parts of the world: license free.

Higher-frequency bands are wider, so the channels in these bands are also wider, allowing higher wireless data rates. For example, licensed channels in the lower frequencies (150–500 MHz) are often regulated to 12.5 KHz, whereas channels in the license-free 2.4 GHz band can be hundreds of times wider.

However, for the same RF power, lower frequency gives greater operating distance. With wireless data, there is always a trade-off between distance and data rate. As radio frequency increases, the possible data rate increases, but operating distance decreases.

20.3.3 Radio Noise

Radio noise is caused by both internal and external factors. Internal factors are those within the wireless circuit boards. There are two internal types; "thermal" EMF noise, which is caused by random electron movements in components, and harmonic noise, which is generated by the RF circuitry. External factors are natural or manufactured and include the following:

- *Radiation from the sun*, or solar flares, can cause noise across the radio spectrum, with variable effect at different times. You might have experienced this type of noise while listening to the radio in your car. Do you remember a time when you stopped at a red light and the radio station seemed to die, but when you moved a little forward the radio started again? This phenomenon was more than likely caused by noise generated by solar flares.
- *Lightning* is another natural source of radio noise. It can strike at any time with a powerful but short-lived effect.
- *Electrical and electronic equipment* produces radio frequency (RF) transmissions; however, the noise amplitude of electrical power equipment at the frequencies used for industrial wireless products is extremely small and usually ignored.
- *Electric motors, pumps, drives, electric power distribution, or welding equipment* provide major sources of electromagnetic interference (EMI). Anywhere electric power is being turned on and off rapidly for large pieces of industrial equipment is a potential source. The spectra for these sources are generally stronger at the lower frequencies and diminishing at the higher frequencies.
- *Radio interference* is noise caused by other wireless transmissions on the same channel or on nearby channels. Sources include walkie-talkies, television, broadcast radio services, and wireless cellular telephones.

The sum of all these interference sources or noise is a volatile mix of varying amplitudes. Over a short period of time, the noise level on any radio channel will have peaks and troughs around an average level, with the amplitude at any instant following a random probability function. Over longer periods of time, the noise pattern will be the same, but the average noise level will change according to what is happening in the surrounding environment.

20.3.4 Radio Signal-to-Noise Ratio (SNR)

A radio receiver tuned to a particular frequency channel will receive whatever is expected or not expected to be transmitted on that channel. Expected transmissions are called *signals*; unexpected transmissions are referred to as *noise*. If the strength of an expected transmission or signal is significantly stronger than the unexpected transmission or noise, the receiver is able to effectively ignore the noise and receive the signal. This is called a good *signal-to-noise ratio* (SNR). If the expected transmission or signal is of similar strength or less than the unexpected transmission or noise, the receiver will not be able to discriminate the actual signal from the noise. This is considered a poor SNR situation.

Poor SNR situations can cause problems in demodulating data from an RF signal. As the noise level gets closer to or exceeds the signal level, the demodulated data have more errors because the noise makes it harder for the receiver to determine whether a demodulated bit of data is a 0 or a 1. This is like trying to talk to your mother on your cell phone in a very busy airport with the loudspeaker above your head announcing all the current flights. *She* can hear *you* because you are speaking directly into the microphone, causing a good SNR situation for her, but the loudspeaker over your head is causing *you* problems because it is louder than the tiny speaker on your cell phone, causing a bad SNR situation for you. You miss words sent by your Mom, just the way a digital receiver would miss the 1s and 0s from a signal. So, in the RF world, a transceiver reorganizes 1s and 0s them into messages to be understood by the host system or application, which is just like you taking words from your Mom and organizing them into sentences you can understand.

Any missed bits or errors are called *bit errors*. The error rate (or errors per total bits) is called the *bit-error ratio*, or BER. The sensitivity of a radio receiver is the lowest RF signal that it can reliably detect reliably—generally quoted at a specified BER. The *data sensitivity* is the lowest RF signal at which the receiver can demodulate a data message with a very low level of external noise. Data sensitivity is normally expressed at a particular BER.

A transmitted signal can vary in strength. During rain or fog, the radio signal is *attenuated* (decreased) by the denser air. In a thunderstorm, the transmitted signal will decrease and the noise level will increase, which could turn a low BER into a high BER.

A SNR of 5 dB means that the average signal measurement is 5 dB stronger than the average noise level. Unfortunately, radio noise is often much less than receiver sensitivity, so SNR is not particularly relevant here.

A more common term that is relevant is the *fade margin*, which is how much a radio signal can decrease (or *fade*) before the receiver can no longer demodulate data. Fade margin is the difference between the transmitted signal and noise or receiver data sensitivity, whichever is greater. It also measures how much the signal needs to fade before it becomes unreliable. The fade margin should be measured on a fine day.

For modern industrial wireless products, a fade margin of 10 dB is adequate; this gives enough margin for loss of signal or increased noise during poor weather or high solar activity. An installation will work reliably with a lower fade margin, but not all the time.

20.3.5 Wireless Reliability

For each data modulation method, the ability of the receiver to demodulate data from the received radio signal decreases as the data rate increases; that is, BER increases as the transmitted data rate increases, due to receiver demodulation operation. However, *demodulation BER* is not the only factor influencing overall data reliability.

Because instantaneous noise closely follows a random probability function, any and every wireless message is vulnerable to random noise "attack" that can corrupt individual bits. Forward-error correction (FEC) techniques can be used to recover the corrupted bits. For messages without FEC, the whole message becomes corrupted if one bit is corrupted. So the probability of a corrupted message relates to its *airtime*—the overall transmission time of the message. This depends on the length of the message (number of bits) and the transmitted data rate. The longer the message, the higher the probability of message corruption, but the higher the data rate, the lower the probability. So, higher data rate has two effects: increasing demodulation BER but also reducing the risk of message corruption.

For reliable operation, all wireless messages need integral error checking to confirm the integrity of the overall message. Message recovery is also required in the event of corruption. This is normally in the form of a message acknowledgment protocol, with corrupted messages being retransmitted or messages being continuously transmitted such that occasional corrupted messages can be ignored.

In summary:

- BER increases with data rate because of receiver demodulation errors.
- The probability of message corruption during transmission decreases with data rate increases.
- The probability of overall message corruption decreases with shorter messages.
- FEC functionality built into the messages improves the rate of successful messages.

As a general rule of thumb, wireless products with short messages, such as wireless I/O transmitting digital data, have a reliability advantage over wireless modems (analog) transmitting much larger amounts of data.

20.3.6 Fixed Frequencies

Fixed frequency, as the name implies, uses a single frequency channel; radios initiate and maintain communications on the same frequency at all times.

Fixed-frequency channels can be license free or licensed. A licensed channel is licensed to the operator of the wireless system by a governing body in each country, such as the Federal Communications Commission (FCC) in the United States. A radio license protects a channel against other users in a specified geographic area, and it specifies the RF power levels that may be used. Generally, licensed channels allow much higher power levels than license-free channels.

In some regions of the world, license-free bands comprise a number of fixed-frequency channels; the user can select one of the channels in the band to use. The problem with license-free fixed-frequency channels is that if more than one user in close vicinity uses a channel, the wireless transmissions can be corrupted. This is much like having multiple people using a single citizen's band (CB) radio channel to communicate. If they all talked at the same time, the words and sentences would become corrupted and no one could understand them.

In industrial situations, most fixed-frequency bands are at a lower frequency than the spread-spectrum bands. Combine the lower frequency with the ability to gain licenses for larger RF power and fixed frequency is a good choice for infrastructure radio over large distances, if licenses are available.

However, there are some administrative and technical drawbacks to consider. The delay associated with obtaining a license varies from country to country, and it is not uncommon for the granting of a license to take long periods. Furthermore, with a limited number of channels from which a license can be granted, licenses can be difficult to come by in areas of great demand. In some regions, they are auctioned off to the highest bidders and have reached prices in the billions of dollars.

20.3.7 Spread Spectrum

Spread-spectrum radios use multiple channels within a continuous band. The frequency is automatically changed or transmissions use multiple frequencies at the same time to reduce the effects of interference.

Spread spectrum allows a large number of wireless systems to share the same band reliably. With a band of fixed-frequency channels, the reliability of a system depends on no other system using the same channel at the same time. Spread spectrum provides a method of interleaving a large number of users into a fixed number of channels. Although there are several types of spread-spectrum techniques, the two most common are frequency hopping spread spectrum (FHSS) and direct sequence spread spectrum (DSSS).

Frequency hopping spread spectrum (FHSS) has the frequency of the transmitted message periodically changed (or *hopped*). The transmitter hops frequencies according to a preset sequence (the *hop sequence*). The receiver either stays synchronized with the transmitter hopping, or is able to detect the frequency of each transmission.

FHSS can hop rapidly several times per message, but generally it transmits a complete message (or data packet) and then hops. Each transmitter hops to a particular hop sequence, which it chooses automatically or is user configured. Because the hop sequences of different transmitters are different, the hopping of a "foreign" transmitter exhibits statistical randomness (though it is not truly random).

FHSS hopping sequences are pseudo-random in that the probability of a foreign transmitter hopping to a particular channel appears to be random.

If more than one system hops onto the same channel, a "hop-clash" event, those radio messages are corrupted. However, when the transmitters hop again, the probability that both transmitters will hop to the same frequency a second time is very remote.

Direct sequence spread spectrum (DSSS) differs from FHSS in that the transmitted data packet is "spread" across a wide channel, effectively transmitting on multiple narrow channels simultaneously. When a data packet is transmitted, the data packet is modulated with a pseudo-random generated key, normally referred to as a *chipping key*, which spreads the transmission across the wideband channel. The receiver decodes and recombines the message using the same chipping key to return the data packet to its original state.

FHSS focuses the transmission power into one frequency channel at any one time. DSSS spreads the power of the transmitter across each channel. As a result, the DSSS distance capabilities as well as the penetrating power are greatly reduced compared to FHSS. However, the possible data rates for DSSS are much higher.

The ability to withstand interference is different between the two techniques. Since the average power level of DSSS transmissions is much lower than FHSS, DSSS cannot tolerate the same level of noise that FHSS can. Furthermore, DSSS has more problems with multipath fading, a phenomenon caused by the mixing of direct and out-of-phase reflected signals. FHSS systems generally transmit signals farther because of the higher effective RF power.

So, generally, you would use DSSS for larger data transmissions with short distances and limited interference and use FHSS for lower data needs with longer distances and interference.

20.3.8 Security

There is heightened awareness of the security risks with use of wireless products, since plant operations are now "in the air." The possibility is very real now for a "bad guy" to sit in the plant parking lot accessing your wireless network for data capture or manipulation, causing operational and safety concerns. Given this fact of modern life, system security is paramount for any industrial wireless system.

Most concerns center around two aspects:

- *Jamming.* A wireless link is deliberately jammed to prevent it from working. This is typically called a *denial-of-service attack*.
- *Hacking.* A wireless link or network is accessed so that wireless data can be stolen or maliciously injected.

20.3.8.1 Jamming

Deliberate or unintentional jamming occurs when another wireless system causes enough interference or noise to prevent your wireless system from operating reliably. The degree of difficulty in doing this depends on the nature of the wireless system.

If the primary wireless system has a high fade margin (or a high signal-to-noise ratio), the jamming signal needs to be very strong. For short-distance applications found in industrial plants and factories, this means that the interfering antenna needs to be fairly close and should be generally easy to locate and eliminate, since the interferer will more than likely be inside the facility.

Fixed-frequency channels are easier to jam than spread-spectrum ones, although some forms of spread spectrum do not provide a large advantage. Direct-sequence devices can be jammed by higher-power direct-sequence devices using the same wide channel. Frequency hopping provides the best protection against jamming, with asynchronous hopping having better performance than synchronous hoppers.

Synchronous frequency hopping is where transmitters continually transmit a radio message for receivers to stay in hop synchronism. Asynchronous transmitters transmit only when there are data to be transmitted, with receivers continually cycling though the top sequence looking for a transmission lead-in signature. Synchronous hoppers can be jammed by a strong fixed-frequency signal, which causes the receivers to lose the transmitter signal in each hopping cycle. However, it is difficult to jam an asynchronous hopper.

Another system factor affecting vulnerability to jamming is the duty cycle of the system. It is difficult to create a jamming signal that is present more than 50 percent of the time. A system that is transmitting continuously (for example, a polling system) can be jammed much more easily than a system that uses a more sophisticated event-driven protocol and transmits with a lower duty cycle. The lower the duty cycle of a system, the less vulnerable it is to jamming.

20.3.8.2 Hacking

The best protection against wireless espionage or hacking is encryption of the wireless data. Although sophisticated modulation techniques and spread spectrum provide a high level of protection, this protection disappears if the offender uses the same type of wireless device (with the same modulation or spread-spectrum technology) as the target system.

Most modern wireless devices provide some degree of security encryption. There has been a lot of publicity about the weakness of the WiFi Wired Equivalent Privacy (WEP) encryption, which transmits the encryption keys as part of the encryption scheme. However, WEP was not intended for secure industrial applications; there are many other encryption methods that provide very secure protection.

Advanced Encryption Standard (AES) is generally recognized as an "unbreakable" scheme. However, this standard requires heavy computing resources and can significantly slow the operation of wireless devices. Many proprietary encryption schemes provide the same degree of protection without the heavy computing resources; the fact that these schemes are not open to continuous scrutiny and testing gives them a higher level of inherent protection.

Security protection does not need to be 100 percent secure; the protection needs to make it so difficult for malicious offenders that they will look for an easier alternative to achieve their goals. Modern wireless systems, if properly engineered, can provide similar or higher-security protection to that of traditional wired systems.

20.3.9 Antennas

An antenna is a circuit element that provides a transition from a guided wave on a transmission line to a free space wave, and vice versa. It is made of a thin electrical conductor to efficiently radiate electromagnetic (EM) energy. Antenna characteristics are the same, whether a transmitting antenna or a receiving antenna. This phenomenon is called *antenna reciprocity*, which implies that an antenna captures EM energy with the same efficiency that it radiates EM energy.

Antennas are designed and built to suit a particular frequency or frequency band. If you use an antenna designed for a different frequency, it will only radiate a small portion of the generated RF power from the transmitter, and it will only absorb a small portion of the RF signal power for the receiver. *Using an antenna of the correct frequency is very important.*

Antennas are compared to a theoretical *isotropic antenna*. This antenna radiates all the power from the transmitter in a three-dimensional spherical pattern—very much like a point source of light without any mirrored reflectors. An isotropic antenna is theoretical only, because in the construction of antennas, the radiation pattern becomes distorted in certain directions.

The term *effective radiated power*, or ERP, is used to measure the power radiated in specific directions. The difference between the effective radiated power and the transmitter power is called the *antenna gain* and is normally expressed in dB:

$$\text{antenna gain in dB} = \log_{10}(P_{ERP}/P_{TX})$$

20.3.9.1 Omnidirectional Antennas

Omnidirectional antennas transmit equally in all directions in the horizontal plane. Normally omnidirectional antennas are used for intraplant applications, where the radio path lies completely within an industrial site or factory. The radio path is often made up of strong reflected signals over relatively short distances (meters versus miles) coming from any direction and that might not always be obvious.

A *dipole* antenna is one type of omnidirectional antenna. It is manufactured with an active "radiator" with a length equal to ½ the wavelength of the design frequency. The RF power envelope radiated by a dipole is distorted by radiating more power in the horizontal plane and less in the vertical plane; that is, there is more power radiated to the sides than up and down. So, the effective RF power of the sides increases and the effective power up or down decreases.

The gain of a dipole antenna to the sides of the antenna is +2.14 dB. This means that the effective radiated power to the sides of the antenna is 2.14 dB more than the power from the transmitter. The gain in the up/down direction will be negative, meaning that the effective radiated power in these directions is less than the power from the transmitter.

A dipole antenna mounted vertically transmits an RF wave in the vertical plane; this is called *vertically polarized*. If the antenna is mounted horizontally, the RF wave will have *horizontal polarity*.

A *collinear* antenna is another type of omnidirectional antenna. Since radiation patterns can be further distorted by connecting multiple dipole elements, these antennas have a higher gain to the sides and a more negative gain up and down.

Collinear antennas are normally manufactured with gains of 5 dB, 8 dB, or 10 dB, compared to the transmitter power. When antenna gains are expressed as a comparison to the transmitter power, it is called *isotropic gain*, or gain compared to an isotropic antenna. Isotropic gain is expressed as dBi.

Another common way to express gain is as compared to a dipole; these gains are expressed as dBd. The difference between dBd and dBi is the intrinsic gain of a dipole, 2.14 dB (normally rounded to 2 dB):

$$\text{DBd} = \text{dBi} - 2$$

Be careful with antenna gain specifications; manufacturers do not always specify whether they are using dBi or dBd. It is best to ask for this clarification from your supplier.

20.3.9.2 Directional Antennas

Directional antennas further distort the radiation patterns and have higher gains in a "forward" or specific direction. These antennas are often used for longer line-of-sight radio paths found in intraplant applications.

A Yagi antenna is an example of a directional antenna that has an active dipole element with "reflector" elements that act to focus power in a forward direction. Before cable communications, most homes had Yagi antennas on their roofs for TV reception. They were pointed in the direction of the transmitting tower, which was usually several miles away.

Yagis are normally available from 2 elements up to 16 elements. The more reflector elements added, the higher gain in the forward direction and the lower gain to the sides and rear. Furthermore, as more elements are added, the directional angle becomes smaller as the gain is more tightly focused.

Yagis are mounted with the central beam horizontal and the orthogonal elements either vertical or horizontal. If the elements are vertical, the antenna is transmitting with vertical polarity; if the elements are horizontal, the polarity is horizontal. Antennas in the same system should have the same polarity.

For higher-frequency Yagi antennas, it is physically possible to add side reflectors to further increase the gain. For 2.4 GHz devices, parabolic reflectors around the dipole element yield extremely high gains and extremely narrow transmission beams.

20.3.9.3 Gains and Losses

Using a high-gain antenna has the following effects at the transmitter:

- Increases the effective transmission power in certain directions and reduces the power in others.
- Gain compensates for loss in coaxial cable.
- Makes the antenna more directional at the transmitter—a good effect for reducing unwanted RF radiation in unrequired directions.

RF power generated by a transmitter is initially reduced by the coaxial cable, then increased by the antenna gain.

Note: Care must be taken that the final effective radiated power is less than the regulated amount. In some countries, there are limits on the gain of antennas as well as the final ERP.

At the receiver, high-gain antennas have the following effects:

- Increases the received signal from certain directions and reduces the signal from others. Gain will also increase the received external noise. If the increased noise exceeds the sensitivity of the receiver, the gain improvement has been negated—that is, in noisy environments antenna gain can have little effect at the receiver, but in no-noise environments, it can have an effect.
- Gain compensates for loss in coaxial cable.
- Makes the antenna more directional—a good effect for reducing unwanted RF noise in unrequired directions but a bad effect if you are relying on reflected signals from various directions.

20.3.9.4 Mounting

If an antenna is not correctly installed, it will not provide optimum performance, reducing RF power radiation.

Whip Antennas Whip antennas (¼ wavelength) are normally connected directly to the wireless device. These antennas are *ground dependent*; that is, they need to be installed onto a metallic plane to "ground" the antenna. Without a ground plane, the antenna will have reduced radiation. For 900 MHz and higher frequencies, the required ground plane is small and the wireless unit itself is normally sufficient.

If the whip antenna is connected via a coaxial connector (as opposed to a fixed connection), some weatherproofing technique is required if the installation is outside. The wireless unit and antenna can be installed inside a weatherproof enclosure.

If the enclosure is metallic, most of the RF energy will be absorbed by the enclosure, although this is still an acceptable installation if the radio path is short. A better installation is to use a nonmetallic enclosure, which will have little blocking effect on the RF signal. *Note:* Make sure the mounting plate in the enclosure is also nonmetallic.

Dipole and Collinear Antennas These antennas are connected to the wireless unit via a length of coaxial cable. If the cable is larger than 6 mm in diameter (¼ inch), do not connect the cable directly to the wireless unit. Thick cables have large-bending radii, and the sideways force on the unit connector can cause a poor connection. Use a short "tail" of RG58 between the thick cable and the wireless unit.

The polarity of these antennas is the same as the main axis, and they are normally installed vertically. They can be mounted horizontally (horizontal polarity), but the antenna at the other end of the wireless link would need to be mounted perfectly parallel for optimum performance. This is very difficult to achieve over distance. If the antenna is mounted vertically, it is only necessary to mount the other antennas vertically for optimum "coupling"; this is easy to achieve.

Dipole and collinear antennas provide best performance when installed with at least 1 to 2 "wavelengths" of clearance for walls or steelwork. The wavelength is based on the frequency:

$$\text{Wavelength in meters} = 300/\text{frequency in MHz}$$

$$\text{Wavelength in feet} = 1{,}000/\text{frequency in MHz}$$

Hence, antennas at 450 MHz need a clearance of approximately 1 meter (~3 feet), 900 MHz requires ~0.3 meters (~1 foot), and needs 2.4 GHz ~15 cm (~½ foot).

Antennas can be mounted with less clearance; however, the radiation from the antenna will be reduced. If the radio path is short, this probably will not matter.

It is important that the antenna mounting bracket be well connected to "earth" or "ground" for good lightning surge protection.

Yagi Antennas Yagi antennas are directional along the central beam of the antenna. The folded element is toward the back of the antenna, and the antenna should be "pointed" in the direction of the transmission.

Yagis should also be mounted with at least 1 to 2 wavelengths of clearance from other objects. The polarity of the antenna is the same as the direction of the orthogonal elements. For example, if the elements are vertical, the Yagi will transmit with vertical polarity.

In networks spread over wide areas, it is common for a central unit to have an omnidirectional antenna and the remote units to have Yagi antennas. In this case, because the omnidirectional antenna will be mounted with vertical polarity, the Yagis need to also have vertical polarity.

Care needs to be taken to ensure that the Yagi is aligned correctly to achieve optimum performance.

Two Yagis can be used for a point-to-point link. In this case, they can be mounted with the elements horizontally to give horizontal polarity. There is a large degree of RF isolation between horizontal and vertical polarity (approx -30 dB), so this installation method is a good idea if there is a large amount of interference from another system close by that is transmitting vertical polarity.

Mounting Near Other Antennas Avoid mounting an antenna near any other antenna. Even if the other antenna is transmitting on a different radio band, the high RF energy of the transmission from a close antenna can "deafen" a receiver. This is a common cause of problems with wireless systems.

Because antennas are designed to transmit parallel to the ground rather than up or down, vertical separation between antennas is a lot more effective than horizontal separation. If mounting near another antenna cannot be avoided, mounting it beneath or above the other antenna is better than mounting them side by side.

Using different polarity to the other antenna (if possible) will also help isolate the RF coupling.

Coaxial Cable If a coaxial cable connects to the antenna via connectors, it is very important to weatherproof the connection using sealing tape. Moisture ingress into a coaxial cable connection is the most common cause of problems with antenna installations. A three-layer sealing process is recommended: an initial layer of adhesive PVC tape followed by a second layer of self-vulcanizing weatherproofing tape (such as 3M 23 tape), with a final layer of adhesive PVC tape.

Allowing a "loop" of cable before the connection is also a good idea. The loop takes any installation strain off the connection and provides spare cable length in case of later moisture ingress into the connectors—in which case the original connectors need to be removed, the cable cut back, and new connectors fitted.

Avoid installing coaxial cables together in long parallel paths. Leakage from one cable to another has an effect similar to mounting an antenna near another antenna.

Lightning Surge Protection Power surges (also known as *voltage surges*) can enter wireless equipment in several ways: via the antenna connection, via the power supply, via connections to equipment connected to the device, and even via the "ground" connection.

Surges are electrical energy following a path to "ground" or "earth." The best way to protect the wireless unit is to remove or minimize the amount of energy entering the unit. This is achieved by "draining" the surge energy to ground via an alternate path.

Wireless devices need to have a solid connection to a ground point, normally an earth electrode, or several earth electrodes if the soil has poor conductivity.

A *solid connection* means a large-capacity conductor, not a small wire. All other devices connected to the wireless unit need to be grounded to the same ground point. There can be significant resistance between different ground points, leading to large voltage differences during surge conditions. Just as many wireless units are damaged by earth potentials as direct surge voltage.

It is very difficult (but not impossible) to protect against direct lightning strikes, but fortunately, the probability of a direct strike at any one location is very small, even in high lightning activity areas. Unfortunately, surges in power lines can occur from lightning strikes up to 5 km away, and electromagnetic energy in the air from lightning activity long distances away can induce high-voltage surges.

20.3.10 Antenna Connection

Electromagnetic energy in the air will be drained to ground via any and every earth path. An earth path exists via an antenna and the wireless unit. To protect against damage, this earth path current must be kept as small as possible. This is achieved by providing better alternate earth paths.

Where an external antenna is used, it is important to ground the antenna to the same ground point as the wireless unit. Antennas are normally mounted to a metal bracket; connecting this bracket to ground is satisfactory. Surge energy induced in antennas will be drained via the grounding connection to the antenna, via the outside shield of the coaxial cable to the ground connection on the wireless unit, and via the internal conductor of the coaxial cable via the radio electronics.

It is the third earth path that causes damage; if the other two paths provide a better earth path, damage can be avoided.

When an antenna is located outside a building or outside in an industrial plant environment, external coaxial surge diverters are recommended to further minimize the effect of surge current in the inner conductor of the coaxial cable.

Coaxial surge diverters have gas-discharge elements that break down in the presence of high surge voltage and divert any current directly to a ground connection. A surge diverter is not normally required when the antenna is within a plant or factory environment, since the plant steelwork provides multiple parallel ground paths and good earthing will provide adequate protection without a surge diverter.

CHAPTER | 20 Industrial Wireless Technology and Planning

20.3.10.1 Connections to Other Equipment

Surges can enter a wireless unit from connected devices, via I/O, serial, or Ethernet connections. Other data devices connected to the wireless unit should be well grounded to the same ground point as the wireless unit.

Special care needs to be taken where the connected data device is remote from the wireless unit, requiring a long data cable. Because the data device and the wireless unit cannot be connected to the same ground point, different earth potentials can exist during surge conditions.

There is also the possibility of surge voltages being induced on long lengths of wire from nearby power cables. Surge diverters can be fitted to the data cable to protect against surges entering the wireless unit.

20.3.10.2 Selecting Antennas/Coaxial Cable

Normally antennas and cable types are selected following a radio test for a proposed installation.

The test will be done with specific antennas and cables, and the results of the test will indicate whether higher-gain antennas or lower-loss cables are required. For example:

- If a test yields a fade margin of 5 dB, you know that by inserting an extra 5 dB antenna gain or removing 5 dB coaxial loss will increase the fade margin to 10 dB. The gain can be inserted at either the transmitter or receiver end, or both.
- If a test shows that net antenna/cable gain of 6 dB is required for reliable operation on a line-of-sight path, and a cable length of 10 meters is required at both ends of the link to give line of sight, you can select which cable to use and then select the antennas to give a net gain of 6 dB.

20.3.10.3 Other Considerations for Cables and Connectors

Coaxial cables have an inner conductor insulated from the surrounding *screen* or *shroud* conductor; the screen is grounded in operation to reduce external interference coupling into the inner conductor. The inner conductor carries the radio signal.

Industrial wireless devices are generally designed to operate with a 50 ohm load—that is, the coaxial cable and antennas are designed to have a 50 ohm impedance to the radio at RF frequencies.

At the high frequencies used in wireless, all insulation appears capacitive, and there is loss of RF signal between the inner conductor and the screen. The quality of the insulation, the frequency of the RF signal, and the length of the cable dictate the amount of loss. Generally, the smaller the outer diameter of the cable, the higher the loss; and loss increases as frequency increases. Cable loss is normally measured in dB per distance—for example, 3 dB per 10 meters, or 10 dB per 100 feet.

Cables need special coaxial connectors fitted. Generally, connectors have a loss of 0.1 to 0.2 dB per connector.

20.3.11 Commissioning

It is much easier to understand commissioning if we break it into parts: a radio test and a bench test.

20.3.11.1 Radio Test

Use portable units to test the radio path. Usually this testing consists of using the same wireless units powered by a couple of batteries and temporarily installed antennas. Testing the radio path prior to installation minimizes the risk of having to change the antenna installation on commissioning.

Most modern wireless devices provide a readout of both received signal strength (normally referred to as RSSI) and background noise level. Record these measurements to assist with future troubleshooting. If the wireless link commissions well but fails later, comparing the RSSI and noise figures with the "as-commissioned" figures gives a good indication of the problem.

A significant drop in signal level indicates that less RF power is being radiated at the transmitter. This could be due to moisture in the cable or antenna, low voltage at the transmitter, transmitter fault, or a change in radio path (trees now have leaves, there's another obstacle in the radio path, antennas have moved, or the like).

A significant increase in noise level indicates that other RF transmitters are active and are close by.

20.3.11.2 Bench Test

Test the configuration of the system on the bench before installing the equipment. If the system works well on the bench but not after installation, the difference is the radio side of the system. Make sure that you bench-test without antennas; connecting antennas will cause the receivers to saturate.

20.3.11.3 Troubleshooting Tips

How to you improve radio performance? Here are some tips:

- If background noise level is high, try to locate the cause of the interference. Is there a nearby antenna causing interference? Is the antenna installed close to noisy equipment such as a variable-frequency motor drive?
- Increase the RF signal level as much as possible.
- Try moving the antenna. If there is multipath fading in an industrial plant or factory, the antenna may only need to be moved a short distance to one side. In longer-distance applications, increase the height of the antenna to clear obstacles.

- Increase transmitter power if this is possible (but do not exceed regulations). Increase antenna gain or decrease cable loss, but again, do not exceed ERP regulations. If the wireless system is a one-way link (transmitter to receiver), you can increase the antenna gain at the receiver without worrying about ERP limits at the transmitter.
- If it is not possible to establish a reliable radio path, can you use an intermediate repeater? Most industrial wireless products have a "store-and-forward" repeater function whereby wireless messages can be received and retransmitted on to other units in the network.

20.3.12 Mesh Technologies

Telecommunications network operators have been building wired networks in "mesh" configurations for decades. A mesh consists of a web of interconnected switches (nodes). The advantage of a mesh wired network is that if a cable connecting two nodes is cut (or a switch malfunctions), traffic between the nodes can be rerouted via other nodes in the web, that is, the network is "self-healing." Similarly, the reliability of a wireless network increases when radio nodes are in a mesh topology, that is, a "web" of nodes is connected wirelessly. In the case of wireless networks, although there are no cables that can get severed, the redundancy offered by a mesh topology is useful in the event of a node malfunctioning or a wireless link getting blocked or a bad signal-to-noise situation.

In a wireless mesh network, the radio nodes work with each other to enable the coverage area of the nodes to work as a single network. Signal strength is sustained by breaking long distances into a series of shorter hops. Intermediate nodes boost the wireless signal and cooperatively make forwarding decisions based on their knowledge of the network.

Wireless mesh networks started out with a single mesh radio providing both client access and backhaul. Network performance was relatively poor as backhaul and service competed for bandwidth. To solve that issue, wireless mesh products began using two radios: one to provide service to clients and one to create a mesh network for backhaul. The two radios can operate in different bands or different channels on the same band. The wireless mesh infrastructure can also be designed to operate with the loss of the central server, system manager, or network manager, allowing the mesh to continue to operate. This is advantageous for industrial applications to keep messages flowing even though the master system is gone.

To get data across the network, messages travel from one node to the next until they arrive at their destinations. Dynamic routing algorithms in each node facilitate this communication. To implement such dynamic routing protocols, each node needs to communicate routing information to other nodes in the network. Each node then determines what to do with the message it receives: either pass it on the next node or keep it, depending on the protocol. The routing algorithms ensure that the message takes the fastest route to its next destination. Depending on the topology, between 10 and 20 mesh routers are mounted per square mile.

Most commercially available wireless mesh networking solutions are proprietary or use WiFi or IEEE 802.11x technologies in the unlicensed band. There are a large number of competing schemes for routing packets across mesh networks. The Institute of Electrical and Electronics Engineers (IEEE) is developing a set of standards under the title 802.11s to define an architecture and protocol for mesh networking. It is expected to be finalized in 2009.

Wireless mesh networks are used for secure, high-bandwidth, scalable access to fixed and mobile applications. Mesh networks can be used for industrial wireless networks, city-wide broadband Internet access, all-wireless offices, and even rural networks. Service can be extended wirelessly to third-party devices such as Internet Protocol (IP) cameras and automated meter readers.

20.3.13 System Management

System management is a specialized function that governs a wireless network, the operation of devices on the network, and the network communications. A system or network manager providing policy-based control of the runtime configuration performs this function.

The system or network manager usually provides the following functions for the network:

- Joining the network and leaving the network.
- Reporting of faults that occur in the network.
- Communication configuration.
- Configuration of clock distribution and the setting of system time.
- Device monitoring.
- Performance monitoring and optimization.
- Security configuration and monitoring.

System management can be executed with a centralized or distributed approach. The centralized approach has all network activities continually coordinated and managed by a single system or network manager. This centralized system manager could be located in the wireless network or on a plant network. In a distributed management approach, the network resources are divided and allocated to distributed system or network managers in the network, allowing them each to manage a pool of devices.

It should be noted that security management of the system is a specialized function that is realized in one entity and that works in conjunction with the system or network management function to enable secure system operation across the network.

20.3.14 System Interfaces

Industrial control networks built over the past few decades use protocols such as Modbus, Fieldbus, OPC, and so on.

These protocols link industrial electronic devices such as sensors, valves, and motors to PLCs, DCSs, SCADA systems, and human/machine interfaces (HMIs).

Industrial wireless local area networks typically use the IEEE 802.11 set of standards for wireless Ethernet. Hence a system interface is required between the wired and wireless portions of industrial control networks to convert between protocols.

System interfaces are required in two locations:

- In wireless transmitters connected to field instrumentation, to convert the signal into Wireless Protocol (for transmission via the wireless network).
- In the gateway between the wireless network and PLC/DCS that manages the plant's control system, to convert from Wireless Protocol to Modbus, Fieldbus, OPC, or other protocols.

There are several suppliers of system interfaces that supply OEMs building wireless industrial control products. When a plant purchases a wireless networking system, the required interfaces are typically a part of the system, invisible to the end user.

20.3.15 Standards and Specifications

20.3.15.1 WiFi 802.11

The 802.11 standard specifies requirements for wireless Ethernet devices using DSSS within the 2.4 GHz and 5 MHz ISM bands. This standard has become popularly known as WiFi (short for *wireless fidelity*). WiFi devices have been in the market since the late 1990s, and several 802.11 standards have emerged, from 802.11a to 802.11n. The most commonly used is 802.11g, which has data rates in the megabits.

Although WiFi is predominantly a commercial wireless standard, WiFi devices have been used extensively in industry for high data rates over very short distance. Industrial WiFi units are available with extended temperature ranges and additional functionality to improve operating reliability.

20.3.15.2 Bluetooth

The Bluetooth standard provides FHSS operation in the 2.4 GHz band with data rates of 800 Kb/s. Distances are very short because of reduced transmitter power. Bluetooth devices are generally used for transferring data between personal computer peripherals and mobile/cellular telephones. Although there are some Bluetooth industrial products, this standard is not commonly used in industry.

20.3.15.3 ZigBee 802.15

ZigBee is a very recent standard that was designed predominantly for battery-powered wireless sensors. ZigBee devices can use either 868/900 MHz or 2.4 GHz ISM bands and have low transmitter power to suit small, embedded batteries in wireless devices. An important feature of this standard is a wireless mesh whereby devices pass information from one unit to another. ZigBee devices are emerging in building management and heating, ventilation and air-conditioning (HVAC) systems and are expected to be used in industrial applications.

20.3.15.4 WiMax

WiMax is a recent wireless standard for fixed broadband data services operating in the 11–66 GHz band. It is being developed with the IEEE 802.16 standards group. WiMax will generally be delivered through commercial supplier networks, although private networks are also envisioned.

20.3.15.5 Cellular Wireless

GSM, GPRS, CDPD, CDMA—these are different cellular standards used for transmitting data. Although they are used in industrial environments for inventory management, they are generally not used for internal plant or factory applications.

20.3.15.6 WirelessHART

WirelessHART was specification developed and released in 2007 by the HART Communications Foundation (HCF). It is a wireless extension to the already popular HART protocol (IEC 61158). It utilizes 802.15.4 radios, channel hopping, spread spectrum, and mesh networking technologies.

20.3.15.7 ISA100.11a

ISA100.11a was a standard developed by the Instruments, Systems, and Automation (ISA) organization. It is protocol independent and designed for plant-wide deployment. It utilizes 802.15.4 radios, channel hopping, spread spectrum, mesh networking technologies, and multiprotocol host interface capabilities.

20.4 PLANNING FOR WIRELESS

The decision to implement wireless technology in an industrial facility is a strategic choice, enabling an infrastructure that will provide significant benefits beyond avoiding the wiring costs. The right decision will help improve safety, optimize the plant, and ensure compliance. Wireless is a complex enabling technology that requires deliberate consideration before broad deployment in an industrial facility. This section outlines an approach for choosing the right industrial wireless network for you.

20.4.1 Imagine the Possibilities

Wireless technology provides a low-cost solution for unlocking value in the plant and for enabling a mobile and more productive workforce. The possibilities are endless. Imagine sensors gathering data where traditional devices cannot reach, providing more real-time data to make knowledgeable decisions. Imagine a wireless network delivering on the promise of lower installed costs. Industrial users expect a secure and reliable network that supports multiple types of wireless-enabled applications that will:

- Keep people, plant, and the environment safe.
- Improve plant and asset reliability.
- Optimize a plant through efficient employees and processes.
- Comply with industrial and environmental standards.

These wireless-enabled solutions will reap operational and safety benefits through the following applications.

Keep people, plants, and the environment safe:

- A real-time location system throughout the facility to monitor employee locations and ensure safe procedural operations.
- Safety shower monitoring.
- An infrastructure that supports emergency responders.
- Wireless leak detection and repair support.
- Integration with existing control and safety systems.

Improve plant and asset reliability:

- Continuous wireless monitoring of equipment and field devices for diagnostic equipment health assessments.
- Mobile worker device commissioning and configuring with automated field operator rounds and access to online data, reports and manuals.
- Equipment health management visualization, such as computerized maintenance management system, inventory management, and document management.
- Utilizing previously unreachable data in existing control system and advanced process control applications.

Optimize a plant through efficient employees and processes:

- Mobile operators operating their desktop applications and their control room displays on handheld computers.
- Input/output modules and sensors to monitor real measurements versus inferred values or for remote control.
- Sensors for upgrading tank instrumentation.
- Voice over IP (VoIP) for communicating among all field workers equipped with WiFi devices.
- Continuous wireless corrosion detection to ensure integrity of piping systems.

Comply with industrial and environmental standards:

- Emissions monitoring.
- Leak detection and repair support.

20.4.2 Getting Ready for Wireless

Many industrial facilities are already deploying wireless networks for targeted requirements. To help plants get ready for the future, the following list presents considerations and questions to ask in preparing for this emerging technology to enable robust business results. It is important to first scope out your current and future wireless needs and make a strategic decision as to the selection of your wireless network based on these needs. This checklist of considerations will not decide for you which strategic direction is best; rather, it will help you explore all the aspects that are important to consider for that decision.

20.4.2.1 Functionality and Applications

Consider how many different functions are more efficient with wireless technology.

- Are you willing to deploy multiple wireless networks to manage and maintain, or do you want just one strategic network? (Many users have multiple uses but want just one wireless network to deploy and manage.)
- Will you consider some simple control applications?
- Do you want to enable your field workers with wireless handheld devices to access data and interact with various servers in the facility?
- Will you want first responders to utilize your wireless network in case of an emergency?

20.4.2.2 Multispeed Support

Do you have requirements for information to reach the control room quickly for some applications and less quickly for others? Can you afford to have your alarms transmitted back at the same rate as monitoring information?

- Will you do low-speed monitoring as well as high-speed monitoring for certain process measurements?

20.4.2.3 Reliability

Can your operations survive without the information conveyed wirelessly? Most can today, but as you look forward and really embrace wireless, your future applications will require a more reliable network. In addition, most wireless solutions are using the unlicensed ISM frequency bands, which provide limited bandwidth for your plant.

- Have you developed a plan for how you are going to use the ISM bands in your industrial facility? This consideration will help a plant ensure solid wireless operations. Suboptimized ISM bands will lead to reduced scalability and reliability, limiting your wireless usage, much like a wiring conduit that is already full.

20.4.2.4 Security

Security is essential to protect against malicious intent and to protect your intellectual property, your bottom line, and your people.

- What security do you need? How much is enough?
- Do you need just one security system, or many?

These are important considerations as you strategically deploy an industrial wireless network. You most likely will want to have just one wireless security approach. This gives you just one system to manage and provides you the opportunity to pick the best available solution to match your wireless uses today and into the future.

20.4.2.5 Self-contained and Predictable Power Management

When most users consider wireless deployments, they understand the upside of no wiring and the cost advantage, but they also envision the downside of having to change many batteries in industrial devices throughout the facility. Device power management is a very important consideration n selecting a wireless network.

- How long do you want your wireless devices to be self-powered? Do your wireless devices require add-on products to maintain and install to meet your reporting rate needs?
- What level of predictable maintenance schedule do you require?

This is a complex question because the answer must consider the power source, the device needs, and how often the device communicates. Most users will require a device that is self-powered for at least three years and, at best, for the lifetime of the device. This is a reasonable demand in selecting a wireless network.

20.4.2.6 Scalability

Planning for future growth and considering what happens when your wireless demands are for several thousand devices must be considerations in selecting your network.

- How many devices can your network handle?
- Will they be enough for the lifetime of your wireless network?
- What happens when you go beyond the limit of your network capacity? Can your network expand?

Many users begin with very limited wireless needs, but as they begin to see the benefits of wireless technology, their needs grow exponentially.

20.4.2.7 Investment Protection and Application Integration

Many plants reap benefits from having previously deployed multiple application interfaces and wireless products throughout the facility. These interfaces can include Modbus, OPC, HART, Foundation Fieldbus, Profibus, Ethernet, and many others. How many of these have you deployed in your facility? Usually plants contain multiple application interfaces driven by different departments. Many users also want information coming from their wireless devices to utilize these existing legacy applications and protocols. When selecting a strategic wireless network, you must have the ability to interface easily with all your legacy applications that will require wireless data. This is very important because this network will service your whole operation, not just one department.

- Can your wireless network serve many application interfaces?
- Will your next choice support your existing wireless devices?

20.4.2.8 Choices

As you select your strategic wireless network, you will need product choices. This opportunity for choice provides ideal pricing alternatives and best-in-class products. When a standard is developed or open solutions are offered, many suppliers adhere to those technology specifications and offer choices for customers in the devices they deploy in the network as well as the applications that can run in the network.

There are many aspects to consider as you deploy wireless in your industrial facility. Deploying this enabling technology is a strategic decision.

Consider all the areas discussed here in creating a comprehensive wireless solution for your facility. Industrial wireless networks that do not address each area satisfactorily might not fit your long-term strategic use of wireless technology.

REFERENCES

http://en.wikipedia.org/wiki/Wireless.
http://en.wikipedia.org/wiki/Electromagnetic_interference.
http://wireless.wilsonmohr.com/wireless.nsf/Content/wirelessresources~industrialwirelessbasics.
www.isa.org/InTechTemplate.cfm?Section=Article_Index1...template=/ContentManagement/ContentDisplay.cfm...ContentID=66911.
ISA100.11 Draft Standard, September 2008.
www.jhsph.edu/wireless/history.html.
Elpro Technologies, *Industrial Wireless Handbook*.
Elpro Technologies, Tech Article No. 1.3, *Wireless Solutions for Process Applications*.
Honeywell White Paper, *Choosing the Right Network*.

Part III

Measurement of Temperature and Chemical Composition

Chapter 21

Temperature Measurement

C. Hagart-Alexander

21.1 TEMPERATURE AND HEAT

21.1.1 Application Considerations

Temperature is one of the most frequently used process measurements. Almost all chemical processes and reactions are temperature dependent. In chemical plants, temperature is not infrequently the only indication of the progress of a process. Where the temperature is critical to the reaction, a considerable loss of product may result from incorrect temperatures. In some cases, loss of control of temperature can result in catastrophic plant failure, with the attendant damage and possible loss of life.

Another area in which accurate temperature measurement is essential is in the metallurgical industries. In the case of many metal alloys, typically steel and aluminum alloys, the temperature of the heat treatment the metal receives during manufacture is a crucial factor in establishing the mechanical properties of the finished product.

There are many other areas of industry in which temperature measurement is essential. Such applications include steam raising and electricity generation, plastics manufacture and molding, milk and dairy products, and many other areas of the food industries.

Where most of us are most aware of temperature is in the heating and air-conditioning systems that make so much difference to people's personal comfort.

Instruments for the measurement of temperature, as with so many other instruments, are available in a wide range of configurations. Everyone must be familiar with the ubiquitous liquid-in-glass thermometer. However, a number of other temperature-dependent effects are used for measurement purposes, and readout can be local to the point of measurement or remote in a control room or elsewhere.

21.1.2 Definitions

To understand temperature measurement, it is essential to have an appreciation of the concepts of temperature and other heat-related phenomena. These are described in the following sections.

21.1.2.1 Temperature

The first recorded temperature measurement was carried out by Galileo at the end of the 16th century. His thermometer depended on the expansion of air. Some form of scale was attached to his apparatus, for he mentions "degrees of heat" in his records.

As with any other measurement, it is necessary to have agreed and standardized units of measurement. In the case of temperature, the internationally recognized units are the Kelvin and the degree Celsius. The definitions of these units are set out in Section 21.2.

Temperature is a measure of stored or potential energy in a mass of matter. It is the state of agitation, both lateral and rotational oscillation, of the molecules of the medium. The higher the temperature of a body, the greater the vibrational energy of its molecules and the greater its potential to transfer this molecular kinetic energy to another body. Temperature is the potential to cause heat to move from a point of higher temperature to one of lower temperature. The rate of heat transfer is a function of that temperature difference.

21.1.2.2 Heat

Heat is thermal energy. The quantity of heat in a body is proportional to the temperature of that body, that is, it is its heat capacity multiplied by its absolute temperature.

Heat is measured in Joules. (Before the international agreements on the SI system of units, heat was measured in calories. One calorie was approximately 4.2 Joules.)

21.1.2.3 Specific Heat Capacity

Various materials absorb different amounts of heat to produce the same temperature rise. The specific heat capacity, or more usually the specific heat, of a substance is the amount of heat that, when absorbed by 1 kg of that substance, will raise its temperature by 1°C.

$$\text{specific heat capacity} = J\ kg^{-1} k^{-1}$$

21.1.2.4 Thermal Conductivity

The rate at which heat is conducted through a body depends on the material of the body. Heat travels very quickly along a bar of copper, for instance, but more slowly through iron. In the case of nonmetals, ceramics, or organic substances, the thermal conduction occurs more slowly still. The heat conductivity is not only a function of the substance but also the form of the substance. Plastic foam is used for heat insulation because the gas bubbles in the foam impede the conduction of heat. Thermal conductivity is measured in terms of:

$$\frac{\text{energy} \times \text{length}}{\text{area} \times \text{time} \times \text{temperatue difference}}$$

$$\text{thermal conductivity} = \frac{J \cdot m}{m^2 \cdot s \cdot K}$$

$$= J \cdot m^{-1} \cdot s^{-1} \cdot K^{-1}$$

21.1.2.5 Latent Heat

When a substance changes state from solid to liquid or from liquid to vapor, it absorbs heat without change of temperature. If a quantity of ice is heated at a constant rate, its temperature will rise steadily until it reaches a temperature of 0°C; at this stage the ice will continue to absorb heat with no change of temperature until it has all melted to water. Now as the heat continues to flow into the water, the temperature will continue to rise but at a different rate from before due to the different specific heat of water compared to ice. When the water reaches 100°C, the temperature rise will again level off as the water boils, changing state from water to steam. Once all the water has boiled to steam, the temperature will rise again, but now at yet another rate dependent on the specific heat of steam. This concept is illustrated in Figure 21.1.

The amount of heat required to convert a kilogram of a substance from solid state to liquid state is the *latent heat of fusion*. Likewise, the *latent heat of evaporation* is the amount of heat required to convert a kilogram of liquid to vapor.

This leveling of temperature rise during change of state accounts for the constant freezing temperatures and constant boiling temperatures of pure materials. The units of measurement of latent heat are Joules per kilogram:

$$\text{latent heat} = J \cdot kg^{-1}$$

21.1.2.6 Thermal Expansion

Expansion of Solids When a solid is heated, it increases in volume. It increases in length, breadth, and thickness. The

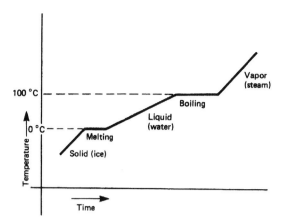

FIGURE 21.1 Increase of temperature during change of state of a mass of water under conditions of constant energy input.

increase in length of any side of a solid will depend on the original length l_0, the rise in temperature t, and the coefficient of linear expansion α.

The coefficient of linear expansion may be defined as the increase in length per unit length when the temperature is raised 1°C. Thus if the temperature of a rod of length l_0 is raised from 0°C to t°C, the new length, l_t, will be given by

$$l_t = l_0 + l_0 \cdot \alpha t = l_0(1 + \alpha t) \qquad (21.1)$$

The value of the coefficient of expansion varies from substance to substance. The coefficients of linear expansion of some common materials are given in Table 21.1.

The increase in area with temperature, that is, the coefficient of superficial expansion, is approximately twice the coefficient of linear expansion. The coefficient of cubic expansion is almost three times the coefficient of linear expansion.

Expansion of Liquids and Gases In dealing with the expansion of liquids and gases, it is necessary to consider the volume expansion or cubical expansion. Both liquids and gases have to be held by a container, which will also expand, so that the apparent expansion of the liquid or gas will be less than the true or absolute expansion. The true coefficient of expansion of a liquid is equal to the coefficient of cubical expansion of the containing vessel. Usually the expansion of a gas is so much greater than that of the containing vessel that the expansion of the vessel may be neglected in comparison with that of the gas.

The coefficient of expansion of a liquid may be defined in two ways. First, there is the zero coefficient of expansion, which is the increase in volume per degree rise in temperature divided by the volume at 0°C, so that volume V_t at temperature t is given by

$$V_t = V_0(l + \beta t) \qquad (21.2)$$

CHAPTER | 21 Temperature Measurement

TABLE 21.1 Coefficients of linear expansion of solids, extracted from *Tables of Physical and Chemical Constants* by Kaye and Laby (Longmans); the values given are per Kelvin and, except where some temperature is specified, for a range of about 20 degrees

Substance	α (ppm)
Aluminum	25.5
Copper	16.7
Gold	13.9
Iron (cast)	10.2
Lead	29.1
Nickel	12.8
Platinum	8.9
Silver	18.8
Tin	22.4
Brass (typical)	18.9
Constantan (Eureka) 60 Cu, 40 Ni	17.0
Duralumin	22.6
Nickel steel,	
10% Ni	13.0
30% Ni	12.0
36% Ni (Invar)	−0.3 to +2.5
40% Ni	6.0
Steel	10.5 to 11.6
Phosphor bronze, 97.6 Cu, 2 Sn, 0.2 P	16.8
Solder, 2 Pb, 1 Sn	25
Cement and concrete	10
Glass (soda)	8.5
Glass (Pyrex)	3
Silica (fused) −80° to 0°C	0.22
Silica (fused) 0° to 100°C	0.50

where V_0 is the volume at 0°C and β is the coefficient of cubical expansion.

There is also the mean coefficient of expansion between two temperatures. This is the ratio of the increase in volume per degree rise of temperature to the original volume. That is,

$$\beta = \frac{V_{t_2} - V_{t_1}}{V_{t_1}(t_2 - t_1)} \quad (21.3)$$

where V_{t_1} is the volume at temperature t_1, and V_{t_2} is the volume at temperature t_2.

This definition is useful in the case of liquids that do not expand uniformly, such as water.

21.1.3 Radiation

There are three ways in which heat may be transferred: conduction, convection, and radiation. *Conduction* is, as already discussed, the direct transfer of heat through matter. *Convection* is the indirect transfer of heat by the thermally induced circulation of a liquid or gas; in "forced convection," the circulation is increased by a fan or pump. *Radiation* is the direct transfer of heat (or other form of energy) across space.

TABLE 21.2 Wavelengths of thermal radiation	
Radiation	Wavelength (μm)
Infrared	100–0.8
Visible light	0.8–0.4
Ultraviolet	0.4–0.01

Thermal radiation is electromagnetic radiation and comes within the infrared, visible, and ultraviolet regions of the electromagnetic spectrum. The demarcation between these three classes of radiation is rather indefinite, but as a guide the wavelength bands are shown in Table 21.2.

So far as the effective transfer of heat is concerned, the wavelength band is limited to about 10 μm in the infrared and to 0.1 μm in the ultraviolet. All the radiation in this band behaves in the same way as light. The radiation travels in straight lines and may be reflected or refracted, and the amount of radiant energy falling on a unit area of a detector is inversely proportional to the square of the distance between the detector and the radiating source.

21.2 TEMPERATURE SCALES

To measure and compare temperatures, it is necessary to have agreed scales of temperature. These temperature scales are defined in terms of physical phenomena that occur at constant temperatures. The temperatures of these phenomena are known as *fixed points*.

21.2.1 Celsius Temperature Scale

The Celsius temperature scale is defined by international agreement in terms of two fixed points: the ice point and the steam point. The temperature of the *ice point* is defined as 0° Celsius and the *steam point* as 100° Celsius.

The ice point is the temperature at which ice and water exist together at a pressure of 1.0132×10^5 N·m^{-2} (originally one standard atmosphere = 760 mm of mercury). The ice should be prepared from distilled water in the form of fine shavings and mixed with ice-cold distilled water.

The steam point is the temperature of distilled water boiling at a pressure of 1.0132×10^5 N·m^{-2}. The temperature at which water boils is very dependent on pressure. At a pressure p, N·m^{-2} the boiling point of water t_p in degrees Celsius is given by

$$t_p = 100 + 2.795 \times 10^{-4}(p - 1.013 \times 10^{-5}) \\ - 1.334 \times 10^{-9}(p - 1.013 \times 10^5)^2 \quad (21.4)$$

The temperature interval of 100°C between the ice point and the steam point is called the *fundamental interval*.

21.2.2 Kelvin, Absolute, or Thermodynamic Temperature Scale

Lord Kelvin defined a scale based on thermodynamic principles that does not depend on the properties of any particular substance. Kelvin divided the interval between the ice and steam points into 100 divisions so that one Kelvin represents the same temperature interval as one Celsius degree. The unit of the Kelvin or thermodynamic temperature scale is called the *Kelvin*. The definition of the Kelvin is the fraction 1/273.16 of the thermodynamic temperature of the triple point of water. This definition was adopted by the 13th meeting of the General Conference for Weights and Measures, in 1967 (13th CGPM, 1967). Note the difference between the ice point (0°) used for the Celsius scale and the triple point of water, which is 0.01°C.

It has also been established that an ideal gas obeys the gas law $PV = RT$, where T is the temperature on the absolute or Kelvin scale and where P is the pressure of the gas, V is the volume occupied, and R is the universal gas constant. Thus the behavior of an ideal gas forms a basis of temperature measurement on the absolute scale. Unfortunately, the ideal gas does not exist, but the so-called permanent gases, such as hydrogen, nitrogen, oxygen, and helium, obey the law very closely, provided that the pressure is not too great. For other gases and for the permanent gases at greater pressures, a known correction may be applied to allow for the departure of the behavior of the gas from that of an ideal gas. By observing the change of pressure of a given mass of gas at constant volume or the change of volume of the gas at constant pressure, it is possible to measure temperatures on the absolute scale.

The constant-volume gas thermometer is simpler in form and is easier to use than the constant-pressure gas thermometer. It is, therefore, the form that is most frequently used. Nitrogen has been found to be the most suitable gas to use for temperature measurement between 500°C and 1500°C, whereas at temperatures below 500°C hydrogen is used. For very low temperatures, helium at low pressure is used.

The relationship between the Kelvin and Celsius scales is such that 0° Celsius is equal to 273.15 K:

$$t = T - 273.15 \quad (21.5)$$

where t represents the temperature in degrees Celsius and T is the temperature Kelvin.

It should be noted that temperatures on the Celsius scale are referred to in terms of degrees Celsius, °C; temperatures on the absolute scale are in Kelvins, K, no degree sign being

CHAPTER | 21 Temperature Measurement

used. For instance, the steam point is written in Celsius as 100°C, but on the Kelvin scale it is 373.15 K.

21.2.3 International Practical Temperature Scale of 1968 (IPTS-68)

The gas thermometer, the final standard of reference, is, unfortunately, rather complex and cumbersome and entirely unsuitable for industrial use. Temperature-measuring instruments capable of a very high degree of repeatability are available. Use of these instruments enables temperatures to be reproduced to a very high degree of accuracy, although the actual value of the temperature on the thermodynamic scale is not known with the same degree of accuracy. To take advantage of the fact that temperature scales may be reproduced to a much higher degree of accuracy than they can be defined, an International Practical Temperature Scale (IPTS) was adopted in 1929 and revised in 1948. The latest revision of the scale occurred in 1968 (IPTS-68). The 1948 scale is still used in many places in industry. The differences between temperatures on the two scales are small, frequently within the accuracy of commercial instruments. Table 21.3 shows the deviation of the 1948 scale from the 1968 revision.

The International Practical Temperature Scale is based on a number of defining fixed points, each of which has been subject to reliable gas thermometer or radiation thermometer observations; these are linked by interpolation using instruments that have the highest degree of reproducibility. In this way the International Practical Temperature Scale is conveniently and accurately reproducible and provides means for identifying any temperature within much narrower limits than is possible on the thermodynamic scale.

The defining fixed points are established by realizing specified equilibrium states between phases of pure substances. These equilibrium states and the values assigned to them are given in Table 21.4.

The scale distinguishes between the International Practical Kelvin Temperature with the symbol T_{68} and the International Practical Celsius Temperature with the symbol t_{68}; the relationship between T_{68} and t_{68} is

$$t_{68} = T_{68} - 273.15 \text{ K} \qquad (21.6)$$

The size of the degree is the same on both scales: 1/273.16 of the temperature interval between absolute zero and the triple point of water (0.01°C). Thus the interval between the ice point 0°C and the boiling point of water 100°C is still 100 Celsius degrees. Temperatures are expressed in Kelvins below 273.15 K (0°C) and in degrees Celsius above 0°C. This differentiation between degrees Celsius and degrees Kelvin is not always convenient, and consequently temperatures below 0°C are usually referred to as *minus degrees Celsius*.

Temperatures between and above the fixed points given in Table 21.4 can be interpolated as follows.

- From 13.81 K to 630.74°C, the standard instrument is the platinum resistance thermometer.
- Above 1,337.58 K (1064.43°C), the scale is defined by Planck's law of radiation, with 1,337.58 K as the reference temperature, and the constant c_2 has a value 0.014 388 meter Kelvin. This concept is discussed in Section 21.6.

In addition to the defining fixed points, the temperatures corresponding to secondary points are given. These points, particularly the melting or freezing points of metals, form convenient workshop calibration points for temperature-measuring instruments (see Table 21.5).

21.2.4 Fahrenheit and Rankine Scales

The Fahrenheit and Rankine temperature scales are now obsolete in Britain and the United States, but because a great deal of engineering data, steam tables, and so on were published using the Fahrenheit and Rankine temperature, a short note for reference purposes is relevant.

Fahrenheit This scale was proposed in 1724. Its original fixed points were the lowest temperature obtainable using ice and water with added salts (ammonium chloride), which was taken as zero. Human blood heat was made 96 degrees (98.4 on the modern scale). On this scale the ice point is 32°F and the steam point is 212°F. There does not appear to be any formal definition of the scale.

To convert from the Fahrenheit to Celsius scale, if t is the temperature in Celsius and f the temperature in Fahrenheit,

$$t = \frac{5}{9}(f - 32) \qquad (21.7)$$

TABLE 21.3 Deviation of IPTS-68 from IPTS-48

t_{68} (°C)	$t_{68} - t_{48}$
−200	0.022
−150	−0.013
0	0.000
50	0.010
100	0.000
200	0.043
400	0.076
600	0.150
1,000	1.24

TABLE 21.4 Defining fixed points of the IPTS-68[1]

Equilibrium state	Assigned value of International Practical temperature	
	T_{68}	t_{68}
Triple point of equilibrium hydrogen	13.81 K	−259.34°C
Boiling point of equilibrium hydrogen at pressure of 33, 330.6 kN · m⁻²	17.042 K	−256.108°C
Boiling point of equilibrium hydrogen	20.28 K	−252.87°C
Boiling point of neon	27.102 K	−246.048°C
Triple point of oxygen	54.361 K	−218.789°C
Boiling point of oxygen	90.188 K	−182.962°C
Triple point of water[3]	273.16 K	0.01°C
Boiling point of water[2][3]	373.15 K	100°C
Freezing point of zinc	692.73 K	419.58°C
Freezing point of silver	1,235.08 K	961.93°C
Freezing point of gold	1,337.58 K	1,064.43°C

1. Except for the triple points and one equilibrium hydrogen point (17.042 K), the assigned values of temperature are for equilibrium states at a pressure P_0 = 1 standard atmosphere (101.325 kN · m⁻²).
   In the realization of the fixed points, small departures from the assigned temperatures will occur as a result of the differing immersion depths of thermometers or the failure to realize the required pressure exactly. If due allowance is made for these small temperature differences, they will not affect the accuracy of realization of the scale.
2. The equilibrium state between the solid and liquid phases of tin (freezing point of tin has assigned value of t_{68} 231.9681°C and may be used an alternative to the boiling point of water).
3. The water used should have the isotopic composition of ocean water.

Rankine The Rankine scale is the thermodynamic temperature corresponding to Fahrenheit. Zero in Rankine is, of course, the same as zero Kelvin. On the Rankine scale the ice point is 491.67°R. Zero Fahrenheit is 459.67°R. To convert temperature from Fahrenheit to Rankine, where R is the Rankine temperature,

$$R = f + 459.67 \qquad (21.8)$$

Table 21.6 illustrates the relationship between the four temperature scales.

21.2.5 Realization of Temperature Measurement

Techniques for temperature measurement are quite varied. Almost any temperature-dependent effect may be used. Sections 21.3–21.6 describe the main techniques for temperature measurement used in industry. However, in laboratories or under special industrial conditions, a wider range of instruments is available. Table 21.7 summarizes the more commonly used measuring instruments in the range quoted. All measuring instruments need to be calibrated against standards. In the case of temperature, the standards are the defining fixed points on the IPTS-68. These fixed points are not particularly easy to achieve in workshop conditions. Although the secondary points are intended as workshop standards, it is more usual, in most instrument workshops, to calibrate against high-grade instruments for which calibration is traceable to the IPTS-68 fixed points.

21.3 MEASUREMENT TECHNIQUES: DIRECT EFFECTS

Instruments for measuring temperature described in this section are classified according to the nature of the change in the measurement probe produced by the change of temperature. They have been divided into four classes: liquid expansion, gas expansion, change of state, and solid expansion.

21.3.1 Liquid-in-Glass Thermometers

The glass thermometer must be the most familiar of all thermometers. Apart from its industrial and laboratory use, it finds application in both domestic and medical fields.

21.3.1.1 Mercury-Filled Glass Thermometer

The coefficient of cubical expansion of mercury is about eight times greater than that of glass. If, therefore, a glass container holding mercury is heated, the mercury will expand more than the container. At a high temperature, the mercury will occupy a greater fraction of the volume of the

TABLE 21.5 Secondary reference points (IPTS-68)

Substance	Equilibrium state	Temperature (K)
Normal hydrogen	TP	13.956
Normal hydrogen	BP	20.397
Neon	TP	24.555
Nitrogen	TP	63.148
Nitrogen	BP	77.342
Carbon dioxide	Sublimation point	194.674
Mercury	FP	234.288
Water	Ice point	273.15
Phenoxy benzine	TP	300.02
Benzoic acid	TP	395.52
Indium	FP	429.784
Bismuth	FP	544.592
Cadmium	FP	594.258
Lead	FP	600.652
Mercury	BP	629.81
Sulphur	BP	717.824
Copper/aluminum eutectic	FP	821.38
Antimony	FP	903.89
Aluminum	FP	933.52
Copper	FP	1,357.6
Nickel	FP	1,728
Cobalt	FP	1,767
Palladium	FP	1,827
Platinum	FP	2,045
Rhodium	FP	2,236
Iridium	FP	2,720
Tungsten	FP	3,660

TP = triple point; FP = freezing point; BP = boiling point.

TABLE 21.6 Comparison of temperature scales

	K	°C	°F	°R
Absolute zero	0	−273.15	−523.67	0
Boiling point O_2	90.19	−182.96	−361.33	162.34
Zero Fahrenheit	255.37	−17.78	0	459.67
Ice point	273.15	0	32	491.67
Steam point	373.15	100	212	671.67
Freezing point of silver	1,235.08	961.93	1,763.47	2,223.14

TABLE 21.7 Temperature measurement techniques

Range (K)	Technique	Application	Resolution (K)
0.01–1.5	Magnetic susceptance of paramagnetic salt	Laboratory	0.001
0.1–50	Velocity of sound in acoustic cavity	Laboratory standard	0.0001
0.2–2	Vapor pressure	Laboratory standard	0.001
1.5–100	Germanium resistance thermometer	Laboratory standard	0.0001
1.5–100	Carbon resistance thermometer	Laboratory	0.001
1.5–1,400	Gas thermometer	Laboratory Industrial	0.002 1.0
210–430	Silicon P-N junction	Laboratory Industrial	0.1 —
4–500	Thermistor	Laboratory Industrial	0.001 0.1
11–550	Quartz crystal oscillator	Laboratory Industrial	0.001 —
15–1,000	Platinum resistance thermometer	Standard Industrial	0.000 01 0.1
20–2,700	Thermocouple	General-purpose	1.0
30–3,000	Sound velocity in metal rod	Laboratory	1%
130–950	Liquid-in-glass	General-purpose	0.1
130–700	Bimetal	Industrial	1–2
270–5,000	Total radiation thermometer	Industrial	10
270–5,000	Spectrally selective radiation thermometer	Industrial	2

container than at a low temperature. If, then, the container is made in the form of a bulb with a capillary tube attached, it can be so arranged that the surface of the mercury is in the capillary tube, its position along the tube will change with temperature and the assembly used to indicate temperature. This is the principle of the mercury-in-glass thermometer.

The thermometer, therefore, consists simply of a stem of suitable glass tubing having a very small but uniform bore. At the bottom of this stem is a thin-walled glass bulb. The bulb may be cylindrical or spherical in shape and has a capacity very many times larger than that of the bore of the stem. The bulb and bore are completely filled with mercury, and the open end of the bore is sealed off either at a high temperature or under vacuum, so that no air is included in the system. The thermometer is then calibrated by comparing it with a standard thermometer in a bath of liquid, the temperature of which is carefully controlled.

When the standard thermometer and the thermometer to be calibrated have reached equilibrium with the bath at a definite temperature, the point on the glass of the thermometer opposite the top of the mercury meniscus is marked. The process is repeated for several temperatures. The intervals between these marks are then divided off by a dividing machine. In the case of industrial thermometers, the points obtained by calibration are transferred to a metal or plastic plate, which is then fixed with the tube into a suitable protecting case to complete the instrument.

The stem of the thermometer is usually shaped in such a way that it acts as a lens, magnifying the width of the mercury column. The mercury is usually viewed against a background of glass that has been enameled white. Figure 21.2 shows the typical arrangement for a liquid-in-glass thermometer.

Mercury-in-glass thermometers are available in three grades: A and B are specified in BS 1041: Part 2.1: 1958; grade C is a commercial-grade thermometer and no limits of accuracy are specified. Whenever possible, thermometers should be calibrated, standardized, and used immersed up to the reading, that is, totally immersed, to avoid errors due to the emergent column of mercury and the glass stem being at a different temperature than the bulb. Errors introduced this way should be allowed for if accurate readings are required. Some thermometers, however, are calibrated for partial immersion and should be used immersed to the specified depth.

When reading a thermometer, an observer should keep his eye on the same level as the top of the mercury column. In this way errors due to parallax will be avoided.

Figure 21.3 shows the effect of observing the thermometer reading from the wrong position. When viewed from

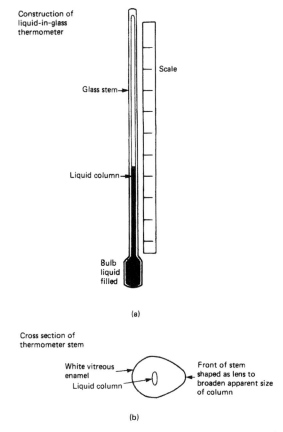

FIGURE 21.2 Mercury-in-glass thermometer. (a) Thermometer and scale. (b) Cross-section of thermometer stem.

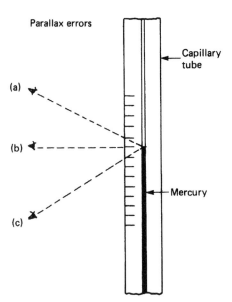

FIGURE 21.3 Parallax errors when reading glass thermometer.

(a) the reading is too high. Taken from (b) the reading is correct, but from (c) it is too low.

A mercury-in-glass thermometer has a fairly large thermal capacity (i.e., it requires quite an appreciable amount of heat to change its temperature by one degree), and glass is not a very good conductor of heat. This type of thermometer will, therefore, have a definite thermal lag. In other words, it will require a definite time to reach the temperature of its surroundings. This time should be allowed for before any reading is taken. If there is any doubt as to whether the thermometer has reached equilibrium with a bath of liquid having a constant temperature, readings should be taken at short intervals of time. When the reading remains constant, the thermometer must be in equilibrium with the bath. If the temperature is varying rapidly, the thermometer may never indicate the temperature accurately, particularly if the tested medium is a gas.

Glass thermometers used in industry are usually protected by metal sheaths. These sheaths may conduct heat to or from the neighborhood of the thermometer bulb and cause the thermometer to read either high or low, according to the actual prevailing conditions. A thermometer should, therefore, be calibrated, whenever possible, under the conditions in which it will be used, if accurate temperature readings are required. If, however, the main requirement is that the temperature indication be consistent for the same plant temperature, an introduced error is not so important, so long as the conditions remain the same and the error is constant.

Errors due to Aging It is often assumed that provided a mercury-in-glass thermometer is in good condition, it will always give an accurate reading. This is not always so, particularly with cheap thermometers. A large error may be introduced by changes in the size of the bulb due to aging. When glass is heated to a high temperature, as it is when a thermometer is made, it does not, on cooling, contract to its original volume immediately. Thus for a long time after it has been made, the bulb continues to contract very slowly so that the original zero mark is too low on the stem and the thermometer reads high. This error continues to increase over a long period and depends on the type of glass used in the manufacture of the thermometer. To reduce to a minimum the error due to this cause, during manufacture thermometers are annealed by baking for several days at a temperature above that which they will be required to measure, and then they are cooled slowly over a period of several days.

Another error due to the same cause is the depression of the zero when a thermometer is cooled rapidly from a high temperature. When cooled, the glass of the thermometer bulb does not contract immediately to its original size, so the reading on the thermometer at low temperature is too low but returns to normal after a period of time. This period depends on the nature of the glass from which the bulb is made.

High-Temperature Thermometers Mercury normally boils at 357°C at atmospheric pressure. To extend the range of a mercury-in-glass thermometer beyond this temperature, the top end of the thermometer bore is enlarged into a bulb with a capacity of about 20 times that of the bore of the stem. This bulb, together with the bore above the mercury, is then filled with nitrogen or carbon dioxide at a sufficiently high pressure to prevent the mercury boiling at the highest temperature at which the thermometer will be used. To extend the range to 500°C, a pressure of about 20 bar is required. In spite of the existence of this gas at high pressure, there is a tendency for mercury to vaporize from the top of the column at high temperatures and to condense on the cooler portions of the stem in the form of minute globules, which will not join up again with the main bulk of the mercury. It is, therefore, inadvisable to expose a thermometer to high temperatures for prolonged periods.

At high temperatures the correction for the temperature of the emergent stem becomes particularly important, and the thermometer should, if possible, be immersed to the top of the mercury column. Where this is not possible, the thermometer should be immersed as far as conditions permit and a correction made to the observed reading for the emergent column. To do this, we need to find the average temperature of the emergent column by means of a short thermometer placed in several positions near to the stem. The emergent column correction may then be found from the formula

$$\text{correction} = 0.0016(t_1 - t_2)n \text{ on Celsius scale}$$

where t_1 is the temperature of the thermometer bulb, t_2 is the average temperature of the emergent column, and n is the number of degrees exposed. The numerical constant is the coefficient of apparent expansion of mercury in glass.

21.3.1.2 Use of Liquids Other than Mercury

In certain industrial uses, particularly in industries where the escape of mercury from a broken bulb might cause considerable damage to the products, other liquids are used to fill the thermometer. These liquids are also used where the temperature range of the mercury-in-glass thermometer is not suitable. Table 21.8 lists some liquids, together with their range of usefulness.

21.3.1.3 Mercury-in-glass Electric Contact Thermometer

A mercury-in-glass thermometer can form the basis of a simple on/off temperature controller that will control the temperature of an enclosure at any value between 40°C and 350°C.

TABLE 21.8 Liquids used in glass thermometers

Liquid	Temperature range (°C)
Mercury	−35 to +510
Alcohol	−80 to +70
Toluene	−80 to +100
Pentane	−200 to +30
Creosote	−5 to +200

Mercury is a good electrical conductor. By introducing into the bore of a thermometer two platinum contact wires, one fixed at the lower end of the scale and the other either fixed or adjustable from the top of the stem, it is possible to arrange for an electrical circuit to be completed when a predetermined temperature is reached. The current through the circuit is limited to about 25 mA. This current is used to operate an electronic control circuit. Contact thermometers find applications in laboratories for the temperature control of water baths, fluidized beds, and incubators. With careful design, temperature control to 0.1°C can be attained.

Formerly, fixed temperature contact thermometers were used for the temperature control of quartz crystal oscillator ovens, but now this duty is more usually performed by thermistors or semiconductor sensors, which can achieve better temperature control by an order of magnitude.

21.3.2 Liquid-Filled Dial Thermometers

21.3.2.1 Mercury-in-Steel Thermometer

Two distinct disadvantages restrict the usefulness of liquid-in-glass thermometers in industry: glass is very fragile, and the position of the thermometer for accurate temperature measurement is not always the best position for reading the scale of the thermometer.

These difficulties are overcome in the mercury-in-steel thermometer, shown in Figure 21.4. This type of thermometer works on exactly the same principle as the liquid-in-glass thermometer. The glass bulb is, however, replaced by a steel bulb and the glass capillary tube by one of stainless steel. As the liquid in the system is now no longer visible, and a Bourdon tube is used to measure the change in its volume. The Bourdon tube, the bulb, and the capillary tube are completely filled with mercury, usually at a high pressure. When suitably designed, the capillary tube may be of considerable length so that the indicator operated by the Bourdon tube

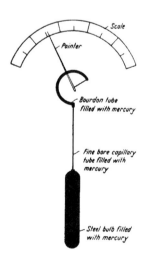

FIGURE 21.4 Mercury-in-steel thermometer.

may be some distance away from the bulb. In this case the instrument is described as being a *distant reading* or *transmitting* type.

When the temperature rises, the mercury in the bulb expands more than the bulb so that some mercury is driven through the capillary tube into the Bourdon tube. As the temperature continues to rise, increasing amounts of mercury are driven into the Bourdon tube, causing it to uncurl. One end of the Bourdon tube is fixed; the motion of the other end is communicated to the pointer or pen arm. Because there is a large force available, the Bourdon tube may be made robust and will give good pointer control and reliable readings.

The Bourdon tube may have a variety of forms, and the method of transmitting the motion to the pointer also varies. Figure 21.5 shows one form of Bourdon tube in which the motion of the free end is transmitted to the pointer by means of a segment and pinion. The free end of the tube forms a trough in which a stainless-steel ball at the end of the segment is free to move. The ball is held against the side of the trough by the tension in the hair-spring. By using this form of construction, lost motion and angularity error are avoided and friction reduced to a minimum. Ambient temperature compensation may be obtained by using a bimetallic strip or by using twin Bourdon tubes in the manner described under the heading of capillary compensation.

Figure 21.6 shows a Bourdon tube with a different form and a different method of transmitting the motion to the pointer. This Bourdon tube is made of steel tube with an almost flat section. A continuous strip of the tubing is wound into two coils of several turns. The coils are arranged one behind the other so that the free end of each is at the center while the other turn of the coils is common to both, as shown in the illustration. One end of the continuous tube—the inner end of the back coil—is fixed and leads to the capillary tube; the other end—the inner end of the front coil—is closed and is attached to the pointer through a small bimetallic coil, which forms a continuation of the Bourdon tube. This bimetallic coil compensates for changes brought about in the elastic properties of the Bourdon tube and in the volume of the mercury within the Bourdon tube due to ambient temperature changes.

This particular formation of the tube causes the pointer to rotate truly about its axis without the help of bearings, but bearings are provided to keep the pointer steady in the presence of vibration. The friction at the bearings will, therefore, be very small, since there is little load on them. Because the end of the Bourdon tube rotates the pointer directly, there will be no backlash.

Thermometer Bulbs The thermometer bulb may have a large variety of forms, depending on the use to which it is put. If the average temperature of a large enclosure is required, the bulb may take the form of a considerable length of tube of smaller diameter, either arranged as a U or wound into a helix. This form of bulb is very useful when the temperature

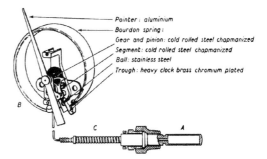

FIGURE 21.5 Construction of mercury-in-steel thermometer. Courtesy of the Foxboro Company.

FIGURE 21.6 Multiturn Bourdon tube.

of a gas is being measured, since it presents a large surface area to the gas and is therefore more responsive than the forms that have a smaller surface area for the same cubic capacity.

In the more usual form, the bulb is cylindrical in shape and has a robust wall; the size of the cylinder depends on many factors, such as the filling medium and the temperature range of the instrument, but in all cases, the ratio of surface area to volume is kept at a maximum to reduce the time lag in the response of the thermometer.

The flange for attaching the bulb to the vessel in which it is placed also has a variety of forms, depending on whether the fitting has to be gas-tight or not and on many other factors. Figure 21.7 shows some forms of bulbs.

The Capillary Tube and Its Compensation for Ambient Temperature The capillary tube used in the mercury-in-steel thermometer is usually made from stainless steel, since mercury will amalgamate with other metals. Changes of temperature affect the capillary and the mercury it contains and, hence, the temperature reading; but if the capillary has a very small capacity, the error owing to changes in the ambient temperature will be negligible.

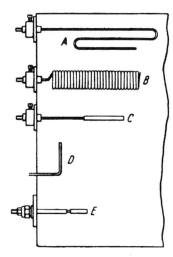

FIGURE 21.7 Forms for bulbs for mercury-in-steel thermometers.

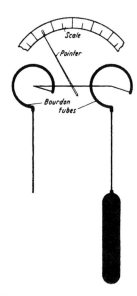

FIGURE 21.8 Ambient temperature compensation of mercury-in-steel thermometer.

Where a capillary tube of an appreciable length is used, it is necessary to compensate for the effects brought about by changes in the temperature in the neighborhood of the tube. This may be done in a number of ways. Figure 21.8 illustrates a method that compensates not only for the changes of temperature of the capillary tube but also for the changes of temperature within the instrument case. To achieve complete temperature compensation, two thermal systems are used that are identical in every respect except that one has a bulb and the other has not. The capillary tubes run alongside each other, and the Bourdon tubes are in close proximity within the same case. If the pointer is arranged to indicate the difference in movement between the free ends of the two Bourdon tubes, it will indicate an effect that is due to the temperature change in the bulb only. If compensation for case temperature only is required, the capillary tube is omitted in the compensating system, but in this case the length of capillary tube used in the uncompensated system should not exceed about 8 meters.

Another method of compensating for temperature changes in the capillary tube is to use a tube of comparatively large bore and to insert into the bore a wire made of Invar or another alloy with a very low coefficient of expansion. Mercury has a coefficient of cubical expansion about six times greater than that of stainless steel. If the expansion of the Invar wire may be regarded as being negligibly small, and the wire is arranged to fill five-sixths of the volume of the capillary bore, then the increase in the volume of the mercury that fills the remaining one-sixth of the bore will exactly compensate for the increase in volume of the containing capillary tube. This method requires the dimensions both of the bore of the capillary tube and of the diameter of the wire insert, to be accurate to within very narrow limits for accurate compensation. The insert may not necessarily be continuous, but it could take the form of short rods, in which case, however, it is difficult to eliminate all trapped gases.

Compensation for changes in the temperature of the capillary tube may also be achieved by introducing compensating chambers of the form shown in Figure 21.9, at intervals along the length of the capillary tube. These chambers operate on exactly the same principle as the Invar-wire-insert type of capillary tube, but the proportion of the chamber occupied by the Invar is now arranged to compensate for the relative increase in volume of the mercury within the chamber and in the intervening length of capillary tube.

21.3.2.2 Other Filling Liquids

Admirable though mercury might be for thermometers, in certain circumstances it has its limitations, particularly at the lower end of the temperature scale. It is also very expensive to weld mercury systems in stainless steel. For these and other reasons, other liquids are used in place of mercury. Details of the liquids used in liquid-in-metal thermometers, with their usual temperature ranges, are given in Table 21.9. Comparison with Table 21.8 shows that liquids are used for different temperature ranges in glass and metal thermometers. In general, in metal thermometers, liquids can be used up to higher temperatures than in glass thermometers because they can be filled to higher pressures.

When liquids other than mercury are used, the bulb and capillary tube need no longer be made of steel. The material of the bulb may, therefore, be chosen from a wide range of metals and alloys and is selected to give the maximum resistance to any corrosive action that may be present where the bulb is to be used.

The capillary tube, too, may be made from a variety of materials, although copper and bronze are the most common. When capillary tubes are made from materials other than stainless steel, it may be necessary to protect them

CHAPTER | 21 Temperature Measurement

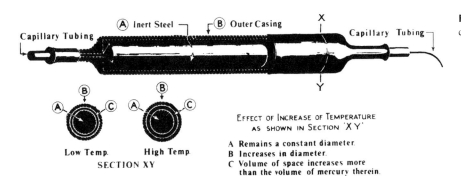

FIGURE 21.9 Ambient temperature compensation chamber.

TABLE 21.9 Liquids used in metal thermometers (expansion type)

Liquid	Temperature range (°C)
Mercury	−39 to +650
Xylene	−40 to +400
Alcohol	−46 to +150
Ether	+20 to +90
Other organic liquids	−87 to +260

from corrosion or mechanical damage. This can be done by covering the tube with thermal insulation material (formerly asbestos was used) and winding the whole in a heavy spiral of bronze. In cases where a bronze outer casing is likely to be damaged, either by acid fumes or mechanically, it may be replaced by a stainless-steel spiral, which results in a much stronger but slightly less flexible construction. For use in damp places or where the tube is liable to be attacked by acid fumes, the capillary and bronze spiral may be protected by a covering of molded rubber, polyvinyl chloride, or rubber-covered woven-fabric hose. For use in chemical plants, such as sulfuric acid plants, both the capillary tube and the bulb are protected by a covering of lead.

The construction of the liquid-in-metal thermometer is the same as that of the mercury-in-steel thermometer, and compensation for changes in ambient temperature may be achieved in the same ways.

Further facts about liquid-in-metal thermometers will be found in Table 21.11, which gives a comparison of the various forms of nonelectrical dial thermometers.

In installations where liquid-filled instruments with very long capillaries are used, care must be taken to see that there is not a significant height difference between the bulb location and that of the instrument. If there is a large height difference, the pressure due to the column of liquid in the capillary will be added to (or subtracted from) the pressure due to the expansion of the liquid in the bulb, resulting in a standing error in the temperature reading. This problem is at its worst with mercury-filled instruments. Instruments with double-capillary ambient temperature compensation (see Figure 21.8) are, of course, also compensated for static head errors.

21.3.3 Gas-Filled Instruments

The volume occupied by a given mass of gas at a fixed pressure is a function of both the molecular weight of the gas and its temperature. In the case of the "permanent gases," provided the temperature is significantly above zero Kelvin, the behavior of a gas is represented by the equation

$$pv = RT \tag{21.9}$$

where p is pressure in $N \cdot m^{-2}$, v is volume in m^3, T is the temperature in K, and R is the gas constant with a value of $8.314 \, J \cdot mol^{-1} \cdot K^{-1}$.

If, therefore, a certain volume of inert gas is enclosed in a bulb, capillary, and Bourdon tube and most of the gas is in the bulb, the pressure as indicated by the Bourdon tube may be calibrated in terms of the temperature of the bulb. This is the principle of the gas-filled thermometer.

Since the pressure of a gas maintained at constant volume increases by 1/273 of its pressure at 0°C for every degree rise in temperature, the scale will be linear, provided that the increase in volume of the Bourdon tube, as it uncurls, can be neglected in comparison with the total volume of gas.

An advantage of the gas-filled thermometer is that the gas in the bulb has a lower thermal capacity than a similar quantity of liquid, so the response of the thermometer to temperature changes will be more rapid than that for a liquid-filled system with a bulb of the same size and shape.

The coefficient of cubical expansion of a gas is many times larger than that of a liquid or solid (air, 0.0037; mercury, 0.00018; stainless steel, 0.00003). It would therefore appear at first sight that the bulb for a gas-filled system would be smaller than that for a liquid-filled system. The bulb must, however, have a cubical capacity many times larger than that of the capillary tube and Bourdon tube if the effects of ambient temperature changes on the system are to be negligible.

It is extremely difficult to get accurate ambient temperature compensation in any other way. The change in dimensions of the capillary tube due to a temperature change is negligible in comparison with the expansion of the gas. Introducing an Invar wire into the capillary bore would not be a solution to the problem because the wire would occupy such a large proportion of the bore that extremely small variations in the dimensions of the bore or wire would be serious.

Further facts about gas expansion thermometers can be found in Table 21.11, in which certain forms of dial thermometers are compared.

21.3.4 Vapor Pressure Thermometers

If a thermometer system similar to that described for gas expansion thermometers is arranged so that the system contains both liquid and vapor and the interface between liquid and vapor is in the bulb, that is, at the temperature for which the value is required, the vapor pressure as measured by the Bourdon tube will give an indication of the temperature. This indication will be completely independent of the volume of the bulb, the capillary, and the Bourdon tube and therefore independent of expansion due to ambient temperature changes.

The saturated vapor pressure of a liquid is not linear with temperature. Figure 21.10 shows the temperature/vapor-pressure relationship for a typical liquid. The form of the vapor pressure graphs for other volatile liquids is similar. It will be seen that pressure versus temperature is nonlinear.

A thermometer based on vapor pressure will have a scale on which the size of the divisions increases with increasing temperature.

The realization of a vapor instrument is essentially the same as a gas-filled instrument except that in the latter the whole instrument is filled with a permanent gas, whereas in

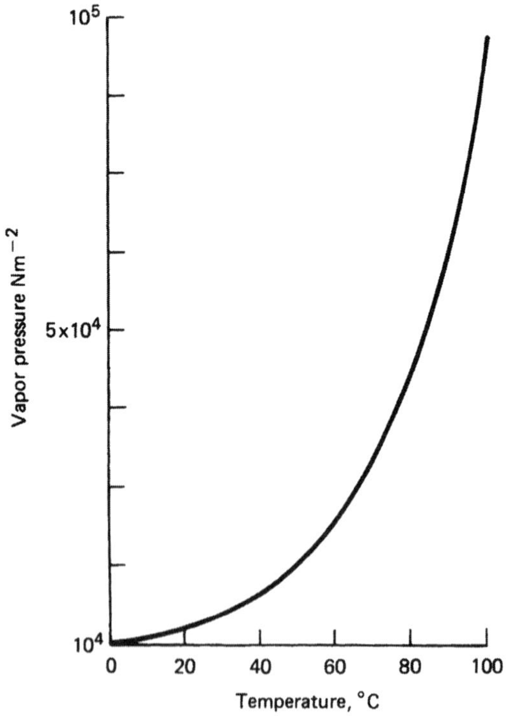

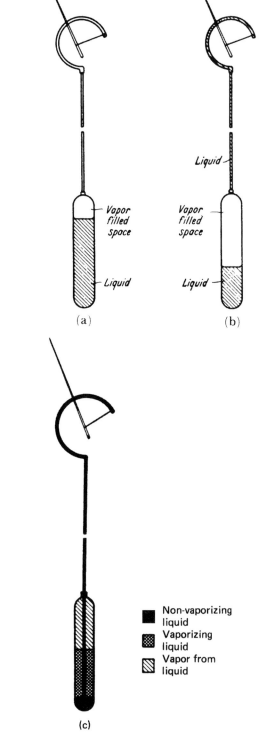

FIGURE 21.10 Saturated vapor pressure of water.

FIGURE 21.11 Vapor pressure thermometer.

CHAPTER | 21 Temperature Measurement

the former the bulb is filled partly with liquid and partly with the vapor of the liquid. This arrangement is shown diagrammatically in Figure 21.11(a).

Many liquids are used for vapor-pressure-actuated thermometers. The liquid is chosen so as to give the required temperature range and so that the usual operating temperature comes within the widely spaced graduations of the instrument. In some forms of the instrument, a system of levers is arranged to give a linear portion to the scale over a limited portion of its range. By suitable choice of filling liquid, a wide variety of ranges is available, but the range for any particular filling liquid is limited. The choice of material for bulb construction is also very wide. Metals such as copper, steel, Monel metal, and tantalummay be used. Table 21.10 shows a number of liquids commonly used for vapor-pressure thermometers, together with their useful operating ranges.

In the instrument, shown diagrammatically in Figure 21.11(a), a quantity of liquid partially fills the bulb. The surface of the liquid in the bulb should be at the temperature that is being measured. The method by which the vapor pressure developed in the bulb is transmitted to the Bourdon tube will depend on whether the temperature of the capillary and Bourdon tube is above or below that of the bulb.

If the ambient temperature of the capillary and Bourdon tube is above that of the bulb, they will be full of vapor, which will transmit the vapor pressure, as shown in Figure 21.11(a). When the ambient temperature increases, it will cause the vapor in the capillary and Bourdon tube to increase in pressure temporarily, but this will cause vapor in the bulb to condense until the pressure is restored to the saturated vapor pressure of liquid at the temperature of the bulb.

Vapor-pressure instruments are not usually satisfactory when the temperature being measured at the bulb is near the ambient temperature of the capillary and the Bourdon tube. In particular, significant measurement delays occur as the measured temperature crosses the ambient temperature. These delays are caused by the liquid distilling into or out of the gauge and capillary, as shown in Figure 21.11(b).

If there is a significant level of difference between the bulb and the gauge, an error will be produced when liquid distills into the capillary due to the pressure head from the column of liquid.

When rapid temperature changes of the bulb occur as it passes through ambient temperature, the movement of the instrument pointer may be quite erratic due to the formation of bubbles in the capillary.

To overcome the defects brought about by distillation of the liquid into and out of the capillary and Bourdon tubes, these tubes may be completely filled with a nonvaporizing liquid, which serves to transmit the pressure of the saturated vapor from the bulb to the measuring system. To prevent the nonvaporizing liquid from draining out of the capillary tube, it is extended well down into the bulb, as shown in Figure 21.11(c), and the bulb contains a small quantity of the non-vaporizing fluid. The nonvaporizing fluid will still tend to leave the capillary tube unless the bulb is kept upright.

Vapor-pressure thermometers are very widely used because they are less expensive than liquid- and gas-filled instruments. They also have an advantage in that the bulb can be smaller than for the other types.

The range of an instrument using a particular liquid is limited by the fact that the maximum temperature for which it can be used must be well below the critical temperature for that liquid. The range is further limited by the nonlinear nature of the scale.

In Table 21.11 the three types of fluid-filled thermometers are compared.

TABLE 21.10 Liquids used in vapor pressure thermometers

Liquid	Critical temperature (°C)	Boiling point (°C)	Typical ranges available (°C)
Argon	−122	−185.7	Used for measuring very low temperatures down to −253°C in connection with the liquefaction of gases
Methyl chloride	143	−23.7	0 to 50
Sulphur dioxide	157	−10	30 to 120
Butane (n)	154	−0.6	20 to 80
Methyl bromide		4.6	30 to 85
Ethyl chloride	187	12.2	30 to 100
Diethyl ether	194	34.5	60 to 160
Ethyl alcohol	243	78.5	30 to 180
Water	375	100	120 to 220
Toluene	321	110.5	150 to 250

TABLE 21.11 Comparison of three types of dial thermometers

	Liquid-in-metal	Gas expansion (constant volume)	Vapor pressure
Scale	Evenly divided.	Evenly divided.	Not evenly divided. Divisions increase in size as the temperature increases. Filling liquid chosen to given reasonably uniform scale in the neighborhood of the operating temperatures.
Range	Wide range is possible with a single filling liquid, particularly with mercury. By choice of suitable filling liquid, temperatures may be measured between −200°C and 57°C, but not with a single instrument.	Usually has a range of at least 50°C between −130°C and 540°C. Can be used for a lower temperature than mercury in steel.	Limited for a particular filling liquid, but with the choice of a suitable liquid almost any temperature between −50°C and used for a lower 320°C may be measured. Instrument is not usually suitable for measuring temperatures near ambient temperatures owing to the lag introduced when bulb temperature crosses ambient temperature.
Power available to operate the indicator	Ample power is available so that the Bourdon tube may be made robust and arranged to give good pointer control.	Power available is very much less than that from liquid expansion.	Power available is very much less than that from liquid expansion.
Effect of difference in level of bulb and Bourdon tube	When the system is filled with a liquid at high pressure, errors due to difference of level between bulb and indicator will be small. If the difference in level is very large, a correction may be made.	No head error, since the pressure due to difference in level is negligible in comparison with the total pressure in the system.	Head error is not negligible, since the pressure in the system is not large. Error may be corrected over a limited range of temperature if the ratio pressure to deflection of the pointer can be considered constant over that range. In this case the error is corrected by resetting the pointer.
Effect of changes in barometric pressure	Negligible.	May produce a large error. Error due to using the instrument at a different altitude from that at which it was calibrated may be corrected by adjusting the zero. Day-to-day variations in barometric pressure may be corrected for in the same way.	Error may be large but may be corrected by resetting the pointer as for head error. Day-to-day errors due to variation in barometric pressure may be corrected by zero adjustment.
Capillary error	Compensation for change in ambient temperature.	Difficult to eliminate.	No capillary error.
Changes in temperature at the indicator	Compensation obtained by means of a bimetallic strip.	Compensation obtained by means of bimetallic strip.	Errors due to changes in the elasticity of the Bourdon tube are compensated for by means of a bimetallic strip.
Accuracy	±½% of range to 320°C ±% of range above 320°C.	±1% of differential range of the instrument if the temperature of the capillary and Bourdon tube does not vary too much.	±1% of differential range, even with wide temperature variation of the capillary and Bourdon tube.

CHAPTER | 21 Temperature Measurement

21.3.5 Solid Expansion

Thermal expansion of solids, usually metals, forms the basis of a wide range of inexpensive indicating and control devices. These devices are not particularly accurate; typically errors of as much as ±5° or more may be expected, but due to their low cost they find wide application, especially in consumer equipment. As indicated earlier in this section, this technique is also used to provide temperature compensation in many instruments.

The temperature-sensitive elements using solid expansion fall into two groups: rodsensing probes and bimetal strips.

There are so many applications that only one or two examples are given here to illustrate the techniques.

21.3.5.1 Rodsensing Probes

The widest application of this technique is for immersion thermostats for use in hot water temperature control. Figure 21.12 shows diagrammatically the operation of an immersion thermostat. The microswitch is operated by the thermal expansion of the brass tube. The reference length is provided by a rod of low thermal expansion, such as Invar. These thermostats, thought not particularly accurate and having a switching differential of several degrees Celsius, provide a very rugged and reliable control system for a noncritical application such as domestic hot water control.

Figure 21.13 shows another rod application. In this case, to achieve greater sensitivity the expanding component is coiled.

21.3.5.2 Bimetal Strip Thermometer

Bimetal strips are fabricated from two strips of different metals with different coefficients of thermal expansion bonded together to form, in the simplest case, a cantilever. Typical metals are brass and Invar. Figure 21.14 illustrates this principle. As the temperature rises, the brass side of the strip expands more than the Invar side, resulting in the strip curling, in this case upward.

In this "straight" form a bimetal strip can form part of a micro-switch mechanism, thus forming a temperature-sensitive switch or thermostat.

To construct a thermometer, the bimetal element is coiled into a spiral or helix. Figure 21.15 shows a typical coiled thermometer element.

A long bimetal strip, consisting of an Invar strip welded to a higher expansion nickel-molybdenum alloy wound around without a break into several compensated helices, arranged coaxially one within the other, forms the temperature-sensitive element of an instrument that may be designed to measure temperature. This method of winding the strip enables a length, sufficient to produce an appreciable movement of the free end, to be concentrated within a small space.

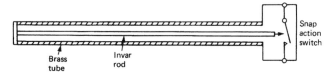

FIGURE 21.12 Rod thermostat.

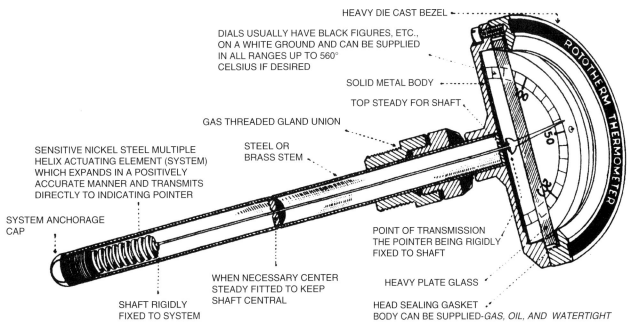

FIGURE 21.13 Dial thermometer.

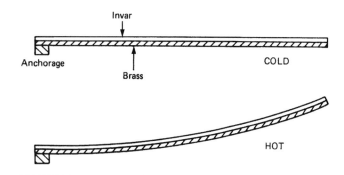

FIGURE 21.14 Action of bimetal strip.

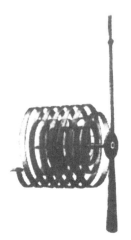

FIGURE 21.15 Helical bimetal strip.

It also makes it possible to keep the thermal capacity of the element and its stem at a low value so that the instrument will respond rapidly to small temperature changes.

The helices in the winding are so compensated that any tendency toward lateral displacement of the spindle in one helix is counteracted by an opposite tendency on the path of one or more of the other helices. Thus the spindle of the instrument is fully floating, retaining its position at the center of the scale, without the help of bearings. The instrument is, therefore, not injured by mechanical shocks that would damage jeweled bearings.

This particular design also results in the angular rotation of the spindle being proportional to the change in temperature for a considerable temperature range. The instrument has a linear temperature scale and can be made to register temperatures up to 300°C to within ±1 percent of the scale range.

Due to its robust construction, this instrument is used in many industrial plants, and a slightly modified form is used in many homes and offices to indicate room temperature. It can be made for a large variety of temperature ranges and is used in many places where the more fragile mercury-in-glass thermometer was formerly used.

21.4 MEASUREMENT TECHNIQUES: ELECTRICAL

21.4.1 Resistance Thermometers

All metals are electrical conductors that, at all but very low temperatures, offer resistance to the passage of electric current. The electrical resistance exhibited by a conductor is measured in ohms. The proportional relationship of electrical current and potential difference is given by Ohm's law:

$$R = \frac{E}{I} \quad (21.10)$$

where R is resistance in Ohms, E is potential difference in volts, and I is current in amperes. Different metals show widely different resistivities. The resistance of a conductor is proportional to its length and inversely proportional to its cross-sectional area, that is,

$$R = \rho \frac{L}{A} \quad (21.11)$$

or

$$\rho = R \frac{A}{L} \quad (21.12)$$

where R is resistance of the conductor, ρ is resistivity of the material, L is length of the conductor, and A is cross-sectional area of the conductor. The units of resistivity are Ohms · meter.

The resistivity of a conductor is temperature dependent. The temperature coefficient of resistivity is positive for metals, that is, the resistance increases with temperature, and for semiconductors the temperature coefficient is negative. As a general guide, at normal ambient temperatures the coefficient of resistivity of most elemental metals lies in the region of 0.35 percent to 0.7 percent per °C.

Table 21.12 shows the resistivity and temperature coefficients for a number of common metals, both elements and alloys.

The metals most used for resistance measurement are platinum, nickel, and copper. These metals have the advantage that they can be manufactured to a high degree of purity, and consequently they can be made with very high reproducibility of resistance characteristics. Copper has the disadvantage of a low resistivity, resulting in inconveniently large sensing elements, and has the further disadvantage of poor resistance to corrosion, resulting in instability of electrical characteristics. The main area of application of copper for resistance thermometers is in electronic instrumentation, where it is in a controlled environment and where an essentially linear temperature characteristic is required.

TABLE 21.12 Resistivities of different metals

Metal	Resistivity at 20°C microhms meter	Temperature coefficient of resistivity (°C^{-1})
Aluminum	282.4	0.0039
Brass (yellow)	700	0.002
Constantan	4,900	10^{-5}
Copper (annealed)	172.4	0.00393
Gold	244	0.0034
Iron (99.98%)	1,000	0.005
Mercury	9,578	0.00087
Nichrome	10,000	0.0004
Nickel	780	0.0066
Platinum (99.85%)	11,060	0.003 927
Silver	159	0.0038
Tungsten	560	0.0045

1. Resistivities of metals dependent on the purity or exact composition of alloys. Some of the figures presented here represent average values.
2. Temperature coefficients of resistivity vary slightly with temperature. The values presented here are for 20°C.

21.4.1.1 Platinum Resistance Thermometers

Platinum is the standard material used in the resistance thermometer, which defines the International Practical Temperature Scale, not because it has a particularly high coefficient of resistivity but because of its stability in use. In fact, a high coefficient is not, in general, necessary for a resistance thermometer material, since resistance values can be determined with a high degree of accuracy using suitable equipment and taking adequate precautions.

Platinum, having the highest possible coefficient of resistivity, is considered the best material for the construction of thermometers. A high value of this coefficient is an indication that the platinum is of high purity. The presence of impurities in resistance thermometer material is undesirable, since diffusion, segregation, and evaporation may occur in service, resulting in a lack of stability of the thermometer. The temperature coefficient of resistivity is also sensitive to internal strains so that it is essential that the platinum should be annealed at a temperature higher than the maximum temperature of service. The combination of purity and adequate annealing is shown by a high value of the ratio of the resistances at the steam and ice points. To comply with the requirements of the International Practical Temperature Scale of 1968, this ratio must exceed 1.39250.

It is essential that the platinum element is mounted in such a way that it is not subject to stress in service.

Platinum is used for resistance thermometry in industry for temperatures up to 800°C. It does not oxidize but must be protected from contamination. The commonest cause of contamination of platinum resistance thermometers is contact with silica or silica-bearing refractories in a reducing atmosphere. In the presence of a reducing atmosphere, silica is reduced to silicon, which alloys with platinum, making it brittle. Platinum resistance thermometers may be used for temperatures down to about 20 K.

For measuring temperatures between 1 K and 40 K, doped germanium sensors are usually used, whereas carbon resistors are used between 0.1 K and 20 K. About 20 K platinum has a greater temperature coefficient of resistivity and has a greater stability. Between 0.35 K and 40 K, a resistance thermometer material (0.5 atomic % iron-rhodium) is also used.

Temperature/Resistance Relationship of Resistance Thermometers BS 1904: 1984 and its internationally harmonized equivalent standard IEC 751: 1983 specify the resistance versus temperature characteristics of industrial platinum resistance thermometers. The standard provides tables of resistance against temperature for 100 Ω resistance thermometers over the temperature range −200°C to 850°C. Two grades of thermometer are specified. The equations are provided from which the tables are derived.

Between −200°C and 0°C, the resistance of the thermometer, R_t, is given by

$$R_t = R_0[1 + At + Bt^2 + C(t-100) \cdot t^3] \qquad (21.13)$$

and for the range 0°C to 850°C:

$$R_t = R_0(1 + At + Bt^2) \qquad (21.14)$$

where $A = 3.90802 \times 10^{-3}\,°C^{-1}$, $B = -5.802 \times 10^{-7}\,°C^{-2}$, and $C = -4.27350 \times 10^{-12}\,°C^{-4}$.

The temperature coefficient is given by

$$\alpha = \frac{R_{100} - R_0}{100 \times R_0} = 0.003850\,°C^{-1}$$

As indicated by Equations (21.14) and (21.15), the value of α is not constant over the temperature range. Figure 21.16 shows the tolerances, in Ohms and degrees Celsius, over the specified temperature range. (Figure 21.16 is based on Figure 2 of BS 104: 1984 and is reproduced with the permission of the BSI. Complete copies of the standard may be obtained by post from BSI Sales, Linford Wood, Milton Keynes, MK14 6LE, UK.) For industrial applications, Class B thermometer sensors are normally used. Class A sensors are available for greater precision, but they are specified over a more restricted temperature range. Although the standard specifies the temperature range down to $-200°C$ (73 K), Class A sensors may be used for temperatures down to 15 K. Resistance thermometers are calibrated to IPTS-68, which is normally done by comparison with a standardized resistance thermometer. For industrial use, most thermometers are made to be 100 Ω at 0°C, but 10 Ω thermometers are manufactured where particularly robust sensors are required.

A wide range of sensor designs is available, the form used depending on the duty and the required speed of response. Some typical forms of construction are illustrated in Figure 21.17. Figure 21.17(a) shows a high-temperature form in which the spiral platinum coil is bonded at one edge of each turn with high-temperature glass inside cylindrical holes in a ceramic rod. In the high-accuracy type, used mainly for laboratory work, the coil is not secured at each turn but is left free to ensure a completely strain-free mounting, as shown in Figure 21.17(b). Where a robust form, suitable for use in aircraft and missiles or any severe vibration condition, is required, the ceramic is in solid rod form and the bifilar wound platinum coil is sealed to the rod by a glass coating, as shown in Figure 21.17(c). Where the sensor is intended for use for measuring surface temperatures, the form shown in Figure 21.17(d) is used. In all forms, the ceramic formers are virtually silica-free and the resistance element is sealed in with high-temperature glass to form an impervious sheath that is unaffected by most gases and hydrocarbons. The external leads, which are silver or platinum of a diameter much larger than the wire of the resistance element, are welded to the fine platinum wire wholly inside the glass seal.

The inductance and capacitance of elements are made as low as possible to allow their use with AC measuring instruments. Typically the elements shown will have self-inductance of 2 μH per 100 Ω and the element self-capacitance will not exceed 5 pF. The current passed

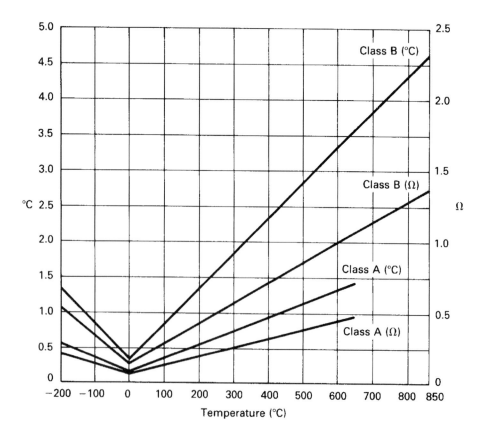

FIGURE 21.16 BS 1904: 1984 specification for tolerances for 100 Ω platinum resistance thermometers. Reproduced by permission of BSI; see text.

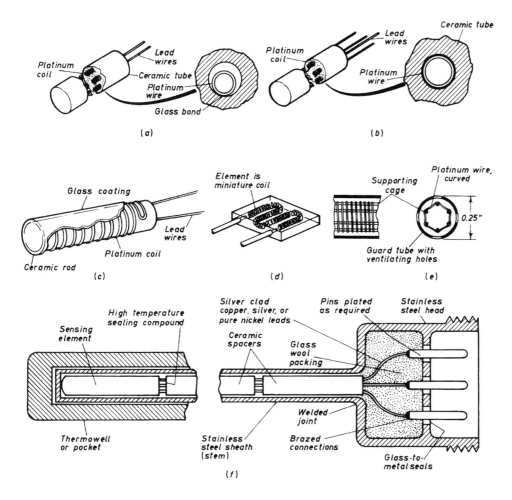

FIGURE 21.17 Construction of resistance thermometers. Courtesy of Fisher-Rosemount Inc.

through a resistance thermometer to measure the resistance must be limited to minimize errors by self-heating of the resistance element. Typical maximum acceptable current is 10 mA for a 100 Ω thermometer.

The rate of response of a resistance thermometer is a function of its construction and encapsulation. A heavy industrial type may have a response of one or two minutes when plunged into water, whereas a naked type, like that shown in Figure 21.17(e), will be only a few milliseconds under the same conditions. Figure 21.17(f) shows the cross-section of a resistance thermometer encapsulated in a metal tube. Figure 21.18 shows a range of typical industrial resistance thermometers.

A more recent development of resistance thermometers has been the replacement of the wire-wound element of the conventional resistance thermometer by a metalized film track laid down on a glass or ceramic substrate. These thermometer elements are made by techniques similar to those used for making hybrid integrated electronic circuits. After the laying down of the metalized film, the film is trimmed by a laser to achieve the required parameters. These metal film devices can be very robust and can be manufactured to a high degree of accuracy.

21.4.1.2 Nickel Resistance Thermometers

Nickel forms an inexpensive alternative to platinum for resistance thermometers. The usable range is restricted to −200°C to +350°C. But the temperature coefficient of resistivity of nickel is 50 percent higher than that of platinum, which is an advantage in some instruments. Nickel resistance thermometers find wide use in water-heating and air-conditioning systems.

As mentioned, the current through a resistance thermometer sensor must be kept low enough to limit self-heating. However, in some applications, such as flowmeters, anemometers, and psychrometers, the self-heating effect is used, the final temperature of the sensor being a function of the flow rate of the process fluid or air. See also Chapter 6.

21.4.1.3 Resistance Thermometer Connections

When resistance thermometers are located at some distance from the measuring instrument, the electrical resistance of the connecting cables will introduce errors of reading. This reading error will, of course, vary as the temperature of the cables changes. However, this error can be compensated by the use

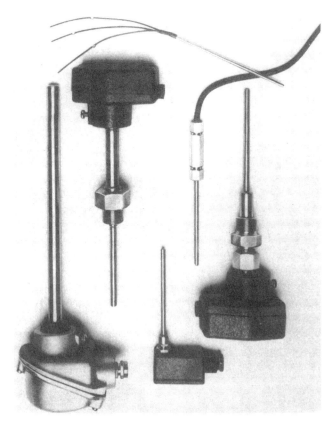

FIGURE 21.18 Typical industrial resistance thermometers. Courtesy of ABB Instrument Group.

of extra conductors. Normally, the change of resistance of a resistance thermometer is measured in a Wheatstone bridge circuit or a modified Wheatstone bridge, so the compensating conductors can be connected in the opposite side of the bridge. In this way bridge unbalance is only a function of the change of resistance of the thermometer element. Figure 21.19(a) shows three-wire compensation. The resistance of Wire 1 is added to that of the resistance thermometer but is balanced by Wire 2 in the reference side of the bridge. Wire 3 supplies the power to the bridge. In Figure 21.19(b), four-wire compensation is shown. The resistance of Wires 1 and 2, which connect to the resistance thermometer, are compensated by the resistance of Wires 3 and 4, which are connected together at the resistance thermometer and are again in the opposite arm of the bridge. A Kelvin double bridge is illustrated in Figure 21.19(c). Resistors R1 and R3 set up a constant current through the resistance thermometer. Resistors R2 and R4 set up a constant current in the reference resistor R5 such that the voltage V_R is equal to the voltage V_t across the resistance thermometer when it is at 0°C. At any other temperature, $V_t = I_t R_t$, and the meter will indicate the difference between V_t and V_R, which will be proportional to the temperature. The indicator must have a very high resistance so that the current in Conductors 1 and 2 is essentially zero. See Part III.

21.4.2 Thermistors

21.4.2.1 Negative Temperature Coefficient Thermistors

An alternative to platinum or nickel for resistance thermometer sensing elements is a semiconductor composed of mixed metal oxides. The composition of these materials depends on the particular properties required. Combinations of two or more of the following oxides are used: cobalt, copper, iron, magnesium, manganese, nickel, tin, titanium, vanadium, and zinc. Devices made of these materials are called *thermistors*. They consist of a piece of the semiconductor to which two connecting wires are attached at opposite sides or ends. Thermistors have a negative temperature coefficient; that is, as the temperature rises the electrical resistance of the device falls. This variation of resistance with temperature is much higher than in the case of metals. Typical resistance values are 10 kΩ at 0°C and 200 Ω at 100°C. This very high sensitivity allows measurement or control to a very high resolution of temperature differences. The accuracy is not as good as for a metallic resistance thermometer owing to the difficulty in controlling the composition of the thermistor material during manufacture. The resolution differs across the usable span of the devices due to their nonlinearity. With the right choice of device characteristics, it is nevertheless possible to control a temperature to within very close limits; 0.001 degree Celsius temperature change is detectable.

The total range that can be measured with thermistors is from −100°C to +300°C. However, the span cannot be covered by one thermistor type; four or five types are needed.

The physical construction of thermistors covers a wide range. The smallest are encapsulated in glass or epoxy beads of 1–2.5 mm diameter; bigger ones come as discs of 5–25 mm diameter or rods of 1–6 mm diameter and up to 50 mm length. The bigger devices are able to pass quite high currents and so operate control equipment directly, without need of amplifiers. Thermistors are also available in metal encapsulations like those used for platinum resistance thermometers.

The big disadvantage of thermistors is that their characteristics are nonlinear. The temperature coefficient of resistivity α at any temperature within the range of a sensor is given by

$$\alpha = -\frac{B}{T^2} \qquad (21.15)$$

where B is the characteristic temperature constant for that thermistor and T is temperature in Kelvin. The units of α are Ohms · K^{-1}.

Most thermistors have a specified resistance at 20°C or 25°C. To determine the resistance at any other temperature, Equation (21.16) is used:

$$R_2 = R_1 \exp\left(\frac{B}{t_2} - \frac{B}{t_1}\right) \qquad (21.16)$$

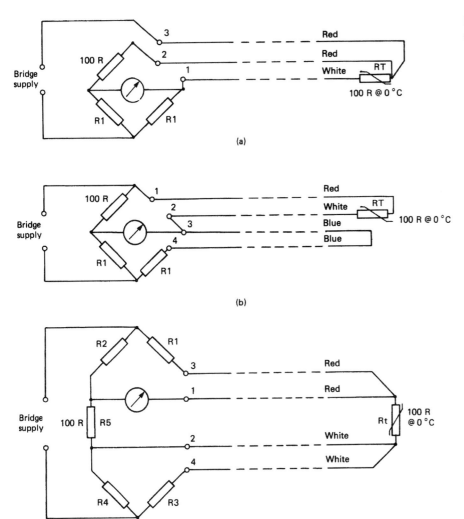

FIGURE 21.19 Connections for compensation of resistance thermometer leads.

where R_1 is resistance of thermistor at temperature $t_1(°C)$ and R_2 is resistance of thermistor at temperature $t_2(°C)$.

Thermistors described as *curvematched* are available. These devices are manufactured to fine tolerances and are interchangeable with an error of less than ±0.2 percent. However, they are expensive and are available in only a limited range of formats.

In general most thermistors are manufactured with tolerances of 10 to 20 percent. Instrumentation for use with these devices must have provision for trimming out the error. Thermistors do not have the stability of platinum resistance thermometers. Their characteristics tend to drift with time. Drifts of up to 0.1°C or more can be expected from some types over a period of some months.

21.4.2.2 Positive Temperature Coefficient Thermistors

Positive temperature coefficient (PTC) thermistors are manufactured from compounds of barium, lead, and strontium titanates. PTC thermistors are primarily designed for the protection of wound equipment such as transformers and motors. The characteristics of these devices have the general shape shown in Figure 21.20. The resistance of PTC thermistors is low and relatively constant, with temperature low. At temperature T_R the increase of resistance with temperature becomes very rapid. T_R is the reference or switching temperature.

In use, PTC thermistors are embedded in the windings of the equipment to be protected. They are connected in series with the coil of the equipment contractor or protection relay. If the temperature of the windings exceeds temperature T_R, the current becomes so small that power is disconnected from the equipment.

21.4.3 Semiconductor Temperature Measurement

21.4.3.1 Silicon Junction Diode

Figure 21.21 shows the forward-bias characteristic of a silicon diode. At voltages below V_F, the forward conduction

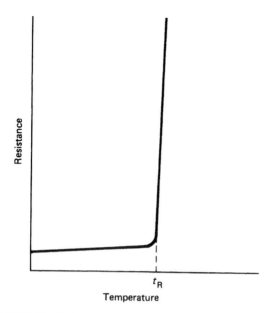

FIGURE 21.20 Resistance temperature characteristic for PTC thermistor.

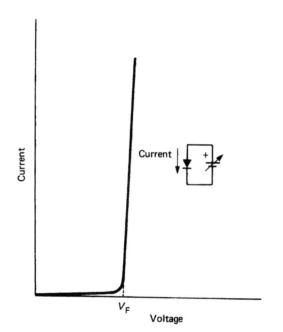

FIGURE 21.21 Forward bias characteristic of silicon diode.

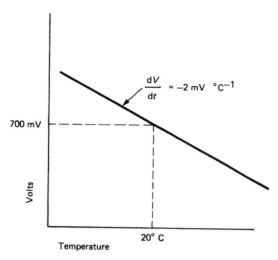

FIGURE 21.22 Temperature characteristic of silicon diode.

voltage, virtually no current flows. Above V_F the diode passes current. The voltage V_F is the energy required by current carriers, either electrons or holes, to cross the junction energy band gap. The value of V_F varies between diode types but is typically 500–700 mV at 20°C. The voltage V_F has a temperature coefficient that is essentially the same for all silicon devices of −2 mV per degree Celsius. The forward voltage against temperature characteristic is linear over the temperature range of 50°C to +150°C. This voltage change with temperature is substantial and as the characteristic is linear it makes a very useful measurement or control signal. There are two principal disadvantages to silicon diodes as control elements. The negative coefficient (see Figure 21.22) is not failsafe. If the control loop is controlling a heater, breakage of the diode wires would be read by the controller as low temperature, and full power would be applied to the heaters. The second disadvantage is the rather limited temperature range. In addition, if a silicon diode is heated above about 200°C, it is completely destroyed, effectively becoming a short circuit.

21.4.3.2 Temperature-Sensing Integrated Circuits

The temperature characteristic of a silicon junction can be improved if the measuring diode is incorporated in an integrated circuit containing an amplifier. Devices are available to provide either an output current proportional to temperature or an output voltage proportional to temperature. Figure 21.23(a) shows the basis of such a device. Figure 21.23(b) shows the circuit of the Analog Devices temperature sensor type AD 590. The operating range of this device is −55°C to +150°C. The temperature is sensed by the emitter-base junctions of two transistors. If two identical transistors are operated at a constant ratio r of collector current densities, the difference in V_t in their base emitter voltages is given by

$$V_t = \frac{KT}{q} \cdot \ln r \qquad (21.17)$$

where K is Boltzmann's constant ($1.380\,66 \times 10^{-23} J \cdot K^{-1}$), q is the electron charge ($1.602\,19 \times 10^{-19}$ coulomb), and T is temperature in Kelvins. It can be seen that V_t is directly proportional to temperature in Kelvins. The voltage is converted to a temperature-dependent current I_t by low-temperature coefficient thin-film resistors R5 and R6. These resistors are laser-trimmed to give the required tolerance at 25°C.

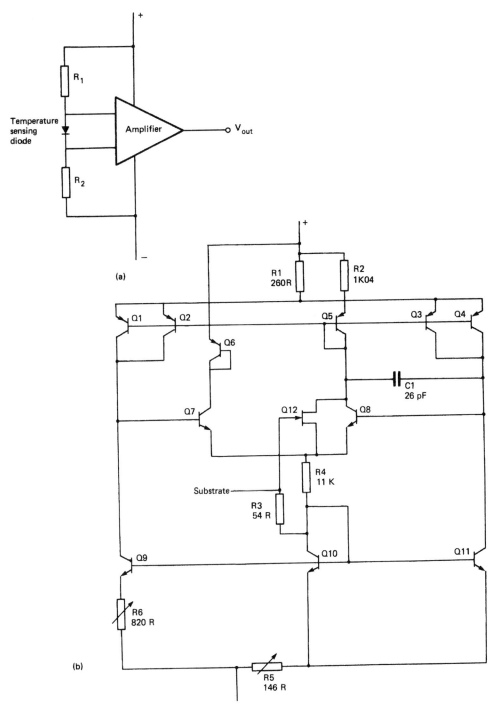

FIGURE 21.23 Semiconductor temperature sensors: (a) diode and amplifier, (b) Analog Devices IC temperature sensor circuit TypeAD 590.

Transistors Q_8 and Q_{11} provide the temperature-dependent voltage V_t. The remaining transistors provide the amplification to give the output current of one microampere per Kelvin. The transistor Q_{10} supplies the bias and substrate leakage currents for the circuit. The device is packaged in a transistor can or ceramic capsule, or it can be supplied as the naked chip for encapsulation into other equipment.

21.5 MEASUREMENT TECHNIQUES: THERMOCOUPLES

21.5.1 Thermoelectric Effects

If an electrical circuit consists of entirely metallic conductors and all parts of the circuit are at the same temperature, there will be no electromotive force in the circuit and therefore no current flows. However, if the circuit consists of more than one metal and if junctions between two metals are at different temperatures, there will be an EMF in the circuit and a current will flow. Figure 21.24 illustrates this effect. The EMF that is generated is called a *thermoelectric EMF* and the heated junction is a *thermocouple*.

21.5.1.1 Seebeck Effect

In 1821 Seebeck discovered that if a closed circuit is formed of two metals and the two junctions of the metals are at different temperatures, an electric current will flow around the circuit. Suppose a circuit is formed, as shown in Figure 21.25, by twisting or soldering together at their ends wires of two different metals such as iron and copper. If one junction remains at room temperature while the other is heated to a higher temperature, a current is produced that flows from copper to iron at the hot junction and from iron to copper at the cold one.

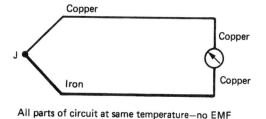

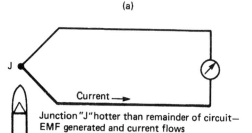

FIGURE 21.24 Basic thermocouple circuit.

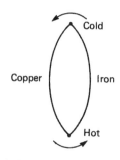

FIGURE 21.25 Simple thermocouple.

Seebeck arranged a series of 35 metals in order of their thermoelectric properties. In a circuit made up of any two of the metals, the current flows across the hot junction from the earlier to the later metal of the series. A portion of his list is as follows: Bi–Ni–Co–Pd–Pt–U–Cu–Mn–Ti–Hg–Pb–Sn–Cr–Mo–Rh–Ir–Au–Zn–W–Cd–Fe–As–Sb–Te.

21.5.1.2 Peltier Effect

In 1834 Peltier discovered that when a current flows across the junction of two metals, heat is absorbed at the junction when the current flows in one direction and is liberated if the current is reversed. Heat is absorbed when a current flows across an iron–copper junction from copper to iron, and it is liberated when the current flows from iron to copper. This heating effect should not be confused with the Joule heating effect, which, being proportional to I^2R, depends only on the size of the current and the resistance of the conductor and does not change to a cooling effect when the current is reversed. The amount of heat liberated or absorbed is proportional to the quantity of electricity that crosses the junction, and the amount liberated or absorbed when unit current passes for a unit time is called the *Peltier coefficient*.

Because heat is liberated when a current does work in overcoming the EMF at a junction and is absorbed when the EMF itself does work, the existence of the Peltier effect would lead one to believe that the junction of the metals is the seat of the EMF produced in the Seebeck effect. It would appear that an EMF exists across the junction of dissimilar metals, its direction being from copper to iron in the couple considered. The EMF is a function of the conduction electron energies of the materials making up the junction. In the case of metals, the energy difference is small, and therefore the EMF is small. In the case of semiconductors the electron energy difference may be much greater, resulting in a higher EMF at the junction. The size of the EMF depends not only on the materials making up the junction but also upon the temperature of the junction. When both junctions are at the same temperature, the EMF at one junction is equal and opposite to that at the second junction, so that the resultant EMF in the circuit is zero. If, however, one junction is heated, the EMF across the hot junction is greater than that across the cold junction, and there will be a resultant EMF in the circuit which is responsible for the current:

$$\text{EMF in the circuit} = P_2 - P_1$$

where P_1 is the Peltier EMF at temperature T_1, and P_2 is the Peltier EMF at temperature T_2, where $T_2 > T_1$. Peltier cooling is used in instrumentation where a small component is required to be cooled under precise control.

Figure 21.26 shows diagrammatically the construction of such a cooler. The conductors and junctions have a big cross-section to minimize IR heating. The warmer face is clamped to a suitable heat sink; the cold face has the component to be cooled mounted in contact with it. Typical size for such a unit is on the order of 5–25 mm. The conductors in Peltier coolers may be either metals or semiconductors; in the latter case they are called *Frigistors*.

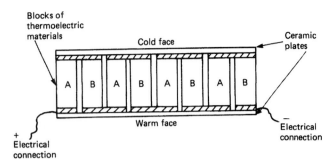

FIGURE 21.26 Peltier cooler.

21.5.1.3 Thomson Effect

Professor William Thomson (later Lord Kelvin) pointed out that if the reversible Peltier effect was the only source of EMF, it would follow that if one junction was maintained at a temperature T_1, and the temperature of the other raised to T_2, the available EMF should be proportional to $(T_2 - T_1)$. This is not true. If the copper–iron thermocouple, already described, is used, it will be found that on heating one junction while the other is maintained at room temperature, the EMF in the circuit increases at first, then diminishes, and, passing through zero, actually becomes reversed. Thomson, therefore, concluded that in addition to the Peltier effects at the junctions, there were reversible thermal effects produced when a current flows along an unequally heated conductor. In 1856, by a laborious series of experiments, he found that when a current of electricity flows along a copper wire of which the temperature varies from point to point, heat is liberated at any point P when the current at P flows in the direction of the flow of heat at P, that is, when the current is flowing from a hot place to a cold place, whereas heat is absorbed at P when the current flows in the opposite direction. In iron, on the other hand, the heat is absorbed at P when the current flows in the direction of the flow of heat at P, whereas heat is liberated when the current flows in the opposite direction from the flow of heat.

21.5.1.4 Thermoelectric Diagram

It will be seen that the Seebeck effect is a combination of the Peltier and Thomson effects and will vary according to the difference of temperature between the two junctions and with the metals chosen for the couple. The EMF produced by any couple with the junctions at any two temperatures may be obtained from a thermoelectric diagram suggested by Professor Tait in 1871. On this diagram the thermoelectric line for any metal is a line such that the ordinate represents the thermoelectric power (defined as the rate of change of EMF acting around a couple with the change of temperature of one junction) of that metal with a standard metal at a temperature represented by the abscissa. Lead is chosen as the standard metal because it does not show any measurable Thomson effect. The ordinate is taken as positive when, for a small difference of temperature, the current

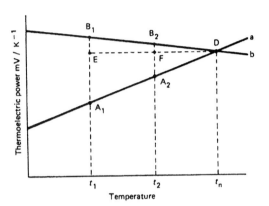

FIGURE 21.27 Thermoelectric diagram of two metals.

flows from lead to the metal at the hot junction. If lines a and b (Figure 21.27) represent the thermoelectric lines for two metals A and B, the EMF around the circuit formed by the two metals, when the temperature of the cold junction is t_1 and that of the hot junction is t_2, will be the difference in the areas of triangle A_1B_1D and A_2B_2D. Now the area of the triangle is

$$A_1B_1D = \frac{1}{2}(A_1B_1 \times ED) \qquad (21.18)$$

and area

$$A_2B_2D = \frac{1}{2}(A_2B_2 \times FD) \qquad (21.19)$$

$$\text{The EMF} = \frac{1}{2}(A_1B_1 \times ED) - \frac{1}{2}(A_2B_2 \times FD) \qquad (21.20)$$

Since triangles A_1B_1D and A_2B_2D are similar triangles, the sides A_1B_1 and A_2B_2 are proportional to ED and FD, respectively. Therefore:

$$\text{EMF} \propto ED^2 - FD^2$$

but

$$ED = t_n - t_1 \text{ and } FD = t_n - t_2$$

so,

$$\text{EMF} \propto (t_n - t_1)^2 - (t_n - t_2)^2$$

$$\propto (t_1 - t_2)\left(\left(\frac{t_1 + t_2}{2}\right) - t_n\right)$$

or

$$\text{EMF} = K(t_1 - t_2)\left(\frac{t_1 + t_2}{2} - t_n\right) \quad (21.21)$$

where K is a constant that, together with t_n must be obtained experimentally for any pair of metals. The temperature t_n is called the *neutral temperature*. Equation (21.21) shows that the EMF in any couple is proportional to the difference of temperature of the junctions and to the difference between the neutral temperature and the average temperature of the junctions. The EMF is zero if either the two junctions are at the same temperature or the average of the temperature of the two junctions is equal to the neutral temperature. Figure 21.28 shows the graph of the EMF of a zinc-iron thermocouple with temperature.

21.5.1.5 Thermoelectric Inversion

This reversal of the thermoelectric EMF is called *thermoelectric inversion*.

Figure 21.29 shows the thermoelectric lines for several common materials. It will be seen that the lines for iron and copper cross at a temperature of 275°C. If the temperature of the cold junction of iron and copper is below 270°C and the temperature of the other junction is raised, the thermoelectric EMF of the circuit (represented by a trapezium) will increase until the temperature of the hot

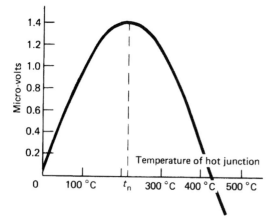

FIGURE 21.28 Temperature/EMF curve for zinc/iron couple.

junction reaches 275°C (when the EMF is represented by a triangle). Further increase in the temperature of the hot junction will result in a decrease in the thermoelectric EMF (the EMF represented by the second triangle will be in the opposite sense). When the average temperature of the two junctions is 275°C or what comes to the same thing, the sum of the two temperatures is 550°C, the areas of the two triangles will be equal, and there will be no thermoelectric EMF: 275°C is the "neutral temperature" for the copper-iron couple. With circuits of other materials, the neutral point will occur at different temperatures. Further increase in the temperature of the hot junction will produce a thermoelectric EMF in the opposite direction: from iron to copper at the hot junction, which will again increase with increasing temperature of the hot junction, as was seen with zinc and iron in Figure 21.28.

In choosing two materials to form a thermocouple to measure a certain range of temperature, it is very important to choose two that have thermoelectric lines that do not cross within the temperature range, that is, the neutral temperature must not fall within the range of temperature to be measured. If the neutral temperature is within the temperature range, there is some ambiguity about the temperature indicated by a certain value of the thermoelectric EMF, because there will be two values of the temperature of the hot junction for which the thermoelectric EMF will be the same. For this reason tungsten-molybdenum thermocouples must not be used at temperatures below 1,250°C.

21.5.1.6 Addition of Thermoelectric EMFs

In measuring the EMF in any circuit due to thermoelectric effects, it is usually necessary to insert some piece of apparatus, such as a millivoltmeter, somewhere in the circuit, and since this generally involves the presence of junctions other than the two original junctions, it is important to formulate the laws according to which the EMFs produced by additional junctions may be dealt with. These laws, discovered originally by experiment, have now been established theoretically.

Law of Intermediate Metals In a thermoelectric circuit composed of two metals A and B with junctions at temperatures t_1 and t_2, the EMF is not altered if one or both junctions are opened and one or more other metals are interposed between metals A and B, provided that all the junctions by which the single junction at temperature t_1 may be replaced are kept at t_1, and all those by which the junction at temperature t_2 may be replaced are kept at t_2.

This law has a very important bearing on the application of thermocouples to temperature measurement, since it means that, provided that all the apparatus for measuring the thermoelectric EMF, connected in the circuit at the cold junction, is kept at the same temperature, the presence of any number of junctions of different metals will not affect

CHAPTER | 21 Temperature Measurement

the total EMF in the circuit. It also means that if another metal is introduced into the hot junction for calibration purposes, it does not affect the thermoelectric EMF, provided that it is all at the temperature of the hot junction.

Law of Intermediate Temperatures The EMF E_{1-3} of a thermocouple with junctions at temperatures t_1 and t_3 is the sum of the EMFs of two couples of the same metals, one with junctions at temperatures t_1 and t_2 (EMF = E_{1-2}) and the other with junctions at t_2 and t_3, (EMF = E_{2-3}) (see Figure 21.30):

$$E_{1-2} + E_{2-3} = E_{1-3} \quad (21.22)$$

This law is the basis on which thermocouple measuring instruments can be manufactured.

21.5.1.7 Cold Junction Compensation

It is not normally practical in industrial applications to have thermocouple cold junctions maintained at 0°C, but with the cold junctions at ambient temperature, cold junction compensation is required. To achieve cold junction compensation, consider a thermocouple with its hot junction at t°C and its cold junction at ambient, its EMF being E_{a-t}. The instrument must indicate an EMF equivalent to having the cold junction at 0°C, that is, an EMF of E_{0-t}. This requires that an EMF must be added at E_{a-t} to provide the required signal:

$$E_{0-t} = E_a + E_{0-a} \quad (21.23)$$

The voltage E_{0-a} is called the *cold junction compensation voltage*.

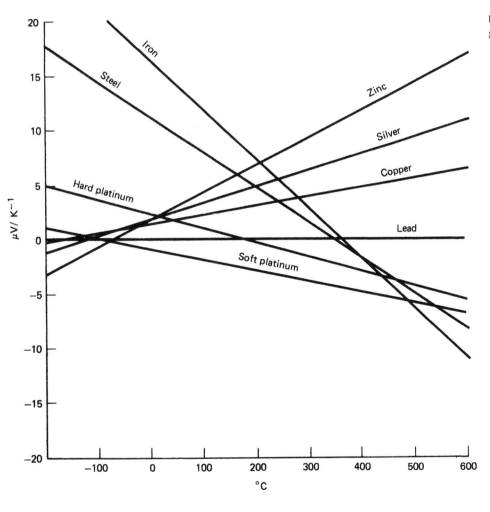

FIGURE 21.29 Thermoelectric diagrams for several metals.

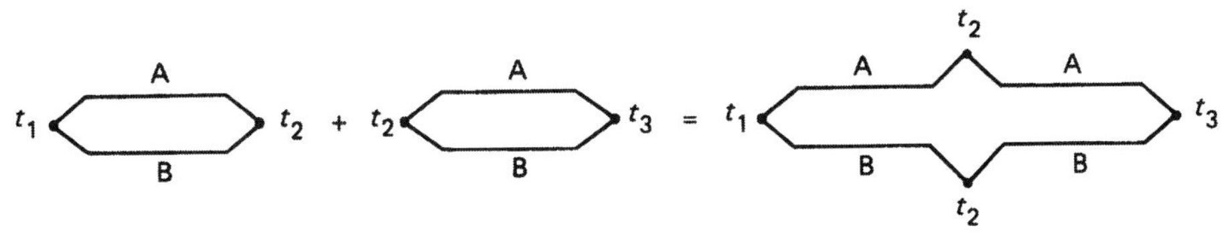

FIGURE 21.30 Law of intermediate metals.

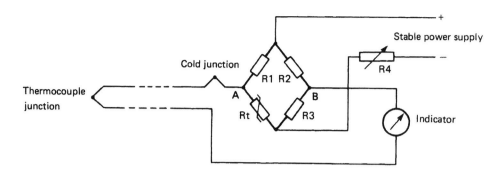

FIGURE 21.31 Bridge circuit to provide cold junction compensation.

This cold junction compensation EMF can be provided automatically by the use of a temperature-sensitive element such as a resistance thermometer, thermistor, or semiconductor sensor in the thermocouple circuit. Figure 21.31 shows such a circuit. In this circuit R_1, R_2 and R_3 are temperature-stable resistors and R_t is a resistance thermometer. The bridge is balanced when all components are at 0°C and the voltage appearing between points A and B is zero. As the temperature changes from 0°C an EMF, which is the unbalance voltage of the bridge, exists across AB. This voltage is scaled by setting R_4 such that the voltage AB is equal to E_{0-a} in Equation (21.23).

Mechanical Cold Junction Compensation An alternative cold junction compensation technique is used when a simple nonelectronic thermometer is required. In this technique the thermocouple is connected directly to the terminals of a moving-coil galvanometer. A bimetal strip is connected mechanically to the mechanical zero adjustment of the instrument in such a way that the instrument zero is offset to indicate the ambient temperature. The EMF E_{a-t} is then sufficient to move the pointer upscale to indicate the true temperature of the thermocouple.

21.5.1.8 Thermocouple Circuit Considerations

Galvanometer Instruments A thermocouple circuit is like any other electrical circuit. There are one or more sources of EMF, which can be batteries, a generator, or in this case the hot and cold junctions. There is a load, the indicator, and there are electrical conductors, which have resistance, to connect the circuit together. The current in this circuit is, as always, governed by Ohm's law:

$$I = \frac{E}{R} \quad (21.24)$$

where I is the current, E is the EMF, and R is the total circuit resistance.

In a practical thermocouple thermometer, the resistance consists of the sum of the resistances of the thermocouple, the compensating cable (see Section 21.5.3.9), and the indicating instrument. Galvanometer-type thermocouple indicators with mechanical cold junction compensation, as described in the

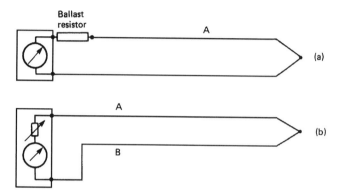

FIGURE 21.32 Use of ballast resistor: (a) external to instrument, (b) adjustable ballast mounted inside instrument.

previous section, are designed either to be used with an external circuit of stated resistance (this resistance value is usually marked on the dial) or they have an internal adjustable resistor. In the latter case the resistance of the external circuit must not exceed a stated maximum value, and the adjustable resistor is adjusted to give the specified total circuit value. Where no internal resistance adjustment is provided, the instrument must be used together with an external ballast resistor; see Figure 21.32(a). This resistor must be mounted as near as possible to the indicating instrument to ensure its being at the same temperature as the cold junction compensating mechanism. The usual practice when installing one of these instruments is to wind the ballast resistor with constantan wire on a small bobbin. The length of constantan wire is chosen to make up the required total resistance. On some instruments the bobbin is made integral with one of the indicator terminals. Figure 21.32(b) shows the arrangement with the ballast resistor integral with the indicating instrument.

Potentiometric Instruments One way to circumvent the critical external resistor is to use a potentiometric indicating device. In a potentiometric device the thermocouple EMF is opposed by an equal and opposite potential from the potentiometer; there is then no current in the circuit and therefore the circuit resistance value is irrelevant.

Potentiometric thermocouple indicators used to be quite common but are now not found so often. However, if the

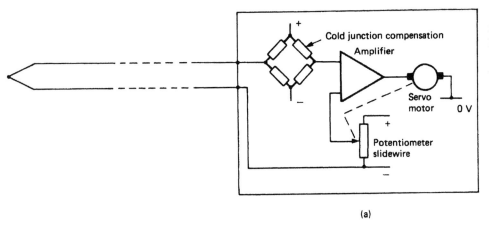

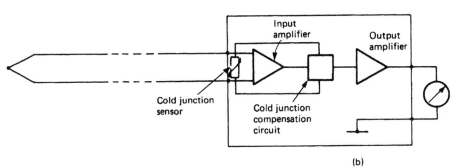

FIGURE 21.33 Cold junction compensation: (a) in conjunction with potentiometric indicating instrument, (b) alternative arrangement for cold junction compensation.

thermocouple indicator is, as it frequently is, a strip chart recorder, it is almost certain to be a potentiometric instrument. Figure 21.33(a) shows the potentiometric arrangement diagrammatically.

Electronic Instruments In modern electronic instruments for thermocouple indication, whether analog or digital devices, the input circuit "seen" by the thermocouple is a high-impedance amplifier. Again, there is negligible current in the thermocouple circuit, and since the resistance of the thermocouple circuit is on the order of 100 Ohms whereas the amplifier input is likely to be a megohm or more, the effect of the external circuit resistance is negligible. Electronic instruments allow their designer much more versatility for cold junction compensation. Instead of the bridge circuit of Figure 21.31, it is possible to arrange the cold junction correction after the input amplifier. This has the advantage that the voltage levels being worked with may be on the order of several volts of amplitude instead of a few millivolts, making it easier to get a higher degree of accuracy for compensation. Figure 21.33(b) shows a block diagram of such an arrangement.

Thermocouple input circuits are available as encapsulated electronic modules. These modules contain input amplifier and cold junction compensation. Since the cold junction consists of the input connections of the module, the connections and the cold junction sensor can be accurately maintained at the same temperature by encapsulation, giving very accurate compensation. These modules can be very versatile. Many are available for use with any of the normal thermocouples. The cold junction compensation is set to the thermocouple in use by connecting a specified value resistor across two terminals of the module. Where the thermocouple instrument is based on a microcomputer, the cold junction compensation can be done by software, the microcomputer being programd to add the compensation value to the thermocouple output. In all electronic equipment for thermocouple signal processing, the location of the sensor for cold junction temperature sensing is critical. It must be very close to the cold junction terminals and preferably in physical contact with them.

21.5.2 Thermocouple Materials

Broadly, thermocouple materials divide into two arbitrary groups based on cost of the materials, namely, base metal thermocouples and precious metal thermocouples.

21.5.2.1 Base Metal Thermocouples

The most commonly used industrial thermocouples are identified for convenience by type letters. The main types, together with the relevant British Standard specification and permitted tolerance on accuracy, are shown in Table 21.13. Also shown are their output EMFs with the cold junction at 0°C. These figures are given to indicate the relative sensitivities of the various couples. Full tables of voltages against hot junction

TABLE 21.13 Thermocouples to British Standards					
Type	Conductors (positive conductor first)	Manufactured to BS 4937 Part No.	Temperature tolerance class 2 thermocouple BS 4937: Part 20: 1991	Output for indicated temperature (cold junction at 0°C)	Service temperature (max. intermittent servicer)
B	Platinum: 30% Rhodium/platinum: 6% Rhodium	Part 7: 1974 (1981)	600–1,700°C ± 3°C	1.241 mV at 500°C	0–1500°C (1700°C). Better life expectancy at high temperature than types R & S.
E	Nickel: chromium/constantan (chromel/constantan) (chromel/advance)	Part 6: 1974 (1981)	−40 + 333°C ± 3°C 333–900°C ± 0.75%	6.317 mV at 100°C	−200 to +850°C (1,100°C). Resistant to oxidizing atmospheres.
J	Iron/constantan	Part 3: 1973 (1981)	−40 to +333°C ± 2.5°C 300–750°C ± 0.75%	5.268 mV at 100°C	−280 to +850°C (1,100°C). Low cost; suitable for general use.
K	Nickel: chromium/nickel: aluminum (chromel/alumel) (C/A) (T1/T2)	Part 4: 1973 (1981)	−40 to +333°C ± 2.5°C 333–1,200°C ± 0.75%	4.095 mV at 100°C	−200 to +1,100°C (1,300°C). Good general-purpose. Best in oxidizing atmosphere.
N	Nickel: chromium: silicon/nickel: silicon: magnesium (nicrosil/nisil)	Part 8: 1986	−40 to + 333°C ± 2.5°C 333–1,200°C ± 0.75%	2.774 mV at 100°C	0–1,100°C (−270°C to +1,300°C). Alternative to type K.
R	Platinum: 13% rhodium/platinum	Part 2: 1973 (1981)	0–600°C ± 1.5°C 600–1,600°C ± 0.25%	4.471 mV at 500°C	0–1,500°C (1.650°C). High temperature. Corrosion resistant.
S	Platinum: 10% rhodium/platinum	Part 1: 1973 (1981)	0–600°C ± 1.5°C 600–1,600°C ± 0.25%	4.234 mV at 500°C	Type R is more stable than type S.
T	Copper/constantan (copper/advance) (Cu/Con)	Part 5: 1974 (1981)	−40 to +375°C ± 1°C	4.277 mV at 100°C	−250 to 400°C (500°C). High resistance to corrosion by water.

temperatures are published in BS 4937. The standard also supplies the equations governing the thermocouple EMFs for convenience for computer programming purposes. These equations are essentially square law; however, provided a thermocouple is used at temperatures remote from the neutral temperature, its characteristic is very nearly linear. Figure 21.34 shows a plot of the characteristic for a type K thermocouple. It can be seen that for temperatures in the range −50°C to 400°C, the characteristic is approximately linear. The commonly used base metal thermocouples are types E, J, K, and T. Of these, J and K are probably the most usual ones. They have a high EMF output, and type K is reasonably resistant to corrosion. Type T has a slight advantage, where the temperature measurement points are very remote from the instrumentation, that because one conductor is copper the overall resistance of the circuit can be lower than for other types. Type N is a newer thermocouple that can be used as an alternative to type K. Table 21.14 shows some commercially available thermocouples that are not currently covered by British Standards.

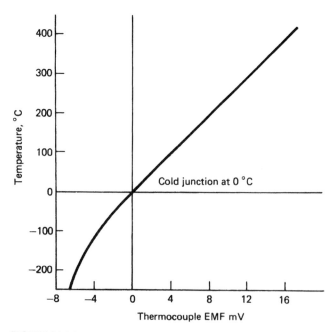

FIGURE 21.34 Type K thermocouple characteristic.

TABLE 21.14 Thermocouples commercially available but not covered by British Standards (composition and accuracy to be agreed with manufacturer)

Type	Conductors (positive conductor first)	Output for indicated temperature (cold junction at 0°C)	Service temperature (max. intermittent service)
W	Tungsten/tungsten: 26% rhenium	34.1 mV at 2,000°C	20–2,300°C (2,600°C)
W_5	Tungsten: 3% Rhenium/tungsten: 26% rhenium	32.404 mV at 2,000°C	
W_3	Tungsten: 3% Rhenium/tungsten: 25% rhenium	35.707 mV at 2000°C	(W_3 suitable for hydrogen atmosphere)
	Tungsten/molybdenum Rhodium: iridium/rhodium	Typically 6.4 mV at 1,200°C	1 to ≥300 K
	Iron/gold: nickel/chromium Iron: gold/silver		1 to ≥300 K

21.5.2.2 Precious Metal Thermocouples

Thermocouple types B, R, and S clearly carry a considerable cost penalty and normally are only used when essential for their temperature range or their relatively high resistance to chemical attack. Their temperature top limit is 1,500°C for continuous use or 1,650°C for intermittent, spot-reading applications. This compares with 1,100°C continuous and 1,300°C intermittent for type K.

Errors in type R and S thermocouple readouts result from strain, contamination, and rhodium drift.

The effect of strain is to reduce the EMF, resulting in low readings. The effect of strain may be removed by annealing the thermocouple. Installations should be designed to minimize strain on the thermocouple wires.

Contamination is by far the most common cause of thermocouple error and often results in ultimate mechanical failure of the wires. Elements such as Si, P, Pb, Zn, and Sn combine with platinum to form low melting point eutectics and cause rapid embrittlement and mechanical failure of the thermocouple wires. Elements such as Ni, Fe, Co, Cr, and Mn affect the EMF output of the thermocouple to a greater or lesser degree, but contamination by these elements does not result in wire breakage and can be detected only by regularly checking the accuracy of the thermocouple. Contamination can be avoided by careful handling of the thermocouple materials before use and by the use of efficient refractory sheathing. Care should be taken to prevent dirt, grease, oil, or soft solder coming into contact with the thermocouple wires before use. If the atmosphere surrounding the thermocouple sheath contains any metal vapor, the sheath must be impervious to such vapors.

Rhodium drift occurs if a rhodium-platinum limb is maintained in air for long periods close to its upper temperature limit. Rhodium oxide will form and volatilize, and some of this oxide can settle on and react with the platinum limb, causing a fall in EMF output. This is a comparatively slow process and is therefore only of significance in installations where the maximum stability and repeatability are required. Type B thermocouples are less susceptible to rhodium drift than types R or S, but type B has a lower EMF than R and S and is subject to higher errors.

Noble metal thermocouples may also be used for measuring cryogenic temperatures. Iron–gold/nickel–chromium or iron–gold/silver (normal silver with 0.37 atomic percent gold) may be used for temperatures from 1 K to above 300 K.

Noble metal thermocouples are often used in the "metal-clad" form, with magnesia or alumina powder as the insulant. This form of construction is described in Section 21.5.3.2.

The following sheath materials are used: nickel, stainless steel, inconel in 1.6 and 3.2 mm sizes, and 5 percent rhodium-plated and 10 percent rhodium-platinum, both in 1.0 mm sizes. For high-temperature work, other special thermocouples have been developed: tungsten 5 percent rhenium/tungsten 20 percent rhenium for use in hydrogen, vacuum, and inert gas atmospheres up to 2,320°C and tungsten/molybdenum and tungsten/iridium for temperatures up to 2,100°C.

Quite a wide range of precious metal thermocouples is available. Types B, R, and S are specified in BS 4937. These three are based only on platinum and rhodium. Gold, iridium, other "platinum metals," and silver are also not uncommonly used. Figure 21.35 shows the characteristics of some of the available options.

21.5.3 Thermocouple Construction

Thermocouples, like resistance thermometers and other temperature sensors, are available in a wide range of mechanical constructions.

21.5.3.1 Plain-Wire Thermocouples

For use in protected environments such as laboratories or inside otherwise enclosed equipment, plain-wire thermocouples can be used. They are also used in plants where the

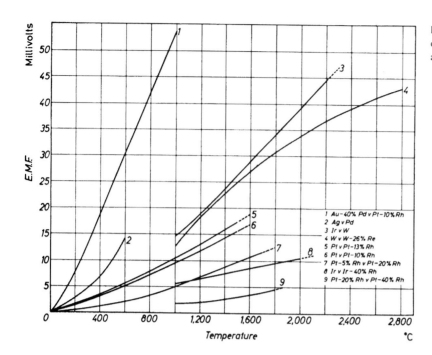

FIGURE 21.35 Summary of thermoelectric properties of precious metal thermocouples. Broken lines indicate areas for intermittent service.

fastest possible response is required. However, they suffer from the obvious disadvantage that they are both fragile and liable to chemical attack. The wires are available insulated with PVC or glass fiber sleeving, or for use with higher temperatures, the wires can be insulated with refractory ceramic beads or sleeves.

21.5.3.2 Sheathed Thermocouples

Thermocouples for use in plant situations, where robust construction is required or where they need to be interchangeable with other types of temperature measurement equipment, are available sheathed in steel or stainless steel designed for direct insertion into process vessels or for use in a thermometer pocket. Figures 21.36(a) and (b) show typical insertion probes. Where thermocouples are to be immersed in very corrosive process fluids or into very high-temperature locations, they are available constructed in ceramic sheaths, as in Figure 21.36(c). Sheathed thermocouples, especially the ceramic ones, suffer from a slow response time, typically a minute or more. However, the locations where they are essential for their mechanical properties are usually in heavy plants where temperatures do not normally move fast in any case.

21.5.3.3 Mineral-Insulated Thermocouples

Probably the most versatile format for thermocouples is the mineral-insulated (MI) construction. In this form the thermocouples are made from mineral-insulated cable similar in concept to the MI cable used for electrical wiring applications. It differs, however, in that the conductors are of thermocouple wire and the sheath is usually stainless steel. The

FIGURE 21.36 Examples of industrial thermocouple probes. Courtesy of ABB.

insulation, however, is similar, being in the form of finely powdered and densely compacted ceramic, usually aluminum oxide or magnesium oxide. Figure 21.36 shows MI thermocouples at (d), (e), and (f).

They are available in diameters from 1 millimeter up to 6 millimeters and can be supplied in any length required. The junction can be either (a) insulated or (b) welded to the tip of the sheath, as shown in Figure 21.37. The latter arrangement

CHAPTER | 21 Temperature Measurement

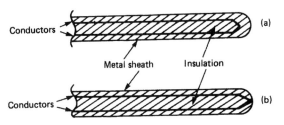

FIGURE 21.37 Mineral insulated thermocouples: (a) insulated junction, (b) junction welded to sheath.

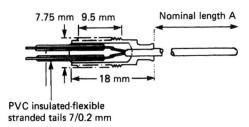

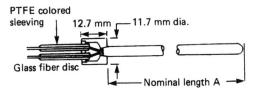

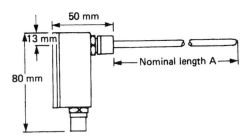

FIGURE 21.38 MI thermocouple terminations. Courtesy of ABB.

has the advantage of very quick response. For some applications the junction being connected to the plant earth via the sheath tip can be unacceptable, so in such cases insulated thermocouples must be used. The principal advantages are their quick response and mechanical flexibility; they can be bent into almost any shape. Care must be taken if reusing MI thermocouples; although they can be straightened or rebent to a new shape, this cannot be done too often. Either the wires break or the insulation gets displaced and the thermocouple becomes short-circuited.

As shown in Figures 21.36 and 21.38, MI thermocouples can be supplied fitted with a variety of terminations. A further useful advantage of MI thermocouples is that the cable can be bought in rolls together with suitable terminations, and the thermocouples can be made up to the required specifications on site. Also, in situations where robust cabling is required, MI thermocouple cable can be used in lieu of compensating cable (see Section 21.5.3.9).

21.5.3.4 Surface Contact Thermocouples

Thermocouples for the measurement of the surface temperature of objects such as pipes or other components or plant items are available. On pipes a surface measurement makes a simple but not very accurate noninvasive temperature measurement. For higher temperatures or more rugged applications, thermocouples are available embedded in a metal plate designed to be clamped or welded to the component to be measured. For lower-temperature applications, below about 200°C, or for use in protected environments, self-adhesive surface thermocouples are supplied. In these probes the thermocouple is embedded in a small plastic pad coated on one face with a suitable contact adhesive.

21.5.3.5 Hot-Metal Thermocouples

Where it is necessary to make spot measurements of the temperature of hot metal billets, very simple test prods are available that consist of a two-pronged "fork." The two prongs are made of the two thermocouple metals with sharpened points. When both prongs are in contact with the hot metal, two junctions are formed, metal A to the billet and the billet to metal B. If the billet is large and enough time is allowed for the tips of the prongs to reach the temperature of the billet, both junctions will be at the same temperature and the error thermal EMFs will cancel. This makes a simple, quick, and very inexpensive way of measuring hot metal temperatures. The points of the prongs are screwed to the main assembly and are expendable. They can be changed as soon as they lose their sharpness or begin to get corroded.

21.5.3.6 Liquid Metal Thermocouples

When measuring the temperature of liquid metals such as steel, it is desirable to use an expendable probe. The cost of a fully protected probe would be very high and the response time slow. A dipstick probe can be used for checking the temperature of liquid steel. The probe itself is robust and constructed with a socket of thermocouple material in the end. A disposable platinum-rhodium/platinum thermocouple itself lasts in the molten metal for a few seconds, long enough to take a temperature measurement. Figure 21.39 shows this arrangement.

21.5.3.7 Thermopiles

Where a very small temperature rise is to be measured, many thermocouples may be connected in series. All the hot junctions are on the object of which the temperature is to be measured, and all the cold junctions are kept at a constant and known temperature. Where a quick temperature response is required, these thermocouples can be of very thin wire

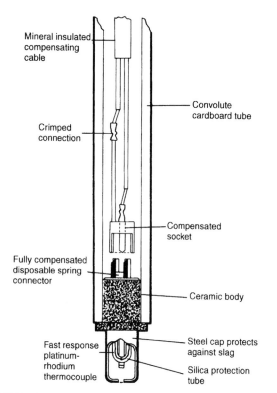

FIGURE 21.39 Liquid metal thermocouple.

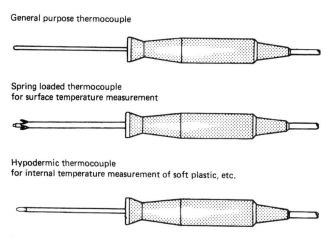

FIGURE 21.40 Handheld thermocouple probes.

of about 25 μm diameter. A speed of response on the order of 10 milliseconds can be achieved. Typical applications of thermopiles are to be found in infrared radiation measurement. This subject is dealt with in Section 21.6.

21.5.3.8 Portable Thermocouple Instruments

With the development over the last decade of microelectronic equipment, portable electrical thermometers have become very popular. They are available with either analog or digital readouts. The analog instruments are about the size of an analog multimeter; the digital instruments are about the size of a pocket calculator. Although most of these instruments use type K thermocouples, they are available for use with other thermocouple materials. There are also portable thermometers available that use resistance thermometer or thermistor sensors. However, the thermocouple instruments are on the whole the most popular. The more sophisticated instruments have the option to use more than one type of thermocouple; a switch on the instrument sets it for the type in use. They are also available with a switched option to read out in Celsius or Fahrenheit. A range of handheld probes are supplied for use with these instruments. Figure 21.40 shows some of the options available. The spring-loaded thermocouples are for surface contact measurements; hypodermic probes are supplied for such applications as temperature measurements in food, such as meat, where it could be an advantage to know the internal temperature of the material.

21.5.3.9 Thermocouple Compensating Cable

Ideally a thermocouple connects back to the reading instrument with cables made of the same metals as the thermocouple. However, this does have two disadvantages in industrial conditions. First, many thermocouple metals have high electrical resistance. This means that on long runs, which on a big plant could be up to 100 meters or more, heavy-gauge conductors must be used. This is not only expensive, it also makes the cables difficult to handle.

Second, in the case of precious metal thermocouples—types B, R, and S, for instance—the cost would be very high indeed. To overcome these problems, compensating cables are used; see Figure 21.41. These cables are made of base metal and are of lower resistivity than the thermocouple material. The alloys they contain have thermoelectric properties that essentially match the thermocouples themselves over a limited ambient temperature range.

Examples of compensating cables are shown in the following table.

Type	Composition	Thermocouples compensated	Temperature limitations
U	Copper/copper-nickel	R and S	0–50°C
Vx	Copper/Constantan	K	0–80°C

Other base metal thermocouples, such as types J and T, comprise relatively inexpensive and low-resistance metals. They are therefore normally installed using cables consisting of the same metals as the thermocouples themselves.

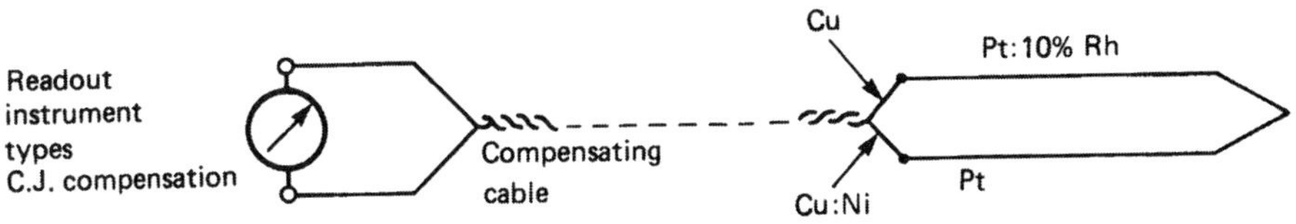

FIGURE 21.41 Thermocouple compensating cable.

21.5.3.10 Accuracy Consideration

The very extensive use of thermocouples stems from their great versatility combined with their low cost. However, as shown in Table 21.13, thermocouples have a fairly wide permitted tolerance. This is due to the fact that most metals used for thermocouples are alloys, and it is not possible to manufacture alloys to the same reproducibility as pure metals. It must be said that, in general, manufacturers do manufacture their thermocouples to better tolerance than BS 4937 demands. But where the highest accuracy is required, it is essential to calibrate thermocouples on installation and to recalibrate them at regular intervals to monitor any deterioration due to corrosion or diffusion of foreign elements into the hot junction.

Where high accuracy is required, it is necessary to first calibrate the thermocouple readout instrument and then the thermocouple itself in conjunction with the instrument.

The calibration of instruments can be done with a precision millivolt source that injects a signal equivalent to the temperature difference between the ambient or cold junction temperature and a temperature in the region in which the thermocouple is to be used.

To calibrate or check thermocouples, the hot junction must be kept at an accurately known temperature. This can be done by inserting it into a heated *isothermal block*. An isothermal block is a block of metal that's large compared with the thermocouple being measured and made of copper or aluminum. The block has provision for heating and in some cases cooling it. It is well insulated from the environment and is provided with suitable holes for inserting various sizes of thermocouple. Where not-so-high precision is required, the thermocouple can be immersed in a heated fluidized sand bath. This consists of an open vessel fitted with a porous bottom (usually made of sintered metal). Heated air is forced up through the bottom. The vessel is filled with carefully graded sand. With the air coming up through it, the sand behaves like a liquid. It takes up the temperature of the air. The sand is a good heat transfer medium. The apparatus makes a most convenient way of calibrating temperature probes. Where maximum accuracy is essential, the thermocouple should be calibrated against one of the IPTS-68 secondary reference points. Table 21.5 shows some of the points.

In carrying out these calibrations, the whole installation needs to be calibrated: thermocouple readout instrument together with compensating cable. In cases where very high accuracy is required, compensating cable should not be used; the conductors should be thermocouple metal for the full length of the installation.

TABLE 21.15 Wavelengths transmitted by lens materials

Lens material	Bandpass (μm)
Pyrex	0.3–2.7
Fused silica	0.3–3.8
Calcium fluoride	0.1–10
Arsenic trisulphide	0.7–12
Germanium	2–12
Zinc selenide	0.5–15

Some very versatile equipment for thermocouple calibration is on the market. Typically, the facilities provided include thermocouple simulation for types E, J, K, R, S, and T; thermocouple output measurement with cold junction compensation; and resistance thermometer simulation. Tests can be static or dynamic using ramp functions.

As with any other type of temperature measurement, the location of the thermocouple junctions is critical. This is just as important for the cold junction as for the hot junction. It must be remembered that there may well be a temperature gradient over quite short distances in an instrument, and unless the cold junction temperature sensor is in close thermal contact with the cold junction itself, a reading error of several degrees Celsius may result. This problem is at its worst with mains electricity-powered measuring instruments, where there is a certain amount of heat liberated by the power unit.

The point to remember is that it is not usually adequate to measure the air temperature in the vicinity of the cold junctions. The sensor should be in good thermal contact with them.

An obvious point, but one which surprisingly often causes trouble, is the mismatch between the thermocouple and the measuring instrument. The obvious mismatch is using the wrong type of thermocouple or compensating cable.

In the case of galvanometric instruments inaccuracies occur if sufficient care has not been taken in the winding of the makeup resistor or if the thermocouple has been changed and the new external circuit resistance not checked. Careless location or makeup of the ballast resistor so that one of the cold junction terminals is too remote from the cold junction compensating element causes variable errors of several degrees as the ambient temperature changes. Where the required ballast resistor is of a low value, 10 ohms or so, the best arrangement may well be to use a coil of compensating cable of the right resistance.

21.6 MEASUREMENT TECHNIQUES: RADIATION THERMOMETERS

21.6.1 Introduction

As mentioned in Section 21.1, thermal energy may be transferred from one body to another by radiation as well as by conduction. The amount of thermal energy or heat leaving a body by radiation and the wavelength of that radiation are functions of the temperature of the body.

This dependence on temperature of the characteristics of radiation is used as the basis of temperature measurement by radiation thermometers. Radiation thermometers are also known as *radiation pyrometers*.

21.6.1.1 Blackbody Radiation

An ideal blackbody is one that, at all temperatures, will absorb all radiation falling on it without reflecting any whatever in the direction of incidence. The absorptive power of the surface, being the proportion of incident radiation absorbed, will be unity. Most surfaces do not absorb all incident radiation but reflect a portion of it. That is, they have an absorptive power of less than unity.

A blackbody is also a perfect radiator. It will radiate more radiation than a body with an absorptive power of less than unity. The emissive power is called the *emissivity* of a surface. The emissivity is the ratio of the radiation emitted at a given temperature compared to the radiation from a perfect blackbody at the same temperature.

The total emissivity of a body is the emissive power over the whole band of thermal radiation wavelengths and is represented by ε_1. When only a small band of wavelengths is considered, the term *spectral emissivity* is used and a subscript is added defining the wavelength band, such as $\varepsilon_{1.5}$ indicating the emissivity at 1.5 μm wavelength.

The emissivity of surfaces is not usually the same over all wavelengths of the spectrum. In general the emissivity of metals is greater at shorter wavelengths and the emissivity of oxides and refractory materials is greater at longer wavelengths. Some materials have a very low emissivity at a particular wavelength band and higher emissivities at shorter and longer wavelength. For instance, glass has an emissivity of almost zero at 0.65 μm.

Realization of a Blackbody Radiator A black-body radiator is achieved in practice by an enclosure (A in Figure 21.42) that has a relatively small orifice B from which black-body radiation is emitted. The inside walls of the enclosure must be at a uniform temperature. To show that the orifice B behaves as a blackbody, consider the ray of radiation C entering the

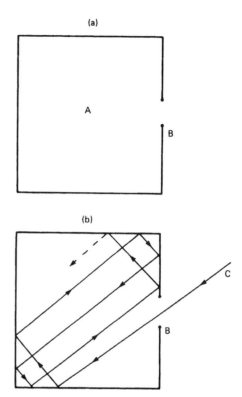

FIGURE 21.42 (a) Blackbody radiator, (b) absorption of ray of radiation by blackbody radiator.

chamber through B. The ray will suffer many reflections on the inside walls of the enclosure before it emerges at B. Provided that the walls of the chamber are not perfectly reflecting, the total energy of the radiation will have been absorbed by the many reflections before the ray can emerge. The orifice is then totally absorbing all radiation that enters it. It is a blackbody.

To show that the orifice must also radiate as a blackbody, first consider a body in a radiant flux at any single wavelength. If that body did not radiate energy at that wavelength as fast as it absorbed it, it would rapidly get warmer than its environment. In practice a body will be at thermal equilibrium with its surroundings, so it must be radiating energy as it receives it.

Therefore, the emissivity ε of a body must equal its absorbance α. The orifice B, which is a black-body absorber, must also be a black-body radiator.

In practice, a sighting hole in a furnace will radiate as a blackbody if the furnace and its contents are in thermal equilibrium and provided that it does not contain a gas or flame that absorbs or radiates preferentially in any wavelength band. However, the radiation from the sighting hole will only be black-body radiation provided that everything in the furnace is at the same temperature. When all objects in the furnace are at the same temperature, all lines of demarcation between them will disappear. If a cold object is introduced to the furnace, it will be absorbing more energy than it is radiating; the rest of the furnace will be losing more radiation than it receives. Under these conditions the radiation will no longer be black-body radiation but will be dependent on the emissivity of the furnace walls.

Prevost's Theory of Exchanges Two bodies A and B in a perfectly heat-insulated space will both be radiating and both be absorbing radiation. If A is hotter than B, it will radiate more energy than B. Therefore B will receive more energy than it radiates, and consequently its temperature will rise. By contrast body A will lose more energy by radiation than it receives, so its temperature will fall. This process will continue until both bodies reach the same temperature. At that stage the heat exchanged from A to B will be equal to that exchanged from B to A.

A thermometer placed in a vessel to measure gas temperature in that vessel will, if the vessel walls are cooler than the gas, indicate a temperature lower than the gas temperature because it will radiate more heat to the vessel walls than it receives from them.

Black-Body Radiation: Stefan–Boltzmann Law The total power of radiant flux of all wavelengths R emitted into the frontal hemisphere by a unit area of a perfectly black body is proportional to the fourth power of the temperature Kelvin:

$$R = \sigma T^4 \qquad (21.25)$$

where σ is the Stefan–Boltzmann constant, having an accepted value of $5.670\,32 \times 10^{-8}\,\text{W} \cdot \text{m}^{-2} K^{-4}$, and T is the temperature Kelvin.

This law is very important, since most total radiation thermometers are based on it. If a receiving element at a temperature T_1 is arranged so that radiation from a source at a temperature T_2 falls on it, it will receive heat at the rate of σT_2^4 and emit it at a rate of σT_1^4. It will, therefore, gain heat at the rate of $\left(\sigma T_2^4 - T_1^4\right)$. If the temperature of the receiver is small in comparison with that of the source, T_1^4 may be neglected in comparison with T_2^4, and the radiant energy gained will be proportional to the fourth power of the temperature Kelvin of the radiator.

21.6.1.2 The Distribution of Energy in the Spectrum: Wien's Laws

When a body is heated, it appears to change color. This is because the total energy and distribution of radiant energy between the different wavelengths is changing as the temperature rises. When the temperature is about 500°C, the body is just visibly red. As the temperature rises, the body becomes dull red at 700°C, cherry red at 900°C, orange at 1,100°C, and finally white-hot at temperatures above 1,400°C. The body appears white-hot because it radiates all colors in the visible spectrum.

It is found that the wavelength of the radiation of the maximum intensity gets shorter as the temperature rises. This is expressed in Wien's displacement law:

$$\lambda_m T = \text{constant}$$
$$= 2898\,\mu\text{m} \cdot \text{K} \qquad (21.26)$$

where λ_m is the wavelength corresponding to the radiation of maximum intensity, and T is the temperature Kelvin. The actual value of the spectral radiance at the wavelength λ_m is given by Wien's second law:

$$L_{\lambda m} = \text{constant} \times T^5 \qquad (21.27)$$

where $L_{\lambda m}$ is the maximum value of the spectral radiance at any wavelength, that is, the value of the radiance at λ_m, and T is the temperature Kelvin. The constant does not have the same value as the constant in Equation (21.26). It is important to realize that it is only the maximum radiance at one particular wavelength that is proportional to T^5; the total radiance for all wavelengths is given by the Stefan-Boltzmann law; in other words, it is proportional to T^4.

Wien deduced that the spectral concentration of radiance, that is, the radiation emitted per unit solid angle per unit area of a small aperture in a uniform temperature enclosure in a direction normal to the area in the range of wavelengths between λ and $\lambda + \delta\lambda$ is $L\lambda \cdot \delta\lambda$, where

$$L_\lambda = \frac{C_1}{\lambda^5 \cdot e^{C_2/\lambda T}} \qquad (21.28)$$

where T is the temperature Kelvin, and C_1 and C_2 are constants. This formula is more convenient to use and applies with less than 1 percent deviation from the more refined Planck's radiation law used to define IPTS-68 provided $\lambda T < 3 \times 10^3\,\text{m} \cdot \text{K}$.

In 1900 Max Planck obtained from theoretical considerations, based on his quantum theory, the expression

$$L_\lambda = \frac{C_1}{\lambda^5 \left(e^{C_2/\lambda T} - 1\right)} \qquad (21.29)$$

where the symbols have the same meaning, and $C_2 = 0.014\,388\,\text{m} \cdot \text{K}$.

These laws also enable the correction to be calculated for the presence of an absorbing medium such as glass in the optical pyrometer as well as the correction required for changes in the spectral emissive power of the radiating surface.

The variation of spectral radiance with wavelength and temperature of a black-body source is given by Figure 21.43.

21.6.2 Radiation Thermometer Types

Since the energy radiated by an object is a function of its absolute temperature, this is a suitable property for the

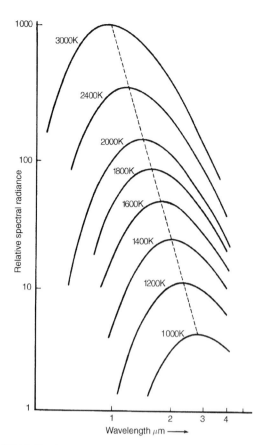

FIGURE 21.43 Spectral energy distribution with temperature.

FIGURE 21.44 General-purpose radiation thermometer. Courtesy of Land Infrared Ltd.

noncontact and nonintrusive measurement of temperature. Instruments for temperature measurement by radiation are called *radiation thermometers*. The terms *pyrometer* or *radiation pyrometer* were formerly used.

There are four principal techniques for the measurement of temperature by the radiation from a hot body: total radiation, pyroelectric, photo-electric, and optical.

Instruments using the first three of these techniques are normally constructed in the same general physical form. Figure 21.44 shows the general format of one of these instruments. It consists of a cylindrical metal body made of aluminum alloy, brass, or plastic. One end of the body carries a lens, which, depending on the wavelength range required, consists of germanium, zinc sulfide, quartz, glass, or sapphire. The opposite end carries the electrical terminations for connecting the sensing head to its signal conditioning module. A typical size of such a sensing head is 250 mm long by 60 mm diameter. A diagrammatic sketch of the construction of the instrument is shown in Figure 21.45. Infrared energy from a target area on the object whose temperature is to be measured is focused by the lens onto the surface of the detector. This energy is converted to an electrical signal that may be amplified by a head amplifier on the circuit board. Power is supplied to the instrument and the output transmitted down a cable, which is connected to terminals in the termination box. In instruments working in the near-infrared region, where the lens is transparent to visible light, a telescope can be provided, built into the instrument, so that it can be focused and aligned by looking through the lens.

A primary advantage of radiation thermometers, especially when they're used to measure high temperatures, is that the instrument measuring head can be mounted remotely from the hot zone in an area cool enough not to exceed the working temperature of the semiconductor electronics, typically about 50–75°C. However, where the instrument has to be near the hot region, such as attached to the wall of a furnace, or where it needs to be of rugged construction, it can be housed in an air- or water-cooled housing. Such a housing is shown in Figure 21.46.

The function of the lens as indicated here is to concentrate the radiation from the source onto the surface of the sensor. This also has the great advantage that the instrument reading is substantially independent of the distance from the source, provided that the source is large enough for its image to fully fill the area of the sensor. The lens material depends on the wavelength to be passed. This will normally be a function of the temperature range for which the instrument is specified. For lower temperatures the lens material will be chosen to give a wide wavelength bandpass. For higher temperatures a narrower bandpass may be acceptable. Of course, the higher the temperature to be measured, the shorter the wavelength that needs to be passed by the lens. Table 21.15 shows the wavelength bandpass of some lens materials.

To achieve a wider wavelength range, the focusing can be achieved with a concave mirror. Figure 21.47 shows diagrammatically the general arrangement of a reflection instrument.

A special application of mirror focusing for radiation thermometry is in the temperature measurement of stars and other astronomic bodies. The thermopile, or more usually a semiconductor detector, is cooled with liquid nitrogen or helium to increase its sensitivity to very small amounts of radiation. It is located at the focus of a reflecting astronomical telescope. The telescope is directed to the body whose temperature is to be measured so that its image

CHAPTER | 21 Temperature Measurement

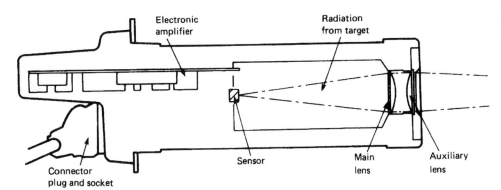

FIGURE 21.45 Diagram of radiation thermometer.

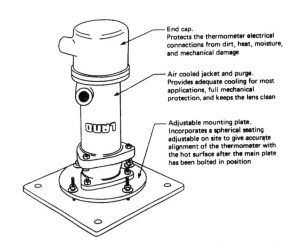

FIGURE 21.46 Air-cooled housing for radiation thermometer. Courtesy of Land Infrared Ltd.

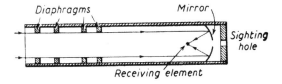

FIGURE 21.47 Mirror-focused radiation thermometer. Courtesy of Land Infrared Ltd.

is focused on the detector. The whole assembly forms a very sensitive radiation thermometer that has the ability to detect temperatures down to a few tens of Kelvins.

21.6.2.1 Total Radiation Thermometer

In this type of instrument, the radiation emitted by the body for which the temperature is required is focused on a suitable thermal-type receiving element. This receiving element may have a variety of forms. It may be a resistance element, which is usually in the form of a very thin strip of blackened platinum, or a thermocouple or thermopile. The change in temperature of the receiving element is then measured as has already been described.

In a typical radiation thermopile, a number of thermocouples made of very fine strips are connected in series and arranged side by side or radially, as in the spokes of a wheel, so that all the hot junctions, which are blackened to increase the energy-absorbing ability, fall within a very small target area. The thermoelectric characteristics of the thermopiles are very stable because the hot junctions are rarely above a few hundred degrees Celsius, and the thermocouples are not exposed to the contaminating atmosphere of the furnace. Stability and the fact that it produces a measurable EMF are the main advantages of the thermopile as a detector. In addition, thermopiles have the same response to incoming radiant energy regardless of wavelength within the range 0.3–20 μm. The main disadvantage of the thermopile is its comparatively slow speed of response, which depends on the mass of the thermocouple elements and the rate at which heat is transferred from the hot to the cold junctions. Increase in this rate of response can be attained only by sacrificing temperature difference, with a resultant loss of output. A typical industrial thermopile of the form shown in Figure 21.48 responds to 98 percent of a step change in incoming radiation in 2 seconds. Special thermopiles that respond within half a second are obtainable, but they have a reduced EMF output.

To compensate for the change in the thermopile output resulting from changes in the cold junction temperature, an ambient temperature sensor is mounted by the cold junctions. Alternative thermal detectors to thermopiles are also used. Thermistors and pyroelectric detectors are currently in use. The advantage of thermistors is that they can be very small and so have a quick speed of response. Their main disadvantage is their nonlinearity, though this is not so great a disadvantage as with a direct measurement of temperature, because provision has to be made to linearize the radiated energy signal anyway.

Correction for Emissivity When the temperature of a hot object in the open is being measured, due regard must be given to the correction required for the difference between the emissivity of the surface of the object and that of a perfect blackbody.

The total radiant flux emitted by the source will be given by

$$R = \varepsilon \sigma A T_a^4 \qquad (21.30)$$

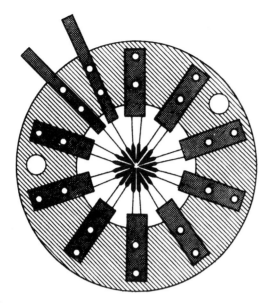

FIGURE 21.48 Thermopile for use in total radiation pyrometer.

where ε is the total emissivity of the body, A is the area from which radiation is received, σ is the Stefan–Boltzmann constant, and T the actual temperature of the body.

This flux will be equal to that emitted by a perfect blackbody at a temperature T_a, the apparent temperature of the body:

$$R = \sigma A T_a^4 \quad (21.31)$$

Equating the value of R in Equations (21.30) and (21.31):

$$\varepsilon \sigma A T^4 = \sigma A T_a^4$$

$$T^4 = \frac{T_a^4}{\varepsilon} \quad (21.32)$$

$$T = \frac{T_a}{\sqrt[4]{\varepsilon}}$$

The actual correction to be applied to the apparent temperature is given in Figure 21.49. Table 21.16 shows the emissivity of some metals at different temperatures.

The radiation from a hot object can be made to approximate much more closely to black-body radiation by placing a concave reflector on the surface. If the reflectivity of the reflecting surface is r, it can be shown that the intensity of the radiation that would pass out through a small hole in the reflector is given by

$$R = \frac{\varepsilon}{1 - r(1 - \varepsilon)} \sigma T^4 \quad (21.33)$$

where R is the radiation intensity through the hole, ε is the emissivity of the surface, σ is the Stefan–Boltzmann constant, and T the temperature in Kelvin. With a gold-plated hemisphere, the effective emissivity of a surface of emissivity 0.6 is increased by this method to a value of 0.97.

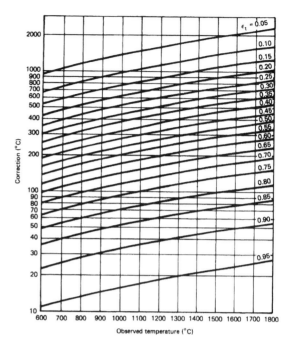

FIGURE 21.49 Emissivity corrections to the readings of a total radiation thermometer.

Surface Radiation Thermometer A surface radiation thermometer manufactured by Land Infrared Ltd. uses the above principle; see Figure 21.50. This instrument uses a thermopile sited on a small hole in a gold-plated hemisphere mounted on the end of a telescopic arm.

Gold is chosen for the reflecting surface because it is the best reflector of infrared radiation known and is not easily tarnished. The hole in the reflector is closed by a fluorite window, which admits a wide range of radiation to the thermopile but excludes dirt and draughts. This pyrometer will give accurate surface temperature readings for most surfaces other than bright or lightly oxidized metals, without any significant error due to surface emissivity changes. The standard instrument covers a temperature range of from 100°C to 1,300°C on three scales. A special low-temperature version is available for the range 0°C to 200°C. The indicator gives a reading in 5 to 6 seconds; the pyrometer should not be left on the hot surface for more than this length of time, particularly at high temperatures. The thermistor bridge provides compensation for changes in the sensitivity of the thermopile at high temperatures, but if the head is too hot to touch, it is in danger of damage to soldered joints, insulation, and the like.

The instrument may be used to measure the mean emissivity of a surface for all wavelengths up to about 10 µm. This value can be used for the correction of total radiation thermometer readings. A black hemispherical insert is provided with the instrument; it can be clipped into the hemispherical reflector to cover the gold. If two measurements are made, one with the gold covered and the other with the gold exposed, the emissivity can readily be deduced from the two measurements. A graph provided with the instrument enables

TABLE 21.16 Total emissivity of miscellaneous materials and total emissivity of unoxidized metals

Material	25°C	100°C	500°C	1,000°C	1,500°C	2,000°C
Aluminum	0.022	0.028	0.060	–	–	–
Bismuth	0.048	0.061	–	–	–	–
Carbon	0.081	0.081	0.079	–	–	–
Chromium	–	0.08	–	–	–	–
Cobalt	–	–	0.13	0.23	–	–
Columbium	–	–	–	–	0.19	0.24
Copper	–	0.02	–	(Liquid 0.15)	–	–
Gold	–	0.02	0.03	–	–	–
Iron	–	0.05	–	–	–	–
Lead	–	0.05	–	–	–	–
Mercury	0.10	0.12	–	–	–	–
Molybdenum	–	–	–	0.13	0.19	0.24
Nickel	0.045	0.06	0.12	0.19	–	–
Platinum	0.037	0.047	0.096	0.152	0.191	–
Silver	–	0.02	0.035	–	–	–
Tantalum	–	–	–	–	0.21	0.26
Tin	0.043	0.05	–	–	–	–
Tungsten	0.024	0.032	0.071	0.15	0.23	0.28
Zinc	(0.05 at 300°C)	–	–	–	–	–
Brass	0.035	0.035	–	–	–	–
Cast iron	–	0.21	–	(Liquid 0.29)	–	–
Steel	–	0.08	–	(Liquid 0.28)	–	–

Total emissivity εdR of miscellaneous materials

Material	Temp. (°C)	ε_t	Material	Temp. (°C)	ε_t
Aluminum (oxidized)	200	0.11	Lead (oxidized)	200	0.63
	600	0.19	Monel (oxidized)	200	0.43
Brass (oxidized)	200	0.61		600	0.43
	600	0.59	Nickel (oxidized)	200	0.37
Calorized copper	100	0.26		1,200	0.85
	500	0.26	Silica brick	1,000	0.80
Calorized copper (oxidized)	200	0.18		1,100	0.85
	600	0.19	Steel (oxidized)	25	0.80
Calorized steel (oxidized)	200	0.52		200	0.79
	600	0.57		600	0.79
Cast iron (strongly oxidized)	40	0.95	Steel plate (rough)	40	0.94
	250	0.95		400	0.97
Cast iron (oxidized)	200	0.64	Wrought iron (dull oxidized)	25	0.94
	600	0.78		350	0.94
Copper (oxidized)	200	0.60	20Ni–25Cr–55Fe (oxidized)	200	0.90
	1,000	0.60		500	0.97
Fire brick	1,000	0.75	60Ni–12Cr–28Fe (oxidized)	270	0.89
Gold enamel	100	0.37		560	0.82
	100	0.74		100	0.87
Iron (oxidized)	500	0.84	80Ni–20Cr (oxidized)	600	0.87
	1,200	0.89		1,300	0.89
Iron (rusted)	25	0.65			

Source: "Temperature: Its measurement and Control" in Science & Industry, American Institute of Physics, Reinhold Publishing Co. (1941).

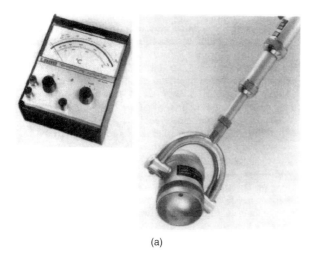

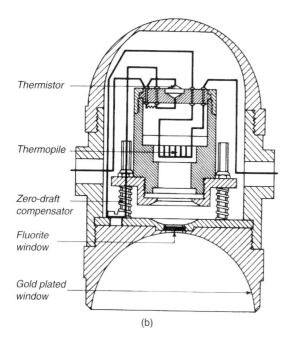

FIGURE 21.50 (a) Surface radiation thermometer. Courtesy of Land Infrared Ltd. (b) Cross-section diagram of Land surface radiation thermometer.

the emissivity to be derived easily from the two readings; a second graph gives an indication of the error involved in the temperature measurement of the hot body.

Calibration of Total Radiation Thermometers A total radiation thermometer may be calibrated by sighting it through a hole into a black-body enclosure of known temperature. A special spherical furnace was developed by the British Iron and Steel Research Association for this purpose. The furnace consisted of a sphere, 0.3 m in diameter, consisting of a diffusely reflecting material. For temperatures up to 1,300°C, stainless steel, 80 Ni 20 Cr alloy, or nickel may be used. For temperatures up to 1,600°C, silicon carbide is necessary; for temperatures up to 3,000°C, graphite may be used, provided that it is filled with argon to prevent oxidation. The spherical core is uniformly wound with a suitable electrical heating element, completely enclosed in a box containing thermal insulation. For calibration of radiation thermometers up to 1,150°C, a hole of 65 mm diameter is required in the cavity, but above this temperature a 45 mm hole is sufficient.

Where the larger hole is used, a correction for the emissivity of the cavity may be required for very accurate work. Two sheathed thermocouples are usually placed in the furnace, one near the back and the other just above the sighting hole. Comparison of the two measured temperatures indicates when the cavity is at a uniform temperature.

Calibration may be carried out by comparing the thermometer and thermocouple temperature, or the test thermometer may be compared with a standard radiation thermometer when both are sighted on to the radiating source, which may or may not be a true blackbody.

Cylindrical furnaces may also be used; here a thermocouple is fitted in the sealed end of the cylinder, which is cut on the inside to form a series of 45° pyramids.

A choice of three aperture sizes is available at the open end. For temperatures up to 1,100°C, the furnace is made of stainless steel, but for higher temperatures refractory materials are used. For further details, see *The Calibration of Thermometers* (HMSO, 1971) and BS 1041: Part 5: 1989. Figure 21.51 shows typical black-body furnaces.

Furnace Temperature by Radiation Thermometer Conditions in a furnace that might otherwise be considered perfect black-body conditions may be upset by the presence of flame, smoke, or furnace gases. In these conditions, a total radiation thermometer generally indicates a temperature between that of the furnace atmosphere and the temperature that would be indicated if such an atmosphere were not present. A thick luminous flame may shield the object almost completely. Nonluminous flames radiate and absorb energy only in certain wavelength bands, principally because of the presence of carbon dioxide and water vapor. The error due to the presence of these gases can be reduced by using a lens of Pyrex, which does not transmit some of these wavelengths, so that the instrument is less affected by variations in quantity of these gases. Where appreciable flame, smoke, and gas are present, it is advisable to use a closed-ended sighting tube or provide a purged sighting path by means of a blast of clean, dry air.

Errors in temperature measurement can also occur owing to absorption of radiation in the cold atmosphere between a furnace and the thermometer. To ensure that the error from this source does not exceed 1 percent of the measured temperature, even on hot, damp days, the distance between thermometer lens and furnace should not exceed 1.5 m if a glass lens is used, 1m if the lens is silica, and 0.6 m if it is of fluorite.

CHAPTER | 21 Temperature Measurement

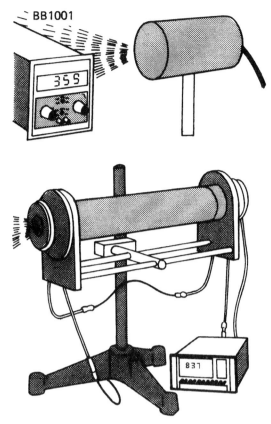

FIGURE 21.51 Blackbody radiators. Courtesy of Polarisers Technical Products.

21.6.2.2 Pyroelectric Techniques

Pyroelectric detectors for thermal radiation are a comparatively recent introduction. Pyroelectric materials, mainly ceramics, are materials for which the molecules have a permanent electric dipole due to the location of the electrons in the molecules. Normally these molecules lie in a random orientation throughout the bulk of the material so that there is no net electrification. Also, at ambient temperatures the orientations of the molecules are essentially fixed. If the temperature is raised above some level characteristic to the particular material, the molecules are free to rotate. This temperature is called the *Curie temperature* by analogy with the magnetic Curie temperature.

If a piece of pyroelectric ceramic is placed between two electrodes at ambient temperature, the molecular dipoles are fixed in a random orientation, as shown in Figure 21.52(a). If it is then heated above its Curie temperature and an electrical potential applied to the electrodes, thus generating an electric field in the ceramic, the molecules will all align themselves parallel to the field, as shown in Figure 21.52(b). On cooling the ceramic back to ambient temperature and then removing the applied potential, the molecules remain aligned, as shown in Figure 21.52(c). The amount of the polarization of the ceramic and therefore the magnitude of

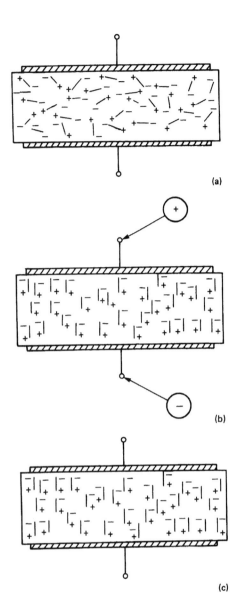

FIGURE 21.52 Pyroelectric effect.

the resulting external electric field is a constant Σ, which is a function of the material. If the field due to the applied voltage was E and the polarization P, then

$$P = \Sigma E \tag{21.34}$$

If the temperature of the polarized pyroelectric ceramic is raised, the molecular dipoles, which are anyway oscillating about their parallel orientation, will oscillate through a greater angle. Figure 21.53 shows one molecular dipole of length x and charge $\pm q$. Its electric moment is qx. If, then, the dipole oscillates through an average angle of $\pm\theta$, the effective length will be z, where

$$z = x \cos\theta \tag{21.35}$$

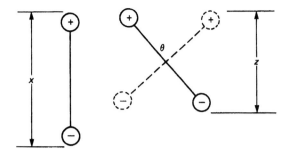

FIGURE 21.53 Mechanism of pyroelectric effect.

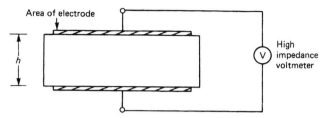

FIGURE 21.54 Pyroelectric detector.

The angle θ will increase with increasing temperature, thus reducing the electric moment of all the molecular dipoles. The electric moment or polarization of the whole piece of pyroelectric ceramic is, of course, the sum of all the molecular dipoles. Thus as the temperature rises, the polarization of the whole piece of material becomes less.

The *Curie point* is the temperature at which the oscillatory energy of the molecular dipoles is such that they can rotate freely into any position, allowing them to return to their random orientation.

As stated earlier, the electric moment M of the whole slice of ceramic is the sum of all the molecular dipole moments:

$$M = PAh \qquad (21.36)$$

where P is the dipole moment per unit volume, h is the thickness of the slice, and A is the electrode area (see Figure 21.54).

If the electric charge at the two surfaces of the slice of pyroelectric ceramic is Q_s, this has a dipole moment of $Q_s \cdot h$, so

$$Q_s = PA \qquad (21.37)$$

If the temperature of the material rises, the polarization is reduced and therefore Q_s becomes less. But if the electrodes are connected by an external circuit to an electrometer or other high-impedance detector, Q_s is normally neutralized by a charge Q on the electrodes. A reduction of Q_s therefore results in an excess charge on the electrodes, and therefore a voltage V is detected.

$$V = \frac{Q}{C} \qquad (21.38)$$

where C is the electrical capacitance of the device; for a temperature change of δT the change of charge δQ is given by

$$\delta Q = \Omega \cdot A \cdot \delta T \qquad (21.39)$$

where Ω is the pyroelectric coefficient of the material. Therefore the voltage change will be

$$\delta V = \frac{\delta Q}{C} = \Omega A \frac{\delta T}{C} \qquad (21.40)$$

where C is the electrical capacitance between the electrodes. The pyroelectric coefficient Ω is a function of temperature reducing with a nonlinear characteristic to zero at the Curie temperature.

When used as a detector in a radiation thermometer, radiation absorbed at the surface of the pyroelectric slice causes the temperature of the detector to rise to a new higher level. At the start the charge on the electrodes will have leaked away through the external electrical circuit, so there will have been zero voltage between the electrodes. As the slice heats up a voltage is detected between the two electrodes. When the device reaches its new temperature, losing heat to its environment at the same rate as it is receiving heat by radiation, the generation of excess charge on the electrodes ceases, the charge slowly leaks away through the electrical circuit, and the detected voltage returns to zero. The device detects the change of incident radiation. To detect a constant flux of radiation, that is, to measure a constant temperature, it is necessary to "chop" the incident radiation with a rotating or oscillating shutter.

The physical construction of a pyroelectric radiation thermometer is essentially identical to a total radiation instrument except for the location of the radiation-chopping shutter just in front of the detector. Figure 21.55(a) shows the location and Figure 21.55(b) a typical profile of the optical chopper in a pyroelectric radiation thermometer.

Figure 21.55(c) shows the graph against time of the chopped radiation, together with the resulting electrical signal.

21.6.2.3 Optical (Disappearing Filament) Thermometer

Optical radiation thermometers provide a simple and accurate means for measuring temperatures in the range 600°C to 3,000°C. Since their operation requires the eye and judgment of an operator, they are not suitable for recording or control purposes. However, they provide an effective way of making spot measurements and for calibration of total radiation thermometers.

In construction an optical radiation thermometer is similar to a telescope. However, a tungsten filament lamp is placed at the focus of the objective lens. Figure 21.56 shows the optical arrangement of an optical radiation thermometer.

CHAPTER | 21 Temperature Measurement

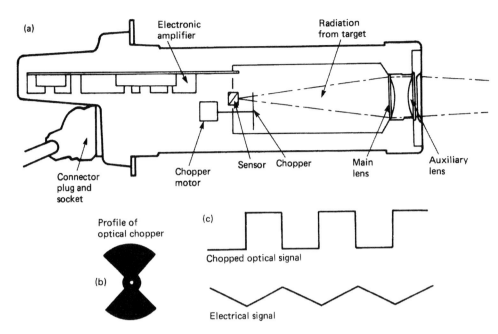

FIGURE 21.55 Diagram of pyroelectric radiation thermometer.

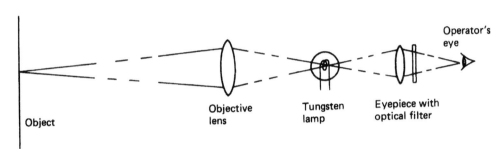

FIGURE 21.56 Optical system of disappearing filament thermometer.

To use the instrument, the point at which the temperature is required to be known is viewed through the instrument. The current through the lamp filament is adjusted so that the filament disappears in the image. Figure 21.57 shows how the filament looks in the eyepiece against the background of the object, furnace, or whatever is to have its temperature measured. At (a) the current through the filament is too high and it looks bright against the light from the furnace; at (c) the current is too low; at (b) the filament is at the same temperature as the background. The temperature of the filament is known from its electrical resistance. Temperature readout is achieved either by a meter measuring the current through the filament or by temperature calibrations on the control resistor regulating the current through the lamp. The filter in the eyepiece shown in Figure 21.56 passes light at a wavelength around 0.65 μm.

Lamps for optical thermometers are not normally operated at temperatures much in excess of 1,500°C. To extend the range of the instrument beyond this temperature, a neutral filter of known transmission factor can be placed in the light path before the lamp. The measurement accuracy of an optical thermometer is typically ±5°C between 800°C and 1,300°C and ±10°C between 1,300°C and 2,000°C.

Corrections for Nonblack-Body Conditions Like the total radiation thermometer, the optical thermometer is affected by the emissivity of the radiation source and by any absorption of radiation, which may occur between the radiation source and the instrument.

The spectral emissivity of bright metal surfaces at 0.65 μm is greater than the total emissivity ε, representing the average emissivity over all wavelengths. The correction required for the departure from black-body conditions is therefore less than in the case of total radiation thermometers.

Due to the fact that a given change of temperature produces a much larger change in radiant energy at 0.65 μm than produced in the average of radiant energy overall wavelengths, the readings of an optical radiation thermometer require smaller corrections than for a total radiation instrument.

The relationship between the apparent temperature T_a and the true temperature T is given by Equation (21.41) which is based on Wien's law,

$$\frac{1}{T} - \frac{1}{T_a} = \frac{\lambda \log_{10} \varepsilon_\lambda}{6245} \tag{21.41}$$

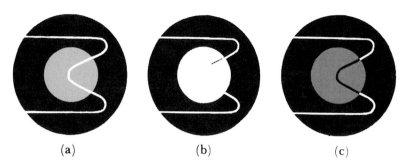

FIGURE 21.57 Appearance of image in optical thermometer.

where λ is the wavelength in micrometers (usually 0.65 μm) and ε_λ is the spectral emissivity at wavelength λ.

21.6.2.4 Photoelectric Radiation Thermometers

The reading obtained with an optical thermometer shows a lower temperature error than a total radiation thermometer. This is because the emissivity error for a given temperature and a known emissivity is proportional to the wavelength of the radiation used to make the measurement. For instance, in the case of oxidized steel at 1,000°C with an emissivity of 0.8, a total radiation thermometer will have an error in excess of 50 degrees, whereas the optical thermometer reading will be within 20 degrees. However, the optical thermometer has two major drawbacks. First, it is only suitable for spot measurements and requires a skilled operator to use it. Second, it is not capable of a quick response and is totally unsuitable for control purposes.

Photoelectric radiation thermometers are ideally suited to the short wavelength application. Structurally they are essentially identical to a total radiation thermometer except that the thermal sensor is replaced by a photodiode.

A photodiode is a semiconductor diode that may be either a silicon or germanium junction diode constructed so that the incident radiation can reach the junction region of the semiconductor. In the case of germanium, the diode will be a plain P–N junction; in the case of silicon it may be either a P–N or P–I–N junction. In service the diodes are operated with a voltage applied in the reverse, that is, nonconduction, direction. Under these conditions the current carriers, that is, electrons, in the semiconductor do not have sufficient energy to cross the energy gap of the junction. However, under conditions of incident radiation, some electrons will gain enough energy to cross the junction. They will acquire this energy by collision with photons. The energy of photons is inversely proportional to the wavelength. The longest wavelength of photons that will, on impact, give an electron enough energy to cross the junction dictates the long wave end of the spectral response of the device.

The short wavelength end of the response band is limited by the transparency of the semiconductor material. The choice of germanium or silicon photodiodes is dictated by the temperature and therefore the wavelength to be measured. Silicon has a response of about 1.1 μm to 0.4 μm. The useful bandpass of germanium lies between 2.5 μm and 1.0 μm. The exact bandpass of photodiodes varies somewhat from type to type, depending on the manufacturing process used, but the preceding figures are typical. Normally the range of wavelengths used is reduced to a narrower bandpass than that detected by the semiconductor sensor. For instance, for general applications above 600°C, a narrow bandpass centered on 0.9 μ is usually used. Wherever possible, silicon is preferred, since it will tolerate higher ambient temperatures than germanium and in general it has the higher speed of response. Small P–I–N photodiodes can have a frequency response up to several hundred megahertz; P–N devices more usually have a response of several kilohertz. Like all other semiconductor devices, the electrical output of photodiodes is temperature dependent. It is therefore necessary to construct these radiation thermometers with thermistors or resistance thermometers in close proximity to the photodiode to provide ambient temperature compensation.

21.6.2.5 Choice of Spectral Wavelength for Specific Applications

It might seem at first sight that apart from optical radiation thermometers, the obvious choice should be to use a total radiation thermometer so as to capture as much as possible of the radiant emission from the target, to achieve the maximum output signal. However, as already mentioned, except at the lowest temperature ranges, there are several reasons for using narrower-wavelength bands for measurement.

Effect of Radiant Emission Against Wavelength One reason relates to the rate at which the radiant emission increases with temperature. An inspection of Figure 21.58 will show that the radiant emission at 2 μm increases far more rapidly with temperature than it does at, say, 6 μm. The rate of change of radiant emission with temperature is always greater at shorter wavelengths. It is clear that the greater this rate of change, the more precise the temperature measurement and the tighter the temperature control. On the other hand, this cannot be carried to extremes, because at a given short wavelength there is a lower limit to the temperature that can be measured. For example, the eye becomes useless below about 600°C. For these reasons alone, we can understand the general rule that the spectral range of the appropriate infrared thermometer shifts to longer wavelengths as the process temperature decreases.

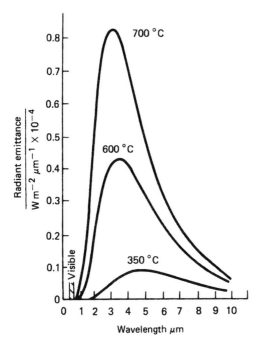

FIGURE 21.58 Black-body radiation characteristics.

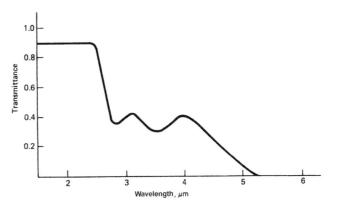

FIGURE 21.59 Transmittance of one millimeter of soda-lime glass.

Emittance, Reflectance, and Transmittance Another important reason for the use of different spectral regions relates to the specific emission characteristics of particular target materials.

The curves of Figure 21.58 show the emission characteristics of the ideal emitter or blackbody. No material can emit more strongly than a blackbody at a given temperature. As discussed previously, however, many materials can and do emit less than a blackbody at the same temperature in various portions of the spectrum. The ratio of the radiant emittance at wavelength λ of a material to that of a blackbody at the same temperature is called *spectral emittance* ($\varepsilon\lambda$). The value of $\varepsilon\lambda$ for the substance can range between 0 and 1 and may vary with wavelength. The emittance of a substance depends on its detailed interaction with radiation. A stream of radiation incident on the surface of a substance can suffer one of three fates: A portion may be reflected. Another portion may be transmitted through the substance. The remainder will be absorbed and degraded to heat.

The sum of the fraction reflected r, the fraction transmitted t, and the fraction absorbed a will be equal to the total amount incident on the substance. Furthermore, the emittance ε of a substance is identical to the absorptance a, and we can write

$$\varepsilon = a = 1 - t - r \qquad (21.42)$$

For the blackbody, the transmittance and reflectance are zero and the emittance is unity. For any opaque substance the transmittance is zero and

$$\varepsilon = 1 - r \qquad (21.43)$$

An example of this case is oxidized steel in the visible and near-infrared, where the transmittance is 0, the reflectance is 0.20, and the emittance is 0.80. A good example of a material for which emittance characteristics change radically with wavelength is glass. Figure 21.59 shows the overall transmission of soda-lime glass. The reflectance of the glass is about 0.03 or less through most of the spectral region shown. At wavelengths below about 2.6 μm, the glass is very highly transparent and the emittance is essentially zero. Beyond 2.6 μm the glass becomes increasingly opaque. From this it is seen that beyond 4 μm, glass is completely opaque and the emittance is above 0.98.

This example of glass clearly illustrates how the detailed characteristics of the material can dictate the choice of the spectral region of measurement. For example, consider the problem of measuring and controlling the temperature of a glass sheet during manufacture at a point where its temperature is 900°C. The rule that suggests a short-wavelength infrared thermometer because of the high temperature obviously fails. To use the region around 1 μm would be useless because the emittance is close to 0. Furthermore, since the glass is highly transparent, the radiation thermometer will "see through" the glass and can give false indications because of a hot wall behind the glass. One can recognize that glass can be used as an effective "window" with a short-wavelength radiation thermometer. By employing the spectral region between 3 and 4 μm, the internal temperature of the glass can be effectively measured and controlled. By operating at 5 μm or more, the surface temperature of the glass is measured. Each of these cases represents a practical application of infrared thermometry.

Atmospheric Transmission A third important consideration affecting the choice of spectral region is that of the transmission of the atmosphere between the target substance and the radiation thermometer. The normal atmosphere always contains a small but definite amount of carbon dioxide and a variable amount of water vapor. Carbon dioxide strongly absorbs radiation between 4.2 and 4.4 μm, and the water vapor absorbs strongly between 5.6 and 8.0 μm and also somewhat in the regions 2.6 to 2.9 μm (see Figure 21.60). It is obvious that these spectral regions should be avoided, particularly in the region of the water bands. If this is not done the temperature calibration will vary with path length and with humidity. If the air temperature is comparable to or higher than the target temperature, the improperly designed infrared thermometer could provide temperature measurements strongly influenced by air temperatures.

21.6.2.6 Signal Conditioning for Radiation Thermometers

Although the output of a radiation thermometer can be used directly in a voltage- or current-measuring instrument, this is unsatisfactory for two prime reasons. First, the energy radiated by a hot body is a function of the fourth power of absolute temperature, resulting in a very nonlinear scale. Second, the radiation detectors are themselves sensitive to ambient temperature. This requires either that the radiation thermometer be maintained at a constant temperature or, alternatively, that an ambient temperature sensor is mounted beside the radiation sensor to provide a signal for temperature correction.

To compensate for these two deficiencies in the signal, suitable electronic circuits must be used to provide linearization of the signal and to provide automatic temperature correction. It is also necessary to provide correction for the emissivity of the target. Typically the instrument itself carries a small "head amplifier" to bring the signal up to a suitable level for transmission to the readout instrument. This head amplifier also provides the required ambient temperature compensation circuits. The linearization and compensation for emissivity are provided at the readout module.

Some modern instruments provide the whole signal conditioning circuitry in the main instrument itself. Figure 21.61 shows such an instrument. In this equipment the output is a 4- to 20-milliamp signal linear with temperature and compensated for ambient temperature.

With the growing use of microprocessors in instrumentation, several manufacturers are introducing instruments in which the linearization and compensation are performed by a microcomputer.

21.6.2.7 Radiation Thermometer Applications

Infrared thermometers are currently used in a wide range of laboratory and industrial temperature control applications. A few low-temperature examples include extrusion, lamination and drying of plastics, paper and rubber, curing of resins, adhesives and paints, and cold-rolling and forming of metals.

Some high-temperature examples include forming, tempering, and annealing of glass; smelting, casting, rolling, forging, and heat treating of metals; and calcining and firing of ceramics and cement.

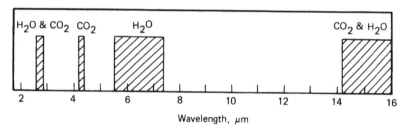

FIGURE 21.60 Atmospheric absorption of infrared radiation.

FIGURE 21.61 Radiation thermometer transmitter. Courtesy of Sirius Instruments Ltd.

In short, the infrared thermometer can be used in almost any application in the range 0 to 3,600°C where its unique capabilities can turn a seemingly impossible measurement and control problem into a practical working process. Many processes now controlled manually can be converted into continuous, automated systems.

21.7 TEMPERATURE MEASUREMENT CONSIDERATIONS

21.7.1 Readout

21.7.1.1 Local Readout

If temperature needs to be measured at a particular point in, say, a chemical plant, what considerations govern the choice of instrument? The obvious first choice to most people is a liquid-in-glass thermometer. However, this requires that one must be able to get close enough to read the instrument accurately. A better solution is a dial thermometer. The type of instrument chosen will of course depend on the accuracy and repeatability required. In general, and especially on bigger plants, local temperature measurement is for general surveillance purposes only; the measurement is probably not essential but is provided as a cross-check on the control instruments to provide operator confidence. An inexpensive bimetal thermometer is probably adequate. If greater accuracy is required, a capillary-type thermometer (see Sections 21.3.2–21.3.4) with a short capillary can be used, or where high accuracy is necessary, an electrical technique may be specified. In the case of furnaces, a portable radiation instrument may be the best choice.

Of course, on small plants not controlled from a separate control room, all measurements will probably be local measurements. It is mainly in this situation that the higher-accuracy local readout is required.

21.7.1.2 Remote Reading Thermometers

The first question to ask in the selection of remote reading instruments is, what is the distance between the measurement point and the readout location? If that distance is less than, say, 100 meters, capillary instruments may well be the best solution. However, if the distance is near the top limit, vapor pressure instruments will probably be ruled out. They may also not be usable if there is likely to be big ambient temperature variation at the readout point or along the length of the capillary.

The next question is, what is the height difference between the thermometer bulb and the readout position? Long vertical runs using liquid-in-metal thermometers can cause measurement offsets due to the liquid head in the vertical capillary adding to (or the subtracting from) the pressure at the instrument Bourdon tube. In the case of height differences greater than, say, 10 meters, liquid thermometers are likely to be unsuitable. This then reduces the choice to gas-filled instruments. A further consideration when specifying instrumentation on a new plant is that it is convenient from itinerary considerations to use as many instruments of the same type as possible. The choice of instrument is then dictated by the most stringent requirement.

On large installations where many different types of instrument are being installed, especially where pneumatic instrumentation is used, capillary instruments can run into an unexpected psychological hazard. Not infrequently a hard-pressed instrument technician, on finding he has too long a capillary, has been known to cut a length out of the capillary and rejoin the ends with a compression coupling. The result is, of course, disaster to the thermometer. Where on installation the capillary tube is found to be significantly too long, it must be coiled neatly in some suitable place. The choice of that place may depend on the type of instrument. In gas-filled instruments the location of the spare coil is irrelevant, but especially with vapor pressure instruments it wants to be in a position where it will receive the minimum of ambient temperature excursions to avoid introducing measurement errors.

For installations with long distances between the point of measurement and the control room, it is almost essential to use an electrical measurement technique. For long runs, resistance thermometers are preferred over thermocouples for two principal reasons. First, the copper cables used for connecting resistance bulbs to their readout equipment are very much less expensive than thermocouple wire or compensating cable. Second, the resistance thermometer signal is a higher level and lower impedance than most thermocouple signals and is therefore less liable to electrical interference.

An added advantage of electrical measurements is that, whether the readout is local or remote, the control engineer is given wider options as to the kinds of readout available. Not only there is a choice of analog or digital readout, there is a wider range of analog readouts, since they are not limited to a rotary dial.

21.7.1.3 Temperature Transmitters

On large installations or where a wide variety of measurements are being made with a wide range of instrumentation, it is more usual to transfer the signal from the measurement point to the control area by means of temperature transmitters. This has the great advantage of allowing standardization of the readout equipment. Also, in the case of electrical transmission by, say, a 4–20 milliamp signal, the measurement is much less liable to degradation from electrical interference. Furthermore, the use of temperature transmitters allows the choice of measurement technique to be unencumbered by considerations of length of run to the readout location.

The choice of electrical or pneumatic transmission is usually dictated by overall plant policy rather than the needs of the particular measurement, in this case temperature. However, where the requirement is for electrical temperature measurement for accuracy or other considerations, the transmission will also need to be electrical. (See Part 4, Chapter 29.)

21.7.1.4 Computer-Compatible Measurements

With the increasing use of computer control of plants, there is a requirement for measurements to be compatible. The tendency here is to use thermocouples, resistance thermometers, or, where the accuracy does not need to be so high, thermistors as the measuring techniques. The analog signal is either transmitted to an interface unit at the control room or to interface units local to the measurement. The latter usually provides for less degradation of the signal.

Because most industrial temperature measurements do not require an accuracy much in excess of 0.5 percent, it is usually adequate for the interface unit to work at 8-bit precision. Higher precision would normally be required only in very special circumstances.

21.7.1.5 Temperature Controllers

Although thermometers, in their widest sense of temperature measurement equipment, are used for readout purposes, probably the majority of temperature measurements in industrial applications are for control purposes. There are therefore many forms of dedicated temperature controllers on the market. As briefly described in Section 21.3.5.1, the simplest of these is a thermostat.

Thermostats A *thermostat* is a device in which the control function, usually electrical contacts but sometimes some other control function such as a valve, is directly controlled by the measurement action. The instrument described in Section 21.3.5.1 uses solid expansion to operate electrical contacts, but any of the other expansion techniques may be used. In automotive applications the thermostat in an engine cooling system is a simple valve directly operated either by vapor pressure or change of state, such as the change of volume of wax when it melts.

Thermostats, however, are very imprecise controllers. In the first place, their switching differential (the difference in temperature between switch-off and switch-on) is usually several Kelvin. Second, the only adjustment is setpoint.

Contact Dial Thermometers A first improvement on a thermostat is the use of a contact dial thermometer. The dial of this instrument carries a second pointer, the position of which can be set by the operator. When the indicating pointer reaches the setpoint pointer, they make electrical contact with one another. The current that then flows between the pointers operates an electrical relay, which controls the load. In this case the switching differential can be very small, typically a fraction of a Kelvin.

Proportional Temperature Controllers Dedicated one-, two-, or three-term temperature controllers are available in either pneumatic or electronic options. The use of such controllers is mainly confined to small plants where there is a cost advantage in avoiding the use of transmitters.

In the case of pneumatic controllers, the input measurement will be liquid, vapor pressure, or gas expansion. The Bourdon tube or bellows used to measure the pressure in the capillary system operates directly on the controller mechanism.

However, in recent years there has been an enormous increase in the number of electronic temperature controllers. The input to these instruments is from either a thermocouple or a resistance thermometer. The functions available in these controllers vary from on/off control to full three-term proportional, integral, and derivative operation. Some of the more sophisticated electronic controllers use an internal microprocessor to provide the control functions. Some units are available with the facility to control several temperature control loops. Of course, the use of an internal microprocessor can make direct computer compatibility a simple matter.

21.7.2 Sensor Location Considerations

To obtain accurate temperature measurement, careful consideration must be given to the siting of temperature-sensing probes. Frequently in industrial applications, temperature-measuring equipment does not live up to the expectations of the plant design engineer. The measurement error is not infrequently 10 or even 20 times the error tolerance quoted by the instrument manufacturer.

Large measurement errors in service may be due to the wrong choice of instrument, but more frequently the error is due to incorrect location of the measurement points. Unfortunately, the location of temperature sensors is dictated by the mechanical design of the plant rather than by measurement criteria.

21.7.2.1 Immersion Probes

To minimize errors in the measurement of the temperature of process fluids, whether liquid or gas, it is preferable to insert the sensor so that it is directly immersed in the fluid. The probe may be directly dipped into liquid in an open vessel, inserted through the wall of the vessel, or inserted into a pipe.

Measurement of Liquid in Vessels Temperature measurement of liquid in a plant vessel may illustrate the dilemma of the control engineer faced with mechanical problems. Consider Figure 21.62, which represents a vessel filled with liquid and stirred by a double-anchor agitator. The ideal place to measure the temperature would be somewhere near the center of the mass at, say, T1. The best arrangement would seem to be a dip probe T2. But even though the design level of the liquid is at A, in operation the liquid level may fall as low as B, leaving probe T2 dry. The only remaining possibility is T3. This is not a very good approach

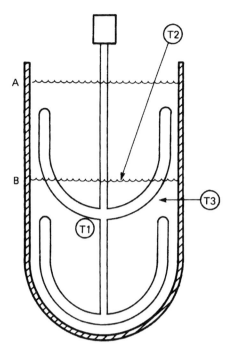

FIGURE 21.62 Problems associated with temperature measurement in a stirred vessel.

to T1 and is subject to error due to conduction of heat from or to the vessel wall.

An approach that can be used if the temperature measurement is critical is to mount a complete temperature-measuring package onto the shaft of the agitator. Wires are then brought up the shaft out of the vessel, from which the temperature signal can be taken off with slip rings, inductively coupled, or radio-telemetered to a suitable receiver. This is, of course, possible only where the temperature of the process is within the operating range of the electronics in the measurement package. The use of slip rings is not very satisfactory because they add unreliability, but in the absence of slip rings the package must also carry its own power supply in the form of batteries.

Probes in Pipes or Ducts There is frequently a requirement to measure the temperature of a fluid flowing in a pipe. This is usually straightforward, but there are still points to watch out for. Figure 21.63 shows three possible configurations for insertion into a pipe. The most satisfactory arrangement is to insert the thermometer probe into the pipe at a bend or elbow. Figure 21.63(a) shows this arrangement. Points to note are:

- To ensure that the probe is inserted far enough for the sensitive length to be wholly immersed and far enough into the fluid to minimize thermal conduction from the sealing coupling to the sensor.
- To insert the probe into the direction of flow as indicated. The reasons for this are to keep the sensor ahead of the turbulence at the bend, which could cause an error due to

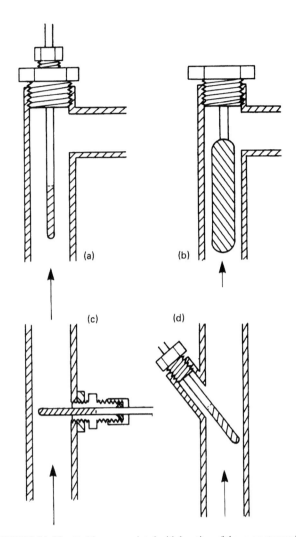

FIGURE 21.63 Problems associated with location of thermometer probe in pipe: (a) preferred arrangement, (b) probe obstructing pipe, (c) sensitive area of probe not fully immersed, (d) alternative preferred arrangement, sensitive portion of probe shaded.

local heating, and to remove the effects of cavitation that could occur at the tip of a trailing probe. Figure 21.63(b) shows the problem that can arise in small pipes where the probe can cause serious obstruction to the flow.

Where it is not possible to put the thermometer at a bend in the pipe, it can be inserted radially, provided that the pipe is big enough. Great care should be taken to ensure complete immersion of the sensitive portion of the probe. Figure 21.63(c) illustrates this problem. A better solution is diagonal insertion as shown at (d). Again the probe should point into the direction of flow.

In measuring temperature in large pipes or ducts, it must be remembered that the temperature profile across the pipe might not be constant. This is especially true for large flue stacks and air-conditioning ducts. The center liquid or gas is usually hotter (or colder, in refrigerated systems) than that at the duct wall. In horizontal ducts carrying slow-moving air or gas, the gas at the top of the duct will be significantly hotter than that at the bottom of the duct. In these circumstances careful consideration must be given as to how a representative measurement can be obtained; it may well be necessary to make several measurements across the duct and average the readings.

21.7.2.2 Radiation Errors

Gas temperature measurements present extra problems compared with temperature measurements in liquids. The difficulties arise from two sources. First, the relatively low thermal conductivity and specific heat of gases result in a poor heat transfer from the gas to the sensing element. This results in a slow response to temperature changes. Second, since most gases are transparent to at least a substantial part of the thermal radiation spectrum, significant measurement errors are likely to occur, as mentioned in Section 21.6. Consider a thermometer bulb inserted into a pipe containing a gas stream. The walls of the pipe or duct are likely to be at a different temperature to the gas—probably, but not necessarily, cooler. This means that although the thermometer is being warmed by receiving heat by contact with the gas, it is also losing heat by radiation to the pipe wall, and if the wall is cooler than the gas, the thermometer will lose more heat than it receives and will therefore register a lower temperature than the true gas temperature. Likewise, if the pipe wall is hotter than the gas, the thermometer reading will be too high.

This error can be reduced by surrounding the sensitive part of the thermometer probe with a cylindrical shield with its axis parallel to the pipe axis. This shield will reach a temperature intermediate between that of the pipe wall and that of the gas (see Figure 21.64). Where more precise measurements are required, an active shield may be employed. In this case a second thermometer is attached to the shield, which is also provided with a small heater. This heater's output is controlled via a controller so that the two thermometers, the one in the gas and the one on the shield, always indicate identical temperatures. In this state the thermometer will be receiving exactly the same amount of radiation from the shield as it radiates back to the shield. Figure 21.65 shows this arrangement.

21.7.2.3 Thermometer Pockets and Thermowells

The direct immersion of temperature-sensing probes into process fluid, although the optimum way to get an accurate measurement, has its disadvantages. First, it has disadvantages from the maintenance point of view: Normally the sensing probe cannot be removed while the plant is on-stream. Second, in the case of corrosive process streams, special corrosion-resistant materials might need to be used. Standard temperature gauges are normally available in only a limited range of materials, typically brass, steel, stainless steel, or ceramic, so a sheath or thermometer pocket or thermowell can be used to protect the temperature-sensing probe.

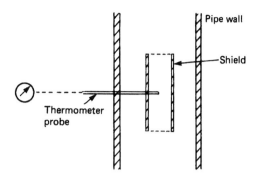

FIGURE 21.64 Radiation shield for gas temperature measurement.

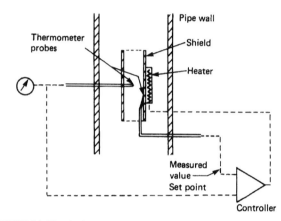

FIGURE 21.65 Active radiation shield.

The use of a thermometer pocket does degrade the measurement accuracy of the instrumentation.

Figure 21.66 shows a thermometer pocket mounted in the wall of a steam-jacketed process vessel. The thermometer probe receives heat from the wall of the pocket by conduction where it touches it and by radiation at other places. The inner wall of the pocket receives heat from the process fluid and by conduction in this case from the steam jacket of the vessel. In the case of a short pocket, the heat conducted along the pocket can cause a significant measurement error, causing too high a reading. In the situation where the outer jacket of the vessel is used for cooling the vessel, for example, a cooling water jacket, the heat flow will be away from the sensing probe and consequently the error will be a low measurement. This conduction error is significant only where the thermometer pocket is short or where the pocket is inserted into a gas stream. To minimize the error, the length of the pocket should be at least three times the length of the sensitive area of the probe.

The use of a thermowell or pocket will also slow the speed of response of an instrument to temperature changes. A directly immersed thermometer probe will typically reach thermal equilibrium within 30 to 90 seconds. However, the same probe in a thermometer pocket may take several minutes to reach equilibrium. This delay to the instrument

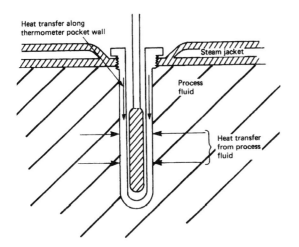

FIGURE 21.66 Thermometer pocket or thermowell.

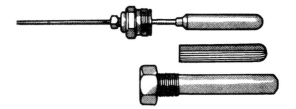

FIGURE 21.67 Taylor thermospeed separable well system. Courtesy of ABB Instrument Group.

response can be improved in those cases where the pocket is mounted vertically pointed downward, or in any position where the closed end is generally lower than the mouth, by filling it with a heat-transfer liquid. This liquid is usually a silicone oil.

An alternative method for improving the rate of heat transfer between the pocket and the bulb is illustrated in Figure 21.67. A very thin corrugated aluminum or bronze sleeve is inserted between the bulb and pocket on one side. This forces the bulb over to the other side, ensuring metal-to-metal contact on this side; on the other side, the sleeve itself, being made of aluminum, which has a high thermal conductivity, provides a reasonable path for the heat. In addition, the bulb should be placed well down the pocket to reduce the possibility of errors due to heat conducted by the pocket to the outside, with consequent reduction of the temperature at the bulb.

The errors associated with thermal conduction along the thermometer pocket are, of course, more critical in the case of gas temperature measurement, since the thermal transfer from gas to thermometer is not nearly as good as it is from liquid.

21.7.2.4 Effect of Process Fluid Flow Rate

Two sources of error in temperature measurement are clearly identified.

Fractional Heating Where the process fluid flows past a probe at high velocity there is, especially in the case of gases, a frictional heating effect. The magnitude of the effect is not easily evaluated, but it is advisable, if possible, to site the probe at a location where the fluid velocity is low.

Conductive Cooling Resistance thermometers and thermistors depend for their operation on an electric current flowing through them. This current causes a small heating effect in the sensor. When such a sensor is used for liquid temperature measurement, the relatively high specific heat of most liquids ensures that this heat is removed and the sensor temperature is that of the liquid. However, in gas measurement the amount of heat removed is a function of the gas velocity and thus a variable source of error can arise dependent on flow rate. In a well-designed instrument, this error should be very small, but it is a potential source of error to be borne in mind.

Cavitation Liquid flowing past a thermometer probe at high speed is liable to cause cavitation at the downstream side of the probe. Apart from any heating effect of the high flow rate, the cavitation will generate noise and cause vibration of the probe. In due course this vibration is likely to cause deterioration or premature catastrophic failure of the probe.

21.7.2.5 Surface Temperature Measurement

Where the temperature of a surface is to be measured, this can be done either with a temperature probe cemented or clamped to the surface or, where a spot measurement is to be made, a sensor can be pressed against the surface. In the former arrangement, which is likely to be a permanent installation, the surface in the region of the sensor itself can be protected from heat loss by lagging with thermally insulating material. Provided that heat losses are minimized, the measurement error can be kept small. Errors can be further reduced where the sensor is clamped to the surface by coating the surface and the sensor with heat-conducting grease. This grease is normally a silicone grease heavily loaded with finely ground alumina. A grease loaded with beryllium oxide has better heat-transfer properties. However, since beryllium oxide is very toxic, this grease must be handled with the greatest of care.

Where spot measurements are to be made, using, for instance, a handheld probe, it is difficult to get accurate readings. The normal practice is to use a probe mounted on a spring so that it can take up any reasonable angle to press flat against the surface to be measured. The mass of the probe tip is kept as small as possible, usually by using a thermocouple or thermistor, to keep the thermal mass of the probe to a minimum. Again, accuracy can be improved somewhat by using thermally conducting grease. Figure 21.40 shows a typical handheld probe.

21.7.3 Miscellaneous Measurement Techniques

Temperature measurement may be the primary measurement required for the control of a plant. There are, however, many cases in which temperature measurement is a tool to get an indication of the conditions in a plant. For instance, in distillation columns it is more convenient and quicker to judge the compositions of the offtake by temperature measurement than to install online analyzers, and as a further bonus the cost of temperature measurement is very significantly less than the cost of analyzers.

The reverse situation, where it is not possible to gain access for a thermometer to the region where the temperature needs to be known, also exists. In this instance some indirect measurement technique must be resorted to. One case of indirect measurement that has already been dealt with at some length is the case of radiation thermometers.

21.7.3.1 Pyrometric Cones

At certain definite conditions of purity and pressure, substances change their state at fixed temperatures. This fact forms a useful basis for fixing temperatures and is the basis of the scales of temperature.

For example, the melting points of metals give a useful method of determining the electromotive force of a thermocouple at certain fixed points on the International Practical Temperature Scale, as has been described.

In a similar way, the melting points of mixtures of certain minerals are used extensively in the ceramic industry to determine the temperature of kilns. These minerals, being similar in nature to the ceramicware, behave in a manner that indicates what the behavior of the pottery under similar conditions is likely to be. The mixtures, which consist of silicate minerals such as kaolin or china clay (aluminum silicate), talc (magnesium silicate), felspar (sodium aluminum silicate), quartz (silica), together with other minerals such as calcium carbonate, are made up in the form of cones known as *Seger cones*. By varying the composition of the cones, a range of temperature between 600°C and 2,000°C may be covered in convenient steps.

A series of cones is placed in the kiln. Those of lower melting point will melt, but eventually a cone is found that will just bend over. This cone indicates the temperature of the kiln. This can be confirmed by the fact that the cone of next-higher melting point does not melt.

Since the material of the cone is not a very good conductor of heat, a definite time is required for the cone to become fluid so that the actual temperature at which the cone will bend will depend to a certain extent on the rate of heating. To obtain the maximum accuracy, which is on the order of ±10°C, the cones must, therefore, be heated at a controlled rate.

21.7.3.2 Temperature-Sensitive Pigments

In many equipment applications it is necessary to ensure that certain components do not exceed a specified temperature range. A typical case is the electronics industry, where it is essential that semiconductor components remain within their rather limited operating range, typically −5°C to 85°C, or for equipment to military specification −40°C to 125°C. These components are too small to fix all but the finest thermocouples to them. To deal with this situation, temperature-sensitive paints can be used. These paints contain pigments that change color at known temperatures with an accuracy of ±1°C. The pigments are available with either a reversible or a nonreversible color change, the latter being the more usually used. In the previous case, a semiconductor component in an electronic machine can have two spots of paint put on its case with color changes at, say, 0°C and 110°C. On subsequent inspection, perhaps after equipment failure, it can be seen at once whether that component has been beyond its temperature tolerance.

As an alternative to paint, these pigments are available on small self-adhesive labels. In either case they are available for temperatures within the range of 0°C to about 350°C in steps of about 5 degrees.

21.7.3.3 Liquid Crystals

A number of liquids, mainly organic, when not flowing, tend to form an ordered structure with, for instance, all the molecules lying parallel to one another. This structure is maintained against the thermal agitation by weak intermolecular bonding such as hydrogen bonding. These bonds hold the structure until the weak bonds between the molecules get broken, as will occur when the liquid begins to flow. The structure can also be changed by electric fields, magnetic fields, or temperature. Different compounds respond to different stimuli. Most people are familiar with the liquid crystal displays on digital watches and pocket calculators; these displays use compounds sensitive to electric fields.

However, in this section we are interested in those liquid crystalline compounds that respond primarily to temperature. The compounds involved are a group of compounds derived from or with molecular structures similar to cholesterol. They are therefore called *cholesteric compounds*. Cholesteric liquids are extremely optically active as a consequence of their forming a helical structure. The molecules have a largely flat form and as a result lie in a laminar arrangement. However, the molecules have side groups that prevent them lying on top of one another in perfect register. The orientation of one layer of molecules lies twisted by a small angle compared to the layer below. This helical structure rotates the plane of polarization of light passing through the liquid in a direction perpendicular to the layers of molecules. Figure 21.68 illustrates this effect diagrammatically. The optical effect is very pronounced, the rotation of polarization being on the order of 1,000° per millimeter of path length. The laminar structure can be enhanced by confining the cholesteric liquid between two parallel sheets of suitable plastic. The choice of polymer for this plastic is based on two prime requirements. First, it is required to be transparent to light, and second, it should be slightly chemically active so that the liquid crystal molecules adjacent to the surface of the polymer are chemically bonded to it, with their axes having the required orientation.

When used for temperature measurement, the liquid crystal is confined between two sheets of transparent plastic a few tens of micrometers apart. The outer surface of one plastic layer is coated with a reflective layer (see Figure 21.69). In (a), a light ray enters the sandwich and travels to the bottom face, where it is reflected. Since the liquid crystal is in its ordered form, it is optically active. The reflected ray interferes destructively with the incident ray and the sandwich looks opaque. In (b), however, the liquid crystal is above the temperature at which the ordered structure breaks up. The material is no longer optically active, and the light ray is reflected in the normal way—the material looks transparent.

The temperature at which the ordered structure breaks up is a function of the exact molecular structure. Using polarized light, a noticeable change in reflected light occurs for a temperature change of 0.001°C. In white light the effect occurs within a temperature range of 0.1°C. Both the appearance of the effect and the exact temperature at which it occurs can be affected by addition of dyes or other materials.

21.7.3.4 Thermal Imaging

In Section 21.6 the measurement of temperature by infrared and visual radiation was discussed in some detail. This technique can be extended to measure surface temperature profiles of objects in a process known as *thermal imaging*. The object to be examined is scanned as for television but at a slower rate and in the infrared region instead of the optical part of the spectrum. The signal so obtained is displayed on a visual display unit. This then builds up an image of the object as "seen" by the infrared radiation from its surface. As well as producing a "picture" of the object, the temperature of the surface is indicated by the color of the image, producing a temperature map of the surface. Surface temperatures can be so imaged to cover a wide range from sub-ambient to very high temperatures. The technique has a very high resolution of temperature on the order of a small fraction of a °C. Applications are to be found in such diverse fields as medicine and geological survey from space.

The technique is dealt with in very much greater detail in Chapter 28.

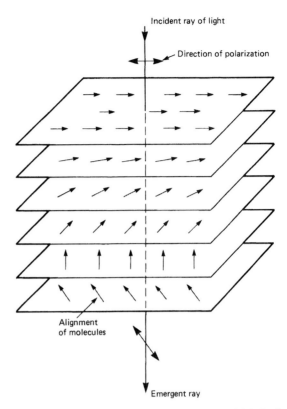

FIGURE 21.68 Rotation of the plane of polarization of light by liquid crystal.

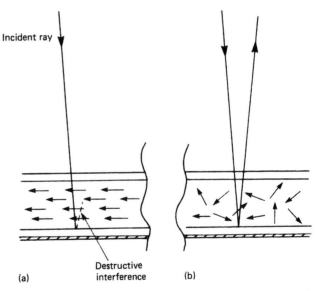

FIGURE 21.69 Destructive interference of reflected ray in liquid crystal.

21.7.3.5 Turbine Blade Temperatures

In the development and design of gas turbines, there is a requirement to measure the temperature and the temperature profile of the turbine rotor blades. This presents some problems, since the turbine may be running at speeds on the

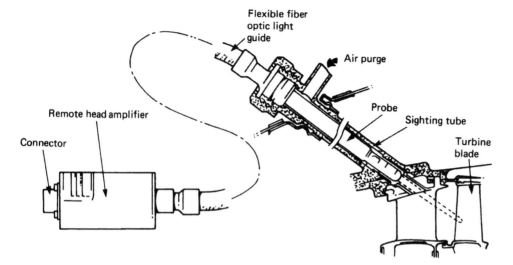

FIGURE 21.70 Radiation thermometer for gas turbine blades.

order of 25,000 revolutions per minute. The rotor may consist of, say, 50 blades, so the time available to measure each blade temperature profile as it passes a point will be about a microsecond.

A technique has been developed by Land Infrared Ltd. to carry out this measurement using fiber optic radiation thermometers. In this arrangement a small optical probe is inserted through the turbine wall and focused onto the rotor blades. The probe is connected by a fiber optic cable to a detector head amplifier unit nearby. Figure 21.70 shows a schematic diagram of focusing a measurement head. By designing the probe so that it focuses on a very small target area, it is possible to "read" a turbine blade temperature profile as it passes the target spot. Figure 21.71 shows the installation arrangement schematically at (a), and at (b) it shows the theoretical and actual signal from the radiation thermometer. The degradation between the theoretical and actual signal is a function of the speed of response of the detector and the frequency bandwidth of the electronics. The theoretical signal consists of a sawtooth waveform. The peak represents the moment when the next blade enters the target area. The hottest part of the blade is its leading edge, the temperature falling toward the trailing edge. The signal falls until the next blade enters the field. The output from the thermometer can be displayed, after signal conditioning, on an oscilloscope or can be analyzed by computer.

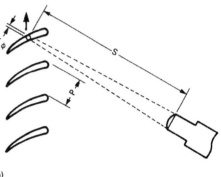

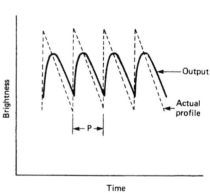

FIGURE 21.71 Measurement of the temperature profile of gas turbine blades: (a) geometry of focusing of thermometer, (b) temperature profile as "seen" by radiation thermometer and electrical output.

REFERENCES

ASTM, *Manual on Use of Thermo-couples in Temperature Measurement*, ASTM Special Technical Publication 470B (1981).

Billing, B. F., and Quinn, T. J. (eds.), *Temperature Measurement 1975*, Adam Hilger, Bristol, U.K. (1975).

Eckert, E. R. G. and Goldstein, R. J. (eds.), *Measurements in Heat Transfer*, McGraw-Hill, New York (1976).

HMSO, *The Calibration of Thermometers* (1971).

Kinzie, P. A., *Thermo-couple Temperature Measurement*, Wiley, Chichester, U.K. (1973).

Quinn, T. J., *Temperature*, Academic Press, New York (1983).

FURTHER READING

Annual Book of ASTM Standards, Vol. 14: Temperature Measurement, American Society for Testing and Materials, Philadelphia, PA (1993).

Dougherty, E. P. (ed.), *Temperature Control Principles for Process Engineers* (1993).

McGhee, T. D., *Principles and Methods of Temperature Measurement*, Wiley, Chichester, U.K. (1988).

Michalski, L., Eckersdorf, K., and McGhee, J., *Temperature Measurement*, Wiley, Chichester, U.K. (1991).

Chapter 22

Chemical Analysis: Introduction

W. G. Cummings; edited by I. Verhappen

22.1 INTRODUCTION TO CHEMICAL ANALYSIS

In the early 20th century, analytical chemistry depended almost entirely on measurements made gravimetrically and by titrimetry, and students were taught that the essential steps in the process were sampling, elimination of interfering substances, the actual measurement of the species of concern, and finally, the interpretation of results. Each step required care, and often substances were analyzed completely so that the components could be checked to total to within an acceptable reach of 100 percent.

Classical analytical methods are still used from time to time, generally for calibrating instruments, but during the past 40 years, the analytical chemistry scene has changed considerably. Spectroscopy and other physical methods of analysis are now widely used, and a comprehensive range of chemical measuring instruments has been developed for specific techniques of analysis. This has meant that chemical analysis is now carried out as a cooperative effort by a team of experts, each having extensive knowledge of their own specialist technique, such as infrared absorption, emission spectrography, electrochemistry, or gas chromatography, while also having considerable knowledge of the capabilities of the methods used by other members of the team.

Thus the analytical chemist has become more than just a chemist measuring the chemical composition of a substance; he is now a problem solver with two more steps in the analytical process—one at the beginning (definition of the problem) and another at the end (solution to the problem). This means that the analytical chemist may measure things other than narrowly defined chemical composition—she may decide, for example, that pH measurements are better than analysis of the final product for controlling a process, or that information on the valency states of compounds on the surface of a metal is more important than determining its composition.

Many elegant techniques have now become available for the analytical chemist's armory, with beautifully constructed electronic instruments, many complete with microprocessors or built-in computers. However, analytical chemists should beware of becoming obsessed solely with the instruments that have revolutionized analytical chemistry and remember that the purpose of their work is to solve problems. They must have an open and critical mind so as to be able to evaluate the analytical instruments available; it is not unknown for instrument manufacturers in their enthusiasm for a new idea to emphasize every advantage of a technique without mentioning major disadvantages. It should also be remembered that, although modern analytical instrumentation can provide essential information quickly, misleading information can equally easily be obtained by inexperienced or careless operators, and chemical measuring instruments must be checked and recalibrated at regular intervals.

Choosing the correct analytical technique or instrument can be difficult because several considerations must be taken into account. First, one must ensure that the required range of concentrations can be covered with an accuracy and precision that is acceptable for the required purpose. Then one must assess the frequency with which a determination must be made to set the time required for an analysis to be made or the speed of response of an instrument. This is particularly important if control of an ongoing process depends on results of an analysis, but it is of less importance when the quality of finished products is being determined whereby ease of handling large numbers of samples may be paramount. Many requirements are conflicting, and decisions have to be made on speed versus accuracy, cost versus speed, and cost versus accuracy; and correct decisions can only be made with a wide knowledge of analytical chemistry and of the advantages and limitations of the many available analytical techniques.

An important consideration is the application of the analytical instrument. This can be in a laboratory, in a rudimentary laboratory or room in a chemical plant area, or working automatically on-stream. It is obvious that automatic on-stream instrumentation will be much more complex and expensive than simple laboratory instruments because the former must withstand the hostile environment of the chemical plant and be capable of coping with temperature changes and plant variables without loss of accuracy. Such instruments have to be constructed to work for long continuous periods without exhibiting untoward drift or being adversely affected by the materials in the plant stream being monitored.

Laboratory instruments, on the other hand, can be much simpler. Here the essential is a robust, easy-to-use instrument for a unique determination. Temperature compensation can be made by manual adjustment of controls at the time of making a determination, and the instrument span can be set by use of standards each time the instrument is used. Thus there is no problem with drift. Laboratory instruments in general-purpose laboratories, however, can be as complex and costly as on-stream instruments but with different requirements. Here flexibility to carry out several determinations on a wide variety of samples is of prime importance, but again, temperature compensation and span adjustment can be carried out manually each time a determination is made. More expensive instruments use microprocessors to do such things automatically, and these are becoming common in modern laboratories. Finally, although the cost of an analytical instrument depends on its complexity and degree of automation, there are other costs that should not be forgotten. Instrument maintenance charges can be appreciable, and there is also the cost of running an instrument. The latter can range from almost nothing in the case of visible and ultraviolet spectrometers to several thousand pounds a year for argon supplies to inductively coupled plasma spectrometers. Many automatic analytical instruments require the preparation of reagent solutions; this, too, can involve an appreciable manpower requirement, also something that should be costed.

More detailed analysis of the factors affecting the costing of analytical chemistry techniques and instrumentation is beyond the scope of this chapter, but other chapters in this reference book give details and comparisons of analytical instrumentation for many applications. For completeness, the remainder of this chapter contains brief descriptions of chromatography, thermal analysis, and polarography.

22.2 CHROMATOGRAPHY

22.2.1 General Chromatography

Around 1900, M. S. Tswett used the adsorbing power of solids to separate plant pigments and coined the term *chromatography* for the method. It was then not used for 20 years, after which the method was rediscovered and used for the separation of carotenes, highly unsaturated hydrocarbons to which various animal and plant substances (e.g., butter and carrots) owe their color.

Chromatography is thus a separating procedure, with the actual measurement of the separated substance made by another method, such as ultraviolet absorption or thermal conductivity, but because it is such a powerful analytical tool, it is dealt with here as an analytical method.

All chromatographic techniques depend on the differing distributions of individual compounds in a mixture between two immiscible phases as one phase (the mobile phase) passes through or over the other (the stationary phase). In practice the mixture of compounds is added to one end of a discrete amount of stationary phase (a tubeful), and the mobile phase is then introduced at the same end and allowed to pass along the stationary phase. The mixture of compounds is eluted, the compound appearing first at the other end of the stationary phase being that which has the smallest distribution into the stationary phase. As the separated compounds appear at the end of the stationary phase, they are detected either by means of unique detectors or by general-purpose detectors that sense the compound only as an impurity in the mobile phase.

The apparatus used varies according to the nature of the two phases. In gas chromatography, the mobile phase is a gas with the stationary phase either a solid or a liquid. This is described in detail in Chapter 25. Liquid chromatography covers all techniques using liquid as a mobile phase; these are column chromatography (liquid/liquid or liquid/solid), paper chromatography, and thin-layer chromatography.

22.2.2 Paper Chromatography and Thin-Layer Chromatography

In paper chromatography the separation is carried out on paper, formerly on ordinary filter papers but more recently on papers specially manufactured for the purpose. These are made free of metallic impurities and have reproducible thickness, porosity, and arrangement of cellulose fibers.

The paper (which must not have been dried) contains adsorbed water, so paper chromatography can be regarded as an absorption process. However, the characteristics of the paper can be changed by applying specific liquids to it. Silicone oils, paraffin oil, petroleum jelly, and rubber latex can be used to produce a paper with nonpolar liquid phases. Specially treated papers are also available, such as those containing ion-exchange resins. Papers for paper chromatography can also be made of glass fibers or nylon as well as cellulose.

In thin-layer chromatography, instead of using paper, a thin layer of an adsorbing substance such as silica gel is coated onto a glass or plastic plate. A very small volume of sample $\sim 30 \mu l$) is transferred onto one end of the plate, which is then placed in a closed tank dipping into a solvent—the mobile phase. As the mobile phase moves along the plate, the components of the sample are separated into a series of spots at different distances from the sample starting position. Figure 22.1 shows alternative arrangements. The location of the spots can be identified by their color or, if colorless,

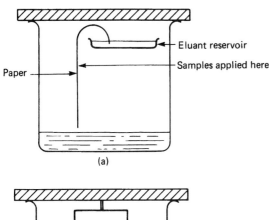

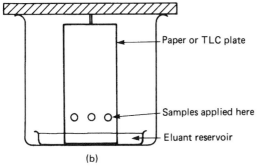

FIGURE 22.1 Apparatus for paper or thin-layer chromatography: (a) descending eluant used with paper chromatography, (b) ascending eluant used with paper chromatography or TLC.

by spraying the plate with a reagent that produces a visible color (or UV-detectable absorbance) with the compounds of interest. The position of the spots identifies the compound, the intensity of the color, and the concentration.

To establish a method for a particular mixture of compounds, one has to select suitable adsorbents, solvents, or mixtures of solvents and a sensitive and selective reagent for detecting the separated compounds. Many textbooks discuss this topic in detail and give applications of the technique.

The apparatus used for measuring the separated substances in both paper and thin-layer chromatography is quite straightforward laboratory-type equipment—for example, visible/ultraviolet spectrometers to determine the color density or the UV absorbance of the spots.

Thin-layer chromatography is generally found to be more sensitive than paper chromatography; development of the chromatogram is faster and it is possible to use a wider range of mobile phases and reagents to detect the position of the spots. Uses include the determination of phenols, carcinogenic polynuclear aromatic hydrocarbons, nonionic detergents, oils, pesticides, amino acids, and chlorophylls.

22.2.2.1 High-Performance Liquid Chromatography

Although liquid chromatography in columns was used by Tswett at the beginning of the 20th century, an improved, quantitative version of the technique, high-performance liquid chromatography (HPLC), has been fully developed more recently. Using precision instruments, determination of trace organic and inorganic materials at concentrations of 10^{-6} to 10^{-12} g is possible. There are also several advantages of HPLC over other chromatographic techniques. HPLC is more rapid and gives better separations than classical liquid chromatography. It also gives better reproducibility, resolution, and accuracy than thin-layer chromatography, although the latter is generally the more sensitive technique. A large variety of separation methods is available with HPLC: liquid/liquid, liquid/solid, ion exchange, and exclusion chromatography—but, again, the sensitivity obtainable is less than with gas chromatography.

Classical column liquid chromatography, in which the mobile liquid passed by gravity through the column of stationary phase, was used until about 1946–1950. In these methods a glass column was packed with a stationary phase such as silica gel and the sample added at the top of the column. Solvent, the mobile phase, was then added at the top of the column, and this flowed through under the force of gravity until the sample components were either separated in the column or were sequentially eluted from it. In the latter case, components were identified by refractive index or absorption spectroscopy. This type of elution procedure is slow (taking several hours), and the identification of the components of the sample is difficult and time consuming.

Modern high-performance liquid chromatography equipment has considerably better performance and is available from many chemical measuring instrument manufacturers. The main parts of a general-purpose HPLC apparatus are shown in Figure 22.2.

The system consists of a reservoir and degassing system, a gradient device, a pump, a pulse dampener, a pre-column, a separating column, and a detector.

Reservoir and Degassing System The capacity of the reservoir is determined by the analysis being carried out; generally, 1 liter is suitable. If oxygen is soluble in the solvent being used, it may need to be degassed. This can be done by distilling the solvent, heating it with stirring, or by applying a reduced pressure.

Gradient Devices If one wants to change the composition of the mobile phase during the separation, this can be done by allowing another solvent to flow by gravity into a stirred mixing vessel that contains the initial solvent and feeds the pump. This change of solvent mix is known as *generating a solvent gradient.*

A better way is to pump the solvents separately into a mixing tube; the desired gradient (composition) can be obtained by programming the pumps. This is elegant but expensive.

Pumps Suitable pumps deliver about 10 ml of solvent per minute at pressures up to 70 bar. These can be pressurized reservoirs, reciprocating pumps, motor-driven syringes, or

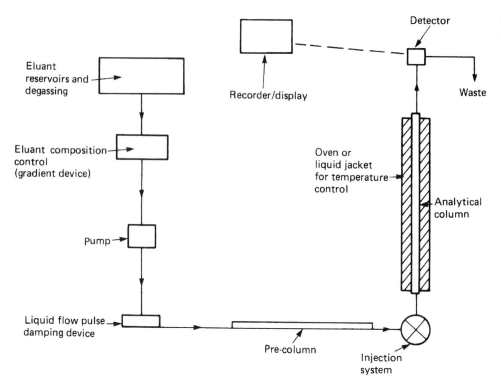

FIGURE 22.2 Line diagram of an HPLC apparatus.

pneumatically operated syringes. It is essential to arrange for pulseless liquid flow, and pulse damping may be required. This can be done using smallbore tubes of small volume or sophisticated constant pressure control equipment.

Pre-Column The solvent (the mobile phase) must be pre-saturated with the stationary liquid phase in the pre-column so that the stationary phase is not stripped off the analytical column.

Sample Introduction Samples can be injected onto the analytical column by injection by syringe through a septum or by means of a sample loop. Injection via a septum can be difficult because of the very high pressures in the column; an alternative is stop-flow injection, where the solvent flow is stopped, the sample injected, and then solvent flow and pressure restored. However, this can cause problems due to the packing in the column shifting its position.

Analytical Columns Very smooth internal walls are necessary for efficient analytical columns, and very thick-walled glass tubing or stainless steel are the preferred materials. Connections between injection ports, columns, and detectors should be of very low volume, and inside diameters of components should be of similar size. Tubing of 2–3 mm internal diameter is most often used, and temperature control is sometimes necessary. This can be done by water-jacketing or by containing the columns within air ovens.

Stationary Phases A very wide variety of materials can be used as solid stationary phases for HPLC—a summary of materials to use has been compiled (R. E. Majors, *Am. Lab.*, 4(5), 27, May 1972). Particle sizes must be small—for example, 35–50 μm and 25–35 μm.

There are various methods of packing the stationary phase into the column. Materials such as ion-exchange resins, which swell when they come into contact with a solvent, must be packed wet as a slurry. Other materials are packed dry, with the column being vibrated to achieve close packing. Packed columns should be evaluated before use for efficiency (a theoretical plate height of about 0.1 mm), for permeability (pressure required), and for speed. (Theoretical plate height is a measure of the separating efficiency of a column analogous to the number of separating plates in a liquid distillation column.) Guidance on column packing materials can be obtained from manufacturers such as Pechiney-St. Gobain, Waters Associates, E. M. Laboratories, Reeve Angel, Restek, Dupont, and Separations Group.

Mobile Phase The mobile phase must have the correct "polarity" for the desired separation, low viscosity, high purity and stability, and compatibility with the detection system. It must also dissolve the sample and wet the stationary phase.

Detectors Commercially available detectors used in HPLC are fluorimetric, conductiometric, heat of absorption detector, Christiansen effect detector, moving wire detector, ultraviolet absorption detector, and the refractive index detector. The last two are the most popular.

Ultraviolet detection requires a UV-absorbing sample and a non-UV-absorbing mobile phase. Temperature regulation is not usually required.

Differential refractometers are available for HPLC, but refractive index measurements are temperature sensitive, and good temperature control is essential if high sensitivity is required. The main advantage of the refractive index detector is wide applicability.

HPLC has been applied successfully to analysis of petroleum and oil products, steroids, pesticides, analgesics, alkaloids, inorganic substances, nucleotides, flavors, pharmaceuticals, and environmental pollutants.

22.3 POLAROGRAPHY AND ANODIC STRIPPING VOLTAMMETRY

22.3.1 Polarography

Polarography is an electrochemical technique; a specific polarographic sensor for the on-stream determination of oxygen in gas streams is described in Chapter 25. However, there are also many laboratory polarographic instruments; these are described briefly here, together with the related technique of anodic stripping voltammetry.

22.3.1.1 Direct Current Polarography

In polarography, an electrical cell is formed with two electrodes immersed in the solution to be analyzed. In the most simple version of the technique (DC polarography), the anode is a pool of mercury in the bottom of the cell (although it is often preferable to use a large-capacity calomel electrode in its place), and the cathode consists of a reservoir of mercury connected to a fine glass capillary with its tip below the surface of the solution. This arrangement allows successive fine drops of mercury to fall through the solution to the anode at the rate of one drop of mercury every 3 or 4 seconds. Figure 22.3 shows the arrangement in practice. The voltage applied across the two electrodes is slowly increased at a constant rate, and the current flowing is measured and recorded.

Figure 22.4 shows the step type of record obtained; the oscillations in the magnitude of the current are due to the changing surface area of the mercury drop during the drop life.

The solutions to be analyzed must contain an "inert" electrolyte to reduce the electrical resistance of the solution and allow diffusion to be the major transport mechanism. These electrolytes can be acids, alkalis, or citrate, tartrate, and acetate buffers, as appropriate. The cells are designed so that oxygen can be removed from the solution by means of a stream of nitrogen, for otherwise the step given by oxygen would interfere with other determinations. The voltage range can run from +0.2 to −2.2 Volts with respect to the calomel electrode. At the positive end the mercury electrode itself oxidizes; at the negative end the "inert" electrolyte is reduced.

The potential at which reduction occurs in a given base electrolyte, conventionally the half-wave potential, is characteristic of the reducible species under consideration, and the polarogram (the record obtained during polarography) thus shows the reducible species present in the solution. The magnitude of the diffusion current is a linear function of the concentration of the ion in solution. Thus, in Figure 22.4,

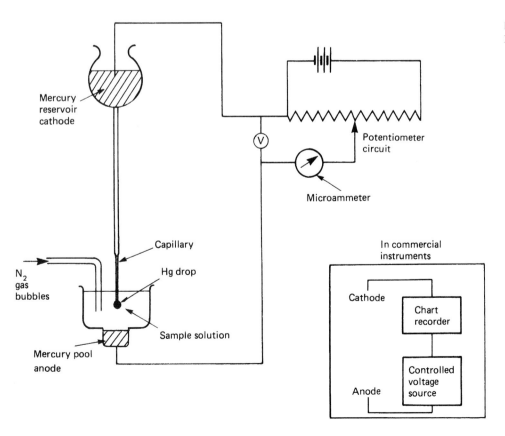

FIGURE 22.3 Arrangement for DC polarography.

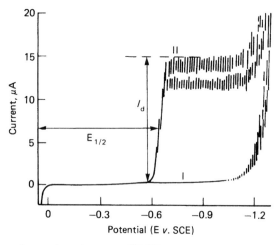

I : d.c. polarogram 1M HCl
II : d.c. polarogram of $5.0 \times 10^{-4}M$ Cd (ii) in 1M HCl
I_d : Diffusion current
$E_{1/2}$: Half-wave potential

FIGURE 22.4 Polarograms of cadmium in hydrochloric acid. Reprinted by courtesy of EG & G Princeton Applied Research and EG & G Instruments Ltd.

$E_{1/2}$ is characteristic of cadmium in a hydrochloric acid electrolyte and I_d is a measure of the amount of cadmium. The limit of detection for DC polarography is about 1 ppm.

22.3.1.2 Sampled DC Polarography

One disadvantage of the simple polarographic technique is that the magnitude of diffusion current has to be measured on a chart showing current oscillations (see Figure 22.4). Because these are caused by the changing surface area of the mercury drop during its lifetime, an improvement can be made by using sampled DC polarography in which the current is measured only during the last milliseconds of the drop life. To do this the mercury drop time must be mechanically controlled. The resulting polarogram has the same shape as the DC polarogram but is a smooth curve without large oscillations.

22.3.1.3 Single-Sweep Cathode Ray Polarography

Another modification to DC polarography is sweep cathode ray polarography. Here an increasing DC potential is applied across the cell but only once in the life of every mercury drop. Drop times of about 7 seconds are used; the drop is allowed to grow undisturbed for 5 seconds at a preselected fixed potential, and a voltage sweep of 0.3 Volt per second is applied to the drop during the last 2 seconds of its life. The sharp decrease in current when the drop falls is noted by the instrument, and the sweep circuits are then automatically triggered back to zero. After the next 5 seconds of drop growing time, another voltage sweep is initiated and is terminated

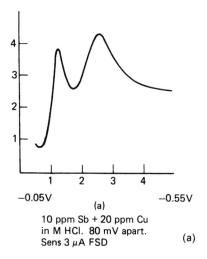

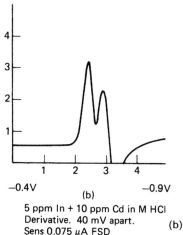

FIGURE 22.5 Single-sweep cathode ray polarograms. (a) Direct. (b) Derivative. Courtesy of R. C. Rooney.

by the drop fall, and so on. The use of a long persistence cathode ray tube enables the rapid current changes to be followed easily with the trace remaining visible until the next sweep. Permanent records can be made by photography.

A characteristic of this technique is the peaked wave (see Figure 22.5(a)), compared with classical DC polarography. This peak is not a polarographic maximum but is due to the very fast voltage sweep past the deposition potential, causing the solution near the drop surface to be completely stripped of its reducible species. The current therefore falls and eventually flattens out at the diffusion current level. The peak height is proportional to concentration in the same way as the diffusion current level, but sensitivity is increased. Resolution between species is enhanced by the peaked waveform, and even this can be improved by the use of a derivative circuit; see Figure 22.5(b). Also, because of the absence of drop growth oscillations, more electronic amplification can be used. This results in the sensitivity of the method being at least 10 times that of conventional DC polarography.

22.3.1.4 Pulse Polarography

The main disadvantage of conventional DC polarography is that the residual current, due mainly to the capacitance effect continually charging and discharging at the mercury drop surface, is large compared with the magnitude of the diffusion current in attempts to determine cations at concentrations of 10^{-5} mol^{-1} or below. Electronic methods have again been used to overcome this difficulty; the most important techniques are pulse and differential pulse polarography.

In normal pulse polarography, the dropping mercury electrode is held at the initial potential to within about 60 milliseconds of the end of the drop life. The potential is then altered in a stepwise manner to a new value and held there for the remainder of the drop life. During the last 20 milliseconds of this process, the current is measured and plotted against the applied potential. Each new drop has the potential increased to enable the whole range of voltage to be scanned. The change in current that occurs when the voltage is stepped comes from the current passed to charge the double-layer capacitance of the electrode to the new potential. This decays very rapidly to zero. There is also a Faradaic current that is observed if the potential is stepped to a value at which an oxidation or reduction reaction occurs. This decays more slowly and is the current that is measured. This technique gives detection limits from 2 to 10 times better than DC polarography (see Figure 22.6), but it is still not as sensitive as differential pulse polarography.

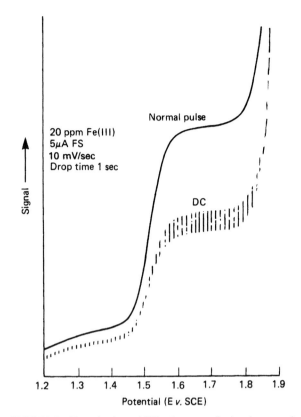

FIGURE 22.6 Normal pulse and DC polarograms for iron in ammonium tartrate buffer, pH 9. Reprinted by courtesy of EG & G Princeton Applied Research and EG & G Instruments Ltd.

22.3.1.5 Differential Pulse Polarography

The most important of modern polarographic techniques is that of differential pulse polarography. Here a 25 or 50 mV amplitude pulse is superimposed at fixed time intervals on the normal linear increasing voltage of 2 or 5 mVs^{-1}, with the mercury drop being dislodged mechanically and so arranged that the pulse occurs once during the lifetime of each drop (see Figure 22.7). The current is measured over a period of about 0.02 second just before the pulse is applied and during 0.02 second toward the end of the drop life. The difference between the two measurements is recorded as a function of the applied DC potential. In practice, a three-electrode potentiostatic arrangement is used (see Figure 22.8). The polarograms obtained in this way are peak shaped (see Figure 22.9); there is increased resolution between any two species undergoing reduction and a great increase in sensitivity, which is mainly a function of the reduction in measured capacitance current. There is a linear relationship between peak height and the concentration of the species being determined, and limits of detection can be as low as 10^{-8} mol^{-1}. The sensitivity of the technique can be varied by varying the pulse height; the peak height increases with increased pulse height but the resolution between peaks suffers (see Figure 22.10). A comparison of the sensitivity

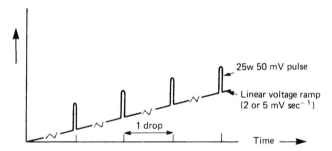

FIGURE 22.7 Voltage waveform for differential pulse polarography.

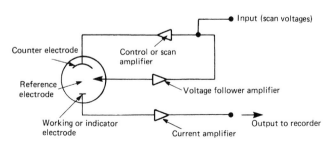

FIGURE 22.8 Practical arrangement for differential pulse polarography.

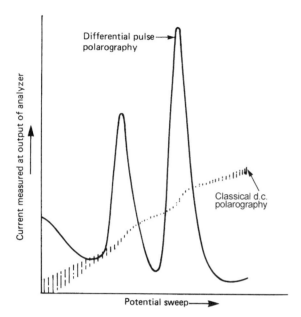

FIGURE 22.9 Differential pulse polarogram.

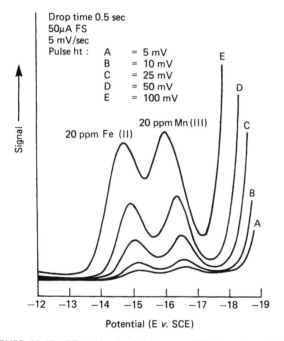

FIGURE 22.10 Effect of pulse height on peak height and resolution. Reprinted by courtesy of EG & G Princeton Applied Research and EG & G Instruments Ltd.

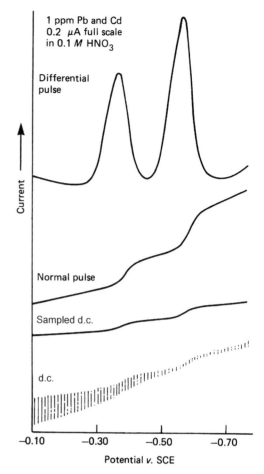

FIGURE 22.11 Comparison of polarographic modes. Reprinted by courtesy of EG & G Princeton Applied Research and EG & G Instruments Ltd.

of DC, sampled DC, normal pulse, and differential pulse polarography is shown in Figure 22.11.

22.3.1.6 Applications of Polarography

Polarographic methods can be used for analyzing a wide range of materials. In metallurgy, Cu, Sn, Pb, Fe, Ni, Zn, Co, Sb, and Bi can be determined in light and zinc-based alloys, copper alloys, and aluminum bronze; the control of effluents is often carried out using polarographic methods. Cyanide concentrations down to ~0.1 ppm can be determined, and sludges and sewage samples as well as fresh and sea waters can be analyzed. Trace and toxic elements can be determined polarographically in foodstuffs and animal feed, in soils, and in pharmaceutical products. In the latter, some compounds are themselves polarographically reducible or oxidizable—for example, ascorbic acid, riboflavin, drugs such as phenobarbitone and ephedrine, and substances such as saccharine. Body fluids, plastics, and explosives can also be analyzed by polarographic techniques.

22.3.2 Anodic Stripping Voltammetry

Anodic stripping voltammetry is really a reversed polarographic method. Metals that are able to form amalgams with mercury, such as Pb, Cu, Cd, and Zn, can be cathodically plated onto a mercury drop using essentially the same instrumentation as for polarography and then the amalgamated metal is stripped off again by changing the potential on the mercury drop linearly with time in an anodic direction. By recording the current as a function of potential, peaks are observed corresponding to the

CHAPTER | 22 Chemical Analysis: Introduction

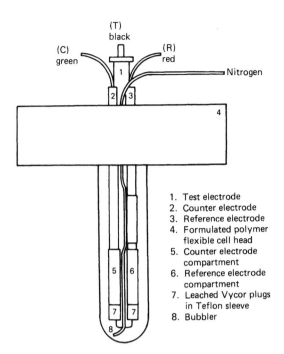

FIGURE 22.12 Cell arrangement for anodic stripping voltammetry. Courtesy of International Laboratory.

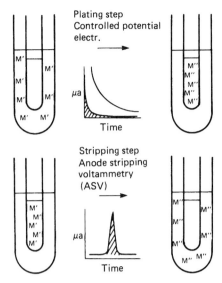

FIGURE 22.13 Plating and stripping steps. Courtesy of International Laboratory.

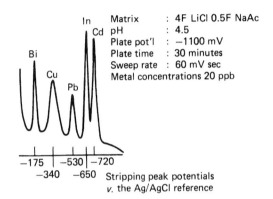

FIGURE 22.14 Stripping peak potentials. Courtesy of International Laboratory.

specific species present in the test solution; the heights of the peaks are proportional to concentration.

In practice, it is not very convenient to use a mercury drop as a cathode; several other types of electrode have been used, including a rotating ring-disc electrode. The most often used, especially for water and environmental analysis, is a wax-treated, mercury-coated graphite rod. This, together with a silver/silver chloride reference electrode and a platinum counter electrode, is immersed in the test solution (see Figure 22.12) and the plating and metal stripping carried out. Figure 22.13 illustrates the plating and stripping steps; Figure 22.14 shows a typical recording of the peak heights of Cd, In, Pb, Cu, and Bi. As with polarography, various electronic modifications have been made to the basic technique, and the stripping step has also been carried out with AC or pulsed voltages superimposed on the linear variation of DC voltage. Details of these systems can be found in reviews of the subject. Equipment for this technique is available at reasonable cost, and units can be obtained for simultaneous plating of up to 12 samples with sequential recording of the stripping stages.

With anodic stripping voltammetry small samples (mg) can be used or very low concentrations of species determined because the plating step can be used as a concentration step. Plating times from 5 to 30 minutes are common depending on the required speed and accuracy of the analysis. Figure 22.14 was obtained using a 30-minute plating time. Good precision and accuracy can be obtained in concentration ranges as low as 0.1 to 10 μg per liter; this, combined with the fact that small samples can be used, means that the technique is most attractive for trace-metal characterization in the analysis of air, water, food, soil, and biological samples.

22.4 THERMAL ANALYSIS

No work on instrumental methods of determining chemical composition would be complete without mention of thermal analysis. This is the name applied to techniques in which a sample is heated or cooled while some physical property of the sample is recorded as a function of temperature. The main purpose in making such measurements is most often not to evaluate the variation of the physical property itself but to use the thermal analysis record to study both the physical and chemical changes occurring in the sample on heating.

There are three main divisions of the technique, depending on the type of parameter recorded on the thermal analysis curve. This can be (a) the absolute value of the measured property, such as sample weight; (b) the difference between some property of the sample and that of a standard material, such as their temperature difference (these are differential measurements); and (c) the rate at which the property is changing with temperature or time, such as the weight loss (these are derivative measurements).

A convention has grown up for thermal analysis nomenclature, and recommendations of the International Confederation for Thermal Analysis are that the terms *thermogravimetry* (TG) be used for measuring sample weight, *derivative thermogravimetry* (DTG) for rate of weight loss, and *differential thermal analysis* (DTA) for measuring the temperature difference between sample and standard. There are also many other terms relating to specific heat measurement, magnetic susceptibility, evolved gases, and so on.

During the past 20 years, a wide choice of commercially available equipment has become available, and thermal analysis is now widely used as a tool in research and product control.

One particular application is to the composition of cast iron in terms of its carbon, silicon, and phosphorus content, which can be calculated from the temperatures at which it freezes. Because it is an alloy, the freezing occurs at two temperatures, the liquidus and the solidus temperatures. At both temperatures, the change of state of the metal releases latent heat. The temperatures at which the liquidus and solidus occur can be measured by the use of equipment made by Kent Industrial Measurements Ltd. To make the measurement, a sample of liquid iron is poured into a special cup made from resin-bonded sand, into which a small type-K thermocouple is mounted (see Figure 22.15). As the iron cools and passes through its two changes of state, its temperature is monitored by the thermocouple. The graph showing the cooling against time (see Figure 22.16) has two plateaus, one at the liquidus and one at the solidus. To complete the analysis, the signal from the thermocouple is processed by a microcomputer that calculates and prints the required analysis.

Figures 22.17–22.22 show other applications of thermogravimetry and derivative thermogravimetry to commercial samples and are largely self-explanatory.

In commercial thermal analysis instruments, the sample is heated at a uniform rate while its temperature and one or more of its physical properties are measured and recorded. A typical arrangement is shown in Figure 22.22(a). The measuring unit has a holder to fix the position of the sample in the furnace, a means of controlling the atmosphere

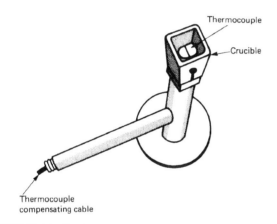

FIGURE 22.15 Cup for thermal analysis of cast iron.

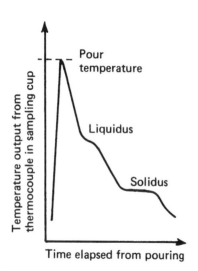

FIGURE 22.16 Cooling profile during cooling of liquid cast iron.

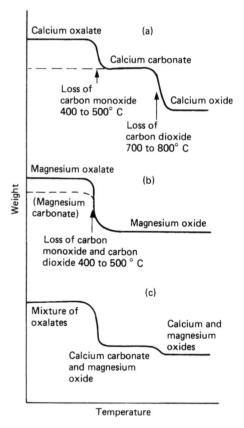

FIGURE 22.17 Weight-loss curves for calcium and magnesium oxalates and a precipitated mixture. Reproduced by permission from *Thermal Analysis* by T. Daniels, published by Kogan Page Ltd.

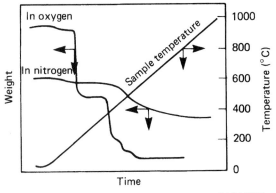

(a) TG CURVES FOR A COAL SAMPLE IN OXYGEN AND NITROGEN (FISHER TG SYSTEM)

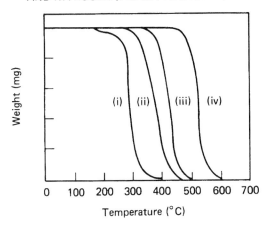

(b) TG CURVES FOR (i) POLYHEXAFLUOROPROPYLENE, (ii) POLYPROPYLENE, (iii) POLYETHYLENE, AND (iv) POLYTETRAFLUOROETHYLENE (Du Pont TG SYSTEM)

FIGURE 22.18 Thermal and thermo-oxidative stability of organic materials. Reproduced by permission from *Thermal Analysis* by T. Daniels, published by Kogan Page Ltd.

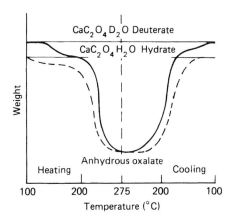

(a) TG PLOTS FOR CALCIUM OXALATE HYDRATE AND DEUTERATE ON HEATING AND COOLING IN A VAPOR ATMOSPHERE

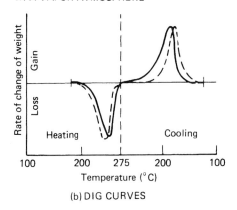

(b) DIG CURVES

FIGURE 22.19 The use of vapor atmospheres in TG. Reproduced by permission from *Thermal Analysis* by T. Daniels, published by Kogan Page Ltd.

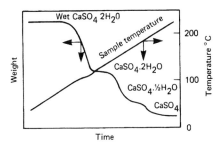

(a) EVALUATION OF THE WATER CONTENT OF GYPSUM

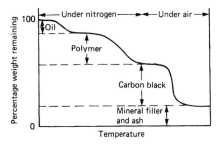

(b) ANALYSIS OF A GUM ELASTOMER (AFTER MAURER,11)

FIGURE 22.20 Analysis of commercial materials by TG. Reproduced by permission from *Thermal Analysis* by T. Daniels, published by Kogan Page Ltd.

around the sample, a thermocouple for measuring the sample temperature and the sensor for the property to be measured, for example, a balance for measuring weight. The design of the property sensor has to be such that it will function accurately over a wide temperature range, and it is most important to ensure that the atmosphere around the sample remains fixed, whether an inert gas, a reactive gas, or a vacuum.

The temperature control unit consists of a furnace and a programming unit, the function of which is to alter the sample temperature (not the furnace temperature) in a predetermined manner. The recording unit receives signals from the property sensor and the sample thermocouple, amplifies them, and displays them as a thermal analysis curve. Figure 22.22(b) shows arrangements for differential instruments where the sample material and a reference material are placed in identical environments with sensors to measure the difference in one of their properties. The differential signal is amplified and recorded as in the basic system. In derivative instruments (see Figure 22.22(c)), a derivative

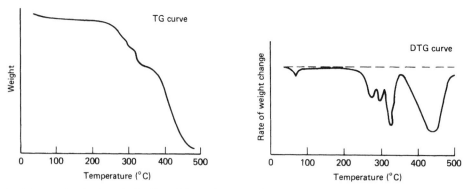

FIGURE 22.21 Dehydration and reduction of $xFe_2O_3 \cdot H_2O$ on heating in hydrogen. Reproduced by permission from *Thermal Analysis* by T. Daniels, published by Kogan Page Ltd.

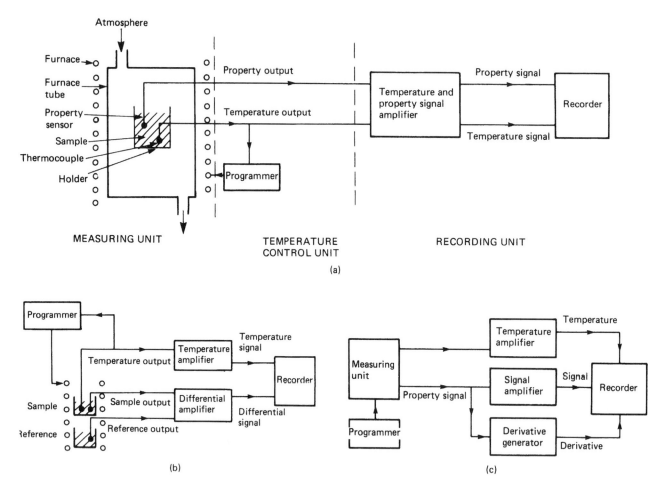

FIGURE 22.22 Construction of thermal analysis instruments: (a) basic thermal analysis system, (b) differential instrument, (c) derivative instrument. Reproduced by permission from *Thermal Analysis* by T. Daniels, published by Kogan Page Ltd.

generator, such as an electro-optical device or an electronic unit, is incorporated to compute the derivative of an input signal. Generally, both the derivative signal and the signal from the property being measured are recorded on the thermal analysis curve. It is, of course, possible to combine both modifications, thereby recording the derivative of a differential signal.

Most measuring units are designed specifically for a particular thermal analysis technique, but furnaces, programmers, amplifiers, and recorders are common to all types of instrument. Instrument manufacturers therefore generally construct a basic control unit containing programming and recording facilities to which can be connected modules designed for specific thermal analysis techniques.

Detailed description of the design of thermal analysis instruments, their applications, and the precautions necessary to ensure good results are beyond the scope of this volume, but there are several well-written books on the topic.

FURTHER READING

Bristow, P. A., *Liquid Chromatography in Practice*, Lab. Data, Florida.

Charsley, E. L. and Warrington, S. B. (eds.), *Thermal Analysis: Techniques and Applications*, Royal Society of Chemistry, London (1992).

Daniels, T., *Thermal Analysis*, Kogan Page, London (1973).

Fried, B. and Sherma, J., *Thin Layer Chromatography: Techniques and Applications*, Marcel Dekker, New York (1982).

Hatakeyma, T. and Quinn, F. X., *Thermal Analysis: Fundamentals and Applications to Polymer Science*, Wiley, New York (1994).

Heyrovsky, J. and Zuman, P., *Practical Polarography*, Academic Press, New York (1968).

Kapoor, R. C. and Aggarwal, B. S., *Principles of Polarography*, Halsted, New York (1991).

Kirkland, J. J. (ed.), *Modern Practice of Liquid Chromatography*, Wiley Interscience, New York (1971).

Lederer, M., *Chromatography for Inorganic Chemistry*, Wiley, New York (1994).

Meites, L., *Polarographic Techniques* (2nd ed.), Interscience, New York (1965).

Perry, S. G., Amos, R., and Brewer, P. I., *Practical Liquid Chromatography*, Plenum, New York (1972).

Snyder, L. R. and Kirkland, J. J., *Introduction to Modern Liquid Chromatography*, Wiley Interscience, New York (1974).

Sofer, G. K. and Nystrom, L. E., *Process Chromatography: A Guide to Validation*, Academic Press, New York (1991).

Speyer, R. F., *Thermal Analysis of Materials*, Marcel Dekker, New York (1993).

Subramanian, G. (ed.), *Preparative and Process-scale Liquid Chromatography*, Ellis Horwood, Chichester, U.K. (1991).

Touchstone, J. C. and Rogers, D. (eds.), *Thin Layer Chromatography Quantitative, Environmental and Clinical Applications*, Wiley, New York (1980).

Wendland, W. W., *Thermal Methods of Analysis*, Interscience, New York (1964).

Wiedemann, H. G. (ed.), *Thermal Analysis*, Vols 1–3, Birkhauser Verlag, Basle and Stuttgart (1972).

Wunderlich, B., *Thermal Analysis*, Academic Press, New York (1990).

Chapter 23

Chemical Analysis: Spectroscopy

A. C. Smith; edited by I. Verhappen

23.1 INTRODUCTION

The analysis of substances by spectroscopic techniques is a rather specialized field and cannot be covered in full depth in a book such as this. However, some 15 techniques are covered here, with the basic principles for each, descriptions of commercial instruments, and, where possible, their use as online analyzers.

Details of other techniques can be found in modern physics textbooks, and greater detail of those techniques that are described here can be found in literature provided by instrument manufacturers such as Pye Unicam, Perkin-Elmer, Rilgers, and Applied Research Laboratories, and in America, ABB Process Analytics, Siemens, Hewlett-Packard (Aligent), Emerson Process (formerly Fisher-Rosemount), and Yokogawa Industrial Automation. There are also many textbooks devoted to single techniques. Some aspects of measurements across the electromagnetic spectrum are dealt with in Chapter 28.

23.2 ABSORPTION AND REFLECTION TECHNIQUES

23.2.1 Infrared

Measurement of the absorption of infrared radiation enables the quantity of many gases in a complex gas mixture to be measured in an industrial environment. Sometimes this is done without restricting the infrared frequencies used (dispersive). Sometimes only a narrow frequency band is used (nondispersive).

23.2.1.1 Nondispersive Infrared Analyzers

Carbon monoxide, carbon dioxide, nitrous oxide, sulfur dioxide, methane, and other hydrocarbons and vapors of water, acetone, ethyl alcohol, benzene, and others may be measured in this way. (Oxygen, hydrogen, nitrogen, chlorine, argon, and helium, being dipolar gases, do not absorb infrared radiation and are therefore ignored.) An instrument to do this is illustrated in Figure 23.1(a). Two beams of infrared radiation of equal energy are interrupted by a rotating shutter, which allows the beams to pass intermittently but simultaneously through an analysis cell assembly and a parallel reference cell and hence into a Luft-pattern detector.

The detector consists of two sealed absorption chambers separated by a thin metal diaphragm. This diaphragm, with an adjacent perforated metal plate, forms an electrical capacitor. The two chambers are filled with the gas to be detected so that the energy characteristic of the gas to be measured is selectively absorbed.

The reference cell is filled with a nonabsorbing gas. If the analysis cell is also filled with a nonabsorbing gas, equal energy enters both sides of the detector. When the sample is passed through the analysis cell, the component to be measured absorbs some of the energy to which the detector is sensitized, resulting in an imbalance of energy, which causes the detector diaphragm to be deflected and thus changes the capacitance. This change is measured electrically and a corresponding reading is obtained on the meter.

Any other gas also present in the sample will not affect the result unless it has absorption bands that overlap those of the gas being determined. In this event, filter tubes containing the interfering gas or gases can be included in one or both optical paths so that the radiation emerging from these tubes will contain wavelengths that can be absorbed by the gas to be detected but will contain very little radiation capable of being absorbed by the interfering gases in the sample, since such radiation has already been removed.

The length of absorption tube to be used depends on the gas being estimated and the concentration range to be covered. The energy absorbed by a column of gas 1 cm long and containing a concentration c of absorbing component is approximately $Ekcl$, where E is the incident energy and k is an absorption constant, provided that kcl is small compared with unity. Thus at low concentrations it is advantageous to

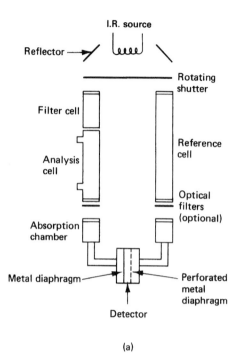

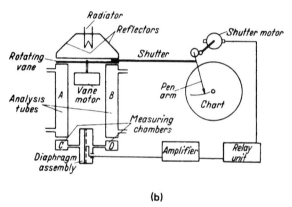

FIGURE 23.1 (a) Luft-type infrared gas analyzer. Courtesy of Grubb Parsons. (b) Infrared gas analyzer of the concentration recorder.

use long absorption paths, provided that kcl remains small and the relationship between energy absorbed and the measured concentration remains reasonably linear. At higher concentrations the energy absorbed is $E[1 - \exp(-kcl)]$, and the relationship between energy absorbed and concentration departs greatly from linearity when absorption exceeds 25 percent. When the absorption reaches this value it is, therefore, necessary to reduce the length of the absorption cell, and the product $c \times 1$ should be kept approximately constant.

The most convenient method of calibrating the instrument is to pass mixtures of the pure gas of known composition through the measuring cell and note the output for each concentration of measured gas. For day-to-day checking, a simple internal calibrating device is fitted, and it is only necessary to adjust the sensitivity control until a standard deflection is obtained.

The instrument is usually run from ac mains through a constant voltage transformer. Where utmost stability is required, an ac voltage stabilizer may be used, since the constant voltage transformer converts frequency variations to voltage changes. Generally the instrument is insensitive to temperature changes, although the gas sensitivity depends on the temperature and pressure of the sample gas in the absorption tube, since it is the number of absorbing molecules in the optical path that determines the meter deflection. For instruments sensitive to water vapor, the detecting condenser has a temperature coefficient of sensitivity of 3 percent per Kelvin, and it is therefore necessary to maintain the detector at a constant temperature.

The approximate maximum sensitivity to certain gases is given in Table 23.1.

Errors due to zero changes may be avoided by the use of a null method of measurement, illustrated in Figure 23.1(b). The out-of-balance signal from the detector is amplified, rectified by a phase-sensitive rectifier, and applied to a servo system that moves a shutter to cut off as much energy from the radiation on the reference side as has been absorbed from the analysis side and so restore balance. The shutter is linked to the pen arm, which indicates the gas concentration.

TABLE 23.1 Sensitivity of nondispersive infrared analyzer

Gas	Minimum concentration for full-scale deflection (Vol. %)	Gas	Minimum concentration for full-scale deflection (Vol. %)
CO	0.05	NO_2	0.1
CO_2	0.01	SO_2	0.02
H_2O	0.1	HCN	0.1
CH_4	0.05	Acetone	0.25
C_2H_4	0.1	Benzene	0.25
N_2O	0.01		

Online Infrared Absorption Meter Using Two Wavelengths To overcome the limitations of other infrared analyzers and provide a rugged, reliable drift-free analyzer for continuous operation on a chemical plant, ICI Mond Division developed an analyzer based on the comparison of the radiation absorbed at an absorption band with that at a nearby wavelength. By use of this comparison method, many of the sources of error, such as the effect of variation in the source intensity, change in the detector sensitivity, or fouling of the measurement cell windows, are greatly reduced.

The absorption at the measurement wavelength (λ_m) is compared with the nearby reference wavelength (λ_{mr}) at which the measured component does not absorb. The two measurements are made alternately using a single absorption path and the same source and detecting system.

The principle of the ICI Mond system is illustrated in Figure 23.2. The equipment consists of two units, the optical unit and the electronics unit, which are connected by a multicore cable. The source unit contains a sealed infrared source, which consists of a coated platinum coil at the focus of a calcium fluoride collimating lens. A chopper motor with sealed bearings rotates a chopper disc, which modulates the energy beam at 600 Hz. The source operates at low voltage and at a temperature well below the melting point of platinum. It is sealed in a nitrogen atmosphere. Energy from the source passes through the absorption cell to the detector unit. A calcium fluoride lens focuses the energy onto an indium antimonide detector. This is mounted on a Peltier cooler in a sealed unit. The temperature is detected by a thermistor inside the sealed module. A preamplifier mounted in the detector unit amplifies the signal to a suitable level for transmission to the electronics unit. Between the lens and the detector module, two interference filters, selected for the measurement and reference wavelengths, are interposed alternately in the beam, at about 6 Hz, so that the detector receives chopped energy at a level corresponding alternately to the measurement and reference transmission levels. Its output is a 600 Hz carrier modulated at 6 Hz.

The two filters are mounted on a counterbalanced arm attached to a stainless-steel torsion band. An iron shoe at the opposite end of the arm moves in and out of the gap in an electromagnet. It also cuts two light beams, which illuminate two silicon phototransistors. The light is provided by two aircraft-type signal lamps, which are underrun to ensure very long life. A drive circuit in the electronics unit causes the system to oscillate at its own natural frequency. One of the photocells provides positive feedback to maintain the oscillation; the other provides negative feedback to control the amplitude. There are no lubricated parts in the detector unit, and the whole can be hermetically sealed if desired.

The absorption cell is a thick-walled tube with heavy flanges. Standard construction is in mild steel, nickel plated, but type 316 stainless-steel construction is available where required. The windows are of calcium fluoride, sealed with

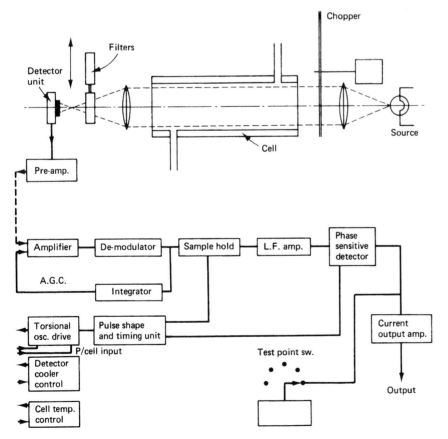

FIGURE 23.2 Dual-wavelength comparison method. Courtesy of Feedback Instruments Ltd.

Viton O-rings and retaining rings. A heater wire is wound on the cell, and the sample gas passes through a tube in thermal contact along the length of the cell before entering it at the end. Provision is made for rodding out tubes and entries in case of blockage. A thermistor embedded in the cell wall detects the cell temperature, which is controlled by a circuit in the electronics unit. The cell is thermally insulated and sealed inside a plastic bellows. The enclosed space is coupled to the purge system. The two end units each have a sealing window, so there is a double seal between the cell and the interior of the detector and source units. Since the source is inside a further sealed module, there is minimal danger of the hot source being exposed to leakage from the sample cell. The gaps between the three units are normally sealed with neoprene gaskets, and the whole device is sufficiently well sealed to maintain a positive purge pressure of at least 2 cm water gauge with a purge gas consumption of 8.3 cm^3/s. For use with highly flammable sample gases, the sealing gaskets at either end of the absorption cell may be replaced by vented gaskets. In this case a relatively large purge flow may be maintained around the cell, escaping to atmosphere across the windows. Thus, any leak at the windows can be flushed out.

To facilitate servicing on site, the source, detector, torsional vibrator, lamps, preamplifier, and source voltage control are all removable without the use of a soldering iron. Since the single-beam system is tolerant to window obscuration and the internal walls of the absorption cell are not polished, cell cleaning will not be required frequently, and in many cases adequate cleaning may be achieved *in situ* by passing solvent or detergent through the measuring cell. There is no need to switch the instrument off while performing this task. If it becomes necessary to do so, the cell can be very quickly removed and disassembled.

The electronics unit contains the power supplies, together with signal processing circuits, temperature control circuits, output and function check meter operating controls, and signal lamps. The housing is of cast-aluminum alloy, designed for flush panel mounting. The circuitry is mostly on plug-in printed circuit boards. The indicating meter, controls, and signal lamps are accessible through a window in the door. The unit is semisealed, and a purge flow may be connected if sealed glands are used at the cable entry. The signal processing circuits are contained on printed circuit boards. Output from the preamplifier is applied to a gain-controlled amplifier that produces an output signal of 3 V peak-to-peak mean. Thus the mean value of $I_r + I_m$ is maintained constantly. The signal is demodulated and smoothed to obtain the 6 Hz envelope waveform. A sample-and-hold circuit samples the signal level near the end of each half-cycle of the envelope, producing a square wave of which the amplitude is related to $I_r - I_m$. Since $I_r + I_m$ is held constant, the amplitude is actually proportional to $(I_r - I_m)/(I_r + I_m)$, which is the required function to give a linearized output in terms of sample concentration. This signal is amplified and passed to a phase-sensitive detector, consisting of a pair of gating transistors that select the positive and negative half-cycles and route them to the inverting and noninverting inputs of a differential amplifier. The output of this amplifier provides the 0–5 V output signal.

The synchronizing signals for the sample-hold and phase-sensitive detector circuits are derived from the torsional oscillator drive circuit via appropriate time delays. The instrument span is governed by selection of feedback resistors in the low-frequency amplifier, and a fine trim is achieved by adjusting the signal level at the gain-controlled amplifier. This is a preset adjustment—no operator adjustment of span is considered necessary or desirable. A front panel zero adjustment is provided. This adds an electrical offset signal at the phase-sensitive detector. The system is normally optically balanced (i.e., $I_r = I_m$) at some specified concentration of the measured variable (usually zero).

The current output and alarm circuits are located on a separate printed circuit board. The voltage output is applied to an operational amplifier, with selected feedback and offset signals to produce 0–10 mA, 5–20 mA, or 10–50 mA output. The required output is obtained by soldered selector links. The output current is unaffected by load resistances up to 1 kΩ at 50 mA or 5 kΩ at 10 mA.

A front panel alarm-setting potentiometer provides a preset signal that is compared with the analyzer output voltage in a differential amplifier. The output of this device opens a relay if the analyzer output exceeds a preset value, which may be either a low or a high analyzer output as required. The alarm condition is indicated by two signal lamps on the panel, and the system can be arranged to operate external alarms or shutdown circuits.

The power to the cell heater and the detector cooler is controlled from a bridge circuit containing thermistors that detect the temperatures of the absorption cell and detector.

The indicating meter on the front panel has a calibrated output scale and is used in conjunction with a selector switch to monitor key points in the circuit, in particular the degree of obscuration in the measuring cell. By choosing the appropriate absorption bands, the analyzer may be made suitable for a wide range of gases or liquids. For gases, it may be used for CO_2, CO, SO_2, CH_4, C_2H_6, C_2H_4, C_6H_6, C_2H_2, NH_3, N_2O, NO, NO_2, $COCl_2$, and H_2O, with ranges of 0–300 ppm and 0–100 percent.

It may also be used for measuring water in ketones, hydrocarbons, organic acids, alcohols, glucols, and oils. The accuracy is ±1 percent and the response time for 90 percent change is 3 s. The instrument is marketed by Anatek Ltd. as the PSA 401 process stream analyzer.

Another instrument based on the same principle is the Miran II Infra Red process analyzer—the chief difference being the sample cell used for gas and liquid streams. These cells are either long path gas cells or multiple internal reflection cells. The gas cells, which are normally manufactured in stainless steel, have a variable path length (see Figure 23.3). Energy passes through the sample gas and reflects one or

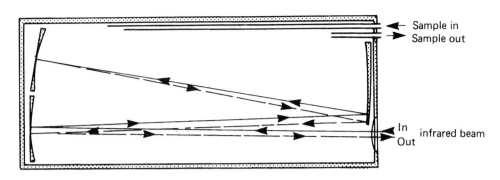

FIGURE 23.3 Internal view of multiple reflections of variable long path cell. Courtesy of Invensys.

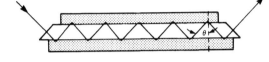

FIGURE 23.4 Principle of MIR sampling technique. Courtesy of Invensys.

more times off the mirrors in the cell before striking the detector. The path length can be adjusted between 0.75 and 20.25 meters by suitable adjustment of the mirrors. These gas cells are used to analyze the presence of low concentrations of components in gases or for those gases requiring a long path length to enhance sensitivity at a weak analytical wavelength.

In a multiple internal reflection (MIR) cell, the infrared beam is directed along or around an optical crystal through which the beam passes (see Figure 23.4). As the beam is reflected on the sample crystal interface, it slightly penetrates the liquid. These penetrations form a path of which the length is dependent on the number of reflections. The energy is absorbed at the analytical wavelength proportionally to concentration, just as in other types of cells. The crystal used is made of KRS (a composite of thallium bromide and iodide). Ordinary transmission cells have limited applicability for high concentrations, viscous or aqueous streams. In many cases, the infrared beam is grossly attenuated or the sample cannot be pumped through such cells. Multiple internal reflection overcomes these problems.

The applications to which this instrument has been put include (a) for gases: the determination of phosgene in methane and plastic production; methane and carbon dioxide in synthetic and natural gases in the range 1 ppm to 100 percent; (b) for liquids: water in acetone distillation, petroleum waste treatments, urea in fertilizer production and isocyanates in urethane and plastic production in the range 50 ppm to 50 percent; (c) for solids: the percentage weight of film coatings such as inks and polymers; and film thickness for nylon and polythene (up to 0.025 mm).

In recent years, there has been much growth in the use of fiber optic probes as a means to nonintrusively introduce infrared and near-infrared wavelengths to the process. Use of these probes negates the need for sample systems, since only the tip of the probe is in contact with the process, typically through a retractable mechanism directly mounted to the stream of interest.

23.2.1.2 Dispersive Infrared Analysis

The previous section was devoted to analysis using only one absorption frequency. However, all organic compounds give rise to a spectrum in the infrared in which there are many absorption frequencies giving a complete fingerprint of that compound. Dispersive infrared can be used, among other things, to identify a substance, for the determination of molecular structure for reaction kinetic studies, and for studies of hydrogen bonding.

In Figure 23.5 is shown a simplified layout of a typical double-beam spectrophotometer. A source provides radiation over the whole infrared spectrum; the monochromator disperses the light and then selects a narrow frequency range, the energy of which is measured by a detector; the latter transforms the energy received into an electrical signal that is then amplified and registered by a recorder or stored in a computer for further processing. The light path and ultimate focusing on the detector is determined by precision-manufactured mirrors.

Light from the radiation source S is reflected by mirrors M_1 and M_2 to give identical sample and reference beams. Each of these focuses on vertical entrance slits S_1 and S_2, the sample and reference cells being positioned in the two beams near their foci. Transmitted light is then directed by a mirror M_3 onto a rotating sector mirror (or oscillating plane mirror) M_4. The latter serves first to reflect the sample beam toward the monochromator entrance slit S_3 and then as it rotates (or oscillates) to block the sample beam and allow the reference beam to pass on to the entrance slit. A collimating mirror M_5 reflects parallel light to a prism P through which it passes, only to be reflected back again through the prism by a rotatable plane mirror M_6. The prism disperses the light beam into its spectrum. A narrow range of this dispersed light becomes focused on a plane mirror M_7, which reflects it out through the exit slit. A further plane mirror M_8 reflects the light to a condenser M_9, which focuses it sharply on the detector D. When the energy of the light transmitted by both sample and reference cells is equal, no signal is produced by the detector. Absorption of radiation by the sample results in an inequality of the two transmitted beams falling

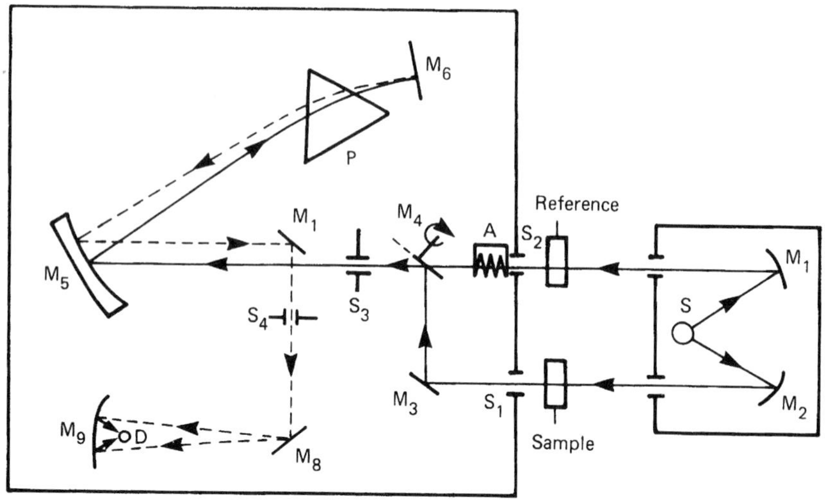

FIGURE 23.5 Simplified spectrophotometer.

on the detector, and a pulsating electrical signal is produced. This is amplified and used to move an attenuator A across the reference beam, cutting down the transmitted light until an energy balance between the two beams is restored. The amount of reference beam reduction necessary to balance the beam energies is a direct measure of the absorption by the sample.

The design and function of the major instrument components now described have a significant influence on its versatility and operational accuracy.

Source IR radiation is produced by electrically heating a Nernst filament (a high-resistance, brittle element composed chiefly of the powdered sintered oxides of zirconium, thorium, and cerium held together by a binding material) or a Globar (SiC) rod. At a temperature in the range 1,100–1,800°C, depending on the filament material, the incandescent filament emits radiation of the desired intensity over the wavelength range 0.4–40 μm.

Monochromator The slit width and optical properties of the components are of paramount importance. The wavelength range covered by various prisms is shown in Table 23.2. Gratings allow better resolution than is obtainable with prisms.

Detector This is usually a bolometer or a thermocouple. Some manufacturers use a Golay pneumatic detector, which is a gas-filled chamber that undergoes a pressure rise when heated by radiant energy. One wall of the chamber functions as a mirror and reflects a light beam directed at it onto a photocell—the output of the photocell bearing a direct relation to the gas chamber expansion.

The infrared spectra of liquids and gases may be obtained by direct study of undiluted specimens. Solids, however, are usually studied after dispersion in one of a number of possible media. These involve reduction of the solid to very small particles, which are then diluted in a mill, pressed into an alkali halide disc at 1500–3300 bar, or spread as pure solid on a cell plate surface.

The interpretation of the spectra—particularly of mixtures of compounds—is a complex problem; readers should consult textbooks on infrared analysis.

23.2.2 Absorption in UV, Visible, and IR

One instrument that uses absorption in the UV, visible, and IR is the Environmental Data Corporation stack-gas monitoring system. It is designed to measure from one to five component gases simultaneously. Depending on requirements, the components may include CO_2, NO, CO, SO_2, H_2, NH_3, hydrocarbons, and opacity or any other gases with selected spectral absorption bands in the UV, visible, or IR. The basis of the system is shown in Figure 23.6. It consists of a light source, receiver, mounting hardware, and recorder. All the gas-monitoring channels are similar in basic operation and calibration. The instrumentation can be mounted on a stack, duct, or other gas stream. A polychromatic beam of light, from a source in an enclosure on one side, is collimated and then passed through the gas to an analyzer on the opposite side. Signals proportional to the gas concentrations are transmitted from analyzer to recorder.

Most gases absorb energy in only certain spectral regions. Their spectra are often quite complex, with interspersed absorbing and nonabsorbing regions. The analyzer section of the instrument isolates the wavelengths characteristic of the gases of interest and measures their individual intensities. Both the intensity at a specific wavelength where the gas uniquely absorbs (A) and the intensity at a nearby region where the gas is nonabsorbing (B) are alternately measured with a single detector 40 times per second. Any light level change, whether due to source variation, darkening of the window, scattering by particulates, water drops, or aerosols in the gas stream, affects both A and B, leaving the ratio unchanged. This ratio gives a reading that is free

CHAPTER | 23 Chemical Analysis: Spectroscopy

TABLE 23.2 Prism frequency ranges

Prism material	Glass	Quartz	CaF$_2$	LiF	NaCl	KBr (CsBr)	CsI
Useful frequency range (cm^{-1})	above 3,500	above 2,860	5,000–1,300	5,000–1,700	5,000–650	1,100–285	1,000–200
Wavelength range (μm)	below 2.86	below 3.5	2.0–7.7	2.0–5.9	2–15.4	9–35	10–5

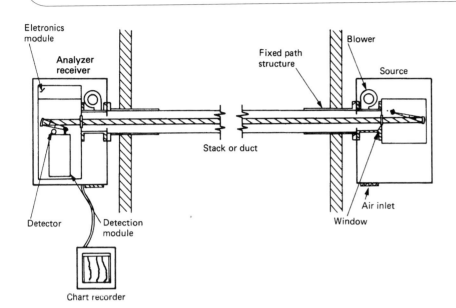

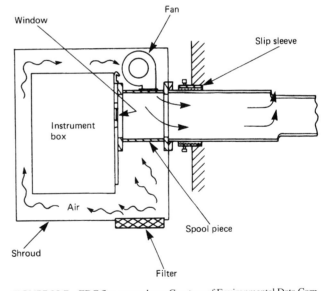

FIGURE 23.6 EDC flue gas analyzer system. Courtesy of Environmental Data Corp.

of interferences, instrumental drift, or the like. Most gases obey approximately Beer's law:

$$B = Ae^{\alpha cl}$$

or

$$\ln\left(\frac{B}{A}\right) = -\alpha cl$$

or

$$c = \frac{\ln(A/B)}{\alpha l}$$

where α is absorption coefficient (known), l is path length (fixed), and c is sample concentration (unknown).

The system response is almost instantaneous and is averaged by damping circuits to typically one second.

The stack gas is separated from the source and analyzer enclosures by means of optical surfaces, such as mirrors or windows. These windows are kept clean by an air curtain system. Self-contained blowers continually renew the air curtains, preventing the gases from contacting the windows directly (see Figure 23.7).

The flow volume and pressure of the purge air is designed for each application to allow a well-defined shear by the flue gas. Thus a known and fixed path length is provided.

When measuring opacity, the instrument measures the reduction in transmission in the visible portion of the spectrum.

FIGURE 23.7 EDC flue gas analyzer. Courtesy of Environmental Data Corp.

Typical ranges covered by the instrument are:

NO   0–25 ppm to 0–5,000 ppm
CO   0–500 ppm to 0–3,000 ppm
CO$_2$ 0–15%
SO$_2$ 0–25 ppm to 0–10,000 ppm
C–H   0–25 ppm to 0–6,000 ppm
H$_2$O   0–1,000 ppm to 0–80%
NH$_3$ 0–100 ppm

23.2.3 Absorption in the Visible and Ultraviolet

Two instruments are worthy of note here.

The first is the Barringer remote sensing correlation spectrometer, designed for the quantitative measurement of gases such as nitrogen oxides or sulfur dioxide in an optical path between the instrument and a suitable source of visible and ultraviolet radiant energy. The sensor is designed for maximum versatility in the remote measurement of gas clouds in the atmosphere, using the day sky or ground-reflected solar illumination as the light source. It may also be used with artificial sources such as quartz-iodine or high pressure Xe lamps.

Very simply, the sensor contains two telescopes to collect light from a distant source, a two-grating spectrometer for dispersion of the incoming light, a disc-shaped exit mask or correlator, and an electronics system (see Figure 23.8). The slit arrays are designed to correlate sequentially in a positive and negative sense with absorption bands of the target gas by rotation of the disc in the exit plane. The light modulations are detected by photomultiplier tubes and processed in the electronics to produce a voltage output that is proportional to the optical depth (expressed in ppm meters) of the gas under observation. The system automatically compensates for changes in average source light intensity in each channel. The basic principle of this method rests on comparison of energy in selected proportions of the electromagnetic spectrum, where absorption by the target gas occurs in accordance with the Beer–Lambert law of absorption.

Typically, this instrument covers the range 1–1,000 ppm or 100–10,000 ppm, this unit being the product of the length of the optical path through the gas and the average concentration (by volume) over that length.

The second instrument that covers absorption in the visible in liquids is the Brinkmann Probe Colorimeter. This instrument is basically a standard colorimeter consisting of a tungsten light source, the output from which passes through one of a series of interchangeable filters covering the wavelength range 420–880 nm, then through a light pipe at the end of which is a probe cell. This cell has a reflecting mirror at one end, so the optical path length is twice the length of the cell. The light then returns to the instrument via a second light pipe to a photomultiplier, the output of which is amplified and fed to a recorder in the usual way. This instrument is ideal for measuring turbidity in liquids and has the advantage that very small volumes of liquid (down to 0.5 ml) may be examined. Its other uses include general quality control, chemical analyses, pollution control, and food processing. Most of these applications make use of the fact that different elements will form colored solutions with reagents. The absorption of these colored solutions is then proportional to the concentration of that particular element.

23.2.4 Measurements Based on Reflected Radiation

Just as measurements of moisture or other components may be made by comparison at two wavelengths of transmitted infrared radiation, the method will work equally well by measuring the attenuation when infrared is reflected or backscattered. The principle is illustrated in Figure 23.9.

For water measurement of paper or granulated material on a conveyor belt, the intensity of the reflected beam at the moisture absorption wavelength of 1.93 μm may be compared with the intensity at a reference wavelength of 1.7 μm. The beams are produced by interposing appropriate filters contained in a rotating disc in front of a lamp producing appropriate radiation. The radiation is then focused onto the measured material and the reflected beam focused onto a lead sulfide photoelectric cell. By measuring the ratio of the intensity of radiation at two wavelengths, the effects of source variation, detector sensitivity, and drift in the electronic circuitry are minimized. Furthermore, calibration has shown that for a number of materials the results are substantially independent of the packing density.

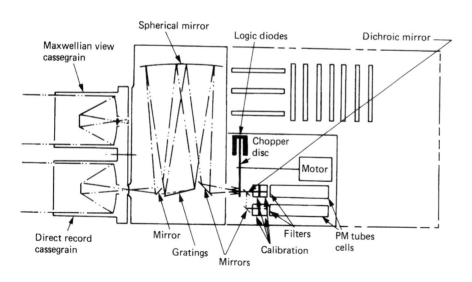

FIGURE 23.8 Barringer remote sensing correlation spectrometer.

CHAPTER | 23 Chemical Analysis: Spectroscopy

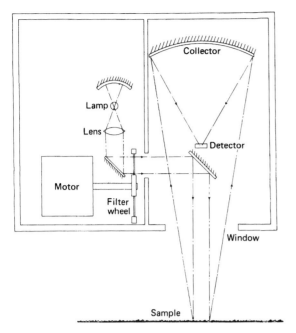

FIGURE 23.9 Backscatter infrared gauge. Courtesy of Infrared Engineering Ltd.

However, if the measured material is a strong absorber of radiation, a powerful source of radiation such as a water-cooled quartz halogen lamp may be necessary.

With this type of instrument, online measurement of the moisture content of sands, clay, dust, or flake, refractory mixtures, paper, textiles, feeding stuffs, and a wide range of other materials may be undertaken with an accuracy of ±1 percent of instrument full scale.

23.2.5 Chemiluminescence

When some chemical reactions take place, energy may be released as light. This phenomenon is known as *chemiluminescence*. Many instruments use this effect for the determination of the concentration of oxides of nitrogen and for ozone. The principles are described in Chapter 25.

23.3 ATOMIC TECHNIQUES: EMISSION, ABSORPTION, AND FLUORESCENCE

23.3.1 Atomic Emission Spectroscopy

Atomic emission spectroscopy is one of the oldest of techniques employed for trace analysis. Because of its relative simplicity, sensitivity, and ability to provide qualitative information quickly, it has been widely used in both industrial and academic analytical problems. It can be used for the analysis of metals, powders, and liquids and is used extensively in the steel and nonferrous alloy industries; the advent of inductively coupled plasma sources for producing spectra has made the technique invaluable for the analysis of some 70 elements in solution—down to concentrations of 1 ppb and less. The basic principles of the technique are as follows.

Each atom consists of a nucleus around which revolves a set of electrons. Normally these electrons follow orbits immediately adjacent to the nucleus. If energy is imparted to the atom by means of a flame or an electric arc or spark, it undergoes excitation and its electrons move into orbits further removed from the nucleus. The greater the energy, the further from the nucleus are the orbits into which the electrons are moved. When sufficient energy is imparted to the electron, it may be torn from the atom, and the atom becomes a positively charged ion. Atoms will not remain in this excited state, especially when removed from the source of energy, and they return to their original states with electrons falling to lower orbits. This electron transition is accompanied by a quantum of light energy. The size of this pulse of light energy and its wavelength depend on the positions of the orbits involved in the transition.

The energy emitted is

$$E = h\nu$$

where h is Planck's constant and ν is the frequency of the radiation. Or

$$E = hc/\lambda$$

where c is the velocity of light and λ the wavelength. Hence the greater the light energy quantum, the shorter is the wavelength of the light emitted.

Only the outer, valence electrons participate in the emission of spectral lines. The number of valence electrons in an atom differs for chemical elements. Thus the alkali elements, sodium, lithium, potassium, and so on, contain only one electron in their outer shell, and these elements have simple spectra. Such elements as manganese and iron have five or six valence electrons, and their spectra are very complex. Generally speaking, the structure of an atom is closely bound up with its optical spectrum. Thus if a mixture of atoms (as found in a sample) are excited by applying energy, then quantities of light are emitted at various wavelengths, depending on the elements present. The intensity of light corresponding to one element bears a relationship to the concentration of that element in the sample.

To sort out the light emitted, a spectroscope is used. Figures 23.10–23.12 show, respectively, the layout of a medium quartz

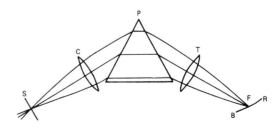

FIGURE 23.10 Optical system of a simple spectroscope. S = slit; C = collimator lens; P = prism; T = telescope lens; F = curve along which the various parts of the spectrum are in focus; B = blue or short wavelength part; R = red or long wavelength part.

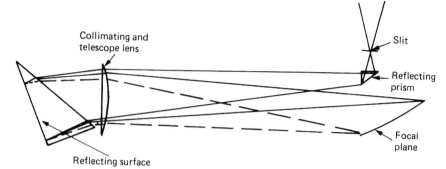

FIGURE 23.11 Diagram of the optical system of a Littrow spectrograph. The lens has been reversed to reduce scattered light.

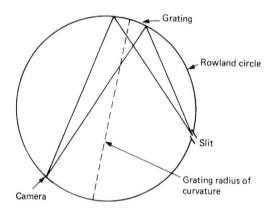

FIGURE 23.12 Elements of Rowland circle.

spectroscope, a Littrow spectrograph, and a spectroscope using a diffraction grating. This last employs the principle, due to Rowland, of having the grating on a concave surface. There are many other configurations. In all cases, each instrument contains three main components: a slit, a dispersive device such as a prism or diffraction grating to separate radiation according to wavelength, and a suitable optical system to produce the spectrum lines, which are monochromatic images of the slit. These images may be recorded on a photographic plate or, by suitable positioning of exit slits, mirrors, and photomultiplier tubes, the light intensity may be recorded electronically.

23.3.1.1 Dispersive Devices

Prisms Prisms are usually made of glass or quartz, and their dispersive ability is based on the variation of the index of refraction with wavelength. As the incident light beam enters the transparent material, it bends toward the normal according to Snell's law:

$$n_1 \sin i = n_2 \sin r$$

where n_1 is the refractive index of air, n_2 is the refractive index of the prism material, i is angle of incidence, and r is angle of refraction. Shorter wavelengths are deviated more than longer ones. The resulting dispersion is greater for the UV than for IR wavelengths.

Gratings Gratings may be considered as a large number of parallel, close, equidistant slits or diffracting lines. The equation $n\lambda = 2d \sin \theta$ shows the dependence of θ on the wavelength of the incident light, where n is an integer, λ is the wavelength of incident light, d is the distance between the lines, and θ is the angle between the diffracted beam and the normal incident beam.

Modern gratings offer the spectroscopist uniform dispersion and coverage of a wide spectral range. Today, nearly all manufacturers have turned almost exclusively to grating instruments.

23.3.1.2 Vacuum Spectrographs

Many elements, particularly the nonmetallic ones, have their most persistent lines in the spectral region 150–220 nm. Light of these wavelengths is absorbed by air, and instruments are manufactured in which the optical paths are evacuated to overcome this problem.

23.3.1.3 Excitation: Spectroscopic Sources

Many factors are considered in the choice of a source. Sample form, necessary sensitivity, and the elements, which must be determined, are the most critical. The main sources used are (1) a dc arc, (2) a high-voltage condensed spark, (3) an arc triggered by a high-voltage spark, (4) flames, (5) plasma jets, and (6) inductively coupled plasmas. A recent form of excitation consists of evaporating a nonconducting sample by means of a laser and exciting the vapor with a high-voltage spark.

23.3.1.4 Standards

To achieve a quantitative estimation of the impurity concentrations, some form of standard sample of known purity must be analyzed under exactly the same conditions as the unknown samples and the intensity of the spectral lines must be compared. Thus a spectrochemical laboratory may have many thousands of standards covering the whole range of materials likely to require analysis.

23.3.1.5 Applications

There are very few online instruments employing atomic emission techniques, but mention should be made of a continuous sodium monitor for boiler/feed water. The water is nebulized into a flame, the sodium emission is isolated by means of a monochromator, and the intensity is measured by means of a photomultiplier and associated electronics. Standard solutions are automatically fed into the instrument from time to time to check the calibration.

In both the steel and nonferrous alloy industries, large grating spectroscopes are used to control the composition of the melts before they are finally poured. A complete analysis for some 30–40 elements can be made within two minutes of a small sample being taken. Suitable additions are then made to the melt to satisfy the required composition specification. In these cases the output from the instrument is fed to a computer, which is programmed to produce actual elemental concentrations as well as the necessary amounts required to be added to known weights of melts in the furnaces for them to be of the correct composition. Analysis of water samples or samples in solution can be carried out using an inductively coupled plasma direct reading spectrometer. Some 60 elements can be determined in each sample every two minutes. The source is ionized argon pumped inductively from an RF generator into which the sample is nebulized. Temperatures of about 8,500°C are achieved. Many instruments of this type are now manufactured and have been of great value to the water industry and to environmental chemists generally—in particular, those instruments manufactured by ARL, Philips, and Jarrell Ash. Limits of detection are on the order of 1 ppb (parts per 10^9) with an accuracy of about 10 percent.

23.3.2 Atomic Absorption Spectroscopy

In emission spectroscopy, as we have already seen, the sample is excited, the emitted radiation dispersed, and the intensities of the selected lines in the emission spectrum measured. If self-absorption and induced emission are neglected, the integrated intensity of emission of a line is given by

$$\int I_\nu d\nu = CN_j F$$

where N_j is the number of atoms in the higher-energy level involved in the transition responsible for the line, F is the oscillation strength of the line, and C is a constant dependent on the dispersing and detecting systems. Assuming that the atoms are in thermal equilibrium at temperature T, the number of atoms in the excited state of excitation energy E_j is given by

$$N_j = N_0 \frac{P_j}{P_0} \exp\left(-\frac{E_j}{KT}\right)$$

where N_0 is the number of atoms in the ground state, P_j and P_0 are statistical weights of the excited and ground states, respectively, and K is Boltzmann's constant. For a spectral term with a total quantum number J_1, P is equal to $2J_1 + 1$. From the preceding equations, it can be seen that the emitted intensity depends on T and E_j. Examples of the variation of N_j/N_0 with temperature are given in Table 23.3.

In nearly all cases, the number of atoms in the lowest excited state is very small compared with the number of atoms in the ground state; the ratio only becomes appreciable at high temperatures. The strongest resonance lines of most elements have wavelengths less than 600 nm, and since temperatures in the flames used are normally less than 3,000°K, the value of N_j will be negligible compared with N_0.

In absorption, consider a parallel beam of radiation of intensity I_0, frequency ν incident on an atomic vapor of thickness 1 cm; if I_ν is the intensity of the transmitted radiation and K_ν is the absorption coefficient of the vapor at frequency ν, then

$$I_\nu = I_0 \exp(-E_\nu l)$$

From classical dispersion theory,

$$\int K_\nu d\nu = \frac{\pi e^2}{mc} N_\nu f$$

where m and e are the electronic mass and charge, respectively, c is the velocity of light, N_ν the number of atoms/cm^3 capable of absorbing radiation of frequency ν, and f the oscillator strength (the average number of electrons per atom capable of being excited by the incident radiation). Thus for a transition initiated from the ground state, where N_ν is for all practical

TABLE 23.3 Values of N_j/N_0 for various resonance lines

Resonance line	Transition	P_j/P_0	N_j/N_0			
			T = 2,000 K	T = 3,000 K	T = 4,000 K	T = 5,000 K
Cs 852.1 nm	$2S_{1/2} - 2P_{3/2}$	2	4.4×10^{-4}	7.24×10^{-3}	2.98×10^{-2}	6.82×10^{-2}
K 766.5 nm	$2S_{1/2} - 2P_{3/2}$	2	2.57×10^{-4}	4.67×10^{-3}	1.65×10^{-2}	3.66×10^{-2}
Na 589.0 nm	$2S_{1/2} - 2P_{3/2}$	2	9.86×10^{-6}	5.88×10^{-4}	4.44×10^{-3}	1.51×10^{-2}
Ca 422.7 nm	$1S_0 - 1P_1$	3	1.21×10^{-7}	3.69×10^{-5}	6.03×10^{-4}	3.33×10^{-6}
Zn 213.8 nm	$1S_0 - 1P_1$	3	7.29×10^{-15}	5.58×10^{-10}	1.48×10^{-7}	4.32×10^{-6}

purposes equal to N_0 (the total number of atoms/cm^3), the integrated absorption is proportional to the concentration of free atoms in the absorbing medium. The theoretical sensitivity is therefore increased because all the atoms present will take part in the absorption, whereas in the emission techniques only a very small number are excited and are used for detection.

In practice, the flame into which the solution is nebulized is treated as though it were the cell of absorbing solution in conventional spectrophotometry. The absorbance in the flame of light of a resonant wavelength of a particular element is a direct measure of the concentration of atoms of that element in solution being nebulized into the flame. A practical system for an atomic absorption spectrometer is shown in Figure 23.13.

When only small volumes of sample are available, the flame may be replaced by a graphite tube or rod furnace. Small volumes (10 μl) are placed on the graphite, and the latter is heated resistively in stages to about 3000°C, and the absorption of a resonant wavelength is measured as a pulse. The sensitivity of this technique is such that very low concentrations of some elements may be determined (~0.001 ppm). The limit of detection using a flame varies from element to element, from less than 1 ppm up to about 50 ppm. The technique has found wide use in analysis of solutions in virtually every industry—from "pure" water analysis to the analysis of plating solutions and from soil extracts to effluent from a steel works.

There are many manufacturers of atomic absorption spectrophotometers, and the modern instruments are very highly automated. The resonant line source is usually a high-intensity hollow cathode lamp; up to 10 of these may be contained in a turret so that each is used in turn. The flames are usually air-propane, air-acetylene, or nitrous oxide–acetylene—the hotter flames being necessary to atomize the more refractory elements. The output from the monochromator and detector is usually handled by a microprocessor so that once the instrument has been calibrated, results are automatically printed out as concentrations. Another instrument based on atomic absorption is the mercury vapor detector. A mercury vapor lamp is the resonant source, and the detector is tuned to the mercury line at 253.6 nm. Air to be sampled is passed through a tube located between source and detector, and the absorption is a measure of the mercury vapor in the air. There are many instruments manufactured for this purpose, and all are very sensitive, with limits of detection of around 0.1 ppm by volume.

23.3.3 Atomic Fluorescence Spectroscopy

This is a technique closely allied to atomic absorption. To initiate atomic fluorescence, neutral atoms in a flame cell are excited as in atomic absorption, that is, by absorption of a characteristic radiation. Fluorescence occurs when these atoms are deactivated by the emission of radiation at the same or a different wavelength. The fluorescent wavelength is characteristic of the atoms in question and its intensity is proportional to the atomic concentration. In practice, initiation is achieved with a high-intensity source, and the fluorescent signal emitted by the atomic vapor is examined at right angles by passing it into a radiation detection system. Very briefly, the basic equation relating the intensity of a fluorescent signal to atomic concentration is

$$F = 2.303\phi\, I_0 e_A lcp$$

where F is the intensity of fluorescent radiation, ϕ the quantum efficiency (which factor has to be used to account for energy losses by processes other than a fluorescence), I_0 is the intensity of the excitation radiation, e_A the atomic absorptivity at the wavelength of irradiation, l the flame path length, c the concentration of the neutral atom absorbing species, and p a proportionality factor relating to the fraction of the total fluorescence observed by the detector. Thus,

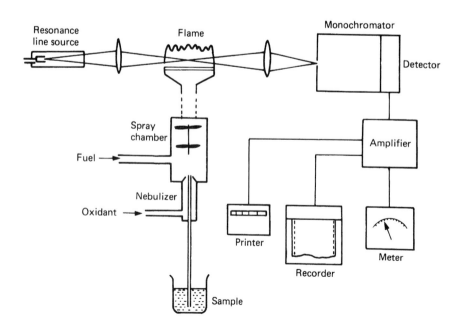

FIGURE 23.13 Practical system for atomic absorption spectrometer.

$F = K\phi I_0 c$ for a particular set of instrumental conditions, and c is proportional to F, and F will increase if the intensity of the irradiating source is increased.

There are four types of atomic fluorescence: resonance, direct-line, stepwise, and sensitized.

Resonance Fluorescence This is the most intense type of fluorescence and most widely used in practice. It occurs when the fluorescent and excitation wavelengths are the same, that is, the atom is excited from the ground state to the first excited state and then emits fluorescent energy on deactivation to the ground state.

Direct-Line Fluorescence Here the valence electron is excited to an energy level above the first excited state. It is then deactivated to a lower energy level (not the ground state), and fluorescent energy is emitted. The wavelength of fluorescence is longer than the excitation wavelength, for example, the initiation of thallium fluorescence at 535 nm by a thallium emission at 377.6 nm.

Stepwise Fluorescence This entails excitation of the atom to a high energy level. The atom is then deactivated to the first excited state. There it emits resonance radiation on returning to the ground state, for example, the emission of sodium fluorescence at 589 nm, following excitation at 330.3 nm.

Sensitized Fluorescence This occurs when the atom in question is excited by collision with an excited atom of another species and normal resonance fluorescence follows. Thallium will fluoresce at 377.6 nm and 535 nm following a collision of neutral thallium atoms with mercury atoms excited at 253.7 nm.

An instrument used to determine trace amounts of elements in solution by atomic fluorescence very simply consists of (a) an excitation source, which can be a high-intensity hollow cathode lamp, a microwave-excited electrodeless discharge tube, some spectral discharge lamps, or, more recently, a tunable dye laser; (b) a flame cell or a graphite rod as in atomic absorption; and (c) a detection system to measure the fluorescence at right angles to the line between source and flame. The detection system is usually a simple monochromator or narrowband filter followed by a photomultiplier tube, amplifier, and recording device. Limits of detection are achieved that are much lower than those obtained by atomic absorption because it is easier to measure small signals against a zero background than to measure small differences in large signals, as is done in atomic absorption. Detection limits as low as 0.0001 ppm are quoted in the literature.

23.4 X-RAY SPECTROSCOPY

23.4.1 X-ray Fluorescence Spectroscopy

Many books have been written about this technique; thus only a brief outline is given here.

The technique is analogous to atomic emission spectroscopy in that characteristic X-radiation arises from energy transferences involved in the rearrangement of orbital electrons of the target element following ejection of one or more electrons in the excitation process. The electronic transitions involved are between orbits nearer to the nucleus (see Figure 23.14).

Thus if an atom is excited by an electron beam or a beam of X-rays, electronic transitions take place and characteristic X-radiation is emitted for that atom. If, after collimation, these X-rays fall onto a crystal lattice—which is a regular periodic arrangement of atoms—a diffracted beam will only result in certain directions, depending on the wavelength of the X-rays λ, the angle of incidence θ, and atomic spacing within the crystal d. Bragg's law for the diffraction of X-rays states that $n\lambda = 2d \sin \theta$. Thus the K_α, K_β, L_α, L_β, M_α, and so on X-radiations will be diffracted at different angles. These fluorescent radiations are then collimated and detected by a variety of detectors. The intensity of these radiations is a measure of the concentration of that particular atom. Thus if a sample containing many elements is subjected to X-radiation, fluorescent radiation for all the elements present will be spread out into a spectrum, depending on the elements present and the crystal being used (see Figure 23.15).

All modern X-ray fluorescence spectrometers use this layout. The source of X-rays is usually an X-ray tube, the anode of which is chromium, tungsten, or rhodium. All types of sample can be analyzed, ranging from metals through powders to solutions. The collimator systems are based on series of parallel plates. Because their purpose is to limit the divergence of the X-ray beam and provide acceptable angular resolution, the distance between the plates must be such that the divergence embraces the width of the diffraction profile of the crystal. In general, this entails a spacing between plates of 200–500 μm.

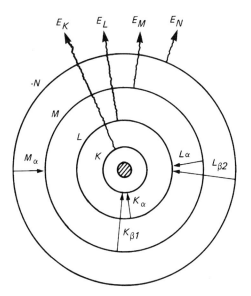

FIGURE 23.14 Transitions giving X-radiation. $(E)_{K_\alpha} = E_K - E_L$; $(E)_{K_\beta} = E_K - E_M$; $(E)_{L_\alpha} = E_L - E_M$; $(E)_{L_\beta} = E_L - E_N$; $(E)_{M_\alpha} = E_M - E_N$.

Most modern instruments can accommodate six analyzing crystals, any one of which can be automatically placed in the fluorescent X-ray beam. Table 23.4 contains a list of the types of crystal used. The detectors are either gas-flow proportional counters or scintillation counters. (See Chapter 29.) The instruments are microprocessor-controlled, and this varies the output of the X-ray source, chooses the correct crystals, and controls the samples going into the instrument. A small computer analyzes the output from the detectors and (having calibrated the instrument for a particular sample type) calculates the concentration of the elements being analyzed—allowing for matrix and inter-element effects. Instruments of this type, made by Philips, Siemens, and ARL, are widely used in the metallurgical industry because the technique—although capable of low limits of detection—is very accurate for major constituents in a sample, such as copper in brass. Analysis of atmospheric particulate pollution is carried out using X-ray fluorescence. The sample is filtered on to a paper and the deposit analyzed.

A portable instrument that uses a radioactive isotope as a source is used to monitor particular elements (depending on settings) in an ore sample before processing. This instrument is now marketed by Nuclear Enterprises. (See Chapter 30.)

Electron probe microanalysis is a technique based on the same principle as X-ray fluorescence, electrons being the exciting source, but by using electronic lenses the electron beam can be focused onto a very small area of a sample, and so analysis of areas as small as 0.1 μm diameter can be carried out. The technique can be used for looking at grain boundaries in metallurgical specimens and plotting elemental maps in suspected heterogeneous alloys. Again, this is a technique that is very specialized.

A further allied technique is photoelectron spectroscopy (PES), or electron spectroscopy for chemical analysis (ESCA). In Figure 23.14, showing the transitions within an atom to produce X-rays, we can see that some electrons are ejected from the various shells in the atom. The energy of these electrons is characteristic of that atom; so by producing an energy spectrum of electrons ejected from a sample when the latter is subjected to X-ray or intense UV radiation, the presence of various elements and their concentrations can

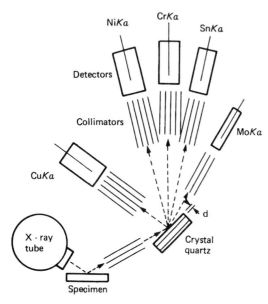

FIGURE 23.15 Multichannel spectrometer having five collimator–detector channels arranged to receive five different analyte lines, each from a different crystallographic plane (*hkil*) from the same quartz crystal.

TABLE 23.4 Analyzing crystals

Crystal	Reflection plane	2d spacing (Å) (1 Å = 0.1 nm)	Lowest atomic number detectable	
			K series	L series
Topaz	(303)	2.712	V (23)	Ce (58)
Lithium fluoride	(220)	2.848	V (23)	Ce (58)
Lithium fluoride	(200)	4.028	K (19)	In (49)
Sodium chloride	(200)	5.639	S (16)	Ru (44)
Quartz	(10$\bar{1}$1)	6.686	P (15)	Zr (40)
Quartz	(10$\bar{1}$10)	8.50	Si (14)	Rb (37)
Penta erythritol	(002)	8.742	Al (13)	Rb (37)
Ethylenediamine tartrate	(020)	8.808	Al (13)	Br (35)
Ammonium dihydrogen phosphate	(110)	10.65	Mg (12)	As (23)
Gypsum	(020)	15.19	Na (11)	Cu (29)
Mica	(002)	19.8	F (9)	Fe (26)
Potassium hydrogen phthalate	(10$\bar{1}$1)	26.4	O (8)	V (23)
Lead stearate		100	B (5)	Ca (20)

be determined. It should be pointed out that this technique is essentially a surface technique and will only analyze a few monolayers of sample. Instruments are manufactured by Vacuum Generators.

23.4.2 X-ray Diffraction

X-ray diffraction is a technique that is invaluable for the identification of crystal structure. In Section 23.4.1 we saw that crystals diffract X-rays according to Bragg's law:

$$n\lambda = 2d \sin \theta$$

Thus if a small crystal of an unidentified sample is placed in an X-ray beam, the X-rays will be diffracted equally on both sides of the sample to produce an X-ray pattern on a film placed behind the sample. The position of the lines on the film (that is, their distance from the central beam) is a function of the crystal lattice structure, and by reference to standard X-ray diffraction data, the crystals in the sample are identified. Again, this is a specialized technique and beyond the scope of this book.

Manufacturers of X-ray fluorescence spectrometers also make X-ray diffraction spectrometers. Typical uses for an instrument are the identification of different types of asbestos, and corrosion deposit studies.

23.5 PHOTO-ACOUSTIC SPECTROSCOPY

An instrument marketed by EDT Research uses this technique to study both liquid and solid samples. Figures 23.16 and 23.17 give schematic diagrams of the instrument and cell. Radiation from an air-cooled high-pressure xenon arc source, fitted with an integral parabolic mirror, is focused onto a variable-speed rotating light chopper mounted at the entrance slit of a high-radiance monochromator. The monochromator has two gratings to enable optical acoustic spectra to be obtained in the UV, visible, and near infrared. The scanning of the monochromator is completely automatic over the spectral range covered, and a range of scan rates can be selected. The exit and entrance slits provide variable-band passes of width 2–16 nm in the UV and 8–64 nm in the IR. A reflective beamsplitter passes a fraction of the dispersed radiation to a pyroelectric detector to provide source compensation and a reference signal. Source radiation is then focused onto the specially designed optoacoustic cell and sample-holder assembly. The sample cell contains a sensitive microphone and preamplifier. Special cells are used for different applications. Absorption of the radiation by the molecular species in the sample occurs and is converted to kinetic energy. The sample temperature fluctuates and causes a variation in the pressure of the gas surrounding the sample. This pressure variation is monitored by the microphone. The amplitude of the microphone signal is recorded as a function of the wavelength of the incident

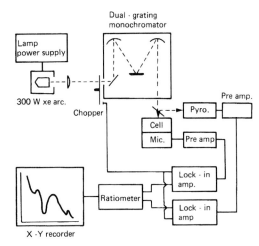

FIGURE 23.16 Photo-acoustic spectrometer layout.

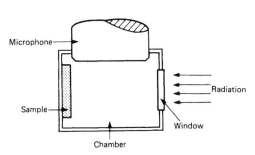

FIGURE 23.17 Schematic representation of a photo-acoustic cell employed for the examination of solid materials.

radiation to give an absorption spectrum of the sample. Typical applications include the identification of foodstuffs, blood and bloodstains, paints and inks, papers and fabrics, and pharmaceutical materials.

23.6 MICROWAVE SPECTROSCOPY

The portion of the electromagnetic spectrum extending approximately from 1 mm (300,000 MHz) to 30 cm (1,000 MHz) is called the *microwave region*. Spectroscopic applications of microwaves consist almost exclusively of absorption work in gaseous samples. With some exceptions, the various types of spectra are distinguished by their energy origins. As mentioned earlier, in the visible and UV regions the transitions between electronic energy states are directly measurable as characteristics of elements, and vibrational and rotational energies of molecules are observed only as perturbation effects. In the infrared region the vibrational spectra are observed directly as characteristic of functional groups, with rotational energies observed as perturbation effects. In the microwave region, transitions between rotational energies of molecules are observed directly as characteristic of absorbing molecules as a whole, with nuclear effects as first-order perturbations. In the radio-frequency (RF) region, the nuclear effects are directly observable. (Especially important today is the observation in the microwave region of paramagnetic resonance absorption, or PMR, as well as nuclear magnetic resonance. Both these techniques will be discussed briefly in a later section.) As in any other type of absorption spectroscopy, the instrument required consists of a source of radiation, a sample cell, and a detector. Unlike optical spectrometers, the microwave spectrometer is a completely electronic instrument requiring no dispersive components, because the source is monochromatic and any frequency can be chosen and measured with very high precision. The most common type of source is the Klystron, a specially designed high-vacuum electron tube. The output is monochromatic under any given set of conditions, and different types are available to cover various parts of the microwave spectrum. The sample cell is usually a waveguide and the detector could be silicon crystal, although bolometers and other heat-type detectors are sometimes used. In addition to the three basic components, a complete spectrometer includes provision for modulation of the absorption spectrum, an ac amplifier for the detector output, a final indicator consisting of a CRT or strip recorder, a sweep generator to vary synchronously the source frequency, a gas sample handling system, and necessary power supplies.

Since the lines in a microwave spectrum are usually completely resolved, it is only necessary to compare these measured frequencies against tables of the frequencies observed for known substances in order to identify molecules. Quantitative analysis is somewhat more complex but is based on the fact that the integrated intensity and the product of the peak height and half-width of a microwave absorption line can be directly related to the concentration of molecules per unit volume. The technique is used extensively in isotopic analysis.

23.6.1 Electron Paramagnetic Resonance (EPR)

This is really a special part of microwave spectroscopy because it usually involves the absorption of microwave radiation by paramagnetic substances in a magnetic field. A typical layout of a spectrometer is given in Figure 23.18. The electromagnet has a homogeneous gap field H that can be swept continuously from near 0 to over 50 microtesla. The sweep generator produces small modulations of the main field H at the center of the air gap. The sample cavity resonates at the Klystron frequency.

The electron, like the proton, is a charged particle; it spins and therefore has a magnetic field. It spins much faster than a proton and so has a much stronger magnetic field. Because of this and since it is lighter than a proton, the electron precesses much more rapidly in a magnetic field. Thus when microwaves travel down a waveguide and produce a rotating magnetic field at any fixed point, it can serve to flip over electron magnets in matter, just as a rotating field in a coil flips protons. If a sample is placed on

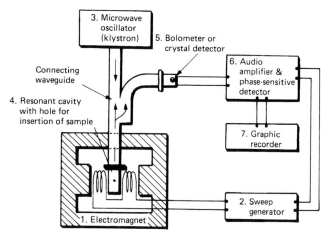

FIGURE 23.18 Block diagram of electron paramagnetic resonance spectrometer.

the sidewall of the waveguide and the microwave radiation, applied to the external magnetic field, causes the electrons to precess, then when the precession rate reaches a resonance value and the electrons flip, they extract energy from the microwaves and the reading on the recorder dips accordingly.

If the electron has not only a magnetic moment along its own spin axis but also one associated with its circulation in an atomic orbit, the electron will possess a total magnetic moment equal to the vector sum of the magnetic moments. The ratio of the total magnetic moment to the spin value is a constant for a given atom in a given environment and is called the *gyromagnetic ratio* or *spectroscopic splitting factor* for that particular electron. The facts that these ratios differ for various atoms and environments and that local magnetic fields depend on the structure of the matter permit spectral separation and EPR spectroscopy. Not all atoms and molecules are susceptible to this technique; in substances in which electrons are paired, magnetism is neutralized. But for unpaired electrons, electronic resonance occurs. This effect is observed in unfilled conduction bands, transition element ions, free radicals, and impurities in semiconductors, and, as might be expected, applications in the biological field are fruitful. The most common use is in the paramagnetic oxygen analyzer.

This same technique is now being applied by a number of companies to measure the water content in hydrocarbon streams. When applying microwave technology to measure oil in water, users must remember that there is an "inflection point" around the 80 percent water content concentration at which it is very difficult to differentiate the two streams. One must also remember that other constituents in the stream, such as silica, will be observed as one or the other phases since the device is unable to discern more than two properties. Manufacturers of this type of equipment include Agar Corporation, Honeywell, Phase Dynamics, and Multifluid Inc.

23.6.2 Nuclear Magnetic Resonance Spectroscopy

When atomic nuclei—the hydrogen proton is the simplest—are placed in a constant magnetic field of high intensity and subjected to a radio-frequency alternating field, a transfer of energy takes place between the high-frequency field and the nucleus to produce a phenomenon known as *nuclear magnetic resonance*.

If a system of nuclei in a magnetic field is exposed to radiation of frequency ν such that the energy of a quantum of radiation $h\nu$ is exactly equal to the energy difference between two adjacent nuclear energy levels, energy transitions may occur in which the nuclei may flip back and forth from one orientation to another. A quantum of energy is equally likely to tip a nucleus in either direction, so there is a net absorption of energy from the radiation only when the number of nuclei in one energy level exceeds the number in another. Under these conditions a nuclear magnetic resonance spectrum is observed. Applications of this technique include such problems as locating hydrogen atoms in solids, measuring bond lengths, crystal imperfections, and determination of crystalline and amorphous fractions in polymers.

23.7 NEUTRON ACTIVATION

Gamma ray spectroscopy is the technique by which the intensities of various gamma energies emanating from a radioactive source are measured. (See Chapter 30.) It can be used for qualitative identification of the components of radionuclide mixtures and for quantitative determination of their relative abundance. Such a situation arises in neutron activation analysis. This is a technique of chemical analysis for extremely minute traces down to ppb (parts per 10^9) of chemical elements in a sample. It employs a beam of neutrons for activation of isotopes that can then be identified, with counters, by the radioactive characteristics of the new nuclear species. This technique has been applied for the trace analysis of many elements in a variety of materials, from coal ash to catalysts, halides in phosphors, and trace impurities in many metals.

Neutron activation is also used as a way to measure level in vessels with very thick walls or high temperature in which normal sensors cannot be placed. The neutron backscatter detector is mounted outside the vessel and measures the gamma radiation "reflected" from the process inside the vessel.

23.8 MASS SPECTROMETERS

The mass spectrometer is capable of carrying out quick and accurate analysis of a wide variety of solids, liquids, and gases and has a wide range of application in process monitoring and laboratory research. Combined with the gas chromatograph, it provides an extremely powerful tool for identifying and quantifying substances that may be present in extremely small quantities.

The optical spectrometer resolves a beam of light into components according to their wavelengths; a mass spectrometer resolves a beam of positive ions into components according to their mass/charge ratio or, if all carry single elementary charges, according to their masses. As with the optical spectrometer, the mass spectrometer may be used to identify substances and to measure the quantity present.

The original mass spectrometer was devised by F. W. Aston around 1919 to measure the mass of individual positive ions. The accuracy of the instrument enabled the different masses of what appeared to be chemically identical atoms to be measured, resulting in the discovery of isotopes. Considerable development has taken place over the years, resulting in very versatile instruments having very high resolving power and sensitivity.

The resolving power of a mass spectrometer is a measure of its ability to separate ions that have a very small difference in mass. If two ions of masses M and M_2 differing in mass by ΔM give adjacent peaks in their spectrum as shown in Figure 23.19, and the height of peak is H above the baseline, then on the 10 percent valley definition the peaks are said to be resolved if the height of the valley h is less than or equal to 10 percent of the peak H, that is,

$$\left(\frac{h}{H}\right) \leq 10\%$$

The resolution is then $M_1/\Delta M$; for example, if the peaks representing two masses 100.000 and 100.005 are separated by a 10 percent valley, the resolution of the instrument is 100.000/0.005, or 20,000. Instruments with a resolution of greater than 150,000 are readily available. The sensitivity, on the other hand, is a measure of the smallest detectable quantity of the substance being identified. An example of the extreme sensitivity of modern instruments is that at a resolution of 1,000, 3 ng/s of a compound, relative molecular mass

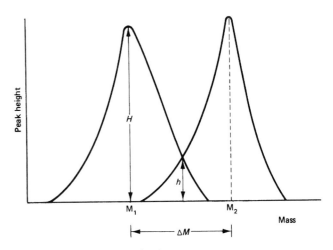

FIGURE 23.19 Peak separation for a mass spectrometer.

300, will give a spectrum with a signal-to-noise ratio of 10:1 for a peak having an intensity of 5 percent of the base peak when a mass range of 10:1 is scanned in 3 s.

The mass spectrometer has a very wide range of use in process monitoring and laboratory research. It is used in refineries for trace element survey, analysis of lubricating oils, and identifying and quantifying the substances in mixtures of organic compounds. Its use in detecting and measuring the concentration of pollutants in air, water, and solids is rapidly increasing, as is its use in biochemical analysis in medicine and other fields, particularly the analysis of drugs in biological extracts.

By means of a double-beam instrument, an unknown sample may be compared with a standard so that the unknown components are readily identified and the concentration measured. By suitable modifications an instrument can be made to provide an energy analysis of electrons released from the surface of a sample by X-radiation, or ultraviolet light.

23.8.1 Principle of the Classical Instrument

There are many different types of mass spectrometers, but the ones described here are the most commonly used.

In all types the pressure is reduced to about 10^5 N/m^2 to reduce collisions between particles in the system. The spectrometer consists of an inlet system by which the sample is introduced into the region in which ions of the sample are produced. The separation of ions according to their mass-to-charge ratio may be achieved by magnetic or electric fields or by a combination of both. The differences between the various types of mass spectrometer lie in the manner in which the separation is achieved. In the instrument illustrated in Figure 23.20, the ions are accelerated by an electrical potential through accelerating and defining slits into the electrostatic analyzer, where ions that have energies within a restricted band are brought to a focus at the monitor slit, which intercepts a portion of the ion beam. They then enter the electromagnetic analyzer, which gives direction and mass focusing. This double focusing results in ions of all masses being focused simultaneously along a given plane. The ions can be recorded photographically on a plate over a period of time to give a very high sensitivity and reduction of the effects of ion-beam fluctuation.

Alternatively, the accelerating or deflecting field may be arranged so that ions of a given mass are focused on a detector, which may consist of a plate or, if initial amplification of the charge is required, onto an electron multiplier or scintillation detector. By arranging the deflecting field to change in a predetermined manner, the instrument may be arranged to scan a range of masses and so record the abundance of ions

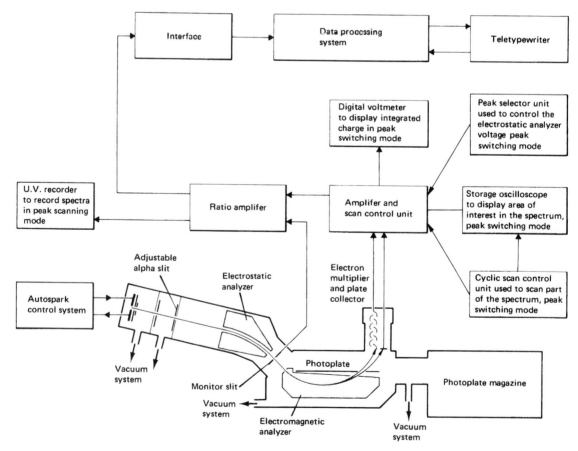

FIGURE 23.20 Schematic diagram of the complete system of a spark source mass spectrometer. Courtesy of Kratos Ltd.

CHAPTER | 23 Chemical Analysis: Spectroscopy

of each particular mass. Such a record is known as a *mass spectrum*, and mathematical analysis of this mass spectrum enables the composition of the sample to be determined. Mass spectra obtained under constant conditions of ionization depend on the structure of the molecules from which the ions originate. Each substance has its own characteristic mass spectrum, and the mass spectrum of a mixture may therefore be analyzed in terms of the spectra of the pure components and the percentage of the different substances in the mixture calculated.

Analysis of the mass spectrum of a mixture may involve the solution of a large number of simultaneous equations, which can be accomplished using a microprocessor or a small computer.

23.8.2 Inlet Systems

The mode of introduction of the sample into the ion source depends on the nature of the sample and, in particular, its volatility.

The simplest system designed to introduce reference compounds into the ion source includes a 35 cm^3 reservoir into which the compound is injected through a septum. Flow into the ion source is through a molecular leak, and a shut-off valve is provided. Facilities for pumping out the system and obtaining temperatures up to 100°C are provided.

Relatively volatile gases and liquids may be introduced by a probe attached to a small reservoir into which the sample is injected and from which it flows to the ion source at a controlled rate. The temperature of the system may be controlled between ambient and 150°C.

For less volatile substances, an all-glass heated system may be used. Glass is used for the system so that catalytic decomposition of the sample is reduced to a minimum. The system can be operated at temperatures up to 350°C and incorporates its own controlled heating and temperature-monitoring facilities. It includes both large and small reservoirs to enable a wide range of quantities of liquid or solid samples to be introduced.

To introduce less volatile and solid samples into the ion chamber, a probe may be used. The sample is loaded onto the tip of the probe, which is inserted into the ion source through a two-stage vacuum lock.

The probe may be heated or cooled independently of the ion chamber as required, from −50 to +350°C. The temperature is measured by a platinum resistance thermometer that forms part of the temperature control system, which enables the temperature to be set from the instrument control panel.

Effluents from a gas chromatograph column usually flow at about 50 cm^3/min and consist mainly of carrier gas. To reduce the flow, the gas is passed through a molecular separator designed to remove as much as possible of the carrier gas but permitting the significant components to pass into the mass spectrometer.

23.8.3 Ion Sources

In the system shown, the ions are produced by a spark passed between electrodes formed from the sample by applying a controlled pulsed RF voltage. Positive ions representative of the sample are produced in the discharge and are accelerated through a simple ion gun. This beam is defined by resolving slits before it passes into the analyzer section.

Other methods may be employed to produce ions of the sample, which are impelled toward the exit slit by a small positive potential in the ion chamber. These methods involve increasing the energy of the sample by some form of radiation. Organic compounds require photons of energy up to 13 eV to produce ionization so that a high-energy beam of short-wavelength radiation is sufficient. Where energies greater than 11 eV are required, window materials become a problem, so the photon source has to emit radiation directly into the ion source. A helium discharge at 21.21 eV provides a convenient source of photons capable of ionizing all organic compounds.

Electrons emitted by a heated filament and accelerated by about 70 eV and directed across the ion chamber may also be used to ionize many substances. Although 70 eV produces the maximum ion yield, any voltage down to the ionization voltage of the compound studied may be used.

The electric field production near a sharp point or edge at a high potential will have a high potential gradient and may be used to produce ions. Ions can also be formed by the collision of an ion and a molecule. This method can produce stable but unusual ions, such as

$$CH_4^+ + CH_4 \rightarrow CH_5^+ + CH_3$$

and is most efficient at pressures of about 10^{-1} N/m^2.

It is most important to realize that the process of producing ions from molecules will in many cases split the original molecule into a whole range of ions of simpler structure, and the peak of maximum height in the spectrum does not necessarily represent the ion of the original molecule. For example, the mass spectrum of *m*-xylene $C_6H_4(CH_3)_2$ may contain 22 peaks of different *m/e* values, and the peak of maximum height represents a *m/e* ratio of 91, whereas the ions having the next highest peak have a *m/e* ratio of 106.

23.8.4 Separation of the Ions

The mass spectrometer shown in Figure 23.20 employs the Mattauch-Herzog geometry, but other forms of geometry achieve a similar result.

The positive ions representative of the sample produced in the ion source are accelerated by a controlled electrostatic field in a simple gun, the spread of the ions being controlled by the resolving slits. If an ion of mass *m* and charge *e* can be regarded as starting from rest, its velocity *v* after falling through a potential V Volts will be represented by the equation

$$\frac{1}{2}mV^2 = eV$$

The ion beam then passes through the electrostatic analyzer, where it passes between two smooth curved plates that are at different potentials, such that an electrostatic field B exists between them and at right angles to the path of the ions. The centrifugal force on the ions will therefore be given by

$$\frac{mV^2}{r} = eB$$

Combining the equations, we see that the radius of curvature r of the path will be given by

$$r = \frac{mV^2}{eB} = \frac{2eV}{eB} = \frac{2V}{B}$$

Thus the curvature of the path of all ions will depend on the accelerating and deflecting fields only and will be independent of the mass/charge ratio. Therefore, if the field B is kept constant, the electrostatic analyzer focuses the ions at the monitor slit in accordance with their translational energies. The monitor slit can be arranged to intercept a given portion of the beam. The energy-focused ion beam is then passed through the electromagnetic analyzer, where a magnetic field at right angles to the electrostatic field is applied (i.e., at right angles to the plane of the diagram). Moving electric charges constitute an electric current, so if each carries a charge e and moves with a velocity v at right angles to a uniform magnetic field H, each particle will be subject to a force F, where $F = Hev$ in a direction given by Fleming's left-hand rule, that is, in a direction mutually at right angles to the magnetic field and the direction of the stream. Thus the ions will move in a curved path radius r such that

$$\frac{mv^2}{r} = Hev$$

or

$$r = \frac{mv^2}{Hev} = \frac{mv}{He}$$

but

$$mv^2 = 2eV \quad \text{or} \quad v = \sqrt{\left(\frac{2eV}{m}\right)}$$

$$\therefore r = \left(\frac{m}{eH}\right)\sqrt{\left(\frac{2eV}{m}\right)}$$

or

$$r^2 = \left(\frac{m^2}{e^2H^2}\right)\left(\frac{2eV}{m}\right)$$

$$= \left(\frac{2V}{H^2}\right)\left(\frac{e}{m}\right)$$

or

$$\frac{m}{e} = \frac{(H^2r^2)}{2V}$$

At constant values of the electrostatic and electromagnetic fields, all ions of the same m/e ratio will have the same radius of curvature. Thus, after separation in the electromagnetic analyzer, ions with a single charge will be brought to a focus along definite lines on the photographic plate according to their mass, starting with the lowest mass on the left-hand edge of the plate and increasing to the highest mass on the right.

The ions will therefore give rise to narrow bands on the photographic plate, and the density of these bands will be a measure of the number of ions falling on the band. The sensitivity range of the plate is limited, and it is necessary to make several exposures for increasing periods of time to record ions that have a large ratio of abundance. Using long exposure, ions that are present in very low abundances may be accurately measured. The intensity of the photographic lines after development of the plate may be compared with a microphotometer similar to that used with optical spectrometers.

Because all ions are recorded simultaneously, ion beam fluctuations affect all lines equally, and the photographic plate also integrates the ions over the whole of the exposure.

The instantaneous monitor current may be measured and used to control the sparking at the electrodes at optimum by adjusting the gap between the electrodes.

The integrated monitor current is a guide to the exposure, and the range of masses falling on the photographic plate may be controlled by adjustment of the value of the electrostatic and magnetic fields.

The plate collector and the electron multiplier detection systems enable quantitative analysis to be carried out with greater speed and precision than with the photographic plate detector. For high sensitivity, the ions may be caused to fall on the first dynode of the electron multiplier and the final current further amplified and recorded on the ultraviolet sensitive strip recorder. The logarithmic ratio of the monitor and collector signals is used in recording spectra to minimize the errors due to variations in the ion beam.

In the peak switching mode, the operator can select the peaks of interest, display them on an oscilloscope, and examine them with greater precision. Increasing the resolving power of the instrument will enable what may initially appear to be a single peak to be split into its components, representing ions differing in mass by a small amount.

Provision is made for changing the amplification in logarithmic steps so that a wide range of abundances may be measured. Where a rapid qualitative and semiquantitative analysis is required for a very wide range of masses, consecutive masses are swept across the multiplier collector by allowing the magnet current to decay from a preset value at a preset rate while the accelerating voltage is kept constant. Values of ion current from the individual ion species received at the detector are amplified and instantaneously compared with a fraction of the total ion current at the

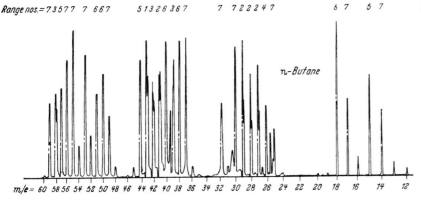

FIGURE 23.21 Ultraviolet-sensitive strip recording.

monitor by means of two logarithmic amplifiers that feed into a summing amplifier. This gives a signal proportional to the relative ion concentrations, which can be recorded on the ultraviolet-sensitive strip recorder and has the form shown in Figure 23.21.

Where large amounts of data are generated, the output from the ratio detector of the electrical detection system can be fed through a suitable interface into a data acquisition and processing system. If necessary this system can be programmed to print out details of the elements present in the sample with an indication of their concentration.

23.8.5 Other Methods of Separation of Ions

23.8.5.1 Time-of-Flight Mass Spectrometer

This type of instrument is shown schematically in Figure 23.22. It has a relatively low resolution but a very fast response time.

In this instrument, the ions are accelerated through a potential V, thus acquiring a velocity v given by:

$$\frac{1}{2}mv^2 = eV \quad \text{or} \quad v = \left[2V\left(\frac{e}{m}\right)\right]^{\frac{1}{2}}$$

If the ions then pass through a field-free (drift) region of length d to the detector, the time of transit t will be d/v. That is,

$$t = \frac{d}{\left[2V\left(\frac{e}{m}\right)\right]^{1/2}} = \left[\left(\frac{e}{m}\right)2d^2V\right]^{1/2}$$

Thus, the ions will arrive at the detector after times proportional to $(m/e)^{1/2}$. The time intervals between the arrival of ions of different mass at the detector are usually very short, and the mass spectrum is most conveniently displayed on a cathode ray tube. The time-of-flight mass spectrometer occupies a unique place in mass spectrometry because it provides a simple, rapid measurement of the abundance of various isotopes or elements comprising a sample. In practice,

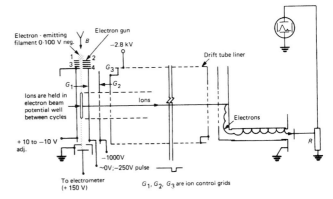

FIGURE 23.22 Time-of-flight spectrometer.

10,000 to 100,000 spectra can be scanned per second. With the aid of suitable electronic circuitry, it is possible to monitor reaction rates and to investigate reaction profiles of only 100 μs duration. Longer-length drift tubes have also contributed to improved mass resolution. It is also possible to scan from 0 to 900 atomic mass units in 1.5 seconds; furthermore, to prevent multiplier saturation when very large ion peaks are present near smaller peaks, appropriate "gating" peaking can be applied to the multiplier. Thus it is possible to suppress mass 40 without interfering with the recording of mass 39 or 41. This has extended the practical range of sensitivity in identifying gas chromatograph effluent by orders of magnitude.

23.8.5.2 Quadrupole Mass Spectrometer

The quadrupole mass spectrometer type of instrument is particularly suited to vacuum system monitoring and to a wide range of gas analysis. Although it has a relatively modest resolving power (about 16,000 maximum) it has the advantages of compactness, robustness, and relatively low cost.

Ions, produced by bombarding the sample with electrons from a filament assembly, are extracted electrostatically from the ionizer and focused by electrostatic lenses into the quadrupole mass filtering system. The latter consists of two pairs of metal rods, precisely aligned and housed in a chamber at a pressure of 2.6×10^{-4} N/m^2. One pair is connected to a source of dc voltage; the other is supplied by an RF voltage. Combination of the dc and RF voltages creates a hyperbolic potential distribution. The applied voltages increase uniformly from zero to a given maximum and then drop to zero again—a voltage sweep that is then repeated. Most ions entering the quadrupole field will undergo an oscillating trajectory of increasing amplitude so that they will eventually be collected on one of the electrodes. However, at any given time, ions of one specific mass/charge ratio are deflected as much to one electrode as to another and are passed by the filter.

As the voltages are swept from zero to their maximum values, the entire mass range is scanned. After passing through the mass filter, the ions impinge on an electron multiplier, and a signal proportional to the collected ion current can be displayed on an oscilloscope or recorder. As the voltages increase, the position of the mass peaks is linearly related to mass, making the spectrum easy to interpret. The instrument covers mass ranges up to about 400 amu. Modern instruments are able to detect partial pressure in the 10^{-13} torr range. They are equipped with variable mass scanning sweeps so that rapidly changing concentrations of gases can be monitored on a continuing basis. There are many other types of ion separators; for details on these, the reader should consult textbooks devoted to mass spectroscopy. Among these types are multiple magnet systems, the cycloidal mass spectrometer, cyclotron resonance types, and RF mass filters.

REFERENCES

Bertin, E. P., *Principles and Practice of X-ray Spectrographic Analysis*, Plenum Press, New York (1970).

Ebdon, L., *An Introduction to Atomic Absorption Spectroscopy: A Self Teaching Approach*, Heyden, London (1982).

Jenkins, R., Gould, R. W., and Gedcke, D., *Quantitative X-ray Spectrometry*, Marcel Dekker, New York (1981).

Price, W. J., *Spectrochemical Analysis by Atomic Absorption*, Heyden, London (1979).

Royal Society of Chemistry, *Annual Reports on Analytical Atomic Spectroscopy*.

Slavin, W., *Atomic Absorption Spectroscopy* (2nd ed.), Wiley, Chichester, U.K. (1978).

Tertian, R. and Claisse, F., *Principles of Quantitative X-ray Fluorescence Analysis*, Heyden, London (1982).

Welvy, E. L. (ed.), *Modern Fluorescence Spectroscopy*, Plenum Press, New York (1981).

White, Fr. A., *Mass Spectrometry in Science and Technology*, Wiley, Chichester, U.K. (1968).

FURTHER READING

Alfassi, Z. B. (ed.), *Activation Analysis*, Vols. I and II, CRC Press, Boca Raton, Fla. (1990).

Izyumov, Y. A. and Chernoplekov, N. A., *Neutron Spectroscopy*, Plenum Publishing, New York (1992).

Hendra, P., et al., *Fourier Transform Raman Spectroscopy: Instrumental and Chemical* Applications, Eelis Horwood, Chichester, U.K. (1991)

Clark, B. J., et al., *UV Spectroscopy: Techniques, Instrumentation and Data Handling*, Chapman & Hall, London (1993).

Parry, S. J., *Activation Spectrometry in Chemical Analysis*, Wiley, Chichester, U.K. (1991).

Chapter 24

Chemical Analysis: Electrochemical Techniques

W. G. Cummings and K. Torrance; edited by I. Verhappen

24.1 ACIDS AND ALKALIS

To appreciate electrochemical techniques of chemical analysis it is necessary to have an understanding of how substances dissociate to form ions.

All acids dissociate when added to water to produce hydrogen ions in the solution, for example, nitric acid:

$$HNO_3 \rightleftharpoons H^+ + NO_3^-$$

The extent to which dissociation takes place varies from acid to acid and increases with increasing dilution until, in very dilute solutions, almost all the acid is dissociated.

According to the ionic theory, the characteristic properties of acids are attributed to the hydrogen ions (H^+) that they produce in solution. Strong acids (nitric, sulfuric, hydrochloric) are those that produce a large concentration of hydrogen ions when added to water. As a result the solutions are excellent conductors of electricity. Weak acids such as carbonic acid (H_2CO_3) and acetic acid (CH_3COOH), when dissolved in water, produce small concentrations of hydrogen ions, and their solutions are poor conductors of electricity.

The strength of a weak acid is indicated by its dissociation constant K, which is defined as

$$K = \frac{[A^-][H^+]}{[HA]}$$

where $[A^-]$ is the molar concentration of the acidic ions, $[H^+]$ is the concentration of hydrogen ions, and $[HA]$ is the concentration of undissociated acid.

The dissociation constant K varies with temperature, but, at a given temperature, if a little more acid is added to the solution, a portion of it dissociates immediately to restore the relative amount of ions and undissociated acid to the original value.

Similarly, the typical properties of alkalis in solution are attributed to hydroxyl ions (OH^-). Strong alkalis such as sodium hydroxide (NaOH) produce large concentrations of hydroxyl ions when added to water, but weak alkalis such as ammonium hydroxide (NH_4OH) are only slightly ionized in water and produce much smaller concentrations of hydroxyl ions.

As with weak acids, the strength of a weak base is indicated by its dissociation constant

$$K = \frac{[B^+][OH^-]}{[BOH]}$$

where $[B^+]$ is the concentration of alkaline ions, $[OH^-]$ is the concentration of hydroxyl ions, and $[BOH]$ is the concentration of undissociated alkali.

Strong electrolytes have no dissociation constant; the expression for strong acids $[A^-][H^+]/[HA]$ and the corresponding expression for alkalis vary considerably with change in concentration. With strong acids and alkalis the apparent degree of ionization can be taken as a measure of the strength of the acid or base.

So far it has been assumed that the effective concentrations or active masses could be expressed by the stoichiometric concentrations, but, according to modern thermodynamics, this is not strictly true. For a binary electrolyte $AB \rightleftharpoons A^+ + B^-$, the correct equilibrium equation is

$$K_a = \frac{a_{A^+} \times a_{B^-}}{a_{AB}}$$

where a_{A^+}, a_{B^-}, and a_{AB} represent the activities of A^+, B^-, and AB, and K_a is the thermodynamic dissociation constant. The thermodynamic quantity "activity" is related to concentration by a factor called the *activity coefficient*, that is, activity = concentration $\times$ activity coefficient.

Using this concept, the thermodynamic activity coefficient is

$$K_a = \frac{[A^+][B^-]}{[AB]} \times \frac{f_{A^+} \times f_{B^-}}{f_{AB}}$$

where f refers to the activity coefficients and the square brackets to the molar concentrations. The activity coefficients of unionized molecules do not differ much from unity, so for weak electrolytes in which the ionic concentration, and therefore the ionic strength, is low, the error introduced by neglecting the difference between the actual values of the activity coefficients of the ions, f_{A^+} and f_{B^-}, and unity is small (less than 5 percent). Hence for weak electrolytes, the constants obtained by using the simpler equation $K = [A^+][B^-]/[AB]$ are sufficiently precise for the purposes of calculation in quantitative analysis. Strong electrolytes are assumed to be completely dissociated, and no correction for activity coefficients needs to be made for dilute solutions.

However, the concept of activity is important in potentiometric techniques of analysis (described later). The activity coefficient varies with concentration, and for ions it varies with the charge and is the same for all *dilute* solutions having the same ionic strength. The activity coefficient depends on the total ionic strength of a solution (a measure of the electrical field existing in the solution) and for ion-selective work it is often necessary to be able to calculate this. The ionic strength I is given by

$$I = 0.5 \sum C_i Z_i^2$$

where C_i is the ionic concentration in moles per liter of solution and Z_i is the charge of the ion concerned. Thus, the ionic strength of 0.1 M nitric acid solution (HNO_3) containing 0.2 M barium nitrate [$Ba(NO_3)_2$] is given by

$$0.5 \big[0.1(\text{for } H^+) + 0.1(\text{for } NO_3^-)$$
$$+ 0.2 \times 2^2 (\text{for } Ba^{++})$$
$$+ 0.4 \times 1 (\text{for } NO_3^-) \big]$$
$$= 0.5[1.4] = 0.7$$

24.2 IONIZATION OF WATER

As even the purest water possesses a small but definite electrical conductivity, water itself must ionize to a very slight extent into hydrogen and hydroxyl ions:

$$H_2O \rightleftharpoons H^+ + OH^-$$

This means that at any given temperature

$$\frac{a_{H^+} \times a_{OH^-}}{a_{H_2O}} = \frac{[H^+] \cdot [OH^-]}{[H_2O]} \times \frac{f_{H^+} \cdot f_{OH^-}}{f_{H_2O}} = K$$

where a_x, $[x]$ and f_x refer to the activity, concentration, and activity coefficient of the species X, and K is a constant.

As water is only slightly ionized, the ionic concentrations are small and the activity coefficients of the ions can therefore be regarded as unity. The activity coefficient of the unionized molecule H_2O may also be taken as unity, and the previous expression therefore reduces to

$$\frac{[H^+] \times [OH^-]}{[H_2O]} = K$$

In pure water, too, because there is only very slight dissociation into ions, the concentration of the undissociated water $[H_2O]$ may also be considered constant and the equation becomes $[H^+] \times [OH^-] = K_w$. The constant K_w is known as the ionic product of water.

Strictly speaking, the assumptions that the activity coefficient of water is constant and that the activity coefficients of the ions are unity are only correct for pure water and for very dilute solutions in which the ionic strength is less than 0.01. In more concentrated solutions, the ionic product for water will not be constant, but because activity coefficients are generally difficult to determine, it is common usage to use K_w.

The ionic product of water, K_w, varies with temperature and is given by the equation

$$\log_{10} K_w = 14.00 - 0.033(t - 25) + 0.00017(t - 25)^2$$

where t is the temperature in °C.

Conductivity measurements show that, at 25°C, the concentration of hydrogen ions in water is 1×10^{-7} mol liter^{-1}. The concentration of hydroxyl ions equals that of the hydrogen ions, therefore, $K_w = [H^+] \times [OH^-] = 10^{-14}$. If the product of $[H^+]$ and $[OH^-]$ in aqueous solution momentarily exceeds this value, the excess ions will immediately recombine to form water. Similarly, if the product of the two ionic concentrations is momentarily less than 10^{-14}, more water molecules will dissociate until the equilibrium value is obtained. Since the concentrations of hydrogen and hydroxyl ions are equal in pure water, it is an exactly neutral solution. In aqueous solutions where the hydrogen ion concentration is greater than 10^{-7}, the solution is acid; if the hydrogen ion concentration is less than 10^{-7} the solution is alkaline.

24.3 ELECTRICAL CONDUCTIVITY

24.3.1 Electrical Conduction in Liquids

As early as 1833, Michael Faraday realized that there are two classes of substances that conduct electricity. In the first class are the metals and alloys and certain nonmetals such as graphite, which conduct electricity without undergoing any chemical change. The flow of the current is due to the motion of electrons within the conductor, and the conduction is described as *metallic* or *electronic*.

In the second class are salts, acids, and bases that, when fused or dissolved in water, conduct electricity owing to the fact that particles known as *ions*, carrying positive or negative electric charges, move in opposite directions through the liquid. It is this motion of electrically charged particles that constitutes the current. Liquids that conduct electricity in this manner are known as *electrolytes*.

24.3.2 Conductivity of Solutions

The passage of current through an electrolyte generally obeys Ohm's law, and the current-carrying ability of any portion of electrolyte is termed its *conductance* and has the units of reciprocal resistance ($1/\Omega$), siemens (S). The specific current-carrying ability of an electrolyte is called its *conductivity* and consequently has the units of $S\ m^{-1}$.

The conductivity of electrolytes varies greatly with their concentration because dilution (a) increases the proportion of the dissolved electrolyte, which forms ions in solution, but (b) tends to reduce the number of these ions per unit of volume. To measure the first effect alone another term, *molar conductivity*, Λ, is defined,

$$\Lambda\left(S\ \frac{m^2}{mol}\right) = \frac{\kappa}{c},$$

where κ is the conductivity and c is the concentration in mol m^{-3}. Although these are the basic SI units, most work is reported using volume units of cm^3, since the liter is a convenient volume for laboratory use and Λ is usually in units of S cm^2/mol.

At infinite dilution the ions of an electrolyte are so widely separated by solvent molecules that they are completely independent and the molar conductivity is equal to the sum of the ionic conductivities, $\lambda°$, of the cation and anion, that is,

$$\Lambda_\infty = \lambda°_- + \lambda°_+$$

The values of $\lambda°$ are the values for unit charge, referred to as *equivalent ionic conductivities* at infinite dilution. The general case is

$$\Lambda_\infty = z_+ n_+ \lambda°_+ + z_- n_- \lambda°_+$$

where z is the charge on the ion and n the number of these ions produced by dissociation of one molecule of the salt, for example,

$$\lambda_\infty(LaCl_3) = 3 \times 1 \times \lambda°_{La} + \times 3 \times \lambda°_{Cl}$$

Since, for example, the ionic conductivity of the chloride ion is the same in all chloride salts, the molar conductivity at infinite dilution of any chloride salt can be calculated if the corresponding value for the cation is known. Values of ionic conductivities at infinite dilution at 25°C are given in Table 24.1.

TABLE 24.1 Limiting ionic conductivities at 25°C

Cation	$\lambda°$ S cm^2/mol	Anion	$\lambda°$ S cm^2/mol
H^+	349.8	OH^-	199.1
Li^+	38.7	F^-	55.4
Na^+	50.1	Cl^-	76.4
K^+	73.5	Br^-	78.1
NH_4^+	73.6	I^-	76.8
$(CH_3)_2NH_2^+$	51.9	NO_3^-	71.5
$\frac{1}{2}Mg^{2+}$	53.1	ClO_4^-	64.6
$\frac{1}{2}Ca^{2+}$	59.5	Acetate	40.9
$\frac{1}{2}Cu^{2+}$	53.6	$\frac{1}{2}SO_4^{2-}$	80.0
$\frac{1}{2}Zn^{2+}$	52.8	$\frac{1}{2}CO_3^{2-}$	69.3

Providing that the concentration of a fully dissociated salt is less than about 10^{-4} mol/l, the conductivity κ at 25°C can be calculated from

$$\kappa(S\ cm^{-1}) = zn(\lambda°_+ + \lambda°_-)c\ 10^{-3}$$

or

$$\kappa(\mu S\ cm^{-1}) = zn(\lambda°_+ + \lambda°_-)c\ 10^3$$

where c is the concentration in mol/l.

Values of limiting ionic conductivities in aqueous solution are highly temperature dependent, and in some cases the value increases five- or six-fold over the temperature range 0–100°C (see Table 24.2). These changes are considered to be due mainly to changes in the viscosity of water and the effect this has on the mobility and hydration of the ions.

24.3.3 Practical Measurement of Electrical Conductivity

From the foregoing, it can be seen that measurement of electrical conductivity enables concentration to be determined.

24.3.3.1 Alternating Current Cells with Contact Electrodes

Conductivity cells provide the means of conducting a small, usually alternating, current through a precise volume of liquid, the conductivity of which we want to know. At its simplest, this process involves the measurement of the resistance between two electrodes of fixed shape and constant distance apart. The relationship between the specific

TABLE 24.2 Ionic conductivities between 0 and 100°C (S cm²/mol)

Ion	0°	5°	15°	18°	25°	35°	45°	55°	100°
H^+	225	250.1	300.6	315	349.8	397.0	441.4	483.1	630
OH^-	105	–	165.9	175.8	199.1	233.0	267.2	301.4	450
Li^+	19.4	22.7	30.2	32.8	38.7	48.0	58.0	68.7	115
Na^+	26.5	30.3	39.7	42.8	50.1	61.5	73.7	86.8	145
K^+	40.7	46.7	59.6	63.9	73.5	88.2	103.4	119.2	195
Cl^-	41.0	47.5	61.4	66.0	76.4	92.2	108.9	126.4	212
Br^-	42.6	49.2	63.1	68.0	78.1	94.0	110.6	127.4	–
I^-	41.4	48.5	62.1	66.5	76.8	92.3	108.6	125.4	–
NO_3^-	40.0	–	–	62.3	71.5	85.4	–	–	195
ClO_4^-	36.9	–	–	58.8	67.3	–	–	–	185
Acetate	20.1	–	–	35	40.9	–	–	–	–
$\frac{1}{2}Mg^{2+}$	28.9	–	–	44.9	53.0	–	–	–	165
$\frac{1}{2}Ca^{2+}$	31.2	–	46.9	50.7	59.5	73.2	88.2	–	180
$\frac{1}{2}SO_4^-$	41	–	–	68.4	80.0	–	–	–	260

conductivity κ of the solution and the resistance R across the electrodes includes a cell constant a such that

$$\kappa = \frac{a}{R}$$

If we express the conductivity in units of S cm⁻¹, the cell constant has the dimension of cm⁻¹. To simplify the electrical circuits of the measuring instruments, it is customary to maintain the resistance of conductivity cells between the limits of 10 and 100,000 Ω. The conductivity of aqueous solutions varies from pure water, with a conductivity of about 5 μ/m, to those of concentrated electrolytes, with conductivities as high as 1000 S/m. To keep within these resistance limits it is necessary, therefore, to have cells with a range of cell constants from 0.01 to 100 cm⁻¹. A working guide to the most appropriate value of cell constant for any given range of conductivity is shown in Table 24.3.

To measure the conductivity accurately it is necessary to know the cell constant accurately. It is usual to determine the cell constant by preferably (a) measuring the conductance when the cell is filled with a solution of which the conductivity is accurately known or, failing that, (b) comparing the measured conductance with that obtained from a cell of known cell constant when both cells contain the same solution at the same temperature.

TABLE 24.3 Guide to cell constant for known conductivity range

Conductivity range μS cm⁻¹	Cell constant cm⁻¹
0.05 to 20	0.01
1 to 200	0.1
10 to 2000	1
100 to 20 000	10
100 to 200 00	50

The only solutions for which conductivities are known with sufficient accuracy to be used for reference purposes are aqueous solutions of potassium chloride. This salt should be of the highest purity, at least analytical reagent grade, and dried thoroughly in an oven at 120°C before preparing solutions by dissolving in deionized water whose conductivity is less than 2 μS/cm at room temperature. The most accurate reference solutions are prepared by weight, and the two most useful solutions are given in Table 24.4.

For many purposes a simpler procedure can be followed. This involves weighing only the potassium chloride and preparing solutions by volume at 20°C; these details are given in Table 24.5.

Calibration of conductivity cells by these solutions requires considerable care if accurate values of cell constants are to be determined. The importance of temperature control cannot be over-emphasized since the conductivity of the potassium chloride solution will change by over 2 percent per Kelvin. Alternatively, the cell constant can be determined by the comparison technique with identical rather than standard conditions in both the "known" and "unknown" cells. Equally important as the effect of temperature is that of polarization in these cells where the electrodes contact the solution and conduct a significant current.

The extent of polarization depends on a number of factors, the most important of which are the nature of the electrode surface and the frequency of the ac signal applied to the cell. The restrictions that polarization errors, arising from electrode material, impose on the choice of cell mean that cells with bright metal electrodes are best suited for measurements of low conductivities, where the proportion of the total resistance due to polarization is very small. Treated or coated electrodes are suitable for low (~0.05 μS cm^{-1}) to intermediate (~0.1 S m^{-1}) conductivities, provided that the frequency of the ac voltage is in the range normally found in commercial instruments (50–1,000 Hz).

Polarization in all the cells we have been discussing can be reduced by increasing the frequency of the applied voltage. This can best be appreciated by considering Figure 24.1, in which the apparent cell constant over a range of conductivities is plotted against three values of ac frequency. The true value of the cell constant was 1 cm^{-1}, and it can be seen that the highest frequency, 3.5 kHz, gave the true value for the cell constant over the widest concentration range. Unfortunately, increase of frequency can introduce capacitative errors into the measurement, particularly from the signal cable, and in many applications the choice of operating frequency is a compromise. Although variable frequency conductivity meters are available as laboratory instruments (e.g., Philips Model PW 9509, High Performance Conductivity Meter), such a facility is not usually found on industrial instruments. In this case it is necessary to consider the range of conductivities to be measured, together with the chemical and physical nature of the solutions to be measured, before specifying the operating frequency. All determinations of cell constant should be carried out at this frequency.

Cell Construction The materials used in cell construction must be unaffected by the electrolyte, and the insulation between the electrodes must be of a high quality and not absorb anything from the process liquid.

A wide range of materials are at present available and cover a wide range of pressures, temperatures, and process

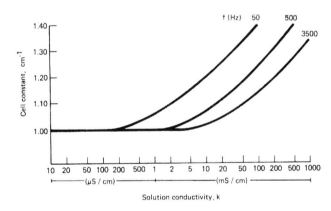

FIGURE 24.1 Effect of frequency on the useful range of a cell with titanium carbide coated stainless steel electrodes. Courtesy of F. Oehme, Polymetron.

TABLE 24.4 Standard solutions for cell calibration

Solution g KCl/ 1000 g solution*	κ at 18°C S m^{-1}	κ at 25°C S m^{-1}
(A) 7.4191	1.1163	1.2852
(B) 0.7453	0.12201	0.14083

*All values are "mass in vacuo."

TABLE 24.5 Standard solutions (volumetric) for cell calibration

Solution	κ at 18°C	κ at 25°C
(A') 7.4365 g KCl/l at 20°C	1.1167 S m^{-1}	1.2856 S m^{-1}
(B') 0.7440 g KCl/l at 20°C	0.1221 S m^{-1}	0.1409 S m^{-1}
(C') 100 ml of solution B' made up to 1 liter at 20°C	–	146.93 μS cm^{-1}

*For the highest accuracy, the conductivity of the dilution water should be added to this value.

fluids. The body may be made of glass, epoxy resins, plastics such as PTFE (pure or reinforced), PVC, Perspex, or any other material suitable for the application, but it must not be deformed in use by temperature or pressure; otherwise, the cell constant will change.

The electrodes may be parallel flat plates or rings of metal or graphite cast in the tube forming the body or in the form of a central rod with a concentric tubular body.

One common form of rod-and-tube conductivity cell consists of a satinized stainless-steel rod electrode surrounded by a cylindrical stainless-steel electrode, with holes to permit the sample to flow freely through the cell. This is surrounded by an intermediate cylinder, also provided with holes, and two O-rings that together with the tapered inner end form a pressure-tight seal onto the outer body when the inner cell is withdrawn for cleaning, so the measured solution can continue to flow and the cell be replaced without interruption of the process. The outer body is screwed into the line through which the measured solution flows. Figure 24.2(a) shows the inserted cell as it is when in use, and (b) shows the withdrawn measuring element with the intermediate sleeve forming a seal on the outer body. The cell may be used at 110°C up to 7 bar pressure.

Many manufacturers offer a type of flow-through conductivity cell with annular graphite electrodes, one form of which is shown in Figure 24.3. It consists of three annular rings of impervious carbon composition material equally spaced within the bore of an epoxy resin molded body. Conduction through the solution within the cell takes place between the central electrode and the two outer rings, which are connected to the earthed terminal of the measuring instrument; thus, electrical conduction is confined entirely within the cell, where it is uninfluenced by the presence of adjoining metal parts in the pipe system. This pattern of cell, having a simple flow path, is ideally suited to the exacting requirements of dialysate concentration monitoring in the artificial kidney machine. Screw-in patterns of this cell are also generally available.

The use of an impervious carbon composition material for the electrodes substantially eliminates polarization error and provides conducting surfaces that do not require replatinization or special maintenance other than periodic but simple and infrequent cleaning by means of a bottle brush. Typical operating temperature and pressure limits for this type of cell are 100°C and 7 bar.

Measuring cells should be installed in positions where they are adequately protected from mechanical shock by passing traffic, dampness, and extremes of temperature. Where a flow-line cell is connected directly in the electrolyte pipe, suitable support should be given to the pipes to ensure that the cell is under no mechanical strain and that the pipe threads in a rigid system are straight and true. Dip pattern cells should be installed so that moving parts in a tank, e.g., agitators, are well clear of the cells.

Where measuring cells are installed in pipework, it is essential that they are positioned in a rising section of the system to ensure that each cell is always full of electrolyte and that pockets of air are not trapped.

Cleaning and Maintenance of Cells Periodic inspection and cleaning of conductivity cells is essential to ensure that the electrode surfaces are free from contamination, which would otherwise alter the electrode area and effective cell constant. The frequency of such procedures is mainly dependent on the nature of the samples, but the design of the cells and the accuracy required for the measurement also must be taken into consideration. All new cells should be thoroughly cleaned before installation; these cleaning procedures depend on the design of the cell and the electrode material.

Platinized Electrodes Cleaning of these electrodes constitutes a major drawback in their application because no form of mechanical cleaning should be attempted. A suitable cleaning solution consists of a stirred mixture of 1 part by volume isopropyl alcohol, 1 part of ethyl ether, and 1 part hydrochloric acid (50 percent). Alternatively, the sensitivity of the electrodes can frequently be restored by immersion in a 10–15 percent solution of hydrochloric or nitric acid for about 2 minutes. The electrodes should be thoroughly rinsed with water before being returned to service.

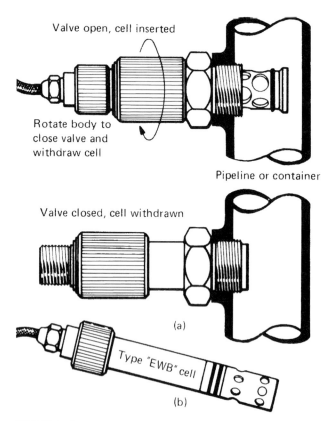

FIGURE 24.2 Retractable conductivity cell. Courtesy of Kent Industrial Measurements Ltd. Analytical Instruments.

CHAPTER | 24 Chemical Analysis: Electrochemical Techniques

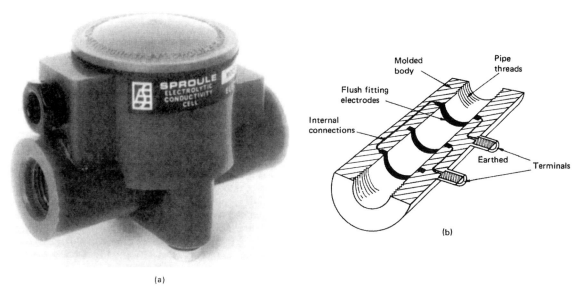

FIGURE 24.3 Flow-through cell. Courtesy of Kent Industrial Measurements Ltd. Analytical Instruments.

Annular Graphitic Electrodes Cleaning should be carried out with a 50 percent solution of water/detergent using a bottle brush. After thorough brushing with this solution, the cell bore should be rinsed several times in distilled water and then viewed. Looking through the bore toward a source of illumination, the surface should be evenly wetted, with no dry patches where the water has peeled away. If dry patches appear rapidly, indicating that a thin film of grease is present, the surface is not clean.

Stainless Steel and Monel A feature of many stainless-steel cells is the frosted appearance of the electrodes, which is essential to reduce polarization. It is most important that this frosting is not polished away by the regular use of abrasive cleaners. This type of cell may be cleaned with a 50 percent water/detergent solution and a bottle brush.

In the case of screw-in cells, the outer electrode may be removed to facilitate cleaning, but on no account should the central electrode be disturbed, because this will impair the accuracy of the electrical constant of the cell. In cases in which metal cells have become contaminated with adherent particulate matter, such as traces of magnetite or other metal oxides, ultrasonic cleaning in the detergent solution has been shown to be effective.

In all cleaning processes care should be taken to keep the external electrical contact, cable entries, and plugs dry.

Instruments for Conventional AC Measurement The conductance of a cell may be measured (a) by Wheatstone bridge methods or (b) by direct measurement of the current through the cell when a fixed voltage is applied.

Wheatstone Bridge Methods The actual conductance of the cell is usually measured by means of a self-balancing

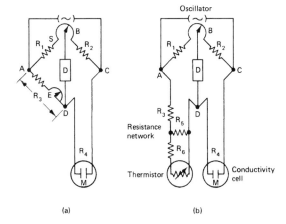

FIGURE 24.4 Measurement of conductance using Wheatstone bridge: (a) simple circuit, (b) thermistor temperature-corrected circuit.

Wheatstone bridge of the form shown in Figure 24.4 and described in detail in Part 3.

Direct Measurement of Cell Conductance The conductance of a cell may be measured directly by the method indicated in Figure 24.5. The current is directly proportional to the conductance, so the output from the current amplifier is applied to the indicator and recorder. Temperature compensation is achieved by connecting a manual temperature compensator in the amplifier circuit, or a resistance bulb may be used to achieve automatic compensation.

Multiple-electrode Cells From the foregoing discussion on errors introduced by polarization, together with the importance of constancy of electrode area, it can be appreciated that two-electrode conductivity cells have their limitations. In circumstances in which accurate measurements of conductivity are required in solutions of moderate or high conductivity or in solutions that can readily contaminate the electrode surfaces, multiple-electrode cells should be considered.

In its simplest form, a multiple-electrode cell has four electrodes in contact with the solution. An outer pair operate similarly to those in a conventional two-electrode cell and an ac current is passed through the solution via these electrodes. The voltage drop across a segment of the solution is measured potentiometrically at a second or inner pair of the electrodes, and this drop will be proportional to the resistivity or inversely proportional to the conductivity of the solution. Four-electrode cells can be operated in either the constant-current or constant-voltage mode, but the latter is the more popular and will be described further. In this form of measurement, the voltage at the inner electrode pair is maintained at a constant value by varying the current passed through the solution via the outer electrodes. The current flowing in the cell will be directly proportional to the conductivity and can be measured as indicated in Figure 24.6.

The circuit shown in the figure is considerably simplified, and there are multiple-electrode cells available from a number of manufacturers that contain additional electrodes, the function of which is to minimize stray current losses in the cell, particularly for solutions flowing through earthed metal pipework.

Since there is imperceptible current flowing through the voltage-sensing electrodes, cells of this type are free from the restrictions imposed by polarization. Therefore multiple-electrode cells can be standardized with any of the potassium chloride solutions given in Tables 24.4 and 24.5. The precaution previously stated about constancy of temperature during any determination of cell constant must still be observed.

Multiple-electrode cells are available with cell constants from 0.1 to 10 cm^{-1} and can therefore be used over a wide range of solution conductivities. However, their most valuable applications are when contamination or polarization is a problem.

Temperature Compensation The conductivity of a solution is affected considerably by change of temperature, and each solution has its own characteristic conductivity-temperature curve. Figure 24.7 shows how different these characteristics can be. When it is required to measure composition rather than absolute conductivity, it is therefore essential to use a temperature compensator to match the solution.

Manual compensators consist of a variable and a fixed resistor in series. The temperature scale showing the position of the contact on the variable resistance is calibrated so that the resistance of the combined resistors changes by the same percentage of the value of conductivity of the solution at 25°C as does the solution. The scale becomes crowded at the upper end, thus limiting the span of the compensator to about 70°C.

Aqueous solutions containing very low (μgl^{-1}) concentrations of electrolytes must have more elaborate compensation

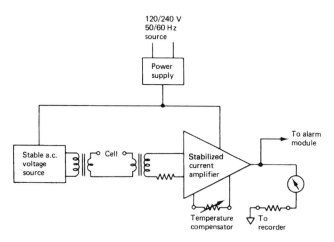

FIGURE 24.5 Direct measurement of cell conductance.

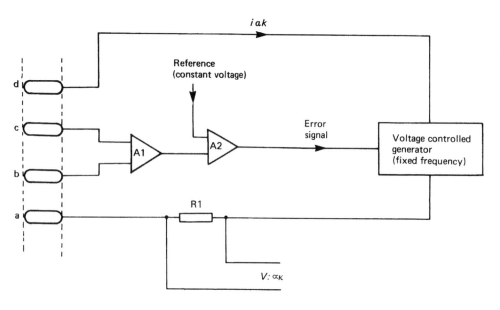

FIGURE 24.6 Four-terminal conductivity measurement. Courtesy of ABB Instrument Group.

CHAPTER | 24 Chemical Analysis: Electrochemical Techniques

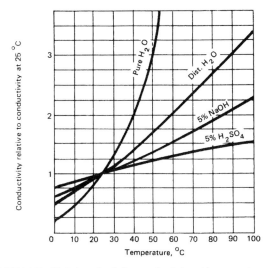

FIGURE 24.7 Variation of solution conductivity with temperature.

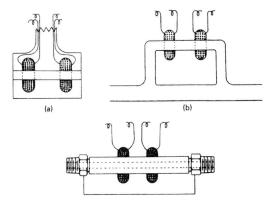

FIGURE 24.8 Electrodeless conductivity cells.

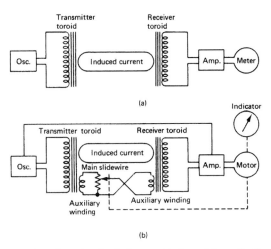

FIGURE 24.9 Measuring circuits for use with electrodeless cells. Courtesy of Beckman Instruments Inc. (a) Direct reading, (b) balanced bridge.

to allow for the nonlinear conductivity-temperature characteristic of pure water. This type of compensation system is applied in all conductivity transmitters (either with two-electrode or multiple-electrode cells) designed for accurate operation in the range up to 0.5 μS cm^{-1}.

24.3.3.2 Electrodeless Method of Measuring Conductivity

The principle of the electrodeless method is to measure the resistance of a closed loop of solution by the extent to which the loop couples two transformer coils. The liquid to be measured is enclosed in a nonconducting pipe or a pipe lined with a nonconducting material. Three forms of measuring units are available, as shown in Figure 24.8. Since the method is most successful with full-scale resistances of 10–1,000 Ω, relatively large-bore pipe may be used, reducing the possible errors due to solid deposition or film formation.

Figure 24.8(a) shows the form used for immersion in a large volume of solution. For measurements on a solution flowing through a pipe, the arrangement shown in Figure 24.8(b) is used. If the liquid contains suspended solids or fibers, wide-bore nonconducting pipe fitted with metallic end pieces connected with a length of wire to complete the circuit may sometimes be used, as shown in Figure 24.8(c).

The principle of the measuring system is shown in Figure 24.9. Figure 24.9(a) shows the simple circuit, which consists of two transformers. The first has its primary winding, the input toroid, connected to an oscillator operating at 3 or 18 kHz; its secondary is the closed loop of solution. The closed loop of solution forms the primary of the second transformer, and its secondary is the output toroid. With constant input voltage, the output of the system is proportional to the conductivity of the solution. The receiver is a high-impedance voltage-measuring circuit that amplifies and rectifies the output and displays it on a large indicator.

To eliminate effects of source voltage and changes in the amplifier characteristics, a null balance system may be provided, as shown in Figure 24.9(b). An additional winding is provided on each toroid, and the position of the contact is adjusted on the main slidewire to restore the system to the original balanced state by means of the balancing motor operated by the amplified out-of-balance signal in the usual way.

The electrodeless measurement of conductivity has obvious advantages in applications where the solution is particularly corrosive or has a tendency to foul or mechanically abrade the electrodes. Typical of these applications are measurements in oleum, hot concentrated sodium hydroxide, and slurries. In addition, this technique is ideal for applications in concentrated electrolytes (not necessarily aggressive) such as estuarine or sea waters where polarization errors would be considerable in a conventional cell. Temperature compensation is normally incorporated.

24.3.4 Applications of Conductivity Measurement

The measurement of electrical conductivity is the simplest and probably the most sensitive method of providing a nonspecific indication of the dissolved solids, or more correctly, the ionic content of a solution. If the number of ionic species in solution is few, it may be possible to use conductivity as a measure of the concentration of a particular component. Undoubtedly the robust nature of conductivity measurements has led to their use in circumstances where their nonspecific response gives rise to errors in interpretation of concentration. Consequently, any successful instrumental application of conductivity as a concentration sensor has to ensure that the species of interest is the dominating ion or the only ion (together with its counter-ion of opposite charge) whose concentration is changing. With these restrictions it can be appreciated that determinations of concentrations by conductivity measurements are often supported by additional analyses or preceded by a physical or chemical separation of the species of interest.

24.3.4.1 Conductivity and Water Purity

Water of the highest purity is increasingly being used for industrial purposes, such as in, for example, the manufacture of electronic components and the preparation of drugs. Other examples of large-scale uses include process steam and feedwater for high-pressure boilers. In all these cases conductivity provides the most reliable measurement of water purity in circumstances in which contamination from nonconducting impurities is considered to be absent. The conductivity of pure water is highly temperature dependent due to the increase in the dissociation of water molecules into hydrogen and hydroxyl ions of water, K_w, with temperature. The extent of this can be seen in Table 24.6.

The conductivity of pure water can be calculated at any temperature, provided that values of $\lambda°_{OH}$, $\lambda°_H$, K_w, the dissociation constant of water, and the density of water d are known at the appropriate temperature.

$$\kappa(\mu S\ cm^{-1}) = (\lambda°_H + \lambda°_{OH})d \cdot \sqrt{K_w} \cdot 10^3$$

In the application under consideration here (i.e., the use of pure water) the exact nature of the ionic species giving rise to a conductivity greater than that of pure water are of no interest, but it is useful to note how little impurity is required to raise the conductivity. For example, at 25°C, only about 10 $\mu g l^{-1}$ of sodium (as sodium chloride) is required to increase the conductivity to twice that of pure water.

24.3.4.2 Condensate Analyzer

The purity of the water used in the steam-water circuit of power stations is particularly important for the prevention of corrosion. An essential component of such a circuit is the

TABLE 24.6 Pure water, conductivity from 0 to 100°C

Temperature (°C)	Conductivity ($\mu S\ cm^{-1}$)	Resistivity (°C)
0	0.0116	86.0
5	0.0167	60.0
10	0.0231	43.3
15	0.0314	31.9
20	0.0418	23.9
25	0.0548	18.2
30	0.0714	14.0
35	0.0903	11.1
40	0.1133	8.82
45	0.1407	7.11
50	0.1733	5.77
60	0.252	3.97
70	0.346	2.89
80	0.467	2.14
90	0.603	1.66
100	0.788	1.27

condenser wherein the steam from the turbines is condensed before returning to the boiler. On one side of the condenser tubes is the highly pure steam and water from the turbines; on the other is cooling water chosen for its abundance (e.g., river water or estuarine water) rather than its chemical purity. Any leakage of this cooling water through the condenser tubes leads to the ingress of unwanted impurities into the boiler and therefore must be immediately detected. Direct measurement of conductivity would detect significant ingress of, say, sodium chloride from estuarine water, but it would not be capable of detecting small leakages, since the conductivity of the condensate would be dominated by the alkaline additives carried over in the steam from the boiler. A better method of detection of leakage is to pass the condensate through a cation exchange column in the H^+ form, then measure the conductivity. Using this procedure, all cations in the condensate are exchanged for hydrogen ions and the solution leaving the column will be weakly acidic if any salts have entered through the condenser. Otherwise, the effluent from the column will ideally be pure water, since the cations of the alkaline boiler water additives (NH_4OH, $NaOH$) will be exchanged and will recombine as

$$H^+ + OH^- \rightleftharpoons H_2O$$

A secondary advantage of such a system is the enhancement of the conductivity due to replacement of cations by hydrogen ions, which gives about a fivefold enhancement in ionic conductance. This is particularly important with very low leak rates.

A schematic diagram of an instrument based on the preceding principles is given in Figure 24.10. The incoming sample flows at about 400 ml min^{-1} through a H^+ form cation exchange column (1), 500 mm deep and 50 mm in diameter, and then to a flow-through conductivity cell (2).

The effluent from the cell flows to waste via an identical column/cell system (3 and 4), which is held in reserve. Since there will be no exchange on this second column, it will not be depleted and the constant flow of water or weak acid keeps it in constant readiness for instant replacement of column (1) when the latter becomes exhausted. The measured conductivity can be recorded and displayed and, where necessary, alarms set for notification of specific salt ingress levels. In the case of power stations using estuarine water for cooling the condensers, the condensate analyzer can be used to give a working guide to the salt going forward to the boiler (see Table 24.7).

24.3.4.3 Conductivity Ratio Monitors

These instruments measure the conductivities at two points in a process system continuously and compare the ratio of the measurements with a preset ratio. When the measured ratio reaches the preset value, a signal from the monitor can either operate an alarm or initiate an action sequence, or both.

One application of this type of dual-conductivity measurement is to control the regeneration frequency of cation exchange units (usually in the H^+-form) in water treatment plants. The conductivity at the outlet of such a unit will be higher than at the inlet since cations entering the ion exchange bed will be replaced by the much more conductive hydrogen ion. For example, an inlet stream containing 10^{-4} mol l^{-1} of sodium chloride will have ratios of 3.5, 3.3, and 2.3 for 100, 90, and 50 percent exchange, respectively. A value corresponding to the acceptable extent of exchange can then be set on the instrument. Reverse osmosis plants use ratio monitors to measure the efficiency of their operation, and these are usually calibrated in percentage rejection or passage.

This type of operational control is most effective when the chemical constituents of the inlet stream do not vary greatly; otherwise the ratio will be subject to errors from unconsidered ionic conductivities.

24.3.4.4 Ion Chromatography

Although conductivity measurements have a nonspecific response, they can, when combined with a separation technique, provide extremely sensitive and versatile detectors of chemical concentration. The best example of this is in ion chromatography, which in recent years has been shown to be an invaluable instrumental technique for the identification and measurement of the concentration of ions, particularly at low levels, in aqueous solution.

The general principles of chromatography are outlined in Chapter 23. In an ion chromatograph, a small volume of sample is injected into a carrier or eluent electrolyte stream. The eluent, together with the sample, is carried forward under high pressure (5–50 bar) to an ion exchange column, where

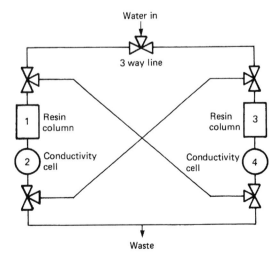

FIGURE 24.10 Condensate analyzer.

TABLE 24.7 Relationship between conductivity and salt fed to the boiler

Conductivity at 25°C ($\mu S\ cm^{-1}$)	Chloride in condensate (ppm)	Salt going forward to boiler (g NaCl/Tonne)
0.137	0.01	0.0165
0.604	0.05	0.0824
1.200	0.10	0.1649
1.802	0.15	0.2473
2.396	0.20	0.3298
6.003	0.50	0.8265

chromatographic separation of either the cations (+ve) or anions (−ve), depending on the nature of the exchanger, takes place. The ion exchange material in these chromatographic separator columns is fundamentally the same as conventional ion exchange resins, but the exchange sites are limited to the surface of very fine resin beads. This form of exchanger has been shown to have the characteristics required for rapid separation and elution of the ionic components in the order expected from the general rules of ion exchange (e.g., Cl^{-1} before Br^{-1} before SO_4^{2-}). At this stage the conductivity can be monitored and the elution peaks corresponding to the separated ionic components measured as increases superimposed on the relatively high background conductivity of the eluent. This is the procedure used in the ion chromatograph manufactured by Wescan Instruments Inc.

In another instrument manufactured by the Dionex Corporation, the eluent from the separator column passes through a second ion exchange column where the ions of opposite charge to those that have been separated chromatographically are all converted to a common form. This second column, termed a *suppressor column*, reduces the background conductivity of the eluent and thus ensures that conductivity changes due to the sample constitute a significant portion of the total measured conductivity. With a system such as this, the retention time identifies the elution peak and the area under the peak is a measure of the concentration of the ionic species giving rise to it. In many cases peak heights rather than areas can be used as the indicator of concentration, thus simplifying the measurement, since an integrator is not required. For most purposes this is adequate, since sharp elution peaks are obtained by keeping mixing to a minimum by use of very narrow-bore transmission tubing combined with a conductivity cell whose volume is of the order of 6 μl. In cells of this size, polarization resistance can be considerable due to the proximity of the electrodes.

A schematic outline of the main features of a typical system for the determination of anions is given in Figure 24.11.

In this particular example the eluent consisting of a mixture of 2.4×10^{-3} mol l^{-1} sodium carbonate

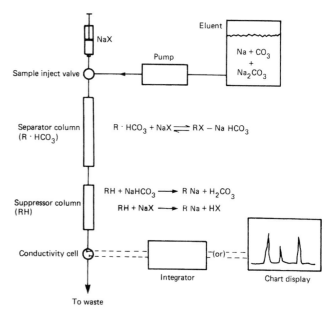

FIGURE 24.11 Flow system for anion chromatography.

and 3×10^{-3} mol l^{-1} sodium bicarbonate has a conductivity of about 700 $\mu S\ cm^{-1}$. The separator column consists of a strong base anion exchanger ($R.HCO_3$), mainly in the bicarbonate form, and the suppressor column is a strong acid cation exchanger in the H^+-form (R.H). After the eluent has passed through the cation exchange, it will be weakly acid carbonic acid (H_2CO_3), having a conductivity level of about 25 $\mu S\ cm^{-1}$, and with this much reduced base conductivity level it is possible to detect quantitatively the small changes due to the acids (H.X) from the sample anions.

A technique used to measure the concentration of sulfur dioxide in air in the parts per hundred million (pphm) range is based on the measurement of the change in the conductivity of a reagent before and after it has absorbed sulfur dioxide. The principle of the measurement is to absorb the sulfur dioxide in hydrogen peroxide solution, thus forming sulfuric acid that increases the electric conductivity of the absorbing reagent.

Continuous measurements can be made by passing air upward through an absorption column down which the hydrogen peroxide absorbing solution is flowing. Provided that flow rates of air and hydrogen peroxide reagent are maintained at a constant rate, the sulfur dioxide concentration is proportional to the measured conductivity of the hydrogen peroxide reagent. Figure 24.12 is a diagram of suitable apparatus.

24.3.4.6 Salt-in-Crude-Oil Monitor

A rapid continuous measurement of the salt in crude oil before and after desalting is based on the measurement of the conductivity of a solution to which a known quantity of crude oil has been added. The sample of crude oil is continuously circulated through a loop in the measurement section of the *salt-in-crude monitor*. When the test cycle is initiated, solvent (xylene) is introduced from a metering cylinder into the analyzer cell. A sample is then automatically diverted from the sample circulating loop into a metering cylinder calibrated to deliver a fixed quantity of crude oil into the analysis cell. A sample is then automatically diverted from the sample circulating loop into a metering cylinder calibrated to deliver a fixed quantity of crude oil into the analysis cell. A solution containing 63 percent *n*-butanol, 37 percent methanol, and 0.25 percent water is then metered into the analysis cell from another calibrated cylinder.

The cell contents are thoroughly mixed by a magnetic stirrer; then the measuring circuit is energized and an ac potential is applied between two electrodes immersed in the liquid. The resulting ac current is displayed on a milliammeter in the electrical control assembly, and a proportional dc millivolt signal is transmitted from the meter to a suitable recorder.

At the end of the measuring period, a solenoid valve is opened automatically to drain the contents of the measuring cell to waste. The minimum cycle time is about 10 minutes.

Provision is made to introduce a standard sample at will to check the calibration of the instrument. Salt concentrations between 1 and 200 kg salt per 1,000 m³ crude oil can be measured with an accuracy of ±5 percent and a repeatability of 3 percent of the quantity being measured.

24.4 THE CONCEPT OF PH

24.4.1 General Theory

Ionic concentrations were discussed in Section 24.2. The range of hydrogen ion concentrations met in practice is very wide; also, when dealing with small concentrations, it is inconvenient to specify hydrogen or hydroxyl concentrations. A method proposed by S. P. L. Sorenson in 1909 is now used universally—this is the concept of a hydrogen ion exponent or pH, defined as:

$$\text{pH} = -\log_{10}[\text{H}^+] = \log_{10}\frac{1}{[\text{H}^+]}$$

Thus pH is the logarithm to base 10 of the reciprocal of the hydrogen ion concentration. The advantage of this nomenclature is that all values of acidity and alkalinity between those of solutions molar with respect to hydrogen and hydroxyl ions can be expressed by a series of positive numbers between 0 and 14. Thus a neutral solution with $[\text{H}^+] = 10^{-7}$ has a pH of 7. If the pH is less than 7, the solution is acid; if greater than 7, the solution is alkaline.

It must be realized that pH measuring devices measure the effective concentration, or activity, of the hydrogen ions and not the actual concentration. In very dilute solutions of electrolyte, the activity and concentration are identical. As the concentration of electrolyte in solution increases above 0.1 mol/liter, however, the measured value of pH becomes a less reliable measure of the concentration of hydrogen ions. In addition, as the concentration of a solution increases, the degree of dissociation of the electrolyte decreases.

A dilute solution of sulfuric acid is completely dissociated and the assumption that pH = $-\log 2\,(\text{H}_2\text{SO}_4)$ is justified. (The 2 occurs because each molecule of acid provides two hydrogen ions.) Anhydrous sulfuric acid is only slightly dissociated, the degree of dissociation rising as the pure acid is diluted.

A maximum hydrogen ion concentration occurs in the neighborhood of 92 percent H_2SO_4, but at this concentration, the difference between actual hydrogen ion concentration

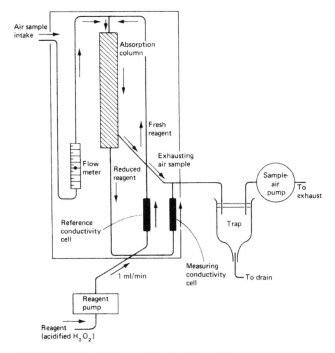

FIGURE 24.12 Continuous sulfur dioxide monitor.

and the activity of the hydrogen ions is large, and the measured pH minimum of about -1.4 occurs at a much lower sulfuric acid content.

A more reliable indication of the ionic behavior of a solution will be obtained if we define pH in terms of the hydrogen ion activity aH^+ so that

$$pH = \log_{10}\left(\frac{1}{aH^+}\right) = -\log_{10} aH^+$$

where aH^+ is related to the hydrogen ion concentration cH^+ by the equation

$$aH^+ = fH^+ cH^+$$

where fH^+ is the activity coefficient; see Section 24.1. The pH values of common acids, bases, and salts are given in Table 24.8.

24.4.2 Practical Specification of a pH Scale

Because the value of pH, defined as $-\log_{10}$ (hydrogen ion activity), is extremely difficult to measure, it is necessary to ensure that when different workers state a pH value, they mean the same thing. An operational definition of pH has been adopted in British Standard 1647:1961. The EMF E_x of the cell

Pt H$_2$/soln. X/conc. KCl soln./ref. electrode

is measured, and likewise the EMF E_s of the cell

Pt H$_2$/soln. S/conc. KCl soln./ref. electrode

both cells being at the same temperature throughout and the reference electrodes and bridge solutions being identical in the two cells.

The pH of the solution X denoted by pH(X) is then related to the pH of the solution S denoted by pH(S) by the definition:

$$pH(X) - pH(S) = \frac{(E_x - E_s)}{(2.3026\, RT/F)}$$

where R is the gas constant, T is temperature in Kelvin, and F is the Faraday constant. Thus defined, pH is a pure number.

To a good approximation, the hydrogen electrodes in both cells may be replaced by other hydrogen-responsive electrodes, such as glass or quinhydrone. The two bridge solutions may be of any molarity not less than 3.5 mol/kg, provided that they are the same.

24.4.3 pH Standards

The difference between the pH of two solutions having been defined as previously, the definition of pH can be completed by assigning at each temperature a value of pH to one or more chosen solutions designated as standards. In BS 1647 the chosen primary standard is a solution of pure potassium hydrogen phthalate that has a concentration of 0.05 mol/liter.

This solution is defined as having a pH value of 4.000 at 15°C and the following values at other temperatures between 0 and 95°C:

Between 0 and 55°C:

$$pH = 4.000 + \frac{1}{2}\left[\frac{(t-15)^2}{100}\right]$$

Between 55 and 95°C:

$$pH = 4.000 + \frac{1}{2}\left[\frac{(t-15)^2}{100}\right] - \frac{(t-55)}{500}$$

Other standard buffer solutions are given in Section 24.4.7.

The EMF E_x is measured and likewise the EMF E_1 and E_2 of similar cells with solution X replaced by standard solutions S_1 and S_2, so that E_1 and E_2 are on either side of and as near as possible to E_x. The pH of the solution X is then obtained by assuming linearity between pH and E, that is,

$$\frac{(pHX - pH\,S_1)}{(pH\,S_2 - pH\,S_1)} = \frac{(E_x - E_1)}{(E_2 - E_1)}$$

24.4.4 Neutralization

When acid and base solutions are mixed, they combine to form a salt and water, for example:

hydrochloric acid		sodium hydroxide		sodium chloride		water
H$^+$Cl	+	Na$^+$ OH	=	Na$^+$Cl	+	HOH
(dissociated)		(dissociated)		(dissociated)		(largely undissociated)

Thus, if equal volume of equally dilute solutions of strong acid and strong alkali are mixed, they yield neither an excess of H$^+$ ions nor of OH$^-$ ions, and the resultant solution is said to be neutral. The pH value of such a solution will be 7.

24.4.5 Hydrolysis

Equivalent amounts of acid and base, when mixed, will produce a neutral solution only when the acids and bases used are strong electrolytes. When a weak acid or base is used, hydrolysis occurs. When a salt such as sodium acetate, formed by a weak acid and a strong base, is present in water, the solution is slightly alkaline because some of the H$^+$ ions from the water are combined with acetic radicals in the relatively undissociated acetic acid, leaving an excess of OH$^-$ ions, thus:

sodium acetate	+	water	→	acetic acid	+	sodium hydroxide
Na$^+$Ac	+	HOH	→	HAc	+	Na$^+$OH
(dissociated)		(largely undissociated)				(dissociated)

TABLE 24.8 pH values of common acids, bases, and salts

Compound	Molarity	pH
Acid benzoic	(Saturated)	2.8
Acid boric	0.1	5.3
Acid citric	0.1	2.1
Acid citric	0.01	2.6
Acid hydrochloric	0.1	1.1
Acid oxalic	0.1	1.3
Acid salicylic	(Saturated)	2.4
Acid succinic	0.1	2.7
Acid tartaric	0.1	2.0
Ammonia, aqueous	0.1	11.3
Ammonium alum	0.05	4.6
Ammonium chloride	0.1	4.6
Ammonium oxalate	0.1	6.4
Ammonium phosphate, primary	0.1	4.0
Ammonium phosphate, secondary	0.1	7.9
Ammonium sulphate	0.1	5.5
Borax	0.1	9.2
Calcium hydroxide	(Saturated)	12.4
Potassium acetate	0.1	9.7
Potassium alum	0.1	4.2
Potassium bicarbonate	0.1	8.2
Potassium carbonate	0.1	11.5
Potassium dihydrogen citrate	0.1	3.7
Potassium dihydrogen citrate	0.02	3.8
Potassium hydrogen oxalate	0.1	2.7
Potassium phosphate, primary	0.1	4.5
Sodium acetate	0.1	8.9
Sodium benzoate	0.1	8.0
Sodium bicarbonate	0.1	8.3
Sodium bisulphate	0.1	1.4
Sodium carbonate	0.1	11.5
Sodium carbonate	0.01	11.0
Sodium hydroxide	0.1	12.9
Sodium phosphate, primary	0.1	4.5
Sodium phosphate, secondary	0.1	9.2
Sodium phosphate, tertiary	0.01	11.7
Sulphamic acid	0.01	2.1

The pH value of the solution will therefore be greater than 7. Experiment shows it to be 8.87 in 0.1 mol/liter solution at room temperature.

Similarly, ammonium chloride (NH_4Cl), the salt of a weak base and a strong acid, hydrolyzes to form the relatively undissociated ammonium hydroxide (NH_4OH), leaving an excess of H^+ ions. The pH value of the solution will therefore be less than 7. Experiment shows it to be 5.13 at ordinary temperatures in a solution having a concentration of 0.1 mol/liter.

A neutralization process therefore does not always produce an exactly neutral solution when one mole of acid reacts with one mole of base.

24.4.6 Common Ion Effect

All organic acids and the majority of inorganic acids are weak electrolytes and are only partially dissociated when dissolved in water. Acetic acid, for example, ionizes only slightly in solution, a process represented by the equation

$$HAc \rightleftharpoons H^+ + Ac^-$$

Its dissociation constant at 25°C is only 1.8×10^{-5}, that is,

$$\frac{([H^+][Ac^-])}{[HAc]} = 1.8 \times 10^{-5} \text{ mol/liter}$$

or

$$[H^+][Ac^-] = 1.8 \times 10^{-5}[HAc]$$

Therefore, in a solution of acetic acid of moderate concentration, the bulk of the acid molecules will be undissociated, and the proportion present as acetic ions and hydrogen ions is small. If one of the salts of acetic acid, such as sodium (NaAc), is added to the acetic acid solution, the ionization of the acetic acid will be diminished. Salts are, with very few exceptions, largely ionized in solution, and consequently when sodium acetate is added to the solution of acetic acid, the concentration of acetic ions is increased. If the preceding equation is to continue to hold, the reaction $H^+ + Ac^{-1}$ HAc must take place, and the concentration of hydrogen ions is reduced and will become extremely small.

Most of the acetic ions from the acid will have recombined; consequently the concentration of unionized acid will be practically equal to the total concentration of the acid. In addition, the concentration of acetic ions in the equilibrium mixture due to the acid will be negligibly small, and the concentration of acetic ions will, therefore, be practically equal to that from the salt. The pH value of the solution may, therefore, be regulated by the strength of the acid and the ratio [salt]/[acid] over a wide range of values.

Just as the ionization of a weak acid is diminished by the addition of a salt of the acid, so the ionization of a weak base will be diminished by the addition of a salt of the base, for example, addition of ammonium chloride to a solution of ammonium hydroxide. The concentration of hydroxyl ions in the mixture will be given by a similar relationship to that obtained for hydrogen ions in the mixture of acid and salt, that is,

$$[OH^-] = K\left[\frac{\text{alkali}}{\text{salt}}\right]$$

24.4.7 Buffer Solutions

Solutions of a weak acid and a salt of the acid such as acetic acid mixed with sodium acetate and solutions of a weak base and one of its salts, such as ammonium hydroxide mixed with ammonium chloride (as explained in Section 24.4.6), undergo relatively little change of pH on the further addition of acid or alkali and the pH is almost unaltered on dilution. Such solutions are called *buffer solutions*; they find many applications in quantitative chemical analysis. For example, many precipitations are made in certain ranges of pH values, and buffer solutions of different values are used for standardizing pH measuring equipment.

Buffer solutions with known pH values over a wide range can be prepared by varying the proportions of the constituents in a buffer solution; the value of the pH is given by

$$pH = \log_{10}\left(\frac{1}{K}\right) + \log_{10}\frac{[\text{salt}]}{[\text{acid}]}$$

The weak acids commonly used in buffer solutions include phosphoric, boric, acetic, phthalic, succinic, and citric acids, with the acid partially neutralized by alkali or the salt of the acid used directly. Their preparation requires the use of pure reagents and careful measurement and weighing, but it is more important to achieve correct proportions of acid to salt than correct concentration. An error of 10 percent in the volume of water present may be ignored in work correct to 0.02 pH units.

The National Institute of Standards and Technology (USA) standard buffer solutions have good characteristics and for pH 4, pH 7, and pH 9.2 are available commercially as preweighed tablets, sachets of powder, or in solution form. Those unobtainable commercially are simple to prepare, provided that analytical-grade reagents are used, dissolved in water with a specific conductance not exceeding 2 µS/cm.

24.5 ELECTRODE POTENTIALS

24.5.1 General Theory

When a metallic electrode is placed in a solution, a redistribution of electrical charges tends to take place. Positive ions of the metal enter the solution, leaving the electrode negatively charged, and the solution will acquire a positive charge. If the solution already contains ions of the metal, there is a tendency for ions to be deposited on the electrode, giving it a positive charge. The electrode eventually reaches an equilibrium potential with respect to the solution, the magnitude and sign of the potential depending on the concentration of metallic ions in the solution and the nature of the metal. Zinc has such a strong tendency to form ions that the metal forms ions in all solutions of its salts, so it is always negatively charged relative to the solution. On the other hand, with copper, the ions have such a tendency to give up their charge that the metal becomes positively charged, even when placed in the most dilute solution of copper salt.

This difference between the properties of zinc and copper is largely responsible for the EMF of a Daniell cell (see Figure 24.13). When the poles are connected by a wire, sudden differences of potential are possible (a) at the junction of the wires with the poles, (b) at the junction of the zinc with the zinc sulfate, (c) at the junction of the zinc sulfate with the copper sulfate, (d) at the junction of the copper with the copper sulfate. The EMF of the cell will be the algebraic sum of these potential differences.

In the measurement of the electrode potential of a metal, a voltaic cell similar in principle to the Daniell cell is used. It can be represented by the scheme

| Metal 1 | Solution containing ions of metal 1 | Solution containing ions of metal 1 | Metal 2 |

Under ordinary conditions, when all the cell is at the same temperature, the thermoelectric EMF at the junctions of wires and electrodes will vanish.

The potential difference that arises at the junction of the solutions, known as the *liquid junction potential*, or *diffusion potential*, is due to the difference in rate of diffusion across the junction of the liquids of the cations and anions. If the cations have a greater rate of diffusion than the anions, the solution into which the cations are diffusing will acquire a positive charge, and the solution that the cations are leaving will acquire a negative charge. Therefore there is a potential gradient across the boundary. If the anions have the greater velocity, the direction of the potential gradient will be reversed. The potential difference at the junction of the two liquids may be reduced to a negligible value either by having present in the two solutions relatively large and equal concentrations of an electrolyte, such as potassium nitrate, which produces ions which diffuse with approximately equal velocities, or by inserting between the two solutions a "salt bridge" consisting of a saturated solution of potassium chloride or of ammonium or potassium nitrate. These salts produce ions of which the diffusion rates are approximately equal.

When salt bridges are used in pH work, the liquid junction potentials are reduced to less than 1 mV unless strong acids or alkalis are involved. If an excess of neutral salt is added to the acid or alkali, the liquid junction potential will be reduced. Thus the error involved is rarely measurable on industrial instruments.

All measurements of the EMF of cells give the potential of one electrode with respect to another. In the Daniell cell, all that can be said is that the copper electrode is 1 Volt positive with respect to the zinc electrode. It is not possible to measure the potential of a single electrode, since it is impossible to make a second contact with the solution without introducing a second metal-solution interface. Practical measurement always yields a difference between two individual electrode potentials.

To assign particular values to the various electrode potentials, an arbitrary zero is adopted; all electrode potentials are measured relative to that of a standard hydrogen electrode (potential taken as zero at all temperatures). By convention, the half-cell reaction is written as a reduction and the potential designated positive if the reduction proceeds spontaneously with respect to the standard hydrogen electrode; otherwise the potential is negative.

The standard hydrogen electrode consists of a platinum electrode coated with platinum black, half immersed in a solution of hydrogen ions at unit activity (1.228 M HCl at 20°C) and half in pure hydrogen gas at one atmosphere pressure. In practice, however, it is neither easy nor convenient to set up a hydrogen electrode, so subsidiary reference electrodes are used, the potential of which relative to the standard hydrogen electrode has previously been accurately determined. Practical considerations limit the choice to electrodes consisting of a metal in contact with a solution that is saturated with a sparingly soluble salt of the metal and that also contains an additional salt with a common anion. Examples of these are the silver/silver chloride electrode ($Ag/AgCl_{(s)}KCl$) and the mercury/mercurous chloride electrode ($Hg/Hg_2Cl_{2s}KCl$) known as the *calomel electrode*. In each case the potential of the reference electrode is governed by the activity of the anion in the solution, which can be shown to be constant at a given temperature.

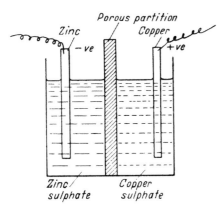

FIGURE 24.13 Daniell cell.

24.5.2 Variation of Electrode Potential with Ion Activity (The Nernst Equation)

The most common measurement of electrode potential is in the measurement of pH, that is, hydrogen ion activity, and selective ion activity, p(ion). The circuit involved is as shown in Figure 24.14.

The measured potential is the algebraic sum of the potentials developed within the system, that is,

$$E = E_{\text{Int.ref.}} + E_s + E_j - E_{\text{Ext.ref.}}$$

where $E_{\text{Int.ref.}}$ is the EMF generated at the internal reference inside the measuring electrode, E_s is the EMF generated at the selective membrane, E_j is the EMF generated at the liquid junction, and $E_{\text{Ext.ref.}}$ is the EMF generated at the external reference electrode.

At a fixed temperature, with the reference electrode potentials constant and the liquid junction potentials zero, the equation reduces to

$$E = E' + E_s$$

where E' is a constant.

The electrode potential generated is related to the activities of the reactants and products that are involved in the electrode reactions.

For a general half-cell reaction

$$\text{oxidized form} + n \text{ electrons} \rightarrow \text{reduced form}$$

or

$$aA + bB + \ldots + ne^- \rightarrow xX + yY + \ldots$$

The electrode potential generated can be expressed by the Nernst equation

$$E = E_0 + \frac{RT}{nF} \ln \frac{\text{OXID}}{\text{RED}} \text{ volts}$$

or

$$E = E_0 + 2.303 \frac{RT}{nF} \log_{10} \frac{[A]^a \cdot [B]^b}{[X]^x \cdot [Y]^y} \text{ volts}$$

where R is the molar gas constant (8.314 Joule. $\text{mol}^{-1} \text{ K}^{-1}$), T is absolute temperature in Kelvin, F is the Faraday constant (96487 coulomb. mol^{-1}), and n is the number of electrons participating in the reaction according to the equation defining the half cell reaction. The value of the term 2.303 RT/nF is dependent on the variables n and T and reduces to 0.059/n volts at 25°C and 0.058/n volts at 20°C.

An ion-selective electrode (say, selective to sodium ions) is usually constructed so that the ion activity of the internal reference solution inside the electrode is constant, and the Nernst equation reduces at constant temperature to

$$E = E_0 + \frac{RT}{nF} \ln a$$

where E_0 includes all the constants and a is the activity of the sodium ion. Because sodium is a positive ion with one charge,

$$E = E_0 + 59.16 \log_{10}(a) \text{ mV at } 25°C$$

This equation shows that a tenfold increase in ion activity will increase the electrode potential by 59.16 mV.

If the ion being measured is doubly charged, the equation becomes

$$E = E_0 + \frac{59.16}{2} \log_{10}(a) \text{ mV at } 25°C$$

The applicability of these equations assumes that the ion-selective electrode is uniquely sensitive to one ion. In most cases, in practice, the electrode will respond to other ions as well but at a lower sensitivity. The equation for electrode potential thus becomes

$$E = E_0 + 59.16 \log_{10}(a_1 + K_2 a_2 + \ldots) \text{ mV}$$

where $K_2 a_2$, etc., represents the ratio of the sensitivity of the electrode of the ion 2 to that of ion 1. The literature on ion-selective electrodes provided by manufacturers usually gives a list of interfering ions and their sensitivity ratios.

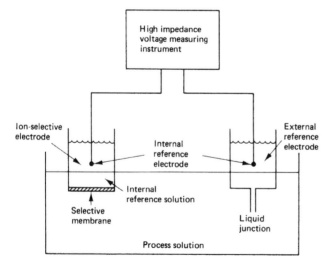

FIGURE 24.14 Method of measuring potential developed at an ion-selective membrane.

24.6 ION-SELECTIVE ELECTRODES

Whereas formerly, ion-selective electrodes were used almost exclusively for measuring hydrogen ion activity (pH), many electrodes have now been developed to respond to a wide range of selected ions. These electrodes are classified into five groups according to the type of membrane used.

24.6.1 Glass Electrodes

The glass electrode used for pH measurement, shown in Figure 24.15(a), is designed to be selective to hydrogen ions, but by choosing the composition of the glass membrane, we can make glass electrodes selective to sodium, potassium, ammonium, silver, and other univalent cations.

24.6.2 Solid-State Electrodes

In these electrodes the membrane consists of a single crystal or a compacted disc of the active material. In Figure 24.15(b) the membrane isolates the reference solution from the solution being measured. In Figure 24.15(c) the membrane is sealed with a metal backing with a solid metal connection. A solid-state electrode selective to fluoride ions employs a membrane of lanthanum fluoride (LaF_3). One that is selective to sulfide ions has a membrane of silver sulfide. There are also electrodes available for measurement of Cl^-, Br^-, I^-, Ag^+, Cu^{2+}, Pb^{2+}, Cd^{2+}, and CN^- ions.

24.6.3 Heterogeneous Membrane Electrodes

These are similar to the solid-state electrodes but differ in having the active material dispersed in an inert matrix. Electrodes in this class are available for Cl^-, Br^-, I^-, S^{2-}, and Ag^+ ions.

24.6.4 Liquid Ion Exchange Electrodes

In this type of electrode, shown in Figure 24.15(d), the internal reference solution and the measured solution are separated by a porous layer containing an organic liquid of low water solubility. Dissolved in the organic phase are large molecules in which the ions of interest are incorporated. The most important of these electrodes is the calcium electrode, but other electrodes in this class are available for the determination of Cl^-, ClO_4^-, NO_3, Cu^{2+}, Pb^{2+}, and BF_4 ions. The liquid ion exchange electrodes have more restricting chemical and physical limitations than the glass or solid-state electrodes, but they may be used to measure ions, which cannot yet be measured with a solid-state electrode.

24.6.5 Gas-Sensing Membrane Electrodes

These electrodes are not true membrane electrodes because no current passes across the membrane. They are complete electrochemical cells, monitored by an ion-selective electrode since the internal chemistry is changed by the ion being

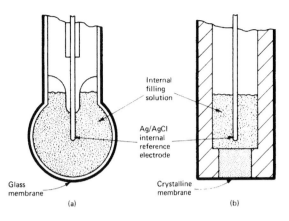

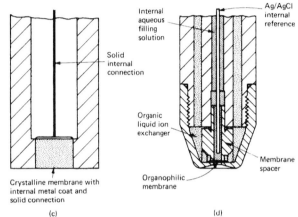

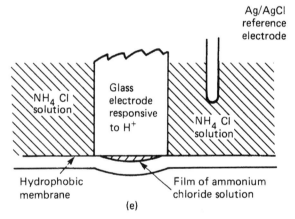

FIGURE 24.15 Ion-selective electrodes: (a) glass, (b) crystalline membrane with internal reference electrode, (c) crystalline membrane with solid connection, (d) liquid ion exchange, (e) gas sensing membrane. Courtesy of Orion Research Inc.

determined passing from the sample solution across the membrane to the inside of the cell.

An example is an ammonia electrode, shown in Figure 24.15(e). The sensing surface of a flat-ended glass pH electrode is pressed tightly against a hydrophobic polymer membrane, which acts as a seal for the end of a tube containing ammonium chloride solution. A silver/silver chloride electrode is immersed in the bulk solution. The membrane permits the diffusion of free ammonia (NH_3), but not ions, between the sample solution and the film of ammonium chloride solution. The introduction of free ammonia changes the pH of the internal ammonium chloride solution, which is sensed by the internal glass pH electrode.

24.6.6 Redox Electrodes

In elementary chemistry, a substance is said to be oxidized when oxygen is combined with it and is said to be reduced when oxygen is removed from it. The definition of oxidation and reduction may, however, be extended. Certain elements, such as iron and tin, can exist as salts in more than one form. Iron, for example, can be combined with sulfuric acid in the form of ferrous iron, valency 2, or ferric iron, valency 3.

Consider the reaction:

ferrous sulphate	+	chlorine	=	ferric chloride	+	ferric sulfate
$6FeSO_4$	+	$3Cl_2$	=	$2FeCl_3$	+	$2Fe_2(SO_4)_3$

The ferrous sulfate is oxidized to ferric sulfate; chlorine is the oxidizing agent. In terms of the ionic theory, the equation may be written

$$6\,Fe^2 + 3\,Cl_2 = 6\,Fe^3 + 6\,Cl^-$$

that is, each ferrous ion loses an electron and so gains one positive charge. When a ferrous salt is oxidized to a ferric salt, each mole of ferrous ions gains one mole (1 Faraday) of positive charges or loses one mole of negative charges, the negative charge so lost being taken up by the oxidizing agent (chlorine). Oxidation, therefore, involves the loss of electrons; reduction, the gain of electrons. Thus the oxidation of a ferrous ion to ferric ion can be represented by the equation

$$Fe^{2+} - e = Fe^{3+}$$

When a suitable electrode, such as an inert metal that is not attacked by the solution and that will not catalyze side reactions, is immersed in a solution containing both ferrous and ferric ions or some other substance in the reduced and oxidized state, the electrode acquires a potential that will depend on the tendency of the ions in the solution to pass from a higher or lower state of oxidation. If the ions in solution tend to become oxidized (i.e., the solution has reducing properties), the ions tend to give up electrons to the electrode, which will become negatively charged relative to the solution. If, on the other hand, the ions in solution tend to become reduced (i.e., the solution has oxidizing properties), the ions will tend to take up electrons from the electrode and the electrode will become positively charged relative to the solution. The sign and magnitude of the electrode potential,

therefore, give a measure of the oxidizing or reducing power of the solution, and the potential is called the *oxidation-reduction* or *redox* potential of the solution, E_h. The potential E_h may be expressed mathematically by the relationship

$$E_h = E_0 + \left(\frac{RT}{nF}\right)\log_{10}\left(\frac{a_o}{a_r}\right)$$

where a_o is the activity of the reduced ion and a_r is the activity of the reduced ion.

To measure the oxidation potential it is necessary to use a reference electrode to complete the electrical circuit. A calomel electrode is often used for this task (see Section 24.7).

The measuring electrode is usually either platinum or gold, but other types are used for special measurements—for example, the hydrogen electrode for use as a primary standard and the quinhydrone electrode for determining the pH of hydrofluoric acid solutions. However, the latter two electrodes do not find much application in industrial analytical chemistry.

24.7 POTENTIOMETRY AND SPECIFIC ION MEASUREMENT

24.7.1 Reference Electrodes

All electrode potential measurements are made relative to a reference electrode, and the EMF generated at this second contact with the solution being tested must be constant. It should also be independent of temperature changes (or vary in a known manner), be independent of the pH of the solution, and remain stable over long periods.

Standard hydrogen electrodes are inconvenient (see the following discussion), and in practice three types of reference are commonly used: silver/silver chloride, mercury/mercurous chloride, or calomel.

Silver/Silver Chloride Electrode This consists of a silver wire or plate coated with silver chloride, in contact with a salt bridge of potassium chloride saturated with silver chloride. The concentration of the potassium chloride may vary from one type of electrode to another, but concentrations of 1.00 or 4.00 mol per liter or a saturated solution are quite common. This saturated type of electrode has a potential of −0.199 V relative to a hydrogen electrode. It has a variety of physical forms, which are discussed in a moment.

Mercury/Mercurous Chloride or Calomel Electrode The metal used is mercury, which has a high resistance to corrosion and, being fluid at ambient temperature, cannot be subject to strain. The mercury is in contact with either mercurous chloride or in some electrodes with mercurous chloride and potassium chloride paste. Contact with the measured solution is through a salt bridge of potassium chloride, the concentration of which may be 3.8 mol per liter or some other concentration appropriate to the application. Contact with the mercury is usually made by means of a platinum wire, which may be amalgamated. The calomel-saturated potassium chloride electrode has a potential relative to the hydrogen electrode of -0.244 V.

Where the use of potassium salt is precluded by the condition of use, it may be replaced by sodium sulfate, the bridge solution having a concentration of 1 mol per liter.

Whatever the type of the reference electrode, contact must be made between the salt bridge and the measured solution. Two common methods are through a ceramic plug whose shape and porosity govern the rate at which the salt bridge solution diffuses out and the process solution diffuses into and contaminates the bridge solution. If the plug is arranged to have a small cross-sectional area relative to its length, the rate of diffusion is very small (say, less than 0.02 cm^3/day), and the electrode can be considered sealed and is used until it becomes unserviceable. It is then replaced by a similar electrode.

Where the application warrants it, a high rate of diffusion from the electrode has to be tolerated (say 1 or 2 cm^3/day), so the relative dimensions and porosity of the plug are changed, or it is replaced by a glass sleeve that permits relatively fast flow of salt bridge solution, thus reducing the rate and degree of fouling of the junction. In these circumstances, the electrode is refilled on a routine basis, or a continuous supply of bridge solution is arranged into the electrode at the appropriate pressure for the application.

A wide range of electrodes is illustrated in Figures 24.16–24.19. The choice of the appropriate reference electrode for the application is vital, and consideration must be given to the pressure, temperature, and nature of the process stream. The accuracy of the measurement and the frequency of maintenance depends on the correct choice of electrode. The EMF of the reference electrode will only remain constant provided satisfactory contact is made by the salt bridge, so the junction must not become plugged by suspended solids, viscous liquids, or reaction products of the process stream. Where this is a danger, the faster flow type of plug must be used. Many routine measurements can, however, be made with the non-flowing electrode, thus avoiding the necessity of refilling, or arranging a pressurized continuous supply. Flowing types of junctions are usually required where an accuracy of ± 0.02 pH units (± 1 or 2 mV) is needed, where frequent or large temperature or composition changes occur, or where the process fluid is such that it is prone to foul the junction.

The temperature of operation will influence the choice of concentration of the filling solutions. Potassium chloride solution with a concentration of 4 mol per liter saturates and starts to precipitate solids at about 19°C and will freeze at -4°C, whereas if the concentration is reduced to 1 mol per liter the solution will freeze at -2°C without becoming

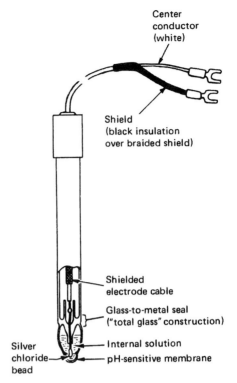

FIGURE 24.16 pH measuring electrode. Courtesy of the Foxboro Company.

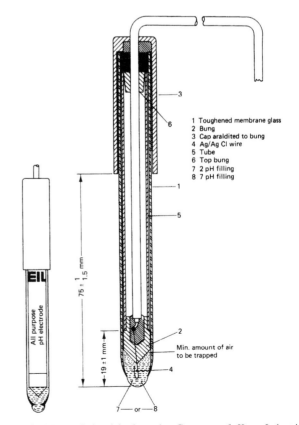

FIGURE 24.17 Industrial electrode. Courtesy of Kent Industrial Measurements Ltd. Analytical Instruments.

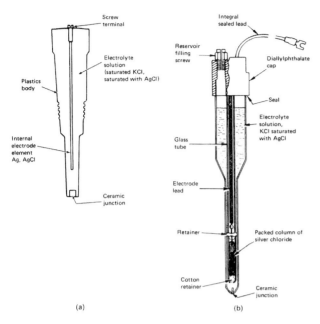

FIGURE 24.18 Reference electrodes (courtesy invensys): (a) sealed electrode, (b) flowing type.

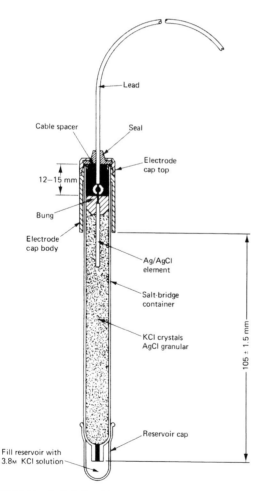

FIGURE 24.19 Sealed silver/silver chloride electrode. Courtesy of Kent Industrial Measurements Ltd. Analytical Instruments.

saturated. Thus no precipitation will take place in the solution of lower concentration. Although not damaging, precipitated potassium chloride and associated silver chloride will tend to clog reference junctions and tubes, decreasing electrolyte flow rate and increasing the risk of spurious potentials. For these reasons, flowing reference electrodes are not recommended for low-temperature applications unless provision is made to prevent freezing or precipitation in the electrode and any associated hardware.

When materials such as sulfides, alkali phosphates, or carbonates, which will react with silver, are present in the process stream, either nonflowing electrodes or electrodes containing potassium chloride at 1 mol per liter should be used. The diffusion rate of silver can be neglected in the nonflowing type, and the solubility of silver chloride in potassium chloride at a concentration of 1 mol per liter is only 1 or 2 percent of that in a solution at 4 mol per liter.

High temperatures with wide fluctuations are best handled by potassium chloride solution at 1 mol per liter.

24.7.2 Measurement of pH

Glass electrode Almost all pH measurements are best made with a glass electrode (the earliest of the ion-selective electrodes), the EMF being measured relative to a reference electrode. The glass electrode can be made to cover practically the whole of the pH scale and is unaffected by most chemicals except hydrofluoric acid. It can also be used in the presence of oxidizing or reducing agents without loss of measuring accuracy.

The electrode consists of a thin membrane of sodium-ion-selective glass sealed onto the end of a glass tube that has no ion-selective properties. The tube contains an internal reference solution in which is immersed the internal reference electrode, and this is connected by a screened lead to the pH meter. The internal reference electrode is almost always a silver/silver chloride electrode although recently, Thalamid electrodes[*] have sometimes been used. The internal reference solution contains chloride ions, to which the internal silver/silver chloride reference electrode responds, and hydrogen ions, to which the electrode as a whole responds. The ion to which the glass electrode responds, hydrogen in the case of pH electrodes, is determined by the composition of the glass membrane.

---

[*]The Thalamid electrode is a metal in contact with a saturated solution of the metallic chloride. Thallium is present as a 40 percent amalgam and the surface is covered with solid thallous chloride. The electrode is immersed in saturated potassium chloride solution. Oxygen access is restricted to prevent the amalgam being attacked. The advantage of the Thalamid electrode is that there is scarcely any time lag in resuming its electrode potential after a temperature change.

A glass pH electrode can be represented as

$$\begin{array}{|c|c|c|c|c|c|} \text{reference electrode} & \text{test solution} \; ^{a}\text{H}^{+} & \text{glass membrane} & \text{internal reference solution} \; ^{a'}\text{H}^{+} \; ^{+a'}\text{Cl} & \text{AgCl} & \text{Ag} \end{array}$$

Glass electrodes for pH measurement are of three main types: (a) general-purpose, for wide ranges of pH over wide ranges of temperature, (b) low-temperature electrodes (less than 10°C), which are low-resistance electrodes and are generally unsuitable for use above pH 9 to 10, and (c) high-pH and/or high-temperature electrodes (greater than 12 pH units). Glass electrodes are manufactured in many forms, some of which are shown in Figures 24.16 and 24.20. Spherical membranes are common, but hemispherical or conical membranes are available to give increased robustness where extensive handling is likely to occur. Electrodes with flat membranes can be made for special purposes, such as measurement of the pH of skin or leather, and microelectrodes are available but at great expense. Combination glass and reference electrodes (see Figure 24.20) can be obtained, and some electrodes can be steam-sterilized.

New electrodes supplied dry should be conditioned before use as the manufacturer recommends or by leaving them overnight in 0.1 mol liter^{-1} hydrochloric acid. Electrodes are best not allowed to dry out, and they should be stored in distilled or demineralized water at temperatures close to those at which they are to be used. The best treatment for pH electrodes for high-pH ranges is probably to condition them and store them in borax buffer solution.

Electrical Circuits for Use with Glass Electrodes For measurement of pH, the EMF in millivolts generated by the glass electrode compared with that of the reference electrode has to be converted to a pH scale, that is, one showing an increase of one unit for a decrease in EMF of approximately 60 mV. The pH scale requires the use of two controls—the calibration control and the slope control. The latter may not always be identified as such on the pH meter, since it often acts in the same way as the temperature compensation control. The slope and temperature compensation controls adjust the number of millivolts equivalent to one pH unit. The calibration control relates the measured EMF to a fixed point on the pH scale.

A typical pH measuring system (glass electrode and reference electrode immersed in a solution) may have a resistance of several hundred megohms. To obtain an accurate measurement of the EMF developed at the measuring electrode, the electrical measuring circuit must have a high input impedance and the insulation resistance of the electrical leads from the electrodes to the measuring circuit must be extremely high ($\sim 10^5$ MΩ—a "Megger" test is useless). The latter is best achieved by keeping the electrode leads as short as possible and using the best moisture-resistant insulating materials available (e.g., polythene or silicone rubber).

The usual method of measurement is to convert the developed EMF into a proportional current by means of a suitable amplifying system. The essential requirements of such a system have been met completely by modern electronic circuits, and one system uses an amplifier with a very high negative feedback ratio. This means that the greater part of the input potential is balanced by a potential produced by passing the meter current through an accurately known resistor, as shown in Figure 24.21. If the PD V_0, developed

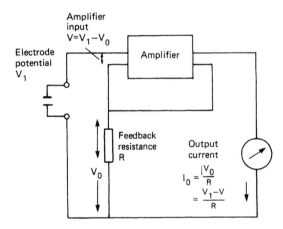

FIGURE 24.21 Principle of d.c. amplifier with negative feedback. Courtesy of Kent Industrial Measurements Ltd.

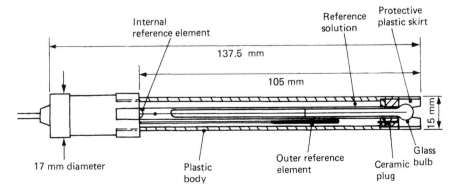

FIGURE 24.20 Combined reference electrode and glass electrode for pH measurement. Courtesy of ABB Instrument Group.

across the feedback resistance is a very large fraction of the measured potential V_1, the input voltage V is a very small fraction of V_1, and

$$I_0 = \frac{(V_1 - V)}{R}, \text{ approaches } \frac{V_1}{R}$$

With modern integrated circuit techniques, it is possible to obtain an amplifier with a very high input impedance and very high gain so that little or no current is drawn from the electrodes.

Such a system is employed in the pH-to-current converter shown in Figure 24.22, which employs zener diode stabilized supplies and feedback networks designed to give a high-gain, high-input impedance diode bridge amplifier.

The dc imbalance signal, resulting from the pH signal, asymmetry-correcting potential, and the feedback voltage, changes the output of a capacity balance diode bridge. This output feeds a transistor amplifier, which supplies feedback and output proportional to the bridge error signal. Zener diode stabilized and potentiometer circuits are used to provide continuous adjustment of span, elevation, and asymmetry potential over the entire operating range of the instrument.

The input impedance of the instrument is about $1 \times 10^{12}\,\Omega$ and the current taken from the electrodes less than 0.5×10^{-12} A.

The principle of another system, which achieves a similar result, is shown in Figure 24.23. It uses a matched pair of field effect transistors (FETs) housed in a single can. Here the EMF produced by the measuring electrode is fed to the gate of one of the pair. The potential that is applied to one side of the high-gain operational amplifier will be governed by the current that flows through the transistor and its corresponding resistance R_3. The potential applied to the gate of the second FET is set by the buffer bias adjustment, which is fed from a zener stabilized potential supply. The potential developed across the second resistance R_4, which is equal in resistance to R_3, will be controlled by the current through the second of the pair of matched FETs. Thus the output of the operational amplifier will be controlled by the difference in the potentials applied to the gates of the FETs, that is, to the difference between the potential developed on the measuring electrode and the highly stable potential set up in the instrument. Thus, the current flowing through the local and remote indicators will be a measure of the change of potential of the measuring electrode.

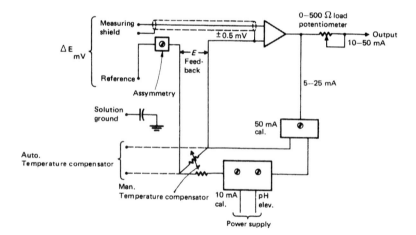

FIGURE 24.22 High gain, high impedance pH-to-current converter. Courtesy of the Foxboro Company.

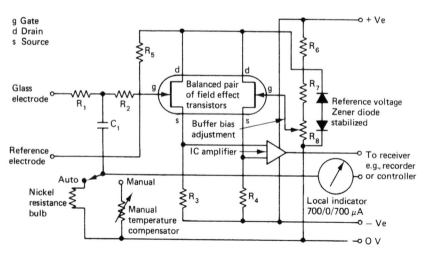

FIGURE 24.23 pH measuring circuit using field effect transistors.

If the EMF given by the glass electrode is plotted against pH for different temperatures, it will be seen that there is a particular value of the pH at which the EMF is independent of temperature. This point is known as the *iso-potential point*.

If the iso-potential point is arranged to be the locus of the slope of the measuring instrument, the pH measuring circuit can be modified to include a temperature sensor arranged to change the negative feedback so that the circuit compensates for the change in slope of the EMF/pH relationship. It is important to realize that the temperature compensation only corrects for the change in the electrode response due to temperature change, and the iso-potential control setting therefore enables pH electrodes calibrated at one temperature to be used at another. The iso-potential control does *not* compensate for the actual change in pH of a solution with temperature. Thus if pH is being measured to establish the composition of a solution, one must carry out the measurements at constant temperature.

A few commercial pH meters have a variable iso-potential control so that they can be used with several different combinations of electrodes, but it is more generally the case that pH meters have fixed iso-potential control settings and can only be used with certain combinations of pH and reference electrodes. It is strongly recommended that, with fixed iso-potential control settings, both the glass and reference electrodes be obtained from the manufacturer of the pH meter. Temperature compensation circuits generally work only on the pH and direct activity ranges of a pH meter and not on the millivolt, expanded millivolt, and relative millivolt ranges.

Modern pH meters with analog displays are scaled 0 to 14 pH units, with the smallest division on the scale equivalent to 0.1 unit, giving the possibility of estimating 0.02 pH units by interpolation. The millivolt scale is generally 0 to 1,400 mV with a polarity switch or -700 to $+700$ mV without one. The smallest division is 10 mV, allowing estimation to 2 mV. Many analog meters have a facility of expanding the scale so that the precision of the reading can be increased up to 10 times. Digital outputs are also available, with the most sensitive ones reading to 0.001 pH unit (unlikely to be meaningful in practice) or 0.1 mV. Instruments incorporating microprocessors are also now available—these can calculate the concentration of substances from pH measurements and give readout in concentration units. Blank and volume corrections can be applied automatically.

Precision and Accuracy Measurements reproducible to 0.05 pH units are possible in well buffered solutions in the pH range 3 to 10. For routine measurements, it is rarely possible to obtain a reproducibility of better than ±0.01 pH units.

In poorly buffered solutions, reproducibility may be no better than ±0.1 pH unit and accuracy may be lost by the absorption of carbon dioxide or by the presence of suspensions, sols, and gels. However, measured pH values can often be used as control parameters, even when their absolute accuracies are in doubt.

Sodium Ion Error Glass electrodes for pH measurement are selective for hydrogen ions, not uniquely responsive to them, and so will also respond to sodium and other ions, especially at alkaline pH values (more than about 11). This effect causes the pH value to be underestimated. Sodium ions produce the greatest error, lithium ions about a half, potassium ions about a fifth, and other ions less than a tenth of the error due to sodium ions. One can either standardize the electrode in an alkaline buffer solution containing a suitable concentration of the appropriate salt, or better, use the special lithium and cesium glass electrodes developed for use in solutions of high alkalinity. These are less prone to interference. For a given glass electrode at a stated measuring temperature, the magnitude of the error can be found from tables provided by electrode manufacturers. An example is shown in Figure 24.24.

Temperature Errors The calibration slope and standard potential of ion-selective electrodes (including glass pH electrodes) are affected by temperature. If the pH is read directly off the pH scale, some form of temperature correction will be available but often only for the calibration slope and not for the standard potential. If measurements are made at a temperature different from that at which the electrode was calibrated there will be an error. This will be small if the meter has an iso-potential setting. For the most accurate work the sample and buffer solutions should be at the same temperature, even if iso-potential correction is possible.

Stirring Factor In well-buffered solutions it may not be necessary to stir when making pH measurements. However, it is essential in poorly buffered solutions.

The Hydrogen Electrode The hydrogen electrode, consisting in practice of a platinum plate or wire coated with platinum block (a finely divided form of the metal), can measure hydrogen ion activity when hydrogen is passed over the electrode. However, this electrode is neither easy nor convenient to use in practice and is now never used in industrial laboratories or on plant.

The Antimony Electrode The antimony electrode is simply a piece of pure antimony rod (~12 mm diameter, 140 mm

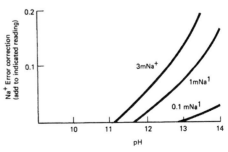

FIGURE 24.24 Relationship of pH and Na ion error. Courtesy of Kent Industrial Measurements Ltd. Analytical Instruments.

long), housed in a protective plastic body resistant to acid attack; see Figure 24.25. The protruding antimony rod, when immersed in a solution containing dissolved oxygen, becomes coated with antimony trioxide Sb_2O_3, and the equilibria governing the electrode potential are

$$Sb \rightarrow Sb^{3+} + 3e^-$$

$$Sb_2O_3 + 6H^+ \rightarrow 2Sb^{3+} + 3H_2O, \; K = \frac{[Sb^{3+}]}{[H^+]^3}$$

However, there are many possible side reactions, depending on the pH and the oxidizing conditions; salt effects are large. There is therefore difficulty in calibrating with buffer solutions; stirring temperature and the amount of oxygen present all have rather large effects. A reproducibility of about 0.1 pH unit is the best that is normally attained; the response is close to Nernstian over the pH range 2 to 7, and the response time can be as short as 3 minutes but is often about 30 minutes.

The outstanding advantage of the antimony electrode is its ruggedness, and for this reason it has been used for determining the pH of soils. Also, of course, it is indispensable for solutions containing hydrofluoric acid, which attack glass. If the electrode becomes coated during use, its performance can be restored by grinding and polishing the active surface and then reforming the oxide film by immersion in oxygenated water before using in deoxygenated solutions. However, there is much more uncertainty to every aspect of behavior of the antimony electrode than with the glass electrode, and even the fragile glass electrodes of years ago with their limited alkaline range displaced the antimony electrode when accurate pH measurements were required. Modern glass electrodes are excellent in respect of robustness and range, and antimony electrodes are not much used apart from the specialized applications already mentioned. In these, the resistance of the measuring system is low, so a simple low-impedance electrical circuit can be used with them—for example, a voltmeter or a potentiometric type of system. Figure 24.26 shows the principle of such a system. Any difference between the electrode EMF and that produced across the potentiometer will be amplified and applied to the servomotor, which moves the slide-wire contact to restore balance.

Industrial pH Systems with Glass Electrodes Two types of electrode systems are in common use: the continuous-flow type of assembly and the immersion, or dip, type of assembly.

Continuous-Flow Type of Assembly The physical form of the assembly may vary a little from one manufacturer to another, but Figure 24.27 illustrates a typical assembly designed with reliability and easy maintenance in mind. Constructed in rigid PVC throughout, it operates at pressures up to 2 bar and temperatures up to 60°C. For higher temperatures and pressures, the assembly may be made from EN 58J stainless steel, flanged, and designed for straight-through flow when pressures up to 3 bar at temperatures up to 100°C can be tolerated. It accommodates the standard measuring electrode, usually of toughened glass.

A reservoir for potassium chloride (or other electrolyte) forms a permanent part of the electrode holder. A replaceable reference element fits into the top of the reservoir and is held in place by an easily detachable clamp nut. A microceramic plug at the lower end of the reservoir ensures slow electrolyte leakage (up to six months' continuous operation without attention is usually obtained). The ceramic junction is housed in a screw-fitting plug and is easily replaceable.

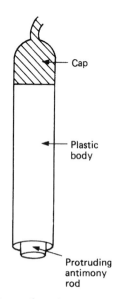

FIGURE 24.25 Antimony electrode.

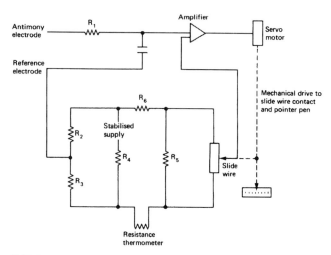

FIGURE 24.26 Low impedance measuring circuit for use with antimony electrodes.

CHAPTER | 24 Chemical Analysis: Electrochemical Techniques

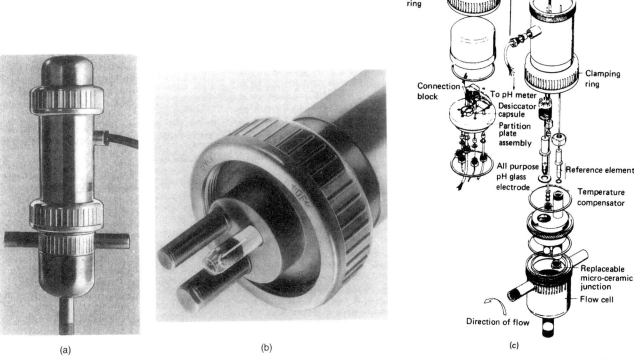

FIGURE 24.27 Flow-type of electrode system (courtesy of Kent Industrial Measurements Ltd. Analytical Instruments): (a) external view, (b) upper section detaches for easy buffering, (c) exploded view showing the components.

The close grouping of electrodes makes possible a small flow cell and hence a fast pH response at low flow rates. An oil-filled reservoir built into the electrode holder houses a replaceable nickel wire resistance element, which serves as a temperature compensator. (This is an optional fitment.)

The flow through the cell creates some degree of turbulence and thus minimizes electrode coating and sedimentation.

The integral junction box is completely weatherproof and easily detachable. Electrode cables and the output cable are taken via individual watertight compression fittings into the base of the junction box. A desiccator is included to absorb moisture, which may be trapped when the cover is removed and replaced.

Two turns of the lower clamp nut allow the entire electrode unit to be detached from the flow cell and hence from the process fluid. The electrodes can be immersed easily in buffer solution.

Immersion Type Basically, this assembly is similar to the flow type except that the flow cell is replaced by a protecting guard, which protects the electrode but allows a free flow of solution to the electrodes. In addition, the upper cap is replaced by a similarly molded tube that supports the electrode assembly but brings the terminal box well above the electrode assembly so that the terminals are clear of the liquid surface when the assembly is in the measured solution. Immersion depths up to 3 m are available.

Electrode assemblies should be designed so that the electrodes can be kept wet when not in use. It is often possible to arrange for the easy removal of the assembly from the process vessel so that it can be immersed in a bucket filled with process liquid, water, or buffer solution during shutdown.

The design of the assembly is often modified to suit the use. For example, in measuring the pH of pulp in a paper beater, the electrodes and resistance bulb are mounted side by side in a straight line and then inclined downstream at about 45°C from the vertical so that they present no pockets to collect pulp and are self-cleaning.

When the assembly is immersed in a tank, care must be taken in the siting to ensure that the instrument is measuring the properties of a representative sample; adequate mixing of the process material is essential. Sometimes it is more convenient to circulate the contents of a tank through a flow type of assembly and then return the liquid to the tank.

The main cause of trouble in electrode assemblies is the fouling of the electrodes. To reduce this issue, two forms of self-cleaning are available, and the choice of method depends on the application. Where the main cause of trouble is deposits on the glass, electrode and mechanical cleaning is required; this may be achieved by the cleaning attachment shown on a dip system in Figure 24.28. The pneumatically driven rubber membrane wipes the electrode, providing a simple, reliable cleaning action. It is driven by compressed air at preset intervals from a controller that incorporates a

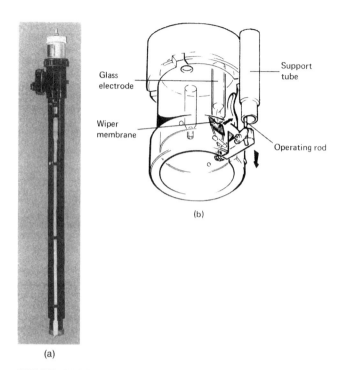

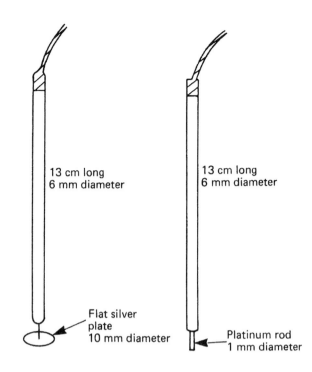

FIGURE 24.28 Electrode cleaning (courtesy of Kent Industrial Measurements Ltd. Analytical Instruments): (a) assembly, (b) detail of cleaning attachment.

FIGURE 24.29 Examples of metal redox electrodes.

programmed timer mechanism that governs the frequency of the wiping action. The cleaning attachment is constructed entirely of polypropylene and 316 stainless steel, except for the rubber wiper, which may be replaced by a polypropylene brush type, should this be more suitable.

Alternatively, an ultrasonic generator operating at 25 kHz can be fitted to the electrode assembly, greatly increasing the periods between necessary electrode cleaning.

24.7.3 Measurement of Redox Potential

When both the oxidized and reduced forms of a substance are soluble in water, the old-fashioned metal redox electrode is useful; an equilibrium is set up between the two forms of the substance and the electrons in the metal electrode are immersed in the solution. Again, a reference electrode, generally calomel, has to be used, and determinations can be made, either by using the redox electrode as an indicator during titrations or by direct potentiometric determination. Arrangements are similar to those for a pH electrode. Redox electrodes, too, can be immersed directly in a liquid product stream when monitoring on plant. The high-impedance EMF measuring circuits as used for pH electrode systems are completely satisfactory, but since metal Redox electrodes are low-resistance systems, low-impedance EMF measuring circuits may also be used as for the antimony pH electrode. (The latter is also a metal redox electrode.)

Apart from the antimony electrode, platinum, silver, and gold electrodes (Figure 24.29) are available commercially, and simple electrodes for use with separate reference and combination electrodes can be obtained both for laboratory and industrial use.

Analytical chemistry applications of redox electrodes include determination of arsenic, cyanides, hydrogen peroxide, hypochlorite or chlorine, ferrous iron, halides, stannous tin, and zinc. The silver electrode is widely used for halide determination. Platinum electrodes are suitable for most other determinations with the exception of occasions in which cyanide is being oxidized with hypochlorite (for example, in neutralizing the toxic cyanide effluent from metal plating baths). In this case a gold electrode is preferable.

24.7.4 Determination of Ions by Ion-Selective Electrodes

General Considerations The measurement of the concentration or the activity of an ion in solution by means of an ion-selective electrode is as simple and rapid as making a pH measurement (the earliest of ion-selective electrodes). In principle it is necessary only to immerse the ion-selective and reference electrodes in the sample, read off the generated EMF by means of a suitable measuring circuit, and obtain the result from a calibration curve relating EMF and concentration of the substance being determined. The difference from pH determinations is that most ion-selective electrode applications require the addition of a reagent to buffer or adjust the ionic strength of the sample before measurement of the potential. Thus, unlike measurement of pH and redox potentials, ion-selective electrodes cannot be immersed directly in a plant stream of liquid product, and a sampling arrangement has to be used. However, this can usually be done quite simply.

pH and PIon Meters High-impedance EMF measuring circuits must be used with most ion-selective electrodes and are basically the same as used for measuring pH with a glass electrode. The pH meters measure EMF in millivolts and are also scaled in pH units. Provided that the calibration control on the pH meter (which relates the measured EMF to a fixed point on the pH scale) has a wide enough range of adjustment, the pH scale can be used for any univalent positive ion; for example, measurement with a sodium-selective electrode can be read on the meter as a pNa scale (or $-\log C_{Na}$). Measurements with electrodes responding to divalent or negative ions cannot be related directly to the pH scale. However, manufacturers generally make some modification to pH meters to simplify measurements with ion-selective electrodes, and the modified meters are called *pIon meters*. Scales are provided, analogous to the pH scale, for ions of various valencies and/or a scale that can be calibrated to read directly in terms of concentration or valency. Meters manufactured as pIon meters generally also have pH and millivolt scales. To date, pIon scales only cover ions with charges of ± 1 and ± 2 because no ion-selective electrodes for determining ions of high charge are yet available commercially. Direct activity scales read in relative units only and so must be calibrated before use in the preferred measurement units.

As with pH meters, pIon meters can be obtained with analog and digital displays, with integral microprocessors, with recorder and printer outputs, and with automatic standardization. Temperature compensation can be incorporated, but although ion-selective and reference electrode combinations have iso-potential points, the facility of being able to set the iso-potential control has so far been restricted to pH measurement. On dual pH/pIon meters, the iso-potential control (if it exists) should be switched out on the pIon and activity scales if one wants to make a slope correction when working with an ion-selective electrode at constant temperature.

For the best accuracy and precision, pIon meters should be chosen that can discriminate 0.1 mV for direct potentiometry; 1 mV discrimination is sufficient when using ion-selective electrodes as indicators for titrimetric methods.

Practical Arrangements For accurate potentiometry, the temperature of the solution being analyzed and the electrode assembly should be controlled and ideally all analyses should be carried out at the same temperature—for example, by using a thermostatically controlled water bath. Solutions must also be stirred; otherwise the EMF developed by the electrode may not be representative of the bulk of the solution. A wide range of stirring speeds is possible, but too slow a speed may give long response times and too high a speed may generate heat in the solution. Precautions must also be taken to minimize contamination.

Taking all these items into account, best results in the laboratory can be obtained by mounting the electrodes in a flow cell through which the test solution is being pumped; see Figure 24.30. This is a mandatory arrangement for on-stream instruments and in the laboratory in cases where the ion concentration being determined is close to the limit of detection of the electrode.

Flow cells should be constructed of a material that will not contaminate a sample with the ion being determined; the flow rates of the solution must be high enough to provide "stirring" but low enough that sample volumes are kept low. There should be good displacement of a previous sample by an incoming one, solution from the reference electrode should not reach the measuring electrode, and when liquid flow through a flow cell stops, the cell must retain liquid

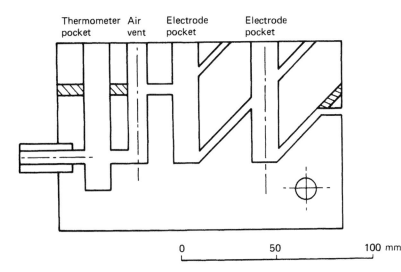

FIGURE 24.30 Flow cell for ion-selective electrodes.

around the electrodes to prevent them drying out. Finally, a flow cell should be water-jacketed so that its temperature can be controlled. Suitable flow cells can be machined out of Perspex and are available commercially.

Pumps used must be capable of pumping at least two channels simultaneously at different rates, the larger volume for the sample and the lesser for the reagent solution. Peristaltic pumps are the most frequently used. It follows that all interconnecting tubing and other components in contact with the sample must be inert with respect to the ion being determined.

As direct potentiometric determination of ions by ion-selective electrodes requires more frequent calibration than the more stable pH systems, industrially developed ion-selective electrode systems often incorporate automatic recalibration. This makes them more expensive than pH measuring systems. A typical scheme for an ion-selective monitor (in this case, for sodium) is shown in Figures 24.31 and 24.32.

Sample water flows to the constant head unit and is then pumped anaerobically at a constant rate into the flow cell, where it is equilibrated with ammonia gas obtained by pumping a stream of air through ammonia solution. (Instead of ammonia gas, a liquid amine could be used, and this would then be the buffer liquid delivered by the second channel of the pump.) The sample then flows through the flow cell to contact the ion-selective and reference electrodes and then to a drain.

Automatic chemical standardization takes place at preset intervals (in this case, once every 24 hours) with provision for manual initiation of the sequence at any time. The standardization sequence commences by activating a valve to stop the sample flow and to allow a sodium ion solution of known strength (the standard sodium solution) to be pumped into the flow cell. When the electrodes have stabilized in the new solution, the amplifier output is compared with a present standard value in the auto-compensation unit, and any error causes a servo-potentiometer to be driven so as to adjust the output signal to the required value. The monitor is then returned to measurement of the sample. The standardization period lasts 30 minutes, a warning lamp shows that standardization is taking place, and any alarm and control contacts are disabled. It is also possible to check the stability of the amplifier and, by a manual introduction of a second sodium standard, to check and adjust the scale length.

Conditioning and Storage of Electrodes The manufacturer's instructions regarding storage and pretreatment of electrodes should be followed closely. The general rules are that (a) glass electrodes should not be allowed to dry out because reconditioning may not be successful, (b) solid-state electrodes can be stored in deionized water for long periods, dry-covered with protective caps, and generally ready for use after rinsing with water, (c) gas-sensing membranes and liquid ion-exchange electrodes must never be allowed to dry out, and (d) reference electrodes are as important as the measuring electrodes and must be treated exactly as specified by the manufacturer. The element must not be allowed to dry out, as would happen if there were insufficient solution in the reservoir.

Ion-selective Electrodes Available and Application areas
A very wide range of electrodes is available. Not only are there many specific ion monitors, but several manufacturers now market standardized modular assemblies that only need different electrodes, different buffer solutions, and minor electrical adjustments for the monitors to cope with many ion determinations.

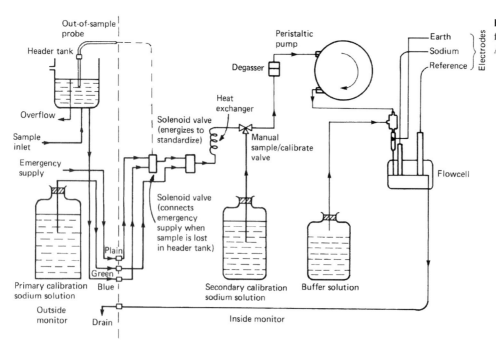

FIGURE 24.31 Schematic diagram for ion-selective monitor. Courtesy ABB Instrument Group.

CHAPTER | 24 Chemical Analysis: Electrochemical Techniques

Thus the fraction K of the total sodium activity will behave as though it were potassium. The *smaller* the value of K, the more selective that electrode is to potassium, that is, the *better* it is. To identify a particular selectivity coefficient, the data are best written in the form:

$$K_{\text{potassium}^+/\text{sodium}^+} = 2.6 \times 10^{-3}$$

This shows that the selectivity of potassium over sodium for the potassium electrode is about 385:1, that is, $1/(2.6 \times 10^{-3})$. It is important to note that selectivity coefficients are not constant but vary with the concentration of both primary and interferent ions, and the coefficients are, therefore, often quoted for a particular ion concentration. They should be regarded as a guide to the effectiveness of an electrode in a particular measurement and not for use in precise calculations, particularly as quoted selectivity coefficients vary by a factor of 10 or more. For accurate work the analyst should determine the coefficient for himself for his own type of solution.

Direct potentiometric determination of ions by means of ion-selective electrodes has many applications. Examples are determination of pH, sodium, and chloride in feed water, condensate, and boiler water in power stations; cyanide, fluoride, sulfide, and chloride in effluents, rivers, and lakes; fluoride, calcium, and chloride in drinking water and sea water; bromide, calcium, chloride, fluoride, iodide, potassium, and sodium in biological samples; calcium, chloride, fluoride, and nitrate in soils; sulfur dioxide in wines and beer; chloride and calcium in milk; sulfide and sulfur dioxide in the papermaking industry; fluoride, calcium, chloride, nitrate, and sulfur dioxide in foodstuffs, pH in water and effluents, papers, textiles, leather, and foodstuffs, and calcium, chloride, fluoride, and potassium in pharmaceuticals.

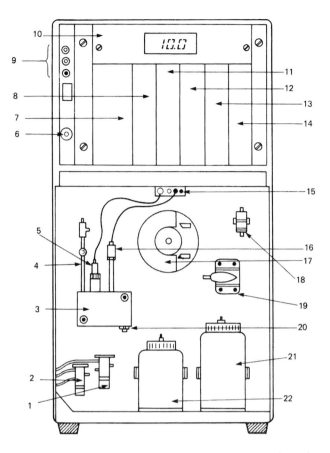

FIGURE 24.32 Diagrammatic arrangement of components for an ion-selective monitor. Courtesy of ABB Instrument Group. 1. solenoid valve (energizes during standardization to connect primary standard solution); 2. solenoid valve (energizes to admit emergency sample supply when sample is lost in the header tank); 3. flow cell; 4. earthing tube; 5. sodium electrode; 6. SUPPLY ON lamp (illuminates when power is connected to the monitor): 7. 8020 100 amplifier; 8. 8033 200 current output module; 9. SERVICE lamp (red) and ONLINE lamp (green) with pushbutton (optional feature); 10. digital display module (linear motor readout optional); 11. 8060 300 compensation module; 12. 8021 400 alarm and temperature control module; 13. 8020 500 power supply; 14. 8020 600 function module; 15. electrodes connection point (junction box); 16. refillable calomel reference electrode; 4. peristaltic pump; 18. gas debubbler; 19. manual SAMPLE/CALIBRATE valve; 20. flow cell drain; 21. secondary standard solution container (1 liter) (heat exchanger located behind the panel at this point); 22. buffer solution container (500 ml).

Table 24.9 shows the ion-selective electrodes available for the more common direct potentiometric determination of ions.

Ion-selective electrodes, as their name implies, are selective rather than specific for a particular ion. A potassium electrode responds to some sodium ion activity as well as to potassium, and this can be expressed as:

$$E_{\text{measured}} = \text{constant} \pm S \log \left(a_{\text{potassium}^+} + K a_{\text{Na}^+} \right)$$

where K is the selectivity coefficient of this electrode to sodium and $0 < K < 1$.

24.8 COMMON ELECTROCHEMICAL ANALYZERS

24.8.1 Residual Chlorine Analyzer

When two dissimilar metal electrodes are immersed in an electrolyte and connected, current will flow due to the buildup of electrons on the more electropositive electrode. The current will soon stop, however, owing to the fact that the cell will become polarized.

If, however, a suitable depolarizing agent is added, a current will continue to flow, the magnitude of which will depend on the concentration and nature of the ions producing the depolarization. Thus, by choice of suitable materials for the electrodes and arranging for the addition of the depolarizing agent, which is in fact the substance for which the concentration is to be measured, amperometric analyzers may be made to measure the concentration of a variety of chemicals. In some instruments a potential difference may be applied to the electrodes, when the current is again a linear function of the concentration of the depolarizing agent.

TABLE 24.9 Available ion-selective electrodes

Solid-state membrane electrodes	Glass membrane electrodes	Liquid ion exchange membrane electrodes	Gas-sensing electrodes
Fluoride	pH	Calcium	Ammonia
Chloride	Sodium	Calcium + magnesium	Carbon dioxide
Bromide	Potassium	(i.e., water hardness)	Sulphur dioxide
Iodide			Nitrous oxide
Thiocyanate			Hydrogen sulphide
Sulphide			Hydrogen fluoride
Silver		Barium	
Copper		Nitrate	
Lead		Potassium	
Cadmium			
Cyanide			
Redox			
pH (antimony)			

The sensitivity of the analyzer is sometimes increased by using buffered water as the electrolyte so that the cell operates at a definite pH. Amperometric instruments are inherently linear in response, but special steps have to be taken to make them specific to the substance for which the concentration is to be measured, because other substances may act as depolarizing agents and so interfere with the measurement. When the interfering substances are known, steps may be taken to remove them.

Where the instrument is intended to measure pollutants in air or gas, the gas to be tested is either bubbled through a suitable cell or arranged to impinge on the surface of the liquid in the cell. In these cases interfering gases can be removed by chemical or molecular filters in the sampling system.

This form of instrument may be used to detect halogens, such as chlorine in air, and instruments with ranges from 0–0.5 to 0–20 ppm are available for measuring with an accuracy of ±2 percent and a sensitivity of 0.01 ppm. By altering the electrolyte, the instrument may be changed to measure the corresponding acid vapors, that is, HCl, HBr, and HF. One type of instrument for measuring chlorine in water is shown in Figure 24.33.

The sample stream is filtered in the tank on the back of the housing and then enters the analyzer unit through the sample flow control valve and up the metering tube into the head control block, where reagent (buffer solution to maintain constant pH) is added by means of a positive displacement feed pump.

Buffered sample flows down tube B, through the flow control block and up tube C to the bottom of the electrode cell assembly. Sample flow rate is adjusted to approximately 150 milliliters per minute. Flow rate is not critical since the relative velocity between the measuring electrode and the sample is established by rotating the electrode at high speed.

In the electrode cell assembly, the sample passes up through the annular space between the concentrically mounted outer (copper) reference electrode and the inner (gold) measuring electrode and out through tube D to the drain. The space between the electrodes contains plastic pellets that are continuously agitated by the swirling of the water in the cell. The pellets keep the electrode surfaces clear of any material, which might tend to adhere. The measuring electrode is coupled to a motor, which operates at 1,550 rev/min. The electrical signal from the measuring electrode is picked up by a springloaded brush on top of the motor and the circuit is completed through a thermistor for temperature compensation, precision resistors, and the instationary copper electrode.

The composition of the electrodes is such that the polarization of the measuring electrode prevents current flow in the absence of a strong oxidizing agent. The presence of the smallest trace of strong oxidizer, such as chlorine (hypochlorous acid), will permit a current to flow by oxidizing the polarizing layer. The amplitude of the self-generated depolarization current is proportional to the concentration of the strong oxidizing agent. The generated

CHAPTER | 24 Chemical Analysis: Electrochemical Techniques

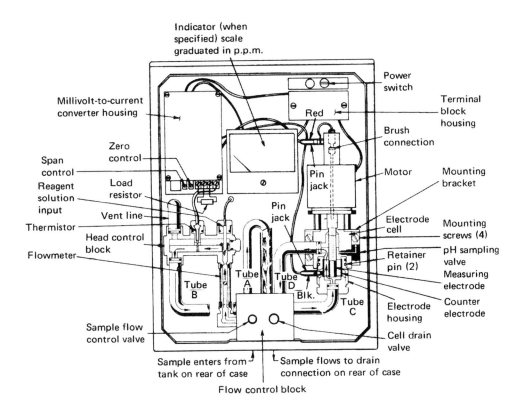

FIGURE 24.33 Residual chlorine analyzer. Courtesy of Capital Controls Division, Severn Trent Ltd.

current is passed through a precision resistor and the millivoltage across the resistor is then measured by the indicating or recording potentiometer. This instrument is calibrated to read in terms of the type (free or total) of residual chlorine measured. In measuring total residual chlorine, potassium iodide is added to the buffer. This reacts with the free and combined chlorine to liberate iodine in an amount equal to the total chlorine. The iodine depolarizes the cell in the same manner as hypochlorous acid, and a current directly proportional to the total residual chlorine is generated.

24.8.2 Polarographic Process Oxygen Analyzer

An instrument using the amperometric (polarographic) method of measurement is an oxygen analyzer used for continuous process measurement of oxygen in flue gas, inert gas monitoring, and other applications.

The key to the instrument is the rugged sensor shown in Figure 24.34. The sensor contains a silver anode and a gold cathode that are protected from the sample by a thin membrane of PTFE. An aqueous KCl solution is retained in the sensor by the membrane and forms the electrolyte in the cell (see Figure 24.35).

Oxygen diffuses through the PTFE membrane and reacts with the cathode according to the equation

$$4e^- + O_2 + 2H_2O \rightarrow 4OH^-$$

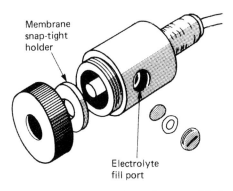

FIGURE 24.34 Process oxygen analyzer. Courtesy of Beckman Instruments Inc.

The corresponding anodic reaction is

$$Ag + Cl^- \rightarrow AgCl + e^-$$

For the reaction to continue, however, an external potential (0.7 volt) must be applied between cathode and anode. Oxygen will then continue to be reduced at the cathode, causing the flow of a current, the magnitude of which is proportional to the partial pressure of oxygen in the sample gas.

The only materials in contact with the process are PVC and PTFE, and the membrane is recessed so that it does not suffer mechanical damage. The cell needs to be recharged with a new supply of electrolyte at three- or six-month intervals, depending on the operating conditions, and the membrane can be replaced easily should it be damaged.

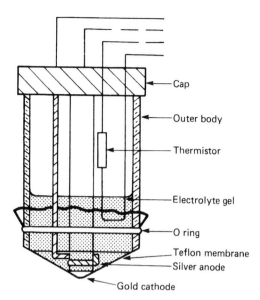

FIGURE 24.35 Diagram of polarographic oxygen sensor. Courtesy of Institute of Measurement and Control.

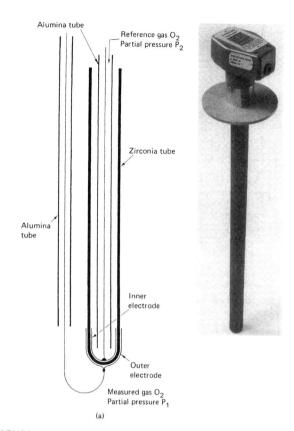

FIGURE 24.36 Oxygen probe. Courtesy of Kent Instruments.

The cell current is amplified by a solid-state amplifier, which gives a voltage output that can be displayed on a selection switch giving ranges of 0–1, 0–5, 0–10 or 0–25 percent oxygen and a calibration adjustment. The calibration is checked by using a reference gas or using air when the instrument should read 20.9 percent oxygen on the 0–25 percent scale. The instrument has an accuracy ±1 percent of scale range at the calibration temperature but an error of ±3 percent of the reading will occur for a 16°C departure in operating temperature.

When in use, the sensor may be housed in an inline-type housing or in a dip type of assembly, usually made of PVC suitable for pressures up to 3.5 bar.

24.8.3 High-temperature Ceramic Sensor Oxygen Probes

Just as an electrical potential can be developed at a glass membrane, which is a function of the ratio of the hydrogen concentrations on either side, a pure zirconia tube maintained at high temperature will develop a potential between its surfaces that is a function of the partial pressure of oxygen that is in contact with its surfaces. This is the principle involved in the oxygen meter shown in Figure 24.36.

The potential developed is given by the Nernst equation:

$$E_S = \left(\frac{RT}{4F}\right)\left\{\ln\left(\frac{\text{internal partial pressure of } O_2^{4-} \text{ ions}}{\text{external partial pressure of } O_2^{4-} \text{ ions}}\right)\right\}$$

Thus, if the potential difference between the surfaces is measured by platinum electrodes in contact with the two surfaces a measure may be made of the ratio of the partial pressure of the oxygen inside and outside the probe. If dry instrument air (20.9 percent oxygen) is fed into the inside of the probe, the partial pressure of oxygen inside the tube may be regarded as constant, so the electrical potential measured in a similar manner to that adopted in pH measurement will be a measure of the concentration of the oxygen in the atmosphere around the measuring probe. Thus by positioning the probe in a stack or flue where the temperature is above 600°C, a direct measurement of the oxygen present may be made. (In another manufacturer's instrument, the probe is maintained at a temperature of 850°C by a temperature-controlled heating element.) The instrument illustrated can operate from 600 to 1,200°C, the reading being corrected for temperature, which is measured by a thermocouple. The probe is protected by a silicon carbide sheath. The zirconia used is stabilized with calcium.

Standard instruments have ranges of oxygen concentration of 20.9–0.1 percent, 1,000–1 ppm, $10^{-5} - 10^{-25}$ partial pressure and can measure oxygen with an accuracy of better than ±10 percent of the reading.

Because temperatures in excess of 600°C must be used, some of the oxygen in the sample will react with any combustible gas present, such as carbon monoxide and hydrocarbons. Thus the measurement will be lower than the correct value but will still afford a rapid means of following changes in the oxygen content of a flue gas caused by changes in combustion conditions.

When using this form of oxygen cell, one must be aware that one is measuring "net oxygen," since any combustible material in the sample stream will be burned or consumed on the outer electrode and in doing so use the stoichiometric amount of oxygen required for combustion.

24.8.4 Fuel Cell Oxygen-measuring Instruments

Galvanic or fuel cells differ from polarographic cells and the high-temperature ceramic sensors in that they are power devices in their own right, that is, they require no external source of power to drive them. One manufacturer's version is shown in Figure 24.37.

A lead anode is made in that geometric form that maximizes the amount of metal available for reaction with a convex disc as the cathode. Perforations in the cathode facilitate continued wetting of the upper surface with electrolyte and ensure minimum internal resistance during the oxygen sensing reaction. The surfaces of the cathode are plated with gold and then covered with a PTFE membrane. Both electrodes are immersed in aqueous potassium hydroxide electrolyte. Diffusion of oxygen through the membrane enables the following reactions to take place:

Cathode anode $\quad 4e^- + O_2 + 2H_2O \rightarrow 4OH^-$
$\quad\quad\quad\quad\quad\quad Pb + 2OH^- \rightarrow PbO + H_2O + 2e$

Overall cell reaction $\quad 2Pb + O_2 \rightarrow PbO$

The electrical output of the cell can be related to the partial pressure of oxygen on the gas side of the membrane in a manner analogous to that described for membrane-covered polarographic cells. In this instance, however, because there is no applied potential and no resultant hydrolysis of the electrolyte, absence of oxygen in the sample corresponds to zero electrical output from the cell. There is a linear response to partial pressure of oxygen, and a single point calibration, for example, on air, is sufficient for most purposes.

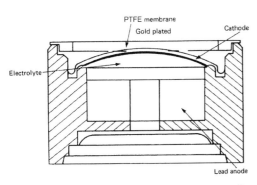

FIGURE 24.37 Diagrammatic microfuel cell oxygen sensor. Courtesy of Analysis Automation.

The main limitation of this type of oxygen sensor is the rate of diffusion of oxygen across the membrane; this determines the speed of response and, at low oxygen partial pressure, this may become unacceptably slow. However, to overcome this, one type of fuel cell oxygen sensor has a completely exposed cathode, that is, not covered with a PTFE membrane.

In common with all membrane cells, the response of the microfuel cell is independent of sample flow rate but the cell has a positive temperature-dependence. This is accommodated by incorporating negative temperature coefficient thermistors in the measuring circuit. These fuel cells have sufficient electrical output to drive readout meters without amplification. However, where dual- or multirange facilities are required some amplification may be necessary.

24.8.5 Hersch Cell for Oxygen Measurement

This galvanic cell differs from fuel cells in that a third electrode is added to the cell and a potential applied to provide anodic protection to the anode. In one manufacturer's cell (see Figure 24.38), the cathode is silver and the anode cadmium. The third electrode is platinum. The anodic protection limits the cadmium current to a few microamperes and extends the life of the cadmium. However, this arrangement gives an electrical output from the cell, which is non-linear with oxygen partial pressure, and it is necessary for the signal to be passed through a "shaping" circuit for the readout to be given in concentration units. Calibration is carried out by generating a predetermined concentration of oxygen in a sample by electrolysis, and electrodes for this are incorporated in the cell. When dry gas samples are being used they must be humidified to prevent the water-based electrolyte in the cell from drying out.

24.8.6 Sensor for Oxygen Dissolved in Water

Electrochemical sensors with membranes for oxygen determination can be applied to measuring oxygen dissolved in water; both polarographic and galvanic sensors can be used.

A most popular type of sensor is the galvanic Mackereth electrode. The cathode is a perforated silver cylinder surrounding a lead anode with an aqueous electrolyte of potassium bicarbonate (see Figure 24.39). The electrolyte is confined by a silicone rubber membrane, which is permeable to oxygen but not to water and interfering ions.

The oxygen that diffuses through the membrane is reduced at the cathode to give a current proportional to the oxygen partial pressure. Equations for the reactions were given in Section 24.8.2.

Accurate temperature control is essential (6 percent error per degree) and thermistor- or resistance-thermometer-controlled compensation circuits are generally used. Working ranges can be from a few μgO_2/liter of water up to 200 percent oxygen saturation. The lead anode is sacrificial, and electrodes therefore have to be refurbished according to the actual design and the total amount of oxygen that has diffused into the cell. Cells are calibrated using water containing known amounts of oxygen. Indicating meters or recorders can be connected, and manufacturers offer both portable instruments and equipment for permanent installation with timing devices, water pumps, and so on. There are also several variations on the basic design of electrodes to cope with oxygen determination in water plant, rivers, lakes, sewage tanks, and so on (see Figure 24.40). One of those shown includes a patented assembly incorporating water sampling by air life—air reversal gives a calibration check and filter clean.

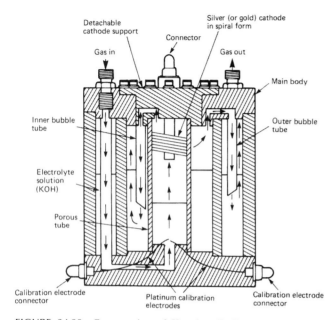

FIGURE 24.38 Cross-section of Hersch cell. Courtesy of Anacon (Instruments) Ltd.

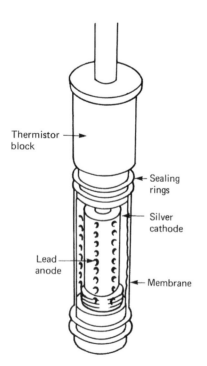

FIGURE 24.39 Diagram of Mackereth oxygen sensor assemblies. Courtesy of ABB Instrument Group.

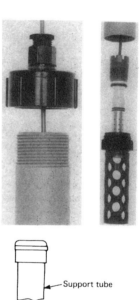

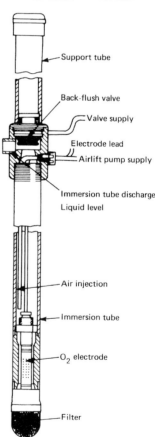

FIGURE 24.40 Varieties of Mackereth oxygen sensor assemblies. Courtesy of Kent Industrial Measurements Ltd. Analytical Instruments.

24.8.7 Coulometric Measurement of Moisture in Gases and Liquids

Moisture from gases (or vaporized from liquids) can be absorbed by a layer of desiccant, generally phosphoric anhydride (P_2O_5), in contact with two platinum or rhodium electrodes. A dc voltage is applied to electrolyze the moisture, the current produced being directly proportional to the mass of moisture absorbed (Faraday's law of electrolysis). The response of such an instrument obviously depends on the flow rate of gas, which is set and controlled accurately at a predetermined rate so that the current measuring meter can be calibrated in vppm moisture. Details are given in Chapter 26.

FURTHER READING

Bailey, P. L., *Analysis with Ion-selective Electrodes*, Heyden, London (1976).

Bates, R. G., *The Determination of pH* (2nd ed.), Wiley Interscience, New York (1973).

Durst, R. A. (ed.), *Ion Selective Electrodes*, National Bureau of Standards Special Publication 314, Dept. of Commerce, Washington, D.C. (1969).

Eisenman, G., *Glass Electrodes for Hydrogen and Other Cations*, Edward Arnold, London/Marcel Dekker, New York (1967).

Freiser, H. (ed.), *Ion-selective Electrodes in Analytical Chemistry*, Vol. I, Plenum Press, New York (1978).

Ives, G. J. and D. J. G. Janz, *Reference Electrodes, Theory and Practice*, Wiley Interscience, New York (1961).

Midgley, D. and K. Torrance, *Potentiometric Water Analysis*, Wiley Interscience, New York (1978).

Perrin, D. D. and B. Dempsey, *Buffers for pH and Metal Ion Control*, Chapman and Hall, London (1974).

Sawyer, D. T. and J. L. Roberts, Jr., *Experimental Electrochemistry for Chemists*, Wiley Interscience, New York (1974).

Chapter 25

Chemical Analysis: Gas Analysis

C. K. Laird; edited by I. Verhappen

25.1 INTRODUCTION

The ability to analyze one or more components of a gas mixture depends on the availability of suitable detectors that are responsive to the components of interest in the mixture and that can be applied over the required concentration range. Gas detectors are now available that exploit a wide variety of physical and chemical properties of the gases detected, and the devices resulting from the application of these detection mechanisms show a corresponding variety in their selectivity and range of response. In a limited number of applications it may be possible to analyze a gas mixture merely by exposure of the sample to a detector that is specific to the species of interest and thus obtain a direct measure of its concentration. However, in the majority of cases no sufficiently selective detector is available, and the gas sample requires some pretreatment, such as drying or removal of interfering components, to make it suitable for the proposed detector. In these cases, a gas analysis system must be used.

A block diagram of the components of a typical gas analyzer is given in Figure 25.1. The sample is taken into the instrument either as a continuous stream or in discrete aliquots and is adjusted as necessary in the sampling unit to the temperature, pressure, and flow-rate requirements of the remainder of the system. Any treatment of the sample—for example, separation of the sample into its components, removal of interfering components, or reaction with an auxiliary gas—is carried out in the processing unit and the sample is passed to the detector. The signal from the detector is amplified if necessary and processed to display or record the concentration of the components of interest in the sample.

In many gas analyzers the time lag between sampling and analysis is reduced to a minimum by taking a continuous stream of sample at a relatively high flow rate and arranging for only a small proportion to enter the analyzer, the remainder being bypassed to waste or returned to the process. Provision is also normally made to check the zero by passing a sample, free of the species to be analyzed, to the detector; the instrument may also include facilities for calibration by means of a "span" switch that feeds a sample of known concentration to the analyzer.

For certain applications, there may be a choice between the use of a highly selective detector, with relatively little

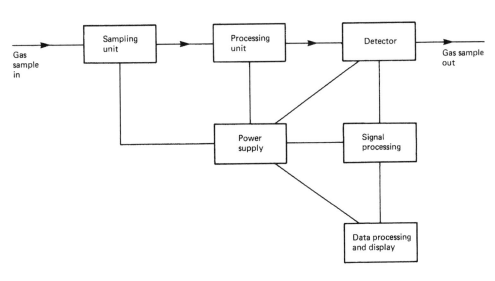

FIGURE 25.1 Schematic diagram of a typical process gas analyzer.

pretreatment of the sample, or use of a detector that responds to a wider range of chemical species, with the sample separated into its components before it reaches the detector. In the special case of gas chromatography, the sample is separated on the basis of the different times taken by each component to pass through a tube or column packed with adsorbent. The outlet gas stream may then be passed through a single detector or through more than one detector in series or switched between detectors to analyze several components of the original sample mixture. By choice of columns, operating conditions, and detectors, a gas-chromatographic analysis system may be built up to be individually tailored to analyze several different preselected components in a single aliquot taken from a gas sample. Because of its importance in process analysis, gas chromatography is given particularly detailed treatment here.

In addition to the analysis techniques described in this chapter, a number of spectroscopic methods are given under that heading in Chapter 23, and some electrochemical methods are outlined in Chapter 24.

25.2 SEPARATION OF GASEOUS MIXTURES

Although detectors have been developed that are specific to particular gases or groups of gases—for example, flammable gases or total hydrocarbons—there is often a need to separate the sample into its components or to remove interfering species before the sample is passed to the detector. A nonspecific detector, such as a katharometer, may also be used to measure one component of a gas mixture by measuring the change in detector response that occurs when the component of interest is removed from the gas mixture.

Methods for separating gaseous mixtures may be grouped under three main headings: chemical reactions, physical methods, and physico-chemical methods.

Chemical Reactions A simple example of chemical separation is the use of desiccants to remove water from a gas stream. The percentage of carbon dioxide in blast furnace gas may be determined by measuring the thermal conductivity of the gas before and after selective removal of the carbon dioxide by passing the gas through soda lime. Similarly, the percentage of ammonia gas in a mixture of nitrogen, hydrogen, and ammonia may be measured by absorbing the ammonia in dilute sulfuric acid or a suitable solid absorbent.

Physical Methods The most powerful physical technique for separation of gases is mass spectrometry, described in Chapter 23—though only minute quantities can be handled in that way. Gases may also be separated by diffusion; for example, hydrogen may be removed from a gas stream by allowing it to diffuse through a heated tube of gold- or silver-palladium alloy.

Physico-Chemical Methods: Chromatography Gas chromatography is one of the most powerful techniques for separation of mixtures of gases or (in their vapor phase) volatile liquids. It is relatively simple and widely applicable. Mixtures of permanent gases, such as oxygen, nitrogen, hydrogen, carbon monoxide, and carbon dioxide, can easily be separated, and when applied to liquids, mixtures such as benzene and cyclohexane can be separated, even though their boiling points differ by only 0.6 K. Separation of such mixtures by other techniques such as fractional distillation would be extremely difficult.

25.2.1 Gas Chromatography

Chromatography is a physical or physico-chemical technique for the separation of mixtures into their components on the basis of their molecular distribution between two immiscible phases. One phase is normally stationary and is in a finely divided state to provide a large surface area relative to volume. The second phase is mobile and transports the components of the mixture over the stationary phase.

The various types of chromatography are classified according to the particular mobile and stationary phases employed in each (see Chapter 22). In gas chromatography, the mobile phase is a gas, known as the *carrier gas*, and the stationary phase is either a granular solid (gas-solid chromatography) or a granular solid coated with a thin film of nonvolatile liquid (gas-liquid chromatography). In gas-solid chromatography, the separation is effected on the basis of the different adsorption characteristics of the components of the mixture on the solid phase; in gas-liquid chromatography, the separation mechanism involves the distribution of the components of the mixture between the gas and stationary liquid phases. Because the components of the mixture are transported in the gaseous phase, gas chromatography is limited to separation of mixtures whose components have significant vapor pressures, which normally means gaseous mixtures or mixtures of liquids with boiling points below approximately 450 K.

The apparatus for gas chromatography, known as the *gas chromatograph*, consists of a tube or column to contain the stationary phase and is itself contained in an environment whose temperature can be held at a constant known value or heated and cooled at controlled rates. The column may be uniformly packed with the granular stationary phase (packed column chromatography), normally used in process instruments. However, it has been found that columns of the highest separating performance are obtained if the column is in the form of a capillary tube, with the solid or liquid stationary phase coated on its inner walls (capillary chromatography). The carrier-gas mobile phase is passed continuously through the column at a constant controlled and known rate. A facility for introduction of known volumes of the mixture to be separated into the carrier-gas stream is provided in the carrier-gas line upstream of the column, and a suitable detector, responsive to changes in the composition of the gas passing through it, is connected to the downstream end of the column.

To analyze a sample, an aliquot of suitable known volume is introduced into the carrier-gas stream, and the output of the detector is continuously monitored. Due to their interaction with the stationary phase, the components of the sample pass through the column at different rates. The processes affecting the separation are complex, but in general, in gas-solid chromatography the component that is least strongly adsorbed is eluted first, whereas in gas-liquid chromatography the dominant process is the solubility of the components in the liquid stationary phase. Thus the separation achieved depends on the nature of the sample and stationary phase, on the length and temperature of the column, and on the flow rate of the carrier gas, and these conditions must be optimized for a particular analysis.

The composition of the gas passing through the detector alternates between pure carrier gas and mixtures of the carrier gas with each of the components of the sample. The output record of the detector, known as the *chromatogram*, is a series of deflections or peaks, spaced in time and each related to a component of the mixture analyzed.

A typical chromatogram of a mixture containing five components is shown in Figure 25.2. The first "peak" (A) at the beginning of the chromatogram is a pressure wave or unresolved peak caused by momentary changes in carrier-gas flow and pressure during the injection of the sample. The recording of the chromatogram provides a visual record of the analysis, but for qualitative analysis each peak must be identified on the basis of the time each component takes to pass through the column by use of single pure compounds or mixtures of known composition. For quantitative analysis the apparatus must be calibrated by use of standard gas mixtures or solutions to relate the detector response to the concentration of the determinand in the initial mixture.

A significant advantage of gas chromatography is that several components of a sample may be analyzed essentially simultaneously in a single aliquot extracted from a process stream. However, sampling is on a regular discrete basis rather than continuous, so the chromatograph gives a series of spot analyses of a sample stream, at times corresponding to the time of sample injection into the instrument. Before a new sample can be analyzed, it is necessary to be certain that all the components of the previous sample have been eluted from the column. It is therefore advantageous to arrange the analytical conditions so that the sample is eluted as quickly as possible, consistent with adequate resolution of the peaks of interest. Two techniques commonly used in process gas chromatography and now finding more application in the laboratory are *heart cut* and *backflush*. Both techniques rely on an understanding of the components being analyzed and their elution times on various phases. With these techniques, the analytical chemist chooses to analyze only those components in which he or she is interested and vents the balance to "waste."

The heart-cut method is the fastest way to separate trace-level concentrations of components when they elute on a tail of a major component. Using two columns, the heart-cut valve diverts the effluent of the heart-cut column either to vent or to the analysis column for further separation. Flow of carrier gas in both columns is maintained the same way using restriction orifices. Normally, the effluent of the heart-cut column is diverted to vent, but when a component of interest appears, it is diverted to the analysis column and then returns to its venting position. In this way a "cut" containing only the component(s) of interest and a narrow band of the background component are introduced to the analytical column.

Reversing the flow of carrier gas in the direction opposite that of the sample injection is called backflushing. Therefore, backflushing a column results in any components still in the column being swept back to the point of injection in approximately the same amount of time it took to flow to their present location in the column. Components

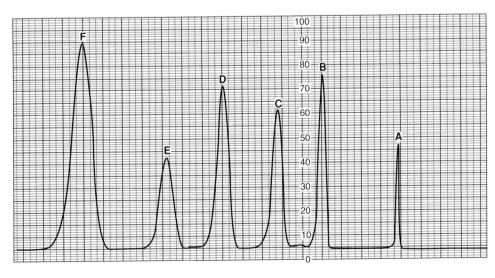

FIGURE 25.2 Chromatogram of a sample containing five components.

will "flush" from the column in the reverse order in which they appear on the column, meaning that in many cases the heavy components will flush back to vent first. This method can be used effectively to, in some cases, reduce flush the light components to vent and direct the heavy components to the detector, thus significantly decreasing the cycle time. Alternately, cycle time can be reduced by not having to flush the entire sample through the column(s), also resulting in increased column life.

25.3 DETECTORS

25.3.1 Thermal Conductivity Detector (TCD)

The thermal conductivity detector is among the most commonly used gas detection devices. It measures the change in thermal conductivity of a gas mixture, caused by changes in the concentration of the species it is desired to detect.

All matter is made up of molecules, which are in constant rapid motion. Heat is the energy possessed by a body by virtue of the motion of the molecules of which it is composed. Raising the temperature of the body increases the energy of the molecules by increasing the velocity of the molecular motion.

In solids the molecules do not alter their position relative to one another but vibrate about a mean position, whereas in a liquid the molecules vibrate about mean positions but may also move from one part of the liquid to another. In a gas the molecular motion is almost entirely translational: The molecules move from one part of the gas to another, impeded only by frequent intermolecular collisions and collisions with the vessel walls. The collisions with the walls produce the pressure of the gas on the walls. In a so-called "perfect gas," the molecules are regarded as being perfectly elastic, so no energy is dissipated by the intermolecular collisions.

Consideration of the properties of a gas that follow as a consequence of the motion of its molecules is the basis of the kinetic theory. Using this theory, Maxwell gave a theoretical verification of laws that had previously been established experimentally. These included Avogadro's law, Dalton's law of partial pressures, and Graham's law of diffusion.

Since heat is the energy of motion of the gas molecules, transfer of heat, or thermal conductivity, can also be treated by the kinetic theory. It can be shown that the thermal conductivity K of component S is given by

$$K_S = \frac{1}{2}\rho\tilde{v}\lambda C_v$$

where ρ is the gas density, $\tilde{v}$ is the mean molecular velocity, λ is the mean free path, and C_v is the specific heat at constant volume. Thus thermal conductivity depends on molecular size, mass, and temperature.

The quantity $\tilde{v}\lambda$ is the diffusion coefficient D of the gas, and the thermal conductivity can be written

$$K_S = \frac{1}{2}D\rho C_v$$

According to this treatment, the thermal conductivity of the gas is independent of pressure. This is found to be true over a wide range of pressures, provided that the pressure does not become so high that the gas may no longer be regarded as being a perfect gas. At very low pressures, the conductivity of the gas is proportional to its pressure, and this is the basis of the operation of the Knudsen hotwire manometer or the Pirani gauge (see Chapter 15).

It can be shown that the conductivity K_T of a pure gas at absolute temperature T varies with temperature according to the equation

$$K_T = K_0\left[b + \frac{273}{b} + T\right]\left[\frac{T}{273}\right]^{3/2}$$

where K_0 is the thermal conductivity at 0° and b is a constant.

The relative thermal conductivities of some gases, relative to air as 1.00, are given in Table 25.1.

It can be shown that the conductivity of a binary mixture of gases is given by

$$K = \frac{K_1}{1 + A\left(\frac{1-x_1}{x_1}\right)P} + \frac{K_2}{1 + B\left(\frac{x_1}{1-x_1}\right)}$$

where A and B are constants known as the Wasil-jewa constants, K_1 and K_2 are the conductivities of the pure gases, and x_1 is the molar fraction of component 1.

In gas analysis, conductivities of pure gases are of limited value; it is much more important to know how the conductivity of a mixture varies with the proportion of the constituent gases. However, as shown, the relationship between the conductivity of a mixture of gases and the proportion of the

TABLE 25.1 Relative thermal conductivities of some common gases

Gas	Conductivity
Air	1.00
Oxygen	1.01
Nitrogen	1.00
Hydrogen	4.66
Chlorine	0.32
Carbon monoxide	0.96
Carbon dioxide	0.59
Sulphur dioxide	0.32
Water vapor	1.30
Helium	4.34

constituents is complicated. When collisions occur between molecules of different gases, the mathematics of the collisions are no longer simple, and the relationship between conductivity and the proportions of the constituents depends on the molecular and physical constants of the gases and on the intermolecular forces during a collision. In practice thermal conductivity instruments are therefore calibrated by establishing the required composition-conductivity curves experimentally.

Several forms of gas sensor based on thermal conductivity have been developed. The majority use the hotwire method of measuring changes in conductivity, with the hotwire sensors arranged in a Wheatstone bridge circuit.

25.3.1.1 Katharometer

A wire, heated electrically and maintained at constant temperature, is fixed along the axis of a cylindrical hole bored in a metal block, which is also maintained at a constant temperature. The cylindrical hole is filled with the gas under test. The temperature of the wire reaches an equilibrium value when the rate of loss of heat by conduction, convection, and radiation is equal to the rate of production of heat by the current in the wire. In practice, conduction through the gas is the most important source of heat loss. End-cooling, convection, radiation, and thermal diffusion effects, though measurable, account for so small a part (less than 1 percent each) of the total loss that they can satisfactorily be taken care of in the calibration. Most instruments are designed to operate with the wire mounted vertically, to minimize losses by convection. Convective losses also increase with the pressure of the gas, so the pressure should be controlled for accurate conductivity measurements in dense gases. The heat loss from the wire depends on the flow rate of gas in the sensor. In some instruments errors due to changes in gas flow are minimized because the gas does not flow through the cell but enters by diffusion; otherwise the gas flow rate must be carefully controlled.

One must also be mindful that in the case of gases, mass flow is also a function of pressure. At pressures typically used in analyzers, the relationship between the change in volume as a function of pressure can be approximated by the ideal gas law, or $P_1 V_1 = P_2 V_2$.

The resistance of the wire depends on its temperature; thus, by measuring the resistance of the wire, we can find its temperature, and the wire is effectively used as a resistance thermometer. The electrical energy supplied to the wire to maintain the excess temperature is a measure of the total heat loss by conduction, convection, and radiation. To measure the effects due to changes in the conductivity of the gas only, the resistance of the hot wire in a cell containing the gas to be tested is compared with the resistance of an exactly similar wire in a similar cell containing a standard gas. This differential arrangement also lessens the effects of changes in the heating current and the ambient temperature conditions. To increase the sensitivity, two measuring and two reference cells, an arrangement usually referred to as a *katharometer*, are often used.

In the katharometer, four filaments with precisely matched thermal and electrical characteristics are mounted in a massive metal block, drilled to form cells and gas paths. A cutaway drawing of a four-filament cell is shown in Figure 25.3. Depending on the specific purpose, the filaments may be made of tungsten, tungsten-rhenium alloy, platinum, or other alloys. For measurements in highly reactive gases, gold-sheathed tungsten filaments may be used. The filaments are connected in a Wheatstone bridge circuit, which may be supplied from either a regulated-voltage or regulated-current power supply. The circuit for a constant-voltage detector is shown in Figure 25.4. The detector is balanced with the same gas in the reference and sample cells. If a gas of different thermal conductivity enters the sample cell, the rate of loss of heat from the sample filaments is altered, so changing their temperature and hence their resistance. The change in resistance unbalances the bridge, and the out-of-balance voltage is recorded as a measure of the change in gas concentration. The katharometer can be calibrated by any binary gas mixture or for a gas mixture that may be regarded as binary, such as carbon dioxide in air.

A theory of the operation of the katharometer bridge follows. This is simplified but is insufficiently rigid for

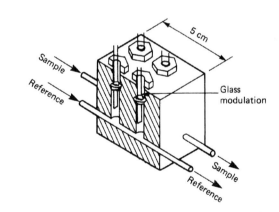

FIGURE 25.3 Cutaway drawing of four-filament diffusion katharometer cell.

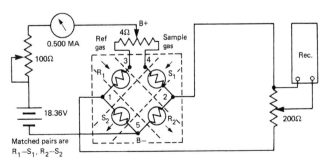

FIGURE 25.4 Circuit for four-filament katharometer cell.

calibrations to be calculated. Small variations in the behavior of individual filaments also mean that each bridge must be calibrated using mixtures of the gas the instrument is to measure.

Assume that the four arms of the bridge (see Figure 25.4) have the same initial resistance R_1 when the bridge current is flowing and the same gas mixture is in the reference and sample cells. Let R_0 be resistance of filament at ambient temperature, R_1 working resistance (i.e., resistance when a current I flows), I current through one filament (i.e., half bridge current), and T wire temperature above ambient.

Then, at equilibrium, energy input is equal to heat loss:

$$I^2 R_1 = K_1 T \quad (25.1)$$

where K_1 is a constant proportional to the thermal conductivity of the gas because most of the heat loss is by conduction through the gas. A simple expression for the working resistance is

$$R_1 = R_0 (1° + \alpha T) \quad (25.2)$$

where α is the temperature coefficient of resistance of the filament material. Then, from Equations (25.1) and (25.2):

$$I^2 R_1 R_0 \alpha = K_1 (R_1 - R_0) \quad (25.3)$$

Then,

$$\begin{aligned} R_1 &= \frac{K_1 R_0}{(K_1 - R_0 I^2 \alpha)} \\ &= R_0 + \frac{K_1 R_0}{(K_1 - R_0 I^2 \alpha)} - R_0 \\ &= R_0 + \frac{K_1 R_0 - K_1 R_0 + I^2 R_0^2}{(K_1 - I^2 R_0 \alpha)} \\ &= R_0 + \frac{I^2 R_0}{(K_1 - I^2 R_0 \alpha)} \end{aligned} \quad (25.4)$$

From Equation (25.3), if $R_1 - R_0$ is small compared with R_1, K_1 must be large compared with $I^2 R_0 \alpha$, and the term $I^2 R_0 \alpha$ can be ignored. Then,

$$R_1 = R_0 + (I^2 R_0^2 \alpha / K_1) \quad (25.5)$$

If the two measurement filaments have a total resistance of R_1 and the reference filaments of R_2, the output voltage of the bridge E is given by

$$E = I(R_1 - R_2) \quad (25.6)$$

Combining Equations (25.5) and (25.6):

$$E = I^3 R_0^2 [(1/K_1) - (1/K_2)] \quad (25.7)$$

where K_1 and K_2 are proportional to the conductivities of the gases in each pair of cells.

Equation (25.7) shows that the output is proportional to the cube of the bridge current, but in practice the index is usually between $I^{2.5}$ and I^3. For accurate quantitative readings, the bridge current must be kept constant.

This equation also shows that the output is proportional to the difference between the reciprocals of the thermal conductivities of the gases in each pair of cells. This is usually correct for small differences in thermal conductivity but does not hold for large differences.

These conditions show that the katharometer has maximum sensitivity when it is used to measure the concentration of binary or pseudo-binary gas mixtures of which the components have widely different thermal conductivities and when the bridge current is as high as possible. The maximum bridge current is limited by the need to avoid overheating and distortion of the filaments, and bridge currents can be highest when a gas of high thermal conductivity is in the cell. When the katharometer is used as the detector in gas chromatography, hydrogen or helium, which have higher thermal conductivities than other common gases, is often used as the carrier gas, and automatic circuits may be fitted to reduce the current to the bridge to prevent overheating.

For maximum sensitivity, especially when it is necessary to operate the detector at low temperatures, the hotwire filaments may be replaced by thermistors. A *thermistor* is a thermally sensitive resistor that has a high negative coefficient of resistance; see Chapter 21. In the same manner as with hot wires, the resistance of the conductor is changed (in this case, lowered) by the passage of current. Thermistor katharometers usually have one sensing and one reference element, the other resistors in the Wheatstone bridge being external resistors.

Many modern katharometers in use today operate on the basis of constant current to the Wheatstone bridge, since this results not only in longer filament or thermistor life but also greater accuracy of measurement.

Except in the case of thermally unstable substances, the katharometer is nondestructive, and it responds universally to all substances. The sensitivity is less than that of the ionization detectors but is adequate for many applications. The detector is basically simple and responds linearly to concentration changes over a wide range. It is used in gas chromatography and in a variety of custom-designed process analyzers.

25.3.2 Flame Ionization Detector (FID)

An extensive group of gas detectors is based on devices in which changes in ionization current inside a chamber are measured. The ionization process occurs when a particle of high energy collides with a target particle, which is thus ionized. The collision produces positive ions and secondary electrons that may be moved toward electrodes by application of an electric field, giving a measurable current, known as the *ionization current*, in the external circuit.

The FID utilizes the fact that, although a hydrogen-oxygen flame contains relatively few ions (10^7 ions/cm^{-3}), it does contain highly energetic atoms. When trace amounts of organic compounds are added to the flame, the number of ions increases (to approximately 10^{11} ions/cm^{-3}) and a measurable ionization current is produced. It is assumed that the main reaction in the flame is

$$CH + O \rightarrow CHO + e$$

However, the FID gives a small response to substances that do not contain hydrogen, such as CCl_4 and CS_2. Hence it is probable that the previously described reaction is preceded by hydrogenation to form CH_4 or CH_3 in the reducing part of the flame. Recombination occurs in addition to the ionization reactions, and the response of the FID is determined by the net overall ionization reaction process.

A schematic diagram of an FID is shown in Figure 25.5; a cross-sectional view of a typical detector is shown in Figure 25.6. The sample gas or effluent from a gas-chromatographic column is fed into a hydrogen-air flame. The jet itself serves as one electrode; a second electrode is placed above the flame. A potential is applied across these electrodes. When sample molecules enter the flame, ionization occurs, yielding a current that, after suitable amplification, may be displayed on a strip chart recorder.

The FID is a mass-sensitive rather than concentration-sensitive detector. This means that it does not respond to the concentration of a component entering it but rather produces a signal proportional to the amount of organic material entering it per unit time. The ion current is effectively proportional to the number of carbon atoms present in the flame, and the sensitivity of the detector may be expressed as the mass of carbon passing through the flame per second required to give a detectable signal. A typical figure is 10^{-11} g C/sec.

The FID is sensitive to practically all organic substances but is insensitive to inorganic gases and water. It has a high sensitivity, good stability, wide range of linear response, and low effective volume. It is widely used as a gas-chromatographic detector and in total hydrocarbon analyzers.

25.3.3 Photo-Ionization Detector (PID)

The photo-ionization detector (see Figure 25.7) has some similarities to the flame ionization detector, and like the FID, it responds to a wide range of organic as well as to some inorganic molecules. An interchangeable sealed lamp

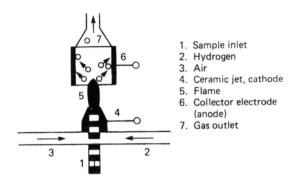

FIGURE 25.5 Flame ionization detector: schematic.

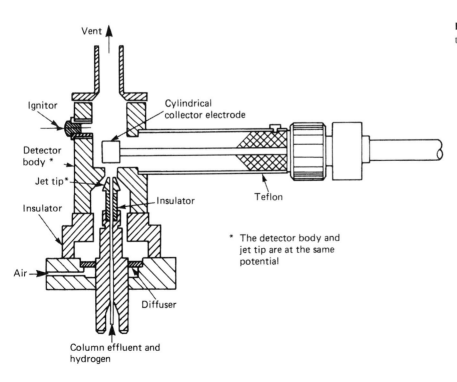

FIGURE 25.6 Cross-section of flame ionization detector.

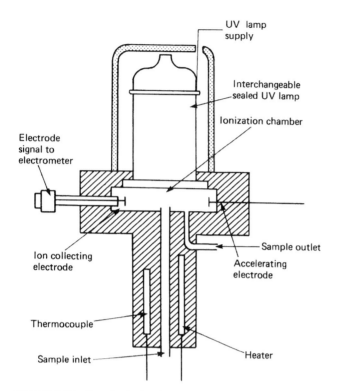

FIGURE 25.7 Photo-ionization detector.

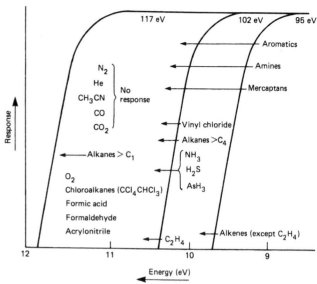

FIGURE 25.8 PID response for the various ultraviolet lamps.

produces monochromatic radiation in the UV region. Molecules having ionization potentials less than the energy of the radiation may be ionized on passing through the beam. In practice, molecules with ionization potentials just above the photon energy of the incident beam may also be ionized due to a proportion being in excited vibrational states. The ions formed are driven to a collector electrode by an electric field and the ion current is measured by an electrometer amplifier.

The flame in the FID is a high-energy ionization source and produces highly fragmented ions from the molecules detected. The UV lamp in the PID is of lower quantum energy, leading to the predominant formation of molecular ions. The response of the PID is therefore determined mainly by the ionization potential of the molecule rather than the number of carbon atoms it contains. In addition, the ionization energy in the PID may be selected by choice of the wavelength of the UV source, and the detector may be made selective in its response. The selectivity obtainable by use of three different UV lamps is shown in Figure 25.8. The ionization potentials of N_2, He, CH_3CN, CO, and CO_2 are above the energy of all the lamps, and the PID does not respond to these gases.

The PID is highly sensitive, typically to picogram levels of organic compounds, and has a wide linear range. It may be used for direct measurements in gas streams or as a gas-chromatographic detector. When it is used as a detector in gas chromatography, any of the commonly used carrier gases is suitable. Some gases, such as CO_2, absorb UV radiation, and their presence may reduce the sensitivity of the detector.

25.3.4 Helium Ionization Detector

Monatomic gases, such as helium or argon, can be raised to excited atomic states by collision with energetic electrons emitted from a β-source. The metastable atomic states are themselves highly energetic and lose their energy by collision with other atomic or molecular species. If the helium contains a small concentration of a gas of which the ionization potential is less than the excitation of the metastable helium atoms, ions will be formed in the collision, so increasing the current-carrying capacity of the gas. This is the basis of the helium ionization detector.

The main reactions taking place can be represented as

$$He + e \rightarrow He^* + e$$
$$He^* + M \rightarrow M^+ + He + e$$

where M is the gas molecule forming ions. However, other collisions can occur—for example, between metastable and ground-state helium atoms or between metastable atoms, which may also result in ion formation.

The helium ionization detector (see Figure 25.9) typically consists of a cylindrical chamber, approximately 1 cm in diameter and a few millimeters long, containing a β-emitting radioactive source. The ends of the cylinder are separated by an insulator and form electrodes. The detector is used as part of a gas-chromatographic system, with helium as the carrier gas.

It can be shown that the ionization mechanism we've described depends on the number of atoms formed in metastable states. It can also be shown that the probability of

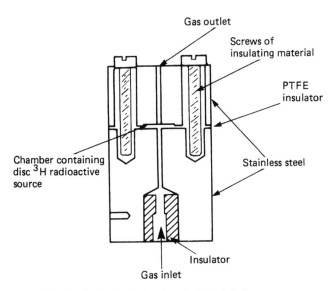

FIGURE 25.9 Helium ionization detector (actual size).

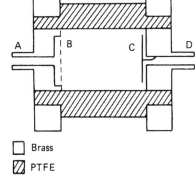

FIGURE 25.10 Electron capture detector.

formation of metastable states depends on the primary electron energy and on the intensity of the applied electric field. The reaction exhibits the highest cross-section for electrons with an energy of about 20 eV and a field strength of 500 V/cm torr. Tritium (^{3}H) sources of 10–10 GBq or ^{63}Ni β-sources of 400–800 MBq activity are usually used, but the free path of the β-particles is very short and the performance of the detector is strongly dependent on its geometry.

The helium ionization detector is used in gas chromatography, when its ability to measure trace levels of permanent gases is useful. However, the carrier gas supply must be rigorously purified.

25.3.5 Electron Capture Detector

The electron capture detector (see Figure 25.10) consists of a cell containing a β-emitting radioactive source, purged with an inert gas. Electrons emitted by the radioactive source are slowed to thermal velocities by collision with the gas molecules and are eventually collected by a suitable electrode, giving rise to a standing current in the cell. If a gas with greater electron affinity is introduced to the cell, some of the electrons are "captured," forming negative ions, and the current in the cell is reduced. This effect is the basis of the electron capture detector. The reduction in current is due both to the difference in mobility between electrons and negative ions and to differences in the rates of recombination of the ionic species and electrons.

The radioactive source may be tritium or ^{63}Ni, with ^{63}Ni usually being preferred, since it allows the detector to be operated at higher temperatures, thus lessening the effects of contamination. A potential is applied between the electrodes that is just great enough to collect the free electrons. Originally, the detector was operated under DC conditions, potentials up to 5 volts being used, but under some conditions space charge effects produced anomalous results. Present detectors use a pulsed supply, typically 25 to 50 volts, 1 microsecond pulses at intervals of 5 to 500 microseconds. Either the pulse interval is selected and the change in detector current monitored, or a feedback system maintains a constant current and the pulse interval is monitored.

The electron capture detector is extremely sensitive to electronegative species, particularly halogenated compounds and oxygen. To obtain maximum sensitivity for a given compound, the choice of carrier gas, pulse interval, or detector current and detector temperature must be optimized.

The electron capture detector is most often used in gas chromatography, with argon, argonmethane mixture, or nitrogen as carrier gas, but it is also used in leak or tracer detectors. The extreme sensitivity of the ECD to halogenated compounds is useful, but high-purity carrier gas and high-stability columns are required to prevent contamination. Under optimum conditions, 1 part in 10^{12} of halogenated compounds, such as Freons, can be determined.

25.3.6 Flame Photometric Detector (FPD)

Most organic and other volatile compounds containing sulfur or phosphorus produce chemiluminescent species when burned in a hydrogen-rich flame. In a flame photometric detector (see Figure 25.11), the sample gas passes into a fuel-rich H_2/O_2 or H_2/air mixture, which produces simple molecular species and excites them to higher electronic states. These excited species subsequently return to their ground states and emit characteristic molecular band spectra. This emission is monitored by a photomultiplier tube through a suitable filter, thus making the detector selective to either sulfur or phosphorus. It may also be sensitive to other elements, including halogens and nitrogen.

The FPD is most commonly used as a detector for sulfur-containing species. In this application, the response is based on the formation of excited S_2 molecules, S_2^*, and their subsequent chemiluminescent emission. The original sulfur-containing molecules are decomposed in the hot inner zone of the flame, and sulfur atoms are formed that combine to form S_2^* in the cooler outer cone of the flame. The exact mechanism of the reaction is uncertain, but it is believed that the excitation energy for the $S_2 \rightarrow S_2^*$ transition may come from the formation of molecular hydrogen or water in the flame, according to the reactions

$$H + H + S_2 \rightarrow S_2^* + H_2 \ (4.5 \text{ eV})$$
$$H + OH + S_2 \rightarrow S_2^* + H_2O \ (5.1 \text{ eV})$$

As the excited S_2 molecule reverts to the ground state, it emits a series of bands in the range 300–450 nm, with the most intense bands at 384.0 and 394.1 nm. The 384.0 nm emission is monitored by the photomultiplier tube.

The FPD is highly selective and sensitive, but the response is not linearly proportional to the mass-flow rate of the sulfur compound. Instead, the relationship is given by:

$$I_{S_2} = I_o [S]^n$$

where I_{S_2} is the observed intensity of the emission (photomultiplier tube output), [S] is the mass-flow rate of sulfur atoms (effectively the concentration of the sulfur compound), and n is a constant, found to be between 1.5 and 2, depending on flame conditions. Commercial analyzers employing the FPD often incorporate a linearizing circuit to give an output that is directly proportional to sulfur mass flow. The detector response is limited to two or three orders of magnitude.

The FPD is highly selective, sensitive (10^{-11} g), and relatively simple but has an extremely nonlinear response. It is used in gas chromatography and in sulfur analyzers.

25.3.7 Ultrasonic Detector

The velocity of sound in a gas is inversely proportional to the square root of its molecular weight. By measuring the speed of sound in a binary gas mixture, its composition can be deduced; this technique is the basis of the ultrasonic detector (see Figure 25.12). A quartz crystal transducer located at one end of the sample cell sound tube acts as the emitter, and an identical crystal located at the other end of the sound tube acts as the receiver. To obtain efficient transfer of sound energy between the gas and the transducers, the detector must be operated at above atmospheric pressure, and the gas in the cell is typically regulated to 1 to 7 bar gauge, depending on the gas. The phase shift of the sound signal traversing the cell between the emitter and receiver is compared to a reference signal to determine the change in speed of sound in the detector.

The detector is most often used in gas chromatography. It has a universal response and the output signal is proportional to the difference in molecular weight between the gaseous species forming the binary mixture. When used as a gas-chromatographic detector, it has good sensitivity (10^{-9}–10^{-10} g) and a wide linear dynamic range (10^6) and allows a wide choice of carrier gas. However, precise temperature control is required, and the electronic circuitry is

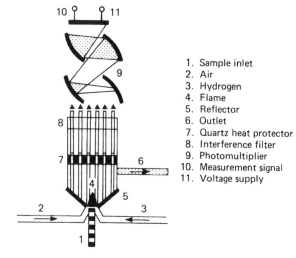

1. Sample inlet
2. Air
3. Hydrogen
4. Flame
5. Reflector
6. Outlet
7. Quartz heat protector
8. Interference filter
9. Photomultiplier
10. Measurement signal
11. Voltage supply

FIGURE 25.11 Flame photometric detector.

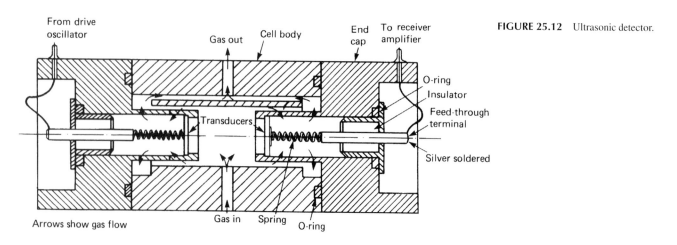

FIGURE 25.12 Ultrasonic detector.

complex. It may be a useful alternative where flames cannot be used or where a katharometer would not respond to all components in a mixture.

25.3.8 Catalytic Detector (Pellistor)

Catalytic gas detectors operate by measuring the heat output resulting from the catalytic oxidation of flammable gas molecules to carbon dioxide and water vapor at a solid surface. By use of a catalyst, the temperature at which the oxidation takes place is much reduced compared with gas phase oxidation. The catalyst may be incorporated into a solid-state sensor containing an electrical heater and temperature-sensing device. A stream of sample gas is fed over the sensor, and flammable gases in the sample are continuously oxidized, releasing heat and raising the temperature of the sensor. Temperature variations in the sensor are monitored to give a continuous record of the flammable-gas concentration in the sample.

The most suitable metals for promoting the oxidation of molecules containing C-H bonds, such as methane and other organic species, are those in Group 8 of the Periodic Table, particularly platinum and palladium. The temperature sensor is usually a platinum resistance thermometer wound in a coil and also used as the electrical heater for the sensor. The resistance is measured by connecting the sensor as one arm of a Wheatstone bridge and measuring the out-of-balance voltage across the bridge.

The construction of a typical catalytic sensing element is shown in Figure 25.13. A coil of 50 μm platinum wire is mounted on two wire supports, which also act as electrical connections. The coil is embedded in porous ceramic material, usually alumina, to form a bead about 1 mm long. The catalyst material is impregnated on the outside of the bead. This type of catalytic sensor is often called a *pellistor*. The choice of catalyst and of the treatment of the outside of the bead—for example, by inclusion of a diffusion layer—influences the overall sensitivity of the sensor and the relative sensitivity to different gases. The sensitivity and selectivity are also influenced by the choice of catalyst and by the temperature at which the sensor is operated. Palladium and its oxides are the most widely used catalysts; they have the advantage that they are much more active than platinum, enabling the sensor to be operated at the lowest possible temperature. The sensor is mounted in a protective open-topped can, as shown in Figure 25.13, so that the gas flow to the sensor is largely diffusion-controlled.

The Wheatstone bridge network commonly used with a catalytic sensor is shown in Figure 25.14. The sensing element forms one arm of the bridge, and the second arm is occupied by a compensator element.

This is a ceramic bead element, identical in construction to the sensor but without the catalytic coating. The sensor and compensator are mounted close together in a suitable housing so that both are exposed to the same sample gas. The pellistor or catalytic sensor is the basis of the majority of portable flammable-gas detectors.

25.3.9 Semiconductor Detector

The electrical conductivity of many metal oxide semiconductors, particularly those of the transition and heavy metals such as tin, zinc, and nickel, is changed when a gas molecule is adsorbed on the semiconductor surface. Adsorption involves the formation of bonds between the gas molecule and the semiconductor by transfer of electrical charge. This charge transfer changes the electronic structure of the semiconductor, changing its conductivity. The conductivity changes are related to the number of gas molecules adsorbed on the surface and hence to the concentration of the adsorbed species in the surrounding atmosphere.

A typical semiconductor detector is shown in Figure 25.15. The semiconducting material is formed as a bead, about 2–3 mm in diameter, between two small coils of platinum wire. One of the coils is used as a heater to raise the temperature of the bead so that the gas molecules it is

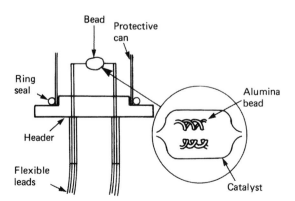

FIGURE 25.13 Catalytic gas-sensing element.

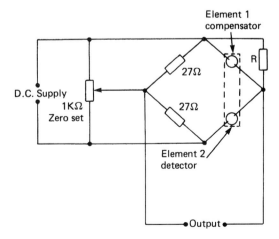

FIGURE 25.14 Wheatstone bridge network used with catalytic detector.

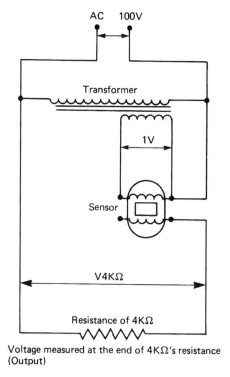

FIGURE 25.16   Measuring circuit for semiconductor sensor.

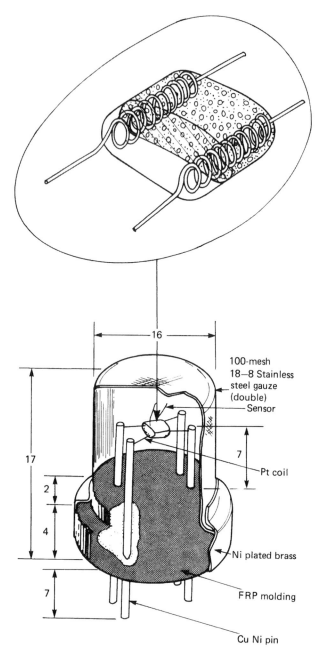

FIGURE 25.15   Semiconductor sensor.

desired to detect are reversibly absorbed on the surface, and the resistance of the bead is measured by measuring the resistance between the two coils. The bead is mounted in a stainless-steel gauze enclosure (see Figure 25.15) to ensure that molecules diffuse to the semiconductor surface, thus ensuring that the device is as free as possible from the effects of changes in the flow rate of the sample gas.

Semiconductor detectors are mainly used as low-cost devices for detection of flammable gases. A suitable power supply and measuring circuit is shown in Figure 25.16. The main defect of the devices at present is their lack of selectivity.

25.3.10 Properties and Applications of Gas Detectors

The properties and applications of the most commonly used gas detectors are summarized in Table 25.2.

25.4 PROCESS CHROMATOGRAPHY

Online or process gas chromatographs are instruments that incorporate facilities to automatically carry out the analytical procedure for chromatographic separation, detection, and measurement of predetermined constituents of gaseous mixtures. Samples are taken from process streams and are presented, in a controlled manner and under known conditions, to the gas chromatograph. Successive analyses may be made on a regular timed basis on aliquots of sample taken from a single stream or, by use of suitable stream-switching valves, a single process chromatograph may carry out automatic sequential analyses on process streams originating from several different parts of the plant.

The main components of a typical process chromatograph system are shown in Figure 25.17. These components are a supply of carrier gas to transport the sample through the column and detector, a valve for introduction of known quantities of sample, a chromatographic column to separate the sample into its components, a detector and associated amplifier to sense and measure the components of the sample in the carrier-gas stream, a programmer to actuate

CHAPTER | 25 Chemical Analysis: Gas Analysis

TABLE 25.2 Properties and applications of gas detectors

Detector	Applicability	Selectivity	Carrier or bulk gas	Lower limit of detection (grams)	Linear range	Typical applications
Thermal conductivity	Universal	Nonselective	He, H_2	10^{-6} 10^{-7}	10^4	Analysis of binary or pseudo-binary mixtures; gas chromatogaphy
Flame ionization	Organic compounds	Nonselective	N_2	10^{-11}	10^6	Gas chromatography; hydrocarbon analyzers
Photo-ionization	Organic compounds except low molecular weight hydrocarbons	Limited	N_2	10^{-11} 10^{-12}	10^7	Gas chromatography
Helium ionization	Trace levels of permanent gases	Nonselective	He	10^{-11}	10^4	Gas chromatography
Electron capture	Halogenated and oxygenated compounds	Response is highly compound-dependent	Ar, N_2, N_2 + 10% CH_4	10^{-12} 10^{-13}	10^3	Gas chromatography, tracer gas detectors, explosive detectors
Flame photometric	Sulphur and phosphorus compounds	Selective to compounds of S or P	N_2, He	10^{-11}	5×10^2 (S) 10^3 (P)	Gas chromatography, sulphur analyzers
Ultrasonic detector	Universal	Nonselective mainly low molecular weight	H_2, He, Ar, N_2, CO_2	10^{-9} 10^{-10}	10^6	Gas chromatography
Catalytic (pellistor)	Flammable gases	Selective to flammable gases	Air	*		Flammable gas detectors
Semiconductor	Flammable gases, other gases	Limited	Air	*		Low-cost flammable gas detectors

*The performance of these detectors depends on the individual design and application.

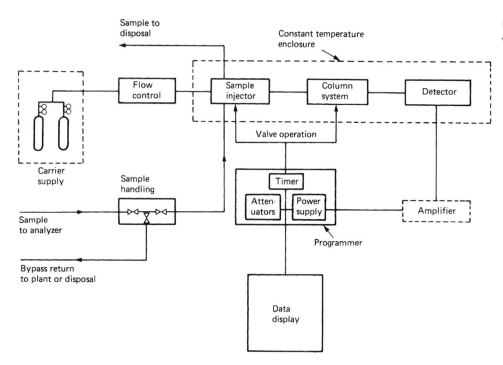

FIGURE 25.17 Functional diagram of process gas chromatograph.

the operations required during the analytical sequence and to control the apparatus, and a display or data-processing device to record the results of the analyses.

25.4.1 Sampling System

The sampling system must present to the gas chromatograph a homogeneous and representative sample of the gas or liquid to be analyzed. In process chromatography, a continuous stream of the sample is taken, usually by means of a fast bypass loop, and treated as necessary—for example, by drying, filtering, or adjusting temperature or pressure. Discrete volumes of the treated sample stream are periodically injected into the carrier gas stream of the chromatograph by means of a gas (or liquid) sampling valve. The chromatograph is normally supplied with the sample from the point or points to be monitored by use of permanently installed sampling lines. However, where the frequency of analysis does not justify the installation of special lines, samples may be collected in suitable containers for subsequent analysis. Gas samples may be collected under pressure in metal (usually stainless steel) cylinders or at atmospheric pressure in gas pipettes, gas sampling syringes, or plastic bags. For analysis of gases at very low concentrations, such as the determination of pollutants in ambient air, the pre-column or adsorption tube concentration technique is often used. Here the sample is drawn or allowed to diffuse through a tube containing a granular solid packing to selectively adsorb the components of interest. The tube is subsequently connected across the sample loop ports of the gas sampling valve on the chromatograph and heated to desorb the compounds to be analyzed into the carrier-gas stream.

It is essential that the sample size should be constant for each analysis and that it is introduced into the carrier gas stream rapidly, as a well-defined slug. The sample should also be allowed to flow continuously through the sampling system to minimize transportation lag. Chromatographic sampling or injection valves are specially designed changeover valves that enable a fixed volume, defined by a length of tubing (the sample loop), to be connected in either one of two gas streams with only momentary interruption of either stream. The design and operation of a typical sampling valve are shown in Figure 25.18. The inlet and outlet tubes terminate in metal (usually stainless steel) blocks with accurately machined and polished flat faces. A slider of soft plastic material, with channels or holes machined to form gas paths, is held against the polished faces and moved between definite positions to fill the loop or inject the sample. The main difference between gas and liquid sampling valves is in the size of sample loop. In the gas sampling valve, the loop is formed externally and typically has a volume in the range 0.1–10 ml. For liquid sampling the volumes required are smaller and the loop is formed in the internal channels of the valve and may have a volume as small as 1 μl. In process chromatography, sampling valves are normally fitted with electric or pneumatic actuators so that they may be operated automatically by the programmer at predetermined times during the analytical sequence.

CHAPTER | 25 Chemical Analysis: Gas Analysis

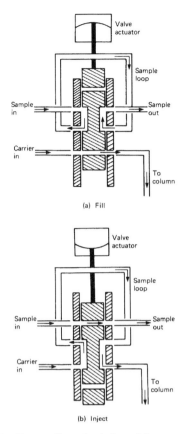

FIGURE 25.18 Gas-sampling valve (schematic).

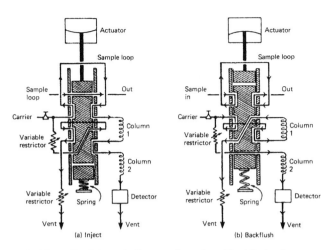

FIGURE 25.19 Schematic diagram of sample and backflush valve.

When it is required to change between columns or detectors during an analysis, similar types of valves are required. The number of ports and the arrangement of the internal channels may be tailored for the individual application. Figure 25.19 shows an arrangement in which a single valve is used for sample injection and backflushing in a chromatograph with two analytical columns in series. The sample is injected onto Column 1, which is chosen so that the components of interest are eluted first, and pass to Column 2. At a predetermined time, the valve is switched to refill the sample loop and to reverse the flow of carrier gas to Column 1 while the forward flow is maintained in Column 2 to effect the final separation of the components of the sample. By this means, components of no interest, such as high-boiling compounds or solvents, can be backflushed to waste before they reach Column 2 or the detector, thus preserving the performance of the columns and speeding the analytical procedure.

As described earlier, another related technique commonly employed is called *heart cut*, in which the component to be analyzed is in the "center" of the sample profile. In this case the light components and heavy components are flushed to vent and with only the components of interest actually being measured by the detector element.

The gas sample must arrive at the sampling valve at or only slightly above atmospheric pressure, at a flow rate typically in the range 10–50 ml min^{-1}, and be free from dust, oil, or abrasive particles. The sampling system may also require filters, pressure or flow controllers, pumps, and shut-off valves for control and processing of the sample stream. All the components of the system must be compatible with the chemical species to be sampled, and must be capable of withstanding the range of pressures and temperatures expected.

Many applications require analysis of two or more process streams with one analyzer. In these instances a sample line from each stream is piped to the analyzer, and sample lines are sequentially switched through solenoid valves to the sampling valve. When multistream analysis is involved, intersample contamination must be prevented. Contamination of samples can occur through valve leakage and inadequate flushing of common lines. To ensure adequate flushing, the capacity of common lines is kept to a minimum and the stream selection valves are timed so that while the sample from one stream is being analyzed, the sample from the next stream is flowing through all common lines.

Prevention of intersample contamination from valve leakage is accomplished by locating valves with respect to pressure drops such that any leakage will flow to vent rather than intermix in common lines. A typical flow arrangement for gas supplies to a chromatograph for multistream application is shown in Figure 25.20. This is designed to ensure that the sample and other supplies are delivered at the correct flow rate and pressure. A pressure-relief valve is fitted to protect the sampling valve from excessive pressure, and shut-off valves are fitted on all services except bottled gas lines.

In some applications additional conditioning of the sample is required. Typical of these are trace-heating of sample lines to maintain a sample in a gaseous state, vaporization to change a liquid to a gas, and elimination of stream contaminants by mechanical or chemical means.

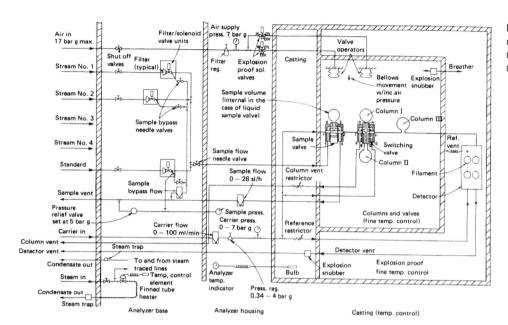

FIGURE 25.20 Flow diagram of multistream chromatograph with thermal conductivity detector. Courtesy of Invensys Foxboro division.

25.4.2 Carrier Gas

The carrier gas transports the components of the sample over the stationary phase in the chromatographic column. The carrier gas must not react with the sample, and for maximum efficiency in using long columns, it is advantageous to use a gas of low viscosity. However, the most important criterion in choosing a carrier gas is often the need to ensure compatibility with the particular detector in use.

The primary factors determining the choice of carrier gas are the effect of the gas on component resolution and detector sensitivity. The carrier gas and type of detector are chosen so that the eluted components generate large signals. For this reason, helium is generally used with thermal conductivity cells because of its high thermal conductivity. Hydrogen has a higher thermal conductivity and is less expensive than helium, but because of precautions necessary when using hydrogen, helium is preferred where suitable.

Specific properties of a particular carrier gas are exploited in other types of detectors—for example, helium in the helium ionization detector. In special instances a carrier gas other than that normally associated with a particular detector may be used for other reasons. For example, to measure hydrogen in trace quantities using a thermal conductivity detector, it is necessary to use a carrier gas other than helium because both helium and hydrogen have high and similar thermal conductivities. Accordingly, argon or nitrogen is used because either has a much lower thermal conductivity than hydrogen, resulting in a larger difference in thermal conductivity and greater output.

The flow rate of carrier gas affects both the retention time of a compound in the column, and the shape of the chromatographic peak and hence the amplitude of the detector signal. It is therefore essential for the flow rate to be readily adjustable to constant known values. The gas is usually supplied from bottles, with pressure-reducing valves to reduce the pressure to a level compatible with the flow control equipment and sufficient to give the required flow rate through the column and detector.

The flow rate of carrier gas may be measured and controlled either mechanically or electronically. Mechanical controllers are either precision pressure regulators, which maintain a constant pressure upstream of the column and detector, or differential pressure regulators, which maintain a constant pressure drop across a variable restriction. The principle of operation of one type of electronic flow controller is shown in Figure 25.21. A proportion of the gas stream is diverted via a narrow tube, fitted with an electric heating coil as shown. Sensor coils of resistance wire are wound on the tube upstream and downstream of the heating coil. Heat at a constant rate is supplied to the heating coil. Gas passing through the tube is heated by the coil, and some heat is transferred to the downstream sensor. The sensor coils are connected in a Wheatstone bridge circuit. The out-of-balance signal from the bridge, caused by the difference in temperatures and hence resistance of the upstream and downstream coils, depends on the mass-flow rate of gas through the tube and on the specific heat of the gas. (See also Chapter 6). The signal, suitably amplified, can be used to give a direct readout of the flow rate of gas through the tube and can be used to control the flow by feeding the signal to open or close a regulating valve in the main gas line downstream of the sensing device.

In cases in which the carrier gas flow rate is controlled mechanically, a rotameter is provided to indicate the flow rate. However, the best indication of correct flow is often the analysis record itself, since the retention times of known components of the sample should remain constant from one injection to the next.

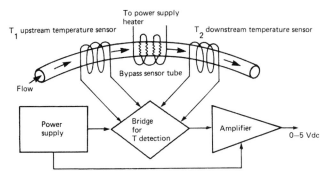

FIGURE 25.21 Principle of operation of electronic mass-flow controller. Courtesy of Brooks Instrument Division, Emerson Process.

25.4.3 Chromatographic Column

The packed separating columns used in process chromatographs are typically 1–2 m lengths of stainless steel tubing, 3–6 mm outer diameter, wound into a helix for convenient housing and packed with a solid absorbent. Separation of permanent gases is normally carried out on columns packed with molecular sieve. These are synthetic zeolites, available in a range of effective pore diameters. Porous polymeric materials have been developed that are capable of separating a wide range of organic and inorganic molecules, and use of these proprietary materials gives much more predictable column performance than when liquid coated solids are used. In addition, the polymeric materials are thermally stable and do not suffer from "bleed" or loss of the liquid stationary phase at high temperatures, which can give rise to detector noise or drift in the baseline of the chromatogram.

One or more columns packed with these materials can normally be tailored to the needs of most process analyses. However, in certain cases it may be necessary to include valves to switch between columns or detectors during the analysis or to divert the carrier gas to waste to prevent a certain component—for example, a solvent present in high concentration—from reaching the detector. These switching operations are referred to as backflushing, or heart cutting if the unwanted peak occurs in the middle of the chromatogram.

Capillary columns are also used for chromatography. They use a solvent coated to walls of the column and can therefore have a high plate count with low pressure drop. The downsides to capillary columns are their fragility and greater tendency to bleed, especially at high pressure or temperature.

25.4.4 Controlled Temperature Enclosures

Many components of the gas chromatograph, including the injection valve, columns, and detectors, are required to be kept at constant temperatures or in environments for which temperature can be altered at known rates, and separate temperature-controlled zones are usually provided in the instrument.

Two general methods are used to distribute heat to maintain the temperature-sensitive components at constant temperatures (± 0.1 K or better) and to minimize temperature gradients. One uses an air bath; the other uses metal-to-metal contact (or heat sink). The former depends on circulation of heated air, the latter on thermal contact of the temperature-sensitive elements with heated metal.

An air bath has inherently fast warmup and comparatively high temperature gradients and offers the advantage of ready accessibility to all components within the temperature-controlled compartment. The air bath is most suitable for temperature programming and is the usual method for control of the temperature of the chromatographic column.

Metal-to-metal contact has a slower warmup but relatively low temperature gradients. It has the disadvantage of being a greater explosion hazard and may require the analyzer to be mounted in an explosion-proof housing, resulting in more limited accessibility and more difficult servicing. The detectors are often mounted in heated metal blocks for control of temperature.

The choice of the method of heating and temperature control may depend on the location in which the instrument is to be used. Instruments are available with different degrees of protection against fire or explosion hazard. For operation in particularly hazardous environments—for example, where there may be flammable gases—instruments are available for which the operation, including temperature control, valve switching, and detector operation, is entirely pneumatic, with the oven being heated by steam.

25.4.5 Detectors

A gas-chromatographic detector should have a fast response, have linear output over a wide range of concentration, be reproducible, and have high detection sensitivity. In addition, the output from the detector must be zero when pure carrier gas from the chromatographic column is passing through the detector.

In process chromatography, the most commonly used detectors are the thermal conductivity and flame ionization types. Both have all the desirable characteristics listed, and one or the other is suitable for most commonly analyzed compounds: The thermal conductivity detector is suitable for permanent gas analysis and responds universally to other compounds, whereas the flame ionization detector responds to almost all organic compounds. In addition, these detectors can be ruggedly constructed for process use and can be used with a wide range of carrier gases. Most other detectors have disadvantages in comparison with these two—for example, fragility, nonlinear response, or a requirement for ultra-pure carrier-gas supplies—and although they are widely used in laboratory chromatographs, their application to process instruments is restricted.

The helium ionization detector may be used for permanent gas analyses at trace levels where the katharometer is insufficiently sensitive, and the ultrasonic detector may be a useful alternative in applications where a flame cannot be used or where a katharometer cannot be used for all components in a mixture. The selective sensitivity of the electron capture detector to halogenated molecules may also find occasional application. A comprehensive list of gas-detecting devices, indicating which are suitable for use in gas chromatography, is given in Table 25.2.

25.4.6 Programmers

Analysis of a sample by gas chromatography requires the execution of a series of operations on or by the instrument at predetermined times after the analytical sequence is initiated by injection of the sample. Other instrumental parameters must also be continuously monitored and controlled. Process gas chromatographs incorporate devices to enable the analytical sequence to be carried out automatically, and the devices necessary to automate a particular instrument are usually assembled into a single module, known as the programmer or controller.

At the most basic level, the programmer may consist of mechanical or electromechanical timers, typically of the cam-timer variety, to operate relays or switches at the appropriate time to select the sample stream to be analyzed, operate the injection valve, and start the data recording process, combined with a facility to correct the output of the chromatograph for baseline drift. Most chromatographs now use built-in microprocessors that incorporate the programmer as part of the central control and data acquisition facility. The programmer itself normally contains a microprocessor and is capable of controlling and monitoring many more of the instrumental parameters as well as acting as a data-logger to record the output of the chromatograph. Computer-type microprocessor-based integrators are available for laboratory use, and in many cases these have facilities to enable them to be used as programmers for the automation of laboratory gas chromatographs. This equipment is then integrated into a laboratory information management system (LIMS) by which analytical results are transmitted directly from the analyzer to a central database for use by customers across the facility.

When the process chromatograph is operated in the automatic mode, all the time-sequenced operations are under programmer control. These will typically include operations to control the gas chromatograph and sampling system, such as sample stream selection, sample injection, column or detector switching, automatic zero and attenuation adjustment, and backflushing. The programmer will also carry out at least some initial processing of the output data, by, for example, peak selection. It is also necessary for a process instrument to incorporate safety devices to prevent damage to itself or to the surroundings in the event of a malfunction, and to give an indication of faults, which may lead to unreliable results. Functions which may be assigned to the programmer include: fault detection and identification, alarm generation, and automatic shutdown of the equipment when a fault is detected.

In addition to the automatic mode of operation the programmer must allow the equipment to be operated manually for startup, maintenance, and calibration.

25.4.7 Data-Processing Systems

The output from a gas chromatograph detector is usually an electrical signal, and the simplest method of data presentation is the chromatogram of the sample, obtained by direct recording of the detector output on a potentiometric recorder. However, the complexity of the chromatograms of typical mixtures analyzed by chromatography means that this simple form of presentation is unsuitable for direct interpretation or display, and further processing is required. The data-processing system of a process chromatograph must be able to identify the peaks in the chromatogram corresponding to components of interest in the sample, and it must measure a suitable parameter of each peak, which can be related to the concentration of that component of the sample. In addition, the system should give a clear indication of faults in the equipment.

Identification of the peaks in the chromatogram is made on the basis of retention time. Provided that instrumental parameters, particularly column temperature and carrier-gas flow rate, remain constant, the retention time is characteristic of a given compound on a particular column. Small changes in operating conditions may change the retention times, so the data-processing system must identify retention times in a suitable "window" as belonging to a particular peak. In addition, retention times may show a long-term drift due to column aging, for which the data-processing system may be required to compensate.

Relation of the output signal to the concentration of the component of interest may be made on the basis of either the height of the peak or the area under it. In both cases a calibration curve must be prepared beforehand by analysis of standard mixtures, and in the more sophisticated systems, this information can be stored and the necessary calculations carried out to give a printed output of the concentrations of the components of interest for each analysis. Automatic updating of the calibration may also be possible. The simplest data-processing systems relate peak height to concentration, but it is usually better to measure peak areas, particularly for complex chromatograms, since this gives some automatic compensation for changes in peak shape caused by adventitious changes in operating conditions. In this case the data-processing system must incorporate an integrator.

25.4.7.1 Display of Chromatographic Data

A refinement of the basic record of the complete chromatogram of the sample is to select and display only peaks corresponding to species of interest, each species being assigned to a separate recorder channel so that successive analyses enable changes in the concentration of each species to be seen. The peaks may be displayed directly or in bar form, as shown in Figure 25.22.

For trend recording, a peak selector accepts the output from the process chromatograph, detects the peak height for each selected measured component, and stores the data. The peak heights are transferred to a memory unit, which holds the value of the height for each peak until it is updated by a further analysis. The output of this unit may be displayed as a chart record of the change in concentration of each measured species. An example of this type of output is shown in Figure 25.22.

25.4.7.2 Gas-Chromatographic Integrators

A variety of gas-chromatographic integrators are available to provide a measure of the areas under the peaks in a chromatogram. The area is obtained by summation of a number of individual measurements of the detector output during a peak, and the number reported by the integrator is typically the peak area expressed in millivolt-seconds. Integrators differ in the method of processing the individual readings of detector output and in the facilities available in the instrument for further processing of the peak area values. In all instruments the analog output signal from the gas chromatograph is first converted to digital form. In simpler integrators an upward change in the baseline level, or in the rate of baseline drift, is taken as the signal to begin the summation process, which continues until the baseline level, or a defined rate of baseline drift, is regained. Since the instrument has to be aware of the baseline change before it can begin integration, a proportion, usually negligibly small, of each peak is inevitably lost, the amount depending on the settings of the slope sensitivity and noise-rejection controls. This difficulty is obviated in the so-called "computing" integrators by storing the digitized detector readings in a memory so that a complete peak, or series of merged peaks, can be stored and integrated retrospectively. Baseline assignment can then also be made retrospectively. In the most sophisticated models the memory is large enough to store data corresponding to a complete chromatogram. Use is also made of the memory to provide facilities for automatic computation of calibration curves, and the integrator may then provide a printed output record giving the concentrations of each component of interest.

25.4.8 Operation of a Typical Process Chromatograph

As an example of process chromatography, the operation of a single-stream instrument designed for high-speed online measurement of the concentration of a single component or group of components is described. The chromatograph is shown schematically in Figure 25.23 and consists of an analyzer, a processor, and a power unit.

The analyzer unit contains those parts of the system required for sample handling and separation and detection of the components. There is a single column and thermal conductivity detector housed in a temperature-controlled zone at the top of the unit, with the associated electronics beneath. The packing and length of the small bore column are chosen to suit the application, and the carrier-gas regulator is designed for high stability under low-flow conditions.

The small-volume thermal conductivity type detector uses thermistor elements to produce the output signals with high speed and stability. The electronic circuit modules mounted in the lower half of the main case control the oven temperature, power the detector, and amplify its output, and provide power pulses to operate the valve solenoids.

The processor contains the electronic circuits that control the sequential operation of the total system. It times

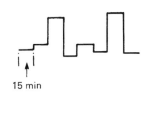

FIGURE 25.22 Methods of display of chromatographic data.

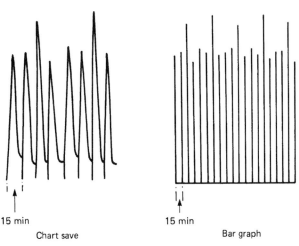

The displays show successive analyses of a single component of the sample.

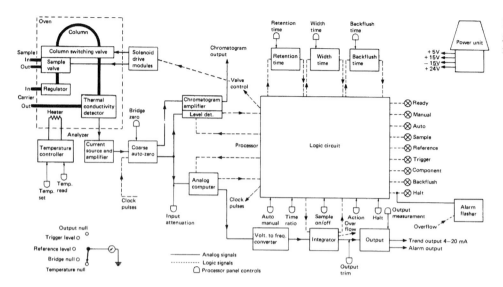

FIGURE 25.23 Schematic diagram of single-channel process chromatograph.

the operation of the simple injection and column switching valves, selects and integrates a chromatographic peak, and updates the trend output signal.

The power unit provides the low voltage regulated supplies for the analyzer and the processor and may be mounted up to 3 meters from the processor.

A typical chromatogram of a sample analyzed by the instrument is shown in Figure 25.24, annotated to show the various switching and logic steps during the analysis.

The operation of the chromatograph can be either on a fixed-time or ratio-time basis. In fixed-time operation the sample injection is the start of the time cycle. At preset times the "integration window" is opened and closed, to coincide with the start and finish of the emergence of the components from the column. While the window is open the detector signal is integrated to give a measure of the concentration of the component. Other operations such as column switching and automatic zeroing are similarly timed from the sample injection. For fixed-time operation to be reliable, pressure and flow rate of carrier gas and temperature and quantity of stationary phase in the column must be closely controlled.

Many of the problems associated with fixed-time operation may be avoided by use of ratio-time operation. In this mode of operation the retention time of components is measured from an early reference peak (corrected retention time; see Figure 25.24) instead of from the time of sample injection. The ratio of two corrected retention times (retention ratio) is less affected by changes in the critical column parameters. The corrected retention time for an early trigger peak is used to predict the time of emergence of the component of interest, that is, the integration window. For the system to be able to operate in the ratio mode, it is necessary to have two specific peaks in the chromatogram in advance of the peak of the component of interest.

Reference Peak The reference peak is due to the first component eluted from the column, with a very low retention time (such as air), and is used as the start point for the ratio timing. If a suitable component is not consistently present in the process sample, one can be injected into the column at the same time as the sample, by using the second loop of the sample valve.

Trigger Peak The trigger peak must appear on the chromatogram between the reference and component peaks. It must be self-evident by virtue of size, and it must be consistent in height and width. As with the reference peak it can be from a component of the process sample or injected separately. Alternatively, it can be a negative peak derived by using a doped carrier gas. The logic circuits measure the time between reference and trigger peaks and use this, together with the preset ratio value, to compute the time for the start of the integration window. Similarly, the trigger peak width is used to define the width of the window. At the start of integration the value of the signal level is stored. The integrator then measures the area under the component peak for the period of the window opening. At this point the signal level is again measured and compared with the stored start value to determine whether any baseline shift has occurred. The integration is corrected for any baseline shifts.

The final value of the integration is stored and used to give an output signal, which represents the concentration of the component. Since this signal is updated after each analysis, the output shows the trend of the concentration.

After the completion of integration, the column is backflushed to remove any later components, the duration of the backflushing being ratioed from the analysis time. Alternatively, for those applications requiring a measurement such as "total heavies," the peak of the total backflushed components can be integrated.

There are some applications in which the ratio-time mode cannot be used. Typically, the measurement of a very early component, such as hydrogen, precludes the existence of

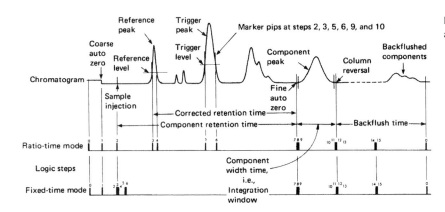

FIGURE 25.24 Chromatogram, showing logic and switching steps.

earlier reference and trigger peaks. Operation of the various functions is then programmed using the fixed-time mode. Selection of the required mode is made using a switch on the processor.

Manual Operation This mode of operation, selected by the "auto/manual" switch on the front panel, provides a single analysis, which is followed by column backflushing and the normal "halt" condition. Single analyses are initiated by operation of the "action" push-button, provided that the previous analysis has been completed. This mode of operation is used during initial programming or servicing.

25.5 SPECIAL GAS ANALYZERS

25.5.1 Paramagnetic Oxygen Analyzers

Many process analyzers for oxygen make use of the fact that oxygen, alone among common gases, is paramagnetic.

25.5.1.1 Basic Principles

The strength of a magnet is expressed as its *magnetic moment*. When a material such as piece of soft iron is placed in a magnetic field, it becomes magnetized by induction, and the magnetic moment of the material divided by its volume is known as the *intensity of magnetization*. The ratio of the intensity of magnetization to the intensity of the magnetic field is called the *volume susceptibility* k of the material. All materials show some magnetic effect when placed in a magnetic field, but apart from elements such as iron, nickel, and cobalt as well as alloys such as steel, all known as *ferromagnetics*, the effect is very small, and intense magnetic fields are required to make it measurable.

Substances that are magnetized in the opposite direction to that of the applied field (so that k is negative) are called *diamagnetics*. Most substances are diamagnetic, and the value of the susceptibility is usually very small. The most strongly diamagnetic substance is bismuth.

The magnetic properties of a substance can be related to its electronic structure. In the oxygen molecule, two of the electrons in the outer shell are unpaired. For this reason, the magnetic moment of the molecule is not neutralized, as is the commoner case, and the permanent magnetic moment is the origin of oxygen's paramagnetic properties.

A ferro- or paramagnetic substance, when placed in a magnetic field in a vacuum or less strongly paramagnetic medium, tries to move from the weaker to the stronger parts of the field. A diamagnetic material in a magnetic field in a vacuum or medium of algebraically greater susceptibility tries—although the effect is very small—to move from the stronger to the weaker parts of the field. Thus when a rod of ferromagnetic or paramagnetic substance is suspended between the poles of a magnet, it will set with its length along the direction of the magnetic field. A rod of bismuth, on the other hand, placed between the poles of a powerful electromagnet, will set at right angles to the field.

It has been shown experimentally that for paramagnetic substances, the susceptibility is independent of the strength of the magnetizing field but decreases with increase of temperature according to the Curie-Weiss law:

$$\text{atomic susceptibility} = \frac{\text{relative atomic mass}}{\text{density}} \times \text{volume susceptibility}$$

$$= \frac{C}{(T - \theta)}$$

where T is the absolute temperature and C and θ are constants.

The susceptibilities of ferromagnetic materials vary with the strength of the applied field, and above a certain temperature (called the *Curie temperature* and characteristic of the individual material) ferromagnetics lose their ability to retain a permanent magnetic field and show paramagnetic behavior. The Curie temperature of iron is 1,000 K.

The susceptibility of diamagnetic substances is almost independent of the magnetizing field and the temperature.

The paramagnetic properties of oxygen are exploited in process analyzers in two main ways: the so-called "magnetic wind" or thermal magnetic instruments and magnetodynamic instruments.

25.5.1.2 Magnetic Wind Instruments

The magnetic wind analyzer, originally introduced by Hartmann and Braun, depends on the fact that oxygen, as a paramagnetic substance, tends to move from the weaker to the stronger part of a magnetic field and that the paramagnetism of oxygen decreases as the temperature is raised:

$$\frac{\text{volume susceptibility}}{\text{density}} = \frac{C}{(T-\theta)} \text{ (Curie-Weiss law)}$$

that is,

$$\text{volume susceptibility} = \frac{C}{(T-\theta)} \times \text{density}$$

But for a gas, the density is proportional to $1/T$, where T is the absolute temperature. Thus,

$$\text{volume susceptibility} = \frac{C}{(T^2 - \theta T)}$$

The principle of the magnetic wind instrument is shown in Figure 25.25. The measuring cell consists of a circular annulus with a horizontal bypass tube, on the outside of which are wound two identical platinum heating coils. These two coils form two arms of a Wheatstone bridge circuit, the bridge being completed by two external resistances. The coils are heated by means of the bridge current, supplied by a DC source of about 12 V. The winding on the left is placed between the poles of a very powerful magnet. When a gas sample containing oxygen enters the cell, the oxygen tends to flow into the bypass tube. Here it is heated so that its magnetic susceptibility is reduced. The heated gas is pushed along the cross-tube by other cold gas entering at the left. This gas flow cools the filaments, the left coil more than the right, and so changes their resistance, as in the flow controller mentioned in Section 25.4.2. The change in resistance unbalances the Wheatstone bridge and the out-of-balance EMF is measured to give a signal, which is proportional to the oxygen content of the gas.

This type of oxygen analyzer is simple and reasonably robust, but it is subject to a number of errors. The instrument is temperature-sensitive: An increase in temperature causes a decrease in the out-of-balance EMF of about 1 percent per Kelvin. This can be automatically compensated by a resistance thermometer placed in the gas stream near the cell. The calibration depends on the pressure of the gas in the cell.

Another error arises from the fact that the analyzer basically depends on the thermal conductivity of the gas passing through the cross-tube. Any change in the composition of the gas mixed with the oxygen changes the thermal balance and so gives an error signal. This is known as the *carrier-gas effect*.

To a first approximation the out-of-balance EMF is given by

$$e = kC_o$$

where e is the EMF, C_o is the oxygen concentration, and k is a factor that varies with the composition of the carrier gas and depends on the ratio of the volumetric specific heat to the viscosity of the carrier gas. For a binary mixture of oxygen with one other gas, k is a constant, and the out-of-balance EMF is directly proportional to the oxygen concentration. For ternary or more complex mixtures, the value of k is constant only if the composition of the carrier gas remains constant.

Values of k for a number of common gases are given in Table 25.3 for an EMF measured in volts and oxygen concentration measured in volume percent. The value of

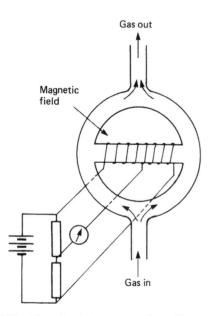

FIGURE 25.25 Magnetic wind oxygen analyzer. Courtesy of Taylor Analytics.

TABLE 25.3 k values for common gases

Gas	k	Gas	k
Ammonia	2.21	Nitrogen	1.00
Argon	0.59	Nitric oxide	0.94
Carbon dioxide	1.54	Nitrous oxide	1.53
Carbon monoxide	1.01	Oxygen	0.87
Chlorine	1.52	Sulphur dioxide	1.96
Helium	0.59	Water vapor	1.14
Hydrogen	1.11		

k for a mixture can be calculated by summing the partial products:

$$k = \frac{(C_A k_A + C_B k_B)}{100}$$

where C_A and C_B are the percentage concentrations of components A and B, and k_A and k_B are the corresponding values of k.

Convective flow or misalignment of the sensor may also change the thermal balance and cause errors. In the case of flammable gases, errors may be caused if they can burn at the temperature in the cross-tube. This type of analyzer is therefore usually considered unsuitable for oxygen measurements in hydrocarbon vapors.

25.5.1.3 Quincke Analyzer

The Quincke analyzer is shown in Figure 25.26. A continuous stream of nitrogen enters the cell and is divided into two streams, which flow over the arms of filaments of a Wheatstone bridge circuit.

The flows are adjusted to balance the bridge to give zero output. One of the nitrogen streams passes the poles of a strong magnet; the other stream passes through a similar volume but without the magnetic field.

The sample gas enters the cell as shown and is mixed with the nitrogen streams immediately downstream of the magnetic field. Oxygen in the sample gas tends to be drawn into the magnetic field, causing a pressure difference in the arms of the cell and changing the flow pattern of the nitrogen over the arms of the Wheatstone bridge. The out-of-balance EMF is proportional to the oxygen concentration of the sample gas.

Because the sample gas does not come into contact with the heated filaments, the Quincke cell does not suffer from the majority of the errors present in magnetic wind instruments, but it does require a separate supply of nitrogen.

25.5.1.4 Magnetodynamic Instruments

Magnetic wind instruments are susceptible to hydrocarbon vapors and to any change in the carrier gas producing a change in its thermal conductivity. These difficulties led to Pauling's development of a measuring cell based on Faraday's work on determination of magnetic susceptibility by measuring the force acting on a diamagnetic body in a nonuniform magnetic field.

25.5.1.5 Magnetodynamic Oxygen Analyzer

In the Pauling cell, two spheres of glass or quartz, filled with nitrogen, which is diamagnetic, are mounted at the ends of a bar to form a dumbbell. The dumbbell is mounted horizontally on a vertical torsion suspension and is placed between the specially shaped poles of a powerful permanent magnet. The gas to be measured surrounds the dumbbell. If oxygen is present, it is drawn into the field and so displaces the spheres of the dumbbell, which are repelled from the strongest parts of the field, so rotating the suspension until the torque produced is equal to the deflecting couple on the spheres (see Figure 25.27). If the oxygen content of the gas in the cell

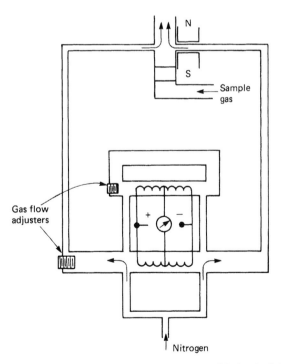

FIGURE 25.26 Quincke oxygen analyzer. Courtesy of Taylor Analytics.

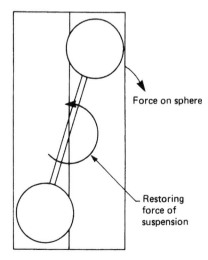

FIGURE 25.27 Magnetodynamic oxygen measuring cell. Courtesy of Taylor Analytics.

changes, there will be a change in the force acting on the spheres, which will take up a new position. The magnitude of the force on the dumbbell may be measured in a number of ways, but a small mirror is commonly attached to the middle of the arm, and the deflection is measured by focusing a beam of light on the mirror. The deflection may either be measured directly, or a force balance system may be used whereby the deflection of the dumbbell is detected but an opposing force is applied to restore it to the null position.

Two different designs of oxygen analyzer, based on the magnetodynamic principle, are shown in Figures 25.28 and 25.29. In the Bendix instrument the suspension is a quartz fiber, and the restoring force is produced electrostatically by the electrodes adjacent to the dumbbell. One electrode is held above ground potential and the other below ground potential by the amplifier controlled from the matched photocells on which the light from the mirror falls. In the Servomex instrument (see Figure 25.29), the suspension is platinum and the restoring force is produced electrically in a single turn of platinum wire connected to the rest of the electronics through the platinum suspension. Electromagnetic feedback is used to maintain the dumbbell in the zero position, and the current required to do this is a measure of the oxygen content of the gas.

The deflecting couple applied to the dumbbell by the magnetic field depends on the magnetic susceptibility of the surrounding gas. The magnetic susceptibilities of all common gases at 20°C are very small (nitrogen, -0.54×10^{-8}; hydrogen, -2.49×10^{-8}; carbon dioxide, -0.59×10^{-8}) compared to that of oxygen ($+133.6 \times 10^{-8}$), and the susceptibility of the gas will depend almost entirely on the concentration of oxygen. This type of analyzer is not influenced by the thermal conductivity of the gas and is unaffected by hydrocarbons. However, the susceptibility of oxygen varies considerably with temperature. This may be overcome by maintaining the instrument at a constant temperature above ambient or the temperature of the measuring cell may be detected and the appropriate temperature correction applied electronically. The reading also depends on the pressure of gas in the cell. This type of analyzer is suitable for measuring the oxygen content of hydrocarbon gases, but paramagnetic gases interfere and must be removed. The most important of these is nitric oxide (susceptibility $+59.3 \times 10^{-8}$), but nitrogen peroxide and chlorine dioxide are also paramagnetic. If the concentration of these gases in the sample is reasonably constant, the instrument may be zeroed on a gas sample washed in acid chromous chloride and the oxygen measured in the usual way.

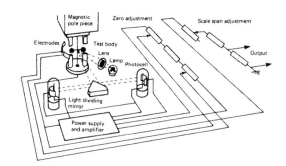

FIGURE 25.28 Bendix oxygen analyzer.

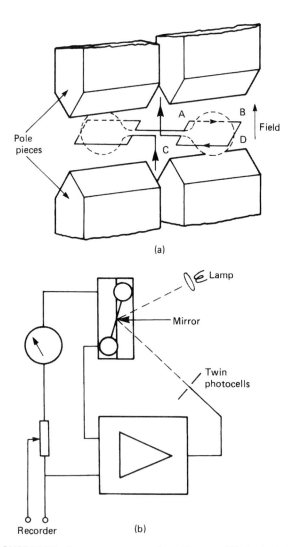

FIGURE 25.29 Servomex oxygen analyzer. Courtesy of Taylor Analytics. (a) Measuring cell. (b) Electronic circuit.

25.5.2 Ozone Analyzer

Continuous analyzers for ozone are based on the chemiluminescent flameless reaction of ozone with ethylene. The light emission from the reaction, centered at 430 nm, is measured by a photomultiplier, and the resulting amplified signal is a measure of the concentration of ozone in the sample stream. The flow diagram and functional block diagram of a typical portable ozone analyzer are given in Figure 25.30. The chemiluminescent light emission from the reaction chamber is a direct function of the ambient temperature, and therefore the temperature is regulated to 50°C. The photomultiplier is contained in a thermo-electrically cooled housing maintained at 25°C to ensure that short- and long-term drift is minimized. The instrument is capable of measuring ozone levels in the range 0.1 to 1,000 ppb.

25.5.3 Oxides of Nitrogen Analyzer

Analyzers for oxides of nitrogen—NO, NO_x (total oxides of nitrogen), NO_2—are based on the chemiluminescent reaction of nitric oxide (NO) and ozone to produce nitrogen dioxide (NO_2). About 10 percent of the NO_2 is produced in an electronically excited state and undergoes a transition to the ground state, emitting light in the wavelength range 590–2600 nm:

$$NO + O_3 \rightarrow NO_2^* + O_2$$
$$NO_2^* \rightarrow NO_2 + h\nu$$

The intensity of the light emission is proportional to the mass-flow rate of NO through the reaction chamber and is measured by a photomultiplier tube.

Analysis of total oxides of nitrogen (NO_x) in the sample is achieved by passing the gases through a stainless-steel tube at 600–800°C. Under these conditions, most nitrogen compounds (but not N_2O) are converted to NO, which is then measured as previously stated. Nitrogen dioxide (NO_2) may be measured directly by passing the air sample over a molybdenum catalyst to reduce it to NO, which is again measured as stated, or the NO_2 concentration may be obtained by automatic electronic subtraction of the NO concentration from the NO_x value.

The flow system of a nitrogen oxides analyzer is shown in Figure 25.31. Ozone is generated from ambient air by the action of UV light

$$3 O_2 \xrightarrow{h\nu} 2 O_3$$

and a controlled flow rate of ozonized air is passed to the reaction chamber for reaction with NO in the air sample, which is passed through the chamber at a controlled flow of 1 l min^{-1}. By selection of a switch to operate the appropriate solenoid valves, a span gas may be directed to the reaction

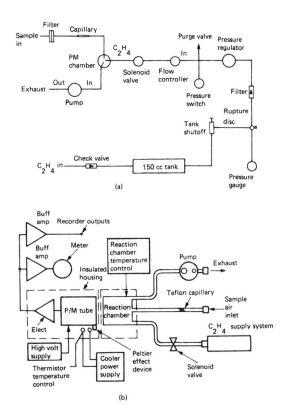

FIGURE 25.30 Ozone analyzer. Courtesy of Columbia Scientific Industries Corp. (a) Flow diagram. (b) Functional block diagram.

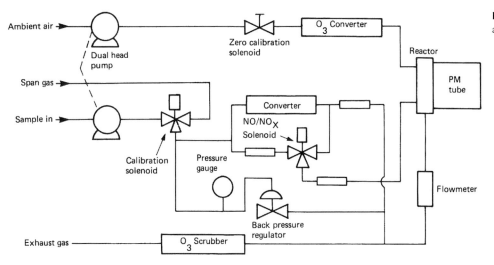

FIGURE 25.31 Oxides of nitrogen analyzer. Courtesy of Beckman.

TABLE 25.4 Measurement principles of special gas analyzers

Gas	Measurement principle
Oxygen	Paramagnetism
	Electrochemical sensor
	Fuel cell
Ozone	Chemiluminescence
	Electrochemical sensor
Nitrogen oxides	Chemiluminescence
Carbon dioxide	Infrared spectrometry
Carbon monoxide	Infrared spectrometry
	Electrochemical sensor
Sulphur oxides	Flame photometry
Hydrocarbons	Flame ionization detector
	Infrared spectrometry
	Catalytic detector
Flammable gases	Catalytic detector
	Semiconductor detector
Hydrogen sulphide	Semiconductor detector
	Flame photometry
	Electrochemical sensor

chamber or a zero calibration may be carried out by shutting off the flow of ozonized air to the reactor. The three-way solenoid valve downstream of the converter is switched to permit NO analysis when bypassing the converter and NO_x analysis when the sample is passed through the converter. The analyzer can measure ozone in air in the range 5 ppb to 25 ppm, with a precision of ±1 percent.

25.5.4 Summary of Special Gas Analyzers

The operating principles of analyzers for the most commonly measured gases are given in Table 25.4.

25.6 CALIBRATION OF GAS ANALYZERS

None of the commonly used gas detectors is absolute; that is, they are devices where the output signal from the detector for the gas mixture under test is compared with that for mixtures of the bulk gas containing known concentrations of the determinand. The use of standard gas mixtures is analogous to the use of standard solutions in solution chemistry, but their preparation and handling present some peculiar problems. As in solution chemistry, the calibration gas mixtures should reflect, as closely as possible, the composition of the samples they are desired to measure. Ideally a number of standard mixtures, whose concentration covers the range of samples to be measured, should be used to establish the response curve of the instrument or detector. However, for routine calibration where the response curve has previously been established or is well known, it is usual to calibrate gas analyzers by use of a "zero" gas mixture that is free of the determinand and establishes the zero of the instrument and one or more "span" gases containing concentrations of the determined close to those it is desired to measure.

The accuracy to which a gas mixture can be prepared depends on the number and nature of the components and on their concentrations. For gas mixtures prepared under pressure in cylinders, it is useful to specify two parameters: the filling and analytical tolerances. The filling tolerance describes the closeness of the final mixture to its original specification and depends mainly on the concentrations of the components. Thus, though it may be possible to fill a cylinder with a component gas at the 50 percent level to a tolerance of ±2.5 percent or ±5 percent of the component (that is, the cylinder would contain between 47.5 and 52.5 percent of the component), at the 10 vpm level the tolerance would typically be ±5 vpm or ±50 percent of the component and the cylinder would contain between 5 and 15 vpm of the component. The analytical tolerance is the accuracy with which the final mixture can be described; it depends on the nature of the mixture and the analytical techniques employed. Accuracies achievable are typically in the range from ±2 percent of component or ±0.2 vpm at the 10 vpm level to ±1 percent of component or ±0.5 percent at the 50 percent level. However, these figures are strongly dependent on the actual gases involved and the techniques available to analyze them.

Gas mixtures may be prepared by either static or dynamic methods. In the static method, known quantities of the constituent gases are admitted to a suitable vessel and allowed to mix, whereas in the dynamic method, streams of the gases, each flowing at a known rate, are mixed to provide a continuous stream of the sample mixture. Cylinders containing supplies of the standard mixtures prepared under pressure are usually most convenient for fixed instruments such as process gas chromatographs, whereas portable instruments are often calibrated by mixtures prepared dynamically. Where mixtures containing low concentrations of the constituents are needed, adsorptive effects may make the static method inapplicable, whereas the dynamic method becomes more complex for mixtures containing large numbers of constituents.

Before any gas mixture is prepared, its properties must be known, particularly if there is any possibility of reaction between the components, over the range of pressures and concentrations expected during the preparation.

25.6.1 Static Methods

Static gas mixtures may be prepared either gravimetrically or by measurement of pressure. Since the weight of gas is usually small relative to the weight of the cylinder required to contain it, gravimetric procedures require balances, which have both high capacity and high sensitivity, and the buoyancy effect of the air displaced by the cylinder may be significant. Measurement of pressure is often a more readily applicable technique.

After preparation gas mixtures must be adequately mixed to ensure homogeneity, usually by prolonged continuous rolling of the cylinder. Once mixed, they should remain homogeneous over long periods of time. Any concentration changes are likely to be due to adsorption on the cylinder walls. This is most likely to happen with mixtures containing vapors near their critical pressures, and use of such mixtures should be avoided if possible.

Another common problem with complex samples is stratification over time, especially if they are stored in cooler ambient temperatures. One method to minimize this effect is to place heating blankets on the cylinders, which introduces thermal currents in the bottles to keep the mixture from "separating."

25.6.2 Dynamic Methods

25.6.2.1 Gas Flow Mixing

Gas mixtures of known concentration may be prepared by mixing streams of two or more components, each of which is flowing at a known rate. The concentration of one gas in the others may be varied by adjustment of the relative flow rates, but the range of concentration available is limited by the range of flows, which can be measured with sufficient accuracy. Electronic mass-flow controllers are a convenient method of flow measurement and control.

25.6.2.2 Diffusion-Tube and Permeation-Tube Calibrators

Standard gas mixtures may be prepared by allowing the compound or compounds of interest to diffuse through a narrow orifice or to permeate through a membrane into a stream of the base gas, which is flowing over the calibration source at a controlled and known rate.

Typical designs of diffusion and permeation tubes are shown in Figure 25.32. In both cases there is a reservoir of the sample, either a volatile liquid or a liquefied gas under pressure, to provide an essentially constant pressure, the saturation vapor pressure, upstream of the diffusion tube or permeation membrane. After an initial induction period it is found that, provided that the tube is kept at constant temperature, the permeation or diffusion rate is constant as long as there is liquid in the reservoir. The tube can then

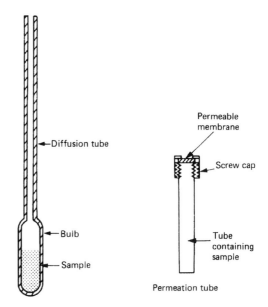

FIGURE 25.32 Cross-sectional diagrams of diffusion and permeation tube calibration sources.

be calibrated gravimetrically to find the diffusion or permeation rate of the sample. The concentration of the sample in the gas stream is then given by

$$C = RK/F$$

where C is the exit gas concentration, R is the diffusion or permeation rate, K is the reciprocal density of the sample vapor, and F is the gas flow rate over the calibration device. The diffusion or permeation rate depends on the temperature of the tube and on the molecular weight and vapor pressure of the sample. Additionally, the diffusion rate depends on the length and inner diameter of the capillary tube, and the permeation rate depends on the nature, area, and thickness of the permeation membrane. Data are available for a large number of organic and inorganic vapors to allow tubes to be designed with the required diffusion or permeation rate, and the exact rate for each tube is then established empirically.

The temperature-dependence of diffusion or permeation means that the tubes must be carefully thermostatted for accurate calibrations. The empirical equation for the temperature-dependence of permeation rate is

$$\log \frac{R_2}{R_1} = 2950 \left(\frac{1}{T_1} - \frac{1}{T_2} \right)$$

where R_1 is permeation rate at T_1 K and R_2 is permeation rate at T_3 K. The permeation rate changes by approximately 10 percent for every 1 K change in temperature. Thus the temperature of the permeation tube must be controlled to within 0.1 K or better if 1 percent accuracy in the permeation rate, and thus the concentration that is being developed, is to be achieved.

The flow diagram of a typical calibrator for use with diffusion or permeation tubes is shown in Figure 25.33. The

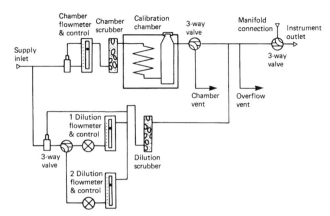

FIGURE 25.33 Flow diagram of gas calibrator.

gas supply is scrubbed before passing through a thermostatted coil and over the calibration source or sources in the calibration chamber. Secondary streams of purified gas may be added to the effluent gas stream to adjust the final concentration to the range required.

The diffusion or permeation technique is especially useful for generating standard mixtures at low concentrations, for example of organic compounds in air for calibration of environmental monitors, air pollution monitors, and so on, and the calibrator can be made portable for field use. The range of compounds that can be used is limited by their saturation vapor pressure; if this is too low, the diffusion or permeation rates, and hence the concentrations available are very small, whereas compounds with high saturation vapor pressures present problems in construction and filling of the calibration tubes.

25.6.2.3 Exponential Dilution

In the exponential dilution technique a volume of gas contained in a vessel in which there is perfect and instantaneous mixing is diluted by passing a stream of a second gas through the vessel at a constant flow rate. It can be shown that, under these conditions, the concentration of any gaseous species in the vessel, and hence the instantaneous concentration in the effluent stream of diluent gas, decays according to the law

$$C = C_0 \exp\left(-\frac{Ut}{V}\right)$$

where C is the concentration of the diluted species at time t, C_0 is the initial concentration, U is the flow rate of diluent gas, and V is the volume of the vessel.

The vessel may either be filled with the gaseous species to be analyzed, in which case the concentration decays from an initial value of 100 percent, or it may be filled with the diluent gas, and a known volume of the gas of interest may be injected into the diluent gas just upstream of the dilution vessel at the start of the experiment. In either case the concentration of the species of interest in the effluent gas stream may be calculated at any time after the start of the dilution.

The exponential dilution vessel is typically a spherical or cylindrical glass vessel of 250–500 ml capacity, fitted with inlet and outlet tubes, and a septum cap or gas sampling valve for introduction of the gas to be diluted. The vessel must be fitted with a stirrer, usually magnetically driven, and baffles to ensure that mixing is as rapid and homogeneous as possible. The diluent gas flows through the vessel at a constant known flow rate, usually in the range 20–30 ml min^{-1}. For a vessel of the dimensions suggested here, this gives a tenfold dilution in approximately 30 minutes.

The exponential dilution technique is a valuable calibration method especially suitable for use at very low concentrations. It is also valuable for studying or verifying the response of a detector over a range of concentrations. However, it should be noted that strict adherence to a known exponential law for the decay of concentrations in the vessel depends on the attainment of theoretically perfect experimental conditions that cannot be achieved in practice.

Changes in the flow rate of the diluent gas or in the temperature or pressure of the gas in the dilution vessel and imperfect or noninstantaneous mixing in the vessel lead to unpredictable deviations from the exponential decay law. Deviations also occur if the determinand is lost from the system by adsorption on the walls of the vessel. Since the technique involves extrapolation from the known initial concentration of the determinand in the diluting gas, any deviations are likely to become more important at the later stages of the dilution. If possible, it is therefore advisable to restrict the range of the dilution to two or three orders of magnitude change in concentration. Where the gas to be diluted is introduced to the dilution vessel by injection with a valve or syringe, the accuracy and precision of the entire calibration curve resulting from the dilution are limited by the accuracy and precision of the initial injection.

FURTHER READING

Cooper, C. J., and A. J. De Rose, "The analysis of gases by chromatography," *Pergamon Series in Analytical Chemistry*, Vol. 7. Pergamon, Oxford (1983).

Cullis, C. F., and J. G. Firth (eds.), *Detection and Measurement of Hazardous Gases*, Heinemann, London (1981).

Grob, R. L. (ed.), *Modern Practice of Gas Chromatography*, Wiley, Chichester, U.K. (1977).

Jeffery, P. F., and P. J. Kipping, *Gas Analysis by Gas Chromatography*, International Series of Monographs in Analytical Chemistry, Vol. 17, Pergamon, Oxford (1972).

Sevcik, J., *Detectors in Gas Chromatography, Journal of Chromatography Library*, Vol. 4, Elsevier, Amsterdam (1976).

Also review articles in *Analytical Chemistry*, and manufacturers' literature.

Chapter 26

Chemical Analysis: Moisture Measurement

D. B. Meadowcroft; edited by I. Verhappen

26.1 INTRODUCTION

The measurement and control of the moisture content of gases, liquids, and solids is an integral part of many industries. Numerous techniques exist, none being universally applicable, and the instrument technologist must be able to choose the appropriate measurement technique for the application. It is particularly important to measure moisture because of its presence in the atmosphere, but it is awkward because it is a condensable vapor that will combine with many substances by either physical adsorption or chemical reaction. Moisture measurement may be needed to ensure that the level remains below a prescribed value or within a specified band, and the range of concentrations involved can be from less than one part per million to percentage values.

A few examples will illustrate the range of applications.

Gases In gas-cooled nuclear reactors, the moisture level of the coolant has to be within a prescribed band (e.g., 250–500 volume parts per million) or below a certain value (e.g., 10 vppm) depending on the type of reactor. Rapid detection of small increases due to leaks from the steam generators is also essential. Moisture must be excluded from semiconductor device manufacture, and glove boxes are fitted with moisture meters to give an alarm at, say, 40 vppm. Environmental control systems need moisture measurement to control the humidity, and even tumble dryers can be fitted with sensors to automatically end the clothes drying cycle.

Liquids The requirement is usually to ensure that the water contamination level is low enough. Examples are the prevention of corrosion in machinery, breakdown of transformer oil, and loss of efficiency of refrigerators or solvents.

Solids Specified moisture levels are often necessary for commercial reasons. Products sold by weight (e.g., coal, ore, tobacco, textiles) can most profitably have moisture contents just below the maximum acceptable limit. Some textiles and papers must be dried to standard storage conditions to prevent deterioration caused by excessive wetness and to avoid the waste of overdrying, since the moisture would be picked up again during storage. Finally, many granulated foods must have a defined moisture content.

The purpose of this chapter is to introduce the reader to the major measurement techniques that are available. The three states—gas, liquid, and solid—will be treated separately. In addition, many commercial instruments measure some parameter that changes reproducibly with moisture concentration, and these instruments must be regularly calibrated by the user. The chapter therefore ends with a discussion of the major calibration techniques that the average user must be willing to employ when using such instruments.

First, it is necessary to clarify a further aspect of moisture measurement that can confuse the newcomer, which is to define the large number of units that are used, particularly for gases, and show how they are interrelated.

26.2 DEFINITIONS

26.2.1 Gases

Although water vapor is not an ideal gas, for most hygrometry purposes—and to gain an understanding of the units involved—it is sufficient to assume that water vapor does behave ideally. The basic unit of moisture in a gas against which other units can readily be referred is *vapor pressure*, and Dalton's law of partial pressures can be assumed to hold if the saturated vapor pressure is not exceeded.

In environmental applications, the unit that's often used is *relative humidity*, which is the ratio in percent of the actual vapor pressure in a gas to the saturation vapor pressure of water at that temperature. It is therefore temperature dependent but is independent of the pressure of the carrier gas.

For chemical measurements, the concentration of moisture is usually required. The *volume concentration* is given by the vapor pressure of moisture divided by the total pressure, often multiplied by 10^6 to give volume parts per million (vppm). The concentration by *weight* in wppm is given by the volume concentration multiplied by the molecular weight of water and divided by that of the carrier gas. Meteorologists often call the weight concentration the *mixing ratio* and express it in g/kg.

When the prime aim is to avoid condensation, the appropriate unit is the *dewpoint*, which is the temperature at which the vapor pressure of the moisture would become saturated with respect to a plane surface. Similarly, the *frostpoint* refers to the formation of ice. The relationship among dewpoints, frostpoints, and saturated vapor pressure is derived from thermodynamic and experimental work and is shown in Figure 26.1.

It should be noted that below 0°C the dewpoint and frostpoint differ. It is possible for supercooled water to exist below 0°C, which can give some ambiguity, but this is unlikely very much below 0°C (certainly not below −40°C). In addition, it can be seen that the saturated vapor pressure increases by an order of magnitude every 15–20 degrees so that in the range −80°C to 50°C dewpoint, there is a vapor pressure change of five orders of magnitude. Table 26.1 lists the vapor pressure for dew- or frostpoint between −90°C and +50°C.

Table 26.2 gives the interrelationships between these various units for some typical values.

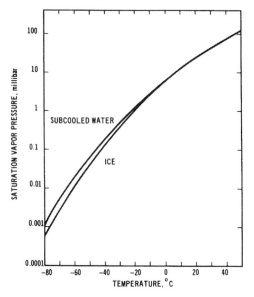

FIGURE 26.1 The relationship between saturation vapor pressure and dewpoint and frostpoint temperatures.

26.2.2 Liquids and Solids

Generally, measurements are made in terms of *concentration*, either as a percentage of the total wet weight of the sample (e.g., in the ceramics industry for clay) or of the dry weight (e.g., in the textile industry, where the moisture concentration is called *regain*). In addition, if a liquid or solid is in equilibrium with the gas surrounding it, the *equilibrium relative humidity* of the gas can be related to the moisture content of the solid or liquid by experimentally derived isotherms (e.g., Figure 26.2) or by Henry's law for appropriate unsaturated liquids. For liquids that obey Henry's law, the partial vapor pressure of the moisture P is related to the concentration of water dissolved in the liquid by $W = KP$, where K is Henry's law constant. K can be derived from the known saturation values of the particular liquid, that is, $K = W_s P_s$, where W_s and P_s are, respectively, saturation concentration and saturation vapor pressure at a given temperature.

TABLE 26.1 The relationship between dew/frostpoint and vapor pressure (μ bar, which is equivalent to vppm at 1 bar total pressure)

Frostpoint (°C)	Saturated vapor pressure (μ bar)	Frostpoint (°C)	Saturated vapor pressure (μ bar)	Dewpoint (°C)	Saturated vapor pressure (μ bar)
−90	0.10	−40	128	0	6,110
−80	0.55	−36	200	4	8,120
−75	1.22	−32	308	8	10,700
−70	2.62	−28	467	12	14,000
−65	5.41	−24	700	16	19,200
−60	10.8	−20	1,030	20	23,400
−56	18.4	−16	1,510	25	31,700
−52	30.7	−12	2,170	30	41,800
−48	50.2	−8	3,100	40	73,000
−44	81.0	−4	4,370	50	120,000

CHAPTER | 26 Chemical Analysis: Moisture Measurement

TABLE 26.2 Some examples of the relationships between the various units for moisture in gases

Dew/frost point (°C)	Vapor pressure (μ bar or vppm at 1 bar)	RH at 20°C ambient (%)	Mixing ratio in air (g/kg)
−70	2.5	0.01	1.5×10^{-3}
−45	72	0.3	0.045
−20	1,030	4.4	0.64
0	6,110	26	3.8
10	12,300	53	7.6
20	23,400	100	14.5

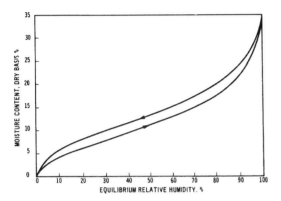

FIGURE 26.2 The relationship between the moisture content of a substance and the equilibrium relative humidity of the surrounding gas, for the example of wool.

26.3 MEASUREMENT TECHNIQUES

Techniques that allow automatic operation have the important advantage that they can be used for process control. Here we therefore concentrate our attention on such techniques. Again, those available for gases, liquids, and solids are discussed separately.

26.3.1 Gases

There is a huge choice of techniques for the measurement of moisture in gases, reflecting the large number of ways in which its presence is manifested. The techniques range from measuring the extension of hair in simple wall-mounted room monitors to sophisticated electronic instruments. To some extent, the choice of technique depends on the property required: dewpoint, concentration, or relative humidity. Only the major techniques are discussed here. More extensive treatments are given in the bibliography.

26.3.1.1 Dewpoint Instruments

The determination of the temperature at which moisture condenses on a plane mirror can be readily estimated (see Figure 26.3) using a small mirror for which the temperature can be controlled by a built-in heater and thermoelectric cooler. The temperature is measured by a thermocouple or platinum resistance thermometer just behind the mirror surface, and the onset of dew is detected by the change of reflectivity measured by a lamp and photocell. A feedback circuit between the cell output and the heater/cooler circuit enables the dewpoint temperature to be followed automatically. Systematic errors can be very small, and such instruments are used as secondary standards, yet with little loss of sophistication they can be priced competitively for laboratory and plant use. Mirror contamination can be a problem in dirty gases, and in some instruments the mirror is periodically heated to reduce the effect of contamination. Condensable carrier gases, which condense at similar temperatures to the moisture, invalidate the technique. It is an ideal method if the dewpoint itself is required, but if another unit is to be derived from it, accurate temperature measurements are essential because of the rapid change in vapor pressure with dewpoint temperature (see Section 26.2.1).

26.3.1.2 Coulometric Instruments

The gas is passed at a constant rate through a sampling tube in which the moisture is absorbed onto a film of partially hydrated phosphoric anhydride (P_2O_5) coated on two platinum electrodes (see Figure 26.4). A dc voltage is applied across the electrodes to decompose the water, the charge produced by the electrolysis being directly proportional to the mass of water absorbed (Faraday's law). Thus the current depends on the flow rate, which must be set and controlled accurately at a predetermined rate (usually 100 ml min^{-1}) so that the current meter can be calibrated directly in ppm. Several points are worth making:

1. The maximum moisture concentration measurable by this technique is in the range 1,000–3,000 vppm, but care must be taken to ensure that surges of moisture level do not wash off the P_2O_5.
2. There is generally a zero leakage current equivalent to a few ppm. To allow for this error, when necessary, the current should be measured at two flow rates and the difference normalized to the flow for 100 ml min^{-1}.

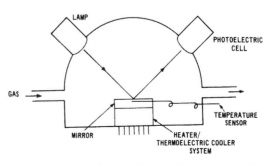

FIGURE 26.3 A schematic diagram of a sensor of a dewpoint mirror instrument.

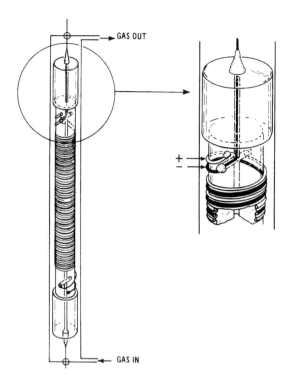

FIGURE 26.4 A schematic diagram of a sensor of a coulometric instrument.

3. Platinum electrodes are not suitable for use in gases containing significant amounts of hydrogen. The platinum can catalyze the recombination of the electrolyzed oxygen, and this water is also electrolyzed, giving inaccurate measurements. Gold or rhodium elements reduce this effect.
4. In the absence of recombination and gas leaks, the response of a coulometric instrument can be regarded as absolute for many purposes.
5. Cells, which work at pressure, can be obtained. This can increase the sensitivity at low moisture levels because it is possible to use a flow rate of 100 ml min^{-1} at the measuring pressure, which does not increase the velocity of gas along the element and hence does not impair the absorption efficiency of the P_2O_5.

26.3.1.3 Infrared Instruments

Water vapor absorbs in the 1–2 μm infrared range, and infrared analyzers (see Chapter 29) can be successfully used as moisture meters. For concentrations in the vppm range, the path length has to be very long, and high sample flow rates of several liters per minute can be necessary to reduce the consequent slow response time. Both single-beam instruments, in which the zero baseline is determined by measuring the absorption at a nearby nonabsorbing wavelength, and double-beam instruments, in which a sealed parallel cell is used as reference, can be used. Single-beam instruments are less affected by deposits on the cell windows and give better calibration stability in polluted gases.

26.3.1.4 Electrical Sensor Instruments

There are many substances for which electrical impedance changes with the surrounding moisture level. If this absorption process is sufficiently reproducible on a thin film, the impedance, measured at either an audio frequency or a radio frequency, can be calibrated in terms of moisture concentration or relative humidity. Materials used in commercial instruments include polymers, tantalum oxide, silicon oxide, chromium oxide, aluminum oxide, lithium chloride mixed with plastic, and carbon-loaded plastics that change length and hence resistance with moisture level. Many such instruments are available commercially, particularly using an anodized aluminum oxide layer that has a very narrow columnar pore structure (see Figure 26.5), but aging and other deterioration processes can occur so that regular calibration is essential. A major advantage of such sensors is that as no imposed gas flow is necessary, they can simply be placed in the gas to be measured—for example, in an environmental chamber. In addition, they can be used at high pressure, they have a wide response range (typically 50°C to −80°C dewpoint for a single aluminum oxide sensor), have a rapid response, and are generally not expensive. These advantages often outweigh any problems of drift and stability and the requirement for regular calibration, but the sensors must be used with care.

26.3.1.5 Quartz Crystal Oscillator Instrument

The oscillation frequency of a quartz crystal coated with hygroscopic material is a very sensitive detector of the weight of absorbed water because very small changes in frequency can be measured. In practice, as shown in Figure 26.6, two quartz crystal oscillators are used, and the wet and a dry gas are passed across them alternately, usually for 30 seconds at a time. The frequency of crystal oscillation is about 9.10^6 Hz and that of the crystal exposed to the wet gas will be lowered and that of the crystal exposed to the dry gas will rise. The resultant audio frequency difference is extracted, amplified, and converted to voltage

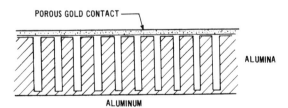

FIGURE 26.5 An idealized representation of the pore structure of anodized alumina. The pores are typically less than 20 nm in diameter and more than 100 μm deep. A porous gold layer is deposited on the alumina for electrical contact when used as a hygrometer sensor.

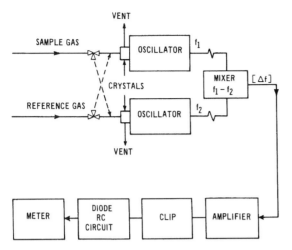

FIGURE 26.6 A block diagram of the arrangement of a piezoelectric humidity instrument. Courtesy of Du Pont Instruments (UK) Ltd.

to give a meter response of which the maximum value on each 30-second cycle is a measure of the moisture level. The range of applicable concentrations is 1–3,000 vppm, and at lower levels the fact that the value after a certain time is measured rather than an equilibrium value means that the instrument can have a more rapid response than alternative methods (sample lines, however, often determine response time). Because the crystals see the sample gas for equal times, contamination of the two crystals should be similar and the frequency difference little affected, resulting in stability. However, regular calibration is still necessary, and the complexity of the instrument makes it expensive.

26.3.1.6 Automatic Psychrometers

The measurement of the temperature difference between a dry thermometer bulb and one surrounded by a wet muslin bag fed by a wick is the classical meteorological humidity measurement. This is called *psychometry*, and automated instruments are available. The rate of evaporation depends on the gas flow as well as on the relative humidity, but generally a flow rate greater than $3\,\mathrm{ms}^{-1}$ gives a constant temperature depression. It is most useful at high relative humidities with accurate temperature measurements.

26.3.2 Liquids

26.3.2.1 Karl Fischer Titration

The Karl Fischer reagent contains iodine, sulfur dioxide, and pyridine (C_5H_5N) in methanol; the iodine reacts quantitatively with water as follows:

$$[3C_5H_5N + I_2 + SO_2] + H_2O \rightarrow 2C_5H_5NHI + C_5H_5NSO_3$$
$$C_5H_5NSO_3 + CH_3OH \rightarrow C_5H_5NHSO_4CH_3$$

If a sample containing water is titrated with this reagent, the endpoint at which all the H_2O has been reacted is indicated by a brown color showing the presence of free iodine. This is the basic standard technique and is incorporated into many commercial instruments with varying levels of automation. In process instruments the endpoint is determined electrometrically by amperometric, potentiometric, or coulometric methods (see Chapter 23). In the amperometric method two platinum electrodes are polarized; when free iodine appears they are depolarized and the resultant current is measured to define the endpoint. Potentiometrically, the potential of an indicator electrode is monitored against a calomel electrode and the endpoint is characterized by a sudden change in potential. Coulometrically, iodine is generated by a constant electrolyzing current from a modified reagent, and the time taken to reach the endpoint gives the mass of water in the sample. This last technique lends itself to automatic operation, with samples injected sequentially or, in one instrument, the moisture in a sample flow is measured continuously by mixing with standardized reagent, and the electrolysis current is a measure of the mass flow of water.

26.3.2.2 Infrared Instruments

The same comments apply as for gases (Section 26.3.1.3), but sample cell lengths are usually shorter, in the range of 1–100 mm. It is an attractive method for online analysis, but care must be taken that other components in the liquid do not interfere with the measurement. Single-beam instruments are most often used.

26.3.2.3 Vapor Pressure Methods

As discussed in Section 26.2.2, the equilibrium relative humidity above a liquid can be used to determine the moisture content in the liquid. Either the relative humidity in a closed volume above the liquid can be measured or a sensor that responds to the moisture vapor pressure in the liquid can be immersed in the liquid. The aluminum oxide sensor (Section 26.3.1.4) can be used either above the liquid because it does not require a gas flow rate or within the liquid because, although the aluminum oxide pores will adsorb water molecules, they will not adsorb the liquid molecules. These techniques are not appropriate if suspended free water is present in the liquid.

One manufacturer has developed a system in which the sensor is a moisture-permeable plastic tube that is immersed in the liquid. A fixed quantity of initially dry gas is circulated through the tube, and the moisture in the gas is measured by an optical dewpoint meter. When equilibrium is reached, the dewpoint measured equals that of the moisture in the liquid.

26.3.2.4 Microwave Instruments

The water molecule has a dipole moment with rotational vibration frequencies, which give absorption in the microwave, S-band (2.6–3.95 GHz), and X-band (8.2–12.4 GHz)

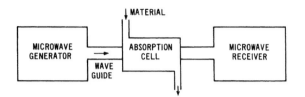

FIGURE 26.7 The basic concept for measuring moisture by microwave absorption.

suitable for moisture measurement (see Figure 26.7). The S-band needs path lengths four times longer than the X-band for a given attenuation, and therefore the microwave band as well as cell dimensions can be chosen to give a suitable attenuation. Electronic developments are causing increased interest in this technique.

26.3.2.5 Turbidity-nephelometer

Undissolved water must be detected in aviation fuel during transfer. After thorough mixing the fuel is divided into two flows; one is heated to dissolve all the water before it passes into a reference cell, and the other passes directly into the working cell. Light beams split from a single source pass through the cells, suspended water droplets in the cell scatter the light, and a differential output is obtained from the matched photoelectric detectors on the two cells. Moisture of 0 to 40 ppm can be detected at fuel temperatures of –30 to 40°C.

26.3.3 Solids

The range of solids in which moisture must be measured commercially is wide, and many techniques are limited to specific materials and industries. In this book just some of the major methods are discussed.

26.3.3.1 Equilibrium Relative Humidity

The moisture level of the air immediately above a solid can be used to measure its moisture content. Electrical probes, as discussed in Section 26.3.1.4, are generally used, and if appropriate they can be placed above a moving conveyor. If a material is being dried, its temperature is related to its equilibrium relative humidity, and a temperature measurement can be used to assess the extent of drying.

26.3.3.2 Electrical Impedance

Moisture can produce a marked increase in the electrical conductivity of a material and, because of water's high dielectric constant, capacitance measurements can also be valuable. Electrical resistance measurements of moisture in timber and plaster are generally made using a pair of sharp pointed probes, as shown in Figure 26.8(a), which are pushed into the material, the meter on the instrument being calibrated directly in percentage moisture. For online measurements of granular materials, electrodes can be rollers, plates (see Figure 26.8(b)), or skids, but uniform density is essential. A difficulty with this and other online methods that require contact between the sensor and the material is that hard materials will cause rapid erosion of the sensor.

26.3.3.3 Microwave Instruments

Most comments appropriate to liquids also apply to solids, but, as mentioned, constant packing density is necessary. For sheet materials such as paper or cloth, measurement is simple, with the sheet passing through a slot in the waveguide. For granular materials, uniform material density is achieved by design of the flow path; alternatively, extruders or compactors can be useful.

26.3.3.4 Infrared Instruments

The basic difference from measurements in gases and liquids is that, for solids, reflectance methods (see Figure 26.9) are usually used rather than transmission methods. Single-beam

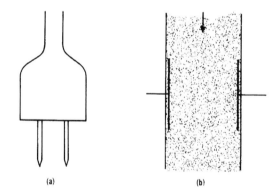

FIGURE 26.8 Two techniques for electrical measurements of moisture in solids: (a) pointed probes for insertion in wood, plaster, etc. to measure resistance, (b) capacitance plates to measure moisture in flowing powder or granules.

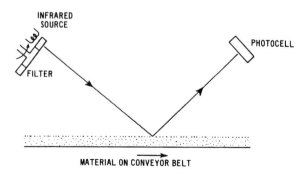

FIGURE 26.9 The principle of infrared reflectance used to measure moisture in a solid on a conveyor.

operation is used with a rotating absorption and reference frequency filter to give regular zero readings. The calibration of a reflectance method can be substantially independent of the packing density as it measures only the surface concentration. For material on a conveyor belt, a plough is often used in front of the sensing position to ensure a measurement more typical of the bulk. The method is not suitable for poorly reflecting materials such as carbon and some metal powders.

26.3.3.5 Neutron Moderation

Hydrogen nuclei slow ("moderate") fast neutrons; therefore, if a fast neutron source is placed over a moist material with a slow neutron detector adjacent, the detector output can be used to indicate the moisture concentration. The concentration of any other hydrogen atoms in the material and its packing density must be known. This technique is described in Chapter 30. Nuclear magnetic resonance can also be used to detect hydrogen nuclei as a means of measuring moisture content.

26.4 CALIBRATION

It will be seen from the preceding sections that many moisture measurement techniques are not absolute and must be calibrated, generally at very regular intervals. It must first be emphasized that the absolute accuracy of moisture measurement, particularly in gases, is not usually high. Though it is possible to calibrate moisture detectors for liquids or solids to 0.1 to 1.0 percent, such accuracies are the exception rather than the rule for gases. Figure 26.10 shows the accuracies of some of the techniques discussed in this chapter compared with the absolute gravimetric standard of the U.S. National Bureau of Standards.

26.4.1 Gases

First, the difficulties of making accurate moisture measurements must be stressed. This is particularly so at low levels, say, less than 100 vppm, because, since all materials absorb moisture to some extent, sample lines must come to equilibrium as well as the detector. At low moisture levels this can take hours, particularly at low-flow rates. A rapid-flow bypass line can be valuable. Patience is mandatory, and if possible the outputs of the instruments should be recorded to establish when stable conditions are achieved. Many plastics are permeable to moisture and must never be used. At high moisture levels copper, Teflon, Viton, glass, or quartz can be satisfactorily used, but at low levels stainless steel is essential. Finally, at high moisture levels it must be remembered that the sample lines and detectors must be at least 10 Kelvins hotter than the dewpoint of the gas.

There are two basic calibration methods that can, with advantage, be combined. Either a sample gas is passed through a reference hygrometer and the instrument under test or a gas of known humidity is generated and passed through the instrument under test. Obviously, it is ideal to double-check the calibration using a known humidity and a reference hygrometer.

The most suitable reference hygrometer is the dewpoint meter, which can be readily obtained with certified calibration traceable to a standard instrument. For many applications, less sophisticated dewpoint instruments would be adequate, and coulometric analyzers are possible for low moisture levels. At high levels gravimetric methods can be used, but they are slow and tedious and difficult to make accurate.

There are a range of possible humidity sources, some of which are available commercially; the choice depends on the facilities available and the application:

1. A plastic tube permeable to moisture and held in a thermostatically controlled water bath will give a constant humidity for a given flow rate. Some manufacturers sell such tubes precalibrated for use as humidity sources, but obviously, the method is not absolute, and the permeation characteristics of the tubes may change with time.
2. Gas cylinders can be purchased with a predetermined moisture level that does not significantly drift because of the internal surface treatment of the cylinder. However, to prevent condensation in the cylinder, the maximum moisture level is limited to about 500 vppm, even with a cylinder pressure of only 10 bar. They are most suitable for spot checks of instruments on site.
3. If an inert gas containing a known concentration of hydrogen is passed through a bed of copper oxide heated to ~350°C, the hydrogen is converted to water vapor. This method relies on the measurement and stability of the hydrogen content, which is better than for moisture. The generated humidity is also independent of flow rate.

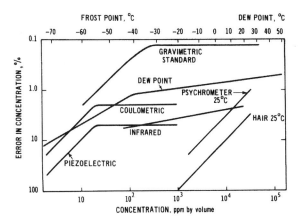

FIGURE 26.10 The accuracy of some of the major techniques for measuring moisture in gases, after Wexler (1970).

4. Water can be continuously injected into a gas stream using either an electrically driven syringe pump or a peristaltic pump. The injection point should be heated to ensure rapid evaporation. The method can be used very successfully, syringes in particular allowing a very wide range to be covered.
5. If a single humidity level can be generated, a range can be obtained using a flow mixing system, but to achieve sufficient accuracy mass-flow meters will probably be necessary.

26.4.2 Liquids

The basic absolute method is that of the Karl Fischer titration, which was described in Section 26.3.2.1.

26.4.3 Solids

There are several methods that allow the absolute moisture level of a solid to be determined, but for all of them, samples of the specific substance being measured by the process technique must be used. The most common technique is, of course, to weigh a sample, dry it, and then weigh again. Drying temperature and time depend on the material; if necessary, the temperature must be limited to avoid decomposition, loss of volatile components, or absorption of gases from the atmosphere.

Balances can be obtained with a built-in heater, which gives a direct reading of moisture content for a fixed initial sample weight. Other favored techniques include measuring the water vapor given off by absorbing it in a desiccant to avoid the effects of volatiles; the Karl Fischer method again; or mixing the substance with calcium carbide in a closed bomb and measuring the pressure of acetylene produced. The method must be carefully chosen to suit the substance and process technique being used. Finally, it is worth noting that rather than an absolute calibration, calibration directly in terms of the desired quality of the substance in the manufacturing process may be the most appropriate.

REFERENCES

Mitchell, J., and Smith, D., *Aquametry. Part I, A Treatise on Methods for the Determination of Water*, Chemical Analysis Series No. 5, Wiley, New York (1977).

Mitchell, J., and Smith, D., *Aquametry. Part 2, The Karl Fischer Reagent*, Wiley, New York (1980).

Verdin, A., *Gas Analysis Instrumentation*, Macmillan, London (1973).

Wexler, A., "Electric hygrometers," National Bureau of Standards Circular No 586 (1957).

Wexler, A. (ed.), *Humidity and Moisture* (3 volumes), papers presented at a conference, Reinhold, New York (1965).

Wexler, A., "Measurement of humidity in the free atmosphere near the surface of the Earth," *Meteorological Monographs*, 11, 262–282 (1970).

Part IV

Electrical and Radiation Measurements

Chapter 27

Electrical Measurements

M. L. Sanderson

27.1 UNITS AND STANDARDS OF ELECTRICAL MEASUREMENT

27.1.1 SI Electrical Units

The *ampere* (A) is the SI base unit (Goldman and Bell, 1982; Bailey, 1982). The Ninth General Conference of Weights and Measures (CGPM), in 1948, adopted the definition of the ampere as that constant current that, if maintained in two straight, parallel conductors of infinite length, of negligible circular cross-section, and placed 1 m apart in vacuum, would produce between these conductors a force equal to 2×10^{-7} newton per meter of length. The force/unit length, F/l, between two such conductors separated by a distance d when each is carrying a current I A is given by:

$$\frac{F}{l} = \frac{\mu_0 I^2}{2\pi d}$$

where μ_0 is the permeability of free space. Thus inherent in this definition of the ampere is the value of μ_0 as exactly $4\pi \times 10^{-7}$ N/A^2.

The derived SI electrical units are defined as follows.

The *volt* (V), the unit of potential difference and electromotive force, is the potential difference between two points of a conducting wire carrying a constant current of 1 A, when the power dissipated between these points is equal to 1 W.

The *ohm* (Ω), the unit of electrical resistance, is the electric resistance between two points of a conductor when a constant potential difference of 1 V, applied to these points, produces in the conductor a current of 1 A, the conductor not being the seat of any electromotive force.

The *coulomb* (C), the unit of quantity of electricity, is the quantity of electricity carried in 1 s by a current of 1 A.

The *farad* (F), the unit of capacitance, is the capacitance of a capacitor between the plates of which there appears a potential difference of 1 V when it is charged by a quantity of electricity of 1C.

The *henry* (H), the unit of electric inductance, is the inductance of a closed circuit in which an electromotive force of 1 V is produced when the electric current varies uniformly at the rate of 1 A/s.

The *weber* (Wb), the unit of magnetic flux, is the flux which linking a circuit of one turn would produce in it an electromotive force of 1 V if it were reduced to zero at a uniform rate in 1 s.

The *tesla* (T) is a flux density of 1 Wb/m^2.

27.1.2 Realization of the SI Base Unit

The definition of the SI ampere does not provide a suitable "recipe" for its physical realization. The realization of the ampere has thus traditionally been undertaken by means of the Ayrton-Jones current balance (Vigoureux, 1965, 1971).

The force, F_x, in a given direction between two electrical circuits carrying the same current I is given by

$$F_x = I^2 \cdot \frac{dM}{dx}$$

where M is the mutual inductance between them.

In the current balance, the force between current-carrying coils is weighed against standard masses. The principle of the balance is shown in Figure 27.1. The balance has two

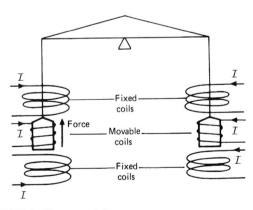

FIGURE 27.1 The current balance.

suspended coils and two pairs of fixed coils through which the same current flows. If the upper and lower coils of the fixed pair carry current in the same direction then the suspended coil experiences no force.

If, however, the currents in the coils of the fixed pair are in opposite directions, the suspended coil experiences an upward or downward force. The force, F_x, is counterbalanced by the weight of a known mass, m, and thus

$$mg = I^2 \cdot \frac{dM}{dx}$$

where g is the acceleration due to gravity.

I can be determined absolutely, that is, in terms of the mechanical base units of mass, length, and time, if dM/dx is known; dM/dx can be calculated from dimensional measurements made on the suspended and fixed coils. Changes in current direction and averaging the masses required to restore the balance condition enable the effects of interactive forces between opposite sides of the balance and of external magnetic fields to be eliminated.

Typically the accuracy of realization of the ampere using the current balance has a probable error of several parts in 10^6. One of the major causes of this inaccuracy is the relative magnitude of the force generated by the coils compared with the mass of the suspended coils. Alternative techniques for the absolute determination of the ampere have been suggested. These include the use of the proton gyromagnetic ratio, γ_p, in conjunction with weak and strong magnetic field measurements (Dix and Bailey, 1975; Vigoureux, 1971) and the measurement of the force on a coil in a magnetic field, together with the measurement of the potential induced when the coil moves in the same magnetic field (Kibble et al., 1983).

27.1.3 National Primary Standards

Because the accuracy of realization of the SI ampere by the current balance is significantly poorer than the precision of intercomparison of standard cells and resistors, and because of the difficulty of storing the realized value of the ampere, most National Standards Laboratories use standard cell banks and resistors as their maintained primary standards. Intercomparison of these national standards is regularly made through the International Bureau of Weights and Measures (BIPM) in Sevres, France. Figure 27.2 (taken from Dix and Bailey, 1975) shows the U.K. primary standards, which are maintained by the National Physical Laboratory (NPL). This figure also shows the relationships of the primary standards to the absolute reference standards; to the national low-frequency AC standards; and to the primary standards of other countries. Table 27.1 lists the U.K. national DC and low-frequency standards apparatus. Radio frequency and microwave standards at NPL are listed in Table 27.2 (Steele et al., 1975). Submillimeter wave measurements and standards are given by Stone et al. (1975).

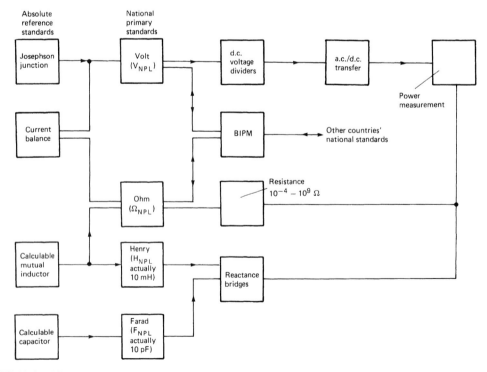

FIGURE 27.2 U.K. National Standards (from Dix and Bailey, 1975).

CHAPTER | 27 Electrical Measurements

TABLE 27.1 U.K. DC and low-frequency standards (from Dix and Bailey, 1975)

Absolute reference standards	National primary standards	Other national standards apparatus		
		Diesselhorst	Volt ratio box	
		Potentiometer (1 in 10^8)	Voltage dividers	Power measurement
Josephson-junction system (1 in 10^7)	Standard cells (3 in 10^8)	Cell comparator (1 in 10^8)	DC/AC thermal transfer Electrostatic voltmeter Inductive dividers	Electrostatic wattmeter Dynamometer wattmeters Calibrated loads
	Standard 1 Ω resistors (1 in 10^7)	Wheatstone bridge (1 in 10^8) Current comparator (1 in 10^8) Build-up resistors (2 in 10^8) Standard resistors	High-current bridge High-resistance bridge Potentiometer Current comparator potentiometer Standard resistors 10^{-4}–10^9 Ω	Electronic sources and amplifiers Rotary generators Reference measurement transformers Transformer measurement systems
Campbell mutual inductor 10 mH (1 in 10^6)		Inductance bridge		Magnetic measurement
		Standard inductors 1 μH to 10 H (2 in 10^5)		Permeameters Vibrating-coil magnetometer
Current balance		Phase-angle standards for L, C and R		Magnetic susceptibility balance Epstein-square magnetic-loss system Local-loss tester
		Capacitance bridge		
Calculate capacitor 0.4 pF (2 in 10^7)	Standard capacitors 10 pF (2 in 10^7)	Standard capacitors 10 pF to 1 nF (5 in 10^7)	Standard capacitors 10 nF to 1 μF	Magnetic-tape calibration

TABLE 27.2 U.K. RF and microwave standards (from Steel et al., 1975)

Quantity	Method	Frequency (GHz)	Level	Uncertainty (95% confidence)
Power in 14 mm coaxial line	Twin calorimeter	0–8.5	10–100 mW	0.2–0.5%
Power in 7 mm coaxial line	Twin calorimeter	0–18	10–100 mW	Under development
Power in WG16 (WR90)	Microcalorimeter	9.0, 10.0, 12.4	10–100 mW	0.2%
Power in WG18 (WR62)	Microcalorimeter	13.5, 15.0, 17.5	10–100 mW	0.2%
Power in WG22 (WR28)	Microcalorimeter	35	10–100 mW	0.5%
Power in WG26 (WR12)	Twin calorimeter	70	10 mW	0.8%
Attenuation	w.b.c.o. piston	0.0306	0–120 dB	0.002 dB
Attenuation in 14 mm coaxial line	w.b.c.o. piston	0–8.5	0–80 dB	0.001 dB/10 dB
Attenuation in WG11A (WR229)	Modulated subcarrier		0–100 dB	From 0.002 dB at low values up to 0.02 dB at 100 dB, for VSWR <1.05
WG15 (WR112)	Modulated subcarrier		0–100 dB	
WG16 (WR90)	Modulated subcarrier		0–100 dB	
WG18 (WR62)	Modulated subcarrier		0–100 dB	
WG22 (WR28)	Modulated subcarrier		0–100 dB	
WG26 (2512)	Modulated subcarrier		0–100 dB	

(Continued)

TABLE 27.2 (Continued)

Quantity	Method	Frequency (GHz)	Level	Uncertainty (95% confidence)
Impedance				
Lumped conductance	RF bridge	1×10^{-3}	10 µS–1S	0.1%
Lumped capacitance	RF bridge	1×10^{-3}	1 pF–10 µF	0.1%
Coaxial conductance	Woods bridge	$5 \times 10^{-3} - 30 \times 10^{-3}$	0–40 mS	0.1% + 0.001 mS
		$30 \times 10^{-3} - 200 \times 10^{-3}$		0.2% + 0.001 mS
Coaxial capacitance	Woods bridge	5×10^{-3}	0–40 pF	0.1% + 0.001 pF
		(these refer to major components only)		
Noise temperature			K	
in 14 mm coaxial line	Thermal	1–2	10^4	
in WG10 (WR284)	Thermal	2.75, 3.0, 3.5	10^4	About 1.5 K
in WG14 (WR137)	Thermal	6.0, 7.0, 8.0	10^4	transfer standards
in WG16 (WR90)	Thermal	9.0, 10.0, 11.2	10^4	calibrate to 110 K
in WG18 (WR62)	Thermal	13.5, 15.0	10^4	0.15 K; transfer
in WG22 (WR28)	Thermal	35	10^4	standards calibrate
in WG11A (WR229)	Cryogenic		77	to 0.6 K
in WG15 (WR112)	Cryogenic		77	

These electrical standards are similar to standards held by other national laboratories—for example, the National Institute for Standards and Testing (NIST) in the United States, Physikalisch-Technische Bundesanstalt (PTB) in West Germany, and others elsewhere.

27.1.3.1 Standard Cells

The U.K. primary standard of voltage is provided by a bank of some 30 Weston saturated mercury cadmium cells, the construction of a single cell being shown in Figure 27.3. The electrodes of the cell are mercury and an amalgam of cadmium and mercury. The electrolyte of cadmium sulphate is kept in a saturated condition over its operating temperature range by the presence of cadmium sulphate crystals. The pH of the electrolyte has a considerable effect on the stability of the EMF of the cell and has an optimal value of 1.4 ± 0.2 (Froelich, 1974). The mercurous sulphate paste over the anode acts as a depolarizer. For details concerning the construction, maintenance, and characteristics of such cells and their use, the reader is directed to the NBS monograph listed in the References.

The nominal value of the EMF generated by the saturated Weston cell is 1.01865 V at 20°C. Cells constructed from the same materials at the same time will have EMFs differing by only a few µV. Cells produced at different times will have EMFs differing by between 10 and 20 µV. The stability of such cells can be on the order of a few parts in 10^7 per year. They can be intercompared by back-to-back measurements to 1 part in 10^8. The internal resistance of the cell is approximately 750 Ω.

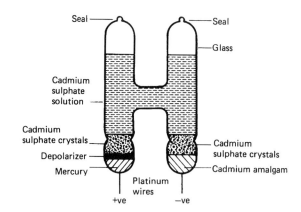

FIGURE 27.3 Weston standard cell.

The variation of the cell EMF with temperature can be described by the equation

$$V_T = V_{20} - 4.06 \times 10^{-5}(T - 20) \\ - 9.07 \times 10^{-7}(T - 20)^2 + 6.6 \times 10^{-9}(T - 20)^3 \\ - 1.5 \times 10^{-10}(T - 20)^4$$

where V_T is the EMF of the cell at a temperature T°C and V_{20} is its EMF at 20°C. For small temperature variations about 20°C, the cell has a temperature coefficient of -40.6 µV/K.

To produce a source of EMF with high stability, it is necessary to maintain the cells in a thermostatically controlled enclosure. At NPL the standard cell enclosure contains up to 54 cells housed in groups of nine in separate copper containers in an air enclosure that has a temperature stability of better than 1 mK/h and a maximum temperature difference between any two points in the enclosure of less than 5 μK. Measurement of the EMFs of the cells is effected under computer control.

27.1.3.2 Monitoring the Absolute Value of National Voltage Standards by Means of the Josephson Effect

Although intercomparison of standard cells can be undertaken to a high degree of precision and such intercomparisons demonstrate that standard cells can be produced that show a high degree of stability with respect to each other, such measurements do not guarantee the absolute value of such cells. The Josephson effect (Josephson, 1962) is now used widely as a means of monitoring the absolute value of national standards of voltage maintained by standard cells to be related to frequency, f, and the Josephson constant $2e/h$; e is the charge on the electron and h is Planck's constant.

The Josephson junction effect shown in Figure 27.4 predicts that if a very thin insulating junction between two superconductors is irradiated with RF energy of frequency f, the voltage-current relationship will exhibit distinct steps, as shown in Figure 27.4(b). The magnitude of one voltage step is given by

$$\Delta V = \frac{h}{2e} \cdot f$$

Thus voltage is related by the Josephson effect to the frequency of the RF radiation and hence to the base unit time. A value of 483,594.0 GHz/V has been ascribed to the Josephson constant, $2e/h$, with an uncertainty of ±5 parts in 10^7.

The insulating junction can be produced in several ways, one of the simplest being to produce a dry solder joint between two conductors. For an irradiation frequency of 10 GHz, the voltage steps are approximately 20 μV. Using the potential difference between a number of steps, it is possible to produce a usable voltage of a few millivolts. Figure 27.5 shows a system employing another application of the Josephson junction known as a superconducting quantum interferometric detector (squid) as the detector in a superconducting potentiometer. This technique enables the comparison of the Josephson junction EMF with the EMF of a standard cell to be made with an accuracy of 1 part in 10^7. Further details of the techniques involved can be found in Dix and Bailey (1975).

27.1.3.3 Standard Resistors

The desirable characteristics of standard resistors are that they should be stable with age, have a low temperature coefficient

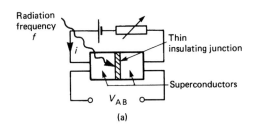

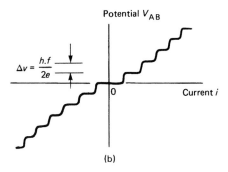

FIGURE 27.4 (a) Josephson junction effect; (b) voltage/current characteristic of Josephson junction.

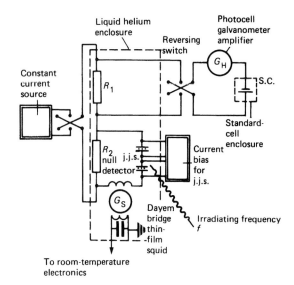

FIGURE 27.5 Voltage comparison system using Josephson junction (from Dix and Bailey, 1975).

of resistance, and be constructed of a material that exhibits only small thermoelectric EMF effects with dissimilar materials. The U.K. national primary standard of resistance consists of a group of standard 1-Ω resistors wound from Ohmal, an alloy with 85 percent copper, 11 percent manganese, and 4 percent nickel, and freely supported by combs on a former. The resistors are immersed in oil. With such resistors it is possible to obtain a stability of 1 part in 10^7/yr.

27.1.3.4 Absolute Determination of the Ohm

The ohm can be determined absolutely by means of the Campbell mutual inductor, whose mutual inductance can

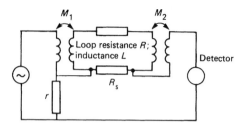

FIGURE 27.6 Campbell bridge.

be determined from geometric measurements made on the coils forming the inductor (Rayner, 1967). When such mutual inductors are used in Campbell's bridge, as shown in Figure 27.6, the balance conditions are

$$R \cdot r + \omega^2 \cdot M_1 \cdot M_2 = 0$$

and

$$M_1 \cdot R_s = L \cdot r$$

where L is the loop inductance and R is its resistance.

Thus the first equation can be used to determine the product $R \cdot r$ in terms of the SI base units of length and time. The ratio of the two resistances R and r can be found using a bridge technique, and thus r can be determined absolutely. This absolute determination has a probable error of 2 parts in 10^6.

An alternative method for the absolute determination of the ohm employs the Thompson-Lampard calculable capacitor (Thompson and Lampard, 1956). This capacitor has a value that can be determined from a knowledge of the velocity of light and a single length measurement.

Consider a cylindrical electrode structure with the symmetrical cross-section shown in Figure 27.7 (a) in which neighboring electrodes are separated only by small gaps; Thompson and Lampard showed that the cross-capacitances per unit length C_1 and C_2 are related by

$$\exp\left(-\frac{\pi C_1}{\varepsilon_0}\right) + \exp\left(-\frac{\pi C_2}{\varepsilon_0}\right) = 1$$

Because of symmetry $C_1 = C_2$ and the cross-capacitance per meter, C, is given by

$$C = \frac{\varepsilon_0 \log_e 2}{\pi} \text{ F/m}$$

Since the velocity of light, c, is given by

$$c^2 = \frac{1}{\varepsilon_0 \mu_0}$$

and the value of μ_0 is, by definition, $4\pi \times 10^{-7}$, then if the velocity of light is known the capacitance per meter of the capacitor can be determined. C has a value of 1.9535485 pF/m.

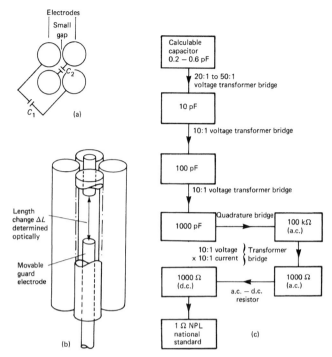

FIGURE 27.7 (a) Cross-section of Thompson-Lampard capacitor; (b) variable capacitor; (c) comparison chain for Thompson-Lampard capacitor.

By inserting a movable guard electrode as shown in Figure 27.7 (b), the position of which can be determined by means of an optical interference technique, it is possible to generate changes in capacitance that can be determined absolutely. The change in capacitance obtained can be compared with the capacitance of a standard 10-pF capacitor and hence, by means of the chain shown in Figure 27.7 (c), used to determine the absolute value of the ohm. The accuracy of this determination is typically 1 part in 10^{-7}.

27.2 MEASUREMENT OF DC AND AC CURRENT AND VOLTAGE USING INDICATING INSTRUMENTS

The most commonly used instruments for providing an analog indication of direct or alternating current or voltage are the permanent magnet-moving coil, moving iron, and dynamometer instruments. Other indicating instruments include thermocouple and electrostatic instruments, the latter based on the attraction between two charged plates. This section provides a description of the basic principles of operation of such instruments. Further details can be found in Golding and Widdis (1963), Harris (1966), Gregory (1973), and Tagg (1974). The accuracy specification and the assessment of influence factors on direct-acting indicating electrical measuring instruments and their accessories are set out in BSI 89:1977 (British Standards Institution, 1977). This is equivalent to IEC 51:1973.

27.2.1 Permanent Magnet-Moving Coil Instruments

Permanent magnet-moving coil instruments are based on the principle of the D'Arsonval moving-coil galvanometer, the movement of which is also used in light spot galvanometers and pen and ultraviolet recorders. A typical construction for a moving-coil instrument is shown in Figure 27.8(a). The current to be measured is passed through a rectangular coil wound on an insulated former, which may be of copper or aluminum, to provide eddy-current damping. The coil is free to move in the gap between the soft iron pole pieces and core of a permanent magnet employing a high-coercivity material such as Columax, Alcomax, or Alnico. The torque produced by the interaction of the current and the magnetic field is opposed by control springs that are generally flat or helical phosphor-bronze springs. These also provide the means by which the current is supplied to the coil. The bearings for the movement are provided by synthetic sapphire jewels and silver-steel or stainless-steel pivots. Alternative means of support can be provided by a taut band suspension, as shown in Figure 27.8(b). This has the advantage of removing the friction effects of the jewel and pivot but is more susceptible to damage by shock loading. The pointer is usually a knife-edge one, and for high-accuracy work it is used in conjunction with a mirror to reduce parallax errors.

The torque, T_g, generated by the interaction of the current, i, and the magnetic field of flux density B is given by

$$T_g = N \cdot B \cdot h \cdot b \cdot i$$

where h and b are the dimensions of the coil having N turns. This is opposed by the restoring torque, T_r, produced by the spring

$$T_r = k\theta$$

where k is the spring constant.

Under static conditions, these two torques are equal and opposite, and thus

$$\theta = \frac{N \cdot B \cdot h \cdot b \cdot i}{k} S \cdot i$$

where S is the sensitivity of the instrument.

Under dynamic conditions, the generated torque, T_g, is opposed by inertial, damping, and spring-restoring forces, and thus

$$T_g = J\frac{d^2\theta}{dt^2} + D \cdot \frac{d\theta}{dt} + k\theta$$

where J is the inertia of the moving system, D is its damping constant, and k is the spring constant.

Damping can be provided by an air damper or by eddy-current damping from the shorted turn of the former or from the coil and external circuit. For eddy-current damping,

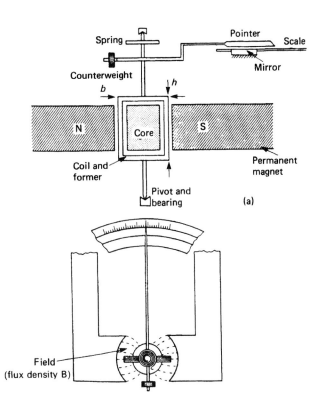

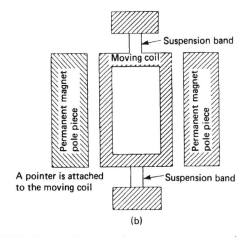

FIGURE 27.8 (a) Elements of a permanent magnet-moving coil instrument; (b) taut band suspension system.

$$D = h^2 b^2 B^2 \left(\frac{N^2}{R} + \frac{1}{R_f}\right)$$

where R represents the resistance of the coil circuit and R_f represents the resistance of the coil making up the former.

The instrument thus has a second-order transfer function given by

$$G(s) = \frac{\theta(s)}{i(s)} = \frac{\left(\frac{k}{J}\right) \cdot S}{s^2 + \left(\frac{D}{J}\right) \cdot s + \left(\frac{k}{J}\right)}$$

Comparing this transfer function with the standard second-order transfer function

$$G(s) = \frac{k\omega_n^2}{s^2 + 2\xi\omega_n s + \omega_n^2}$$

The natural frequency of the instrument is given by $\omega_n = \sqrt{(k/j)}$, and its damping factor $\xi = D/2\sqrt{(J/k)}$.

If $D^2 > 4kJ$, then $\xi > 1$ and the system is overdamped. The response to a step input of current magnitude I at $t = 0$ is given by

$$\theta(t) = S \cdot I \left\{ 1 - \frac{\xi + \sqrt{(\xi^2 - 1)}}{2\sqrt{(\xi^2 - 1)}} e^{[-\xi + \sqrt{(\xi^2-1)}]\omega_n t} \right.$$

$$\left. + \frac{\xi - \sqrt{(\xi^2 - 1)}}{2\sqrt{(\xi^2 - 1)}} e^{[-\xi + \sqrt{(\xi^2-1)}]\omega_n t} \right\}$$

If $D^2 = 4kJ$, then $\xi = 1$ and the system is critically damped. The response to the step input is given by

$$\theta(t) = S \cdot I[1 - (1 + \omega_n t)e^{-\omega_n t}]$$

If $D^2 < 4kJ$, then $\xi < 1$ and the system is underdamped. The response to the step input is given by

$$\theta(t) = S \cdot I \cdot \left\{ 1 - \frac{e^{-\xi \omega_n t}}{\sqrt{(1 - \xi^2)}} \cdot \sin\left[\sqrt{(1 - \xi^2)} \cdot \omega_n t + \phi\right] \right\}$$

and $\phi = \cos^{-1} \xi$.

These step responses are shown in Figure 27.9.

27.2.1.1 Range Extension

The current required to provide full-scale deflection (FSD) in a moving-coil instrument is typically in the range of 10 µA to 20 mA. DC current measurement outside this range is provided by means of resistive shunts, as shown in Figure 27.10(a). The sensitivity of the shunted ammeter, S_A, is given by

$$S_A = \frac{\theta}{I} = \frac{R_p}{R_p + R_m} \cdot S$$

where R_p is the resistance of the shunt, R_m is the resistance of the coil and swamping resistance, and S is the sensitivity of the unshunted movement.

High-current ammeters usually employ a movement requiring 15 mA for FSD. The shunts are usually four-terminal devices made of manganin. The voltage drop across the instrument is 0.075 V and thus the power dissipated in the shunt is approximately 0.075 IW. Table 27.3 gives the power dissipation in the shunt for various current ratings.

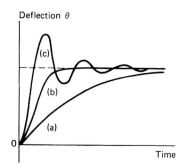

FIGURE 27.9 Second-order system responses. (a) Overdamped; (b) critically damped; (c) underdamped.

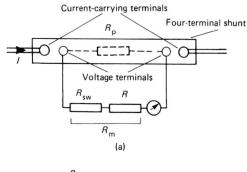

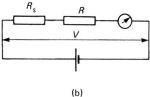

FIGURE 27.10 (a) Ammeter using a moving-coil instrument; (b) voltmeter using a moving-coil instrument.

TABLE 27.3 Power dissipated in shunt for various current ratings

Current (A)	Power dissipated (W)
1	0.075
2	0.150
5	0.375
10	0.75
20	1.50
50	3.75
100	7.50
200	15.00
500	37.50
1,000	75.00

CHAPTER | 27 Electrical Measurements

For use as a DC voltmeter the sensitivity, S_v, is given by

$$S_v = \frac{\theta}{V} = \frac{S}{R_s + R}$$

where R_s is the series resistance, R is the resistance of the coil, and S is the sensitivity of the movement (see Figure 27.10(b)).

The value of the series resistance depends on the sensitivity of the moving coil. For a movement with an FSD of 10 mA, it is 100 Ω/V. If FSD requires only 10 µA, the resistance has a value of 100,000 Ω/V. Thus for a voltmeter to have a high input impedance, the instrument movement must have a low current for FSD.

27.2.1.2 Characteristics of Permanent Magnet-Moving Coil instruments

Permanent magnet-moving coil instruments have a stable calibration, low power consumption, and a high torque-to-weight ratio and can provide a long uniform scale. They can have accuracies of up to 0.1 percent of FSD. With the use of shunts or series resistors, they can cover a wide range of current and voltage. The errors due to hysteresis effects are small, and they are generally unaffected by stray magnetic fields. It is possible to adjust the damping in such instruments to any required value. The major errors are likely to be caused by friction in the bearings and changes in the resistance of the coil with temperature. Copper wire, which is used for the coil, has a temperature coefficient of +0.4%/K. When used as a voltmeter, this temperature variation is usually swamped by the series resistance. When used as an ammeter with manganin shunts, it is necessary to swamp the coil resistance with a larger resistor, usually manganin, as shown in Figure 27.10(a). This has the effect of more closely matching the temperature coefficient of the coil/swamp resistance combination to that of the shunt, thus effecting a constant current division between the instrument and the shunt over a given temperature range.

27.2.1.3 AC Voltage and Current Measurement Using Moving-Coil Instruments

The direction of the torque generated in a moving-coil instrument is dependent on the instantaneous direction of the current through the coil. Thus an alternating current will produce no steady-state deflection.

Moving-coil instruments are provided with an AC response by the use of a full-wave bridge rectifier, as shown in Figure 27.11. The bridge rectifier converts the AC signal into a unidirectional signal through the moving-coil instrument, which then responds to the average DC current through it. Such instruments measure the mean absolute value of the waveform and are calibrated to indicate the RMS value of the wave on the assumption that it is a sinusoid.

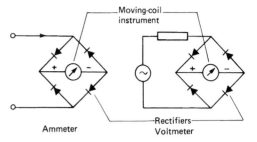

FIGURE 27.11 AC current and voltage measurement using a rectifier-moving-coil instrument.

For a periodic current waveform $I(t)$ through the instrument, the mean absolute value, I_{mab}, is given by

$$I_{mab} = \frac{1}{T}\int_0^T |I(t)| \cdot dt$$

and its RMS value is given by

$$I_{rms} = \sqrt{\left[\frac{1}{T}\int_0^T I^2(t) \cdot dt\right]}$$

where T is the period of the wave.

The form factor (FF) for the current waveform is defined as

$$FF = \frac{I_{rms}}{I_{mab}} \sqrt{\frac{\left[\left(\frac{1}{T}\right)\int_0^T I^2(t) \cdot dt\right]}{\left(\frac{1}{T}\right)\int_0^T |I(t)| \cdot dt}}$$

For a sinusoid $I(t) = \hat{I}.\sin \omega t$ the rms value is $\hat{I}/\sqrt{2}$ and its mean absolute value is $2\hat{I}/\pi$. The FF for a sinusoid is thus 1.11. Rectifier instruments indicate 1.11. I_{mab}. For waveforms that are not sinusoidal, rectifier instruments will provide an indication; they will have an error of

$$\left(\frac{1.11 - FF}{FF}\right) \times 100\%$$

Figure 27.12 shows several waveforms with their form factors and the errors of indication that occur if they are measured with mean absolute value measuring-RMS scaled instruments. This FF error also occurs in the measurement of AC current and voltage using digital voltmeters that employ rectification for the conversion from AC to DC.

As current-measuring devices, the diodes should be selected for their current-carrying capability. The nonlinear characteristics of the diodes make range extension using shunts impractical, and therefore it is necessary to use rectifier instruments with current transformers (see Section 27.2.3). The forward diode drop places a lower limit on the voltage, which can be measured accurately and gives such instruments a typical minimum FSD of 10 V. When used as a voltmeter, the variation of the diode forward drop with temperature can provide the instrument with a sensitivity to ambient temperature. It is possible to design such instruments to provide an accuracy of 1 percent of FSD from 50 Hz to 10 kHz.

Waveshape	Form Factor	Percentage error in measurement using mean sensing-rms indicating instruments
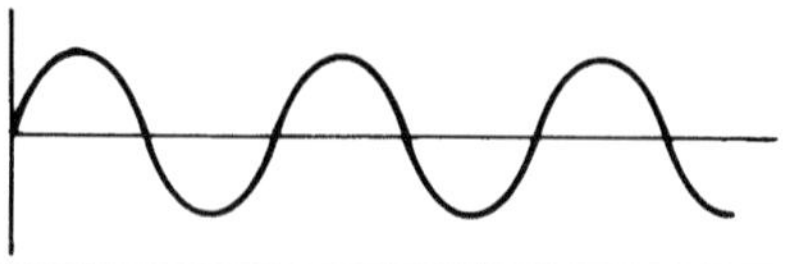 (sine wave)	1.11	0
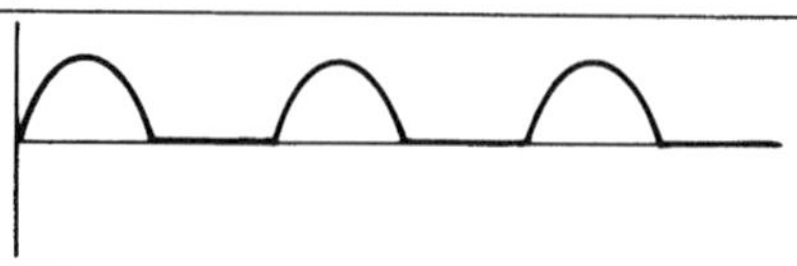 (half-wave rectified)	1.57	−29.3
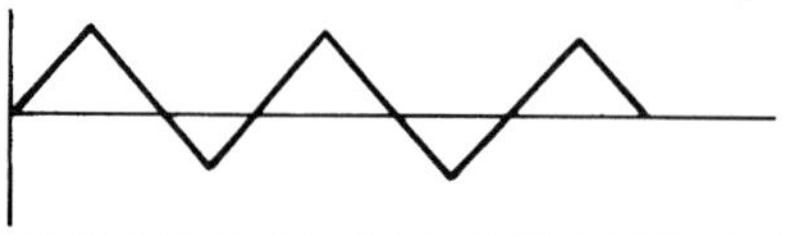 (triangular)	1.15	−3.96
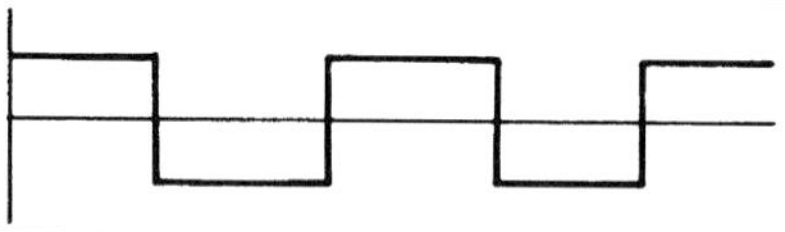 (square wave)	1	+11.1
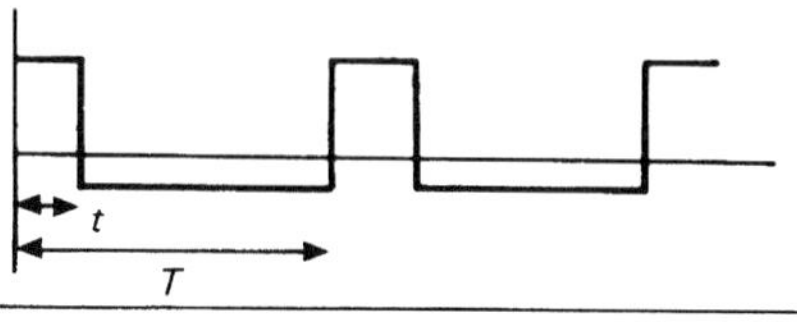 (pulse, K = t/T)	$\frac{1}{2}\sqrt{\left[\frac{1}{K(1-K)}\right]}$	$\{2.22\sqrt{[K(1-K)]} - 1\} \times 100$

FIGURE 27.12 Waveform form factors and errors of indication for rectifier instruments.

27.2.1.4 Multimeters

These are multirange devices using a permanent magnet-moving coil instrument. They enable the measurement of DC and AC current and voltage and resistance. One of the most common instruments of this type is the AVO-Biddle Model 8 Mark 6 (Thorn-EMI). Table 27.4 gives the specification for this instrument and Figure 27.13 shows the circuit diagram. The basic movement has a full-scale deflection of 50 μA; therefore, this gives the instrument a sensitivity of 20,000 Ω/V on its DC voltage ranges. The three ranges of resistance operate by measuring the current passing through the resistance on applying a DC voltage supplied from internal batteries. A zero control on these ranges, used with the instrument probes shorted together, enables compensation for changes in the EMF of the internal batteries to be made.

27.2.1.5 Electronic Multimeters

By using the electronic input in Figure 27.14(a) it is possible to achieve a high input impedance irrespective of the voltage range. This is used as shown in Figure 27.14(b) to measure current, resistance, and AC quantities. For current measurement the maximum input voltage can be made to be the same on all ranges. Resistance measurements can be made with lower voltage drops across the resistors and with a linear indication. AC quantities are measured using rectification and mean or peak sensing. Table 27.5 gives the specification of such an instrument (Hewlett-Packard HP410C General Purpose Multi-Function Voltmeter).

27.2.2 Moving-Iron Instruments

There are two basic types of moving-iron instrument: the attraction and repulsion types, both shown in Figure 27.15. In the attraction type a piece of soft iron in the form of a disc is attracted into the coil, which is in the form of a flat solenoid. Damping of the instrument is provided by the air-damping chamber. The shape of the disc can be used to control the scale shape. In the repulsion instrument two pieces of iron, either in the form of rods or vanes, one fixed and the other movable, are magnetized by the field current to be measured. In both instruments the torque, T_g, generated by the attraction or repulsion is governed by

$$T_g = \frac{1}{2} \cdot \frac{dL}{d\theta} \cdot i^2$$

where L is the inductance of the circuit. The restoring torque, T_r, is produced by a spring:

$$T_r = k\theta$$

CHAPTER | 27 Electrical Measurements

TABLE 27.4 Multimeter specification

DC voltage	8 ranges: 100 mV, 3, 10, 30, 100, 300, 600 V, 1 kV
DC current	7 ranges: 50 µA, 300 µA, 1, 10, 100 mA, 1A, and 10 A
AC voltage	7 ranges: 3, 10, 30, 100, 300, 600 V, 1 kV
AC current	4 ranges: 10 mA, 100 mA, 1 A, and 10 A
Resistance	3 ranges: ×1:0–2 kΩ ×100:0–200 kΩ ×10k:0–20 MΩ
Source for resistance measurement	One 15 V type B 121 battery (for ×10k range) One 1.5 V type SP2 single cell (for x1, ×100 range)
Accuracy	DC ±% fsd AC (150Hz) ±2% fsd Resistance ±3% center scale
Sensitivity	DC 20,000 Ω/V all ranges AC 100 Ω/V 3 V range 1,000 Ω/V 10 V range 2,000 Ω/V all other ranges
Overload protection	High-speed electromechanical cut-out with a fuse on the two lower resistance ranges
Decibels	−10 to +55 using AC voltage scale
Voltage drop at terminals	DC 100 mV on 50 µA range, approx. 400 mV on other ranges AC less than 450 mV at 10 A
Frequency response AC voltage range (up to 300 V)	< ±3% discrepancy between 50 Hz reading and readings taken between 15 Hz and 15 Hz

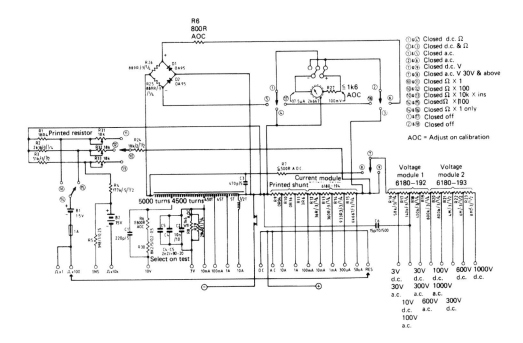

FIGURE 27.13 Multimeter. Courtesy of Thorn EMI Instruments Ltd.

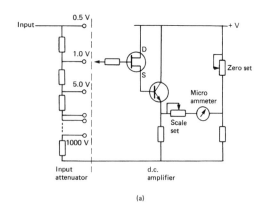

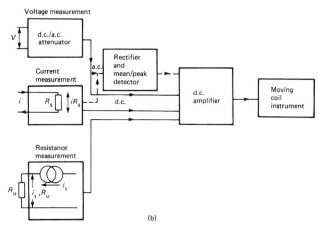

FIGURE 27.14 Electronic multimeter, (a) Electronic input; (b) schematic of electronic multimeter.

and thus

$$\theta = \frac{1}{2} \cdot \frac{1}{k} \cdot \frac{dL}{d\theta} \cdot i^2$$

The deflection of the instrument is proportional to the mean square of the current and thus the instrument provides a steady-state deflection from an AC current. The scales of such instruments are usually calibrated in terms of RMS values and they tend to be nonlinear, being cramped at the lower end.

Friction in the bearings of the instrument causes error. Hysteresis effects in the iron of the instrument give rise to different indications for increasing and decreasing current. Errors can also be caused by the presence of stray magnetic fields. Variation in ambient temperature causes changes in the mechanical dimensions of the instrument, alters the permeability of the iron, and changes the resistance of the coil. This last effect is the most important. Used as an ammeter, the change in resistance causes no error, but when used as a voltmeter the change in resistance of the copper winding of +0.4%/K causes the sensitivity of the voltmeter to change. This effect is usually reduced by using a resistance in series with the coil wound with a wire having a low temperature coefficient. The inductance of the instrument can also cause changes in its sensitivity with frequency when used as a voltmeter. This is shown in Figure 27.16(a). At a given angular frequency w, the error of reading of the voltmeter is given by $(\omega^2 L^2)/(2R^2)$, where L is its inductance and R its

TABLE 27.5 Electronic multimeter specification

DC voltmeter
Voltage ranges: ±15 mV to ±1500 V full scale in 15, 50 sequence (11 ranges) Accuracy: ±2% of full scale on any range Input resistance: 100 MΩ ± 1% on 500 mV range and above, 10 MΩ ± 3% on 150 mV range and below
AC voltmeter
Voltage ranges: 0.5 V to 300 V full scale in 0.5, 1.5, 5 sequence (7 ranges) Frequency range: 20 Hz to 700 MHz Accuracy: ±3% of full scale at 400 Hz for sinusoidal voltages from 0.5 V–300 V rms. The AC probe responds to the positive peak-above-average value of the applied signal. The meter is calibrated in RMS Frequency response: ±2% from 100 Hz to 50 MHz (400 Hz ref.); 0 to −4% from 50 MHz to 100 MHz; ±10% from 20 Hz to 100 Hz and from 100 MHz to 700 MHz Input impedance; input capacitance 1.5 pF, input resistance >10 MΩ at low frequencies. At high frequencies, impedance drops off due to dielectric loss Safety: the probe body is grounded to chassis at all times for safety. All AC measurements are referenced to chassis ground
DC ammeter
Current ranges: ±1.5 μA to ±150 mA full scale in 1.5, 5 sequence (11 ranges) Accuracy: ±3% of full scale on any range Input resistance: decreasing from 9 kΩ on 1.5 μA range to approximately 0.3 Ω on the 150 mA range Special current ranges: ±1.5, ±5 and ±15 μA may be measured on the 15, 50, and 150 mV ranges using the DC voltmeter probe, with ±5% accuracy and 10 MΩ input resistance

(Continued)

CHAPTER | 27 Electrical Measurements

> **TABLE 27.5** (Continued)
>
> Ohmmeter
>
> Resistance range: resistance from 10 Ω to 10 MΩ center scale (7 ranges)
> Accuracy: zero to midscale: ±5% of reading of ±2% of midscale, whichever is greater; ±7% from midscale to scale value of 2; ±8% from scale value of 2 to 3; ±9% from scale value to 3 to 5; ±10% from scale value of 5 to 10
> Maximum input: DC: 100 V on 15, 50 and 150 mV ranges, 500 V on 0.5 to 15 V ranges, 1,600 V on higher ranges
> AC: 100 times full scale or 450 V p, whichever is less

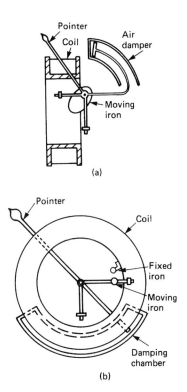

FIGURE 27.15 Moving-iron instrument, (a) attraction; (b) repulsion (from Tagg, 1974).

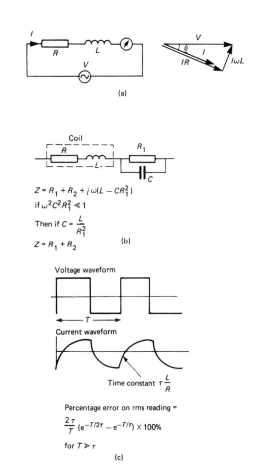

FIGURE 27.16 (a) Inductance effects in moving-iron voltmeters (b) compensation for effect of inductance; (c) errors in measurement of nonsinusoidal waveforms.

resistance. Figure 27.16(b) shows a compensation method for this error.

Although the moving-iron instrument is a mean square indicating instrument errors can be introduced in measuring the RMS value of a nonsinusoidal voltage waveform. These errors are caused by the peak flux in the instrument exceeding the maximum permitted flux and by attenuation of the harmonic current through the instrument by the time constant of the meter, as shown in Figure 27.16(c).

Moving-iron instruments are capable of providing an accuracy of better than 0.5 percent of FSD. As ammeters they have typical FSDs in the range of 0.1–30 A without shunts. The minimum FSD, when they are used as voltmeters, is typically 50 V with a low input impedance of order 50 Ω/V. Their frequency response is limited by their high inductance and stray capacitance to low frequencies, although instruments are available that will measure at frequencies up to 2,500 Hz. Moving-iron instruments have relatively high power requirements and therefore they are unsuitable for use in high-impedance AC circuits.

27.2.3 AC Range Extension Using Current and Voltage Transformers

In Section 27.2.1.1, extension of the range of permanent magnet-moving coil instruments using current shunt and resistive voltage multipliers was described. The same techniques can be applied in AC measurements. However, in power measurements with large currents, the power dissipated in the shunt becomes significant (see Table 27.3). For high voltage measurements, the resistive voltage multiplier provides no isolation for the voltmeter. For these reasons, range extension is generally provided by the use of current and voltage transformers. These enable single-range ammeters and voltmeters, typically with FSDs of 5 A and 110 V, respectively, to be used.

The principle of the current transformer (CT) is shown in Figure 27.17(a) and its equivalent circuit is shown in Figure 27.17(b). The load current being measured flows through the primary winding while the ammeter acts as a secondary load. The operation of the CT depends on the balance of the ampere turns (the product of current and turns) produced by the primary and secondary windings. If the transformer is ideal with no magnetizing current or iron loss. then

$$\frac{I_P}{I_S} = n_{ct}$$

where n_{ct} is the current transformer turns ratio given by

$$n_{ct} = \frac{n_S}{n_P}$$

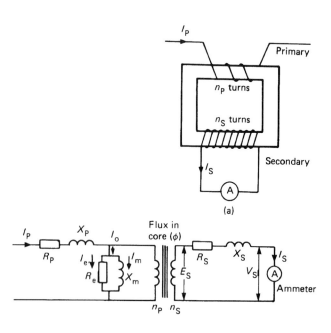

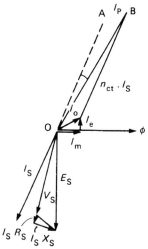

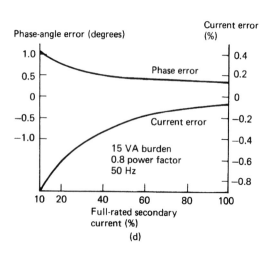

FIGURE 27.17 (a) Current transformer; (b) equivalent circuit; (c) phasor diagram of current transformer; (d) current ratio and phase-angle errors for current transformer.

CHAPTER | 27 Electrical Measurements

The CT is generally constructed with a toroidal core of a high-permeability, low-loss material such as mumetal or strip-wound silicon steel. This construction minimizes the magnetizing current, iron loss, and leakage flux, ensuring that the actual primary to secondary current ratio is close to the inverse-turns ratio.

Figure 27.17(c) shows the effect of magnetizing current and iron loss on the relative magnitudes and phases of the primary and secondary currents.

Two errors of CTs can be identified in Figure 27.17(c). These are the current or ratio error and the phase angle error or phase displacement. The current or ratio error is defined as

$$\frac{\text{Rated ratio } (I_P/I_S) - \text{actual ratio } (I_P/I_S)}{\text{Actual ratio } (I_P/I_S)} \times 100\%$$

The phase-angle error or phase displacement is the phase angle between the primary and secondary current phasors, drawn in such a way (as in Figure 27.17(c)) that for a perfect transformer there is zero phase displacement. When the secondary current leads the primary current, the phase displacement is positive.

These errors are expressed with respect to a particular secondary load that is specified by its burden and power factor. The burden is the VA rating of the instrument at full load current. A typical burden may be 15 VA with a power factor of 0.8 lagging. Figure 27.17(d) shows typical current and phase angle errors for a CT as a function of secondary load current. BS 3938:1973 sets limits on ratio and displacement errors for various classes of CT (British Standards Institution, 1973).

The ampere turn balance in the current transformer is destroyed if the secondary circuit is broken. Under these circumstances a high flux density results in the core, which will induce a high voltage in the secondary winding. This may break down the insulation in the secondary winding and prove hazardous to the operator. It is therefore important not to open-circuit a current transformer while the primary is excited.

Voltage transformers (VTs) are used to step down the primary voltage to the standard 110 V secondary voltage. Figure 27.18(a) shows the connection of such a transformer; Figure 27.18(b) shows its equivalent circuit. For an ideal transformer,

$$\frac{V_P}{V_S} = n_{vt}$$

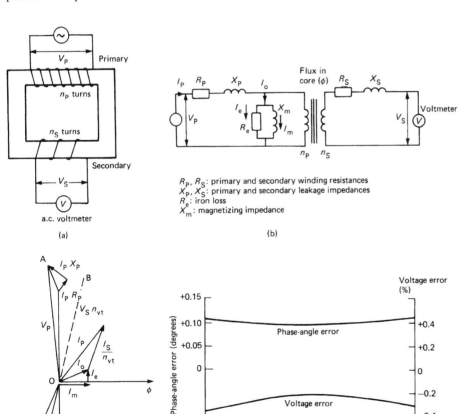

FIGURE 27.18 (a) Voltage transformer; (b) equivalent circuit; (c) phasor diagram of voltage transformer; (d) voltage and phase-angle errors for voltage transformer.

where n_{vt} is the voltage transformer turns ratio given by

$$n_{vt} = \frac{n_P}{n_S}$$

Figure 27.18(c) shows the phasor diagram of an actual voltage transformer. The two errors of voltage transformers are the voltage or ratio error and the phase-angle error or phase displacement.

The voltage error is defined to be

$$\frac{\text{Rated voltage ratio } (V_P/V_S) - \text{actual ratio } (V_P/V_S)}{\text{Actual voltage ratio } (V_P/V_S)} \times 100\%$$

The phase displacement is the phase displacement between the primary and secondary voltages, as shown in Figure 27.18(c), and is positive if the secondary voltage leads the primary voltage. Figure 27.18(d) shows typical curves for the voltage ratio and phase angle errors for a VT as a function of secondary voltage. BS 3941:1974 sets out specifications for voltage transformers (British Standards Institution, 1974).

Ratio errors are significant in CTs and VTs when they are used in current and voltage measurement. Both ratio errors and phase-angle errors are important when CTs and VTs are used to extend the range of wattmeters (see Section 27.4).

27.2.4 Dynamometer Instruments

The operation of the dynamometer instrument is shown in Figure 27.19. The instrument has two air- or iron-cored coil systems—one fixed and the other pivoted and free to rotate. The torque, T_g, generated by the interaction of the two currents is given by

$$T_g = \frac{dM}{d\theta} \cdot i_1 \cdot i_2$$

and the restoring torque produced by the control springs is given by

$$T_r = k \cdot \theta$$

Thus the deflection, θ, is given by

$$\theta = \frac{1}{k} \cdot \frac{dM}{d\theta} \cdot i_1 \cdot i_2$$

Now, if the same current flows through both coils, the steady-state deflection is proportional to the mean square of the current. Alternatively, if swamping resistances are employed, the instrument can be used as a voltmeter. The scale of such instruments is usually calibrated in RMS quantities and thus is nonlinear. Air-cored instruments have no errors due to hysteresis effects, but the absence of an iron core requires the coils to have a large number of ampere turns to provide the necessary deflecting torque. This results in a high power loss to the circuit to which the instrument is connected. The torque-to-weight ratio is small and therefore friction effects are more serious, and the accuracy of these instruments can be affected by stray magnetic fields. Dynamometer instruments tend to be more expensive than other types of ammeters and voltmeters. The most important use of the dynamometer principle is in the wattmeter (see Section 27.4.1).

27.2.5 Thermocouple Instruments

Figure 27.20 shows the elements of a thermocouple instrument. These are a heating element that usually consists of a fine wire or a thin-walled tube in an evacuated glass envelope, a thermocouple having its hot junction in thermal contact with the heating element, and a permanent magnet-moving coil millivoltmeter. Thermocouple instruments respond to the heating effect of the current passing through the heating element and are thus mean-square sensing devices and provide an indication which is independent of the current waveshape. They are capable of operating over a wide frequency range. At low frequencies (less than 10 Hz) their operation is limited by pointer vibration caused by the thermal response of the wire. At high frequencies (in excess of 10 MHz) their operation is limited by the skin effect, altering the resistance of the heating element.

Thermocouple instruments have FSDs typically in the range 2–50 mA and are usually calibrated in RMS values.

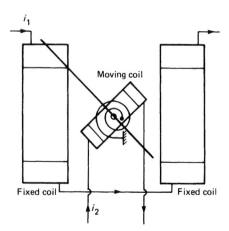

FIGURE 27.19 Dynamometer instrument.

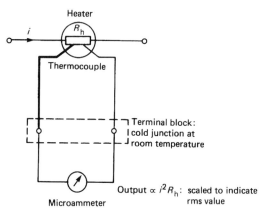

FIGURE 27.20 Thermocouple instrument.

The scale is thus nonlinear. They are fragile and have only a limited overrange capability before the heating element is melted by overheating. The frequency range of the instrument as a voltmeter is limited by the ability to produce nonreactive series resistors.

27.2.6 Electrostatic Instruments

Electrostatic instruments that may be used as voltmeters and wattmeters depend for their operation on the forces between two charged bodies. The torque between the fixed and moving vane in Figure 27.21(a) is given by

$$T = \frac{1}{2} \cdot \frac{dC}{d\theta} \cdot V^2$$

where C is the capacitance between the plates.

The usual form of the electrostatic voltmeter is the four-quadrant configuration shown in Figure 27.21(b). There are two possible methods of connection for such a voltmeter. These are the heterostatic and idiostatic connections shown in Figure 27.21(c). Commercial instruments usually employ the idiostatic connection, in which the needle is connected to one pair of quadrants. In this configuration the torque produced is proportional to the mean-square value of the voltage. If the instrument is scaled to indicate the RMS value, the scale will be nonlinear. The torques produced by electrostatic forces are small, and multicellular devices of the form shown in Figure 27.21(d) are used to increase the available torque. Multicellular instruments can be used for voltages in the range 100–1,000 V. Electrostatic instruments have the advantage of a capacitive high input impedance. They are fragile and expensive, and therefore their use is limited to that of a transfer standard between AC and DC quantities.

27.3 DIGITAL VOLTMETERS AND DIGITAL MULTIMETERS

Analog indicating instruments provide a simple and relatively cheap method of indicating trends and changes in measured quantities. As voltmeters, direct indicating instruments have low input impedance. At best they provide only limited accuracy, and this is achieved only with considerable skill

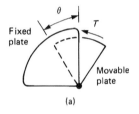

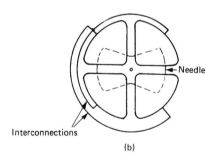

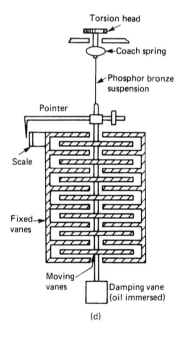

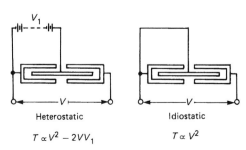

FIGURE 27.21 (a) Principle of electrostatic voltmeter; (b) four-quadrant electrostatic voltmeter; (c) heterostatic and idiostatic connections; (d) multicellular electrostatic voltmeter.

on the part of the observer. Their speed of response is also slow. Digital instruments, in contrast, can provide high input impedance, high accuracy and resolution, and a high speed of measurement. They provide an indication to the observer that is free from ambiguity and requires no interpolation.

27.3.1 Analog-to-Digital Conversion Techniques

Fundamental to both digital voltmeters (DVMs), the functions of which are limited to the measurement of DC and AC voltage, and digital multimeters (DMMs), the functions of which may include voltage, current, and resistance measurement, is an analog-to-digital converter (ADC). ADCs are dealt with in detail in Part 4 as well as in Owens (1983), Arbel (1980), and Sheingold (1977). In this section consideration is limited to the successive-approximation, dual-ramp, and pulse-width techniques.

ADCs take an analog signal of which the amplitude can vary continuously and convert it into a digital form that has a discrete number of levels. The number of levels is fixed by the number of bits employed in the conversion; this sets the resolution of the conversion. For a binary code with N bits, there are 2^N levels. Since the digital representation is discrete, there is a range of analog values that all have the same digital representation. Thus there is a quantization uncertainty of $\pm 1/2$ least significant bit (LSB), and this is in addition to any other errors that may occur in the conversion itself. ADCs used in DVMs and DMMs are either sampling ADCs or integrating ADCs, as shown in Figure 27.22. Sampling ADCs provide a digital value equivalent to the voltage at one time instant. Integrating ADCs provide a digital value equivalent to the average value of the input over the period of the measurement. The successive-approximation technique is an example of a sampling ADC. The dual-ramp and pulse-width techniques described here are examples of integrating ADCs. Integrating techniques require a longer time to perform their measurement but have the advantage of providing noise- and line-frequency signal rejection.

27.3.1.1 Successive-Approximation ADCs

This technique is an example of a feedback technique that employs a digital-to-analog converter (DAC) in such a way as to find the digital input for the DAC for which the analog output voltage most closely corresponds to the input voltage that is to be converted. Detailed consideration of DACs is found in Part 4.

Figure 27.23(a) shows an N-bit R-$2R$ ladder network DAC. The output of this device is an analog voltage given by

$$V_{out} = \frac{V_{ref}}{2^N} \sum_{n=0}^{N-1} a_n 2^n$$

where the a_n take values of either 1 or 0, dependent on the state of the switches, and $-V_{ref}$ is the reference voltage.

The successive-approximation technique shown in Figure 27.23(b) employs a decision-tree approach to the conversion problem. The control circuitry on the first cycle of the conversion sets the most significant bit of the DAC (MSB), bit a_{N-1}, to 1 and all the rest of the bits to 0. The output of the comparator is examined. If it is a 0, implying that the analog input is greater than the output, then the MSB is maintained at a 1; otherwise it is changed to a 0. The next cycle determines whether the next most significant bit is a 1 or a 0. This process is repeated for each bit of the DAC. The conversion period for the successive-approximation ADC technique is fixed for a given ADC irrespective of the signal level and is equal to $N\tau$, where N is the number of bits and τ is the cycle time for determining a single bit. Integrated

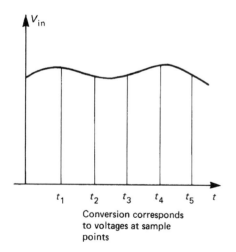

Conversion corresponds to voltages at sample points

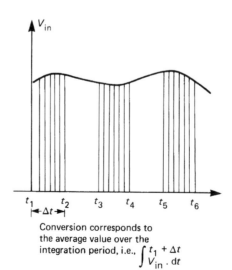

Conversion corresponds to the average value over the integration period, i.e., $\int_{t_1}^{t_1 + \Delta t} V_{in} \cdot dt$

FIGURE 27.22 Sampling and integrating ADCs.

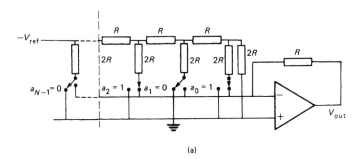

FIGURE 27.23 (a) $R\ 2R$ ladder network DAC; (b) successive-approximation ADC.

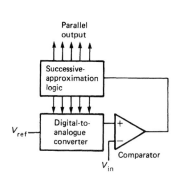

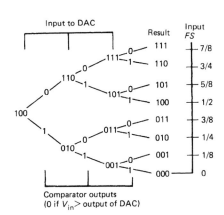

circuit successive-approximation logic-generating chips are available to be used in conjunction with standard DACs and comparators to produce medium-speed ADCs. A typical 8-bit ADC will have a conversion time of 10 μs. Successive-approximation ADCs are limited to 16 bits, equivalent to a five-decade conversion.

27.3.1.2 Dual-Ramp ADCs

The dual-ramp conversion technique is shown in Figure 27.24 and operates as follows:

The input voltage, V_{in}, is switched to the input of the integrator for a fixed period of time t_1, after which the integrator will have a value of

$$\frac{-V_{in} \cdot t_1}{RC}$$

The reference voltage $-V_{ref}$ is then applied to the integrator and the time is then measured for the output of the integrator to ramp back to zero. Thus

$$\frac{V_{in} \cdot t_1}{RC} = \frac{V_{ref} \cdot t_2}{RC}$$

from which

$$\frac{t_2}{t_1} = \frac{V_{in}}{V_{ref}}$$

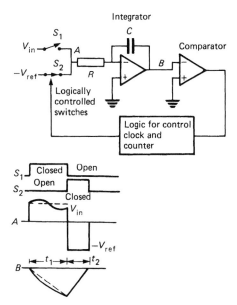

FIGURE 27.24 Dual-slope ADC.

If t_1 corresponds to a fixed number of counts, n_1, of a clock having a period τ and t_2 is measured with the same clock, say, n_2 counts, then

$$n_2 = \frac{V_{in}}{V_{ref}} \cdot n_1$$

The values of the R and C components of the integrator do not appear in the defining equation of the ADC; neither does the frequency of the reference clock. The only variable that appears explicitly in the defining equation is the reference voltage. The effect of the offset voltage on the comparator will be minimized as long as its value remains constant over the cycle and also providing it exhibits no hysteresis. Modifications of the technique employing quad-slope integrators are available which reduce the effects of switch leakage current and offset voltage and bias current in the integrator to second-order effects (Analog Devices, 1984). Errors caused by nonlinearity of the integrator limit the conversion using dual-ramp techniques to five decades.

The dual-ramp conversion technique has the advantage of line-frequency signal rejection (Gumbrecht, 1972). If the input is a DC input with an AC interference signal superimposed on it,

$$V_{in} = V_{d.c.} + V_{a.c.} \sin(\omega t + \phi)$$

where ϕ represents the phase of the interference signal at the start of the integration, then the value at the output of the integrator, V_{out}, at the end of the period t_1, is given by

$$V_{out} = -\frac{V_{d.c.} t_1}{RC} - \frac{1}{RC} \int_0^{t_1} V_{a.c.} \sin(\omega t + \phi)$$

If the period t_1 is made equal to the period of the line frequency, then the integral of a line-frequency signal or any harmonic of it over the period will be zero, as shown in Figure 27.24. At any other frequency it is possible to find a value of ϕ such that the interference signal gives rise to no error. It is also possible to find a value ϕ_{max} such that the error is a maximum. It can be shown that the value of ϕ_{max} is given by

$$\tan \phi_{max} = \frac{\sin \omega t_1}{(1 - \cos \omega t_1)}$$

The series or normal mode rejection of the ADC is given as the ratio of the maximum error produced by the sine wave to the peak magnitude of the sine wave. It is normally expressed (in dBs) as series mode rejection (SMR):

$$= -20 \log_{10} \frac{\omega t_1}{\cos \phi_{max} - \cos(\omega t_1 + \phi_{max})}$$

A plot of the SMR of the dual-slope ADC is shown in Figure 27.25. It can be seen that ideally it provides infinite SMR for any frequency given by n/t_1, $n = 1, 2, 3 \ldots$ Practically, the amount of rejection such an ADC can provide is limited because of nonlinear effects, due to the fact that the period t_1 can only be defined to a finite accuracy and that the frequency of the signal to be rejected may drift. However, such a technique can easily provide 40 dB of line-frequency rejection.

Figure 27.26 shows a schematic diagram of a commercially available dual-slope integrated-circuit chip set.

27.3.1.3 Pulse-Width ADCs

A simple pulse-width ADC is shown in schematic form in Figure 27.27. The ADC employs a voltage-controlled monostable to produce a pulse the width of which is proportional to the input voltage. The width of the pulse is then measured by means of a reference clock. Thus the counter has within it at the end of the conversion period a binary number that corresponds to the analog input. The accuracy of the technique depends on the linearity and stability of the voltage to pulse-width converter and the stability of the reference clock. High-speed conversion requires the use of a high-frequency clock. By summing the counts over longer periods of time, the effect of line frequency and noise signals can be integrated out.

A modified pulse-width technique for use in precision voltmeters is shown in Figure 27.28 (Pitman, 1978; Pearce, 1983). Precision pulses generated by chopping +ve and −ve reference voltages are fed into the input of an integrator that is being forced to ramp up and down by a square wave. The ramp waveform applied to the two comparators generates two pulse trains, which are used to gate the reference voltages. In the absence of an input voltage, feedback ensures that the width of the +ve and −ve pulses will be equal. The outputs of the comparators are fed to an up-down counter. For the duration of the +ve pulse the counter counts up, and during the −ve pulses it counts down. Thus ideally with no input the count at the end of the integration period will be zero.

If an input is applied to the integrator, the width of the +ve and −ve pulse widths are adjusted by the feedback mechanism, as shown in Figure 27.28. If the period of the square wave is approximately 312 μs and the clock runs at approximately 13 MHz, it is possible to provide a reading with a resolution of 1 part in 4,000 over a single period. Figure 27.29 shows the variation of the pulse widths for a time-varying input. By extending the integration period to 20 ms, the resolution becomes 1 part in 260,000, and

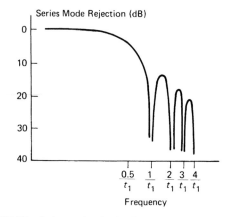

FIGURE 27.25 Series mode rejection for dual-slope ADC.

CHAPTER | 27 Electrical Measurements

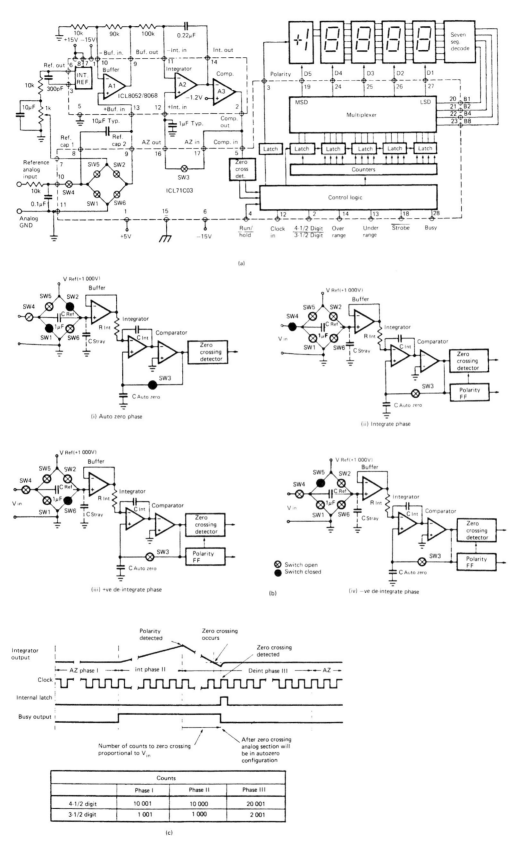

FIGURE 27.26 Dual-slope integrated-circuit chip set. (a) Function block diagram. (b) Operating phase of converter. Courtesy of Intersil Datel (UK) Ltd. (c) Timing.

significant rejection of 50 Hz line frequency is achieved. The method allows trading to occur between resolution and speed of measurement.

27.3.1.4 Voltage References in ADCs

In all ADCs a comparison is made with some reference voltage. Therefore, to have accurate conversion it is necessary to have an accurate and stable voltage source. Most digital voltmeters use Zener diodes to generate their reference voltages, although high-precision devices often have facilities to employ standard cells to provide the reference voltage. Two types of Zener devices are commonly used. The compensated Zener diode is a combination of a Zener junction and a forward-biased junction in close proximity to the Zener junction so that the temperature coefficient can be set to a few ppm by choosing the correct Zener current. Active Zener diodes have a temperature controller built into the silicon chip around the Zener diode. Table 27.6 (taken from Spreadbury, 1981) compares the characteristics of these two types of Zener devices with bandgap devices and the Weston standard cell.

27.3.2 Elements in DVMs and DMMs

The ADC is the central element of a DVM or DMM. The ADC is, however, a limited input range device operating usually on unipolar DC signals. Figure 27.30 shows the elements of a complete DVM or DMM.

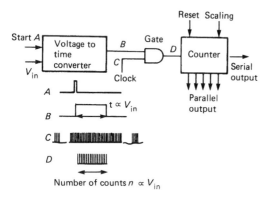

FIGURE 27.27 Pulse-width ADC.

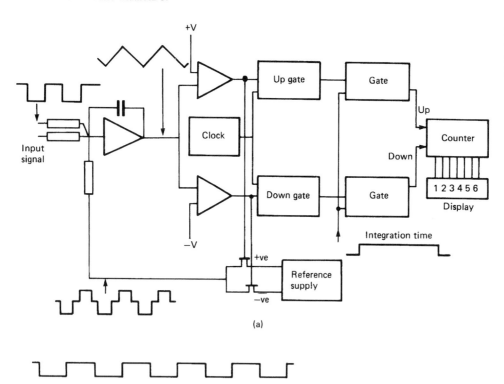

FIGURE 27.28 Precision pulse-width ADC. (a) Circuit; (b) timing. Courtesy of Solartron Instruments Ltd.

27.3.2.1 DC Input Stage and Guarding

The DC input stage provides high input impedance together with attenuation/amplification and polarity sensing of the signal to ensure that the voltage applied to the ADC is of the correct magnitude and polarity.

DVMs and DMMs are often used to measure small DC or AC signals superimposed on much larger common-mode signals. For example, in measuring the output signal from a DC Wheatstone bridge, as shown in Figure 27.31(a), the common-mode voltage is half the bridge supply. If a transducer is situated some distance away from its associated DVM, the common-mode signal may be generated by line-frequency ground currents, as shown in Figure 27.31(b), and thus the potential to be measured may be superimposed on an AC line-frequency common-mode signal. Figure 27.31(c) shows the equivalent circuit for the measurement circuit and the input of the DVM or DMM. R_A and R_B represent the high and low side resistances of the measurement circuit, R_{in} the input resistance of the DVM or DMM, and R_i and C_i the leakage impedance between the low terminal of the instrument and power ground. The leakage impedance between the high terminal and the instrument ground can be neglected because the high side is usually a single wire whereas the low side often consists of a large metal plate or plane. The divider consisting of R_B and R_i and C_i converts common-mode signals to input signals. Typically R_i is 10^9 Ω and C_i may be as high as 2.5 nF. For specification purposes R_A is taken as zero and R_B is taken as 1 kΩ. Thus at DC the common-mode rejection is −120 dB and at 50 Hz it is −62 dB.

The common-mode rejection can be improved by the addition of an input guard. This is shown in Figure 27.31(d) and can be considered as the addition of a metal box around the input circuit. This metal box is insulated both from the input low and the power ground. It is available as a terminal of the input of the instrument. If the guard is connected to the low of the measurement circuit, the effect of current flow between the low terminal and guard is eliminated, since they are at the same potential. The potential dividing action now occurs between the residual leakage impedance between low and power ground in the presence of the guard. The value of these leakage impedances are of order 10^{11} Ω and 2.5 pF. The DC common-mode rejection has now been increased to −160 dB and the 50 Hz common-mode rejection to −122 dB. Thus a DC common-mode signal of 100 V will produce an input voltage of 1 μV, and a 20 V, 50 Hz common-mode signal will produce an input of less than 20 μV.

In situations in which there is no common-mode signal, the guard should be connected to the signal low; otherwise unwanted signals may be picked up from the guard.

27.3.2.2 AC/DC Conversion

Two techniques are commonly used in AC voltage and current measurement using digital instruments. Low-cost DVMs and DMMs employ a mean absolute value measurement-RMS indicating technique similar to that employed in AC current and voltage measurement using a permanent magnet-moving coil instrument. By the use of operational techniques, as shown in Figure 27.32, the effect of the forward diode drop can be reduced and thus precision rectification

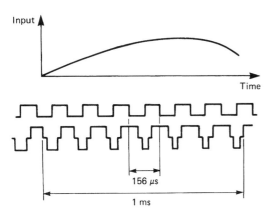

FIGURE 27.29 Effect of time-varying input on pulse-width ADC. Courtesy of Solartron Instruments Ltd.

TABLE 27.6 Reference voltage sources

	Weston cell	Compensated Zener	Active Zener	Bandgap device
Stable level, V	1.018	6.4	7	1
Temperature coefficient parts in 10^6 per deg C	−40	1	0.2	30
Internal resistance	500 Ω	15 Ω at 7.5 mA	½ Ω at 1 mA	½ Ω
	(in all cases, with op. amp. can be reduced to 0.001 Ω)			
Aging, parts in 10^6 per year	0.1 to 3	2 to 10	20	100
Noise, μV rms	0.1	1	7	6

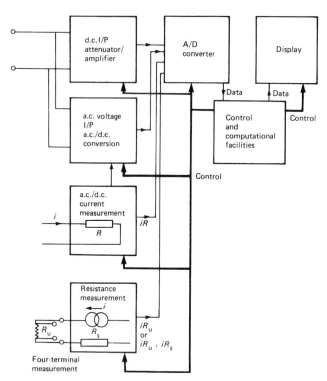

FIGURE 27.30 Elements of DVM/DMM.

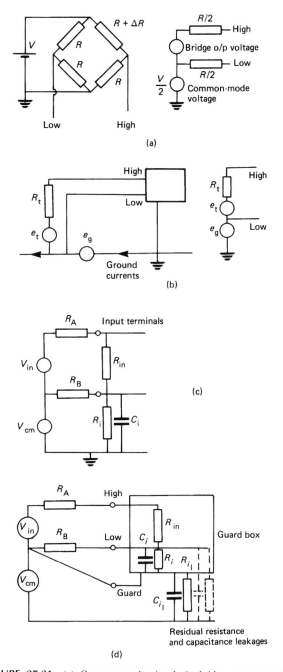

FIGURE 27.31 (a) Common-mode signals in bridge measurements; (b) ground current-generated common-mode signals; (c) input equivalent circuit; (d) input guarding.

can be achieved. However, because the instrument is then not RMS sensing but relies on the waveform being sinusoidal for correct indication, this technique suffers from the form factor errors shown in Section 27.2.1.3.

True RMS measurement can be obtained either by use of analog electronic multipliers and square-root extractors, as shown in Figure 27.33(a), or by the use of thermal converters, as shown in Figure 27.33(b). High-precision instruments employ vacuum thermocouples to effect an AC/DC transfer. Brodie (1984) describes an AC voltmeter using such a technique, which provides a measurement accuracy of 160 ppm for any signal level from 100 mV to 125 V in a frequency band from 40 Hz to 20 kHz. This voltmeter is capable of measuring over a range from 12.5 mV to 600 V in a frequency band from 10 Hz to 1 MHz with reduced accuracy.

In true RMS sensing instruments, the manufacturer often specifies the maximum permissible crest factor for the instrument. The crest factor is the ratio of the peak value of the periodic signal to its RMS value. Typically the maximum permissible crest factor is 5.

27.3.2.3 Resistance and Current Measurement

Resistance measurement is provided by passing a known current through the resistor and measuring the voltage drop across it. Four-terminal methods, as shown in Figure 27.30, enable the effect of lead resistance to be reduced. High-precision DMMs employ ratiometric methods in which the same current is passed through both the unknown resistance and a standard resistance, and the unknown resistance is computed from the ratio of the voltages developed across the two resistances and the value of the standard resistor. AC and DC current measurements use a shunt across which a voltage is developed. This voltage is then measured by the ADC.

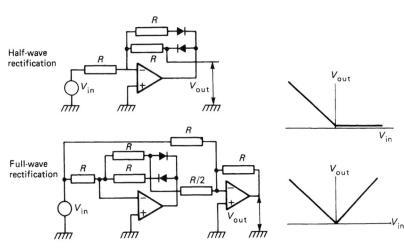

FIGURE 27.32 AC signal precision rectification.

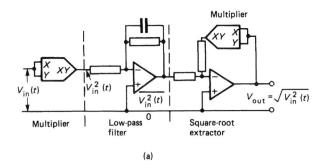

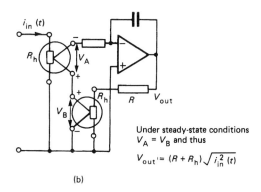

FIGURE 27.33 (a) RMS evaluation by analog multiplication; (b) AC/DC conversion using thermal techniques.

27.3.2.4 Control and Post-Measurement Computational Facilities

The control element in DVMs and DMMs is increasingly provided by a microprocessor. The use of the microprocessor also enables the digital instrument to provide the user with a large range of post-measurement storage and computational facilities. These may include:

1. Collection and storage of a set of readings, with a given time interval between readings.
2. The application of scaling and offset calculations to the readings to provide an output of the form $y = mx + c$, where x is the reading and m and c are constants input by the operator. This enables the measured value to be output in engineering units.
3. Testing readings to ascertain whether they are within preset limits. In this mode the instrument may either display "hi-lo-pass" or may count the number of readings in each category.
4. The calculation and display of the percentage deviation from a given set point input by the operator.
5. Calculation of the ratio of the measured value to some value input by the operator.
6. Storing the maximum and minimum value of the measured variable.
7. Generating statistical data from a given set of measurements to provide the sample average, standard deviation, variance, or RMS value.
8. Digital filtering of the measured variable to provide a continuous average, an average over n readings, or a walking window average over n readings.

27.3.2.5 Output

The visual display of DVMs and DMMs is commonly provided by light-emitting diodes (LEDs) or liquid crystal displays (LCDs). The relative merits of each of these displays is considered in Chapter 30 in Part 4. If the results are to be communicated to further digital systems, the output may be provided as either a parallel binary or binary-coded decimal (BCD) output. Many DVMs and DMMs are fitted with the standard IEEE-488 or RS232 parallel or serial interfaces, which allow data and control to pass between the instrument and a host control computer. The characteristics of IEEE-488 and RS232 interfaces are considered in Part 4.

27.3.3 DVM and DMM Specifications

DVMs and DMMs cover a wide range of instruments, from handheld, battery-operated multimeters through panel meters and bench instruments to standards laboratory instruments. These digital instruments are specified primarily by their resolution, accuracy, and speed of reading. The resolution of the instrument, which may be higher than its accuracy, corresponds to the quantity indicated by a change in the least significant digit of the display. Typically, digital instruments have displays that are between 8½ and 8½ digits. The half-digit indicates that the most significant digit can only take the value 1 or 0. Thus a 3½ digit instrument has a resolution of 1 part in 2,000 and an 8½-digit one has a resolution of 1 part in 2×10^8. The accuracy of the instrument is specified as $\pm$ (x percent of reading (R) + y per cent of scale (S) + n digits). Table 27.7 gives condensed specifications for comparison of a handheld 3½-digit DMM, a 5½-digit intelligent multimeter, and an 8½-digit standards laboratory DVM. The accuracies quoted in Table 27.7 are for

TABLE 27.7 Comparison of digital voltmeter specifications

	3½-digit multimeter (Fluke 8026B)	5½-digit intelligent multimeter (Thurlby 1905A)	8½-digit Standards Laboratory DVM (Solartron 7081)
DC voltage ranges	199.9 mV–1000 V	210.000 mV–1100.00 V	0.1 V–1000 V
Typical accuracy	±(0.1% R + 1 digit)	±(0.015% R + 0.0015% S + 2 digits)	Short-term stability ±(1.2 ppm R + 0.3 ppm S)
Input impedance	10 MΩ on all ranges	>1 GΩ on lowest two ranges 10 MΩ on remainder	>10 GΩ on 3 lowest ranges 10 MΩ on remainder
AC voltage ranges	199.9 mV–750 RMS	210.00 mV–750 V RMS	0.1 V–1000 V rms
Type	True RMS sensing crest factor 3:1 ±(1% R + 3 digits)	Mean sensing/RMS calibrated for sinusoid	True rms sensing crest factor 5:1 short-term stability
Typical accuracy	45 Hz–10 kHz	±(2% R + 10 digits)	±(0.05% R + 0.03% S)
Frequency range	10 MΩ∥100 pF	45 Hz–20 kHz	10 Hz–100 kHz
Input impedance		10 MΩ∥47 pF	1 MΩ∥100 pF
DC current ranges	1.999 mA–1.999 A	210.000 μA–2100.00 mA	
Typical accuracy	±(0.75% R + 1 digit)	±(0.1% R + 0.0015% S + 2 digits)	
Voltage burden	0.3 V max. on all ranges except 1.999 A range. Max. burden on 1.999 A range 0.9 V	0.25 V max. on all ranges except 2100 mA range. Max. burden on 2100 mA range 0.75 V	Not applicable
AC current ranges	1.999 mA–1.999 A	210.00 μA–2100.0 mA	
Type	True rms sensing crest factor 3:1	Mean sensing/rms calibrated for sinusoid	
Typical accuracy	±(1.5% R + 2 digits)	±(0.3% R + 5 digits)	Not applicable
Frequency range	45 Hz–1 kHz	45 Hz–500 Hz	
Voltage burden	0.3 V max. on all ranges except 1.999 A range. Max. burden on 1.999 A range 0.9 V	0.25 V max. on all ranges except 2100.0 mA range. Max. burden on 2100.0 mA range 0.75 V	
Resistance ranges	199.9 Ω–19.99 MΩ	210.000 Ω–21.000 MΩ	0.1 kΩ–1,000 MΩ
Typical accuracy	±(0.1% R + 1 digit)	±(0.04% R + 0.0015% S + 2 digits)	Short-term stability (2 ppm R + 0.4 ppm S)
Current employed	Max. current 0.35 mA on 199.9 Ω range	Max. current 1 mA on 210.000 Ω range	Max. current 1 mA on 0.1, 1 and 10 kΩ ranges
Speed of reading		3 per second	100 per second to 1 per 51.2 s
Common-mode rejection ratio	>100 dB at DC, 50 Hz, and 60 Hz with 1 kΩ unbalance for DC ranges >60 dB for 50 and 60 Hz with 1 kΩ unbalance on AC ranges	>120 dB at DC or 50 Hz	Effective CMR [CMR + SMR] with 1 kΩ unbalance 5½–8½ digit >140 dB at 50(60) Hz >120 dB at 400 Hz for AC measurement
Series mode rejection	>60 dB at 50 Hz or 60 Hz	>60 dB at 50 Hz	>40 dB at 50(60) Hz 5½–8½ digits >70 dB at 50(60) or 400 Hz
Additional notes	Battery operated with LCD display Also provides conductance measurement and continuity testing	LED display Intelligent functions include: scaling and offsetting, percentage deviation, low-hi-pass, max-min, filtering, averaging, and data logging RS232 and IEEE-488 interfaces True rms option available	LED display Intelligent functions include: ratio, scaling and offsetting, digital filtering, statistics, limits, time: real or elapsed, history file with 1500 numeric readings or 500 readings with time and channel mode RS232 and IEEE-488 interfaces

CHAPTER | 27 Electrical Measurements

guidance only; for complete specifications the reader should consult the specification provided by the manufacturer.

27.4 POWER MEASUREMENT

For a two-terminal passive network, if the instantaneous voltage across the network is $v(t)$ and the instantaneous current through it is $i(t)$, the instantaneous power, $p(t)$, taken or returned to the source is given by

$$p(t) = v(t) \cdot i(t)$$

For a linear network, if $v(t)$ is sinusoidal, that is,

$$v(t) = \overline{V} \cdot \sin \omega t$$

then $i(t)$ must be of the form

$$i(t) = \hat{I} \sin(\omega t + \phi)$$

and the instantaneous power, $p(t)$, is given by

$$p(t) = v(t) \cdot i(t) = \overline{V}\hat{I} \sin \omega t \sin(\omega t + \phi)$$

The average power dissipated by the network is given by

$$P = \frac{1}{T} \int_0^T p(t) \cdot dt$$

where T is the period of the waveform and thus

$$P = \frac{\omega}{2\pi} \int_0^{2\pi/\omega} \overline{V}\hat{I} \sin(\omega t + \phi) \cdot dt$$

Therefore, P is given by

$$P = \frac{\overline{V}\hat{I}}{2} \cdot \cos \phi$$

The RMS voltage, V, is given by

$$V = \frac{\overline{V}}{\sqrt{2}}$$

and the RMS current, I, is given by

$$I = \frac{\hat{I}}{\sqrt{2}}$$

Thus the average power dissipated by the network is given by

$$P = VI \cos \phi$$

($\cos \phi$ is known as the power factor).

27.4.1 The Three-Voltmeter Method of Power Measurement

By using a noninductive resistor and measuring the three voltages shown in Figure 27.34(a), it is possible to measure the power dissipation in the load without using a wattmeter. Figure 27.34(b) shows the phasor diagram for both leading and lagging power factors.

From the phasor diagram by simple trigonometry,

$$V_A^2 = V_B^2 + V_C^2 + 2V_B V_C \cos \phi$$

and

$$V_B = IR$$

Since the average power dissipated in the load is given by

$$P = V_C I \cos \phi$$

then

$$P = \frac{V_A^2 - V_B^2 - V_C^2}{2R}$$

and the power factor $\cos \phi$ is given by

$$\cos \phi = \frac{V_A^2 - V_B^2 - V_C^2}{2V_B V_C}$$

27.4.2 Direct-Indicating Analog Wattmeters

Direct-indicating analog wattmeters employ the dynamometer, induction, electrostatic, or thermocouple principles.

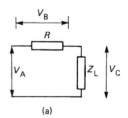

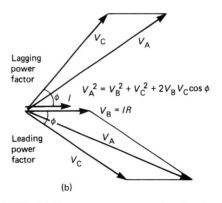

FIGURE 27.34 (a) Power measurement using the three-voltmeter method; (b) phasor diagram for the three-voltmeter method.

These are shown in Figures 27.35 and 27.36. Of these, the dynamometer is the most commonly used. In the dynamometer wattmeter shown in Figure 27.35(a), the current into the network is passed through the fixed coils while the moving coil carries a current that is proportional to the applied voltage. The series resistance in the voltage coil is noninductive. The series torque is provided by a spring; thus the mean deflection of the wattmeter from Section 27.2.4 is given by

$$\theta = \frac{1}{k} \cdot \frac{1}{R_s} \cdot \frac{dM}{d\theta} \cdot V \cdot I \cdot \cos\phi$$

The primary errors in dynamometer wattmeters occur as a consequence of magnitude and phase errors in the voltage coil and power loss in the wattmeter itself. Other errors are caused by the capacitance of the voltage coil and eddy currents.

If the resistance and inductance of the voltage coil are R_V and L_V, respectively, and if R_S is the resistance in series with the voltage coil, the current through the voltage coil at an angular frequency Ω has a magnitude given by

$$I_V = \frac{V}{\sqrt{\left[(R_V + R_S)^2 + \omega^2 L_V^2\right]}}$$

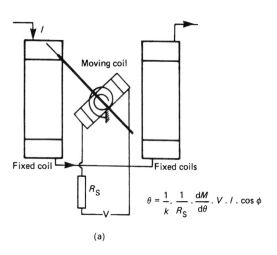

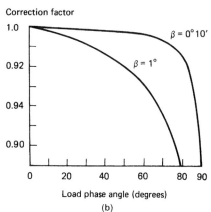

FIGURE 27.35 (a) Dynamometer wattmeter; (b) wattmeter correction factors.

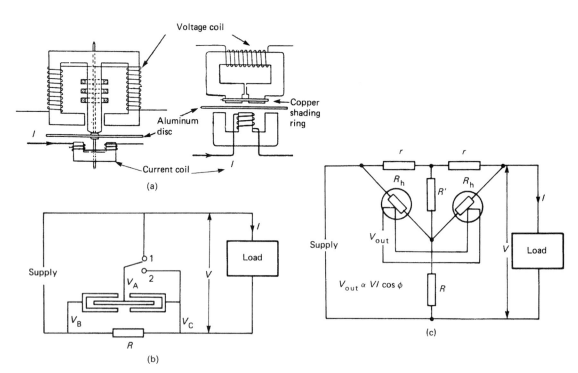

FIGURE 27.36 (a) Induction wattmeter; (b) electrostatic wattmeter; (c) thermocouple wattmeter.

with a phase angle, β, given by

$$\beta = \tan^{-1}\frac{\omega L_V}{(R_V + R_S)} \simeq \frac{\omega L_V}{(R_V + R_S)}$$

Thus altering the frequency alters both the sensitivity and phase angle of the voltage coil.

If the load circuit has a lagging power factor, $\cos\phi$, the wattmeter true indication will be

$$\frac{\cos\phi}{\cos\beta \cdot \cos(\phi - \beta)} \times \text{actual indication}$$

and the error as a percentage of actual indication will be

$$\frac{\sin\beta}{(\cos\phi + \sin\beta)} \times 100\%$$

The wattmeter reads high on lagging power factors. Figure 27.35(b) shows the correction factors for $\beta = 1°$ and $\phi = 0°10'$.

The induction wattmeter in Figure 27.36(a) operates on a principle similar to the shaded pole induction watt-hour meter described in Section 27.5 in that the torque is generated by the interaction of eddy currents induced in a thin aluminum disc with the imposed magnetic fields. The average torque generated on the disc is proportional to the average power. In the induction wattmeter, the generated torque is opposed by a spring and thus it has a scale that can be long and linear.

In the electrostatic wattmeter shown in Figure 27.36(b) with the switch in position 1 the instantaneous torque is given by

$$T \propto (V_A - V_B)^2 - (V_A - V_C)^2$$

and thus

$$T \propto 2R\left(V \cdot i + \frac{i^2 R}{2}\right)$$

where V and i are the instantaneous load voltage and current, respectively.

If this torque is opposed by a spring then the average deflection will be given by

$$\theta \propto 2R\left(VI\cos\phi + \frac{I^2 R}{2}\right)$$

that is, the average power dissipated in the load plus half the power dissipated in R.

With the switch in position 2 the instantaneous torque is given by

$$T \propto (V_A - V_B)^2$$

and the average deflection will be given by

$$\theta \propto R(I^2 R)$$

that is, the power dissipated in R.

Thus from these two measurements the power in the load can be computed.

In the compensated thermal wattmeter employing matched thermocouples, as shown in Figure 27.36(c), the value of the resistance R is chosen such that

$$R' = \frac{(R_h + r) \cdot R}{r}$$

The output of the wattmeter can then be shown to be given by

$$V_{out} = \frac{k \cdot r}{(r + R_h)(r + R)} \cdot V \cdot I \cdot \cos\phi$$

where k is a constant of the thermocouples.

In the compensated thermal wattmeter there are no errors due to the power taken by either the current or voltage circuits.

Dynamometer wattmeters are capable of providing an accuracy of order 0.25 percent of FSD over a frequency range from DC to several kHz. Induction wattmeters are suitable only for use in AC circuits and require constant supply voltage and frequency for accurate operation. The electrostatic wattmeter is a standards instrument having no waveform errors and suitable for measurements involving low power factors, such as the measurement of iron loss, dielectric loss, and the power taken by fluorescent tubes. Thermocouple wattmeters are capable of providing measurements up to 1 MHz with high accuracy.

27.4.3 Connection of Wattmeters

There are two methods of connecting a dynamometer wattmeter to the measurement circuit. These are shown in Figures 27.37(a) and (b). In the connection shown in Figure 27.37(a), the voltage coil is connected to the supply side of the current coil. The wattmeter therefore measures the power loss in the load plus the power loss in the current coil. With the wattmeter connected as in Figure 27.37(b), the current coil takes the current for both the load and the voltage coil. This method measures the power loss in the load and in the voltage coil. For small load currents the voltage drop in the current coil will be small; therefore, the power loss in this coil will be small and the first method of connection introduces little error. For large load currents, the power loss in the voltage coil will be small compared with the power loss in the load, and the second method of connection is preferred.

Compensated wattmeters of the type shown in Figure 27.38 employ a compensating coil in series with the voltage

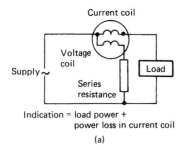

(a)

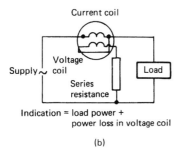

(b)

FIGURE 27.37 Wattmeter connection.

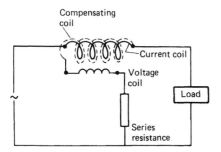

FIGURE 27.38 Compensated dynamometer wattmeter.

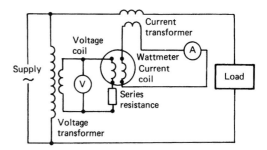

FIGURE 27.39 Wattmeter used with instrument transformers.

windings. This compensating coil is identical to the current coil and tightly wound with it to produce a magnetic field in opposition to the main magnetic field due to the load current. Thus the effect of the voltage coil current is eliminated, and therefore the wattmeter connected in the manner shown in Figure 27.38 shows no error due to the power consumption in the voltage coil.

For electronic wattmeters the power loss in the voltage detection circuit can be made to be very small, and thus the second method of connection is preferred.

The current and voltage ranges of wattmeters can be extended by means of current and voltage transformers, as shown in Figure 27.39. These transformers introduce errors in the measurement, as outlined in Section 27.2.3.

27.4.4 Three-Phase Power Measurement

For an n conductor system, the power supplied can be measured by n wattmeters if they are connected with each wattmeter having its current coil in one of the conductors and its potential coil between the conductor and a single common point. This method of measurement is shown for both star- and delta-connected three-phase systems in Figures 27.40(a) and (b). The power dissipated in the three-phase system is given by

$$P = W_1 + W_2 + W_3$$

Blondel's theorem states that if the common point for the potential coil is one of the conductors, the number of wattmeters is reduced by one. Thus it is possible to measure the power in a three-phase system using only two wattmeters, irrespective of whether the three-phase system is balanced. This method is shown in Figures 27.41(a) and (b). The phasor diagram for a star-connected balanced load is shown in Figure 27.41(c). The total power dissipated in the three-phase system is given by

$$P = W_1 + W_2$$

That is, the power dissipated is the algebraic sum of the indications on the wattmeters.

It should be noted that if the voltage applied to the voltage coil of the wattmeter is more than 90 degrees out of phase with the current applied to its current coil, the wattmeter will indicate in the reverse direction. It is necessary under such circumstances to reverse the direction of the voltage winding and to count the power measurement as negative. If the power factor of the load is 0.5 so that I_1 lags 60° behind V_{10}, the phase angle between V_{12} and I_1 is 90° and wattmeter W_1 should read zero.

It is also possible in the case of a balanced load to obtain the power factor from the indication on the two wattmeters, since

$$W_2 - W_1 = \sqrt{3} \cdot V \cdot I \cdot \sin \phi$$

and therefore

$$\tan \phi = \sqrt{3 \frac{(W_2 - W_1)}{(W_1 + W_2)}}$$

If the three-phase system is balanced, it is possible to use a single wattmeter in the configuration shown in Figure 27.42.

CHAPTER | 27 Electrical Measurements

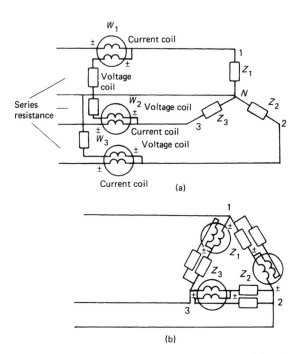

FIGURE 27.40 (a) Power measurement in a star-connected three-phase load using three wattmeters; (b) power measurement in a delta-connected three-phase load using three wattmeters.

With the switch in position 1, the indication on the wattmeter is given by

$$W_1 = \sqrt{3} \cdot V \cdot I \cos(30 + \phi)$$

With the switch in position 2, the wattmeter indicates

$$W_2 = \sqrt{3} \cdot V \cdot I \cos(30 - \phi)$$

The sum of these two readings is therefore

$$W_1 + W_2 = 3VI \cos \phi = P$$

that is, the total power dissipated in the system.

27.4.5 Electronic Wattmeters

The multiplication and averaging process involved in wattmetric measurement can be undertaken by electronic means, as shown in Figure 27.43. Electronic wattmeters fall into two categories, depending on whether the multiplication and averaging is continuous or discrete.

In the continuous method the multiplication can be by means of a four-quadrant multiplier, as shown in Figure 27.44(a) (Simeon and McKay, 1981); time-division multiplication, as in Figure 27.44(b) (Miljanic et al., 1978); or by the use of a Hall-effect multiplier, as in Figure 27.44(c) (Bishop and Cohen, 1973).

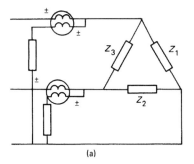

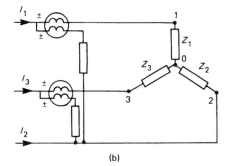

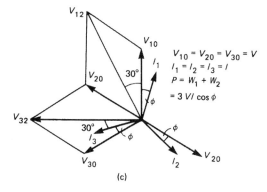

FIGURE 27.41 (a) Two-wattmeter method of power measurement in a three-phase delta-connected load; (b) two-wattmeter method of power measurement in a three-phase star-connected load; (c) phasor diagram for two-wattmeter method in a balanced star-connected load.

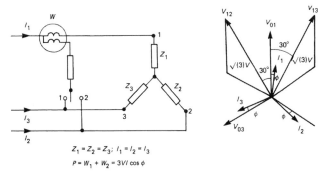

FIGURE 27.42 One-wattmeter method for balanced three-phase systems.

The sampling wattmeter shown in Figure 27.45 takes simultaneous samples of both the voltage and current waveforms, digitizes these values, and provides multiplication and averaging using digital techniques (Dix, 1982; Matouka, 1982).

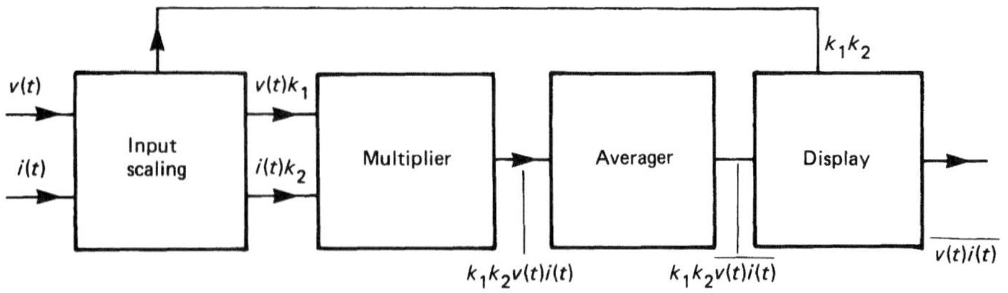

FIGURE 27.43 Electronic wattmeter.

If the voltage and current waveforms have fundamental and harmonic content with a fundamental period T, the instantaneous power can be written as a Fourier series:

$$p(t) = P + \sum_{k=1}^{\infty} p_k \cdot \sin\left(\frac{2\pi k t}{T} + \rho_k\right)$$

where P is the average power.

If the waveforms are uniformly sampled n times over m periods, the time t_j of the jth sample is given by

$$t_j = j \cdot \frac{m}{n} \cdot T$$

and the measured average power W is given by

$$W = \frac{1}{n} \sum_{j=0}^{n-1} p(t_j)$$

The error between the measured and true mean values is given by

$$W - P = \frac{1}{n} \sum_{k=1}^{\infty} p_k \sum_{j=0}^{n-1} \sin\left(\frac{2\pi k \cdot j \cdot m}{n} + \rho_k\right)$$

It can be shown (Clarke and Stockton, 1982; Rathore, 1984), that the error of measurement is given by

$$|W - P| = \left|\sum_{k<0}^{*} p_k \sin(\rho_k)\right| \leq \sum_{k>0}^{*} |p_k|$$

where Σ^* indicates summation over those terms where $k \cdot m/n$ is an integer, that is, those harmonics of the power signal for which the frequencies are integer multiples of the sampling frequency.

Matouka (1982) has analyzed other sources of error in sampling wattmeters, including amplifier, offset, sampled data, amplitude and time quantization, and truncation errors.

Continuous analog methods employing analog multipliers are capable of providing measurement of power, typically up to 100 kHz. The Hall-effect technique is capable of measurement up to the region of several GHz and can be used in power measurement in a waveguide. Using currently available components with 15-bit A/D converters, the sampling wattmeter can achieve a typical uncertainty of 1 part in 10^4 at power frequencies. Table 27.8 gives the characteristics of an electronic wattmeter providing digital display.

27.4.6 High-Frequency Power Measurement

At high frequencies, average power measurement provides the best method of measuring signal amplitude because power flow, unlike voltage and current, remains constant along a loss-less transmission line. Power measurements are made by measuring the thermal effects of power or by the use of a square-law device such as a diode (Hewlett-Packard, 1978; Fantom, 1985).

Static calorimetric techniques employ a thermally insulated load and a means for measuring the rise in temperature caused by the absorbed RF power. Flow calorimeters consist of a load in which absorbing liquid such as water converts the RF power into heat, together with a circulating system and a means for measuring the temperature rise of the circulating liquid. Because of their potentially high accuracy, calorimetric methods are used as reference standards. However, because of the complexity of the measurement systems, they are not easily portable.

Commercially available thermal techniques employ either thermistors or thermocouple detectors. Figure 27.46 shows the equivalent circuit of a thermistor system. The detecting thermistor is in either a coaxial or waveguide mount. The compensating thermistor is in close thermal contact with the detecting thermistor but shielded from the RF power.

Figure 27.47 shows a thermistor power meter employing two self-balancing DC bridges. The bridges are kept in balance by adjusting their supply voltages. With no applied RF power, V_c is made equal to V_{rf0}, that is, the value of V_{rf} with no applied RF energy. After this initialization process, ambient temperature changes in both bridges track each other.

If RF power is applied to the detecting thermistor, V_{rf} decreases such that

$$P_{rf} = \frac{V_{rf0}^2}{4R} - \frac{V_{rf}^2}{4R}$$

CHAPTER | 27 Electrical Measurements

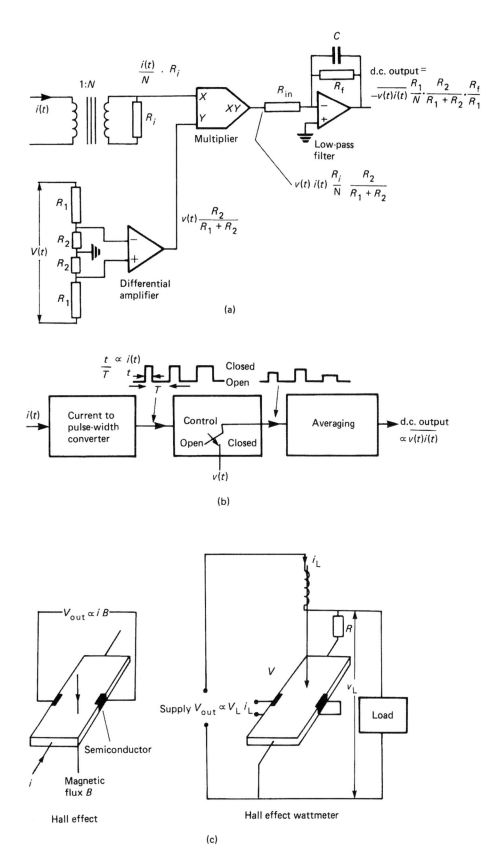

FIGURE 27.44   (a) Four-quadrant analog multiplier wattmeter; (b) time-division multiplication wattmeter; (c) Hall-effect wattmeter.

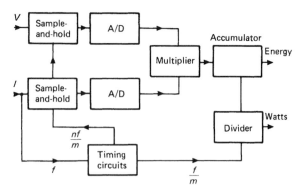

FIGURE 27.45 Sampling wattmeter (from Dix, 1982).

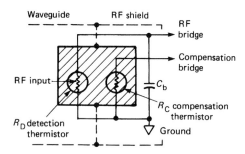

FIGURE 27.46 Equivalent circuit of a thermistor rf power detector (from Hewlett-Packard, 1978).

TABLE 27.8 Electronic Wattmeter Specification

Valhalla Scientific Digital Power Analyzer Model 2100 range/resolution table

True RMS voltage ranges	True RMS current ranges		
	0.2000 A	2.000 A	20.00 A
150.00 V	30.00 W	300.0 W	3,000 W
300.0 V	60.00 W	600.0 W	6,000 W
600.0 V	120.00 W	1,200.0 W	12,000 W
	True watts ranges		

Performance specifications:

AC/DC current (true RMS):
Crest factor response: 50:1 for minimum rms input, linearly decreasing to 2.5:1 for full-scale rms input
Peak indicator: Illuminates at 2.5 × full scale
Minimum input: 5% of range
Maximum input: 35 A peak, 20 A DC or RMS; 100 A DC or RMS for 16 mS without damage
Overrange: 150% of full scale for DC up to maximum input

AC/DC voltage (true RMS):
Crest factor response: 50:1 for minimum RMS input, linearly decreasing to 2.5:1 for full-scale RMS input
Minimum input: 5% of range
Maximum input: 600 V DC or RMS AC, 1,500 V peak
Maximum common mode: 1,500 V peak, neutral to earth
Peak indicator: Illuminates at 2.5 × full scale

Watts (true power – VI cos ϕ):
Power factor response: Zero to unity leading or lagging
Accuracy: (V-A-W 25°C ± 5°C, 1 year)
DC and 40 Hz to 5 kHz: 0.25% of reading ±6 digits
5 Hz to 10 kHz: ±0.5% of reading ±0.5% of range
10 kHz to 20 kHz: ±1% of reading ±1% of range (2A range only)

General specifications:
Displays: Dual 4½-digit large high-intensity 7-segment LED
Operating temperature range: 0–50°C
Temperature coefficient: ±0.025% of range per °C from 0°C to 20°C and 30–50°C
Conversion rate: Approximately 600 mS
Power: 115/230 V AC ±10%, 50–60 Hz 5 W

where R is the resistance of the thermistor, and since

$$V_{rf0} = V_c$$

then the RF power can be calculated from

$$P_{rf} = \frac{1}{4R}(V_c - V_{rf})(V_c - V_{rf})$$

The processing electronics perform this computation on the output signals from the two bridges.

27.5 MEASUREMENT OF ELECTRICAL ENERGY

The energy supplied to an electrical circuit over a time period T is given by

$$E = \int_0^T p(t) \cdot dt$$

The most familiar instrument at power frequencies for the measurement of electrical energy is the watt-hour meter used to measure the electrical energy supplied to consumers by electricity supply undertakings. The most commonly used technique is the shaded pole induction watt-hour meter, shown in schematic form in Figure 27.48(a). This is essentially an induction motor for which the output is absorbed by its braking system and dissipated in heat. The rotating element is an aluminum disc, and the torque is produced by the interaction of the eddy currents induced in the disc with the imposed magnetic fields. The instantaneous torque is proportional to

$$(\phi_v i_i - \phi_i i_v)$$

where ϕ_v is the flux generated by the voltage coil, ϕ_i is the flux generated by the current coil, i_v is the eddy current generated in the disc by the voltage coil, and i_i is the eddy current generated in the disc by the current coil.

CHAPTER | 27 Electrical Measurements

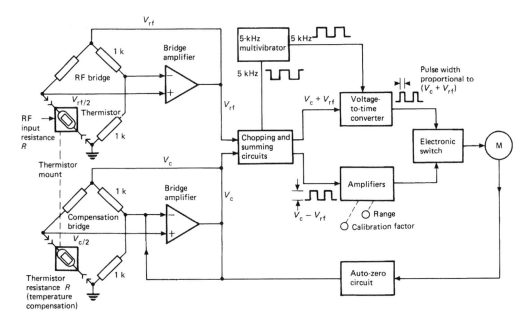

FIGURE 27.47 Thermistor RF power meter (from Hewlett-Packard, 1978).

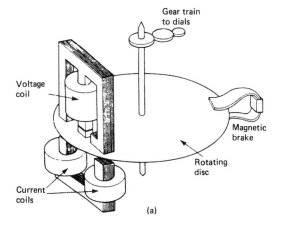

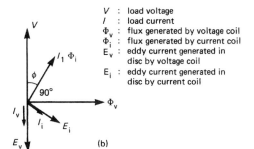

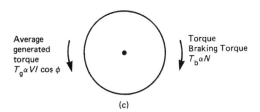

FIGURE 27.48 (a) Watt-hour meter; (b) phasor diagram of fluxes and eddy currents in watt-hour meter; (c) torque balance in a watt-hour meter.

The relative phases of these quantities are shown in Figure 27.48(b). The flux generated by the current coil is in phase with the current and the flux generated by the voltage coil is adjusted to be exactly in quadrature with the applied voltage by means of the copper shading ring on the voltage magnet.

The average torque, T_g, can be shown to be proportional to the power

$$T_g \propto VI\cos\phi$$

The opposing torque, T_b, is provided by eddy-current braking and thus is proportional to the speed of rotation of the disc, N, as shown in Figure 27.48(c). Equating the generated and braking torques,

$$T_b = T_g; \quad \text{and} \quad N \propto VI\cos\phi$$

and therefore the speed of rotation of the disc is proportional to the average power and the integral of the number of revolutions of the disc is proportional to the total energy supplied. The disc is connected via a gearing mechanism to a mechanical counter that can be read directly in watt-hours.

27.6 POWER-FACTOR MEASUREMENT

Power-factor measurement is important in industrial power supply, since generating bodies penalize users operating on poor power factors, because this requires high current-generating capacity but low energy transfer. It is possible to employ the dynamometer principle to provide an indicating instrument for power factor. This is shown in Figure 27.49.

The two movable coils are identical in construction but orthogonal in space. The currents in the two coils are equal in magnitude but time-displaced by 90°. There is no restoring torque provided in the instrument, and the movable coil system aligns itself so that there is no resultant torque. Thus:

$$VI\cos\phi \cdot \frac{dM_1}{d\theta} + VI\cos(\phi - 90°)\frac{dM_2}{d\theta} = 0$$

If the mutual inductance between the current carrying coil and the voltage coil 1 is given by

$$M_1 = k_1 \cos\theta$$

and if the mutual inductance between the current-carrying coil and the voltage coil 2 is given by

$$M_2 = k_1 \sin\theta$$

then the rest position of the power factor instrument occurs when

$$\theta = \phi$$

The dial of the instrument is usually calibrated in terms of the power factor, as shown in Figure 27.49. The method can also be applied to power-factor measurement in balanced three-phase loads (Golding and Widdis, 1963).

27.7 THE MEASUREMENT OF RESISTANCE, CAPACITANCE, AND INDUCTANCE

The most commonly used techniques for the measurement of these quantities are those of bridge measurement. The word *bridge* refers to the fact that in such measurements two points in the circuit are bridged by a detector, which detects either a potential difference or a null between them. Bridges are used extensively by National Standards Laboratories to maintain electrical standards by facilitating the calibration and intercomparison of standards and substandards. They are used to measure the resistance, capacitance, and inductance of actual components, and they do this by comparison with standards of these quantities. For details of the construction of standard resistors, capacitors, and inductors, readers should consult Hague and Foord (1971) and Dix and Bailey (1975). In a large number of transducers, nonelectrical quantities are converted into corresponding changes in resistance, capacitance, or inductance, which has led to the use of bridges in a wide variety of scientific and industrial measurements.

27.7.1 DC Bridge Measurements

The simplest form of a DC four-arm resistance bridge is the Wheatstone bridge, which is suitable for the measurement of resistance typically in the range from 1 Ω to 10 MΩ and is shown in Figure 27.50. The bridge can be used in either

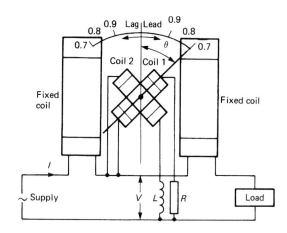

FIGURE 27.49 Power-factor instrument.

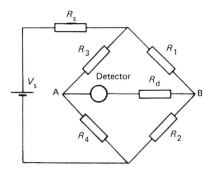

FIGURE 27.50 Wheatstone bridge.

a balanced, that is, null, mode or a deflection mode. In the balanced mode the resistance to be measured is R_1, and R_3 is a variable standard resistance. R_2 and R_4 set the ratio. The detector, which may be either a galvanometer or an electronic detector, is used to detect a null potential between the points A and B of the bridge. A null occurs when

$$R_1 = \frac{R_2}{R_4} \cdot R_3$$

The bridge is balanced either manually or automatically using the output signal from the detector in a feedback loop to find the null position. The null condition is independent of the source resistance, R_s, of the voltage source supplying the bridge or the sensitivity or input resistance, R_d, of the detector. These, however, determine the precision with which the balance condition can be determined. The sensitivity, S, of the bridge can be expressed as

$$S = \frac{\text{Bridge output voltage, } V_{\text{out}}, \text{ for a change } \Delta R_1 \text{ in } R_1}{\text{Bridge supply voltage}}$$

Near the balance condition for a given fractional change, δ, in R_1 given by

$$\delta = \frac{\Delta R_1}{R_1}$$

the sensitivity is given by

$$S = \frac{\delta R_d}{\sum_{i=1}^{d} R_i + R_d\left[2 + \left(\frac{R_2}{R_4}\right)\left(\frac{R_4}{R_3}\right)\right]} + R_s\left[2 + \left(\frac{R_3}{R_1}\right) + \left(\frac{R_1}{R_3}\right)\right] + R_d R_s \sum_{i=1}^{4}\left(\frac{1}{R_i}\right)$$

With an electronic detector, R_d can be made large, and if R_s is small, then S is given by

$$S = \frac{\delta}{\left[2 + \left(\frac{R_3}{R_4}\right)\left(\frac{R_4}{R_3}\right)\right]}$$

which has a maximum value of $\delta/4$ when $(R_3/R_4) = 1$.

The unbalanced mode is shown in Figure 27.51(a) and is often used with strain gauges (Chapter 4). R_1 is the active strain gauge and R_2 is the dummy gauge subject to the same temperature changes as R_1 but no strain. The output from the bridge is given by

$$V_{out} = \frac{V_s}{2}\left\{1 - \frac{1}{\left[1 + \left(\frac{\delta}{2}\right)\right]}\right\}$$

where

$$\delta = \frac{\Delta R}{R}$$

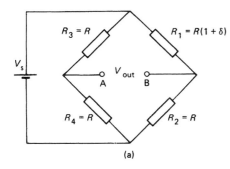

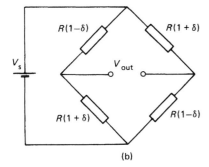

FIGURE 27.51 (a) Unbalanced Wheatstone bridge; (b) unbalanced Wheatstone bridge with increased sensitivity.

For $\delta \ll 1$ the output of the bridge is linearly related to the change in resistance, that is,

$$V_{out} = \frac{V_s}{4} \cdot \delta$$

Self-heating generally limits the bridge supply voltage and hence the output voltage. Amplification of the bridge output voltage has to be undertaken with an amplifier that has a high common-mode rejection ratio (CMRR), since the output from the bridge is in general small, and the common-mode signal applied to the amplifier is $V_s/2$. Further details of amplifiers suitable for use as bridge detectors can be found in Part 4.

The output from a strain-gauge bridge can be increased if four gauges are employed, with two in tension and two in compression, as shown in Figure 27.51(b). For such a bridge the output is given by

$$V_{out} = V_s \cdot \delta$$

Strain gauges and platinum resistance thermometers may be situated at a considerable distance from the bridge, and the long leads connecting the active element to the bridge will have a resistance that will vary with temperature. Figure 27.52 shows the use of the Wheatstone bridge in three-lead resistance measurement where it can be seen that close to balance the effect of the lead resistance and its temperature variation is approximately self-canceling and that the canceling effect deteriorates the further the bridge condition departs from balance. Figure 27.53 shows the use of Smith and Muller bridges to eliminate the lead resistance of a four-lead platinum resistance thermometer. (See also Chapter 1.)

27.7.1.1 Low-Resistance Measurement

Contact resistance causes errors in the measurement of low resistance, and therefore, to accurately define a resistance, it is necessary to employ the four-terminal technique shown in Figure 27.54. The outer two terminals are used to supply the current to the resistance, and the inner two, the potential terminals, determine the precise length of conductor over which the resistance is defined.

Measurement of low resistance is undertaken using the Kelvin double bridge shown in Figure 27.55(a). R_1 is the resistance to be measured and R_2 is a standard resistance on the same order of magnitude as R_1. The link between them, which is sometimes referred to as the *yoke,* has resistance r. The current through R_1 and R_2 is regulated by R. R_3, R_4, r_3, and r_4 are four resistances of which either R_3 and r_3 or R_4 and r_4 are variable and for which

$$\frac{R_3}{R_4} = \frac{r_3}{r_4}$$

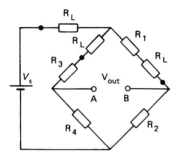
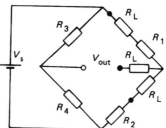
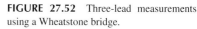

FIGURE 27.52 Three-lead measurements using a Wheatstone bridge.

R_1 is unknown resistance
R_L represents lead resistances
● Connections to R_1

$R_4 = R_2$

Balance condition:

$$\frac{R_1 + R_L}{R_2} = \frac{R_3 + R_L}{R_4}$$

$$\therefore \quad R_1 = R_3$$

Out-of-balance condition:

$R_2 = R_3 = R_4 = R$

$R_1 = R(1 + \delta)$

$$V_{out} = \frac{V_s \delta}{4}(1 - \beta)$$

$$\beta = \frac{R_L[R(4 + \delta) + R_L]}{R^2(3 + 2\delta) + RR_L(4 + \delta) + R_L^2}$$

R_1 is unknown resistance
R_L represents lead resistances
● Connections to R_1

$R_3 = R_4$

Balance condition:

$$\frac{R_1 + R_L}{R_2 + R_L} = \frac{R_3}{R_4}$$

$$\therefore \quad R_1 = R_2$$

Out-of-balance condition:

$R_2 = R_3 = R_4 = R$

$R_1 = R(1 + \delta)$

$$V_{out} = \frac{V_s \delta}{4}(1 - \beta')$$

$$\beta' = \frac{R_L}{R_L + R[1 + (\delta/2)]}$$

The delta star transformation applied to the bridge as shown in Figure 27.55(b) apportions the yoke resistance between the two sides of the bridge. The balance condition is given by

$$\frac{R_1 + r_a}{R_2 + r_c} = \frac{R_3}{R_4}; \quad r_a = \frac{r_3 \cdot r}{(r_3 + r_4 + r)} \quad r_c = \frac{r_4 \cdot r}{(r_3 + r_4 + r)}$$

and thus the unknown resistance R_1 is given by

$$R_1 = \frac{R_3}{R_4}(R_2 + r_c) - r_a$$

$$= \frac{R_3}{R_4} \cdot R_2 + \frac{r_4 \cdot r}{r_3 + r_4 + r}\left(\frac{R_3}{R_4} - \frac{r_3}{r_4}\right)$$

The term involving the yoke resistance r can be made small by making r small and by making

$$\frac{R_3}{R_4} = \frac{r_3}{r_4}$$

The bridge can be used to measure resistances typically from 0.1 μΩ to 1 Ω. For high precision the effect of thermally generated EMFs can be eliminated by reversing the current in R_1 and R_2 and rebalancing the bridge. The value of R_1 is then taken as the average of the two measurements.

27.7.1.2 High-Resistance Measurement

Modified Wheatstone bridges can be used to measure high resistance up to 10^{15} Ω. The problems in such measurements arise from the difficulty of producing stable, high-value standard resistors and errors caused by shunt-leakage resistance.

The problem of stable high-resistance values can be overcome by using the bridge with lower value and therefore more stable resistances. This leads to bridges that have larger ratios and hence reduced sensitivity. By operating the bridge with R_4 as the variable element then as $R_1 \to R_4 \to 0$.

The shunt leakage is made up of leakage resistance across the leads and the terminals of the bridge and across the unknown resistor itself. High-value standard resistors are constructed with three terminals. In the bridge arrangement shown in Figure 27.56(a), R_{sh1} shunts R_3; thus if $R_1 \gg R_3$, this method of connection decreases the effect of the leakage resistance. The only effect of R_{sh2} is to reduce the sensitivity of the balance condition.

Figure 27.56(b) shows a DC form of the Wagner grounding arrangement used to eliminate the effect of leakage resistance. The bridge balance then involves balancing the bridge with the detector across BC by adjusting R_6 and then balancing the bridge with the detector across AB by adjusting R_4. The procedure is then repeated until a balance is achieved under both conditions. The first balance condition ensures

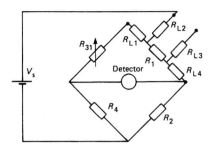
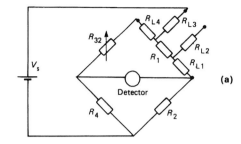

FIGURE 27.53 (a) Smith bridge for four-lead platinum resistance thermometer measurement; (b) Muller bridge for four-lead platinum resistance thermometer measurement.

Balance condition:

$R_1 + R_{L4} = R_{31} + R_{L1}$

Balance condition:

$R_1 + R_{L1} = R_{32} + R_{L4}$

Thus $R_1 = \dfrac{R_{31} + R_{32}}{2}$

R_1 is unknown resistance; $R_3 = R_4$; R_{L1}, R_{L2}, R_{L3}, R_{L4} are lead resistances

Bridge connections for first balance

Bridge connections for second balance

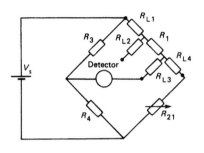
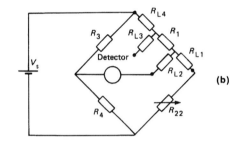

Balance condition:

$R_1 + R_{L1} = R_{21} + R_{L4}$

Balance condition:

$R_1 + R_{L4} = R_{22} + R_{L1}$

Thus $R_1 = \dfrac{R_{21} + R_{22}}{2}$

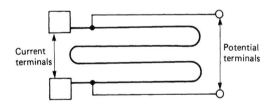

FIGURE 27.54 A four-terminal resistance.

that there is no potential drop across R_{sh2} and thus no current flows through it.

27.7.2 AC Equivalent Circuits of Resistors, Capacitors, and Inductors

Resistors, capacitors, and inductors do not exist as pure components. They are in general made up of combinations of all three impedance elements. For example, a resistor may have both capacitive and inductive parasitic elements. Figure 27.57 shows the complete equivalent circuits for physical realizations of the three components together with simplified equivalent circuits, which are commonly used. Further details of these equivalent circuits can be found in Oliver and Cage (1971).

At any one frequency, any physical component can be represented by its complex impedance $Z = R \pm jX$ or its admittance $Y = G \pm jB$. Since $Y = 1/Z$ and $Z = 1/Y$, then

$$R = \frac{G}{G^2 + B^2}; \qquad X = \frac{-B}{G^2 + B^2}$$

and

$$G = \frac{R}{R^2 + X^2}; \qquad B = \frac{-X}{R^2 + X^2}$$

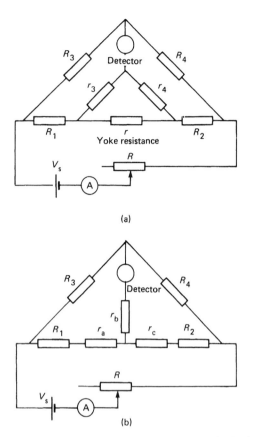

FIGURE 27.55 (a) Kelvin double bridge; (b) equivalent circuit of Kelvin double bridge.

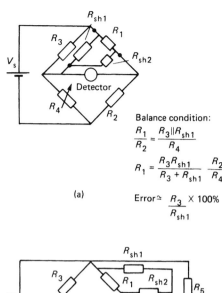

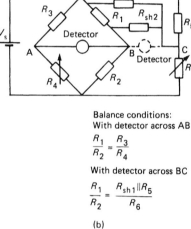

FIGURE 27.56 (a) Wheatstone bridge for use with three-terminal high resistances; (b) DC Wagner earthing arrangement.

These two representations of the component correspond to series and parallel equivalent circuits. If at a given frequency the impedance is $Z = R + jX$, the equivalent circuit at that frequency in terms of ideal components is a resistor in either series or parallel with an inductor, as shown in Figure 27.58(a). This figure also gives the conversion formulae between the two representations. For components for which the impedance at any given frequency is given by $Z = R - jX$, the equivalent circuits are series or parallel combinations of a resistor and a capacitor, as in Figure 27.58(b).

The quality factor, Q, is a measure of the ability of a reactive element to act as a pure storage element. It is defined as

$$Q = \frac{2\pi \times \text{maximum stored energy in the cycle}}{\text{Energy dissipated per cycle}}$$

The dissipation factor, D, is given by

$$D = \frac{1}{Q}$$

The Q and D factors for the series and parallel inductive and capacitive circuits are given in Figure 27.58. From this figure it can be seen that Q is given by $\tan\theta$ and D by $\tan\delta$, where δ is the loss angle. Generally, the quality of an inductance is measured by its Q factor and the quality of a capacitor by its D value or loss angle.

27.7.3 Four-Arm AC Bridge Measurements

If the resistive elements of the Wheatstone bridge are replaced by impedances and the DC source and detector are replaced by their AC equivalents, as shown in Figure 27.59, then if Z_1 is the unknown impedance, the balance condition is given by

$$Z_1 = \frac{Z_2 Z_3}{Z_4}; \quad R_1 + jX_1 = \frac{(R_2 + jX_2)(R_3 + jX_3)}{(R_4 + jX_4)}$$

or

$$|Z_1| = \frac{|Z_2||Z_3|}{|Z_4|} \text{ and } \angle Z_1 = \angle Z_2 + \angle Z_3 - \angle Z_4$$

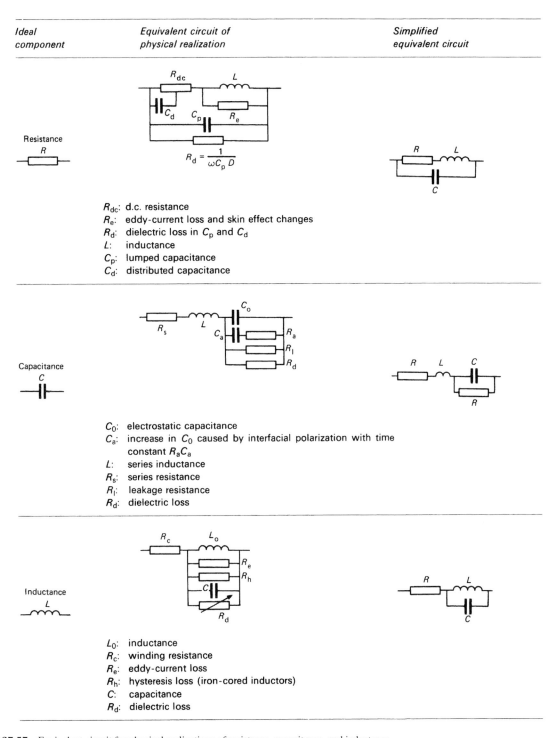

FIGURE 27.57 Equivalent circuit for physical realizations of resistance, capacitance, and inductance.

There are, therefore, a very large number of possible bridge configurations. The most useful can be classified according to the following scheme due to Ferguson. Since the unknown impedance has only two parameters R_1 and X_1, it is therefore sufficient to adjust only two of the six available parameters on the right-hand side of the balance equation. If the adjustment for each parameter of the unknown impedance is to be independent, the variables should be adjusted in the same branch. Adjusting the parameters R_2, X_2, is the same as adjusting parameters R_3, X_3, and thus four-arm bridges can be classified into one of two types, either ratio bridges or product bridges.

In the ratio bridge, the adjustable elements in either Z_2 or Z_3 are adjacent to the unknown impedance and the

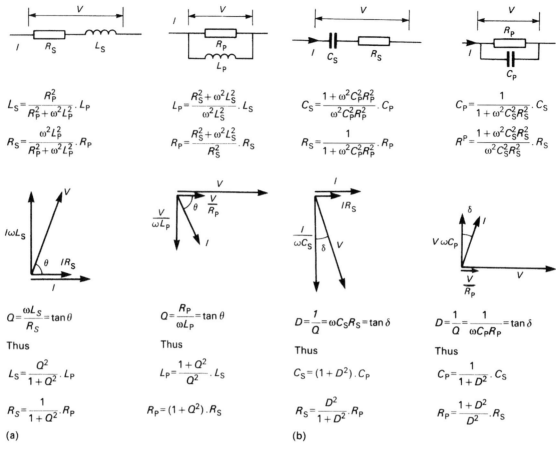

FIGURE 27.58 (a) Equivalent series/parallel resistor and inductor circuits; (b) equivalent series/parallel resistor and capacitor circuits.

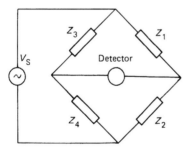

FIGURE 27.59 AC four-arm bridge.

ratio, either Z_3/Z_4 or Z_2/Z_4, must be either real or imaginary but not complex if the two elements in the balance condition are to be independent. In product bridges the balance is achieved by adjusting the elements in Z_4, which is opposite the unknown. For the adjustments to be independent requires $Z_2 \cdot Z_3$ to be real or imaginary but not complex.

Figure 27.60 gives examples of a range of commonly used four-arm bridges for the measurement of C and L. For further details concerning the application of such bridges, the reader should consult Hague and Foord (1971).

27.7.3.1 Stray Impedances in AC bridges

Associated with the branches, source, and detector of an AC bridge are distributed capacitances to ground. The use of shields around these elements enables the stray capacitances to be defined in terms of their location, magnitude, and effect. Figure 27.61(a) shows these capacitances, and Figure 27.61(b) shows the equivalent circuit with the stray capacitances transformed to admittances across the branches of the bridge and the source and detector. The stray admittances across the source and detector do not affect the balance condition. The balance condition of the bridge in terms of the admittances of the branches and the admittances of the stray capacitances across them is given by

$$(Y_1 + Y_{AB})(Y_4 + Y_{CD}) = (Y_3 + Y_{AD})(Y_2 + Y_{CB})$$

where, for example,

$$Y_{AB} = \frac{Y_A Y_B}{Y_A + Y_B + Y_C + Y_D} = \frac{Y_A Y_B}{\Delta};$$

$$\Delta = Y_A + Y_B + Y_C + Y_D$$

CHAPTER | 27 Electrical Measurements 481

Bridge	Circuit	Balance conditions	Notes
Maxwell	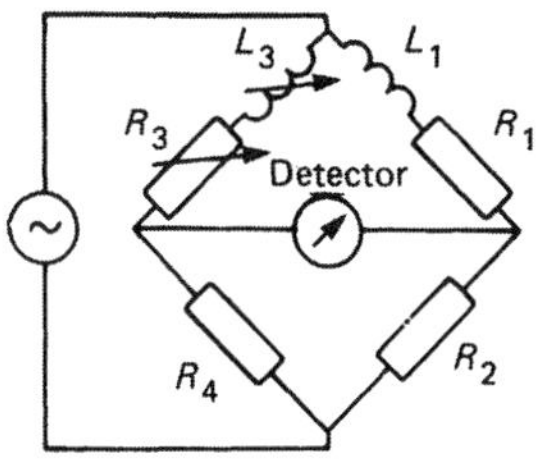	$L_1 = \dfrac{R_2}{R_4} L_3$ $R_1 = \dfrac{R_2}{R_4} R_3$	Ratio bridge with inductive and resistive standards for the measurement of the series inductance and resistance of an unknown inductor; balance condition is frequency independent and therefore purity of source is unimportant; a parallel form of the bridge can be used to measure the parallel components of an unknown inductance
Maxwell–Wien	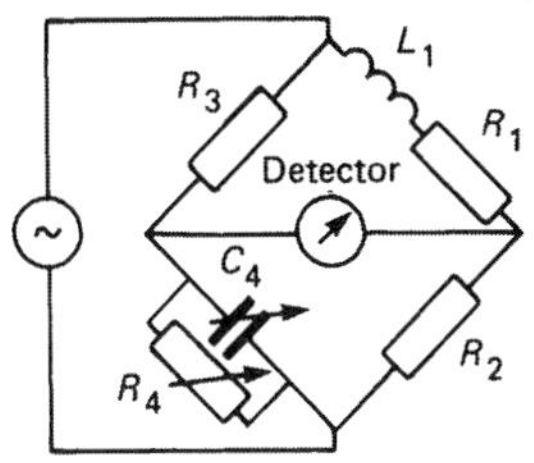	$L_1 = R_2 R_3 C_4$ $R_1 = \dfrac{R_2 R_3}{R_4}$ $Q_1 = \omega C_4 R_4$	Product bridge employing capacitive and resistive standards for the measurement of the series inductance and resistance of an unknown inductor; widely used for the measurement of inductance; if C_4 and R_4 are variable bridge measures L_1 and R_1; if R_4 and R_2 or R_3 are variable bridge measures L_1 and Q_1
Hay	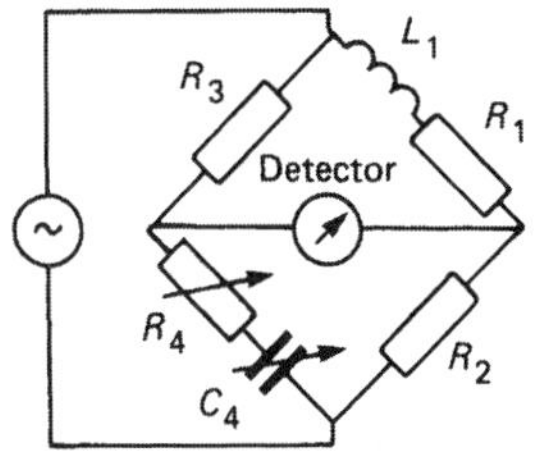	$L_1 = \dfrac{R_2 R_3 C_4}{1 + \omega^2 C_4^2 R_4^2}$ $R_1 = \dfrac{R_2 R_3 \omega^2 C_4^2 R_4^2}{(1 + \omega^2 C_4^2 R_4^2)}$ $Q_1 = \dfrac{1}{\omega C_4 R_4}$	Product bridge employing capacitive and resistive standards for the measurement of the series inductance and resistance of an unknown inductor; suitable for the measurement of a.c. inductance in the presence of d.c. bias current; used for the measurement of inductances with high L and Q
Owen	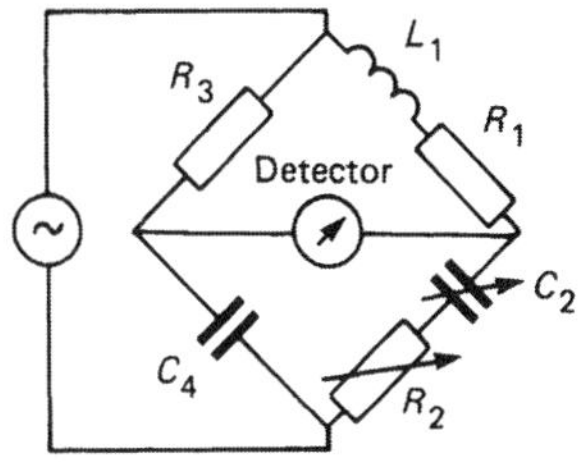	$L_1 = C_4 R_3 \cdot R_2$ $G_1 = \dfrac{1}{R_1} = \dfrac{1}{C_4 R_3} \cdot C_2$	Ratio bridge employing capacitive and resistive standards for the measurement of the series inductance and conductance of an unknown inductor; used as a high-precision bridge
Series capacitance component bridge	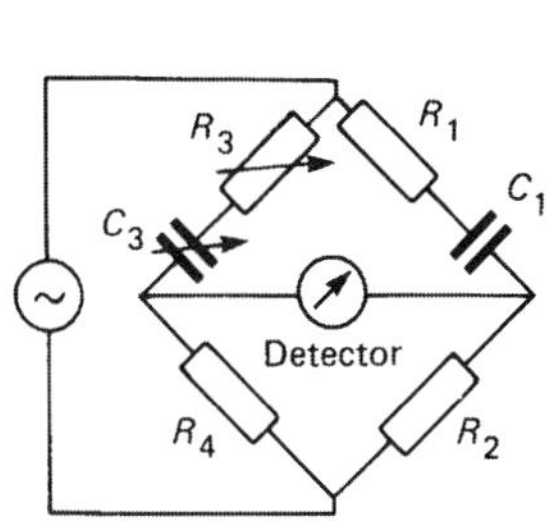	$C_1 = \dfrac{R_4}{R_2} \cdot C_3$ $R_1 = \dfrac{R_2}{R_4} \cdot R_3$ $D_1 = \omega C_3 R_3$	Ratio bridge employing capacitive and resistive standards for the measurement of the series capacitance and resistance of an unknown capacitor; widely used for the measurement of capacitance; if C_3 and R_3 are variable bridge measures C_1 and R_1; if R_3 and R_4 are variable bridge measures C_1 and D_1

FIGURE 27.60 AC four-arm bridges for the measurement of capacitance and inductance.

(Continued)

FIGURE 27.60 (Continued)

Bridge	Circuit	Balance conditions	Notes
Parallel capacitance component bridge	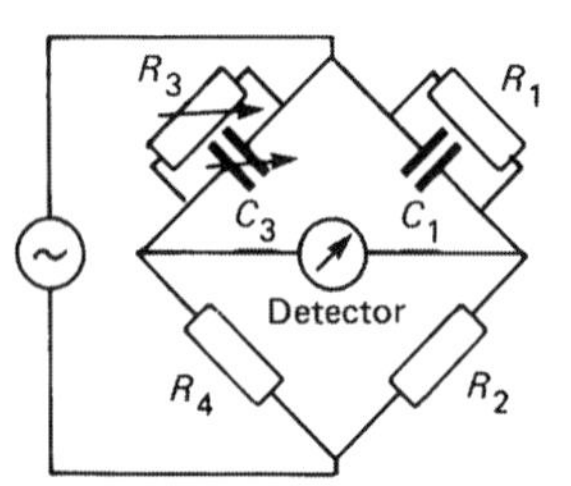	$C_1 = \dfrac{R_4}{R_2} \cdot C_3$ $R_1 = \dfrac{R_4}{R_2} \cdot R_3$ $D_1 = \dfrac{1}{\omega C_3 R_3}$	Ratio bridge employing capacitive and resistive standards for the measurement of the parallel capacitance and resistance of an unknown capacitor; used particularly for high D capacitor measurement
Maxwell–Wien	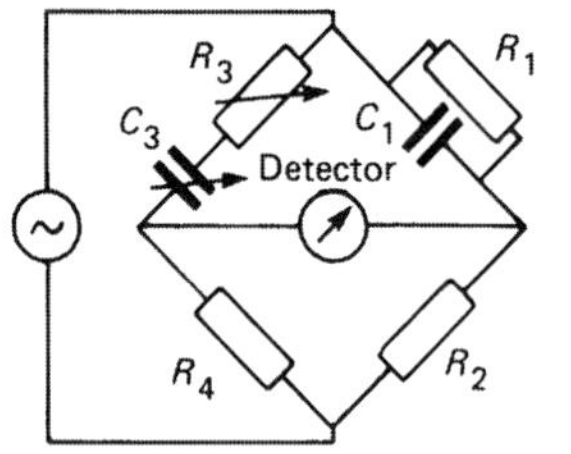	$C_1 = \dfrac{R_4}{R_2} \cdot \dfrac{C_3}{1+\omega^2 C_3^2 R_3^2}$ $R_1 = \dfrac{R_2}{R_4} \cdot \dfrac{1+\omega^2 C_3^2 R_3^2}{\omega^2 C_3^2 R_3}$ $D_1 = \omega C_3 R_3$	Ratio bridge employing capacitive and resistive standards for the measurement of the parallel capacitance and resistance of an unknown capacitor; used as a frequency-dependent circuit in oscillators
Schering	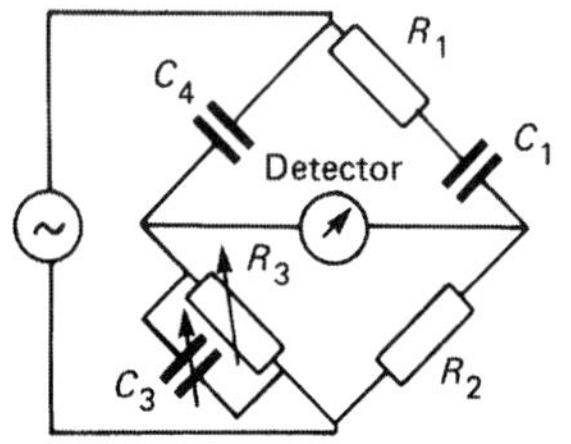	$C_1 = \dfrac{C_4}{R_2} \cdot R_3$ $R_1 = \dfrac{R_2}{C_4} \cdot C_3$ $D_1 = \omega C_3 R_3$	Product bridge employing capacitive and resistive standards for the measurement of the parallel capacitance and resistance of an unknown capacitor; used for measuring dielectric losses at high voltage and r.f. measurements

and thus the balance condition is given by

$$(Y_1 Y_4 - Y_2 Y_3)$$
$$+ \frac{1}{\Delta}(Y_1 Y_C Y_D + Y_4 Y_A Y_B - Y_3 Y_C Y_B - Y_2 Y_A Y_D) = 0$$

If the stray capacitances are to have no effect on the balance condition, this must be given by

$$Y_1 Y_4 = Y_2 Y_3$$

and the second term of the balance condition must be zero. It can be easily shown that this can be achieved by either

$$\frac{Y_A}{Y_C} = \frac{Y_1}{Y_2} = \frac{Y_3}{Y_4} \quad \text{or} \quad \frac{Y_B}{Y_D} = \frac{Y_1}{Y_3} = \frac{Y_2}{Y_4}$$

Thus, the stray impedances to ground have no effect on the balance condition if the admittances at one opposite pair of branch points are in the same ratio as the admittances of the pairs of branches shunted by them.

The Wagner earthing arrangement shown in Figure 27.62 ensures that Points D and B of the balanced bridge are at ground potential; thus the effect of stray impedances at these points is eliminated. This is achieved by means of an auxiliary arm of the bridge, consisting of the elements Y_5 and Y_6. The bridge is first balanced with the detector between D and B by adjusting Y_3. The detector is moved between B and E and the auxiliary bridge balanced by adjusting Y_5 and Y_6. This ensures that Point B is at earth potential. The two balancing processes are repeated until the bridge balances with the detector in both positions. The balance conditions for the main bridge and the auxiliary arm are then given by

$$Y_1 Y_4 = Y_2 Y_3 \quad \text{and} \quad Y_3(Y_6 + Y_C) = Y_4(Y_5 + Y_A)$$

27.7.4 Transformer Ratio Bridges

Transformer ratio bridges, which are also called *inductively coupled bridges*, largely eliminate the problems associated with stray impedances. They also have the advantage that only a small number of standard resistors and capacitors is needed. Such bridges are therefore commonly used as universal bridges to measure the resistance, capacitance, and inductance of components having a wide range of values at frequencies up to 250 MHz.

The element that is common to all transformer ratio bridges is the tapped transformer winding, shown in Figure 27.63. If the transformer is ideal, the windings have zero leakage flux, which implies that all the flux from one winding links with the other, and zero winding resistance. The core material on which the ideal transformer is wound has zero eddy-current and hysteresis losses. Under these circumstances, the ratio of the voltages V_1 to V_2 is identical to the ratio of the turns n_1 to n_2, and this ratio is independent of the loading applied to either winding of the transformer.

In practice the transformer is wound on a tape-wound toroidal core made from a material such as supermalloy or supermumetal, which has low eddy-current and hysteresis loss as well as high permeability. The coil is wound as a multistranded rope around the toroid, with individual strands in the rope joined in series, as shown in Figure 27.64. This configuration minimizes the leakage inductance of the windings. The windings are made of copper with the largest cross-sectional area to minimize their resistance. Figure 27.65 shows an equivalent circuit of such a transformer. L_1 and L_2 are

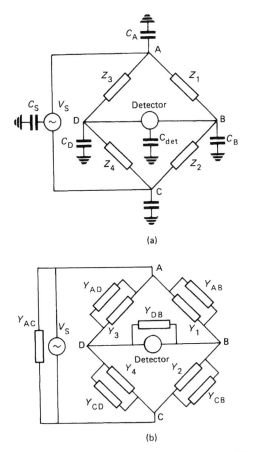

FIGURE 27.61 (a) Stray capacitances in a four-arm AC bridge; (b) equivalent circuit of an AC four-arm bridge with stray admittances.

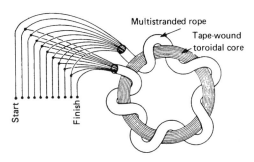

FIGURE 27.64 Construction of a toroidal tapped transformer.

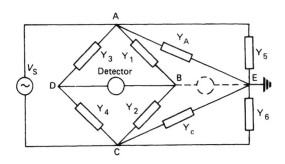

FIGURE 27.62 Wagner earthing arrangement.

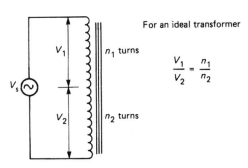

FIGURE 27.63 Tapped transformer winding.

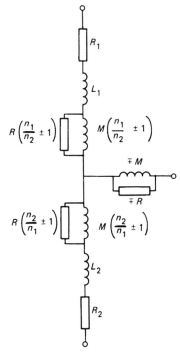

FIGURE 27.65 Equivalent circuit of a tapped transformer.

the leakage inductances of the windings; R_1 and R_2 are the winding resistances; M is the mutual inductance between the windings; and R represents hysteresis and eddy-current loss in the core.

The ratio error from the ideal value of n_1/n_2 is given approximately by

$$\frac{n_2(R_1 + j\omega L_1) - n_1(R_2 + j\omega L_2)}{(n_1 + n_2)} \cdot \left(\frac{1}{R} + \frac{1}{j\omega M}\right) \times 100\%$$

and this error can be made to be less than 1 part in 10^6.

The effect of loading is also small. An impedance Z applied across the n_2 winding gives a ratio error of

$$\frac{(n_1/n_2)(R_2 + j\omega L_2)(n_2/n_1)(R_1 + j\omega L_1)}{[(n_1 + n_2)/n_2] \cdot Z} \times 100\%$$

For an equal bridge, with $n_1 = n_2$, this is

$$\frac{R_2 + j\omega L_2}{Z} \times 100\%$$

which is approximately the same error as if the transformer consisted of a voltage source with an output impedance given by its leakage inductance and the winding resistance. These can be made to be small, and thus the effective output impedance of the transformer is low; therefore, the loading effect is small. The input impedance of the winding seen by the AC source is determined by the mutual inductance of the windings (which is high) and the loss resistance (which is also high).

Multidecade ratio transformers, as shown in Figure 27.66, use windings either with separate cores for each decade or all wound on the same core. For the multicore transformer, the input for the next decade down the division chain is the output across a single tap of the immediately higher decade. For the windings on a single core, the number of decades that can be accommodated is limited by the need to maintain the volts/turn constant over all the decades, and therefore the number of turns per tap at the higher decade becomes large. Generally a compromise is made between the number of cores and the number of decades on a single core.

27.7.4.1 Bridge Configurations

There are three basic bridge configurations, as shown in Figure 27.67. In Figure 27.67(a) the detector indicates a null when

$$\frac{Z_1}{Z_2} = \frac{V_1}{V_2}$$

and for practical purposes

$$\frac{V_1}{V_2} = \frac{n_1}{n_2} = n$$

Thus

$$Z_1 = nZ_2; \quad |Z_1| = n|Z_2| \quad \text{and} \quad \angle Z_1 = \angle Z_2$$

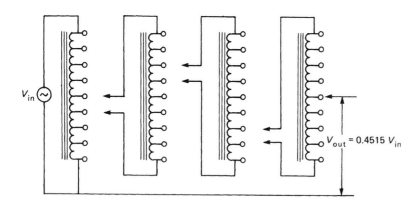

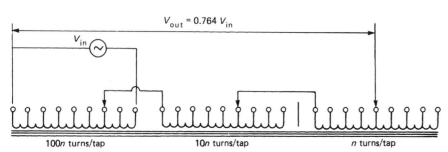

FIGURE 27.66 Multidecade ratio transformers.

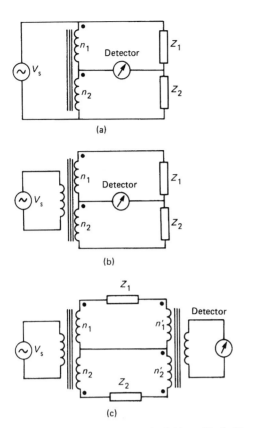

FIGURE 27.67 (a) Autotransformer ratio bridge; (b) double-wound transformer ratio bridge; (c) double ratio bridge.

The bridge can therefore be used for comparing like impedances.

The three-winding voltage transformer shown in Figure 27.67(b) has the same balance condition as the bridge in Figure 27.67(a). However, in the three-winding bridge the voltage ratio can be made more nearly equal to the turns ratio. The bridge has the disadvantage that the leakage inductance and winding resistance of each section is in series with Z_1 and Z_2; therefore the bridge is most suitable for the measurement of high impedances.

Figure 27.67(c) shows a double ratio transformer bridge in which the currents I_1 and I_2 are fed into a second double-wound transformer. The detector senses a null condition when there is zero flux in the core of the second transformer. Under these conditions for an ideal transformer

$$I_1 n'_1 = I_2 n'_2; \quad \frac{I_1}{I_2} = \frac{n'_2}{n'_1} = \frac{1}{n'}$$

and the second transformer presents zero input impedance. Therefore, since

$$\frac{I_1}{I_2} = \frac{Z_2}{Z_1} \cdot \frac{n_1}{n_2} = \frac{Z_2}{Z_1} n$$

then

$$Z_1 = nn' Z_2; \quad |Z_1| = nn' |Z_2| \quad \text{and} \quad \angle Z_1 = \angle Z_2$$

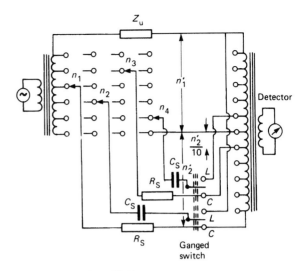

FIGURE 27.68 Universal bridge.

By using the two ratios, this bridge extends the range of measurement that can be covered by a small number of standards.

Figure 27.68 shows a universal bridge for the measurement of R, C, and L. In the figure only two decades of the inductive divider that control the voltages applied to the bank of identical fixed capacitors and resistors are shown. The balance condition for the bridge, when connected to measure capacitance, is given by

$$C_u = \frac{n'_2}{n'_1} \left(\frac{n_2}{10} + \frac{n_4}{100} \right) \cdot C_s$$

and

$$\frac{1}{R_u} = \frac{n'_2}{n'_1} \left(\frac{n_1}{10} + \frac{n_3}{100} \right) \cdot \frac{1}{R_s}$$

When measuring inductance, the current through the capacitor and inductor are summed into the current transformer and the value of capacitance determined is the value that resonates with the inductance. For an unknown inductance, its measured values in terms of its parallel equivalent circuit are given by

$$L_{up} = \frac{1}{\omega^2 C_u}; \quad \frac{1}{R_{up}} = \frac{1}{R_u}$$

where the values of C_u and R_u are given in these equations. The value of w is chosen such that it is a multiple of 10, and therefore the values of L_{up} and C_u are reciprocal. The values of L_{up} and R_{up} can be converted to their series equivalent values using the equations in Section 27.7.2.

The transformer ratio bridge can also be configured to measure low impedances, high impedances, and network and amplifier characteristics. The ampere turn balance used in ratio bridges is also used in current comparators employed in the calibration of current transformers and for intercomparing four-terminal impedances. Details of these applications can be found in Gregory (1973), Hague and Foord (1971), and Oliver and Cage (1971). The current comparator principle can also be extended to enable current comparison to be made at DC (Dix and Bailey, 1975).

Transformer ratio bridges are often used with capacitive and inductive displacement transducers because they are immune to errors caused by earth-leakage impedances and since they offer an easily constructed, stable, and accurately variable current or voltage ratio (Hugill, 1983; Neubert, 1975).

27.7.4.2 The Effect of Stray Impedances on the Balance Condition of Inductively Coupled Bridges

Figure 27.69 shows the unknown impedance with its associated stray impedances Z_{sh1} and Z_{sh2}. The balance condition of the bridge is unaffected by Z_{sh1} since the ratio of V_1 to V_2 is unaffected by shunt loading. At balance the core of the current transformer has zero net flux. There is no voltage drop across its windings; hence there is no current flow through Z_{sh2}. Z_{sh2} has therefore no effect on the balance condition. Thus the bridge rejects both stray impedances. This enables the bridge to measure components *in situ* while still connected to other components in a circuit. In practice, if the output impedance of the voltage transformer has a value Z_{vt} and the current transformer has an input impedance of Z_{ct}, the error on the measurement of Z_1 is given approximately by

$$\left(\frac{Z_{vt}}{Z_{sh1}} + \frac{Z_{ct}}{Z_{sh2}}\right) \times 100\%$$

27.7.4.3 The Use of Inductively Coupled Bridges in an Unbalanced Condition

The balance condition in inductively coupled bridges is detected as a null. The sensitivity of the bridge determines the output under unbalance conditions and therefore the precision with which the balance can be found. Figure 27.70 shows the two-winding voltage and current transformers and their equivalent circuits. Figure 27.71 shows the sensitivities of the two bridges when used with capacitive and inductive

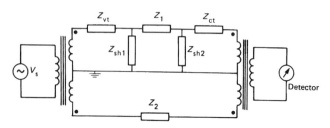

FIGURE 27.69 Effect of stray impedances on balance condition.

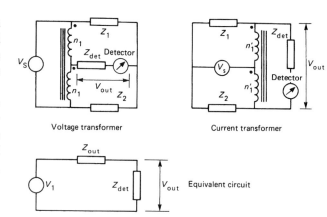

	V_1	Z_{out}
Voltage transformer	$\dfrac{V_S}{2} \dfrac{(Z_2 - Z_1)}{(Z_2 + Z_1)}$	$Z_1 \| Z_2$
Current transformer	$\dfrac{2V_S (Z_2 - Z_1)}{(Z_2 + Z_1 + Z_1 Z_2 / Z_c)}$	$Z_1 \| 2Z_c + Z_2 \| 2Z_c$

$Z_c = j\omega L_c$: L_c is inductance of ratio arms
$L_c = M_c$ mutual inductance of ratio arms

FIGURE 27.70 Unbalanced inductively coupled bridge.

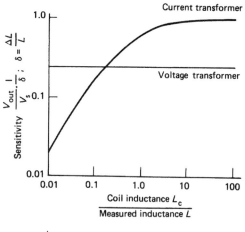

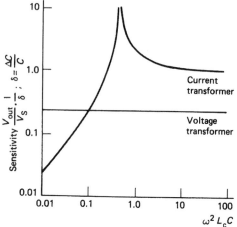

FIGURE 27.71 Sensitivity of current and voltage transformer bridges.

elements. The capacitors form a resonant circuit with the current transformer, and for frequencies below the resonant frequency, the sensitivity of the bridge is dependent on both w, the angular excitation frequency of the bridge, and L_c, the self-inductance of the winding, as shown in Figure 27.71. The dependence of the sensitivity on w and L_c can be reduced at the cost of reduced sensitivity (Neubert, 1975).

27.7.4.4 Autobalancing Ratio Bridges

By employing feedback as shown in Figure 27.72, the transformer ratio bridge can be made to be self-balancing. The high-gain amplifier ensures that at balance the current from the unknown admittance Y_u is balanced by the current through the feedback resistor. Thus at balance

$$V_1 Y_u n'_1 = \frac{V_{out}}{R} \cdot n'_2$$

with

$$V_1 = \overline{V}_1 \sin \omega t$$

$$V_{out} = \overline{V}_{out} \sin(\omega t + \phi)$$

and

$$Y_u = G_u + jB_u$$

$$G_u = \frac{n'_2}{n'_1} \cdot \frac{1}{R} \cdot \frac{V_{out}}{V_1} \cos \phi;$$

$$B_u = \frac{n'_2}{n'_1} \cdot \frac{1}{R} \cdot \frac{V_{out}}{V_1} \cdot \sin \phi$$

The amplifier output and a signal 90° shifted from that output are then passed into two phase-sensitive detectors. These detectors employ reference voltages that enable the resistive and reactive components of the unknown to be displayed.

Windings can be added to the bridge to enable the bridge to measure the difference between a standard and the unknown.

27.7.5 High-Frequency Impedance Measurement

As the frequency of measurement is increased, the parasitic elements associated with real components begin to dominate the measurement. Therefore, RF bridges employ variable capacitors (typically less than 1,000 pF) as the adjustable elements in bridges and fixed resistors whose physical dimensions are small. A bridge that can be constructed using these elements is the Schering bridge, shown in Figure 27.60. Great care has to be taken with shielding and wiring layout in RF bridges, to avoid large coupling loops. The impedance range covered by such bridges decreases as the frequency is raised. At microwave frequencies all the wiring is coaxial, discrete components are no longer used, and impedance measurements can only be undertaken for impedances close to the characteristic impedance of the system. Further details of high-frequency measurements can be found in Oliver and Cage (1971) and Somlo and Hunter (1985).

The bridged T and parallel T circuits (shown in Figure 27.73 together with their balance conditions) can be used for measurements at RF frequencies. The parallel T measurement technique has the advantage that the balance can be achieved using two grounded variable capacitors.

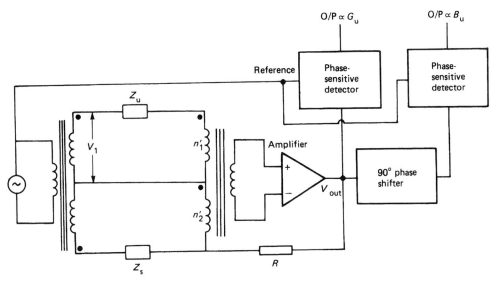

FIGURE 27.72 Autobalancing ratio bridge.

If $Z_s = \infty$ output gives conductance and susceptance of unknown impedance Z_u
If $Z_s \neq \infty$ output gives deviation of conductance and susceptance from the values of the standard impedance Z_s

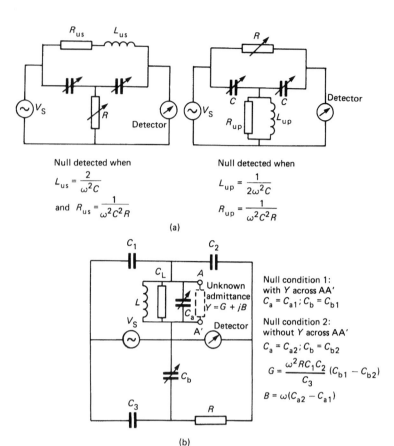

FIGURE 27.73 Bridged T (a) and parallel T (b) circuits for the measurement of impedance at high frequencies.

Resonance methods can also be used for the measurement of components at high frequencies. One of the most important uses of resonance in component measurement is the Q meter, shown in Figure 27.74. In measuring inductance as shown in Figure 27.74(a), the variable capacitor C, which forms a series-resonant circuit with L_{us}, is adjusted until the detector detects resonance at the frequency f. The resonance is detected as a maximum voltage across C. At resonance, Q is given by

$$Q = \frac{V_c}{V_{in}} = \frac{V_L}{V_{in}}$$

and L_{us} is given by

$$L_{us} = \frac{1}{4\pi^2 f^2 C}$$

The value of R_{us} is given by

$$R_{us} = \frac{1}{2\pi f C Q}$$

The self-capacitance of an inductor can be determined by measuring the value of C—say, C_1—which resonates with it at a frequency f together with value of C—say, C_2—which resonates with the inductance at $2f$. Then C_0, the self-capacitance of the coil, is given by

$$C_0 = \frac{C_1 - 4C_2}{3}$$

In Figure 27.74(b), the use of the Q meter to measure the equivalent parallel capacitance and resistance of a capacitor is shown. Using a standard inductor at a frequency f, the capacitor C is adjusted to a value C_1C_c at which resonance occurs. The unknown capacitor is connected across C, and the value of C is adjusted until resonance is found again. If this value is C_2, the unknown capacitor C_{up} has a value given by

$$C_{up} = C_1 - C_2$$

Its dissipation factor, D, is given by

$$D = \frac{Q_1 - Q_2}{Q_1 Q_2} \cdot \frac{C_1}{C_1 - C_2}$$

where Q_1 and Q_2 are the measured Q values at the two resonances. Its parallel resistance, R_{up}, is given by

$$R_{up} = \frac{Q_1 Q_2}{Q_1 - Q_2} \cdot \frac{1}{2\pi f C_1}$$

The elements of the high-frequency equivalent circuit of a resistance in Figure 27.74(c) can also be measured. At a

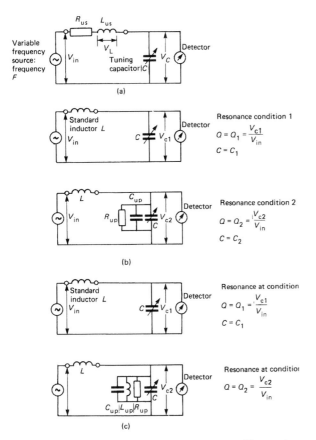

FIGURE 27.74 Q meter, (a) Inductance measurement; (b) capacitance measurement; (c) resistance measurement.

given frequency, f, the capacitor C is adjusted to a value C_1 such that it resonates with L. The resistor is then connected across the capacitor and the value of C adjusted until resonance is reestablished. Let this value of C be C_2. If the values of Q at the resonances are Q_1 and Q_2, respectively, values of the unknown elements are given by

$$R_{up} = \frac{Q_1 Q_2}{(Q_1 - Q_2)} \cdot \frac{1}{2\pi f C_1}$$

$$C_{up} = C_1 - C_2$$

and

$$L_{up} = \frac{1}{(2\pi f)^2 \cdot C_{up}}$$

27.8 DIGITAL FREQUENCY AND PERIOD/TIME-INTERVAL MEASUREMENT

These measurements, together with frequency ratio, phase difference, rise and fall time, and duty-factor measurements, employ digital counting techniques and are all fundamentally related to the measurement of time.

The SI unit of time is defined as the duration of 9,192,631,770 periods of the radiation corresponding to the transition between the $F = 4$, $m_f = 0$ and $F = 3$, $m_f = 0$ hyperfine levels of the ground state of the cesium-133 atom. The unit is realized by means of the cesium-beam atomic clock in which the cesium beam undergoes a resonance absorption corresponding to the required transition from a microwave source. A feedback mechanism maintains the frequency of the microwave source at the resonance frequency. The SI unit can be realized with an uncertainty of between 1 part in 10^{13} and 10^{14}. Secondary standards are provided by rubidium gas cell resonator-controlled oscillators or quartz crystal oscillators. The rubidium oscillator uses an atomic resonance effect to maintain the frequency of a quartz oscillator by means of a frequency-lock loop. It provides a typical short-term stability (averaged over a 100-s period) of five parts in 10^{13} and a long-term stability of one part in 10^{11}/month. Quartz crystal oscillators provide inexpensive secondary standards with a typical short-term stability (averaged over a 1-s period) of five parts in 10^{12} and a long-term stability of better than one part in 10^8/month. Details of time and frequency standards can be found in Hewlett-Packard (1974).

Dissemination of time and frequency standards is also undertaken by radio broadcasts. Radio stations transmit waves for which the frequencies are known to an uncertainty of a part in 10^{11} or 10^{12}. Time-signal broadcasting on a time scale known as Coordinated Universal Time (UTC) is coordinated by the Bureau International de L'Heure (BIH) in Paris. The BIH annual report details the national authorities responsible for time-signal broadcasts, the accuracies of the carrier frequencies of the standard frequency broadcasts, and the characteristics of national time-signal broadcasts. Table 27.9 provides details of time broadcast facilities in the United Kingdom.

27.8.1 Frequency Counters and Universal Timer/Counters

Frequency measurements are undertaken by frequency counters, the functions of which (in addition to frequency measurement) may also include frequency ratio, period measurement, and totalization. Universal timer/counters provide the functions of frequency counters with the addition of time-interval measurement. Figure 27.75 shows the elements of a microprocessor-controlled frequency counter. The input signal conditioning unit accepts a wide range of input signal levels, typically with a maximum sensitivity corresponding to a sinusoid having an RMS value of 20 mV and a dynamic range from 20 mV rms to 20 V rms. The trigger circuit has a trigger level that is either set automatically with respect to the input wave or can be continuously adjusted over some range. The trigger circuit generally employs hysteresis to reduce the effect of noise on the waveform, as shown in

TABLE 27.9 U.K. time broadcasts

GBR 16 kHz radiated from Rugby (52° 22′ 13″ N 01 ° 10′ 25″ W)

Power: ERP 65 kW
Transmission modes: A1, FSK (16.00 and 15.95 kHz), and MSK (future)

Time signals: Schedule (UTC)	Form of the time signals
0255 to 0300 0855 to 0900 1455 to 1500 2055 to 2100 There is an interruption for maintenance from 1000 to 1400 every Tuesday	A 1 type second pulses lasting 100 ms, lengthened to 500 ms at the minute The reference point is the start of carrier rise Uninterrupted carrier is transmitted for 24 s from 54 m 30 s and from 0 m 6 s DUTI: CCIR code by double pulses

MSF 60 kHz radiated from Rugby

Power: ERP 27 kW

Schedule (UTC)	Form of the time signals
Continuous except for an interruption for maintenance from 1000 to 1400 on the first Tuesday in each month	Interruptions of the carrier of 100 ms for the second pulses and of 500 ms for the minute pulses. The epoch is given by the beginning of the interruption BCD NRZ code, 100 bits/s (month, day of month, hour, minute), during minute interruptions BCD PWM code, 1 bit/s (year, month, day of month, day of week, hour, minute) from seconds 17 to 59 in each minute DUT1: CCIR code by double pulses

The MSF and GBR transmission are controlled by a cesium beam frequency standard. Accuracy $\pm 2 \times 10^{-12}$

Figure 27.76(a), although this can cause errors in time measurement, as shown in Figure 27.76(b).

The quartz crystal oscillator in a frequency counter or universal counter timer can be uncompensated, temperature compensated, or oven stabilized. The frequency stability of quartz oscillators is affected by aging, temperature, variations in supply voltage, and changes in power supply mode, that is, changing from line-frequency supply to battery supply. Table 27.10 gives comparative figures for the three types of quartz oscillator. The uncompensated oscillator gives sufficient accuracy for five- or six-digit measurement in most room-temperature applications. The temperature-compensated oscillator has a temperature-dependent compensating network for frequency correction and can give sufficient accuracy for a six- or seven-digit instrument. Oven-stabilized oscillators maintain the temperature of the crystal typically at $70 \pm 0.01°C$. They generally employ higher mass crystals with lower resonant frequencies and operate at an overtone of their fundamental frequency. They have better aging performance than the other two types of crystal and are suitable for use in seven- to nine-digit instruments.

The microprocessor provides control of the counting operation and the display and post-measurement computation.

Conventional frequency counters count the number of cycles, n_i, of the input waveform of frequency, f_i, in a gating period, t_g, which corresponds to a number of counts, n_{osc}, of the 10 MHz crystal oscillator. They have an uncertainty corresponding to ±1 count of the input waveform. The relative resolution is given by

Relative resolution
$$= \frac{\text{Smallest measurable change in measurement value}}{\text{Measurement value}}$$

and for the measurement of frequency is thus

$$\pm \frac{1}{\text{Gating period} \times \text{input frequency}} = \pm \frac{1}{t_g \cdot f_i}$$

To achieve measurements with good relative resolution for low-frequency signals, long gating times are required. Reciprocal frequency counters synchronize the gating time to the input waveform, which then becomes an exact number of cycles of the input waveform. The frequency of the input waveform is thus calculated as

$$f_i = \frac{\text{Number of cycles of input waveform}}{\text{Gating period}}$$
$$= \frac{n_i}{n_{osc}} \times 10^{-7} \text{ Hz}$$

The relative resolution of the reciprocal method is

$$\pm \frac{10^{-7}}{\text{Gating time}} = \pm \frac{10^{-7}}{t_g} = \pm \frac{1}{n_{osc}}$$

independent of the input frequency, and thus it is possible to provide high-resolution measurements for low-frequency signals. Modern frequency counters often employ both methods, using the conventional method to obtain the high resolution at high frequencies.

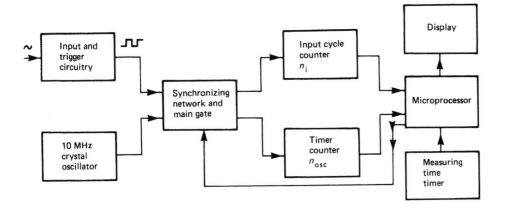

FIGURE 27.75 Digital frequency counter.

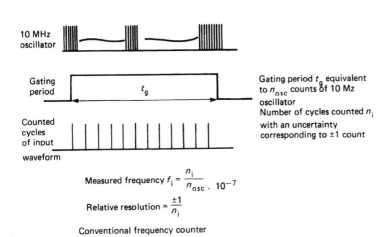

Conventional frequency counter

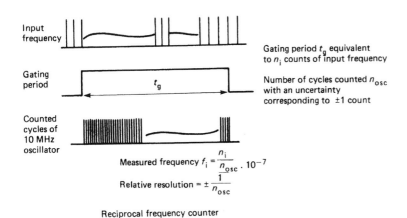

Reciprocal frequency counter

The period, T_1, of the input wave is calculated from

$$T_i = \frac{1}{f_i} = \frac{\text{Gating period}}{\text{Number of cycles of input waveform}}$$

$$= \frac{n_{osc} \times 10^{-7}}{n_i}$$

with a relative resolution of ± 1 in n_{osc}.

The accuracy of frequency counters is limited by four factors. These are the system resolution and the trigger, systematic, and time-base errors. Trigger error (TE) is the absolute measurement error due to input noise causing triggering that is too early or too late. For a sinusoidal input waveform it is given by

$$TE = \pm \frac{1}{\pi f_i} \text{ (input signal to noise ratio)}$$

and for a nonsinusoidal wave

$$TE = \pm \frac{\text{Peak-to-peak noise voltage}}{\text{Signal slew rate}}$$

Systematic error (SE) is caused by differential propagation delays in the start and stop sensors or amplifier channels of the counter or by errors in the trigger level settings of the start and stop channels. These errors can be removed by calibration. The time-base error (TBE) is caused by deviation on the frequency of the crystal frequency from its calibrated value. The causes of the deviation have been considered previously.

The relative accuracy of frequency measurement is given by

$$\pm \frac{\text{Resolution of } f_i}{f_i} \pm \frac{TE}{t_g} \pm \text{Relative TBE}$$

and the relative accuracy of period measurement is given by

$$\pm \frac{\text{Resolution of } T_i}{T_i} \pm \frac{TE}{t_g} \pm \text{Relative TBE}$$

Figure 27.77 shows the techniques employed in single-shot time interval measurement and frequency ratio measurement.

Table 27.11 gives the characteristics of a series of 200 MHz universal timer/counters (Racal-Dana 9902/9904/9906).

27.8.2 Time-Interval Averaging

Single-shot time interval measurements using a 10 MHz clock have a resolution of ± 100 ns. However, by performing repeated measurements of the time interval, it is possible to significantly improve the resolution of the measurement (Hewlett-Packard, 1977b). As shown in Figure 27.78, the number of counts of the digital clock in the time interval, T, will be either n or $n + 1$. It can be shown that if the measurement clock and the repetition rate are asynchronous the best estimate of the time interval, T, is given by

$$T = \bar{n} T_{osc}$$

where $\bar{n}$ is the average number of counts taken over N repetitions and T_{osc} is the period of the digital clock.

The standard deviation, σ_T, which is a measure of the resolution in time-interval averaging (TIA) for large N is given by

$$\sigma_T = \frac{T_{osc}}{\sqrt{N}} \sqrt{[F(F-1)]}$$

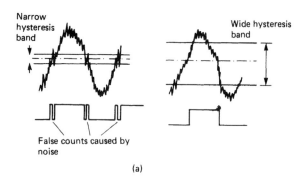

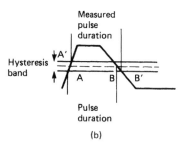

FIGURE 27.76 (a) The use of hysteresis to reduce the effects of noise; (b) timing errors caused by hysteresis.

Table 27.10 Quartz oscillator characteristics

Stability against	Uncompensated	Temperature compensated	Oven stabilized
Aging: /24 h	n.a.	n.a.	$<5 \times 10^{-10}$*
/month	$<5 \times 10^{-7}$	$<1 \times 10^{-7}$	$<1 \times 10^{-8}$
/year	$<5 \times 10^{-6}$	$<1 \times 10^{-7}$	$<5 \times 10^{-8}$
Temperature: 0–50°C ref. to +23°C	$<1 \times 10^{-5}$	$<1 \times 10^{-6}$	$<5 \times 10^{-9}$
Change in measuring and supply mode: line/int. battery/ext. D.C. 12–26 V	$<3 \times 10^{-7}$	$<5 \times 10^{-8}$	$<3 \times 10^{-9}$
Line voltage: ±10%	$<1 \times 10^{-8}$	$<1 \times 10^{-9}$	$<5 \times 10^{-10}$
Warm-up time to reach within 10^{-7} of final value	n.a.	n.a.	<15 min

*After 48 h of continuous operation.

CHAPTER | 27 Electrical Measurements

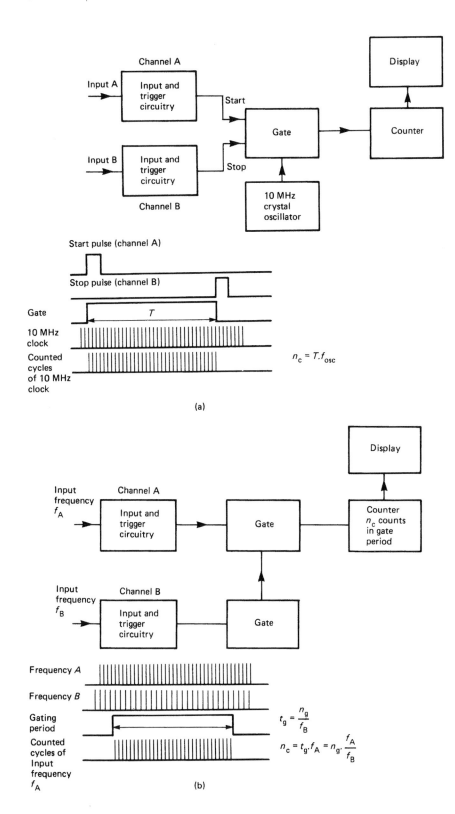

FIGURE 27.77 (a) Single-shot time-interval measurement; (b) frequency ratio measurement.

where F lies between 1 and 0, dependent on the time interval being measured. Thus the maximum standard deviation on the time estimate is $T_{osc}/(2\sqrt{N})$. By employing repeated measurements using a 10 MHz clock, it is possible to obtain a resolution of 10 ps. Repeated measurements also reduce errors due to trigger errors caused by noise. The relative accuracy of TIA measurements is given by

$$\pm \frac{\text{Resolution of } T}{T} \pm \frac{TE}{\sqrt{(N)} \cdot T} \pm \frac{SE}{T} \pm \text{Relative TBE}$$

TABLE 27.11 Universal timer/counter specifications

	Measuring functions
Modes of operation	Frequency
	Single and multiple period
	Single and multiple ratio
	Single and double-line time interval
	Single and double-line time interval averaging
	Single and multiple totalizing

	Frequency measurement
Input	Channel A
Coupling	AC or DC
Frequency range	DC to 50 MHz (9902 and 9904)
	HF DC to 30 MHz
	VHF 10 MHz to 200 MHz prescaled by 4 (9906)
Accuracy	±1 count ± timebase accuracy
Gate times (9000 and 9902)	Manual: 1 ms to 100 s
	Automatic: gate times up to 1 s are selected automatically to avoid overspill
	Hysteresis avoids undesirable range changing for small frequency changes
	1 ms to 100 s in decade steps (9904)
	HF: 1 ms to 100 s
	VHF: 4 ms to 400 s

	Single- and multiple-period measurement
Input	Channel A
Range	1 µs to 1 s single period 100 ns to 1 s multiple period (9902 and 9904)
Clock unit	1 µs to 100 s single period 100 ns to 100 s multiple period (9906)
Coupling	1 µs
Periods averaged	AC or DC
Resolution	1 to 10^5 in decade steps
Accuracy	10 ps maximum
	$\dfrac{\pm 0.3\%}{\text{Number of periods averaged}} \pm \text{count} \pm \text{timebase accuracy}$
	(measured at 50 mV rms input with 40 dB S/N ratio)
Bandwidth	Automatically reduced to 10 MHz (3 dB) when period selected

	Time interval single and double input
Input	Single input: channel B
Time range	Double input: start channel B stop channel A
	100 ns to 10^4 s (2.8 h approx.) (9902)
	100 ns to 10^5 s (28 h approx.) (9904)
	100 ns to 10^6 s (280 h approx.) (9906)
	±1 count ± trigger error ± timebase accuracy
Accuracy	Trigger error $= \dfrac{5}{\text{Signal slope at the trigger point (V/}\mu\text{s)}}$ ns
Clock units	100 ns to 10 ms in decade steps
Start/stop signals	Electrical or contact
Manual start/stop	By single push button on front panel
Trigger slope selection	Positive or negative slope can be selected on both start and stop
Manual start/stop (9900)	By single push button on front panel
	N.B. Input socket automatically biased for contact operation (1 mA current sink)
Trigger slope selection (9900)	Electrical-positive or negative slopes can be selected on both start and stop signals
	Contact-opening or closure can be selected on both start and stop signals
Bounce protection (9900)	A 10 ms dead time is automatically included when contact operation is selected

	Time-interval averaging single and double input
Input	Single input: channel B
	Double input: Start channel
	B Stop channel A

CHAPTER | 27 Electrical Measurements

TABLE 27.11 (*Continued*)

	Time-interval averaging single and double input
Time range	150 ns to 100 ms (9902)
	150 ns to 1 s 9904
	150 ns to 10 s (9906)
Dead time between intervals	150 ns
Clock unit	100 ns
Time intervals averaged	1 to 10^5 in decade steps
Resolution	100 ns to 1 ps
Accuracy	± Timebase accuracy ± system error ± averaging error
	System error: 10 s per input channel. This is the difference in delays between start and stop signals and can be minimized by matching externally
	Averaging error = $\dfrac{\text{Trigger error} \pm 100}{\sqrt{\text{Intervals averaged}}}$ ns
	Trigger error = $\dfrac{5}{\text{Signal slope at the trigger point}(V/\mu s)}$ ns
	Ratio
Higher-frequency input	Channel A
Higher-frequency range	10 Hz to 30 MHz (9900)
	DC to 50 MHz (9902, 9904)
Lower-frequency input	Channel B
Lower-frequency range reads	DC to 10 MHz
	$\dfrac{\text{Frequency } A}{\text{Frequency } B} \times n$
Multiplier n	1 to 10^5 in decade steps
Accuracy	$\dfrac{\pm 1 \text{count} \pm \text{trigger error on Channel B}}{\text{No. of gated periods}}$
	Trigger error = $\dfrac{5}{\text{Signal slope at the trigger point}(V/\mu s)}$ ns
	Totalizing
Input	Channel A (10 MHz max.)
Max. rate	10^7 events per second
Pulse width	50 ns minimum at trigger points
Pre-scaling	Events can be pre-scaled in decade multiples (n) from 1 to 10^5
Reads	$\dfrac{\text{No. of input events} \pm 1 \text{ count} - 0}{n}$
Manual start/stop	By single push button on front panel
Electrical start/stop	By electrical signal applied to Channel B

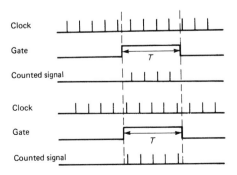

FIGURE 27.78 Resolution of one-shot time-interval measurement.

With a high degree of confidence this can be expressed as

$$\pm \frac{1}{N} \cdot \frac{1}{n} \pm \frac{TE}{\sqrt{(N)} \cdot n \cdot T} \pm \frac{SE}{n \cdot T_{\text{osc}}} \pm \text{Relative TBE}$$

27.8.3 Microwave-Frequency Measurement

By the use of pre-scaling, as shown in Figure 27.79, in which the input signal is frequency divided before it goes into the gate to be counted, it is possible to measure frequencies up

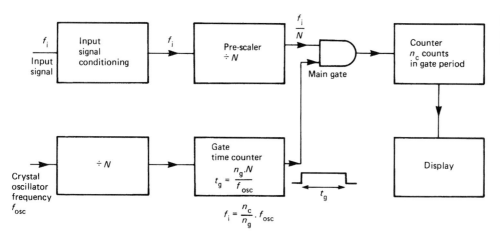

FIGURE 27.79 Frequency measurement range extension by input pre-scaling.

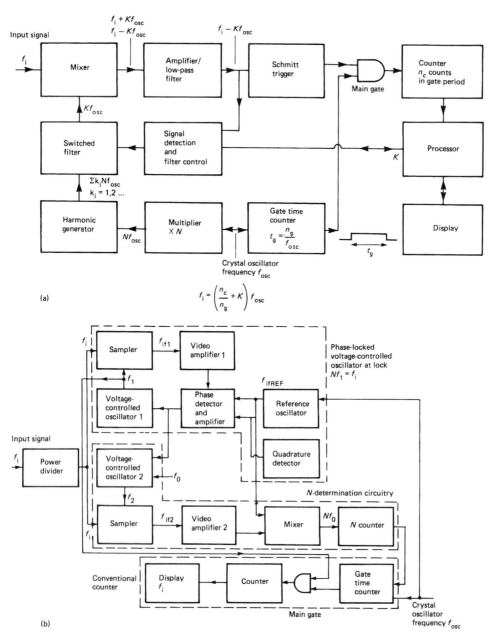

FIGURE 27.80 (a) Heterodyne converter counter; (b) transfer oscillator counter (from Hewlett-Packard, 1977a).

CHAPTER | 27 Electrical Measurements

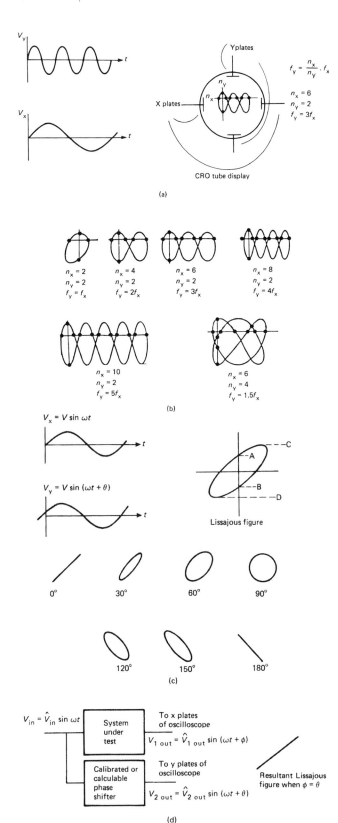

FIGURE 27.81 (a) Frequency measurement using Lissajous figures; (b) Lissajous figures for various ratios of f_x to f_y; (c) phase measurement using Lissajous figures; (d) improved phase measurement using Lissajous figures.

to approximately 1.5 GHz. Higher frequencies typically up to 20 GHz can be measured using the heterodyne converter counter shown in Figure 27.80(a), in which the input signal is down-mixed by a frequency generated from a harmonic generator derived from a crystal-controlled oscillator. In the transfer oscillator technique shown in Figure 27.80(b), a low-frequency signal is phase-locked to the microwave input signal. The frequency of the low-frequency signal is measured, together with its harmonic relationship to the microwave signal. This technique typically provides measurements up to 23 GHz. Hybrid techniques using both heterodyne down-conversion and transfer oscillator extend the measurement range, typically to 40 GHz. Further details of these techniques can be found in Hewlett-Packard (1977a).

27.9 FREQUENCY AND PHASE MEASUREMENT USING AN OSCILLOSCOPE

Lissajous figures can be used to measure the frequency or phase of a signal with respect to a reference source, though with an accuracy much lower than for other methods described. Figure 27.81 (a) shows the technique for frequency measurement. One signal is applied to the X plates of the oscilloscope and the other to the Y plates. Figure 27.81(b) shows the resulting patterns for various ratios of the frequency f_x applied to the X plates to the frequency f_x applied to the Y plates. If f_x is the known frequency, it is adjusted until a stationary pattern is obtained. f_y is then given by

$$f_y = \frac{f_x \cdot n_x}{n_y}$$

where n_x is the number of crossings of the horizontal line and n_y the number of crossings of the vertical line, as shown in Figure 27.81(b).

If the two signals are of the same frequency, their relative phases can be determined from the figures shown in Figure 27.81(c). The phase angle between the two signals is given by

$$\sin \theta = \frac{AB}{CD}$$

The accuracy of the method can be significantly increased by the use of a calibrated or calculable phase shift introduced to ensure zero phase shift between the two signals applied to the plates of the oscilloscope, as shown in Figure 27.81(d).

REFERENCES

Analog Devices, *Data Acquisition Databook*, Analog Devices, Norwood, Mass., **10**, 123–125 (1984).

Arbel, A. F., *Analog Signal Processing and Instrumentation*, Cambridge University Press, Cambridge (1980).

Bailey, A. E., "Units and standards of measurement," *J. Phys. E: Sci. Instrum.*, **15**, 849–856 (1982).

Bishop, J. and E. Cohen, "Hall effect devices in power measurement," *Electronic Engineering (GB)*, **45**, No. 548, 57–61 (1973).

British Standards Institution, BS 3938: 1973: Specification for Current Transformers, BSI, London (1973).

British Standards Institution, BS 3941: 1974: Specification for Voltage Transformers, BSI, London (1974).

British Standards Institution, BS 89: 1977: Specification for Direct Acting Electrical Measuring Instruments and their Accessories, BSI, London (1977).

Brodie, B., "A 160ppm digital voltmeter for use in ac calibration," *Electronic Engineering (GB)*, **56**, No. 693, 53–59 (1984).

Clarke, F. J. J., and Stockton, J. R., "Principles and theory of wattmeters operating on the basis of regularly spaced sample pairs," *J. Phys. E: Sci. Instrum.*, **15**, 645–652 (1982).

Dix, C. H., "Calculated performance of a digital sampling wattmeter using systematic sampling," *Proc. I.E.E.*, **129**, Part A, No. 3, 172–175 (1982).

Dix, C. H., and Bailey, A. E. "Electrical Standards of Measurement – Part 1 D. C. and low frequency standards," *Proc. I.E.E.*, **122**, 1018–1036 (1975).

Fantom, A. E., *Microwave Power Measurement*, Peter Peregrinus for the IEE, Hitchin (1985).

Froelich, M., "The influence of the pH value of the electrolyte on the e.m.f. stability of the international Weston cell," *Metrologia*, **10**, 35–39 (1974).

Goldman, D. T. and R. J. Bell, *SI: The International System of Units*, HMSO, London (1982).

Golding, E. W. and Widdis, F. C., *Electrical Measurements and Measuring Instruments*, 5th ed., Pitman, London (1963).

Gregory, B. A., *Electrical Instrumentation*, Macmillan, London (1973).

Gumbrecht, A. J., *Principles of Interference Rejection*, Solartron DVM Monograph No. 3, Solartron, Farnborough, U.K. (1972).

Hague, B. and Foord, T. R., *Alternating Current Bridge Methods*, Pitman, London (1971).

Harris, F. K., *Electrical Measurements*, John Wiley, New York (1966).

Hewlett-Packard, *Fundamentals of Time and Frequency Standards*, Application Note 52-1, Hewlett-Packard, Palo Alto, Calif. (1974).

Hewlett-Packard, *Fundamentals of Microwave Frequency Counters*, Application Note 200-1, Hewlett-Packard, Palo Alto, Calif. (1977a).

Hewlett-Packard, *Understanding Frequency Counter Specifications*, Application Note 200-4, Hewlett-Packard, Palo Alto, Calif. (1977b).

Hewlett-Packard, *Fundamentals of RF and Microwave Power Measurements*, Application Note 64-1, Hewlett-Packard, Palo Alto, Calif. (1978).

Hugill, A. L., "Displacement transducers based on reactive sensors in transformer ratio bridge circuits," in *Instrument Science and Technology*, Volume 2, ed. Jones, B. E., Adam Hilger, Bristol (1983).

Josephson, B. D., "Supercurrents through barriers," *Phys. Letters*, **1**, 251 (1962).

Kibble, B. P., Smith, R. C., and Robinson, I. A., "The NPL moving coil-ampere determination," *I.E.E.E. Trans.*, **IM-32**, 141–143 (1983).

Matouka, M. F., "A wide-range digital power/energy meter for systems with non-sinusoidal waveforms," *I.E.E.E. Trans.*, **IE-29**, 18–31 (1982).

Miljanic, P. N., Stojanovic, B. and Petrovic, V. "On the electronic three-phase active and reactive power measurement," *I.E.E.E. Trans.*, **IM-27**, 452–455 (1978).

NBS, NBS Monograph 84: Standard Cells—Their Construction, Maintenance, and Characteristics, NBS, Washington, D.C. (1965).

Neubert, H. P. K., *Instrument Transducers*, 2nd ed., Oxford University Press, London (1975).

Oliver, B. M. and Cage, J. M., *Electronic Measurements and Instrumentation*, McGraw-Hill, New York (1971).

Owens, A. R., "Digital signal conditioning and conversion," in *Instrument Science and Technology*, Volume 2 (ed. B. E. Jones), Adam Hilger, Bristol (1983).

Pearce, J. R., "Scanning, A-to-D conversion and interference," *Solartron Technical Report Number 012183*, Solartron Instruments, Farnborough, U.K. (1983).

Pitman, J. C., "Digital voltmeters: a new analog to digital conversion technique," *Electronic Technology (GB)*, **12**, No. 6, 123–125 (1978).

Rathore, T. S., "Theorems on power, mean and RMS values of uniformly sampled periodic signals," *Proc. I.E.E.*, **131**, Part A, No. 8, 598–600 (1984).

Rayner, G. H., "An absolute determination of resistance by Campbell's method," *Metrologia*, **3**, 8–11 (1967).

Simeon, A. O. and McKay, C. D., "Electronic wattmeter with differential inputs," *Electronic Engineering (GB)*, **53**, No. 648, 75–85 (1981).

Sheingold, D. H., *Analog/digital Conversion Notes*, Analog Devices, Norwood, Mass. (1977).

Somlo, P. I. and Hunter, J. D., *Microwave Impedance Measurement*, Peter Peregrinus for IEE, Hitchin (1985).

Spreadbury, P. J., "Electronic voltage standards," *Electronics and Power*, **27**, 140–142 (1981).

Steele, J. A., Ditchfield, C. R., and Bailey, A. E., "Electrical standards of measurement Part 2: RF and microwave standards," *Proc. I.E.E.*, **122**, 1037–1053 (1975).

Stone, N. W. B., et al., "Electrical standards of measurement Part 3: submillimeter wave measurements and standards," *Proc. I.E.E.*, **122**, 1053–1070 (1975).

Tagg, G. F., *Electrical Indicating Instruments*, Butterworths, London (1974).

Thompson, A. M. and Lampard, D. G., "A new theorem in electrostatics with applications to calculable standards of capacitance," *Nature (GB)*, **177**, 888 (1956).

Vigoureux, P., "A determination of the ampere," *Metrologia*, **1**, 3–7 (1965).

Vigoureux, P., *Units and Standards for Electromagnetism*, Wykenham Publications, London (1971).

FURTHER READING

Carr, J. J., *Elements of Electronic Instrumentation and Measurement*, Prentice Hall, Englewood Cliffs, N.J. (1986).

Coombs, F., Electronic Instrument Handbook (1994).

Fantom, A. E., Bailey, A. E., and Lynch, A. C., (eds.), *Radio Frequency and Microwave Power Measurement*, Institution of Electrical Engineers (1990).

O'Dell, *Circuits for Electronic Instrumentation*, Cambridge University Press, Cambridge (1991).

Schnell, L. (ed.), *Technology of Electrical Measurements*, Wiley.

Chapter 28

Optical Measurements

A. W. S. Tarrant

28.1 INTRODUCTION

A beam of light can be characterized by its spectral composition, its intensity, its position and direction in space, its phase, and its state of polarization. If something happens to it to alter any of those quantities and the alterations can be quantified, a good deal can usually be found out about the "something" that caused the alteration. Consequently, optical techniques can be used in a huge variety of ways, but it would be quite impossible to describe them all here.

This chapter describes a selection of widely used instruments and techniques. Optical instruments can be conveniently thought of in two categories: those basically involving image formation (for example, microscopes and telescopes) and those that involve intensity measurement (for example, photometers). Many instruments (for example, spectrophotometers) involve both processes, and it is convenient to regard these as falling in the second, intensity measurement category. For the purposes of this book we are almost entirely concerned with instruments in this second category. Image-formation instruments are well described in familiar textbooks such as R. S. Longhurst's *Geometrical and Physical Optics*.

The development of optical fibers and light guides has enormously broadened the scope of optical techniques. Rather than take wires to some remote instrument to obtain a meaningful signal from it, we can often now take an optical fiber, an obvious advantage where rapid response times are involved or in hazardous environments.

In all branches of technology, it is quite easy to make a fool of oneself if one has no previous experience of the particular techniques involved. One purpose of this book is to help the nonspecialist to find out what is and what is not possible. The author would like to pass on one tip to people new to optical techniques: When we consider what happens in an optical system, we must consider what happens in the *whole optical system*; putting an optical system together is not quite like putting an electronic system together. Take, for example, a spectrophotometer, which consists of a light source, a monochromator, a sample cell, and a detector. An alteration in the position of the lamp will affect the distribution of light on the detector and upset its operation, despite the fact that the detector is several units "down the line." Optical systems must be thought of as a whole, not as a group of units acting in series.

It should be noted that the words *optical* and *light* are used in a very loose sense when applied to instruments. Strictly, these terms should only refer to radiation within the visible spectrum, that is, the wavelength range 380–770 nm. The techniques used often serve equally well in the near-ultraviolet and near-infrared regions of the spectrum, and naturally we use the same terms. It is quite usual to hear people talking about ultraviolet light when they mean *ultraviolet radiation*, and many "optical" fibers are used with infrared radiation.

28.2 LIGHT SOURCES

Light sources for use in instruments may be grouped conveniently under two headings: (1) conventional or incoherent sources and (2) laser or coherent sources. Conventional sources are dealt with in Sections 28.2.1 to 28.2.3 and laser sources in Section 28.2.4.

The principal characteristics of a conventional light source for use in instruments are:

1. Spectral power distribution.
2. Luminance or radiance.
3. Stability of light output.
4. Ease of control of light output.
5. Stability of position.

Other factors that may have to be taken into account in the design of an instrument system are heat dissipation, the nature of auxiliary equipment needed, source lifetime, cost, and ease of replacement. By the *radiance* of a light source, we mean the amount of energy per unit solid angle radiated from a unit area of it. Very often this quantity is much more important than the actual power of the lamp. For example, if a xenon arc is to be used with a spectroscopic system, the quantity of interest is the amount of light that can be got through the slit of the spectrometer. A low-power lamp with a high radiance can be focused down to give a small, intense image at the slit and thus get a lot of light through; but if the lamp has a low radiance, it does not matter how powerful it is, it cannot be refocused to pass the same amount of radiation through the system. It can be easily shown in this case that the radiance is the only effective parameter of the source.

If light output in the visible region only is concerned, *luminance* is sometimes used instead of *radiance*. There is a strict parallel between units and definitions of "radiant" quantities and "luminous" quantities (BS Spec. 4727; IEC–CIE, *International Lighting Vocabulary*). The unit of light, the lumen, can be thought of as a unit of energy weighted with regard to wavelength according to its ability to produce a visible sensation of light (Walsh, 1958, p. 138).

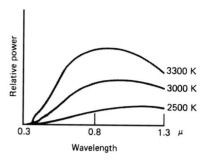

FIGURE 28.1 Spectral power distribution of tungsten lamp.

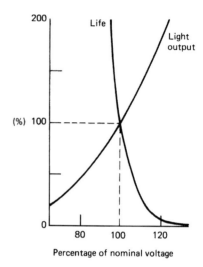

FIGURE 28.2 Variation with voltage of life and light output of a tungsten lamp (after Henderson and Marsden).

28.2.1 Incandescent Lamps

Incandescent sources are those in which light is generated by heating material electrically until it becomes white hot. Normally this material is tungsten, but if only infrared radiation is wanted it may be a ceramic material. In a tungsten lamp the heating is purely resistive, and the use of various filament diameters enables lamps to be made of similar power but different voltage ratings. The higher the voltage, the finer and more fragile is the filament. For instrument purposes, small and compact filaments giving the highest radiance are usually needed, so low-voltage lamps are often used. For lamps used as radiation standards it is customary to use a solid tungsten ribbon as a filament, but these require massive currents at low voltage (for example, 18 A, 6 V).

The spectral power distribution of a tungsten lamp corresponds closely to that of a Planckian radiator, as shown in Figure 28.1, and the enormous preponderance of red energy will be noted. Tungsten lamps have a high radiance, are very stable in light output provided that the input power is stabilized, and are perfectly stable in position. The light output can be precisely controlled by varying the input power from zero to maximum, but because the filament has a large thermal mass, there is no possibility of deliberately modulating the light output. If a lamp is run on an AC supply at mains frequency, some modulation at twice that frequency invariably occurs and may cause trouble if other parts of the instrument system use mains–frequency modulation. The modulation is less marked with low-voltage lamps, which have more massive filaments, but can only be overcome by using either a smoothed dc supply or a high-frequency power supply (10 kHz).

The main drawback to tungsten lamps is the limited life. The life depends on the voltage (see Figure 28.2), and it is common practice in instrument work to under-run lamps to get a longer life.

Longer lamp lives are obtained with tungsten halogen lamps. These have a small amount of a halogen—usually bromine or iodine—in the envelope that retards the deterioration of the filament. It is necessary for the wall temperature of the bulb to be at least 300°C, which entails the use of a small bulb made of quartz. However, this allows a small amount of ultraviolet radiation to escape, and with its small size and long life, the tungsten halogen lamp is a very attractive light source for instrument purposes.

28.2.1.1 Notes on Handling and Use

After lengthy use troubles can arise with lamp holders, usually with contact springs weakening. Screw caps are preferable to bayonet caps. The envelopes of tungsten halogen lamps should not be touched by hand—doing so will leave grease on them, which will burn into the quartz when hot and ruin the surface.

28.2.2 Discharge Lamps

Discharge lamps are those in which light is produced by the passage of a current through a gas or vapor, hence producing mostly line spectra. Enclosed arcs of this kind have negative temperature/resistance characteristics, so current-limiting devices are necessary. Inductors are often used if the lamp is to be run on an ac mains supply. Many types of lamp are

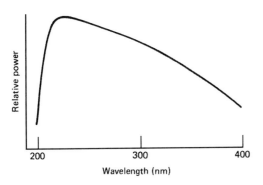

FIGURE 28.3 Typical spectral power distribution for a deuterium lamp.

available (Henderson and Marsden, 1972); some commonly met with in instrument work are mentioned here.

28.2.2.1 Deuterium Lamps

The radiation is generated by a low current density discharge in deuterium. Besides the visible line spectrum, a continuous spectrum (see Figure 28.3) is produced in the ultraviolet. The radiance is not high, but provided that the input power is stabilized, these lamps are stable in both output and position. To obtain the necessary ultraviolet transmission, either the whole envelope is made of silica or a silica window is used. These lamps are used as sources in ultraviolet spectrophotometers and are superior to tungsten lamps for that purpose at wavelengths below 330 nm.

28.2.2.2 Compact Source Lamps

Some types of discharge lamp offer light sources of extremely high luminance. These involve discharges of high current density in a gas at high pressure. The lamp-filling gases may be xenon, mercury, mercury plus iodine, or a variety of other "cocktails." The light emitted is basically a line spectrum plus some continuous radiation, but in many cases the spectrum is, in effect, continuous. The spectrum of xenon, for example, has over 4,000 lines and extends well down into the ultraviolet region. Xenon lamps are widely used in spectrofluorimeters and other instruments for which a very intense source of ultraviolet radiation is required.

Many lamps of this kind require elaborate starting arrangements and are particularly difficult to restart if switched off when hot. "Igniter" circuits are available, but since these involve voltages up to 50 kV, special attention should be given to the wiring involved. All these lamps contain gas under high pressure, even when cold, and the maker's safety instructions should be rigidly followed.

Xenon arcs produce quite dangerous amounts of ultraviolet radiation—dangerous both in itself and in the ozone that is produced in the atmosphere. They must never be used unshielded, and to comply with the Occupational Safety and Health Administration (OSHA) rules, interlocks should be arranged so that the lamp is switched off if the instrument case is opened. Force-ducted ventilation should be used with the larger sizes, unless "ozone-free" lamps are used. These are lamps with envelopes that do not transmit the shorter ultraviolet wavelengths.

28.2.3 Electronic Sources: Light-emitting Diodes

By applying currents to suitably doped semiconductor junctions, it is possible to produce a small amount of light. The luminance is very low, but the light output can be modulated, by modulating the current, up to very high frequencies, and thus the light-emitting diode (LED) is a good source for a fiber optic communication system. LEDs are also commonly used in display systems. The spectral power distribution depends on the materials used in the junction. For electro-optical communication links, there is no need to keep to the visible spectrum, and wavelengths just longer than visible (for example, 850 nm) are often used. There are no serious operating problems and only low voltages are needed, but the light output is miniscule compared with, say, a tungsten lamp.

28.2.4 Lasers

Light from lasers differs from that from conventional sources by virtue of being *coherent*, whereas conventional sources produce *incoherent* light. In an incoherent beam, there is no continuous phase relationship between light at one point of the beam and any other. The energy associated with any one quantum or wave packet can be shown to extend over a finite length—somewhere around 50 cm—as it travels through space. For that reason no interference effects can be observed if the beam is divided and subsequently superimposed if the path difference is longer than 50 cm or so. The same effect is responsible for the fact that monochromatic light from a conventional source in fact has a measurable bandwidth.

However, in a laser, the light is produced not from single events occurring randomly within single atoms but from synchronized events within a large number of atoms—hence the "finite length of wave train" is not half a meter but can be an immense distance. Consequently, laser light is much more strictly monochromatic than that from conventional sources; it is also very intense and is almost exactly unidirectional. Thus it is easy to focus a laser beam down to a very small spot at which an enormous density of energy can be achieved.

Lasers are valuable in applications in which (1) the extended length of wave train is used (for example, holography and surveying), (2) a high energy density is needed (for example, cutting of sheet metal, ophthalmic surgery), and (3) the narrowness of the beam is used (for example, optical alignment techniques in engineering or building construction).

The operating principle of the laser is the stimulated emission of radiation. In any normal gas the number of electrons in atoms in each of the possible energy levels is determined by the temperature and other physical factors. In a laser this normal distribution is deliberately upset so as to overpopulate one of the higher levels. The excited atoms then not only release their excess energy as radiation, they do so *in phase*, so the emissions from vast numbers of atoms are combined in a single wave train. Lasing action can also be produced in solid and liquid systems; hundreds of atomic systems are now known that can be used in lasers, so a wide range of types is available for either continuous or pulsed operation. A simple explanation of the principle is given in Heavens (1971) and numerous specialist textbooks (Dudley, 1976; Koechner, 1976; Mooradian et al., 1976).

Although lasers are available in many types and powers, by far the most commonly used in laboratory work is the helium–neon laser operating at a wavelength of 632.8 nm. The power output is usually a few milliwatts. For applications in which a high-energy density is needed (for example, metal cutting), CO_2 lasers are often used. Their wavelength is in the infrared range (about 10.6 μm), and the output power may be up to 500 W. Their advantage in industrial work is their relatively high efficiency—about 10 percent of the input power appears as output power.

For other wavelengths in the visible region, krypton, argon, or "tuneable dye" lasers can be used. The krypton and argon types can be made to operate at a variety of fixed wavelengths. Tuneable dye lasers use a liquid system involving organic dyes. In such systems the operating frequency can be altered within a limited range by altering the optical geometry of the system.

28.2.4.1 Laser Safety

Lasers are, by their nature, dangerous. The foremost risk is that of damage to eyesight caused by burning of the retina. Even a moment's exposure may be catastrophic. Consequently, safety precautions must be taken and strictly maintained.

It is an OSHA requirement that all due precautions are taken. In practice this means that, among other things, all rooms in which lasers are used must be clearly marked with approved warning notices. The best precautions are to use the lowest laser powers that are possible and to design equipment using lasers to be totally enclosed. A full description of safety requirements is given in *Standards for the Safe Use of Lasers*, published by the American National Standards Institute, which should be read and studied before any work with lasers is started. Useful guidance may also be obtained from BS 4803 and *Safety in Universities: Notes for Guidance*.

28.3 DETECTORS

The essential characteristics of a radiation detector are:

1. The spectral sensitivity distribution;
2. The response time.
3. The sensitivity.
4. The smallest amount of radiation that it can detect.
5. The size and shape of its effective surface.
6. Its stability over a period of time.

Other factors to be borne in mind in choosing a detector for an instrument are the precision and linearity of response, physical size, robustness, and the extent of auxiliary equipment needed. Detectors can be used in three ways:

1. Those in which the detector is used to effect an actual measurement of light intensity.
2. Those in which it is used to judge for equality of intensity between two beams.
3. Those in which it is required to establish the presence or absence of light.

In case (1), there needs to be an accurately linear relationship between the response and the intensity of radiation incident on the detector. Many detectors do not have this property, and this fact may determine the design of an instrument; for example, many infrared spectrophotometers have elaborate optical arrangements for matching the intensity in order that a nonlinear detector can be used.

It should be noted that in almost all detectors the sensitivity varies markedly from point to point on the operative surface. Consequently, if the light beam moves with respect to the detector, the response will be altered. This effect is of critical importance in spectrophotometers and like instruments; if a solution cell with faces that are not perfectly flat and parallel is put into the beam, it will act as a prism, move the beam on the detector, and produce an erroneous result. Precautions should be taken against this effect by ensuring that imperfect cells are not used.

Some detectors are sensitive to the direction of polarization of incident light. Although in most optical instruments the light is randomly polarized, the distribution of intensities between the polarization directions is by no means uniform, especially after passage through monochromators. This effect often causes no trouble, but it can be an extremely abstruse source of error.

28.3.1 Photomultipliers

Photomultipliers rely on the photoemissive effect. The light is made to fall on an emitting surface (the photocathode; see Figure 28.4) with a very low work function within a vacuum tube and causes the release of electrons. These electrons are attracted to a second electrode at a strongly positive voltage where each causes the emission of several secondary electrons, which are attracted to a third electrode, and so on. By repeating this process at a series of "dynodes," the original electron stream is greatly multiplied and results in a current of up to 100 μA or so at the final anode.

The spectral sensitivity is determined by the nature of the photocathode layer. The actual materials need not concern us here; different types of cathodes are referred to by a series of numbers from S1 upward. The response is linear, response time rapid, and the sensitivity is large. The sensitivity may be widely varied by varying the voltage applied to the dynode chain. Thus

$$S \propto V^{n/2}$$

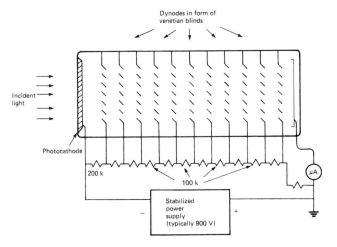

FIGURE 28.4 Construction of photomultiplier and typical circuit.

where S = sensitivity, V = voltage applied, and n = number of dynode stages. Conversely, where accurate measurements are needed, the dynode chain voltage must be held extremely stable, since between 8 and 14 stages are used. A variety of cathode shapes and multiplier configurations are available. There is no point in having a cathode of a much larger area than that actually to be used, because this will add unnecessarily to the noise.

Emission from the photocathode also occurs as a result of thermionic emission, which produces a permanent "dark current." It is random variations in this dark current—noise—that limit the ultimate sensitivity.

Photomultipliers are also discussed in Chapter 29, and excellent information on their use is given in the makers' catalogues, to which readers are referred. It should be noted that photomultipliers need very stable high-voltage power supplies, are fragile, and are easily damaged by overloads. When they are used in instruments, it is essential that interlocks are provided to remove the dynode voltage before any part of the case is opened. Moreover, photomultipliers must never be exposed to sunlight, even when disconnected.

28.3.2 Photovoltaic and Photoconductive Detectors (Photodiodes)

When light falls on a semiconductor junction, there is nearly always some effect on the electrical behavior of that junction, and such effects can be made use of in light detectors. There are two main categories: (1) those in which the action of the light is used to generate an EMF, and (2) those in which the action of light is used to effectively alter the resistance of the device. Those of the first type are referred to as *photovoltaic detectors*—sometimes called *solar cells*; those of the second type are called *photoconductive detectors*. There are some materials that show photoconductive effects but that are not strictly semiconductors (for example, lead sulfide), but devices using these are included in the category of photoconductive detectors.

In photovoltaic detectors the energy of the light is actually converted into electrical energy, often to such effect that no external energy source is needed to make a measurement; the solar cell uses this principle. The more sensitive photovoltaic detectors need an external power source, as do all photoconductive detectors.

28.3.2.1 Simple Photovoltaic Detectors

In many applications there is sufficient light available to use a photovoltaic cell as a self-powered device. This might be thought surprising, but it is perfectly feasible. If we consider a detector of 15 mm diameter, illuminated to a level of 150 lux (average office lighting), the radiant power received in the visible spectrum will be about 100 μW. If the conversion efficiency is only 5 percent, that gives an output power of 5 μW, and if that is fed to a galvo or moving coil meter of 50 Ω resistance, a current of around 300 μA will be obtained—more than enough for measurement purposes.

Detectors of this type were in use for many years before the advent of the semiconductor era in the 1950s and were then called *rectifier cells* or *barrier-layer cells*. The simplest type (see Figure 28.5) consists of a steel plate on which a layer of selenium is deposited. A thin transparent film of gold is deposited on top of that to serve as an electrode. In earlier models an annular contact electrode was sputtered on to facilitate contact with the gold film. Nowadays a totally encapsulated construction is used for cells of up to 30 mm diameter.

The semiconductor action takes place at the steel–selenium junction; under the action of light there is a buildup of electrons in the selenium, and if an external circuit is made via the gold electrode, a current will flow. If there is zero resistance in that external circuit, the current will be proportional to the light intensity. Normally there will be some resistance, and the EMF developed in it by the current will oppose the current-generation process. This means that the response will not be linear with light intensity (see Figure 28.6). If the external resistance is quite high (for example, 1,000 Ω), the response will approximate to a logarithmic one, and it is this feature that enables this type of detector to be used over a very wide

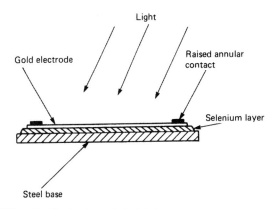

FIGURE 28.5 Simple photovoltaic detector.

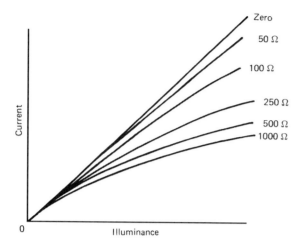

FIGURE 28.6 Nonlinear response of photovoltaic detectors.

range of intensities in the photographic exposure meter and various daylight recording instruments.

A circuit known as the Campbell–Freeth circuit was devised to present zero resistance to the detector (Thewlis, 1961) so as to obtain a truly linear response. It is not widely used in current practice since if high accuracy is sought, it is unlikely that a photovoltaic detector will be used.

Detectors of this variety, based on steel plates, are perhaps best described as "cheap and cheerful." The accuracy is not high—1 or 2 percent at best; they are not particularly stable with temperature changes, and they show marked fatigue effects. They are also extremely noisy and have a relatively long time response compared with more modern versions. However, they are cheap and robust, may be fabricated into a variety of shapes, and have the outstanding advantage that no external power supply is required. Consequently, they are widely used in exposure meters, street-lighting controls, photometers, "abridged" spectrophotometers and colorimeters, flame photometers, simple densitometers, and so on—all those cases in which high accuracy is not needed but price is an important consideration.

The spectral response of this type of detector is broader than that of the human eye, but a filter can be used to produce a spectral response that is a good enough match for most photometric purposes. Such filters are usually encapsulated with the detectors into a single unit.

28.3.2.2 The Silicon Diode

The advent of doped semiconductor materials has enabled many other "barrier-layer" systems to be used. One such is the diffused silicon photodiode; in this case the barrier layer is provided by a p–n junction near the surface of a silicon layer. This is very much less noisy than the detector described previously and can thus be used for much lower light levels. External amplification is needed, but rapid response can be obtained (50 ns in some cases) so that the silicon diode is not far short of the performance obtained from the photomultiplier, with the advantages of a wider spectral range, less bulk, and much less complexity.

28.3.2.3 Photoconductive Detectors

Almost all semiconductor materials are light sensitive, since light falling on them produces an increased number of current carriers and hence an increase in conductivity; semiconductor devices have to be protected from light to prevent the ambient lighting upsetting their operation. Consequently, it is possible to make a wide variety of light-sensitive devices using this principle.

These devices may be in the form of semiconductor diodes or triodes. In the former case the material is usually deposited as a thin film on a glass plate with electrodes attached; under the action of light the resistance between the electrodes drops markedly, usually in a nonlinear fashion. Since there are dozens of semiconductor materials available, these devices can be made with many different spectral responses, covering the visible and near infrared spectrum up to 5 μm or so. The response time is also dependent on the material, and though many are fast, some are very slow—notably cadmium sulfide, which has a response time of about 1 s.

Photoconductive detectors of the triode type are, in effect, junction transistors that are exposed to light.[*] They offer in-built amplification, but again, usually produce a nonlinear signal. Devices are now available in which a silicon diode is combined with an amplifier within a standard transistor housing.

All photoconductive devices are temperature sensitive, and most drift quite badly with small changes of temperature. For that reason they are used with "chopped" radiation (see Section 28.4.3 on detector techniques) in all critical applications. They are commonly used as detectors in spectrophotometers in the near-infrared range. A common technique is to use them as null-balance detectors comparing the sample and reference beams (see Section 28.6.1 on spectrophotometers) so that a nonlinear response does not matter.

28.3.3 Pyroelectric Detectors

Although they are strictly "thermal" detectors, pyroelectric detectors are used very widely as light detectors. They rely on the use of materials that have the property of temperature-dependent spontaneous electric polarization. These may be in the form of crystals, ceramics, or thin plastic films. When radiation falls on such a material, thermal expansion takes place, minutely altering the lattice spacing in the crystal, which alters the electrical polarization and results in an

---

[*]The joke is often made that the light-detecting properties of one famous type of phototransistor were discovered only when one batch of transistors was accidentally left unpainted. The author regrets that he cannot confirm this story.

EMF and a charge being developed between two faces of the crystal. These faces are equipped with electrodes to provide connection to an external circuit and to an appropriate amplifier.

Pyroelectric detectors using this principle are extremely sensitive, and, being basically thermal detectors, they are sensitive to an enormous range of wavelengths. In practice the wavelength range is limited by the transmission of windows used, the absorption characteristics of the pyroelectric material, and the reflection characteristics of its surfaces. The last-mentioned quality can be adjusted by the deposition of thin films so that relatively wide spectral responses may be obtained; alternatively, the sensitivity to a particular wavelength can be considerably enhanced.

It is important to remember that pyroelectric detectors respond in effect only to *changes* in the radiation falling on them and not to steady-state radiation. Their response speed is extremely fast, but because they inherently have some capacity, there is a compromise between sensitivity and response speed that can be determined by an appropriate choice of load resistor. The ability to respond only to changes in the radiation field enables them to be used for laser pulse measurements, and at a more mundane level they make excellent sensors for burglar alarms. They are usually made in quite small sizes (for example, 1×2 mm), but they can be made up to 1 cm in diameter with a lower response speed. They can also be made in the form of linear and two-dimensional arrays. If they are required to measure steady-state radiation, as in a spectrophotometer, the usual technique of beam chopping (see Section 28.4.3) can be used; the pyroelectric detector will then respond only to the chopped beam and nothing else.

A variety of materials are used, notably lithium tantalate and doped lead zirconate titanate. The pyroelectric detector, with its wide spectral range, fast response, and relatively low cost, is probably capable of further development and application than any other and will be seen in a very wide range of applications in the next few years.

28.3.4 Array Detectors

The devices described thus far are suitable for making a single measurement of the intensity of a beam of light at any one instant. However, often the need arises to measure the intensity of many points in an optical image, as in a television camera. In television camera tubes the image is formed on a photocathode resembling that of a photomultiplier, which is scanned by an electron beam. Such tubes are outside the scope of this book, but they are necessarily expensive and require much supporting equipment.

In recent years array detectors using semiconductor principles have been developed, enabling measurements to be made simultaneously at many points along a line or, in some cases, over a whole area comprising many thousands of image points. All these devices are based on integrated-circuit technology, and they fall into three main categories:

1. Photo-diode arrays.
2. Charge-coupled devices.
3. Charge injection devices.

It is not possible to go into their operation in detail here, but more information can be found in the review article by Fry (1975) and in the book by Beynon (1979).

1. A photodiode array consists of an array of photodiodes of microscopic dimensions, each capable of being coupled to a signal line in turn through an associated transistor circuit adjacent to it on the chip. The technique used is to charge all the photodiode elements equally; on exposure to the image, discharging takes place, those elements at the points of highest light intensity losing the most charge. The array is read by connecting each element in turn to the signal line and measuring the amount of charge needed to restore each element to the original charge potential. This can be carried out at the speeds normally associated with integrated circuits, the scan time and repetition rate depending on the number of elements involved—commonly one or two thousand. It should be noted that since all array detectors are charge-dependent devices, they are in effect time integrating over the interval between successive readouts.

2. Charge-coupled devices (CCDs) consist of an array of electrodes deposited on a substrate, so that each electrode forms part of a metal-oxide-semiconductor device. By appropriate voltage biasing of electrodes to the substrate it is possible to generate a potential well under each electrode. Furthermore, by manipulating the bias voltages on adjacent electrodes, it is possible to transfer the charge in any one potential well to the next, and with appropriate circuitry at the end of the array the original charge in each well may be read in turn. These devices were originally developed as delay lines for computers, but since all semiconductor processes are affected by incident light, they make serviceable array detectors. The image is allowed to fall on the array, and charges will develop at the points of high intensity. These are held in the potential wells until the reading process is initiated at the desired interval.

3. Charge-injection devices (CIDs) use a similar principle but one based on coupled pairs of potential wells rather than a continuous series. Each pair is addressed by using an X–Y coincident voltage technique. Light is measured by allowing photogenerated charges to build up on each pair and sensing the change as the potential well fills. Once reading has been completed, this charge is injected into the substrate by removing the bias from both electrodes in each pair simultaneously.

Recent developments have made photodiode arrays useful. One vendor, Hamamatsu, offers silicon photodiode arrays consisting of multiple photodiode elements, formed in a linear or matrix arrangement in one package. Some arrays are supplied coupled with a CMOS multiplexer. The multiplexer simplifies design and reduces the cost of the output electronic circuit. Hamamatsu's silicon photodiode arrays are used in a wide range of applications such as laser beam position detection, color measurement and spectrophotometry.

CCD sensors are now widely used in cameras. The recent NASA Kepler spacecraft, for example, has a CCD camera with a 0.95-meter aperture, wide field-of-view Schmidt telescope, and a 1.4-meter primary mirror. With more than 95 megapixels, Kepler's focal plane array of 42 backside illuminated CCD90s from E2v Technologies forms the largest array of CCDs ever launched into space by NASA. New CCDs are ideal for applications in low light level and high-speed industrial inspection, particularly where UV and NIR sensitivity are required.

By the 1990s, as CMOS sensors were gaining popularity, CIDs were adapted for applications demanding high dynamic range and superior antiblooming performance (Bhaskaran, et al). CID-based cameras have found their niche in applications requiring extreme radiation tolerance and the high dynamic range scientific imaging. CID imagers have progressed from passive pixel designs using proprietary silicon processes to active pixel devices using conventional CMOS processing. Scientific cameras utilizing active pixel CID sensors have achieved a factor of 7 improvement in read noise (30 electrons (rms) versus 225 electrons (rms)) at vastly increased pixel frequencies (2.1 MHz versus 50 kHz) when compared to passive pixel devices. Radiation-hardened video cameras employing active pixel CIDs is the enabling technology in the world's only solid-state radiation-hardened color camera, which is tolerant to total ionizing radiation doses of more than 5 Mega-rad. Performance-based CID imaging concentrates on leveraging the advantages that CIDs provide for demanding applications.

28.4 DETECTOR TECHNIQUES

In nearly all optical instruments that require a measurement of light intensity, we rely on the detector producing a signal (usually a current) that is accurately proportional to the light intensity. However, all detectors, by their nature, pass some current when in total darkness, and we have to differentiate between the signal that is due to "dark current" and that due to "dark current plus light current." A further problem arises if we try to measure very small light currents or larger light currents very accurately. In all detectors there are small random variations of the dark current—that is, noise—and it is this noise that limits the ultimate sensitivity of any detector.

The differentiation between "dark and dark plus light" signals is achieved by taking the difference of the signals when the light beam falls on the detector and when it is obscured. In some manually operated instruments, two

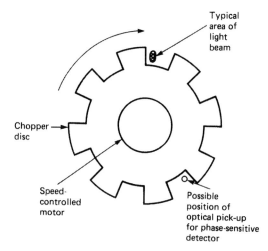

FIGURE 28.7 Chopper disc.

settings are made, one with and one without a shutter in the beam. In most instruments a chopper disc (see Figure 28.7) is made to rotate in the light beam so that the beam is interrupted at a regular frequency and the signal is observed by AC circuits that do not respond to the continuous dark current. This technique is called *beam chopping*.

The effects of noise can be reduced in three ways:

1. Prolonging the time constant of the detector circuitry.
2. Cooling the detector.
3. Using synchronized techniques.

28.4.1 Detector Circuit Time Constants

Dark current noise is due to random events on the atomic scale in the detector. Hence if the time constant of the detector circuit is made sufficiently long, the variations in the output current will be smoothed out, and it will be of a correct average value. The best choice of time constant will depend on the circumstances of any application, but clearly long time constants are not acceptable in many cases. Even in manually read instruments, a time constant as long as 1 s will be irritating to the observer.

28.4.2 Detector Cooling

It is not possible to go into a full discussion of detector noise here but in many detectors the largest contribution to the noise is that produced by thermionic emission within the detector. Thermionic emission from metals follows Langmuir's law:

$$i \propto e^{3T/2}$$

where T is absolute temperature and i represents the emission current. Room temperature is around 295 K in terms of absolute temperature, so a significant reduction in emission current and noise can be achieved by cooling the detector.

Detectors may be cooled by the use of liquid nitrogen (77 K, –196°C), solid CO_2 (195 K, –79°C), or by Peltier effect cooling devices. The use of liquid nitrogen or solid CO_2 is cumbersome; it usually greatly increases the complexity of the apparatus and requires recharging. Another problem is that of moisture from the air condensing or freezing on adjacent surfaces. Peltier-effect cooling usually prevents these problems but does not achieve such low temperatures. Unless special circumstances demand it, detector cooling is less attractive than other methods of noise reduction.

28.4.3 Beam Chopping and Phase-Sensitive Detection

Random noise may be thought of as a mixture of a large number of signals, all of different frequencies. If a beam chopper is used and running at a fixed frequency and the detector circuitry is made to respond preferentially to that frequency, a considerable reduction in noise can be achieved. Although this can be effected by the use of tuned circuits in the detector circuitry, it is not very reliable, since the chopper speed usually cannot be held precisely constant. A much better technique is to use a phase sensitive detector, phase locked to the chopper disc by means of a separate pickup.

Such a system is illustrated in Figure 28.8. The light beam to be measured is interrupted by a chopper disc, A, and the detector signal (see Figure 28.9) is passed to a phase-sensitive detector. The gating of the detector is controlled by a separate pickup, B, and the phase relationship between A and B is adjusted by moving B until the desired synchronization is achieved (see Figure 28.10); a double-beam oscilloscope is useful here.

The chopper speed should be kept as uniform as possible. Except when thermal detectors are used, the chopping frequency is usually in the range 300–10,000 Hz. The phase-locking pickup may be capacitive or optical, the latter being greatly preferable. Care must be taken that the chopper disc is not exposed to room light or this may be reflected from the back of the blades into the detector and falsely recorded as a "dark" signal. It should be noted that because the beam to be measured has finite width, it is not possible to "chop square."

By making the light pulses resemble the shape and phase of the gating pulses, a very effective improvement in signal-to-noise ratio can be obtained—usually at least 1,000:1. This technique is widely used, but stroboscopic trouble can occur when pulsating light sources are involved, such as fluorescent lamps or CRT screens, when the boxcar system can be used.

28.4.4 The Boxcar Detector

A further improvement in signal-to-noise ratio can be obtained with the boxcar system. In some ways this is similar to the phase-sensitive detector system, but instead of the chopping being done optically, it is carried out electronically. When this method is used with a pulsating source (for example, a fluorescent lamp), the detector signal is sampled at intervals and phase-locked with the source, the phase locking being provided from the power supply feeding the

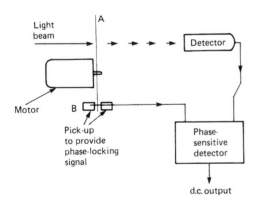

FIGURE 28.8 Beam chopper used with a phase-sensitive detector.

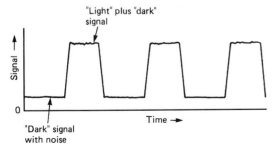

FIGURE 28.9 Typical output from a beam-chopped detector.

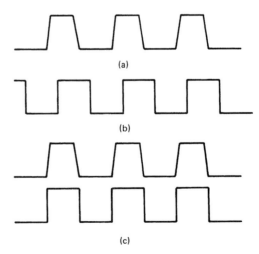

FIGURE 28.10 Input signals to phase-sensitive detector: (a) signal from detector; (b) gating signal derived from pickup, incorrectly phased; (c) detector and gating signals, correctly phased.

source (see Figure 28.11). The sampling period is made very narrow, and the position of the sampling point within the phase is adjusted by a delay circuit (see Figure 28.12). It is necessary to establish the dark current in a separate experiment. If a steady source is to be measured, an oscillator is used to provide the "phase" signal.

This system both offers a considerable improvement in noise performance and enables us to study the variation of light output of pulsating sources within a phase. Although more expensive than the beam chopping-PSD system, it is very much more convenient to use, since it eliminates all the mechanical and optical problems associated with the chopper. (The American term *boxcar* arises from the fact that the gating signal, seen upside down on a CRT, resembles a train of boxcars.)

28.4.5 Photon Counting

When measurement of a very weak light with a photomultiplier is required, the technique of photon counting may be used. In a photomultiplier, as each electron leaves the photocathode, it gives rise to an avalanche of electrons at the anode of very short duration (see Section 28.3.1). These brief bursts of output current can be seen with the help of a fast CRT. When very weak light falls on the cathode, these bursts can be counted over a given period of time, giving a measure of the light intensity.

Although extremely sensitive, this system is by no means easy to use. It is necessary to discriminate between pulses due to photoelectrons leaving the cathode and those that have originated from spurious events in the dynode chain; this is done with a pulse-height discriminator, which has to be very carefully adjusted. Simultaneous double pulses are also a problem. Another problem in practice is the matching of photon counting with electrometer (i.e., normal) operation. It also has to be remembered that it is a counting and not a measuring operation, and since random statistics apply, if it is required to obtain an accuracy of ±1 percent on a single run, then at least 10,000 counts must be made.

This technique is regularly used successfully in what might be termed research laboratory instrumentation—Raman spectrographs and the like—but is difficult to set up and its use can really only be recommended for those cases where the requirement for extreme sensitivity demands it.

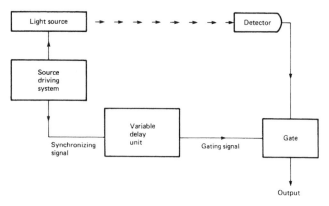

FIGURE 28.11 Boxcar detector system applied to a pulsating source.

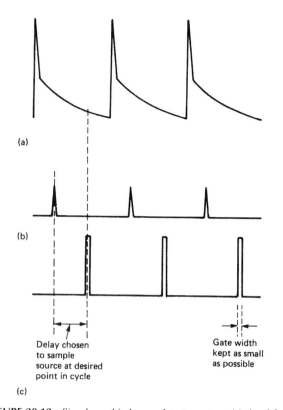

FIGURE 28.12 Signals used in boxcar detector system: (a) signal from detector produced by pulsating source (e.g., CRT); (b) synchronizing signal derived from source; (c) gating signal.

28.5 INTENSITY MEASUREMENT

The term *light-intensity measurement* can be used to refer to a large range of different styles of measurement. These can be categorized loosely into (1) those where the spectral sensitivity of the detector is used unmodified and (2) those where the spectral sensitivity of the detector is modified deliberately to match some defined response curve. Very often we are concerned with the comparison of light intensities, where neither the spectral power distribution of the light falling on the detector nor its geometrical distribution with respect to the detector will change. Such comparative measurements are clearly in category (1), and any appropriate detector can be used. However, if, for example, we are concerned with the purposes of lighting engineering where we have to accurately measure "illuminance" or "luminance" with sources of any spectral or spatial distribution, we must use a photometer with a spectral response accurately matched to that of the human eye and a geometrical response accurately following a cosine law; that is, a category (2) measurement.

Some "measurements" in category (1) are only required to determine the presence or absence of light, as in a very large number of industrial photoelectric controls and counters, burglar alarms, street-lighting controls, and so on. The only critical points here are that the detector should be arranged so that it receives only light from the intended source and that it shall have sufficient response speed. Photodiodes and phototransistors are often used and are robust, reliable, and cheap. Cadmium sulfide photoconductive cells are often used in lighting controls, but their long response time (about 1 s) restricts their use in many other applications.

28.5.1 Photometers

A typical photometer head for measurements of illuminance (the amount of visible light per square meter falling on a plane) is shown in Figure 28.13. Incident light falls on the opal glass cylinder A, and some reaches the detector surface C after passing through the filter layer B. The detector may be of either the photoconductive or photovoltaic type. The filter B is arranged to have a spectral transmission characteristic such that the detector-filter combination has a spectral sensitivity matching that of the human eye. In many instruments the filter layer is an integral part of the detector. The cosine response is achieved by careful design of the opal glass A in conjunction with the cylindrical protuberance D in its mounting. The reading is displayed on an appropriately calibrated meter connected to the head by a meter or two of a thin cable, so that the operator does not "get in his own light" when making a reading.

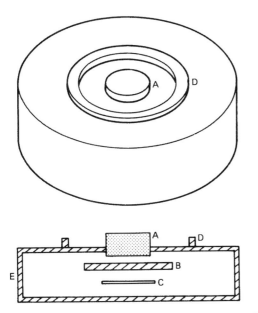

FIGURE 28.13 Cosine response photometer head. A: opal glass cylinder; B: filter; C: detector; D: light shield developed to produce cosine response; E: metal case.

Instruments of this kind are available at a wide range of prices, depending on the accuracy demanded; the best instruments of this type can achieve an accuracy of 1 or 2 percent of the illuminance of normal light sources. Better accuracy can be achieved with the use of a Dresler filter, which is built up from a mosaic of different filters rather than a single layer, but this is considerably more expensive.

A widely used instrument is the Hagner photometer, which is a combined instrument for measuring both illuminances and luminances. The "illuminance" part is as described previously, but the measuring head also incorporates a telescopic optical system focused on to a separate, internal detector. A beam divider enables the operator to look through the telescope and point the instrument at the surface whose luminance is required; the internal detector output is indicated on a meter, also arranged to be within the field of view.

The use of silicon detectors in photometers offers the possibility of measuring radiation of wavelengths above those of the visible spectrum, which is usually regarded as extending from 380 to 770 nm. Many silicon detectors will operate at wavelengths up to 1170 nm. In view of the interest in the 700–1100 nm region for the purposes of fiber optic communications, a variety of dual-function "photometer/radiometer" instruments are available. Basically these are photometers that are equipped with two interchangeable sets of filters: (1) to modify the spectral responsivity of the detector to match the spectral response of the human eye and (2) to produce a flat spectral response so that the instrument responds equally to radiation of all wavelengths and thus produces a reading of radiant power. In practice this "flat" region cannot extend above 1,170 nm, and the use of the phrase *radiometer* is misleading because that implies an instrument capable of handling *all* wavelengths; traditional radiometers operate over very much wider wavelength ranges.

In construction, these instruments resemble photometers, as described, except that the external detector head has to accommodate the interchangeable filters and the instrument has to have dual calibration. These instruments are usually restricted to the measurement of illuminance and irradiance, unlike the Hagner photometer, which can measure both luminance and illuminance. The effective wavelength operating range claimed in the radiometer mode is usually 320–1,100 nm.

28.5.2 Ultraviolet Intensity Measurements

In recent years, a great deal of interest has developed in the nonvisual effects of radiation on humans and animals, especially ultraviolet radiation. Several spectral response curves for photobiological effects (for example, erythema and photokeratitis) are now known (Steck, 1982). Ultraviolet photometers have been developed accordingly, using the same general principles as visible photometers but with appropriate detectors and filters. Photomultipliers are usually used as detectors, but the choice of filter materials is restricted; to date, nothing like the exact correlation of visible response to the human eye has been achieved.

An interesting development has been the introduction of ultraviolet film badges—on the lines of X-ray film badges—for monitoring exposure to ultraviolet radiation. These use photochemical reactions rather than conventional detectors (Young et al., 1980).

28.5.3 Color-Temperature Meters

The color of incandescent light sources can be specified in terms of their *color temperature*, that is, the temperature at which the spectral power distribution of a Planckian black-body radiator most closely resembles that of the source concerned. (*Note:* This is *not* the same as the actual temperature.) Since the spectral power distribution of a Planckian radiator follows the law

$$E_\lambda = c_1 \left\{ \lambda^s \left(e^{c_2/\lambda T} - 1 \right) \right\}^{-1}$$

(see Section 28.8), it is possible to determine T by determining the ratio of the E_λ values at two wavelengths. In practice, because the Planckian distribution is quite smooth, broadband filters can be used.

Many photometers (see Section 28.6.1) are also arranged to act as color-temperature meters, usually by the use of a movable shade arranged so that different areas of the detectors can be covered by red and blue filters; thus the photometer becomes a "red-to-blue ratio" measuring device.

Devices of this kind work reasonably well with incandescent sources (which include sunlight and daylight) but will give meaningless results if presented to a fluorescent or discharge lamp.

28.6 WAVELENGTH AND COLOR

28.6.1 Spectrophotometers

Instruments that are used to measure the optical transmission or reflection characteristics of a sample over a range of wavelengths are termed *spectrophotometers*. This technique is widely used for analytical purposes in all branches of chemistry and is the physical basis of all measurement of color; thus it is of much interest in the consumer industries. Transmission measurements, usually on liquid samples, are the most common.

All spectrophotometers contain four elements:

1. A source of radiation.
2. An optical system, or monochromator, to isolate a narrow band of wavelengths from the whole spectrum emitted by the source.
3. The sample (and its cell if it is liquid or gaseous).
4. A detector of radiation and its auxiliary equipment.

Note that theoretically it does not matter whether the light passes first through the monochromator and then the sample, or vice versa; the former is usual in visible or ultraviolet instruments, but for infrared work the latter arrangement offers some advantages.

Spectrophotometers may be either single beam, in which the light beam takes a single fixed path and the measurements are effected by taking measurements with and without the sample present, or double beam, in which the light is made to pass through two paths, one containing the sample and the other a reference; the intensities are then compared. In work on chemical solutions the reference beam is usually passed through a cell identical to that of the sample containing the same solvent so that solvent and cell effects are cancelled out; in reflection work the reference sample is usually a standard white reflector. The single-beam technique is usual in manually operated instruments and the double-beam one in automatic instruments, nowadays the majority.

Two main varieties of double-beam techniques are used (see Figure 28.14). That shown in Figure 28.14(a) relies for accuracy on the linearity of response of the detector and is sometimes called the *linearity method*. The light beam is made to follow alternate sample and reference paths, and the detector is used to measure the intensity of each in turn; the ratio gives the transmission or reflection factor at the wavelength involved. The other method, shown in Figure 28.14(b), is called the *optical-null method*. Here the intensity of the reference beam is reduced to equal that of the sample beam by some form of servocontrolled optical attenuator, and the detector is called on only to judge for equality between the two beams. The accuracy thus depends on the optical attenuator, which may take the form of a variable aperture or a system of polarizing prisms.

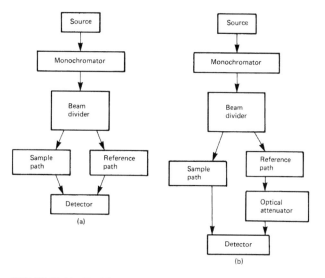

FIGURE 28.14 Double-beam techniques in spectrophotometry: (a) linearity method; (b) optical null method.

Since spectrophotometric results are nearly always used in extensive calculations and much data can easily be collected, spectrophotometers are sometimes equipped with microprocessors. These microprocessors are also commonly used to control a variety of automatic functions—for example, the wavelength-scanning mechanism, automatic sample changing, and so on.

For chemical purposes it is nearly always the absorbance (i.e., optical density) of the sample rather than the transmission that is required:

$$A = \log_{10} \frac{1}{T}$$

where A is the absorbance or optical density and T is the transmission. The relation between transmission and absorbance is shown in Table 28.1.

Instruments for chemical work usually read in absorbance only. Wave number—the number of waves in one centimeter—is also used by chemists in preference to wavelength. The relation between the two is shown in Table 28.2.

It is not economic to build all-purpose spectrophotometers in view of their use in widely different fields. The most common varieties are:

1. Transmission/absorbance, ultraviolet, and visible range (200–600 nm).
2. Transmission and reflection, near-ultraviolet, and visible (300–800 nm).
3. Transmission/absorbance, infrared (2.5–25 μm).

The optical parts of a typical ultraviolet-visible instrument are shown in Figure 28.15. Light is taken either from a tungsten lamp A or deuterium lamp B by moving mirror C to the appropriate position. The beam is focused by a mirror on to the entrance slit E of the monochromator. One of a series of filters is inserted at F to exclude light of submultiples of the desired wavelength. The light is dispersed by the diffraction grating G, and a narrow band of wavelengths is selected by the exit slit K from the spectrum formed. The wavelength is changed by rotating the grating by a mechanism operated by a stepper motor.

The beam is divided into two by an array of divided mirror elements at L, and images of the slit K are formed at R and S, the position of the reference and sample cells. A chopper disc driven by a synchronous motor M allows light to pass through only one beam at a time. Both beams are directed to T, a silica diffuser, so that the photomultiplier tube is presented alternately with sample and reference beams; the diffuser is necessary to overcome nonuniformities in the photomultiplier cathode (see Section 28.4).

TABLE 28.1 Relation between transmission and absorbance

Transmission (%)	Absorbance
100	0
50	0.301
10	1.0
5	1.301
1	2.0
0.1	3.0

TABLE 28.2 Relation between wavelength and wavenumber

Wavelength	Warenumber (no. of waves/cm)
200 mm	50,000
400 nm	25,000
500 nm	20,000
1 μ	10,000
5 μ	2,000
10 μ	1,000
50 μ	200

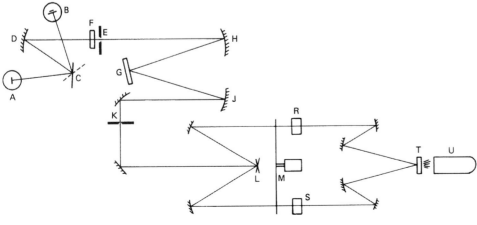

FIGURE 28.15 Typical ultraviolet–visible spectrophotometer.

The signal from U is switched appropriately to sample or reference circuits by the signal driving the chopper M and the magnitudes compared. Their ratio gives the transmission, which may be recorded directly, or an absorbance figure may be calculated from it and recorded. A microprocessor is used to control the functions of wavelength scanning, slit width, filter, and lamp selection and to effect any desired calculations on the basic transmission results.

28.6.2 Spectroradiometers

The technique of measuring the spectral power distribution (SPD) of a light source is termed *spectroradiometry*. We may be concerned with the SPD in relative or absolute terms. By "relative" we refer to the power output per unit waveband at each wavelength of a range, expressed as a ratio of that at some specified wavelength. (For the visible spectrum this is often 560 nm.) By "absolute" we mean the actual power output per steradian per unit waveband at each wavelength over a range. Absolute measurements are much more difficult than relative ones and are not often carried out except in specialized laboratories.

Relative SPD measurements are effected by techniques similar to those of spectrophotometry (see Section 28.6.1). The SPD of the unknown source is compared with that of a source for which the SPD is known. In the single-beam method, light from the source is passed through a monochromator to a detector, the output of which is recorded at each wavelength of the desired range. This is repeated with the source for which the SPD is known, and the ratio of the readings of the two at each wavelength is then used to determine the unknown SPD in relative terms. If the SPD of the reference source is known in absolute terms, the SPD of the unknown can be determined in absolute terms.

In the double-beam method, light from the two sources is passed alternately through a monochromator to a detector, enabling the ratio of the source outputs to be determined wavelength by wavelength. This method is sometimes offered as an *alternative mode* of using double-beam spectrophotometers. The experience of the author is that with modern techniques, the single-beam technique is simpler, more flexible, more accurate, and as rapid as the double-beam one. It is not usually worthwhile to try to modify a spectrophotometer to act as a spectroradiometer if good accuracy is sought.

A recent variation of the single-beam technique is found in the optical multichannel analyzer. Here light from the source is dispersed and made to fall not on a slit but on a multiple-array detector; each detector bit is read separately with the aid of a microprocessor. This technique is not as accurate as the conventional single-beam one but can be used with sources that vary rapidly with time (for example, pyrotechnic flares).

Both single-beam and double-beam methods require the unknown source to remain constant in intensity while its spectrum is scanned—which can take up to 3 minutes or so. Usually this is not difficult to arrange, but if a source is inherently unstable (for example, a carbon arc lamp) the whole light output or output at a single wavelength can be monitored to provide a reference (Tarrant, 1967).

Spectroradiometry is not without its pitfalls, and the worker intending to embark in the field should consult suitable texts (Forsythe, 1941; Commission Internationale de l'Eclairage, 1984).

Particular problems arise (1) where line and continuous spectra are present together and (2) where the source concerned is time modulated (for example, fluorescent lamps, cathode ray tubes, and so on). When line and continuous spectra are present together, they are transmitted through the monochromator in differing proportions, and this must be taken into account or compensated for (Henderson, 1970; Moore, 1984). The modulation problem can be dealt with by giving the detector a long time constant (which implies a slow scan speed) or by the use of phase-locked detectors (Brown and Tarrant, 1981; see Section 28.4.3).

Few firms offer spectroradiometers as stock lines because they nearly always have to be custom built for particular applications. One instrument—the Surrey spectroradiometer—is shown in Figure 28.16. This instrument was developed for work on cathode ray tubes but can be used on all steady sources (Brown and Tarrant, 1981). Light from the source is led by the front optics to a double monochromator, which allows a narrow waveband to pass to a photomultiplier. The output from the photomultiplier tube is fed to a boxcar detector synchronized with the tube-driving signal so that the tube output is sampled only over a chosen period, enabling the initial glow or afterglow to be studied separately. The output from the boxcar detector is recorded by a desktop computer, which also controls the wavelength-scanning mechanism by means of stepper motor. The known source is scanned first and the readings are also held in the computer, so that the SPD of the unknown source can be printed out as soon as the wavelength scan is completed. The color can also be computed and printed out. The retro-illuminator is a retractable unit used when setting up. Light can be passed backward through the monochromator to identify the precise area of the CRT face viewed by the monochromator system.

This instrument operates in the visible range of wavelengths 380–760 nm and can be used with screen luminances as low as 5 cd/m^2. When used for color measurement, an accuracy of ± 0.001 in x and y (CIE 1931 system) can be obtained.

28.6.3 The Measurement of Color

28.6.3.1 Principles

The measurement of color is not like the measurement of physical quantities such as pressure or viscosity because color is not a physical object. It is a visual phenomenon, a part of the process of vision. It is also a *psychophysical phenomenon*, and if we attempt to measure it we must not lose sight of that fact.

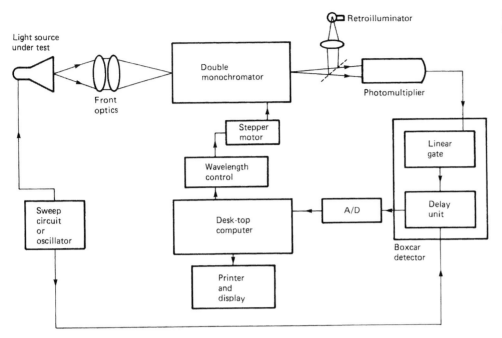

FIGURE 28.16 The Surrey spectroradiometer.

The nature of color is discussed briefly by the author elsewhere (Tarrant, 1981). The newcomer to the subject should consult the excellent textbooks on the subject by Wright (1969) and by Judd and Wysecki (1975). Several systems of color measurement are in use; for example, the CIE system and its derivatives, the Lovibond system (Chamberlin, 1979), and the Munsell system. It is not possible to go into details of these systems here; we shall confine ourselves to a few remarks on the CIE 1931 system. (The letters CIE stand for Commission Internationale de l'Eclairage, and the 1931 is the original fundamental system. About six later systems based on it are in current use.)

There is strong evidence to suggest that in normal daytime vision, our eyes operate with three sets of visual receptors, corresponding to red, green, and blue (a very wide range of colors can be produced by mixing these together), and that the responses to these add in a single arithmetical way. If we could then make some sort of triple photometer (each channel having spectral sensitivity curves corresponding to those of the receptor mechanisms), we should be able to make physical measurements to replicate the functioning of eyes. This cannot in fact be done, since the human visual mechanisms have *negative* responses to light of certain wavelengths. However, it is possible to produce photocell filter combinations that correspond to red, green, and blue and that can be related to the human color-matching functions (within a limited range of colors) by simple matrix equations.

This principle is used in photoelectric colorimeters, or *tristimulus colorimeters*, as they are sometimes called. One is illustrated in Figure 28.17. The sample is illuminated by a lamp and filter combination, which has the SPD of one of the defined Standard Illuminants. Light diffusely reflected from the sample is passed to a photomultiplier through a set of filters, carefully designed so that the three filter/

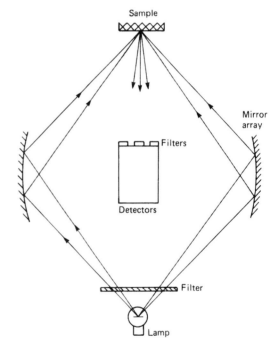

FIGURE 28.17 Tristimulus colorimeter.

photomultiplier spectral sensitivity combinations can be related to the human color-matching functions. By measuring each response in turn and carrying out a matrix calculation, the color specification in the CIE 1931 system (or its derivatives) can be found. Most instruments nowadays incorporate microprocessors, which remove the labor from these calculations so that the determination of a surface color can be carried out rapidly and easily.

Consequently, colorimetric measurements are used on a large scale in all consumer industries, and the colors of manufactured products can now be very tightly controlled.

It is possible to achieve a high degree of precision so that the minimum color difference that can be measured is slightly smaller than that which the human eye can perceive. It should be noted that nearly all surfaces have markedly directional characteristics, and hence if a sample is measured in two instruments that have different viewing/illuminating geometry, different results must be expected.

Diagrams of these instruments, such as Figure 28.17, make them look very simple. In fact, the spectral sensitivity of the filter/photomultiplier combination has to be controlled and measured to a very high accuracy; it certainly is not economic to try to build a do-it-yourself colorimeter. It is strongly emphasized that the foregoing remarks are no substitute for a proper discussion of the fascinating subject of colorimetry, and no one should embark on color measurement without reading at least one of the books mentioned.

28.7 MEASUREMENT OF OPTICAL PROPERTIES

Transparent materials affect light beams passing through them, notably changing the speed of propagation; refractive index is, of course, a measure of this. In this section we describe techniques for measuring the material properties that control such effects.

28.7.1 Refractometers

The precise measurement of the refractive index of transparent materials is vital to the design of optical instruments but is also of great value in chemical work. Knowledge of the refractive index of a substance is often useful in both identifying and establishing the concentration of organic substances, and by far the greatest use of refractometry is in chemical laboratories. Britton has used the refractive index of gases to determine concentration of trilene in air, but this involves an interferometric technique that will not be discussed here.

When light passes from a less dense to a denser optical medium—for example, from air into glass—the angle of the refracted ray depends on the angle of incidence and the refractive indices of the two media (see Figure 28.18(a)), according to Snell's law:

$$\frac{n_2}{n_1} = \frac{\sin i_1}{\sin i_2}$$

In theory, then, we could determine the refractive index of an unknown substance in contact with air by measuring these angles and assuming that the refractive index of air is unity (in fact, it is 1.000 27).

In practice, for a solid sample we have to use a piece with two nonparallel flat surfaces; this involves also measuring the angle between them. This method can be used with the aid of a simple table spectrometer. Liquid samples can be measured in this way with the use of a hollow prism, but it is a laborious method and requires a considerable volume of the liquid.

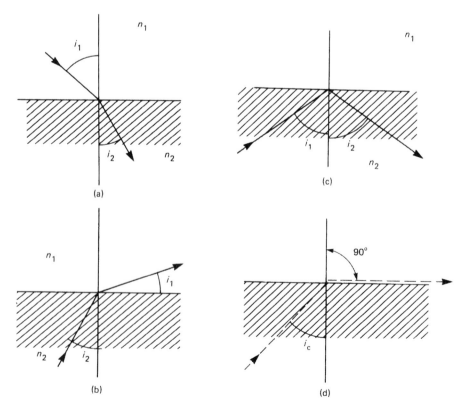

FIGURE 28.18 (a) Refraction of ray passing from less dense to more dense medium; (b) refraction of ray passing from more dense to less dense medium; (c) total internal reflection; (d) the critical angle case, where the refracted ray can just emerge.

Most refractometers instead make use of the critical angle effect. When light passes from a more dense to a less dense medium, it may be refracted, as shown in Figure 28.18(b), but if the angle i_2 becomes so large that the ray cannot emerge from the dense medium, the ray is totally internally reflected, as illustrated in Figure 28.18(c).

The transition from refraction to internal reflection occurs sharply, and the value of the angle of i_2 at which this occurs is called the *critical angle*, illustrated in Figure 28.18(d). If we call that angle i_c, then

$$\frac{n_2}{n_1} = \frac{1}{\sin i_c}$$

Hence by determining i_c we can find n_1, if n_2 is known.

28.7.1.1 The Abbé Refractometer

The main parts of the Abbé refractometer which uses this principle are shown in Figure 28.19. The liquid under test is placed in the narrow space between prisms A and B. Light from a diffuse monochromatic source (L), usually a sodium lamp, enters prism A, and thus the liquid layer, at a wide variety of angles. Consequently, light will enter prism B at a variety of angles, sharply limited by the critical angle. This light then enters the telescope (T), and on moving the telescope around, a sharp division is seen at the critical angle; one half of the field is bright and the other is almost totally dark. The telescope is moved to align the crosswires on the light/dark boundary and the refractive index can be read off from a directly calibrated scale attached to it. This calibration also takes into account the glass/air refraction that occurs when the rays leave prism B.

Although simple to use, this instrument suffers from all the problems that complicate refractive index measurements. It should be noted that:

1. In all optical materials, the refractive index varies markedly with wavelength in a nonlinear fashion. Hence either monochromatic sources or "compensating" devices must be used. For high-accuracy work, monochromatic sources are invariably used.
2. The refractive index of most liquids also varies markedly with temperature, and for accurate work temperature control is essential.
3. Since the refractive index varies with concentration, difficulties may be encountered with concentrated solutions, especially of sugars, which tend to become inhomogeneous under the effects of surface tension and gravity.
4. The range of refractive indices that can be measured in critical-angle instruments is limited by the refractive index of the prism A. Commercial instruments of this type are available for measuring refractive indices up to 1.74.
5. In visual instruments the light/dark field boundary presents such severe visual contrast that it is sometimes difficult to align the crosswires on it.

28.7.1.2 Modified Version of the Abbé Refractometer

If plenty of liquid is available as a sample, prism A of Figure 28.19 may be dispensed with and prism B simply dipped in the liquid. The illuminating arrangements are as shown in Figure 28.20. Instruments of this kind are usually called *dipping refractometers*.

When readings are required in large numbers or continuous monitoring of a process is called for, an automatic type of Abbé refractometer may be used. The optical system is essentially similar to that of the visual instrument except that instead of moving the whole telescope to determine the position of the light/dark boundary, the objective lens is kept fixed and a differentiating detector is scanned across its image plane under the control of a stepper motor. The detector's response to the sudden change of illuminance at the boundary enables its position, and thus the refractive index, to be determined. To ensure accuracy, the boundary is scanned in both directions and the mean position is taken.

28.7.1.3 Refractometry of Solid Samples

If a large piece of the sample is available, it is possible to optically polish two faces at an angle on it and then to measure the deviation of a monochromatic beam that it produces with a table spectrometer. This process is laborious and expensive. The Hilger–Chance refractometer is designed for the determination of the refractive indices of optical glasses

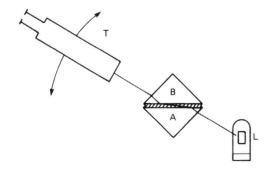

FIGURE 28.19 Abbé refractometer.

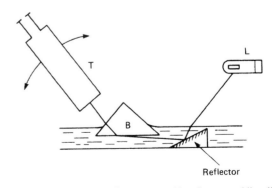

FIGURE 28.20 Dipping refractometer with reflector enabling light to enter prism B at near-grazing incidence.

and requires that only two roughly polished surfaces at right angles are available. It can also be used for liquids.

The optical parts are shown in Figure 28.21. Monochromatic light from the slit (A) is collected by the lens (B) and passes into the V-shaped prism block (C), which is made by fusing two prisms together to produce a very precise angle of 90 degrees between its surfaces. The light emerges and enters the telescope (T). When the specimen block is put in place, the position of the emergent beam will depend on the refractive index of the sample. If it is greater than that of the V block, the beam will be deflected upward; if lower, downward. The telescope is moved to determine the precise beam direction, and the refractive index can be read off from a calibrated scale.

In the actual instrument the telescope is mounted on a rotating arm with the reflecting prism, so that the axis of the telescope remains horizontal. A wide slit with a central hairline is used at A, and the telescope eyepiece is equipped with two lines in its focal plane so that the central hairline may be set between them with great precision (see Figure 28.22). Since it is the bulk of the sample, not only the surfaces, that is responsible for the refraction, it is possible to place a few drops of liquid on the V-block so that perfect optical contact may be achieved on a roughly polished specimen. The V-block is equipped with side plates so that it forms a trough suitable for liquid samples.

When this device used for measuring optical glasses, an accuracy of 0.0001 can be obtained. This is very high indeed, and the points raised about accuracy in connection with the Abbé refractometer should be borne in mind. Notice that the instrument is arranged so that the rays pass the air/glass interfaces at normal or near-normal incidence so as to reduce the effects of changes in the refractive index of air with temperature and humidity.

28.7.1.4 Solids of Irregular Shape

The refractive index of irregularly shaped pieces of solid materials may theoretically be found by immersing them in a liquid of identical refractive index. When this happens, rays traversing the liquid are not deviated when they encounter the solid but pass straight through, so that the liquid–solid boundaries totally disappear. A suitable liquid may be made up by using liquids of different refractive index together. When a refractive index match has been found, the refractive index of the liquid may be measured with an Abbé refractometer. Suitable liquids are given by Longhurst (1974):

	n
Benzene	1.504
Nitrobenzene	1.553
Carbon bisulphide	1.632
α-monobromonaphthalene	1.658

The author cannot recommend this process. Granted, it can be used for a magnificent lecture-room demonstration, but in practice these liquids are highly toxic, volatile, and have an appalling smell. Moreover, the method depends on

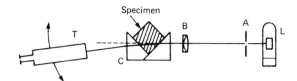

FIGURE 28.21 Hilger–Chance refractometer.

FIGURE 28.22 Appearance of field when the telescope is correctly aligned.

both the solid and the liquid being absolutely colorless; if either has any trace of color it is difficult to judge the precise refractive index match at which the boundaries disappear.

28.7.2 Polarimeters

Some solutions and crystals have the property that when a beam of plane-polarized light passes through them, the plane is rotated. This phenomenon is known as *optical activity*, and in liquids it occurs only with those molecules that have no degree of symmetry. Consequently, few compounds show this property, but one group of compounds of commercial importance does: sugars. Hence the measurement of optical activity offers an elegant way of determining the sugar content of solutions. This technique is referred to as *polarimetry* or occasionally by its old-fashioned name of *saccharimetry*.

Polarimetry is one of the oldest physical methods applied to chemical analysis and has been in use for almost 100 years. The original instruments were all visual, and though in recent years photoelectric instruments have appeared, visual instruments are still widely used because of their simplicity and low cost.

If the plane of polarization in a liquid is rotated by an angle θ on passage through a length of solution l, then

$$\theta = \alpha c l$$

where c is the concentration of the optically active substance and α is a coefficient for the particular substance called the *specific rotation*. In all substances the specific rotation increases rapidly with decreasing wavelength (see Figure 28.23), and for that reason monochromatic light sources are always used—very often a low-pressure sodium lamp. Some steroids show anomalous behavior of the specific rotation with wavelength, but the spectropolarimetry involved is beyond the scope of this book.

28.7.2.1 The Laurent Polarimeter

The main optical parts of this instrument are shown in Figure 28.24. The working is best described if the solution

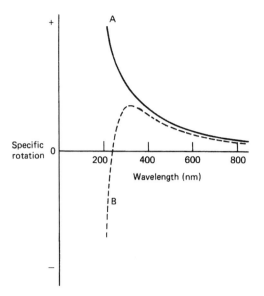

FIGURE 28.23 Variation of specific rotation with wavelength. A: typical sugar; B: steroid showing reversion.

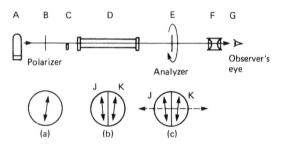

FIGURE 28.24 Laurent polarimeter. (a) Plane-polarized light after passage through B; (b) polarization directions in field after passage through C; (c) broken arrow shows plane of analyzer at position of equal brilliance with no sample present.

tube is at first imagined not to be present. Light from a monochromatic source (A) passes through a sheet of Polaroid (B) so that it emerges plane-polarized. It then encounters a half-wave plate (C), which covers only half the area of the beam. The effect of the half-wave plate is to slightly alter the plane of polarization of the light that passes through it so that the situation is as shown in Figure 28.24(b). If the solution tube (D) is not present, the light next encounters a second Polaroid sheet at E. This is mounted so that it can be rotated about the beam. On looking through the eyepiece (F), the two halves of the field will appear of unequal brilliance until E is rotated and the plane it transmits is as shown in Figure 28.24(c). Since the planes in J and K differ only by a small angle, the position of equal brilliance can be judged very precisely. If the position of the analyzer (E) is now read, the solution tube (D) can be put in position and the process repeated so that the rotation θ may be determined. Since the length of the solution tube is known, the concentration of the solution may be determined if the specific rotation is known.

28.7.2.2 The Faraday-Effect Polarimeter

Among the many effects that Faraday discovered was the fact that glass becomes weakly optically active in a magnetic field. This discovery lay unused for over 100 years until its employment in the Faraday-effect polarimeter. The main optical parts are shown schematically in Figure 28.25.

A tungsten lamp, filter, and Polaroid are used to provide plane-polarized monochromatic light. This is then passed through a Faraday cell (a plain block of glass situated within a coil), which is energized from an oscillator at about 380 Hz, causing the plane of polarization to swing about 3° either side of the mean position. If we assume for the time being that there is no solution in the cell and no current in the second

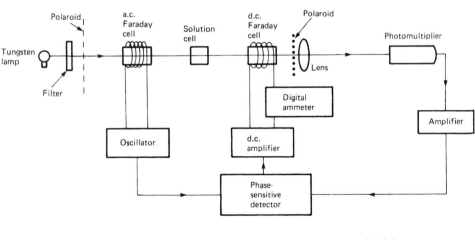

FIGURE 28.25 Faraday-effect polarimeter. (a) Photomultiplier signal with no sample present; (b) photomultiplier signal with uncompensated rotation.

Faraday cell, this light will fall unaltered on the second Polaroid. Since this is crossed on the mean position, the photomultiplier will produce a signal at *twice* the oscillator frequency, because there are two pulses of light transmission in each oscillator cycle, as shown in Figure 28.25(a). If an optically active sample is now put in the cell, the rotation will produce the situation in Figure 28.25(b) and a component of the same frequency as the oscillator output. The photomultiplier signal is compared with the oscillator output in a phase-sensitive circuit so that any rotation produces a dc output. This is fed back to the second Faraday cell to oppose the rotation produced by the solution, and, by providing sufficient gain, the rotation produced by the sample will be completely restored. In this condition the current in the second cell will in effect give a measure of the rotation, which can be indicated directly with a suitably calibrated meter.

This arrangement can be made highly sensitive and a rotation of as little as 1/10,000th of a degree can be detected, enabling a short solution path length to be used—often 1 mm. Apart from polarimetry, this technique offers a very precise method of measuring angular displacements.

28.8 THERMAL IMAGING TECHNIQUES

Much useful information about sources and individual objects can be obtained by viewing them not with the visible light they give off but by the infrared radiation they emit. We know from Planck's radiation law:

$$P_\lambda = \frac{C_1}{\lambda^5 \left[e^{C_2/\lambda T} - 1 \right]}$$

(where P_λ represents the power radiated from a body at wavelength λ, T the absolute temperature, and C_1 and C_2 are constants) that objects at the temperature of our environment radiate significantly, but we are normally unaware of this because all that radiation is well out in the infrared spectrum. The peak wavelength emitted by objects at around 20°C (293 K) is about 10 μm, whereas the human eye is only sensitive in the range 0.4–0.8 μm. By the use of an optical system sensitive to infrared radiation, this radiation can be studied, and since its intensity depends on the surface temperature of the objects concerned, the distribution of temperatures over an object or source can be made visible. It is possible to pick out variations in surface temperature of less than 1 K in favorable circumstances, and so this technique is of great value; for example, it enables the surface temperature of the walls of a building to be determined and so reveals the areas of greatest heat loss. Figure 28.26 illustrates this concept: (a) is an ordinary photograph of a house; (b) gives the same view using thermal imaging techniques. The higher temperatures associated with heat loss from the windows and door are immediately apparent. Thermal imaging can be used in medicine to reveal variations of surface temperature on a patient's body and thus reveal failures of circulation; it is of great military value since it will

(a)

(b)

FIGURE 28.26 Thermal imaging techniques applied to a house. Courtesy of Agema Infrared Systems.

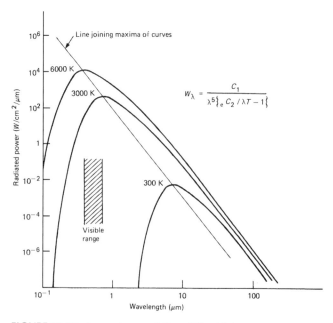

FIGURE 28.27 Log scale presentation of Planck's law.

function in darkness, and it has all manner of applications in the engineering field.

An excellent account of the technique is given by Lawson (1979) in *Electronic Imaging*; Figure 28.27 and Table 28.3

Table 28.3 Solar and black-body radiation at various wavelengths (after Lawson, 1979)

Wavelength band (μm)	Typical value of solar radiation (Wm^{-2})	Emission from black body at 300 K (Wm^{-2})
0.4–0.8	750	0
3–5	24	6
8–13	1.5	140

Note: Complete tables of data relating to the spectral power distribution of blackbody radiators are given by M. Pivovonsky and M. Nagel.

are based on this work. Although the spectrum of a body around 300 K has its peak at 10 μm, the spectrum is quite broad. However, the atmosphere is effectively opaque from about 5–8 μm and above 13 μm, which means that in practice the usable bands are 3–5 μm and 8–13 μm. Although there is much more energy available in the 8–13 μm band (see Table 28.3) and if used outdoors there is much less solar radiation, the sensitivity of available detectors is much better in the 3–5 μm band, and both are used in practice. Nicholas (1968) has developed a triple waveband system using visible 2–5 μm and 8–13 μm ranges, but this technique has not so far been taken up in commercially available equipment.

In speaking of black-body radiations we must remember that, in practice, no surfaces have emissivities of 1.0 and often they have much less, so there is not a strict relationship between surface temperatures and radiant power for all the many surfaces in an exterior source. In daylight, reflected solar power is added to the emitted power and is a further complication. However, the main value of the technique is in recognizing differences of temperature rather than temperature in absolute terms.

REFERENCES

American National Standards Institute, Standards for the Safe Use of Lasers.
Beynon, J. D. E., *Charge Coupled Devices and their Applications*, McGraw-Hill, New York (1979).
Bhaskaran, S., Chapman, T., Pilon, M., VanGorden, S., *Performance Based CID Imaging—Past, Present and Future*, Thermo Fisher Scientific, Liverpool, NY. http://www.thermo.com/eThermo/CMA/PDFs/Product/productPDF_8997.pdf
BS 4803, *Guide on Protection of Personnel against Hazards from Laser Radiation* (1983).
Brown, S. and Tarrant, A. W. S., *A Sensitive Spectroradiometer using a Boxcar Detector*, Association International de la Couleur (1981).
Chamberlin, G. J. and Chamberlin, D. G., *Color: its Measurement, Computation and Application*, Heyden, London (1979).
Commission Internationale de I'Eclairage, *The Spectroradiometric Measurement of Light Sources*, CIE Pub. No. 63 (1984).
Dudley, W. W., *Carbon Dioxide Lasers, Effects and Applications*, Academic Press, London (1976).
Dresler, A. and Frühling, H. S., "Uber ein Photoelektrische Dreifarben Messgerät," *Das Licht*, **11**, 238 (1938).
Forsythe, W. E., *The Measurement of Radiant Energy*, McGraw-Hill, New York (1941).
Fry, P. W., "Silicon photodiode arrays," *J. Sci. Inst.*, **8**, 337 (1975).
Geutler, G., *Die Farbe*, **23**, 191 (1974).
Heavens, O. S., *Lasers*, Duckworth, London (1971).
Henderson, S. T., *J. Phys. D.*, **3**, 255 (1970).
Henderson, S. T. and Marsden, A. M., *Lamps and Lighting*, Edward Arnold, London (1972).
IEC-CIE, *International Lighting Vocabulary*.
Judd, D. B. and Wysecki, G., *Color in Business, Science and Industry*, Wiley, New York (1975).
Koechner, W., *Solid State Laser Engineering*, Springer, New York (1976).
Lawson, W. D., "Thermal imaging," in *Electronic Imaging* (eds., T. P. McLean and P. Schagen), Academic Press, London (1979).
Longhurst, R. S., *Geometrical and Physical Optics*, Longman, London (1974).
Mooradian, A., Jaeger, T., and Stokseth, P., *Tuneable Lasers and Applications*, Springer, New York (1976).
Moore, J., "Sources of error in spectro-radiometry," *Lighting Research and Technology*, **12**, 213 (1984).
Nichols, L. W. and Laner, J., *Applied Optics*, **7**, 1757 (1968).
Pivovonsky, M. and Nagel, M., *Tables of Blackbody Radiation Functions*, Macmillan, London (1961).
Safety in Universities—Notes for Guidance, Part 2: 1, Lasers, Association of Commonwealth Universities (1978).
Steck, B., "Effects of optical radiation on man," *Lighting Research and Technology*, **14**, 130 (1982).
Tarrant, A. W. S., *Some Work on the SPD of Daylight*, PhD thesis, University of Surrey (1967).
Tarrant, A. W. S., "The nature of color – the physicist's viewpoint," in *Natural Colors for Food and other Uses* (ed. J. N. Counsell), Applied Science, London (1981).
Thewlis, J. (ed.), *Encyclopaedic Dictionary of Physics*, Vol. 1, p. 553, Pergamon, Oxford (1961).
Walsh, J. W. T., *Photometry*, Dover, London (1958).
Wright, W. D., *The Measurement of Color*, Hilger, Bristol (1969).
Young, A. R., Magnus, I. A., and Gibbs, N. K., "Ultraviolet radiation radiometry of solar simulation," *Proc. Conf. on Light Measurement*, SPIE. Vol. 262 (1980).

Chapter 29

Nuclear Instrumentation Technology

D. Aliaga Kelly and W. Boyes

29.1 INTRODUCTION

Nuclear gauging instruments can be classified as those that measure the various radiations or particles emitted by radioactive substances or nuclear accelerators, such as alpha particles, beta particles, electrons and positrons, gamma- and X-rays, neutrons, and heavy particles such as protons and deuterons. A variety of other exotic particles also exist, such as neutrinos, mesons, muons, and the like, but their study is limited to high-energy research laboratories and uses special detection systems; they will not be considered in this book.

An important factor in the measurements to be made is the energy of the particles or radiations. This is expressed in electron-volts (eV) and can range from below 1 eV to millions of eV (MeV). Neutrons of very low energies (0.025 eV) are called *thermal neutrons* because their energies are comparable to those of gas particles at normal temperatures. However, other neutrons can have energies of 10 MeV or more; X-rays and gamma-rays can range from a few eV to MeV and sometimes GeV (10^9 eV).

The selection of a particular detector and detection system depends on a large number of factors that have to be taken into account in choosing the optimum system for a particular project. One must first consider the particle or radiation to be detected, the number of events to be counted, whether the energies are to be measured, and the interference with the measurement by background radiation of similar or dissimilar types. Then the selection of the detector can be made, bearing in mind cost and availability as well as suitability for the particular problem. Choice of electronic units will again be governed by cost and availability as well as the need to provide an output signal with the information required. It can be seen that the result will be a series of compromises, since no detector is perfect, even if unlimited finance is available.

The radioactive source to be used must also be considered, and a list of the more popular types is given in Tables 29.1 and 29.2. Some other sources, used particularly in X-ray fluorescence analysis, are given in Table 30.1 in the next chapter.

29.1.1 Statistics of Counting

The variability of any measurement is measured by the standard deviation σ, which can be obtained from replicate determinations by well-known methods. There is an inherent variability in radioactivity measurements because the disintegrations occur in a random manner, described by the Poisson distribution. This distribution is characterized by the property that the standard deviation σ of a large number of events, N, is equal to its square root, that is,

$$\sigma(N) = \sqrt{N} \qquad (29.1)$$

For ease in mathematical application, the normal (Gaussian) approximation to the Poisson distribution is ordinarily used. This approximation, which is generally valid for numbers of events, N equal to or greater than 20, is the particular normal distribution whose mean is N and whose standard deviation is $\sqrt{N}$.

Generally, the concern is not with the standard deviation of the number of counts but rather with the deviation in the rate (= number of counts per unit time):

$$R' = \frac{N}{t} \qquad (29.2)$$

where t is the time of observation, which is assumed to be known with such high precision that its error may be neglected. The standard deviation in the counting rate, $\sigma(R')$, can be calculated by the usual methods for propagation of error:

$$\sigma(R') = \frac{\sqrt{N}}{t} = \left(\frac{R'}{t}\right)^{\frac{1}{2}} \qquad (29.3)$$

In practice, all counting instruments have a background counting rate, B, when no radioactive source is present.

TABLE 29.1 Radiation Sources in General Use

Isotope	Half-life	Emissions (and energies–MeV)				
		Beta		Gamma	Alpha	
		E_{max}	E_{av}			
^{3}H (Tritium)	12.26 yr	0.018	0.006	Nil	Nil	
^{14}C	5730 yr	0.15	0.049	Nil	Nil	
^{22}N	2.6 yr	0.55	0.21	1.28	Nil	Also 0.511 MeV annihilation
^{24}Na	15 h	1.39	1.37	2.75		
^{32}P	14.3 d	1.71	0.69	Nil	Nil	Pure beta emitter
^{36}Cl	3×10^5 yr	0.71	0.32	Nil	Nil	Betas emitted simulate fission products
^{60}Co	5.3 yr	0.31	0.095	1.17 (100)	Nil	Used in radiography, etc.
				1.33 (100)	Nil	
^{90}Sr	28 yr	0.54	0.196	Nil	Nil	Pure beta emitter
^{90}Y	64.2 h	2.25	0.93	Nil	Nil	
^{131}I	8.04 d	0.61+	0.18+	0.36 (79)+	Nil	Used in medical applications
^{137}Cs	30 yr	0.5+	0.19+	0.66 (86)	Nil	Used as standard gamma calibration source
^{198}Au	2.7 d	0.99+	0.3+	0.41 (96)	Nil	Gammas adopted recently as universal standard
^{226}Ra	1600 yr	–	–	0.61 (22)		Earliest radioactive source isolated by Mme Curie, still used in medical applications
				1.13 (13)		
				1.77 (25)		
				+ others		
^{241}Am	457 yr	–	–	0.059 (35)	5.42 (12)	
				+ others	5.48 (85)	Alpha X-ray calibration source

Note: Figures in parentheses show the percentage of primary disintegration that goes into that particular emission (i.e., the abundance). + Indicates other radiation of lower abundance.

TABLE 29.2 Neutron sources

Source and type	Neutron emission n/s per unit activity or mass	Half-life	energies of neutrons emitted		
^{124}Sb–Be (γ, n)	5×10^{-5}/Bq	60 dyas	Low: 30 keV		For field assay of beryllium ores
^{226}Ra–Be (α, n)	3×10^{-4}/Bq	1622 yr	Max: 12 MeV		Early n source, now replaced by Am/Be
			Av: 4 MeV		
^{210}Po–Be (α, n)	7×10^{-5}/Bq	138 days	Max: 10.8 MeV		Short life is disadvantage
			Av: 4.2 MeV		
^{241}Am–Be (α, n)	6×10^{-5}/Bq	433 yr	Max: 11 MeV		Most popular neturon source
			Av: 3–5 MeV		
^{252}Cf fission	2.3×10^6/µg	2.65 yr	Simulates reactor neutron spectrum		Short life and high cost

CHAPTER | 29 Nuclear Instrumentation Technology

When a source is present, the counting rate increases to R_0. The counting rate R due to the source is then

$$R = R_0 - B \qquad (29.4)$$

By propagation-of-error methods, the standard deviation of R can be calculated as follows:

$$\sigma(R') = \left(\frac{R_0}{t_1} + \frac{B}{t_2}\right)^{\frac{1}{2}} \qquad (29.5)$$

where t_1 and t_2 are the times over which source-plus-background and background counting rates were measured, respectively. Practical counting times depend on the activity of the source and of the background. For low-level counting, one has to reduce the background by the use of massive shielding, careful material selection for the components of the counter, the use of such devices as anticoincidence counters, and as large a sample as possible.

The optimum division of a given time period for counting source and background is given by

$$\frac{t_2}{t_1} = \frac{1}{1 + (R_0/B)^{\frac{1}{2}}} \qquad (29.6)$$

TABLE 29.3 Limits of the quantity χ^2 for sets of counts with random errors

Number of observations	Lower limit for χ^2	Upper limit for χ^2
3	0.103	5.99
4	0.352	7.81
5	0.711	9.49
6	1.14	11.07
7	1.63	12.59
8	2.17	14.07
9	2.73	15.51
10	3.33	16.92
15	6.57	23.68
20	10.12	30.14
25	13.85	36.42
30	17.71	42.56

29.1.1.1 Nonrandom Errors

These may be due to faults in the counting equipment, personal errors in recording results or operating the equipment, or errors in preparing the sample. The presence of such errors may be revealed by conducting a statistical analysis on a series of repeated measurements. For errors in the equipment or in the way it is operated, the analysis uses the same source, and for source errors a number of sources are used.

Although statistical errors will follow a Gaussian distribution, errors that are not random will not follow such a distribution, so that if a series of measurements cannot be fitted to a Gaussian distribution curve, nonrandom errors must be present. The chi-squared test allows one to test the goodness of fit of a series of observations to the Gaussian distribution. If nonrandom errors are not present, the values of χ^2 as determined by the relation given in Equation (29.7) should lie between the limits quoted in Table 29.3 for various groups of observations:

$$\chi^2 = \frac{\sum_{i=1}^{i=q}(\bar{n} - n_i)^2}{\bar{n}} \qquad (29.7)$$

where $\bar{n}$ is the average count observed, n_i is the number counted in the ith observation, and q is the number of observations. If a series of observations fits a Gaussian distribution, there is a 95 percent probability that χ^2 will be greater than or equal to the lower limit quoted in Table 29.1 but only a 5 percent probability that χ^2 will be greater than or equal to the upper limit quoted. Thus, for 10 observations, if χ^2 lies outside the region of 3.33–16.92 it is very probable that errors of a nonrandom kind are present.

In applying the chi-squared test, the number of counts recorded in each observation should be large enough to make the statistical error less than the accuracy required for the activity determination. Thus if 10,000 counts are recorded for each observation and for a series of observations χ^2 lies between the expected limits, it can be concluded that nonrandom errors of a magnitude greater than about ± 2 percent are not present.

29.1.1.2 Radioactive Decay

Radioactive sources have the property of disintegrating in a purely random manner, and the rate of decay is given by the law

$$\frac{dN}{dt} = -\lambda N \qquad (29.8)$$

where λ is called the decay constant and N is the total number of radioactive atoms present at a time t. This may be expressed as

$$N = N_0 \exp(-\lambda t) \qquad (29.9)$$

where N_0 is the number of atoms of the parent substance present at some arbitrary time 0.

Combining these equations, we have

$$\frac{dN}{dt} = -\lambda N_0 \exp(-\lambda t) \qquad (29.10)$$

showing that the rate of decay falls off exponentially with time. It is usually more convenient to describe the decay in terms of the "half-life" $T_{1/2}$ of the element. This is the time required for the activity, dN/dt, to fall to half its initial value and $\lambda = 0.693/T_{1/2}$.

When two or more radioactive substances are present in a source, the calculation of the decay of each isotope becomes more complicated and will not be dealt with here.

The activity of a source is a measure of the frequency of disintegration occurring in it. Activity is measured in Becquerels (Bq), one Becquerel corresponding to one disintegration per second. The old unit, the Curie (Ci), is still often used, and 1 Megabecquerel = 0.027 millicuries.

It is also often important to consider the radiation that has been absorbed—the *dose*. This is quoted in grays, the gray being defined as that dose (of any ionizing radiation) that imparts 1 joule of energy per kilogram of absorbing matter at the place of interest. So 1 Gy = 1 J kg^{-1}. The older unit, the rad, 100 times smaller than the gray, is still often referred to.

29.1.2 Classification of Detectors

Various features of detectors are important and have to be taken into account in deciding the choice of a particular system, notably:

1. Cost.
2. Sizes available.
3. Complexity in auxiliary electronics needed.
4. Ability to measure energy and/or discriminate between various types of radiations or particles.
5. Efficiency, defined as the probability of an incident particle being recorded.

Detectors can be grouped generally into the following classes. Most of these are covered in more detail later, but Cherenkov detectors and cloud chambers are specialized research tools and are not discussed in this book.

29.1.2.1 Gas Detectors

Gas detectors include ionization chambers, gas proportional counters, Geiger counters, multiwire proportional chambers, spark counters, drift counters, and cloud chambers.

29.1.2.2 Scintillation Counters

Some substances have the property that, when bombarded with nuclear particles or ionizing radiation, they emit light that can be picked up by a suitable highly sensitive light detector that converts the light pulse into an electronic pulse that can be amplified and measured.

29.1.2.3 Cherenkov Detectors

When a charged particle traverses a transparent medium at a speed greater than that of light within the medium, Cherenkov light is produced. Only if the relative velocity $\beta = v/c$ and the refractive index n of the medium are such that $n\beta > 1$ will the radiation exist. When the condition is fulfilled, the Cherenkov light is emitted at the angle given by the relation

$$\cos \theta = \frac{1}{n\beta} \qquad (29.11)$$

where θ is the angle between the velocity vector for the particle and the propagation vector for any portion of the conical radiation wavefront.

29.1.2.4 Solid-State Detectors

Some semiconductor materials have the property that when a potential is applied across them and an ionizing particle or ionizing radiation passes through the volume of material, ions are produced just as in the case of a gas-ionization chamber, producing electronic pulses in the external connections that can be amplified, measured, or counted. A device can thus be made, acting like a solid ionization chamber. The materials that have found greatest use in this application are silicon and germanium.

29.1.2.5 Cloud Chambers

These were used in early research work and are still found in more sophisticated forms in high-energy research laboratories to demonstrate visually (or photographically) the actual paths of ionizing particles or radiation by means of trails of liquid droplets formed after the passage of such particles or radiation through supersaturated gas.

Photographic film can be used to detect the passage of ionizing particles or radiation, since they produce latent images along their paths in the sensitive emulsion. On development, the grains of silver appear along the tracks of the particles or ions.

29.1.2.6 Plastic Film Detectors

Thin (5 μm) plastic films of polycarbonate can be used as detectors of highly ionizing particles which can cause radiation damage to the molecules of the polycarbonate film. These tracks may be enlarged by etching with a suitable chemical and visually measured with a microscope.

Alternatively, sparks can be generated between two electrodes, one of which is an aluminized mylar film, placed on either side of a thin, etched polycarbonate detector. The sparks that pass through the holes in the etched detector can be counted using a suitable electronic scaler.

29.1.2.7 Thermoluminescent Detectors

For many years it was known that if one heated some substances, particularly fluorites and ceramics, they could be made to emit light photons and, in the case of ceramics, could be made incandescent. When ionizing radiation is absorbed in matter, most of the absorbed energy goes into heat, whereas a small fraction is used to break chemical bonds. In some materials a very minute fraction of the energy is stored in metastable energy states. Some of the energy thus stored can be recovered later as visible light photons if the material is heated, a phenomenon known as *thermoluminescence* (TL).

In 1950 Daniels proposed that this phenomenon could be used as a measurement of radiation dose, and in fact it was used to measure radiation after an atom-bomb test. Since then interest in TL as a radiation dosimeter has progressed to the stage that it could now well replace photographic film as the approved personnel radiation badge carried by people who may be involved with radioactive materials or radiation.

29.1.2.8 Materials for TL Dosimetry

The most popular phosphor for dosimetric purposes is lithium fluoride (LiF). This can be natural LiF or with the lithium isotopes ^{6}Li and ^{7}Li enriched or depleted, as well as variations in which an activator such as manganese (Mn) is added to the basic LiF. The advantages of LiF are:

1. Its wide and linear energy response from 30 KeV up to and beyond 2 MeV.
2. Its ability to measure doses from the mR to 10^5R without being affected by the rate at which the dose is delivered; this is called *dose-rate independence*.
3. Its ability to measure thermal neutrons as well as X-rays, gamma rays, beta rays, and electrons.
4. Its dose response is almost equivalent to the response of tissue, that is, it has almost the same response as the human body.
5. It is usable in quite small amounts, so it can be used to measure doses to the fingers of an operator without impeding the operator's work.
6. It can be reused many times, so it is cheap.

Another phosphor that has become quite popular in recent years is calcium fluoride with manganese (CaF_2:Mn), which has been found to be more sensitive than LiF for low-dose measurements (some 10 times) and can measure a dose of 1 mR yet is linear in doserate response up to 10^5 R. However, it exhibits a large energy dependence and is not linear below 300 KeV.

Thermoluminescence has also been used to date ancient archaeological specimens such as potsherds, furnace floors, ceramic pots, and so on. This technique depends on the fact that any object heated to a high temperature loses inherent thermoluminescent powers and, if left for a long period in a constant radioactive background, accumulates an amount of TL proportional to the time it has lain undisturbed in that environment.

29.1.3 Health and Safety

Anyone who works with radioactive materials must understand clearly the kinds of hazards involved and their magnitude. Because radioactivity is not directly observable by the body's senses, it requires suitable measuring equipment and handling techniques to ensure that any exposure is minimized; because of this, suitable legislation governs the handling and use of all radioactive material. In Part 4 an outline of the regulations is given as well as advice on contacting the local factory inspector before the use of any radioactive source is contemplated.

Because everyone in the world already receives steady radiation (from the natural radiopotassium in the human body and from the general background radiation to which all are subjected), the average human body acquires a dose of about 300 micro-grays (μGy) (equivalent to 30 millirads) per year. Hence, though it is almost impossible to reduce radiation exposure to zero, it is important to ensure that using a radioactive source does not increase the dose to a level greater than many other hazards commonly met in daily life.

There are three main methods for minimizing the hazards due to the use of a radioactive source:

1. *Shielding*. A thickness of an appropriate material, such as lead, should be placed between the source and the worker.
2. *Distance*. An increase in distance between source and worker reduces the radiation intensity.
3. *Time*. The total dose to the body of the worker depends on the length of time spent in the radiation field. This time should be reduced to the minimum necessary to carry out the required operation.

These notes are for sources that are contained in sealed capsules. Those that are in a form that might allow them to enter the body's tissues must be handled in ways that prevent such an occurrence (for example, by operating inside a "glove box," which is a box allowing open radioactive sources to be dealt with while the operator stays outside the enclosure). Against internal exposure the best protection is good housekeeping, and against external radiation the best protection is good instrumentation, kept in operating condition and *used*.

Instruments capable of monitoring radioactive hazards depend on the radiation or particles to be monitored. For gamma rays, emitted by the most usual radioactive sources to be handled, a variety of instruments is available. Possibly the cheapest yet most reliable monitor contains a Geiger counter, preferably surrounded by a suitable metal covering to modify the counter's response to the varied energies from gamma-emitting radioisotopes to make it similar to the response of the human body. There are many such instruments available from commercial suppliers. More elaborate ones are based on ionization chambers, which are capable of operating over a much wider range of intensities and are correspondingly more expensive. For beta emitters, a Geiger counter with a thin window to allow the relatively easily absorbed beta particles to enter the counter is, again, the cheapest monitor. More expensive monitors are based on scintillation counters, which can have large window areas, useful for monitoring extended sources or accidents where radioactive beta emitters have been spilt. Alpha detection is particularly difficult, since most alphas are absorbed in extremely thin windows. Geiger counters with very thin windows can be used, or an ionization chamber with an open front, used in air at normal atmospheric pressure. More expensive scintillation counters or semiconductor detectors can also be used. Neutrons require much more elaborate and expensive monitors, ranging from ionization or proportional counters containing BF_3 or ^3He, to scintillation counters using ^6LiI, ^6Li-glass, or plastic scintillators, depending on the energies of the neutrons.

29.2 DETECTORS

29.2.1 Gas Detectors

Gas-filled detectors may be subdivided into those giving a current reading and those indicating the arrival of single particles. The first class comprises the current ionization chambers and the second the counting or pulse-ionization chambers, proportional counters, and Geiger counters. The object of the ionization chamber is always the same: to measure the rate of formation of ion pairs within the gas. One must therefore be certain that the voltage applied to the electrodes is great enough to give saturation, that is, to ensure that there will be no appreciable recombination of positive and negative ions.

To understand the relation between the three gas-filled detectors, we can consider a counter of very typical geometry: two coaxial cylinders with gas between them. The inner cylinder, usually a fine wire (the anode), is at a positive potential relative to the outer cylinder (the cathode). Let us imagine ionization to take place in the gas, from a suitable radioactive source, producing, say, 10 electrons. The problem is to decide how many electrons (n) will arrive at the anode wire.

Figure 29.1 shows the voltage applied across the counter V, plotted against the logarithm of n, that is, $\log_{10} n$. When V is very small, on the order of volts or less, all 10 electrons do not arrive at the anode wire, because of recombination. At V_1 the loss has become negligible because saturation has been achieved and the pulse contains 10 electrons. As V is increased, n remains at 10 until V_2 is reached, usually some tens or hundreds of volts. At this point the electrons begin to acquire sufficient energy between collisions at the end of

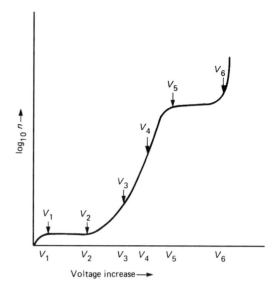

FIGURE 29.1 Response of gas counter to increase in voltage.

their paths for ionization by collision to occur in the gas and this multiplication causes n to rise above 10, more or less exponentially with V as each initial electron gives rise to a small avalanche of secondary electrons by collision close to the wire anode. At any potential between V_2 and V_3 the multiplication is constant, but above this the final number of electrons reaching the wire is no longer proportional to the initial ionization. This is the region of limited proportionality, V_4. Above this, from V_5 to V_6 the region becomes that of the Geiger counter, where a single ion can produce a complete discharge of the counter. It is characterized by a spread of the discharge throughout the whole length of the counter, resulting in an output pulse size independent of the initial ionization. Above V_6 the counter goes into a continuous discharge. The ratio of n, the number of electrons in the output pulse, to the initial ionization at any voltage is called the gas-amplification factor A and varies from unity in the ionization chamber region to 10^3 to 10^4 in the proportional region, reaching 10^5 just below the Geiger region.

Ionization-chamber detectors can be of a variety of shapes and types. Cylindrical geometry is the most usual one adopted, but parallel plate chambers are often used in research. These chambers are much used in radiation-protection monitors, since they can be designed to be sufficiently sensitive for observing terrestrial radiation yet will not overload when placed in a very high field of radiation such as an isotopic irradiator. They can also be used, again in health physics, to integrate over a long period the amount of ionizing radiation passing through the chamber. An example is the small integrating chamber used in X-ray and accelerator establishments to observe the amount of radiation produced in personnel who carry the chambers during their working day.

Proportional counters are much more sensitive than ionization chambers; this allows weak sources of alpha and beta particles and low energy X-rays to be counted. The end-window proportional counter is particularly useful for counting flat sources, because it exhibits nearly 2π geometry—that is, it counts particles entering the counter over a solid angle of nearly 2π. Cylindrical proportional counters are used in radiocarbon dating systems because of their sensitivity for the detection of low-energy ^{14}C beta particles (E_{max} = 156 keV) and even tritium ^{3}H beta particles (E_{max} = 156 keV).

29.2.1.1 Geiger–Mueller Detectors

The Geiger counter has been and is the most widely used detector of nuclear radiation. It exhibits several very attractive features, some of which are:

1. Its cheapness. Manufacturing techniques have so improved the design that Geiger–Mueller tubes are a fraction of the cost of solid-state or scintillation detectors.
2. The output signal from a Geiger–Mueller tube can be on the order of 1 V, much higher than that from proportional, scintillation, or solid-state detectors. This means that the cost of the electronic system required is a fraction of that of other counters. A Geiger–Mueller tube with a simple high-voltage supply can drive most scaler units directly, with minimal or no amplification.
3. The discharge mechanism is so sensitive that a single ionizing particle entering the sensitive volume of the counter can trigger the discharge.

With these advantages there are, however, some disadvantages which must be borne in mind. These include:

1. The inability of the Geiger–Mueller tube to discriminate between the energies of the ionizing particles triggering it.
2. The tube has a finite life, though this has been greatly extended by the use of halogen fillings instead of organic gases. The latter gave lives of only about 10^{10} counts, whereas the halogen tubes have lives of 10^{13} or more counts.
3. There is a finite period between the initiation of a discharge in a Geiger–Mueller counter and the time when it will accept a new discharge. This, called the *dead time*, is on the order of 100 μs.

It is important to ensure that with Geiger counters the counting rate is such that the dead-time correction is only a small percentage of the counting rate observed. This correction can be calculated from the relation:

$$R' = \frac{R}{1 - R\tau} \qquad (29.12)$$

where R is the observed counting rate per unit time, R' is the true counting rate per unit time, and τ is the counter dead time.

Dead time for a particular counter may be evaluated by a series of measurements using two sources. Geiger tube manufacturers normally quote the dead time of a particular counter, and it is customary to increase this time electronically in a unit often used with Geiger counters so that the total dead time is constant but greater than any variations in the tube's dead time, since between individual pulses the dead time can vary.

The counting rate characteristics can be understood by reference to Figure 29.2. The starting voltage V_s is the voltage that, when applied to the tube viewing a fixed radioactive source, makes it just start to count. As the high voltage is increased, the counting rate rapidly increases until it reaches what is called the *plateau*. Here the increase in counting rate from V_A to V_B is small, on the order of 1–5 percent per 100 V of high voltage. Above V_B the counting rate rises rapidly and the tube goes into a continuous discharge, which will damage the counter. An operating point is selected (V_{op}) on the plateau so that any slight variation of the high-voltage supply has a minimal effect on the counting rate.

To count low-energy beta or alpha particles with a Geiger counter, a thin window must be provided to allow the particles to enter the sensitive volume and trigger the counter. Thin-walled glass counters can be produced, with wall thicknesses on the order of 30 mg/cm^2, suitable for counting high-energy beta particles. (In this context, the mass per unit area is more important than the linear thickness: for glass, 25 mg/cm^2 corresponds to about 0.1 mm.) For low-energy betas or alphas, a very thin window is called for, and these have been made with thicknesses as low as 1.5 mg/cm^2. Figure 29.3 gives the transmission through windows of

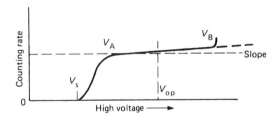

FIGURE 29.2 Geiger counter characteristic response.

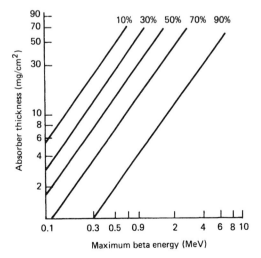

FIGURE 29.3 Transmission of thin windows.

TABLE 29.4 Transmission of thin windows

Nuclide	Max. energy E_{max} (MeV)	Percentage transmission for window thickness of			
		30mg/cm²	20mg/cm²	7mg/cm²	3mg/cm²
^{14}C	0.15	0.01	0.24	12	40
^{32}P	1.69	72	80.3	92	96

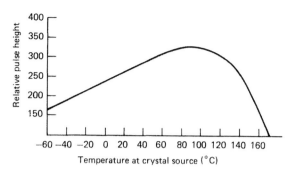

FIGURE 29.4 Relative output of CsI (Na) as a function of temperature.

different thickness, and Table 29.4 shows how it is applied to typical sources.

Alternatively, the source can be introduced directly into the counter by mixing as a gas with the counting gas, or if a solid source, by placing it directly inside the counter and allowing the flow of counting gas to continuously pass through the counter. This is the flow-counter method.

29.2.2 Scintillation Detectors

Scintillation counters comprise three main items: the scintillator, the light-to-electrical pulse converter (generally a photomultiplier), and the electronic amplifier. Here we consider the wide variety of each of these items that are available today.

The scintillator can consist of a single crystal of organic or inorganic material, a plastic fluor, an activated glass, or a liquid. Shapes and sizes can vary enormously, from the small NaI (Tl) (this is the way of writing "sodium iodide with thallium additive") crystal used in medical probes up to tanks of liquid scintillator of thousands of cubic meters used in cosmic ray research.

The selection of a suitable material—some of the characteristics of which are given in Tables 29.5 and 29.6—depends on a number of competing factors. No material is perfect, so compromises must be made in their selection for particular purposes.

29.2.2.1 Inorganic Scintillators

NaI (Tl) is still after many years the best gamma- and X-ray detector actually available, yet it is very hygroscopic and must be completely sealed from moisture. It is sensitive to shock, except in the new form of extruded NaI (Tl) called Polyscin, developed by Harshaw, and it is expensive, especially in large sizes. Its light output is, in general, proportional to the energy of the photon absorbed, so the pulse height is a measure of the energy of the incident photon and, when calibrated with photons of known energy, the instrument can be used as a spectrometer. The decay lifetime of NaI (Tl) is relatively slow, being about 230 ns, although NaI without any thallium when operated at liquid-nitrogen temperatures (77 K) has a decay lifetime of only 65 ns.

The next most used inorganic scintillator is CsI. When activated with thallium as CsI (Tl), it can be used at ambient temperatures with a light output some 10 percent lower than NaI (Tl) but with considerable resistance to shock. Its absorption coefficient is also greater than that of NaI (Tl), and these two characteristics have resulted in its use in space vehicles and satellites, since less mass is necessary and its resistance to the shock of launch is valuable. In thin layers it can be bent to match a circular light guide and has been used in this manner on probes to measure excited X-rays in soil. When activated with sodium instead of thallium, the light output characteristics are changed. The light output is slightly higher than CsI (Tl), and the temperature/light output relation is different (see Figure 29.4). The maximum light output is seen to occur, in fact, at a temperature of about 80°C. This is of advantage in borehole logging, where increased temperatures and shock are likely to occur as the detector is lowered into the drill hole.

CsI (Tl) has been a popular detector for alpha particles, since it is not very much affected by moisture from the air and so can be used in windowless counters. CsI (Na), on the other hand, quickly develops a layer impervious to alpha particles of 5–10 MeV energies when exposed to ambient air and thus is unsuitable for such use. CaF and CaF (Eu) are scintillators that have been developed to give twice the light output of NaI (Tl), but only very small specimens have been grown and production difficulties make their use impossible at present.

$B_4Ge_3O_{12}$ (bismuth germanate, or BGO) was developed to meet the requirements of the medical tomographic scanner, which calls for large numbers of very small scintillators capable of high absorption of photons of about 170 keV energy yet able to respond to radiation changes quickly without exhibiting "afterglow," especially when the detector has to integrate the current output of the photomultiplier. Its higher density than NaI (Tl) allows the use of smaller crystals, but its low light output (8 percent of NaI (Tl)) is a disadvantage. Against this it is nonhygroscopic, so it can be used with only light shielding.

CsF is a scintillator with a very fast decay time (about 5 ns) but a low light output (18 percent of NaI (Tl)). It also has been used in tomographic scanners.

CHAPTER | 29 Nuclear Instrumentation Technology

TABLE 29.5 Inorganic scintillator materials

Material[1]	Density (g/cm^3)	Refractive index (n)	Light output[2] (% anthracene)	Decay constant (s)	Wavelength of maximum emission (nm)	Operating temperature (°C)	Hygroscopic
NaI (Tl)	3.67	1.775	230	0.23×10^{-6}	413	Room	Yes
NaI (pure)	3.67	1.775	440	0.06×10^{-6}	303		Yes
CsI (Tl)	4.51	1.788	95	1.1×10^{-6}	580	Room	No
CsI (Tl)	4.51	1.787	150–190	0.65×10^{-6}	420	Room	Yes
CsI (pure)	4.51	1.788	500	0.6×10^{-6}	400		No
CaF$_2$ (EU)	3.17	1.443	110	1×10^{-6}	435	Room	No
LiI (EU)	4.06	1.955	75	1.2×10^{-6}	475	Room	Yes
CaWO$_4$	6.1	1.92	36	6×10^{-6}	430	Room	No
ZnS (Ag)	4.09	2.356	300	0.2×0^{-6}	1450	Room	No
ZnO (Ga)	5.61	2.02	90	0.4×10^{-9}	385	Room	No
CdWO$_4$	7.90			$0.9\text{–}20 \times 10^{-6}$	530	Room	No
Bi$_4$Ge$_3$O$_{12}$	7.13	2.15		0.3×10^{-6}	480	Room	No
CsF	4.64	1.48	40	5×10^{-12}	390	Room	No

1. *The deliberately added impurity is given in parentheses.*
2. *Light output is expressed as a percentage of that of a standard crystal of anthracene used in the same geometry.*

Table 29.6 Properties of organic scintillators

Material	Scintillator	Density (g/cm³)	Refractive index (n)	Boiling melting or softening point (°C)	Light output (% anthracene)	Decay constant (ns)	Wavelength of max. emission (nm)	Loading content (% by weight)	H/C Number of H atoms/ C atoms	Attenuation length (l/e m)	Principal applications
Plastic	NE102A	1.032	1.581	75	65	2.4	423	–	1.104	2.5	γ, α, β fast n
	NE104	1.032	1.581	75	68	1.9	406	–	1.100	1.2	Ultra-fast counting
	NE104B	1.032	1.58	75	59	3.0	406	–	1.107	1.2	Ditto with BBQ[1] light guides
	NE105	1.037	1.58	75	46	–	423	–	1.098	–	Air-equivalent for dosimetry
	NE110	1.032	1.58	75	60	3.3	434	–	1.104	4.5	γ, α, β fast n, etc.
	NE111A	1.032	1.58	75	55	1.6	370	1.103		–	Ultra-fast timing
	NE114	1.032	1.58	75	50	4.0	434	1.109		–	Cheaper for large arrays
	NE160	1.032	1.58	80	59	2.3	423	–	1.105	–	For use at higher temperatures—usable up to 150°C
	Pilot U	1.032	1.58	75	67	1.36	391	–	1.100	–	Ultra-fast timing
	Pilot 425	1.19	1.49	100	–	425	1.6				Cherenkov detector
Liquid	NE213	0.874	1.508	141	78	3.7	425	1.213			Fast n (PSD)[2] α, β (internal counting)
	Ne216	0.885	1.523	141	78	3.5	425	1.171			Internal counting, dosimetry
	NE220	1.306	1.442	104	65	3.8	425	29% O	1.669		α, β (internal counting)
	NE221	1.08	1.442	104	55	4	425	Gel	1.669		γ, fast n
	NE224	0.887	1.505	169	80	2.6	425		1.330		γ, insensitive to n
	NE226	1.61	1.38	80	20	3.3	430		0		n (heptane-based)
	NE228	0.71	1.403	99	45	–	385		2.11		Deuterated
	NE230	0.945	1.50	81	60	3.0	425	14.2% D	0.984		Deuterated
	NE232	0.89	1.43	81	60	4	430	24.5% D	1.96		α, β (internal counting)
	NE233	0.874	1.506	117	74	3.7	425		1.118		For large tanks
	NE235	0.858	1.47	350	40	4	420		2.0		Internal counting, dosimetry
	NE250	1.035	1.452	104	50	4	425	32% O	1.760		

1. BBQ is wavelength-shifter.
2. PSD means pulse shape discrimination.

LiI (Eu) is a scintillator particularly useful for detecting neutrons, and since the lithium content can be changed by using enriched ^{6}Li to enhance the detection efficiency for slow neutrons or by using almost pure ^{7}Li to make a detector insensitive to neutrons, it is a very versatile, if expensive, neutron detector. When it is cooled to liquid-nitrogen temperature the detection efficiency for fast neutrons is enhanced and the system can be used as a neutron spectrometer to determine the energies of the neutrons falling on the detector. LiI (Eu) has the disadvantage that it is extremely hygroscopic, even more so than NaI (Tl).

Cadmium tungstate ($CdWO_4$) and calcium tungstate ($CaWO_4$) single crystals have been grown, with some difficulty, and can be used as scintillators without being encapsulated, since they are nonhygroscopic. However, their refractive index is high, and this causes 60–70 percent of the light emitted in scintillators to be entrapped in the crystal.

CaF_2 (Eu) is a nonhygroscopic scintillator that is inert toward almost all corrosives and thus can be used for beta detection without a window or in contact with a liquid, such as the corrosive liquids used in fuel-element treatment. It can also be used in conjunction with a thick NaI (Tl) or CsI (Tl) or CsI (Na) crystal to detect beta particles in a background of gamma rays, using the Phoswich concept, where events occurring only in the thin CaF_2 (Eu), unaccompanied by a simultaneous event in the thick crystal, are counted. That is, only particles that are totally absorbed in the CaF_2 (Eu) are of interest—when both crystals display coincident events, these are vetoed by coincidence and pulse-shape discrimination methods. Coincidence counting is discussed further in Section 29.3.6.4.

29.2.2.2 Organic Scintillators

The first organic scintillator was introduced by Kallman in 1947 when he used a naphthalene crystal to show that it would detect gamma rays. Later, anthracene was shown to exhibit improved detection efficiency and stilbene was also used. The latter has proved particularly useful for neutron detection. Mixtures of organic scintillators such as solutions of anthracene in naphthalene, liquid solutions, and plastic solutions were also introduced, and now the range of organic scintillators available is very great. The plastic and liquid organic scintillators are generally cheaper to manufacture than inorganic scintillators and can be made in relatively large sizes. They are generally much faster in response time than most inorganic scintillators and, being transparent to their own scintillation light, can be used in very large sizes. Table 29.6 gives the essential details of a large range of plastic and liquid scintillators.

The widest use of organic scintillators is probably in the field of liquid-scintillation counting, where the internal counting of tritium, ^{14}C, ^{55}Fe, and other emitters of low-energy beta particles at low activity levels is being carried out on an increasing scale. Biological samples of many types have to be incorporated into the scintillator, and it is necessary to do this with the minimum quenching of the light emitted, minimum chemiluminescence, as well as minimum effort—the last being an important factor where large numbers of samples are involved.

At low beta energies the counting equipment is just as important as the scintillator. Phosphorescence is reduced to a minimum by the use of special vials and reflectors. Chemiluminescence, another problem with biological and other samples, is not completely solved by the use of two photomultipliers in coincidence viewing the sample. This must be removed or reduced, for example, by removing alkalinity and/or peroxides by acidifying the solution before mixing with the scintillator.

29.2.2.3 Loaded Organic Scintillators

To improve the detection efficiency of scintillators for certain types of particles or ionizing or nonionizing radiations, small quantities of some substances can be added to scintillators without greatly degrading the light output. It must be borne in mind that in nearly all cases there is a loss of light output when foreign substances are added to an organic scintillator, but the gain in detection efficiency may be worth a slight drop in this output. Suitable loading materials are boron, both natural boron and boron enriched in ^{10}B and gadolinium—these are to increase the detection efficiency for neutrons. Tin and lead have been used to improve the detection efficiency for gamma rays and have been used in both liquid and plastic scintillators.

29.2.2.4 Plastic Scintillators

Certain plastics such as polystyrene and polyvinyltoluene can be loaded with small quantities of certain substances such as p-terphenyl that cause them to scintillate when bombarded by ionizing particles or ionizing radiation. An acrylic such as methyl methacrylate can also be doped to produce a scintillating material but not with the same high light output as the polyvinyltoluene-based scintillators. It can be produced, however, much more cheaply and it can be used for many high-energy applications.

Plastic scintillators have the ability to be molded into the most intricate shapes to suit a particular experiment, and their inertness to water, normal air, and many chemical substances allows their use in direct contact with the activity to be measured. Being of low atomic number constituents, the organic scintillators are preferred to inorganics such as NaI (Tl) or CsI (Tl) for beta counting, since the number of beta particles scattered out of an organic scintillator without causing an interaction is about 8 percent, whereas in a similar NaI (Tl) crystal the number scattered out would be 80–90 percent.

When used to detect X- or gamma rays, organic scintillators differ in their response compared with inorganic scintillators. Where inorganic scintillators in general have basically three main types of response, called photoelectric, Compton, and pair production, because of the high Z (atomic weight) of the materials of the inorganic scintillators, the low-Z characteristics of the basic carbon and similar components in organic scintillators lead only to Compton reactions, except at very low energies, with the result that for a monoenergetic gamma emitter, the spectrum produced is a Compton distribution.

For study of the basic interactions between gamma and X-rays and scintillation materials, see Price (1964), since these reaction studies are beyond the scope of this chapter.

The ability to produce simple Compton distribution spectra has proved of considerable use in cases in which one or two isotopes have to be measured at low intensities and a large inorganic NaI (Tl) detector might be prohibitively expensive. Such is the case with whole-body counters used to measure the ^{40}K and ^{137}Cs present in the human body—the ^{40}K being the natural activity present in all potassium, the ^{137}Cs the result of fallout from the many atomic-bomb tests. Similarly, in measuring the potassium content of fertilizer, a plastic scintillator can carry this out more cheaply than an inorganic detector. The measurement of moisture in soil by gamma-ray transmission through a suitable sample is also performed more easily with a plastic scintillator, since the fast decay time of the organics compared with, say, NaI (Tl) allows higher counting rates to be used, with consequent reduction of statistical errors.

29.2.2.5 Scintillating Ion-exchange Resins

By treating the surfaces of plastic scintillating spheres in suitable ways, the extraction and counting of very small amounts of beta-emitting isotopes may be carried out from large quantities of carrier liquid, such as rainwater, cooling water from reactors, effluents, or rivers, rather than having to evaporate large quantities of water to obtain a concentrated sample for analysis.

29.2.2.6 Flow Cells

It is often necessary to continuously monitor tritium, ^{14}C, and other beta-emitting isotopes in aqueous solution, and for this purpose the flow cell developed by Schram and Lombaert and containing crystalline anthracene has proved valuable. A number of improvements have been made to the design of this cell, resulting in the NE806 flow cell. The standard flow cell is designed for use on a single 2-in.-diameter low-noise photomultiplier and can provide a tritium detection efficiency of 2 percent at a background of 2 c/s and a ^{14}C detection efficiency of 30 percent at a background of 1 c/s.

29.2.2.7 Photomultipliers

The photomultiplier is the device that converts the light flash produced in the scintillator into an amplified electrical pulse. It consists generally of an evacuated tube with a flat glass end onto the inner surface of which is deposited a semitransparent layer of metal with a low "work function," that is, it has the property of releasing electrons when light falls on it. The most usual composition of this photocathode, as it is called, is cesium plus some other metal. A series of dynodes, also coated with similar materials, form an electrical optical system that draws secondary electrons away from one dynode and causes them to strike the next one with a minimum loss of electrons. The anode finally collects the multiplied shower of electrons, which forms the output pulse to the electronic system. Gains of 10^6 to 10^7 are obtained in this process.

Depending on the spectrum of the light emitted from the scintillator, the sensitivity of the light-sensitive photocathode can be optimized by choice of the surface material. One can detect single electrons emitted from the photocathode with gallium arsenide (GaAs) as the coating.

Silicon photodiodes can also be used to detect scintillation light, but because the spectral sensitivity of these devices is in the red region (~500–800 nm) as opposed to the usual scintillator light output of ~400–500 nm, a scintillator such as CsI (T1) must be used for which the output can match the spectral range of a silicon photodiode. Light detectors are discussed further in Chapter 21.

29.2.3 Solid-state Detectors

It was observed earlier that the operation of a solid-state or semiconductor detector could be likened to that of an ionization chamber. A study of the various materials that were thought to be possible for use as radiation detectors has been carried out in many parts of the world, and the two materials that proved most suitable were silicon and germanium. Both these materials were under intense development for the transistor industry, so the detector researchers were able to make use of the work in progress. Other materials tested, which might later prove valuable, were cadmium telluride (CdTe), mercuric iodide (HgI_2), gallium arsenide (GaAs), and silicon carbide. CdTe and HgI_2 can be used at room temperature, but to date, CdTe has been produced in only relatively small sizes and with great difficulty. Mercuric iodide (HgI) has been used successfully in the measurement of X-ray fluorescence in metals and alloy analysis.

In this book no attempt has been made to go deeply into the physics of semiconductor detectors, but it is useful to give a brief outline of their operation, especially compared with the gas-ionization chamber, of which they are often regarded as the solid-state equivalent.

First, a much smaller amount of energy is required to release electrons (and therefore holes) in all solids than in gases. An average energy of only 3 eV is required to produce an electron–hole pair in germanium (and 3.7 eV in silicon), whereas about 30 eV is required to produce the equivalent in gases. This relatively easy production of free holes and electrons in solids results from the close proximity of atoms, which causes many electrons to exist at energy levels just below the conduction band. In gases the atoms are isolated and the electrons much more tightly bound. As a result, a given amount of energy absorbed from incident radiation produces more free charges in solids than in gas, and the statistical fluctuations become a much smaller fraction of the total charge released. This is the basic reason for semiconductor detectors producing better energy resolution than gas detectors, especially at high energies. At low energies, because the signal from the semiconductor is some 10 times larger than that from the gas counter, the signal/noise ratio is enhanced.

To obtain an efficient detector from a semiconductor material, we may consider what occurs should we take a slice of silicon 1 cm² in area and 1 mm thick and apply a potential across the faces. If the resistivity of this material is 2,000 Ω cm, with ohmic contacts on each side, the slice would behave like a resistor of 200 Ω, and if 100 V is applied across such a resistor, Ohm's law states that a current of 0.5 A would pass. If radiation now falls on the silicon slice, a minute extra current will be produced, but this would be so small compared with the standing 0.5 A current that it would be undetectable. This is different from the gas-ionization chamber, where the standing current is extremely small.

The solution to this problem is provided by semiconductor junctions. The operation of junctions depends on the fact that a mass action law compels the product of electron and hole concentrations to be constant for a given semiconductor at a fixed temperature. Therefore, heavy doping with a donor such as phosphorus not only increases the free electron concentration, it also depresses the hole concentration to satisfy the relation that the product must have a value dependent only on the semiconductor. For example, silicon at room temperature has the relation $n \times p \approx 10^{20}$, where n is the number of holes and p is the number of electrons. Hence in a region where the number of donors is doped to a concentration of 10^{18}, the number of holes will be reduced to about 10^2. McKay, of Bell Telephone Laboratories, first demonstrated in 1949 that if a reverse-biased p–n junction is formed on the semiconductor, a strong electric field may be provided across the device, sweeping away free holes from the junction on the p-side (doped with boron; see Figure 29.5) and electrons away from it on the n-side (doped with phosphorus). A region is produced that is free of holes or electrons and is known as the *depletion region*. However, if an ionizing particle or quantum of gamma energy passes through the region, pairs of holes and electrons are produced that are collected to produce a current in the external circuit. This is the basic operation of a semiconductor detector. The background signal is due to the collection of any pairs of holes plus electrons produced by thermal processes. By reducing the temperature of the detector by liquid nitrogen to about 77 K, most of this background is removed, but in practical detectors the effect of surface contaminants on those surfaces not forming the diode can be acute. Various

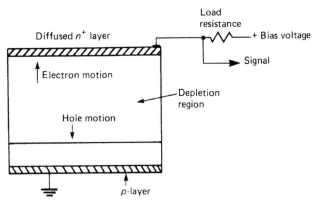

FIGURE 29.5 Schematic diagram of semiconductor detector.

methods of avoiding these problems, such as the use of guard rings, have reduced much of this problem. However, the effects of very small amounts of oxygen and the like can have devastating results on a detector, and most are enclosed in a high-vacuum chamber.

By doping a germanium or silicon crystal with lithium (an interstitial donor), which is carried out at moderate temperatures using an electric field across the crystal, the acceptors can be almost completely compensated in p-type silicon and germanium. This allows the preparation of relatively large detectors suitable for high-energy charged particle spectroscopy. By this means coaxial detectors with volumes up to about 100 cm³ have been made, and these have revolutionized gamma-ray spectroscopy, since they can separate energy lines in a spectrum that earlier NaI (Tl) scintillation spectrometers could not resolve.

New work on purifying germanium and silicon has resulted in the manufacture of detectors of super-pure quality such that lithium drifting is not required. Detectors made from such material can be cycled from room temperature to liquid-nitrogen temperature and back when required without the permanent damage that would occur with lithium-drifted detectors. Surface-contamination problems, however, still require them to be kept *in vacuo*. Such material is the purest ever produced—about 1 part in 10^{13} of contaminants.

29.2.4 Detector Applications

In all radiation-measuring systems there are several common factors that apply to measurements to be carried out. These are:

1. Geometry.
2. Scattering.
3. Backscattering.
4. Absorption.
5. Self-absorption.

Geometry Since any radioactive source emits its products in all directions (in 4π geometry), it is important to be able

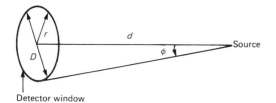

FIGURE 29.6 Geometry of radiation collection.

to calculate how many particles or quanta may be collected by the active volume of the counter. If we consider a point source of radiation, as in Figure 29.6, all the emitted radiation from the source will pass through an imaginary sphere with the source at center, providing there is no absorption. Also, for any given sphere size, the average radiation flux in radiations per unit time per unit sphere surface area is constant over the entire surface. The geometry factor, G, can be therefore written as the fraction of total 4π solid angle subtended by source and detector. For the case of the point source at a distance d from a circular window of radius r, we have the following relation:

$$G = 0.5(1 - \cos\phi) \qquad (29.13)$$

$$= 0.5\left\{1 - \frac{1}{[1 + (r^2/d^2)]^{1/2}}\right\} \approx \frac{D^2}{16d^2} \qquad (29.14)$$

Scattering Particles and photons are scattered by material through which they pass, and this effect depends on the type of particle or photon, its energy, its mass, the type of material traversed, its mass, and density. What we are concerned with here is not the loss of particle energy or particles themselves as they pass through a substance but the effects caused by particles deflected from the direct path between radioactive source and detector. It is found that some particles are absorbed into the material surrounding the source. Others are deflected away from it but are later rescattered into the detector, so increasing the number of particles in the beam. Some of these deflected particles can be scattered at more than 90° to the beam striking the detector—these are called *backscattered*, since they are scattered back in the direction from which they came. Scattering also occurs with photons, as is particularly demonstrated by the increase in counting rate of a beam of gamma rays when a high-Z material such as lead is inserted into the beam.

Backscattering Backscattering increases with increasing atomic number Z and with decreasing energy of the primary particle. For the most commonly used sample planchets (platinum), the backscattering factor has been determined as 1.04 for an ionization chamber inside which the sample is placed, known as a *50 percent chamber*.

Absorption Because the particles to be detected may be easily absorbed, it is preferred to mount source and detector in an evacuated chamber or insert the source directly into the gas of an ionization chamber or proportional counter. The most popular method at present uses a semiconductor detector, which allows the energies to be determined very accurately when the detector and source are operated in a small evacuated cell. This is especially important for alphas.

Self-absorption If visible amounts of solid are present in a source, losses in counting rate may be expected because of self-absorption of the particles emitted from the lower levels of the source that are unable to leave the surface of the source. Nader et al. give an expression for the self-absorption factor for alpha particles in a counter with 2π geometry (this is the gas counter with source inside the chamber):

and
$$f_s = 1 - \frac{s}{2pR} \text{ for } s < pR \qquad (29.15)$$

$$f_s = \frac{0.5R}{s} \text{ for } s > pR \qquad (29.16)$$

where s is the source thickness, R the maximum range of alpha particles in source material, and p the maximum fraction of R that particles can spend in source and still be counted.

Radiation Shield The detector may be housed in a thick radiation shield to reduce the natural background that is found everywhere to a lower level where the required measurements can be made. The design of such natural radiation shields is a subject in itself and can range from a few centimeters of lead to the massive battleship steel or purified lead used for whole-body monitors, where the object is to measure the natural radiation from a human body.

29.2.4.1 Alpha-detector Systems

The simplest alpha detector is the air-filled ionization chamber, used extensively in early work on radioactivity but now only used for alpha detection in health physics surveys of spilled activities on benches, and so on. Even this application is seldom used, since more sensitive semiconductor or scintillation counters are usual for this purpose. Thin-window ionization or gas-proportional counters can also be used, as can internal counters in which the sample is inserted into the active volume of a gas counter. Due to the intense ionization produced by alpha particles, it is possible to count them in high backgrounds from other radiation such as betas and gamma rays by means of suitable discrimination circuits.

Ionization Chambers Because alpha particles are very readily absorbed in a small thickness of gas or solid, an early method used in counting them involved the radioactive source being placed inside a gas counter with no intermediate window. Figure 29.7 shows the schematic circuit of such a counter in which the gas, generally pure methane (CH_4), is allowed to flow through the counting volume. The counter

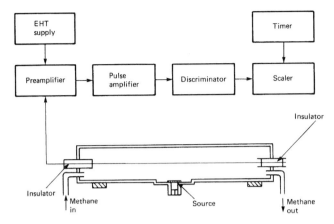

FIGURE 29.7 Gas flow-type proportional counter system.

can be operated in the ionization counter region, but more usually the high-voltage supply is raised so that it operates in the proportional counter region, allowing discrimination to be made between alpha particles and beta particles, which may be emitted from the same source, and improving the ratio of signal to noise.

Ionization chambers are used in two different ways. Depending on whether the time constants of the associated electronic circuits are small or large, they are either *counting*, that is, responding to each separate ionizing event, or *integrating*, that is, collecting the ionization over a relatively long period of time. The counting mode is little used nowadays except for alpha particles.

Gas Proportional Counters Thin-window proportional counters allow alphas to be counted in the presence of high beta and gamma backgrounds, since these detectors can discriminate between the high ionization density caused by alpha particles passing through the gas of the counter and the relatively weak ionization produced by beta particles or gamma-ray photons.

With counters using flowing gas, the source may be placed inside the chamber and the voltage across the detector raised to give an amplification of 5–50 times. In this case pure methane (CH_4) is used with the same arrangement as shown in Figure 29.7.

The detection efficiency of this system is considerably greater than that provided by the thin-window counter with the source external to the counter. This is due (1) to the elimination of the particles lost in penetrating the window and (2) to the improved geometry whereby all particles that leave the source and enter the gas are counted.

Geiger Counters The observations made about alpha-particle detection by gas proportional counters apply also to Geiger counters. That is, thin entrance windows or internally mounted sources are usable and the system of Figure 29.7 is also applicable to Geiger-counter operation. However, some differences have to be taken into account. Since operation in the Geiger region means that no differences in particle energy are measured, all particles are able to trigger the counter and no energy resolution is possible. On the other hand, the counter operating in the Geiger region is so sensitive that high electronic amplification is not required, and in many cases the counter will produce an output sufficient to trigger the scaler unit directly. To operate the system in its optimum condition, some amplification is desirable and should be variable. This is to set the operating high-voltage supply and the amplifier gain on the Geiger plateau (see Section 29.2.1.1) so that subsequent small variations in the high-voltage supply do not affect the counting characteristics of the system.

Scintillation Counters Since alpha particles are very easily absorbed in a very short distance into any substance, problems are posed if alphas are to enter a scintillator through a window, however thin, which would also prevent light from entering and overloading the photomultiplier. The earliest scintillation detector using a photomultiplier tube had a very thin layer of zinc sulfide powder sprinkled on the glass envelope of a side-viewing photomultiplier of the RCA 931-A type, on which a small layer of adhesive held the zinc sulfide in place. Due to the low light transmission of powdered zinc sulfide, a layer 5–10 mg/cm^2 is the optimum thickness. The whole system was enclosed in a light-tight box, source and detector together.

Later experimenters used a layer of aluminum evaporated onto the zinc sulfide to allow the alphas in but to keep the light out. This was not successful, since an adequate thickness of aluminum to keep the light out also prevented the alphas from entering the zinc sulfide. When photomultipliers with flat entrance windows began to become available, the zinc sulfide was deposited on the face of a disc of transparent plastic such as Perspex/Lucite. This was easier to evaporate aluminum on than directly onto the photomultiplier, but the light problem was still present, and even aluminized mylar was not the answer. Operation in low- or zero-light conditions was still the only satisfactory solution.

Thin scintillating plastics were also developed to detect alphas, and in thin layers, generally cemented by heat to a suitable plastic light guide, proved cheaper to use than zinc sulfide detectors. The light output from plastic scintillators was, however, less than that from zinc sulfide, but both scintillators, in their very thin layers, provide excellent discrimination against beta particles or gamma rays, often present as background when alpha-particle emitters are being measured. One major advantage of the plastic scintillator over the inorganic zinc sulfide detector is its very fast response, on the order of 4×10^{-9} s for the decay time of a pulse, compared with $4-10 \times 10^{-6}$ s for zinc sulfide.

Another scintillator that has been used for alpha detection is the inorganic crystal CsI(Tl). This can be used without any window (but in the dark) because it is nonhygroscopic, and, if suitably beveled around the circumference, will produce an output proportional to the energy of the ionizing particle incident on it, thus acting as a spectrometer. Extremely thin CsI (T1) detectors have been made by vacuum deposition and used where the beta and gamma background must be reduced to very small proportions, yet the heavy particles of interest must be detected positively.

Inorganic single crystals are expensive, and when large areas of detector must be provided, either the zinc sulfide powder screen or a large-area very thin plastic scintillator is used. The latter is much cheaper to produce than the former, since the thin layer of zinc sulfide needs a great deal of labor to produce and it is not always reproducible. In health physics applications, for monitors for hand and foot contamination or bench-top contamination, various combinations of plastic scintillator, zinc sulfide, or sometimes anthracene powder scintillators are used.

If the alpha emitter of interest is mixed in a liquid scintillator, the geometrical effect is almost completely eliminated and, provided that the radioactive isotope can be dissolved in the liquid scintillator, maximum detection efficiency is obtained. However, not all radioactive sources can be introduced in a chemical form suitable for solution, and in such cases the radioactive material can be introduced in a finely divided powder into a liquid scintillator with a gel matrix; such a matrix would be a very finely divided extremely pure grade of silica. McDowell (1980) has written a useful review of alpha liquid scintillation counting from which many references in the literature may be obtained.

29.2.4.2 Detection of Beta Particles

Ionization Chambers Although ionization chambers were much used for the detection of beta particles in the early days, they are now used only for a few special purposes:

1. *The calibration of radioactive beta sources for surface dose rate.* An extrapolation chamber varies the gap between the two electrodes of a parallel plate ionization chamber, and the ionization current per unit gap versus air gap is plotted on a graph as the gap is reduced, which extrapolates to zero gap with an uncertainty that is seldom as much as 1 percent, giving a measure of the absolute dose from a beta source.
2. *Beta dosimetry.* Most survey instruments used to measure dose rate incorporate some sort of device to allow beta rays to enter the ionization chamber. This can take the form of a thin window with a shutter to cut off the betas and allow gamma rays only to enter when mixed beta-gamma doses are being measured. However, the accuracy of such measurements leaves a great deal to be desired, and the accurate measurement of beta dose rates is a field where development is needed.

Proportional Counters Beta particles may originate from three different positions in a proportional counter; first, from part of the gaseous content of the counter, called *internal counting*; second, from a solid source inside the counter itself; and third, from an external source with the beta particles entering the counter by means of a thin window.

The first method, internal counting, involves mixing the radioactive source in the form of a gas with a gas suitable for counting; or the radioactive source can be transformed directly into a suitable gaseous form. This is the case when detection of ^{14}C involves changing solid carbon into gaseous carbon dioxide (CO_2), acetylene (CH), or methane (CH_4), any of which can be counted directly in the proportional counter. This method is used in measurement of radiocarbon to determine the age of the carbon: radiocarbon dating (see Section 4.4.2.1).

The second method involves the use of solid sources introduced by means of a gas-tight arm or drawer so that the source is physically inside the gas volume of the counter. This method was much used in early days but is now used only in exceptional circumstances. The third method, with the source external to the counter, is now the most popular, since manufacturers have developed counters with extremely thin windows that reduce only slightly the energies of the particles crossing the boundary between the gas of the counter and the air outside. Thin mica or plastic windows can be on the order of a few mg/cm^2, and they are often supported by wire mesh to allow large areas to be accommodated.

A specialized form of proportional counter is the $2\pi/4\pi$ type used in the precise assay of radioactive sources. This is generally in the form of an open-ended pillbox, and the source is deposited on a thin plastic sheet, either between two counters face to face (the 4π configuration) or with only a single counter looking at the same source (the 2π configuration). Such counters are generally made in the laboratory, using plastic as the body onto which a conducting layer of suitable metal is deposited to form the cathode. One or more wires form the anode. A typical design is shown in Figure 29.8.

Geiger Counters The Geiger counter has been the most popular detector for beta particles for a number of reasons. First, it is relatively cheap, either when manufactured in the laboratory or as a commercially available component. Second, it requires little in the way of special electronics to make it work, as opposed to scintillation or solid-state detectors. Third, the later halogen-filled versions have quite long working lives. To detect beta particles, some sort of window must be provided to allow the particles to enter the detector.

The relations between the various parameters affecting the efficiency with which a particular radioactive source is detected by a particular Geiger counter have been studied extensively. Zumwalt (1950), in an early paper, provided the basic results that have been quoted in later books and papers, such as Price (1964) and Overman and Clark (1960). In general, the observed counting rate R and the corresponding disintegration rate A can be related by the equation

$$R = YA \qquad (29.17)$$

CHAPTER | 29 Nuclear Instrumentation Technology

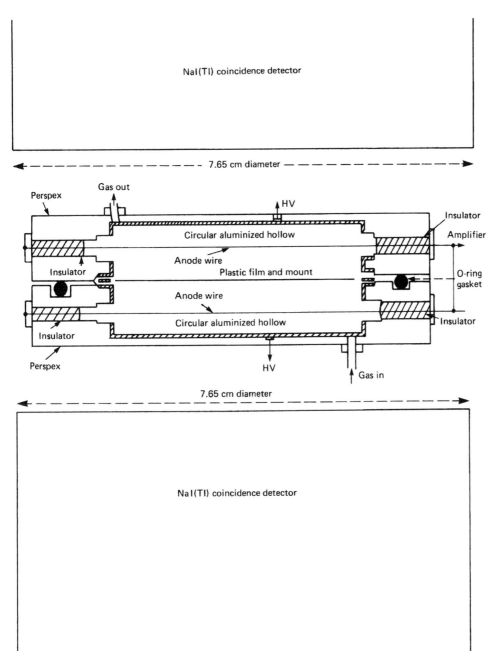

FIGURE 29.8 Typical design of 4π counter.

where Y is a factor representing the counting efficiency. Some 11 factors are contained in Y, and these are set out in those books we mentioned.

29.2.4.3 Detection of Gamma Rays (100 keV Upward)

Ionization Chambers Dosimetry and dose-rate measurement of gamma rays are the main applications in which ionization chambers are used to detect gamma rays today. They are also used in some industrial measuring systems although scintillation, and other counters are now generally replacing them.

A device called the *free air ionization chamber* (see Figure 29.9) is used in X-ray work to calibrate the output of X-ray generators. This chamber directly measures the charge liberated in a known volume of air, with the surrounding medium being air. In Figure 29.9 only ions produced in the volume $V_1 + V + V_2$ defined by the collecting electrodes are collected and measured; any ionization outside this volume is not measured. However, due to the physical size of the chamber and the amount of auxiliary equipment required, this device is only used in standardizing laboratories, where small Bragg–Gray chambers may be calibrated by comparison with the free air ionization chamber.

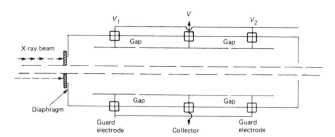

FIGURE 29.9 Free air ionization chamber.

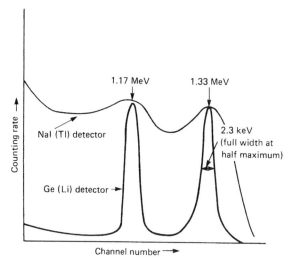

FIGURE 29.10 Comparison of energy resolution by different detectors.

The Bragg–Gray chamber depends on the principle that if a small cavity exists in a solid medium, the energy spectrum and the angular distribution of the electrons in the cavity are not disturbed by the cavity, and the ionization occurring in the cavity is characteristic of the medium, provided that:

1. The cavity dimensions are small compared with the range of the secondary electrons.
2. Absorption of the gamma radiation by the gas in the cavity is negligible.
3. The cavity is surrounded by an equilibrium wall of the medium.
4. The dose rate over the volume of the medium at the location of the cavity is constant.

The best-known ionization chamber is that designed by Farmer; this has been accepted worldwide as a substandard to the free air ionization chamber for measuring dose and dose rates for human beings.

Solid-state Detectors The impact of solid-state detectors on the measurement of gamma-rays has been dramatic; the improvement in energy resolution compared with a NaI (Tl) detector can be seen in Figure 29.10, which shows how the solid-state detector can resolve the 1.17 and 1.33 MeV lines from ^{60}Co with the corresponding NaI (Tl) scintillation detector result superimposed. For energy resolution the solid-state detector provides a factor of 10 improvement on the scintillation counter. However, there are a number of disadvantages which prevent this detector superseding the scintillation detector. First, the present state of the art limits the size of solid-state detectors to some 100 cm^3 maximum, whereas scintillation crystals of NaI (Tl) may be grown to sizes of 76 cm diameter by 30 cm, while plastic and liquid scintillators can be even larger. Second, solid-state detectors of germanium have to be operated to liquid-nitrogen temperatures and *in vacua*. Third, large solid-state detectors are very expensive. As a result, the present state of the art has tended towards the use of solid-state detectors when the problem is the determination of energy spectra, but scintillation counters are used in cases where extremely low levels of activity are required to be detected. A very popular combined use of solid-state and scintillation detectors is embodied in the technique of surrounding the relatively small solid-state detector in its liquid nitrogen-cooled and evacuated cryostat with a suitable scintillation counter, which can be an inorganic crystal such as NaI (Tl) or a plastic or liquid scintillator. By operating the solid-state detector in anticoincidence with the annular scintillation detector, Compton-scattered photons from the primary detector into the anticoincidence shield (as it is often called) can be electronically subtracted from the solid-state detector's energy response spectrum.

29.2.4.4 The Detection of Neutrons

The detection of nuclear particles usually depends on the deposition of ionization energy and, since neutrons are uncharged, they cannot be detected directly. Neutron sensors therefore need to incorporate a conversion process by which the incoming particles are changed to ionizing species and nuclear reactions such as ^{235}U (fission) + ~200 MeV or ^{10}B(n, α)Li^7 + ~2 MeV are often used. Exothermic reactions are desirable because of the improved signal-to-noise ratio produced by the increased energy per event, but there are limits to the advantages that can be gained in this way. In particular, these reactions are not used for neutron spectrometry because of uncertainty in the proportion of the energy carried by the reaction products, and for that purpose the detection of proton recoils in a hydrogenous material is often preferred.

There are also many ways of detecting the resultant ionization. It may be done in real time with, for example, solid-state semiconductor detectors, scintillators, or ionization chambers, or it may be carried out at some later, more convenient time by measuring the activation generated by the neutrons in a chosen medium (such as ^{56}M from ^{55}Mn). The choice of technique depends on the information required and the constraints of the environment. The latter are often imposed by the neutrons themselves. For example, boron-loaded scintillators can be made very sensitive and are convenient for detecting low fluxes, but scintillators and photomultipliers are vulnerable to damage by neutrons and are sensitive to the gamma fluxes which tend to accompany them. They are not suitable for the high-temperature, high-flux applications found in nuclear reactors.

Neutron populations in reactors are usually measured with gas-filled ionization chambers. Conversion is achieved by fission in ^{235}U oxide applied as a thin layer (~1 mg cm^{-2}) to the chamber electrode(s). The ^{10}B reaction is also used, natural or enriched boron being present either as a painted layer or as BF$_3$ gas. Ionization chambers can be operated as pulse devices, detecting individual events; as DC generators, or in which events overlap to produce current; or in the so-called "current fluctuation" or "Campbell" mode in which neutron flux is inferred from the magnitude of the noise present on the output as a consequence of the way in which the neutrons arrive randomly in time. Once again the method used is chosen to suit operational requirements. For example, the individual events due to neutrons in a fission chamber are very much larger than those due to gammas, but the gamma photon arrival rate is usually much greater than that of the neutrons. Thus, pulse counters have good gamma rejection compared with DC chambers at low fluxes. On the other hand, the neutron-to-gamma flux ratio tends to improve at high reactor powers while counting losses increase and gamma pulse pile-up tends to simulate neutron events. DC operation therefore takes over from pulse measurement at these levels. The current fluctuation mode gives particularly good gamma discrimination because the signals depend on the mean square charge per event, i.e., the initial advantage of neutrons over gammas is accentuated. Such systems can work to the highest fluxes, and it is now possible to make instruments which combine pulse counting and current fluctuation on a single chamber and which cover a dynamic range of more than 10 decades.

The sensitivity of a detector is proportional to the probability of occurrence of the expected nuclear reaction and can conveniently be described in terms of the cross-section of a single nucleus for that particular reaction. The unit of area is the barn, that is, 10^{-24}cm^2. ^{10}B has a cross-section of order 4,000 b to slow (thermal) neutrons whilst that of ^{235}U for fission is only ~550 b. In addition, the number of reacting atoms present in a given thickness of coating varies inversely with atomic weight so that, in principle, ^{10}B sensors are much more sensitive than those that depend on fission. This advantage is offset by the lower energy per event and by the fact that boron is burnt up faster at a given neutron flux, that is, that such detectors lose sensitivity with time.

Neutrons generate activation in most elements, and if detector constructional materials are not well chosen, this activation can produce residual signals analogous to gamma signals and seriously shorten the dynamic range. The electrodes and envelopes of ion chambers are therefore made from high-purity materials that have small activation cross-sections and short daughter half-lives. Aluminum is usually employed for low-temperature applications, but some chambers have to operate at ~550°C (a dull red), and these use titanium and/or special low-manganese, low-cobalt stainless steels. Activity due to fission products from fissile coatings must also be considered and is a disadvantage to fission chambers. The choice of insulators is also influenced by radiation and temperature considerations. Polymers deteriorate at high fluences, but adequate performance can be obtained from high-purity, polycrystalline alumina and from artificial sapphire, even at 550°C. Analogous problems are encountered with cables and special designs with multiple coaxial conductors insulated with high-purity compressed magnesia have been developed. Electrode/cable systems of this type can provide insulation resistances of order 10^9 Ω at 550°C and are configured to eliminate electrical interference even when measuring microamp signals in bandwidths of order 30 MHz under industrial plant conditions. Figure 29.11 shows

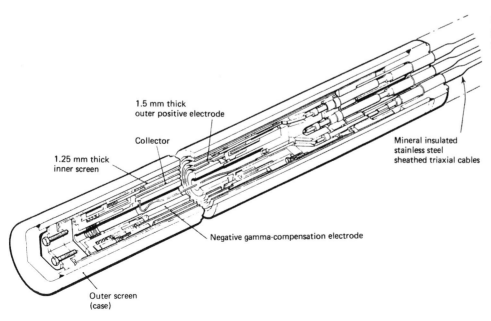

FIGURE 29.11 Typical reactor control ionization chamber. Courtesy of UKAEA, Winfrith.

the construction of a boron-coated gamma-compensated dc chamber designed to operate at 550°C and 45 bar pressure in the AGR reactors. The gas filling is helium and β activity from the low-manganese steel outer case is screened by thick titanium electrodes. The diameter of this chamber is 9 cm; it is 75 cm long and weighs 25 kg. By contrast, Figure 29.12 shows a parallel plate design, some three hundred of which were used to replace fuel "biscuits" to determine flux distributions in the ZEBRA experimental fast reactor at AEE Winfrith.

Boron trifluoride (BF_3) proportional counters are used for thermal neutron detection in many fields; they are convenient and sensitive, and are available commercially with sensitivities between 0.3 and 196 S^{-1} (unit flux)$^{-1}$. They tend to be much more gamma sensitive than pulse-fission chambers because of the relatively low energy per event from the boron reaction, but this can be offset by the larger sensitivity. A substantial disadvantage for some applications is that they have a relatively short life in terms of total dose (gammas plus neutrons), and in reactor applications it may be necessary to provide withdrawal mechanisms to limit this dose at high power.

Proton-recoil counters are used to detect fast neutrons. These depend on the neutrons interacting with a material in the counter in a reaction of the (n, p) type, in which a proton is emitted that, being highly ionizing, can be detected. The material in the counter can be either a gas or a solid. It must have low-Z nuclei, preferably hydrogen, to allow the neutron to transfer energy. If a gas, the most favored are hydrogen or helium, because they contain the greatest number of nuclei per unit volume. If a solid, paraffin, polyethylene, or a similar low-Z material can be used to line the inside of the counter.

This type of counter was used by Hurst et al. to measure the dose received by human tissue from neutrons in the range 0.2–10 MeV. It was a three-unit counter (the gas being methane) and two individual sections contained a thin (13.0 mg/cm^2) layer of polyethylene and a thick (100 mg/cm^2) layer of polyethylene. The energy responses of the three sources of protons combine in such a way as to give the desired overall response, which matches quite well the tissue-dose curve over the energy range 0.2–10 MeV. This counter also discriminates well against the gamma rays which nearly always accompany neutrons, especially in reactor environments.

Improvements in gas purification and counter design have led to the development of ^{3}He-filled proportional counters. ^{3}He pressures of 10–20 atm allow the use of these counters as direct neutron spectrometers to measure energy distributions, and in reactor neutron spectrum analysis they are found in most reactor centers all over the world.

As explained, it is necessary for the measurement of neutrons that they shall interact with substances that will then emit charged particles. For thermal energy neutrons (around 0.025 eV in energy), a layer of fissionable material such as ^{235}U will produce reaction products in the form of alpha particles, fission products, helium ions, and so on. However, more suitable materials for producing reaction products are ^{6}Li and ^{10}B. These have relatively high probability of a neutron producing a reaction corresponding to a cross-section of 945 barns for ^{6}Li and for ^{10}B of 3770 barns. By mixing ^{6}Li or ^{10}B with zinc sulfide powder and compressing thin rings of the mixture into circular slots in a methyl methacrylate disc, the reaction products from the ^{6}Li or ^{10}B atom disintegration when a neutron is absorbed strike adjacent ZnS particles and produce light flashes, which can be detected by the photomultiplier to which the detector is optically coupled. Figure 29.13 shows such a neutron-detector system.

Neutron-proton reactions allow the detection of neutrons from thermal energies up to 200 MeV and higher. For this reaction to take place a hydrogen-type material is required with a high concentration of protons. Paraffin, polyethylene

FIGURE 29.12 Pulse-fission ionization chamber for flux-distribution measurement. Courtesy of UKAEA, Winfrith.

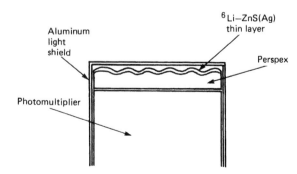

FIGURE 29.13 Thermal neutron scintillation counter.

or gases such as hydrogen, ^{3}H, methane, etc., provide good sources of protons, and by mixing a scintillator sensitive to protons (such as ZnS) with such a hydrogenous material, the protons produced when a neutron beam interacts with the mixed materials can be counted by the flashes of light produced in the ZnS. Liquids can also be used, in which the liquid scintillator is mixed with boron, gadolinium, cadmium, and so on, in chemical forms which dissolve, in the scintillator. Large tanks of 500-1000-1 sizes have been made for high-energy studies, including the study of cosmic ray neutrons. ^{6}Li can also be used dissolved in a cerium-activated glass, and this has proved a very useful neutron detector as the glass is inert to many substances. This ^{6}Li glass scintillator has proved very useful for studies in neutron radiography, where the neutrons are used in the manner of an X-radiograph to record on a photographic film the image produced in the glass scintillator by a beam of neutrons.

Another neutron detector is the single crystal of lithium iodide activated with europium ^{6}Li (Eu). When this crystal is cooled to liquid-nitrogen temperature it can record the spectrum of a neutron source by pulse-height analysis. This is also possible with a special liquid scintillator NE213 (xylene-based), which has been adopted internationally as the standard scintillator for fast neutron spectrometry from 1 to 20 MeV.

One of the problems in neutron detection is the presence of a gamma-ray background in nearly every practical case. Most of the detectors described here do have the useful ability of being relatively insensitive to gamma rays. That is with the exception of LiI (Eu), which, because of its higher atomic number due to the iodine, is quite sensitive to gamma rays. By reducing the size of the scintillator to about 4 mm square by 1 mm thick and placing it on the end of a long thin light guide placed so that the detector lies at the center of a polyethylene sphere, a detector is produced that can measure neutron dose rate with a response very close to the response of human tissue. Figure 29.14 shows a counter of this kind that is nearly isotropic in its response (being spherical) and with much reduced sensitivity to gamma rays. This is known as the *Bonner sphere*, and the diameter of the sphere can be varied between 10 and 30 cm to cover the whole energy range.

For thermal neutrons ($E = 0.025$ eV), an intermediate reaction such as ^{6}Li(n, α) or ^{10}B(n, α) or fission can be used, a solid-state detector being employed to count the secondary particles emitted in the reaction. For fast neutrons, a radiator in which the neutrons produce recoil protons can be mounted close to a solid-state detector, and the detector counts the protons. By sandwiching a thin layer of ^{6}LiF between two solid-state detectors and summing the coincident alpha and tritium pulses to give an output signal proportional to the energy of the incident neutron plus 4.78 MeV, the response of the assembly with respect to incident neutron energy was found to be nearly linear.

29.3 ELECTRONICS

A more general treatment of the measurement of electrical quantities is given in Chapter 20. We concentrate here on aspects of electronics that are particularly relevant to nuclear instrumentation.

29.3.1 Electronics Assemblies

Although it is perfectly feasible to design a set of electronics to perform a particular task—and indeed this is often done for dedicated systems in industry—the more usual system is to incorporate a series of interconnecting individual circuits into a common frame. This permits a variety of arrangements to be made by plugging in the required elements into this frame, which generally also provides the necessary power supplies. This "building-block" system has become standardized worldwide under the title of NIM and is based on the U.S. Nuclear Regulatory Commission (USNRC) Committee on Nuclear Instrument Modules, presented in *USNRC Publication TID-20893*. The basic common frame is 483 mm (19 in.) wide and the plug-in units are of standard dimensions, a single module being 221 mm high × 34.4 mm wide × 250 mm deep (excluding the connector). Modules can be in widths of one, two, or more multiples of 34.4 mm. Most standard units are single or double width. The rear connectors that plug into the standard NIM bin have a standardized arrangement for obtaining positive and negative stabilized supplies from the common power supply, which is mounted at the rear of the NIM bin. The use of a standardized module system allows the use of units from a number of different manufacturers in the same bin, since it may not be possible or economic to obtain all the units required from one supplier. Some 70 individual modules are available from one manufacturer, which gives some idea of the variety.

A typical arrangement is shown in Figure 29.15, where a scintillation counter is used to measure the gamma-ray energy spectrum from a small source of radioactivity. The detector could consist of a NaI (T1) scintillator optically coupled to the photocathode of a photomultiplier, the whole contained in a light-tight shielded enclosure of metal, with the dynode

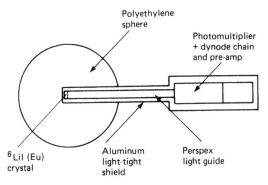

FIGURE 29.14 Sphere fast-neutron detector.

resistor chain feeding each of the dynodes (see Section 21.3.1) located in the base of the photomultiplier, together with a suitable preamplifier to match the high-impedance output of the photomultiplier to the lower input impedance of the main pulse amplifier, generally on the order of 50 Ω. This also allows the use of a relatively long coaxial cable to couple the detector to the electronics if necessary.

The main amplifier raises the amplitudes of the input pulses to the range of about 0.5–10 V. The single-channel analyzer can then be set to cover a range of input voltages corresponding to the energies of the source to be measured. If the energy response of the scintillator-photomultiplier is linear—as it is for NaI (T1)— the system can be calibrated using sources of known energies, and an unknown energy source can be identified by interpolation.

29.3.2 Power Supplies

The basic power supplies for nuclear gauging instruments are of two classes: the first supplies relatively low dc voltages at high currents (e.g., 5–30 V at 0.5–50 A) and the second high voltages at low currents (e.g., 200–5000 V at $200 \mu A - 5mA$). Alternatively, batteries, both primary and secondary (i.e., rechargeable) can be used for portable instruments. In general, for laboratory use dc supplies are obtained by rectifying and smoothing the mains ac supply. In the United Kingdom and most European countries, the mains ac power supply is 50 Hz, whereas in the United States and South America it is 60 Hz, but generally a supply unit designed for one frequency can be used on the other. However, mains-supply voltages vary considerably, being 240 V in the United Kingdom and 220 V in most of the EEC countries, with some countries having supplies of 110, 115, 120, 125, 127, and so on. The stability of some of these mains supplies can leave much to be desired, and fluctuations of plus and minus 50 V have been measured. As nuclear gauging instruments depend greatly on a stable mains supply, the use of special mains-stabilizing devices is almost a necessity for equipment that may have to be used on such varying mains supplies.

Two main types of voltage regulator are in use at present. The first uses a saturable inductor in the form of a transformer with a suitably designed air gap in the iron core. This is a useful device for cases where the voltage swing to be compensated is not large. The second type of stabilizer selects a portion of the output voltage, compares it with a standard, and applies a suitable compensation voltage (plus or minus) to compensate. Some of these units use a motor-driven tapping switch to vary the input voltage to the system—this allows for slow voltage variations. A more sophisticated system uses a semiconductor-controlled voltage supply to add to or subtract from the mains voltage.

The simplest power supply is obtained by a transformer and a rectifier. The best results are obtained with a full-wave rectifier (see Figure 29.16) or a bridge rectifier (see Figure 29.17). A voltage-doubling circuit is shown in Figure 29.18. The outputs from either system are then smoothed using a suitable filter, as shown in Figure 29.19. A simple stabilizer may be fitted in the form of a Zener diode, which has the characteristic that the voltage drop across it is almost independent of the current through it. A simple stabilizer is shown in Figure 29.20. Zeners may be used in series to allow quite high voltages to be stabilized.

An improved stabilizer uses a Zener diode as a reference element rather than an actual controller. Such a circuit is

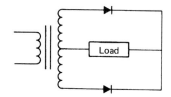

FIGURE 29.16 Full-wave rectifier.

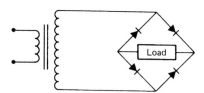

FIGURE 29.17 Bridge rectifier.

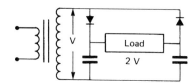

FIGURE 29.18 Voltage-doubling circuit.

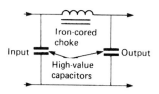

FIGURE 29.19 Smoothing filter.

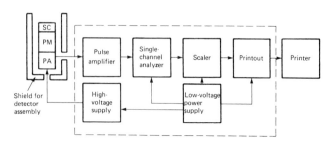

FIGURE 29.15 Typical arrangement of electronics. SC = scintillator; PM = photomultiplier; PA = preamplifier.

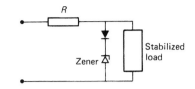

FIGURE 29.20 Simple stabilizer.

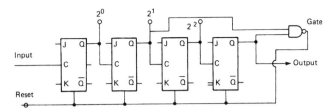

FIGURE 29.22 Decade-counting circuit using binary units (flip-flops).

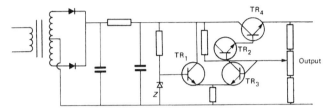

FIGURE 29.21 Improved stabilizer.

29.3.4 Sealers

From the earliest days it has been the counting of nuclear events which has been the means of demonstrating the decay of radioactive nuclei. Early counters used thermionic valves in scale-of-two circuits, which counted in twos. By using a series of these, scale-of-10 counters may be derived. However, solid-state circuits have now reduced such scale-of-10 counters to a single semiconductor chip, which is far more reliable than the thermionic valve systems and a fraction of the size and current consumption.

As sealers are all based on the scale-of-two, Figure 29.22 shows the arrangement for obtaining a scale-of-10, and with present technology, many scales-of-10 may be incorporated on a single chip. The basic unit is called a J-K binary because of the lettering on the original large-scale version of the binary unit.

Rates of 150–200 MHz may be obtained with modern decade counters which can be incorporated on a single chip together with the auxiliary units, such as standard oscillator, input and output circuits, and means of driving light displays to indicate the count achieved.

shown in Figure 29.21, where the sensing element, the transistor TR3, compares a fraction of the output voltage that is applied to its base with the fixed Zener voltage. Through the series control transistor TR$_4$, the difference amplifier TR$_1$, TR$_3$, and TR$_2$ corrects the rise or fall in the output voltage that initiated the control signal.

29.3.2.1 High-voltage Power Supplies

High voltages are required to operate photomultipliers, semiconductor detectors, multi- and single-wire gas proportional counters, and the like, and their output must be as stable and as free of pulse transients as possible. For photomultipliers the stability requirements are extremely important, since a variation in overall voltage across a photomultiplier of, say, 0.1 percent can make a 1 percent change in output. For this reason stabilities on the order of 0.01 percent over the range of current variation expected and 0.001 percent over mains ac supply limits are typically required in such power supplies.

29.3.5 Pulse-Height Analyzers

If the detector of nuclear radiation has a response governed by the energy of the radiation to be measured, the amplitude of the pulses from the detector is a measure of the energy. To determine the energy, therefore, the pulses from the detector must be sorted into channels of increasing pulse amplitude.

Trigger circuits have the property that they can be set to trigger for all pulses above a preset level. This is acting as a discriminator. By using two trigger circuits, one set at a slightly higher triggering level than the other, and by connecting the outputs to an anticoincidence circuit (see Section 29.3.6.4), the output of the anticoincidence circuit will be only pulses that have amplitudes falling within the voltage difference between the triggering levels of the two discriminators. Figure 29.23 shows a typical arrangement; Figure 29.24 shows how an input pulse 1, below the triggering level V, produces no output, nor does pulse 3, above the triggering level $V + \triangle V$. However, pulse 2, falling between the two triggering levels, produces an output pulse. $\triangle V$ is called the channel width.

29.3.3 Amplifiers

29.3.3.1 Preamplifiers

Detectors often have to be mounted in locations quite distant from the main electronics, and if the cable is of any appreciable length, considerable loss of signal could occur. The preamplifier therefore serves more as an impedance transformer, to convert the high impedance of most detector outputs to a sufficiently low impedance which would match a 50- or 70-Ω connecting cable. If the detector is a scintillating counter, the output impedance of the photomultiplier is on the order of several thousand ohms, and there would be almost complete loss of signal if one coupled the output of the counter directly to the 50-Ω impedance of the cable–hence the necessity of providing a suitable impedance matching device.

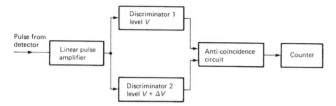

FIGURE 29.23 Single-channel pulse-height analyzer.

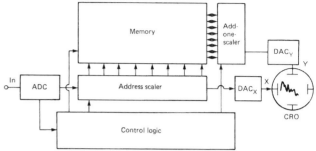

FIGURE 29.25 Block diagram of multichannel analyzer.

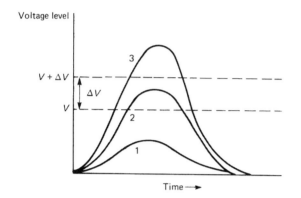

FIGURE 29.24 Waveforms and operation of Figure 3.23.

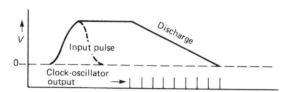

FIGURE 29.26 Principle of Wilkinson ADC using linear discharge of a capacitor.

A multichannel analyzer (MCA) allows the separation of pulses from a detector into channels determined by their amplitudes. Early analyzers used up to 20 or more single-channel analyzers set to successively increasing channels. These, however, proved difficult to stabilize, and the introduction of the Hutchinson–Scarrott system of an analog-to-digital converter (ADC) combined with a computer memory enabled more than 8,000 channels to be provided with good stability and adequate linearity. The advantages of the MCA are offset by the fact that the dead time (that is, the time during which the MCA is unable to accept another pulse for analysis) is longer than that of a single-channel analyzer and so it has a lower maximum counting rate. A block diagram of a typical multichannel analyzer is shown in Figure 29.25. The original ADC was that of Wilkinson, in which a storage capacitor is first charged up so that it has a voltage equal to the peak height of the input pulse. The capacitor is then linearly discharged by a constant current, so producing a ramp waveform, and during this period a high-frequency clock oscillator is switched on (see Figure 29.26). Thus the period of the discharge and the number of cycles of the clock are proportional to the magnitude of the input pulse. The number of clock pulses recorded during the ramp gives the channel number, and after counting these in a register the classification can be recorded, usually in a ferrite-core memory.

A later development is the use of the successive approximation analog-to-digital converter, due to Gatti, Kandiah, and so on, which provides improved channel stability and resolution. ADCs are further discussed in Chapter 20.

29.3.6 Special Electronic Units

29.3.6.1 Dynode Resistor Chains

Each photomultiplier requires a resistor chain to feed each dynode an appropriate voltage to allow the electrons ejected by the scintillator light flash to be accelerated and multiplied at each dynode. For counting rates up to about 10^5 per second, the resistors are usually equal in value and high resistance, so that the total current taken by the chain of resistors is of the order of a few hundred microamperes. Figure 29.27 shows a typical dynode resistor chain. As has already been pointed out, the high voltage supplying the dynode chain must be extremely stable and free from voltage variations, spurious pulses, etc. A 0.1 percent change in voltage may give nearly 1 percent change of output. The mean current taken by the chain is small, and the fitting of bypass capacitors allows pulses of current higher in value than the standing current to be supplied to the dynodes, particularly those close to the anode where the multiplied electron cascade, which started from the photocathode, has now become a large number due to the multiplication effect. However, as the number of pulses to be counted becomes more than about 15,000–50,000 per second, it is necessary to increase the standing current through the dynode chain. Otherwise space charge effects cause the voltages on the dynodes to drop, so reducing the gain in the photomultiplier. When the counting rate to be measured is high, or very fast rise times have to be counted, then the dynode current may have to be increased from a few hundred microamperes to some 5–10 mA, and a circuit as shown in Figure 29.28 is used. Photomultipliers are also discussed in Chapter 21.

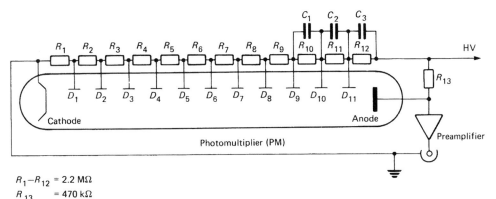

FIGURE 29.27 Dynode resistor chain for counting rates up to about 15,000 c/s.

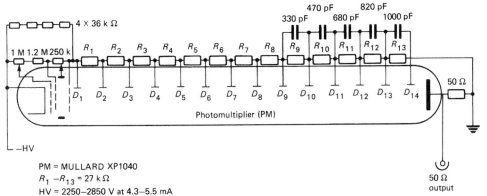

FIGURE 29.28 Dynode resistor chain for high counting rates ~106 cs.

29.3.6.2 Adders/Mixers

When a number of detector signals have to be combined into a single output, a mixer (or fan-in) unit is used. This sums the signals from up to eight preamplifiers or amplifiers to give a common output. This is used, for example, when the outputs from several photomultipliers mounted on one large scintillator must be combined. Such a unit is also used in whole-body counters, where a number of separate detectors are distributed around the subject being monitored. Figure 29.29 shows the circuit of such a unit.

29.3.6.3 Balancing Units

These units are, in effect, potentiometer units which allow the outputs from a number of separate detectors, often scintillation counters with multiple-photomultiplier assemblies on each large crystal, to be balanced before entering the main amplifier of a system. This is especially necessary when pulse-height analysis is to be performed, since otherwise the variations in output would prevent equal pulse heights being obtained for given energy of a spectrum line.

29.3.6.4 Coincidence and Anticoincidence Circuits

A coincidence circuit is a device with two or more inputs that gives an output signal when all the inputs occur at the same time. These circuits have a finite *resolving time*—that is, the greatest interval of time τ that may elapse between signals for the circuit still to consider them coincident. Figure 29.30 shows a simple coincidence circuit where diodes are used as switches. If either or both diodes are held at zero potential, the relevant diode or diodes conduct and the output of the circuit is close to ground. However, if both inputs are caused to rise to the supply voltage level V_c by simultaneous application of pulses of height V_c, both diodes cease to conduct, and the output rises to the supply level for as long as the input pulses are present. An improved circuit is shown in Figure 29.31.

The use of coincidence circuits arose as a result of studies of cosmic rays, since it allowed a series of counters to be used as a telescope to determine the direction of path of such high-energy particles. By the use of highly absorbing slabs between counters, the nature and energies of these cosmic particles and the existence of showers of simultaneous

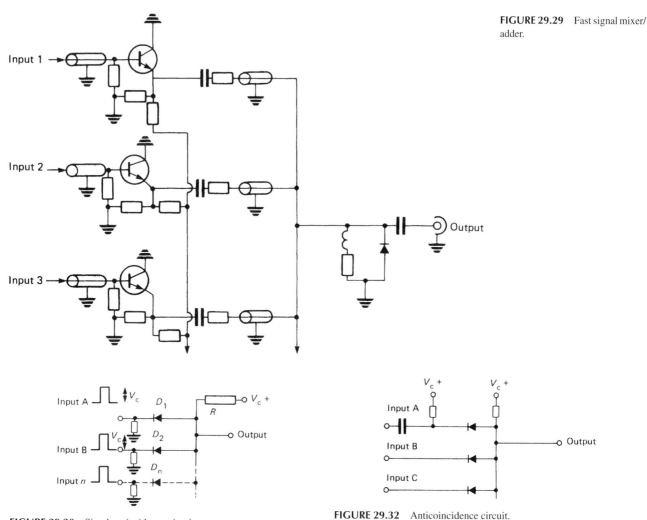

FIGURE 29.29 Fast signal mixer/adder.

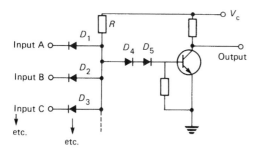

FIGURE 29.30 Simple coincidence circuit.

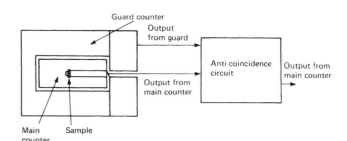

FIGURE 29.32 Anticoincidence circuit.

FIGURE 29.31 Improved coincidence circuit.

FIGURE 29.33 Use of anticoincidence circuit.

particles were established. The anticoincidence circuit was used in these measurements to determine the energies of particles which were absorbed in dense material such as lead, having triggered a telescope of counters before entering the lead, but not triggering counters below the lead slab.

Nowadays coincidence circuits are used with detectors for the products of a nuclear reaction or particles emitted rapidly in cascade during radioactive decay or the two photons emitted in the annihilation of a positron. The latter phenomenon has come into use in the medical scanning and analysis of human living tissue by computer-activated tomography (CAT scanning).

Anticoincidence circuits are used in low-level counting by surrounding a central main counter, such as would be used in radiocarbon-dating measurements, with a guard counter such that any signal occurring simultaneously in the main and guard counters would *not* be counted but only signals originating solely in the main central counter. An anticoincidence circuit is shown in Figure 29.32; Figure 29.33 gives a block diagram of the whole system.

REFERENCES

Birks, J. B., *The Theory and Practice of Scintillation Counting*, Pergamon Press, Oxford (1964).

Dearnaley, G., and Northrup, D. C., *Semiconductor Counters for Nuclear Radiations*, Spon, London (1966).

Eichholz, G. G., and Poston, J. W., *Principles of Nuclear Radiation Detection*, Wiley, Chichester, U.K. (1979).

Fremlin, J. H., *Applications of Nuclear Physics*, English Universities Press, London (1964).

Heath, R. L., *Scintillation Spectrometry Gamma-Ray Spectrum Catalogue*, Vols. I and II: USAEC Report IDO-16880, Washington, D.C. (1964).

Hoffer, P. B., Beck, R. N., and Gottschalk, A., (eds.), *Semiconductor Detectors in the Future of Nuclear Medicine,* Society of Nuclear Medicine, New York (1971).

Knoll, G. F., *Radiation Detection and Measurement*, Wiley, Chichester, U.K. (1979).

McDowell, W. J., *In-Liquid Scintillation Counting, Recent Applications and Developments* (ed. C. T. Peng,), Academic Press, London (1980).

Overman, R. T., and Clark, H. M., *Radioisotope Techniques*, McGraw-Hill New York (1960).

Price, W. J., *Nuclear Radiation Detection*, McGraw-Hill, New York (1964).

Segre, E., *Experimental Nuclear Physics*, Vol. III, Wiley, Chichester, U.K. (1959).

Sharpe, J., *Nuclear Radiation Detection*, Methuen, London (1955).

Snell, A. H. (ed.), *Nuclear Instruments and their Uses,* Vol. I, Ionization Detectors, Wiley, Chichester, U.K. (1962) (Vol. II was not published).

Taylor, D., *The Measurement of Radio Isotopes*, Methuen, London (1951)

Turner, J. C., *Sample Preparation for Liquid Scintillation Counting*, The Radiochemical Center, Amersham, U.K. (1971, revised).

Watt, D. E., and Ramsden, D., *High Sensitivity Counting Techniques*, Pergamon Press, Oxford (1964).

Wilkinson, D. H., *Ionization Chambers and Counters,* Cambridge University Press, Cambridge (1950).

Zumwalt, L. R., Absolute Beta Counting using End-Window Geiger-Muller Counters and Experimental Data on Beta-Particle Scattering Effects, USAEC Report AECU-567 (1950).

FURTHER READING

Eichholz, G. G., and Poston, J. W., *Principles of Nuclear Radiation Detection*, Lewis Publishers (1986).

Chapter 30

Measurements Employing Nuclear Techniques

D. Aliaga Kelly and W. Boyes

30.1 INTRODUCTION

There are two important aspects of using nuclear techniques in industry which must be provided for before any work takes place. These are:

1. Compliance with the many legal requirements when using or intending to use radioactive sources;
2. Adequate health physics procedures and instruments to ensure that the user is meeting the legal requirements

The legal requirements cover the proposed use of the radioactive source, the way in which it is delivered to the industrial site, the manner in which it is used, where and how it is stored when not in use and the way it is disposed of, either through waste disposal or return to the original manufacturer of the equipment or the source manufacturer. Each governing authority, such as the U.S. Nuclear Regulatory Commission, the Atomic Energy Commission of Canada, and other national bodies, has set requirements for use of nuclear gauging instrumentation. When using these instruments, rely on the manufacturer to guide you through the requirements, procedures, and documentation.

There are differences in regulatory requirements from country to country that make it impossible to list here. It must be noted that in some applications, operators must be badged and use monitoring devices, while in other applications, particularly in the chemical and hydrocarbon processing industries, these procedures are not necessarily required.

It is impossible to cover exhaustively all the applications of radioisotopes in this book; here we deal with the following:

1. Density;
2. Thickness;
3. Level;
4. Flow;
5. Tracer applications;
6. Material analysis.

Before considering these applications in detail we discuss some general points that are relevant to all or most of them.

One of the outstanding advantages of radioisotopes is the way in which their radiation can pass through the walls of a container to a suitable detector without requiring any contact with the substance being measured. This means that dangerous liquids or gases may be effectively controlled or measured without risk of leakage, either of the substance itself out from the container or of external contamination to the substance inside. Thus in the chemical industry many highly toxic materials may be measured and controlled completely without risk of leakage from outside the pipes, etc., conveying it from one part of the factory to another. (See also Section 30.1.4.)

Another important advantage in the use of radioisotopes is that the measurement does not affect, for example, the flow of liquid or gas through a pipe, and such flow can be completely unimpeded. Thus the quantity of tobacco in a cigarette-making machine can be accurately determined as the continuous tube of paper containing the tobacco moves along, so that each cigarette contains a fixed amount of tobacco.

Speed is another important advantage of the use of radioisotopes. Measurements of density, level, etc., may be carried out continuously so that processes may be readily controlled by the measuring signal derived from the radioisotope system. Thus, a density gauge can control the mixing of liquids or solids, so that a constant density material is delivered from the machine. Speed in determining flow allows, for example, measurement of the flow of cooling gas to a nuclear reactor and observation of small local changes in flow due to obstructions which would be imperceptible using normal flow-measuring instruments.

The penetrating power of radiations from radio-isotopes is particularly well known in its use for gamma radiography. A tiny capsule of highly radioactive cobalt, for example, can be used to radiograph complex but massive metal castings where a conventional X-ray machine would be too bulky to fit. Gamma radiography is discussed further in Chapter 24. Also, leaks in pipes buried deep in the ground can be found by passing radioactive liquid along them and afterwards monitoring the line of the pipe for radiation from the active liquid which has soaked through the leak into the surrounding earth. If one uses a radioisotope with a very short half-life (half-life is the time taken for a particular radioisotope to decay to half its initial activity) the pipeline will be free of radioactive contamination in a short time, allowing repairs to be carried out without hazard to the workmen repairing it or to the domestic consumer when the liquid is, for example, the local water supply.

30.1.1 Radioactive Measurement Relations

When radiations from radioactive isotopes pass through any material, they are absorbed according to (1) their energy and (2) the density and type of material. This absorption follows in general the relationship

$$I = I_0 B \exp -(\mu_L x) \quad (30.1)$$

where I_0 is the intensity of the incident radiation, I the intensity of the radiation after passing through the material, x the thickness of material (cm), μ_L the linear absorption coefficient (cm^{-1}) and B the build-up factor.

The *absorption coefficient* is a factor which relates the energy of the radiation and the density type of material, and suitable tables are available (Hubbell) from which this factor may be obtained for the particular conditions of source and material under consideration. As the tables are usually given in terms of the *mass absorption coefficient* (μ_L) (generally in cm^2/g), it is useful to know that the *linear absorption coefficient* (μ_L) (in cm^{-1}) may be derived by multiplying the *mass absorption coefficient* (μ_L) (in cm^2/g) by the density (ρ) of the material (in g/cm^3). It must be borne in mind that in a mixture of materials each will have a different *absorption coefficient* for the same radiation passing through the mixture.

The *build-up factor, B*, is necessary when dealing with gamma- or X-radiation, where scattering of the incident radiation can make the intensity of the radiation which actually falls on the detector different from what it would be if no scattering were to take place. For electrons or beta particles this factor can be taken equal to 1. The complication of gamma ray absorption is illustrated by the non-linearity of the curves in Figure 30.1, which gives the thickness of different materials needed to effect a ten-fold attenuation in a narrow beam.

From Equation (30.1) we can obtain some very useful information in deciding on the optimum conditions for making a measurement on a particular material, or, conversely, if we have a particular radioactive source, we

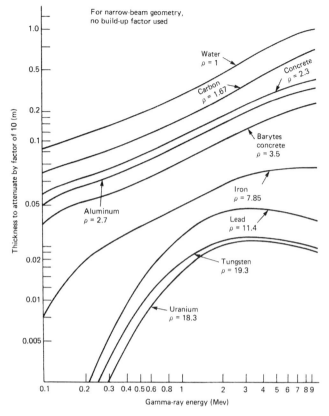

FIGURE 30.1 Thickness needed to attenuate a narrow gamma-ray beam by a factor of 10.

can determine what are its limits for measuring different materials.

First, it can be shown that the maximum sensitivity for a density measurement is obtained when

$$x = \frac{1}{\mu_L} \quad (30.2)$$

that is, when the thickness of material is equal to the reciprocal of the *linear absorption coefficient*. This reciprocal is also called the *mean-free path*, and it can be shown that for any thickness of a particular substance there is an optimum type and intensity of radioactive source. For very dense materials, and for thick specimens, a source emitting high-energy radiation will be required. Therefore ^{60}Co, which emits gamma rays of 1.33 and 1.17 MeV, is frequently used. At the other end of the scale, for the measurement of very thin films such as, for example, *Melinex*, a soft beta or alpha particle emitter would be chosen. For measurement of thickness it is generally arranged to have two detectors and one source. The radiation from the source is allowed to fall on one detector through a standard piece of the material to be measured, while the other detector measures the radiation from the source which has passed through the sample under test. The signal from each of the pair of detectors is generally combined as two dc levels, the difference between which drives a potentiometer-type pen recorder.

30.1.2 Optimum Time of Measurement

The basic statistics of counting were outlined in the previous chapter (Section 22.1.1). We now consider how that is applied to particular measurements.

Suppose that the number of photons or particles detected per second is n and that the measurement is required to be made in a time t. The number actually recorded in t will be $nt \pm \sqrt{nt}$, where $\sqrt{nt}$ is the standard deviation of the measurement according to Poisson statistics and is a measure of the uncertainty of the true value of nt. The relative uncertainty (coefficient of variation) is given by

$$\frac{\sqrt{(nt)}}{(nt)} = \frac{1}{\sqrt{(nt)}}$$

A radioisotope instrument is used to measure some quality X of a material in terms of the output I of a radiation detector. The instrument sensitivity, or relative sensitivity, S, is defined as the ratio of the fractional change $\delta I/I$ in detector output which results from a given fractional change $\delta X/X$ in the quality being measured, i.e.,

$$S = \frac{\delta I}{I} \bigg/ \frac{\delta X}{X} \qquad (30.3)$$

If in a measurement, the only source of error is the statistical variation in the number of recorded events, the coefficient of variation in the value of the quality measures

$$\frac{\delta X}{X} = \frac{1}{S} \cdot \sqrt{\frac{(nt)}{nt}} = \frac{1}{S\sqrt{(nt)}} \qquad (30.4)$$

To reduce this to as small a value as possible, then S, n, or t or all three of these variables should be increased to as high a value as possible. In many cases, however, the time available for measurement is short. This is particularly true on high-speed production lines of sheet material, where only a few milliseconds may be available for the measurement.

It can now be seen how measurement time, collimation, detector size, and absorber thickness may affect the error in the measurement. The shorter the measurement time, the greater the degree of collimation, the thicker the absorber and the smaller the detector, the greater will be the source activity required to maintain a constant error. A larger source will be more expensive, and in addition its physical size may impose a limit on the activity usable. Bearing in mind that a source radiates equally in all directions, only a very small fraction can be directed by collimation for useful measurement; the rest is merely absorbed in the shielding necessary to protect the source.

30.1.3 Accuracy/Precision of Measurements

The precision or reproducibility of a measurement is defined in terms of the ability to repeat measurements of the same quantity. Precision is expressed quantitatively in terms of the standard deviation, σ, from the average value obtained by repeated measurements. In practice it is determined by statistical variations in the rate of emission from the radioactive source, instrumental instabilities, and variations in measuring conditions.

The accuracy of a measurement is an expression of the degree of correctness with which an actual measurement yields the true value of the quantity being measured. It is expressed quantitatively in terms of the deviation from the true value of the mean of repeated measurements. The accuracy of a measurement depends on the precision and also on the accuracy of calibration. If the calibration is exact, then in the limit, accuracy and precision are equal. When measuring a quantity such as thickness it is relatively easy to obtain a good calibration. In analyzing many types of samples, on the other hand, the true value is often difficult to obtain by conventional methods and care may have to be taken in quoting the results.

In general, therefore, a result is quoted along with the calculated error in the result and the confidence limits to which the error is known. Confidence limits of both one standard deviation, 1σ (68 percent of results lying within the quoted error), and two standard deviations, 2σ (95 percent of results lying within the quoted error), are used.

In analytical instruments, when commenting on the smallest quantity or concentration which can be measured, the term "limit of detection" is preferred. This is defined as the concentration at which the measured value is equal to some multiple of the standard deviation of the measurement.

In practice, the accuracy of radioisotope instruments used to measure the thickness of materials is generally within ± 1 percent, except for very lightweight materials, when it is about ± 2 percent. Coating thickness can usually be measured to about the same accuracy. Level gauges can be made sensitive to a movement of the interface of ± 1 mm. Gauges used to measure the density of homogeneous liquids in closed containers generally can operate to an accuracy of about ± 0.1 percent, though some special instruments can reduce the error to ± 0.01 percent. The accuracy of bulk density gauges is in the range ± 0.5 to ± 5 percent depending on the application and on the measuring conditions.

30.1.4 Measurements on Fluids in Containers

Nuclear methods may be used to make certain measurements on fluids flowing in pipes from 12.7 mm to 1 m in diameter. For plastics or thin-walled pipes up to 76 mm in diameter the combined unit of source and detector shown in Figure 30.2(a) is used, while for larger pipes the system consisting of a separate holder and detector shown in Figure 30.2(b) is used.

The gamma-ray source is housed in a shielded container with a safety shutter so that the maximum dose rate is less than 7.5 μGy/h, and is mounted on one side of the pipe or tank. A measuring chamber containing argon at 20 atm is fitted on the other side of the pipe or tank. It is fitted with

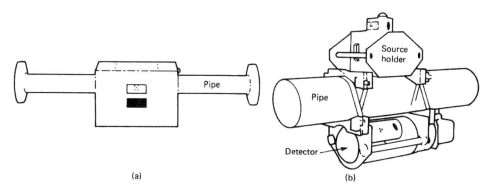

FIGURE 30.2 Fluid-density measuring systems. Courtesy of Nuclear Enterprises Ltd. (a) Combined detector/source holder; (b) separate units for larger pipes.

a standardizing holder and has a detection sensitivity of ±0.1 percent.

The system may be used to measure density over a range of 500–4000 kg/m² with a sensitivity of 0.5 kg/m³, specific gravity with an accuracy of ±0.0005, percentage solids ±0.05 percent, and moisture content of slurries of a constant specific gravity to within ±0.25 percent.

The principle of the measurement is that the degree of absorption of the gamma rays in the flowing fluid is measured by the ionization chamber, where output is balanced against an adjustable source which is set by means of the calibrated control to the desired value of the material being measured. Deviations from this standard value are then shown on the calibrated meter mounted on the indicator front panel. Standardization of the system for the larger pipes is performed manually and a subsidiary source is provided for this purpose. The selection of the type of source depends on (1) the application, (2) the wall thickness and diameter of the pipe and (3) the sensitivity required. Sources in normal use are ^{137}Cs (source life 30 yr), ^{241}Am (460 yr) and ^{60}CO (5 yr).

The measuring head has a temperature range of –10 to +55°C and the indicator 5–40°C, and the response time minimum is 0.1 s, adjustable to 25 s.

30.2 MATERIALS ANALYSIS

Nuclear methods of analysis, particularly neutron activation analysis, offer the most sensitive methods available for most elements in nature. However, no one method is suitable for all elements, and it is necessary to select the techniques to be used with due regard to the various factors involved, some of which may be listed as follows:

1. Element to be detected;
2. Quantities involved;
3. Accuracy of the quantitative analysis required;
4. Costs of various methods available;
5. Availability of equipment to carry out the analysis to the statistical limits required;
6. Time required;
7. Matrix in which the element to be measured is located;
8. Feasibility of changing the matrix mentioned in (7) to a more amenable substance.

For example, the environmental material sample may be analyzed by many of the methods described, but the choice of method must be a compromise, depending on all the factors involved.

30.2.1 Activation Analysis

When a material is irradiated by neutrons, photons, alpha or beta particles, protons, etc. a reaction may take place in the material depending on a number of factors. The most important of these are:

1. The type of particle or photon used for the irradiation;
2. The energy of the irradiation;
3. The flux in the material;
4. The time of irradiation.

The most useful type of particle has been found to be neutrons, since their neutral charge allows them to penetrate the high field barriers surrounding most atoms, and relatively low energies are required. In fact, one of the most useful means of irradiation is the extremely low energy neutrons of thermal energy (0.025 eV) which are produced abundantly in nuclear reactors. The interactions which occur cause some of the various elements present to become radioactive, and in their subsequent decay into neutral atoms again, to emit particles and radiation which are indicative of the elements present. Neutron activation analysis, as this is called, has become one of the most useful and sensitive methods of identifying certain elements in minute amounts without permanently damaging the specimen, as would be the case in most chemical methods of analysis. A detector system can be selected which responds uniquely to the radiation emitted as the excited atoms of the element of interest decay with emission of gamma rays or beta particles, while not responding to other types of radiation or to different energies from other elements which may be present in the sample. The decay half-life of the radiation is also used to identify the element of interest, while the actual magnitude of the response at the particular energy involved is a direct measure of the amount of the element of interest in the sample. Quantitatively the basic relation between A, the induced activity present at the end of the irradiation in Becquerels, i.e., disintegrations per second, is given by

$$A = N\sigma\phi \left[1 - \exp\left(-\frac{0.693 t_i}{T_{\frac{1}{2}}}\right)\right] \quad (30.5)$$

where N is the number of target atoms present, σ the cross-section (cm^2), ϕ the irradiation flux (neutrons cm^{-2} s^{-1}), t_i the irradiation time and $T_{1/2}$ the half-life of product nuclide.

From this we may calculate N, the number of atoms of the element of interest present, after appropriate correction factors have been evaluated.

Beside the activation of the element and measurement of its subsequent decay products one may count directly the excitation products produced whilst the sample is being bombarded. This is called "prompt gamma-ray analysis," and it has been used, for example, to analyze the surface of the moon.

Most elements do not produce radioactivity when bombarded with electrons, beta particles, or gamma rays. However, most will emit characteristic X-rays when bombarded and the emitted X-rays are characteristic of each element present. This is the basis of X-ray fluorescence analysis, discussed below.

Electrons and protons have also been used to excite elements to emit characteristic X-rays or particles, but this technique requires operation in a very high vacuum chamber.

High-energy gamma rays have the property of exciting a few elements to emit neutrons—this occurs with beryllium when irradiated with gamma rays of energy greater than 1.67 MeV and deuterium with gamma rays of energy greater than 2.23 MeV. This forms the basis of a portable monitor for beryllium prospecting in the field, using an ^{124}Sb source to excite the beryllium. Higher-energy gamma rays from accelerators, etc., have the property of exciting many elements but require extremely expensive equipment.

30.2.2 X-ray Fluorescence Analysis

30.2.2.1 Dispersive X-ray Fluorescence Analysis

In dispersive X-ray fluorescence analysis the energy spectrum of the characteristic X-ray emitted by the substance when irradiated with X-rays is determined by means of a dispersive X-ray spectrometer, which uses as its analyzing element the regular structure of a crystal through which the characteristic X-rays are passed. This property was discovered by Bragg and Bragg in 1913, who produced the first X-ray spectrum by crystal diffraction through a crystal of rock salt.

Figure 30.3 shows what happens when an X-ray is diffracted through a crystal of rock salt. The Braggs showed that the X-rays were reflected from the crystal, meeting the Bragg relationship

$$n\lambda = 2a \sin\theta \quad (30.6)$$

where λ is the wavelength of the incident radiation, a the distance between lattice planes and n the order of the reflection. θ is the angle of incidence and of reflection, which are equal.

To measure the intensity distribution of an X-ray spectrum by this method the incident beam has to be collimated and the detector placed at the corresponding position on the opposite side of the normal (see Figure 30.4). The system effectively selects all rays which are incident at the appropriate angle for Bragg reflection. If the angle of incidence is varied (this will normally involve rotating the crystal through a controllable angle and the detector by twice this amount) then the detector will receive radiation at a different wavelength and a spectrum will be obtained. If the system is such that the angular range $d\theta$ is constant over the whole range of wavelength investigated, as in the geometry illustrated in Figure 30.4, we can write

$$n\,d\lambda = 2a\,d(\sin\theta) = 2a\cos\theta\,d\theta \quad (30.7)$$

The intensity thus received will be proportional to $\cos\theta$ and to $d\theta$. After dividing by $\cos\theta$ and correcting for the variation with angle of the reflection coefficient of the crystal and the variation with wavelength of the detector efficiency, the recorded signal will be proportional to the intensity per unit wavelength interval, and it is in this form that continuous X-ray spectra are traditionally plotted.

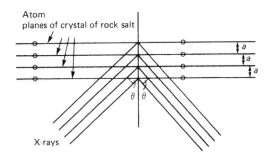

FIGURE 30.3 Diffraction of X-rays through crystal of rock salt.

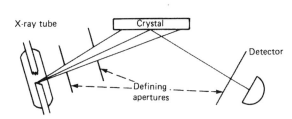

FIGURE 30.4 Dispersive X-ray spectrometer.

30.2.2.2 X-ray Fluorescence Analysis (Non-dispersive)

When a substance is irradiated by a beam of X-rays or gamma rays it is found that the elements present fluoresce, giving out X-rays of energies peculiar to each element present. By selecting the energy of the incident X-rays selection may be made of particular elements required. As an example, if a silver coin is irradiated with X-rays, the silver present emits X-rays of energies about 25.5 keV, and if other elements are also present, such as copper or zinc, "characteristic" X-rays, as they are called, will be emitted with energies of 8.0 and 8.6 keV, respectively.

The Si (Li) or Ge (Li) detector, cooled to the temperature of liquid nitrogen, will separate the various spectral lines and a multichannel analyzer will allow their intensity to be evaluated electronically. However, it must be pointed out that as the incident X-rays and the excited, emergent, characteristic X-rays have a very short path in metals the technique essentially measures the elemental content of the metal surface. The exciting X-rays are selected for their energy to exceed what is called the "K-absorption edge" of the element of interest, so that elements of higher atomic weight will not be stimulated to emit their characteristic X-rays. Unfortunately, the range of exciting sources is limited, and Table 30.1 lists those currently used. Alternatively, special X-ray tubes with anodes of suitable elements have been used to provide X-rays for specific analyses as well as intensities greater than are generally available from radioisotope sources.

TABLE 30.1 Exciting sources for X-ray fluorescence analysis

Isotope	Half-life	Principal photon energies (keV)	Emission (%)
^{241}Am	433 yr	11.9–22.3	~40
		59.5	35.3
^{109}Cd	453 d	22.1, 25.0	102.3
		2.63–3.80	~10
		88.0	3.6
^{57}Co	270.5 d	6.40, 7.06	~55
		14.4	9.4
		122.0	85.2
		136.5	11.1
^{55}Fe	2.7 yr	5.89, 6.49	~28
^{153}Gd	241.5 d	41.3, 47.3	~110
		69.7	2.6
		97.4	30
		103.2	20
^{125}I	60.0 d	27.4, 31.1	138
		35.5	7
^{210}Pb	22.3 yr	9.42–16.4	~21
		46.5	~4
		+ Bremsstrahlung to 1.16 MeV	
^{147}Pm	2.623 yr	Characteristic X-rays of target	~0.4
+ target		+ Bremsstrahlung to 225 keV	
^{238}Pu	87.75 yr	11.6–21.7	~13
^{125m}Te	119.7 d	27.4, 31.1	~50
		159.0	83.5

Continued

TABLE 30.1 *Continued*

^{170}Tm	128 d	52.0, 59.7	~5
		84.3	3.4
		+ Bremsstrahlung to 968 keV	
Tritium (^{3}H)	12.35 yr		
+ Ti target		4.51, 4.93	~10^{-2}
		+ Bremsstrahlung to 18.6 keV	
+ Zr target		1.79–2.5	~10^{-2}
		+ Bremsstrahlung to 18.6 keV	

Gas proportional counters and NaI (Tl) scintillation counters have also been used in non-dispersive X-ray fluorescence analysis, but the high-resolution semiconductor detector has been the most important detector used in this work. While most systems are used in a fixed laboratory environment, due to the necessity of operating at liquid-nitrogen temperatures, several portable units are available commercially in which small insulated vessels containing liquid nitrogen, etc., give a period of up to 8 h use before requiring to be refilled.

The introduction of super-pure Ge detectors has permitted the introduction of portable systems which can operate for limited periods at liquid-nitrogen temperatures, but as long as they remain in a vacuum enclosure they can be allowed to rise to room temperature without damage to the detector, as would occur with the earlier Ge (Li) detector, where the lithium would diffuse out of the crystal at ambient temperature. As Si (Li) detectors are really only useful for X-rays up to about 30 keV, the introduction of the Ge (HP) detector allows X-ray non-dispersive fluorescence analysis to be used for higher energies in non-laboratory conditions.

In the early 1980s with the availability of microprocessors as well as semiconductor detectors such as HgI (mercuric iodide), it became possible to build small, lightweight non-dispersive devices that could be operated on batteries and taken into the field. These devices were able to operate at near-room temperature, due to the incorporation of Peltier cooling circuitry in their designs, and made *in-situ* non-destructive elemental analysis possible. Typically their use has been for positive material identification, especially specialty metal alloys, but they have also been used for coal ash analysis, lead in paint, and other nonspecific elemental analysis.

30.2.3 Moisture Measurement: By Neutrons

If a beam of fast neutrons is passed through a substance any hydrogen in the sample will cause the fast neutrons to be slowed down to thermal energies, and these slow neutrons can be detected by means of a BF$_3$- or ^{3}He-filled gas proportional counter system. As the major amount of hydrogen present in most material is due to water content, and the slowing down is directly proportional to the hydrogen density, this offers a way of measuring the moisture content of a great number of materials. Some elements such as cadmium, boron, and the rare-earth elements chlorine and iron, however, can have an effect on the measurement of water content since they have high thermal-neutron capture probabilities. When these elements are present this method of moisture measurement has to be used with caution. On the other hand, it provides a means of analyzing substances for these elements, provided the actual content of hydrogen and the other elements is kept constant.

Since the BF$_3$ or ^{3}He counter is insensitive to fast neutrons it is possible to mount the radioactive fast neutron source close to the thermal neutron detector. Then any scattered thermal neutron from the slowing down of the fast neutrons, which enters the counter, will be recorded.

This type of equipment can be used to measure continuously the moisture content of granular materials in hoppers, bins, and similar vessels. The measuring head is mounted external to the vessel, and the radiation enters by a replaceable ceramic window (Figure 30.5).

The transducer comprises a radioisotope source of fast neutrons and a slow neutron detector assembly, mounted within a shielding measuring head. The source is mounted on a rotatable disc. It may thus be positioned by an electropneumatic actuator either in the center of the shield or adjacent to the slow neutron detector and ceramic radiation "window" at one end of the measuring head, which is mounted externally on the vessel wall. Fast neutrons emitted from the source are slowed down, mainly by collisions with atoms of hydrogen in the moisture of material situated near the measuring head. The count rate in the detector increases with increasing hydrogen content and is used as an indication of moisture content.

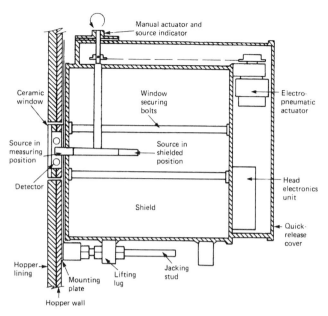

FIGURE 30.5 Moisture meter for granular material in hoppers, surface mounting. Courtesy of Nuclear Enterprises Ltd.

The slow neutron detector is a standard, commercially available detector and is accessible without dismantling the measuring head. The electronics, housed in a sealed case with the measuring head, consist of an amplifier, pulse-height analyzer, and a line-driver module, which feed preshaped 5 V pulses to the operator unit.

The pulses from the head electronic unit are processed digitally in the operator's unit to give an analog indication of moisture content on a 100-mm meter mounted on the front panel, or an analog or digital signal to other equipment. The instrument has a range of 0–20 percent moisture and a precision of ±0.5 percent moisture, with a response time of 5–25 s. The sample volume is between 100 and 300 l, and the operating temperature of the detector is 5–70°C and of the electronics 5–30°C. Figure 30.6 shows such a moisture gauge in use on a coke hopper.

A similar arrangement of detector and source is used in borehole neutron moisture meters. The fast neutron source is formed into an annulus around the usually cylindrical thermal neutron detector and the two components are mounted in a strong cylinder of steel, which can be let down a suitable borehole and will measure the distribution of water around the borehole as the instrument descends. This system is only slightly sensitive to the total soil density (provided no elements of large cross-section for absorbing neutrons are present), so only a rough estimate is needed of the total soil density to compensate for its effect on the measured value of moisture content. The water content in soil is usually quoted as percentage by weight, since the normal gravimetric method of determining water content measures the weight of water and the total weight of the soil sample. To convert the water content as measured by the neutron gauge

FIGURE 30.6 Coke moisture gauge in use on hoppers. Courtesy of Nuclear Enterprises Ltd.

measurement into percentage by weight one must also know the total density of the soil by some independent measurement. This is usually performed, in the borehole case, by a physically similar borehole instrument, which uses the scattering of gamma-rays from a source in the nose of the probe, around a lead plug to a suitable gas-filled detector in the rear of the probe. The lead plug shields the detector from any direct radiation from the source. Figure 30.7 shows the response of the instrument.

At zero density of material around the probe the response of the detector is zero, since there is no material near the detector to cause scattering of the gamma rays and subsequent detection. In practice, even in free air, the probe will show a small background due to scattering by air molecules. As the surrounding density increases, scattering into the detector begins to occur, and the response of the instrument increases linearly with density but reaches a maximum for the particular source and soil material. The response then decreases until theoretically at maximum density the response will be zero. Since the response goes through a maximum with varying density, the probe parameters should be adjusted so that the density range of interest is entirely on one side of the maximum. Soil-density gauges are generally designed to operate on the negative-slope portion of the response.

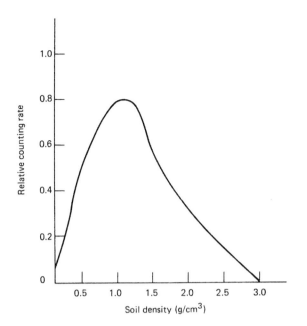

FIGURE 30.7 Response of a scattered-gamma-ray gauge to soil density.

30.2.3.1 Calibration of Neutron-moisture Gauges

Early models of neutron gauges used to be calibrated by inserting them into concrete blocks of known densities, but this could lead to serious error, since the response in concrete is quite different from that in soil. It has been suggested by Ballard and Gardner that in order to eliminate the sensitivity of the gauge to the composition of the material to be tested, one should use non-compactive homogeneous materials of known composition to obtain experimental gauge responses. This data is then fitted by a "least-squares" technique to an equation which enables a correction to be obtained for a particular probe used in soil of a particular composition.

Dual Gauges Improved gauges have been developed in which the detector is a scintillation counter with the scintillator of cerium-activated lithium glass. As such a detector is sensitive to both neutrons and gamma rays, the same probe may be used to detect both slowed-down neutrons and gamma rays from the source—the two can be distinguished because they give pulses of different shapes. This avoids the necessity of having two probes, one to measure neutrons and the other to measure scattered gammas, allowing the measurement of both moisture and density by pulse-shape analysis.

Surface-neutron gauges are also available, in which both radioactive source and detector are mounted in a rectangular box which is simply placed on a flat soil surface. It is important to provide a smooth flat surface for such measurements, as gaps cause appreciable errors in the response.

30.2.4 Measurement of Sulfur Contents of Liquid Hydrocarbons

Sulfur occurs in many crude oils at a concentration of up to 5 percent by weight and persists to a lesser extent in the refined product. As legislation in many countries prohibits the burning of fuels with a high sulfur content to minimize pollution, and sulfur compounds corrode engines and boilers and inhibit catalysts, it is essential to reduce the concentration to tolerable levels. Thus rapid measurement of sulfur content is essential, and the measurement of the absorption of appropriate X-rays provides a suitable online method. In general, the mass absorption coefficient of an element increases with increase of atomic number (Figure 30.8) and decreases with shortening of the wavelength of the X-rays. In order to make accurate measurement the wavelength chosen should be such that the absorption will be independent of changes in the carbon-hydrogen ratio of the hydrocarbon. When the X-rays used have an energy of 22 keV the mass attenuation for carbon and hydrogen are equal. Thus by using X-rays produced by allowing the radiation from the radio-element ^{241}Am to produce fluorescent excitation in a silver target which gives X-rays having an energy of 23 keV, the absorption is made independent of the carbon–hydrogen ratio. As this source has a half-life of 450 yr, no drift occurs owing to decay of the source. The X-rays are passed through a measuring cell through which the hydrocarbon flows and, as the absorption per unit weight of sulfur is many times greater than the absorption of carbon and hydrogen, the fraction of the X-rays absorbed is a measure of the concentration of the sulfur present. Unfortunately the degree of absorption of X-rays is also affected by the density of the sample and by the concentration of trace elements of high atomic weight and of water.

The concentration of X-rays is measured by a high-resolution proportional counter so the accuracy will be a function of the statistical variation in the count rate and the stability of the detector and associated electronics, which can introduce an error of ±0.01 per cent sulfur. A water content of 650 ppm will also introduce an error of 0.01 percent sulfur. Compensation for density variations may be achieved by measuring the density with a non-nucleonic meter and electronically correcting the sulfur signal.

Errors caused by impurities are not serious, since the water content can be reduced to below 500 ppm and the only serious contaminant, vanadium, seldom exceeds 50 ppm.

The stainless steel flow cell has standard flanges and is provided with high-pressure radiation windows and designed so that there are no stagnant volumes. The flow cell may be removed for cleaning without disturbing the source. Steam tracing or electrical heating can be arranged for samples likely to freeze. The output of the high-resolution proportional counter, capable of high count rates for statistical accuracy, is amplified and applied to a counter and digital-to-analog converter when required. Thus both digital

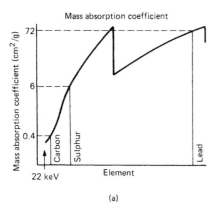
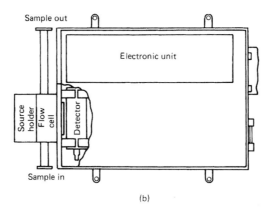

FIGURE 30.8 Online sulfur analyzer. Courtesy of Nuclear Enterprises Ltd. (a) Mass absorption coefficient; (b) arrangement of instrument.

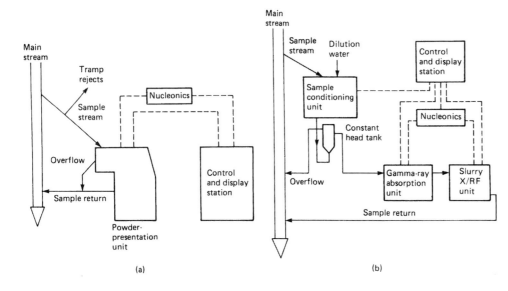

FIGURE 30.9 Block diagrams of calcium in cement raw material measuring instrument. Courtesy of Nuclear Enterprises Ltd. (a) Dry powder form of instrument; (b) slurry form.

and analog outputs are available for display and control purposes.

The meter has a range of up to 0–6 percent sulfur by weight and indicates with a precision of ±0.01 percent sulfur by weight or ±1 percent of the indicated weight, whichever is the larger. It is independent of carbon-hydrogen ratio from 6:1 to 10:1 and the integrating times are from 10 to 200 s. The flow cell is suitable for pressures up to 15 bar and temperatures up to 150°C, the temperature range for the electronics being −10 to +45°C.

The arrangement of the instrument is as shown in Figure 30.8.

30.2.5 The Radioisotope Calcium Monitor

The calcium content of raw material used in cement manufacture may be measured on-line in either dry powder or slurry form. The basis of the method is to measure the intensity of the characteristic K X-rays emitted by the flowing sample using a small ^{55}Fe radio-isotope source as the means of excitation. This source is chosen because of its efficient excitation of Ca X-rays in the region which is free from interference by iron.

In the form of instrument shown in Figure 30.9(a), used for dry solids, a sample of material in powder form is extracted from the main stream by a screw conveyor and fed into a hopper. In the powder-presentation unit the powder is extracted from the hopper and fed on to a continuously weighted sample presenter at a rate which is controlled so as to maintain a constant mass of sample per unit area on the latter within very close limits. After measurement, the sample is returned to the process. A system is fitted to provide an alarm if the mass per unit area of sample wanders outside preset limits, and aspiration is provided to eliminate airborne dust in the vicinity of the sample presenter and measuring head.

Under these conditions it is possible to make precise reproducible X-ray fluorescence measurements of elements from atomic number 19 upward without pelletizing.

The signal from the X-ray detector in the powder-presentation unit is transmitted via a head amplifier and a standard nucleonic counting chain to a remote display and control station. Analog outputs can be provided for control purposes.

In the form used for slurries shown in Figure 30.9(b) an additional measurement is made of the density and hence the solids content of the slurry. The density of the slurry is measured by measuring the absorption of a highly collimated 660 ke V gamma-ray beam. At this energy the measurement is independent of changes in solids composition.

The dry powder instrument is calibrated by comparing the instrument readings with chemical analysis carried out under closely controlled conditions with the maximum care taken to reduce sampling errors. This is best achieved by calibrating while the instrument is operating in closed loop with a series of homogeneous samples recirculated in turn. This gives a straight line relating percentage of calcium carbonate to the total X-ray count.

With slurry, a line relating percentage calcium carbonate to X-ray count at each dilution is obtained, producing a nomogram which enables a simple special-purpose computer to be used to obtain the measured value indication or signal.

In normal operation the sample flows continuously through the instrument and the integrated reading obtained at the end of 2–5 min, representing about 2 kg of dry sample or 30 liters of slurry, is a measure of the composition of the sample.

An indication of CaO to within ±0.15 percent for the dry method and ±0.20 percent for slurries should be attainable by this method.

30.2.6 Wear and Abrasion

The measurement of the wear experienced by mechanical bearings, pistons in the cylinder block or valves in internal combustion engines is extremely tedious when performed by normal mechanical methods. However, wear in a particular component may be easily measured by having the component irradiated in a neutron flux in a nuclear reactor to produce a small amount of induced radioactivity. Thus the iron in, for example, piston rings, which have been activated and fitted to the pistons of a standard engine, will perform in an exactly similar way to normal piston rings, but when wear takes place, the active particles will be carried around by the lubrication system, and a suitable radiation detector will allow the wear to be measured, as well as the distribution of the particles in the lubrication system and the efficiency of the oil filter for removing such particles.

To measure the wear in bearings, one or other of the bearing surfaces is made slightly radioactive, and the amount of activity passed to the lubricating system is a measure of the wear experienced.

30.2.7 Leak Detection

Leakage from pipes buried in the ground is a constant problem with municipal authorities, who may have to find leaks in water supplies, gas supplies, or sewage pipes very rapidly and with the minimum of inconvenience to people living in the area, as essential supplies may have to be cut off until the leak is found and the pipe made safe again.

To find the position of large leaks in water distribution pipes two methods have been developed. The first uses an inflatable rubber ball with a diameter nearly equal to that of the pipe and containing 100 or so MBq of ^{24}Na which is inserted into the pipe after the leaking section has been isolated from the rest of the system. The only flow is then towards the leak and the ball is carried as far as the leak, where it stops. As ^{24}Na emits a high-energy gamma ray its radiation can be observed on the surface through a considerable thickness of soil, etc., by means of a sensitive portable detector. Alternatively, radioactive tracer is introduced directly into the fluid in the pipe.

After a suitable period the line of the pipe can be monitored with a sensitive portable detector, and the buildup of activity at the point of the leak can be determined. ^{24}Na is a favored radioactive source for leak testing, especially of domestic water supply or sewage leaks, since it has a short half-life (15 h), emits a 2.7 MeV gamma ray and is soluble in water as ^{24}Na Cl. Thus, leak tests can be rapidly carried out and the activity will have decayed to safe limits in a very short time.

30.3 MECHANICAL MEASUREMENTS

30.3.1 Level Measurement

30.3.1.1 Using X- or Gamma Rays

Level measurements are usually made with the source and detector fixed in position on opposite sides of the outer wall of the container (Figure 30.10). Because many containers in the chemical engineering and oil-refining industries, where most level gauges are installed, have large dimensions, high-activity sources are required and these have to be enclosed in thick lead shields with narrow collimators to reduce scattered

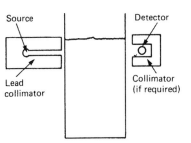

FIGURE 30.10 Level gauge (fixed).

radiation which could be a hazard to people working in the vicinity of such gauges. Because of cost, Geiger counters are the most usual detectors used, though they are not as efficient as scintillation counters. The important criterion in the design of a level gauge is to select a radioactive source to give the optimum path difference signal when the material or liquid to be monitored just obscures the beam from source to detector. The absorption of the beam by the wall of the container must be taken into account, as well as the absorption of the material being measured. A single source and single detector will provide a single response, but by using two detectors with a single source (Figure 30.11), it may be able to provide three readings: low (both detectors operating), normal (one detector operating), and high (neither detector operating). This system is used in papermaking systems to measure the level of the hot pulp.

Another level gauge to give a continuous indication of material or liquid level has the detector and the source mounted on a servocontrolled platform which follows the level of the liquid or material in the container (Figure 30.12). This provides a continuous readout of level, and it is possible to use this signal to control the amount of material or liquid entering or leaving the container in accordance with some preprogrammed schedule.

Another level gauge uses a radioactive source inside the container but enclosed in a float which rises and falls with the level of the liquid inside the container. An external detector can then observe the source inside the container, indicate the level and initiate refilling procedure to keep the level at a predetermined point.

Portable level gauges consisting of radioactive source and Geiger detector in a hand-held probe, with the electronics and counting display in a portable box, have been made to detect the level of liquid CO_2 in high-pressure cylinders. This is a much simpler method than that of weighing the cylinder and subtracting this value from the weight when the cylinder was originally received.

30.3.1.2 Using Neutrons

Some industrial materials have a very low atomic number (Z), such as oils, water, plastics, etc., and by using a beam of fast neutrons from a source such as ^{241}Am/Be or ^{252}Cf, and a suitable thermal-neutron detector such as cerium-activated lithium glass in a scintillation counter, or ^{10}BF$_3$ or ^{3}He gas-filled counters, it is possible to measure the level of such material (Figure 30.13). Fast neutrons from the source are moderated or slowed down by the low-Z material and some are scattered back into the detector. By mounting both the source of fast neutrons and the slow neutron detector at one side of the vessel the combination may be used to follow a varying level using a servocontrolled mechanism, or in a fixed position to control the level in the container to a preset position. This device is also usable as a portable detector system to find blockages in pipes, valves, etc., which often occur in plastics-manufacturing plants.

30.3.2 Measurement of Flow

There are several methods of measuring flow using radioactive sources, as follows.

30.3.2.1 Dilution Method

This involves injection of a liquid containing radioactivity into the flow line at a known constant rate: samples are taken further down the line where it is known that lateral mixing has been completed. The ratio of the concentrations of the radioactive liquid injected into the line and that of the samples allows the flow rate to be computed.

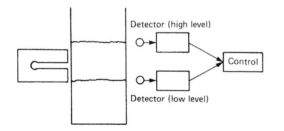

FIGURE 30.11 Dual detector level gauge.

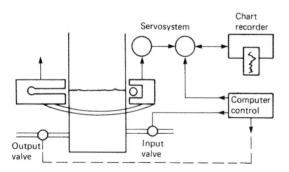

FIGURE 30.12 Level gauge (continuous) with automatic control.

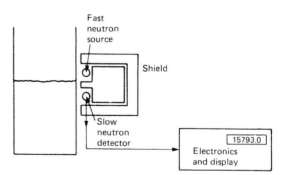

FIGURE 30.13 Level measurement by moderation of fast neutrons.

30.3.2.2 The "Plug" Method

This involves injecting a radioactive liquid into the flow line in a single pulse. By measuring the time this "plug" of radioactive liquid takes to pass two positions a known distance apart, the flow can be calculated.

A variation of the "plug" method again uses a single pulse of radioactive liquid injected into the stream, but the measurement consists of taking a sample at a constant rate from the line at a position beyond which full mixing has been completed. Here the flow rate can be calculated by measuring the average concentration of the continuous sample over a known time.

30.3.3 Mass and Thickness

Since the quantitative reaction of particles and photons depends essentially on the concentration and mass of the particles with which they are reacting it is to be expected that nuclear techniques can provide means for measuring such things as mass. We have already referred to the measurement of density of the material near a borehole. We now describe some other techniques having industrial uses.

30.3.3.1 Measurement of Mass, Mass Per Unit Area, and Thickness

The techniques employed in these measurements are basically the same. The radiation from a gamma ray source falls on the material and the transmitted radiation is measured by a suitable detector. In the nucleonic belt weigher shown in Figure 30.14, designed to measure the mass flow rate of granular material such as iron ore, limestone, coke, cement, fertilizers, etc., the absorption across the total width of the belt is measured. The signal representing the total radiation falling on the detector is processed with a signal representing the belt speed by a solid-state electronic module and displayed as a mass flow rate and a total mass. The complete equipment comprises a C frame assembly housing the source, consisting of ^{137}Cs enclosed in a welded steel capsule mounted in a shielding container with a radiation shutter, and the detector, a scintillation counter whose sensitive length matches the belt width, housed in a cylindrical flame-proof enclosure suitable for Groups 11 A and B gases, with the preamplifier. A calibration plate is incorporated with the source to permit a spot check at a suitable point within the span. In addition, there is a dust- and moisture-proof housing for the electronics which may be mounted locally or up to 300m from the detector.

The precision of the measurement is better than ±1 percent, and the operating temperature of the detector and electronics is −10 to +40°C. The detector and preamplifier may be serviced by unclassified staff, as the maximum dose rate is less than 7.5 μGy/h.

Similar equipment may be used to measure mass per unit area by restricting the area over which the radiation falls to a finite area, and if the thickness is constant and known the reading will be a measure of the density.

30.3.3.2 Measurement of Coating Thickness

In industry a wide variety of processes occur where it is necessary to measure and sometimes automatically control the thickness of a coating applied to a base material produced in strip form. Examples of such processes are the deposition of tin, zinc, or lacquers on steel, or adhesives, wax, clay bitumen, or plastics to paper, and many other processes.

By nucleonic methods measurement to an accuracy of ±1 percent of coating thickness can be made in a wide variety of circumstances by rugged equipment capable of a high reliability. Nucleonic coating-thickness gauges are based on the interaction of the radiation emitted from a radioisotope source with the material to be measured. They consist basically of the radioisotope source in a radiation shield and a radiation detector contained in a measuring head, and an electric console.

When the radiation emitted from the source is incident on the subject material, part of this radiation is scattered, part is absorbed, and the rest passes through the material. A part of the absorbed radiation excites characteristic fluorescent X-rays in the coating and/or backing.

Depending on the measurement required, a system is used in which the detector measures the intensity of scattered, transmitted, or fluorescent radiation. The intensity of radiation monitored by the detector is the measure of the thickness (mass per unit area) of the coating. The electric console contains units which process the detector signal and indicate total coating thickness and/or deviation from the target thickness. The measuring head may be stationary

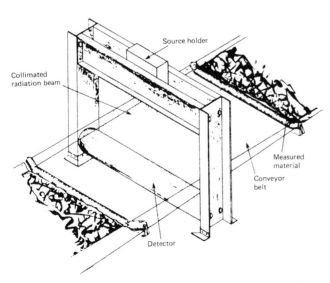

FIGURE 30.14 Nucleonic belt weigher. Courtesy of Nuclear Enterprises Ltd.

or programmed to scan across the material. Depending on the type and thickness of coating and base materials, and machine details, one of four gauge types is selected: differential beta transmission, beta backscatter, X-ray fluorescence, and preferential absorption.

Differential Beta-transmission Gauge (Figure 30.15) The differential beta-transmission gauge is used to measure coating applied to base materials in sheet form when the coating has a total weight of not less than about one-tenth of the weight of the base material, when both sides of the base and coated material are accessible, and when the composition of coating and base is fairly similar. Here the thickness (mass per unit area) of the coating is monitored by measuring first the thickness of the base material before the coating is applied, followed by the total thickness of the material with its coating, and then subtracting the former from the latter. The difference provides the coating thickness. The readings are obtained by passing the uncoated material through one measuring head and the coated material through the other, the coating being applied between the two positions. The intensity of radiation transmitted through the material is a measure of total thickness. Separate meters record the measurement determined by each head, and a third meter displays the difference between the two readings, which corresponds to the coating thickness.

Typical applications of this gauge are the measurements of wax and plastics coatings applied to paper and aluminum sheet or foil, or abrasives to paper or cloth.

Beta-backscatter Gauge (Figure 30.16) The beta-backscatter gauge is used to measure coating thickness when the process is such that the material is only accessible from one side and when the coating and backing material are of substantially different atomic number. The radioisotope source and the detector are housed in the same enclosure. Where radiation is directed, for example, on to an uncoated calender roll it will be backscattered and measurable by the detector. Once a coating has been applied to the roll the intensity of the backscattered radiation returning to the detector will change. This change is a measure of the thickness of the coating. Typical applications of this gauge are the measurement of rubber and adhesives on calenders, paper on rollers, or lacquer, paint, or plastics coatings applied to sheet steel.

Measurement of Coating and Backing by X-ray Fluorescence X-ray fluorescent techniques, employing radioisotope sources to excite the characteristic fluorescent radiation, are normally used to measure exceptionally thin coatings. The coating-fluorescence gauge monitors the increase in intensity of an X-ray excited in the coating as the coating thickness is increased. The backing-fluorescence gauge excites an X-ray in the backing or base material and measures the decrease in intensity due to attenuation in the coating as the coating thickness is increased. The intensity of fluorescent radiation

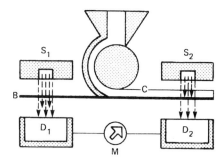

FIGURE 30.15 Differential beta-transmission gauge. Courtesy Nuclear Enterprises Ltd. S_1: first source; D_1: first detector; S_2: second source; D_2: second detector; B: base material; C: coating; M: differential measurement indicator.

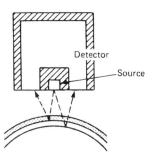

FIGURE 30.16 Beta-backscatter gauge. Courtesy of Nuclear Enterprises Ltd.

is normally measured with an ionization chamber, but a proportional or scintillation counter may sometimes be used.

By the use of compact geometry, high-efficiency detectors, and a fail-safe radiation shutter, the dose rates in the vicinity of the measuring head are kept well below the maximum permitted levels, ensuring absolute safety for operators and maintenance staff.

Figure 30.17(a) illustrates the principle of the coating-fluorescence gauge, which monitors the increase in intensity of an X-ray excited in the coating as the coating thickness is increased.

The instrument is used to measure tin, zinc, aluminum, and chromium coatings applied to sheet steel, or titanium coatings to paper or plastics sheet.

The Preferential Absorption Gauge (Figure 30.18) This gauge is used when the coating material has a higher mean atomic number than the base material. The gauge employs low-energy X-rays from a sealed radioisotope source which are absorbed to a much greater extent by materials with a high atomic number, such as chlorine, than by materials such as paper or textiles, which have a low atomic number. It is thus possible to monitor variations in coating thickness by measuring the degree of preferential X-ray absorption by the coating, using a single measuring head. The instrument is used to measure coatings which contain clay, titanium, halogens, iron or other substances with a high atomic number which have been applied to plastics, paper, or textiles.

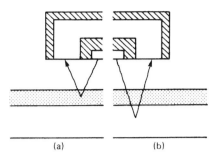

FIGURE 30.17 X-ray fluorescence gauge. Courtesy of Nuclear Enterprises Ltd. (a) Coating-fluorescence gauge which monitors the increase in intensity of X-rays excited in coating as its thickness increases; (b) backing-fluorescence gauge monitors the decrease in intensity of radiation excited in the backing materials as the coating thickness increases.

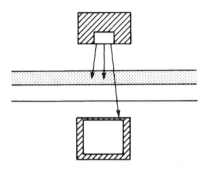

FIGURE 30.18 Preferential absorption gauge. Courtesy Nuclear Enterprises Ltd.

30.4 MISCELLANEOUS MEASUREMENTS

30.4.1 Field-survey Instruments

In prospecting for uranium, portable instruments are used (1) in aircraft, (2) in trucks or vans, (3) hand-held and (4) for undersea surveys. Uranium is frequently found in the same geological formations as oil, and uranium surveys have been used to supplement other more conventional methods of surveying, such as seismic analyses. The special case of surveying for beryllium-bearing rocks was discussed in Section 4.2.1.

As aircraft can lift large loads, the detectors used in such surveys have tended to be relatively large NaI (Tl) groups of detectors. For example, one aircraft carried four NaI (Tl) assemblies, each 29.2 cm diameter and 10 cm thick, feeding into a four- or five-channel spectrometer which separately monitored the potassium, uranium, thorium, and the background. Simultaneously a suitable position-finding system such as Loran-C is in operation, so that the airborne plot of the radioactivities, as the aircraft flies over a prescribed course, is printed with the position of the aircraft onto the chart recorder. In this way large areas of land or sea, which may contain ground-based survey teams with suitable instruments can survey the actual areas pinpointed in the aerial survey as possible sources of uranium.

30.4.1.1 Land-based Radiometrical Surveys

Just as the aircraft can carry suitable detector systems and computing equipment, so can land-based vehicles, which can be taken to the areas giving high-uranium indications. While similar electronics can usually be operated from a motor vehicle, the detectors will have to be smaller, especially if the terrain is very rugged, when manually portable survey monitors will be called for. These also can now incorporate a small computer, which can perform the necessary analyses of the signals received from potassium, uranium, thorium, and background.

30.4.1.2 Undersea Surveys

Measurement of the natural gamma radiation from rocks and sediments can be carried out using a towed seabed gamma-ray spectrometer. The spectrometer, developed by UKAEA Harwell in collaboration with the Institute of Geological Sciences, has been used over the last ten years and has traversed more than 10,000 km in surveys of the United Kingdom continental shelf. It consists of a NaI (Tl) crystal-photo-multiplier detector assembly, containing a crystal 76 mm diameter × 76 mm or 127 mm long, with an EMI type 9758 photomultiplier together with a preamplifier and high-voltage generator which are potted in silicone rubber. The unit is mounted in a stainless steel cylinder, which in turn is mounted in a 30-m long flexible PVC hose 173 mm diameter, which is towed by a cable from the ship, and also contains suitable ballast in the form of steel chain to allow the probe to be dragged over the surface of the seabed without becoming entangled in wrecks or rock outcrops.

The electronics on board the ship provide four channels to allow potassium, uranium, and thorium, as well as the total gamma radioactivity to be measured and recorded on suitable chart recorders and teletypes, and provision is also made to feed the output to a computer-based multichannel analyzer.

30.4.2 Dating of Archaeological or Geological Specimens

30.4.2.1 Radiocarbon Dating by Gas-proportional or Liquid-scintillation Counting

The technique of radiocarbon dating was discovered by W. F. Libby and his associates in 1947, when they were investigating the radioactivity produced by the interaction of cosmic rays from outer space with air molecules. They discovered that interactions with nitrogen in the atmosphere produced radioactive ^{14}C which quickly transformed into $^{14}CO_2$, forming about 1 percent of the total CO_2 in the world. As cosmic rays have been bombarding the earth at a steady rate over millions of years, forming some two atoms of ^{14}C per square centimeter of the earth's surface per second, then an equilibrium stage should have been reached, as the half-life (time for a radioactive substance to decay to half its original value) of ^{14}C is some 5,000 years. As CO_2 enters all living matter the distribution should be uniform. However, when the human being, animal, tree, or plant dies, CO_2 transfer ceases, and the carbon already present in the now-dead object is fixed. The ^{14}C in this carbon will therefore start to decay with a half-life of some 5,000 years, so that measurement of the ^{14}C present in the sample allows one to determine, by the amount of ^{14}C still present, the time elapsed since the death of the person, animal, tree, or plant. We expect to find two disintegrations per second for every 8 g of carbon in living beings or dissolved in sea water or in the atmosphere CO_2 for the total carbon in these three categories adds to 8 (7.5 in the oceans, 0.125 in the air, 0.25 in life forms, and perhaps 0.125 in humus).

There are several problems associated with radiocarbon dating which must be overcome before one can arrive at an estimated age for a particular sample. First, the sample must be treated so as to release the ^{14}C in a suitable form for counting. Methods used are various, depending on the final form in which the sample is required and whether it is to be counted in a gas counter (Geiger or proportional) or in a liquid-scintillation counter.

One method is to transform the carbon in the sample by combustion into a gas suitable for counting in a gas-proportional counter. This can be carbon dioxide (CO_2), methane (CH_4), or acetylene (C_2H_2). The original method used by Libby, in which the carbon sample was deposited in a thin layer inside the gas counter, has been superseded by the gas-combustion method. In this the sample is consumed by heating in a tube furnace or, in an improved way, in an oxygen bomb. The gas can be counted directly in a gas proportional counter, after suitable purification, as CO_2 or CH_4, or it can be transformed into a liquid form such as benzene, when it can be mixed with a liquid scintillator and measured in a liquid-scintillation counter.

Counting Systems When one considers that there are, at most, only two ^{14}C disintegrations per second from each 8 g of carbon, producing two soft beta particles (E_{max} = 0.156 MeV), one can appreciate that the counting is, indeed, very "low-level." The natural counting rate (unshielded) of a typical gas proportional counter 15 cm diameter $\times$ 60 cm long would be some 75.6 counts per second, whereas the signal due to 1 atm of live modern CO_2 or CH_4 filling the counter would be only 0.75 count per second. In order to achieve a standard deviation of 1 percent in the measurement, 10,000 counts would have to be measured, and at the rate of 0.75 count per second the measurement would take 3.7 days.

It is immediately apparent that the natural background must be drastically reduced and, if possible, the sample size increased to improve the counting characteristics (Watt and Ramsden, 1964).

Background is due to many causes, some of the most important being:

1. Environmental radioactivity from walls, air, rocks, etc.;
2. Radioactivities present in the shield itself;
3. Radioactivities in the materials used in the manufacture of the counters and associated devices inside the shield;
4. Cosmic rays;
5. Radioactive contamination of the gas or liquid scintillator itself;
6. Spurious electronic pulses, spikes due to improper operation or pick-up of electromagnetic noise from the electricity mains supply, etc.

Calculation of a Radiocarbon Date Since the measurement is to calculate the decay of ^{14}C, we have the relation

$$I = I_0 \exp-(\lambda t) \tag{30.8}$$

where I is the activity of the sample when measured, I_0 the original activity of the sample (as reflected by a modern standard), λ the decay constant = $0.693/T_{1/2}$ (where $T_{1/2}$ = half-life) and t the time elapsed. If $T_{1/2}$ = 5,568 yr (the latest best value found for the half-life of ^{14}C is 5,630 yr, but internationally it has been agreed that all dates are still referred to 5,568 yr to avoid the confusion which would arise if the volumes of published dates required revision) then Equation (30.8) may be rewritten as

$$t = 8033 \log_e \frac{S_s - B}{S_0 - B} \tag{30.9}$$

where S_s is the count rate of sample, S_0 the count rate of modern sample, and B the count rate of dead carbon. A modern carbon standard of oxalic acid, 95 percent of which is equivalent to 1890 wood, is used universally, and is available from the National Institute of Standards and Testing (NIST) in Gaithersburg, Maryland.

Anthracite coal, with an estimated age of 2×10^9 yr, can be used to provide the dead carbon background. Corrections must also be made for isotopic fractionation which can occur both in nature and during the various chemical procedures used in preparing the sample.

Statistics of Carbon Dating The standard deviation σ of the source count rate when corrected for background is given by

$$\sigma = \left[\frac{S}{T - t_b} + \frac{B}{t_b}\right]^{\frac{1}{2}} \tag{30.10}$$

where S is the gross count of sample plus background, B the background counted for a time t_b and T the total time available.

In carbon dating $S \approx 2B$ and the counting periods for sample and background are made equal, usually of the order of 48 h. Thus if $t_1 = t_b = T_{1/2}$ and $S = D + B$ we have

$$\sigma_D = \left(\frac{D + 2B}{t}\right)^{\frac{1}{2}} \tag{30.11}$$

The maximum age which can be determined by any specific system depends on the minimum sample activity which can be detected. If the "2σ criterion" is used, the minimum sample counting rate detectable is equal to twice the standard deviation and the probability that the true value of D lies within the region $\pm 2\sigma_D$ is 95.5 percent. Some laboratories prefer to use the "4σ criterion," which gives a 99.99 percent probability that the true value is within the interval $\pm 4\sigma$. If $D_{min} = 2\sigma_D$, then

$$D_{min} = 2\sqrt{\left(\frac{D_{min} + 2B}{t}\right)} \quad (30.12)$$

then the maximum dating age T_{max} which can be achieved with the system can be estimated as follows. From Equation (30.9)

$$T_{max} = \frac{T_{\frac{1}{2}}}{\log_e 2} \log_e \frac{D_0}{D_{min}}$$

$$= \frac{T_{\frac{1}{2}}}{\log_e 2} \log_e \left[\frac{D_0 \sqrt{t}}{2\sqrt{(D_{min} + 2B)}}\right] \quad (30.13)$$

As $D_m \ll 2B$ the equation can be simplified to

$$T_{max} = \frac{T_{\frac{1}{2}}}{\log_e 2} \log_e \left[\frac{D_0}{\sqrt{B}} \sqrt{\frac{t}{8}}\right] \quad (30.14)$$

where D_0 is the activity of the modern carbon sample, corrected for background. The ratio $D_0\sqrt{B}$ is considered the factor of merit for a system.

For a typical system such as that of Libby (1985)

$$t = 48\,\text{h} \quad \text{and} \quad T_{\frac{1}{2}} = 5568\,\text{yr}$$
$$D_0 = 6.7\,\text{cpm} \quad \text{and} \quad B = 5\,\text{cpm}, \quad \text{so}$$
$$D_0/\sqrt{B} \approx 3$$

Hence

$$T_{max} = \frac{5568}{\log_e 2} \log_e \left[3\sqrt{\left(\frac{48 \times 62}{8}\right)}\right]$$
$$= 8034 \times 4.038$$
$$= 32442\,\text{yr}$$

Calibration of the Radiocarbon Time Scale A number of corrections have to be applied to dates computed from the radiocarbon decay measurements to arrive at a "true" age. First, a comparison of age by radiocarbon measurement with the age of wood known historically initially showed good agreement in 1949. When the radiocarbon-dating system was improved by the use of CO_2 proportional counters, in which the sample was introduced directly into the counter, closer inspection using many more samples showed discrepancies in the dates calculated by various methods. Some of these discrepancies can be accounted for, such as the effect on the atmosphere of all the burning of wood and coal since the nineteenth century—Suess called this the "fossil-fuel effect." Alternative methods of dating, such as dendrochronology (the counting of tree rings), thermoluminescent dating, historical dating, etc., have demonstrated that variations do occur in the curve relating radiocarbon dating and other methods. Ottaway (1983) describes in more detail the problems and the present state of the art.

30.4.3 Static Elimination

Although not strictly instrumentation, an interesting application of radioactive sources is in the elimination of static electricity. In a number of manufacturing processes static electricity is produced, generally, by friction between, for example, a sheet of paper and the rollers used to pass it through a printing process. This can cause tearing at the high speeds that are found in printing presses. In the weaving industry, when a loom is left standing overnight, the woven materials and the warp threads left in the loom remain charged for a long period. It is found that dust particles become attracted to the cloth, producing the so-called "fog-marking," reducing thereby the value of the cloth. In rubber manufacture for cable coverings, automobile tires, etc., the rubber has to pass through rollers, where the static electricity so generated causes the material to remain stuck to the rollers instead of moving to the next processing stage.

All these static problems can be overcome, or at least reduced by mounting a suitable radioactive source close to the place where static is produced, so that ions of the appropriate sign are produced in quantities sufficient to neutralize the charges built up by friction, etc.

A wide variety of sources are now available in many shapes to allow them to be attached to the machines to give optimum ionization of the air at the critical locations. Long-strip sources of tritium, $^{90}Sr/^{90}Y$, and ^{241}Am are the most popular. The importance of preventing static discharges has been highlighted recently in two fields. The first is in the oil industry, where gas-filled oil tankers have exploded due to static discharges in the empty tanks. The second is in the microchip manufacturing industry, where static discharges can destroy a complete integrated circuit.

REFERENCES

Clayton, C. G., and Cameron, J. F., *Radioisotope Instruments*, Vol. 1, Pergamon Press, Oxford (1971) (Vol. 2 was not published). This has a very extensive bibliography for further study.

Gardner, R. P., and Ely, R. L., Jr, *Radioisotope Measurement Applications in Engineering*, Van Nostrand, New York (1967).

Hubbell, J. H., *Photon Cross-sections, Attenuation Coefficients, and Energy Absorption Coefficients from 10 keV to 100 GeV*, NS RDS NBS 29 (1969).

Libby, W. F., *Radiocarbon Dating*, University of Chicago Press (1955).

Ottaway, B. S. (ed.), "Archaeology, dendrochronology and the radiocarbon calibration curve," *Edinburgh University Occasional Paper No. 9* (1983).

Shumilovskii, N. N., and Mel'ttser, L. V., *Radioactive Isotopes in Instrumentation and Control*, Pergamon Press, Oxford (1964).

Sim, D. F., *Summary of the Law Relating to Atomic Energy and Radioactive Substances*, UK Atomic Energy Authority (yearly). Includes (a) *Radioactive Substances Act 1960*, (b) Factories Act—The Ionizing Radiations (Sealed Sources) Regulations 1969. (HMSO Stat. Inst. 1969, No. 808); (c) *Factories Act—The Ionizing Radiations Regulations*, HMSO (1985).

Watt, D. E., and Ramsden, D., *High Sensitivity Counting Techniques*, Pergamon Press, Oxford (1964).

Chapter 31

Non-Destructive Testing

Scottish School of Non-Destructive Testing

31.1 INTRODUCTION

The driving force for improvements and developments in non-destructive testing instrumentation is the continually increasing need to demonstrate the integrity and reliability of engineering materials, products, and plant. Efficient materials manufacture, the assurance of product quality, and re-assurance of plant at regular intervals during use represent the main need for non-destructive testing (NDT). This "state-of-health" knowledge is necessary for both economic and safety reasons. Indeed, in the UK the latter reasons have been strengthened by legislation such as the Health and Safety at Work Act 1974, and in the United States by the Occupational Safety and Health Administration (OSHA).

Failures in engineering components generally result from a combination of conditions, the main three being inadequate design, incorrect use, or the presence of defects in materials. The use of non-destructive testing seeks to eliminate the failures caused predominantly by defects. During manufacture these defects may, for example, be shrinkage and porosity in castings, laps and folds in forgings, laminations in plate material, lack of penetration and cracks in weldments. Alternatively, with increasing complexity of materials and conditions of service, less obvious factors may require control through NDT. For example, these features may be composition, microstructure, and homogeneity.

Non-destructive testing is not confined to manufacture. The designer and user may find application to on-site testing of bridges, pipelines in the oil and gas industries, pressure vessels in the power-generation industry, and in-service testing of nuclear plant, aircraft, and refinery installations. Defects at this stage may be deterioration in plant due to fatigue and corrosion.

The purpose of non-destructive testing during service is to look for deterioration in plant to ensure that adequate warning is given of the need to repair or replace. Periodic checks also give confidence that "all is well."

In these ways, therefore, non-destructive testing plays an important role in the manufacture and use of materials. Moreover, as designs become more adventurous and as new materials are used, there is less justification for relying on past experience. Accordingly, non-destructive testing has an increasingly important role.

Methods of non-destructive testing are normally categorized in terms of whether their suitability is primarily for the examination of the surface features of materials (Figure 31.1) or the internal features of materials (Figure 31.2). Closely allied to this is the sensitivity of each method, since different situations invariably create different levels of quality required. Consequently the range of applications is diverse. In the following account the most widely used methods of nondestructive testing are reviewed, together with several current developments.

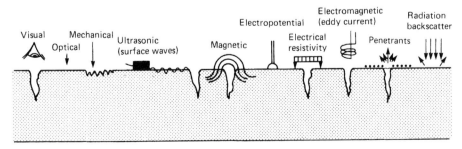

FIGURE 31.1 NDT methods for surface inspection.

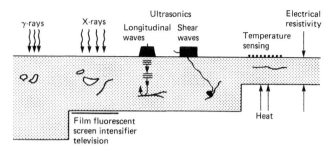

FIGURE 31.2 NDT methods for sub-surface inspection.

31.2 VISUAL EXAMINATION

For many types of components, integrity is verified principally through visual inspection. Indeed, even for components that require further inspection using ultrasonics or radiography visual inspection still constitutes an important aspect of practical quality control.

Visual inspection is the most extensively used of any method. It is relatively easy to apply and can have one or more of the following advantages:

1. Low cost;
2. Can be applied while work is in progress;
3. Allows early correction of faults;
4. Gives indication of incorrect procedures;
5. Gives early warning of faults developing when item is in use.

Equipment may range from that suitable for determining dimensional non-conformity, such as the Welding Institute Gauges (Figure 31.3), to illuminated magnifiers (Figure 31.4) and the more sophisticated fiberscope (Figure 31.5). The instrument shown in Figure 31.5 is a high-resolution flexible fiberscope with end tip and focus control. Flexible lengths from 1 m to 5 m are available for viewing inaccessible areas in boilers, heat exchangers, castings, turbines, interior welds, and other equipment where periodic or troubleshooting inspection is essential.

31.3 SURFACE-INSPECTION METHODS

The inspection of surfaces for defects at or close to the surface presents great scope for a variety of inspection techniques. With internal-flaw detection one is often limited to radiographic and ultrasonic techniques, whereas with surface-flaw detection visual and other electromagnetic methods such as magnetic particle, potential drop, and eddy current become available.

31.3.1 Visual Techniques

In many instances defects are visible to the eye on the surface of components. However, for the purposes of recording or

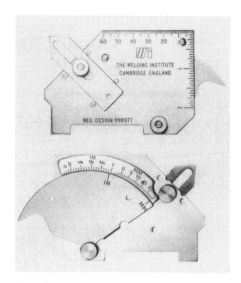

FIGURE 31.3 Gauges for visual inspection. Courtesy the Welding Institute.

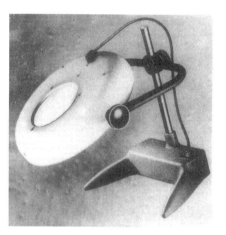

FIGURE 31.4 Illuminated magnifiers for visual inspection. Courtesy P. W. Allen & Co.

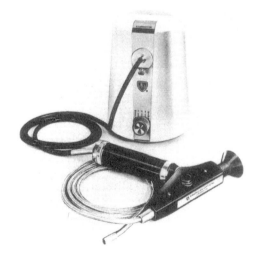

FIGURE 31.5 High-resolution flexible fiberscope. Courtesy P. W. Allen & Co.

gaining access to difficult locations, photographic and photomicrographic methods can be very useful. In hazardous environments, as encountered in the nuclear and offshore fields, remote television cameras coupled to video recorders allow inspection results to be assessed after the test. When coupled to remote transport systems these cameras can be used for pipeline inspection, the cameras themselves being miniaturized for very narrow pipe sections.

When surface-breaking defects are not immediately apparent, their presence may be enhanced by the use of dye penetrants. A penetrating dyeloaded liquid is applied to a material surface where, due to its surface tension and wetting properties, a strong capillary effect exists, which causes the liquid to penetrate into fine openings on the surface. After a short time (about 10 minutes), the surface is cleaned and an absorbing powder applied which blots the dye penetrant liquid, causing a stain around the defects. Since the dye is either a bright red or fluorescent under ultraviolet light, small defects become readily visible. The penetrant process itself can be made highly automated for large-scale production, but still requires trained inspectors for the final assessment. To achieve fully automated inspection, scanned ultraviolet lasers which excite the fluorescent dye and are coupled to photodetectors to receive the visible light from defect indications are under development.

31.3.2 Magnetic Flux Methods

When the material under test is ferromagnetic the magnetic properties may be exploited to provide testing methods based on the localized escape of flux around defects in magnetized material. For example, when a magnetic flux is present in a material such as iron, below magnetic saturation, the flux will tend to confine itself within the material surface. This is due to the continuity of the tangential component of the magnetic field strength, H, across the magnetic boundary. Since the permeability of iron is high, the external flux density, B_{ext}, is small (Figure 31.6(a)). Around a defect, the presence of a normal component of B incident on the defect will provide continuity of the flux to the air, and a localized flux escape will be apparent (Figure 31.6(b)). If only a tangential component is present, no flux leak occurs, maximum leakage conditions being obtained when B is normal to the defect.

31.3.2.1 Magnetization Methods

To detect flux leakages, material magnetization levels must be substantial (values in excess 0.72 Tesla for the magnetic flux density). This, in turn, demands high current levels. Applying Ampere's current law to a 25-mm diameter circular bar of steel having a relative permeability of 240, the current required to achieve a magnetic field strength of 2400 A/m at the surface giving the required flux value is 188 A peak current.

Such current levels are applied either as ac current from a step-down transformer whose output is shorted by the specimen or as halfwave rectified current produced by diode rectification. Differences in the type of current used become apparent when assessing the "skin depth" of the magnetic field. For ac current in a magnetic conductor, magnetic field penetration, even at 50 Hz, is limited; the dc component present in the half-wave rectified current produces a greater depth of penetration.

Several methods of achieving the desired flux density levels at the material surface are available. They include the use of threading bars, coils, and electromagnets (Figure 31.7).

31.3.2.2 Flux-Leakage Detection

One of the most effective detection systems is the application of finely divided ferric-oxide particles to form an indication by their accumulation around the flux leakage. The addition of a fluorescent dye to the particle enables the indication to be easily seen under ultraviolet light. Such a system, however, does not sustain recording of defect information except through photographic and replication techniques such as strippable magnetic paint and rubber. Alternative flux-detection

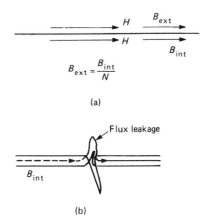

FIGURE 31.6 Principle of magnetic flux test.

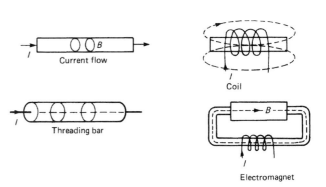

FIGURE 31.7 Ways of inducing flux.

techniques are becoming available, such as the application of modified magnetic recording tape which is wrapped around the material before magnetization. After the test, the tape can be unwound and played through a tape unit, which detects the presence of the recorded flux leakages.

For more dynamic situations, such as in the online testing of tube and bar materials, a faster detection technique for the recording of indications is required. A small detector head, comprising a highly permeable yoke on which a number of turns of wire are wrapped, can be used to detect small flux leakages by magnetic induction as the flux leak passes under the detector (Figure 31.8). The material motion can be linked to a chart recorder showing impulses as they occur.

31.3.3 Potential Drop Techniques

The measurement of material resistance can be related to measurements of the depth of surfacebreaking cracks. A four-point probe head (Figure 31.9) is applied to a surface and current passed between the outer probes. The potential drop across the crack is measured by the two inner probes and, as the crack depth increases, the greater current path causes an increasing potential drop. By varying probe spacing, maximum sensitivity to changes in crack depth can be obtained. In addition, the application of ac current of varying frequency permits the depth of current penetration beneath the surface to be varied due to the "skin effect" (see below).

31.3.4 Eddy-Current Testing

A powerful method of assessing both the material properties and the presence of defects is the eddycurrent technique. A time-changing magnetic field is used to induce weak electrical currents in the test material, these currents being sensitive to changes in both material conductivity and permeability. In turn, the intrinsic value of the conductivity depends mainly on the material composition but is influenced by changes in structure due to crystal imperfections (voids or interstitial atoms); stress conditions; or work hardening dependent upon the state of dislocations in the material. Additionally, the presence of discontinuities will disturb the eddy-current flow patterns giving detectable changes.

The usual eddy-current testing system comprises a coil which due to the applied current produces an ac magnetic field within the material. This, in turn, excites the eddy currents which produce their own field, thus altering that of the current (Figure 31.10). This reflects also in the impedance of the coil, whose resistive component is related to eddy-current losses and whose inductance depends on the magnetic circuit conditions. Thus, conductivity changes will be reflected in changes in coil resistance, whilst changes in permeability or in the presentation of the coil to the surface will affect the coil inductance.

The frequency of excitation represents an important test parameter, due to the "skin effect" varying the depth of current penetration beneath the surface. From the skin-depth formula

$$\delta = \frac{1}{\sqrt{(\pi f \mu \sigma)}}$$

where f is the frequency, μ the permeability, and σ the conductivity, it can be seen that in ferromagnetic material the skin depth, δ, is less than in non-magnetic materials by the large factor of the square root of the permeability.

By selection of the appropriate frequency, usually in the range 100 kHz to 10 MHz, the detection of discontinuities and other subsurface features can be varied. The higher the frequency, the less the depth of penetration. In addition, in ferromagnetic material the ability of an ac magnetic field to bias the material into saturation results in an incremental permeability close to that of the non-magnetic material.

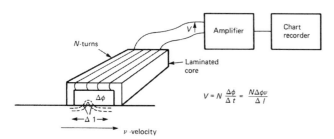

FIGURE 31.8 Detection of flux leakage. Δl: flux-leakage width; $\Delta \phi$: flux-leakage magnitude; V: induced voltage.

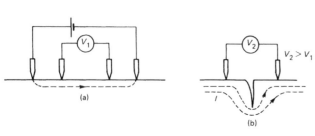

FIGURE 31.9 Probe for potential drop technique.

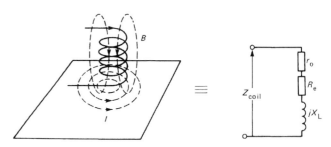

FIGURE 31.10 Principle of eddy-current testing. $Z_{coil} = (r_0 + R_e) + jX_L$; R_e are the additional losses due to eddy-current flow.

The eddy-current method therefore represents a very general testing technique for conducting materials in which changes in conductivity, permeability, and surface geometry can be measured.

31.3.4.1 Eddy-Current Instrumentation

In eddy-current testing the coils are incorporated into a balanced-bridge configuration to provide maximum detection of small changes in coil impedance reflecting the material changes. The simplest type of detector is that which measures the magnitude of bridge imbalance. Such units are used for material comparison of known against unknown and for simple crack detection (Figure 31.11).

A more versatile type of unit is one in which the magnitude and phase of the coil-impedance change is measured (Figure 31.12), since changes in inductance will be 90° out of phase with those from changes in conductivity. Such units as the vector display take the bridge imbalance voltage $V_0 e^{j(\omega t + \phi)}$ and pass it through two quadrature phase detectors. The 0° reference detector produces a voltage proportional to $V_0 \cos \phi$, whilst the 90° detector gives a voltage of $V_0 \sin \phi$. The vector point displayed on an X–Y storage oscilloscope represents the magnitude and phase of the voltage imbalance and hence the impedance change.

To allow positioning of the vector anywhere around the screen a vector rotator is incorporated using sine and cosine potentiometers. These implement the equation

$$V'_x = V_x \cos \phi - V_y \sin \phi$$
$$V'_y = V_x \sin \phi + V_y \cos \phi$$

where V'_x and V'_y are the rotated X and Y oscilloscope voltages and V_x and V_y are the phasedetector outputs.

In setting up such a unit the movement of the spot during the lift-off of the probe from the surface identifies the magnetic circuit or permeability axis, whereas defect deflection and conductivity changes will introduce a component primarily along an axis at right angles to the permeability axis (Figure 31.13).

Additional vector processing can take place to remove the response from a known geometrical feature. Compensation probes can produce a signal from the feature, such as support members, this signal being subtracted from that of the defect plus feature. Such cancellation can record small defect responses which are being masked by the larger geometrical response. Also, a number of different frequencies can be applied simultaneously (multi-frequency testing) to give depth information resulting from the response at each frequency.

The third type of testing situation for large-scale inspection of continuous material comprises a testing head through which the material passes. A detector system utilizing the phase detection units discussed previously is set up to respond to known defect orientations. When these defect signals

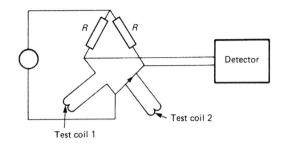

FIGURE 31.11 Simple type of eddy-current detector.

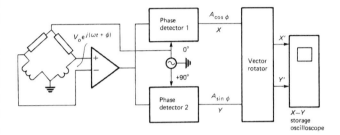

FIGURE 31.12 Eddy-current detection with phase discrimination.

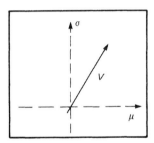

FIGURE 31.13 CRT display for eddy-current detection.

exceed a predetermined threshold it is recorded along with the tube position on a strip chart, or the tube itself is marked with the defect position.

31.4 ULTRASONICS

31.4.1 General Principles of Ultrasonics

Ultrasonics, when applied to the non-destructive testing of an engineering component, relies on a probing beam of energy directed into the component interacting in an interpretable way with the component's structural features. If a flaw is present within the metal, the progression of the beam of energy is locally modified and the modification is detected and conveniently displayed to enable the flaw to be diagnosed. The diagnosis largely depends on a knowledge of the nature of the probing energy beam, its interaction with the structural features of the component under test, and the manufacturing history of the component.

The ultrasonic energy is directed into the material under test in the form of mechanical waves or vibrations of very high frequency. Although its frequency may be anything in excess of the upper limit of audibility of the human ear, or 20 kHz, ultrasonic non-destructive testing frequencies normally lie in the range 0.5–10 MHz. The equation

$$\lambda = \frac{V}{f}$$

where λ is the wavelength, V the velocity, and f the frequency, highlights this by relating wavelength and frequency to the velocity in the material.

The wavelength determines the defect sensitivity in that any defect dimensionally less than half the wavelength will not be detected. Consequently the ability to detect small defects increases with decreasing wavelength of vibration and, since the velocity of sound is characteristic of a particular material, increasing the frequency of vibration will provide the possibility of increased sensitivity. Frequency selection is thus a significant variable in the ability to detect small flaws.

The nature of ultrasonic waves is such that propagation involves particle motion in the medium through which they travel. The propagation may be by way of volume change, the compression wave form, or by a distortion process, the shear wave form. The speed of propagation thus depends on the elastic properties and the density of the particular medium.

Compression wave velocity

$$V_c = \sqrt{\left(\frac{E}{\rho} \cdot \frac{1-\mu}{(1+\mu)(1-2\mu)}\right)}$$

Shear wave velocity

$$V_s = \sqrt{\left(\frac{E}{\rho} \cdot \frac{1}{(2+\mu)}\right)} = \sqrt{\frac{G}{\rho}}$$

where E is the modulus of elasticity, ρ the density, μ Poisson's ratio, and G the modulus of shear.

Other properties of ultrasonic waves relate to the results of ultrasound meeting an interface, i.e., a boundary wall between different media. When this occurs, some of the wave is reflected, the amount depending on the acoustic properties of the two media and the direction governed by the same laws as for light waves. If the ultrasound meets a boundary at an angle, the part of the wave that is not reflected is refracted, suffering a change of direction for its progression through the second medium. Energy may be lost or attenuated during the propagation of the ultrasound due to energy absorption within the medium and to scatter which results from interaction of the waves with microstructural features of size comparable with the wavelength. This is an important factor, as it counteracts the sensitivity to flaw location on the basis of frequency selection. Hence high frequency gives sensitivity to small flaws but may be limited by scatter and absorption to short-range detection.

The compression or longitudinal wave is the most common mode of propagation in ultrasonics. In this form, particle displacement at each point in a material is parallel to the direction of propagation. The propagating wavefront progresses by a series of alternate compressions and rarefactions, the total distance occupied by one compression and one rarefaction being the wavelength. Also commonly used are shear or transverse waves, which are characterized by the particle displacement at each point in a material being at right angles to the direction of propagation. In comparing these wave motions it should be appreciated that for a given material the shear waves have a velocity approximately five-ninths of that of compressional waves. It follows that for any frequency, the lower velocity of shear waves corresponds to a shorter wavelength. Hence, for a given frequency, the minimum size of defect detectable will be less in the case of shear waves.

Other forms of shear motion may be produced. Where there is a free surface a Rayleigh or surface wave may be generated. This type of shear wave propagates on the surface of a body with effective penetration of less than a wavelength. In thin sections bounded by two free surfaces a Lamb wave may be produced. This is a form of compressional wave which propagates in sheet material, its velocity depending not only on the elastic constant of the material but also on plate thickness and frequency. Such waveforms can be used in ultrasonic testing. A wave of a given mode of propagation may generate or transform to waves of other modes of propagation at refraction or reflection, and this may give rise to practical difficulties in the correct interpretation of test signals from the material.

Ultrasonic waves are generated in a transducer mounted on a probe. The transducer material has the property of expanding and contracting under an alternating electrical field due to the piezoelectric effect. It can thus transform electrical oscillations into mechanical vibrations and vice versa. Since the probe is outside the specimen to be tested, it is necessary to provide a coupling agent between probe and specimen. The couplant, a liquid or pliable solid, is interposed between probe surface and specimen surface, and assists in the passage of ultrasonic energy. The probe may be used to transmit energy as a transmitter, receive energy as a receiver, or transmit and receive as a transceiver. A characteristic of the transceiver or single-crystal probe is the dead zone, where defects cannot be resolved with any accuracy due to the transmission-echo width. Information on the passage of ultrasonic energy in the specimen under test is provided by way of the transducer, in the form of electrical impulses which are displayed on a cathode ray tube screen. The most commonly used presentation of the information is A-scan, where the horizontal base line represents distance or time intervals and the vertical axis gives signal amplitude or intensities of transmitted or reflected signals.

The basic methods of examination are transmission, pulse-echo, and resonance. In the transmission method an ultrasonic beam of energy is passed through a specimen and investigated by placing an ultrasonic transmitter on one face and a receiver on the other. The presence of internal flaws is indicated by a reduction in the amplitude of the received signal or a loss of signal. No indication of defect depth is provided.

Although it is possible with the pulse echo method to use separate probes it is more common to have the transmitter and receiver housed within the one probe, a transceiver. Hence access to one surface only is necessary. This method relies on energy reflected from within the material, finding its way back to the probe. Information is provided on the time taken by the pulse to travel from the transmitter to an obstacle, backwall, or flaw and return to the receiver. Time is proportional to the distance of the ultrasonic beam path, hence the cathode ray tube may be calibrated to enable accurate thickness measurement or defect location to be obtained (Figure 31.14(a)). For a defect suitably orientated to the beam direction an assessment of defect size can be made from the amplitude of the reflected signal. Compression probes are used in this test method to transmit compressional waves into the material normal to the entry surface. On the other hand, shear probes are used to transmit shear waves where it is desirable to introduce the energy into the material at an angle, and a reference called the skip distance may be used. This is the distance measured over the surface of the body between the probe index or beam exit point for the probe and the point where the beam axis impinges on the surface after following a double traverse path. Accurate defect location is possible using the skip distance and the beam path length (Figures 31.14(b) and (c)).

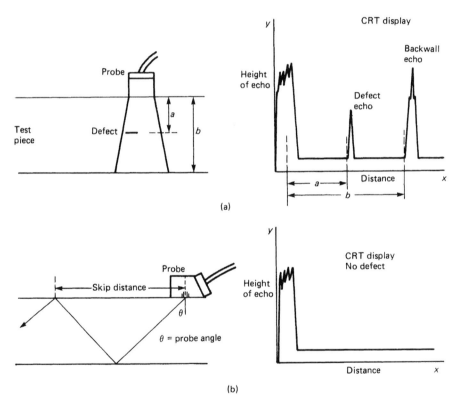

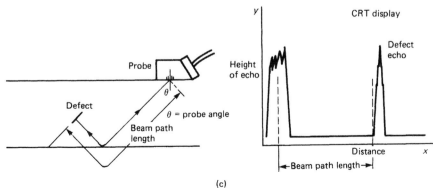

FIGURE 31.14 Displays presented by different ultrasonic probes and defects, (a) Distance (time) of travel–compression-wave examination; (b) skip distance–shear-wave examination; (c) beam path length (distance of travel)–shear-wave examination.

A condition of resonance exists when the thickness of a component is exactly equal to half the wavelength of the incident ultrasonic energy, i.e. the component will vibrate with its natural frequency. The condition causes an increase in the amplitude of the received pulse which can readily be identified. The condition of resonance can also be obtained if the thickness is a multiple of the half-wavelength. The resonance method consequently involves varying the frequency of ultrasonic waves to excite a maximum amplitude of vibration in a body or part of a body, generally for the purpose of determining thickness from one side only.

31.4.2 The Ultrasonic Test Equipment Controls and Visual Presentation

In most ultrasonic sets, the signal is displayed on a cathode ray tube (CRT). Operation of the instrument is similar in both the through transmission and pulse-echo techniques, and block diagrams of the test equipment are shown in Figures 31.15(a) and (b).

The master timer controls the rate of generation of the pulses or pulse repetition frequency (PRF) and supplies the timebase circuit giving the base line whilst the pulse generator controls the output of the pulses or pulse energy which is transmitted to the probe. At the probe, electrical pulses are converted via the transducer into mechanical vibrations at the chosen frequency and directed into the test material. The amount of energy passing into the specimen is very small. On the sound beam returning to the probe, the mechanical vibrations are reconverted at the transducer into electrical oscillations. This is known as the piezoelectric effect. In the main, transmitter and receiver probes are combined. At the CRT the timebase amplifier controls the rate of sweep of the beam across the face of the tube which, by virtue of the relationship between the distance traveled by ultrasonic waves in unit time, i.e., velocity, can be used as a distance depth scale when locating defects or measuring thickness. Signals coming from the receiver probe to the signal amplifier are magnified, this working in conjunction with an incorporated attenuator control. The signal is produced in the vertical axis. Visual display is by CRT with the transmitted and received signals in their proper time sequence with indications of relative amplitude.

A-, B-, and C-scan presentations can be explained in terms of CRT display (Figure 31.16). Interposed between the cathode and anode in the CRT is a grid which is used to limit the electron flow and hence control ultimate screen brightness as it is made more positive or negative. There is also a deflector system immediately following the electron gun which can deflect the beam horizontally and vertically over the screen of the tube. In A-scan presentation, deflector plate X and Y coordinates represent timebase and amplified response. However, in B-scan the received and amplified echoes from defects and from front and rear surfaces of the test plate are applied, not to the Y deflector plate as in normal A-scan, but to the grid of the CRT in order to increase the

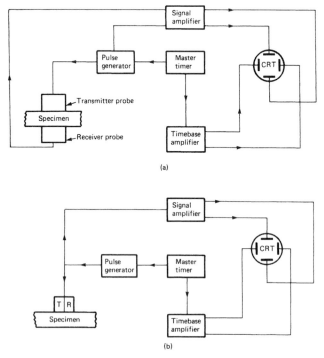

FIGURE 31.15 Block diagram of ultrasonic flaw detectors using (a) through transmission and (b) pulse-echo techniques.

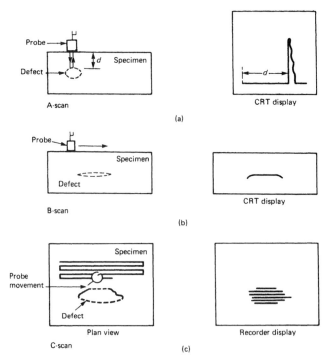

FIGURE 31.16 A-, B-, and C-scan presentations. Courtesy the Welding Institute.

brightness of the trace. If a signal proportional to the movement of the probe along the specimen is applied to the X deflector plates, the whole CRT display can be made to represent a cross section of the material over a given length of the sample (Figure 31.16(b)). A further extension to this type of presentation is used in C-scan, where both X and Y deflections on the tube follow the corresponding coordinates of probe traverse. Flaw echoes occurring within the material are gated and used to modulate the grid of the CRT and produce brightened areas in the regions occupied by the defects. A picture is obtained very like a radiograph but with greater sensitivity (Figure 31.17(c)). Figure 31.17 is a block diagram for equipment that can give an A- or B-scan presentation, and Figure 31.18 does the same for C-scan.

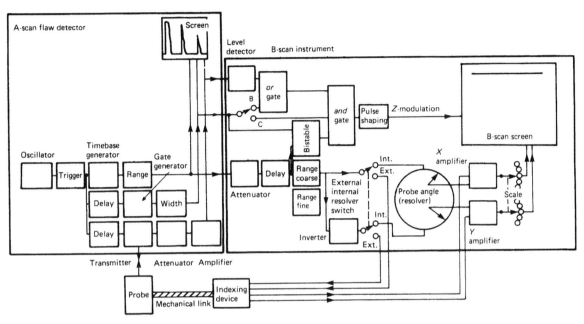

FIGURE 31.17 A- and B-scan equipment. Courtesy the Welding Institute.

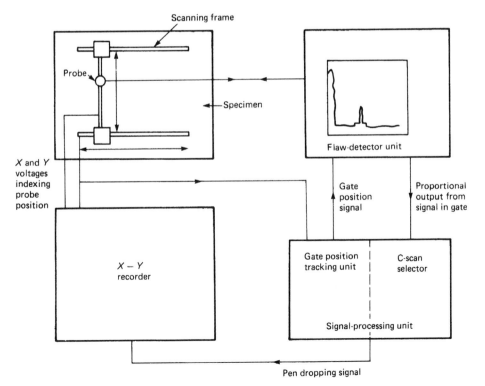

FIGURE 31.18 C-scan equipment. Courtesy the Welding Institute.

31.4.3 Probe Construction

In ultrasonic probe construction the piezoelectric crystal is attached to a non-piezoelectric front piece or shoe into which an acoustic compression wave is emitted (Figure 31.19(a)). The other side of the crystal is attached to a material which absorbs energy emitted in the backward direction. These emitted waves correspond to an energy loss from the crystal and hence increase the crystal damping. Ideally, one would hope that the damping obtained from the emitted acoustic wave in the forward direction would be sufficient to give the correct probe performance. Probes may generate either compression waves or angled shear waves, using either single or twin piezoelectric crystals. In the single-crystal compression probe the zone of intensity variation is not confined within the Perspex wear shoe. However, twin-crystal probes (Figure 31.19(b)) are mainly used since the dead zone or zone of non-resolution of the single form may be extremely long. In comparison, angle probes work on the principle of mode conversion at the boundary of the probe and workpiece. An angled beam of longitudinal waves is generated by the transducer and strikes the surface of the specimen under test at an angle. If the angle is chosen correctly, only an angled shear wave is transmitted into the workpiece. Again a twin crystal form is available as for compression probes, as shown in Figure 31.19(b).

Special probes of the focused, variable-angle, crystal mosaic, or angled compression wave type are also available (Figure 31.20). By placing an acoustic lens in front of a transducer crystal it is possible to focus the emitted beam in a similar manner to an optical lens. If the lens is given a cylindrical curvature, it is possible to arrange that, when testing objects in immersion testing, the sound beam enters normal to a cylindrical surface. Variable-angle probes can be adjusted to give a varying angle of emitted sound wave, these being of value when testing at various angles on the same workpiece. In the mosaic type of probe a number of crystals are laid side by side. The crystals are excited in phase such that the mosaic acts as a probe of large crystal size equal to that of the mosaic. Finally, probes of the angled compression type have been used in the testing of austenitic materials where, due to large grain structure, shear waves are rapidly attenuated. Such probes can be made by angling the incident compressional beam until it is smaller than the initial angle for emission of shear waves only. Although a shear wave also occurs in the test piece it is rapidly lost due to attenuation.

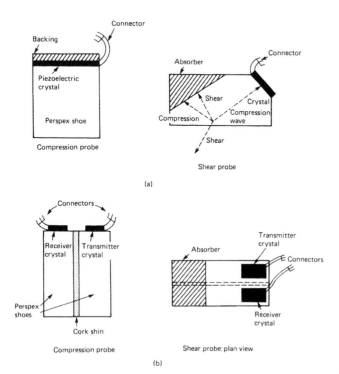

FIGURE 31.19 Single (a) and twin-crystal (b) probe construction.

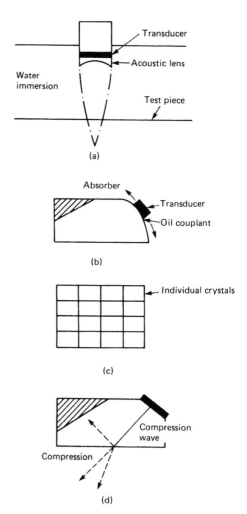

FIGURE 31.20 Special probe construction, (a) Focused probe; (b) variable-angle probe; (c) mosaic probe; (d) angled compression probe.

31.4.4 Ultrasonic Spectroscopy Techniques

Ultrasonic spectroscopy is a technique used to analyze the frequency components of ultrasonic signals. The origins of spectroscopy date back to Newton, who showed that white light contained a number of different colors. Each color corresponds to a different frequency, hence white light may be regarded as the sum of a number of different radiation frequencies, i.e., it contains a spectrum of frequencies. Short pulses of ultrasonic energy have characteristics similar to white light, i.e., they carry a spectrum of frequencies. The passage of white light through a material and subsequent examination of its spectrum can yield useful information regarding the atomic composition of the material. Likewise the passage of ultrasound through a material and subsequent examination of the spectrum can yield information about defect geometry, thickness, transducer frequency response, and differences in microstructure. The difference in the type of information is related to the fact that ultrasonic signals are elastic instead of electromagnetic waves.

Interpretation of ultrasonic spectra requires a knowledge of how the ultrasonic energy is introduced into the specimen. Two main factors determine this:

1. The frequency response characteristic of the transducer;
2. The output spectrum of the generator that excites the transducer.

The transducer response is important. Depending on the technique, one or two transducers may be used. The response depends on the mechanical behavior, and elastic constants and thickness normal to the disc surface are important. In addition, damping caused by a wear plate attached to the front of the specimen and transducer-specimen coupling conditions can be significant factors. Crystals which have a low Q value give a broad, flat response satisfactory for ultrasonic spectroscopy but suffer from a sharp fall-off in amplitude response, i.e., sensitivity.

Results indicate that crystal response is flat at frequencies well below the fundamental resonant frequency, f_o, suggesting that higher f_o values would be an advantage. However, there is a limit to the thickness of crystal material which remains robust enough to be practical. In general, some attempt is made to equalize the transducer response's effect, for instance, by changing the amplitude during frequency modulation.

A straightforward procedure can be adopted of varying frequency continuously and noting the response. Using such frequency-modulation techniques (Figure 31.21) requires one probe for transmission and one for reception. If a reflection test procedure is used, then both transducers may be mounted in one probe. Figure 31.21 shows a block outline of the system. The display on the CRT is then an amplitude versus frequency plot. Some modifications to the system include:

1. A detector and suitable filter between the wide band amplifier and oscilloscope to allow the envelope function to be displayed; and
2. Substitution of an electronically tuned rf amplifier to suppress spurious signals.

Alternatively, a range of frequencies may be introduced, using the spectra implicit in pulses. The output spectra for four types of output signal are shown in Figure 31.22. The types of output are:

1. Single dc pulse with reactangular shape;
2. Oscillating pulse with rectangular envelope and carrier frequency f_o;
3. dc pulse with an exponential rise and decay;
4. Oscillating pulse with exponential rise and decay and carrier frequency f_o.

From these results it can be seen that the main lobe of the spectrum contains most of the spectral energy, and its width

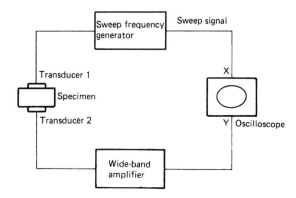

FIGURE 31.21 Frequency-modulation spectroscope.

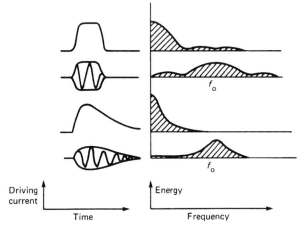

FIGURE 31.22 Pulse shape and associated spectra.

is inversely proportional to the pulse duration. In order to obtain a broad ultrasonic spectrum of large amplitude the excitation pulse must have as large an amplitude and as short a duration as possible. In practice, a compromise is required, since there is a limit to the breakdown strength of the transducer and the output voltage of the pulse generator. Electronic equipment used in pulse spectroscopy (Figure 31.23) differs considerably from that for frequency modulation, including a time gate for selecting the ultrasonic signals to be analyzed, and so allowing observations in the time and frequency domains.

A pulse frequency-modulation technique can also be used which combines both the previous procedures (Figure 31.24). The time gate opens and closes periodically at a rate considerably higher than the frequency sweep rate, hence breaking the frequency sweep signals into a series of pulses. This technique has been used principally for determining transducer response characteristics.

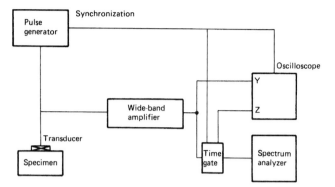

FIGURE 31.23 Pulse-echo spectroscope.

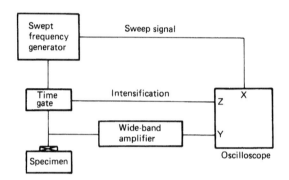

FIGURE 31.24 Pulsed frequency-modulation spectroscope.

31.4.5 Applications of Ultrasonic Spectroscopy

31.4.5.1 Transducer Response

It is known that the results of conventional ultrasonic inspection can vary considerably from transducer to transducer even if element size and resonant frequency remain the same. To avoid this difficulty it is necessary to control both the response characteristics and the beam profile during fabrication. To determine the frequency response both the pulse and pulsed frequency-modulation techniques are used. The test is carried out by analyzing the first backwall echo from a specimen whose ultrasonic attenuation has a negligible dependence on frequency and which is relatively thin, to avoid errors due to divergence of the ultrasonic beam. Analysis of the results yields response characteristics, typical examples of which are shown in Figure 31.25.

31.4.5.2 Microstructure

The attenuation of an ultrasonic signal as it passes through an amorphous or polycrystalline material will depend on the microstructure of the material and the frequency content of the ultrasonic signal. Differences in microstructure can therefore be detected by examining the ultrasonic attenuation spectra. The attenuation of ultrasound in polycrystalline materials with randomly oriented crystallites is caused mainly by scattering. The elastic properties of the single crystal and the average grain size determine the frequency dependence of the ultrasonic attenuation. Attenuation in glass, plastics, and other amorphous materials is characteristic of the material, although caused by mechanisms other than grain-boundary scattering.

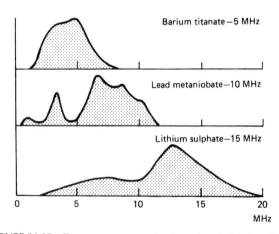

FIGURE 31.25 Frequency response of various piezoelectric transducers.

31.4.5.3 Analyzing Defect Geometry

The assessment of the shape of a defect is often made by measuring changes in the ultrasonic echo height. Intepretation on this basis alone may give results that are less than accurate, since factors such as orientation, geometry, and acoustic impedance could affect the echo size at least as much as the defect size. One technique that allows further investigation is to vary the wavelength of the ultrasonic beam by changing the frequency, hence leading to ultrasonic spectroscopy.

The pulse echo spectroscope is ideally suited for this technique. Examination of test specimens has shown that the use of defect echo heights for the purpose of size assessment is not advisable if the spectral "signatures" of the defect echoes show significant differences.

31.4.6 Other Ways of Presenting Information from Ultrasonics

Many other techniques of processing ultrasonic information are available, and in the following the general principles of two main groups: (1) ultrasonic visualization or (2) ultrasonic imaging will be outlined. More detailed information may be found in the references given at the end of this chapter.

Ultrasonic visualization makes use of two main techniques: (1) Schlieren methods and (2) photoelastic methods. A third method combining both of these has been described by D. M. Marsh in *Research Techniques in Non-destructive Testing* (1973). Schlieren methods depend on detecting the deviation of light caused by refractive index gradients accompanying ultrasonic waves. In most cases this technique has been used in liquids. Photoelastic visualization reveals the stresses in an ultrasonic wave using crossed Polaroids to detect stress birefringence of the medium. Main uses have been for particularly intense fields, such as those in solids in physical contact with the transducer. Many other methods have been tried.

The principle of the Schlieren system is shown in Figure 31.26. In the absence of ultrasonics, all light is intercepted by E. Light diffracted by ultrasound and hence passing outside E forms an image in the dark field. Considerable care is required to make the system effective. Lenses or mirrors may be used in the system as focusing devices; mirrors are free from chromatic aberration and can be accurately produced even for large apertures. Problems may exist with layout and off-axis errors. Lens systems avoid these problems but can be expensive for large apertures.

The basic principles of photoelastic visualization (Figure 31.27) are the same as those used in photoelastic stress analysis (see Chapter 4, Section 4.9). Visualization works well in situations where continuous ultrasound is being used. If a pulsed system is used then collimation of the light beam crossing the ultrasound beam and a pulsed light source are required.

The principal advantage of a photoelastic system is its compactness and lack of protective covering in contrast to the Schlieren system.

Photoelastic systems are also cheaper. However, the Schlieren system's major advantage lies in its ability to visualize in fluids as well as in solids. The sensitivity of photoelastic techniques to longitudinal waves is half that for shear waves. Measurements with Schlieren systems have shown that the sensitivity of ultrasound visualization systems is particularly affected by misalignment of the beam. For example, misalignment of the beam in water is

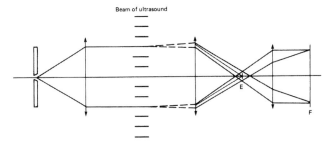

FIGURE 31.26 Schlieren visualization.

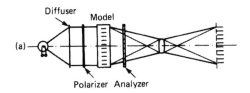

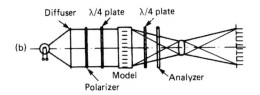

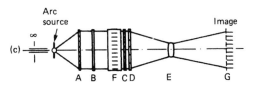

FIGURE 31.27 Diagrams of photoelastic visualization methods. (a) Basic anisotropic system; (b) circular polarized system for isotropic response; (c) practical large-aperture system using Fresnel lenses.

magnified by refraction in the solid. It is therefore essential to ensure firm securing of transducers by clamps capable of fine adjustment.

An alternative approach to visualization of ultrasonic waves is the formation of an optical image from ultrasonic radiation. Many systems exist in this field and, once again, a few selected systems will be outlined.

Ultrasonic cameras of various types have been in existence for many years. In general, their use has been limited to the laboratory due to expense and limitations imposed by design. Most successful cameras have used quartz plates forming the end of an evacuated tube for electron-beam scattering, and the size of these tubes is limited by the need to withstand atmospheric pressure. This can be partly overcome by using thick plates and a harmonic, although there is a loss in sensitivity. A new approach to ultrasonic imaging is the use of liquid crystals to form color images.

All these systems suffer from the need to have the object immersed in water, where the critical angle is such that radiation can only be received over a small angle.

Increasingly, acoustic holography is being introduced into ultrasonic flaw detection and characterization. A hologram is constructed from the ultrasonic radiation and is used to provide an optical picture of any defects. This type of system is theoretically complex and has practical problems associated with stability, as does any holographic system, but there are an increasing number of examples of its use (for example, medical and seismic analysis). Pasakomy has shown that real-time holographic systems may be used to examine welds in pressure vessels. Acoustic holography with scanned hologram systems has been used to examine double-"V" notch welds. Reactor pressure vessels were examined using the same technique and flaw detection was within the limits of error.

31.4.7 Automated Ultrasonic Testing

The use of microcomputer technology to provide semi-automatic and automatic testing systems has increased, and several pipe testers are available which are controlled by microprocessors. A typical system would have a microprocessor programmed to record defect signals which were considered to represent flaws. Depending on the installation, it could then mark these flaws, record them, or transmit them via a link to a main computer. Testing of components online has been developed further by British Steel and by staff at SSNDT.

Allied to the development of automatic testing is the development of data-handling systems to allow flaw detection to be automatic. Since the volume of data to be analyzed is large, the use of computers is essential. One danger associated with this development is that of over-inspection. The number of data values collected should be no more than is satisfactory to meet the specification for flaw detection.

31.4.8 Acoustic Emission

Acoustic emission is the release of energy as a solid material undergoes fracture or deformation. In non-destructive testing two features of acoustic emission are important: its ability to allow remote detection of crack formation or movement at the time it occurs and to do this continuously.

The energy released when a component undergoes stress leads to two types of acoustic spectra: continuous or burst type. Burst-type spectra are usually associated with the leading edge of a crack being extended; i.e., crack growth. Analysis of burst spectra is carried out to identify material discontinuities. Continuous spectra are usually associated with yield rather than fracture mechanisms. Typical mechanisms which would release acoustic energy are: (1) crack growth, (2) dislocation avalanching at the leading edge of discontinuities, and (3) discontinuity surfaces rubbing.

Acoustic spectra are generated when a test specimen or component is stressed. Initial stressing will produce acoustic emissions from all discontinuities, including minor yield in areas of high local stress. The most comprehensive assessment of a structure is achieved in a single stress application; however, cyclic stressing may be used for structures that may not have their operating stresses exceeded. In large structures which have undergone a period of continuous stress a short period of stress relaxation will allow partial recovery of acoustic activity.

Acoustic emission inspection systems have three main functions: (1) signal detection, (2) signal conditioning, and (3) analysis of the processed signals. It is known that the acoustic emission signal in steels is a very short pulse. Other studies have shown that a detecting system sensitive mainly to one mode gives a better detected signal. Transducer patterns may require some thought, depending on the geometry of the item under test. Typical transducer materials are PZT-5A, sensitive to Rayleigh waves and shear waves, and lithium niobate in an extended temperature range. Signal conditioning basically takes the transducer output and produces a signal which the analysis system can process. Typically a signal conditioning system will contain a low-noise pre-amplifier and a filter amplification system.

The analysis of acoustic emission spectra depends on being able to eliminate non-recurring emissions and to process further statistically significant and predominant signals. A typical system is shown in Figure 31.28. The main operating functions are: (1) a real-time display and (2) a source analysis computer. The real-time display gives the operator an indication of all emissions while significant emissions are processed by the analysis system. A permanent record of significant defects can also be produced.

In plants where turbulent fluid flow and cavitation are present, acoustic emission detection is best carried out in the low-megahertz frequency range to avoid noise interference.

Acoustic emission detection has been successfully applied to pipe rupture studies, monitoring of known flaws in pressure vessels, flaw formation in welds, stress corrosion cracking, and fatigue failure.

31.5 RADIOGRAPHY

Radiography has long been an essential tool for inspection and development work in foundries. In addition, it is widely used in the pressure vessel, pipeline, offshore drilling platform, and many other industries for checking the integrity of welded components at the in-process, completed, or in-service stages. The technique also finds application in the aerospace industry.

The method relies on the ability of high-energy, short-wavelength sources of electromagnetic radiation such as

CHAPTER | 31 Non-Destructive Testing

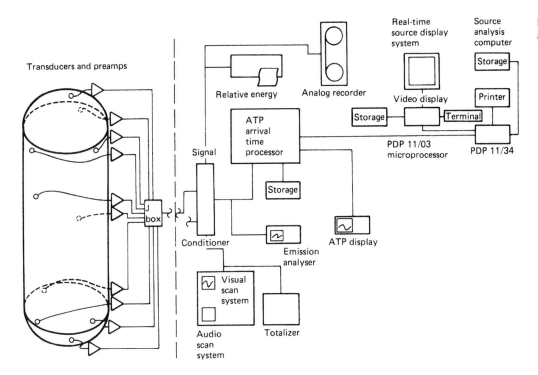

FIGURE 31.28 System for acoustic emission detection.

X-rays, gamma rays (and neutron sources) to penetrate solid materials. By placing a suitable recording medium, usually photographic film, on the side of the specimen remote from the radiation source and with suitable adjustment of technique, a shadowgraph or two-dimensional image of the surface and internal features of the specimen can be obtained. Thus radiography is one of the few non-destructive testing methods suitable for the detection of internal flaws, and has the added advantage that a permanent record of these features is directly produced.

X- and gamma rays are parts of the electro-magnetic spectrum (Figure 31.29). Members of this family are connected by the relationship velocity equals frequency times wavelength, and have a common velocity of 3×10^8 m/s. The shorter the wavelength, the higher the degree of penetration of matter. Equipment associated with them is also discussed in Chapter 22.

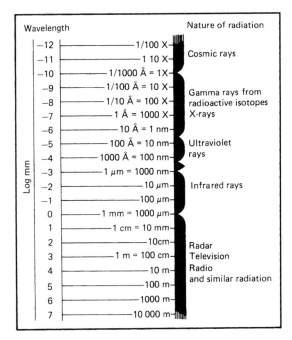

FIGURE 31.29 Electromagnetic spectrum.

31.5.1 Gamma Rays

The gamma-ray sources used in industrial radiography are artificially produced radioactive isotopes. Though many radioactive isotopes are available, only a few are suitable for radiography. Some of the properties of the commonly used isotopes are shown in Table 31.1. Since these sources emit radiation continuously they must be housed in a protective container which is made of a heavy metal such as lead or tungsten. When the container is opened in order to take the radiograph it is preferable that this be done by remote control in order to reduce the risk of exposing the operator to the harmful rays. Such a gamma-ray source container is shown in Figure 31.30.

TABLE 31.1 Isotopes commonly used in gamma radiography

Source	Effective equiv. energy (MeV)	Half-life	Specific emission (R/h/Ci)	Specific activity (Ci/l)	Used for steel thickness (mm) up to
Thulium^{170}	0.08	117 days	0	0.0025	9
Iridium^{192}	0.4	75 days	0.48	25	75
Cesium^{137}	0.6	30 yr	0.35	25	80
Cobalt^{60}	1.1; 1.3	5.3 yr	1.3	120	140

FIGURE 31.30 Gamma-ray source container. Saddle is attached to the work piece. The operator slips the source container into the saddle and moves away. After a short delay the source moves into the exposed position. When the preset exposure time expires the source returns automatically into its safe shielding. Courtesy Pantatron Radiation Engineering Ltd.

31.5.2 X-rays

X-rays are produced when high-speed electrons are brought to rest by hitting a solid object. In radiography the X-rays are produced in an evacuated heavy-walled glass vessel or "tube." The typical construction of a tube is shown in Figure 31.31. In operation a dc voltage in the range 100 kV to 2 MV is applied between a heated, spirally wound filament (the cathode) and the positively charged fairly massive copper anode. The anode has embedded in it a tungsten insert or target of effective area 2–4 mm^2 and it is on to this target that the electrons are focused. The anode and cathode are placed about 50 mm apart, and the tube current is kept low (5–10 mA) in order to prevent overheating of the anode. Typical voltages for a range of steel thicknesses are shown in Table 31.2.

The high voltage needed to produce the X-rays is obtained by relatively simple circuitry consisting of suitable combinations of transformers, rectifiers, and capacitors. Two of the more widely used circuits are the Villard, which produces a pulsating output, and the Greinacher, producing an almost constant potential output. These are shown in Figure 31.32 along with the form of the resulting voltage. A voltage stabilizer at the 240 V input stage is desirable.

Exposure (the product of current and time) varies from specimen to specimen, but with a tube current of 5 mA exposure times in the range 2–30 min are typical.

31.5.3 Sensitivity and IQI

Performance can be optimized by the right choice of type of film and screens, voltage, exposure, and film focal (target) distance. The better the technique, the higher the sensitivity of the radiograph. Sensitivity is a measure of the smallness

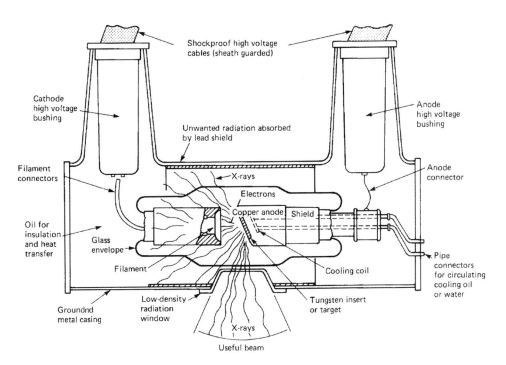

FIGURE 31.31 Single section, hot-cathode, high vacuum, oil-cooled, radiographic X-ray tube.

TABLE 31.2 Maximum thickness of steel which can be radiographed with different X-ray energies

X-ray energy (KeV)s	High-sensitivity technique thickness (mm)	Low-sensitivity technique thickness (mm)
100	10	25
150	15	50
200	25	75
250	40	90
400	75	110
1000	125	160
2000	200	250
5000	300	350
30000	300	380

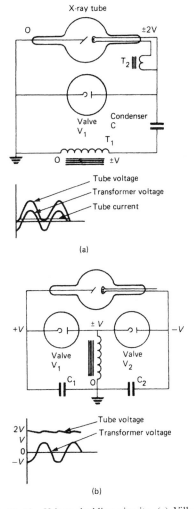

FIGURE 31.32 Voltage-doubling circuits. (a) Villard; (b) Greinacher constant-potential.

of flaw which may be revealed on a radiograph. Unfortunately, this important feature cannot be measured directly, and so the relative quality of the radiograph is assessed using a device called an image indicator (IQI).

A number of different designs of IQI are in use, none of which is ideal. After extensive experimentation, the two types adopted by the British Standards Institution and accepted by the ISO (International Standards Organization) are the wire type and the step-hole type. In use, the IQI should, wherever possible, be placed in close contact with the source side of the specimen, and in such a position that it will appear near the edge of whichever size of film is being used.

31.5.3.1 Wire Type of IQI

This consists of a series of straight wires 30 mm long and diameters as shown in Table 31.3. Five models are available, the most popular containing seven consecutive wires. The wires are laid parallel and equidistant from each other and mounted between two thin transparent plastic sheets of low absorption for X- or gamma rays. This means that although the wires are fixed in relation to each other, the IQI is flexible and thus useful for curved specimens. The wires should be of the same material as that being radiographed. IQIs for steel, copper, and aluminum are commercially available. For other materials it may not be possible to obtain an IQI of matching composition. In this case, a material of similar absorptive properties may be substituted.

Also enclosed in the plastic sheet are letters and numbers to indicate the material type and the series of wires (Figure 31.33). For example, 4Fe10 indicates that the IQI is suitable for iron and steel and contains wire diameters from 0.063 to 0.25 mm.

For weld radiography, the IQI is placed with the wires placed centrally on the weld and lying at right angles to the length of the weld.

The percentage sensitivity, S, of the radiograph is calculated from

$$S = \frac{\text{Diameter of thinnest wire visible on the radiograph} \times 100}{\text{Specimen thickness (mm)}}$$

Thus the better the technique, the more wires will be imaged on the radiograph. It should be noted that the smaller the value of S, the better the quality of the radiograph.

31.5.3.2 Step or Hole Type of IQI

This consists of a series of uniform thickness metal plaques each containing one or two drilled holes. The hole diameter is equal to step thickness. The step thickness and hole diameters are shown in Table 31.4.

For steps 1 to 8, two holes are drilled in each step, each hole being located 3 mm from the edge of the step. The remaining steps, 9 to 18, have only a single hole located in the center of the step.

For convenience, the IQI may be machined as a single step wedge (Figure 31.34). For extra flexibility each plaque is mounted in line but separately between two thin plastic or rubber sheets. Three modules are available, the step series in each being as shown in Table 31.5.

In this type also, a series of letters and numbers identifies the material and thickness range for which the IQI is suitable. The percentage sensitivity, S, is calculated from

$$S = \frac{\text{Diameter of smallest hole visible on the radiograph} \times 100}{\text{Specimen thickness (mm)}}$$

For the thinner steps with two holes it is important that both holes are visible before that diameter can be included in the calculation. In use, this type is placed close to, but not on, a weld.

TABLE 31.3 Voltage-doubling circuits: (a) Villard; (b) Greinacher constant-potential

Wire no.	Diameter (mm)	Wire no.	Diameter (mm)	Wire no.	Diameter (mm)
1	0.032	8	0.160	15	0.80
2	0.010	9	0.200	16	1.00
3	0.050	10	0.250	17	1.25
4	0.063	11	0.320	18	1.00
5	0.080	12	0.400	19	2.00
6	0.100	13	0.500	20	2.50
7	0.125	14	0.63	21	3.20

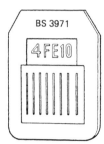

FIGURE 31.33 Wire-type IQI.

TABLE 31.4 Hole and step dimensions for IQI

Step no.	Diameter and step thickness (mm)	Step no.	Diameter and step thickness (mm)
1	0.125	10	1.00
2	0.160	11	1.25
3	0.200	12	1.60
4	0.250	13	2.00
5	0.320	14	2.50
6	0.400	15	3.20
7	0.500	16	4.00
8	0.630	17	5.00
9	0.800	18	6.30

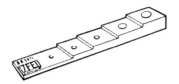

FIGURE 31.34 Step/hole type IQI.

TABLE 31.5 Step and holes included in different models of IQI

Model	Step and hole sizes
A	1–6 inclusive
B	7–12 inclusive
C	13–18 inclusive

31.5.4 Xerography

Mainly because of the high price of silver, attempts have been made to replace photographic film as the recording medium. Fluoroscopy is one such method but is of limited application due to its lack of sensitivity and inability to cope with thick sections in dense materials. An alternative technique which has been developed and which can produce results of comparable sensitivity to film radiography in the medium voltage range (100–250 kV) is xerography. In this process, the X-ray film is replaced by an aluminum plate coated on one side with a layer of selenium 30–100 μm thick. A uniform positive electric charge is induced in the selenium layer. Since selenium has a high resistivity, the charge may be retained for long periods provided the plate is not exposed to light, ionizing radiations, or unduly humid atmospheres.

Exposure to X- or gamma rays causes a leakage of the charge from the selenium to the earthed aluminum backing plate—the leakage at any point being proportional to the radiation dose falling on it. The process of forming the image is shown in Figure 31.35. This latent image is developed by blowing a fine white powder over the exposed plate. The powder becomes charged (by friction) and the negatively charged particles are attracted and adhere to the positively charged selenium, the amount attracted being proportional to the amount of charge at each point on the "latent" selenium image.

The image is now ready for viewing, which is best done in an illuminator using low-angle illumination, and the image will withstand vibration, but it must not be touched.

The process is capable of a high degree of image sharpness since the selenium is virtually free of graininess, and sharpness is not affected by the grain size of the powder.

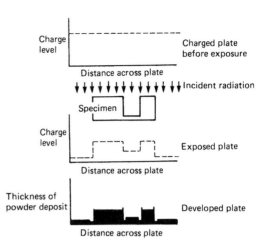

FIGURE 31.35 Diagrammatic representation of the process of xerography.

An apparent drawback is the low inherent contrast of the image, but there is an outlining effect on image detail which improves the rendering of most types of weld and casting flaws thus giving an apparent high contrast. The overall result is that sensitivity is similar to that obtained with film.

31.5.5 Fluoroscopic and Image-Intensification Methods

In fluoroscopy the set-up of source, specimen, and recording medium is similar to that for radiography. However, instead of film a specially constructed transparent screen is used which fluoresces, i.e., emits light when X-rays fall on it. This enables a positive image to be obtained since greater amounts of radiation, for example that passing through thinner parts of the specimen, will result in greater brightness.

Fluoroscopy has the following advantages over radiography:

1. The need for expensive film is eliminated
2. The fluorescent screen can be viewed while the specimen is moving, resulting in:
   A. Easier image interpretation
   B. Faster throughput.

Unfortunately, the sensitivity possible with fluoroscopy is considerably inferior to that obtained with film radiography. It is difficult to obtain a sensitivity better than 5 percent whereas for critical work a sensitivity of 2 percent or better is required. Therefore, although the method is widely used in the medical field, its main use in industry is for applications where resolution of fine detail is not required. There are three reasons for the lack of sensitivity:

1. Fluoroscopic images are usually very dim. The characteristics of the human eye are such that, even when fully dark adapted, it cannot perceive at low levels of brightness the small contrasts or fine detail which it can at higher levels.

2. In an attempt to increase image brightness the fluorescent screens are usually constructed using a zinc sulphide-cadmium sulphide mixture which, although increasing brightness, gives a coarser-grained and hence a more blurred image.
3. The image produced on a fluoroscopic screen is much less contrasty than that on a radiograph.

31.5.5.1 Image-Intensification Systems

In fluoroscopy the main problem of low screen brightness is due mainly to:

1. The low efficiency—only a fraction of the incident X-rays are converted into light.
2. The light which is produced at the screen is scattered in all directions, so that only a small proportion of the total produced is collected by the eye of the viewer.

In order to overcome these limitations, a number of image-intensification and image-enhancement systems have been developed.

The electron tube intensifier is the commonest type. Such instruments are commonly marketed by Philips and Westinghouse. In this system use is made of the phenomenon of photoelectricity, i.e., the property possessed by some materials of emitting electrons when irradiated by light.

The layout of the Philips system is shown in Figure 31.36. It consists of a heavy-walled glass tube with an inner conducting layer over part of its surface which forms part of the electron-focusing system. At one end of the tube there is a two-component screen comprising a fluorescent screen and, in close contact with it, a photoelectric layer supported on a thin curved sheet of aluminum. At the other end of the tube is the viewing screen and an optical system.

The instrument operates as follows. When X-rays fall on the primary fluorescent screen they are converted into light which, in turn, excites the photoelectric layer in contact with it and causes it to emit electrons: i.e., a light image is converted into an electron image. The electrons are accelerated across the tube by a dc potential of 20–30 kV and focused on the viewing screen. Focusing of the electron image occurs largely because of the spherical curvature of the photocathode. However, fine focusing is achieved by variation of a small positive voltage applied to the inner conducting layer of the glass envelope. As can be seen from Figure 31.36, the electron image reproduced on the viewing screen is much smaller and hence much brighter than that formed at the photo-cathode. Further increase in brightness is obtained because the energy imparted to the electrons by the accelerating electric field is given up on impact.

Although brighter, the image formed on the final screen is small and it is necessary to increase its size with an optical magnifier. In the Philips instrument an in-line system provides a linear magnification of nine for either monocular or binocular viewing.

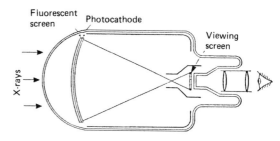

FIGURE 31.36 Diagram of 5-inch Philips image-intensifier tube.

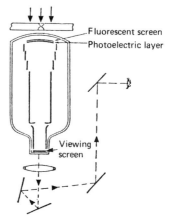

FIGURE 31.37 Diagram of Westinghouse image-intensifier tube and optical system.

As can be seen from Figure 31.37, the Westinghouse instrument is somewhat similar to that of Philips, but there are two important differences:

1. Westinghouse uses a subsidiary electron lens followed by a final main electron lens, fine focusing being achieved by varying the potential of the weak lens.
2. Westinghouse uses a system of mirrors and lenses to prepare the image for final viewing.

The advantages of this system are that viewing is done out of line of the main X-ray beam and the image can be viewed simultaneously by two observers.

31.6 UNDERWATER NON-DESTRUCTIVE TESTING

The exploration and recovery of gas and oil offshore based on large fabricated structures has created a demand for non-destructive testing capable of operation at and below the surface of the sea. Because of high capital costs, large operating costs, and public interest the structures are expected to operate all the year round and to be re-certified by relevant authorities and insurance companies typically on a five-year renewal basis.

The annual inspection program must be planned around limited opportunities and normally covers:

1. Areas of previous repair;
2. Areas of high stress;
3. Routine inspection on a five-year cycle.

The accumulated inspection records are of great importance. The inspection is designed to include checks on:

1. Structural damage caused by operational incidents and by changes in seabed conditions;
2. Marine growth accumulation both masking faults and adding extra mass to the structure;
3. Fatigue cracking caused by the repeated cyclic loading caused by wind and wave action;
4. Corrosion originating in the action of salt water.

The major area of concern is within the splash zone and just below, where incident light levels, oxygen levels, repeated wetting/drying, and temperature differentials give rise to the highest corrosion rates. The environment is hostile to both equipment and operators in conditions where safety considerations make demands on inspection techniques requiring delicate handling and high interpretative skills.

31.6.1 Diver Operations and Communication

Generally, experienced divers are trained as non-destructive testing operators rather than inspection personnel being converted into divers. The work is fatiguing and inspection is complicated by poor communications between diver and the supervising surface inspection engineer. These constraints lead to the use of equipment which is either robust and provides indications which are simple for first-line interpretation by the diver or uses sophisticated data transmission to the surface for interpretation, and relegates the diver's role to one of positioning the sensor-head. Equipment requiring a high degree of interaction between diver and surface demands extensive training for optimum performance.

The communication to the diver is speech based. The ambient noise levels are high both at the surface, where generators, compressors, etc., are operating, and below it, where the diver's microphone interacts with the breathing equipment. The microphone picks up the voice within the helmet. Further clarity is lost by the effect of air pressure altering the characteristic resonance of the voice, effectively increasing the frequency band so that, even with frequency shifters, intelligible information is lost. Hence any communication with the diver is monosyllabic and repetitive.

For initial surveys and in areas of high ambient danger the mounting of sensor arrangements on remote-controlled vehicles (RCV) is on the increase.

Additional requirements for robust techniques and equipments are imposed by 24-hour operation and continual changes in shift personnel. Equipment tends to become common property and not operated by one individual who can undertake maintenance and calibration.

Surface preparation prior to examination is undertaken to remove debris, marine growth, corrosion products, and, where required, the existing surface protection. The preparation is commonly 15 cm on either side of a weld and down to bare metal. This may be carried out by hand scraping, water jetting, pneumatic/hydraulic needle descaler guns, or hydraulic wire brushing. Individual company requirements vary in their estimation of the effects of surface-cleaning methods which may peen over surface-breaking cracks.

31.6.2 Visual Examination

The initial and most important examination is visual in that it provides an overall and general appreciation of the condition of the structure, the accumulation of marine growth and scouring. Whilst only the most obvious cracks will be detected, areas requiring supplementary non-destructive testing will be highlighted. The examination can be assisted by closed-circuit television (CCTV) and close-up photography. CCTV normally uses low-light level silicon diode cameras, capable of focusing from 4 inches to infinity.

In order to assist visual examination, templates, mimics, and pit gauges are used to relay information on the size of defects to the surface. Where extensive damage has occurred, a template of the area may be constructed above water for evaluation by the inspection engineer who can then design a repair technique.

31.6.3 Photography

Light is attenuated by scatter and selective absorption in seawater and the debris held in suspension so as to shift the color balance of white light towards green. Correction filters are not normally used as they increase the attenuation. Correct balance and illumination levels are achieved with flood- or flashlights. The photography is normally on 35 or 70 mm format using color film with stereoscopic recording where later analysis and templating will assist repair techniques.

Camera types are normally waterproofed versions of land-based equipment. When specifically designed for underwater use, the infrequent loading of film and power packs is desirable with built-in flash and single-hand operation or remote firing from a submersible on preset or automatic control.

CCTV provides immediate and recordable data to the surface. It is of comparatively poor resolution and, in black and white, lacks the extra picture contrast given by color still photography.

31.6.4 Magnetic Particle Inspection (MPI)

MPI is the most widely accepted technique for underwater non-destructive testing. As a robust system with wide operator latitude and immediate confirmation of a successful application by both diver and surface, the technique is well suited to cope in the hostile environment.

Where large equipment cannot gain access, magnetization is by permanent magnets with a standard pull-off strength, although use will be limited away from flat material, and repeated application at 6-inch intervals is required.

When working near to the air–sea boundary, the magnetization is derived from flexible coils driven from surface transformers and the leakage flux from cracks disclosed by fluorescent ink supplied from the surface. Ac is used to facilitate surface crack detection. The flexible cables carrying the magnetization current are wrapped around a member or laid in a parallel conductor arrangement along the weld. At lower levels the primary energy is taken to a subsea transformer to minimize power losses. The transformer houses an ink reservoir which dilutes the concentrate 10:1 with seawater. Ink illumination is from hand-held ultraviolet lamps which also support the ink dispenser (Figure 31.38). At depth the low ambient light suits fluorescent inspection whilst in shallow conditions inspection during the night is preferred. Photographic recording of indications is normal along with written notes and any CCTV recording available.

31.6.5 Ultrasonics

Ultrasonic non-destructive testing is mainly concerned with simple thickness checking to detect erosion and corrosion. Purpose-built probes detect the back-wall ultrasonic echo and will be hand-held with rechargeable power supplies providing full portability. The units may be self-activating, switching on only when held on the test material, and are calibrated for steel although alternative calibration is possible. Ideally the unit therefore has no diver controls and can display a range of steel thickness from 1 to 300 mm on a digital read-out. Where detailed examination of a weld is required, either a conventional surface A-scan set housed in a waterproof container with external extensions to the controls is used, or the diver operates the probes of a surface-located set under the direction of the surface non-destructive testing engineer. In either case, there is a facility for monitoring of the master set by a slave set. The diver, for instance, may be assisted by an audio feedback from the surface set which triggers on a threshold gate open to defect signals. The diver's hand movements will be monitored by a helmet video to allow monitoring and adjustment by the surface operator.

31.6.6 Corrosion Protection

Because seawater behaves as an electrolytic fluid, corrosion of the steel structures may be inhibited by providing either sacrificial zinc anodes or impressing on the structure a constant electrical potential to reverse the electrolytic action. In order to check the operation of the impressed potential or the corrosion liability of the submerged structure, surface voltage is measured with reference to silver/silver chloride cell (Figure 31.39). Hand-held potential readers containing a reference cell, contact tip, and digital reading volt-meter along with internal power supply will allow a survey to be completed by a diver which may be remotely monitored from the surface.

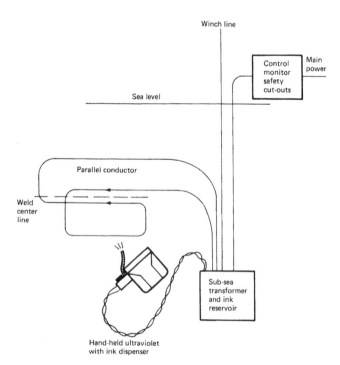

FIGURE 31.38 Magnetic particle inspection in deep water.

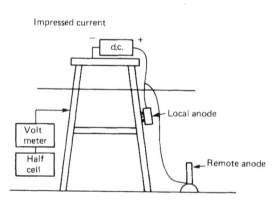

FIGURE 31.39 Measuring cathodic protection.

31.6.7 Other Non-Destructive Testing Techniques

There are a variety of other non-destructive techniques which have not yet gained common acceptance within the oil industry but are subject to varying degrees of investigation and experimental use. Some of these are described below.

31.6.7.1 Eddy Current

Eddy-current techniques are described in Section 31.3.4. The method can, with suitable head amplification, be used for a search-and-follow technique. Whilst it will not detect sub-surface cracks, the degree of surface cleaning both on the weld and to each side is not critical. Thus substantial savings in preparation and reprotection can be made.

31.6.7.2 AC Potential Difference (ACIPD)

As mentioned in Section 31.3.3, changes in potential difference can be used to detect surface defects. This is particularly valuable under water.

An ac will tend to travel just under the surface of a conductor because of the skin effect. The current flow between two contacts made on a steel specimen will approximately occupy a square with the contact points on one diagonal. In a uniform material there will be a steady ohmic voltage drop from contact to contact which will map out the current flow. In the presence of a surface crack orientated at right angles to the current flow there will be a step change in the potential which can be detected by two closely spaced voltage probes connected to a sensitive voltmeter (Figure 31.40). Crack penetration, regardless of attitude, will also influence the step voltage across the surface crack and allow depth estimation. The method relies upon the efficiency of the contact made by the current driver and the voltage probe tips, and limitations occur because of the voltage safety limitations imposed on electrical sources capable of producing the constant current required. The voltages are limited to those below the optimum required to break down the surface barriers. The technique, however, is an obvious choice to first evaluate MPI indications in order to differentiate purely surface features (for example, grinding marks) from cracks.

31.6.7.3 Bulk Ultrasonic Scanning

The alternative to adapting surface non-destructive testing methods is to use the surrounding water as a couplant to transfer a large amount of ultrasonic energy into the structure and then monitor the returning energy by scanning either a single detector or the response from a tube from which acoustic energy can be used to construct a visual image (Figure 31.41). Such techniques are experimental but initial results indicate that very rapid inspection rates can be obtained with the diver entirely relegated to positioning the sensor. Analysis of the returning information is made initially by microcomputers, which attempt to remove the background variation and highlight the signals which alter the sensor position or scan. The devices do not attempt to characterize defects in detail, and for this other techniques are required.

31.6.7.4 Acoustic Emission

Those developments which remove the need for a diver at every inspection are attractive but, as yet, are not fully proven. Acoustic emission is detected using probes fixed to the structure, "listening" to the internal noise. As described in Section 31.4.8, the system relies upon the stress concentrations and fatigue failures to radiate an increasing amount

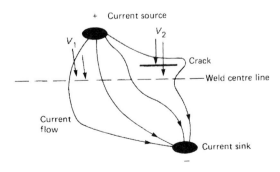

FIGURE 31.40 AC potential difference (AC/PD).

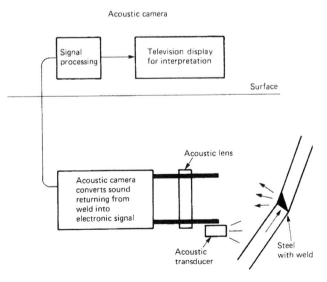

FIGURE 31.41 Bulk ultrasonic scanning.

of energy as failure approaches, and this increased emission is detected by the probes.

31.7 DEVELOPMENTS

Many of the recent development in non-destructive testing, and in ultrasonic testing in particular, have been in the use of computers to control inspections and analyze results.

The most widespread use of computer technology is in digital flaw detectors. These instruments digitize the incoming data signal, allowing it to be stored on disc, recalled, and printed. Digital flaw detectors are also able to simplify the task of taking inspections by providing functions such as automatic calibration and DAC curve plotting. The use of B-, C-, and D-scans to produce clear images of defects is well established and commonly available at a relatively low cost.

Many advances in instrumentation are in the off-line processing of information, both before and after the actual inspection (Carter and Burch, 1986). The most common use is to enhance C-scans by color coding defect areas. Other data processing techniques include SAFT (Software Aperture Focusing Technique) (Doctor et al., 1986), TOFD (Time of Flight Diffraction) (Carter, 1984), expert systems (Moran and Bower, 1987), and neural networks (Windsor et al., 1993).

One of the most interesting innovations in this field is the concept of an integrated NDT work-bench (McNab and Dunlop, 1993). These work-benches provide the hardware and software systems required to design, carry out, and analyze an ultrasonic inspection, all operated from a single computer. In theory, such a system would include a CAD package to provide the exact geometry of the part under inspection, controllers to move a mechanical probe over the inspection area, digital signal processing software to enhance the incoming data, and an expert system to help assess the results.

The use of mechanical devices to take inspections is becoming more common (Mudge, 1985), especially in the nuclear and pressure vessel industries. Due to the inflexibility of these mechanical systems, however, the majority of inspections are still performed by a manual operator. Designers of NDT equipment are now turning their attention to how computer technology can improve the reliability. For this type of equipment, ease of use is a prime consideration, since the operator must be able to concentrate on taking the inspection without the distraction of a complicated computer system. Work is under way at the University of Paisley to determine the optimum structure and design of system interface to produce the best aid to manual operators. The use of speech recognition as a form of remote control for operators is being examined with great interest.

31.8 CERTIFICATION OF PERSONNEL

No overview of non-destructive testing would be complete without some reference to the various operator-certification schemes currently in use worldwide. The range of products, processes and materials to which such methods have been applied has placed increasing demands on the skills and abilities of its practitioners. Quality assurance requires not only the products to have fitness for purpose but also the relevant personnel.

Developments in various countries in operator certification for non-destructive testing are given by Drury (1979).

APPENDIX 31.1 Fundamental standards used in ultrasonic testing

British Standards	BS	
BS 2704	(78)	Calibration Blocks
BS 3683 (Part 4)	(77)	Glossary of Terms
BS 3889 (Part 1A)	(69)	Ultrasonic Testing of Ferrous Pipes
BS 3923		Methods of Ultrasonic Examination of Welds
BS 3923 (Part 1)	(78)	Manual Examination of Fusion Welds in Ferritic Steel
BS 3923 (Part 2)	(72)	Automatic Examination of Fusion Welded Butt Joints in Ferritic Steels
BS 3923 (Part 3)	(72)	Manual Examination of Nozzle Welds
BS 4080	(66)	Methods of Non-Destructive Testing of Steel Castings
BS 4124	(67)	Non-Destructive Testing of Steel Forgings—Ultrasonic Flaw Detection
BS 4331		Methods for Assessing the Performance Characteristics of Ultrasonic Flaw Detection Equipment
BS 4331 (Part 1)	(78)	Overall Performance—Site Methods

(Continued)

CHAPTER | 31 Non-Destructive Testing

APPENDIX 31.1 (Continued)

BS 4331 (Part 2)	(72)	Electrical Performance
BS 4331 (Part 3)	(74)	Guidance on the In-service Monitoring of Probes (excluding Immersion Probes)
BS 5996	(81)	Methods of Testing and Quality Grading of Ferritic Steel Plate by Ultrasonic Methods
M 36	(78)	Ultrasonic Testing of Special Forgings by an Immersion Technique
M 42	(78)	Non-Destructive Testing of Fusion and Resistance Welds in Thin Gauge Material (ASTM)
American Standards		
E 114	(75)	Testing by the Reflection Method using Pulsed Longitudinal Waves Induced by Direct Contact
E 127	(75)	Aluminum Alloy Ultrasonic Standard Reference Blocks
E 164	(74)	Ultrasonic Contact Inspection of Weldments
E 213	(68)	Ultrasonic Inspection of Metal Pipe and Tubing for Longitudinal Discontinuities
E 214	(68)	Immersed Ultrasonic Testing
E 273	(68)	Ultrasonic Inspection of Longitudinal and Spiral Welds of Welded Pipes and Tubings
E 317	(79)	Performance Characteristics of Pulse Echo Ultrasonic Testing Systems
E 376	(69)	Seamless Austenitic Steel Pipe for High Temperature Central Station Service
E 388	(71)	Ultrasonic Testing of Heavy Steel Forgings
E 418	(64)	Ultrasonic Testing of Turbine and Generator Steel Rotor Forgings
E 435	(74)	Ultrasonic Inspection of Steel Plates for Pressure Vessels
E 50	(64)	Ultrasonic Examination of Large Forged Crank Shafts
E 531	(65)	Ultrasonic Inspection of Turbine-Generator Steel Retaining Rings
E 557	(73)	Ultrasonic Shear Wave Inspection of Steel Plates
E 578	(71)	Longitudinal Wave Ultrasonic Testing of Plain and Clad Steel Plates for Special Applications
E 609	(78)	Longitudinal Beam Ultrasonic Inspection of Carbon and Low Alloy Steel Casting
West German Standards (DIN)—Translations available		
17175		Seamless Tubes of Heat Resistant Steels: Technical Conditions of Delivery
17245		Ferritic Steel Castings Creep Resistant at Elevated Temperatures: Technical Conditions of Delivery
54120		Non-Destructive Testing: Calibration Block 1 and its Uses for the Adjustment and Control of Ultrasonic Echo Equipment
54122		Non-Destructive Testing: Calibration Block 2 and its Uses for the Adjustment and Control of Ultrasonic Echo Equipment

REFERENCES

BS 3683, Glossary of Terms used in Non-destructive Testing: Part 3, Radiological Flaw Detection (1964).

BS 3971, Image Quality Indications (1966).

BS 4094, Recommendation for Data on Shielding from Ionising Radiation (1966).

Blitz, J., Ultrasonic Methods and Applications, Butterworths, London (1971).

Carter, P., "Experience with the time-of-flight diffraction technique and accompanying portable and versatile ultrasonic digital recording system," *Brit. J. NDT*, **26**, 354 (1984).

Carter, P. and S. F., Burch, "The potential of digital processing for ultrasonic inspection," NDT-86: Proceedings of the 21st Annual British Conference on Non-Destructive Testing, (eds.) J. M. Farley and P. D. Hanstead (1986).

Doctor, S. R., T. E., Hall, and L. D., Reid, "SAFT—the evolution of a signal processing technology for ultrasonic testing," NDT International (June 1986).

Drury, J., "Developments in various countries in operator certification for non-destructive testing," in *Developments in Pressure Vessel Technology*, Vol. 2, Applied Science, London (1979).

Electricity Supply Industry, Standards published by Central Electricity Generating Board.

Ensminger, D., *Ultrasonics*, Marcel Dekker, New York (1973).

Erf, R. K., *Holographic Non-destructive Testing*, Academic Press, London (1974).

Farley, J. M., et al., "Developments in the ultrasonic instrumentation to improve the reliability of NDT of pressurised components," Conf. on In-service Inspection, Institution of Mechanical Engineers (1982).

Filipczynski, L., et al., *Ultrasonic Methods of Testing Materials*, Butterworths, London (1966).

Greguss, P., *Ultrasonic Imaging*, Focal Press, London (1980).

Halmshaw, R., *Industrial Radiology: Theory and Practice*, Applied Science, London (1982).

HMSO, The Ionising Radiations (Sealed Sources) Regulations (1969).

Institute of Welding, *Handbook on Radiographic Apparatus and Techniques*.

Krautkramer, J. and H. Krautkramer, *Ultrasonic Testing of Materials*, 3rd ed., Springer-Verlag, New York (1983).

Marsh, D. M., "Means of visualizing ultrasonics," in *Research Techniques in Non-destructive Testing*, Vol. 2 (ed., R. S. Sharpe). Academic Press, London (1973).

Martin, A. and S. Harbison, *An Introduction to Radiation Protection*, Chapman and Hall, London (1972).

McGonnigle, W. J., *Non-destructive Testing*, Gordon and Breach, London (1961).

McNab, A. and Dunlop, I., "Advanced visualisation and interpretation techniques for the evaluation of ultrasonic data: the NDT workbench," *British Journal of NDT* (May 1993).

Moran, A. J. and K. J. Bower, "Expert systems for NDT—hope or hype?," Non-Destructive Testing: Proceedings of the 4th European Conference 1987, Vol. 1 (eds.) J. M. Farley and R. W. Nichols (1987).

Mudge, P. J., "Computer-controlled systems for ultrasonic testing," *Research Techniques in Non-Destructive Testing*, Vol. VII, R. S. Sharp, ed. (1985).

Sharpe, R. S. (ed.), *Research Techniques in Non-destructive Testing*, Vols. 1–5, Academic Press, London (1970–1982).

Sharpe, R. S., *Quality Technology Handbook*, Butterworths, London (1984).

Windsor, C. G., F. Anelme, L. Capineri, and J. P. Mason, "The classification of weld defects from ultrasonic images: a neural network approach," *British Journal of NDT* (January 1993).

FURTHER READING

Gardner, W. E. (ed.), *Improving the Effectiveness and Reliability of Non-destructive Testing*, Pergamon Press, Oxford (1992).

Chapter 32

Noise Measurement

J. Kuehn

32.1 SOUND AND SOUND FIELDS

32.1.1 The Nature of Sound

If any elastic medium, whether it be gaseous, liquid, or solid, is disturbed, then this disturbance will travel away from the point of origin and be propagated through the medium. The way in which the disturbance is propagated and its speed will depend upon the nature and extent of the medium, its elasticity, and density.

In the case of air, these disturbances are characterized by very small fluctuations in the density (and hence atmospheric pressure) of the air and by movements of the air molecules. Provided these fluctuations in pressure take place at a rate (frequency) between about 20 Hz and 16 kHz, then they can give rise to the sensation of audible sound in the human ear. The great sensitivity of the ear to pressure fluctuations is well illustrated by the fact that a peak-to-peak fluctuation of less than 1 part in 10^9 of the atmospheric pressure will, at frequencies around 3 kHz, be heard as audible sound. At this frequency and pressure the oscillating molecular movement of the air is less than 10^{-7} of a millimeter.

The magnitude of the sound pressure at any point in a sound field is usually expressed in terms of the rms (root-mean-square) value of the pressure fluctuations in the atmospheric pressure. This value is given by

$$P_{ms} = \sqrt{\left(\frac{1}{T}\int_0^T P^2(k)dt\right)} \qquad (32.1)$$

where $P(t)$ is the instantaneous sound pressure at time t and T is a time period long compared with the periodic time of the lowest frequency present in the sound.

The SI unit for sound pressure is Newton/m² (N/m²), which is now termed Pascal, that is, 1 Newton/m² = 1 Pascal. Atmospheric pressure is, of course, normally expressed in bars (1 bar = 10^5 Pascal).

The great sensitivity of the human hearing mechanism, already mentioned, is such that at frequencies where it is most sensitive it can detect sound pressures as small as 2×10^{-5} Pascals. It can also pick up sound pressures as high as 20 or even 100 Pascals. When dealing with such a wide dynamic range (pressure range of 10 million to 1) it is inconvenient to express sound pressures in terms of Pascals and so a logarithmic scale is used. Such a scale is the decibel scale. This is defined as ten times the logarithm to the base 10 of the ratio of two powers. When applying this scale to the measurement of sound pressure it is assumed that sound power is related to the square of the sound pressure. Thus

$$\text{Sound pressure level} = 10\log_{10}\left(\frac{W_1}{W_0}\right)$$
$$= 10\log_{10}\left(\frac{P^2}{P_0^2}\right)$$
$$= 20\log_{10}\left(\frac{P}{P_0}\right) \text{dB} \qquad (32.2)$$

where P is the sound pressure being measured and P_0 is a "reference" sound pressure (standardized at 2×10^{-5} Pascals).

It should be noted that the use of the expression "sound-pressure *level*" always denotes that the value is expressed in decibels. The reference pressure will, of course, be the 0 dB value on the decibel scale. The use of a reference pressure close to that representing the threshold of hearing at the frequencies where the ear is most sensitive means that most levels of interest will be positive. (A different reference pressure is used for underwater acoustics and for the 0 dB level on audiograms.) A good example of the use of the decibel scale of sound-pressure level and of the complicated response of the human ear to pure tones is given in Figure 32.1, showing equal loudness contours for pure tones presented under binaural, free-field listening conditions.

The equal loudness level contours of Figure 32.1 (labeled in Phons) are assigned numerical values equal to

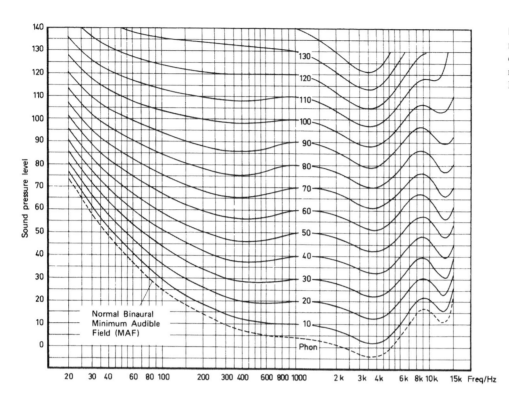

FIGURE 32.1 Normal equal loudness contours for pure tones. (Most of the figures in this chapter are reproduced by courtesy of Bruel & Kjaer (UK) Ltd.)

that of the sound-pressure level at 1 kHz through which they pass. This use of 1 kHz as a "reference" frequency is standardized in acoustics. It is the frequency from which the audible frequency range is "divided up" when choosing octave-band widths or one-third octave-band widths. Thus the octave band centered on 1 kHz and the third octave-band width centered on 1 kHz are included in the "standardized" frequency bands for acoustic measurements.

In daily life, most sounds encountered (particularly those designated as "noise") are not the result of simple sinusoidal fluctuations in atmospheric pressure but are associated with pressure waveforms which vary with time both in frequency and magnitude. In addition, the spatial variation of sound pressure (i.e., the sound field) associated with a sound source is often complicated by the presence of sound-reflecting obstacles or walls.

32.1.2 Quantities Characterizing a Sound Source or Sound Field

The definition of noise as "unwanted sound" highlights the fact that the ultimate measure of a sound, as heard, involves physiological and often psychological measurements and assessments. Objective measurements can be used to evaluate the noise against predetermined and generally acceptable criteria.

When planning the installation of machinery and equipment it is often necessary to be able to predict the magnitude and nature of the sound fields at a distance from the noise source, which means using noise data associated only with the noise source and not with the environment in which the measurements were made.

Although the sound-pressure level, at a point, is the most common measure of a sound field it is far from being a comprehensive measure. As a scalar quantity it provides no information on the direction of propagation of the sound and, except in the case of very simple sound fields (such as that from a physically small "point" source or a plane wave; see BS 4727), it is not directly related to the sound power being radiated by the source. The importance, in some cases, of deducing or measuring the sound power output of a source and its directivity has received much attention in recent years, and equipment is now available (as research equipment and also commercially available for routine measurements) for measuring particle velocity or the sound intensity (W/m^2) at a point in a sound field.

In addition to some measure of the overall sound-pressure level at a point it is also usually necessary to carry out some frequency analysis of the signal to find out how the sound pressure levels are distributed throughout the audible range of frequencies.

32.1.3 Velocity of Propagation of Sound Waves

As mentioned earlier, sound-pressure fluctuations (small compared with the atmospheric pressure) are propagated through a gas at a speed which is dependent on the elasticity and density of the gas. An appropriate expression for the velocity of propagation is

$$c = \sqrt{\left(\frac{\gamma P_0}{\rho_0}\right)} \qquad (32.3)$$

where γ is the ratio of the specific heat of the gas at constant pressure to that at constant volume (1.402 for air), P_0 is the gas pressure (Newtons/m^2) and ρ_0 is the density of the gas (kg/m^3). This expression leads to a value of 331.6 m/s for air at 0°C and a standard barometric pressure of 1.013×10^5 Newtons/m^2 (1.013 bar).

The use of the general gas law also leads to an expression showing that the velocity is directly proportional to the square root of the absolute temperature in K:

$$c = c_0 \sqrt{\left(\frac{t + 273}{273}\right)} \frac{\text{m}}{\text{s}} \quad (32.4)$$

where c_0 is the velocity of sound in air at 0°C (331.6 m/s) and t is the temperature (°C).

The speed of propagation of sound in a medium is an extremely important physical property of that medium, and its value figures prominently in many acoustic expressions. If sound propagates in one direction only then it is said to propagate as a "plane free progressive wave" and the ratio of the sound pressure to the particle velocity, at any point, is always given by $\rho_0 c$.

Knowledge of the velocity of sound is important in assessing the effect of wind and of wind and temperature gradients upon the bending of sound waves. The velocity of sound also determines the wavelength (commonly abbreviated to λ) at any given frequency and it is the size of a source *in relation to the wavelength of the sound radiation* that greatly influences the radiating characteristics of the source. It also determines the extent of the near-field in the vicinity of the source. The screening effect of walls, buildings, and obstacles is largely dependent on their size in relation to the wavelength of the sound. Thus a vibrating panel 0.3 m square or an open pipe of 0.3 m diameter would be a poor radiator of sound at 50 Hz [$\lambda = c/f \simeq 6.6$ m] but could be an excellent, efficient radiator of sound at 1 kHz [$\lambda \simeq 0.33$ m] and higher frequencies. The relation between frequency and wavelength is shown in Figure 32.2.

32.1.4 Selecting the Quantities of Interest

It cannot be emphasized too strongly that before carrying out any noise measurements the ultimate objective of those measurements must be clearly established so that adequate, complete, and appropriate measurements are made.

The following examples indicate how the purpose of a noise measurement influences the acoustic quantities to be measured, the techniques to be employed, and the supporting environmental or operational measurements/observations which may be needed.

32.1.4.1 Measurements to Evaluate the Effect of the Noise on Human Beings

The ultimate aim might be to assess the risk of permanent or temporary hearing loss as a result of exposure to the noise or perhaps the ability of the noise to mask communication or to assess the likely acceptability of the noise in a residential area.

In these cases frequency analysis of the noise may be required, i.e., some measure of its fluctuation in level with time and a measure of the likely duration of the noise. Other factors of importance could be the level of other (background) noises at various times of the day and night, weather conditions, and the nature of the communication or other activity being undertaken. A knowledge of the criteria against which the acceptability or otherwise of the noise is to be judged will also usually indicate the appropriate bandwidths to be used in the noise measurements. In some cases, octave band or one-third octave band measurements could suffice whilst in other cases narrow-band analyses might be required with attendant longer sampling times.

32.1.4.2 Measurements for Engineering Design or Noise-Control Decisions

Most plant installations and many individual items of machinery contain more than one source of noise, and when the ultimate objective is plant noise, control the acoustic measurements have to:

1. Establish the operational and machine installation conditions which give rise to unacceptable noisy operation;
2. Identify and quantify the noise emission from the various sources;
3. Establish an order of priority for noise control of the various sources;
4. Provide adequate acoustic data to assist the design and measure the performance of noise control work.

These requirements imply that detailed noise analyses are always required and that these acoustic measurements must be fully supported by other engineering measurements to

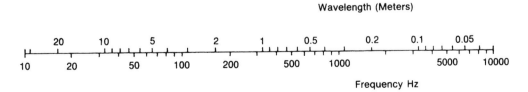

FIGURE 32.2 Relation between wavelength and frequency.

establish the plant operating and installation conditions and, where possible, assist in the design of the noise-control work that will follow. Such measurements could include measurements of temperatures, pressures, gas flows, and vibration levels. All too often the value of acoustic measurements is much reduced because of the absence of these supporting measurements.

32.1.4.3 Noise Labeling

In some cases "noise labeling" of machines may be required by government regulations and by purchasers of production machinery, vehicles, plant used on construction sites, etc. Regulations, where they exist, normally specify the measurement procedure and there are U.S., British, or international standards which apply (see Appendix 32.1).

32.1.4.4 Measurements for Diagnostic Purposes

Investigation of the often complex sources of noise of machines and of ways and means of reducing them may require very detailed frequency analysis of both the noise and vibrations and the advanced features offered by modern analyzing and computing instrumentation.

In recent years, the use of acoustic measurements for "fingerprinting" machinery to detect, in the early stages, changes in the mechanical condition of machinery has been further developed in conjunction with measurements of vibration for the same purpose. These measurements are also based on detailed frequency analyses.

32.2 INSTRUMENTATION FOR THE MEASUREMENT OF SOUND-PRESSURE LEVEL

The basic instrumentation chain consists of a microphone, signal-conditioning electronics, some form of filtering or weighting, and a quantity indicator, analog or digital. This is shown schematically in Figure 32.3.

32.2.1 Microphones

The microphone is the most important item in the chain. It must be physically small so as not to disturb the sound field and hence the sound pressure which it is trying to measure, it must have a high acoustic impedance compared with that of the sound field at the position of measurement, it must be stable in its calibration, have a wide frequency response, and be capable of handling without distortion (amplitude or phase distortion) the very wide dynamic range of sound-pressure levels to which it will be subjected. When used in instruments for the measurement of low sound-pressure levels it must also have an inherent low self-generated noise level. To permit the use of closed cavity methods of calibration it should also have a well-defined diaphragm plane.

All these requirements are best met by the use of condenser or electret microphones, and with such high electrical impedance devices it is essential that the input stage for the microphone—which might take the form of a 0 dB gain impedance transforming unit—be attached directly to the microphone so that long extension cables may be used between the microphone and the rest of the measuring instrument.

32.2.1.1 Condenser Microphone

The essential components of a condenser microphone are a thin metallic diaphragm mounted close to a rigid metal backplate from which it is electrically isolated. An exploded view of a typical instrument is given in Figure 32.4. A stabilized dc polarization voltage E_0 (usually around 200 V) is applied between the diaphragm and the backplate, fed from a high-resistance source to give a time constant $R(C_t + C_s)$ (see Figure 32.5) much longer than the period of the lowest frequency sound-pressure variation to be measured.

If the sound pressure acting on the diaphragm produces a change in the value of C_t—due to movement of the diaphragm—of $\Delta C(t)$ then the output voltage V_0 fed from the microphone to the preamplifier will be

$$V_0(t) = \frac{\Delta C(t) \cdot E_0}{C_t + C_s}$$

since $C_t \gg \Delta C(t)$. It should be noted that the microphone sensitivity is proportional to the polarization voltage but inversely proportional to the total capacitance $C_t + C_s$. Moreover, if we wish the microphone to have a pressure sensitivity independent of frequency then the value of $\Delta C(t)$ (and hence the deflection of the diaphragm) for a given sound pressure must be independent of frequency, that is, the diaphragm must be "stiffness" controlled. This requires a natural frequency above that of the frequency of the sounds to be measured.

Condenser microphones, which are precision instruments, can be manufactured having outstanding stability of calibration and sensitivities as high as 50 mV per Pascal. The selection of a condenser microphone for a specific application is determined by the frequency range to be covered, the dynamic range of sound-pressure levels, and the likely incidence of the sound.

A condenser microphone is a pressure microphone, which means that the electrical output is directly proportional to the sound pressure acting on the diaphragm. At higher frequencies, where the dimensions of the diaphragm become a significant fraction of the wavelength, the presence of the diaphragm in the path of a sound wave creates

CHAPTER | 32 Noise Measurement

APPENDIX 32.1

Standards play a large part in noise instrumentation. Therefore we give a list of international and British documents that are relevant. BS 4727, part 3, Group 08 (1985), *British Standards Glossary of Acoustics and Electroacoustics Terminology particular to Telecommunications and Electronics*, is particularly helpful. This standard covers a very wide range of general terms, defines a wide variety of levels, and deals with transmission and propagation, oscillations, transducers, and apparatus, as well as psychological and architectural acoustics.

International standards	British standard	Title
IEC 225	BS 2475	Octave and third-octave filters intended for the analysis of sound and vibration
IEC 327	BS 5677	Precision method of pressure calibration of one inch standard condenser microphones by the reciprocity method
IEC 402	BS 5678	Simplified method for pressure calibration of one inch condenser microphone by the reciprocity method
IEC 486	BS 5679	Precision method for free field calibration of one inch standard condenser microphones
IEC 537	BS 5721	Frequency weighting for the measurement of aircraft noise
IEC 651	BS 5969	Sound level meters
IEC 655	BS 5941	Values for the difference between free field and pressure sensitivity levels for one inch standard condenser microphones
IEC 704		Test code for determination of airborne acoustical noise emitted by household and similar electrical appliances
IEC 804	BS 6698	Integrating/averaging sound level meters
ISO 140	BS 2750	Field and laboratory measurement of airborne and impact sound transmission in buildings
ISO 226	BS 3383	Normal equal loudness contours for pure tones and normal threshold of hearing under free field listening conditions
ISO 266	BS 3593	Preferred frequencies for acoustic measurements
ISO 362	BS 3425	Method for the measurement of noise emitted by motor vehicles
ISO 717	BS 5821	Rating of sound insulation in buildings and of building elements
ISO 532	BS 4198	Method of calculating loudness
ISO R.1996	BS 4142	Method of rating industrial noise affecting mixed residential and industrial areas
ISO 1999	BS 5330	Acoustics—assessment of occupational noise exposure for hearing conservation purposes
ISO 2204		Acoustics—guide to the measurement of airborne noise and evaluation of its effects on man
ISO 3352		Acoustics—assessment of noise with respect to its effects on the intelligibility of speech
ISO 3740–3746	BS 4196 (Parts 1–6)	Guide to the selection of methods of measuring noise emitted by machinery
ISO 3891	BS 5727	Acoustics—procedure for describing aircraft noise heard on ground
ISO 4871		Acoustics—noise labeling of machines
ISO 4872		Acoustics—measurement of airborne noise emitted by construction equipment intended for outdoor use–method for determining compliance with noise limits
ISO 5130		Acoustics—measurement of noise emitted by stationary vehicles—survey method
ISO 6393		Acoustics—measurement of airborne noise emitted by earth moving machinery, method for determining compliance, with limits for exterior noise—stationary test conditions
	BS 5228	Noise control on construction and demolition sites
	BS 6402	Specification of personal sound exposure meters

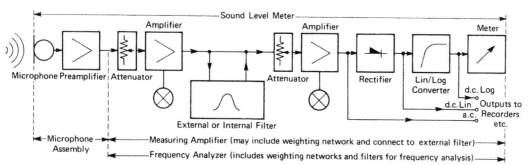

FIGURE 32.3 Block diagram of a noise-measuring system.

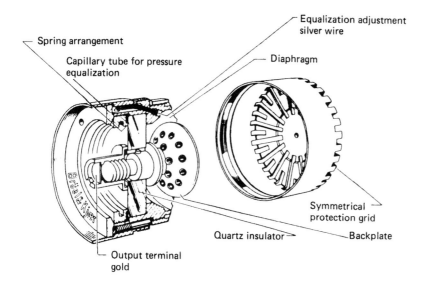

FIGURE 32.4 Exploded view of a condenser microphone.

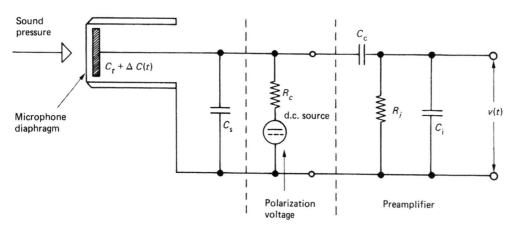

FIGURE 32.5 Equivalent circuit of a condenser microphone and microphone preamplifier.

a high-impedance obstacle, from which the wave will be reflected. When this happens the presence of the microphone modifies the sound field, causing a higher pressure to be sensed by the diaphragm. A microphone with flat pressure response would therefore give an incorrect reading. For this reason, special condenser microphones known as free field microphones are made. They are for use in free field conditions, with perpendicular wave incidence on the diaphragm. Their pressure frequency response is so tailored as to give a flat response to the sound waves which would exist if they were not affected by the presence of the microphone. Free field microphones are used with sound level meters.

Effective pressure increases due to the presence of the microphone in the sound field are shown as free-field corrections in Figure 32.6 for the microphone whose outline is given. These corrections are added to the pressure response of the microphone. Figure 32.7 shows the directional response of a typical ½-inch condenser microphone. Microphones

CHAPTER | 32 Noise Measurement

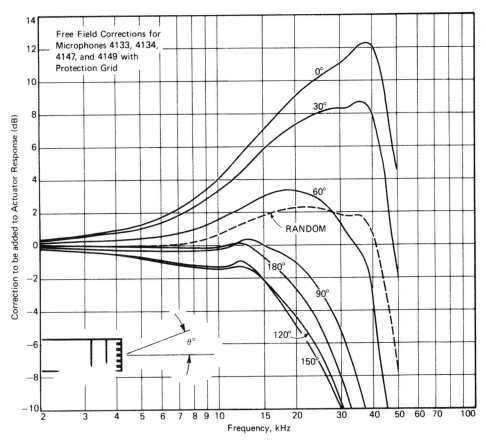

FIGURE 32.6 Directional characteristics of a half-inch microphone mounted on a sound-level meter.

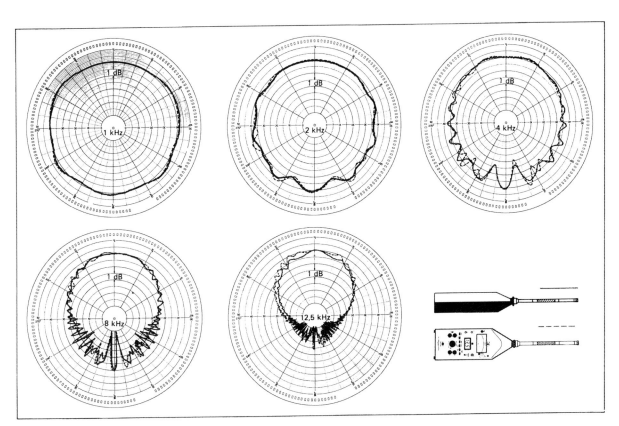

FIGURE 32.7 Free-field corrections to microphone readings for a series of B and K half-inch microphones.

with flat pressure responses are mainly intended for use in couplers and should be used with grazing incidence in free-field conditions.

A complete condenser microphone consists of two parts, a microphone cartridge and a preamplifier, which is normally tubular in shape, with the same diameter as the cartridge. The preamplifier may be built into the body of a sound-level meter.

Generally, a larger diameter of microphone means higher sensitivity, but frequency-range coverage is inversely proportional to diaphragm dimensions. Most commonly used microphones are standardized 1-inch and ½-inch but ¼-inch and even ⅛-inch are commercially available. The small-diaphragm microphones are suitable for the measurement of very high sound-pressure levels.

32.2.1.2 Electret Microphone

Although requirements for precision measurement are best met by condenser microphones, the need to have a source of dc voltage for the polarization has led to the search for microphones which have an inherent polarized element. Such an element is called an electret. In the last fifteen years electrets have been produced which have excellent long-term stability and have been used to produce prepolarized condenser microphones. Many of the design problems have been overcome by retaining the stiff diaphragm of the conventional condenser microphone and applying the electret material as a thin coating on the surface of the backplate. The long-term stability of the charge-carrying element (after artificially aging and stabilizing) is now better than 0.2 dB over a year.

The principle of operation of the electret condenser microphone is, of course, precisely the same as that of the conventional condenser microphone. At present, its cost is slightly in excess of that of a conventional condenser microphone of similar performance but it has application where lower power consumption and simplified associated electronics are at a premium.

32.2.1.3 Microphones for Low-Frequency Work

The performance of condenser microphones at low frequencies is dependent on static pressure equalization and on the ratio of input impedance of the microphone preamplifier to the capacitive impedance of the microphone cartridge. For most measurements, good response to about 10 Hz is required and is provided by most microphones. Extension of performance to 1 or 2 Hz is avoided in order to reduce the sensitivity of a microphone to air movement (wind) and pressure changes due to opening and closing of doors and windows.

Microphones for low-frequency work exist, including special systems designed for sonic boom measurement, which are capable of excellent response down to a small fraction of 1 Hz.

32.2.2 Frequency Weighting Networks and Filters

The perception of loudness of pure tones is known and shown in Figure 32.1. It can be seen that it is a function of both the frequency of the tone and of the sound pressure. Perception of loudness of complete sounds has attracted much research, and methods have been devised for the computation of loudness or loudness level of such sounds.

Many years ago it was thought that an instrument with frequency-weighting characteristics corresponding to the sensitivity of the ear at low, medium, and high sound intensities would give a simple instrument capable of measuring loudness of a wide range of sounds, simple and complex. Weightings now known as A, B, and C were introduced, but it was found that, overall, the A weighting was most frequently judged to give reasonable results and the measurement of A-weighted sound pressure became accepted and standardized throughout the world. Their response curves are shown in Figure 32.8. Environmental noise and noise at work are measured with A weighting, and many sound-level measuring instruments no longer provide weightings other than A.

It cannot be claimed that A weighting is perfect, that the result of A-weighted measurements is a good measure of loudness or nuisance value of all types of noise, but at least they provide a standardized, commonly accepted method.

Where it is acknowledged that A weighting is a wrong or grossly inadequate description, and this is particularly likely when measuring low-frequency noise, sounds are analyzed by means of octave or third-octave filters and the results manipulated to give a desired value.

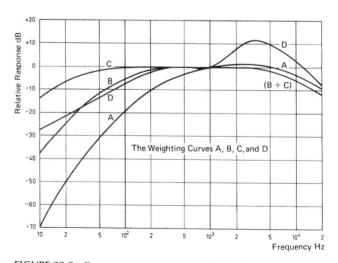

FIGURE 32.8 Frequency response curves of the A, B, C, and D weighting networks.

CHAPTER | 32 Noise Measurement

Aircraft noise is a special case, where the noise is analyzed simultaneously in third-octave bands and individual band levels are sampled every 0.5 s, each one weighted according to a set of tables and summed, to give the answer in forms of a perceived noise level in decibels. An approximation to this figure may be obtained by the use of a simpler system, a sound-level meter with standardized "D" weighting, also shown in Figure 32.8.

32.2.3 Sound-Level Meters

These are most widely used for the measurement of sound and can be purchased in forms ranging from simple instruments fitted with one frequency weighting network only and a fixed microphone to versions capable of handling a whole range of microphone probes and sophisticated signal processing with digital storage of data. The instrument therefore deserves special mention.

As with most other instruments, its cost and complication is related to the degree of precision of which it is capable. IEC Publication 651 and BS 5969 classify sound-level meters according to their degree of precision. The various classes (or types) and the intended fields of application are given in Table 32.1.

The performance specifications for all these types have the same "center value" requirements, and they differ only in the tolerances allowed. These tolerances take into account such aspects as (1) variation of microphone response with the angle of incidence of the sound, (2) departures from design objectives in the frequency-weighting networks (tolerances allowed) and (3) level linearity of the amplifiers.

While it is not possible to state the precision of measurement for any type in any given application—since the overall accuracy depends on care with the measurement technique as well as the calibration of the measurement equipment—it is possible to state the instrument's absolute accuracy at the reference frequency of 1 kHz when sound is incident from the reference direction at the reference sound-pressure level (usually 94 dB), and these values are given in Table 32.1.

Apart from the absolute accuracy at the reference frequency the important differences in performance are:

1. Frequency range over which reliable measurements may be taken—with Type 2 and 3 instruments having much wider tolerances at both the low and the high ends of the audio frequency range (see Figure 32.9).
2. Validity of rms value when dealing with high crest factor (high ratio of peak to rms) signals, i.e., signals of an impulsive nature such as the exhaust of motorcycles and diesel engines.
3. Validity of rms value when dealing with short-duration noises.
4. Linearity of readout over a wide dynamic range and when switching ranges.

Prior to 1981, sound-level meters were classified either as "Precision grade" (see IEC Publication 179 and BS 4197) or "industrial grade" (see IEC 123 or BS 3489) with very wide tolerances. The current Type 1 sound-level meter is equivalent, broadly, to the old precision grade whilst Type 3 meters are the equivalent of the old industrial grade with minor modifications. A typical example of a simple sound-level meter which nevertheless conforms to Type 1 class of accuracy is shown in Figure 32.10. It has integrating facilities as discussed below. Figure 32.11 shows a more sophisticated instrument which includes filters but is still quite compact.

32.2.3.1 Integrating Sound-Level Meters

Standard sound-level meters are designed to give the reading of sound level with fast or slow exponential time averaging. Neither is suitable for the determination of the "average" value if the character of the sound is other than steady or is composed of separate periods of steady character. A special category of "averaging" or "integrating" meters is standardized and available, which provides the "energy average," measured over a period of interest.

The value measured is known as L_{eq} (the equivalent level), which is the level of steady noise which, if it persisted over the whole period of measurement, would give the same total energy as the fluctuating noise measured over the period.

As in the case of sound-level meters, Types 0 and 1 instruments represent the highest capability of accurately measuring noises which vary in level over a very wide dynamic range, or which contain very short bursts or impulse, for example gunshots, hammer-blows, etc. Type 3 instruments are not suitable for the accurate measurement of such events but may be used in cases where level variation is not wide or sudden.

TABLE 32.1 Classes of sound-level meter

Type	Intended field of application	Absolute accuracy (at reference frequency) in reference direction at the reference sound-pressure level (dB)
0	Laboratory reference standards	±0.4
1	Laboratory use and field use where the acoustical environment can be closely specified or controlled	±0.7
2	General field work	±1.0
3	Field noise surveys	±1.5

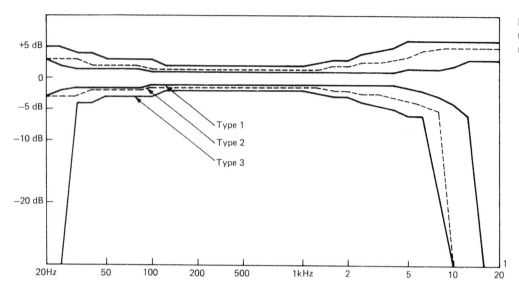

FIGURE 32.9 Tolerances in the response of sound-level meters at different frequencies.

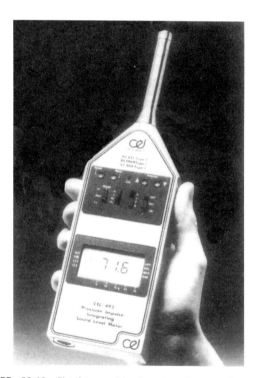

FIGURE 32.10 Simple sound-level meter. Courtesy Lucas CEL Instruments Ltd.

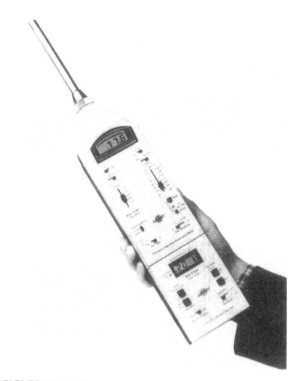

FIGURE 32.11 More sophisticated sound-level meter.

The precise definition of the L_{eq} value of a noise waveform over a measurement period T is given by

$$L_{eq} = 10 \log \frac{1}{T} \int_0^T \frac{[p(t)]^2}{p_0^2} dt$$

where L_{eq} is the equivalent continuous sound level, T the measurement duration, $p(t)$ the instantaneous value of the (usually A-weighted) sound pressure and p_0 the reference rms sound pressure of 20 µPa.

A measure of the A-weighted L_{eq} is of great significance when considering the possibility of long-term hearing loss arising from habitual exposure to noise (see Barns and Robinson 1970).

Most integrating sound-level meters also provide a value known as sound energy level or SEL, sometimes also described as single-event level. This represents the sound energy of a short-duration event, such as an aircraft flying overhead or an explosion, but expressed as if the event lasted only 1 s. Thus

$$\text{SEL} = 10 \log \int \frac{[p(t)]^2}{p_0^2} dt$$

Many integrating sound-level meters allow the interruption of the integrating process or the summation of energy in separated periods.

32.2.3.2 Statistical Sound-Level Meters

In the assessment of environmental noise, for example, that produced by city or motorway traffic, statistical information is often required. Special sound-level meters which analyze the sound levels are available. They provide probability and cumulative level distributions and give percentile levels, for example L_{10}, the level exceeded for 10 percent of the analysis time as illustrated in Figures 32.12 and 32.13. Some of these instruments are equipped with built-in printers and can be programmed to print out a variety of values at preset time intervals.

32.2.4 Noise-Exposure Meters/Noise-Dose Meters

The measurement of noise exposure to which a worker is subjected may be carried out successfully by an integrating sound-level meter if the worker remains in one location throughout the working day. In situations where he or she moves into and out of noisy areas or performs noisy operations and moves at the same time a body-worn instrument is required.

BS 6402 describes the requirements for such an instrument. It may be worn in a pocket but the microphone must be capable of being placed on the lapel or attached to a safety helmet. Figure 32.14 shows one in use with the microphone mounted on an earmuff and the rest of the equipment in a breast pocket. The instrument measures A-weighted exposure according to

$$E = \int_0^T \text{Pa}^2(t) dt$$

where E is the exposure (Pa^2 · h) and T is the time (h).

Such an instrument gives the exposure in Pascal-squared hours, where 1 Pa^2 · h is the equivalent of 8 h exposure to a steady level of 84.9 dB. It may give the answers in terms of percentage of exposure as allowed by government regulations. For example, 100 percent may mean 90 dB for 8 h, which is the maximum allowed with unprotected ears by the current regulations.

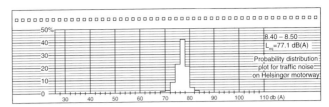

FIGURE 32.12 Probability distribution plot for motorway noise.

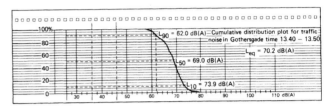

FIGURE 32.13 Cumulative distribution plot for traffic noise in a busy city street.

FIGURE 32.14 Noise-exposure meter in use.

The actual performance requirements are similar to those of Type 2 sound-level meters and integrating sound-level meters, as accuracy of measurement is much more dependent on the clothing worn by the worker, the helmet or hat, and the direction of microphone location in relation to noise source than on the precision of the measuring instrument.

32.2.5 Acoustic Calibrators

It is desirable to check the calibration of equipment from time to time, and this can be done with acoustic calibrators, devices which provide a stable, accurately known sound pressure to a microphone which is part of an instrument system. Standards covering the requirements for such calibrators are in existence, and they recommend operation at a level of 94 dB (1 Pa) or higher and a frequency of operation between 250 and 1000 Hz. The latter is often used, as it is a frequency at which the A-weighting network provides 0 dB insertion loss, so the calibration is valid for both A-weighted and unweighted measurement systems.

One type of calibrator known as a pistonphone offers extremely high order of stability, both short and long term, with known and trusted calibration capability to a tolerance within 0.1 dB. In pistonphones the level is normally 124 dB and the frequency 250 Hz.

The reason for the relatively high sound-pressure levels offered by calibrators is the frequent need to calibrate in the field in the presence of high ambient noise levels. Most calibrators are manufactured for operation with one specific microphone or a range of microphones. Some calibrators are so designed that they can operate satisfactorily with microphones of specified (usually standardized) microphone diameters but with different diaphragm constructions and protective grids (see Figures 32.15 and 32.16).

In a pistonphone the sound pressure generated is a function of total cavity volume, which means dependence on the diaphragm construction. In order to obtain the accuracy of calibration offered by a pistonphone the manufacturer's manual must be consulted regarding corrections required for specific types of microphones. The pressure generated by a pistonphone is also a function of static pressure in the cavity, which is linked with the atmosphere via a capillary tube. For the highest accuracy, atmospheric pressure must be measured and corrections applied.

Some calibrators offer more than one level and frequency. Calibrators should be subjected to periodic calibrations in approved laboratories. Calibration is discussed further in Section 32.6.

32.3 FREQUENCY ANALYZERS

Frequently the overall A-weighted level, either instantaneous, time-weighted, or integrated over a period of time, or even statistical information on its variations are insufficient descriptions of sound. Information on frequency content of sounds may be required. This may be in terms of content in standardized frequency bands, of octave or thirdoctave bandwidth. There are many standards which call for presentation of noise data in these bands.

Where information is required for diagnostic purposes very detailed, high-resolution analysis is performed. Some analog instruments using tuned filter techniques are still in use—the modern instrument performs its analyzing function by digital means.

32.3.1 Octave Band Analyzers

These instruments are normally precision sound-level meters which include sets of filters or allow sets of filters to be

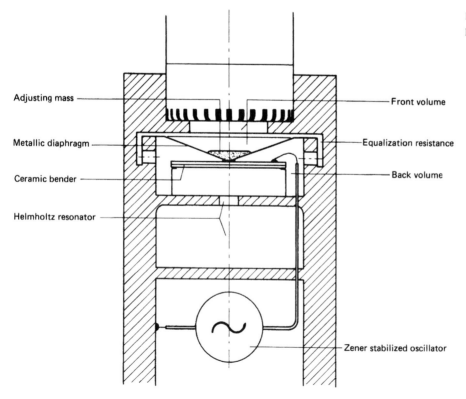

FIGURE 32.15 Principle of operation of a portable sound-level calibrator.

CHAPTER | 32 Noise Measurement

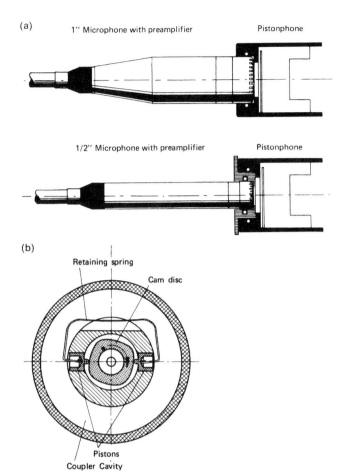

FIGURE 32.16 Pistonphone. (a) Mounting microphones on a pistonphone; (b) cross-sectional view showing the principle of operation.

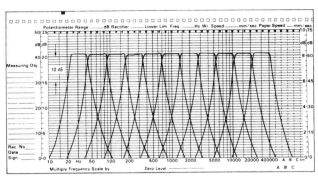

FIGURE 32.17 Frequency characteristics of octave filters in a typical analyzer.

connected, so that direct measurements in octave bands may be made. The filters are bandpass filters, whose passband encompasses an octave, i.e., the upper band edge frequency equals twice the lower band edge one. The filters should have smooth response in the passband, preferably offering 0 dB insertion loss. The center frequency fm is the geometric mean

$$fm = \sqrt{(f_2 \cdot f_1)}$$

where f_2 is the upper band edge frequency and f_1 the lower band edge frequency.

Filters are known by their nominal center frequencies. Attenuation outside the passband should be very high. ISO Standard 225 and BS 3593 list the "preferred" frequencies while IEC 266 and BS 2475 give exact performance requirements for octave filters. A typical set of characteristics is given in Figure 32.17.

If a filter set is used with a sound-level meter then such a meter should possess a "linear" response function, so that the only filtering is performed by the filters. Answers obtained in such analyses are known as "octave band levels" in decibels, re 20 μPa.

From time to time, octave analysis is required for the purpose of assessing the importance of specific band levels in terms of their contribution to the overall A-weighted level. Octave band levels may then be corrected by introducing A-weighting attenuation at center frequencies. Some sound-level meters allow the use of filters with A weighting included, thus giving answers in A-weighted octave band levels.

A simple portable instrument is likely to contain filters spanning the entire audio frequency range. The analyzer performs by switching filters into the system in turn and indicating results for each individual band. If noises tend to fluctuate, "slow" response of the sound-level meter is used, but this may still be too fast to obtain a meaningful average reading.

Special recorders, known as level recorders, may be used with octave band filters/sound-level meters which allow a permanent record to be obtained in the field. Parallel operation of all the filters in a set is also possible with special instruments, where the answers may be recorded in graphical form or in digital form, and where the octave spectrum may be presented on a CRT.

32.3.2 Third-Octave Analyzers

Octave band analysis offers insufficient resolution for many purposes. The next step is thirdoctave analysis by means of filters meeting ISO 266, IEC 225, and BS 2475. These filters may again be used directly with sound-level meters. As there are three times as many filters covering the same total frequency span of interest and the bandwidths are narrower, requiring longer averaging to obtain correct answers in fluctuating noise situations, the use of recording devices is almost mandatory. Typical characteristics are shown in Figure 32.18.

Apart from the field instruments described above, high-quality filter sets with automatic switching and recording facilities exist, but the modern trend is for parallel sets of filters used simultaneously with graphical presentation of third-octave spectra on CRT.

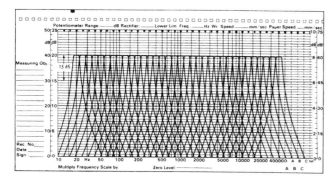

FIGURE 32.18 Frequency characteristics of third-octave filters in a typical analyzer.

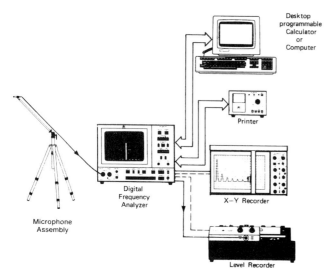

FIGURE 32.19 Output options with a typical digital frequency analyzer.

Digital filter instruments are also available, in which the exact equivalent of conventional, analog filtering techniques is performed. Such instruments provide time coincident analysis, exponential averaging in each band or correct integration with time, presentation of spectra on a CRT and usually XY recorder output plus interfacing to digital computing devices (see Figure 32.19).

32.3.3 Narrow-Band Analyzers

Conventional analog instruments perform the analysis by a tunable filter, which may be of constant bandwidth type (Figure 32.20) or constant percentage (Figure 32.21). Constant percentage analyzers may have a logarithmic frequency sweep capability, while constant bandwidth would usually offer a linear scan. Some constant bandwidth instruments have a synchronized linkup with generators, so that very detailed analysis of harmonics may be performed. The tuning may be manual, mechanical, and linked with a recording device, or electronic via a voltage ramp. In the last case computer control via digital-to-analog conversion may be achieved. With narrow band filtering of random signals, very

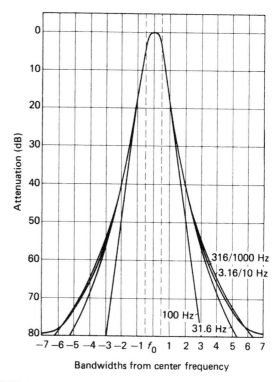

FIGURE 32.20 Characteristics of a typical constant bandwidth filter.

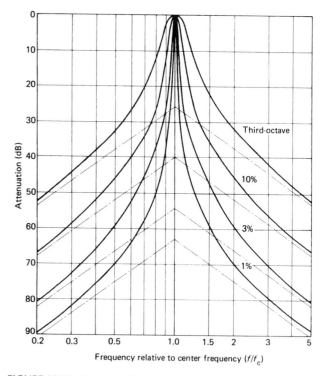

FIGURE 32.21 Bandpass filter selectivity curves for four constant percentage bandwidths.

long averaging times may be required, making analysis very time consuming.

Some narrow-band analyzers attempt to compute octave and third-octave bands from narrow-band analysis, but at present such techniques are valid only for the analyses of stationary signals.

32.3.4 Fast Fourier Transform Analyzers

These are instruments in which input signals are digitized and fast Fourier transform (FFT) computations performed. Results are shown on a CRT and the output is available in analog form for XY plotters or level recorders or in digital form for a variety of digital computing and memory devices.

Single-channel analyzers are used for signal analysis and two-channel instruments for system analysis, where it is possible to directly analyze signals at two points in space and then to perform a variety of computations in the instruments. In these, "real-time" analysis means a form of analysis in which all the incoming signal is sampled, digitized, and processed. Some instruments perform real-time analysis over a limited frequency range (for example, up to 2 KHz). If analysis above these frequencies is performed, some data are lost between the data blocks analyzed. In such a case, stationary signal analyses may be performed but transients may be missed.

Special transient analysis features are often incorporated (for example, sonic boom) which work with an internal or external trigger and allow the capture of a data block and its analysis. For normal operation, linear integration or exponential averaging are provided. The analysis results are given in 250–800 spectral lines, depending on type.

In a 400-line analyzer, analysis would be performed in the range of, say, 0–20 Hz, 0–200 Hz, 0–2000 Hz and 0–20 000 Hz. The spectral resolution would be the upper frequency divided by number of spectral lines, for example, 2000/400, giving 5 Hz resolution.

Some instruments provide a "zoom" mode, allowing the analyzer to show results of analysis in part of the frequency range, focusing on selected frequency in the range and providing much higher resolution (see Figure 32.22).

Different forms of cursor are provided by different analyzers, which allow the movement of the cursor to a part of the display to obtain a digital display of level and frequency at the point.

The frequency range of analysis and spectral resolution are governed by the internal sampling rate generator, and sampling rate may also be governed by an external pulse source. Such a source may be pulses proportional to a rotating speed of a machine, so that analysis of the fundamental and some harmonic frequencies may be observed, with the spectral pattern stationary in the frequency domain in spite of rotating speed changes during run-up tests.

Most analyzers of the FFT type will have a digital interface, either parallel or serial, allowing direct use with desktop computers in common use. In two-channel analyzers, apart from the features mentioned above, further processing can provide such functions as transfer function (frequency response) correlation functions, coherence function, cepstrum, phase responses, probability distribution, and sound intensity.

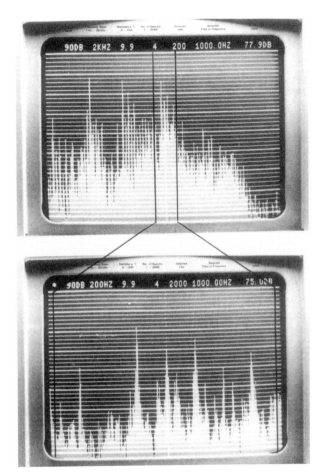

FIGURE 32.22   Scale expansion in presentation of data on spectra.

32.4 RECORDERS

32.4.1 Level Recorders

Level recorders are recording voltmeters, which provide a record of dc or the rms value of ac signal on linear or logarithmic scale; some provide synchronization for octave and third-octave analyzers, so that filter switching is timed for frequency-calibrated recording paper or synchronization with generators for frequency-response plotting. Such recorders are also useful for recording noise level against time.

32.4.2 XY Plotters

These are recorders in which the Y plot is of dc input and X of a dc signal proportional to frequency. They are frequently used with FFT and other analyzers for the recording of memorized CRT display.

32.4.3 Digital Transient Recorders

Digital transient recorders are dedicated instruments specifically designed for the capture of transients such as gunshot, sonic boom, mains transient, etc. The incoming signals are digitized at preset rates and results captured in a memory when commanded by an internal or external trigger system, which may allow pre- and post-trigger recording. The memorized data may be replayed in digital form or via a D/A converter, usually at a rate hundreds or thousands of times slower than the recording speed, so that analog recordings of fast transients may be made.

32.4.4 Tape Recorders

These are used mainly for the purpose of gathering data in the field. Instrumentation tape recorders tend to be much more expensive than domestic types, as there are stringent requirements for stability of characteristics. Direct tape recorders are suitable for recording in the AF range, but where infrasonic signals are of interest FM tape recorders are used. The difference in range is shown in Figure 32.23, which also shows the effect of changing tape speed.

A direct tape recorder is likely to have a better signal-to-noise ratio, but a flatter frequency response and phase response will be provided by FM type.

IRIG standards are commonly followed, allowing recording and replaying on different recorders. The important standardization parameters are head geometry and track configuration, tape speeds, center frequencies, and percentage of frequency modulation. Tape reels are used as well as tape cassettes, and the former offer better performance.

Recent developments in digital tape recorders suitable for field-instrumentation use enhance performance significantly in the dynamic range.

32.5 SOUND-INTENSITY ANALYZERS

Sound-intensity analysis may be performed by two-channel analyzers offering this option. Dedicated sound-intensity analyzers, based on octave or third-octave bands, are available, mainly for the purpose of sound-power measurement and for the investigation of sound-energy flow from sources.

The sound-intensity vector is the net flow of sound energy per unit area at a given position, and is the product of sound pressure and particle velocity at the point. The intensity vector component in a given direction r is

$$I_r = \overline{p \times u_r}$$

where the horizontal bar denotes time average.

To measure particle velocity, current techniques rely on the finite difference approximation by means of integrating over time the difference in sound pressure at two points, A and B, separated by ΔR, giving

$$u_r = -\rho_0 \int \frac{(P_B - P_A)}{\Delta R} dt$$

where P_A and P_B are pressures at A and B and ρ_0 is the density of air. An outline of the system is given in Figure 32.24. This two-microphone technique requires the two transducers and associated signal paths to have very closely matched phase and amplitude characteristics. Only a small fraction of a degree in phase difference may be tolerated. Low

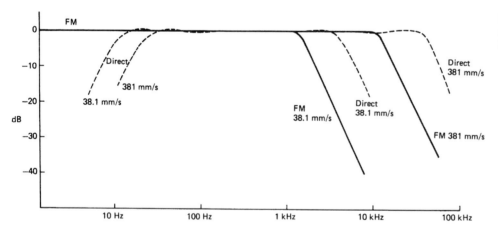

FIGURE 32.23 Typical frequency response characteristics of AM and FM recording.

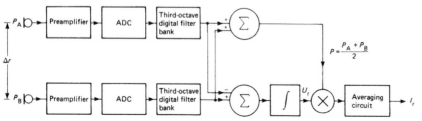

FIGURE 32.24 Sound-intensity analysis using two channels.

frequencies require relatively large microphone separation (say, 50 mm) while at high frequencies only a few millimeters must be used. Rapid developments are taking place in this area.

32.6 CALIBRATION OF MEASURING INSTRUMENTS

32.6.1 Formal Calibration

This is normally performed by approved laboratories, where instrumentation used is periodically calibrated and traceable to primary standards; the personnel are trained in calibration techniques, temperature is controlled, and atmospheric pressure and humidity accurately measured.

Instruments used for official measurements, for example by consultants, environmental inspectors, factory inspectors, or test laboratories, should be calibrated and certified at perhaps one- or two-year intervals. There are two separate areas of calibration, the transducers and the measuring, analyzing, and computing instruments.

Calibration of transducers, i.e., microphones, requires full facilities for such work. Condenser microphones of standard dimensions can be pressure calibrated by the absolute method of reciprocity. Where accurate free-field corrections are known for a specific type of condenser microphone, these can be added to the pressure calibration. Free-field reciprocity calibration of condenser microphones (IEC 486 and BS 5679) is very difficult to perform and requires a first-class anechoic chamber.

Microphones other than condenser (or nonstandard size condenser) can be calibrated by comparison with absolutely calibrated standard condenser microphones. This again requires an anechoic chamber.

Pressure calibration of condenser microphones may be performed by means of electrostatic actuators. Although this is not a standardized method, it offers very good results when compared with the reciprocity method, is much cheaper to perform, and gives excellent repeatability. It is not a suitable method for frequencies below about 50 Hz. Calibration methods for two-microphone intensity probes are in the stage of development and standardization.

Sound-level meter calibration is standardized with a recommendation that acoustic tests on the complete unit are performed. This is a valid requirement if the microphone is attached to the body of the sound-level meter. Where the microphone is normally used on an extension or with a long extension cable it is possible to calibrate the microphone separately from the electronics. Filters and analyzers require a purely electronic calibration, which is often time consuming if not computerized.

Microphone calibrators (see Section 32.2.5) are devices frequently calibrated officially. This work is normally done with traceable calibrated microphones or by means of direct comparison with a traceable calibrator of the same kind. Sophisticated analyzers, especially those with digital interfaces, lend themselves to computer-controlled checks and calibrations of all the complicated functions. The cost of this work by manual methods would be prohibitive.

32.6.2 Field Calibration

This normally implies the use of a calibrated, traceable microphone calibrator (for example, a pistonphone) which will provide a stable, accurately known sound pressure, at a known frequency, to a microphone used in the field instrument set-up (for example, a sound-level meter).

Although such "calibration" is only a calibration at one point in frequency domain and at one level, it is an excellent check, required by most measurement standards. Its virtue is in showing departures from normal operation, signifying the need for maintenance and recalibration of instrumentation.

32.6.3 System Calibration

Coomplete instrument systems—for example, a microphone with an analyzer or a microphone with a sound-level meter and a tape recorder—may be calibrated by a good-quality field calibrator. Again, the calibration would be at one frequency and one level, but if all instrument results line up within a tight tolerance a high degree of confidence in the entire system is achieved.

32.6.4 Field-System Calibration

Some condenser microphone preamplifiers are equipped with a facility for "insert voltage calibration." This method is used mainly with multimicrophone arrays, where the use of a calibrator would be difficult and time consuming to perform. Permanently installed outdoor microphones may offer a built-in electrostatic actuator which allows regular (for example, daily) microphone and system checks.

32.7 THE MEASUREMENT OF SOUND-PRESSURE LEVEL AND SOUND LEVEL

The measurement of sound-pressure level (spl) implies measurement without any frequency weighting, whereas sound level is frequency weighted spl. All measurement systems have limitations in the frequency domain, and, even though many instruments have facilities designated as "linear," these limitations must be borne in mind. Microphones have low-frequency limitations, as well as complex responses at the high-frequency end, and the total response of the measurement system is the combined response of the transducer and the following electronics.

When a broad-band signal is measured and spl quoted, this should always be accompanied by a statement defining the limits of the "linear" range, and the response outside it. This is especially true if high levels of infrasound and/or ultrasound are present.

In the measurement of sound level, the low-frequency performance is governed by the shape of the weighting filter, but anomalies may be found at the high-frequency end. The tolerances of weighting networks are very wide in this region, the microphones may show a high degree of directionality (see Section 32.2.1.1), and there may also be deviations from linearity in the response of the microphone.

Sound-pressure level or sound level is a value applicable to a point in space, the point at which the microphone diaphragm is located. It also applies to the level of sound pressure at the particular point in space, when the location of sound source or sources is fixed and so is the location of absorbing or reflective surfaces in the area as well as of people.

In free space, without reflective surfaces, sound radiated from a source will flow through space, as shown in Figure 32.25. However, many practical sources of noise have complex frequency content and also complex radiation patterns, so that location of the microphone and distance from the source become important as well as its position relating to the radiation pattern. In free space sound waves approximate to plane waves.

32.7.1 Time Averaging

The measurement of sound-pressure level or sound level (weighted sound-pressure level), unless otherwise specified, means the measurement of the rms value and the presentation of results in decibels relative to the standardized reference pressure of 20 μPa.

The rms value is measured with exponential averaging, that is, temporal weightings or time constants standardized for sound-level meters. These are known as F and S (previously known as "Fast" and "Slow").

In modern instruments, the rms value is provided by wide-range log • mean square detectors. The results are presented on an analog meter calibrated in decibels, by a digital display, or by a logarithmic dc output which may easily be recorded on simple recorders.

Some instruments provide the value known as "instantaneous" or "peak" value. Special detectors are used for this purpose, capable of giving the correct value of signals lasting for microseconds. Peak readings are usually taken with "linear" frequency weighting.

In addition to the above, "impulse"-measuring facilities are included in some sound-level measuring instruments. These are special detectors and use averaging circuits which respond very quickly to very short-duration transient sounds but which give very slow decay from the maximum value reached. At the time of writing, no British measurement standards or regulations require or allow the use of "impulse" value other than for determination as to whether the sound is of an impulsive character or not.

One specific requirement for maximum "fast" level is in vehicle drive-by test, according to BS 3425 and ISO 362. The "slow" response gives slightly less fluctuating readout than the "fast" mode—its main use is in "smoothing" out the fluctuations.

"Impulse" time weighting may be found in some sound-level meters, according to IEC 651 and BS 5969. This form of time weighting is mainly used in Germany, where it originated. The value of meters incorporating this feature is in their superior rms detection capability when used in fast or slow mode.

Occasionally, there is a requirement for the measurement of maximum instantaneous pressure, unweighted. This can be performed with a sound-level meter with a linear response or a microphone and amplifier. The maximum instantaneous value, also known as peak value, may be obtained by displaying ac output on a storage oscilloscope or by using an instrument which faithfully responds to very short rise times and which has a "hold" facility for the peak value. These hold circuits have limitations, and

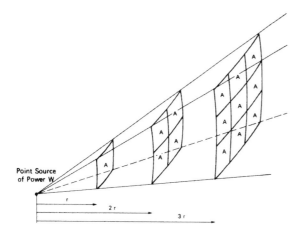

FIGURE 32.25 Flow of sound from a source.

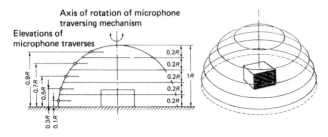

FIGURE 32.26 Coaxial circular paths for microphone traverses.

their specifications should be studied carefully. The peak value may be affected by the total frequency and phase response of the system.

32.7.2 Long Time Averaging

In the measurement of total sound energy at a point over a period of time (perhaps over a working day) as required by environmental regulations, such as those relating to noise on construction and demolition sites or hearing-conservation regulations, special rules apply. The result required is in the form of L_{eq}, or equivalent sound level, which represents the same sound energy as that of the noise in question, however, fluctuating with time.

The only satisfactory way of obtaining the correct value is to use an integrating/averaging sound-level meter to IEC 804 of the correct type. Noises with fast and wide-ranging fluctuations demand the use of Type 0 or Type 1 instruments. Where fluctuations are not so severe, Type 2 will be adequate.

The value known as SEL (sound-exposure level), normally available in integrating sound-level meters, is sometimes used for the purpose of describing short-duration events, for example, an aircraft flying overhead or an explosion. Here the total energy measured is presented as though it occurred within 1 s. At present, there are no standards for the specific use of SEL in the UK.

32.7.3 Statistical Distribution and Percentiles

Traffic noise and environmental noise may be presented in statistical terms. Requirements for the location of microphones in relation to ground or facades of buildings are to be found in the relevant regulations. Percentile values are the required parameter. L_{10} means a level present or exceeded for 10 percent of the measurement time and L_{50} the level exceeded for 50 percent of the time. The value is derived from a cumulative distribution plot of the noise, as discussed in Section 32.2.3.2.

32.7.4 Space Averaging

In the evaluation of sound insulation between rooms a sound is generated and measured in one room and also measured in the room into which it travels. In order to obtain information of the quality of insulation relating to frequency and compile an overall figure in accordance with relevant ISO and BS Standards, the measurements are carried out in octave and third-octave bands.

The sound is generated normally as a broadband (white) noise or bands of noise. As the required value must represent the transfer of acoustic energy from one room to another, and the distribution of sound in each room may be complex, "space averaging" must be carried out in both rooms. Band-pressure levels (correctly time averaged) must be noted for each microphone location and the average value found.

The space average for each frequency band is found by

$$L_p = 10 \log \left\{ \frac{1}{N} \left[\sum_{i=1}^{N} 10^{0.1 L_{pi}} \right] \right\}$$

L_p = space average level

where L_{pi} is the sound-pressure level at the ith measurement point and N the total number of measurement points. If large differences are found over the microphone location points in each band, a larger number of measurement points should be used. If very small variations occur, it is quite legitimate to use an arithmetic average value.

This kind of measurement may be carried out with one microphone located at specified points and levels noted and recorded, making sure that suitable time averaging was used at each point. It is possible to use an array of microphones and use a scanning (multiplexing) arrangement, feeding the outputs into an averaging analyzer. The space average value may also be obtained by using one microphone on a rotating boom, feeding its output into an analyzer with averaging facility. The averaging should cover a number of complete revolutions of the boom.

In integrating/averaging a sound-level meter may be used to obtain the space average value without the need for any calculations. It requires the activation of the averaging function for identical periods of time at each predetermined location and allowing the instrument to average the values as total L_{eq}. With the same time allocated to each measurement point, L_{eq} = space average.

32.7.5 Determination of Sound Power

As sound radiation from sources may have a very complex pattern, and sound in the vicinity of a source is also a function of the surroundings, it is not possible to describe or quantify the sound emission characteristics in a simple way by making one or two measurements some distance away, nor even by making a large number of such measurements. The only method is to measure the total sound power radiated by the source. This requires the measurement to be carried out in specific acoustic conditions. ISO 3740–3746 and BS 4196 specify the acoustic conditions suitable for such tests and assign uncertainties which apply in the different conditions.

These standards describe the measurements of pressure, over a theoretical sphere or hemisphere or a "box" which encloses the sound source, the computation of space average, and finally, the computation of sound-power level in decibels relative to 1 μW or sound power level in watts. The results are often given in octave or third-octave bands or are

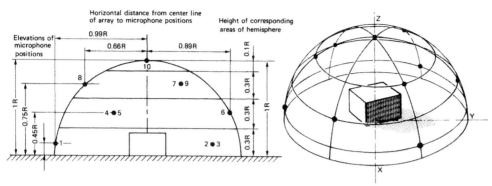

FIGURE 32.27 Microphone positions on equal areas of the surface of a hemisphere.

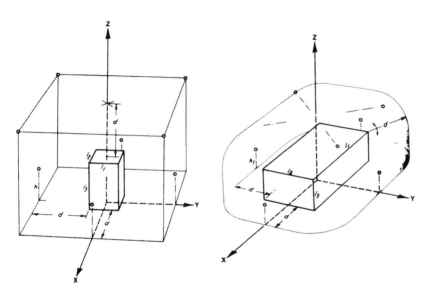

FIGURE 32.28 Microphone positions on a parallelepiped and a conformal surface.

A weighted. Figures 32.27 and 32.28 show the appropriate positions for a microphone.

The great disadvantage of obtaining sound power from the measurement of sound pressure is that special, well defined acoustic conditions are required, and the achievement of such conditions for low uncertainties in measurements is expensive and often impossible. Large, heavy machinery would require a very large free space or large anechoic or reverberation chambers and the ability to run the machinery in them, i.e., provision of fuel or other source of energy, disposal of exhausts, coupling to loads, connection to gearboxes, etc.

With smaller sound sources it may be possible to perform the measurement by substitution or juxtaposition method *in situ*. These methods are based on measurements of noise of a source in an enclosed space, at one point, noting the results and then removing the noise source and substituting it by a calibrated sound source and comparing the sound-pressure readings.

If the unknown sound source gives a reading of X dB at the point and the calibrated one gives $X + N$ dB, then the sound-power level of the unknown source is N dB lower than that of the calibrated source.

The calibrated source may have variable but calibrated output. In situations where it is not possible to stop the unknown noise source its output may be adjusted so that the two sources together, working from the same or nearly the same location, give a reading 3dB higher than the unknown source. When this happens, the power output of both sources is the same.

The two methods described above are simple to use but must be considered as approximations only.

32.7.6 Measurement of Sound Power by Means of Sound Intensity

Accurate and reliable sound-power measurement by means of pressure has the great disadvantage of requiring either free-field conditions in an open space or anechoic chamber or a well diffused field in a reverberation chamber. Open spaces are difficult to find as are rain-free and wind-free conditions.

Anechoic chambers are expensive to acquire, and it is difficult to achieve good echo-free conditions at low frequencies; neither is it often convenient or practical to take machinery and install it in a chamber. Reverberation rooms are also expensive to build, and diffuse field conditions in them are not very easy to achieve.

In all above conditions machines require transporting, provision of power, gears, loads, etc., and this may mean that other sources or noise will interfere with the measurement.

The technique of measuring sound intensity, as discussed in Section 32.5, allows the measurement of sound power radiated by a source under conditions which would be considered adverse for pressure-measurement techniques. It allows the source to be used in almost any convenient location, for example, on the factory floor where it is built or assembled or normally operating, in large or small enclosures, near reflecting surfaces and, what is extremely important, in the presence of other noises which would preclude the pressure method. It allows the estimation of sound power output of one item in a chain of several.

At present the standards relating to the use of this technique for sound-power measurement are being prepared but, due to its many advantages, the technique is finding wide acceptance. In addition, it has great value in giving the direction of energy flow, allowing accurate location of sources within a complex machine.

32.8 EFFECT OF ENVIRONMENTAL CONDITIONS ON MEASUREMENTS

32.8.1 Temperature

The first environmental condition most frequently considered is temperature. Careful study of instrument specifications will show the range of temperatures in which the equipment will work satisfactorily, and in some cases the effect (normally quite small) on the performance. Storage temperatures may also be given in the specifications. It is not always sufficient to look at the air temperature; metal instrument cases, calibrators, microphones, etc., may be heated to quite high temperatures by direct sunlight. Instruments stored in very low temperatures—for example, those carried in a boot of a car in the middle of winter—when brought into warm and humid surroundings may become covered by condensation and their performance may suffer. In addition, battery life is significantly shortened at low temperatures.

32.8.2 Humidity and Rain

The range of relative humidity for which the instruments are designed is normally given in the specifications. Direct rain on instruments not specifically housed in showerproof (NEMA 4) cases must be avoided. Microphones require protective measures recommended by the manufacturers.

32.8.3 Wind

This is probably the worst enemy. Wind impinging on a microphone diaphragm will cause an output, mainly at the lower end of the spectrum (see Figure 32.29). Serious measurements in very windy conditions are virtually impossible, as are measurements of low-level noises even in slight wind conditions.

In permanent noise-monitoring systems it is now common practice to include an anemometer so that high-wind conditions may be identified and measurements discarded or treated with suspicion. Windscreens on microphones offer some help but do not eliminate the problems. These also have the beneficial effect of protecting the diaphragm from dust and chemical pollution, perhaps even showers. All windscreens have some effect on the frequency response of microphones.

32.8.4 Other Noises

The measurement of one noise in the presence of another does cause problems if the noise to be measured is not at least 10 dB higher than other noise or noises. Providing the two noise sources do not have identical or very similar spectra, and both are stable with time, corrections may be made as shown in Figure 32.30. Sound-intensity measuring techniques may be used successfully in cases where other noises are as high as the noise to be measured, or even higher.

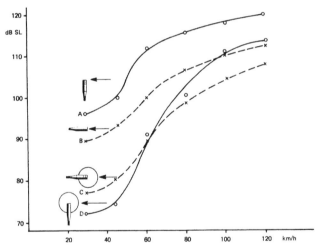

FIGURE 32.29 Noise levels induced in a half-inch free-field microphone fitted with a nosecone. A: With standard protection grid, wind parallel to diaphragm; B: as A but with wind at right angle to diaphragm; C: as B with windscreen; D: as A with windscreen.

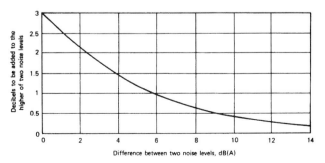

FIGURE 32.30 Noise-level addition chart.

REFERENCES

Burns, W. and D. W. Robinson, *Hearing and Noise in Industry*, HMSO, London (1970).

Hassall, J. R. and Zaveri, Z., *Acoustic Noise Measurements*, Bruel and Kjaer, Copenhagen (1979).

ISO Standards Handbook 4, *Acoustics, Vibration and Shock* (1985).

Randall, R. B., *Frequency Analysis*, Bruel and Kjaer, Copenhagen (1977).

FURTHER READING

Anderson, J. and M. Bratos-Anderson, *Noise: Its Measurement, Analysis, Rating and Control*, Ashgate Publishing (1993).

Harris, C. M., *Handbook of Acoustical Measurements and Noise Control*, McGraw-Hill, New York (1991).

Wilson, C. E., *Noise Control: Measurement, Analysis and Control of Sound and Vibration*, Krieger (1994).

Part V

Controllers, Actuators, and Final Control Elements

Chapter 33

Field Controllers, Hardware and Software

W. Boyes

33.1 INTRODUCTION

In this section, we look at the hardware and software that take the measurements from the field sensors and the instructions from the control system and provide the instructions to the final control elements that form the end points of the control loops in a plant.

In the past 30 years, there have been amazing changes in the design and construction of field controllers and the software that operates on them, while there have been many fewer changes in the design of the final control elements themselves.

It all started with Moore's law, of course. In the April 1965 issue of *Electronics Magazine*, Intel cofounder Gordon E. Moore described the doubling of electronic capabilities. "The complexity for minimum component costs has increased at a rate of roughly a factor of two per year …" he wrote. "Certainly over the short term this rate can be expected to continue, if not to increase."

Even though there were and are pundits who believe that Moore's law will finally be exceeded, the cost and power of electronic products continues to follow his law. Costs drop by half and power increases by a factor of two every two years. This has now been going on for over 43 years. Every time it looks like there will be a slowdown, new processes are developed to continue to make more and more powerful electronics less expensively.

So, what has this meant? Manufacturing, indeed all of society, has been radically changed by the applications of Moore's law. In 1965, manufacturing was done with paper routers and instructions. Machining was done by hand, according to drawings. Drawings themselves were done with pencil or India ink by draftspeople who did nothing else all day. Engineers used slide rules.

Machines were controlled by electromechanical relays and mechanical timers and human operators. If a new product was required, the production lines needed to be shut down, redesigned, rewired, and restarted, often at the cost of months of lost production.

The computers that put a man on the moon in 1969 had far less processing capability than the average inexpensive cell phone does in 2009.

33.2 FIELD CONTROLLERS, HARDWARE, AND SOFTWARE

In 1968, working on parallel paths, Richard Morley, of Bedford Associates (later Modicon; see Figure 33.1), and Otto Struger, of Allen-Bradley Co., created the first programmable logic controllers (PLCs). These devices were developed

FIGURE 33.1 Richard Morley with the Modicon. Courtesy of Richard Morley.

to replace hardwired discrete relay logic control systems in discrete manufacturing scenarios such as automotive assembly lines. The first PLCs used dedicated microprocessors running proprietary real-time operating systems (RTOSs) and were programmed using a special programming language called "ladder logic," created by Morley and his associates. Ladder logic came about as a digital adaptation of the ladder diagrams electricians used to create relay logic prior to the advent of PLCs. The first generation PLCs had 4 kilobytes of memory, maximum. They revolutionized industrial production, in both the discrete and the process fields.

In 1976, Robert Metcalfe, of Xerox, and his assistant, David Boggs, published *Ethernet: Distributed Packet-Switching For Local Computer Networks*. Most computers and PLCs today use some form of Ethernet to move data. The patents, interestingly, were not on the software but on the chips to produce hubs, routers, and the like, which had become practical because of Moore's law.

In 1981, Moore's law permitted IBM to release its first Personal Computer, or PC. It ran with 16 kilobytes of RAM on an Intel 4.88 MHz 8088 chip. Each of the original PCs was more powerful than the triple modular redundant computers that still (in 2009) drive the space shuttle.

By 1983, Moore's law had progressed to the point where a joint venture of Yamatake and Honeywell produced the first "smart transmitter." This was a field device: a pressure transmitter that had an onboard microprocessor transmitter—a computer inside a field instrument that could communicate digitally and be programmed like a computer. Other companies quickly followed suit.

In 1996, Fisher-Rosemount Inc., now Emerson Process Management, changed the definition of a distributed control system by combining a commercial off-the-shelf (COTS) PC made by Dell with a proprietary field controller and a suite of integrated proprietary software, running over standard Ethernet networks, and called it the DeltaV. This device was possible only because Moore's law had made the PC powerful enough to replace the "big iron" proprietary computers used in previous DCS designs, both from Fisher-Rosemount and other vendors.

In 2002, Craig Resnick, an analyst with ARC Advisory Group, coined the name *programmable automation controller* (PAC) for an embedded PC running either a version of Windows or a proprietary RTOS (see Figure 33.2).

In 1968, process field controllers were of the analog type, standalone single-loop controllers. In 2009, even standalone single-loop controllers are digital microprocessor-based special-purpose computers. They may even be PACs.

The time since 1968 has seen the convergence of the discrete PLC controller and the process loop controller. Products like the Control Logix platform from Rockwell Automation (the successor to Allen-Bradley Co.), the Simatic S7 platform from Siemens, the C200 platform from Honeywell, or any number of other PAC platforms now combine the discrete digital input/output features of the original PLCs and the advanced loop control functions of the analog single-loop controller.

FIGURE 33.2 A programmable automation controller. Courtesy of Advantech Inc.

Chapter 34

Advanced Control for the Plant Floor

Dr. James R. Ford, P. E.

34.1 INTRODUCTION

Advanced process control (APC) is a fairly mature body of engineering technology. Its evolution closely mirrors that of the digital computer and its close cousin, the modern microprocessor. APC was born in the 1960s, evolved slowly and somewhat painfully through its adolescence in the 1970s, flourished in the 1980s (with the remarkable advances in computers and digital control systems, or DCSs), and reached maturity in the 1990s, when model predictive control (MPC) ascended to the throne of supremacy as the preferred approach for implementing APC solutions.

As Zak Friedman[1] dared to point out in a recent article, the current decade has witnessed tremendous APC industry discontent, self-examination, and retrenchment. He lists several reasons for the malaise, among them: "cutting corners" on the implementation phase of the projects, poor inferred property models (these are explained later), tying APC projects to "optimization" projects, and too-cozy relationships between APC software vendors/implementers and their customers. In a more recent article[2] in the same magazine, Friedman interviewed me because I offer a different explanation for the APC industry problems.

This chapter traces the history of the development of process control, advanced process control, and related applied engineering technologies and discusses the reasons that I think the industry has encountered difficulties. The chapter presents some recommendations to improve the likelihood of successful APC project implementation and makes some predictions about the future direction of the technology.

34.2 EARLY DEVELOPMENTS

The discovery of oil in Pennsylvania in 1859 was followed immediately by the development of processes for separating and recovering the main distillable products, primarily kerosene, heating oil, and lubricants. These processes were initially batch in nature. A pot of oil was heated to boiling, and the resulting vapor was condensed and recovered in smaller batches.

The first batch in the process was extremely light (virgin naphtha), and the last batch was heavy (fuel oil or lubricating oil). Eventually, this process was transformed from batch to continuous, providing a means of continuously feeding fresh oil and recovering all distillate products simultaneously. The heart of this process was a unit operation referred to as *countercurrent, multicomponent, two-phase fractionation*.

Whereas the batch process was manual in nature and required very few adjustments (other than varying the heat applied to the pot), the continuous process required a means of making adjustments to several important variables, such as the feed rate, the feed temperature, the reflux rate, and so on, to maintain stable operation and to keep products within specifications.

Manually operated valves were initially utilized to allow an operator to adjust the important independent variables. A relatively simple process could be operated in a fairly stable fashion with this early "process control" system.

Over the next generation of process technology development, process control advanced from purely manual, open-loop control to automatic, closed-loop control. To truly understand this evolution, we should examine the reasons that this evolution was necessary and how those reasons impact the application of modern process control technology to the operation of process units today.

34.3 THE NEED FOR PROCESS CONTROL

Why do we need process control at all? The single most important reason is to respond to process disturbances. If process disturbances did not occur, the manual valves mentioned here would suffice for satisfactory, stable operation of process plants. What, then, do we mean by a process disturbance?

---

1. "Has the APC Industry Completely Collapsed?," *Hydrocarbon Processing*, January 2005, p. 15.
2. "Jim Ford's Views on APC," *Hydrocarbon Processing*, November 2006, p. 19.

A process disturbance is a change in any variable that affects the flow of heat and/or material in the process. We can further categorize disturbances in two ways: by time horizon and by measurability.

Some disturbances occur slowly over a period of weeks, months, or years. Examples of this type of disturbance are:

- *Heat exchanger fouling.* Slowly alters the rate of heat transfer from one fluid to another in the process.
- *Catalyst deactivation.* Slowly affects the rate, selectivity, and so on of the reactions occurring in the reactor.

Automatic process control was *not* developed to address long time-horizon disturbances.

Manual adjustment for these types of disturbances would work almost as well. So, automatic process control is used to rectify disturbances that occur over a much shorter time period of seconds, minutes, or hours. Within this short time horizon, there are really two main types of disturbances: measured and unmeasured.

34.4 UNMEASURED DISTURBANCES

Automatic process control was initially developed to respond to unmeasured disturbances. For example, consider the first automatic devices used to control the level of a liquid in a vessel. (See Figure 34.1.) The liquid level in the vessel is sensed by a float. The float is attached to a lever.

A change in liquid level moves the float up or down, which mechanically or pneumatically moves the lever, which is connected to a valve. When the level goes up the valve opens, and vice versa. The control loop is responding to an unmeasured disturbance, namely, the flow rate of material into the vessel.

The first automatic controllers were not very sophisticated. The float span, lever length, connection to the valve, and control valve opening had to be designed to handle the full range of operation. Otherwise, the vessel could overflow or drain out completely. This type of control had no specific target or "set point." At constant inlet flow, the level in the vessel would reach whatever resting position resulted in the proper valve opening to make the outflow equal to the inflow.

Level controllers were probably the first type of automatic process controller developed because of the mechanical simplicity of the entire loop. Later on, it became obvious that more sophisticated control valves were needed to further automate other types of loops. The pneumatically driven, linear-position control valve evolved over the early 20th century in all its various combinations of valve body and plug design to handle just about any type of fluid condition, pressure drop, or the like. The development of the automatic control valve ushered in the era of modern process control.

34.5 AUTOMATIC CONTROL VALVES

The first truly automatic control valves were developed to replace manual valves to control flow. This is the easiest type of variable to control, for two reasons. First, there is essentially no dead time and very little measurement lag between a change in valve opening and a change in the flow measurement. Second, a flow control loop is not typically subjected to a great deal of disturbance. The only significant disturbance is a change in upstream or downstream pressure, such as might occur in a fuel gas header supplying fuel gas through a flow controller for firing a heater or boiler. Other less significant disturbances include changes in the temperature and density of the flowing fluid. The flow control loop has become the foundation of all automatic process control, for several reasons.

Unlike pressure and temperature, which are intensive variables, flow is an extensive variable. Intensive variables are key control variables for stable operation of process plants because they relate directly to composition. Intensive variables are usually controlled by adjusting flows, the extensive variables. In this sense, intensive variables are higher in the control hierarchy. This explains why a simple cascade almost always involves an intensive variable as the master, or primary, in the cascade and flow as the slave, or secondary, in the cascade. When a pressure or temperature controller adjusts a control valve directly (rather than the flow in a cascade), the controller is actually adjusting the flow of material through the valve.

What this means, practically speaking, is that, unlike the intensive variables, a flow controller has no predetermined or "best" target for any given desired plant operation. The flow will be wherever it needs to be to maintain the higher-level intensive variable at its "best" value. This explains why "optimum" unit operation does not require accurate flow measurement. Even with significant error in flow measurement, the target for the measured flow will be adjusted

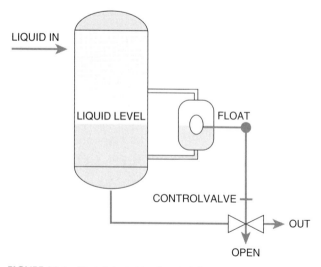

FIGURE 34.1 Float-Actuated level control diagram.

CHAPTER | 34 Advanced Control for the Plant Floor

(in open or closed loop) to maintain the intensive variable at its desired target. This also explains why orifices are perfectly acceptable as flow controller measurement devices, even though they are known to be rather inaccurate.

These comments apply to almost all flow controllers, even important ones like the main unit charge rate controller. The target for this control will be adjusted to achieve an overall production rate, to push a constraint, to control the inventory of feed (in a feed drum), and so on. What about additional feed flows, such as the flow of solvent in an absorption or extraction process? In this case, there is a more important, higher-level control variable, an intensive variable—namely, the ratio of the solvent to the unit charge rate. The flow of solvent will be adjusted to maintain a "best" solvent/feed ratio. Again, measurement accuracy is not critical; the ratio target will be adjusted to achieve the desired higher-level objective (absorption efficiency, etc.), regardless of how much measurement inaccuracy is present.

Almost all basic control loops, either single-loop or simple cascades, are designed to react, on feedback, to unmeasured disturbances. Reacting to unmeasured disturbances is called *servo*, or *feedback control*. Feedback control is based on reacting to a change in the process variable (the PV) in relation to the loop target, or set point (the SP). The PV can change in relation to the SP for two reasons: either because a disturbance has resulted in an unexpected change in the PV or because the operator or a higher-level control has changed the SP. Let's ignore SP changes for now.

34.6 TYPES OF FEEDBACK CONTROL

So, feedback control was initially designed to react to unmeasured disturbances. The problem in designing these early feedback controllers was figuring out how much and how fast to adjust the valve when the PV changed. The first design was the float-type level controller described earlier. The control action is referred to as *proportional* because the valve opening is linearly proportional to the level. The higher the level, the more open the valve.

This type of control may have been marginally acceptable for basic levels in vessels, but it was entirely deficient for other variables. The main problem is that proportional-only control cannot control to a specific target or set point. There will always be offset between the PV and the desired target. To correct this deficiency, the type of control known as *integral*, or *reset*, was developed. This terminology is based on the fact that, mathematically, the control action to correct the offset between SP and PV is based on the calculus operation known as *integration*. (The control action is based on the area under the SP-PV offset curve, integrated over a period of time.) The addition of this type of control action represented a major improvement in feedback control. Almost all flow and pressure loops can be controlled very well with a combination of proportional and integral control action. A less important type of control action was developed to handle situations in which the loop includes significant measurement lag, such as is often seen in temperature loops involving a thermocouple, inside a thermowell, which is stuck in the side of a vessel or pipe. Control engineers noted that these loops were particularly difficult to control, because the measurement lag introduced instability whenever the loops were tuned to minimize SP-PV error. For these situations, a control action was developed that reacts to a change in the "rate of change" of the PV. In other words, as the PV begins to change its "trajectory" with regard to the SP, the control action is "reversed," or "puts on the brakes," to head off the change that is coming, as indicated by the change in trajectory. The control action was based on comparing the rate of change of the PV over time, or the derivative of the PV. Hence the name of the control action: derivative.

In practice, this type of control action is utilized very little in the tuning of basic control loops. The other type of change that produces an offset between SP and PV is an SP change. For flow control loops, which are typically adjusted to maintain a higher-level intensive variable at its target, a quick response to the SP change is desirable; otherwise, additional response lag is introduced. A flow control loop can and should be tuned to react quickly and equally effectively to both PV disturbances and SP changes.

Unfortunately, for intensive variables, a different closed-loop response for PV changes due to disturbances vs. SP changes is called for. For temperature, and especially pressure, these loops are tuned as tightly as possible to react to disturbances. This is because intensive variables are directly related to composition, and good control of composition is essential for product quality and yield. However, if the operator makes an SP change, a much less "aggressive" control action is preferred.

This is because the resulting composition change will induce disturbances in other parts of the process, and the goal is to propagate this disturbance as smoothly as possible so as to allow other loops to react without significant upset.

Modern DCSs can provide some help in this area. For example, the control loop can be configured to take proportional action on PV changes only, ignoring the effect of SP changes. Then, following an SP change, integral action will grind away on correcting the offset between SP and PV. However, these features do not fully correct the deficiency discussed earlier. This dilemma plagues control systems to this very day and is a major justification for implementation of advanced controls that directly address this and other shortcomings of basic process control.

34.7 MEASURED DISTURBANCES

The other major type of disturbance is the measured disturbance. Common examples are changes in charge rate, cooling water temperature, steam header pressure, fuel gas header pressure, heating medium temperature and ambient air temperature, where instruments are installed to measure those variables. The first 20 years of the development of APC technology focused primarily on using measured disturbance information for improving the quality of control. Why? Modern process units are complex and highly interactive. The basic control system, even a modern DCS, is incapable of maintaining fully stable operation when disturbances occur. APC was developed to mitigate the destabilizing effects of disturbances and thereby to reduce process instability. This is still the primary goal of APC. Any other claimed objective or direct benefit is secondary.

Why is it so important to reduce process instability? Process instability leads to extremely conservative operation so as to avoid the costly penalties associated with instability, namely, production of off-spec product and violation of important constraints related to equipment life and human safety. Conservative operation means staying well away from constraint limits. Staying away from these limits leaves a lot of money on the table in terms of reduced yields, lower throughput, and greater energy consumption. APC reduces instability, allowing for operation much closer to constraints and thereby capturing the benefits that would otherwise be lost.

As stated earlier, early APC development work focused on improving the control system's response to measured disturbances. The main techniques were called *feed-forward*, *compensating*, and *decoupling*. In the example of a fired heater mentioned earlier, adjusting the fuel flow for changes in heater charge rate and inlet temperature is feed-forward. The objective is to head off upsets in the heater outlet temperature that are going to occur because of these feed changes. In similar fashion, the fuel flow can be "compensated" for changes in fuel gas header pressure, temperature, density, and heating value, if these measurements are available. Finally, if this is a dual-fuel heater (fuel gas and fuel oil), the fuel gas flow can be adjusted when the fuel oil flow changes so as to "decouple" the heater from the firing upset that would otherwise occur. This decoupling is often implemented as a heater fired duty controller.

A second area of initial APC development effort focused on controlling process variables that are not directly measured by an instrument. An example is reactor conversion or severity.

Hydrocracking severity is often measured by how much of the fresh feed is converted to heating oil and lighter products. If the appropriate product flow measurements are available, the conversion can be calculated and the reactor severity can then be adjusted to maintain a target conversion. Work in this area of APC led to the development of a related body of engineering technology referred to as *inferred properties* or *soft sensors*.

A third area of APC development work focused on pushing constraints. After all, if the goal of APC is to reduce process instability so as to operate closer to constraints, why not implement APCs that accomplish that goal? (Note that this type of APC strategy creates a measured process disturbance; we are going to move a major independent variable to push constraints, so there had better be APCs in place to handle those disturbances.) Especially when the goal was to increase the average unit charge rate by pushing known, measured constraints, huge benefits could often be claimed for these types of strategies. In practice, these types of strategies were difficult to implement and were not particularly successful.

While all this development work was focused on reacting to changes in measured disturbances, the problems created by unmeasured disturbances continued to hamper stable unit operation (and still do today). Some early effort also focused on improving the only tool available at the time, the proportional-integral-derivative (PID) control algorithm, to react better to unmeasured disturbances.

One of the main weaknesses of PID is its inability to maintain stable operation when there is significant dead time and/or lag between the valve movement and the effect on the control variable. For example, in a distillation column, the reflux flow is often adjusted to maintain a stable column tray temperature. The problem arises when the tray is well down the tower. When an unmeasured feed composition change occurs, upsetting the tray temperature, the controller responds by adjusting the reflux flow. But there may be dead time of several minutes before the change in reflux flow begins to change the tray temperature. In the meantime, the controller will have continued to take more and more integral action in an effort to return the PV to SP. These types of loops are difficult (or impossible) to tune. They are typically detuned (small gain and integral) but with a good bit of derivative action left in as a means of "putting on the brakes" when the PV starts returning toward SP.

Some successes were noted with algorithms such as the Smith Predictor, which relies on a model to predict the response of the PV to changes in controller output. This algorithm attempts to control the predicted PV (the PV with both dead time and disturbances included) rather than the actual measured PV. Unfortunately, even the slightest model mismatch can cause the controller using the Smith Predictor to become unstable.

We have been particularly successful in this area with development of our "smart" PID control algorithm. In its simplest form, it addresses the biggest weakness of PID, namely, the overshoot that occurs because the algorithm continues to take integral action to reduce the offset between SP and PV, even when the PV is returning to SP. Our algorithm turns the integral action on and off according to a proven decision process made at each controller execution. This algorithm excels in loops with significant dead time and lag. We use this algorithm on virtually all APC projects.

34.8 THE NEED FOR MODELS

By the mid-1980s, many consulting companies and in-house technical staffs were involved in the design and implementation of the types of APC strategies described in the last few paragraphs. A word that began to appear more and more associated with APC was *model*.

For example, to implement the constraint-pushing APC strategies we've discussed, a "dynamic" model was needed to relate a change in the independent variable (the charge rate) to the effect on each of the dependent, or constraint, variables. With this model, the adjustments that were needed to keep the most constraining of the constraint variables close to their limits could be determined mathematically.

Why was it necessary to resort to development of models? As mentioned earlier, many of the early constraint-pushing efforts were not particularly successful. Why not? It's the same problem that plagues feedback control loops with significant dead time and lag.

There are usually constraints that should be honored in a constraint-pushing strategy that may be far removed (in time) from where the constraint-pushing move is made. Traditional feedback techniques (PID controllers acting through a signal selector) do not work well for the constraints with long dead time. We addressed this issue by developing special versions of our smart PID algorithm to deal with the long dead times, and we were fairly successful in doing so.

34.9 THE EMERGENCE OF MPC

In his Ph.D. dissertation work, Dr. Charles Cutler developed a technique that incorporated normalized step-response models for the constraint (or control) variables, or CVs, as a function of the manipulated variables, or MVs. This allowed the control problem to be "linearized," which then permitted the application of standard matrix algebra to estimate the MV moves to be made to keep the CVs within their limits. He called the matrix of model coefficients the *dynamic matrix* and developed the *dynamic matrix control* (DMC) control technique. He also incorporated an objective function into the DMC algorithm, turning it into an "optimizer." If the objective function is the sum of the variances between the predicted and desired values of the CVs, DMC becomes a minimum variance controller that minimizes the output error over the controller time horizon.

Thus was ushered in the control technology known in general as multivariable, model-predictive control (MVC or MPC). Dr. Cutler's work led eventually to formation of his company, DMC Corporation, which was eventually acquired by AspenTech. The current version of this control software is known as DMCPlus. There are many competing MPC products, including Honeywell RMPCT, Invensys Connoisseur, and others.

MPC has become the preferred technology for solving not only multivariable control problems but just about any control problem more complicated than simple cascades and ratios. Note that this technology no longer relies on traditional servo control techniques, which were first designed to handle the effect of unmeasured disturbances and which have done a fairly good job for about 100 years. MPC assumes that our knowledge of the process is perfect and that all disturbances have been accounted for. There is no way for an MPC to handle unmeasured disturbances other than to readjust at each controller execution the bias between the predicted and measured value of each control variable. This can be likened to a form of integral-only control action. This partially explains MPC's poor behavior when challenged by disturbances unaccounted for in the controller.

DMCPlus uses linear, step-response models, but other MPC developers have incorporated other types of models. For example, Pavilion Technologies has developed a whole body of modeling and control software based on neural networks. Since these models are nonlinear, they allow the user to develop nonlinear models for processes that display this behavior. Polymer production processes (e.g., polypropylene) are highly nonlinear, and neural net-based controllers are said to perform well for control of these processes. GE (MVC) uses algebraic models and solves the control execution prediction problem with numerical techniques.

34.10 MPC VS. ARC

There are some similarities between the older APC techniques (feed-forward, etc.) and MPC, but there are also some important differences. Let's call the older technique *advanced regulatory control* (ARC). To illustrate, let's take a simplified control problem, such as a distillation column where we are controlling a tray temperature by adjusting the reflux flow, and we want feed-forward action for feed rate changes. The MPC will have two models: one for the response of the temperature to feed rate changes and one for the response of the temperature to reflux flow changes. For a feed rate change, the controller knows that the temperature is going to change over time, so it estimates a series of changes in reflux flow required to keep the temperature near its desired target.

The action of the ARC is different. In this case, we want to feed-forward the feed rate change directly to the reflux flow. We do so by delaying and lagging the feed rate change (using a simple dead time and lag algorithm customized to adjust the reflux with the appropriate dynamics), then adjusting the reflux with the appropriate steady-state gain or sensitivity (e.g., three barrels of reflux per barrel of feed). The ultimate sensitivity of the change in reflux flow to a change in feed rate varies from day to day; hence, this type of feed-forward control is adaptive and, therefore, superior to MPC (the MPC models are static). *Note:* Dr. Cutler recently formed a new corporation, and he is now offering adaptive DMC, which includes real-time adjustment of the response models.

How does the MPC handle an unmeasured disturbance, such as a feed composition change? As mentioned earlier, it can do so only when it notices that the temperature is not where it's supposed to be according to the model prediction from the series of recent moves of the feed rate and reflux. It resets the predicted vs. the actual bias and then calculates a reflux flow move that will get the temperature back where it's supposed to be, a form of integral-only feedback control.

On the other hand, the ARC acts in the traditional feedback (or servo) manner with either plain or "smart" PID action.

34.11 HIERARCHY

Prior to MPC, most successful APC engineers used a process engineering-based, hierarchical approach to developing APC solutions. The bottom of the control hierarchy, its foundation, is what we referred to earlier as "basic" process control, the single loops and simple cascades that appear on P&IDs and provide the operator with the first level of regulatory control.

Simple processes that are not subject to significant disturbances can operate in a fairly stable fashion with basic process control alone. Unfortunately, most process units in refineries and chemical plants are very complex, highly interactive, and subject to frequent disturbances. The basic control system is incapable of maintaining fully stable operation when challenged by these disturbances—thus the emergence of APC to mitigate the destabilizing effects of disturbances.

The hierarchical approach to APC design identifies the causes of the disturbances in each part of the process, then layers the solutions that deal with the disturbances on top of the basic control system, from the bottom up. Each layer adds complexity and its design depends on the disturbance(s) being dealt with.

As an example of the first layer of the APC hierarchy, consider the classic problem of how to control the composition of distillation tower product streams, such as the overhead product stream. The hierarchical approach is based on identifying and dealing with the disturbances. The first type of disturbance that typically occurs is caused by changes in ambient conditions (air temperature, rainstorms, etc.), which lead to a change in the temperature of the condensed overhead vapor stream, namely the reflux (and overhead product). This will cause the condensation rate in the tower to change, leading to a disturbance that will upset column separation and product qualities. Hierarchical APC design deals with this disturbance by specifying, as the first level above basic control, internal reflux control (IRC).

The IRC first calculates the net internal reflux with an equation that includes the heat of vaporization, heat capacity and flow of the external reflux, the overhead vapor temperature, and the reflux temperature. The control next back-calculates the external reflux flow required to maintain constant IR and then adjusts the set point of the external reflux flow controller accordingly. This type of control provides a fast-responding, first-level improvement in stability by isolating the column from disturbances caused by changes in ambient conditions. There are multiple inputs to the control (the flow and temperatures which contribute to calculation of the internal reflux), but typically only one output—to the set point of the reflux flow controller.

Moving up the hierarchy to the advanced supervisory control level, the overhead product composition can be further stabilized by controlling a key temperature in the upper part of the tower, since temperature (at constant pressure) is directly related to composition. And, since the tower pressure could vary (especially if another application is attempting to minimize pressure), the temperature that is being controlled should be corrected for pressure variations. This control adjusts the set point of the IRC to maintain a constant pressure-corrected temperature (PCT). Feed-forward action (for feed rate changes) and decoupling action (for changes in reboiler heat) can be added at this level.

Further improvement in composition control can be achieved by developing a correlation for the product composition, using real-time process measurements (the PCT plus other variables such as reflux ratio, etc.), then using this correlation as the PV in an additional APC that adjusts the set point of the PCT. This type of correlation is known as an *inferred property* or *soft sensor*.

Thus, the hierarchical approach results in a multiple-cascade design, an inferred property control adjusting a PCT, adjusting an IRC, adjusting the reflux flow.

In general, APCs designed using hierarchical approaches consist of layers of increasingly complex strategies. Some of the important advantages of this approach are:

- Operators can understand the strategies; they appeal to human logic because they use a "systems" approach to problem solving, breaking a big problem down into smaller problems to be solved.
- The control structure is more suitable for solutions at a lower level in the control system; such solutions can often be implemented without the requirement for additional hardware and software.
- The controls "degrade gracefully"; when a problem prohibits a higher-level APC from being used, the lower-level controls can still be used and can capture much of the associated benefit.

How would we solve the preceding control problem using MPC? There are two approaches. The standard approach is to use the inferred property as a CV and the external reflux flow controller as the MV. In this case, then, how does the MPC deal with the other disturbances such as the reflux temperature, reboiler heat, and feed rate? These variables must be included in the model matrix as additional independent variables. A step-response model must then be developed for the CV (the inferred property) as a function of each of the additional independent variables. This is the way most of the APC industry designs an MPC.

The hierarchical approach would suggest something radically different. The CV is the same because the product composition is the variable that directly relates to profitability. However, in the hierarchical design, the MV is the PCT. The lower-level controls (the PCTC and IRC) are implemented at a lower level in the control hierarchy, typically in the DCS.

Unfortunately, the industry-accepted approach to MPC design violates the principles of hierarchy. Rarely, if ever, are intermediate levels of APC employed below MPC. There is no hierarchy—just one huge, flat MPC controller on top of the basic controllers, moving all of them at the same time in seemingly magical fashion. Operators rarely understand them or what they are doing. And they do not degrade gracefully. Consequently, many fall into disuse.

34.12 OTHER PROBLEMS WITH MPC

A nonhierarchical design is only one of the many problems with MPC. The limitations of MPC have been thoroughly exposed, though probably widely ignored.[3] Here are a few other problems.

In many real control situations, even mild model mismatch introduces seriously inappropriate controller moves, leading to inherent instability as the controller tries at each execution to deal with the error between where it thinks it is going and where it really is.

MPC is not well suited for processes with noncontinuous phases of operation, such as delayed cokers. The problem here is how to "model" the transitory behavior of key control variables during coke drum prewarming and switching. Unfortunately, every drum-switching operation is different. This means that both the "time to steady state" and ultimate CV gain are at best only approximations, leading to model mismatch during every single drum operation. No wonder they are so difficult to "tune" during these transitions.

Correcting model mismatch requires retesting and remodeling, a form of very expensive maintenance. MPC controllers, particularly complex ones implemented on complex units with lots of interactions, require a lot of "babysitting"—constant attention from highly trained (and highly paid) control engineers. This is a luxury that few operating companies can afford.

The licensors of MPC software will tell you that their algorithms "optimize" operation by operating at constraints using the least costly combination of manipulated variable assets. That is certainly correct, mathematically; that is the way the LP or QP works. In practice, however, any actual "optimizing" is marginal at best. This is due to a couple of reasons. The first is the fact that, in most cases, one MV dominates the relationships of other MVs to a particular CV. For example, in a crude oil distillation tower, the sensitivity between the composition of a side-draw distillate product and its draw rate will be much larger than the sensitivity with any other MV. The controller will almost always move the distillate draw rate to control composition. Only if the draw rate becomes constrained will the controller adjust a pump-around flow or the draw rate of another distillate product to control composition, regardless of the relative "cost" of these MVs. The second reason is the fact that, in most cases, once the constraint "corners" of the CV/MV space are found, they tend not to change. The location of the corners is most often determined either by "discretionary" operator-entered limits (for example, the operator wants to limit the operating range of an MV) or by valve positions. In both situations, these are constraints that would have been pushed by any kind of APC, not just a model-predictive controller. So, the MPC has not "optimized" any more than would a simpler ARC or ASC.

When analyzing the MPC controller models that result from plant testing, control engineers often encounter CV/MV relationships that appear inappropriate and that are usually dropped because the engineer does not want that particular MV moved to control that particular CV. Thus, the decoupling benefit that could have been achieved with simpler ASC is lost.

If every single combination of variables and constraint limits has not been tested during controller commissioning (as is often the case), the controller behavior under these untested conditions is unknown and unpredictable. For even moderately sized controllers, the number of possible combinations becomes unreasonably large such that all combinations cannot be tested, even in simulation mode.

Control engineers often drop models from the model matrix to ensure a reasonably stable solution (avoiding approach to a singular matrix in the LP solution). This is most common where controllers are implemented on fractionation columns with high-purity products and where the controller is expected to meet product purity specifications on both top and bottom products. The model matrix then reduces to one that is almost completely decoupled. In this case, single-loop controllers, rather than an MPC, would be a clearly superior solution.

What about some of the other MPC selling points? One that is often mentioned is that, unlike traditional APC, MPC eliminates the need for custom programming. This is simply not true. Except for very simple MPC solutions, custom code is almost always required—for example, to calculate a variable used by the controller, to estimate a flow from a valve position, and so on.

---

3. The emperor is surely wearing clothes! See the excellent article by Alan Hugo, "Limitations of Model-Predictive Controllers," that appeared in the January 2000 issue of *Hydrocarbon Processing*, p. 83.

The benefits of standardization are often touted. Using the same solution tool across the whole organization for all control problems will reduce training and maintenance costs. But this is like saying that I can use my new convertible to haul dirt, even though I know that using my old battered pickup would be much more appropriate. Such an approach ignores the nature of real control problems in refineries and chemical plants and relegates control engineers to pointers and clickers.

34.13 WHERE WE ARE TODAY?

Perhaps the previous discussion paints an overly pessimistic picture of MPC as it exists today. Certainly many companies, particularly large refining companies, have recognized the potential return of MPC, have invested heavily in MPC, and maintain large, highly trained staffs to ensure that the MPCs function properly and provide the performance that justified their investment.

But, on the other hand, the managers of many other companies have been seduced by the popular myth that MPC is easy to implement and maintain—a misconception fostered at the highest management levels by those most likely to benefit from the proliferation of various MPC software packages.

So, what can companies do today to improve the utilization and effectiveness of APC in general and MPC in particular? They can do several things, all focused primarily on the way we analyze operating problems and design APC solutions to solve the problems and thereby improve productivity and profitability. Here are some ideas.

As mentioned earlier, the main goal of APC is to isolate operating units from process disturbances. What does this mean when we are designing an APC solution? First, identify the disturbance and determine its breadth of influence. Does the disturbance affect the whole unit? If so, how? For example, the most common unit-encompassing disturbance is a feed-rate change. But how is the disturbance propagated? In many process units, this disturbance merely propagates in one direction only (upstream to downstream), with no other complicating effects (such as those caused by recycle or interaction between upstream and downstream unit operations). In this case, a fairly straightforward APC solution involves relatively simple ARCs for inventory and load control—adjusting inter-unit flows with feed-forward for inventory control, and adjusting load-related variables such as distillation column reflux with ratio controls or with feed-forward action in the case of cascades (for example, controlling a PCT by adjusting an IRC). No MVC is required here to achieve significantly improved unit stability. If the effect of the disturbance can be isolated, design the APC for that isolated part of the unit and ignore potential second-order effects on other parts of the unit. A good example is the IR control discussed earlier. The tower can be easily isolated from the ambient conditions disturbance by implementing internal reflux control. The internal reflux can then be utilized as an MV in a higher-level control strategy (for example, to control a pressure-compensated tray temperature as an indicator of product composition) or an inferred property.

What about unmeasured disturbances, such as feed composition, where the control strategy must react on feedback alone? As mentioned earlier, we have had a great deal of success with "intelligent" feedback-control algorithms that are much more effective than simple PID. For example, our Smart PID algorithm includes special logic that first determines, at each control algorithm execution, whether or not integral action is currently advised based on the PV's recent trajectory toward or away from set point. Second, it determines the magnitude of the integral move based on similar logic. This greatly improves the transient controller response, especially in loops with significant dead time and/or lag.

Another successful approach is to use model-based, adaptive control algorithms, such as those incorporated in products like Brainwave. Here are some additional suggestions:

Using a hierarchical approach suggests implementing the lower-level APC strategies as low in the control system as possible. Most modern DCSs will support a good bit of ARC and ASC at the DCS level. Some DCSs such as Honeywell Experion (and earlier TDC systems) have dedicated, DCS-level devices designed specifically for implementation of robust APC applications. We are currently implementing very effective first-level APC in such diverse DCSs as Foxboro, Yokogawa, DeltaV, and Honeywell.

When available, and assuming the analyses are reliable, use lab data as much as possible in the design of APCs. Develop inferred property correlations for control of key product properties and update the model correlations with the lab data.

Track APC performance by calculating, historizing, and displaying a variable that indicates quality of control. We often use the variance (standard deviation), in engineering units, of the error between SP and PV of the key APC (for example, an inferred property). Degradation in the long-term trend of this variable suggests that a careful process analysis is needed to identify the cause of the degraded performance and the corrective action needed to return the control to its original performance level.

34.14 RECOMMENDATIONS FOR USING MPC

If MPC is being considered for solution of a control problem, apply it intelligently and hierarchically.

Intelligent application of MPC involves asking important questions, such as:

- Is the control problem to be solved truly multivariable? Can (and should) several MVs be adjusted to control one CV? If not, don't use MPC.

CHAPTER | 34 Advanced Control for the Plant Floor

- Are significant dead time and lag present such that the model-based, predictive power of MPC would provide superior control performance compared to simple feed-forward action?
- Look at the model matrix. Are there significant interactions between *many* dependent variables and *many* independent variables, or are there just isolated "islands" of interactions where each island represents a fairly simple control problem that could just as easily be solved with simpler technology?
- How big does the controller really have to be? Could the problem be effectively solved with a number of small controllers without sacrificing the benefits provided by a unit-wide controller?

Applying MPC hierarchically means doing the following:

- Handle isolated disturbance variables with lower-level ARC. For example, stabilize and control a fired heater outlet temperature with ARC, not MPC. The ARC can easily handle disturbance variables such as feed rate, inlet temperature, and fuel gas pressure.
- Use the lower-level ARCs, such as the fired heater outlet temperature just mentioned, as MVs in the MPC controller.
- Use intensive variables, when possible, as MVs in the controller. For example, use a distillate product *yield*, rather than flow rate, as the MV in a controller designed to control the quality of that product; this eliminates unit charge rate as a disturbance variable.

34.15 WHAT'S IN STORE FOR THE NEXT 40 YEARS?

A cover story that appeared in *Hydrocarbon Processing* about 20 years ago included an interview with the top automation and APC guru of a major U.S. refining company. The main point proffered by the interviewee was that the relentless advances in automation technology, both hardware- and software-wise (including APC), would lead one day to the "virtual" control room in which operating processes would no longer require constant human monitoring—no process operators required. I'm not sure how this article was received elsewhere, but it got a lot of chuckles in our office.

Process control, whether continuous or discrete, involves events that are related by cause and effect. For example, a change in temperature (the cause event) needs a change in a valve position (the effect event) to maintain control of the temperature. Automation elevates the relationships of the cause and effect events by eliminating the requirement for a human to connect the two.

With the huge advances in automation technology accomplished in the period of the 1970s and 1980s, one might imagine a situation in which incredibly sophisticated automation solutions would eventually eliminate the need for direct human involvement.

That's not what this author sees for the next 40 years. Why not? Three reasons. First, the more sophisticated automation solutions require process models. For example, MPC requires accurate MV/CV or DV/CV response models. Real-time rigorous optimization (RETRO) requires accurate steady-state process models. Despite the complexity of the underlying mathematics and the depth of process data analysis involved in development of these models, the fact is that these will always remain just "models." When an optimizer runs an optimization solution for a process unit, guess what results? The optimizer spits out an optimal solution for the model, not the process unit! Anybody who believes that the optimizer is really optimizing the process unit is either quite naïve or has been seduced by the same type of hype that accompanied the advent of MPCs and now seems to pervade the world of RETRO.

When well-designed and -engineered (and -monitored), RETRO can move operations in a more profitable direction and thereby improve profitability in a major way. But the point is that models will never be accurate enough to eliminate the need for human monitoring and intervention.

The second important reason involves disturbances. Remember that there are two types of process disturbances, measured and unmeasured. Despite huge advances in process measurement technology (analyzers, etc.), we will never reach the point (at least in the next 40 years) when *all* disturbances can be measured. Therefore, there will always be target offset (deviation from set point) due to unmeasured disturbances and model mismatch. Furthermore, it is an absolute certainty that some unmeasured disturbances will be of such severity as to lead to process "upsets," conditions that require human analysis and intervention to prevent shutdowns, unsafe operation, equipment damage, and so on. Operators may get bored in the future, but they'll still be around for the next 40 years to handle these upset conditions.

The third important reason is the "weakest link" argument. Control system hardware (primarily field elements) is notoriously prone to failure. No automation system can continue to provide full functionality, in real time, when hardware fails; human intervention will be required to maintain stable plant operation until the problem is remedied.

Indeed, the virtual control room is still more than 40 years away. Process automation will continue to improve with advances in measurement technology, equipment reliability, and modeling sophistication, but process operators, and capable process and process-control engineers, will still maintain and oversee its application. APC will remain an "imperfect" technology that requires for its successful application experienced and quality process engineering analysis.

The most successful APC programs of the next 40 years will have the following characteristics:

- The APC solution will be designed to solve the operating problem being dealt with and will utilize the appropriate control technology; MPC will be employed only when required.
- The total APC solution will utilize the principle of "hierarchy"; lower-level, simpler solutions will solve lower-level problems, with ascending, higher-level, more complex solutions achieving higher-level control and optimization objectives.
- An intermediate level of model-based APC technology, somewhere between simple PID and MPC (Brainwave, for example), will receive greater utilization at a lower cost (in terms of both initial cost and maintenance cost) but will provide an effective means of dealing with measured and unmeasured disturbances.
- The whole subfield of APC known as *transition management* (e.g., handling crude switches) will become much more dominant and important as an adjunct to continuous process control.
- APC solutions will make greater use of inferred properties, and these will be based on the physics and chemistry of the process (not "artificial intelligence," like neural nets).
- The role of operating company, in-house control engineers will evolve from a purely technical position to more of a program manager. Operating companies will rely more on APC companies to do the "grunt work," allowing the in-house engineers to manage the projects and to monitor the performance of the control system.

APC technology has endured and survived its growing pains; it has reached a relatively stable state of maturity, and it has shaken out its weaker components and proponents. Those of us who have survived this process are wiser, humbler, and more realistic about how to move forward. Our customers and employers will be best served by a pragmatic, process engineering-based approach to APC design that applies the specific control technology most appropriate to achieve the desired control and operating objectives at lowest cost and with the greatest probability of long-term durability and maintainability.

Chapter 35

Batch Process Control

W. H. Boyes

35.1 INTRODUCTION

"Batch manufacturing has been around for a long time," says Lynn Craig, one of the founders of WBF, formerly World Batch Forum, and one of the technical chairs of ISA88, the batch manufacturing standard. "Most societies since prehistoric times have found a way to make beer. That's a batch process. It's one of the oldest batch processes. Food processing and the making of poultices and simples (early pharmaceutical manufacturing) are so old that there is no record of who did it first."

"The process consisted of putting the right stuff in the pot and then keeping the pot just warm enough, without letting it get too warm. Without control, beer can happen, but don't count on it. Ten years ago or a little more, batch was a process that, generally speaking, didn't have a whole lot of instrumentation and control, and all of the procedures were done manually. It just didn't fit into the high-tech world."[1]

In many cases, the way food is manufactured and the way pharmaceuticals and nutraceuticals are made differs hardly at all from the way they were made 100 years ago. In fact, this is true of many processes in many different industries.

What is different since the publication of the ISA S88 batch standard is that there's now a language and a way of describing batch processes so that they can be repeated precisely and reliably, anywhere in the world.

The S88 model describes any process (including continuous processes) in terms of levels or blocks (see Figure 35.1). This is similar to the ISO's Computer Integrated Manufacturing (CIM) standard, which uses a set of building blocks that conceptually allow any process to be clearly elaborated.

The highest three levels—enterprise, site, and area—aren't, strictly speaking, controlled by the batch standard but are included to explain how the batch standard's language can interface with the business systems of the area,

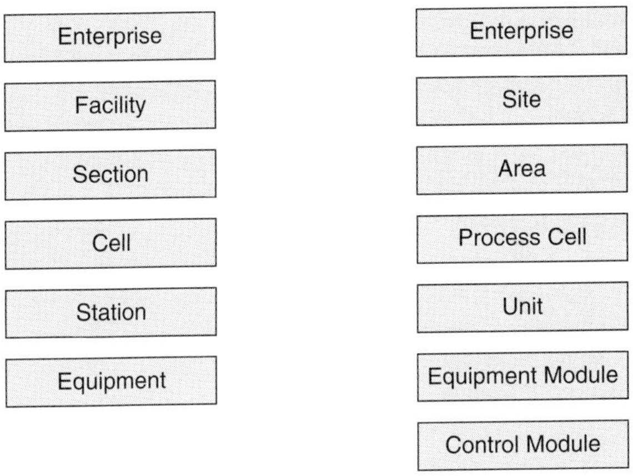

FIGURE 35.1 Comparing standard models. ISA's S88.01 model (right) is similar to the ISO's CIM model (left).

plant site, and business enterprise as a whole. The enterprise is a common term for a complete business entity; a site is whatever the enterprise defines it to be, commonly a place where production occurs.

The second edition of Tom Fisher's seminal book on the batch standard, *Batch Control Systems*, which was completely rewritten by William Hawkins after Fisher's death, states that, "Areas are really political subdivisions of a plant, whose borders are subject to the whims of management … and so it (the area) joins Site and Enterprise as parts of the physical model that are defined by management, and not by control engineers."

The next two levels, process cell and unit, are the building blocks of the manufacturing process. Process cell is used in a way intentionally similar to the widely used discrete manufacturing cell concept. One or more units are contained in a process cell. A unit is a collection or set of controlled equipment, such as a reactor vessel and the ancillary equipment necessary to operate it.

Within the unit are the equipment module and the control module. The equipment module is the border around a minor group of equipment with a process function. An equipment

---

1. Conversation with the author at WBF North America Conference, 2006.

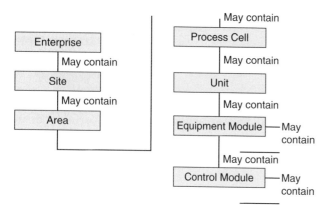

FIGURE 35.2 Getting physical: S88 physical model.

module may contain control module(s) and even subsidiary equipment modules. See Figure 35.2. The control module contains the equipment and systems that perform the actual control of the process.

Since the introduction of the S88 batch standard, every major supplier of automation equipment and software has introduced hardware and software designed to work in accordance with the standard. Many suppliers now build the software required to implement the S88 standard directly into their field controllers. Many of those suppliers are corporate members of WBF, formerly known as World Batch Forum. WBF was founded as a way to promulgate the use of the S88 standards worldwide. Currently it calls itself the Organization for Production Technology, since the S88 and S95 standards are in extremely common use in industry.

FURTHER READING

Hawkins, William M., and Thomas G. Fisher, *Batch Control Systems*, ISA Press, 2006.

Parshall, Jim, and Lamb, L. B., *Applying S88: Batch Control from a User's Perspective*, ISA Press, 2000.

WBF Body of Knowledge, www.wbf.org.

Chapter 36

Applying Control Valves[1]

B. G. Liptak; edited by W. H. Boyes

36.1 INTRODUCTION

Control valves modulate the flows of process or heat-transfer fluids and stabilize variations in material and heat balance of industrial processes. They manipulate these flows by changing their openings and modifying the energy needed for the flow to pass through them. As a control valve closes, the pressure differential required to pass the same flow increases, and the flow is reduced. Figure 36.1 illustrates the pump curve of a constant-speed pump and the system curve of the process that the pump serves. The elevation (static) head of the process is constant, whereas the friction loss increases with flow. The pressure generated by the pump is the sum of the system curves (friction and static) and the pressure differential (ΔP) required by the control valve. As the valve throttles, the pump travels on its curve while delivering the required valve pressure drop.

The pumping energy invested to overcome the valve differential is wasted energy and is the difference between the pressure required to "push" (transport) the fluid into the process (system curve, at bottom left in Figure 36.1) and the pump curve of the constant-speed pump. Pumps are selected to meet the maximum possible flow demand of the process, so they tend to be oversized during normal operation. Consequently, using control valves to manipulate the flow generated by constant-speed pumps wastes energy and increases plant-operation costs.

Therefore, when designing a control system, a process-control engineer must first decide whether a control valve or a variable-speed pump should be used to throttle the flow. Variable-speed pumps reduce flow by reducing pump speed. So, instead of burning energy unnecessarily introduced by the pump head, that energy isn't introduced in the first place. This lessens operating costs but increases capital investment because variable and constant-speed pumps usually cost more than control valves.

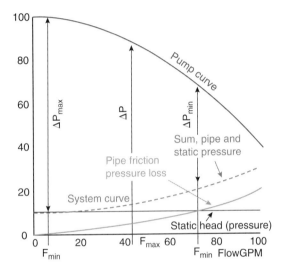

FIGURE 36.1 Pump curve vs. process. Courtesy of Putman Media Inc.

When several users are supplied by the same variable-speed pump, its speed can be automatically adjusted by a valve-position controller (VPC), which detects the opening of the most-open user valve (MOV). The MOV isn't allowed to open beyond 80 to 90 percent because when the set point of the VPC is reached, this integral-only controller starts increasing the pump speed. This increases the available pressure drop for all the valves, which in turn reduces their openings.

36.2 VALVE TYPES AND CHARACTERISTICS

If the cost/benefit analysis comparing constant and variable-speed pumping systems favors using throttling valves, the next task is to select the right valve type for the application. Figure 36.2 shows that various valve designs have different pressure and temperature ratings, costs, capacities ($Cd = Cv/d2$), and so on. Once the valve type is selected, the next task is to select the valve characteristics and size the valve.

---

1. This section originally appeared in different form in *Control* magazine and is reprinted by permission of Putman Media Inc.

WWW.CONTROLGLOBAL.COM
CONTROL
FOR THE PROCESS INDUSTRIES

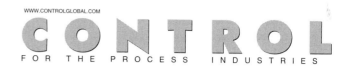

PICKING THE RIGHT CONTROL VALVE

Features and Applications	Ball: Conventiona	Ball: Characterized	Butterfly:Conventional	Butterfly: High-performance	Digital	Globe: Single-ported	Globe: Double-ported	Globe: Angle	Globe: Eccentric disc	Pinch	Plug: Conventional	Plug: Characterized	Saunders Sliding gate: V-Insert	Sliding gate: V-Insert	Sliding gate: Positioned disc	Special: Dynamically balanced
ANSI class pressure rating (max.)	2500	600	300	600	2500	2500	2500	2500	600	150	2500	300	150	150	2500	1500
Max. capacity (Cd)	45	25	40	25	14	12	15	12	13	60	35	25	20	30	10	30
Characteristics	F	G	P	F, G	E	E	E	E	G	P	P	F, G	P,F	F	F	F,G
Corrosive Service	E	E	G	G	F, G	G, E	G, E	G, E	F, G	G	G, E	G	G	F, G	G	G,E
Cost (relative to single-port globe)	0.7	0.9	0.6	0.9	3.0	1.0	1.2	1.1	1.0	0.5	0.7	0.9	0.6	1.0	2.0	1.5
Cryogenic service	A	S	A	A	A	A	A	A	A	NA	A	S	NA	A	NA	NA
High pressure drop (over 200 PSI)	A	A	NA	A	E	G	G	E	A	NA	A	A	NA	NA	E	E
High temperature (over 500°F)	Y	S	E	G	Y	Y	Y	Y	Y	NA	S	S	NA	NA	S	NA
Leakage (ANSI class)	V	IV	I	IV	V	IV	II	IV	IV	IV	IV	IV	V	I	IV	II
Liquids: Abrasive service	C	C	NA	NA	P	G	G	E	G	G, E	F, G	F, G	F, G	NA	E	G
Cavitation resistance	L	L	L	L	M	H	H	H	M	NA	L	L	NA	L	H	M
Dirty service	G	G	F	G	NA	F, G	F	G	F, G	E	G	G	G, E	G	F	F
Flashing applications	P	P	P	F	F	G	G	E	G	F	P	P	F	P	G	P
Slurry including fibrous service	G	G	F	F	NA	F, G	F, G	G,E	F, G	E	G	G	E	G	P	F
Viscous service	G	G	G	G	F	G	F, G	G, E	F, G	G, E	G	G	G,E	F	F	F
Gas/Vapor: Abrasive, erosive	C	C	F	F	P	G	G	E	F, G	G, E	F, G	F, G	G	NA	E	E
Dirty	G	G	G	G	NA	G	F, G	G	F, G	G	G	G	G	G	F	G

Abbreviations:
A = Available
C = All-ceramic design available
F = Fair
G = Good
E = Excellent
H = High
L = Low
M = Medium
NA = Notavailable
P = Poor
S = Special designs only
Y = Yes

FIGURE 36.2 Picking the right control valve. Courtesy of Putman Media Inc.

These characteristics determine the relationship between valve stroke (control signal received) and the flow through the valve; size is determined by maximum flow required.

After startup, if the control loop tends to oscillate at low flows but is sluggish at high flows, users should consider switching the valve trim characteristics from linear to equal-percentage trim. Inversely, if oscillation is encountered at high and sluggishness at low flows, the equal-percentage trim should be replaced with a linear one. Changing the valve characteristics can also be done (sometimes more

CHAPTER | 36 Applying Control Valves

easily) by characterizing the control signal leading to the actuator rather than by replacing the valve trim.

Once the valve type is selected, the next task is to choose the valve characteristics and size the valve. Three of the most common valve characteristics are described in Figure 36.3, and the following tabulation lists the recommended selections of valve characteristics for some of the most common process applications.

The characteristic curves were drawn, assuming that the pressure drop through the valves remain constant while the valves throttle. The three valve characteristics differ from each other in their gain characteristics.

The gain of a control valve is the ratio between the change (=%) in the control signal that the valve receives and the resulting change (=%) in the flow through the valve. Therefore, the valve gain (Gv) can be expressed as GPM/% stroke. The gain (Gv = GPM/%) of a linear valve is constant, the gain of an equal percentage (=%) valve is increasing at a constant slope, and the gain of a quick opening (QO) valve is dropping as the valve opens.

The valve characteristics are called *linear* (straight line in Figure 36.3) if the gain is constant (Gv = 1) and a 1 percent change in the valve lift (control signal) results in the same amount (GPM) of change in the flow through the valve, no matter how open the valve is. This change is the slope of the straight line in Figure 36.3, and it can be expressed as a percentage of maximum flow per a 1 percent change in lift, or as a flow quantity of, say, 5 GPM per percent lift, no matter how open the valve is.

If a 1 percent change in the valve stroke results in the same percentage change (not quantity, but percent of the flow that is occurring!), the valve characteristic is called *equal percentage* (=%). If the valve characteristic is =%, the amount of change in flow is a small quantity when the valve is nearly closed, and it becomes larger and larger as the valve opens.

As shown in Figure 36.3, in case of quick opening (QO) valves, the opposite is the case; at the beginning of the stroke, the valve gain is high (the flow increases at a fast slope) and towards full opening, the slope is small.

The recommended choice of the valve characteristic is a function of the application. For common applications, the recommendations are tabulated at the bottom of Figure 36.2. It should be noted that Figure 36.3's valve characteristics assume that the valve pressure drop is constant. Unfortunately, in most applications it isn't constant but drops off as the load (flow) increases. This is why the valve characteristics recommended in Figure 36.2 are different if the ratio of maximum to minimum pressure differential across the valve is above or below 2:1.

One approach to characterizing an analog control signal is to insert either a divider or a multiplier into the signal line. By adjusting the zero and span, a complete family of curves can be obtained. A divider is used to convert an air-to-open, equal-percentage valve into a linear one or an air-to-close linear valve into an equal-percentage one. A multiplier is used to convert an air-to-open linear valve into an equal-percentage or an air-to-close equal-percentage valve into linear one.

36.3 DISTORTION OF VALVE CHARACTERISTICS

Figure 36.4 shows the effect of the distortion coefficient (DC, defined in Figure 36.4) on the characteristics of an =% valve. As the ratio of the minimum to maximum pressure

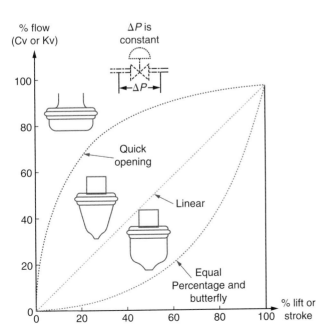

FIGURE 36.3 Valve characteristics. Courtesy of Putman Media.

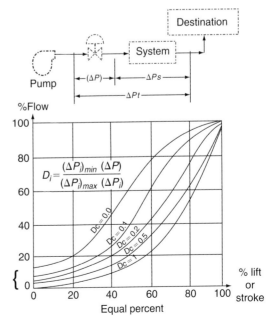

FIGURE 36.4 Distortion coefficient. Courtesy of Putman Media Inc.

drop increases, the DC coefficient drops and the =% characteristics of the valve shifts toward linear.

Similarly, under these same conditions, the characteristics of a linear valve would shift toward quick opening (QO, not shown in the figure). In addition, as the DC coefficient drops, the controllable minimum flow increases and therefore the "rangeability" of the valve also drops.

36.4 RANGEABILITY

The conventional definition of rangeability is the ratio between the maximum and minimum "controllable" flows through the valve. Minimum controllable flow (Fmin) is not the leakage flow (which occurs when the valve is closed) but the minimum flow that is still controllable and can be changed up or down as the valve is throttled.

Using this definition, manufacturers usually claim a 50:1 rangeability for equal-percentage valves, 33:1 for linear valves, and about 20:1 for quick-opening valves. These claims suggest that the flow through these valves can be controlled down to 2, 3, and 5 percent of maximum. However, these figures are often exaggerated. In addition, as shown in the figure, the minimum controllable flow (Fmin) rises as the distortion coefficient (DC) drops. Therefore, at a DC of 0.1, the 50:1 rangeability of an equal-percentage valve drops to about 10:1.

Consequently, the rangeability should be defined as the flow range over which the actual installed valve gain stays within ±25 percent of the theoretical (inherent) valve gain (in the units of GPM per % stroke). To illustrate the importance of this limitation, Figure 36.3 shows the actual gain of an equal percentage valve starts to deviate from its theoretical gain by more than 25 percent, when the flow reaches about 65 percent.

Therefore, in determining the rangeability of such a valve, the maximum allowable flow should be 65 percent. Actually, if one uses this definition, the rangeability of an =% valve is seldom more than 10:1. In such cases, the rangeability of a linear valve can be greater than that of an =% valve. Also, the rangeability of some rotary valves can be higher because their clearance flow tends to be lower and their body losses near the wide-open position also tend to be lower than those of other valve designs.

To stay within ±25 percent of the theoretical valve's gain, the maximum flow should not exceed 60 percent of maximum in a linear valve or 70 percent in an =% valve. In terms of valve lift, these flow limits correspond to 85 percent of maximum lift for =% and 70 percent for linear valves.

36.5 LOOP TUNING

In the search for the most appropriate control valve, gain and stability are as crucial as any other selection characteristics.

The gain of any device is its output divided by its input. The characteristic range and gain of control valves are interrelated. The gain of a linear valve is constant. This gain (Gv) is the maximum flow divided by the valve stroke in percentage (Fmax/100 percent).

Most control loops are tuned for quarter-amplitude damping. This amount of damping (reduction in the amplitude of succeeding peaks of the oscillation of the controlled variable) is obtained by adjusting the controller gain (Gc = 100/%PB) until the total loop gain (the product of the gains of all the control loop components) reaches 0.5 (see Figure 36.5).

The gains of a linear controller (Gv = plain proportional) and a linear transmitter (if it is a temperature transmitter, its gain is Gs = 100%/°F) are both constant. Therefore, if the process gain (Gp = °F/GPM) is also constant, a linear valve is needed to maintain the total loop gain at 0.5 (Gv = 0.5/GvGcGs = constant, meaning linear).

If the transmitter is nonlinear, such as in the case of a d/p cell (sensor gain increases with flow), one can correct for that nonlinearity by using a nonlinear valve the gain of which drops with flow increases (quick opening). In case of heat transfer over a fixed area, the efficiency of heat transfer (process gain Gp) drops as the amount of heat to be transferred rises. To compensate for this nonlinearity (drop in process gain = Gp), the valve gain (Gv) must increase with load. Therefore, an equal-percentage valve should be selected for all heat-transfer temperature control applications.

In case of flow control, one effective way of keeping the valve gain (Gv) perfectly constant is to replace the control valve with a linear cascade slave flow control loop. The limitation of this cascade configuration (in addition to its higher cost) is that if the controlled process is faster than the flow loop, cycling will occur. This is because the slave in any cascade system must be faster than its master. The only way

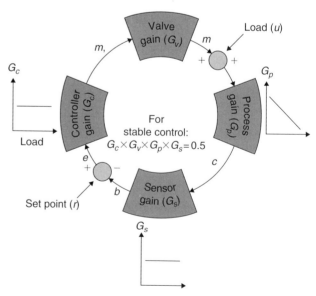

FIGURE 36.5 Well-tuned loops. Courtesy of Putman Media Inc.

CHAPTER | 36 Applying Control Valves

to overcome this cycling is to slow (detune) the master by lowering its gain (increasing its proportional band), which in turn degrades its control quality. Therefore, this approach should only be considered on slow or secondary temperature control loops.

36.6 POSITIONING POSITIONERS

A valve positioner is a high-gain (0.5 to 10 percent proportional band), sensitive, proportional-only, valve-stroke position controller. Its set point is the control signal from the controller. The main purpose of having a positioner is to guarantee that the valve does in fact move to the position that corresponds to the value of the controller output.

The addition of a positioner can correct for such maintenance-related effects as variations in packing friction due to dirt buildup, corrosion, or lack of lubrication; variations in the dynamic forces of the process; or nonlinearity in the valve actuator. In addition, the positioner can allow for split-ranging the controller signal between valves or can increase the actuator speed or actuator thrust by increasing the pressure and/or volume of the actuator air signal. In addition, it can modify the valve characteristics by the use of cams or function generators.

A positioner will improve performance on most slow loops, such as the control of analytical properties, temperature, liquid level, blending, and large-volume gas flow. A controlled process can be considered slow if its period of oscillation is three or more times the period of oscillation of the positioned valve.

Positioners also are useful to overcome the "dead band" of the valve, which can be caused by valve-stem friction. The result of this friction is that whenever the direction of the control signal is reversed, the stem remains in its last position until the dead band is exceeded. Positioners will eliminate this limit cycle by closing a loop around the valve actuator. Integrating processes, such as liquid level, volume (as in digital blending), weight (not weight-rate), and gas-pressure control loops, are prone to limit cycling and will usually benefit from the use of positioners.

In the case of fast loops (fast flow, liquid pressure, small-volume gas pressure), positioners are likely to degrade loop response and cause limit cycling because the positioner (a cascade slave) is not faster than the speed at which its set point (the control signal) can change. A controlled process is considered fast if its period of oscillation is less than three times that of the positioned valve.

Split ranging of control valves does not necessarily require the use of positioners, because one can also split-range the valves through the use of different spring ranges in the valve actuators.

If the need is only to increase the speed or the thrust of the actuator, it is sufficient to install an air volume booster or a pressure amplifier relay, instead of using a positioner. If the goal is to modify the valve characteristics on fast processes, this should not be done by the use of positioners but instead by installing dividing or multiplying relays in the controller output.

36.7 SMARTER SMART VALVES

Much improvement has occurred and more is expected in the design of intelligent and self-diagnosing positioners and control valves. The detection and correction for the wearing of the trim, hysteresis caused by packing friction, air leakage in the actuator, and changes in valve characteristics can all be automated. If the proper intelligence is provided, the valve can compare its own behavior with its past performance and, when the same conditions result in different valve openings, it can conclude, for example, that its packing is not properly lubricated or the valve port is getting plugged. In such cases, the valve can automatically request and schedule its own maintenance.

A traditional valve positioner serves only the purpose of keeping the valve at the opening that corresponds to the control signal. Digital positioners can also collect and analyze valve-position data, valve operating characteristics, and performance trends and can enable diagnostics of the entire valve assembly. The control signals into smart positioners can be analog (4–20 mA) or digital (via bus systems). The advantages of digital positioners relative to their analog counterparts include increased accuracy (0.1 to 1 percent versus 0.3 to 2 percent for analog), improved stability (about 0.1 percent compared to 0.175 percent), and wider range (up to 50:1 compared to 10:1).

Smart valves should also be able to measure their own inlet, outlet, and vena contracta pressures, flowing temperature, valve opening (stem position) and actuator air pressure. Valve performance monitoring includes the detection of "zero" position and span of travel, actuator air pressure versus stem travel, and the ability to compare these against their values when the valve was new. Major deviations from the "desired" characteristic can be an indication of the valve stuffing box being too tight, a corroded valve stem or a damaged actuator spring.

Additional features offered by smart valves include the monitoring of packing box or bellows leakage by "sniffing" (using miniaturized chemical detectors), checking seat leakage by measuring the generated sound frequency, or by comparing the controller output signal at "low flow" with the output when the valve was new. Another important feature of digital positioners is their ability to alter the inherent characteristics of the valve.

36.8 VALVES SERVE AS FLOWMETERS

A control valve can also be viewed as a variable area flowmeter. Therefore, smart valves can measure their own flow by solving their valve-sizing equation. For example, in case of turbulent liquid flow applications, where the valve capacity coefficient Cv can be calculated as $Cv = \dfrac{q}{F_p}\sqrt{\dfrac{G_f}{\Delta p}}$, the valve data can be used to calculate flow. This is done by inserting the known values of Cv, Gf, Δp, and the piping geometry coefficient (Fp) into the applicable equation for Cv. Naturally, in order for the smart valves of the future to be able to accurately measure their own flow, they must be provided with sufficient intelligence to identify the applicable sizing equation for the particular process. (See Section 6.15 in Volume 2 of the *Instrument Engineers' Handbook* for valve sizing equations.)

FURTHER READING

Liptak, Bela G., *Instrument Engineer's Handbook*, Vol. 2, Chilton's and ISA Press, 2002.

Emerson Process Management, *Control Valve Handbook*, 4th ed., 2006; downloadable from www.controlglobal.com/whitepapers/2006/056.html.

Bauman, Hans D., *Control Valve Primer: A User's Guide,* 4th ed., ISA Press, 2009.

Part VI

Automation and Control Systems

Chapter 37

Design and Construction of Instruments

C. I. Daykin and W. H. Boyes

37.1 INTRODUCTION

The purpose of this chapter is to give an insight into the types of components and construction used in commercial instrumentation.

To the designer, the *technology* in Instrument Technology depends on the availability of components and processes appropriate to his task. Being aware of what is possible is an important function of the designer, especially with the rapidity with which techniques now change. New materials such as ceramics and polymers have become increasingly important, as have semicustom (ASICs) and large-scale integrated (LSI and VLSI) circuits. The need for low-cost automatic manufacture is having the greatest impact on design techniques, demanding fewer components and suitable geometries. Low volume instruments and one-offs are now commonly constructed in software, using "virtual instrumentation" graphical user interfaces such as Labview and Labtech, or "industrial strength" HMI programs such as Wonderware or Intellution.

The distinction between computer and instrument has become blurred, with many instruments offering a wide range of facilities and great flexibility. Smart sensors, which interface directly to a computer, and Fieldbus interconnectivity have shifted the emphasis to software and mechanical aspects.

Historical practice, convention, and the emergence of standards also contribute significantly to the subject. Standards, especially, benefit the designer and the user, and have made the task of the author and the reader somewhat simpler.

Commercial instruments exist because there is a market, and so details of their design and construction can only be understood in terms of a combination of commercial as well as technical reasons. A short section describes these trade-offs as a backdrop to the more technical information.

37.2 INSTRUMENT DESIGN

37.2.1 The Designer's Viewpoint

Many of the design features found in instruments are not obviously of direct benefit to the user. These can best be understood by also considering the designer's viewpoint.

The instrument designer's task is to find the best compromise between cost and benefit to the users, especially when competition is fierce. For a typical medium-volume instrument, its cost as a percentage of selling price is distributed as follows:

Purchase cost
- Design cost 20%
- Manufacturing cost 30%
- Selling cost 20%
- Other overheads 20%
- Profit 10%
- 100%

Operating/maintenance cost may amount to 10 percent per annum. Benefits to the user can come from many features, for example:

1. Accuracy
2. Speed
3. Multipurpose
4. Flexibility
5. Reliability
6. Integrity
7. Maintainability
8. Convenience

Fashion, as well as function, is very important, since a smart, pleasing, and professional appearance is often essential when selling instruments on a commercial basis.

For a particular product the unit cost can be reduced with higher volume production, and greater sales can be achieved with a lower selling price. The latter is called its "market elasticity." Since the manufacturer's objective is to maximize return on investment, the combination of selling price and volume which yields the greatest profit is chosen.

37.2.2 Marketing

The designer is generally subordinate to marketing considerations, and consequently, these play a major role in determining the design of an instrument and the method of its manufacture. A project will only go ahead if the anticipated return on investment is sufficiently high and commensurate with the perceived level of risk. It is interesting to note that design accounts for a significant proportion of the total costs and, by its nature, involves a high degree of risk.

With the rapid developments in technology, product lifetimes are being reduced to less than three years, and market elasticity is difficult to judge. In some markets, especially in test and measurement systems, product lifetime has been reduced to one single production run. In this way design is becoming more of a process of evolution, continuously responding to changes in market conditions. Initially, therefore, the designer will tend to err on the cautious side, and make cost reductions as volumes increase.

The anticipated volume is a major consideration and calls for widely differing techniques in manufacture, depending on whether it is a low-, medium-, or high-volume product.

37.2.3 Special Instruments

Most instrumentation users configure systems from low-cost standard components with simple interfaces. Occasionally the need arises for a special component or system. It is preferable to modify a standard component wherever possible, since complete redesigns often take longer than anticipated, with an uncertain outcome.

Special systems also have to be tested and understood in order to achieve the necessary level of confidence. Maintenance adds to the extra cost, with the need for documentation and test equipment making specials very expensive. This principle extends to software as well as hardware.

37.3 ELEMENTS OF CONSTRUCTION

37.3.1 Electronic Components and Printed Circuits

Electronic circuitry now forms the basis of most modern measuring instruments. A wide range of electronic components are now available, from the simple resistor to complete data acquisition subsystems. Table 37.1 lists some of the more commonly used types.

Computer-aided design makes it possible to design complete systems on silicon using standard cells or by providing the interconnection pattern for a standard chip which has an array of basic components. This offers reductions in size and cost and improved design security.

The most common method of mounting and interconnecting electronic components at one time was by double-sided, through-hole plated fiberglass printed circuit board (PCB) (Figure 37.1 (a)). Component leads or pins are pushed through holes and soldered to tinned copper pads (Figure 37.1 (b)). This secures the component and provides the connections. The solder joint is thus the most commonly used component and probably the most troublesome. In the past eight years, new techniques for surface-mount assembly have made through-hole circuit boards obsolete for anything but the most low-volume production.

Tinning the copper pads with solder stops corrosion and makes soldering easier. Fluxes reduce oxidation and surface tension, but a temperature-controlled soldering iron is indispensable. Large-volume soldering can be done by a wave-soldering machine, where the circuit board is passed over a standing wave of molten solder. Components have to be on one side only, although this is also usually the case with manual soldering.

The often complicated routing of connections between components is made easier by having two layers of printed "tracks," one on each surface, permitting cross-overs. Connections between the top and bottom conductor layers are provided by plated-through "via" holes. It is generally considered bad practice to use the holes in which components are mounted as via holes, because of the possibility of damaging the connection when components are replaced. The through-hole plating (see Figure 37.1(c)) provides extra support for small pads, reducing the risk of them peeling off during soldering. They do, however, make component removal more difficult, often requiring the destruction of the component so that its leads can be removed individually. The more expensive components are therefore generally provided with sockets, which also makes testing and servicing much simpler.

The PCB is completed by the addition of a solder mask and a printed (silk-screened) component identification layer. The solder mask is to prevent solder bridges between adjacent tracks and pads, especially when using an automatic soldering technique such as wave soldering. The component identification layer helps assembly, testing and servicing.

For very simple low-density circuits a single-sided PCB is often used, since manufacturing cost is lower. The pads and tracks must be larger, however, since without through-hole plating they tend to lift off more easily when soldering.

For very high-density circuits, especially digital, multilayer PCBs are used, as many as nine layers of printed circuits being laminated together with through-hole plating providing the interconnections.

Most electronic components have evolved to be suitable for this type of construction, along with machines for automatic handling and insertion. The humble resistor is an interesting example; this was originally designed for wiring

CHAPTER | 37 Design and Construction of Instruments

TABLE 37.1 Electronic components

Component and symbol	Main types	Appearance	Range	Character
Resistors	Metal oxide		0 Ω–100 MΩ	1% general purpose
	Metal film		0.1 Ω 100 MΩ	0.001% low drift
	Wire wound		0.01 ΩMΩ	0.01% high power
Capacitors	Air dielectric		0.01 pF–100 pF	0.01% high stability
	Ceramics		1 pF–10 μF	5% small size
	Polymer		1 pF–10 μF	1% general purpose
	Electrolytic		0.1 μF–1 F	10% small size
Inductors	Cylindrical core		0.1 μH–10 mH	10% general purpose
	Pot core		1 μH–1 H	0.1% high stability
	Toroidal core		100 μH–100 H	20% high values
				10^{-7} high accuracy ratios
Transformers	Cylindrical core		RF, IF types	
	Pot core		0.1 μH–1 mH	0.1% mutual inductor
	Toroidal core		0.1 H–10 H	20% high inductance
Diodes	PN junction		1 pA(on)–10^3 A(off)	Wide range
Transistors	Bipolar		10^{-17} W(input)–10^3 W(output)	high freq., low noise
				high power, etc.
	FETs		10^{-23}(input)–10^3 W(output)	as above
Integrated circuits	Analog		operational amplifiers	wide range
			function blocks	multiply, divide
			Amplifiers	high frequency
			switches	high accuracy
			semi-custom	
	Digital		small scale integ. (SSI)	logic elements
			medium scale integ. (MSI)	function blocks
			large scale integ. (LSI)	major functions
			v. large scale integ. (VLSI)	complete systems
			semi-custom	
	Monolithic hybrid		A/D D/A conversion	
			special functions	
			semi-custom	
Others	Thyristors		1 A–10^3 A	high power switch
	Triacs		1 A–10^3 A	high power switch
	Opto-couplers		singles, duals, quads	
	Thermistors		+or – temp. coefficient	
	Relay sys		0.01 Ω(on) – 10^{12} Ω(off)	

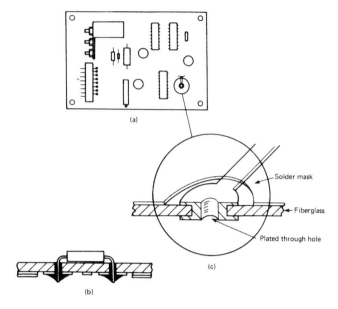

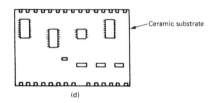

FIGURE 37.1 Printed electronic circuits. (a) Printed circuit board (PCB); (b) traditional axial component; (c) through-hole plating (close-up); (d) surface-mounted assemblies.

between posts or tag-strips in valve circuits. Currently, they are supplied on long ribbons, and machines or hand tools are used for bending and cropping the leads ready for insertion (see Figure 37.1(b)).

Important principles relevant to the layout of circuits are also discussed in Chapter 35.

37.3.2 Surface-Mounted Assemblies

As noted earlier, surface-mounting took over as the design of choice in the 1990s. Sometimes, the traditional fiberglass rigid board itself has been replaced by a flexible sheet of plastic with the circuits printed on it. Semiconductors, chip resistors and chip capacitors are available in very small outline packages, and are easier to handle with automatic placement machines.

Surface mounting eliminates the difficult problem of automatic insertion and, in most cases, the costly drilling process as well. Slightly higher densities can be achieved by using a ceramic substrate instead of fiberglass (Figure 37.1(d)). Conductors of palladium silver, insulators, and resistive inks are silk-screened and baked onto the substrate to provide connections, cross-overs, and some of the components. These techniques have been developed from the older "chip and wire" hybrid thick film integrated circuit technique, used mainly in high-density military applications. In both cases, reflow soldering techniques are used due to the small size. Here, the solder is applied as a paste and silk-screened onto the surface bonding pads. The component is then placed on its pads and the solder made to reflow by application of a short burst of heat which is not enough to damage the component. The heat can be applied by placing the substrate onto a hot vapor which then condenses at a precise temperature above the melting point of the solder. More simply, the substrate can be placed on a temperature-controlled hot plate or passed under a strip of hot air or radiant heat.

The technique is therefore very cost effective in high volumes, and with the increasing sophistication of silicon circuits results in "smart sensors" where the circuitry may be printed onto any flat surface.

37.3.2.1 Circuit Board Replacement

When deciding servicing policy it should be realized that replacing a whole circuit board is often more cost effective than trying to trace a faulty component or connection. To this end, PCBs can be mounted for easy access and provided with a connector or connectors for rapid removal. The faulty circuit board can then be thrown away or returned to the supplier for repair.

37.3.3 Interconnections

There are many ways to provide the interconnection between circuit boards and the rest of the instrument, of which the most common are described below.

Connectors are used to facilitate rapid making and breaking of these connections and simplify assembly test and servicing. Conventional wiring looms are still used because of their flexibility and because they can be designed for complicated routing and branching requirements. Termination of the wires can be by soldering, crimp, or wire-wrap onto connector or circuit board pins. This, however, is a labor-intensive technique and is prone to wiring errors. Looms are given mechanical strength by lacing or sleeving wires as a tight bunch and anchoring to the chassis with cable ties.

Ribbon cable and insulation displacement connectors are now replacing conventional looms in many applications. As many as sixty connections can be made with one simple loom with very low labor costs. Wiring errors are eliminated since the routing is fixed at the design stage (see Figure 37.2).

Connectors are very useful for isolating or removing a subassembly conveniently. They are, however, somewhat expensive and a common source of unreliability.

Another technique, which is used in demanding applications where space is at a premium, is the "flexy" circuit. Printed circuit boards are laminated with a thin, flexible sheet of Kapton which carries conductors. The connections

CHAPTER | 37 Design and Construction of Instruments

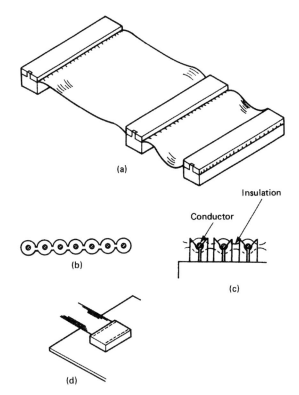

FIGURE 37.2 Ribbon cable interconnection. (a) Ribbon cable assembly; (b) ribbon cable cross-section; (c) insulation displacement terminator; (d) dual in-line header.

are permanent, but the whole assembly can be folded up to fit into a limited space.

It is inappropriate to list here the many types of connectors. The connector manufacturers issue catalogs full of different types, and these are readily available.

37.3.4 Materials

A considerable variety of materials are available to the instrument designer, and new ones are being developed with special or improved characteristics, including polymers and superstrong ceramics. These materials can be bought in various forms, including sheet, block, rod, and tube, and processed in a variety of ways.

37.3.4.1 Metals

Metals are usually used for strength and low cost as structural members. Aluminum for low weight and steel are the most common. Metals are also suitable for machining precise shapes to tight tolerances.

Stainless steels are used to resist corrosion, and precious metal in thin layers helps to maintain clean electrical contacts. Metals are good conductors and provide electrical screening as well as support. Mumetal and radiometal have high permeabilities and are used as very effective magnetic screens or in magnetic components. Some alloys–notably beryllium-copper—have very good spring characteristics, improved by annealing, and this is used to convert force into displacement in load cells and pressure transducers. Springs made of nimonic keep their properties at high temperatures, which is important in some transducer applications.

The precise thermal coefficient of the expansion of metals makes it possible to produce compensating designs, using different metals or alloys, and so maintain critical distances independent of temperature. Invar has the lowest coefficient of expansion at less than 1 ppm per K over a useful range, but it is difficult to machine precisely.

Metals can be processed to change their characteristics as well as their shape; some can be hardened after machining and ground or honed to a very accurate and smooth finish, as found in bearings.

Metal components can be annealed, i.e., taken to a high temperature, in order to reduce internal stresses caused in the manufacture of the material and machining. Heat treatments can also improve stability, strength, spring quality, magnetic permeability, or hardness.

37.3.4.2 Ceramics

For very high temperatures, ceramics are used as electrical and heat insulators or conductors (e.g., silicon carbide). The latest ceramics (e.g., zirconia, sialon, silicon nitride, and silicon carbide) exhibit very high strength, hardness, and stability even at temperatures over 1,000°C. Processes for shaping them include slip casting, hot isostatic pressing (HIP), green machining, flame spraying, and grinding to finished size. Being hard, their grinding is best done by diamond or cubic boron nitride (CBN) tools. Alumina is widely used, despite being brittle, and many standard mechanical or electrical components are available.

Glass-loaded machinable ceramics are very convenient, having very similar properties to alumina, but are restricted to lower temperatures (less than 500 °C). Special components can be made to accurate final size with conventional machining and tungsten tools.

Other compounds based on silicon include sapphires, quartz, glasses, artificial granite, and the pure crystalline or amorphous substance. These have well behaved and known properties (e.g., thermal expansion coefficient, conductivity and refractive index), which can be finely adjusted by adding different ingredients. The manufacture of electronic circuitry, with photolithography, chemical doping, and milling, represents the ultimate in materials technology. Many of these techniques are applicable to instrument manufacture, and the gap between sensor and circuitry is narrowing—for example, in chemfets, in which a reversible chemical reaction produces a chemical potential that is coupled to one or more field-effect transistors. These transistors give amplification and possibly conversion to digital form before transmission to an indicator instrument with resulting higher integrity.

37.3.4.3 Plastics and Polymers

Low-cost, lightweight, and good insulating properties make plastics and polymers popular choices for standard mechanical components and enclosures. They can be molded into elaborate shapes and given a pleasing appearance at very low cost in high volumes. PVC, PTFE, polyethylene, polypropylene, polycarbonates, and nylon are widely used and available as a range of composites, strengthened with fibers or other ingredients to achieve the desired properties. More recently, carbon composites and Kevlar have exhibited very high strength-to-weight ratio, useful for structural members. Carbon fiber is also very stable, making it suitable for dimensional calibration standards. Kapton and polyamides are used at higher temperatures and radiation levels.

A biodegradable plastic, poly 3-hydroxy-buty-rate, or PHB, is also available which can be controlled for operating life. Manufactured by cloned bacteria, this material represents one of many new materials emerging from advances in biotechnology.

More exotic materials are used for special applications, and a few examples are:

1. Mumetal: very high magnetic permeability.
2. PVDF: polyvinylidene fluoride, piezoelectric effect.
3. Samarium/cobalt: very high magnetic remanence (fixed magnet).
4. Sapphire: very high thermal conductivity.
5. Ferrites: very stable magnetic permeability, wide range available.

37.3.4.4 Epoxy Resins

Two-part epoxy resins can be used as adhesives, as potting material, and as paint. Parameters such as viscosity, setting time, set hardness, and color can be controlled. Most have good insulating properties, although conducting epoxies exist, and all are mechanically strong, some up to 300°C. The resin features in the important structure material: epoxy bonded fiberglass. Delicate assemblies can be ruggedized or passivated by a prophylactic layer of resin, which also improves design security.

Epoxy resin can be applied to a component and machined to size when cured. It can allow construction of an insulating joint with precisely controlled dimensions. Generally speaking, the thinner the glue layer, the stronger and more stable the joint.

37.3.4.5 Paints and Finishes

The appearance of an instrument is enhanced by the judicious use of texture and color in combination with its controls and displays. A wide range of British Standard coordinated colors are available, allowing consistent results (BS 5252 and 4800). In the United States, the Pantone color chart is usually used, and colors are generally matched to a PMS (Pantone Matching System) color. For example, PMS 720 is a commonly used front panel color, in a royal blue shade.

Anodized or brushed aluminum panels have been popular for many years, although the trend is now back toward painted or plastic panels with more exotic colors. Nearly all materials, including plastic, can be spray-painted by using suitable preparation and curing. Matte, gloss, and a variety of textures are available.

Despite its age, silk-screen printing is used widely for lettering, diagrams, and logos, especially on front panels.

Photosensitive plastic films, in one or a mixture of colors, are used for stick-on labels or as complete front panels with an LED display filter. The latter are often used in conjunction with laminated pressure pad-switches to provide a rugged, easy-to-clean, splash-proof control panel.

37.3.5 Mechanical Manufacturing Processes

Materials can be processed in many ways to produce the required component. The methods chosen depend on the type of material, the volume required, and the type of shape and dimensional accuracy.

37.3.5.1 Bending and Punching

Low-cost sheet metal or plastic can be bent or pressed into the required shape and holes punched with standard or special tools (Figure 37.3). Simple bending machines and a fly press cover most requirements, although hard tooling is more cost effective in large volumes. Most plastics are thermosetting and require heating, but metals are normally worked cold. Dimensional accuracy is typically not better than 0.5 mm.

37.3.5.2 Drilling and Milling

Most materials can be machined, although glass (including fiberglass), ceramics, and some metals require specially hardened tools. The hand or pillar drill is the simplest tool, and high accuracy can be achieved by using a jig to hold the work-piece and guide the rotating bit.

A milling machine is more complex, where the workpiece can be moved precisely relative to the rotating tool. Drills, reamers, cutters, and slotting saws are used to create complex and accurate shapes. Tolerances of 1 μm can be achieved.

37.3.5.3 Turning

Rotating the workpiece against a tool is a method for turning long bars of material into large numbers of components at low unit cost. High accuracies can be achieved for internal and external diameters and length, typically 1 μm, making cylindrical components a popular choice in all branches of engineering.

37.3.5.5 Lapping

A fine sludge of abrasive is rubbed onto the work-piece surface to achieve ultra-high accuracy, better than 10 nm if the metal is very hard. In principle, any shape can be lapped, but optical surfaces such as flats and spherics are most common, since these can be checked by sensitive optical methods.

37.3.5.6 Chemical and Electrochemical Milling

Metal can be removed or deposited by chemical and electrochemical reactions. Surfaces can be selectively treated through masks. Complex shapes can be etched from sheet material of most kinds of metal using photolithographic techniques. Figure 37.3(b) shows an example where accuracies of 0.1 mm are achieved.

Gold, tin, copper, and chromium can be deposited for printed circuit board manufacture or servicing of bearing components. Chemical etching of mechanical structures into silicon in combination with electronic circuitry is a process currently under development.

37.3.5.7 Extruding

In extruding, the material, in a plastic state and usually at a high temperature, is pushed through an orifice with the desired shape. Complex cross-sections can be achieved, and a wide range of standard items are available, cut to length. Extruded components are used for structural members, heat sinks, and enclosures (Figure 37.4). Initial tooling is, however, expensive for non-standard sections.

37.3.5.8 Casting and Molding

Casting, like molding, makes the component from the liquid or plastic phase but results in the destruction of the mold. It usually refers to components made of metals such as

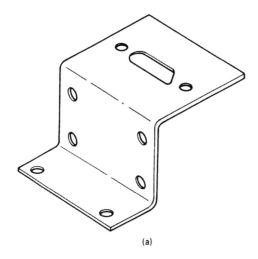
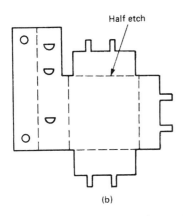

FIGURE 37.3 Sheet metal. (a) Bent and drilled or punched; (b) chemical milling.

A fully automatic machining center and tool changer and component handler can produce vast numbers of precise components of many different types under computer control.

37.3.5.4 Grinding and Honing

With grinding, a hard stone is used to remove a small but controlled amount of material. When grinding, both tool and component are moved to produce accurate geometries including relative concentricity and straightness (e.g., parallel) but with a poor surface finish. Precise flats, cylinders, cones, and spherics are possible. The material must be fairly hard to get the best results and is usually metal or ceramic.

Honing requires a finer stone and produces a much better surface finish and potentially very high accuracy (0.1 μm). Relative accuracy (e.g., concentricity between outside and inside diameters) is not controllable, and so honing is usually preceded by grinding or precise turning.

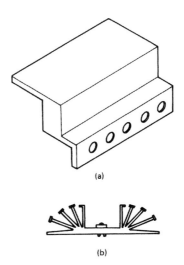

FIGURE 37.4 Extrusion. (a) Structural member; (b) heat sink.

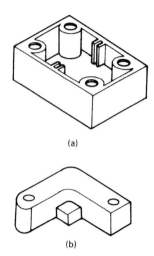

FIGURE 37.5 Examples of casting and molding. (a) Molded enclosure; (b) cast fixing.

aluminum alloys and sand casts made from a pattern. Very elaborate forms can be made, but further machining is required for accuracies better than 0.5 mm. Examples are shown in Figure 37.5. Plastics are molded in a variety of ways, and the mold can be used many times. Vacuum forming and injection molding are used to achieve very low unit cost, but tooling costs are high. Rotational low-pressure molding (rotomolding) is often used for low-volume enclosure runs.

37.3.5.9 Adhesives

Adhesive technology is advancing at a considerable rate, finding increasing use in instrument construction. Thin layers of adhesive can be stable and strong and provide electrical conduction or insulation. Almost any material can be glued, although high-temperature curing is still required for some applications. Metal components can be recovered by disintegration of the adhesive at high temperatures. Two-part adhesives are usually best for increased shelf life.

Jigs can be used for high dimensional accuracies, and automatic dispensing for high volume and low-cost assembly.

37.3.6 Functional Components

A wide range of devices is available, including bearings, couplings, gears, and springs. Figure 37.6 shows the main types of components used and their principal characteristics.

37.3.6.1 Bearings

Bearings are used when a controlled movement, either linear or rotary, is required. The simplest bearing consists of rubbing surfaces, prismatic for linear, cylindrical for rotation, and spherical for universal movement. Soft materials such as copper and bronze and PTFE are used for reduced friction, and high precision can be achieved. Liquid or solid lubricants are sometimes used, including thin deposits of PTFE,

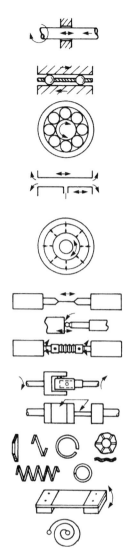

FIGURE 37.6 Mechanical components.

graphite, and organic and mineral oils. Friction can be further reduced by introducing a gap and rolling elements between the surfaces. The hardened steel balls or cylinders are held in cages or a recirculating mechanism. Roller bearings can be precise, low friction, relatively immune to contamination, and capable of taking large loads.

The most precise bearing is the air bearing. A thin cushion of pressurized air is maintained between the bearing surfaces, considerably reducing the friction and giving a position governed by the average surface geometry. Accuracies of 0.01 μm are possible, but a source of clean, dry, pressurized air is required. Magnetic bearings maintain an air gap and have low friction but cannot tolerate side loads.

With bearings have evolved seals to eliminate contamination. For limited movement, elastic balloons of rubber or metal provide complete and possibly hermetic sealing. Seals made of low-friction polymer composites exclude larger particles, and magnetic liquid lubricant can be trapped between magnets, providing an excellent low-friction seal for unlimited linear or rotary movement.

37.3.6.2 Couplings

It is occasionally necessary to couple the movement of two bearings, which creates problems of clashing. This can be overcome by using a flexible coupling which is stiff in the required direction and compliant to misalignment of the bearings. Couplings commonly used include:

1. Spring wire or filaments.
2. Bellows.
3. Double hinges.

Each type is suitable for different combinations of side load, misalignment, and torsional stiffness.

37.3.6.3 Springs

Springs are used to produce a controlled amount of force (e.g., for preloaded bearings, force/pressure transducers or fixings). They can take the form of a diaphragm, helix, crinkled washer, or shaped flat sheet leaf spring. A thin circular disc with chemically milled Archimedes spinal slots is an increasingly used example of the latter. A pair of these can produce an accurate linear motion with good sideways stiffness and controllable spring constant.

37.4 CONSTRUCTION OF ELECTRONIC INSTRUMENTS

Electronic instruments can be categorized by the way they are intended to be used physically, resulting in certain types of construction:

1. Site mounting.
2. Panel mounting.
3. Bench mounting.
4. Rack mounting.
5. Portable instruments.

37.4.1 Site Mounting

The overriding consideration here is usually to get close to the physical process which is being measured or controlled. This usually results in the need to tolerate harsh environmental conditions such as extreme temperature, physical shock, muck, and water. Signal conditioners and data-acquisition subsystems, which are connected to transducers and actuators, produce signals suitable for transmission over long distances, possibly to a central instrumentation and control system some miles away. Whole computerized systems are also available with ruggedized construction for use in less hostile environments.

The internal construction is usually very simple, since there are few, if any, controls or displays. Figure 37.7 shows an interesting example which tackles the common problem of wire terminations. The molded plastic enclosure is sealed

FIGURE 37.7 Instrument for site mounting. Courtesy of Solartron Instruments.

at the front with a rubber "O" ring and is designed to pass the IPC 65 "hosepipe" test (see BS 5490 or the NEMA 4 and 4X standard). The main electronic circuit is on one printed circuit board mounted on pillars, connected to one of a variety of optional interface cards. The unit is easily bolted to a wall or the side of a machine.

37.4.2 Panel Mounting

A convenient way for an instrument engineer to construct a system is to mount the various instruments which require control or readout on a panel with the wiring and other system components protected inside a cabinet. Instruments designed for this purpose generally fit into one of a number of DIN standard cut-outs (see DIN 43 700). Figure 37.8 is an example illustrating the following features:

1. The enclosure is an extruded aluminum tube.
2. Internal construction is based around five printed circuit boards, onto which the electronic displays are soldered. The PCBs plug together and can be replaced easily for servicing.
3. All user connections are at the rear, for permanent or semi-permanent installation.

37.4.3 Bench-mounting Instruments

Instruments which require an external power source but a degree of portability are usually for benchtop operation. Size is important, since bench space is always in short supply.

Instruments in this category often have a wide range of controls and a display requiring careful attention to ergonomics. Figure 37.9(a) shows a typical instrument, where the following points are worth noting:

1. The user inputs are at the front for easy access.
2. There is a large clear display for comfortable viewing.

FIGURE 37.8 Panel-mounting instrument. Courtesy of Systemteknik AB.

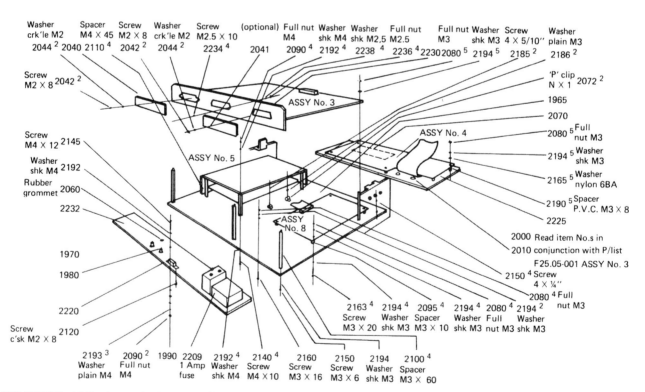

FIGURE 37.9 (a) Bench-mounting instrument. Courtesy of Automatic Systems Laboratories Ltd. (b) General assembly drawing of instrument shown in Figure 37.9(a).

3. The carrying handle doubles up as a tilt bar.
4. It has modular internal construction with connectors for quick servicing.

The general assembly drawing for this instrument is included as Figure 37.9(b), to show how the parts fit together.

37.4.4 Rack-mounting Instruments

Most large electronic instrumentation systems are constructed in 19-inch wide metal cabinets of variable height (in units of 1.75 inch 1U). These can be for bench mounting, free standing, or wall mounting. Large instruments are normally designed for bench operation or rack mounting for which optional brackets are supplied. Smaller modules plug into subracks which can then be bolted into a 19-inch cabinet.

Figure 37.10 shows some of the elements of a modular instrumentation system with the following points:

1. The modules are standard Eurocard sizes and widths (DIN 41914 or IEC 297).
2. The connectors are to DIN standard (DIN 41612).
3. The subrack uses standard mechanical components and can form part of a much larger instrumentation system.

The degree of modularity and standardization enables the user to mix a wide range of instruments and components from a large number of different suppliers worldwide.

37.4.5 Portable Instruments

Truly portable instruments are now common, due to the reduction in size and power consumption of electronic circuitry. Figure 37.11 shows good examples which incorporate the following features:

1. Lightweight, low-cost molded plastic case.
2. Low-power CMOS circuitry and liquid crystal display (LCD).
3. Battery power source gives long operating life.

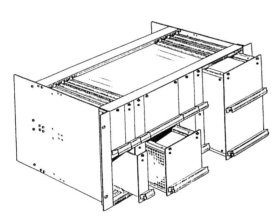

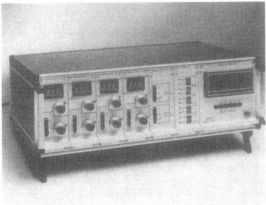

FIGURE 37.10 Rack-based modular instruments. Courtesy of Schroff (U.K.) Ltd. and Automatic Systems Laboratories Ltd.

FIGURE 37.11 Portable instruments. Courtesy of Solomat SA.

Size reduction is mainly from circuit integration onto silicon and the use of small outline components.

37.4.6 Encapsulation

For particularly severe conditions, especially with regard to vibration, groups of electronic components are sometimes *encapsulated* (familiarly referred to as "potting"). This involves casting them in a suitable material, commonly epoxy resin. This holds the components very securely in position, and they are also protected from the atmosphere to which the instrument is exposed. To give further protection against stress (for instance, from differential thermal expansion), a complex procedure is occasionally used, with compliant silicone rubber introduced as well as the harder resin.

Some epoxies are strong up to 300°C. At higher temperatures (450°C) they are destroyed, allowing encapsulated components to be recovered if they are themselves heat resistant. Normally an encapsulated group would be thrown away if any fault developed inside it.

37.5 MECHANICAL INSTRUMENTS

Mechanical instruments are mainly used to interface between the physical world and electronic instrumentation. Examples are:

1. Displacement transducers (linear and rotary).
2. Force transducers (load cells).
3. Accelerometers.

Such transducers often have to endure a wide temperature range, shock, and vibration, requiring careful selection of materials and construction.

Many matters contribute to good mechanical design and construction, some of which are brought out in the devices described in other chapters of this book. We add to that here by showing details of one or two instruments where particular principles of design can be seen. Before that, however, we give a more general outline of kinematic design, a way of proceeding that can be of great value for designing instruments.

37.5.1 Kinematic Design

A particular approach sometimes used for high-precision mechanical items is called kinematic design. When the relative location of two bodies must be constrained, so that there is either no movement or a closely controlled movement between them, it represents a way of overcoming the uncertainties that arise from the impossibility of achieving geometrical perfection in construction. A simple illustration is two flat surfaces in contact. If they can be regarded as ideal geometrical planes, then the relative movement of the two bodies is accurately defined. However, it is expensive to approach geometrical perfection, and the imperfections of the surfaces mean that the relative position of the two parts will depend upon contact between high spots, and will vary slightly if the points of application of the forces holding them together are varied. The points of contact can be reduced, for instance, to four with a conventional chair, but it is notorious that a four-legged chair can rock unless the bottoms of the legs match the floor perfectly. *Kinematic design* calls for a three-legged chair, to avoid the *redundancy* of having its position decided by four points of contact. More generally, a rigid solid body has 6 *degrees of freedom* which can be used to fix its position in space. These are often thought of as three Cartesian coordinates to give the position of one point of the body, and when that has been settled, rotation about three mutually perpendicular axes describes the body's attitude in space. The essence of kinematic design is that each degree of freedom should be constrained in an identifiable localized way. Consider again the three-legged stool on a flat surface. The Z-coordinate of the tip of the leg has been constrained, as has rotation about two axes in the flat surface. There is still freedom of X- and- Y-coordinates and for rotation about an axis perpendicular to the surface: 3 degrees of freedom removed by the three constraints between the leg-tips and the surface.

A classical way of introducing six constraints and so locating one body relative to another is to have three V-grooves in one body and three hemispheres attached to the other body, as shown in Figure 37.12. When the hemispheres enter the grooves (which should be deep enough for contact to be made with their sides and not their edges), each has two constraints from touching two sides, making a total of six.

If one degree of freedom, say, linear displacement, is required, five spheres can be used in a precise groove as in Figure 37.13. Each corresponds to a restricted movement.

For the required mating, it is important that contact should approximate to point contact and that the construction materials should be hard enough to allow very little deformation perpendicular to the surface under the loads normally encountered. The sphere-on-plane configuration

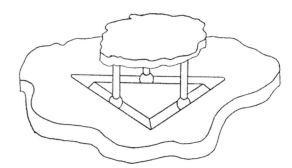

FIGURE 37.12 Kinematic design: three-legged laboratory stands, to illustrate that six contacts fully constrain a body. (Kelvin clamp, as also used for theodolite mounts.)

described is one possible arrangement: crossed cylinders are similar in their behavior and may be easier to construct.

Elastic hinges may be thought of as an extension of kinematic design. A conventional type of door hinge is expensive to produce if friction and play are to be greatly reduced, particularly for small devices. An alternative approach may be adopted when small, repeatable rotations must be catered for. Under this approach, some part is markedly weakened, as in Figure 37.14, so that the bending caused by a turning moment is concentrated in a small region. There is elastic resistance to deformation but very little friction and high repeatability.

The advantages of kinematic design may be listed as:

1. Commonly, only simple machining operations are needed at critical points.
2. Wide tolerances on these operations should not affect repeatability, though they may downgrade absolute performance.
3. Only small forces are needed. Often gravity is sufficient or a light spring if the direction relative to the vertical may change from time to time.
4. Analysis and prediction of behavior is simplified.

The main disadvantage arises if large forces have to be accommodated. Kinematically designed constraints normally work with small forces holding parts together, and if these forces are overcome—even momentarily under the inertia forces of vibration—there can be serious malfunction.

Indeed, the lack of symmetry in behavior under kinematic design can prove a more general disadvantage (for instance, when considering the effects of wear).

Of course, the small additional complexity often means that it is not worth changing to kinematic design. Sometimes a compromise approach is adopted, such as localizing the points of contact between large structures without making them literal spheres on planes. In any case, when considering movements and locations in instruments it is helpful to bear the ideas of kinematic design in mind as a background to analysis.

37.5.2 Proximity Transducer

This is a simple device which is used to detect the presence of an earthed surface which affects the capacitance between the two electrodes E1 and E2 in Figure 37.15. In a special application it is required to operate at a temperature cycling between 200°C and 400°C in a corrosive atmosphere and survive shocks of 1,000 g. Design points to note are:

1. The device is machined out of the solid to avoid a weak weld at position A.
2. The temperature cycling causes thermal stresses which are taken up by the spring washer B (special nimonic spring material for high temperatures).
3. The ceramic insulator blocks are under compression for increased strength.

37.5.3 Load Cell

As discussed in Chapter 7, a load cell converts force into movement against the reaction of a spring. The movement is then measured by a displacement transducer and converted into electrical form.

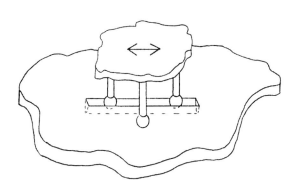

FIGURE 37.13 Kinematic design: five constraints allow linear movement.

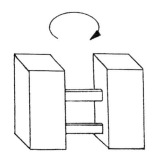

FIGURE 37.14 Principle of elastic hinge.

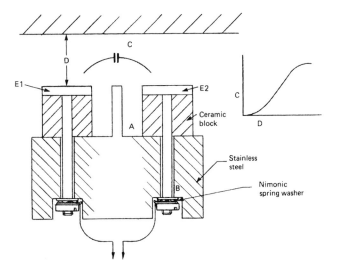

FIGURE 37.15 Rugged proximity transducer.

The load cell in Figure 37.16 consists of four stiff members and four flexures, machined out of a solid block of high-quality spring material in the shape of a parallelogram. The members M1, M2 and M3, M4 remain parallel as the force F bends the flexures at the thin sections (called hinges).

Any torque, caused by the load being offset from the vertical, will result in a small twisting of the load cell, but this is kept within the required limit by arranging the rotational stiffness to be much greater than the vertical stiffness. This is determined by the width.

The trapezoidal construction is far better in this respect than a normal cantilever, which would result in a nonlinear response.

37.5.4 Combined Actuator Transducer

Figure 37.17 illustrates a more complex example, requiring a number of processing techniques to fabricate the complete item. The combined actuator transducer (CAT) is a low-volume product with applications in automatic optical instruments for mirror positioning. The major bought-in components are a torque motor and a miniature pre-amplifier produced by specialist suppliers. The motor incorporates the most advanced rare-earth magnetic materials for compactness and stability, and the pre-amplifier uses small outline electronic components, surface mounted on a copper/fiberglass printed circuit board.

The assembled CAT is shown in Figure 37.17(a). It consists of three sections: the motor in its housing, a capacitive

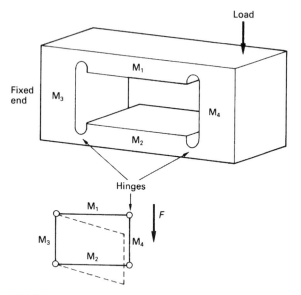

FIGURE 37.16 Load cell spring mechanism.

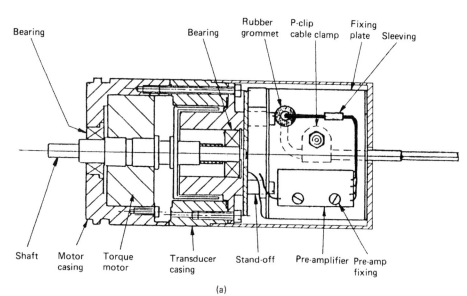

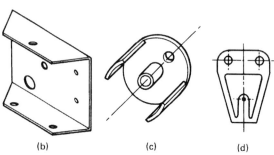

FIGURE 37.17 Combined actuator/transducer. (a) Whole assembly; (b) fixing plate; (c) transducer rotor; (d) spring contact.

angular transducer in its housing, and a rear-mounted plate (Figure 37.17(b)) for pre-amplifier fixing and cable clamping. The motor produces a torque which rotates the shaft in the bearings. Position information is provided by the transducer, which produces an output via the pre-amplifier. The associated electronic servocontrol unit provides the power output stage, position feedback, and loop-stabilization components to control the shaft angle to an arc second. This accuracy is attainable by the use of precise ball bearings which maintain radial and axial movement to within 10 μm.

The shaft, motor casing, and transducer components are manufactured by precise turning of a nonmagnetic stainless steel bar and finished by fine bead blasting. The motor and transducer electrodes (not shown) are glued in place with a thin layer of epoxy resin and in the latter case finished by turning to final size.

The two parts of the transducer stator are jigged concentric and held together by three screws in threaded holes A, leaving a precisely determined gap in which the transducer rotor (Figure 37.17(c)) rotates. The transducer rotor is also turned, but the screens are precision ground to size, as this determines the transducer sensitivity and range.

The screens are earthed, via the shaft and a hardened gold rotating point contact held against vibration by a spring (Figure 37.17(d)). The spring is chemically milled from thin beryllium copper sheet.

The shaft with motor and transducer rotor glued in place, motor stator and casing, transducer stator and bearings are then assembled and held together with three screws as B. The fixing plate, assembled with cable, clamp, and pre-amplifier separately, is added, mounted on standard stand-offs, the wires made off, and then finally the cover put on.

In addition to the processes mentioned, the manufacture of this unit requires drilling, reaming, bending, screen printing, soldering, heat treatment, and anodizing. Materials include copper, PTFE, stainless steel, samarium cobalt, epoxy fiberglass, gold, and aluminum, and machining tolerances are typically 25 μm for turning, 3 μm for grinding and 0.1 mm for bending.

The only external feature is the clamping ring at the shaft end for axial fixing (standard servo type size 20). This is provided because radial force could distort the thin wall section and cause transducer errors.

REFERENCES

Birbeck, G., "Mechanical Design," in *A Guide to Instrument Design*, SIMA and BSIRA, Taylor and Francis, London (1963).

Clayton, G. B., *Operational Amplifiers*, Butterworths, London (1979).

Furse, J. E., "Kinematic design of fine mechanisms in instruments," in *Instrument Science and Technology*, Volume 2, ed. E. B. Jones, Adam Hilger, Bristol, U.K. (1983).

Horowitz, P., and Hill, W., *The Art of Electronics*, Cambridge University Press, Cambridge (1989).

Kibble, B. P., and Rayner, G. H., *Co-Axial AC Bridges*, Adam Hilger, Bristol, U.K. (1984).

Morrell, R., *Handbook of Properties of Technical and Engineering Ceramics* Part 1, An introduction for the engineer and designer, HMSO, London (1985).

Oberg, E., and Jones, F. D., *Machinery Handbook*, The Machinery Publishing Company (1979).

Shields, J., *Adhesives Handbook*, Butterworths, London (revised 3rd ed., 1985).

Smith, S. T., and Chetwynd, J., *Foundations of Ultraprecision Mechanism Design*, Gordon & Breach, London (1992).

The standards referred to in the text are:

BS 5252 (1976) and 4800 (1981): Framework for color coordination for building purposes

BS 5490 (1977 and 1985): Environmental protection provided by enclosures.

DIN 43 700 (1982): Cutout dimensions for panel mounting instruments.

DIN 41612: Standards for Eurocard connectors.

DIN 41914 and IEC 297: Standards for Eurocards.

Chapter 38

Instrument Installation and Commissioning

A. Danielsson

38.1 INTRODUCTION

Plant safety and continuous effective plant operability are totally dependent upon correct installation and commissioning of the instrumentation systems. Process plants are increasingly becoming dependent upon automatic control systems, owing to the advanced control functions and monitoring facilities that can be provided in order to improve plant efficiency, product throughput, and product quality.

The instrumentation on a process plant represents a significant capital investment, and the importance of careful handling on site and the exactitude of the installation cannot be overstressed. Correct installation is also important in order to ensure long-term reliability and to obtain the best results from instruments which are capable of higher-order accuracies due to advances in technology. Quality control of the completed work is also an important function.

Important principles relevant to installing instrumentation are also discussed in Chapter 35.

38.2 GENERAL REQUIREMENTS

Installation should be carried out using the best engineering practices by skilled personnel who are fully acquainted with the safety requirements and regulations governing a plant site. Prior to commencement of the work for a specific project, installation design details should be made available which define the scope of work and the extent of material supply and which give detailed installation information related to location, fixing, piping, and wiring. Such design details should have already taken account of established installation recommendations and measuring technology requirements. The details contained in this chapter are intended to give general installation guidelines.

38.3 STORAGE AND PROTECTION

When instruments are received on a job site it is of the utmost importance that they are unpacked with care, examined for superficial damage, and then placed in a secure store which should be free from dust and suitably heated. In order to minimize handling, large items of equipment, such as control panels, should be programmed to go directly into their intended location, but temporary anti-condensation heaters should be installed if the intended air-conditioning systems have not been commissioned.

Throughout construction, instruments and equipment installed in the field should be fitted with suitable coverings to protect them from mechanical abuse such as paint spraying, etc. Preferably, after an installation has been fabricated, the instrument should be removed from the site and returned to the store for safe keeping until ready for precalibration and final loop checking. Again, when instruments are removed, care should be taken to seal the ends of piping, etc., to prevent ingress of foreign matter.

38.4 MOUNTING AND ACCESSIBILITY

When instruments are mounted in their intended location, either on pipe stands, brackets, or directly connected to vessels, etc., they should be vertically plumbed and firmly secured. Instrument mountings should be vibration free and should be located so that they do not obstruct access ways which may be required for maintenance to other items of equipment. They should also be clear of obvious hazards such as hot surfaces or drainage points from process equipment.

Locations should also be selected to ensure that the instruments are accessible for observation and maintenance.

Where instruments are mounted at higher elevations, it must be ensured that they are accessible either by permanent or temporary means.

Instruments should be located as close as possible to their process tapping points in order to minimize the length of impulse lines, but consideration should be paid to the possibility of expansion of piping or vessels which could take place under operating conditions and which could result in damage if not properly catered for. All brackets and supports should be adequately protected against corrosion by priming and painting.

When installing final control elements such as control valves, again, the requirement for maintenance access must be considered, and clearance should be allowed above and below the valve to facilitate servicing of the valve actuator and the valve internals.

38.5 PIPING SYSTEMS

All instrument piping or tubing runs should be routed to meet the following requirements:

1. They should be kept as short as possible.
2. They should not cause any obstruction that would prohibit personnel or traffic access.
3. They should not interfere with the accessibility for maintenance of other items of equipment.
4. They should avoid hot environments or potential fire-risk areas.
5. They should be located with sufficient clearance to permit lagging which may be required on adjacent pipework.
6. The number of joints should be kept to a minimum consistent with good practice.
7. All piping and tubing should be adequately supported along its entire length from supports attached to firm steelwork or structures (not handrails).

(*Note*: Tubing can be regarded as thin-walled seamless pipe that cannot be threaded and which is joined by compression fittings, as opposed to piping, which can be threaded or welded.)

38.5.1 Air Supplies

Air supplies to instruments should be clean, dry, and oil free. Air is normally distributed around a plant from a high-pressure header (e.g., 6–7 bar g), ideally forming a ring main. This header, usually of galvanized steel, should be sized to cope with the maximum demand of the instrument air users being serviced, and an allowance should be made for possible future expansion or modifications to its duty.

Branch headers should be provided to supply individual instruments or groups of instruments. Again, adequate spare tappings should be allowed to cater for future expansion.

Branch headers should be self draining and have adequate drainage/blow-off facilities. On small headers this may be achieved by the instrument air filter/regulators.

Each instrument air user should have an individual filter regulator. Piping and fittings installed after filter regulators should be non-ferrous.

38.5.2 Pneumatic Signals

Pneumatic transmission signals are normally in the range of 0.2–1.0 bar (3–15psig), and for these signals copper tubing is most commonly used, preferably with a PVC outer sheath. Other materials are sometimes used, depending on environmental considerations (e.g., alloy tubing or stainless steel). Although expensive, stainless steel tubing is the most durable and will withstand the most arduous service conditions.

Plastic tubing should preferably only be used within control panels. There are several problems to be considered when using plastic tubes on a plant site, as they are very vulnerable to damage unless adequately protected, they generally cannot be installed at subzero temperatures, and they can be considerably weakened by exposure to hot surfaces. Also, it should be remembered that they can be totally lost in the event of a fire.

Pneumatic tubing should be run on a cable tray or similar supporting steelwork for its entire length and securely clipped at regular intervals. Where a number of pneumatic signals are to be routed to a remote control room they should be marshaled in a remote junction box and the signals conveyed to the control room via multitube bundles. Such junction boxes should be carefully positioned in the plant in order to minimize the lengths of the individually run tubes. (See Figure 38.1 for typical termination of pneumatic multitubes.)

38.5.3 Impulse Lines

These are the lines containing process fluid which run between the instrument impulse connection and the process tapping point, and are usually made up from piping and pipe fittings or tubing and compression fittings. Piping materials must be compatible with the process fluid.

Generally, tubing is easier to install and is capable of handling most service conditions provided that the correct fittings are used for terminating the tubing. Such fittings must be compatible with the tubing being run (i.e., of the same material).

Impulse lines should be designed to be as short as possible, and should be installed so that they are self-draining for liquids and self-venting for vapors or gases. If necessary, vent plugs or valves should be located at high points in liquid-filled lines and, similarly, drain plugs or valves should be fitted at low points in gas or vapor-filled lines. In

CHAPTER | 38 Instrument Installation and Commissioning

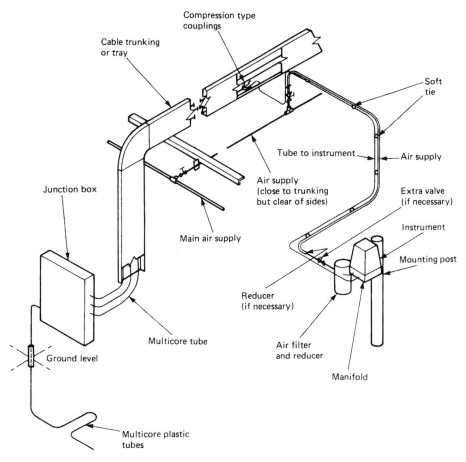

FIGURE 38.1 Typical field termination of pneumatic multitubes.

any case, it should be ensured that there are provisions for isolation and depressurizing of instruments for maintenance purposes. Furthermore, filling plugs should be provided where lines are to be liquid sealed for chemical protection and, on services which are prone to plugging, rodding-out connections should be provided close to the tapping points.

38.6 CABLING

38.6.1 General Requirements

Instrument cabling is generally run in multicore cables from the control room to the plant area (either below or above ground) and then from field junction boxes in single pairs to the field measurement or actuating devices.

For distributed microprocessor systems the inter-connection between the field and the control room is usually via duplicate data highways from remote located multiplexers or process interface units. Such duplicate highways would take totally independent routes from each other for plant security reasons.

Junction boxes must meet the hazardous area requirements applicable to their intended location and should be carefully positioned in order to minimize the lengths of individually run cables, always bearing in mind the potential hazards that could be created by fire.

Cable routes should be selected to meet the following requirements:

1. They should be kept as short as possible.
2. They should not cause any obstruction that would prohibit personnel or traffic access.
3. They should not interfere with the accessibility for maintenance of other items of equipment.
4. They should avoid hot environments or potential fire-risk areas.
5. They should avoid areas where spillage is liable to occur or where escaping vapors or gases could present a hazard.

Cables should be supported for their whole run length by a cable tray or similar supporting steelwork. Cable trays should preferably be installed with their breadth in a vertical plane. The layout of cable trays on a plant should be carefully selected so that the minimum number of instruments in the immediate vicinity would be affected in the case of a local fire. Cable joints should be avoided other than in approved junction boxes or termination points. Cables entering junction boxes from below ground should be specially protected by fire-resistant ducting or something similar.

38.6.2 Cable Types

There are three types of signal cabling generally under consideration, i.e.,

1. Instrument power supplies (above 50 V).
2. High-level signals (between 6 and 50 V). This includes digital signals, alarm signals, and high-level analog signals (e.g., 4–20 mAdc).
3. Low-level signals (below 5V). This generally covers thermocouple compensating leads and resistance element leads.

Signal wiring should be made up in twisted pairs. Solid conductors are preferable so that there is no degradation of signal due to broken strands that may occur in stranded conductors. Where stranded conductors are used, crimped connectors should be fitted. Cable screens should be provided for instrument signals, particularly low-level analog signals, unless the electronic system being used is deemed to have sufficient built-in "noise" rejection. Further mechanical protection should be provided in the form of singlewire armor and PVC outer sheath, especially if the cables are installed in exposed areas, e.g., on open cable trays. Cables routed below ground in sand-filled trenches should also have an overall lead sheath if the area is prone to hydrocarbon or chemical spillage.

38.6.3 Cable Segregation

Only signals of the same type should be contained within any one multicore cable. In addition, conductors forming part of intrinsically safe circuits should be contained in a multicore reserved solely for such circuits.

When installing cables above or below ground they should be separated into groups according to the signal level and segregated with positive spacing between the cables. As a general rule, low-level signals should be installed furthest apart from instrument power supply cables with the high-level signal cables in between. Long parallel runs of dissimilar signals should be avoided as far as possible, as this is the situation where interference is most likely to occur.

Cables used for high-integrity systems such as emergency shutdown systems or data highways should take totally independent routes or should be positively segregated from other cables. Instrument cables should be run well clear of electrical power cables and should also, as far as possible, avoid noise-generating equipment such as motors. Cable crossings should always be made at right angles.

When cables are run in trenches, the routing of such trenches should be clearly marked with concrete cable markers on both sides of the trench, and the cables should be protected by earthenware or concrete covers.

38.7 GROUNDING

38.7.1 General Requirements

Special attention must be paid to instrument grounding, particularly where field instruments are connected to a computer or microprocessor type control system. Where cable screens are used, ground continuity of screens must be maintained throughout the installation with the grounding at one point only, i.e., in the control room. At the field end the cable screen should be cut back and taped so that it is independent from the ground. Intrinsically safe systems should be grounded through their own ground bar in the control room. Static grounding of instrument cases, panel frames, etc., should be connected to the electrical common plant ground. (See Figure 38.2 for a typical grounding system.)

Instrument grounds should be wired to a common bus bar within the control center, and this should be connected to a remote ground electrode via an independent cable (preferably duplicated for security and test purposes). The resistance to ground, measured in the control room, should usually not exceed 1 Ω unless otherwise specified by a system manufacturer or by a certifying authority.

38.8 TESTING AND PRE-COMMISSIONING

38.8.1 General

Before starting up a new installation the completed instrument installation must be fully tested to ensure that the equipment is in full working order. This testing normally falls into three phases, i.e., pre-installation testing; piping and cable testing; loop testing or pre-commissioning.

38.8.2 Pre-installation Testing

This is the testing of each instrument for correct calibration and operation prior to its being installed in the field. Such testing is normally carried out in a workshop which is fully equipped for the purpose and should contain a means of generating the measured variable signals and also a method of accurately measuring the instrument input and output (where applicable). Test instruments should have a standard of accuracy better than the manufacturer's stated accuracy for the instruments being tested and should be regularly certified.

Instruments are normally calibration checked at five points (i.e., 0, 25, 50, 75, and 100 percent) for both rising and falling signals, ensuring that the readings are within the manufacturer's stated tolerance.

After testing, instruments should be drained of any testing fluids that may have been used and, if necessary, blown through with dry air. Electronic instruments should be energized for a 24-hour warm-up period prior to the calibration

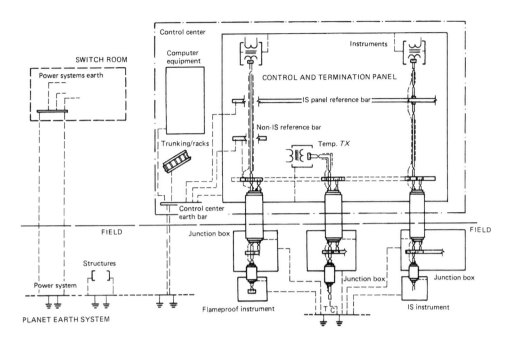

FIGURE 38.2 A typical control center gounding system.

test being made. Control valves should be tested *in situ* after the pipework fabrication has been finished and flushing operations completed. Control valves should be checked for correct stroking at 0, 50, and 100 percent open, and at the same time the valves should be checked for correct closure action.

38.8.3 Piping and Cable Testing

This is an essential operation prior to loop testing.

38.8.3.1 Pneumatic Lines

All air lines should be blown through with clean, dry air prior to final connection to instruments, and they should also be pressure tested for a timed interval to ensure that they are leak free. This should be in the form of a continuity test from the field end to its destination (e.g., the control room).

38.8.3.2 Process Piping

Impulse lines should also be flushed through and hydrostatically tested prior to connection of the instruments. All isolation valves or manifold valves should be checked for tight shutoff. On completion of hydrostatic tests, all piping should be drained and thoroughly dried out prior to reconnecting to any instruments.

38.8.3.3 Instrument Cables

All instrument cables should be checked for continuity and insulation resistance before connection to any instrument or apparatus. The resistance should be checked core to core and core to ground.

Cable screens must also be checked for continuity and insulation. Cable tests should comply with the requirements of Part 6 of the IEE Regulation for Electrical Installations (latest edition), or the rules and regulations with which the installation has to comply. Where cables are installed below ground, testing should be carried out before the trenches are back filled. Coaxial cables should be tested using sine-wave reflective testing techniques. As a prerequisite to cable testing it should be ensured that all cables and cable ends are properly identified.

38.8.4 Loop Testing

The purpose of loop testing is to ensure that all instrumentation components in a loop are in full operational order when interconnected and are in a state ready for plant commissioning.

Prior to loop testing, inspection of the whole installation, including piping, wiring, mounting, etc., should be carried out to ensure that the installation is complete and that the work has been carried out in a professional manner. The control room panels or display stations must also be in a fully functional state.

Loop testing is generally a two-person operation, one in the field and one in the control room who should be equipped with some form of communication, e.g., field telephones or radio transceivers. Simulation signals should be injected at the field end equivalent to 0, 50, and 100 percent of the instrument range, and the loop function should be checked for correct operation in both rising and falling modes. All results should be properly documented on calibration or loop check sheets. All ancillary components in the loop should be checked at the same time.

Alarm and shutdown systems must also be systematically tested, and all systems should be checked for "fail-safe" operation, including the checking of "burn-out" features on thermocouple installations. At the loop-checking stage all ancillary work should be completed, such as setting zeros, filling liquid seals, and fitting of accessories such as charts, ink, fuses, etc.

38.9 PLANT COMMISSIONING

Commissioning is the bringing "on-stream" of a process plant and the tuning of all instruments and controls to suit the process operational requirements. A plant or section thereof is considered to be ready for commissioning when all instrument installations are mechanically complete and all testing, including loop testing, has been effected.

Before commissioning can be attempted it should be ensured that all air supplies are available and that all power supplies are fully functional, including any emergency standby supplies. It should also be ensured that all ancillary devices are operational, such as protective heating systems, air conditioning, etc. All control valve lubricators (when fitted) should be charged with the correct lubricant.

Commissioning is usually achieved by first commissioning the measuring system with any controller mode overridden. When a satisfactory measured variable is obtained, the responsiveness of a control system can be checked by varying the control valve position using the "manual" control function. Once the system is seen to respond correctly and the required process variable reading is obtained, it is then possible to switch to "auto" in order to bring the controller function into action. The controller responses should then be adjusted to obtain optimum settings to suit the automatic operation of plant.

Alarm and shutdown systems should also be systematically brought into operation, but it is necessary to obtain the strict agreement of the plant operation supervisor before any overriding of trip systems is attempted or shutdown features are operated.

Finally, all instrumentation and control systems would need to be demonstrated to work satisfactorily before formal acceptance by the plant owner.

REFERENCES

BS 6739, British Standard Code of Practice for Instrumentation in Process Control Systems: Installation Design and Practice (1986).

Regulations for electrical installations 15th ed. (1981) as issued by the Institution of Electrical Engineers.

The reader is also referred to the National Electrical Code of the United States (current edition) and relevant ANSI, IEC, and ISA standards.

Chapter 39

Sampling

J. G. Giles

39.1 INTRODUCTION

39.1.1 Importance of Sampling

Any form of analysis instrument can only be as effective as its sampling system. Analysis instruments are out of commission more frequently due to trouble in the sampling system than to any other cause. Therefore time and care expended in designing and installing an efficient sampling system is well repaid in the saving of servicing time and dependability of instrument readings. The object of a sampling system is to obtain a truly representative sample of the solid, liquid, or gas which is to be analyzed, at an adequate and steady rate, and transport it without change to the analysis instrument, and all precautions necessary should be taken to ensure that this happens. Before the sample enters the instrument it may be necessary to process it to the required physical and chemical state, i.e., correct temperature, pressure, flow, purity, etc., without removing essential components. It is also essential to dispose of the sample and any reagent after analysis without introducing a toxic or explosive hazard. For this reason, the sample, after analysis, is continuously returned to the process at a suitable point, or a sample-recovery and disposal system is provided.

39.1.2 Representative Sample

It is essential that the sample taken should represent the mean composition of the process material. The methods used to overcome the problem of uneven sampling depend on the phase of the process sample, which may be in solid, liquid, gas, or mixed-phase form.

39.1.2.1 Solids

When the process sample is solid in sheet form it is necessary to scan the whole sheet for a reliable measurement of the state of the sheet (e.g., thickness, density, or moisture content). A measurement at one point is insufficient to give a representative value of the parameter being measured.

If the solid is in the form of granules or powder of uniform size, a sample collected across a belt or chute and thoroughly mixed will give a reasonably representative sample. If measurement of density or moisture content of the solid can be made while it is in a vertical chute under a constant head, packing density problems may be avoided.

In some industries where the solids are transported as slurries, it is possible to carry out the analysis directly on the slurry if a method is available to compensate for the carrier fluid and the velocities are high enough to ensure turbulent flow at the measurement point.

Variable-size solids are much more difficult to sample, and specialist work on the subject should be consulted.

39.1.2.2 Liquids

When sampling liquid it is essential to ensure that either the liquid is turbulent in the process line or that there are at least 200 pipe diameters between the point of adding constituents and the sampling point. If neither is possible, a motorized or static mixer should be inserted into the process upstream of the sample point.

39.1.2.3 Gases

Gas samples must be thoroughly mixed and, as gas process lines are usually turbulent, the problem of finding a satisfactory sample point is reduced. The main exception is in large ducts such as furnace or boiler flues, where stratification can occur and the composition of the gas may vary from one point to another. In these cases special methods of sampling may be necessary, such as multiple probes or long probes with multiple inlets in order to obtain a representative sample.

39.1.2.4 Mixed-Phase Sampling

Mixed phases such as liquid/gas mixtures or liquid/solids (i.e., slurries) are best avoided for any analytical method that involves taking a sample from the process. It is always preferable to use an in-line analysis method where this is possible.

39.1.3 Parts of Analysis Equipment

The analysis equipment consists of five main parts:

1. Sample probe.
2. Sample-transport system.
3. Sample-conditioning equipment.
4. The analysis instrument.
5. Sample disposal.

39.1.3.1 Sample Probe

This is the sampling tube that is used to withdraw the sample from the process.

39.1.3.2 Sample-Transport System

This is the tube or pipe that transports the sample from the sample point to the sample-conditioning system.

39.1.3.3 Sample-Conditioning System

This system ensures that the analyzer receives the sample at the correct pressure and in the correct state to suit the analyzer. This may require pressure increase (i.e., pumps) or reduction, filtration, cooling, drying, and other equipment to protect the analyzer from process upsets. Additionally, safety equipment and facilities for the introduction of calibration samples into the analyzer may also be necessary.

39.1.3.4 The Analysis Instrument

This is the process analyzer complete with the services such as power, air, steam, drain vents, carrier gases, and signal conditioning that are required to make the instrument operational. (Analysis techniques are described in Part 2 of this book.)

39.1.3.5 Sample Disposal

The sample flowing from the analyzer and sample conditioning system must be disposed of safely. In many cases it is possible to vent gases to atmosphere or allow liquids to drain, but there are times when this is not satisfactory. Flammable or toxic gases must be vented in such a way that a hazard is not created. Liquids such as hydrocarbons can be collected in a suitable tank and pumped back into the process, whereas hazardous aqueous liquids may have to be treated before being allowed to flow into the drainage system.

39.1.4 Time Lags

In any measuring instrument, particularly one which may be used with a controller, it is desirable that the time interval between occurrence of a change in the process fluid and its detection at the instrument should be as short as possible consistent with reliable measurement. In order to keep this time interval to a minimum, the following points should be kept in mind.

39.1.4.1 Sample-Transport Line Length

The distance between the sampling point and the analyzer should be kept to the minimum. Where long sample transport lines are unavoidable a "fast loop" may be used. The fast loop transports the sample at a flow rate higher than that required by the analyzer, and the excess sample is either returned to the process, vented to atmosphere, or allowed to flow to drain. The analyzer is supplied with the required amount of sample from the fast loop through a short length of tubing.

39.1.4.2 Sampling Components

Pipe, valves, filter, and all sample-conditioning components should have the smallest volume consistent with a permissible pressure drop.

39.1.4.3 Pressure Reduction

Gaseous samples should be filtered, and flow in the sample line kept at the lowest possible pressure, as the mass of gas in the system depends on the pressure of the gas as well as the volume in the system.

When sampling high-pressure gases the pressure reducing valve must be situated at the sample point. This is necessary, because for a fixed mass flow rate of gas the response time will increase in proportion to the absolute pressure of the gas in the sample line (i.e., gas at 10 bar A will have a time lag five times that of gas at 2 bar A). This problem becomes more acute when the sample is a liquid that has to be vaporized for analysis (e.g., liquid butane or propane).

The ratio of volume of gas to volume of liquid can be in the region of 250:1, as is the case for propane. It is therefore essential to vaporize the liquid at the sample point and then treat it as a gas sample from then on.

39.1.4.4 Typical Equations

1. $t = \dfrac{L}{S}$

    t = time lag

    S = velocity (m/s)

    L = line length (m)

2. General gas law for ideal gases:

$$\frac{pv}{T} = \frac{8314 \times W}{10^5 \times M}$$

p = pressure
T = abs. temperature (K)
v = volume (1)
W = mass (g)
M = molecular weight

3. Line volume:

$$\frac{\pi d^2}{4} = V_I$$

d = internal diameter of tube (mm)
V_I = volume (ml/m)

4. $t = \dfrac{6L \times V_I}{100F}$

L = line length (m)
V_I = internal volume of line (ml/m)
F = sample flow rate (1/min)
t = time lag (s)

(For an example of a fast loop calculation see Section 39.3.2.2, Table 39.1.)

39.1.4.5 Useful Data

Internal volume per meter (V_I) of typical sample lines:

⅛ in OD × 0.035 wall = 1.5 ml/m
¼ in OD × 0.035 wall = 16.4 ml/m
⅜ in OD × 0.035 wall = 47.2 ml/m
½ in OD × 0.065 wall = 69.4 ml/m
½ in nominal bore steel pipe (extra strong)(13.88 mm ID)
   = 149.6 ml/m
3 mm OD × 1 mm wall = 0.8 ml/m
6 mm OD × 1 mm wall = 12.6 ml/m
8 mm OD × 1 mm wall = 28.3 ml/m
10 mm OD × 1 m wall = 50.3 ml/m
12 mm OD × 1.5 mm wall = 63.6 ml/m

39.1.5 Construction Materials

Stainless steel (Type 316 or 304) has become one of the most popular materials for the construction of sample systems due to its high resistance to corrosion, low surface adsorption (especially moisture), wide operating temperature range, high-pressure capability, and the fact that it is easily obtainable. Care must be taken when there are materials in the sample which cause corrosion, such as chlorides and sulfides, in which case it is necessary to change to more expensive materials such as Monel.

When atmospheric sampling is carried out for trace constituents, Teflon tubing is frequently used, as the surface adsorption of the compounds is less than stainless steel, but it is necessary to check that the compound to be measured does not diffuse through the wall of the tubing.

For water analysis (e.g., pH and conductivity) it is possible to use plastic (such as PVC or ABS) components, although materials such as Kunifer 10 (copper 90 percent, nickel 10 percent) are increasing in popularity when chlorides (e.g., salt water) are present, as they are totally immune to chloride corrosion.

TABLE 39.1 Fast-loop calculation for gas oil sample

Customer	Demo.
Date	Jan. 1986
Order No.	ABC 123
Tag number	Gas oil (sample line)
Density	825.00 kg/m³
Viscosity	3.00 centipoise
Response time	39.92 s
Flow rate	16.60 l/min (1 m³/h)
Length	73.00 m
Diameter (ID)	13.88 mm (1/2 in nominal bore (extra strong))
Velocity	1.83 m/s
RE =	6979
Flow	TURB
Friction factor	0.038
Delta P	278.54 kPa (2.7854 bar)
Customer	Demo.
Date	Jan. 1986
Order No.	ABC 123
Tag number	Gas oil (return line)
Density	825.00 kg/m³
Viscosity	3.00 centipoise
Response time	71.31 s
Flow rate	16.60 l/min (1 m³/h)
Length	73.00 m
Diameter (ID)	18.55 mm (1/4 in nominal bore (extra strong))
Velocity	1.02 m/s
RE =	5222
Flow	TURB
Friction factor	0.040
Delta P.	67.85 kPa (0.6785 bar)

39.2 SAMPLE SYSTEM COMPONENTS

39.2.1 Probes

The most important function of a probe is to obtain the sample from the most representative point (or points) in the process line.

39.2.1.1 Sample Probe

A typical probe of this type for sampling all types of liquid and gases at low pressure is shown in Figure 39.1. It can be seen that the probe which is made of 21 mm OD (11.7 mm ID) stainless steel pipe extends to the center of the line being sampled. However, if the latter is more than 500mm OD the probe intrusion is kept at 250 mm to avoid vibration in use.

39.2.1.2 Small-Volume Sample Probe

This probe is used for sampling liquids that must be vaporized or for high-pressure gases (Figure 39.2). Typically, a 6mm OD × 2 mm ID tube is inserted through the center of a probe of the type described in Section 39.2.1.1. The probe may be withdrawn through the valve for cleaning.

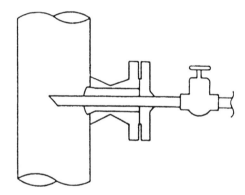

FIGURE 39.1 Sample probe. Courtesy of Ludlam Sysco.

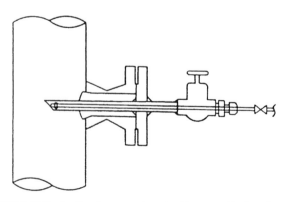

FIGURE 39.2 Small-volume sample probe. Courtesy of Ludlam Sysco.

39.2.1.3 Furnace Gas Probes

Low-Temperature Probe Figure 39.3 shows a gas-sampling probe with a ceramic outside filter for use in temperatures up to 400°C.

Water-Wash Probe This probe system is used for sampling furnace gases with high dust content at high temperatures (up to 1600°C) (Figure 39.4). The wet gas-sampling probe is water cooled and self-priming. The water/gas mixture passes from the probe down to a water trap, where the gas and water are separated. The gas leaves the trap at a pressure of approximately 40 mbar, and precautions should be taken

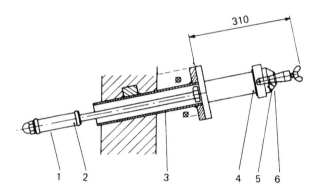

FIGURE 39.3 Gas-sampling probe. Courtesy of ABB Hartmann and Braun. 1. Gas intake; 2. ceramic intake filter; 3. bushing tube with flange; 4. case with outlet filter; 5. internal screwthread; 6. gas outlet.

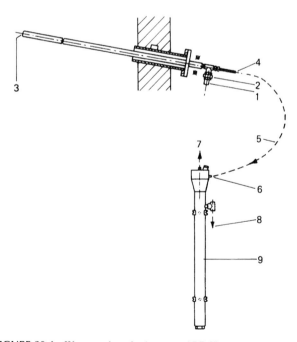

FIGURE 39.4 Water-wash probe (courtesy ABB Hartmann and Braun). 1. Water intake; 2. water filter; 3. gas intake; 4. gas-water outlet; 5. connecting hose; 6. gas-water intake; 7. gas outlet; 8. water outlet; 9. water separator.

to avoid condensation in the analyzer, either by ensuring that the analyzer is at a higher temperature than the water trap or by passing the sample gas through a sample cooler at 5°C to reduce the humidity.

Note that this probe is not suitable for the measurement of water-soluble gases such as CO_2, SO_2, or H_2S.

Steam Ejector The steam ejector illustrated in Figure 39.5 can be used for sample temperatures up to 180°C and, because the condensed steam dilutes the condensate present in the flue gas, the risk of corrosion of the sample lines when the steam/gas sample cools to the dew point is greatly reduced.

Dry steam is supplied to the probe and then ejected through a jet situated in the mouth of a Venturi. The flow of steam causes sample gas to be drawn into the probe. The steam and gas pass out of the probe and down the sample line to the analyzer system at a positive pressure. The flow of steam through the sample line prevents the build-up of any corrosive condensate.

39.2.2 Filters

39.2.2.1 "Y" Strainers

"Y" strainers are available in stainless steel, carbon steel, and bronze. They are ideal for preliminary filtering of samples before pumps or at sample points to prevent line scale from entering sample lines. Filtration sizes are available from 75 to 400 μm (200 to 40 mesh). The main application for this type of filter is for liquids and steam.

39.2.2.2 In-Line Filters

This design of filter is normally used in a fast loop configuration and is self-cleaning (Figure 39.6). Filtration is through a stainless steel or a ceramic element. Solid particles tend to be carried straight on in the sample stream so that maintenance time is very low. Filtration sizes are available from 150 μm (100 mesh) down to 5 μm. It is suitable for use with liquids or gases.

39.2.2.3 Filters with Disposable Glass Microfiber Element

These filters are available in a wide variety of sizes and porosities (Figure 39.7). Bodies are available in stainless

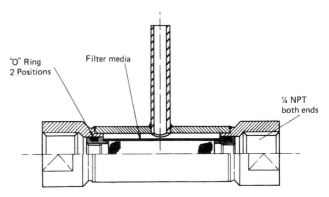

FIGURE 39.6 In-line filter. Courtesy of Microfiltrex.

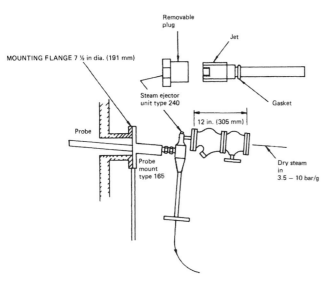

FIGURE 39.5 Steam ejection probe. Courtesy of Servomex.

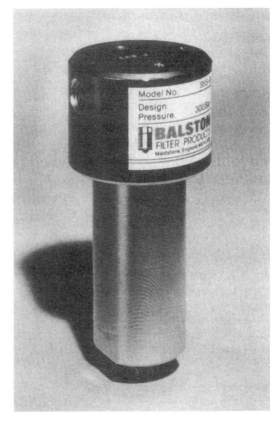

FIGURE 39.7 Filter with disposable element. Courtesy of Balston.

steel, aluminum, or plastic. Elements are made of glass microfiber and are bonded with either resin or fluorocarbon. The fluorocarbon-bonded filter is particularly useful for low-level moisture applications because of the low adsorption/desorption characteristic.

The smallest filter in this range has an internal volume of only 19 ml and is therefore suitable when a fast response time is required.

39.2.2.4 Miniature in-Line Filter

These are used for filtration of gases prior to pressure reduction and are frequently fitted as the last component in the sample system to protect the analyzer (Figure 39.8).

39.2.2.5 Manual Self-Cleaning Filter

This type of filter works on the principle of edge filtration using discs usually made of stainless steel and fitted with a manual cleaning system (Figure 39.9). The filter is cleaned by rotating a handle which removes any deposits from the filter element while the sample is flowing. The main uses are for filtration of liquids where filter cleaning must be carried out regularly without the system being shut down; it is especially suitable for waxy material which can rapidly clog up normal filter media.

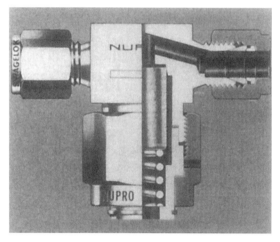

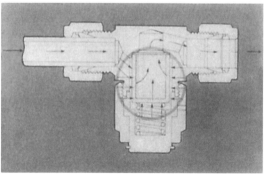

FIGURE 39.8 Miniature in-line filter. Courtesy of Nupro.

39.2.3 Coalescers

Coalescers are a special type of filter for separating water from oil or oil from water (Figure 39.10). The incoming sample flows from the center of a specially treated filter element through to the outside. In so doing, the diffused water is slowed down and coalesced, thus forming droplets which, when they reach the outer surface, drain downwards as the water is denser than the hydrocarbon. A bypass stream is taken from the bottom of the coalescer to remove the water. The dry hydrocarbon stream is taken from the top of the coalescer.

39.2.4 Coolers

39.2.4.1 Air Coolers

These are usually used to bring the sample gas temperature close to ambient before feeding into the analyzer.

39.2.4.2 Water-Jacketed Coolers

These are used to cool liquid and gas samples and are available in a wide range of sizes (Figure 39.11).

39.2.4.3 Refrigerated Coolers

These are used to reduce the temperature of a gas to a fixed temperature (e.g., +5°C) in order to condense the water out of a sample prior to passing the gas into the analyzer. Two types are available: one with an electrically driven compressor type refrigerator and another using a Peltier cooling element. The compressor type has a large cooling capacity whereas the Peltier type, being solid state, needs less maintenance.

39.2.5 Pumps, Gas

Whenever gaseous samples have to be taken from sample points which are below the pressure required by the analyzer, a sample pump of some type is required. The pumps that are available can broadly be divided into two groups:

1. The eductor or aspirator type
2. The mechanical type.

39.2.5.1 Eductor or Aspirator Type

All these types of pump operate on the principle of using the velocity of one fluid which may be liquid or gas to induce the flow in the sample gas. The pump may be fitted before or after the analyzer, depending on the application. A typical application for a water-operated aspirator (similar to a laboratory vacuum pump) is for taking a sample of flue gas for oxygen measurement. In this case the suction port of the aspirator is connected directly to the probe via

CHAPTER | 39 Sampling

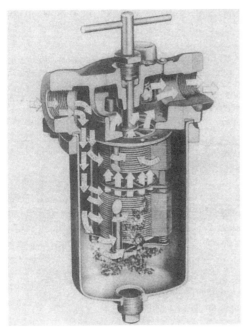

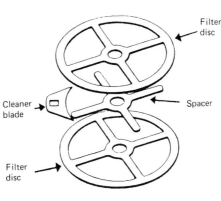

FIGURE 39.9 Manual self-cleaning filter. Courtesy of AMF CUNO.

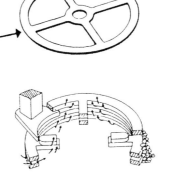

FIGURE 39.10 Coalescer. Courtesy of Fluid Data.

FIGURE 39.11 Water-jacketed cooler. Courtesy of George E. Lowe. Dimensions are shown in mm.

a sample line and the water/gas mixture from the outlet feeds into a separator arranged to supply the sample gas to the analyzer at a positive pressure of about 300 mm water gauge.

In cases where water will affect the analysis it is sometimes possible to place the eductor or aspirator after the analyzer and draw the sample through the system. In these cases the eductor may be supplied with steam, air, or water to provide the propulsive power.

39.2.5.2 Mechanical Gas Pumps

There are two main types of mechanical gas pump available:

1. Rotary pump
2. Reciprocating piston or diaphragm pump.

Rotary Pumps Rotary pumps can be divided into two categories, the rotary piston and the rotating fan types, but the latter is very rarely used as a sampling pump.

The rotary piston pump is manufactured in two configurations. The Rootes type has two pistons of equal size which rotate inside a housing with the synchronizing carried out by external gears. The rotary vane type is similar to those used extensively as vacuum pumps. The Rootes type is ideal where very large flow rates are required and, because there is a clearance between the pistons and the housing, it is possible to operate them on very dirty gases.

The main disadvantage of the rotary vane type is that, because there is contact between the vanes and the housing, lubrication is usually required, and this may interfere with the analysis.

Reciprocating Piston and Diaphragm Pump Of these two types the diaphragm pump has become the most popular. The main reason for this is the improvement in the types of material available for the diaphragms and the fact that there are no piston seals to leak. The pumps are available in a wide variety of sizes, from the miniature units for portable personnel protection analyzers to large heavy-duty industrial types.

A typical diaphragm pump (Figure 39.12) for boosting the pressure of the gas into the analyzer could have an all-stainless-steel head with a Terylene reinforced Viton diaphragm and Viton valves. This gives the pump a very long service life on critical hydrocarbon applications.

Many variations are possible; for example, a Teflon-coated diaphragm can be fitted where Viton may not be compatible with the sample, and heaters may be fitted to the head to keep the sample above the dew point.

The piston pump is still used in certain cases where high accuracy is required in the flow rate (for example, gas blending) to produce specific gas mixtures. In these cases the pumps are usually operated immersed in oil so that the piston is well lubricated, and there is no chance of gas leaks to and from the atmosphere.

39.2.6 Pumps, Liquid

There are two situations where pumps are required in liquid sample systems:

1. Where the pressure at the analyzer is too low because either the process line pressure is too low, or the pressure drop in the sample line is too high, or a combination of both.
2. When the process sample has to be returned to the same process line after analysis.

The two most common types of pumps used for sample transfer are:

1. Centrifugal (including turbine pump)
2. Positive displacement (e.g., gear, peristaltic, etc.).

25.2.6.1 Centrifugal

The centrifugal and turbine pumps are mainly used when high flow rates of low-viscosity liquids are required. The turbine pumps are similar to centrifugal pumps but have a special impeller device which produces a considerably higher pressure than the same size centrifugal. In order to produce high pressures using a centrifugal pump there is a type available which has a gearbox to increase the rotor speed to above 20,000 rev/min.

39.2.6.2 Positive-Displacement Pumps

Positive-displacement pumps have the main characteristic of being constant flow devices. Some of these are specifically designed for metering purposes where an accurate flow rate must be maintained (e.g., process viscometers). They can take various forms:

1. Gear pump
2. Rotary vane pump
3. Peristaltic pump.

Gear Pumps Gear pumps are used mainly on high-viscosity products where the sample has some lubricating properties.

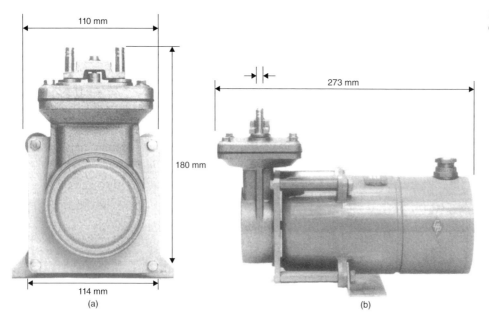

FIGURE 39.12 Diaphragm pump. Courtesy of Charles Austen Pumps.

They can generate high pressures for low flow rates and are used extensively for hydrocarbon samples ranging from diesel oil to the heaviest fuel oils.

Rotary Vane Pumps These pumps are of two types, one having rigid vanes (usually metal) and the other fitted with a rotor made of an elastomer such as nitrile (Buna-*n*) rubber or Viton. The metal vane pumps have characteristics similar to the gear pumps described above, but can be supplied with a method of varying the flow rate externally while the pump is operating.

Pumps manufactured with the flexible vanes (Figure 39.13) are particularly suitable for pumping aqueous solutions and are available in a wide range of sizes but are only capable of producing differential pressures of up to 2 bar.

Peristaltic Pumps Peristaltic pumps are used when either accurate metering is required or it is important that no contamination should enter the sample. As can be seen from Figure 39.14, the only material in contact with the sample is the special plastic tubing, which may be replaced very easily during routine servicing.

39.2.7 Flow Measurement and Indication

Flow measurement on analyzer systems falls into three main categories:

1. Measuring the flow precisely where the accuracy of the analyzer depends on it
2. Measuring the flow where it is necessary to know the flow rate but it is not critical (e.g., fast loop flow)
3. Checking that there is flow present but measurement is not required (e.g., cooling water for heat exchangers).

It is important to decide which category the flowmeter falls into when writing the specification, as the prices vary over a wide range, depending on the precision required.

The types of flowmeter available will be mentioned but not the construction or method of operation, as this is covered in Chapter 1.

39.2.7.1 Variable-Orifice Meters

The variable-orifice meter is extensively used in analyzer systems because of its simplicity, and there are two main types.

Glass Tube This type is the most common, as the position of the float is read directly on the scale attached to the tube, and it is available calibrated for liquids or gases. The high-precision versions are available with an accuracy of ±1 percent full-scale deflection (FSD), whereas the low-priced units have a typical accuracy of ±5 percent FSD.

Metal Tube The metal tube type is used mainly on liquids for high-pressure duty or where the liquid is flammable or hazardous. A good example is the fast loop of a hydrocarbon analyzer. The float has a magnet embedded in it, and the position is detected by an external follower system. The accuracy of metal tube flowmeters varies from ±10 percent FSD to ±2 percent FSD, depending on the type and whether individual calibration is required.

39.2.7.2 Differential-Pressure Devices

On sample systems these normally consist of an orifice plate or preset needle valve to produce the differential pressure, and are used to operate a gauge or liquid-filled manometer when indication is required or a differential pressure switch when used as a flow alarm.

39.2.7.3 Spinner or Vane-Type Indicators

In this type the flow is indicated either by the rotation of a spinner or by the deflection of a vane by the fluid. It is ideal

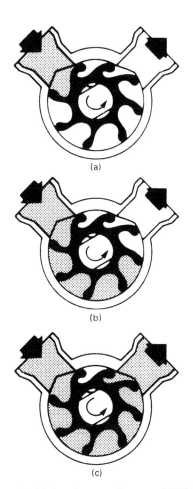

FIGURE 39.13 Flexible impeller pump. Courtesy of ITT Jabsco. (a) Upon leaving the offset plate the impeller blade straightens and creates a vacuum, drawing in liquid–instantly priming the pump, (b) As the impeller rotates it carries the liquid through the pump from the intake to outlet port, each successive blade drawing in liquid, (c) When the flexible blades again contact the offset plate they bend with a squeezing action which provides a continuous, uniform discharge of liquid.

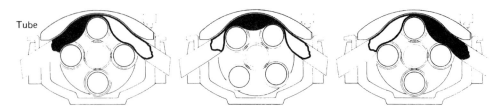

FIGURE 39.14 Peristaltic pump. Courtesy of Watson-Marlow. The advancing roller occludes the tube which, as it recovers to its normal size, draws in fluid which is trapped by the next roller (in the second part of the cycle) and expelled from the pump (in the third part of the cycle). This is the peristaltic flow-inducing action.

for duties such as cooling water flow, where it is essential to know that a flow is present but the actual flow rate is of secondary importance.

39.2.8 Pressure Reduction and Vaporization

The pressure-reduction stage in a sample system is often the most critical, because not only must the reduced pressure be kept constant, but also provision must be made to ensure that under faulty conditions dangerously high pressures cannot be produced. Pressure reduction can be carried out in a variety of ways.

39.2.8.1 Simple Needle Valve

This is capable of giving good flow control if upstream and downstream pressures are constant.

Advantage: Simplicity and low cost.
Disadvantage: Any downstream blockage will allow pressure to rise.

They are only practical if downstream equipment can withstand upstream pressure safely.

39.2.8.2 Needle Valve with Liquid-Filled Lute

This combination is used to act as a pressure stabilizer and safety system combined. The maintained pressure will be equal to the liquid head when the needle value flow is adjusted until bubbles are produced (Figure 39.15).

It is essential to choose a liquid that is not affected by sample gas and also does not evaporate in use and cause a drop in the controlled pressure.

39.2.8.3 Diaphragm-Operated Pressure Controller

These regulators are used when there is either a very large reduction in pressure required or the downstream pressure must be accurately controlled (Figure 39.16). They are frequently used on gas cylinders to provide a controllable low-pressure gas supply.

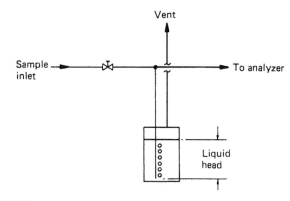

FIGURE 39.15 Lute-type pressure stabilizer. Courtesy of Ludlam Sysco.

39.2.8.4 Vaporization

There are cases when a sample in the liquid phase at high pressure has to be analyzed in the gaseous phase. The pressure reduction and vaporization can be carried out in a specially adapted diaphragm-operated pressure controller as detailed above, where provision is made to heat the complete unit to replace the heat lost by the vaporization.

39.2.9 Sample Lines, Tube and Pipe Fitting

39.2.9.1 Sample Lines

Sample lines can be looked at from two aspects: first, the materials of construction, which are covered in Section 39.1.5, and, second, the effect of the sample line on the process sample, which is detailed below.

The most important consideration is that the material chosen must not change the characteristics of the sample during its transportation to the analyzer. There are two main ways in which the sample line material can affect the sample.

Adsorption and Desorption Adsorption and desorption occur when molecules of gas or liquid are retained and discharged from the internal surface of the sample line material at varying rates. This has the effect of delaying the transport of the adsorbed material to the analyzer and causing erroneous results.

CHAPTER | 39 Sampling

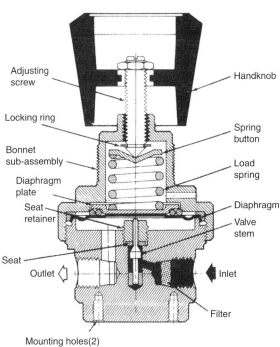

FIGURE 39.16 Diaphragm-operated pressure controller. Courtesy of Tescom.

Water and hydrogen sulfide at low levels are two common measurements where this problem is experienced. An example is when measuring water at a level of 10 ppm in a sample stream, where copper tubing has an adsorption/desorption which is twenty times greater than stainless steel tubing, and hence copper tubing would give a very sluggish response at the analyzer.

Where this problem occurs it is possible to reduce the effects in the following ways:

1. Careful choice of sample tube material
2. Raising the temperature of the sample line
3. Cleaning the sample line to ensure that it is absolutely free of impurities such as traces of oil
4. Increasing the sample flow rate to reduce the time the sample is in contact with the sample line material.

Permeability Permeability is the ability of gases to pass through the wall of the sample tubing. Two examples are:

1. Polytetrafluoroethylene (PTFE) tubing is permeable to water and oxygen.
2. Plasticized polyvinyl chloride (PVC) tubing is permeable to the smaller hydrocarbon molecules such as methane.

Permeability can have two effects on the analysis:

1. External gases getting into the sample such as when measuring low-level oxygen using PTFE tubing. The results would always be high due to the ingress of oxygen from the air.
2. Sample gases passing outwards through the tubing, such as when measuring a mixed hydrocarbon stream using plasticized PVC. The methane concentration would always be too low.

39.2.9.2 Tube, Pipe, and Method of Connection Definition

1. Pipe is normally rigid, and the sizes are based on the nominal bore.

   Typical materials:
   Metallic: carbon steel, brass, etc.
   Plastic: UPVC, ABS, etc.

2. Tubing is normally bendable or flexible, and the sizes are based on the outside diameter and wall thickness.

   Typical materials:
   Metallic: carbon steel, brass, etc.
   Plastic: UPVC, ABS, etc.

Methods of joining
Pipe (metallic):
   1. Screwed
   2. Flanged
   3. Welded
   4. Brazed or soldered

Pipe (plastic):
   1. Screwed
   2. Flanged
   3. Welded (by heat or use of solvents)

Tubing (metallic):
1. Welding
2. Compression fitting
3. Flanged

Tubing (plastic):
1. Compression
2. Push-on fitting (especially for plastic tubing) with hose clip where required to withstand pressure.

General The most popular method of connecting metal tubing is the compression fitting, as it is capable of withstanding pressures up to the limit of the tubing itself and is easily dismantled for servicing as well as being obtainable manufactured in all the most common materials.

39.3 TYPICAL SAMPLE SYSTEMS

39.3.1 Gases

39.3.1.1 High-Pressure Sample to a Process Chromatograph

The example taken is for a chromatograph analyzing the composition of a gas which is in the vapor phase at 35 bar (Figure 39.17). This is the case described in Section 39.1.4.3, where it is necessary to reduce the pressure of the gas at the sample point in order to obtain a fast response time with a minimum wastage of process gas.

The sample is taken from the process line using a low-volume sample probe (Section 39.2.1.2) and then flows immediately into a pressure reducing valve to drop the pressure to a constant 1.5 bar, which is measured on a local

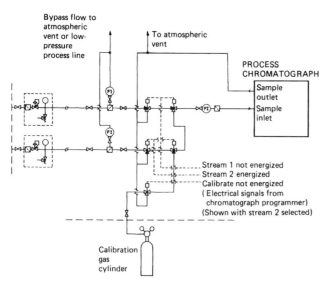

FIGURE 39.17 Schematic: high-pressure gas sample to chromatograph. Courtesy of Ludlam Sysco.

pressure gauge. A pressure relief valve set to relieve at 4 bar is connected at this point to protect downstream equipment if the pressure-reducing valve fails.

After pressure reduction the sample flows in small-bore tubing (6 mm OD) to the main sample system next to the analyzer, where it flows through a filter (such as shown in Section 39.2.2.3) to the sample selection system.

The fast loop flows out of the bottom of the filter body, bypassing the filter element, and then through a needle valve and flowmeter to an atmospheric vent on a low-pressure process line.

The stream selection system shown in Figure 39.17 is called a block-and-bleed system, and always has two or more three-way valves between each stream and the analyzer inlet. The line between two of the valves on the stream which is not in operation is vented to atmosphere, so guaranteeing that the stream being analyzed cannot be contaminated by any of the other streams. A simple system without a block-and-bleed valve is described in Section 39.3.2.1 below.

After the stream-selection system the sample flows through a needle valve flowmeter and a miniature in-line filter to the analyzer sample inlet. The analyzer sample outlet on this system flows to the atmospheric vent line.

39.3.1.2 Furnace Gas Using Steam-Injection Probe Inside the Flue

This system utilizes a Venturi assembly located inside the flue (Figure 39.18). High-pressure steam enters the Venturi via a separate steam tube, and a low-pressure region results inside the flue at the probe head. A mixture of steam and sample gas passes down the sample line. Butyl or EDPDM rubber-lined steam hose is recommended for sample lines, especially when high-sulfur fuels are used. This will minimize the effects of corrosion.

At the bottom end of the sample line the sample gas is mixed with a constant supply of water. The gas is separated from the water and taken either through a ball valve or a solenoid valve towards the sample loop, which minimizes the dead volume between each inlet valve and the analyzer inlet.

The water, dust, etc., passes out of a dip leg (A) and to a drain. It is assumed that the gas leaves the separator saturated with water at the temperature of the water. In the case of a flue gas system on a ship operating, for example, in the Red Sea, this could be at 35°C. The system is designed to remove condensate that may be formed because of lower temperatures existing in downstream regions of the sample system.

At the end of the loop there is a second dip leg (B) passing into a separator. A 5 cm water differential pressure is produced by the difference in depth of the two dip legs, so there is always a continuous flow of gas round the loop and out to vent via dip-leg (B).

CHAPTER | 39 Sampling

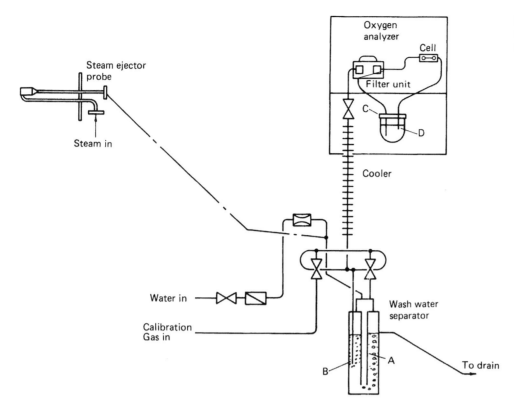

FIGURE 39.18 Schematic: furnace gas sampling. Courtesy of Servomex.

The gas passes from the loop to a heat exchanger, which is designed so that the gas leaving the exchanger is within 1 K of the air temperature. This means that the gas leaving the heat exchanger can be at 36°C and saturated with water vapor. The gas now passes into the analyzer which is maintained at 60°C.

The gas arrives in the analyzer and enters the first chamber, which is a centrifugal separator in a stainless steel block at 60°C. The condensate droplets will be removed at this point and passed down through the bottom end of the separator into a bubbler unit (C). The bubbles in this tube represent the bypass flow. At the same time the gas is raised from 36°C to 60°C inside the analyzer.

Gas now passes through a filter contained in the second chamber, the measuring cell, and finally to a second dip leg (D) in the bubbler unit. The flow of gas through the analyzer cell is determined by the difference in the length of the two legs inside the bubbler unit and cannot be altered by the operator.

This system has the following operating advantages:

1. The system is under a positive pressure right from inside the flue, and so leaks in the sample line can only allow steam and sample out and not air in.
2. The high-speed steam jet scours the tube, preventing build-up.
3. The steam maintains the whole of the sample probe above the dew point and so prevents corrosive condensate forming on the outside of the probe.
4. The steam keeps the entire probe below the temperature of the flue whenever the temperature of the flue is above the temperature of the steam.
5. The actual sampling system is intrinsically safe, as no electrical pumps are required.

39.3.1.3 Steam Sampling for Conductivity

The steam sample is taken from the process line by means of a special probe and then flows through thick-wall 316 stainless steel tubing to the sample system panel (Figure 39.19). The sample enters the sampling panel through a high-temperature, high-pressure isolating valve and then flows into the cooler, where the steam is condensed and the condensate temperature is reduced to a suitable temperature for the analyzer (typically 30°C).

After the cooler, the condensate passes to a pressure-control valve to reduce the pressure to about 1 bar gauge. The temperature and pressure of the sample are then measured on suitable gauges and a pressure-relief valve (set at 2 bar) is fitted to protect downstream equipment from excess pressure if a fault occurs in the pressure control valve. The constant-pressure, cooled sample passes through a needle valve, flowmeter, and three-way valve into the conductivity cell and then to drain.

Facilities are provided for feeding water of known conductivity into the conductivity cell through the three-way valve for calibration purposes. The sample coolers are normally supplied with stainless steel cooling coils which are suitable where neither the sample nor the coolant contain appreciable chloride which can cause stress corrosion cracking.

When chlorides are known to be present in the sample or cooling water, cooling coils are available, made of alternative materials which are resistant to chloride-induced stress corrosion cracking.

39.3.2 Liquids

39.3.2.1 Liquid Sample to a Process Chromatograph

The example taken is for a chromatograph measuring butane in gasoline (petrol) (Figure 39.20). The chromatograph in this case would be fitted with a liquid inject valve so that the sample will remain in the liquid phase at all times within the sample system.

In a liquid inject chromatograph the sample flow rate through the analyzer is very low (typically 25 ml/min), so that a fast loop system is essential.

The sample flow from the process enters the sample system through an isolating valve, then through a pump (if required) and an in-line filter, from which the sample is taken to the analyzer. After the filter the fast loop flows through a flowmeter followed by a needle valve, then through an isolating valve back to the process. Pressure gauges are fitted, one before the in-line filter and one after the needle valve, so that it is possible at any time to check that the pressure differential is sufficient (usually 1 bar minimum) to force the sample through the analyzer.

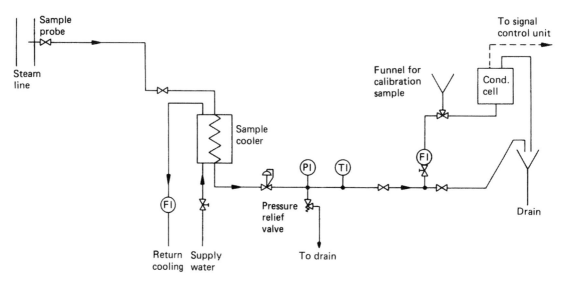

FIGURE 39.19 Schematic: steam sampling for conductivity. Courtesy of Ludlam Sysco.

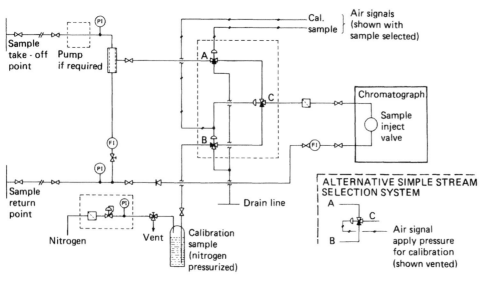

FIGURE 39.20 Schematic: liquid sample to process chromatograph. Courtesy of Ludlam Sysco.

The filtered sample flows through small-bore tubing (typically, 3 mm OD) to the sample/calibration selection valves. The system shown is the block-and-bleed configuration as described in Section 39.3.1.1. Where there is no risk of cross contamination the sample stream-selection system shown in the inset of Figure 39.20 may be used.

The selected sample flows through a miniature in-line filter (Section 39.2.2.4) to the analyzer, then through the flow control needle valve and non-return valve back to the fast loop return line.

When the sample is likely to vaporize at the sample return pressure it is essential to have the flow control needle valve after the flowmeter in both the fast loop and the sample through the analyzer. This is done to avoid the possibility of any vapor flashing off in the needle valve and passing through the flowmeter, which would give erroneous readings.

The calibration sample is stored in a nitrogen-pressurized container and may be switched either manually or automatically from the chromatograph controller.

39.3.2.2 Gas Oil Sample to a Distillation Point Analyzer

In this case the process conditions are as follows (Figure 39.21):

Sample tap:
 Normal pressure: 5 bar g
 Normal temperature: 70°C
 Sample line length: 73 m

Sample return:
 Normal pressure: 5 bar g
 Return line length: 73 m

This is a typical example of an oil-refinery application where, for safety reasons, the analyzer has to be positioned at the edge of the process area and consequently the sample and return lines are relatively long. Data to illustrate the fast loop calculation, based on equations in Crane's Publication No. 410M, are given in Table 39.1.

An electrically driven gear pump is positioned immediately outside the analyzer house which pumps the sample round the fast loop and back to the return point. The sample from the pump enters the sample system cabinet and flows through an in-line filter from which the sample is taken to the analyzer, through a needle valve and flowmeter back to the process. The filtered sample then passes through a water-jacketed cooler to reduce the temperature to that required for the coalescer and analyzer. After the cooler the sample is pressure reduced to about 1 bar with a pressure-control valve.

The pressure is measured at this point and a relief valve is fitted so that, in the event of the pressure control valve failing open, no downstream equipment will be damaged.

The gas oil sample may contain traces of free water and, as this will cause erroneous readings on the analyzer, it is removed by the coalescer. The bypass sample from the bottom of the coalescer flows through a needle valve and flow-meter to the drain line. The dry sample from the coalescer flows through a three-way ball valve for calibration purposes and then a needle valve and flowmeter to control the sample flow into the analyzer.

The calibration of this analyzer is carried out by filling the calibration vessel with the sample which had previously been accurately analyzed in the laboratory. The vessel is then pressurized with nitrogen to the same pressure as that set on the pressure-control valve and then the calibration sample is allowed to flow into the analyzer by turning the three-way ball valve to the calibrate position.

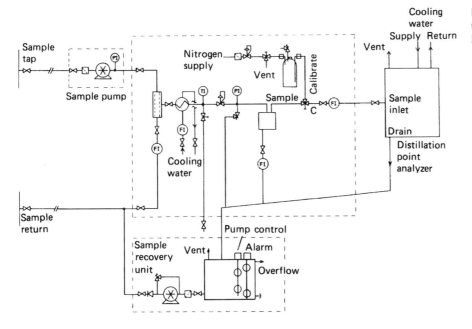

FIGURE 39.21 Schematic: gas oil sample to distillation point analyzer. Courtesy of Ludlam Sysco.

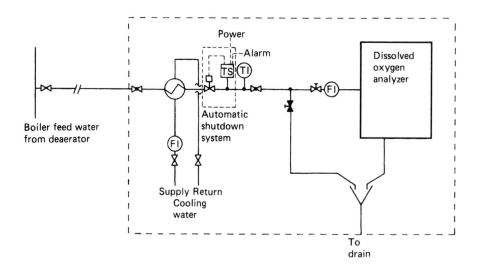

FIGURE 39.22 Schematic: water sample system for dissolved oxygen analyzer. Courtesy of Ludlam Sysco.

The waste sample from the analyzer has to flow to an atmospheric drain and, to prevent product wastage, it flows into a sample recovery unit along with the sample from the coalescer bypass and the pressure relief valve outlet.

The sample recovery unit consists of a steel tank from which the sample is returned to the process intermittently by means of a gear pump controlled by a level switch. An extra level switch is usually fitted to give an alarm if the level rises too high or falls too low.

A laboratory sample take-off point is fitted to enable a sample to be taken for separate analysis at any time without interfering with the operation of the analyzer.

39.3.2.3 Water-Sampling System for Dissolved Oxygen Analyzer

Process conditions:

Sample tap:

Normal pressure: 3.5 bar
Normal temperature: 140°C

Sample line length: 10 m

This system (Figure 39.22) illustrates a case where the sample system must be kept as simple as possible to prevent degradation of the sample. The analyzer is measuring 0–20 μg/l oxygen and, because of the very low oxygen content, it is essential to avoid places in the system where oxygen can leak in or be retained in a pocket. Wherever possible ball or plug valves must be used, as they leave no dead volumes which are not purged with the sample.

The only needle valve in this system is the one controlling the flow into the analyzer, and this must be mounted with the water flowing vertically up through it so that all air is displaced.

The sample line, which should be as short as possible, flows through an isolating ball valve into a water-jacketed cooler to drop the temperature from 140°C to approximately 30°C, and then the sample is flow controlled by a needle valve and flowmeter. In this case, it is essential to reduce the temperature of the water sample before pressure reduction; otherwise, the sample would flash off as steam. A bypass valve to drain is provided so that the sample can flow through the system while the analyzer is being serviced.

When starting up an analyzer such as this it may be necessary to loosen the compression fittings a little while the sample pressure is on to allow water to escape, fill the small voids in the fitting, and then finally tighten them up again to give an operational system.

An extra unit that is frequently added to a system such as this is automatic shutdown on cooling-water failure to protect the analyzer from hot sample. This unit is shown dotted on Figure 39.22, and consists of an extra valve and a temperature detector, either pneumatic or electronically operated, so that the valve is held open normally but shuts and gives an alarm when the sample temperature exceeds a preset point. The valve would then be reset manually when the fault has been corrected.

REFERENCES

Cornish, D. C., et al., *Sampling Systems for Process Analyzers*, Butterworths, London (1981).
Flow of Fluids, Publication No. 410M, Crane Limited (1982).
Marks, J. W., *Sampling and Weighing of Bulk Solids*, Transtech Publications, Clausthal-Zellerfeld (1985).

Chapter 40

Telemetry

M. L. Sanderson

40.1 INTRODUCTION

Within instrumentation there is often a need for telemetry in order to transmit data or information between two geographical locations. The transmission may be required to enable centralized supervisory data logging, signal processing, or control to be exercised in large-scale systems which employ distributed data logging or control subsystems. In a chemical plant or power station these subsystems may be spread over a wide area. Telemetry may also be required for systems which are remote or inaccessible, such as a spacecraft, a satellite, or an unmanned buoy in the middle of the ocean. It can be used to transmit information from the rotating sections of an electrical machine without the need for slip rings. By using telemetry-sensitive signal processing and recording, an apparatus can be physically remote from hazardous and aggressive environments and can be operated in more closely monitored and controlled conditions.

Telemetry has traditionally been provided by either pneumatic or electrical transmission. Pneumatic transmission, as shown in Figure 40.1, has been used extensively in process instrumentation and control. The measured quantity (pressure, level, temperature, etc.) is converted to a pneumatic pressure, the standard signal ranges being 20–100 kPa gauge pressure (3–15 lb/in^2/g) and 20–180 kPa (3–27 lb/in^2/g). The lower limit of pressure provides a live zero for the instrument which enables line breaks to be detected, eases instrument calibration and checking, and provides for improved dynamic response since, when venting to atmospheric pressure, there is still sufficient driving pressure at 20 kPa. The pneumatic signals can be transmitted over distances up to 300 m in 6.35 mm or 9.5 mm OD plastic or metal tubing to a pneumatic indicator, recorder, or controller. Return signals for control purposes are transmitted from the control element. The distance is limited by the speed of response, which quadruples with doubling the distance. Pneumatic instrumentation generally is covered at greater length in Chapter 31.

Pneumatic instruments are intrinsically safe, and can therefore be used in hazardous areas. They provide protection against electrical power failure, since systems employing air storage or turbine-driven compressors can continue to provide measurement and control during power failure.

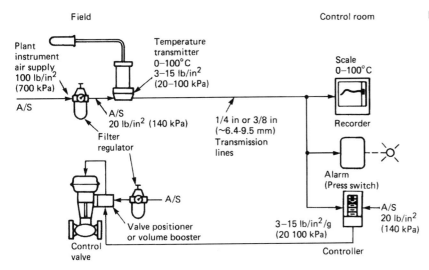

FIGURE 40.1 Pneumatic transmission.

Pneumatic signals also directly interface with control valves which are pneumatically operated and thus do not require the electrical/pneumatic converters required by electrical telemetry systems, although they do suffer from the difficulty of being difficult to interface to data loggers. Pneumatic transmission systems require a dry, regulated air supply. Condensed moisture in the pipework at subzero temperatures or small solid contaminants can block the small passages within pneumatic instruments and cause loss of accuracy and failure. Further details of pneumatic transmission and instrumentation can be found in Bentley (1983) and Warnock (1985).

Increasingly, telemetry in instrumentation is being undertaken using electrical, radio frequency, microwave, or optical fiber techniques. The communication channels used include transmission lines employing two or more conductors which may be a twisted pair, a coaxial cable, or a telephone line physically connecting the two sites; radio frequency (rf) or microwave links which allow the communication of data by modulation of an rf or microwave carrier; and optical links in which the data are transmitted as a modulation of light down a fiber-optic cable. All of these techniques employ some portion of the electromagnetic spectrum, as shown in Figure 40.2.

Figure 40.3 shows a complete telemetry system. Signal conditioning in the form of amplification and filtering normalizes the outputs from different transducers and restricts their bandwidths to those available on the communication channel. Transmission systems can employ voltage, current, position, pulse, or frequency techniques in order to transmit analog or digital data. Direct transmission of analog signals as voltage, current, or position requires a physical connection between the two points in the form of two or more wires and cannot be used over the telephone network. Pulse and frequency telemetry can be used for transmission over both direct links and also for telephone, rf, microwave, and optical links. Multiplexing either on a time or frequency basis enables more than one signal to be transmitted over the same channel. In pulse operation the data are encoded as the amplitude, duration, or position of the pulse or in a digital form. Transmission may be as a baseband signal or as an amplitude, frequency, or phase modulation of a carrier wave.

In the transmission of digital signals the information capacity of the channel is limited by the available bandwidth, the power level, and the noise present on the channel. The Shannon–Hartley theorem states that the information capacity, C, in bits/s (bps) for a channel having a bandwidth B Hz and additative Gaussian band-limited white noise is given by

$$C = B \cdot \log_2\left(1 + \frac{S}{N}\right)$$

where S is the average signal power at the output of the channel and N is the noise power at the output of the channel.

FIGURE 40.2 Electromagnetic spectrum.

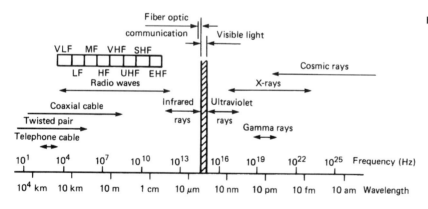

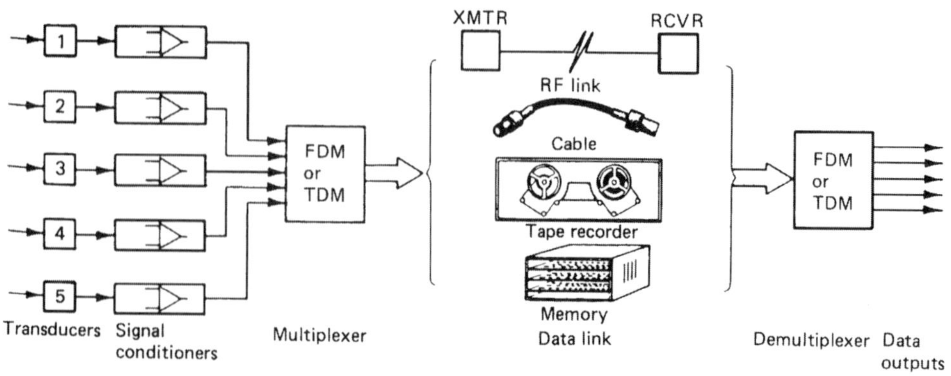

FIGURE 40.3 Telemetry system.

This capacity represents the upper limit at which data can be reliably transmitted over a particular channel. In general, because the channel does not have the ideal gain and phase characteristics required by the theorem and also because it would not be practical to construct the elaborate coding and decoding arrangements necessary to come close to the ideal, the capacity of the channel is significantly below the theoretical limit.

Channel bandwidth limitations also give rise to bit rate limitations in digital data transmission because of intersymbol interference (ISI), in which the response of the channel to one digital signal interferes with the response to the next. The impulse response of a channel having a limited bandwidth of B Hz is shown in Figure 40.4(a). The response has zeros separated by $1/2B$ s. Thus for a second impulse transmitted across the channel at a time $1/2B$ s later there will be no ISI from the first impulse. This is shown in Figure 40.4(b). The maximum data rate for the channel such that no ISI occurs is thus $2B$ bps. This is known as the Nyquist rate. Figure 40.4(c) shows the effect of transmitting data at a rate in excess of the Nyquist rate.

40.2 COMMUNICATION CHANNELS

40.2.1 Transmission Lines

Transmission lines are used to guide electromagnetic waves, and in instrumentation these commonly take the form of a twisted pair, a coaxial cable, or a telephone line. The primary constants of such lines in terms of their resistance, leakage conductance, inductance, and capacitance are distributed as shown in Figure 40.5. At low frequencies, generally below 100 kHz, a medium-length line may be represented by the circuit shown in Figure 40.6, where R_L is the resistance of the wire and C_L is the lumped capacitance of the line. The line thus acts as a low-pass filter. The frequency response can be extended by loading the line with regularly placed lumped inductances.

Transmission lines are characterized by three secondary constants. These are the characteristic impedance, Z_0; the attenuation, α, per unit length of line which is usually expressed in dB/unit length; and the phase shift, β, which is measured in radians/unit length. The values of Z_0, α, β are related to the primary line constants by:

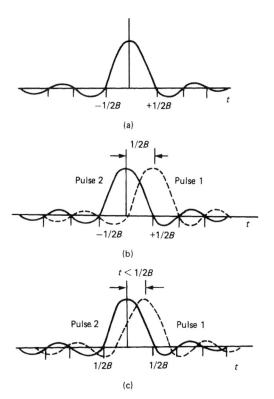

$$Z_0 = \sqrt{\left(\frac{R + j\omega L}{G + j\omega C}\right)}\, \Omega$$

$$\alpha = 8.68\left[0.5\left(\left\{(R^2 + \omega^2 L^2)(G^2 + \omega^2 C^2)\right\}^{1/2}\right.\right.$$
$$\left.\left. + (RG - \omega^2 LC)\right)\right]^{1/2} \text{ dB/unit length}$$

$$\beta = \left[0.5\left(\left\{(R^2 + \omega^2 L^2)(G^2 + \omega^2 C^2)\right\}^{1/2}\right.\right.$$
$$\left.\left. - (RG - \omega^2 LC)\right)\right]^{1/2} \text{ radians/unit length}$$

where R is the resistance per unit length, G is the leakage conductance per unit length, C is the capacitance per unit length, and L is the inductance per unit length.

FIGURE 40.4 (a) Impulse response of a bandlimited channel; (b) impulse responses delayed by $1/2B$ s; (c) impulse responses delayed by less than $1/2B$ s.

FIGURE 40.5 Distributed primary constants of a transmission line.

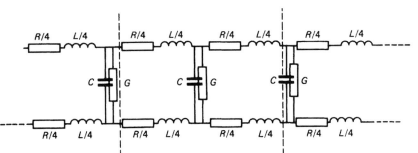

It is necessary to terminate transmission lines with their characteristic impedance if reflection or signal echo is to be avoided. The magnitude of the reflection for a line of characteristic impedance Z_0 terminated with an impedance Z_T is measured by the reflection coefficient, ρ, given by:

$$\rho = \frac{Z_T - Z_0}{Z_T + Z_0}$$

Twisted pairs are precisely what they say they are, namely, two insulated conductors twisted together. The conductors are generally copper or aluminum, and plastic is often used as the insulating material. The twisting reduces the effect of inductively coupled interference. Typical values of the primary constants for a 22 gauge copper twisted pair are $R = 100$ Ω/km, $L = 1$ mH/km, $G = 10^{-5}$ S/km and $C = 0.05$ μF/km. At high frequencies the characteristic impedance of the line is approximately 140 Ω. Typical values for the attenuation of a twisted pair are 3.4 dB/km at 100 kHz, 14 dB/km at 1 MHz, and 39 dB/km at 10 MHz. The high-frequency limitation for the use of twisted pairs at approximately 1 MHz occurs not so much as a consequence of attenuation but because of crosstalk caused by capacitive coupling between adjacent twisted pairs in a cable.

Coaxial cables which are used for data transmission at higher frequencies consist of a central core conductor surrounded by a dielectric material which may be either polyethylene or air. The construction of such cables is shown in Figure 40.7. The outer conductor consists of a solid or braided sheath around the dielectric. In the case of the air dielectric the central core is supported on polyethylene spacers placed uniformly along the line. The outer conductor is usually covered by an insulating coating. The loss at high frequencies in coaxial cable is due to the "skin effect," which forces the current in the central core to flow near to its surface and thus increases the resistance of the conductor. Such cables have a characteristic impedance of between 50 and 75 Ω. The typical attenuation of a 0.61 cm diameter coaxial cable is 8 dB/100 m at 100 MHz and 25 dB/100 m at 1 GHz.

Trunk telephone cables connecting exchanges consist of bunched twisted conductor pairs. The conductors are insulated with paper or polyethylene, the twisting being used to reduce the crosstalk between adjacent conductor pairs. A bunch of twisted cables is sheathed in plastic, and the whole cable is given mechanical strength by binding with steel wire or tape which is itself sheathed in plastic. At audio frequencies the impedance of the cable is dominated by its capacitance and resistance. This results in an attenuation that is frequency dependent and phase delay distorted, since signals of different frequencies are not transmitted down the cable with the same velocity. Thus a pulse propagated down a cable results in a signal which is not only attenuated (of importance in voice and analog communication) but which is also phase distorted (of importance in digital signal transmission). The degree of phase delay distortion is measured by the group delay $d\beta/d\Omega$. The bandwidth of telephone cables is restricted at low frequencies by the use of ac amplification in the repeater stations used to boost the signal along the line. Loading is used to improve the high-frequency amplitude response of the line. This takes the form of lumped inductances which correct the attenuation characteristics of the line. These leave the line with a significant amount of phase delay distortion and give the line attenuation at high frequencies. The usable frequency band of the telephone line is between 300 Hz and 3 kHz. Figure 40.8 shows typical amplitude and phase or group delay distortions

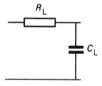

FIGURE 40.6 Low-frequency lumped approximation for a transmission line.

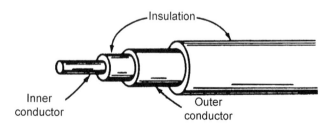

FIGURE 40.7 Coaxial cable.

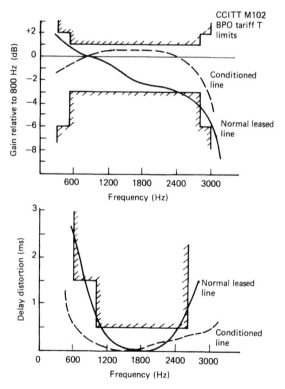

FIGURE 40.8 Gain and delay distortion on telephone lines.

relative to 800 Hz for a typical leased line and a line to which equalization or conditioning has been applied.

In order to transmit digital information reliably the transmission equipment has to contend with a transmission loss which may be as high as 30 dB; a limited bandwidth caused by a transmission loss which varies with frequency; group delay variations with frequency; echoes caused by impedance mismatching and hybrid crosstalk; and noise which may be either Gaussian or impulsive noise caused by dial pulses, switching equipment, or lightning strikes. Thus it can be seen that the nature of the telephone line causes particular problems in the transmission of digital data. Devices known as modems (Modulators/DEModulators) are used to transmit digital data along telephone lines. These are considered in Section 40.9.1.

40.2.2 Radio Frequency Transmission

Radio frequency (rf) transmission is widely used in both civilian and military telemetry and can occur from 3 Hz (which is referred to as very low frequency (VLF)) up to as high as 300 GHz (which is referred to as extremely high frequency (EHF)). The transmission of the signal is by means of line-of-sight propagation, ground or surface wave diffraction, ionospheric reflection or forward scattering (Coates 1982). The transmission of telemetry or data signals is usually undertaken as the amplitude, phase, or frequency modulation of some rf carrier wave. These modulation techniques are described in Section 40.5. The elements of an rf telemetry system are shown in Figure 40.9.

The allocation of frequency bands has been internationally agreed upon under the Radio Regulations of the International Telecommunication Union based in Geneva. These regulations were agreed to in 1959 and revised in 1979 (HMSO, 1980). In the UK the Radio Regulatory Division of the Department of Trade approves equipment and issues licenses for the users of radio telemetry links. In the United States, the Federal Communications Commission (FCC) serves the same purpose. In other countries, there is an analogous office. For general-purpose low-power telemetry and telecontrol there are four bands which can be used. These are 0–185 kHz and 240–315 kHz, 173.2–173.35 MHz and 458.5–458.8 MHz. For high-power private point systems the allocated frequencies are in the UHF band 450–470 MHz. In addition, systems that use the cellular telephony bands are becoming common.

For medical and biological telemetry there are three classes of equipment. Class I are low-power devices operating between 300 kHz and 30 MHz wholly contained within the body of an animal or human. Class II is broad-band equipment operating in the band 104.6–105 MHz. Class III equipment is narrow-band equipment operating in the same frequency band as the Class II equipment. Details of the requirements for rf equipment can be found in the relevant documents cited in the References (HMSO, 1963, 1978, 1979).

40.2.3 Fiber-Optic Communication

Increasingly, in data-communication systems there is a move toward the use of optical fibers for the transmission of data. Detailed design considerations for such systems can be found in Keiser (1983), Wilson and Hawkes (1983), and Senior (1985). As a transmission medium fiber-optic cables offer the following advantages:

1. They are immune to electromagnetic interference.
2. Data can be transmitted at much higher frequencies and with lower losses than twisted pairs or coaxial cables. Fiber optics can therefore be used for the multiplexing of a large number of signals along one cable with greater distances required between repeater stations.
3. They can provide enhanced safety when operating in hazardous areas.
4. Ground loop problems can be reduced.
5. Since the signal is confined within the fiber by total internal reflection at the interface between the fiber and the cladding, fiber-optic links provide a high degree of data security and little fiber-to-fiber crosstalk.
6. The material of the fiber is very much less likely to be attacked chemically than copper-based systems, and it can be provided with mechanical properties which will make such cables need less maintenance than the equivalent twisted pair or coaxial cable.
7. Fiber-optic cables can offer both weight and size advantages over copper systems.

40.2.3.1 Optical Fibers

The elements of an optical fiber, as shown in Figure 40.10, are the core material, the cladding, and the buffer coating. The core material is either plastic or glass. The cladding is a material whose refractive index is less than that of the core. Total internal reflection at the core/cladding interface confines the light to travel within the core. Fibers with plastic cores also have plastic cladding. Such fibers exhibit high

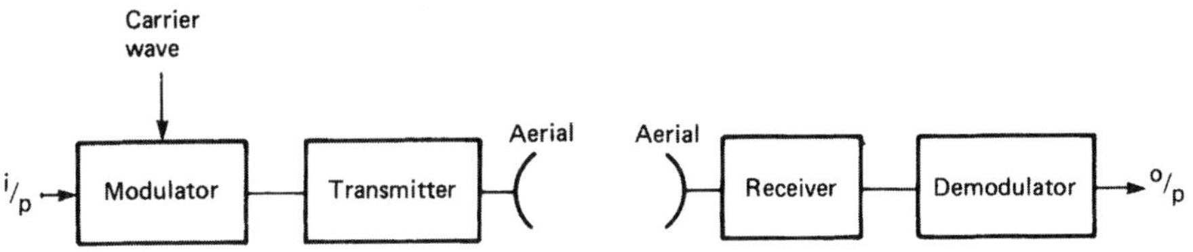

FIGURE 40.9 RF telemetry system.

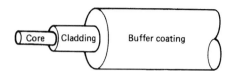

FIGURE 40.10 Elements of an optical fiber.

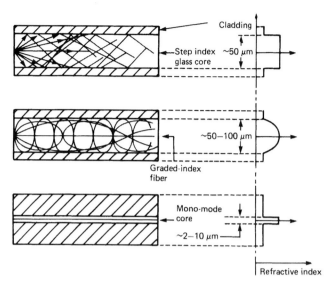

FIGURE 40.11 Propagation down fibers.

losses but are widely used for short-distance transmission. Multicomponent glasses containing a number of oxides are used for all but the lowest-loss fibers, which are usually made from pure silica. In low- and medium-loss fibers the glass core is surrounded by a glass or plastic cladding. The buffer coating is an elastic, abrasion-resistant plastic material which increases the mechanical strength of the fiber and provides it with mechanical isolation from geometrical irregularities, distortions, or roughness of adjacent surfaces which could otherwise cause scattering losses when the fiber is incorporated into cables or supported by other structures.

The numerical aperture (NA) of a fiber is a measure of the maximum core angle for light rays to be reflected down the fiber by total internal reflection.

By Snell's Law:

$$NA = \sin\theta = \sqrt{\left(\mu_1^2 - \mu_2^2\right)}$$

where μ_1 is the refractive index of the core material and μ_2 is the refractive index of the cladding material.

Fibers have NAs in the region of 0.15–0.4, corresponding to total acceptance angles of between 16 and 46 degrees. Fibers with higher NA values generally exhibit greater losses and low bandwidth capabilities.

The propagation of light down the fibers is described by Maxwell's equations, the solution of which gives rise to a set of bounded electromagnetic waves called the "modes" of the fiber. Only a discrete number of modes can propagate down the fiber, determined by the particular solution of Maxwell's equation obtained when boundary conditions appropriate to the particular fiber are applied. Figure 40.11 shows the propagation down three types of fiber. The larger-core radius multimode fibers are either step index or graded index fibers. In the step index fibers there is a step change in the refractive index at the core/cladding interface. The refractive index of the graded index fiber varies across the core of the fiber. Monomode fibers have a small-core radius, which permits the light to travel along only one path in the fiber.

The larger-core radii of multimode fibers make it much easier to launch optical power into the fiber and facilitate the connecting of similar fibers. Power can be launched into such a fiber using light-emitting diodes (LEDs), whereas single-mode fibers must be excited with a laser diode.

Intermodal dispersion occurs in multimode fibers because each of the modes in the fibers travels at a slightly different velocity. An optical pulse launched into a fiber has its energy distributed among all its possible modes, and therefore as it travels down the fiber the dispersion has the effect of spreading the pulse out. Dispersion thus provides a bandwidth limitation on the fiber. This is specified in MHz · km. In graded index fibers the effect of intermodal dispersion is reduced over that in step index fibers because the grading bends the various possible light rays along paths of nominal equal delay. There is no intermodal dispersion in a single-mode fiber, and therefore these are used for the highest-capacity systems. The bandwidth limitation for a plastic clad step index fiber is typically 6–25 MHz · km. Employing graded index plastic-clad fibers this can be increased to the range of 200–400 MHz · km. For monomode fibers the bandwidth limitation is typically 500–1500 MHz · km.

Attenuation within a fiber, which is measured in dB/km, occurs as a consequence of absorption, scattering, and radiative losses of optical energy. Absorption is caused by extrinsic absorption by impurity atoms in the core and intrinsic absorption by the basic constituents of the core material. One impurity which is of particular importance is the OH (water) ion, and for low-loss materials this is controlled to a concentration of less than 1 ppb. Scattering losses occur as a consequence of microscopic variations in material density or composition, and from structural irregularities or defects introduced during manufacture. Radiative losses occur whenever an optical fiber undergoes a bend having a finite radius of curvature.

Attenuation is a function of optical wavelength. Figure 40.12 shows the typical attenuation versus wavelength characteristics of a plastic and a monomode glass fiber. At 0.8 μm the attenuation of the plastic fiber is 350 dB/km and that of the glass fiber is approximately 1 dB/km. The minimum attenuation of the glass fiber is 0.2 dB/km at 1.55 μm. Figure 40.13 shows the construction of the light- and medium-duty optical cables.

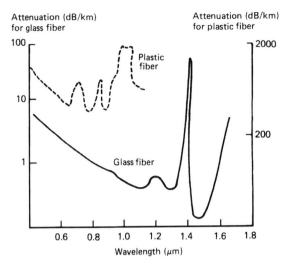

FIGURE 40.12 Attenuation characteristics of optical fibers.

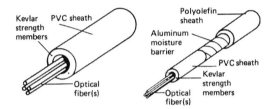

FIGURE 40.13 Light- and medium-duty optical cables.

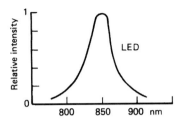

FIGURE 40.14 Spectral output from a LED.

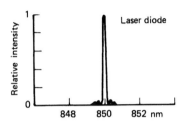

FIGURE 40.15 Spectral output from a laser diode.

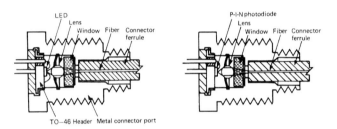

FIGURE 40.16 LED and *p-i-n* diode detector for use in a fiber-optic system.

40.2.3.2 Sources and Detectors

The sources used in optical fiber transmission are LEDs and semiconductor laser diodes. LEDs are capable of launching a power of between 0.1 and 10 mW into the fiber. Such devices have a peak emission frequency in the near infrared, typically between 0.8 and 1.0 µm. Figure 40.14 shows the typical spectral output from a LED. Limitations on the transmission rates using LEDs occur as a consequence of rise time, typically between 2 and 10 ns, and chromatic dispersion. This occurs because the refractive index of the core material varies with optical wavelength, and therefore the various spectral components of a given mode will travel at different speeds.

Semiconductor laser diodes can provide significantly higher power, particularly with low duty cycles, with outputs typically in the region of 1 to 100 mW. Because they couple into the fiber more efficiently, they offer a higher electrical to optical efficiency than do LEDs. The lasing action means that the device has a narrower spectral width compared with a LED, typically 2 nm or less, as shown in Figure 40.15. Chromatic dispersion is therefore less for laser diodes, which also have a faster rise time, typically 1 ns.

For digital transmissions of below 50 Mbps LEDs require less complex drive circuitry than laser diodes and require no thermal or optical power stabilization.

Both *p-i-n* (*p* material-intrinsic-*n* material) diodes and avalanche photodiodes are used in the detection of the optical signal at the receiver. In the region 0.8–0.9 µm silicon is the main material used in the fabrication of these devices. The *p-i-n* diode has a typical responsivity of 0.65 A/W at 0.8 µm. The avalanche photodiode employs avalanche action to provide current gain and therefore higher detector responsivity. The avalanche gain can be 100, although the gain produces additional noise. The sensitivity of the photodetector and receiver system is determined by photodetector noise which occurs as a consequence of the statistical nature of the production of photoelectrons, and bulk and dark surface current, together with the thermal noise in the detector resistor and amplifier. For *p-i-n* diodes the thermal noise of the resistor and amplifier dominates, whereas with avalanche photodiodes the detector noise dominates.

Figure 40.16 shows a LED and *p-i-n* diode detector for use in a fiber-optic system.

40.2.3.3 Fiber-Optic Communication Systems

Figure 40.17 shows a complete fiber-optic communications system. In the design of such systems it is necessary to compute the system insertion loss in order that the system can

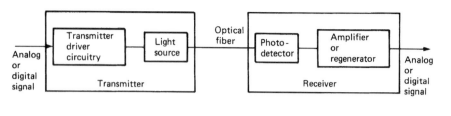

FIGURE 40.17 Fiber-optic communication system.

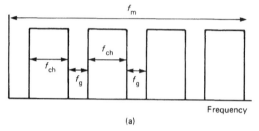

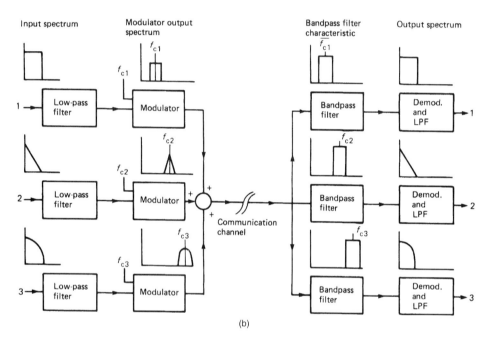

FIGURE 40.18 (a) Frequency-division multiplexing: (b) transmission of three signals using FDM.

be operated using the minimum transmitter output flux and minimum receiver input sensitivity. In addition to the loss in the cable itself, other sources of insertion loss occur at the connections between the transmitter and the cable and the cable and the receiver; at connectors joining cables; and at points where the cable has been spliced. The losses at these interfaces occur as a consequence of reflections, differences in fiber diameter, NA, and fiber alignment. Directional couplers and star connectors also increase the insertion loss.

40.3 SIGNAL MULTIPLEXING

In order to enable several signals to be transmitted over the same medium it is necessary to multiplex the signals. There are two forms of multiplexing: frequency-division multiplexing (FDM) and time-division multiplexing (TDM). FDM splits the available bandwidth of the transmission medium into a series of frequency bands and uses each of the frequency bands to transmit one of the signals. TDM splits the transmission into a series of time slots and allocates certain time slots, usually on a cyclical basis, for the transmission of one signal.

The basis of FDM is shown in Figure 40.18(a). The bandwidth of the transmission medium f_m is split into a series of frequency bands, having a bandwidth f_{ch}, each one of which is used to transmit one signal. Between these channels there are frequency bands, having bandwidth f_g, called "guard bands," which are used to ensure that there is adequate separation and minimum cross-talk between any two adjacent channels. Figure 40.18(b) shows the transmission of three band-limited signals having spectral characteristics as shown, the low-pass filters at the input to the modulators being used to bandlimit the signals. Each of the signals then modulates a carrier. Any form of carrier modulation can be used, although it is desirable to use a modulation which requires minimum bandwidth. The modulation shown in Figure 40.18(b) is amplitude modulation (see Section 40.5).

The individually modulated signals are then summed and transmitted. Bandpass filters after reception are used to separate the channels, by providing attenuation which starts in the guard bands. The signals are then demodulated and smoothed.

TDM is shown schematically in Figure 40.19. The multiplexer acts as a switch connecting each of the signals in turn to the transmission channel for a given time. In order to recover the signals in the correct sequence it is necessary to employ a demultiplexer at the receiver or to have some means inherent within the transmitted signal to identify its source. If N signals are continuously multiplexed then each one of them is sampled at a rate of $1/N$ Hz. They must therefore be band-limited to a frequency of $1/2N$ Hz if the Shannon sampling theorem is not to be violated.

The multiplexer acts as a multi-input-single-output switch, and for electrical signals this can be done by mechanical or electronic switching. For high frequencies electronic multiplexing is employed, with integrated circuit multiplexers which use CMOS or BIFET technologies.

TDM circuitry is much simpler to implement than FDM circuitry, which requires modulators, band-pass filters, and demodulators for each channel. In TDM only small errors occur as a consequence of circuit non-linearities, whereas phase and amplitude non-linearities have to be kept small in order to limit intermodulation and harmonic distortion in FDM systems. TDM achieves its low channel crosstalk by using a wideband system. At high transmission rates errors occur in TDM systems due to timing jitter, pulse accuracy, and synchronization problems. Further details of FDM and TDM systems can be found in Johnson (1976) and Shanmugan (1979).

40.4 PULSE ENCODING

Pulse code modulation (PCM) is one of the most commonly used methods of encoding analog data for transmission in instrumentation systems. In PCM the analog signal is sampled and converted into binary form by means of an ADC, and these data are then transmitted in serial form. This is shown in Figure 40.20. The bandwidth required for the transmission of a signal using PCM is considerably in excess of the bandwidth of the original signal. If a signal having a bandwidth f_d is encoded into an N-bit binary code the minimum bandwidth required to transmit the PCM encoded signal is $f_d \cdot N$ Hz, i.e., N times the original signal bandwidth.

Several forms of PCM are shown in Figure 40.21. The non-return to zero (NRZ-L) is a common code that is easily interfaced to a computer. In the non-return to zero mark (NRZ-M) and the non-return to zero space (NRZ-S) codes level transitions represent bit changes. In bi-phase level (BIΦ-L) a bit transition occurs at the center of every period. One is represented by a "1" level changing to a "0" level at the center transition point, and zero is represented by a "0" level changing to a "1" level. In bi-phase mark and space code (BIΦ-M) and (BIΦ-S) a level change occurs at the beginning of each bit period. In BIΦ-M one is represented by a mid-bit transition; a zero has no transition. BIΦ-S is the converse of BIΦ-M. Delay modulation code DM-M and DM-S have transitions at mid-bit and at the end of the bit time. In DM-M a one is represented by a level change at mid-bit; a zero followed by a zero is represented by a level change after the first zero. No level change occurs if a zero precedes a one. DM-S is the converse of DM-M. Bi-phase codes have a transition at least every bit time which can be used for synchronization, but they require twice the bandwidth of the NRZ-L code. The delay modulation codes offer the greatest bandwidth saving but are more susceptible to error, and are used if bandwidth compression is needed or high signal-to-noise ratio is expected.

Alternative forms of encoding are shown in Figure 40.22. In pulse amplitude modulation (PAM) the amplitude of the

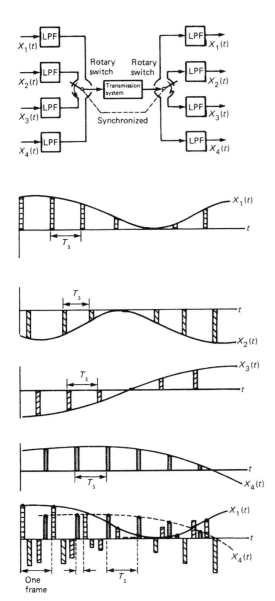

FIGURE 40.19 Time-division multiplexing.

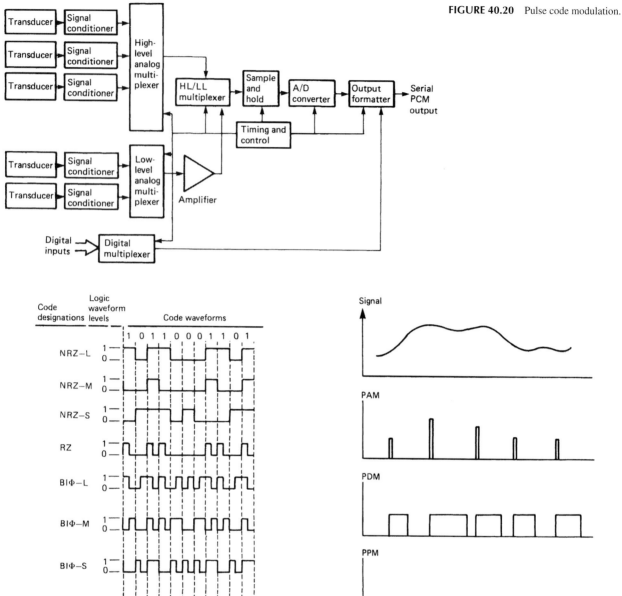

FIGURE 40.20 Pulse code modulation.

FIGURE 40.21 Types of pulse code modulation.

FIGURE 40.22 Other forms of pulse encoding.

signal transmitted is proportional to the magnitude of the signal being transmitted, and it can be used for the transmission of both analog and digital signals. The channel bandwidth required for the transmission of PAM is less than that required for PCM, although the effects of ISI are more marked. PAM as a means of transmitting digital data requires more complex decoding schemes, in that it is necessary to discriminate between an increased number of levels.

Other forms of encoding which can be used include pulse-width modulation (PWM), otherwise referred to as pulse-duration modulation (PDM), which employs a constant height variable width pulse with the information being contained in the width of the pulse. In pulse-position modulation (PPM) the position of the pulses corresponds to the width of the pulse in PWM. Delta modulation and sigma-delta modulation use pulse trains, the frequencies of which are proportional to either the rate of change of the signal or the amplitude of the signal itself. For analyses of the above systems in terms of the bandwidth required for transmission, signal-to-noise ratios, and error rate, together with practical details, the reader is directed to Hartley et al. (1967), Cattermole (1969), Steele (1975), and Shanmugan (1979).

40.5 CARRIER WAVE MODULATION

Modulation is used to match the frequency characteristics of the data to be transmitted to those of the Transmission channel, to reduce the effects of unwanted noise and interference, and to facilitate the efficient radiation of the signal. These are all effected by shifting the frequency of the data into some frequency band centered around a carrier frequency. Modulation also allows the allocation of specific frequency bands for specific purposes, such as in a FDM system or in rf transmission systems, where certain frequency bands are assigned for broadcasting, telemetry, etc. Modulation can also be used to overcome the limitations of signal-processing equipment in that the frequency of the signal can be shifted into frequency bands where the design of filters or amplifiers is somewhat easier, or into a frequency band that the processing equipment will accept. Modulation can be used to provide the bandwidth against signal-to-noise trade-offs which are indicated by the Hartley–Shannon theorem.

Carrier-wave modulation uses the modulation of one of its three parameters, namely amplitude, frequency, or phase, and these are all shown in Figure 40.23. The techniques can be used for the transmission of both analog and digital signals.

In amplitude modulation the amplitude of the carrier varies linearly with the amplitude of the signal to be transmitted. If the data signal $d(t)$ is represented by a sinusoid $d(t) = \cos 2\pi f_d t$ then in amplitude modulation the carrier wave $c(t)$ is given by:

$$c(t) = C\left(1 + m \cdot \cos^2 \pi f_d t\right)\cos 2\pi f_c t$$

where C is the amplitude of the unmodulated wave, f_c its frequency, and m is the depth of modulation which has a value lying between 0 and 1. If $m = 1$ then the carrier is said to have 100 percent modulation. The above expression for $c(t)$ can be rearranged as:

$$c(t) = C \cdot \cos^2 \pi f_c t + \frac{Cm}{2}\left[\cos 2\pi (f_c + f_d)t + \cos 2\pi (f_c - f_d)t\right]$$

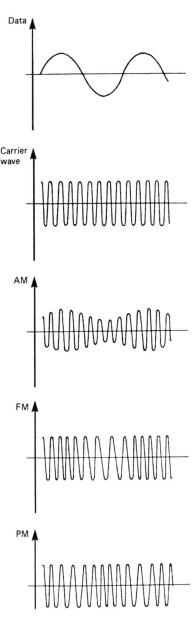

FIGURE 40.23 Amplitude, frequency, and phase modulation of a carrier wave.

showing that the spectrum of the transmitted signal has three frequency components at the carrier frequency f_c and at the sum and difference frequencies $(f_c + f_d)$ and $(f_c - f_d)$. If the signal is represented by a spectrum having frequencies up to f_d then the transmitted spectrum has a bandwidth of $2f_d$ centered around f_c. Thus in order to transmit data using AM a bandwidth equal to twice that of the data is required. As can be seen, the envelope of the AM signal contains the information, and thus demodulation can be effected simply by rectifying and smoothing the signal.

Both the upper and lower sidebands of AM contain sufficient amplitude and phase information to reconstruct the data, and thus it is possible to reduce the bandwidth requirements of the system. Single side-band modulation (SSB) and vestigial side-band modulation (VSM) both transmit the data using amplitude modulation with smaller bandwidths than straight AM. SSB has half the bandwidth of a simple AM system; the low-frequency response is generally poor. VSM transmits one side band almost completely and only a trace of the other. It is very often used in high-speed data transmission, since it offers the best compromise between bandwidth requirements, low-frequency response, and improved power efficiency.

In frequency modulation consider the carrier signal $c(t)$ given by:

$$c(t) = C\cos\left(2\pi f_c t + \phi(t)\right)$$

Then the instantaneous frequency of this signal is given by:

$$f_i(t) = f_c + \frac{1}{2\pi} \cdot \frac{d\phi}{dt}$$

The frequency deviation

$$\frac{1}{2\pi} \cdot \frac{d\phi}{dt}$$

of the signal from the carrier frequency is made to be proportional to the data signal. If the data signal is represented by a single sinusoid of the form:

$$d(t) = \cos 2\pi f_d t$$

then:

$$\frac{d\phi(t)}{dt} = 2\pi k_f \cdot \cos 2\pi f_d t$$

where k_f is the frequency deviation constant which has units of Hz/V. Thus:

$$c(t) = C \cos\left(2\pi f_c t + 2\pi k_f \int_{-\infty}^{t} d(\tau) \cdot d\tau\right)$$

and assuming zero initial phase deviation, then the carrier wave can be represented by:

$$c(t) = C \cos(2\pi f_c t + \beta \sin 2\pi f_d t)$$

where β is the modulation index and represents the maximum phase deviation produced by the data. It is possible to show that $c(t)$ can be represented by an infinite series of frequency components $f \pm nf_d$, $n = 1, 2, 3,...$, given by:

$$c(t) = C \sum_{n=-\infty}^{\infty} J_n(\beta) \cos(2\pi f_c + n2\pi f_d)t$$

where $J_n(\beta)$ is a Bessel function of the first kind of order n and argument β. Since the signal consists of an infinite number of frequency components, limiting the transmission bandwidth distorts the signal, and the question arises as to what is a reasonable bandwidth for the system to have in order to transmit the data with an acceptable degree of distortion. For $\beta \ll 1$ only J_0 and J_1 are important, and large β implies large bandwidth. It has been found in practice that if 98 percent or more of the FM signal power is transmitted, then the signal distortion is negligible. Carson's rule indicates that the bandwidth required for FM transmission of a signal having a spectrum with components up to a frequency of f_d is given by $2(f_\Delta + f_d)$, where f_Δ is the maximum frequency deviation. For narrow-band FM systems having small frequency deviations the bandwidth required is the same as that for AM. Wide-band FM systems require a bandwidth of $2f_\Delta$. Frequency modulation is used extensively in rf telemetry and in FDM.

In phase modulation the instantaneous phase deviation ϕ is made proportional to the data signal. Thus:

$$\phi = k_p \cdot \cos 2\pi f_d t$$

and it can be shown that the carrier wave $c(t)$ can be represented by:

$$c(t) = C \cos(2\pi f_c t + \beta \cos 2\pi f_d t)$$

where β is now given by k_p. For further details of the various modulation schemes and their realizations the reader should consult Shanmugan (1979) and Coates (1982).

40.6 ERROR DETECTION AND CORRECTION CODES

Errors occur in digital data communications systems as a consequence of the corruption of the data by noise. Figure 40.24 shows the bit error probability as a function of signal-to-noise ratio for a PCM transmission system using NRZ-L coding.

In order to reduce the probability of an error occurring in the transmission of the data, bits are added to the transmitted message. These bits add redundancy to the transmitted data, and since only part of the transmitted message is now the actual data, the efficiency of the transmission is reduced.

There are two forms of error coding, known as forward error detection and correction coding (FEC), in which the transmitted message is coded in such a way that errors can

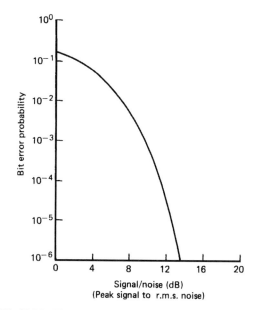

FIGURE 40.24 Bit-error probability for PCM transmission using a NRZ-L code.

be both detected and corrected continuously, and automatic repeat request coding (ARQ), in which if an error is detected then a request is sent to repeat the transmission. In terms of data-throughput rates FEC codes are more efficient than AQR codes because of the need to retransmit the data in the case of error in an ARQ code, although the equipment required to detect errors is somewhat simpler than that required to correct the errors from the corrupted message. ARQ codes are commonly used in instrumentation systems.

Parity-checking coding is a form of coding used in ARQ coding in which $(n - k)$ bits are added to the k bits of the data to make an n-bit data system. The simplest form of coding is parity-bit coding in which the number of added bits is one, and this additional bit is added to the data stream in order to make the total number of ones in the data stream either odd or even. The received data are checked for parity. This form of coding will detect only an odd number of bit errors. More complex forms of coding include linear block codes such as Hamming codes, cyclic codes such as Bose–Chanhuri–Hocquenghen codes, and geometric codes. Such codes can be designed to detect multiple-burst errors, in which two or more successive bits are in error. In general the larger the number of parity bits, the less efficient the coding is, but the larger are both the maximum number of random errors and the maximum burst length that can be detected. Figure 40.25 shows examples of some of these coding techniques. Further details of coding techniques can be found in Shanmugan (1979), Bowdell (1981), and Coates (1982).

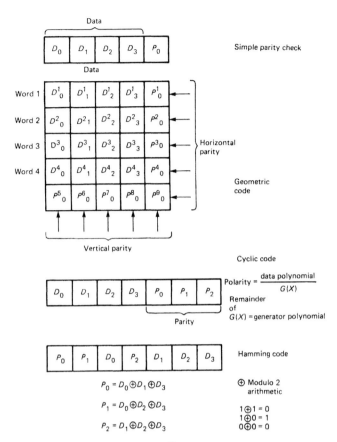

FIGURE 40.25 Error-detection coding

40.7 DIRECT ANALOG SIGNAL TRANSMISSION

Analog signals are rarely transmitted over transmission lines as a voltage since the method suffers from errors due to series and common mode inductively and capacitively coupled interference signals and those due to line resistance. The most common form of analog signal transmission is as current.

Current transmission as shown in Figure 40.26 typically uses 0–20 or 4–20 mA. The analog signal is converted to a current at the transmitter and is detected at the receiver either by measuring the potential difference developed across a fixed resistor or using the current to drive an indicating instrument or chart recorder. The length of line over which signals can be transmitted at low frequencies is primarily limited by the voltage available at the transmitter to overcome voltage drop along the line and across the receiver. With a typical voltage of 24 V the system is capable of transmitting the current over several kilometers. The percentage error in a current transmission system can be calculated as 50 × the ratio of the loop resistance in ohms to the total line insulation resistance, also expressed in ohms. The accuracy of current transmission system systems is typically ±0.5 percent.

The advantage of using 4–20 mA instead of 0–20 mA is that the use of a live zero enables instrument or line faults to be detected. In the 4–20 mA system zero value is represented by 4 mA and failure is indicated by 0 mA. It is possible to use a 4–20 mA system as a two-wire transmission system in which both the power and the signal are transmitted along the same wire, as shown in Figure 40.27. The 4 mA standing current is used to power the remote instrumentation and the transmitter. With 24 V drive the maximum power available to the remote station is 96 mW. Integrated-circuit devices such as the Burr–Brown XTR 100 are available for providing two-wire transmission. This is capable of providing a 4–20 mA output span for an input voltage as small as 10 mV, and is capable of transmitting at frequencies up to 2 kHz over a distance of 600 m. Current transmission cannot be used over the public telephone system because it requires a dc transmission path, and telephone systems use ac amplifiers in the repeater stations.

Position telemetry transmits an analog variable by reproducing at the receiver the positional information available at the transmitter. Such devices employ null techniques with either resistive or inductive elements to achieve the position telemetry. Figure 40.28 shows an inductive "synchro."

The ac power applied to the transmitter induces EMF in the three stator windings, the magnitude of which are dependent upon the position of the transmitter rotor. If the receiver

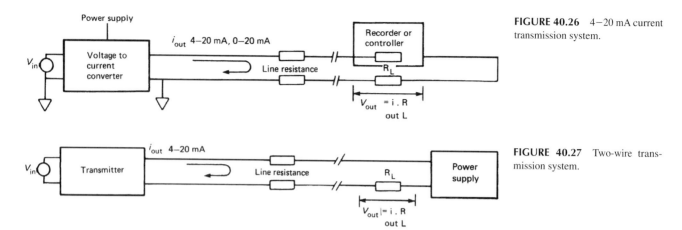

FIGURE 40.26 4–20 mA current transmission system.

FIGURE 40.27 Two-wire transmission system.

rotor is aligned in the same direction as the transmitter rotor then the EMF induced in the stator windings of the receiver will be identical to those on the stator windings of the transmitter. There will therefore be no resultant circulating currents. If the receiver rotor is not aligned to the direction of the transmitter rotor then the circulating currents in the stator windings will be such as to generate a torque which will move the receiver rotor in such a direction as to align itself with the transmitter rotor.

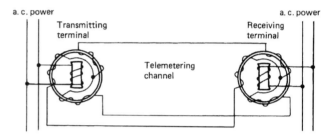

FIGURE 40.28 Position telemetry using an inductive "synchro."

40.8 FREQUENCY TRANSMISSION

By transmitting signals as frequency the characteristics of the transmission line in terms of amplitude and phase characteristics are less important. On reception the signal can be counted over a fixed period of time to provide a digital measurement. The resolution of such systems will be one count in the total number received. Thus for high resolution it is necessary to count the signal over a long time period, and this method of transmission is therefore unsuitable for rapidly changing or multiplexed signals but is useful for such applications as batch control, where, for example, a totalized value of a variable over a given period is required. Figure 40.29(a) shows a frequency-transmission system.

Frequency-transmission systems can also be used in two-wire transmission systems, as shown in Figure 40.29(b), where the twisted pair carries both the power to the remote device and the frequency signal in the form of current modulation. The frequency range of such systems is governed by the bandwidth of the channel over which the signal is to be transmitted, but commercially available integrated circuit V-to-f converters, such as the Analog Devices AD458, convert a 0–10 V dc signal to a frequency in the range 0–10 kHz or 0–100 kHz with a maximum non-linearity of ±0.01 percent of FS output, a maximum temperature coefficient of ±5 ppm/K, a maximum input offset voltage of ±10 mV, and a maximum input offset voltage temperature coefficient of 30 μV/K. The response time is two output pulses plus 2 μ. A low-cost f to V converter, such as the Analog Devices AD453, has an input frequency range of 0–100 kHz with a variable threshold voltage of between 0 and ±12 V, and can be used with low-level signals as well as high-level inputs from TTL and CMOS. The converter has a full-scale output of 10 V and a non-linearity of less than ±0.008 percent of FS with a maximum temperature coefficient of ±50 ppm/K. The maximum response time is 4 ms.

40.9 DIGITAL SIGNAL TRANSMISSION

Digital signals are transmitted over transmission lines using either serial or parallel communication.

For long-distance communication serial communication is the preferred method. The serial communication may be either synchronous or asynchronous. In synchronous communication the data are sent in a continuous stream without stop or start information. Asynchronous communication refers to a mode of communication in which data are transmitted as individual blocks framed by start and stop bits. Bits are also added to the data stream for error detection. Integrated circuit devices known as universal asynchronous receiver transmitters (UARTS) are available for converting parallel data into a serial format suitable for transmission over a twisted pair or coaxial line and for reception of the data in serial format and reconversion to parallel format with parity-bit checking. The schematic diagram for such a device is shown in Figure 40.30.

Because of the high capacitance of twisted-pair and coaxial cables the length of line over which standard 74 series TTL can transmit digital signals is limited typically to a length of

CHAPTER | 40 Telemetry

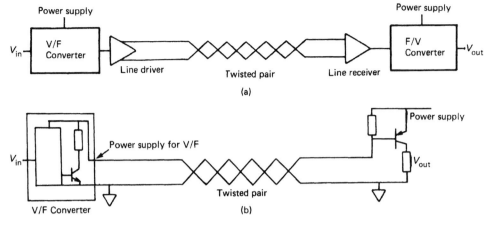

FIGURE 40.29 (a) Frequency-transmission system; (b) two-wire frequency-transmission system.

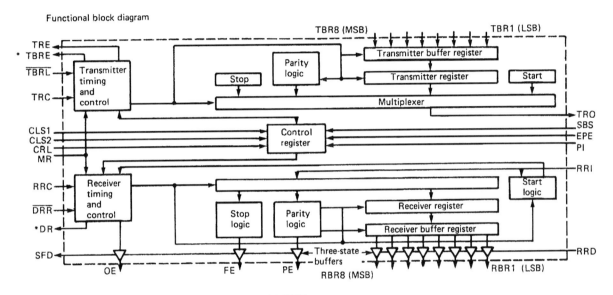

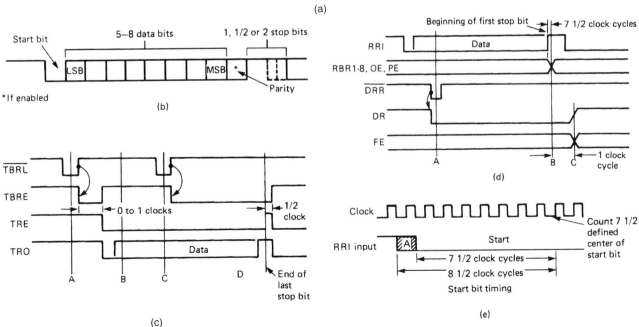

FIGURE 40.30 (a) Universal asynchronous receiver transmitter (UART). (b) Serial data format. (c) Transmitter timing (not to scale). (d) Receiver timing (not to scale). (e) Start bit timing.

3 m at 2 Mbit/s. This can be increased to 15 m by the use of open-collector TTL driving a low-impedance terminated line.

In order to drive digital signals over long lines special-purpose line driver and receiver circuits are available. Integrated circuit Driver/Receiver combinations such as the Texas Instruments SN75150/SN75152, SN75158/SN75157, and SN75156/SN75157 devices meet the internationally agreed EIA Standards RS-232C, RS-422A, and RS-423A, respectively (see Section 40.9.2).

40.9.1 Modems

In order to overcome the limitations of the public telephone lines digital data are transmitted down these lines by means of a modem. The two methods of modulation used by modems are frequency-shift keying (FSK) and phase-shift keying (PSK). Amplitude-modulation techniques are not used because of the unsuitable response of the line to step changes in amplitude. Modems can be used to transmit information in two directions along a telephone line.

Full-duplex operation is transmission of information in both directions simultaneously; half-duplex is the transmission of information in both directions but only in one direction at any one time; and simplex is the transmission of data in one direction only.

The principle of FSK is shown in Figure 40.31. FSK uses two different frequencies to represent a 1 and a 0, and this can be used for data transmission rates up to 1200 bits/s. The receiver uses a frequency discriminator whose threshold is set midway between the two frequencies. The recommended frequency shift is not less than 0.66 of the modulating frequency. Thus a modem operating at 1200 bits/s has a recommended central frequency of 1700 Hz and a frequency deviation of 800 Hz, with a 0 represented by a frequency of 1300 Hz and a 1 by a frequency of 2100 Hz. At a transmission rate of 200 bits/s it is possible to operate a full-duplex system. At 600 and 1200 bits/s half-duplex operation is used incorporating a slow-speed backward channel for supervisory control or low-speed return data.

At bit rates above 2400 bits/s the bandwidth and group delay characteristics of telephone lines make it impossible to transmit the data using FSK. It is necessary for each signal to contain more than one bit of information. This is achieved by a process known as phase shift keying (PSK), in which the phase of a constant-amplitude carrier is changed. Figure 40.32(a) shows the principle of PSK and

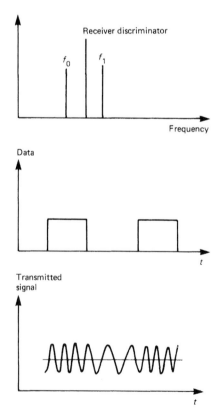

FIGURE 40.31 Frequency-shift keying.

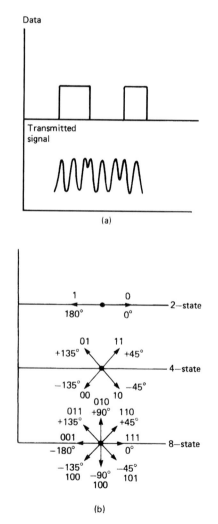

FIGURE 40.32 (a) Principle of phase-shift keying; (b) two-, four-, and eight-state shift keying.

Figure 40.32(b) shows how the information content of PSK can be increased by employing two-, four-, and eight-state systems. It should now be seen that the number of signal elements/s (which is referred to as the Baud rate) has to be multiplied by the number of states to obtain the data transmission rate in bits/s. Thus an eight-state PSK operating at a rate of 1200 baud can transmit 9600 bits/s. For years, 9600 bps was the fastest transmission over telephone cables using leased lines, i.e., lines which are permanently allocated to the user as opposed to switched lines. At the higher data transmission rates it is necessary to apply adaptive equalization of the line to ensure correct operation and also to have in-built error-correcting coding. Details of various schemes for modem operation are to be found in Coates (1982) and Blackwell (1981). The International Telephone and Telegraph Consultative Committee (CCITT) has made recommendations for the mode of operation of modems operating at different rates over telephone lines. These are set out in recommendations V21, V23, V26, V27, and V29. These are listed in the References. Since the publication of the second edition of this book, much improvement in telephone modems has been made. Typically, today's modems utilize the V90 protocol and operate at maximum speeds of up to 56 Kbps.

Figure 40.33 shows the elements and operation of a typical modem. The data set ready (DSR) signal indicates to the equipment attached to the modem that it is ready to transmit data. When the equipment is ready to send the data it sends a request to send (RTS) signal. The modem then starts transmitting down the line. The first part of the transmission is to synchronize the receiving modem. Having given sufficient time for the receiver to synchronize, the transmitting modem sends a clear to send (CTS) signal to the equipment, and the data are then sent. At the receiver the detection of the transmitted signal sends the data carrier detected (DCD) line high, and the signal transmitted is demodulated.

40.9.2 Data Transmission and Interfacing Standards

To ease the problem of equipment interconnection various standards have been introduced for serial and parallel data transmission. For serial data transmission between data terminal equipment (DTE), such as a computer or a piece of peripheral equipment, and data communication equipment (DCE), such as modem, the standards which are currently being used are the RS-232C standard produced in the USA by the Electronic Industries Association (EIA) in 1969 and their more recent RS-449 standard with its associated RS-422 and RS-423 standards.

The RS-232C standard defines an electro-mechanical interface by the designation of the pins of a 25-pin plug and socket which are used for providing electrical ground, data interchange, control, and clock or timing signals between the

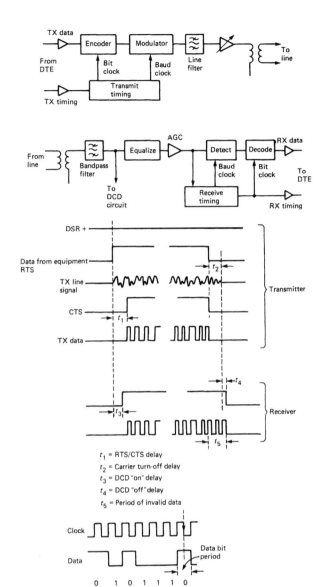

FIGURE 40.33 Modem operation.

two pieces of equipment. The standard also defines the signal levels, conditions, and polarity at each interface connection. Table 40.1 gives the pin assignments for the interface, and it can be seen that only pins 2 and 3 are used for data transmission. Logical 1 for the driver is an output voltage between −5 and −15 V, with logical zero being between +5 and +15 V. The receiver detects logical 1 for input voltages <-3 V and logical 0 for input voltages >3 V, thus giving the system a minimum 2 V noise margin. The maximum transmission rate of data is 20,000 bits/s, and the maximum length of the interconnecting cable is limited by the requirement that the receiver should not have more than 2500 pF across it. The length of cable permitted thus depends on its capacitance/unit length.

The newer RS-449 interface standard which is used for higher data-transmission rates defines the mechanical characteristics in terms of the pin designations of a 37-pin interface. These are listed in Table 40.2. The electrical

TABLE 40.1 Pin assignments for RS-232

Pin number	Signal nomenclature	Signal abbreviation	Signal description	Category
1	AA	–	Protective ground	Ground
2	BA	TXD	Transmitted data	Data
3	BB	RXD	Received data	Data
4	CA	RTS	Request to send	Control
5	CB	CTS	Clear to send	Control
6	CC	DSR	Data set ready	Ground
7	AB	–	Signal ground	
8	CF	DCD	Received line signal detector	Control
9	–	–	–	Reserved for test
10	–	–	–	Reserved for test
11	–	–	–	Unassigned
12	SCF	–	Secondary received line signal detector	Control
13	SCB	–	Secondary clear to send	Control
14	SBA	–	Secondary transmitted data	Data
15	DB	–	Transmission signal element timing	Timing
16	SBB	–	Secondary received data	Data
17	DD	–	Received signal element timing	Timing
18	–	–	–	Unassigned
19	SCA	–	Secondary request to send	Control
20	CD	DTR	Data terminal ready	Control
21	CG	–	Signal quality detector	Control
22	CE	–	Ring indicator	Control
23	CH/CI	–	Data signal rate selector	Control
24	DA	–	Transmit signal element timing	Timing
25	–	–	–	Unassigned

characteristics of the interface are specified by the two other associated standards, RS-422, which refers to communication by means of a balanced differential driver along a balanced interconnecting cable with detection being by means of a differential receiver, and RS-423, which refers to communication by means of a single-ended driver on an unbalanced cable with detection by means of a differential receiver. These two systems are shown in Figure 40.34.

The maximum recommended cable lengths for the balanced RS-422 standard are 4000 ft at 90 kbits/s, 380 ft at 1 Mbits/s and 40 ft at 10 Mbits/s. For the unbalanced RS-423 standard the limits are 4000 ft at 900 bits/s, 380 ft at 10 kbits/s and 40 ft at 100 kbits/s. In addition, the RS-485 standard has been developed to permit communication between multiple addressable devices in a ring format, with up to 33 addresses per loop.

For further details of these interface standards the reader is directed to the standards produced by the EIA and IEEE. These are listed in the References. Interfaces are also discussed in Part 5.

The IEEE-488 Bus (IEEE 1978), often referred to as the HPIB Bus (the Hewlett-Packard Interface Bus or the GPIP, General Purpose Interface Bus), is a standard which specifies a communications protocol between a controller and instruments connected onto the bus. The instruments typically connected on to the bus include digital voltmeters, signal generators, frequency meters, and spectrum and impedance analyzers. The bus allows up to 15 such instruments to be connected onto the bus. Devices talk, listen, or do both, and at least one device on the bus must provide control, this usually being a computer. The bus uses 15 lines, the pin connections for which are shown in Table 40.3. The signal

TABLE 40.2 Pin assignments for RS-449

Circuit mnemonic	Circuit name	Circuit direction	Circuit type
SG	Signal ground	–	Common
SC	Send common	To DCE	
RC	Received common	From DCE	
IS	Terminal in service	To DCE	Control
IC	Incoming call	From DCE	
TR	Terminal ready	To DCE	
DM	Data mode	From DCE	
SD	Send data	To DCE	Primary channel data
RD	Receive data	From DCE	
TT	Terminal timing	To DCE	Primary channel timing
ST	Send timing	From DCE	
RT	Receive timing	From DCE	
RS	Request to send	To DCE	Primary channel control
CS	Clear to send	From DCE	
RR	Receiver ready	From DCE	
SQ	Signal quality	From DCE	
NS	New signal	To DCE	
SF	Select frequency	To DCE	
SR	Signal rate selector	To DCE	
SI	Signal rate indicator	From DCE	
SSD	Secondary send data	To DCE	Secondary channel data
SRD	Secondary receive data	From DCE	
SRS	Secondary request to send	To DCE	Secondary
SCS	Secondary clear to send	From DCE	channel
SRR	Secondary receiver ready	From DCE	control
LL	Local loopback	To DCE	Control
RL	Remote loopback	To DCE	
TM	Test mode	From DCE	
SS	Select standby	To DCE	Control
SB	Standby indicator	From DCE	

levels are TTL and the cable length between the controller and the device is limited to 2 m. The bus can be operated at a frequency of up to 1 MHz. The connection diagram for a typical system is shown in Figure 40.35. Eight lines are used for addresses, program data, and measurement data transfers, three lines are used for the control of data transfers by means of a handshake technique, and five lines are used for general interface management.

CAMAC (which is an acronym for Computer Automated Measurement and Control) is a multiplexed interface system which not only specifies the connections and the communications protocol between the modules of the system which act as interfaces between the computer system and peripheral

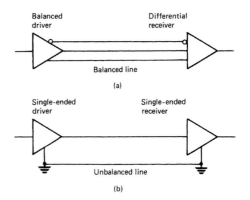

FIGURE 40.34 RS-422 and RS-423 driver/receiver systems.

TABLE 40.3 Pin assignment for IEEE-488 interface

Pin no.	Function	Pin no.	Function
1	DIO 1	13	DIO 5
2	DIO 2	14	DIO 6
3	DIO 3	15	DIO 7
4	DIO 4	16	DIO 8
5	EOI	17	REN
6	DAV	18	GND twisted pair with 6
7	NRFD	19	GND twisted pair with 7
8	NDAC	20	GND twisted pair with 8
9	IFC	21	GND twisted pair with 9
10	SRQ	22	GND twisted pair with 10
11	ATN	23	GND twisted pair with 11
12	Shield (to earth)	24	
	Signal ground		
	DIO = Data Input-Output		
	EOI = End Or Identify		
	REN = Remote Enable		
	DAV = Data Valid		
	NRFD = Not Ready For Data		
	NDAC = Not Data Accepted		
	IFC = Interface Clear		
	SRQ = Service Request		
	ATN = Attention		
	GND = Ground		

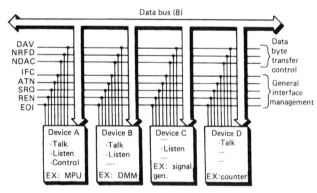

FIGURE 40.35 IEEE-488 bus system.

devices but also stipulates the physical dimensions of the plug-in modules. These modules are typically ADCs, DACs, digital buffers, serial to parallel converters, parallel to serial converters, and level changers. The CAMAC system offers a 24-bit parallel data highway via an 86-way socket at the rear of each module. Twenty-three of the modules are housed in a single unit known as a "crate," which additionally houses a controller. The CAMAC system was originally specified for the nuclear industry, and is particularly suited for systems where a large multiplexing ratio is required. Since each module addressed can have up to 16 subaddresses a crate can have up to 368 multiplexed inputs/outputs. For further details of the CAMAC system the reader is directed to Barnes (1981) and to the CAMAC standards issued by the Commission of the European Communities, given in the References.

The S100 Bus (also referred to as the IEEE-696 interface (IEEE 1981)) is an interface standard devised for bus-oriented systems and was originally designed for interfacing microcomputer systems. Details of this bus can be found in the References.

Two more bus systems are in common usage: USB or Universal Serial Bus, and IEEE 1394 "FireWire." These are high-speed serial buses with addressable nodes and the ability to pass high bandwidth data. USB is becoming the preferred bus for microcomputer peripherals, while Fire-Wire is becoming preferred for high bandwidth data communications over short distances, such as testbed monitoring.

In addition to telephone modems, new methods of information transfer that have reached widespread use since 1990 include ISDN, the digital subscriber line ADSL (asynchronous) and SDSL (synchronous), and high-bandwidth cable modems.

REFERENCES

Barnes, R. C. M., "A standard interface: CAMAC," in *Minicomputers: A Handbook for Engineers, Scientists, and Managers* (ed. Y. Parker), Abacus, London (1981), pp. 167–187.

Bentley, J., *Principles of Measurement Systems*, Longman, London (1983).

Bowdell, K., "Interface data transmission," in *Microcomputers: A Handbook for Engineers, Scientists, and Managers* (ed. Y. Parker), Abacus, London (1981), pp. 148–166.

Blackwell, J., "Long distance communication," in *Minicomputers: A Handbook for Engineers, Scientists, and Managers* (ed. Y. Parker), Abacus, London (1981), pp. 301–316.

Cattermole, K. W., *Principles of Pulse Code Modulation*, Iliffe, London (1969).

CCITT, Recommendation V 24. List of definitions for interchange circuits between data-terminal equipment and data circuit-terminating equipment, in CCITT, Vol. 8.1, *Data Transmission over the Telephone Network*, International Telecommunication Union, Geneva (1977).

Coates, R. F. W., *Modern Communication Systems*, 2d ed., Macmillan, London (1982).

EEC Commission: CAMAC, *A Modular System for Data Handling. Revised Description and Specification*, EUR 4100e, HMSO, London (1972).

EEC Commission: CAMAC, *Organisation of Multi-Crate Systems. Specification of the Branch Highway and CAMAC Crate Controller Type A*, EUR 4600e, HMSO, London (1972).

EEC Commission: CAMAC, *A Modular Instrumentation System for Data Handling. Specification of Amplitude Analog Signals*, EUR 5100e, HMSO, London (1972).

EIA, *Standard RS-232C Interface between Data Terminal Equipment and Data Communications Equipment Employing Serial Binary Data Interchange*, EIA, Washington, D.C. (1969).

EIA, *Standard RS-449 General-purpose 37-position and 9-position Interface for Data Terminal Equipment and Data Circuit-terminating Equipment Employing Serial Binary Data Interchange*, EIA, Washington, D.C. (1977).

Hartley, G., P. Mornet, F. Ralph, and D. J. Tarron, *Techniques of Pulse Code Modulation in Communications Networks*, Cambridge University Press, Cambridge (1967).

HMSO, *Private Point-to-point Systems Performance Specifications (Nos W. 6457 and W. 6458) for Angle-Modulated UHF Transmitters and Receivers and Systems in the 450–470 Mcls Band*, HMSO, London (1963).

HMSO, *Performance Specification: Medical and Biological Telemetry Devices*, HMSO, London (1978).

HMSO, *Performance Specification: Transmitters and Receivers for Use in the Bands Allowed to Low Power Telemetry in the PMR Service*, HMSO, London (1979).

HMSO, International Telecommunication Union World Administrative Radio Conference, 1979, *Radio Regulations. Revised International Table of Frequency Allocations and Associated Terms and Definitions*, HMSO, London (1980).

IEEE, *IEEE-488–1978 Standard Interface for Programmable Instruments*, IEEE, New York (1978).

IEEE, *IEEE-696–1981 Standard Specification for S-100 Bus Interfacing Devices*, IEEE, New York (1981).

Johnson, C. S., "Telemetry data systems," *Instrument Technology*, 39–53, Aug. (1976); 47–53, Oct. (1976).

Keiser, G., *Optical Fiber Communication*, McGraw-Hill International, London (1983).

Senior, J., *Optical Fiber Communications, Principles and Practice*, Prentice-Hall, London (1985).

Shanmugan, S., *Digital and Analog Communications Systems*, John Wiley, New York (1979).

Steele, R., *Delta Modulation Systems*, Pentech Press, London (1975).

Warnock, J. D., Section 16.27 in *The Process Instruments and Controls Handbook* 3rd ed., (ed. by D. M. Considine), McGraw-Hill, London (1985).

Wilson, J., and J. F. B. Hawkes, *Optoelectronics: An Introduction*, Prentice-Hall, London (1983).

FURTHER READING

Strock, O. J., *Telemetry Computer Systems: An Introduction*, Prentice-Hall, London (1984).

Chapter 41

Display and Recording

M. L. Sanderson

41.1 INTRODUCTION

Display devices are used in instrumentation systems to provide instantaneous but non-permanent communication of information between a process or system and a human observer. The data can be presented to the observer in either an analog or a digital form. Analog indicators require the observer to interpolate the reading when it occurs between two scale values, which requires some skill on the part of the observer. They are, however, particularly useful for providing an overview of the process or system and an indication of trends when an assessment of data from a large number of sources has to be quickly assimilated. Data displayed in a digital form require little skill from the observer in reading the value of the measured quantity, though any misreading can introduce a large observational error as easily as a small one.

Using digital displays it is much more difficult to observe trends within a process or system and to quickly assess, for example, the deviation of the process or system from its normal operating conditions. Hybrid displays incorporating both analog and digital displays combine the advantages of both.

The simplest indicating devices employ a pointer moving over a fixed scale; a moving scale passing a fixed pointer; or a bar graph in which a semi-transparent ribbon moves over a scale. These devices use mechanical or electromechanical means to effect the motion of the moving element. Displays can also be provided using illuminative devices such as light-emitting diodes (LEDs), liquid crystal displays (LCDs), plasma displays, and cathode ray tubes (CRTs). The mechanisms and configurations of these various display techniques are identified in Table 41.1.

Recording enables hard copy of the information to be obtained in graphical or alphanumeric form or the information to be stored in a format which enables it to be retrieved at a later stage for subsequent analysis, display, or conversion into hard copy. Hard copy of graphical information is made by graphical recorders; x–t recorders enable the relationships between one or more variables and time to be obtained while x–y recorders enable the relationship between two variables to be obtained. These recorders employ analog or digital drive mechanisms for the writing heads, generally with some form of feedback. The hard copy is provided using a variety of techniques, including ink pens or impact printing on normal paper, or thermal, optical, or electrical writing techniques on specially prepared paper. Alphanumeric recording of data is provided by a range of printers, including impact printing with cylinder, golf ball, daisywheel, or dot matrix heads; or non-impact printing techniques including ink-jet, thermal, electrical, electrostatic, electromagnetic, or laser printers.

Recording for later retrieval can use either magnetic or semiconductor data storage. Magnetic tape recorders are used for the storage of both analog and digital data. Transient/waveform recorders (also called waveform digitizers) generally store their information in semiconductor memory. Data-logger systems may employ both semiconductor memory and magnetic storage on either disc or tape. Table 41.2 shows the techniques and configurations of the commonly used recording systems.

The display and recording of information in an instrumentation system provides the human/machine interface (HMI) between the observer and the system or process being monitored. It is of fundamental importance that the information should be presented to the observer in as clear and unambiguous a way as possible and in a form that is easily assimilated. In addition to the standard criteria for instrument performance such as accuracy, sensitivity, and speed of response, ergonomic factors involving the visibility, legibility, and organization and presentation of the information are also of importance in displays and recorders.

41.2 INDICATING DEVICES

In moving-pointer indicator devices (Figure 41.1) the pointer moves over a fixed scale which is mounted vertically or horizontally. The scale may be either straight or an

TABLE 41.1 Commonly used display techniques

Display technique	Mechanism	Configurations
Indicating devices		
Moving pointer	Mechanical/electromechanical movement of pointer over a fixed scale	Horizontal/vertical, straight, arc circular, or segment scales with edgewise strip, hairline, or arrow-shaped pointers
Moving scale	Mechanical/electromechanical movement of scale; indication given by position of scale with respect to fixed pointer	Moving dial or moving drum analog indicators, digital drum indicators
Bar graph	Indication given by height or length of vertical or horizontal column	Moving column provided by mechanically driven ribbon or LED or LCD elements
Illuminative displays		
Light emitting diodes	Light output provided by recombination electroluminescence in a forward-biased semiconductor diode	Red, yellow, green displays configured as lamps, bar graphs, 7- and 16-segment alphanumeric displays, dot matrix displays
Liquid crystal displays	The modulation of intensity of transmitted-reflected light by the application of an electric field to a liquid crystal cell	Reflective or transmissive displays, bar graph, 7-segment, dot matrix displays, alphanumeric panels
Plasma displays	Cathode glow of a neon gas discharge	Nixie tubes, 7-segment displays, plasma panels
CRT displays	Conversion into light of the energy of scanning electron beam by phosphor	Monochrome and color tubes, storage tubes, configured as analog, storage, sampling, or digitizing oscilloscopes, VDUs, graphic displays

arc. The motion is created by mechanical means, as in pressure gauges, or by electromechanical means using a moving coil movement. (For details of such movements the reader is directed to Chapter 20 in Part 3 of this book.) The pointer consists of a knife edge, a line scribed on each side of a transparent member, or a sharp, arrow-shaped tip. The pointers are designed to minimize the reading error when the instrument is read from different angles. For precision work such "parallax errors" are reduced by mounting a mirror behind the pointer. Additional pointers may be provided on the scale. These can be used to indicate the value of a set point or alarm limits.

The pointer scales on such instruments should be designed for maximum clarity. BS 3693 sets out recommended scale formats. The standard recommends that the scale base length of an analog indicating device should be at least $0.07 D$, where D is the viewing distance. At a distance of 0.7 m, which is the distance at which the eye is in its resting state of accommodation, the minimum scale length should be 49 mm. The reading of analog indicating devices requires the observer to interpolate between readings. It has been demonstrated that observers can subdivide the distance between two scale markings into five, and therefore a scale which is to be read to within 1 percent of full-scale deflection (FSD) should be provided with twenty principal divisions. For electromechanical indicating instruments accuracy is classified by BS 89 into nine ranges, from ±0.05 percent to ±5 percent of FSD. A fast response is not required for visual displays since the human eye cannot follow changes much in excess of 20 Hz. Indicating devices typically provide frequency responses up to 1–2 Hz.

Moving-scale indicators in which the scale moves past a fixed pointer can provide indicators with long scale lengths. Examples of these are also shown in Figure 41.1.

In the bar graph indicator (Figure 41.2) a semitransparent ribbon moves over a scale. The top of the ribbon indicates the value of the measured quantity and the ribbon is generally driven by a mechanical lead. Arrays of LEDs or LCDs can be used to provide the solid state equivalent of the bar graph display.

41.3 LIGHT-EMITTING DIODES (LEDs)

These devices, as shown in Figures 41.3(a) and 41.3(b), use recombination (injection) electroluminescence and consist of a forward-biased p–n junction in which the majority carriers

TABLE 41.2 Commonly used recording techniques

Recording system	Technique	Configurations
Graphical recorders	Provide hard copy of data in graphical form using a variety of writing techniques, including pen-ink, impact printing, thermal, optical, and electric writing	Single/multichannel x–t strip chart and circular chart recorders, galvanometer recorders, analog and digital x–y recorders, digital plotters
Printers	Provide hard copy of data in alphanumeric form using impact and non-impact printing techniques	Serial impact printers using cylinder, golf ball, daisywheel, or dot matrix heads. Line printers using drum, chain-belt, oscillating bar, comb, and needle printing heads. Non-impact printers using thermal, electrical, electrostatic, magnetic, ink-jet, electrophotographic, and laser printing techniques
Magnetic recording	Use the magnetization of magnetic particles on a substrate to store information	Magnetic tape recorders using direct, frequency modulation, or pulse code modulation technique for the storage of analog or digital data. Spool to spool or cassette recorders. Floppy or hard discs for the storage of digital data
Transient recorders	Use semiconductor memory to store high-speed transient waveforms	Single/multichannel devices using analog-to-digital conversion techniques. High-speed transient recorders using optical scanning techniques to capture data before transfer to semiconductor memory
Data loggers	Data-acquisition system having functions programmed from the front panel	Configured for a range of analog or digital inputs with limited logical or mathematical functions. Internal display using LED, LCD, CRT. Hard copy provided by dot matrix or thermal or electrical writing technique. Data storage provided by semiconductor or magnetic storage using tape or disc

from both sides of the junction cross the internal potential barrier and enter the material on the other side, where they become minority carriers, thus disturbing the local minority carrier population. As the excess minority carriers diffuse away from the junction, recombination occurs. Electroluminescence takes place if the recombination results in radiation. The wavelength of the emitted radiation is inversely proportional to the band gap of the material, and therefore for the radiation to be in the visible region the band gap must be greater than 1.8 eV.

$GaAs_{0.6}P_{0.4}$ (gallium arsenide phosfide) emits red light at 650 nm. By increasing the proportion of phosphorus and doping with nitrogen the wavelength of the emitted light is reduced. $GaAs_{0.15}P_{0.65}N$ provides a source of orange light, whilst $GaAs_{0.15}P_{0.85}N$ emits yellow light at 589 nm. Gallium phosfide doped with nitrogen radiates green light at 570 nm.

Although the internal quantum efficiencies of LED materials can be high, the external quantum efficiencies may be much lower. This is because the materials have a high refractive index and therefore a significant proportion of the emitted radiation strikes the material/air interface beyond the critical angle and is totally internally reflected. This is usually overcome by encapsulating the diode in a hemispherical dome made of epoxy resin, as shown in Figure 41.3(c). The construction of an LED element in a seven-segment display is shown in Figure 41.3(d).

The external quantum efficiencies of green diodes tend to be somewhat lower than those of red diodes, but, for the same output power, because of the sensitivity of the human eye the green diode has a higher luminous intensity.

Typical currents required for LED elements are in the range of 10–100 mA. The forward diode drop is in the range

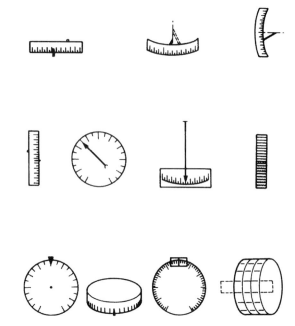

FIGURE 41.1 Moving-pointer and moving-scale indicators.

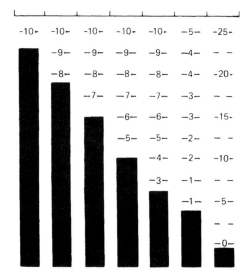

FIGURE 41.2 Bar-graph indicator.

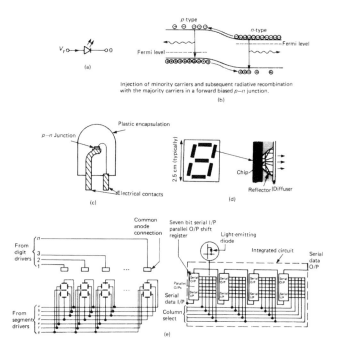

FIGURE 41.3 Light-emitting diodes.

of 1.6–2.2 V, dependent on the particular device. The output luminous intensities of LEDs range from a few to over a hundred millicandela, and their viewing angle can be up to ±60°C. The life expectancy of a LED display is twenty years, over which time it is expected that there will be a 50 percent reduction in output power.

LEDs are available as lamps, seven- and sixteen-segment alphanumeric displays, and in dot matrix format (Figure 41.3(e)). Alphanumeric displays are provided with on-board decoding logic which enables the data to be entered in the form of ASCII or hexadecimal code.

41.4 LIQUID CRYSTAL DISPLAYS (LCDs)

These displays are passive and therefore emit no radiation of their own but depend upon the modulation of reflected or transmitted light. They are based upon the optical properties of a large class of organic materials known as liquid crystals. Liquid crystals have molecules which are rod shaped and which, even in their liquid state, can take up certain defined orientations relative to each other and also with respect to a solid interface. LCDs commonly use nematic liquid crystals such as p-azoxyanisole, in which the molecules are arranged with their long axes approximately parallel, as shown in Figure 41.4(a). They are highly anisotropic—that, is, they have different optical or other properties in different directions. At a solid-liquid interface the ordering of the crystal can be either homogeneous (in which the molecules are parallel to the interface) or homeotropic (in which the molecules are aligned normal to the interface), as shown in Figure 41.4(b). If a liquid crystal is confined between two plates which without the application of an electric field is a homogeneous state, then when an electric field is applied the molecules will align themselves with this field in order to minimize their energy. As shown in Figure 41.4(c), if the E field is less than some critical value E_c, then the ordering is not affected. If $E > E_c$ then the molecules farthest away from the interfaces are realigned. For values of the electric field $E \gg E_c$ most of the molecules are realigned.

A typical reflective LCD consists of a twisted nematic cell in which the walls of the cell are such as to produce a homogeneous ordering of the molecules but rotated by 90° (Figure 41.4(d)). Because of the birefringent nature of the crystal, light polarized on entry in the direction of alignment

CHAPTER 41 Display and Recording

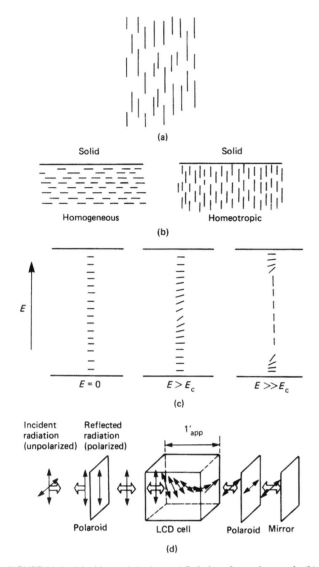

FIGURE 41.4 Liquid crystal displays. (a) Ordering of nematic crystals; (b) ordering at liquid crystal/solid interface; (c) application of an electric field to a liquid crystal cell; (d) reflective LCD using a twisted nematic cell.

employed with the cell having a response corresponding to the rms value of the applied voltage. The frequency of the ac is in the range 25 Hz to 1 kHz. Power consumption of LCD displays is low, with a typical current of 0.3–0.5 µA at 5 V. The optical switching time is typically 100–150 ms. They can operate over a temperature range of $-10°C$ to $+70°C$.

Polarizing devices limit the maximum light which can be reflected from the cell. The viewing angle is also limited to approximately $\pm 45°$. This can be improved by the use of cholesteric liquid crystals used in conjunction with dichroic dyes which absorb light whose direction of polarization is parallel to its long axis. Such displays do not require polarizing filters and are claimed to provide a viewing angle of $\pm 90°$.

LCDs are available in bar graph, seven-segment, dot matrix, and alphanumeric display panel configurations. The availability of inexpensive color LCD panels for laptop computers has made possible a whole new class of recording instruments, in which the LCD replaces the pointer-and-paper chart recorder as a virtual instrument.

41.5 PLASMA DISPLAYS

In a plasma display as shown in Figure 41.5(a) a glass envelope filled with neon gas, with added argon, krypton, and mercury to improve the discharge properties of the display, is provided with an anode and a cathode. The gas is ionized by applying a voltage of approximately 180 V between the anode and cathode. When ionization occurs there is an orange yellow glow in the region of the cathode. Once ionization has taken place the voltage required to sustain the ionization can be reduced to about 100 V.

The original plasma display was the Nixie tube, which had a separate cathode for each of the figures 0–9. Plasma displays are now available as seven-segment displays or plasma panels as shown in Figures 41.5(a) and 41.5(b). The seven-segment display used a separate cathode for each of the segments. The plasma panel consists, for example, of 512×512 vertical and horizontal x and y electrodes. By applying voltages to specific x and y electrode pairs (such that the sum of the voltages at the spot specified by the (x, y) address exceeds the ionization potential) then a glow will be produced at that spot. The display is operated by applying continuous ac voltages equivalent to the sustaining potential to all the electrodes. Ionization is achieved by pulsing selected pairs. The glow at a particular location is quenched by applying antiphase voltages to the electrodes.

The display produced by plasma discharge has a wide viewing angle capability, does not need back lighting, and is flicker free. A typical 50 mm seven-segment display has a power consumption of approximately 2 W.

of the molecules at the first interface will leave the cell with its direction of polarization rotated by 90°. On the application of an electric field in excess of the critical field (for a cell of thickness 10 µm a typical critical voltage is 3 V) the molecules align themselves in the direction of the applied field. The polarized light does not then undergo any rotation. If the cell is sandwiched between two pieces of polaroid with no applied field, both polarizers allow light to pass through them, and therefore incident light is reflected by the mirror and the display appears bright. With the application of the voltage the polarizers are now crossed, and therefore no light is reflected and the display appears dark. Contrast ratios of 150:1 can be obtained.

DC voltages are generally not used in LCDs because of the electromechanical reactions. AC waveforms are

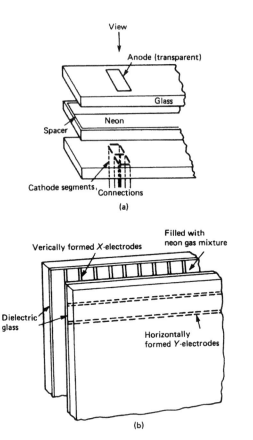

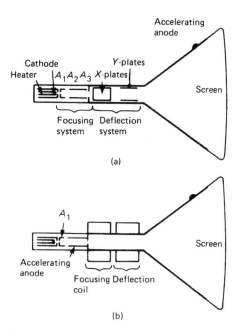

FIGURE 41.6 Cathode ray tube. (a) Electrostatic focusing and deflection; (b) electromagnetic focusing and deflection.

FIGURE 41.5 Plasma display. (a) Seven-segment display; (b) plasma panel.

41.6 CATHODE RAY TUBES (CRTs)

CRTs are used in oscilloscopes which are commonly employed for the display of repetitive or transient waveforms. They also form the basis of visual display units (VDUs) and graphic display units. The display is provided by using a phosphor which converts the energy from an electron beam into light at the point at which the beam impacts on the phosphor.

A CRT consists of an evacuated glass envelope in which the air pressure is less than 10^{-4} Pascal (Figure 41.6). The thermionic cathode of nickel coated with oxides of barium, strontium, and calcium is indirectly heated to approximately 1100K and thus gives off electrons. The number of electrons which strike the screen and hence control the brightness of the display is adjusted by means of the potential applied to a control grid surrounding the cathode. The control grid which has a pin hole through which the electrons can pass is held at a potential of between 0 and 100 V negative with respect to the cathode. The beam of electrons pass through the first anode A_1 which is typically held at a potential of 300 V positive with respect to the cathode before being focused, accelerated, and deflected.

Focusing and deflection of the beam can be by either electrostatic or magnetic means. In the electrostatic system shown in Figure 41.6(a) the cylindrical focusing anode A_2, which consists of disc baffles having an aperture in the center of them, is between the first anode A_1 and the accelerating anode A_3, which is typically at a potential of 3–4 kV with respect to the cathode. Adjusting the potential on A_2 with respect to the potentials on A_1 and A_3 focuses the beam such that the electrons then travel along the axis of the tube. In a magnetic focusing system magnetic field coils around the tube create a force on the electrons, causing them to spiral about the axis and also inwardly. By employing magnetic focusing it is possible to achieve a smaller spot size than with electrostatic focusing. Deflection of the electron beam in a horizontal and vertical direction moves the position of the illuminated spot on the screen. Magnetic deflection provides greater deflection capability and is therefore used in CRTs for television, alphanumeric, and graphical displays. It is slower than electrostatic deflection, which is the deflection system commonly used in oscilloscopes.

Acceleration of the beam is by either the use of the accelerating electrode A_3 (such tubes are referred to as monoaccelerator tubes) or by applying a high potential (10–14 kV) on to a post-deflection anode situated close to the CRT screen. This technique, which is known as post-deflection acceleration (PDA), gives rise to tubes with higher light output and increased deflection sensitivity.

The phosphor coats the front screen of the CRT. A range of phosphors are available, the choice of which for a particular situation depends on the color and efficiency of the luminescence required and its persistence time, that is, the time for which the afterglow continues after the electron beam has been removed. Table 41.3 provides the characteristics of some commonly used phosphors.

TABLE 41.3 Characteristics of commonly used phosphors (Courtesy of Tektronix)

Phosphor	Fluorescence	Relative[a] luminance (%)	Relative[b] photographic writing speed (%)	Decay	Relative burn resistance	Comments
PI	Yellow–green	50	20	Medium	Medium	In most applications replaced by P31
P4	White	50	40	Medium/short	Medium/high	Television displays
P7	Blue	35	75	Long	Medium	Long-decay, double-layer screen
P11	Blue	15	100	Medium/short	Medium	For photographic applications
P31	Green	100	50	Medium/short	High	General purposes, brightest available phosphor

[a] Measured with a photometer and luminance probe incorporating a standard eye filter. Representative of 10 kV aluminized screens with P31 phosphor as reference.
[b] P11 as reference with Polaroid 612 or 106 film. Representative of 10 kV aluminized screens.

41.6.1 Color Displays

Color displays are provided using a screen that employs groups of three phosphor dots. The material of the phosphor for each of the three dots is chosen such that it emits one of the primary colors (red, green, blue). The tube is provided with a shadow mask consisting of a metal screen with holes in it placed near the screen and three electron guns, as shown in Figure 41.7(a). The guns are inclined to each other so that the beams coincide at the plane of the shadow mask. After passing through the shadow mask the beams diverge and, on hitting the screen, energize only one of the phosphors at that particular location. The effect of a range of colors is achieved by adjusting the relative intensities of the three primary colors by adjusting the electron beam currents. The resolution of color displays is generally lower than that of monochrome displays because the technique requires three phosphors to produce the effect of color. Alignment of the shadow mask with the phosphor screen is also critical, and the tube is sensitive to interfering magnetic fields.

An alternative method of providing a color display is to use penetration phosphors which make use of the effect that the depth of penetration of the electron beam is dependent on beam energy. By using two phosphors, one of which has a non-luminescent coating, it is possible by adjusting the energy of the electron beam to change the color of the display. For static displays having fixed colors this technique provides good resolution capability and is reasonably insensitive to interfering magnetic fields.

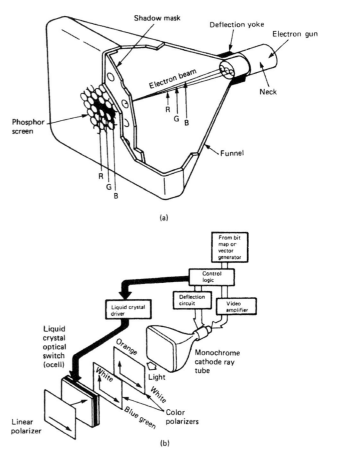

FIGURE 41.7 Color displays. (a) Shadow mask color tube; (b) liquid crystal color display.

Liquid crystal color displays employ a single phosphor having two separate emission peaks (Figure 41.7(b)). One is orange and the other blue-green. The color polarizers orthogonally polarize the orange and blue-green components of the CRT's emission, and the liquid crystal cell rotates the polarized orange and blue-green information into the transmission axis of the linear polarizer and thus selects the color of the display. Rotation of the orange and blue-green information is performed in synchronism with the information displayed by the sequentially addressed CRT. Alternate displays of information viewed through different colored polarizing filters are integrated by the human eye to give color images. This sequential technique can be used to provide all mixtures of the two primary colors contained in the phosphor.

41.6.2 Oscilloscopes

Oscilloscopes can be broadly classified as analog, storage, sampling, or digitizing devices. Figure 41.8 shows the elements of a typical analog oscilloscope. The signals to be observed as a function of time are applied to the vertical (Y) plates of the oscilloscope. The input stage of the vertical system matches the voltage levels of these signals to the drive requirements of the deflection plate of the oscilloscope which will have a typical deflection sensitivity of 20 V/cm. The coupling of the input stage can be either dc or ac.

The important specifications for the vertical system include bandwidth and sensitivity. The bandwidth is generally specified as the highest frequency which can be displayed with less than 3 dB loss in amplitude compared with its value at low frequencies. The rise time, T_r, of an oscilloscope to a step input is related to its bandwidth, B, by $T_r = 0.35/B$. In order to measure the rise time of a waveform with an accuracy of better than 2 percent it is necessary that the rise time of the oscilloscope should be less than 0.2 of that of the waveform. Analog oscilloscopes are available having bandwidths of up to 1 GHz.

The deflection sensitivity of an oscilloscope, commonly quoted as mV/cm, mV/div, or μV/cm, μV/div, gives a measure of the smallest signal the oscilloscope can measure accurately. Typically, the highest sensitivity for the vertical system corresponds to 10 μV/cm. There is a trade-off between bandwidth and sensitivity since the noise levels generated either by the amplifier itself or by pickup by the amplifier are greater in wideband measurements. High-sensitivity oscilloscopes may provide bandwidth-limiting controls to improve the display of low-level signals at moderate frequencies.

For comparison purposes, simultaneous viewing of multiple inputs is often required. This can be provided by the use of dual-trace or dual-beam methods. In the dual-trace method the beam is switched between the two input signals. Alternate sweeps of the display can be used for one of the two signals, or, in a single sweep, the display can be chopped between the two signals. Chopping can occur at frequencies up to 1 MHz. Both these methods have limitations for measuring fast transients since in the alternate method the transient may occur on one channel whilst the other is being displayed. The chopping rate of the chopped display limits the frequency range of the signals that can be observed.

Dual-beam oscilloscopes use two independent electron beams and vertical deflection systems. These can be provided with either a common horizontal system or two independent horizontal systems to enable the two signals to be displayed at different sweep speeds. By combining a dual-beam system with chopping it is possible to provide systems with up to eight inputs.

Other functions which are commonly provided on the Y inputs include facilities to invert one channel or to take the

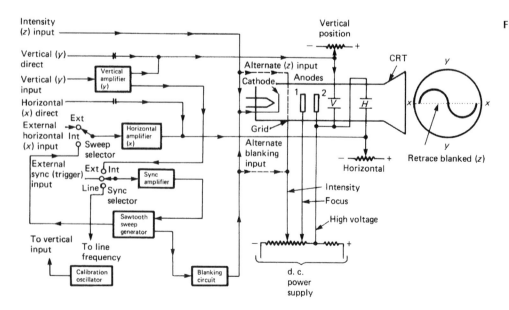

FIGURE 41.8 Analog oscilloscope.

difference of two input signals and display the result. This enables unwanted common mode signals present on both inputs to be rejected.

For the display of signals as a function of time the horizontal system provides a sawtooth voltage to the X plates of the oscilloscope together with the blanking waveform necessary to suppress the flyback. The sweep speed required is determined by the waveform being observed. Sweep rates corresponding to as little as 200 ps/div can be obtained. In time measurements the sweep can either be continuous, providing a repetitive display, or single shot, in which the horizontal system is triggered to provide a single sweep. To provide a stable display in the repetitive mode the display is synchronized either internally from the vertical amplifier or externally using the signal triggering or initiating the signal being measured. Most oscilloscopes provide facilities for driving the X plates from an external source to enable, for example, a Lissajous figure to be displayed.

Delayed sweep enables the sweep to be initiated sometime after the trigger. This delayed sweep facility can be used in conjunction with the timebase to provide expansion of one part of the waveform.

The trigger system allows the user to specify a position on one or more of the input signals in the case of internal triggering or on a trigger signal in the case of external triggering where the sweep is to initiate. Typical facilities provided by trigger level controls are auto (which triggers at the mean level of the waveform) and trigger level and direction control, i.e., triggering occurs at a particular signal level for positive-going signals.

41.6.3 Storage Oscilloscopes

Storage oscilloscopes are used for the display of signals that are either too slow or too fast and infrequent for a conventional oscilloscope. They can also be used for comparing events occurring at different times.

The techniques which are used in storage oscilloscopes include bistable, variable persistence, fast transfer storage, or a combination of fast transfer storage with bistable or variable persistence storage. The storage capability is specified primarily by the writing speed. This is usually expressed in cm/µs or div/µs.

The phosphor in a bistable CRT has two stable states: written and unwritten. As shown in Figure 41.9(a), when writing the phosphor is charged positive in those areas where it is written on. The flood gun electrons hit the unwritten area but are too slow to illuminate it. However, in the written areas the positive charge of the phosphor attracts the electrons and provides them with sufficient velocity to keep the phosphor lit and also to knock out sufficient secondaries to keep the area positive. A bistable tube displays stored data at one level of intensity. It provides a bright, long-lasting display (up to 10 h), although with less contrast than other techniques. It has a slow writing speed with a typical value

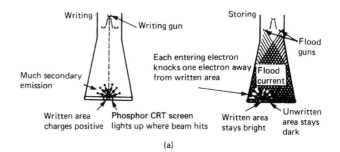

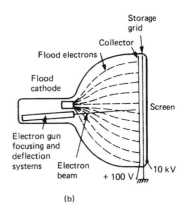

FIGURE 41.9 Storage CRTs. (a) Bistable; (b) variable persistence.

of 0.5 Cm/µs. Split-screen operation, in which the information on one part of the screen is stored whilst the remainder has new information written onto it, can be provided using a bistable CRT.

The screen of the variable persistence CRT, shown in Figure 41.9(b), is similar to that of a conventional CRT. The storage screen consists of a fine metal mesh coated with a dielectric. A collector screen and ion-repeller screen are located behind the storage screen. When the writing gun is employed a positive trace is written out on the storage screen by removing electrons from its dielectric. These electrons are collected by the collector screen. The positively charged areas of the dielectric are transparent to low-velocity electrons. The flood gun sprays the entire screen with low-velocity electrons. These penetrate the transparent areas but not other areas of the storage screen. The storage screen thus becomes a stencil for the flood gun. The charge on the mesh can be controlled, altering the contrast between the trace and the background and also modifying how long the trace is stored. Variable persistence storage provides high contrast between the waveform and the background. It enables waveforms to be stored for only as long as the time between repetitions, and therefore a continuously updated display can be obtained. Time variations in system responses can thus be observed. Integration of repetitive signals can also be provided since noise or jitter not common to all traces will not be stored or displayed. Signals with low repetition rates and fast rise times can be displayed by allowing successive repetitions to build up the trace brightness. The typical writing rate for variable persistence storage is 5 cm/μs. A typical storage time would be 30 s.

Fast transfer storage uses an intermediate mesh target that has been optimized for speed. This target captures the waveform and then transfers it to another mesh optimized for long-term storage. The technique provides increased writing speeds (typically up to 5500 cm/μs). From the transfer target storage can be by either bistable or variable persistence. Oscilloscopes are available in which storage by fast variable persistence, fast bistable, variable persistence, or bistable operation is user selectable. Such devices can provide a range of writing speeds, with storage time combinations ranging from 5500 cm/μs and 30 s using fast variable persistence storage to 0.2 cm/μs and 30 min using bistable storage.

41.6.4 Sampling Oscilloscopes

The upper frequency limit for analog oscilloscopes is typically 1 GHz. For the display of repetitive signals in excess of this frequency, sampling techniques are employed. These extend the range to approximately 14 GHz. As shown in Figure 41.10, samples of different portions of successive waveforms are taken. Sampling of the waveform can be either sequential or random. The samples are stretched in time, amplified by relatively low bandwidth amplifiers,

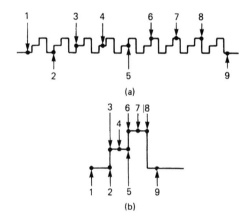

FIGURE 41.10 Sampling oscilloscope.

and then displayed. The display is identical to the sampled waveform. Sampling oscilloscopes are typically capable of resolving events of less than 5 mV in peak amplitude that occur in less than 30 ps on an equivalent timebase of less than 20 ps/cm.

41.6.5 Digitizing Oscilloscopes

Digitizing oscilloscopes are useful for measurements on single-shot or low-repetition signals. The digital storage techniques they employ provide clear, crisp displays. No fading or blooming of the display occurs and since the data are stored in digital memory the storage time is unlimited.

Figure 41.11 shows the elements of a digitizing oscilloscope. The Y channel is sampled, converted to a digital form, and stored. A typical digitizing oscilloscope may provide dual-channel storage with a 100 MHz, 8-bit ADC on each channel feeding a 1 K × 8 bit store, with simultaneous sampling of the two channels. The sample rate depends on the timebase range but typically may go from 20 samples/s at 5 s/div to 100 M samples/s at 1 μs/div. Additional stores are often provided for comparative data.

Digitizing oscilloscopes can provide a variety of display modes, including a refreshed mode in which the stored data and display are updated by a triggered sweep; a roll mode in which the data and display are continually updated, producing the effect of new data rolling in from the right of the screen; a fast roll mode in which the data are continually updated but the display is updated after the trigger; an arm and release mode which allows a single trigger capture; a pre-trigger mode which in roll and fast roll allocates 0, 25, 75, and 100 percent of the store to the pre-trigger signal; and a hold display mode which freezes the display immediately.

Communication to a computer system is generally provided by a IEEE-488 interface. This enables programming of the device to be undertaken for stored data to be sent to an external controller, and, if required, to enable the device to receive new data for display. Digitizing oscilloscopes may

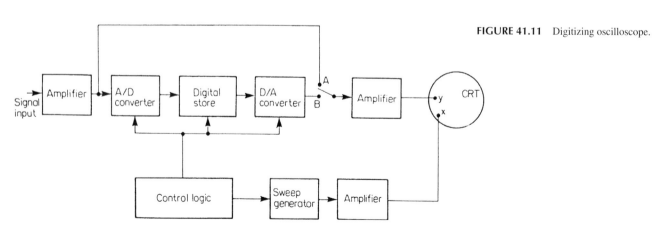

FIGURE 41.11 Digitizing oscilloscope.

also provide outputs for obtaining hard copy of the stored data on an analog *x–y* plotter.

41.6.6 Visual Display Units (VDUs)

A VDU comprising a keyboard and CRT display is widely used as a HMI in computer-based instrumentation systems. Alphanumeric VDU displays use the raster scan technique, as shown in Figure 41.12. A typical VDU will have 24 lines of 80 characters. The characters are made up of a 7×5 dot matrix. Thus seven raster lines are required for a single line of text and five dots for each character position. The space between characters is equivalent to one dot and that between lines is equivalent to one or two raster scans.

As the electron beam scans the first raster line the top row of the dot matrix representation for each character is produced in turn. This is used to modulate the beam intensity. For the second raster scan the second row of the dot matrix representation is used. Seven raster scans generate one row of text. The process is repeated for each of the 24 lines and the total cycle repeated at a typical rate of 50 times per second.

The characters to be displayed on the screen are stored in the character store as 7-bit ASCII code, with the store organized as 24×80 words of 7 bits. The 7-bit word output of the character store addresses a character pattern ROM. A second input to the ROM selects which particular row of the dot matrix pattern is required. This pattern is provided as a parallel output which is then converted to a serial form to be applied to the brightness control of the CRT. For a single scan the row-selection inputs remain constant whilst the character pattern ROM is successively addressed by the ASCII codes corresponding to the 80 characters on the line. To build up a row of character the sequence of ASCII codes remains the same but on successive raster scans the row address of the character pattern ROM is changed.

41.6.7 Graphical Displays

A graphical raster-scan display is one in which the picture or frame is composed of a large number of raster lines, each one of which is subdivided into a large number of picture elements (pixels). Standard television pictures have 625 lines, consisting of two interlaced frames of 313 lines. With each 313-line frame scanned every 20 ms the line scan rate is approximately 60 µs/line.

A graphic raster-scan display stores every pixel in random access or serial store (frame), as shown in Figure 41.13. The storage requirements in such systems can often be quite large. A system using 512 of the 625 lines for display with 1024 pixels on each line having a dual intensity (on/off) display requires a store of 512 Kbits. For a display having the same resolution but with either eight levels of brightness or eight colors the storage requirements then increase to 1.5 Mbits.

Displays are available which are less demanding on storage. Typically, these provide an on/off display having a resolution of 360 lines, each having 720 pixels. This requires a $16 \text{ K} \times 16$ bit frame store. Limited graphics can be provided by alphanumeric raster scan displays supplying additional symbols to the ASCII set. These can then be used to produce graphs or pictures.

41.7 GRAPHICAL RECORDERS

Hard copy of data from a process or system can be displayed in a graphical form either as the relationships between one or more variables and time using an *x–t*

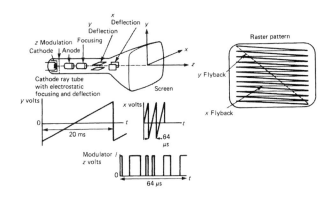

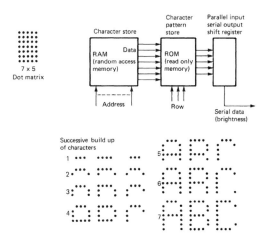

FIGURE 41.12 Visual display unit.

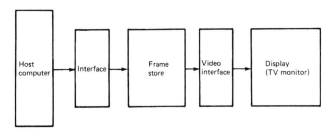

FIGURE 41.13 Elements of a graphical display.

recorder or as the relationship between two variables using an x–y recorder. x–t recorders can be classified as either strip chart recorders, in which the data are recorded on a continuous roll of chart paper, or circular chart recorders, which, as their name implies, record the data on a circular chart.

41.7.1 Strip Chart Recorders

Most strip chart recorders employ a servo-feedback system (as shown in Figure 41.14) to ensure that the displacement of the writing head across the paper tracks the input voltage over the required frequency range. The position of the writing head is generally measured by a potentiometer system. The error signal between the demanded position and the actual position of the writing head is amplified using an ac or dc amplifier, and the output drives either an ac or dc motor. Movement of the writing head is effected by a mechanical linkage between the output of the motor and the writing head. The chart paper movement is generally controlled by a stepping motor.

The methods used for recording the data onto the paper include:

1. *Pen and ink.* In the past these have used pens having ink supplied from a refillable reservoir. Increasingly, such systems use disposable fiber-tipped pens. Multichannel operation can be achieved using up to six pens. For full-width recording using multiple pens, staggering of the pens is necessary to avoid mechanical interference.
2. *Impact printing.* The "ink" for the original impact systems was provided by a carbon ribbon placed between the pointer mechanism and the paper. A mark was made on the paper by pressing the pointer mechanism onto the paper. Newer methods can simultaneously record the data from up to twenty variables. This is achieved by having a wheel with an associated ink pad which provides the ink for the symbols on the wheel. By rotating the wheel different symbols can be printed on the paper for each of the variables. The wheel is moved across the paper in response to the variable being recorded.
3. *Thermal writing.* These systems employ thermally sensitive paper which changes color on the application of heat. They can use a moving writing head which is heated by an electric current or a fixed printing head having a large number of printing elements (up to several hundred for 100 mm wide paper). The particular printing element is selected according to the magnitude of the input signal. Multichannel operation is possible using such systems and time and date information can be provided in alphanumeric form.
4. *Optical writing.* This technique is commonly used in galvanometer systems (see Section 41.7.3). The source of light is generally ultra-violet to reduce unwanted effects from ambient light. The photographic paper used is sensitive to ultraviolet light. This paper develops in daylight or artificial light without the need for special chemicals. Fixing of the image is necessary to prevent long-term degradation.
5. *Electric writing.* The chart paper used in this technique consists of a paper base coated with a layer of a colored dye—black, blue, or red—which in turn is coated with a very thin surface coating of aluminum. The recording is effected by a tungsten wire stylus moving over the aluminum surface. When a potential of approximately 35 V is applied to this stylus an electric discharge occurs which removes the aluminum, revealing the dye. In multichannel recording the different channels are distinguished by the use of different line configurations (for example, solid, dashed, dotted). Alphanumeric information can also be provided in these systems.
6. *Inkjet writing.* In this method, a standard chart paper is used and a monochrome or color inkjet printhead, adapted from microcomputer printers, is used to draw the alphanumeric traces. The adaptability of this method makes possible many graphical formats from a single printhead.

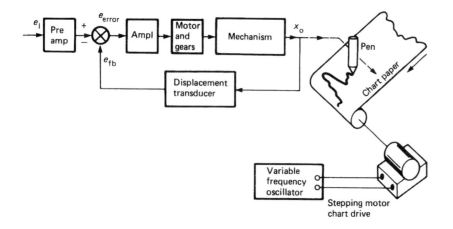

FIGURE 41.14 Strip chart recorder.

The major specifications for a strip chart recorder include:

1. *Number of channels.* Using a numbered print wheel, up to 30 channels can be monitored by a single recorder. In multichannel recorders, recording of the channels can be simultaneous or on a time-shared basis. The channels in a multichannel recording can be distinguished either by color or by the nature of line marking.
2. *Chart width.* This can be up to 250 mm.
3. *Recording technique.* The recorder may employ any of the techniques described above.
4. *Input specifications.* These are given in terms of the range of the input variable which may be voltage, current, or from a thermocouple, RTD, pH electrode, or conductivity probe. Typical dc voltage spans are from 0.1 mV to 100 V. Those for dc current are typically from 0.1 mA to 1 A. The zero suppression provided by the recorder is specified by its suppression ratio, which gives the number of spans by which the zero can be suppressed. The rejection of line frequency signals at the input of the recorder is measured by its common and normal mode rejection ratios.
5. *Performance specifications.* These include accuracy, deadband, resolution, response time, and chart speed. A typical voltage accuracy specification may be $\pm(0.3 + 0.1 \times$ suppression ratio) percent of span, or 0.20 mV, whichever is greater. Deadband and resolution are usually expressed as a percentage of span, with 0.1 and ± 0.15 percent, respectively, being typical figures. Chart recorders are designed for use with slowly varying inputs (<1 Hz for full-scale travel). The response time is usually specified as a step response time and is often adjustable by the user to between 1 and 10 s. The chart speed may vary between mm/h and m/h, depending on the application.

41.7.2 Circular Chart Recorders

These recorders, as shown in Figure 41.15, generally use a 12-in diameter chart. They are particularly well suited for direct actuation by a number of mechanical sensors without the need for a transducer to convert the measured quantity into an electrical one. Thus they can be configured for a temperature recorder using a filled bulb thermometer system or as an absolute, differential, or gauge pressure recorder employing a bellows or a bourdon tube. Up to four variables can be recorded on a single chart. The rotation of the circular scale can be provided by a clockwork motor, which makes the recorder ideally suited in remote locations having no source of power.

41.7.3 Galvanometer Recorders

The D'Arsonval movement used in the moving-coil indicating instruments can also provide the movement in an optical galvanometer recorder, as shown in Figure 41.16. These devices can have bandwidths in excess of 20 KHz. The light source in such devices is provided by either an ultraviolet or tungsten lamp. Movement of the light beam is effected by the rotation of a small mirror connected to the galvanometer movement. The light beam is focused into a spot on the light-sensitive paper. Positioning of the trace may also be achieved by mechanical means. A recorder may have several galvanometers and, since it is light that is being deflected, overlapping traces can be provided without the need for time staggering. Identification of individual traces is by sequential trace interruption in which the light from each trace in turn is interrupted by a series of pins passing in front of the galvanometers. Amplitude and time reference grids may also be provided.

Trade-offs are available between the sensitivity and frequency response of the galvanometers with high-sensitivity devices having low bandwidths. Manufacturers generally provide a range of plug-in galvanometers, having a range of sensitivities and bandwidths for different applications, which enables them to be quickly removed from the magnet block and replaced. Galvanometer systems are available which enable up to 42 channels to be recorded on a 300 mm

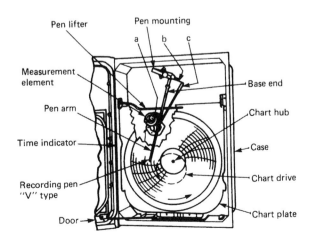

FIGURE 41.15 Circular chart recorder.

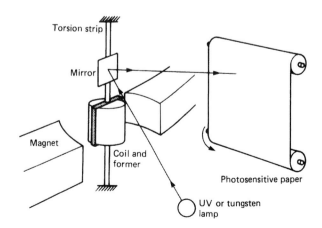

FIGURE 41.16 Galvanometer recorder.

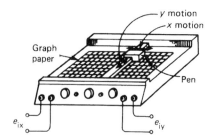

FIGURE 41.17 x–y recorder.

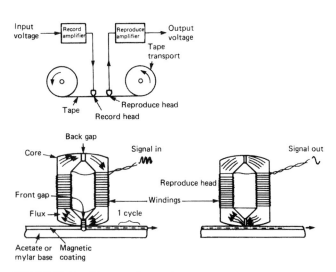

FIGURE 41.18 Elements of magnetic tape recorder system.

wide chart drive. The drive speed of the paper can be up to 5000 mm/s.

41.7.4 X–Y Recorders

In an analog x–y recorder (Figure 41.17) both the x and y deflections of the writing head are controlled by servo feedback systems. The paper, which is usually either A3 or A4 size, is held in position by electrostatic attraction or vacuum. These recorders can be provided with either one or two pens. Plug-in timebases are also generally available to enable them to function as x–t recorders. A typical device may provide x and y input ranges which are continuously variable between 0.25 mV/cm and 10 V/cm with an accuracy of ±0.1 percent of full scale. Zero offset adjustments are also provided. A typical timebase can be adjusted between 0.25 and 50 s/cm with an accuracy of 1 percent.

The dynamic performance of x–y recorders is specified by their slewing rate and acceleration. A very high-speed x–y recorder capable of recording a signal up to 10 Hz at an amplitude of 2 cm peak to peak will have a slewing rate of 97 cm/s and a peak acceleration of 7620 cm/s^2. Remote control of such functions as sweep start and reset, pen lift, and chart hold can be provided.

Digital x–y plotters replace the servo feedback with an open-loop stepping motor drive. These can replace traditional analog recorders and provide increased measurement and graphics capabilities. A digital measurement plotting system can provide, for example, simultaneous sampling and storage of a number of input channels; a variety of trigger modes, including the ability to display pre-trigger data; multi-pen plotting of the data; annotation of the record with date, time, and set-up conditions; and an ability to draw grids and axes. Communication with such devices can be by means of the IEEE-488 or RS-232 interfaces.

For obtaining hard copy from digital data input, graphics plotters are available. With appropriate hardware and software these devices can draw grids, annotate charts, and differentiate data by the use of different colors and line types. They are specified by their line quality, plotting speed, and paper size. Intelligence built into the plotter will free the system's CPU for other tasks. The availability of a graphics language to control the plotter functions and graphics software packages also simplifies the user's programming tasks.

41.8 MAGNETIC RECORDING

Using magnetic tape recording for data which are to be subsequently analyzed has the advantage that, once recorded, the data can be replayed almost indefinitely. The recording period may vary from a few minutes to several days. Speed translation of the captured data can be provided in that fast data can be slowed down and slow data speeded up by using different record and reproduce speeds. Multichannel recording enables correlations between one or more variables to be identified.

The methods employed in recording data onto magnetic tape include direct recording, frequency modulation (FM), and pulse code modulation (PCM). Figure 41.18 shows the elements of a magnetic tape recorder system. Modulation of the current in the recording head by the signal to be recorded linearly modulates the magnetic flux in the recording gap. The magnetic tape, consisting of magnetic particles on an acetate or mylar base, passes over the recording head. As they leave from under the recording head the particles retain a state of permanent magnetization proportional to the flux in the gap. The input signal is thus converted to a spatial variation of the magnetization of the particles on the tape. The reproduce head detects these changes as changes in the reluctance of its magnetic circuit, which induces a voltage in its winding. This voltage is proportional to the rate of change of flux. The reproduce head amplifier integrates the signal to provide a flat frequency characteristic.

Since the reproduce head generates a signal that is proportional to the rate of change of flux the direct recording method cannot be used down to dc. The lower limit is typically 100 Hz. The upper frequency limit occurs when the induced variation in magnetization varies over distances smaller than the gap in the reproduce head. This sets an upper limit for direct recording of approximately 2 MHz, using gaps and tape speeds commonly available.

In FM recording systems the carrier signal is frequently modulated by the input signal (frequency modulation is discussed in Chapter 29). The central frequency is chosen with respect to the tape speed, and frequency deviations of up to ±40 percent of the carrier frequency are used. FM systems provide a dc response but at the expense of the high-frequency recording limit.

PCM techniques, which are also described in Chapter 29, are used in systems for the recording of digital data. The coding used is typically Non-Return to Zero Level or Delay Modulation.

A typical portable instrumentation tape recorder can provide up to 14 data channels using ½-in tape. Using direct record/reproduce, such a system can record signals in a frequency band from 100 Hz to 300 KHz with an input sensitivity of between 0.1 to 2.5 V rms and a signal-to-noise ratio of up to 40 dB. FM recording with such a system provides bandwidths extending from dc to as high as 40 KHz. A typical input sensitivity for FM recording is 0.1 to 10 V rms with a signal-to-noise ratio of up to 50 dB. TTL data recording using PCM can be achieved at data transfer rates of up to 480 kbits/s using such a system.

Magnetic tape recorders are becoming obsolete as the cost and durability of systems with hard discs become lower and better. In addition, the much lower cost of RAM memory because of the proliferation of PCs, and especially laptops with very low power requirements, has made possible high volume RAM-based data storage units.

41.9 TRANSIENT/WAVEFORM RECORDERS

These devices are used for the high-speed capture of relatively small amounts of data. The specifications provided by general-purpose transient recorders in terms of sampling rate, resolution, number of channels, and memory size are similar to those of the digitizing oscilloscope. Pre-trigger data capture facilities are also often provided. High-speed transient recorders, using scan conversion techniques in which the signal is written onto a siliconidiode target array, can provide sampling rates of 10^{11} samples/s (10 ps/point) with record lengths of 512 points and are capable of recording single-shot signals having bandwidths of up to 500 MHz.

41.10 DATA LOGGERS

Data loggers can be broadly defined as data-acquisition systems whose functions are programmed from the front panel. They take inputs from a mixture of analog and digital signals, perform limited mathematical and logical operations on the inputs, and provide storage either in the form of semiconductor memory or magnetic tape or disc systems. Many data loggers are provided with an integral alphanumeric display and printer. Figure 41.19 shows the elements of a typical data-logging system. These systems can be characterized by:

1. *The number and type of analog or digital inputs which can be accepted.* The analog inputs may be ac or dc current, or from thermocouples or RTDs.
2. *The scanning rate and modes of scanning available.* These can include single-scan, monitor, interval scan, and continuous scan.
3. *The method of programming.* This can be by one button per function on the front panel; by the use of a menu-driven approach by presenting the information to the user on a built-in CRT screen; or by using a high-level language such as BASIC or C.
4. *The mathematical functions provided by the logger.* These may include linearization for thermocouple inputs, scaling to provide the user with an output in engineering units, and data averaging over a selectable number of scans.
5. *The nature and capacity of the internal memory.* If the memory is semiconductor memory, is it volatile? If so, is it provided with battery back-up in case of power failure?
6. *The printer technique and the width of the print output.* Data loggers typically use dot matrix ink printers or thermal or electric writing technique with continuous strip paper.
7. *The display.* Data loggers generally provide display using LEDs or LCDs, although some incorporate CRTs.
8. *Communication with other systems.* This is generally provided by RS 232/422/423 interfaces. Some data loggers are capable of being interrogated remotely using the public telephone network.

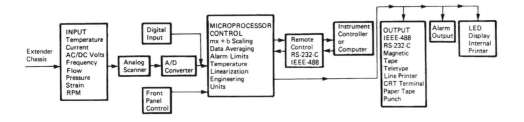

FIGURE 41.19 Data logger.

REFERENCES

Agard, P. J., et al., *Information and Display Systems in Process Instruments and Control Handbook* (eds D. M. Considine and G. Considine), McGraw-Hill, New York (1985).

Bentley, J., *Principles of Measurement Systems*, Longman, London (1983).

Bosman, D., "Human factors in display design," in *Handbook of Measurement Science*, Volume 2, *Practical Fundamentals* (ed. P. Sydenham), John Wiley, Chichester (1983).

British Standards Institution, BS 89: 1977, Direct Acting Indicating Electrical Instruments and their Accessories (1977).

British Standards Institution, BS 3693 (Part 1: 1964 and Part 2: 1969), The Design of Scales and Indexes (1964 and 1969).

Doebelin, E. O., *Measurement Systems Application and Design*, 3rd ed., McGraw-Hill, London (1983).

Lenk, J. D., *Handbook of Oscilloscopes, Theory and Application*, Prentice-Hall, Englewood Cliffs, N.J. (1982).

Wilkinson, B. and D. Horrocks, *Computer Peripherals.* Hodder and Stoughton, London (1980).

Wilson, J. and J. F. B. Hawkes, *Optoelectronics: An Introduction.* Prentice-Hall, Englewood Cliffs, N.J. (1983).

Chapter 42

Pneumatic Instrumentation

E. H. Higham; edited by W. L. Mostia Jr., PE

42.1 BASIC CHARACTERISTICS

The early evolution of process control was centered on the petroleum and chemical industries, where the process materials gave rise to hazardous environments and therefore the measuring and control systems had to be intrinsically safe. Pneumatic systems were particularly suitable in this connection, and, once the flapper/nozzle system and its associated pneumatic amplifier (customarily called relay) had been developed for the detection of small movements, sensors to measure temperature, flow, pressure, level, and density were devised. These were followed by pneumatic mechanisms to implement the control functions whilst air cylinders or diaphragm motors were developed to actuate the final control elements.

Although it is clear that pneumatic systems are easily made intrinsically safe, the evolution of instruments began with mechanical systems, and because pneumatics allowed remote locating of sensors and final control elements, they were used widely. As electronic instruments became available, they quickly replaced pneumatics except in the most rugged petrochemical environments. Pneumatic plants were still common in the 1960s and 1970s and were interfaced with computer control systems using P/I and P/E converters as well as Scanivalves. As digital instruments became more widely available in the 1990s, the use of pneumatics, even in refinery applications, declined.

The pneumatic systems had further advantages in that the installation did not involve special skills, equipment or materials. The reservoir tanks, normally provided to smooth out the pulsations of the air compressor, also stored a substantial volume of compressed air which, in the event of failure of the compressor or its motive power, stored sufficient energy to enable the control system to remain in operation until an orderly shutdown had been implemented. The provision of comparable features with electronic or other types of control systems involves a great deal of additional equipment. In addition, another advantage of pneumatic systems is that you do not necessarily require electrical power. Prior to the advent of battery-operated intrinsically safe devices, even some instruments on pipeline systems were pneumatic, running on natural gas.

Although pneumatic control systems have many attractive features they compare unfavorably with electronic systems in two particular respects, namely signal transmission and signal conditioning or signal processing.

For the majority of process applications a distance/velocity lag of about 1 second is quite acceptable. This corresponds typically to a pneumatic transmission distance of about 100 meters and is not a limitation for a great many installations, but in large plants, where measurement signals may have to be transmitted over distances of a kilometer or more, electrical methods for transmission have to be adopted. When the successful operation of a process depends on conditioning the signal (e.g., taking the square root, multiplying, dividing, summing or averaging, differentiating or integrating the signals) the pneumatic devices are undoubtedly more complex, cumbersome, and very often less accurate than the corresponding electronic devices.

Pneumatic systems are virtually incompatible with digital ones and cannot compete at all with their advantages for signal transmission, multiplexing, computation, etc. On the other hand, when all these opportunities have been exploited and implemented to the full, the signal to the final control element is nearly always converted back to pneumatics so that the necessary combination of power, speed, and stroke/movement can be achieved with a pneumatic actuator.

The majority of diagrams in this chapter are taken from the technical publications of the Foxboro Company (now an Invensys division) and are reproduced with their permission. It must be said that the most common pneumatic instruments seen today are Fisher (Fisher-Rosemount Inc.).

42.2 PNEUMATIC MEASUREMENT AND CONTROL SYSTEMS

Pneumatic systems are characterized by the simplicity of the technology on which they are based and the relative ease with which they can be installed, operated, and maintained. They are based on the use of a flapper/nozzle in conjunction with a pneumatic relay to detect a very small relative movement (typically, <0.01 mm) and to control a supply of compressed air so that a considerable force can be generated under precise control.

A typical flapper/nozzle system is shown in Figure 42.1. It comprises a flat of metal attached to the device or member of which the relative motion is to be detected. This flapper is positioned so that when moved it covers or uncovers a 0.25 mm diameter hole located centrally in the 3 mm diameter flat surface of a truncated 90° cone. A supply of clean air, typically at 120 kPa, is connected via a restrictor and a "T" junction to the nozzle. A pressure gauge connected at this "T" junction would show that, with the nozzle covered by the flapper, the pressure approaches the supply pressure, but when the flapper moves away from the nozzle the pressure falls rapidly to a value determined by the relative values of the discharge characteristics of the nozzle and the restrictor, as shown in Figure 42.2.

For measurement purposes this change in pressure needs to be amplified, and this is effected by means of a pneumatic relay (which is equivalent to a pneumatic amplifier). In practice, it is convenient to incorporate the restrictor associated with the nozzle in the body of the relay, as shown in Figure 42.3. However, typically in latter day practice, due to maintenance constraints, this restrictor is normally mounted on the relay mounting assembly, and is removable via a screwdriver. This figure also shows that the relay comprises two chambers isolated from each other by a flexible diaphragm that has a conical seat and stem that act as a valve to cover or uncover the exhaust port. The stem acts against a small ball retained by the leaf spring so that it functions as a second valve which controls the flow of air from the supply to the output port.

In operation, when the nozzle is covered, the pressure in the associated chamber builds up, causing the conical valve to close the exhaust port and the ball valve to allow air to

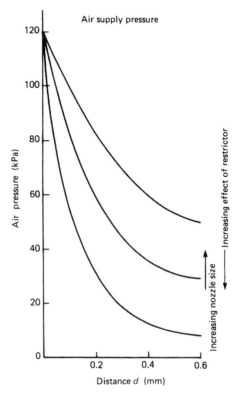

FIGURE 42.2 Relation between nozzle back pressure and flapper/nozzle distance.

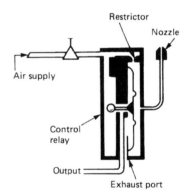

FIGURE 42.3 Pneumatic relay (leaf spring).

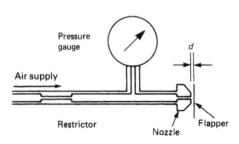

FIGURE 42.1 Typical flapper/nozzle system.

flow from the supply to the output port so that the output pressure rises.

When the nozzle is uncovered by movement of the flapper the flexible diaphragm moves so that the ball valve restricts the flow of air from the supply. At the same time the conical valve moves off its seat, opening the exhaust so that the output pressure falls.

With such a system the output pressure is driven from 20 to 100 kPa (3 to 15 psig) as a result of the relative movement between the flapper and nozzle of about 0.02 mm. Although the device has a non-linear character, when it is incorporated

CHAPTER | 42 Pneumatic Instrumentation

into a sensor or final control element, this is taken into account so that the output or response is essentially linear.

There are two basic schemes for utilizing the flapper/nozzle/relay system, namely, the motion-balance and the force-balance systems. These are illustrated in Figures 42.4 and 42.5. Figure 42.4 shows a motion-balance system in which the input motion is applied to point A on the lever AB. The opposite end (B) of this lever is pivoted to a second lever BCD which in turn has point D pivoted in a lever positioned by movement of the feedback bellows. At the center (C) of the lever BD there is a stem on which one end of the lever CEF is supported while it is pivoted at point F and has a flapper nozzle located at E. A horizontal displacement which causes A to move to the left is transmitted via B to C, and as a result the flapper at E moves off the nozzle so that the back pressure falls. This change is amplified by the relay so that the pressure in the bellows falls and the lever carrying the pivot D moves down until equilibrium is re-established. The output pressure is then proportional to the original displacement. By changing the inclination of the lever CEF sensitivity or gain of the system may be changed.

Figure 42.5 illustrates a force-balance system. The measurement signal, in the form of a pressure, is applied to a bellows which is opposed by a similar bellows for the reference signal. The force difference applied to the lever supported on an adjustable pivot is opposed by a spring/bellows combination. Adjacent to the bellows is a flapper/nozzle sensor. In operation, if the measurement signal exceeds the reference signal the resultant force causes the force bar to rotate clockwise about the adjustable pivot so that the flapper moves closer to the nozzle, with the result that the pressure in the output bellows increases until equilibrium is re-established. The change in output pressure is then proportional to the change in the measurement signal.

42.3 PRINCIPAL MEASUREMENTS

42.3.1 Introduction

Virtually all pneumatic measuring systems depend on a primary element such as an orifice plate, Bourdon tube, etc., to convert the physical parameter to be measured into either a force or a displacement which, in turn, can be sensed by some form of flapper/nozzle system or used directly to operate a mechanism such as an indicator, a recorder pen, or a switch.

The measurements most widely used in the process industries are temperature, pressure, flow, level, and density. In the following sections a description is given of the methods for implementing these measurements pneumatically, as opposed to describing the characteristics of the primary elements themselves. Also described are the pneumatic controllers which were evolved for the process industries and which are still very widely used.

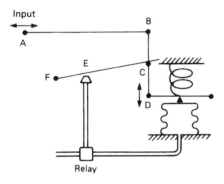

FIGURE 42.4 Basic pneumatic motion-balance system.

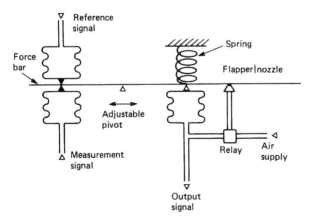

FIGURE 42.5 Basic pneumatic force-balance system.

42.3.2 Temperature

Filled thermal systems are used almost exclusively in pneumatic systems for temperature measurement and control. The sensing portion of the system comprises a bulb connected via a fine capillary tube to a spiral or helical Bourdon element or a bellows. When the bulb temperature rises, the increased volume of the enclosed fluid or its pressure is transmitted to the Bourdon element, which responds to the change by movement of its free end. This movement can be used directly to position a recording pen or an indicating pointer, either of which could also be coupled to a flapper/nozzle mechanism by a system of links and levers to actuate a pneumatic controller.

When a bellows is used to terminate the capillary the change in bulb temperature is converted into a force which, in turn, can be used to actuate a force-balance mechanism. Details of the materials used for filling the bulbs and their characteristics are given in Chapter 14.

In the motion-balance systems the free end of the Bourdon tube is connected via adjustable links to a lever pivoted about the axis A in Figure 42.6. The free end of this lever bears on a second lever system pivoted about the axis B and driven by the feedback bellows. The free end of this second lever system is held in contact with a stem on a further lever system by a light spring. This latter lever is pivoted about the axis C and its free end is shaped to form the flapper of a flapper/nozzle system. The control relay associated with the flapper/nozzle system generates the output signal which is also applied to the feedback bellows.

In operation, the links and levers are adjusted so that with the bulb held at the temperature corresponding to the lower range value, the output signal is 20 kPa. If the measured temperature then rises, the lever pivoted at A moves in a clockwise direction (viewed from the left). In the absence of any movement of the bellows, this causes that associated lever pivoted about the axis C to rotate clockwise so that the flapper moves towards the nozzle. This, in turn, causes the nozzle back pressure to rise, a change which is amplified by the control relay, and this is fed back to the bellows so that the lever pivoted about the axis B moves until balance is restored. In this way, the change in the sensed temperature is converted into a change in the pneumatic output signal.

An example of the force-balance system is shown in Figure 42.7. There are two principal assemblies, namely the force-balance mechanism and the thermal system, which comprises the completely sealed sensor and capillary assembly filled with gas under pressure. A change in temperature at the sensor causes a change in gas pressure. This change is converted by the thermal system capsule into a change in the force applied at the lower end of the force bar which, being pivoted on cross-flexures, enables the force due to the thermal system to be balanced by combined forces developed by the compensating bellows, the feedback bellows, and the base elevation spring. If the moment exerted by the thermal system bellows for the force developed by the compensating bellows (which compensates the effect of ambient temperature at the transmitter), the feedback bellows (which develops a force proportional to the output signal), and the elevation spring (which provides means for adjusting the lower range value), the gap between the top of the force bar and the nozzle is reduced, causing the nozzle back pressure to rise. This change is amplified by the relay and applied to the feedback bellows so that the force on the force bar increases until balance is re-established. In this way, the forces applied to the force bar are held in balance so that the output signal of the transmitter is proportional to the measured temperature.

This type of temperature transmitter is suitable for measuring temperatures between 200 and 800K, with spans from 25 to 300K.

42.3.3 Pressure Measurement

As explained in Chapter 9, the majority of pressure measurements utilize a Bourdon tube, a diaphragm, or a bellows (alone or operated in conjunction with a stable spring) as the primary elements which convert the pressure into a motion or a force. Both methods are used in pneumatic systems; those depending on motion use a balancing technique similar to that described in the previous section for temperature measurements, while those which operate on the force-balance principle utilize a mechanism that has wider application. Force-balance was initially developed for differential pressure measurements which, because they provide a common basis for measuring flow, level, and density, have very wide application in industry. By selecting alternative primary elements the same force-balance mechanism can be used for measurement of gauge or absolute pressures. The basic arrangement of the mechanism is shown in Figure 42.8.

The force developed by the primary element is applied via a flexure to one end of the force bar which comprises two sections threaded so that they can be joined together with a thin circular diaphragm of cobalt-nickel alloy clamped in the joint. This diaphragm serves as the pivot or flexure for the force bar as well as the seal between the process fluid and the force-balance mechanism, particularly in the differential pressure measuring instruments.

The outer edge of the diaphragm seal is clamped between the body enclosing the primary element and the framework

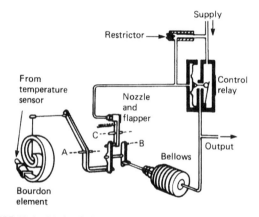

FIGURE 42.6 Motion-balance system for temperature measurement.

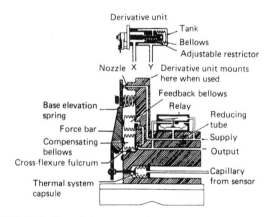

FIGURE 42.7 Force-balance system for temperature measurement.

that supports the force-balance mechanism. This framework carries the zero adjustment spring, the feedback bellows, the pneumatic relay, and the nozzle of the flapper/nozzle system.

At the lower end of the range bar are mounted the feedback bellows and zero adjustment spring, whilst at its upper end it carries the flapper and a flexure that connects it to the upper end of the force bar. The range bar is threaded so that the position of the range wheel (and hence the sensitivity of the system) can be varied. Figure 42.9 shows the diagram of forces for the mechanism.

In operation, force (F_1) from the primary element applied to the lower end of the force bar produces a moment ($F_1 a$) about the diaphragm seal pivot which is transmitted to the upper end of the force bar where it becomes a force (F_2) applied via the flexural pivots and a transverse flexure connector to the upper end of the range bar. This force produces a moment ($F_2 c$) about the range wheel, acting as a pivot, which is balanced by the combined moments produced by the forces F_3 and F_4 of the feedback bellows and zero spring, respectively.

Thus, at balance

$$F_1 a = F_2 b$$

and

$$F_2 c = F_3 d + F_4 e$$

from which it follows that:

$$F_1 = \frac{bd}{ac} \cdot F_3 + \frac{be}{ac} \cdot F_4$$

By varying the position of the pivot on the range bar the ratio of F_1 to F_3 can be adjusted through a range of about 10 to 1. This provides the means for adjusting the span of an instrument.

The feedback loop is arranged so that when an increased force is generated by the primary element the resultant slight movement of the force bar causes the flapper to move closer to the nozzle so that its back pressure increases. This change is amplified by the relay and is applied to the bellows. As a result, the force that it applies to the range bar increases until a new equilibrium position is established and the various forces are balanced. Figure 42.10 illustrates the use of this

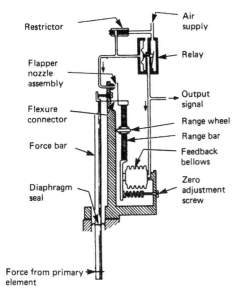

FIGURE 42.8 Basic arrangement of pneumatic force-balance mechanism.

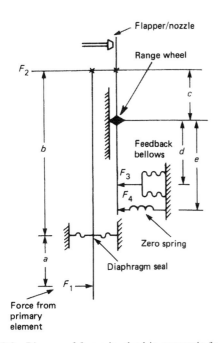

FIGURE 42.9 Diagram of forces involved in pneumatic force-balance mechanism.

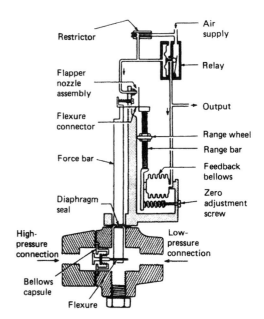

FIGURE 42.10 Force-balance mechanism incorporated into gauge pressure transmitter.

force-balance mechanism in a gauge pressure transmitter if the low-pressure connection is open to the atmosphere or a high-range differential pressure transmitter if both high- and low-pressure signals are connected.

Figure 42.11 shows essentially the same instrument but as an absolute pressure transmitter that uses a bellows as the primary element. In this case the bellows is evacuated and sealed on what was the low pressure side in the previous configuration, so that the instrument measures absolute pressure.

Figure 42.12 shows how the mechanism is used in a high-pressure transmitter in which the primary element is a Bourdon tube.

Figure 42.13 shows the use of the same mechanism attached to the differential pressure sensor which has become the most widely used form of pneumatic sensor.

The principal feature of the differential pressure sensor is the diaphragm capsule shown in more detail in Figure 42.14.

It comprises two identical corrugated diaphragms, welded at their perimeters to a back-up plate on which matching contours have been machined. At the center, the two diaphragms are connected to a stem carrying the C-flexure by which the diaphragm is connected to the force bar. The small cavity between the diaphragms and the back-up plate is filled with silicone oil to provide damping.

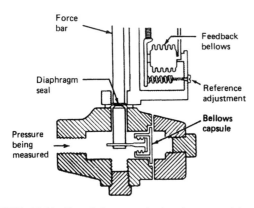

FIGURE 42.11 Force-balance mechanism incorporated into absolute pressure transmitter.

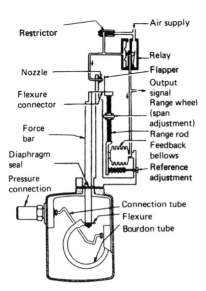

FIGURE 42.12 Force-balance mechanism incorporated into high-pressure transmitter incorporating a Bourdon tube.

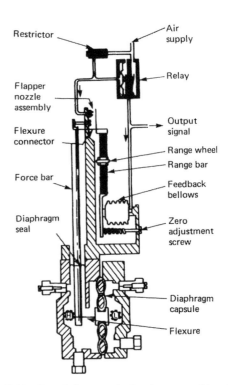

FIGURE 42.13 Force-balance mechanism incorporated into differential pressure transmitter.

CHAPTER | 42 Pneumatic Instrumentation

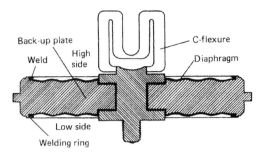

FIGURE 42.14 Construction of differential pressure sensor capsule.

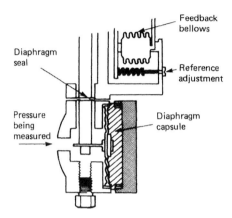

FIGURE 42.15 Modification of sensor capsule for absolute pressure measurements.

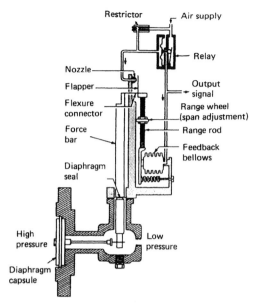

FIGURE 42.16 Force-balance mechanism incorporated into a flange-mounted level transmitter.

If equal pressures are applied on either side of the capsule there is no resultant movement of the central stem. If, however, the force on one side is greater than that on the other, the resultant force is transmitted via the central stem to the C-flexure. If the force on one side greatly exceeds that on the other, then the diaphragm exposed to the higher pressure moves onto the matching contours of the back-up plate and is thereby protected from damage. This important feature has been a major factor in the successful widespread application of the differential pressure transmitter for measurement of many different process parameters, the most important being the measurement of flow in conjunction with orifice plates and the measurement of level in terms of hydrostatic pressures.

Figure 42.15 shows how the diaphragm capsule of the differential pressure transmitter is modified to convert it into an absolute pressure transmitter.

42.3.4 Level Measurements

The diaphragm capsule and force-balance mechanism can be adapted for measurement of liquid level. Figure 42.16 shows how it is mounted in a flange which, in turn, is fixed to the side of the vessel containing the liquid whose level is to be measured. If the tank is open then the low-pressure side of the capsule is also left open to the atmosphere. If, on the other hand, the measurement is to be made on a liquid in an enclosed tank, then the pressure of the atmosphere above the liquid must be applied to the rear of the capsule via the low-pressure connection. In operation, the hydrostatic pressure applied to the diaphragm being proportional to the head of liquid produces a force which is applied to the lower end of the force bar.

The same type of pneumatic force-balance mechanism is used to measure this force and so generate an output signal which is directly proportional to the liquid level.

42.3.5 Buoyancy Measurements

The same mechanism can be adapted to function as a buoyancy transmitter to measure either liquid density or liquid level, according to the configuration of the float which is adopted. This is shown in Figure 42.17, and further details of this method are given in Chapter 6.

42.3.6 Target Flow Transmitter

This transmitter combines in a single unit both a primary element and a force-balance transmitter mechanism. As shown in Figure 42.18, the latter is essentially the same force-balance mechanism that is used in the other pneumatic transmitters described previously. The primary element is a disc-shaped target fixed to the end of the force bar and located centrally in the pipe carrying the process liquid.

As explained in Chapter 26, the force on the target is the difference between the upstream and downstream surface pressure integrated over the area of the target. The square root of this force is proportional to the flow rate.

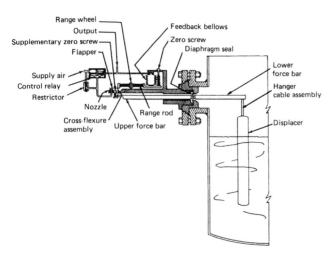

FIGURE 42.17 Force-balance mechanism incorporated into a buoyancy transmitter.

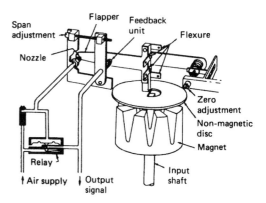

FIGURE 42.19 Modified force-balance mechanism incorporated into a speed transmitter.

The output pressure which establishes the force-balance is the transmitted pneumatic signal and is proportional to the speed of rotation. It may be used to actuate the pneumatic receiver in an indicator, recorder, or controller.

42.4 PNEUMATIC TRANSMISSION

In relatively simple control systems it is usually possible to link the primary sensing element directly to the controller mechanism and to locate this reasonably close to the final control element. However, when the process instrumentation is centralized, the primary sensing elements are arranged to operate in conjunction with a mechanism that develops a pneumatic signal for transmission between the point of measurement and the controller, which is usually mounted in a control room or sheltered area.

A standard for these transmission signals has been developed and is applied almost universally, with only small variations that arise as a result of applying different units of measure. If the lower range value is represented by a pressure P then the upper range value is represented by a pressure $5P$ and the span by $4P$. Furthermore, it is customary to arrange the nominal pressure of the air supplied to the instrument to be between $6.5P$ and $32.0P$.

In SI units, the zero or lower range value of the measurement is represented by a pressure of 20 kPa, the upper range value by a pressure of 100 kPa, and the span therefore by a change in pressure of 80 kPa. The corresponding Imperial and ANSI units are 3 psi, 15 psi, and 12 psi, whilst the corresponding metric units are 0.2 kg/cm² (0.2 bar), 1.0 kg/cm² (1.0 bar), and 0.8 kg/cm² (0.8 bar). (1 bar = 100 = kPa = 14.7 psi.)

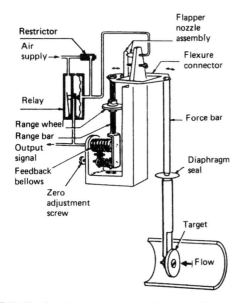

FIGURE 42.18 Force-balance mechanism incorporated into a target flow transmitter.

42.3.7 Speed

A similar force-balance technique can be used to generate a pneumatic signal proportional to speed, as shown in Figure 42.19. As the transmitter input shaft rotates, the permanent magnet attached to it generates a magnetomotive force. This force tends to cause the non-magnetic alloy disc to rotate in the same direction. Since the disc is connected to the flapper by means of flexures, any rotation of the disc causes a corresponding movement of the flapper, thus charging the clearance between the flapper and the nozzle.

As the flapper/nozzle relationship changes, the output pressure from the relay to the feedback unit changes until the force at the feedback unit balances the rotational force.

If a change in measured value occurs, this is sensed by the transmitter, and its output pressure increases accordingly. When the transmitter is remote from the controller and connected to it by an appreciable length of tubing, there is a finite delay before the change of pressure generated at the transmitter reaches the controller. In the case of an increased signal, this delay is governed by the magnitude of the change, the capability of the transmitter to supply the necessary compressed air to raise the pressure throughout the transmission line, the rate at which the change can be propagated along the line, and finally, the capacity of the receiving element.

A rigorous description of the overall characteristics of a pneumatic transmission system is beyond the scope of this book, but Figures 42.20–42.22 show the magnitude of the

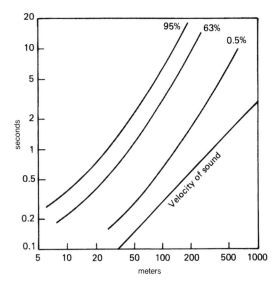

FIGURE 42.22 Time delay for 63.2 percent response to step change of applied pressure for various lengths of tube.

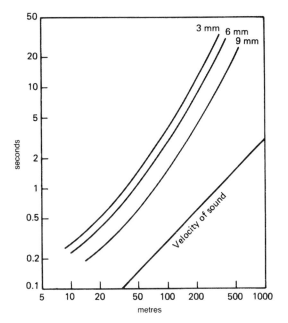

FIGURE 42.20 Time delay in response to step change of pneumatic pressure applied to various lengths of tube.

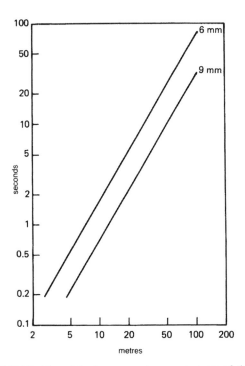

FIGURE 42.21 Time delay in response to a constant rate of change of pneumatic pressure applied to various lengths of tube.

effects for reasonably representative conditions. It is worth noting, however, that too-short tubing (capacity) can cause problems (oscillations), and four- and five-pipe systems were developed to overcome transmission delays. Booster relays were also often used in conjunction with final control elements that required larger capacities.

Figure 42.20 shows the time taken for the output to reach 95 percent, 63 percent, and 0.5 percent of its ultimate value following a step change from 20 kPa to 100 kPa at the input according to the length of tube. Figure 42.21 shows the time lag versus tubing length when the input is changing at a constant rate. Figure 42.22 shows the time delay for a 63.2 percent response to a step change of 20 to 100 kPa versus tubing length for 9 mm, 6 mm, and 3 mm tubing.

42.5 PNEUMATIC CONTROLLERS

42.5.1 Motion-Balance Controllers

Many of the early pneumatic controllers were based on the principle of motion balance because the sensor associated with the primary element produced a mechanical movement, rather than a force, to provide the signal proportional to the measured quantity. The technique is still widely used, and the Foxboro Model 43A, an example of a modern implementation of the concept embodying several variations of the control actions, is described in the following sections.

In most of these controllers a spring-loaded bellows or a Bourdon element converts the incoming signal into a motion. This motion is used directly to drive a pointer so that the measured value is displayed on a scale. On the same scale there is a second adjustable pointer, the position of which identifies the set point or desired value for the controller action. For the majority of applications the required action is proportional plus integral, and is derived as follows.

As shown in Figure 42.23, a mechanical linkage between the two pointers is arranged so that movement of its midpoint is proportional to the difference between the measured value and the set point. This difference is applied horizontally at one end of the proportioning lever which is pivoted at its center to a flat spring. A bellows connected to the pneumatic output of the controller and opposed by the integral action bellows applies a force to the flat spring, with the result that the proportioning lever is positioned virtually according to the magnitude of the output signal combined with the integral action.

The lower free end of the proportioning lever bears on the striker bar of the flapper/nozzle system, and in so doing causes the flapper to cover or uncover the nozzle. Thus if the measurement exceeds the set point, the proportioning lever moves to the left (in the figure) and, being pivoted centrally, its lower free end moves to the right, off the striker bar, with the result that the flapper covers the nozzle. This causes the nozzle back pressure to increase, a change that is amplified by the control relay to a value that becomes the instrument output signal and is also applied to the proportional bellows, causing the proportioning lever to move downwards until it impinges again on the striker bar which, in turn, moves the flapper off the nozzle. Consequently the nozzle back pressure falls and the change, after being amplified by the control relay, is applied to the proportioning bellows as well as to the instrument output and via the integral tank and restrictor to the integral bellows. In this way any residual error between the measured value and the desired value is integrated and the output gradually adjusted until the error is eliminated.

This type of control action meets the majority of operational requirements. However, in a few instances proportional action alone is sufficient, in which case the integral bellows are replaced by a spring and the integral tank and restrictor are omitted.

On the other hand, there are a few instances where it is desirable to enhance the performance further by including derivative action, i.e., adding to the output signal a component that is proportional to the rate of change of the error signal. This is implemented by modifying the proportional bellows system as shown in Figure 42.24. In operation, a sudden change in the measurement signal actuates the flapper nozzle mechanism so that the output signal changes. This

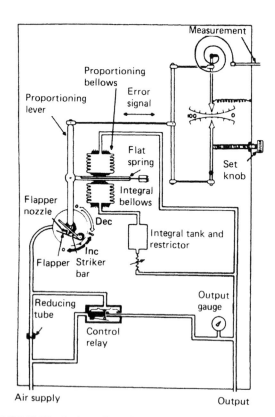

FIGURE 42.23 Basic configuration of pneumatic motion-balance two-term controller.

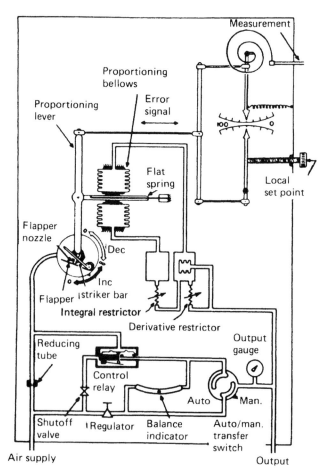

FIGURE 42.24 Basic configuration of pneumatic motionbalance three-term controller with auto/manual balance feature.

change is fed back to the proportional bellows directly by the derivative mechanism, but the derivative restrictor causes the transient signal to die away so that the pressure in the proportioning bellows becomes equal to the output pressure. In the meantime, the integral action builds up to reduce any difference between the measured value and the set point.

A further feature available in the instrument is the switch for transferring the controller from manual to automatic operation and vice versa. The switch itself is a simple two-way changeover device, but associated with it are a regulator and a sensitive balance indicator, comprising a metal ball mounted in a curved glass tube.

When it is desired to transfer the control loop from automatic to manual operation the regulator is adjusted until the ball in the balance indicator is positioned centrally to show that the regulator pressure has been set equal to the controller output. The process can then be transferred from automatic to manual control without imposing any transient disturbance. The reverse procedure is applied when the process has to be transferred from manual to automatic control, except that in this instance the set point control is adjusted to balance the controller output signal against the manually set value.

Other modes of operation which are available include differential gap action (also known as on/off control with a neutral zone), automatic shutdown (which is used to shut down a process when the measured value reaches a predetermined limit), on/off control (which is the simplest configuration of the instrument), remote pneumatic set (in which the manual set point control is replaced by an equivalent pneumatically driven mechanism) and batch operation (in which provision is made for avoiding saturation of the integral action circuit during the interval between successive batch process sequences). However, these configurations are rare compared with the simple proportional and proportional-plus-integral controllers.

42.5.2 Force-Balance Controllers

The Foxboro Model 130 series of controllers serves to illustrate the method of operation of modern pneumatic force-balance controllers. The basic mechanism of the control unit is shown in Figure 42.25. There are four bellows, one each for the set point, measurement, proportional feedback, and integral ("reset") feedback, which bear on a floating disc with two bellows on each side of a fulcrum whose angular position can be varied. In operation, the forces applied via each bellows multiplied by the respective distances from the fulcrum are held in balance by operation of a conventional flapper/nozzle system which generates the pneumatic output signal.

If the angular position of the fulcrum is set so that it is directly above the proportional feedback bellows and the reset bellows (as shown in Figure 42.26(a)), then the unit

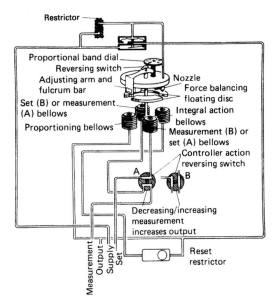

FIGURE 42.25 Basic mechanism of force-balance pneumatic controller.

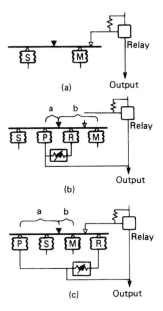

FIGURE 42.26 Force diagram for various controller settings. Adjustable fulcrum set for (a) on-off control action, (b) 25 percent proportional band or (c) 400 percent proportional band.

functions as an on/off controller. The slightest increase in measurement signal above the set point signal causes the nozzle to be covered so that output signal rises to the supply pressure. Any decrease in the measurement signal below the set point signal uncovers the nozzle so that the output signal falls to the zero level.

If the adjustable fulcrum is moved to the position shown in Figure 42.26(b), the proportional band is 25 percent (or the gain is 4) ($a/b = 1/4$). If the measurement signal increases, the flapper/nozzle will be covered so that the output pressure rises until balance is restored. This will occur when the output pressure has increased by a factor of four (if the integral action is disregarded).

If the fulcrum is adjusted to the position shown in Figure 42.26(c), the proportional band is 400 percent (or the gain is 0.25) ($a/b = 1/4$). Then an increase in the measurement signal causes the flapper/nozzle to be covered so that the output pressure also rises until balance is restored. This will occur when the change in output pressure is one quarter that of the change in measurement signal.

Referring again to Figure 42.26 (b), if the measurement signal increases, so does the pressure in the output bellows by a factor of four. However, this change of output signal is applied via a restrictor to the reset bellows, which acts on the same side of the fulcrum. Thus as the pressure in the reset bellows rises, the output signal must rise proportionally more to restore balance. A difference in pressure then exists between the proportional bellows and the reset bellows until the process reacts to the controller output and reduces the measurement signal. This, in turn, reduces the pressure in the proportional bellows until the difference is eliminated. The reset action continues so long as there is a difference between the measurement and set point, and hence a difference between the reset and proportional bellows.

The configuration of the complete controller shown in Figure 42.27 is centered around the automatic control unit. The manual set point is adjusted by a knob in the front panel of the instrument. It drives a pointer on the scale and operates a mechanism that generates a proportional pressure to be applied to the automatic control unit.

The local/remote set point mechanism includes a receiver bellows which drives a second pointer to indicate the received value of the set point. It includes a switch which allows either the local or remote signal to be selected (Figure 42.27). The derivative unit is actuated from the measurement signal only and is not affected by changes of set point. "Derivative" is

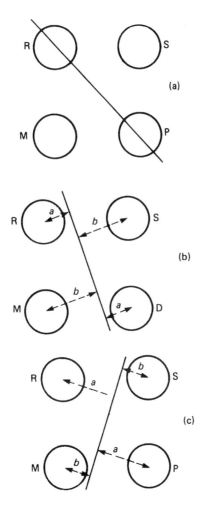

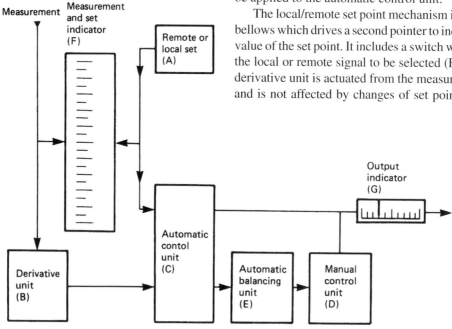

FIGURE 42.27 Configuration of complete pneumatic controller.

also known as "rate" and was initially known as "pre-act." As shown in Figure 42.28, it comprises a further force balance mechanism in which the moment generated by the signal to the measurement bellows A is balanced by the combined moments generated by two further bellows, one (C) arranged to generate a substantially larger moment than the other (B).

In steady conditions, the moment due to bellows A is equal to the combined moments of bellows B and C. Any increase in the measurement signal disturbs the balance, with the result that the relative position of the flapper/nozzle is reduced and the output of the linear aspirator relay rises. This signal is applied not only to the measurement bellows in the automatic control unit but also to bellows B and C.

However, the supply to bellows C passes through a restrictor so that, initially, balance is restored by the force generated by the bellows B alone. Subsequently, the force generated by bellows C rises and the effect is compensated by a fall in the output signal. After a time determined by the value of the restrictor, the pressures in bellows B and C become equal and they are then also equal to the pressure in bellows A. Thus, as a result of the transient, the effective measurement signal to the controller is modified by an amount proportional to the rate of change of the incoming signal, thereby providing the derivative controller action.

At the same time the full air supply pressure is applied via restrictors to chambers C and B so that the diaphragm takes up a position where the air vented to the atmosphere through the nozzle is reduced to a very low value.

42.5.2.1 Automatic Manual Transfer Switch

The design of previous pneumatic controllers has been such that, before transferring the control function from automatic to manual operation or vice versa, it is necessary to balance the two output signals manually; otherwise, the process would be subjected to a "bump," which may well cause unacceptable transient conditions.

To avoid this manual balancing operation, a unit has been devised and incorporated as a basic feature of the instrument. It is known as the automatic balancing unit, and, as shown in Figures 42.29 and 42.30, consists essentially of a chamber divided into four separate compartments isolated from each other by a floppy diaphragm pivoted at the center. One of the chambers includes a nozzle through which air can escape slowly in normal operation as a result of the relative proximity of the diaphragm to the nozzle. When the position of the diaphragm moves away from the nozzle the flow of air increases and vice versa.

The interconnection of the unit when the controller is set in the manual operation mode is shown in Figure 42.29. In this mode, the manual controller generates the output signal which is also applied to the "A" chamber in the automatic balancing unit. If the controller were in the automatic mode this same signal would be generated by the force balance unit operating in conjunction with the relay, in which case it would pass via the reset restrictor and tank to the reset bellows. But in the manual mode both these connections are closed by pneumatic logic switches.

Referring now to chamber D in the automatic balancing unit, this is connected directly to the proportional bellows (P) in the force-balance unit and so is held at that pressure. Any difference between the pressure in chambers A and D causes the floppy diaphragm to move with respect to the nozzle and so modify the pressure in chamber B which (in manual mode) is supplied via a restrictor. Thus as the output signal is changed by adjustment of the manual controller, the pressure in chamber A varies accordingly, causing the position of the floppy diaphragm to change. This, in turn, alters the rate at which air escapes via the nozzle from chamber B. This change is transferred via the restrictor to both chamber C and the reset bellows R in the force-balance unit so that the latter causes the pressure in its output bellows P to change until balance is restored through its normal modus operandi.

However, this same signal is also applied to chamber D, so that, in effect, the pressure in the proportional bellows P continuously follows that set by the manual controller. Consequently, when the overall operation is transferred from manual to automatic, the force-balance control unit will have previously driven its own output signal to the value set by the manual controller so that the process is not subjected to a transient disturbance.

Before the controller action is transferred from automatic to manual operation the mechanism in the manual controller will have been receiving the output signal. Therefore the manual control lever is continually driven to a position corresponding to the output signal so that it is always ready for the controller to be transferred from automatic to manual control.

When the transfer switch is operated, the thumbwheel becomes engaged so that the output signal is now generated by the manual controller instead of the force-balance automatic controller. Prior to this, the automatic controller output signal is applied via the pneumatic logic switch to both chambers A and D in the automatic balancing unit so that half of the unit does not apply any moment.

42.5.2.2 Batch Operation

When the controller is used to control a batch process or one in which the controller is held in the quiescent condition for an appreciable time it is necessary to include an additional feature which prevents saturation of the integral action that would otherwise occur because the measured value is held below the set point. (This is sometimes known as reset or integral wind-up.) When the process is restarted after being held in the quiescent state, the measured value overshoots

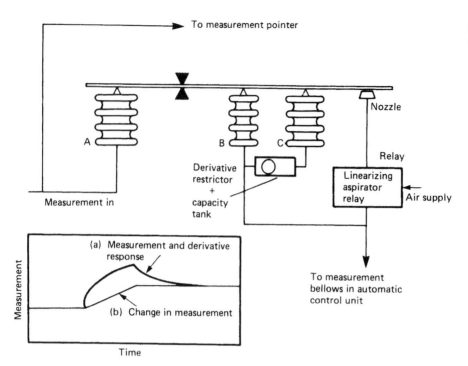

FIGURE 42.28 Basic mechanism of derivative function generation.

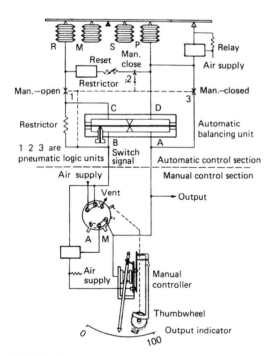

FIGURE 42.29 Manual-to-automatic transfer.

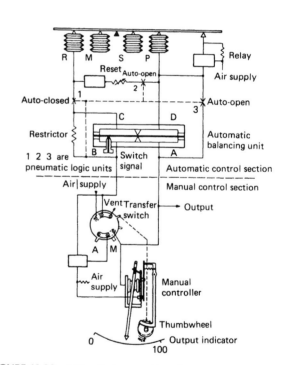

FIGURE 42.30 Automatic-to-manual transfer.

the set point unless the normal modus operandi of the reset action is modified. Reset windup does not just occur in batch processes. It can occur when a signal selector is used on the output, i.e., to select the highest or lowest of two controllers or when a shutdown system has operated and the controller winds up. The signal to the valve is used as an external signal that is fed back to the reset bellows of the controllers involved, or an external limiter is put on the reset signal.

The modification involves inclusion of a batch switch, which, as shown in Figure 42.31, is essentially a pressure switch actuated by the output signal from the controller but with the trip point set by a spring or by an external pneumatic

CHAPTER | 42 Pneumatic Instrumentation

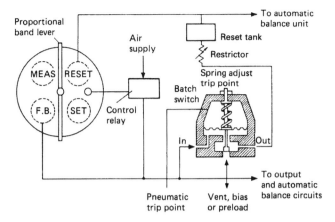

FIGURE 42.31 Functional diagram of batch-switch system. *Notes*: (1) Bellows are beneath circular plate, nozzle is above; (2) an increase in measurement causes a decrease in output.

signal. While the controller is functioning as a controller, the batch switch has no effect, but when the process is shut down and the measurement signal moves outside the proportional band, the batch switch is tripped, whereupon it isolates the controller output from the reset bellows which is vented to atmosphere.

When the measurement returns within the proportional band the batch switch resets and the integral action immediately becomes operative in the normal manner.

42.6 SIGNAL CONDITIONING

42.6.1 Integrators

The most frequent requirement for integrators in pneumatic systems is to convert the signal from head-type flowmeters into direct reading or count of flow units. Since head-type flowmeters generate signals that are proportional to the square of flow rate, the integrator must extract the square root before totalizing the flow signal. In operation, the Foxboro Model 14A Integrator accepts a standard pneumatic signal, proportional to 0–100 percent of differential pressure from a flow transmitter, which is applied to the integrator receiver bellows A in Figure 42.32. The force exerted by the bellows positions a force bar B in relation to a nozzle C. With an increase in differential pressure, the force bar approaches the nozzle, and the resulting back pressure at the relay regulates the flow of air to drive the turbine rotor E. As the rotor revolves, the weights F, which are mounted on a cross-flexure assembly G on top of the rotor, develop a centrifugal force. This force feeds back through the thrust pin H to balance the force exerted on the force bar by the bellows.

The centrifugal force is proportional to the square of the turbine speed. This force balances the signal pressure which, for head-type flowmeters, is proportional to the square root of the flow rate. Therefore, turbine speed is directly proportional to flow and the integrator count, which is a totalization of the number of revolutions of the turbine rotor, is directly proportional to the total flow.

The turbine rotor is geared directly to counter J through gearing K. Changes in flow continuously produce changes in turbine speed to maintain a continuous balance of forces.

42.6.2 Analog Square Root Extractor

The essential components of the square root extractor are shown in Figure 42.33, while Figure 42.34 shows the functional diagram. The input signal is applied to the A bellows and creates a force which disturbs the position of the nozzle with respect to the force arm. If the signal increases, the force arm is driven closer to the nozzle so that the back pressure increases. This change is amplified by the relay and its output is applied to both the C bellows and the B diaphragm.

Because the flexure arm is restrained so that the end remote from the flexure can only move along an arc, the combined effect of forces B and C is to drive the force arm away from the nozzle. This causes the nozzle back pressure to fall, which, in turn, reduces forces B and C until balance is restored.

Consideration of the vector force diagram shows that

$$\tan\theta = \frac{A}{B} \text{ or } B\tan\theta = A \qquad (42.2)$$

For small changes, $\tan\theta$ is approximately equal to θ. Hence $B\theta = A$. Because (1) force A is proportional to the input signal, (2) force B is proportional to the output signal and (3) the position of the flexure arm to establish the angle θ is proportional to the force C, it follows that $B \times C \propto A$. As the bellows C is internally connected to B it follows that $B^2 \propto A$ or $B \propto \sqrt{A}$.

42.6.3 Pneumatic Summing Unit and Dynamic Compensator

There are numerous process control systems that require two or more analog signals to be summed algebraically, and feed-forward control systems require lead/lag and impulse functions to be generated. The force-balance mechanism of the controller can be adapted to serve these functions. Figure 42.35 represents the arrangement of the bellows with respect to the fulcrum. It can be seen that with the gain set at unity (i.e., $a = b$) the two signals to be summed applied to the A and B bellows, respectively, the P and C bellows connected in parallel and supplied with air from the relay (which also supplies the output signal), then the output signal will be the average of the two input signals. If summing of the input signals is required, then only either the P or C bellows should be used. Similarly, signals may be subtracted by applying them to the A or B and C bellows, in which case the output is taken from the P bellows. In all these arrangements, the

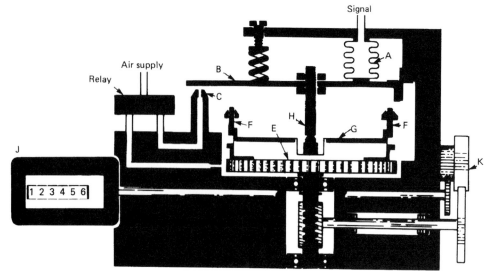

FIGURE 42.32 Functional diagram of pneumatic integrator.

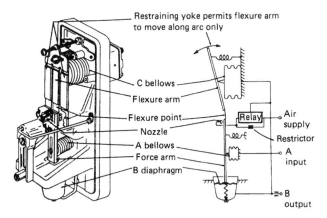

FIGURE 42.33 Functional diagram of pneumatic analog square root extractor.

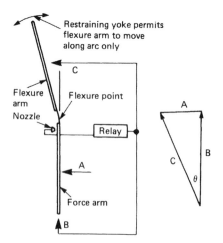

FIGURE 42.34 Force diagram for analog square root extractor.

positioning of the fulcrum allows a constant to be introduced into the equation.

To generate a lead/lag, the mechanism is arranged as shown in Figure 42.36. The input signal is applied to the A bellows, and the B and C bellows are connected together and supplied via a restriction and capacity tank from the P bellows, which, in common with the output, is supplied by the relay. As shown in Figure 42.37, the response of the system is determined by the setting of the gain and the restrictor.

Figure 42.38 shows the arrangement of the mechanism which generates an impulse response. The input signal is applied to bellows B, bellows P is connected to bellows B via a restrictor and a capacity tank, a plug valve is included so that one or the other of these two bellows can be vented, whilst the relay generates the signal for bellows C and the output from the relay. For reverse action, the roles of B and P are interchanged.

As shown in Figure 42.39, the output signal includes a positive- or negative-going impulse, according to the setting of the fulcrum and whether bellows P or B is vented. The recovery time is determined by the setting of the restrictor.

42.6.4 Pneumatic-to-Current Converters

It is sometimes necessary to provide an interface between pneumatic and electronic control systems. This is particularly true when an existing pneumatic measurement and control system is being extended in a manner that involves the transmission of signals over long distances. In general, pneumatic transmission systems are more costly to install than the equivalent electronic current transmission systems, and they suffer from a transmission lag which detracts from the system performance.

CHAPTER | 42 Pneumatic Instrumentation

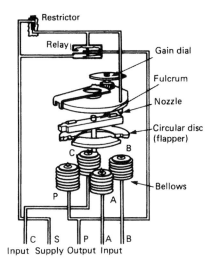

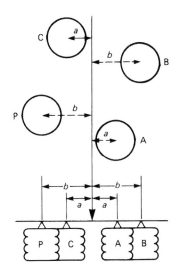

FIGURE 42.35 Basic configuration of pneumatic summing unit and dynamic compensator and force diagram.

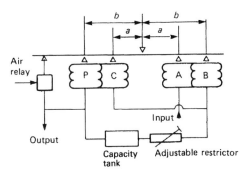

FIGURE 42.36 Force diagram for lead/lag unit.

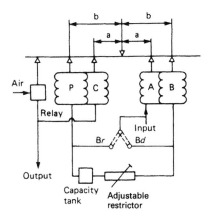

FIGURE 42.38 Force diagram for impulse generator.

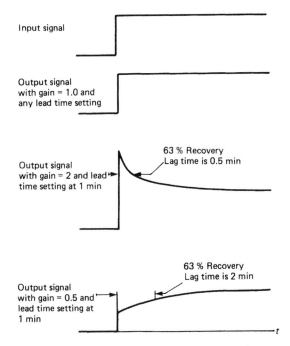

FIGURE 42.37 Response characteristic of lead/lag unit.

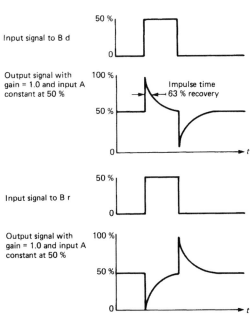

FIGURE 42.39 Response characteristic of impulse generator.

Various methods have been used in the past to convert the pneumatic signals into a proportional electric current; the majority of these units are now based on the piezoresistive sensors described in Part 1. These are fabricated on a wafer of silicon where a circular portion at the center is etched away or machined to form a thin diaphragm. One side of the diaphragm is exposed to the pressure to be measured while, on the reverse side, the strain and temperature sensors are formed by ion implantation or similar techniques which have been developed for semiconductor manufacture. The strain sensors are connected in the arms of a Wheatstone bridge from which the out-of-balance signal provides the measurement signal, but, because the gauge factor is affected by any change in the temperature of the silicon wafer, the temperature sensors are used to compensate for this by varying the excitation applied to the bridge network. There are several proprietary ways of implementing this compensation, but most require selection or adjustment of some components to optimize the compensation over a useful temperature range. Compared with most other pressure transmitters, these units are only intended for converting the output signal from a pneumatic transmitter into a proportional current, and, although they may be subjected to equally hostile environmental conditions, the sensor itself is only exposed to "instrument type" air pressures. In the majority of cases, this is a pressure between 20 and 100 kPa or the equivalent in metric or imperial units of measure. Occasionally there is a requirement to measure signals in the range 20–180 kPa, or the equivalent in other units, when power cylinders or positioners are involved.

A typical converter is shown in Figure 42.40. Its accuracy, including linearity, hysteresis, and repeatability, is ±0.25 percent of span and the ambient temperature effect is less than ±0.75 percent per 30° change within an overall temperature range from 0 to 50°C and ±1 percent per 30° change within the range −30 to +80°C. In some instances the converters can be located in a clean or protected environment such as a room in which other instrumentation is accommodated. For these applications, the robust enclosure is not required and the sensors, together with the associated electronic circuits, can be assembled into smaller units so that up to 20 of them can be stacked together and mounted in a standard 19-inch rack, as shown in Figure 42.40(b).

42.7 ELECTROPNEUMATIC INTERFACE

42.7.1 Diaphragm Motor Actuators

The pneumatic diaphragm actuator remains unchallenged as the most effective method of converting the signal from a controller into a force that can be used to adjust the setting of the final operator. In the majority of process plants, the final control element is a control valve, and the pneumatic diaphragm actuator is particularly well suited to provide the necessary force and stroke.

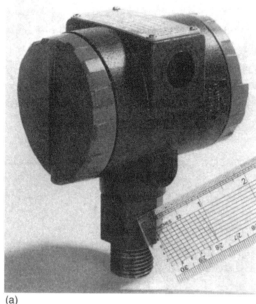

FIGURE 42.40 (a) Pneumatic-to-current converter; (b) pneumatic-to-current converters stacked in rack.

A typical actuator is shown in Figure 42.41. A flexible diaphragm that separates the upper and lower sections of the airtight housing is supported against a backing plate mounted on the shaft. A powerful spring, selected according to the required stroke and the air-operating pressure, opposes motion of the stem when compressed air is admitted to the upper section of the actuator housing. The stroke is limited by stops operating against the diaphragm backing plate.

The arrangement of the actuator and valve shown in Figure 42.41 is the customary configuration in which the spring forces the valve stem to its upper limit if the air supply or the air signal fails. A three-way valve (similar to a Fisher YD valve) is shown in the figure, and in this case a spring would force the plug to direct the flow from the input port C to the exit port L in the event of an air failure. If the actuator were to be attached to a straight-through valve, as would be the case if the exit port were blanked off, then the valve would close in the event of an air failure. In many instances this would represent a "fail-safe" mode of operation (see Part 4, Chapter 8), but in some plants the safe mode following failure of the air signal would be for the valve to be driven fully open. To deal with this eventuality, the actuator body can be inverted so that instead of the usual position shown in Figure 42.42(a) it is assembled as shown in Figure 42.42(b), in which case the spring would drive the valve stem downwards in the event of an air failure.

Pneumatic valve stem positioners, electropneumatic converters, or electropneumatic positioners, mounted on the valve yoke, can be used to enhance the precision or speed of response of the system.

42.7.2 Pneumatic Valve Positioner

A valve positioner is used to overcome stem friction and to position a valve accurately in spite of unbalanced forces in the valve body. One such unit is shown in Figure 42.43. It is usually mounted on the yoke of the valve and, in its normal mode of operation, a peg mounted on the valve stem and located in a slot in the feedback arm converts the stem movement into a shaft rotation.

Within the instrument, a flexure in the form of a "U" with one leg extended is mounted on the shaft. The incoming pneumatic signal is fed to a bellows which applies a proportional force to the flexure at a point in line with the shaft. The free end of the flexure carries a steel ball which bears on a flapper which, together with the associated nozzle, is mounted on a disc that can be rotated about an axis perpendicular to that of the shaft.

In the normal mode of operation, an increase in the input signal causes the bellows to apply a force to the flexure so that the ball at its free end allows the flapper to move closer to the nozzle. This causes the back pressure to rise, and the change after amplification by the relay becomes the output signal, which is applied to the pneumatic actuator and so causes the valve stem to move until balance is re-established. The rotatable disc provides the means for adjusting the sensitivity as well as the direct or reverse action when used with either "air-to-lift" or "air-to-lower" actuators.

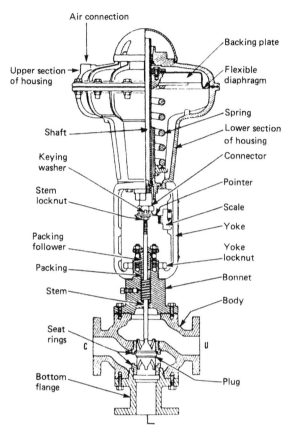

FIGURE 42.41 Pneumatic diaphragm motor actuator (mounted on three-way valve).

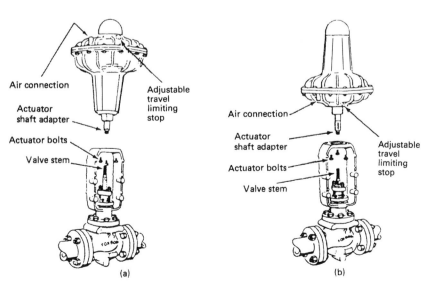

FIGURE 42.42 (a) Mounting for direct action; (b) mounting for reverse action.

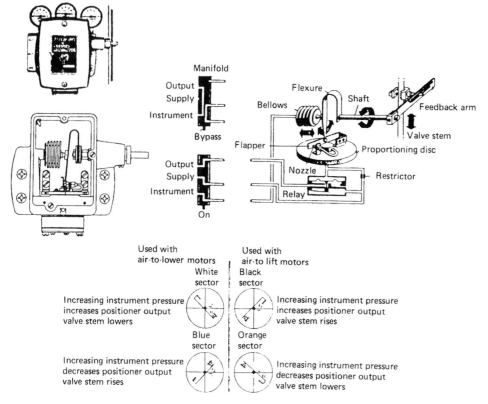

FIGURE 42.43 Pneumatic valve positioner.

42.7.3 Electropneumatic Converters

Although individual control loops may utilize electronic devices for sensing the primary measurement and deriving the control function, in the great majority of instances the final operator is required to provide a force or motion or both which can most readily be implemented pneumatically. Hence there is a requirement for devices which provide an interface between the electronic and pneumatic systems either to convert a current into a proportional air pressure or to convert a current into a valve setting by controlling the air supply to a diaphragm actuator. In most instances the devices have to be mounted in the process plants, where they are likely to be subjected to severe environmental conditions. Therefore it is important for them to be insensitive to vibration and well protected against mechanical damage and extremes of temperatures as well as corrosive atmospheres.

The Foxboro E69F illustrates one method of achieving these requirements. It involves a motion-balance system based on a galvanometer movement, as shown in Figure 42.44. The system comprises a coil assembly mounted on cross-flexures which allow it relative freedom to rotate about its axis but prevent any axial movement. The coil is suspended in a powerful magnetic field so that a current passing through the coil causes it to rotate. In the case of the electropneumatic converter, this rotation is sensed by a flapper/nozzle system, as shown in Figure 42.45.

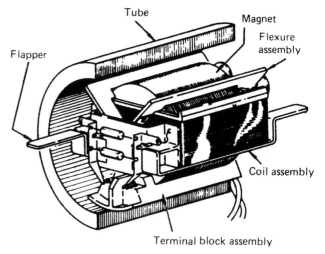

FIGURE 42.44 Galvanometer mechanism from electropneumatic converter.

The pneumatic section of the instrument comprises a feedback bellows and a bias spring which apply a force to a lever pivoted at one end by a flexure and carrying a nozzle at the free end. The nozzle is supplied via a restrictor and its back pressure is applied to a pneumatic relay whose output is applied to the feedback bellows and is also used as the pneumatic output from the unit.

CHAPTER | 42 Pneumatic Instrumentation

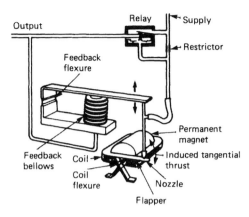

FIGURE 42.45 Basic configuration of electropneumatic converter.

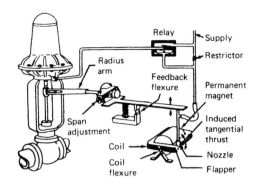

FIGURE 42.46 Basic configuration of electropneumatic valve-stem positioner.

With the current corresponding to the lower range value (e.g., 4 mA for a 4 to 20 mA converter) the relative position of the flapper and nozzle are adjusted so that the pneumatic output pressure is equal to the required value (e.g., 20 kPa for a 20 to 100 kPa system).

An increase in the input current causes the coil to rotate, and in so doing moves the flapper towards the nozzle so that the back pressure is increased. The change is amplified by the relay and applied to the feedback bellows, so that the lever moves the nozzle away from the flapper until a new balance position is established.

The system is arranged so that when the coil current reaches the upper-range value (e.g., 20 mA) the pneumatic output signal reaches its corresponding upper-range value (e.g., 100 kPa). A limited range of adjustment is available by radial movement of the nozzle with respect to the axis of the coil.

42.7.4 Electropneumatic Positioners

The Foxboro E69P shown in Figure 42.46 uses the same sensing mechanism in a positioner that drives the stem of a valve to a setting which is proportional to the incoming current signal. The positioner is mounted on the valve yoke and the valve stem is mechanically linked to it via a peg that is held by a spring against one side of a slot in the radius arm. As shown in Figure 42.47, movement of the valve stem causes the radius arm to rotate a shaft which passes through the instrument housing and the span pivot assembly. This comprises a roller whose axis is parallel to that of the shaft, but the offset can be adjusted (to vary the span). The roller bears on the spring-loaded lever which carries the nozzle at its free end.

In operation, an increase in the current signal causes the coil to rotate so that the flapper moves closer to the nozzle. This increases the nozzle back pressure and the change, after being amplified by the relay, is applied to the valve actuator, causing the valve stem to move. This movement is transmitted via the radius arm to the shaft carrying the roller, with the result that the springloaded lever is repositioned until the valve stem has moved to a position corresponding to the new input current.

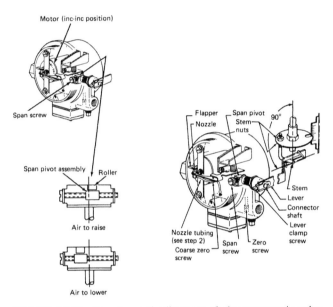

FIGURE 42.47 Mounting and adjustment of electropneumatic valve stem positioner.

Compared with straightforward air-operated actuators, positioners have the advantage of virtually eliminating the effects of friction and stiffness as well as the reaction of the process pressure on the valve stem.

REFERENCES

Anderson, N. A., Instrumentation for Process Measurement and Control, Chilton, London (1980).

Considine, D. M. (ed.), *Process Instruments and Controls Handbook*, 3rd ed., McGraw-Hill, New York (1985).

Foxboro, Introduction to Process Control (PUB 105B), Foxboro (1986).

Miller, J. T. (ed.), *The Instrument Manual*, United Trades Press (1971).

Fisher Bulletin 62.1:546 has an excellent picture of a modern I/P, while Fisher Bulletin 62.1–3582 is an excellent representation of a positioner. The reader is referred to the Fisher-Rosemount website (www.frco.com) for these bulletins, as they represent the most commonly used pneumatic instruments today.

Chapter 43

Reliability in Instrumentation and Control

J. Cluley

43.1 RELIABILITY PRINCIPLES AND TERMINOLOGY

43.1.1 Definition of Reliability

Reliability is generally defined as the probability that a component or assembly will operate without failure for a prescribed period under specified conditions. In order to ensure a meaningful result the conditions of operation, both physical and electrical, must be specified in detail, as must the standard of performance required.

Typical conditions which need to be considered are:

1. Variations in power supply voltage and size of voltage transients.
2. With a.c. supplies, the variations in frequency and harmonic content.
3. The level of unwanted rf energy radiated by the equipment must not cause interference to radio communications.
4. The equipment must be able to tolerate some rf radiation if it is to be used near high-power radio or radar transmitters.
5. Equipment for satellites and nuclear plants may need shielding from the ionizing radiations which they may experience.
6. Maximum and minimum ambient temperatures.
7. Maximum and minimum humidity.
8. Vibration and shock levels.
9. External conditions such as exposure to sand and dust storms, rainstorms, saltwater spray or solar radiation.
10. Air pressure.
11. Variations in loading (where relevant).

When quoting component reliability we need to consider only some of the above factors; for example, with small ceramic dielectric capacitors used in laboratory conditions only working voltage and temperature may be specified. When quoting the reliability of an electronic controller for the jet engine of an airliner, however, we may need to specify nearly all the factors, since the operating conditions are much more severe.

Where a range of variation is shown, for example with temperature, the equipment may be subject to rapid changes or regular cycling, and facilities must be provided to implement these changes when performing life tests. Also, to ensure testing under the worst conditions likely to be encountered it may be necessary to correlate variations of two parameters. For example, the worst conditions for producing condensation are minimum temperature and maximum humidity.

For continuously operating equipment the factor which indicates the amount of use is simply the elapsed time, but for intermittently used systems some deterioration may occur even though it is not energized. Thus the usage is indicated by adding to the switched-on time a fraction of the stand-by time. For some devices such as switches and relays usage is related to the number of operations rather than the switched-on time.

43.1.2 Reliability and MTBF

Although reliability as defined above is an important parameter to equipment users, it has the disadvantage that its value depends upon the operating period so that the manufacturer cannot specify a value for reliability which applies to all applications. It is convenient to look for such a value; the one usually adopted is the mean time between failures (MTBF). This applies to a maintained system in which each failure is repaired and the equipment is then restored to service. For systems which cannot be maintained, such as satellite control systems, the value of interest is the mean time to failure (MTTF).

For maintained systems the MTBF can be measured by operating them for a period T hours long enough to produce a number N of faults. The MTBF is then $M = T/N$ hours. This expression assumes that the time between faults does not change significantly with time, a reasonable assumption for much electronic equipment. It is not possible to test every system for all of its life, so any value for MTBF is subject to a sampling error.

Another measure of reliability which is often quoted is failure rate, that is, the number of failures per unit time. This is particularly useful in relation to individual components since it is the most convenient figure to use when estimating the reliability of a complete system in terms of the performance of its components. It is equal to the reciprocal of the MTBF, that is, the failure rate $\lambda = 1/M$.

For electronic components which have inherently high reliability the time interval is usually taken as 10^6 or 10^9 hours; for smaller periods the failure rate will be a small fraction.

Equipment manufacturers generally quote MTBF as an indication of reliability since the value does not depend upon the operating period which is often different for each user. However, reliability is the factor of prime concern to the user, and the relation between this and MTBF is thus important. The simplest case occurs when the failure rate and thus MTBF is constant with time. This is a somewhat drastic assumption but failure records of many maintained electronic systems show it to be justified. It is in most cases a slightly pessimistic assumption since the components used to replace faulty ones have been manufactured later after more experience with the production process has been acquired and are thus usually more reliable. Also those components which fail early are less reliable and in time are replaced by better versions.

Mechanical components can initially be assigned a constant failure rate but they will ultimately show an increasing failure rate as they reach the wear-out phase. Thus the maintenance program will usually involve changing or refurbishing these items well before they reach wear-out.

The relation between MTBF and reliability can be established using the Poisson probability distribution. This applies to a random process where the probability of an event does not change with time and the occurrence of one event does not affect the probability of another event. The distribution depends on only one parameter, the expected number of events μ in the period of observation T.

The series

$$S = \exp(-\mu)\left(1 + \mu + \frac{\mu^2}{2!} + \frac{\mu^3}{3!} + \cdots + \frac{\mu^r}{r!} \cdots \right)$$

gives the probability that 0, 1, 2, 3, r events will occur in the period T. If we take an event to be a system failure the probability we require is that of zero failures, that is, the reliability. This is then

$$R = \exp(-\mu) \quad (43.1)$$

43.1.3 The Exponential Failure Law

If the system MTBF is M we expect one fault in the period M and thus T/M in the period $T \cdot T/M$ is thus the expected number of events μ and the reliability is

$$R = \exp\left(\frac{-T}{M}\right) \quad (43.2)$$

The two quantities T and M must be expressed in the same units. Graphs of this relation on linear and logarithmic scales are shown in Figures 43.1 and 43.2.

As an example we consider the instrumentation module of a research satellite which is designed to have a working life of 5 years. If it has an MTBF of 40,000 hours what is its probability of surviving 5 years without failure? Here T must be expressed in hours as $5 \times 8760 = 43,800$.

Thus the probability of zero failure is

$$\exp\left(\frac{-T}{M}\right) = \exp\left(\frac{-43,800}{40,000}\right) = 0.335$$

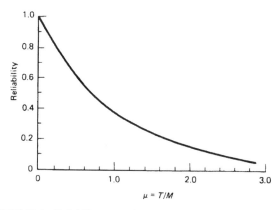

FIGURE 43.1 Reliability versus time graph: linear scale.

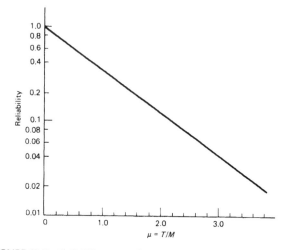

FIGURE 43.2 Reliability versus time graph: logarithmic scale.

This is not an acceptable figure. We can also use equation (43.2) to calculate what MTBF would be required to afford a more acceptable reliability, say 0.8. This gives

$$0.8 = \exp\left(\frac{-43,800}{M}\right)$$

Thus

$$\frac{-43,800}{M} = \log_e(0.8) = -\log_e(1.25)$$
$$= -0.2231$$

hence

$$M = \frac{43,800}{0.2231} = 196,000 \text{ hours}$$

If the module contains 500 components, their mean MTBF must be

$$500 \times 196,000 = 98 \times 10^6 \text{ hours}$$

The corresponding mean failure rate is the reciprocal of this, or $1/(98 \times 10^6) = 10.2 \times 10^{-9}$ per hour or 10.2 per 10^9 hours.

This is an extremely low failure rate, but not unattainable; the maintenance records of early American electronic telephone exchanges show component replacement rates of 10 per 10^9 hours for low-power silicon transfers, 5–10 per 10^9 hours for small fixed capacitors and only 1 per 10^9 hours for carbon film and metal film resistors. These exchanges are expected to have a working life of some 25 years, so high-reliability components are mostly used to build them.

43.1.4 Availability

Many instrumentation and control systems are used in situations where repairs cannot be started as soon as a fault occurs. The factor which indicates the probability of successful operation for a certain time is the reliability, which is the most important parameter for the user. This figure is critical, for example, for aircraft electronics, where there are no facilities for in-flight maintenance, although when the aircraft returns to the base the faulty unit is normally replaced and sent for subsequent repair. It is also the relevant factor for all satellite electronics which are considered non-repairable, although recent activities from the American Space Shuttle suggest that it may be possible in the near future to make some repairs to satellite equipment.

The manager of, for example, an automatic test facility is not interested directly in the MTBF of his or her equipment if he or she can get a fault repaired quickly. The manager's concern is the amount of work that can be put through in a given time; this involves both the time between failures and the repair time. The quantity usually quoted is the availability, defined as the proportion of the switched-on time during which the equipment is available for work.

This can be determined from the running log of the equipment by dividing the total switched-on time into the up-time U during which it is working and the down-time D during which it is faulty or being repaired. The switched-on time is then $U + D$ and the availability is

$$A = \frac{U}{U + D} \quad (43.3)$$

The unavailability or down-time ratio is $1 - A$ or $D/(D + U)$.

Where the equipment is used for some sort of batch operation there may be an extra period which must be included in D to allow for restarting the equipment. Thus if automatic test equipment fails part-way through a test, that test must be abandoned and after repair it must be repeated. To take this into account, the definition of down-time must include the time to find and repair the fault and the time needed to rerun any aborted test.

In this case there may be a regularly scheduled maintenance period and this is normally excluded from the down-time. However, for continuously running equipment any time devoted to regular maintenance must be classed as down-time.

The availability can also be expressed in terms of the MTBF M and the mean time to repair R as

$$A = \frac{M}{M + R} \quad (43.4)$$

This is the asymptotic value to which A converges; it is initially 1 and it decays to within 2 percent of the final value after a period of about $4/R$ from switching on.

The assumption of constant failure rate is an acceptable simplification for most purposes, but experience suggests that the mean repair time tends to fall somewhat with time as the maintenance staff acquire experience in diagnosing and repairing faults. However, the staff maintaining some very reliable equipment, such as electronic telephone exchanges with a design life of some 25 years, encounter faults so rarely that they have little opportunity to gain experience in a working environment. The usual way of providing practice is then to use a simulated environment into which various faults can be deliberately inserted.

43.1.5 Choosing Optimum Reliability

When the designer of electronic equipment asks the purchaser what level of reliability he or she requires in the product the initial answer may be, "As high as possible." In an ideal world this answer may be acceptable, but in the real world economic factors generally intrude. A widely used criterion is the total cost of ownership; the design which leads to the lowest total cost is then selected. The cost of ownership includes the initial purchase price, the cost of repairing faults and replacing components, and the cost of stand-by equipment. In some cases, for example, telephone exchanges or power generating equipment which is expected to be available 24 hours a day, there is an extra cost involved whenever it is inoperable as some income is lost during the down-time.

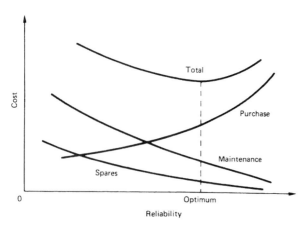

FIGURE 43.3 Relation between total life cost and reliability.

As a higher reliability is demanded the costs of design, manufacture, and testing all increase rapidly; this is reflected in the purchase price. The costs of repairs and maintenance and the standby equipment, however, all fall as reliability increases, as indicated in Figure 43.3. The total cost of ownership over the life of the equipment generally has a minimum value and the design which corresponds to this value is often chosen if no other criteria are involved. For high-reliability systems where failure may imperil human lives, there is usually a statutory requirement for a specified minimum reliability, and this becomes the overriding factor. The designer's aim is then directed towards achieving this reliability at minimum cost. Such requirements are stipulated for blind-landing control systems for passenger aircraft and for the safety circuits of nuclear power reactors.

The choice of the optimum reliability for some products may be affected by the existence of a guarantee period after purchase during which the manufacturer agrees to repair any fault (not caused by misuse) free of charge. Many domestic products fall into this category, generally having a guarantee period of a year. It is generally reckoned that any service call which has to be made during the guarantee period will largely cancel out the manufacturer's profit. The target MTBF in this case must be considerably greater than the guarantee period (usually 1 year) so that very few of the products fail during this period. Any change which increases product reliability must have its cost compared with the savings in the cost of free service calls which it produces. One change which appears to be advantageous is the substitution of the mechanical timer in a washing machine by an electronic controller with an embedded microprocessor. This gives a marked improvement in reliability at little additional cost and also provides more flexible control.

43.1.6 Compound Systems

In most assessments of reliability we need to evaluate overall system reliability in terms of the reliabilities of the system components.

This depends upon two rules for combining probabilities. The first is the product rule which gives the combined probability that two independent events will both occur as the product of the separate probabilities. Since reliability is the probability of zero failures, if the reliabilities of two components of a system are R_1 and R_2 this gives the system reliability as

$$R = R_1 \times R_2 \tag{43.5}$$

The assumption here is that both components must be working for the system to work.

This result can easily be extended to systems with more than two components as a product of all the component reliabilities. For example, if we consider a telemetry transmitting system consisting of a transducer and signal processing unit, a radio transmitter and aerial, and a power unit with reliabilities for a given duty of 0.95, 0.89, and 0.91, the overall system reliability is

$$R = 0.95 \times 0.89 \times 0.91 = 0.77$$

This kind of system in which all units must be working for the system to work is often called a series system. For some purposes we need to examine an alternative situation in which a number of different events may occur and we need to estimate the probability of any one occurring. This is simplest when the events are mutually exclusive, so that only one can occur at a time. This situation is covered by the addition rule which states that if p_1 and p_2 are the probabilities of the two events occurring separately, the probability of either one occurring is

$$P = p_1 + p_2 \tag{43.6}$$

Such a situation could be applied to the failure of a diode or transistor, for example. If the probability of a short-circuit fault is 9.5 and of an open-circuit fault is 2.4, both per million hours, the probability of either fault occurring is the sum, that is, 11.9 per million hours. This situation is one in which the two events are mutually exclusive since a device cannot be simultaneously open circuit and short circuit.

A more frequent case is that in which the two events can both occur. Here we have four possible outcomes:

1. Neither event occurs.
2. Event 1 only occurs.
3. Event 2 only occurs.
4. Both events occur.

The probability of one or more events occurring is the sum of the probabilities of (2), (3), and (4) since these are mutually exclusive. If the probabilities of the two events occurring are p_1 and p_2 the probabilities of these events are:

2. $p_1 \times (1 - p_2)$
3. $p_2 \times (1 - p_1)$
4. $p_1 \times p_2$

Here we use another axiom of probability theory: that is, if the probability of an event occurring is p, the probability of its not occurring is $1 - p$. Thus, in (2), p_2 is the probability that event (2) will occur, so the probability that it will not occur is $1 - p_2$.

Thus the combined probability required is the sum of these:

$$P = p_1 - p_1 p_2 + p_2 - p_1 p_2 + p_1 p_2$$
$$= p_2 + p_2 - p_1 p_2 \qquad (43.7)$$

When estimating the probability of failure of a complex assembly we generally simplify equation (43.7) since the individual probabilities are very small in a system which has an acceptable reliability. Thus the product terms will be negligible and we can estimate the overall probability of failure as the sum of the separate probabilities of all the components.

The probability of system failure is then

$$P = p_1 + p_2 + p_3 + \cdots + p_n \qquad (43.8)$$

We assume that the failure of any components will result in system failure. For example, if the individual component failure rates are of the order of 10^{-5} over the operating period, the product terms will be of the order of 10^{-10} and will be vastly less than the likely error in the assumed component failure rates. They can thus be safely ignored.

We can apply equation (43.7) to the calculation of the combined reliability of a system comprising two parallel units in which the system operates when one unit or both are working. Here the probabilities p_1 and p_2 are the reliabilities of the two units and the overall reliability is given by

$$R_T = R_1 + R_2 - R_1 R_2 \qquad (43.9)$$

For two identical units the combined reliability is

$$R_T = 2R - R^2 \qquad (43.10)$$

In a similar manner the combined reliability of a triplicate parallel system in which any one unit can provide service can be shown to be

$$R_T = 3R - 3R^2 + R^3 \qquad (43.11)$$

which is often called a 1-out-of-3 system.

The probability of failure of one unit is $1 - R$, so that the probability of failure of all three is $(1 - R)^3$. This is the only outcome which gives system failure. The reliability is thus $1 - (1 - R)^3$ which reduces to $3R - 3R^2 + R^3$.

This result assumes that any combining or switching operation required to maintain service has a negligible chance of failure. If this is not the case R_T should be multiplied by the reliability of the switching unit.

Equation (43.11) can be applied to a high-reliability power supply which incorporates three independent units each feeding the common power line through an isolating diode. If the diodes are heavily derated they should have a very low failure rate and the system reliability can be expressed in terms of the reliability of the three units. Using equation (43.11) involves the assumption that any one unit can supply the load on the system.

An alternative method of organizing a triplicate system is to follow it by a majority voting circuit. The system will then operate correctly if any two or all three of the units are working. Assuming identical units, the probability of all three units working is R^3 where the reliability of each unit is R. The probability of two units working and one faulty is $R^2(1 - R)$. This can occur in three ways since each of the three units can be the faulty one. Thus the overall reliability is

$$R_T = R^3 + 3R^2(1 - R) = 3R^2 - 2R^3 \qquad (43.12)$$

By combining series and parallel arrangements the reliability of more complex systems can be calculated. Some of these systems and their overall reliability R_T are shown in Figure 43.4.

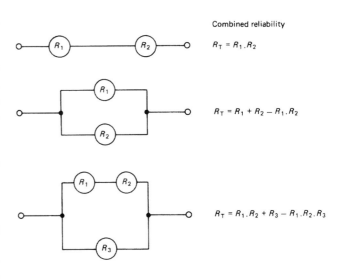
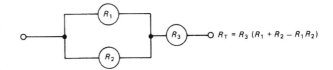
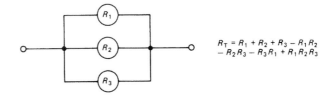

FIGURE 43.4 The reliability of compound systems.

43.2 RELIABILITY ASSESSMENT

43.2.1 Component Failure Rates

An essential part of the design of reliable instrumentation and control systems is a regular assessment of reliability as the design proceeds. The design can then be guided towards the realization of the target MTBF as the finer details are decided and refined where necessary. If reliability assessment is delayed until the design is almost completed there is a danger that the calculated MTBF will be so wide of the target that much design work has to be scrapped and work restarted almost from the beginning.

In the preliminary design phase only an estimation of component numbers will be available and an approximation to the expected MTBF is all that can be computed. As the design proceeds more detail becomes available, more precise component numbers are available, and a more accurate calculation is warranted. In the final assessment we can examine the stress under which each component operates and allocate to it a suitable failure rate.

The basis of all reliability assessment of electronic equipment is thus the failure rates of the components from which it is assembled. These rates depend upon the electrical stresses imposed on each component, the environment in which it functions, and the period of operation. The effect of the environment and the stress will vary widely, depending upon the type of component involved, but all components tend to show similar trends when failure rates are plotted against time.

43.2.2 Variation of Failure Rate with Time

The general behavior of much electronic equipment is shown in Figure 43.5 in which failure rate is plotted against time. In view of its shape this is often called the "bathtub" curve. The graph can be divided into three areas: an initial period when the failure rate is falling, usually called "infant mortality;" a longer period with an approximately constant failure rate corresponding to the normal working life; and a final period with an increasing failure rate, usually called the "wear-out" phase. This kind of behavior was first observed in equipment which used thermionic valves, but similar characteristics are also found in transistorized equipment, with a somewhat longer period of infant mortality. The failure rate is subsequently approximately constant. The fault statistics from maintained equipment, however, sometimes show a slowly decreasing failure rate which is usually attributed to an improvement in the quality of the replacement components. The argument is that as time passes the component manufacturer learns from failed components what are the more common causes of failure and can take steps to eliminate them. The manufacturer is thus ascending a "learning curve," and the reliability of the components is expected to increase slowly with time. The evidence from equipment fault recording and life tests thus either supports the assumption of a constant failure rate, or suggests that it is a slightly pessimistic view.

In a complex item of equipment such as a digital computer there are inevitably a few components with less than average reliability and these are the ones that fail initially. Their replacements are likely to be more reliable, so giving an improvement in system reliability. As these weaker items are eliminated we reach the end of the infant mortality phase and the system failure rate settles down.

Where high reliability is important, for example, in aircraft control systems, the customer usually tries to avoid the higher failure rate associated with the infant mortality phase and may specify that the equipment should undergo a "burn-in" period of operation at the manufacturer's factory before dispatch. The burn-in time is then chosen to cover the expected infant mortality phase.

The wear-out phase is never normally reached with modern transistorized equipment, as it is almost always scrapped as obsolete long before any sign of wear-out occurs. Even in equipment designed for a long life, typically 25 years, such as submarine repeaters and electronic telephone exchanges, the wear-out phase is not expected to appear.

Any component which has moving parts such as a switch or a relay must, however, experience a wear-out phase which

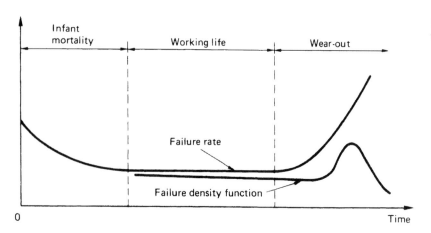

FIGURE 43.5 Failure rate and failure density function variation with time.

must somehow be avoided where high reliability is needed. Where no repair is possible, for example, in satellite control systems, bearings and other points of wear must be of sufficient size and their loading must be suitably low. This will enable the designer to ensure that wear will not degrade performance until the operating period is well above the expected life of the system. Long-life components of this kind need special design and testing and are thus expensive. Where regular maintenance can be undertaken, it may be better to use generally available components and replace them at regular intervals, before any wear causes a decrease in reliability.

43.2.3 Failure Modes

A component can be classed as faulty when its characteristics change sufficiently to prevent the circuit in which it is used from working correctly. The change may be gradual or sudden, classed as degradation or catastrophic failure. Catastrophic failure is due almost invariably to a short-circuit or open-circuit condition; typically diodes and capacitors may develop short circuits and resistors and relay coils may develop open circuits. Generally catastrophic faults are permanent, but some short-circuit faults may be only temporary if they are caused by small conducting pieces of wire which can be moved if the equipment is subject to vibration. Some years ago a major repair effort was needed on a number of avionic units into which a batch of faulty diodes had been built. In a defective manufacturing process some conducting whiskers had been sealed inside a batch of semiconductor diodes. They lay dormant during the normal testing procedures but as soon as the equipment was subjected to severe vibration in service the whiskers migrated and caused intermittent short circuits.

The effect of degradation failure depends upon the type of circuit into which the component is built. Generally analog circuits are more critical than digital circuits in that a smaller drift in characteristics will cause a fault. In a narrow band filter a change in the value of a capacitor by only 1 percent may be unacceptable; for example, in an *LC* tuned circuit resonant at 10 MHz it would shift the resonant frequency by 0.5 percent or 50 kHz. This would not be acceptable in a system with 25 kHz channel spacing. On the other hand, capacitors used for decoupling or power supply filtering may change value to a much greater extent before the circuit will fail. For example, to reduce size and cost, electrolytic capacitors are often used to perform these circuit functions and with conservative design the circuit can cope with typical manufacturing tolerances on capacitance of −10 percent to +50 percent without failure. For reliable circuit design it is important to calculate the change which each capacitor can undergo without causing failure and select the most suitable type. Fortunately there are many types of capacitors available from which to choose, each with its own characteristics and tolerances.

The situation is quite different with transistors in that an important characteristic, the common emitter current gain, always has a wide variation. It may be specified, for example, as within the range 200–800. Some manufacturers divide transistors of a given type into grades A, B, and C with different ranges of current gain, but even then the range of gain for a given type is quite wide and the designer has to accept this and produce circuits which will work correctly with any expected value of gain. With analog circuits this usually involves stabilizing the circuit gain by using a substantial amount of negative feedback.

Digital circuits are usually more tolerant of variations in current gain and can thus work satisfactorily with transistors having a wider range of current gain than can most analog circuits. Thus when considering degradation failure where transistor parameters drift slowly, it may be appropriate to allocate a higher failure rate to transistors used in analog circuits than to the same transistors used in digital circuits. Catastrophic faults will of course affect both types of circuit equally, causing complete failure. The relative frequencies of degradation and catastrophic faults will determine how much allowance must be made for the type of circuit in which the transistor is used. This is an example of one of the complications attending reliability assessment; the failure rate to be used for a particular component may depend much upon the type of circuit in which it is used.

43.2.4 The Effect of Temperature on Failure Rates

Almost all components exhibit a failure rate which increases with the temperature at which they operate. In many cases, particularly with semiconductor devices, the rate of increase with temperature agrees with that predicted theoretically by a law long used by chemists to relate the speed of a chemical reaction to the temperature at which it occurs. There are two failure modes which can be envisaged in which the speed of a reaction determines the onset of failure; the first is that in which a component has a slightly permeable sealing and is operated in an atmosphere containing contaminants, or despite good sealing a small amount of contaminant is trapped inside the component during manufacture. This may occur even if the component is fabricated in a so-called "clean room" with well-filtered air. No filter is able to remove all suspended matter and there is always the possibility of a minute amount of contaminant being present which can find its way into a component. In either case the contaminant diffuses into or reacts with the semiconductor and if present in sufficient quantity it will ultimately impair the device characteristics enough to cause failure. The faster the reaction proceeds the sooner will the device fail; consequently, if a number of the devices are used in an instrument its failure rate will be proportional to the speed of the reaction (Cluley 1981).

Another failure mechanism is dependent upon the nature of the encapsulation of a component and was a problem with early plastic moldings. If these were subject to electrical stress and humidity and were in a dirty atmosphere the film of moisture could conduct. The flow of current then damages the surface and leaves a carbonized track which provides a permanent leakage path across the insulator. This fault is more likely to occur if there is some defect, such as a small crack which provides a preferred leakage path.

There are two expressions used by chemists relating reaction speed and temperature. The simplest of these, which nevertheless agrees well with experimental results, is that due to Arrhenius:

$$\text{Reaction speed } k = c \times \exp\left(\frac{-b}{T}\right) \quad (43.13)$$

where b and c are constants and T is the absolute temperature (Klassen and van Peppen 1989). This law was originally developed empirically to fit the experimental results and it can be modified by taking logarithms of both sides of the equation to give

$$\log k = \log c \frac{-b}{T} \quad (43.14)$$

Thus if the component failure rate λ which is proportional to k is plotted against $1/T$ on log-linear paper the result should be a straight line with a slope of $-b$. If the failure mechanism is known the value of b can be calculated.

This expression is useful to enable the failure rate at one temperature to be calculated, knowing the failure rate at another temperature. It is particularly applied to estimate the failure rate of components at working temperatures from data accumulated in accelerated life tests conducted at high temperatures.

The form of equation (43.14) shows that the failure rate will change by a fixed ratio for a given small change in T. For reliability assessment it is convenient to use a 10°C increment in T. The corresponding factor for the increase in failure rate varies between about 1.2 and 2 for semiconductors, with somewhat larger values for other components. Equation (43.14) can be used to deduce how component failure rates depend upon working temperature; for example, if the factor for failure rate increase is 1.6 for a 10°C rise, and tests have established a failure rate of 15 per 10^9 hours for a working temperature of 30°C, the effect of increasing the working temperature to 55°C will be to increase the failure rate to $15 \times (1.6)\ 25/10 = 48.6$ per 10^9 hours. If accelerated life tests have been conducted at several high temperatures a plot of failure rates against $1/T$ can be used to extrapolate the results down to working temperature, so saving years of testing time.

43.2.5 Estimating Component Temperature

We have seen that component failure rates are markedly dependent upon temperature; it is thus important in any reliability assessment to establish the temperature limits that various components experience. This is relatively easy for components such as low-loss capacitors which do not dissipate any appreciable amount of heat and thus assume the same temperature as their surroundings. However, any component which absorbs electrical energy will dissipate this as heat and will thus be at a higher temperature than its surroundings. The maximum internal temperature for reliable operation is normally specified by the manufacturer and so as the ambient temperature increases the power dissipated must be reduced. This process is called "derating." Typical circuit design standards for carbon and metal film resistors suggest using the recommended power rating for ambient temperatures up to 45°C and derating linearly down to zero dissipation at 90°C. This is for normal free-air conditions; if forced cooling is used, higher dissipation can be allowed.

Some degree of derating is suggested for electronic equipment in commercial aircraft, as follows:

Tantalum capacitors	$T_C = T_M \times 0.67$
Aluminum electrolytic, paper, ceramic, glass, and mica capacitors	$T_C = T_M \times 0.72$
All resistors	$T_C = T_M \times 0.68$ or 120°C, whichever is lower
Semiconductors, ICs, diodes	Max. junction temp. = $T_{MJ} \times 0.6$
Relays and switches	$T_C = T_M \times 0.75$
Transformers, coils, and chokes	T_C to be at least 35°C below the manufacturer's hot spot temperature

Here T_C is the external surface or case temperature of a component, T_M the component manufacturer's maximum permissible body temperature at zero dissipation, and T_{MJ} the maximum junction temperature specified by the manufacturer with zero dissipation.

Where extremely high reliability is required the derating factors 0.67, 0.72, etc., may be reduced, but this will involve using larger resistors and heat sinks, so increasing equipment volume.

A critical factor in determining transistor reliability is the junction temperature; consequently it is important that the circuit designer should be able to estimate this fairly accurately. The usual basis for this is the analogy between the flow of heat and the flow of electric current. If we have a current source of strength I amperes at a potential of V_1 connected via a resistance of R to a sink at a potential of V_2 the relation between the potentials is

$$V_2 = V_1 + I \times R$$

The analogs of potential and current are temperature and thermal power (usually expressed in watts), so giving the equation

$$T_2 = T_1 + \theta \times W \quad (43.15)$$

Here temperatures are in degrees Celsius. θ is thermal resistance in units of Kelvin per Watt, and W is the power generated by the source in watts. In the case of a transistor T_2 is the junction temperature and T_1 the ambient temperature. The thermal resistance has two components: θ_1 the internal resistance between the junction and the transistor case and θ_E the external resistance between the transistor case and the surroundings. θ_1 is fixed by the design of the transistor and cannot be changed by the user, but θ_E depends upon the transistor mounting. In order to minimize the junction temperature θ_E can be decreased by mounting the transistor on a heat sink, a block of metal which is a good heat conductor (usually aluminum) with fins to help to dissipate heat and painted black.

For a BFY50, a metal-cased low-power transistor rated at 800 mW maximum power, the value of θ_1 is 35 K/W, and with no heat sink θ_E is 185 K/W. If we have a transistor dissipation of 800 mW and an ambient temperature of 30°C the junction temperature without heat sink would be

$$T_2 = 30 + 0.8(35 + 185) = 195°C$$

This is much too high for reliable operation, so a heat sink will be needed to improve the cooling. If we specify that the junction temperature must not exceed 100°C we can use equation (43.15) to calculate the thermal resistance of a suitable heat sink. Thus

$$T_2 = 100 = 30 + 0.8(35 + \theta_E)$$

whence $\theta_E = 52.5$ K/W. A corrugated press-on heat sink is available for transistors similar to the BFY50 (TO39 outline) with a thermal resistance of 48 K/W, which would provide adequate natural cooling. If the dissipation or the ambient temperature were higher forced air cooling would be needed.

The above calculation assumes that the transistor dissipation is constant. If the transistor is pulsed the situation depends upon the relation between the pulse duration and its thermal time constant. If the pulse duration is much smaller than the time constant the important factor is the average dissipation which can be used in equation (43.15) to determine the junction temperature. This will not fluctuate appreciably. However, if the pulse duration becomes comparable with the time constant it will be necessary to investigate the heating and cooling of the junction and determine its maximum temperature.

43.2.6 The Effect of Operating Voltage on Failure Rates

The main effect of a change in the working voltage on resistor reliability is that caused by the change in dissipation and so in the working temperature. Where resistors are used in low-power applications so that their temperature rise above ambient is small, the operating voltage will not significantly affect their failure rate.

There is some evidence that using voltage derating reduces the failure rate of semiconductors, and some users ensure that the manufacturer's maximum rated voltage is at least twice the operating voltage in order to improve reliability. This procedure can be applied to discrete transistors and diodes, but not to most digital devices. Packages such as transistor-transistor and emitter-coupled logic and most storage devices are designed to operate at fixed voltages, with tolerances of typically ±5 percent, so that any attempt to improve reliability by reducing the supply voltage will prevent circuit operation. CMOS digital circuits, on the other hand, are typically specified for operation with a supply voltage of 3–15 V. This allows scope for some voltage derating but there is a disadvantage in that the speed of operation is roughly proportional to the supply voltage. Thus voltage derating will result in longer switching times and slower operation.

Analog integrated circuits, such as operational amplifiers, on the other hand, can operate with a range of supply voltage, typically ±5 to ±18 V or ±3 to ±18 V. Some degree of voltage derating can be applied, but the maximum output voltage swing is approximately proportional to the supply voltage so that the consequence of derating is a reduction in the maximum output voltage.

Experience with paper dielectric capacitors showed a marked dependence of failure rates on operating voltage and an empirical law which fitted the results showed a fifth power relation with the failure rate being proportional to $(V/V_R)_5$. Here V is the peak operating voltage and V_R the rated voltage. The same relation can also be applied to polyester capacitors which are widely used in sizes up to about $1\mu F$. These are made with dc voltage ratings from 63 to 400 V so that substantial voltage derating can be used in typical transistor circuits with supply potentials of 20 V or less.

For example, if a polyester capacitor rated at 40 Vac has a failure rate of 12.5 per 10^9 hours at its working temperature, derating it to 25 Vac should give a failure rate of

$$12.5 \times \left(\frac{25}{40}\right)^5 = 1.19 \text{ per } 10^9 \text{ hours}$$

43.2.7 Accelerated Life Tests

The failure rates of modern components are so low at typical operating temperatures that any life tests under these circumstances will involve many thousands of components on test for long periods. For example, if we wish to test components which have an expected failure rate of 20 per 10^9 hours we need to put, say, 5000 on test for just over a year before a failure is likely. We need more than one fault to produce useful data so that a test such as this is bound to be very lengthy. We can obtain useful data in a much shorter time by artifically increasing the failure rate by a known amount, generally by operating the components at a high temperature. Transistors for submarine cable repeaters are required to have a working life of some 20 years and are tested typically at junction temperatures of 250°C and 280°C. Integrated circuits for electronic telephone exchanges are also required to have a 20-year life, but the reliability is not quite so important, as the cost of repairing a fault in an office or an exchange is much less than in a submarine repeater. Typical test temperatures for these are 125°C, 150°C, and 160°C. The acceleration factor for the test at 160°C compared with a working temperature of 70°C is calculated as 555 so that a test of 320 hours' duration represents a working life of 20 years. If we make a somewhat gross approximation that a 10°C rise in temperature causes an increase of failure rate by a fixed ratio, the ratio in this case is about 2.02.

Although accelerated life tests of semiconductors at elevated temperatures are widely used they are sometimes deceptive, as the failure mechanism changes at high temperatures and the straight-line extrapolation down to working temperatures fails as the graph changes slope. Despite this the high-temperature tests are valuable for comparing one component design with another, and also for producing faults in a reasonably short time for failure mode analysis.

43.2.8 Component Screening

Since the failure rate of nearly all components shows an initial value greater than that measured after the "infant mortality" period has elapsed, it would improve system reliability if these initial failures could be weeded out. Some can be eliminated by a few hundred hours of "soak test" at the maximum working temperature but this will not find all early semiconductor failures. It is thus common to give these separate screening tests at typically 125°C for 168 hours. The components are tested at ambient temperature, usually 25°C, before and after the screening, and any item which shows a significant change in characteristics is rejected. For transistors the most sensitive parameter is common emitter current gain; for operational amplifiers open-loop gain and cut-off frequency f_T. In addition to electrical tests, thermal cycling between temperature limits, vibration testing, and acceleration testing in a centrifuge may be used for screening out mechanically weak items. The bonding of the leads in current semiconductors is so effective that accelerations of 30,000 g are used for screening.

Data collected by AT&T from telephone exchanges suggest that the infant mortality phase can last for up to a year for equipment used in such benign surroundings. Thus high-temperature screening with an acceleration factor of, say, 100 will require some 87 hours to eliminate all likely early failures.

43.2.9 Confidence Limits and Confidence Level

A problem which arises in evaluating system reliability from the failure rates of its components is the relation between the figures to be used in design calculations and the results of life tests and equipment fault records. All life tests are carried out on only a sample of the components manufactured, and their validity depends upon the degree to which the sample is representative of the full population. The sample should be selected randomly, but even so its characteristics will not be exactly the same as those of the whole population, and will vary from sample to sample. Thus any data from life tests cannot be applied directly to the general population, but we can assign a probability, called the confidence level, that its failure rate lies between prescribed limits, called the confidence limits. Thus failure rates are stated in the form, "The failure rate of this component lies between 3.6×10^{-9} and 7.5×10^{-9} per hour with a confidence level of 90 percent." This implies that the probability that the failure rate lies between the limits quoted is 90 percent. Figures given in BS 4200: Part 7: 1982 show the number of component hours of testing needed to establish a specific failure rate with a 60 percent confidence level. Here we are interested only in the upper confidence limit, giving us an upper bound for the failure rate, so the lower limit is ignored. For a demonstrated failure rate of 10^{-8} per hour (10 FITs) the testing times are in millions of component hours:

No faults	91.7
1 fault	203
2 faults	311
3 faults	418
4 faults	524
5 faults	630

This is a considerably shorter test than would be required for a confidence level of 90 percent, in which for zero faults the test time would be 230 million hours and 389 million for one fault.

43.2.10 Assembly Screening

In addition to component screening, circuit boards are often subjected to screening. This is unnecessary if the complete system is screened, but any boards destined for spares must be subjected to the same treatment as the system. The usual tests may include temperature cycling, a soak test, and vibration. For American military and commercial equipment a range of test programs are used. For equipment used in an air-conditioned room with little temperature variation, temperature limits of 20 to 30°C are specified, whereas equipment used in a fighter aircraft is subject to rapid and wide-ranging temperature variations and would be cycled between −54 and 71°C whilst undergoing vibration and with the power supply periodically switched on and off. The test conditions are designed to represent the most arduous environment that the equipment is expected to encounter in service.

43.2.11 Dealing with the Wear-out Phase

The usual method of estimating system failure rate is to add together the individual failure rates of all the components in the system. This assumes that we are dealing with a minimum system with no redundant items, and that every component must be in working order if the system is to work properly. This simplified calculation ignores the probability of multiple faults, but modern components have such a low failure rate that this is a justified assumption for any normal working period.

In addition it is assumed that component failure rates are constant over the period considered. This means that none of them will have reached the "wear-out" phase of their working life. In repairable systems it may be necessary to replace some components with moving parts at regular intervals to ensure that they do not reach the "wear-out" phase. Otherwise, as their working life is less than that of most electronic components, they may severely limit the system reliability.

This procedure may be applied to small switches which may be expected to survive a million operations before failure. If such a switch is operated 100 times per day it should survive for just over 27 years. If, however, it is used in a hostile environment which reduces its expected life by a factor of 10, or it is operated 1000 times per day, the survival time is reduced to only 2.7 years, which may be much less than the expected operating period. It may then be necessary to replace the switches at, say, yearly intervals to ensure that they do not become a significant factor in reducing system reliability.

If test data on the time to failure of a batch of components are available a more precise estimate of the appropriate replacement policy can be made. This is based upon the failure rate or hazard rate curve shown in Figure 43.5 and a curve derived from it called the failure density function, also shown in Figure 43.5. The failure rate is defined as the ratio of the number of failures Δn occurring in the interval from t to $t + \Delta t$ to the number of survivors $n(t)$ at time t, divided by the interval Δt (Klaasen and van Peppern, 1989). It is thus

$$Z(t) = \frac{\Delta n}{[\Delta t \times n(t)]}$$

The failure density function $f(t)$ is defined as the ratio of the number of failures Δn occurring in the interval from t to $t + \Delta t$ to the size of the original test batch $n(0)$ at time $t = 0$, hence

$$f(t) = \frac{\Delta n}{[\Delta t \times n(0)]}$$

In many situations we expect very few failures during the operating period of reliable equipment, so $z(t)$ will be almost identical, but if we continue well into the wear-out phase during life tests we expect most of the items to fail. Then $f(t)$ and $z(t)$ will diverge as shown in Figure 43.5. If possible, we need to acquire enough data to plot approximately the wear-out peak in the failure density function $f(t)$. This is often a bell-shaped curve, as shown in Figure 43.6, the "normal" distribution. This arises when the variations in the time to failure are due to a number of causes, each producing a small change. The data used to plot the curve can be used to estimate the mean life T and the standard deviation σ. Tables of the distribution show that only 0.13 percent of items will have a time to failure less than $T - 3\sigma$, so this is often taken as the working life of the items. In a system expected to operate for a longer period, the items should be replaced at this time.

43.2.12 Estimating System Failure Rate

We take as an example for estimating unit reliability a small assembly consisting of the components in Table 43.1. The MTBF is thus $10^9/102.7 = 9.74 \times 10^6$ hours.
If five of these assemblies are used in a non-redundant satellite control system which is expected to have a 7-year life the total failure rate is $5 \times 102.7 = 513.5$ per 10^9 hours.

The MTBF is thus $10^9/513.5 = 1.95 \times 10^6$ hours and the reliability is

$$R = \exp\left(\frac{-T}{M}\right) = \exp\left(\frac{-7 \times 8760}{1.95 \times 10^6}\right) = 0.969$$

TABLE 43.1 Components of a small assembly

Component	Failure rate per 10^9 hours	No.	Total failure rate per 10^9 hours
Tantalum capacitor	3.0	2	6.0
Ceramic capacitor	2.8	8	22.4
Resistor	0.42	11	4.62
Transistor	8.3	3	24.9
Integrated circuit	12.4	2	24.8
Diods	4.0	5	20.0
			102.7

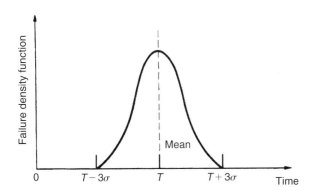

FIGURE 43.6 Normally distributed failure density function.

This calculation assumes that each assembly has no redundancy and it will thus not work satisfactorily if any component fails.

The expected number of faults in the 7-year operating period is $513.5 \times 7 \times 8760/10^9 = 0.0315$. Since $1/0.0315 = 31.8$ we may expect that if we had 32 satellites in orbit we would expect one of them to fail during its planned life.

43.2.13 Parallel Systems

We have seen that in series systems the overall reliability is the product of the reliabilities of the units which comprise the system. Thus the overall reliability must be less than that of any of the units. Using a parallel configuration we can construct systems which have a greater reliability than that of any of the units. This arrangement is thus of interest to designers of high-reliability systems. The system is usually designed so that any one of the parallel units can provide satisfactory service, and some switching mechanism is used to isolate a faulty unit. A simple example is a triple power supply for a critical control system in which the load can be supplied either by a regulated power supply connected to the mains, a float-charged battery, or a stand-by diesel generator. Each source is connected to the load through an isolating diode. If the reliabilities of the three sources are $R_1 = 0.92$, $R_2 = 0.95$, and $R_3 = 0.90$ for a particular operating period, the overall reliability will be: $R_T = R_1 + R_2 + R_3 - R_1R_2 - R_2R_3 - R_3R_1 + R_1R_2R_3 = 2.77 - 0.874 - 0.855 - 0.828 + 0.7866 = 0.9996$.

The expression for R_T is derived easily by considering the probability that the system will not work. This can happen only if all three units are faulty, the probability of this event being $R_F = (1 - R_1)(1 - R_2)(1 - R_3)$. The overall reliability (the probability of the system working) is the complement of this or $(1 - R_F)$, which gives the expression for R_T shown above. In this calculation we have assumed that the switching action of the isolating diodes cannot fail. This is clearly an optimistic assumption, although in some cases the diodes are duplicated to improve the reliability of the switching action. If they are also derated their failure probability will be very much less than the unit failure probabilities implicit in the above figures, and so it can justifiably be neglected. On the other hand if the diode failure rate is likely to be a significant factor it can be incorporated in the unit reliability figures. Thus if the diode reliability is estimated as R_D, the reliabilities of the three separate power sources must be changed to $0.92 \times R_D$, $0.95 \times R_D$ and $0.90 \times R_D$.

In some situations we may have a combination of series and parallel units; for example, in an instrumentation system involving a telemetry link from a moving vehicle there may be a power supply unit, a signal converter and multiplexor unit, and twin radio transmitters and aerials. The aerials may be separated to allow one to illuminate the shadows caused by buildings in the field of the other. Generally the output of one aerial alone will give a useful service, so we can evaluate the overall reliability by assuming that the system gives an acceptable performance with only one transmitter operating. The reliability diagram is then as shown in Figure 43.7.

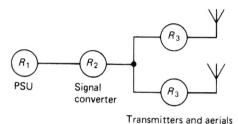

FIGURE 43.7 Reliability diagram of telemetry transmitting system.

This can be analyzed as a series system of three items, the PSU, the signal converter, and the transmitter/aerial combination. The respective reliabilities are: R_1, R_2, and $2R_3 - R_3^2$. The overall reliability is thus $R_1 \times R_2(2R_3 - R_3^2)$.

Other compound systems can be analyzed by using the expressions given in Figure 43.4.

43.2.14 Environmental Testing

Although estimates of MTBF are needed during the design and development phase of electronic systems, users often expect some demonstration of claimed figures. This is particularly the case with military and avionic systems where the specification generally calls for a minimum demonstrated MTBF.

This requires samples of the production run to be tested under the most arduous working conditions likely to be encountered in service. These may vary widely, depending upon the working environment. Thus instrumentation systems to be installed in power station control rooms will be working in a benign environment with no vibration, comparatively small temperature and humidity variations, a stable mains supply and free from dust and water spray. They can thus be tested adequately in a normal laboratory without extra facilities. The transducers and their associated electronics, however, which supply data to the rest of the system, are attached to boilers, turbines, alternators, etc., and so will be subjected to some vibration and high temperatures. These conditions must be incorporated into any realistic test program. The vibration may be a random waveform or a sinusoidal signal swept over a prescribed range of frequencies, and some specifications may also call for a number of shocks to be applied. Even more severe conditions are experienced by flight test equipment which will be subject to considerable shock and vibration, with rapid and large changes in temperature and rapid changes in humidity. It is also likely to encounter changes in power supply voltage and frequency considerably larger than those typical of a ground environment. All

CHAPTER | 43 Reliability in Instrumentation and Control

these factors must be represented in a full environmental test which requires the use of an environmental test chamber. This allows the user to regulate all of the conditions mentioned above, and also to subject the equipment under test to fresh- and saltwater spray and simulated solar radiation. Its electrical performance should also be monitored during the test.

Where the equipment may be subjected to electromagnetic fields provision must be made to generate such fields of known strength and frequency during the test to simulate service conditions. Also, where the equipment is to be used near sensitive radio apparatus, there may be a specified limit to the field strength it generates. To check this property, a field strength measuring set which covers the appropriate frequency is required. Where the power supply may be contaminated by short high-voltage transients such as those caused when switching inductive loads, provision must be made to inject transients into the power supplied to the equipment on test and to check that no malfunctions are caused by them.

Control equipment in satellites is subject to ionizing radiation not experienced by ground-based apparatus, and its resistance to this radiation must be checked in detailed environmental tests.

Generally the testing of equipment to be used in benign surroundings such as computer rooms and laboratories requires a minimum of environmental control, but as the environment to be simulated becomes more severe, with greater excursions of temperature, humidity, etc., and increasing vibration levels, the complexity and cost of simulating the working environment becomes much greater.

43.3 SYSTEM DESIGN

43.3.1 Signal Coding

All but the most elementary control and instrumentation systems involve several units, such as transducers, amplifiers, recorders, and actuators, which are connected together to perform the task required. An important feature of the system is the manner in which the signals passed between units are coded.

All early systems used analog coding in which the signal which represents a system variable has a magnitude proportional to it. For example, an oil lamp invented by Philon in about 250 BC used a float regulator in which the movement of the regulator was proportional to the level of oil.

One of the earliest applications of an analog controller for industrial purposes was the rotating ball governor used to control the speed of steam engines, developed in 1769 by James Watt. The geometry ensures that whenever the engine speed increases above the desired value the steam supply is reduced, so ensuring that any changes in speed are kept to a minimum.

Analog techniques continued to dominate control and instrumentation systems until the early 1950s, all the feedback schemes developed during the Second World War for the precise positioning of anti-aircraft guns, radar aerials, etc., being analog systems.

Analog coding is simple and in many ways convenient since most transducers deliver analog outputs, and many analog indicators such as pointer meters and chart recorders are simple and relatively cheap. However, it is difficult to store analog signals for later display and processing, and to transmit them over long distances without introducing noise and errors.

Many analog systems use electrical voltage as the quantity which represents the physical variables since this is the output coding of most transducers and a wide range of voltage measuring devices are available to display output quantities.

The accuracy of the display is dependent partly upon the degree to which the voltage across the display device equals the output of the signal processing element. In circuit terms, this means a negligible voltage drop in the wiring between them. Such a condition can easily be satisfied in a compact system, but installations which cover a large area such as power station instrumentation schemes may include some wiring runs which are many hundreds of meters long. In these circumstances the voltage drop in the wiring can no longer be neglected. However, if we change the analog quantity used from voltage to current the voltage drop, if not too large, will not affect the current in the line, and we can preserve the system accuracy. Accurate current generators are not difficult to design, and this arrangement is widely used in many boiler house and process control systems which involve long runs of cable. Although the use of current as the analog quantity almost eliminates the effect of conductor resistance on display accuracy, it requires the maintenance of a high insulation resistance between the pair of wires feeding a display, as any current leakage between them will degrade the accuracy of the display. In the hostile environment of a typical boiler house there may be heat and damp to contend with, and the avoidance of leakage requires great care with both the installation and the maintenance of all the system wiring.

The advent in the 1950s of transistors which were fast enough to be used in small digital computers enabled such machines to be made smaller, lighter, cheaper, and much more reliable than the valve machines which preceded them. They could consequently be used as the nucleus of a control or instrumentation system without incurring any major cost penalty. Since they could also complete complex calculations in a very short time and could easily store masses of data for as long as needed, they soon became the natural choice for major control and instrumentation systems.

When microcomputers were developed in the early 1970s it became possible to fabricate the central components of a small digital computer on a few silicon chips, each one only a few millimeters square. The manufacturing process was a high-volume operation, and so the cost and size of the computing element fell by a factor of some hundreds. This immensely widened the scope of digital control and measuring systems since they could be used in much cheaper products without incurring extra cost. For example, they were built into the weighing machines used in retail shops, pumps gasoline, and domestic appliances such as washing machines and microwave ovens. By this time microprocessors had been developed to include storage for both program and data, interface registers, counter/timers, and DACs and ADCs. Thus a working system needed only a few packages in addition to the microprocessor itself.

43.3.2 Digitally Coded Systems

The availability of fast arithmetic operations enables much signal processing to be performed in real time so that results can be displayed quickly and, generally more important, can be used in a feedback control system. Examples of typical calculations are fast Fourier analysis and the rms value of a sampled waveform. Even simple 8-bit microprocessors such as the Motorola 6809 can multiply two 8-bit numbers together in only microseconds.

A further advantage of digital coding is its ability to withstand signal attenuation and noise. For instance, many digital systems use signal levels consistent with those of TTL circuits. In the original versions of these circuits, binary 1 is coded as a voltage between 2.4 and 3.5 when transmitted. The receiving device will accept this as a 1 if it exceeds 2.0 V. The interconnection can thus tolerate a negative induced noise transient of 0.4 V for a 1 signal. The maximum transmitted level for a logic 0 is 0.4 V, and an input will be recognized as a 0 if it does not exceed 0.8 V. Thus in this logic state the circuit will also tolerate a positive induced transient of 0.4 V (the noise margin) before an error is introduced.

Later versions of TTL with Schottky diodes incorporated to reduce transistor saturation had a somewhat larger noise margin in the 1 state but the same in the 0 state.

If we compare this with an analog system with a full-scale signal of 10 V, the same induced voltage of 0.4 V will give an error of 0.4/10 or 4 percent. If the full-scale voltage is 5 V the error will be 8 percent. TTL packages are used where high currents are needed, for example, to drive the solenoids in a printer, but much less current is needed to drive other logic packages, and MOS logic suffices for this. These circuits will operate on any power supply voltage from 3 to 15 V, but take longer to switch than TTL. As they have complementary transistors in the output stage their noise margin is about 40 percent of the supply voltage, that is, 2 V for a 5 V supply and 4 V for a 10 V supply. With a 10 V analog system, a 4 V noise component would give an error of 40 percent, which would be quite unacceptable. Although these figures show that digital systems can tolerate much more induced noise than analog systems, the comparison is not quite as marked as indicated since the analog system will usually require a much smaller signal bandwidth than digital systems. Thus short noise pulses which would cause an error in a digital system can be much reduced in analog systems by passing the signals through a low-pass filter which can be designed to pass the relatively low-frequency analog signals but provide considerable attenuation to the noise pulses. However, low-frequency noise such as mains interference will not be affected by such filtering.

A further advantage of digital coding is that by adding extra digits to a byte, or a longer packet of data, error detection or error correction can be provided. The simpler schemes can cope only with a single error so the size of a block to which the redundant check bits are added must be small enough to make the probability of two errors insignificant.

43.3.3 Performance Margins in System Design

Although when describing how a control or instrumentation system operates it is customary to start with the system inputs and follow their path to the output devices, the reverse process is needed for system design. Thus, if we are designing a ship's automatic steering control, the first data needed are the maximum torque which must be exerted on the rudder to maintain a given course and the maximum rate at which the rudder angle must be altered. These two items enable the power level of the steering motor and its gearing to be calculated, as well as the electrical input power needed by the motor. This must be controllable in magnitude and polarity by low-voltage circuits, often incorporating a microprocessor, which sense the ship's actual heading and compare it with the requested course. A large degree of power amplification is needed to supply the steering motor, since the control input may be only 25 mW (5 V and 5 mA) and the motor may require many kilowatts.

An important factor in designing a reliable system is the provision of adequate performance margins in its various components. In this case reserve motor power should be provided to allow for such factors as wear, the increasing frictional load caused by corrosion, and the extra rudder angular velocity needed for emergency maneuvers. An equal reserve of both output power and gain must be provided in the power amplifier which supplies the motor.

Safety margins of this kind are mainly decided by engineering judgment and previous experience since there is no exact method of deciding their size. However, where we need to allow for known tolerances on various system parameters we can make a more exact estimate.

43.3.4 Coping with Tolerance

In order to illustrate the way in which parameter tolerances can be handled we take as an example an instrumentation system with transducers, signal amplifiers, signal processors such as phase-sensitive detectors, and some indicating or recording device.

Each of these will have a characteristic such as gain, or in the case of a transducer, the output expected for a given stimulus. In the case of a force transducer this might be millivolt output per kilogram of force exerted on it. This is specified by the manufacturer with an associated tolerance. When the system is first installed its overall performance will normally be calibrated, in this case by placing a known weight on the transducer and adjusting the system gain until the output device reads correctly.

The calibration test will normally be repeated at regular intervals, their spacing being determined by the rate at which the performance drifts and the maximum permissible system error.

When considering system design it is important to allow an adequate performance margin overall to allow the system to be restored to its specified performance during the whole of its expected life. This means that the maximum expected variation in all system characteristics must be estimated to allow the likely variations in the end-to-end characteristic to be assessed. The usual procedure is the "worst-case" design in which all parameters are assigned values, usually at the extreme bounds of their tolerance, which will cause the greatest effect on the overall characteristic. The variation in the overall performance can then be calculated and provision made to cope with it. If we start with all parameters at their minimum value we can determine the least value of gain needed in the signal amplifier to enable the system to be calibrated. We assume that a number of systems are being constructed, all to the same design. We must then consider a system with all the parameters at their maximum values, and calculate its performance. This will be well in excess of the specified performance, and we can determine how much adjustable attenuation must be included in the system to enable it to be correctly calibrated.

As a simple example we will consider a weighing system consisting of a strain gauge transducer (sensitivity 2 mV ± 5 percent per kg) followed by an amplifier/phase-sensitive detector (PSD) output $G \pm 10$ percent Vdc for 1 V rms input) which drives a digital display requiring 10 V ± 3 percent for the full-scale indication of 10 kg. The arrangement is shown in Figure 43.8.

The minimum transducer output for the full-scale load of 10 kg is 20 mV − 5 percent = 19 mV. A display having minimum sensitivity will require an input of 10 V + 3 percent = 10.3 V.

Thus the minimum amplifier gain needed is $10.3/(19 \times 10^{-3}) = 542$. This must be the minimum gain, i.e., $G - 10$ percent, whence the nominal gain $G = 542/0.9 = 602$. The maximum gain is 10 percent above this, i.e., $602 \times 1.1 = 662$.

The maximum transducer output is 20 mV + 5 percent = 21 mV for full-scale load. This will produce a display input of 662×0.021 V = 13.9 V.

The display will require 10 V − 3 percent = 9.7 V at maximum sensitivity. Thus the attenuation ratio needed to enable the system to be calibrated in all circumstances is 13.9/9.7 = 1.43. If the input resistance of the display is large compared with R_1 and R_2, this gives the conditions $(R_1 + R_2)/R_2 > 1.43$.

43.3.5 Component Tolerances

In order to design reliable electronic equipment it is essential to allow for changes in device characteristics during their working life as well as the tolerances in their initial values. Generally tolerances are specified as measured just before components are delivered to the user. There is usually some delay before the components are assembled, so tolerances at assembly will be a little wider to allow for small changes during storage. For example, carbon film resistors typically have a maximum drift during storage of 2 percent per year, whereas metal film resistors which are much more stable are quoted as having a maximum drift of only 0.1 percent per year. The assembly process usually involves soldering the component to a printed circuit board and so heating it momentarily. This is also likely to cause a small change in the component value. Some equipment will be screened, which usually involves a short period of operation at high temperature and may cause a small change in value. Finally, the value will drift during the working life of the equipment. The end-of-life tolerance which the designer must allow for is thus significantly greater than that measured immediately after the component is manufactured.

Both passive and semiconductor components will experience a drift in characteristics, but as semiconductor tolerances in parameters such as current gain are so large the comparatively small changes during assembly and normal life pose little problem to the designer. For example, if a transistor is specified to have a current gain in the range 100–300, any

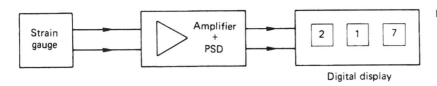

FIGURE 43.8 Strain gauge instrumentation.

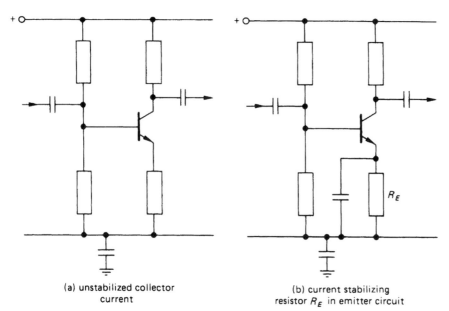

FIGURE 43.9 Stabilizing collector current by use of an emitter resistor.

circuit which can accept this wide variation can easily cope with a 5 percent drift during assembly and service.

A circuit block often required in instrumentation and control systems is a voltage amplifier having a closely specified gain. In view of the wide variation in open-loop gain caused by transistor tolerances, the customary way of meeting the requirement is to use overall negative feedback. As this is increased the closed-loop gain depends increasingly upon the attenuation in the passive feedback path. At its simplest this will be the ratio of two fixed resistors. Thus we can cope with wide variations in the characteristics of the active amplifier components if we can ensure constant resistance values in the feedback path. This is a much easier task, since we can obtain metal oxide resistors which at low power levels will drift less than 0.1 percent (film temperature 30°C) during a 25-year life. This application is for submarine repeaters, but a similar requirement for long life in an environment which precludes repair arises in the control systems of commercial satellites. The same resistor is estimated to have a drift of just over 1 percent in 25 years at 70°C.

Many low-power amplifier requirements are conveniently met using integrated circuits. These generally have even wider gain tolerances than discrete transistor amplifiers, and only the minimum gain is usually specified. They are usually operated with a high degree of feedback to stabilize the overall gain and reduce distortion; again, the performance is dependent upon resistor stability.

43.3.6 Temperature Effects

Some environments in which control and instrumentation systems operate, such as manned control rooms, have a measure of temperature regulation, and the equipment they house is subject to only small temperature variations. At the other extreme, electronic engine controllers used in aircraft are mounted near jet engines and may thus suffer wide temperature variations. For reliable performance the equipment designer must investigate the component changes caused by temperature variations and ensure that they will not prevent system operation.

For example, metal film resistors typically have a temperature coefficient of +50 parts per million (ppm). A temperature change of 80°C will cause a change in resistance of only 0.4 percent, which is less than the manufacturer's tolerance of ±1 percent and should not prevent most circuits from operating correctly. In many cases, particularly feedback amplifiers, the important factor is the ratio of two resistors rather than their absolute value. If the same type of resistor is used, both resistors should change by nearly the same proportion, and their ratio will change very little. Where high precision is important, wire-wound resistors having a very low temperature coefficient of around +5 ppm are available, but owing to their inductance they are not suitable for use at high frequencies.

The most stable capacitors for values up to 10 nF are silvered mica types which have a typical temperature coefficient of +35 ppm, so that for most purposes the change due to temperature variations can be neglected.

Inductors also have significant temperature coefficients which can be minimized by using a single-layer air-cored coil. This results in coefficients of 5 to 15 ppm. Low-inductance coils wound on ceramic formers, or better, with the low-expansion conductor deposited in a groove on the surface of the former, yield coefficients of around 1 ppm.

These low-temperature coefficients of inductance and capacitance cause designers few problems except when both components are connected together in LC oscillator circuits where frequency stability is important. One method of reducing frequency drift is to split the tuning capacitor into two sections, one of which has a negative coefficient. If a suitable combination of negative and positive coefficients is used, the frequency drift can be reduced to well below 1 ppm.

Where the potentials in a circuit are determined by a resistor chain we have seen that the effect of temperature changes on these potentials will be very small. Matters are quite different, however, if the circuit includes semiconductor junctions. For silicon devices the current increases about 15 percent for a 1°C rise in temperature at constant voltage. This means that a 20°C rise would cause an increase in current by a factor of just over 16. Since in an adverse environment the temperature may change much more than this, constant voltage operation of diode and transistor junctions is quite unacceptable. The usual method of stabilizing the transistor current against temperature changes is to connect a resistor in series with the emitter, chosen to ensure a voltage drop across it of least 2 V. The base voltage is held almost constant by a resistive potential divider. If the junction current is held constant, the base-emitter voltage falls by about 1.5 mV for each °C junction temperature rise. Thus for an 80°C rise, V_{be} falls by about 120 mV. For a fixed base voltage, the voltage across the emitter resistor will rise by the same amount, so causing an increase in emitter current by 0.12/2 or 6 percent. This is a rather crude calculation, but it is adequate to show the effectiveness of the method which is widely adopted. An alternative method of stabilizing emitter current is to make the base bias voltage fall with temperature at the required rate of about 2 mV per °C. This is often done by deriving the bias voltage from the voltage across a diode supplied with constant current. To obtain effective stabilization the diode and the transistor it regulates must be at the same temperature, generally arranged by mounting them on the same heat sink. This form of biasing is usually adopted for high-power amplifiers which have comparatively low-voltage supplies (typically 12 V for mobile operation). The 2 V dropped across an emitter resistor would then represent a significant power loss and a reduction in the effective voltage available for the transistor.

A final method of removing the effect of temperature changes is to isolate the circuit from them by enclosing it in an oven maintained at a constant temperature which must, of course, be above the maximum ambient temperature. The cost and power drain needed for this scheme means that it can be used in practice for only a small circuit package, typically a tuned circuit or a crystal used to determine the frequency of an oscillator. We would expect an improvement in frequency stability by an order of magnitude or more when using a constant temperature oven.

43.3.7 Design Automation

Although some degree of automation is generally used in the design of electronic systems, it is largely confined to detailed activity such as the analysis of analog and digital circuits and simulating their behavior and assistance to the manufacturing process by helping the layout of printed circuit boards and integrated circuits. It is also used in the testing of the product at various stages.

Most of the programs used have been available for some years, and nearly all of the faults in them have been discovered and removed. Despite this, devices designed with their help still reveal occasional unexpected errors. The problem is that nearly all design aids involve at some stage computer programs which cannot at present be generated without some human effort, which in turn is likely to introduce errors. Thus all design activity should assume that errors will be present initially and some procedure for finding and correcting them is necessary.

The usual recommendation is to hold regular audits or reviews of the design, preferably conducted by engineers not involved directly in the design process. Experience shows that the designer is likely to overlook a mistake if he or she conducts the review him- or herself. This follows the advice given to authors that they should ask someone else to proofread their work.

The problem of eliminating design errors has become of increasing interest as hardware has become more reliable and more faults are attributed to design and fewer to components. Although formal methods of designing systems are being developed (Diller 1990) they are as yet unable to tackle complex logical devices and are not in use commercially. The problem of exhaustive testing of intricate devices such as microprocessors lies in the large number of combinations of data, instructions, and storage locations to be investigated. For example, even a small 8-bit microprocessor containing perhaps 70,000 transistors has many instructions and can address over 60,000 storage locations; a multiplier handling two 16-bit integers will have over 4 billion different input combinations. If every instruction is to be tested with all possible data values and all storage locations the test will take some hundreds of years. Thus only limited testing is practicable, and it is important to design the tests to cover as much of the device logic as possible.

As microprocessors are now embodied in most military equipment, there is much interest in producing reliable devices without design errors. This can be largely overcome by using formal mathematical methods to specify and verify the processor. A team at RSRE Malvern has been working on this project, using a formalism called LCF-LSM (logic of computable functions-logic of sequential machines) and leading to a device called VIPER (Verifiable Integrated Processor for Enhanced Reliability) intended for safety-critical applications (Dittmar 1986). This has a 32-bit data bus and a 20-bit address bus, and to avoid possible timing problems there is no provision for interrupts. All external requests for service are dealt with by polling; this can take longer than an interrupt if many devices are connected to the microprocessor, but with a fast processor and the moderate response time acceptable for servicing mechanical systems, no problems arise. The only commercial use of the VIPER device reported is in signaling equipment for the Australian railway network.

43.3.8 Built-in Test Equipment

Where equipment can be maintained availability can be increased by conducting regular system checks so that any fault is discovered as soon as possible. This enables repairs to be started as soon as possible, so minimizing the downtime. Two methods have been used: initial testing and periodic checking. Initial testing is usually included in single instruments such as high-bandwidth oscilloscopes and logic-state analyzers. These generally incorporate microprocessors to control their functions and are configured so that each time the equipment is switched on, an interrupt is created which starts a test routine. This checks the calibration of the system and as many of its functions as possible. As this type of apparatus is generally used intermittently and hardly ever left running continuously, it is tested often enough to ensure that faults cannot give incorrect readings for very long.

Periodic testing is necessary for critical systems which are normally energized continuously and so would only have an initial test very occasionally. At regular intervals the system is diverted from its normal task and enters a test routine which conducts a quick system check, reporting any fault discovered. In large installations further diagnostic tests can then be carried out which will investigate the fault in more detail and give more information about its location.

Built-in tests of this kind are used in non-maintained systems only if some redundancy is provided and there are facilities for disconnecting faulty equipment and switching in alternative units.

43.3.9 Sneak Circuits

A problem in some situations is the occurrence of what has been called "sneak" circuits. These have been defined as latent paths or conditions in an electrical system which inhibit desired conditions or initiate unintended or unwanted actions (Arsenault and Roberts 1980). The conditions are not caused by component failures but have been inadvertently designed into the system. They are liable to occur at interfaces where different designers have worked on two packages but there has not been sufficient analysis of the combined system. They are also liable to occur after design modifications have been introduced when the new configuration has not been tested exhaustively.

One frequent source of sneak errors is the arrival of several signals required for particular action in an unexpected order. A similar source of potential error was recognized some years ago with the development of electronic logic circuits. If an input change causes more than one signal to propagate through a logic network, and two or more of these are inputs to the same gate, the resulting action can depend upon the order in which the various inputs arrive. If the output of the logic gate should not change it may nevertheless emit a short unwanted pulse which could advance a counter and cause an error.

This phenomenon is called a "race hazard" and it must be avoided if the system is to operate reliably. It can be tackled in two main ways. The first generates what is called a "masking" signal as an extra input to the gate which prevents a false output regardless of the timing of the input signals. This is satisfactory where the race can occur at only very few gates. In more complex systems such as digital computers it may occur many times, and the effort of analyzing these and introducing the extra logic is prohibitive. The solution adopted in this case is to inhibit the output of the gate until one can be certain that all inputs have arrived. The inhibiting signal is usually a train of constant frequency or clock pulses which is applied to all storage elements. Any inputs which arrive between clock pulses are not allowed to alter the state of the storage device until the next clock pulse arrives.

The procedures adopted to deal with race hazards in logic circuits can to some degree be applied to the prevention of the unwanted consequences of sneak circuits. Some of these may occur through the incorrect state of logic elements when power is applied to a package. Most logic devices which include some storage may set themselves in either logic condition when power is first applied. In order to ensure that they all start operation from some known condition an initializing pulse is usually sent to them a short time after power is applied to set them into the desired state.

The same process can be used to avoid sneak circuits by delaying any action until one can be sure that all changes have occurred.

Extensive computer programs are now available to analyze systems and discover any sneak paths. They were originally written to handle electrical control circuits, including relays, switches, lamps, etc., and were subsequently extended to apply to digital logic circuits. One event which helped to stimulate NASA to invest in sneak circuit analysis programs occurred at the launch of a Redstone booster in 1986. After 50 successful launches, a launch sequence was started, but after lifting several inches off the pad the engine cut out. The Mercury capsule separated and ejected its parachutes, leaving a very explosive rocket on the pad with no means of control. It was left for just over 24 hours until the liquid oxygen had evaporated and the batteries had run down before being approached. Subsequent investigations showed a timing error had occurred, in that the tail plug cable had disconnected 29 milliseconds before the control plug cable and the sneak circuit caused the engine to cut out. The cables were intended to disconnect in the reverse order, and the cable arrangements were later altered to ensure this.

An unwanted digital input can occur if unused inputs to logic gates are not connected to either a logic 1 or logic 0 potential. Manufacturers always advise users to do this to prevent the inputs from picking up stray noise pulses. In one recorded case (Brozendale, 1989) an unused input in an interface unit of a chemical plant was not earthed as intended and picked up an induced voltage. This caused an incorrect address to be sent to the controlling computer which gave the output commands for the wrong device. The result was that a number of valves were opened wrongly, breaking a gas line and releasing a toxic gas. Since plant safety depends critically upon correctly identifying the device which needs attention, a safe system design should include more than one means of identification. The design principle is that all information exchanges between the processor and peripheral devices should have some degree of redundancy in the interests of reliable operation. Thus in addition to checking items such as addresses, it is desirable to read back into the computer all data sent to the peripheral devices so that they can be checked.

43.4 BUILDING HIGH-RELIABILITY SYSTEMS

43.4.1 Reliability Budgets

Few electronic systems are designed for which no reliability target exists. This may vary from "no worse than the opposition" for a mass-produced domestic article to a closely specified minimum MTBF for an avionic or military system, with perhaps a financial bonus for exceeding the minimum.

In the past some designs have been largely completed before an estimate of reliability was started. If this diverges significantly from the target, a major redesign is required involving much extra time and cost. Consequently it is now accepted as a principle of good design that where reliability is a significant item in the specification, it should be a major consideration at all stages of the design.

In a system which can be regarded as a number of separate units, each of which must operate correctly if the system is to deliver its required output, it is useful to establish a reliability budget. In this the required overall reliability is partitioned between the various units so that the designer of each unit has his or her own reliability target.

A simple initial procedure which can be applied where the units have roughly the same complexity is an equal division. Thus if the overall reliability figure specified is R, the reliability target for each of n units is $\sqrt[n]{R}$. Thus for four units the target would be $\sqrt[4]{R}$. If the units vary in complexity the allocation should be unequal; a value can be assigned to each unit based upon previous experience or, given some preliminary design data, on a count of components, semiconductors, or integrated circuit packages. The overall specification requires a relation between the unit reliabilities R_1, R_2, R_3, R_4, and R of

$$R = R_1 \times R_2 \times R_3 \times R_4 \quad (43.16)$$

It is more likely that the system will be specified as having a particular MTBF of M hours. In this case a system comprising four similar units will require each unit to have an MTBF of $4M$. Where the units have differing complexities and are expected to have differing MTBFs M_1, M_2, M_3, and M_4, the relation between them must be

$$\frac{1}{M} = \frac{1}{M_1} + \frac{1}{M_2} + \frac{1}{M_3} + \frac{1}{M_4} \quad (43.17)$$

An initial estimate of the various MTBFs can be made using a simple parts count and refined later as the design proceeds.

43.4.2 Component Selection

Electronic components have been developed over many years to improve their performance, consistency, and reliability, and consequently, the less demanding reliability targets can often be attained by using widely available commercial components. A modest improvement in reliability can in these cases be obtained by derating the components. There still remain, however, many applications for which high reliability is demanded and which thus need components of higher and well-established reliability.

Attempts to improve the reliability of electronic equipment were first tackled in a systematic way towards the end of the Second World War, when the American services discovered that some of their equipment cost during its lifetime at least ten times more to maintain than to purchase initially. An early outcome of this was a program to develop more reliable thermionic valves, which were responsible for many failures.

Work started in the U.K. some 40-plus years ago when an increasing number of agencies, such as the Post Office, commercial airlines, the armed services, and the railways required very reliable electronic systems and attempted to issue their own specifications for component performance and reliability. The manufacturers found great difficulty in coping with all these different requirements and the long testing programs needed for them. In consequence a committee chaired by Rear-Admiral G. F. Burghard was established to develop a set of common standards for electronic parts of assessed reliability suitable for both military and civilian applications. The committee's final report in 1965 was accepted by industry and government, and the British Standard Institution (BSI) accepted responsibility for publishing the appropriate documents. The basic document is BS 9000 which prescribes a standard set of methods and procedures by which electronic components are specified and their conformance to specification is assessed. The system is implemented by the BSI and operated under the monitoring of the National Supervising Inspectorate.

BS 9001 gives tables and rules for sampling component production, and BS 9002 gives details of all components which have been approved and their manufacturers. There is such variety in the items used to manufacturer electronic equipment that separate specifications are needed for each family such as:

BS 901X	Cathode-ray and camera tubes, valves, etc.
BS 907X	Fixed capacitors
BS 9090	Variable capacitors
BS 9093	Variable preset capacitors
BS 911X	Fixed resistors
BS 913X	Variable resistors, etc.

In some cases, for example discrete semiconductors (BS 93XX), these have been divided into subfamilies such as signal diodes, switching diodes, voltage reference diodes, voltage regulator diodes, etc.

BS 9301	General-purpose silicon diodes
BS 9305	Voltage regulator diodes
BS 9320	Microwave mixer diodes (CW operation)
BS 9331	Medium-current rectifier diodes
BS 9364	Low-power switching transistors, etc.

There is now a European dimension to the BS 9000 scheme in that many British Standards are now harmonized with the standards of the European CECC (CENELEC Electronic Components Committee), and constitute the BS E9000 series.

Also, many CECC standards have been adopted as British Standards, for example:

BS CECC 00107	Quality assessment procedures
BS CECC 00108	Attestation of conformity
BS CECC 00109	Certified test records, etc.

The BS 9000 scheme includes provision for the collection of the results of life tests so as to build up a data bank of component performance.

43.4.3 The Use of Redundancy

Although greatly increased reliability can be obtained by using specially developed components, derating them, and keeping them as cool and vibration free as possible, there is a limit to the benefit this can bring. There are many critical applications where yet higher reliability is required, and the usual method of coping with this requirement is to introduce some degree of redundancy.

In general terms this means providing more than one way of producing the desired output. The assumption is that if one path is inoperative due to a fault, another path will provide the correct output. This may involve sending copies of the same information along different paths (spatial redundancy) or sending copies of the information along the same path at different times (temporal redundancy). The former is much more powerful, as it can cope with a permanent fault in one path, whereas the latter is generally simpler to implement but copes best with transient errors such as impulsive noise on a telephone line or a radio circuit. As only one path is provided, it cannot cope with a permanent fault.

Many control and instrumentation systems are confined to a restricted area and so need no measures to cope with transmission faults. Thus the technique most applicable is that of spatial redundancy which requires extra equipment and, if fully implemented, the replication of the entire system. The simplest form of this is duplication. We postulate two identical channels, each of which can deliver the outputs needed, with some provision for switching to the spare channel when the working channel fails. If the probability of failure for a single channel is p, the probability of system failure is p^2 since the system will fail only when both channels fail. The reliability is thus

$$R = 1 - p^2 \qquad (43.18)$$

This result assumes that both systems are independent so that the failure of one channel makes no difference to the probability of failure of the second channel. This assumption is not always valid, since any common item such as a power supply which feeds both channels will invalidate it. Even if two independent power supplies are provided, they will normally be connected to the same mains supply, and the result given in equation (43.18) should be multiplied by the reliability of the mains supply.

We can generalize this result for n identical channels, on the assumption that only one working channel will provide the required output, to give

$$R = 1 - p^n \qquad (43.19)$$

This, again, is not a realistic calculation, since a multi-channel system will need some mechanism for checking the working channel and switching to the next channel when it fails. A better figure for overall reliability is given by multiplying R by the reliability of the checking and switching mechanism.

In some installations the checking may be done by a short program module in the computer which controls the system. This, however, may not be acceptable if it means suspending the computer's normal operation at regular intervals to run the test.

Where we have some reserve of data handling power and data appear in bursts we may be able to test a channel by injecting test signals between the bursts. This does not involve suspending the computer's operation, but is practicable only when we can be certain that there will be intervals in the demands for service which are long enough and frequent enough to allow adequate testing.

The program in such cases is usually divided into modules with varying degrees of importance which are executed, some at regular intervals of time, others when particular patterns of data occur. At any moment the program being executed is that which has the highest priority, and when that has ended the module having the next highest priority is invoked. At any time the current program module can have its execution interrupted by another module of higher priority which becomes active. In the priority list the system test program is often put at the bottom so that when there is no other call for the computer's services it continues execution of the test program rather than idling.

43.4.4 Redundancy with Majority Voting

If we ignore the possibility of failure in the switching mechanism, and assume a constant failure rate for each channel of λ, the MTBF for a duplicate system is $M = 3/2\lambda$. For a triplicate system it increases to $11/6\lambda$.

The scheme mentioned above means that the normal operation of the system must be halted for a short period to permit a channel to be tested. If it is found to be faulty it is disconnected and another channel switched into operation. This is bound to involve some interruption to the output, which cannot always be tolerated. For example, in a real-time vehicle control system any interruption can mean a major deviation from the desired path. In such circumstances some mechanism which does not involve a break in output is needed; one of these involves three identical channels fed with the same input signal with a majority voting circuit at the output. This is easiest to implement with a digital system where the required voting circuit has to implement the logical function $X = A \cdot B + B \cdot C + C \cdot A$, where A, B, and C are the outputs of three channels. This requires only four gates as shown in Figure 43.10. The MTBF for this arrangement is $5/6\lambda$.

It is important in maintained systems to provide an indication that a fault has occurred even though the redundancy has prevented this from causing a system failure. This enables corrective action to be started as soon as possible; without this the system is less reliable than a single system since a fault on either of the two remaining working channels will cause a system failure. The logic expression which must be implemented is derived easily by considering the outputs before a fault occurs. In this case the outputs must be either three ones or three zeros. The logic expression is thus $Y = A \cdot B \cdot C + \overline{A} \cdot \overline{B} \cdot \overline{C}$. A fault condition is indicated when this expression has a value zero. Thus to deliver a logic 1 signal when a fault has occurred we need the complement of this, that is,

$$Z = \overline{A \cdot B \cdot C + \overline{A} \cdot \overline{B} \cdot \overline{C}} = B \cdot \overline{C} + C \cdot \overline{A} + A \cdot \overline{B}$$

Although we have discussed majority voting in its simplest and most widely used form of triplicated channels, the same voting procedure can be used with any odd number of channels. The logic expression for the voting circuit is a little more complicated; for example, if we have five channels the terms comprise all of the combinations of three items selected from five, that is,

$$\begin{aligned}X = {} & A \cdot B \cdot C + A \cdot B \cdot D + A \cdot B \cdot E + A \cdot C \cdot D \\ & + A \cdot C \cdot E + A \cdot D \cdot E + B \cdot C \cdot D \\ & + B \cdot C \cdot E + B \cdot D \cdot E + C \cdot D \cdot E\end{aligned}$$

This requires 10 three-input AND gates and one 10-way OR gate.

As an example of the benefit of triplication we take the control circuits of a recent optical fiber submarine cable. This has four separate channels, three working and one spare. At each repeater location the repeater inputs and outputs of each working channel can be switched to the spare channel in the event of a fault. If we assume that there are 200 components with an average failure rate of 0.2×10^{-9} per hour in the monitoring and switching operation at each repeater housing and we look for a working life of 20 years, the expected number of faults per housing is $200 \times 0.2 \times 10^{-9} \times 20 \times 8760 = 0.007008$.

The failure rate is a somewhat crude assessment as it is a weighted average over all the components, but the value is in line with the target failure rates quoted for a long-haul submarine repeater which vary from 1.0 FIT for transistors to 0.1 FIT for capacitors and resistors (1 FIT is a failure rate of 10^{-9} per hour). The corresponding reliability is

$$\exp(-0.007008) = 0.992945$$

If there are 16 repeaters in the cable, the overall reliability of the switching operation will be

$$0.992945^{16} = 0.8929$$

If we introduce a triplicate redundancy scheme with majority voting at each repeater site the reliability will be

$$R_T = 3R^2 - 2R^3$$

where R is the reliability of each channel. This can be shown by considering the circumstances in which the system fails,

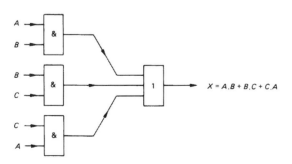

FIGURE 43.10 Majority voting logic circuit.

that is, when two channels are faulty and one working, or all three are faulty. If $p = (1 - R)$ is the probability of a channel failing, the probability of two or three failing is

$$P_T = 3p^2(1 - p) + p^3$$
$$= 3p^2 - 2p^3$$

since $(1 - p)$ is the probability of one channel working and p^2 is the probability of two being faulty; there are three ways in which this can occur. Expressing this in terms of reliability gives

$$P_T = 3(1 - R)^2 - 2(1 - R)^3$$

Finally, the overall reliability is given by

$$R_T = 1 - P_T$$
$$= 3R^2 - 2R^3 \qquad (43.20)$$

Returning to the repeater calculation, the reliability of a triplicated version with majority voting is given by putting $R = 0.992945$ in equation (43.20). The overall reliability then becomes $R_T = 0.9976$. Thus the probability of a failure has been reduced from 10.7 percent to 0.24 percent. This is a somewhat optimistic calculation since the reliability of the majority voting element has not been included. However, it should require far fewer components than the 200 we have assumed for each repeater station and thus should be much more reliable than the switching units.

43.4.5 The Level of Redundancy

The scheme shown in Figure 43.10 uses only one voting circuit, as the final element in the system. The overall reliability can be improved by subdividing the system, replicating each subsystem, and following it by a majority voting circuit. It can be shown by using a somewhat simplified system model that the optimum scheme is one in which the system is subdivided so that the reliability of the subsystem is equal to the reliability of the voting circuit (Cluley, 1981). Since in a digital system the same logic hardware is used in both the working channels and the voting circuit, the conclusion is that the subsystem and voting circuit should be of similar sizes. This is a practicable arrangement where discrete components are used, but most current equipment, both analog and digital, makes much use of integrated circuits which generally have a much greater complexity than a voting circuit. We are thus forced to conclude that optimum redundancy is impracticable in current equipment, and the number of voting circuits which can be introduced is limited by the system design. However, we can still obtain improved reliability by subdividing each channel and replicating the sub-assemblies; we can also

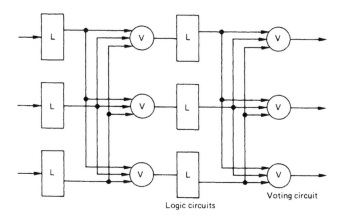

FIGURE 43.11 Triplicated logic and voting circuits.

ensure a further improvement by replicating the voting circuits so that each subassembly has its own voting circuit, as shown in Figure 43.11. The ultimate limit to the overall reliability is set by the final voting circuit, which cannot be replicated, although it could have some degree of component redundancy.

43.4.6 Analog Redundancy

The same increase in system reliability due to triplication which is obtained in digital systems can also be obtained in analog systems. A practical difficulty is the design of suitable majority voting elements. One circuit which was developed for a triplicated analog autopilot system for aircraft will give a majority vote if any two channels have the same output. If the outputs are all different, it will follow whichever output has a value intermediate between the other two. Thus it will give the desired output if one output is hardover to zero and a second is hardover to full scale; we assume that the third output is correct. If, however, two outputs both give either zero or full scale the circuit gives an incorrect output (Cluley 1981). Another arrangement was used in a later aircraft control system in which the three channels drove servo motors which rotated the shaft on which the control surface was mounted. The shaft summed the three torques generated by the motors and so achieved an approximate majority vote. To avoid damage to the motors or their driving amplifiers the motor current and hence the torque was strictly limited. The effective voting element was the control shaft which could easily be made large enough to ensure that the probability of its failure was negligible.

An alternative arrangement which is convenient to use with integrated circuit amplifiers is to operate them in parallel pairs, with provision for disconnecting a faulty amplifier. As these devices are directly coupled, any fault will almost certainly disturb the potential at the output. This is fairly simple to arrange if the amplifier output is restricted. For

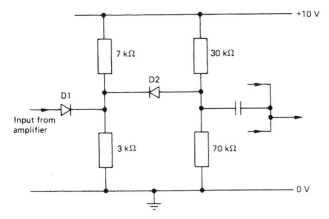

FIGURE 43.12 Circuit to disconnect redundant amplifier when its output is greater than 7 V or less than 3 V.

example, if it does not exceed 5 V ± 2 V with a 10 V supply a two-diode circuit as shown in Figure 43.12 will disconnect the amplifier from the output when its potential falls below 3 V or above 7 V. In the first case D1 disconnects, and in the second case D2 disconnects. In practice the disconnection is not abrupt owing to the diode characteristic, but when the amplifier output is hardover to earth or the positive supply there will be nearly 3 V reverse bias on one of the diodes, which is ample to ensure complete disconnection of the faulty amplifier.

Where a data signal can vary widely in amplitude in a random fashion much advantage can be obtained by combining several versions of the signal provided that the amplitude fluctuations are largely independent. This is the technique used in radio receivers for long-distance circuits. The received signals arrive after one or more reflections from the ionosphere and are liable to fluctuate in amplitude because of interferences between waves which have traveled along different paths. Experiment shows that the fluctuations in signal level received by aerials spaced 10 wavelengths apart have very little correlation, and if they are combined on the basis that the largest signal is always used, the result will show much less fluctuation than any component signal. This technique is called diversity reception and is often used to combat fading in long-distance radio reception, particularly with amplitude-modulated transmissions. The amplitude of the received carrier can be used to indicate signal strength, and is used as a feedback signal to control receiver gain. In triple diversity the three gain-control signals are taken to three diodes with a common output which automatically selects the largest signal. This is connected to all three receivers, and the audio outputs are also commoned. The gain of the receivers handling the weaker signals will be reduced and so their contribution to the common audio output will also be reduced. Although this technique was first used for HF reception it was also found to improve VHF reception from moving vehicles in telemetry links for instrumentation.

43.4.7 Common Mode Faults

A crucial factor in designing redundant systems is ensuring that all of the replicated channels are independent, so that the existence of a fault in one channel makes no difference to the probability of a fault occurring in another channel. Any fault which will affect all channels is called a "common mode" fault, and we can only obtain the full improvement in reliability which redundancy promises if common mode faults are extremely unlikely. Two likely causes of common mode faults are power supplies and common environmental factors.

Where all the channels of a redundant system are driven from a common power supply, the reliability of this supply will be a limiting factor in the overall system reliability. Should this be inadequate, the power unit can be replicated in the same way as the data channel, and its reliability can be included in the estimate of channel reliability. There is still a common mode hazard since all the power units are connected to the same source of energy. Past records give a useful indication of the probable reliability of the supply; if this is considered to be inadequate there are several ways of coping. The first is the use of uninterrupted power supply (UPS). This consists of a motor-generator set supplying the load which has a large flywheel mounted on the common shaft. When the main power supply fails, a standby diesel engine is started, and as soon as it has attained full speed it is connected via a clutch to the motor-generator shaft and takes up the load. While the diesel engine is starting, the flywheel supplies the energy needed by the generator. The shaft speed will fall somewhat, but this can usually be tolerated. Where long supply interruptions may occur a second diesel generator can be provided.

There are some variants of this: for low-power applications the supply may be obtained from a battery during a mains failure, the battery otherwise being trickle charged from the mains. For higher power loads the main supply can be from a mains-driven rectifier, with a standby generator in the event of a mains failure. To allow time for the generator to run up to speed a stand-by battery is normally provided, sufficient to supply the load for 10–20 minutes. The scheme depends upon the switching generally performed by diodes which automatically connect the load to the highest voltage supply available. Thus the normal working supply voltage and the generator voltage must both be a little greater than that of the standby battery. The most likely failure mode of the power diodes used for switching is to a short circuit, so the reliability can be improved using twin diodes in series as shown in Figure 43.13.

Other common factors which need addressing are the environment which is likely to affect all channels of a replicated system, and secondary damage. Since all adverse environmental conditions, such as large temperature fluctuations, excessive vibration, or a corrosive atmosphere reduce the reliability of electronic equipment their effect

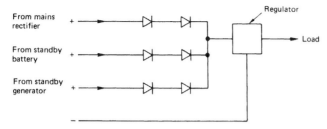

FIGURE 43.13 Redundant power supply switching.

will be particularly severe if they affect all the channels of a redundant system. The effect of the environment may be diminished by enclosing the complete system in an insulating housing, but this will not completely remove the chance of common mode faults unless there is some segregation between the channels. For example, a fault in one channel may cause it to overheat or emit toxic fumes; without segregation between the channels this may cause a fault in the other channels.

Common mode faults of this kind have occurred in power stations where, despite some degree of redundancy, all of the control cables were routed along the same duct as some power cables. A cable fault once caused a fire which damaged all of the control cables, so nullifying the benefit of redundancy. The only way to avoid such common mode faults is to ensure complete physical separation not only between power and control cables, but also between the cables incorporated in each of the redundant channels.

With the increasing reliability of electronic hardware, a greater proportion of faults caused by design errors appear in maintenance records. These are likely to occur as common mode failures since the same error will arise in all channels of a replicated system. Although such systems are normally subjected to exhaustive testing, it is very difficult to ensure that a fault which occurs only with a particular set of data has been eliminated. The complexity of current computer-based systems means that they would be obsolete and out of production long before all possible combinations of data, instructions, and storage locations had been tried.

One expensive but apparently worthwhile way of reducing the consequence of design faults in redundant systems is to use different teams to design the nominally identical channels. The extra expense has been thought beneficial in high-reliability applications such as satellite control systems, both for hardware and software design.

43.5 THE HUMAN OPERATOR IN CONTROL AND INSTRUMENTATION

43.5.1 The Scope for Automation

Although for economic reasons control and instrumentation systems are increasingly being automated there remain many situations in which decisions have to be made with complex and sometimes incomplete data. In many of these situations a human operator is still included in the loop on account of his or her flexibility and ability to accept a wide range of data. Thus although it has been technically possible for many years to apply the brakes automatically on U.K. mainline trains when they pass a danger signal, railway managers have hitherto insisted on leaving this critical decision to the driver.

In some situations the control information is provided by a number of different sources which are not directly detectable by a human controller. For example, in the blind landing system used for civil aircraft the normal visual data the pilot uses is absent due to adverse weather, and information about height, course, and position relative to the runway is provided by radio beams and the magnetic fields of cables in the ground.

Although it would be possible to convey this information to the pilot in a visual form the quantity and rate of change of the data would impose a very severe load, and international regulations do not permit manual landing in these circumstances. The last few minutes of the landing are then controlled automatically.

In other situations where automation is technically possible a human operator is retained in the interest of safety on account of greater flexibility and the ability to cope with quite unexpected conditions. The postal service has run driverless underground trains across London for some decades, but they carry only freight. As yet there are very few miles of track which carry driverless trains for public use.

Most commercial passenger ships have automatic steering gear, but international safety legislation requires a suitably qualified person to be in charge of the ship whenever it is under way.

43.5.2 Features of the Human Operator

One way of analyzing the dynamic behavior of a control system is to determine its overall transfer function, by combining the transfer functions of all its component parts. This type of analysis thus requires a knowledge of the transfer function of a human operator, when he or she is an essential part of the control loop, for example, as a car driver or an aircraft pilot. Unfortunately, it is very difficult to define a satisfactory model of human behavior as there are so many possible transfer functions which may be exhibited. Most operator errors are noticed and corrected by the operator, who is often unwilling to admit mistakes. Thus any attempt to measure performance must be sufficiently sensitive to detect these corrected errors and must at the same time be inconspicuous so as not to distract the operator. A further area of uncertainty is that performance depends upon the physical and mental state of the operator; it is very difficult to determine this and to specify it in any meaningful way.

Some attempts to estimate the reliability of a human operator have been made by breaking down his or her task

TABLE 43.2 Error rates

Type of error	Rate
Process involving creative thinking, unfamiliar operations where time is short, high-stress situation	10^{-0} to 10^{-1}
Errors of omission where dependence is placed on situation cues and memory	10^{-2}
Errors of commission, e.g., operating the wrong button, reading the wrong dial, etc.	10^{-3}
Errors in regularly performed, commonplace task	10^{-4}
Extraordinary errors: difficult to conceive how they could occur; stress-free situation, powerful cues helping success	$<10^{-5}$

into a series of simple operations and assigning a probability of error to each of these. Some estimates of error rates are given in Table 43.2. As some confirmation of these figures it is generally accepted that the average person will make an error when dialing a telephone number once in about 20 attempts. The error rate is, however, somewhat lower if pushbuttons are used instead of a circular dial.

One factor which seriously affects the effectiveness of the human operator is the duration of the control task. In 1943 the RAF asked for tests to determine the optimum length of a watch for radar operators on antisubmarine patrol, as it was believed that some targets were being missed. Tests showed that a marked deterioration in performance occurred after about 30 minutes, and this conclusion has often been confirmed since.

It is important not to regard this particular result as directly applicable to other circumstances. A specific feature of the radar operator's task on antisubmarine patrol is the low data rate and the use of only one of the senses. Many control tasks involve much higher data rates and data input to several senses; for example, an activity which many people find comparatively undemanding, driving a car, involves visual and audible input and sense feedback from the pedals and the steering wheel.

There is some evidence that a human operator can handle a greater data rate if several senses are involved, rather than using a single sense. For reliable control action it is important to choose the appropriate sense to convey data to the operator. Visual communication is best where the message is long, complex, or contains technical information, where it may need referring to later, or where it may contain spatial data.

On the other hand, audible communication is more effective for very simple, short messages, particularly warnings and alarms, or where the operator may move around. In a crowded environment the operator may easily move to a position where a display cannot be seen, whereas an audible warning would be heard. Also, audible information is preferable where precise timing is required to synchronize various actions; for example, an audible countdown is always used when spacecraft are launched although ample visual displays are always available.

Despite its many advantages, speech communication is liable to errors; the average human operator has a vocabulary of some tens of thousands of words and it is easy to confuse one with another. A method often used to make the communication more reliable is to diminish the number of words which can be used by confining the message as far as possible to a set of standard phrases. This is a procedure used successfully by the services and air traffic control. Ideally each message should be read back to enable the originator to check its contents, but the rate at which information must be handled often precludes this. Sometimes there are alternative sources of information; for example, in air traffic control the radar display enables the controller to check that the pilot has understood his or her message and is flying on the requested bearing. Mistakes do still occur, however, and these may have serious consequences. In 1977 two Boeing 747 aircraft collided on the ground at Tenerife with the loss of 583 lives and a cost of some $150 million. A number of human errors contributed to this accident, a major one being a misunderstanding between the air traffic controller and the pilot of one aircraft. The pilot thought that he had been cleared to take off, whereas he had been cleared only to taxi to the end of the runway and was expected to request further clearance when he reached there.

Tests on the efficiency with which simple tasks are performed show that this depends upon the rate at which data are presented to the operator. If the data rate is too slow, attention slackens and performance suffers. Thus the designer of a control system which involves a human operator should ensure that the rate at which information is presented to the operator must be high enough to keep him or her alert, but not too high to overload his or her capacity to accept and comprehend it. The period for which a control task can be continued before performance begins to deteriorate is of course dependent to some degree upon the task and the individual, but experiments suggest that regular breaks will improve the reliability of the operator.

One outstanding feature of the human operator is his or her adaptability; an operator is able to alter his or her working pattern fairly quickly when presented with a new machine or working environment, and can cope with quite complex data. For example, an operator can read handwritten information comparatively easily despite its variety of size and character; computers cannot yet perform this task reliably. An operator can also learn to handle new machines and procedures comparatively quickly. However, if the old procedure has been in place for some time and has become very familiar, there are many cases in which, under stress, the operator will revert to his or her earlier behavior. Records of aircraft accidents and near misses contain many examples of this behavior.

When a human operator forms part of a control process it is important to have some data about the operator's response time. Measurements of this have been made when various tasks are being performed. A relation quoted between response time t and the display information H in bits is

$$t = a + bH \text{ seconds} \qquad (43.21)$$

Here a is the lower limit of human response time, equal to 0.2 second, b is the reciprocal of the information handling rate, typically 15 bits per second. The task performed was monitoring a set of instruments and operating toggle switches when certain readings occurred.

H is derived from the display by calculating

$$H = \log_2 n$$

where n is the number of equiprobable, independent, alternative readings of the display.

Although one particular advantage of including a human operator in a control system is adaptability, this may in some cases cause problems. In any complex system the operator needs a degree of autonomy to enable him or her to cope with unexpected situations, but it is important that this is not used to cut corners and thus to allow dangerous situations to arise. It is a natural human characteristic that, although an operator may start a new task carrying out all the instructions correctly, without careful supervision the operator tends to relax and discover ways of easing the task and perhaps defeating some of the built-in safety provisions.

43.5.3 User-Friendly Design

In many early control panels much of the designer's work was directed towards producing a balanced and often symmetrical layout of the instruments and controls. Too often the needs of the operator were ignored; the operator had to stretch out to reach some switches, and meters were not always in a convenient position to read quickly and accurately. The result was that in times of stress or emergency operators made errors, and in order to minimize these a systematic study was made of the human frame to determine where switches and dials should be placed to be within convenient reach, the position and intensity of lighting, and other similar matters which would ease the operator's task. Many of the arrangements proposed as a result of the study were contrary to previous ideas for a balanced and symmetrical layout.

For example, to help pilots to identify the particular control knob they needed to operate when these were grouped closely together, each knob was given a different shape and surface texture. Also, in control panels used for chemical plant and power stations a simplified line diagram of the plant was sometimes printed on the panel, with the controls and instruments connected to each unit of the plant near to the symbol for that unit.

Many control activities require the operator to turn a knob or move a lever to bring the reading on a meter to a target value. In this situation it is important to ensure that the meter and knob are mounted close to one another and that there is proper coordination between the direction in which the knob is turned and the direction in which the meter pointer moves. Normally a movement of the pointer from left to right should be caused by a clockwise rotation of the control knob. This is satisfactory with the knob beneath the meter or to its left or right. Another factor needing attention in this situation is the amount of pointer movement caused by a given rotation of the knob. If this is too small, control action is delayed because the knob may need several turns; on the other hand, this arrangement assists the exact setting of the control. Several schemes are used to minimize the delay: two gear ratios may be provided—fast for rapid movement and slow for fine adjustment—or a flywheel may be mounted on the control shaft so that a rapid spin will continue to cause rotation for some seconds.

On the other hand, if a small knob rotation causes a large deflection of the pointer, it will be difficult to set its position accurately, and this may take longer than necessary.

In situations where rapid, almost instinctive action may be needed in an emergency, for example, driving a car or piloting an aircraft, it is important to have the most frequently used controls in standard positions. For some 70 years the positions of brake, clutch, and throttle have been standardized on most of the world's manually operated cars, so preventing many potential accidents which could arise when drivers use unfamiliar cars. Unfortunately the same is not true for cars with automatic gearboxes, and a number of accidents have been caused by variations in the placing of the controls and their labeling.

Such variations were much more liable to cause errors in flying aircraft, and some research was conducted soon after the Second World War into events which had been classed as "pilot errors." This research revealed that in many of these events the awkward layout of the controls and the confusing displays were a contributing factor to the accident. A particular example of this is the position of three important controls which are mounted on the throttle quadrant in three American military aircraft then in common use (see Table 43.3). Clearly pilots who move from one aircraft to another are very likely to confuse these controls in an emergency, and several cases of this were uncovered.

An unusual case of failure to interpret instrument readings correctly occurred in the crash of a Boeing 737-400 at Kegworth in January 1989, which killed 47 people. The official accident report said that a fan blade fracture had caused a fire in the left engine, causing smoke and fumes to reach the flight deck. Together with heavy engine vibration, noise, and shuddering the resulting situation was outside the pilots' experience, as they had not received any flight training for the recognition of engine failure on the electronic engine instrument system. The captain later told

CHAPTER | 43 Reliability in Instrumentation and Control

TABLE 43.3 The position of controls in American military aircraft

Aircraft	Position on throttle quadrant		
	Left	Center	Right
B-25	Throttle	Propeller	Mixture
C-47	Propeller	Throttle	Mixture
C-82	Mixture	Throttle	Propeller

investigators that he had not obtained any clear indication of the source of the problem when he looked at the instrument panel, and the copilot had no recollection of what he had seen. However, evidence from the flight data recorder and the remains of the aircraft show clearly that the crew shut down the right engine instead of the left engine, so depriving the aircraft of all power. They mistakenly assumed that they had behaved correctly since after shutting down the right engine the noise and shuddering from the left engine ceased. The complete loss of power meant that the aircraft could not reach the runway at which the captain hoped to land, and it crashed on the edge of the M1 motorway. Although many people on board, including the cabin crew, had seen flames coming from the left engine, this information did not reach the pilots.

It is difficult to understand how two experienced pilots could both misinterpret the engine instruments so fatally as to shut down the wrong engine, but subsequent analysis of the engine instrument panel revealed major deficiencies in its layout. The primary instruments for both engines were in a panel on the left and the secondary instruments in a panel on the right. Each panel was aligned with a throttle control as shown in Figure 43.14.

The clearest warning of the engine failure was given by the vibration indicator, which is a secondary instrument and thus in the right group and nearest to the right throttle. In the confusion the right engine was throttled down. The accident report suggested an alternative layout also shown in Figure 43.14. Here the primary instruments are aligned with the throttle controls and the secondary instruments are on either side of them. The Royal Air Force Institute of Aviation Medicine (IAM) carried out tests after the crash on the 737–400 instrument layout to find out how it was possible for the pilots to misinterpret the information presented on them. The IAM found that the actual combination of layout and dial design was the least efficient of the four combinations tested and gave 60 percent more reading errors than the best and took 25 percent longer to read. Also pointers pivoted at their centers were found to be much more effective than the cursors fitted. These were small light-emitting diode displays which moved round the edge of a circular scale.

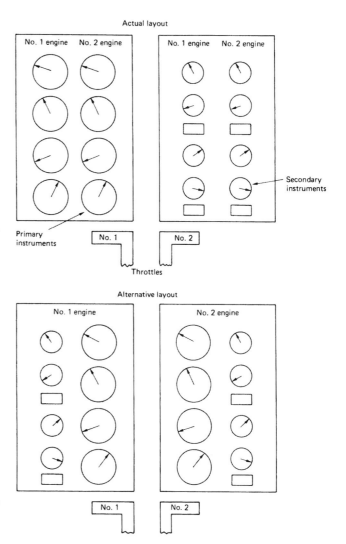

FIGURE 43.14 Layout of engine instruments in Boeing 737–400 aircraft.

The main display showed fan speed, exhaust gas temperature, core speed, and fuel flow. In addition to the analog display given by three LEDs a digital display was given in the center of the device. The secondary displays were a little smaller and showed oil pressure, oil temperature, and A and B system hydraulic pressures. These were analog only. The main display LEDs have 81 different positions, and the secondary displays have 31.

A useful indication of normal conditions is the same reading for each quantity of the two engines. With the earlier pointer display this was clearly shown by the parallelism of the two pointers. It is considerably more difficult to detect this from the position of three small LEDs outside the marked scale of the electronic display.

In order to assist pilots to recognize warning indications correctly, all modern cockpits now adopt the dark/quiet principle, meaning that if all systems are working as they should be there will be no lights or sound. In consequence any warning given by light or sound is instantly recognized.

This arrangement is used in all the Airbus models and is preferred by pilots.

The instrument panel in the Airbus 320 differs considerably from that of earlier aircraft in that many of the data are presented on cathode-ray tube (CRT) displays. This allows much more information to be available to the pilot since he or she can switch the CRT to several different sources of data. This facility is a valuable help in fault finding. There is, however, one point in which the cockpit arrangements differ from those of previous aircraft: when the engines are under automatic control the throttle levers do not move when the engine power is altered, although they are operated as normal when under manual control. In other aircraft the throttle controls move when under automatic control so giving the pilot an indication of the power demanded. Pilots have become accustomed to having this information and find that it takes longer to find by looking at the instrument display.

43.5.4 Visual Displays

An important factor in the design of warning systems is the amount of information which is presented to the operator. This is important both in total and in each separate display. It is usually recommended that in a text display not more than 25 percent of the available space should be used. If the text is packed more tightly it becomes more difficult to read and understand. The same general principle applies to diagrams; if these are too crowded they become much more difficult to interpret.

The eye is most sensitive at illumination levels typical of electronic displays to light of about 550 nanometers wavelength, a yellow/green color, so this is advised for monochrome displays. This view is supported by tests in which the time the subject took when asked to read the first character in a four-digit display was measured using light of various colors. Other tests on the use of passive displays which reflect or scatter incident light and active displays which emit light, such as LEDs, were conducted. These indicated that both types were equally readable in high levels of illumination, but the active display was much easier to read in low light levels such as exist in the less well lit parts of the cockpit.

There is a limit to the rate at which humans can read and assimilate information, and if this is presented by too many warnings and indicators the result is confusion and often incorrect responses. In some complex control environments the designers attempt to provide a warning for almost all abnormal conditions, and in an emergency the operator may be overwhelmed with visual and audible signals. Current civil airliners are in danger of suffering from this excess; for example, there are 455 warnings and caution alerts on the Boeing 747.

A notable example of operators being overwhelmed by an excess of alarms occurred in the American nuclear power station on Three Mile Island in March 1979. At 4 a.m., a pump feeding water to the steam generators failed. The emergency pumps started but could not function properly because outlet valves which should be open were closed. As a result, water backed up into a secondary loop, causing a rise of pressure in the reactor. A relief valve then opened automatically, but stuck open when it should have reclosed as the pressure fell. The emergency core cooling system was activated and water poured into the reactor and out through the relief valve. A further fault in the instrumentation caused the operators to assume that the reactor was covered with water and so safe, whereas the water was running out. The operators then took over and shut down the emergency cooling system, so depriving the reactor of all cooling. The fuel rods then soon reached a temperature of 2500°C. This complex set of events produced so many alarms and flashing lights that the operators were completely confused and it was nearly 2 hours before they realized the real problem.

43.5.5 Safety Procedures

Although the benefits of including a human operator in any complex control system are well recognized—the greater flexibility of action and the ability to tackle situations not previously encountered, for example—there may well be accompanying hazards. One feature of humans is the change in attitude caused by carrying out a fairly well defined control task many times. Initially a new task tends to be undertaken carefully with due attention to all the prescribed checks and data recording. With time, however, there is a temptation to cut corners and look for the easiest way to achieve a marginally acceptable result, and unless strict supervision and monitoring are insisted upon, some safety procedures may be neglected. In the design of high-voltage switching stations provision must be made to isolate some parts of the system to permit routine maintenance and extensions, and it is essential that the power should not be reconnected until work has finished and the staff have left the scene. Although there are administrative procedures to ensure this, such as "permit to work" cards, there is also a physical interlock system. This requires the power to be switched off before a key is released which gives access to the working area. When the gate to the working area is open, the key is trapped and cannot be used to turn on the power.

The high-voltage areas of radio transmitters and nuclear installations are normally fenced off, and the interlock is usually a switch fitted to the access doors. When these are opened the power is automatically disconnected and cannot be reconnected without moving outside the area. Unfortunately there are occasionally undisciplined operators who try to make adjustments with the power on by climbing the protective fence with fatal consequences.

A particular accident in which human errors were a major factor was the explosion in the Chernobyl nuclear reactor in 1986. Several of the people in senior positions in the power station had little experience in running a large nuclear plant, and the prevailing attitude of secrecy prevented the details and lessons of earlier accidents from reaching plant operators and managers. Even the information about the Three Mile Island accident was withheld, although the rest of the world knew all about it. In addition, those in charge of the industry put much more emphasis on production than on safety. The experiment which was being conducted when the accident occurred was to test the ability of the voltage regulator to maintain the busbar voltage when the turbines were slowing down after the reactor was shut down in an emergency. The voltage needed to be held near to normal for 45–50 seconds when supplying the essential load to ensure that safety mechanisms were effective, including the emergency core cooling system. The regulators on other similar stations had given some trouble, and many proposals had been made to conduct the same kind of test. However, this was considered risky, and the plant managers had refused to allow it. The management at Chernobyl had nevertheless agreed to the test; it is unlikely that they fully realized the possible consequences of their decision.

43.6 SAFETY MONITORING

43.6.1 Types of Failure

Hitherto we have considered that a fault in a component occurs when its characteristics change enough to prevent the system in which it is incorporated from producing the expected output. The only categories which have been recognized are those dealing with changes with time—permanent or intermittent faults—and those dealing with the rate of change—degradation or catastrophic failures. For example, both open-circuit and short-circuit failures are regarded equally as catastrophic failures. This classification is valid for all data handling systems, but needs further refinement for some control systems and safety monitoring systems.

In these we can envisage two consequences of failure, depending upon the state in which the system is left afterwards. In safety systems which are used to monitor some process, plant, or machinery, the important process parameters are measured continuously or at frequent intervals. When any parameter departs from its target value so far as to constitute a safety hazard, the safety system should shut the process down or ensure that it moves to a safe condition.

The safety system then has only two possible output states: it can shut the system down, as it is required to do if a fault exists, or it can do nothing and allow the process to continue, as it should do if there is no fault. A fault in the monitoring equipment can then have two consequences: what is called a "fail-safe" error occurs when the system is shut down although no fault exists, and a "fail-dangerous" error occurs when the monitoring function fails so that the system will not be shut down when a plant fault occurs.

Generally the consequence of a fail-safe error is much less than that of a fail-dangerous error. A fail-safe error normally interrupts plant operation, and so will reveal itself immediately when the plant is operating. If the plant is shut down, the fault will be detected only when an attempt is made to bring the plant into operation. The shut-down process is usually designed so that no damage will be caused to the plant or its operators.

A fail-dangerous error may cause serious damage to the plant and operators. Without routine maintenance a fail-dangerous error on the safety monitoring equipment will manifest itself only when a plant fault occurs; the expected automatic shut-down will then not occur.

The most severe consequence of a fail-dangerous error lies probably in the safety monitoring circuits of a nuclear power reactor. A fail-safe error will cause an unscheduled shut-down of the station, costing possibly some hundreds of thousands of pounds per hour of downtime. A fail-dangerous error may at worst cause a calamity on the scale of the Chernobyl disaster with a cost which is very difficult to quantify, but which will certainly be very large. An American investigation in 1957 led to the Brookhaven Report (WASH-740) which estimated that the property loss in a worst-case release of radioactivity could be $7 billion. A later survey in 1964–1965 took into account price rises and inflation and revised the total cost to some $40.5 billion.

Generally, the design philosophy will be to equalize approximately the damage potentials of safe and dangerous failures. The damage potential is defined as the product of the probability of the failure and its effect (Klassen and van Peppen, 1989). The most convenient measure of the effect is its cost. This is often difficult to estimate, particularly if loss of life is involved, but is necessary to establish a justifiable design. Some international agreements may be helpful; for example, there is an agreed maximum payment for loss of life in accidents involving civil airliners.

The cost of a major disaster at a nuclear power station will certainly be much greater than the cost of a safe failure, so the probability of a dangerous failure must be reduced far below that of the probability of a safe failure. This usually involves a high degree of redundancy in the safety system. A proposed target figure for the average probability of failing to meet a demand to trip is 10^{-7} over 5000 hours.

43.6.2 Designing Fail-Safe Systems

Any attempt at designing an electronic monitoring system which is much more likely to fail safely than fail dangerously immediately raises the question of fault data. Ideally, we need to discover the most likely failure mode of all components and design our circuits so that this type of failure leaves the system being monitored in a safe condition. For many components the two basic types of failure are open circuit and short circuit. Thus we need to know not just the raw component failure rate, but what proportion of failures are to open and short circuit. Unfortunately there is little information of this kind readily available; MIL-HDBK-217B, a major reference text of failure rates, makes no distinction between open-and short-circuit faults.

Despite the scarcity of data there are some conclusions that can be reached by examining the nature of certain components; for example, it is difficult to envisage a mechanism which will cause a short-circuit failure in a low-power carbon film or metal film resistor so that the predominant failure mode is open circuit. To take care of the remote possibility of a short circuit it is sometimes given a failure probability of 1 percent of the total failure rate.

Relay contacts also have a predominant failure mode. If they are open and are required to close when the relay coil is energized there is a very small probability that they will not do so because a speck of contamination has landed on one of the contacts and forms an insulating barrier between them, or because the coil is open circuit. In order to diminish the probability of this occurrence twin contacts are provided on a widely used telephone relay so that correct operation is assured even if one of the two contact pairs is non-conducting. An alternative approach is to seal the contacts into a gas-tight enclosure, typically filled with dry nitrogen, so that no dust or contamination can reach the contacts and they cannot oxidize. This is the method adopted in the reed relay, which has high reliability but cannot handle such heavy currents as the standard telephone relay nor drive as many contact pairs from one coil.

If a set of relay contacts have already closed and are in contact and the coil is then de-energized it is difficult to postulate a mechanism other than gross mechanical damage to the relay which will prevent the contacts from opening and breaking the circuit. Consequently in analyzing failure modes in relay circuits the open-circuit failure rate is often assumed to be some fifty times greater than the short-circuit rate. Here the short-circuit failure rate is associated with the failure of the contacts to open when the current in the relay coil is interrupted.

43.6.3 Relay Tripping Circuits

Many alarm and monitoring systems have a variety of sensing elements and are required to operate alarms and in critical situations shut down the plant they control. The sensing elements may be those used for system control or in critical cases a separate set. In both arrangements the alarm action must operate when any one plant variable goes out of range. The corresponding circuit may be a parallel or a series connection. If all the alarm outputs are connected in parallel they must normally be open circuit and must close to signal a fault. As this mode has a failure rate much higher than that where the relay contacts are normally closed and open to signal a fault it is not generally used. The preferred circuit is that with all the relay contacts in series and with all of them closed, as shown in Figure 43.15.

A fault detected by any unit will open the contact chain and can be made to energize an alarm to take emergency action. The contact opening could be caused either by energizing or de-energizing the relay coil. To minimize dangerous failures the coil is designed to be normally energized and the current is interrupted to indicate a fault. This means that a failure of the relay power supply or an open-circuit fault in the relay coil will be a fail-safe rather than a fail-dangerous event.

43.6.4 Mechanical Fail-Safe Devices

The same design principle can be applied where objects have to be moved when an alarm is signalled. For example, when a nuclear reactor develops excess power or overheats it must be shut down as soon as possible. This involves inserting control rods made of neutron-absorbing material such as boron into the reactor. One widely used scheme uses vertical channels into which the rods can drop. They are normally held out of the reactor by electromagnets attached to movable rods, and any condition calling for shutdown will interrupt the current in the electromagnets, so causing the rods to drop into the reactor under gravity. This means that any failure in the power supply or a break in the cabling or the electromagnet winding is a safe failure.

Passenger lifts comprise another group of control systems in which mechanical fail-safe mechanisms find a place. After some early accidents caused by failure of the hoist cable or the drive gearbox, public agitation stimulated a search for a safety device which would prevent this kind of accident. All failures in the drive or the cable result in a loss of tension in the connection between the cable and the lift cage. One fail-safe mechanism based upon this fact uses hinged spring-loaded cams attached to the top of the lift which are held back by the tension in the cable. When this tension disappears and the lift begins free fall, the cams fly outwards and jam against the lift guides, so stopping the descent of the lift.

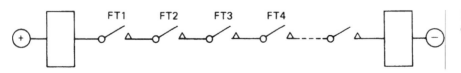

FIGURE 43.15 Series connection of relay contacts in guard line.

43.6.5 Control System Faults

Classifying errors in a control system is more difficult than in an alarm system; in some cases where there is no provision for reversion to manual control any failure may be dangerous. Manned spacecraft are generally controlled by on-board computers, but the crew have on occasions had to take over control. This was possible because critical landing operations are usually carried out in fair weather and in daylight. However, if there is no time to change over to manual control the fault must be classed as dangerous.

The way in which faults in aircraft autopilot controls are classified depends upon the particular phase of the flight in which they occur. For example, in mid-flight a fault which deprives the autopilot of power but leaves the aircraft flying a straight course may be comparatively safe as there should be time for the pilot to take over. On the other hand, a fault which sends the aircraft into a full-power dive and prevents manual control may cause structural failure and a major accident.

When a so-called "blind" landing is made in bad weather, the pilot has few of the visual indications normally available to show the aircraft's position relative to the runway. The aircraft is guided to the vicinity of the airfield by a radio beacon and then aligned with the runway by further short-range radio beams. A vertical beam gives a guide to the end of the runway and these signals together with the aircraft's terrain clearance measuring radio give all the information needed for an automatic landing. When still several hundred feet above the runway, the pilot can decide to abort the landing and either make another attempt to land or divert to another airport. Beyond this point the pilot must hand control over to the automatic control system, and no reversion to manual control is possible.

The redundant control system is equipped with a fault-detection system which warns the pilot if there is a disagreement between the channels which operate each control surface, so that he or she would normally commit the aircraft to an automatic landing only if there were no fault. The duration of this final phase during which the aircraft must be controlled automatically is about 30 seconds, and for this time any failure of the control system must be classed as a dangerous failure.

In today's traffic conditions steering a car requires continual vigilance, and any failure of the steering mechanism would clearly be a dangerous fault. Consequently, attempts to reduce the effort needed by the driver using power derived from the engine are designed to be fail-safe, so that if the hydraulic power mechanism fails, the driver retains control and reverts to manual steering. In one form of the system, a spring-loaded, spool-type valve is fitted in the steering rod which connects the steering box to the front wheels, as shown in Figure 43.16. With no torque exerted on the steering wheel, the springs hold the valve centrally, and the oil from the pump is returned directly to it. When the driver turns the wheel the springs deflect and allow oil to pass through the valve to a piston and cylinder which are attached to the drag link, so turning the road wheels in the direction the driver desires. When the road wheels have turned as much as the driver wishes, the force in the drag link ceases, the spool valve is returned to the neutral position, and the oil supply to the cylinder is cut off. End-stops fitted to the spool valve allow only a small degree of movement between the valve and its housing and transmit the steering force when the hydraulic power assistance fails.

43.6.6 Circuit Fault Analysis

In order to estimate the probability of fail-safe and fail-dangerous faults in a circuit it is necessary to examine the consequence of each component sustaining a short- or an open-circuit fault. As an example, we take a relay-driving circuit which is driven from an open-circuit collector TTL gate such as the 7403. The circuit is shown in Figure 43.17. The input is normally low, holding T1 and T2 in conduction and the relay energized. To indicate an alarm the input becomes high impedance so that the current in R1 and R2 ceases. T1 and T2 are cut off, and the relay current falls to zero. A fail-safe fault will occur when the relay releases with the input held low. This could be caused by the following faults:

R2	Open circuit (O/C)	T1	O/C
R3	O/C	T2	O/C
D1	Short circuit (S/C)	RL	Coil O/C or S/C Contacts O/C

Although there is a small probability of the resistors suffering a short-circuit failure, this is much less than other probabilities which also could cause a fail-safe failure, and it has been neglected. A short circuit of R1 and R4 comes into this category.

A fail-dangerous condition would be caused by any fault which would hold the relay contacts closed when the input was open circuit. The obvious faults are T1 or T2 S/C, but R1 or R4 O/C might also give trouble if the ambient temperature were so high as to increase the leakage current in T1 or the input driver. An electrically noisy environment might also cause trouble since R2 would be at a high impedance to earth and would thus be liable to pick up noise voltages which could hold T1 partly conducting. There is also a small probability that the relay contacts will develop a short circuit.

Using the failure rate data in Table 43.4 we can estimate the probabilities of fail-safe and fail-dangerous faults. The probability of a fail-safe fault is then 45.8 FIT, and the probability of a fail-dangerous fault between 5.2 and 15.2 FIT. The lower figure is perhaps a little optimistic since it assumes that neither R1 nor R4 O/C will cause a fault. This is likely to be so if the unit is not subjected to high temperature and is not in a noisy environment. If we cannot be certain of this we must use the figure of 15.2 FIT.

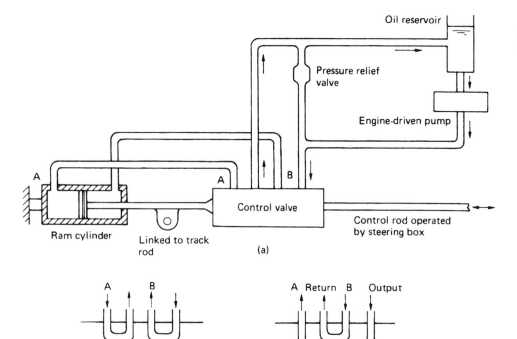

FIGURE 43.16 Power-assisted steering: (a) layout; (b) flow of hydraulic fluid.

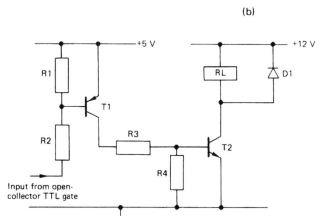

FIGURE 43.17 Relay-driving circuit.

TABLE 43.4 Failure rates* used in the example (see text)

Component	O/C failure rate (FIT)	S/C failure rate (FIT)
Transistor	2.5	2.5
Resistor	5	–
Relay coil	15	5
Diode	1.2	0.8
Contacts	10	0.2

*1 FIT is a failure rate of 10^{-9} per hour.

A further possible cause of a safe failure is the collapse of either power supply voltage. If the probability of this occurring is comparable with the component failure rates it must be included in the total. Often, however, several alternative power sources are available with automatic switching, and the probability of a power failure may then become much smaller than the other failure probabilities; it can thus be neglected.

A difficulty with any alarm system is that a failsafe error immediately reveals itself, as the plant it is connected to will shut down automatically. A fail-dangerous error, on the other hand, will not reveal itself until the plant develops a potentially dangerous fault and the expected automatic shutdown does not occur. In order to minimize the likelihood of this it is normal practice to conduct routine maintenance of the system with particular emphasis on checking for fail-dangerous faults. Such maintenance reduces the period during which any particular sensing unit can remain in a fail-dangerous state and so reduces the chance of a serious accident, which can occur only when the sensing unit is in this state and a plant failure also occurs.

43.7 SOFTWARE RELIABILITY

43.7.1 Comparison with Hardware Reliability

During the last two decades, computers and microprocessors have undergone steady development, with particular improvement in reliability. Consequently, an increasing proportion of failures arise not from hardware malfunction but from errors in computer programs. These, unfortunately, are more difficult to locate and remove than are hardware faults.

We can use the same definition of reliability as we use for hardware, namely, the probability that the program will operate for a specified duration in a specified environment without a failure. We can also take advantage of redundancy to improve performance and use the same methods of calculation to estimate the overall reliability of a compound program in terms of the reliabilities of the separate sections of the program. The same procedure can also be used to estimate the reliability of a complete system in terms of the separate reliabilities of the hardware and software.

43.7.2 The Distinction between Faults and Failures

When discussing software reliability it is important to distinguish between faults and failures. A failure occurs when a program fails to produce the output expected (we assume that the hardware is operating correctly). This arises from some error or fault in the program.

Most control and measurement programs comprise a large number of routines, only some of which are executed during each run of the program, so that a fault may be dormant for some time until the routine which contains it happens to be executed. The large number of different paths through a typical program mean that the chance of any particular execution revealing a fault is very small; for example, a program which contains 10 two-way branches has about 10^6 paths through it. Some measurements quoted by Musa et al. (1987) on a mixture of programs showed that the number of failures per fault was in the region of 10^{-6} to 10^{-7}.

These facts have important implications for program-testing strategy. If we test only complete programs we need to undertake a very large number of tests with a wide range of data to ensure that all paths through the program are traversed. On the other hand, if the program is divided into small modules, each of which can be tested separately, we are much more likely to find faults since there are far fewer paths to test. In off-line computer operations such as data processing, the time required for the execution of a program segment is not usually critical; the important factor is the total time needed to process a suite of programs such as, for example, those involved in payroll calculations. The situation may be different in on-line operations such as vehicle or process control. Here the calculations involved in a control loop must be performed in a certain time; otherwise the system becomes unstable. Thus there is an additional failure mode which must be considered; the right answer is no use if it is not available in time. This point may be a restriction in the use of software redundancy schemes that involve extra computation to mask a fault.

43.7.3 Typical Failure Intensities

Although the user is generally interested in the failure intensity, that is, the number of faults per hour of operation, this is not a factor relating solely to the program, since it depends upon the number of instructions executed per second. Thus the same program run on a faster machine will have a higher failure intensity. It is useful to have a measure which is independent of execution rate; this is the failure density, or the number of faults in a given size of module (often 1000 lines of code).

Computer folklore has it that many commercial programs contain about one fault per 1000 instructions when first released. Measurements reported (Musa et al. 1987) for programs of about 100,000 lines of source code reveal failure densities of this order.

43.7.4 High-Reliability Software

Failure densities which can be tolerated in commercial software are quite unacceptable for safety-critical situations, such as aircraft controls and the monitoring of nuclear power reactors. The cost of a failure in terms of money or human lives is so great in these cases that extreme efforts are required to produce error-free programs. For example, the error rate quoted for the "fly-bywire" controls of the A320 Airbus is 10^{-9} per hour. For earlier passenger aircraft, error rates of 10^{-9} per mission were proposed, a mission being a flight of 1–10 hours' duration. In order to construct software to this standard it is necessary to specify the requirement in very precise terms, perhaps using a language specially designed for the purpose. In these applications natural languages such as English include far too many ambiguities.

Two examples of job specification languages are Z, developed at RSRE (Sennett 1989), and Gipsy, developed at the University of Texas at Austin. The use of a formal language to specify a problem raises the possibility of using the computer to translate the specification into a computer program automatically. At present, however, this is practicable only for comparatively trivial programs.

43.7.5 Estimating the Number of Faults

In order to monitor the task of eliminating faults from a program it would be useful to have some estimate of the number of faults in the program being tackled. One method of doing this is called "seeding" and is based upon a technique used to estimate the number of fish of a certain species living in a large pond. The pond is seeded by introducing into the pond a number of similar tagged fish. After a suitable interval to allow the tagged fish to become evenly distributed throughout the pond, a sample of fish is removed from the pond and the number of tagged and unmarked fish is counted. If there are 10 tagged fish and 80 unmarked fish of the species being investigated in the sample and 150 tagged fish were added to the pond, we conclude that the population of the pond is 80 × 150/10 = 1200 fish.

In the case of a computer program the seeding process consists of adding errors to the program before the debugging activity is started. The debugging team is not aware of this and reports all faults found. The supervisor then counts the number of deliberate faults found (say, 10) and the number of genuine faults found (say, 50). If 40 faults were introduced deliberately, the estimated total number of faults is 50 × 40/10 = 200.

This technique will predict the number of faults accurately only if they are all equally likely to be found in any particular debugging session, including those planted. In practice, some faults will be easier to find than others, so the result of the process should be given a margin of error.

43.7.6 Structured Programming

Many techniques have been proposed to improve program reliability and to make programs easier to write and check. One of these which has gained almost universal acceptance is structured programming. This has been defined in several ways, but essentially it involves dividing the program into modules which can be compiled and tested separately, and which should have only one entry and one exit. Consequently, GOTO statements should be avoided, and only three types of program block are required. The first is the sequence in which statements are executed in turn, without branching. The second is the loop in which a set of statements are executed while a condition is satisfied (while… do). The third is the selection in which one set of statements is executed if a condition is satisfied, if not another set is executed (if… then… else…). Flow charts for these constructs are shown in Figure 43.18. Some languages allow for a number of alternatives in the selection process, but the same result can be obtained by repeated use of "if… then… else…."

The size of the modules is a matter of some compromise; if they are too large they become too complex to be easily understood and checked, and if they are too small the number required for any significant project becomes excessive. For many purposes a module of 40–50 lines seems appropriate.

One of the early approaches to the design of structured programs was the Michael Jackson or data structure design method. The basic feature of this is that the structure of the program should correspond to the structure of the files that the program is intended to process. The data are analyzed to produce a data structure diagram in which the components of each data item are placed in a box below it and connected to it by lines. Thus in designing a program to produce a report the item "line" would be below the item "page" which in turn would be below "report." The program structure can then be produced by replacing each item by a module which processes it. Thus "line" is replaced by a module which generates a line, and this is used by the module above which generates a page. Modules on the same level are normally processed as a sequence, but two of them may be alternatives and they are marked to denote this. The modules defined in this way are all of the three types mentioned above (sequences, loops, or selections).

A later development which has been promoted by the National Computing Center is Structured Systems Analysis and Design Methodology (SSADM) which is now claimed to be the most widely used structured method in Europe. It is also being submitted as the basis for a European software development methodology.

Experience has shown that the use of structured methods in programming makes for fewer errors, needs less time in writing and testing programs, and makes the subsequent maintenance and enhancement of programs much easier. Its advantages are particularly evident in large projects where many programmers are each working on a small part of the program and need precisely defined interfaces between their various segments. It is often convenient to make each module a subroutine or procedure; this ensures that any variables used in the subroutine cannot be accessed from outside the subroutine unless they have been declared as "global." We can then use the same variable names in a number of different modules without confusion. This is particularly useful

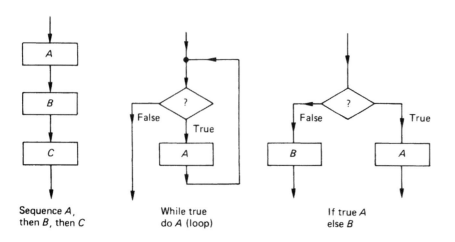

FIGURE 43.18 Flowcharts for structured programming.

in developing large programs involving many programmers who do not then need to keep a constantly changing list of all variable names used in the project.

Since structured programming was first introduced many versions of it have been proposed, some of which depend upon the use of particular languages. Early high-level languages such as FORTRAN were not designed with structured programming in mind. Later languages such as ALGOL, PASCAL, and ADA were specifically given features which simplified the division of the program into independent modules with a well defined start and finish, as required for structured programming. ADA was developed specifically for high-integrity applications; it was adopted by the U.S. Department of Defense in 1979 and has also been used in the U.K. for defense projects. An ANSI standard version of ADA was issued in 1983. Although ADA addresses many of the problems which arise with earlier languages it is large and complex. Consequently, it is difficult to ensure that all compilers produce the same object code from a given source program, since the language contains some ambiguities.

Another language which has been developed at RSRE specifically for high-integrity applications is Newspeak (Sennett 1989). This addresses the problem of limited word and register sizes by incorporating into the compiler facilities for checking that arithmetic operations cannot cause overflow. This is done when declaring each variable by stating the range of values which it can take; thus, for example, a variable of type byte must lie within the range 0–255 (decimal). This information allows the compiler to check whether arithmetic operations will cause overflow or other errors by evaluating them with extreme values of the variables. Any expression which could cause overflow is rejected by the compiler. This arrangement eliminates some run-time errors which could be disastrous in a system controlling an aircraft or a nuclear reactor, which is required to operate for long periods without a failure.

Another possible source of error which is eliminated by Newspeak is the possibility of confusion caused by collecting data in different units. Almost all the input signals handled by a computer embedded in a control or instrumentation system represent some physical quantity and are consequently meaningless unless the units of measurement are known. Newspeak allows the unit to be associated with each data item and thus errors due to combining data items expressed in different units can be avoided, since only variables measured in the same units can be included in the same calculation. If this condition is not satisfied, the compiler will reject the program. Mixed units could occur, for example, in maritime navigation where chart depths are shown in meters and sonar depth sounders may indicate feet or fathoms.

43.7.7 Failure-Tolerant Systems

One method of coping with software faults, as with hardware faults, is to use redundancy. However, multiple copies of the same program cannot be used since they will all contain the same faults. We must use different programs, written by different teams, which are designed to perform the same task. For example, in the A310 Airbus slat and flap control system two independently developed versions of the software are used. Also, in the American Space Shuttle a single back-up computer runs in parallel with four primary computers. All critical functions can be performed by either of two separate completely independent programs, one in the primary computers and the other in the back-up computer.

43.8 ELECTRONIC AND AVIONIC SYSTEMS

43.8.1 Radio Transmitters

Where the reliability of a radio link using diversity reception is a critical part of a control or instrumentation system it may be necessary to provide some degree of redundancy in the transmitting end of the link. The same procedure as used for any complex system can be adopted, that is, the division of the system into smaller assemblies, each of which can be replicated. The exact arrangement depends upon the type of transmitter and whether it can be maintained.

Generally, duplicate redundancy is adequate and in a maintained system both of the duplicate units may normally be active. When a fault appears, the faulty unit is disconnected, and a reduction in the quality of service is accepted for the short time during which the repair is undertaken.

For an AM transmitter the master oscillator, RF amplifier, modulator, and power supply may be duplicated. To avoid synchronizing problems only one master oscillator would normally be used at a time, the second being switched in when needed. Where a single aerial such as a rhombic is used, the outputs of the two amplifiers can be combined and fed to the aerial. Where an aerial array is used this can be split into two sections, each fed from one amplifier. This gives some redundancy in the aerial system, which may be an advantage if it is sited in an exposed position where inclement weather might cause damage.

The general arrangement is shown in Figure 43.19. To cope with more than one failure, facilities can be included to cross-connect, for example, RFA1 to MOD2, MOD1 to A2, or PSU1 to either RFA2 or MOD2.

Another technique which has been used to improve reliability is to duplicate the main transmitter components but operate them so that each chain produces half of the rated output. This will give a substantial degree of derating and so will enhance reliability. When one unit fails the other is switched to full power.

A source of unwanted shutdown in radio transmitters is transients which can produce over-voltages on valves, transmission lines, and aerials. These generally cause protective devices such as circuit breakers to trip. The overvoltages can be caused by overmodulation in AM transmitters or lightning

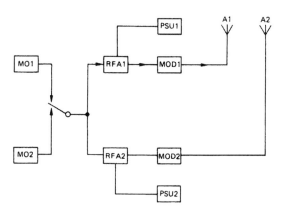

FIGURE 43.19 Block diagram of duplicate AM radio transmitter.

discharges to lines or aerials. Although limiters are nearly always fitted to the modulation input chain, short transient overloads do occasionally occur. At maintained stations the first action after such an interruption is to reclose the circuit breaker; in most cases the transient will be over before this and normal operation can be resumed. However, to reduce costs many stations are now unattended, so to prevent lengthy periods of shutdown the circuit breakers are fitted with automatic reclose mechanisms. These will reclose the breaker, say, 5 seconds after it has opened and repeat the process once or twice more. Thus unless the transient fault is repeated several times, the transmitter will only be out of action for a few seconds. This arrangement requires circuit breakers for protection rather than fuses, but gives much better service when short transient overloads are encountered.

Where frequency modulation is used a fault in a duplicate system as described above will not cause a reduction in the received demodulated signal, only a 3 dB reduction in the signal-to-noise ratio. If the link has a reasonable margin of signal strength this should give a very small change in its performance.

One use of radio transmitters which requires very high reliability is as radio beacons used by aircraft for navigating along prescribed tracks between airports. These transmitters are the major source of navigational data for aircraft in flight and usually include some form of redundancy.

Another need for highly reliable communication has recently arisen in the provision of UHF and microwave links between offshore oil platforms and the onshore terminals and refineries. These systems are required to handle telemetry data and control signals and are usually equipped with error-correcting facilities and double checking of control commands. Use is also made of satellite links for the longer links.

43.8.2 Satellite Links

In non-maintainable systems such as satellite links it is customary to have the stand-by item unpowered until a fault occurs. It can then be energized and switched into operation, so extending the satellite's life. Typical units duplicated in this way are the low-noise amplifier and the power amplifier. The latter originally used a traveling-wave tube (TWT) which had a limited life and required a high-voltage power supply. Solid-state amplifiers are now available for the 4–6 GHz band which are more reliable, but are also duplicated.

The earliest satellites provided one wideband channel which handled a number of separate signals, including television. Unfortunately, the TWT is non-linear, particularly at maximum power output, so causing intermodulation, that is, the mixing together of the various signals being handled. To avoid this, current satellites such as INTELSAT V have some 30 different transponders, most of which receive 6 GHz signals and transmit back on 4 GHz. Each transponder handles only one signal, so avoiding intermodulation. This kind of problem arises only when analog signals are transmitted; when using digital coding time-division multiplexing can be used so that each input signal has access to the channel in turn for a short interval. As only one signal at a time is handled there is no opportunity for intermodulation, and the TWT can be used at maximum power output and so at maximum efficiency. This multiplicity of channels enables a switch matrix to be included in the system between the outputs of the receivers and the inputs of the TWTs. It is thus possible to connect any receiver to any TWT so that by providing a few spare channels, these can act as redundant standby units for the remaining 20 or so working channels. It is also possible to cross-connect these to the 14/11 GHz channels. The switching matrix can be controlled from the earth station, so giving the satellite the ability to continue operation even with several units faulty.

Although early satellites were used mainly to expand the provision of telephone, radio and television links, and thus could not be regarded as of the highest priority, later satellites included some intended solely to aid navigation, controlled by an international organization called INMARSAT. These now supply navigational data to some thousands of ships, and a failure would put many lives at risk. To reduce the chance of this there are now a number of navigational satellites, and channels are leased from INTELSAT and other satellite operators.

43.8.3 Aircraft Control Systems

An important application of complex control systems is in the guidance of aircraft. The earliest systems which took over from the pilot for a short time were used for landing in low visibility. Since a failure could involve major loss of life the U.K. Air Regulation Board laid down a stringent safety requirement. This was based upon statistics of the accident rate for manually controlled landings in good weather conditions, with a safety factor of 10, leading to a requirement that the probability of a major failure should not exceed 1 in 10^7 landings.

If we assume that the last landing phase in which the aircraft has been committed to the automatic control and the pilot cannot intervene lasts for 30 seconds, the system MTBF needed is 83,500 hours or nearly 10 years. This can be derived from the failure probability which is 10^{-7} for 30 seconds or 1.2×10^{-5} per hour. To demonstrate this MTBF with a confidence level of only 60 percent requires a test period of 76,500 hours, or nearly 9 years (BS 4200). The extremely high MTBF required for the complete guidance system and the very lengthy testing time make this an impractical scheme. The only way to reduce both of these times is to introduce some degree of redundancy, for example, by a triplicated majority voting system. In this scheme the system will operate correctly if any two of the three replicated channels are working. The probability of system failure is the probability that two or three channels will fail. By far the largest of these is the failure of two channels which has a probability of $3p^2$ where p is the probability of one channel failing. Thus $3p^2 = 10^{-7}$, whence $p = 1.83 \times 10^{-4}$ for a 30-second period, or 2.19×10^{-2} per hour. This corresponds to a channel MTBF of only $10^2/2.19 = 45.6$ hours.

This is a much simplified calculation, but it indicates that this form of redundancy can provide an adequate system with a much less onerous target for the channel MTBF and for the time needed to demonstrate it. An alternative scheme which was used in the VC-10 aircraft involved two autopilots, each with a monitoring channel which could detect most errors. The general arrangement of the triplicate scheme is shown in Figure 43.20, and an example of the use of two pairs of processors in the Lockheed L 1011–500 airliner is shown in Figure 43.21.

In practice what is classed above as a single channel consists of one complete control system, comprising separate sensing and servo mechanisms to control the elevator, aileron, and rudder. Each of these may comprise some six units, so the complete channel may involve some 18 units. Making the rather simple assumption that they have similar MTBFs, the requirement for each unit becomes about 822 hours. This is not a difficult design parameter for current, mainly solid-state, equipment.

Where the automatic control is required only for landing, the rudder control is required only to provide a final "kick-off," that is, a corrective action needed in a cross-wind to align the aircraft along the runway. Prior to this the aircraft must point a little into the wind to maintain its track along the center line of the runway. The small use of the rudder allows a duplicate form of redundancy to be used without prejudice to the overall system reliability. Where in-flight control is needed the rudder controls will have the same degree of redundancy as the other parts of the guidance system.

There is a further assumption implicit in this calculation which needs examination: the nature of the final voting element. There can be only one of these, and its reliability must be taken into account, since it cannot be replicated and it may determine the complete system reliability. The first systems used three torque-limited motors coupled to the same shaft, which operated the control surface. The shaft was of ample size so that its shearing strength was much greater than the maximum motor torque, and its failure was consequently extremely unlikely.

Later systems replicated the control surfaces, so that the "voting" operation is the summation of the three forces exerted on the control surfaces by the airflow over them. It is difficult to imagine any way in which this summation could fail so long as the aircraft's speed is maintained: thus this is one case in which the reliability of the voting operation can be taken as 100 percent.

Later control systems included a requirement that automatic control should be available throughout the flight, with a typical duration of 2 hours. This means a much greater MTBF for each channel, of the order of 11,000 hours.

To diminish the MTBF required for each channel, some use has been made of quadruplex redundancy. In this arrangement several modes of operation are possible; one involves comparing all the channels. So long as they all agree, no action is needed. When a disparity occurs the channel whose output disagrees is disabled. If the fault is in the earlier stages of the channel it may be possible

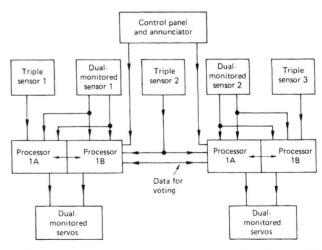

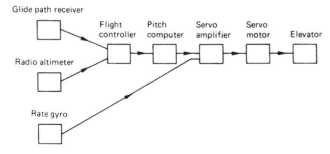

FIGURE 43.20 One channel of a triplicate elevator control system.

FIGURE 43.21 Two pairs of processors used in the Lockheed L1011–500 aircraft.

to couple the input to the final hydraulic drive to that of another channel; otherwise the control surface can, if possible, be driven to a central position. The three remaining channels can then be configured as a majority voting triplicate system.

The Lockheed L 1011–500 airliner is an example of four processors being used in pairs as shown in Figure 43.21. The sensors are connected in either monitored pairs or 2/3 majority logic.

In such a high-reliability application it is essential that great care should be taken to avoid the possibility of common mode faults which could imperil all the control channels. The most probable of these is a failure of the power supply. To counteract this each engine has an electrical generator, and there is in addition a separate generator independent of the main engines. In addition to electrical supplies a high-pressure oil supply is needed for the actuators which drive the control surfaces. Again each engine has an oil pump, and a final standby is provided by an air-turbine-driven pump which can be lowered into the airstream in an emergency.

Some current aircraft which rely heavily on electronic controls rather then mechanical links require very high reliability in the essential flight control systems. For example, the manufacturers of the latest A 320 Airbus claim a failure rate of 10^{-9} per hour of flight for the control systems. In the few years it has been in service, however, there were three fatal accidents involving A 320s by January 1992. In the crash at Habsheim, France, it could be argued that pilot error contributed to the outcome, but this has not been established for the other two.

In the 1980s NASA adopted a failure rate criterion for aircraft controls of 10^{-9} for a 10-hour civilian flight, for the succeeding decade. This is a target which can only be attained by a high degree of redundancy. In practice on a long-haul flight the pilot can act as a back-up during much of the flight; the only time fully automatic operation is needed is during the landing phase. For example, on transatlantic flights of the Boeing 747 it is customary to switch on only one autopilot during most of the flight; the other two are switched on only for the landing to give full redundancy.

No redundancy of controllers will give the required reliability unless the probability of common mode faults such as power failures is very small. To this end it is usual to provide a high degree of redundancy in the supply of electrical power and high-pressure oil for hydraulic servos.

The plans for the Boeing 777 which came into service in 1995 provide for a 120 kVA generator on each engine and in the auxiliary power unit. There is also a 30 kVA back-up generator on each engine. The captain's primary flight display, navigational display, and some engine displays can be supported from only one back-up generator. Extra back-up is provided for the flaps which are driven electrically if the hydraulic drive fails. Other aircraft use batteries for emergency power supplies.

In addition to the triplicated flight control system, modern aircraft such as the Boeing 767 have a duplicate flight management computer which optimizes the total cost, including crew time, maintenance and fuel, a thrust management computer which regulates the automatic throttle and fuel flow, and a dual redundant control display which provides computer-generated flight plans, performance data, and advice to the pilot. All of these are connected to a digital bus which also carries the outputs of all subsystems and sensors.

43.8.4 Railway Signaling and Control

When railways were first built engine drivers had a large degree of autonomy; there were no time-tables and the absence of modern electrical signaling meant that there was no nationwide standard of time. However, as traffic grew it became clear that much more discipline imposed centrally was essential if a safe service was to be offered to the public.

Early accidents caused public alarm and subsequent legislation requiring safety measures such as signal interlocking and improved braking. Braking on early trains was by modern standards extremely poor; brakes were fitted only to the engine and to the guard's van. After an accident in 1889, which killed 78 people when part of a train rolled backwards down a hill and collided with an oncoming train, an Act was passed requiring continuous braking for passenger trains. This involved redundancy in the braking system by fitting brakes to every carriage, so providing a much greater braking force than previously available.

Unfortunately, the first brake controls were operated from the engine, and when a coupling failed and the train split into two parts the rear portion had no brake power. The need was for some fail-safe mechanism which would automatically apply the brakes if two carriages became uncoupled. After tests it became clear that the Westinghouse vacuum brake was by far the most satisfactory arrangement, and this was adopted for all U.K. trains. The brakes in this scheme are applied under spring loading, and a piston fitting inside a cylinder can pull off the brakes when the air is pumped out of the cylinder. The engine provides the necessary vacuum through a pipe which passes along all the carriages.

As soon as two carriages become uncoupled, the pipe is disconnected, the vacuum is broken, and the brakes applied automatically under spring pressure. This simple arrangement has proved very reliable over some 100 years of use. Later versions used compressed air for brake actuation to provide greater operating force, but a second sensing pipe was used to ensure that the brakes were applied if two carriages became uncoupled.

As train speeds increased with the development of more powerful engines, the reliability of the signaling system became more important. The most likely failure of the mechanical system was a break in the steel cable used to operate the signal semaphore arm. This was originally built so that in the safe condition the cable was not in tension and the arm pointed downwards; to indicate danger the cable was tensioned to bring the arm up to the horizontal position. It was later realized that for a fault involving a break in the wire or the failure of a joint or link (by far the most likely type of failure) this was a "fail-dangerous" condition, since if the signalman set the signal to danger it would revert to clear, so allowing a train to pass. The signal aspect was then changed so that the clear position was with the arm pointing upwards and the cable in tension. A break in the cable would then allow the arm to fall to the horizontal position, meaning danger. The break would then cause a "fail-safe" fault which would stop any train arriving at the signal.

Even greater train speeds caused difficulties in seeing the signals, and four-aspect color signals were introduced. These do not involve any movement of a signal arm, and there is no possibility of a fail-safe design based on the operation of a mechanical link. The signals use electric filament lamps, and the reliability of the signaling depends mainly upon the probability that the lamps will not fail. Even when derated with the penalty of reduced light output the life of a filament lamp is only some 4000 hours, so that used alone, it would be unacceptable. To give a much better life redundancy is used in the form of a twin-filament lamp. The current taken by the main filament is monitored, and when it fails the second filament is automatically switched on, and a relay gives a fault indication in the signal box so that the lamp can be changed as soon as possible. If the second filament fails, the nearby signals are automatically set to danger. Thus as well as using redundancy, the final back-up is the fail-safe action of stopping all oncoming trains.

Despite the redundancy and the fail-safe features of this form of signaling its effectiveness depends on the driver's ability to interpret correctly the information given by the signals. Thus any further progress requires some back-up for the driver. This was first introduced by the GWR in the 1930s by using a movable ramp between the rails which gave an audible warning in the cab when the train passed a distant signal set at danger. This had to be cancelled manually by the driver within a short period; otherwise the brakes would be applied automatically.

A similar scheme was later adopted nationally, but with induction loop signaling between the track and the train. The London Underground system has a much greater traffic density than the mainline railways, trains following one another at intervals of 2 minutes or less. A back-up system which relies on the driver to apply the brakes may involve unnecessary delay, so on the Victoria line the track signals control both the train speed and the brakes.

As it is likely that further developments in automatic train control will involve microprocessors, some development work was initiated by British Rail into high-reliability systems. The scheme they selected used twin processors which had identical inputs from axle-driven tachometers and voltages induced by track-mounted cables and magnets. The processors are cross-connected and have identical inputs. Periodically each one compares its output states with those input from the other processor, and any disagreement is interpreted as a fault and the check result line is energized. This shuts down the system permanently by blowing a fuse which also removes the supply from any displays. This scheme can easily be extended to a triplicate scheme in which any disagreement will shut down only the faulty processor. The remaining two can continue operation so long as their outputs agree.

43.8.5 Robotic Systems

The increasing use of automation in manufacturing processes has led to the use of a large number of industrial robots. By 1985 some 100,000 were in service, mainly in Europe, the United States, and Japan. In the interest of safety they should be segregated from human operators, but in many cases this is not completely possible since they have to be loaded with workpieces and the finished product removed. A further problem arises with robots which have to be "taught" what to do. These are used in work such as paint spraying in which it is impossible to calculate the path along which the robot hand should move. The robot is provided with internal storage capacity and the hand is moved under manual control to simulate the action of a human operator by using a small hand-held control box. The trajectory of the hand is then stored within the robot and can be repeated automatically as often as required.

Clearly this means that someone must be inside the robot enclosure when the power is applied, and it is essential that this person's safety is ensured at all times. Also, it may not be possible to maintain and test the robot properly without touching it.

Anyone approaching the robot will have no knowledge of the path the arm is programmed to follow, and there is no external indication of this. It is thus liable to make sudden and unexpected moves. Furthermore, if the power suddenly fails it may drop its load.

All these factors make the reliability of the robot control system (generally based on a small computer) of extreme importance. Redundancy can be used with either two or three channels. With two channels the two output signals can be continuously compared; any disparity immediately stops the robot's motion. Three-channel redundancy can be arranged as a majority voting system. Other safety features which can be included are:

1. In the "teaching" mode, the linear speed of the arm can be held down to, say, 20 cm/s to avoid danger to the operator.
2. An area extending at least a meter beyond the robot can be fenced off to prevent access.

3. The robot arm and other moving parts can be equipped with touch-sensitive pads which sense when anything is touched and so halt the robot.
4. Emergency buttons can be fitted which stop the pump and dump the high-pressure hydraulic supply.
5. Built-in test equipment can be fitted which checks most of the robot control system each time power is switched on. Any fault will prevent the robot's motion.

43.9 NUCLEAR REACTOR CONTROL SYSTEMS

43.9.1 Requirements for Reactor Control

In the design of many control systems the importance of reliability depends upon the consequences of a failure. The cost of a failure in the control circuits of a domestic appliance such as a washing machine is largely restricted to the cost of repair; this will fall on the manufacturer if it occurs within the guarantee period, after this on the owner. The manufacturer is interested in making a product with sufficient reliability to ensure a very low level of returns under guarantee and to keep up with the competition.

At the other end of the spectrum, the cost of a failure of the control or safety systems of a nuclear reactor is extremely large but not easy to quantify. However, the consequences of the Chernobyl accident have been very great in the destruction of equipment, the sterilization of land, and illness and loss of life. In 1957 an American report (WASH-740) predicted that the cost to property alone in a worst-case accident could be $7 billion; this was updated in 1964–1965 to a figure of $40.5 billion and would be much larger at today's prices. The accident at Three Mile Island in the United States in 1979 caused a loss of coolant and a partial meltdown of the core. Although there were no casualties, and no one off the site received excessive radiation, the cost to the owners was considerable. The TMI-2 reactor was disabled, and the Nuclear Regulatory Commission would not allow the twin TMI-1 reactor to restart until the accident had been investigated.

In consequence of this extremely high cost the reliability specified for reactor safety systems is also very high; the U.S. Nuclear Regulatory Commission in 1986 made a policy statement about future advanced reactors, which were expected to show a considerably better safety record than existing reactors and have a probability of less than 10^{-6} per year (1.14×10^{-10} per hour) for excessive radio-nuclide releases following a core meltdown.

France and Switzerland have safety targets similar to that of the United States and a similar target was set in the U.K. for the core-melt frequency of the Sizewell B station. Sweden has a similar target for core melt, but the probability of severe radio-active contamination of large land areas is required to be even lower. The Swedish firm ASEA-Atom has proposed an inherently safe reactor (PIUS) which can be abandoned at any time by its operators and will then automatically shut itself down and stay safe for about a week. The core and the primary coolant system is surrounded by a large pool of borated water, and any system upset will cause the borated water to flood into the reactor coolant system. As boron is a strong neutron absorber, this will shut the reactor down, and it is also designed to establish natural convection cooling. Another automatic shut-down mechanism has been proposed for a Power Reactor Inherently Safe Module (PRISM) in which any upset which causes a considerable increase in the coolant temperature would automatically cause safety rods to fall into the reactor core. The unit consists of nine reactor modules feeding three turbine generators. Each module has six neutron-absorbing rods for power control and three articulated rods suspended by Curie point magnets. These are for emergency shut-down. When the magnets reach their Curie temperature, they lose their magnetism, and the rods fall into the core.

43.9.2 Principles of Reactor Control

The reliability required for a nuclear power station control system is almost certainly greater than that required in other systems, as can be seen from the specified failure rates quoted in the previous section. This reliability involves a high degree of redundancy in the control and safety systems and calls for particular features in the design of the reactor itself.

It is interesting to recall that the first reactor which demonstrated self-sustaining nuclear fission, the Enrico Fermi pile built at the University of Chicago, had primitive but redundant safety measures. A large neutron-absorbing boron rod was suspended by a clothes line above the pile and a graduate student was given an axe and told to cut the line, so releasing the rod, in the event of an accident. Also another group of students were stationed above the pile and given bottles of gadolinium (another neutron absorber). If a problem arose, they were told to throw the bottles and run away.

Two major concerns in the design of the reactor are to prevent the escape of radioactive material into the surroundings and to prevent a core melt-down. As far as possible the reactor is designed to provide at least two methods of coping with any failure. To prevent the escape of radioactive material, the fuel is contained in closed cans, and the entire reactor core is enclosed in a large enclosure—concrete in the early gas-cooled reactors and steel in light-water reactors.

Also, to prevent the failure of a coolant pump causing overheating of the core several pumps are used, and the reactor can survive the failure of any one. For example, the early Magnox stations had six gas circulators, and the later AGRs, such as Hinkley, have eight. Also, the loss of the power supply for the pumps will automatically shut down the reactor.

CHAPTER | 43 Reliability in Instrumentation and Control

The choice and location of the sensors used to monitor the state of the reactor is decided largely by the principle that any conceivable failure of the reactor system should be detected by at least two types of instrument. Thus excess activity within the reactor will be shown by an increase in fuel can temperature and also by excess neutron flux. A rupture in the coolant circulation pipes of a gas-cooled reactor would be revealed by a large rate of change of gas pressure (dp/dt) and by changes in the channel outlet gas temperature.

One design factor which will make the control task much easier is to choose a combination of fuel, coolant, and moderator which has a degree of inherent stability. This is usually a matter of minimizing fluctuations of fuel temperature, by selecting a combination which has an overall negative temperature coefficient of reactivity. This means that if a disturbance causes an increase in neutron flux, and thus in core temperature, the result is a decrease in reactivity which will limit the disturbance. This is a form of negative feedback.

The early Magnox gas-cooled reactors had a variety of sensors for detecting faults, mostly arranged in 2/3 majority voting logic, as shown in Table 43.5. The later AGR reactors had 27 different parameters measured, all arranged in 2/4 logic, and proposed PWRs will have a few more parameters in the primary system with nine additional parameters being measured in the secondary protection system. All of these are connected in 2/4 logic.

The main limitation on the reactor output is the maximum permitted fuel can temperature. In order to regulate this the neutron-absorbing control rods are adjusted, and this requires a large number of thermocouples which measure can temperature. Because these deteriorate with high temperature and irradiation they are usually replicated. In some reactors they are fitted in groups of 16. The signal amplifier scans them and selects the element which shows the highest temperature to indicate the spot temperature.

TABLE 43.5 Fault sensors in early Magnox gas-cooled reactors

Quantity sensed	No. of channels	Logic scheme
Fuel can temperature	9	$(2/3)^2$
Rate of change of pressure	6	$2 \times 2/3$
Channel outlet gas temperature	36	$12 \times 2/3$
High-power excess flux	9	$3 \times 2/3$
Low log excess flux	3	2/3
High log doubling time	3	2/3
Low log doubling time	3	2/3
Loss of blower supply	6	3/6

To ensure the necessary high reliability in the control function, and particularly in the provision for automatically shutting the reactor down under fault conditions, a high degree of redundancy is used, generally triplicated with majority voting. This is used either singly or in two stages, starting with nine separate inputs, as shown in Figure 43.22. For higher reliability 2/4 schemes have also been used, for example, in the Heysham 2 AGR.

For a 2/3 majority voting scheme the output will be incorrect if either two or three of the inputs are incorrect. If the probability of a false input is p, the probability of a false output is

$$p_1 = 3p^2 + p^3 \qquad (43.22)$$

Since p is normally small the second term is much smaller than the first, and it can be neglected in view of the uncertainty associated with the value of p. For a two-stage version, often called $(2/3)^2$ we can use Equation (43.22) again to give

$$P = 3p_1^2 = 3(3p^2)^2 = 27p^4 \qquad (43.23)$$

This result ignores the failure probability of the voting circuit which should be added if it is significant.

The use of 2/4 logic improves the reliability of the redundant system with regard to dangerous failures, but is rather worse than 2/3 logic with regard to safe failures. Thus in order to prevent the correct response to a reactor fault three or four channels of a group must fail. If the probability of one channel failing is p, the probability of a dangerous failure is then $P_D = 4p^3 + p^4$. Generally, the second term is much less than the first, so that the value can be taken as $4p^3$, compared with $3p^2$ for 2/3 logic.

However, if only two of the channels give an incorrect "trip" indication, the reactor will be shut down, giving a safe failure. The probability of this is $P_s = 6p_1^2$, where p_1 is the probability of a channel failing safely.

The failure rates of solid-state equipment in nuclear power stations have been found to lie in the range of 0.01 to 0.1 failures per year. Taking an average value of 0.05 gives a failure probability of 0.0125 for a 3-month interval. This is of interest as the license for the early reactors required all monitoring equipment to be tested regularly every 3 months.

This figure is for all failures; the critical ones are the dangerous failures which on average form about one-third of the total. Thus the dangerous failure rate is 0.0042 over 3 months. In a $(2/3)^2$ arrangement this gives a failure probability of $27(0.042)^4 = 8.40 \times 10^{-9}$.

A dangerous failure can occur at any time during the maintenance cycle, and it will persist until the next test/calibrate/repair action. Thus the fault can remain for a period varying from zero to the maintenance interval, in our case 3 months. The average time is thus 1½ months.

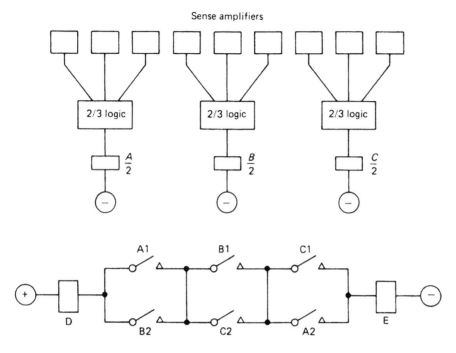

FIGURE 43.22 Double 2/3 logic.

In order to assess the consequent likelihood of a severe accident we need also to know the chance of a reactor having an upset which would call for the action of the safety system. This is not an easy figure to discover, but some guidance is available from American sources; these show that 52 significant events occurred between 1969 and 1979 during about 400 reactor years of operation. This is a frequency of about 1 in 8 reactor years, or 0.0156 in ½ months.

Some later information published in 1989 deals with seven events during some 940 reactor years of BWR operation; this corresponds to a frequency of 1 in 134 reactor years, but deals with only one type of reactor. There is a considerable discrepancy between this and the previous figure, but at least it suggests that 1 even in 8 reactor years is a conservative assumption.

The probability of a simultaneous reactor upset and a dangerous failure of the monitoring equipment is then

$$8.4 \times 10^{-9} \times 0.0156 = 1.31 \times 10^{-10}$$

This calculation relates to the interval between maintenance actions: 3 months. We assume that after service the equipment is as good as new, so that each 3-monthly period has the same probability of failure. Thus the probability of a reactor upset which will find the monitoring equipment in a fail-dangerous state is

$$4 \times 1.31 \times 10^{-10} = 5.24 \times 10^{-10} \text{ per year}$$

This result relates to only one group of sensors, but even with 100 groups the probability is well below the target figure of 10^{-6} per year and allows for some common mode failures.

Using 2/4 logic with the same fail-dangerous probability per channel per 3 months of $p = 0.0042$ gives a group failure rate of

$$p_1 = 4p^3 = 4 \times (0.0042)^3 = 2.96 \times 10^{-7}$$

Combining this with the probability of a reactor upset of 0.0156 per 3 months gives an overall probability of a dangerous fault as 4.62×10^{-9} over 3 months, or for a year the figure of 1.85×10^{-8}. This configuration of double 2/4 logic was used for the later AGR stations which included solid-state monitoring units, and these are likely to have a somewhat lower failure rate than is assumed in the above calculation. In the double 2/4 or $(2/4)^2$ configuration the first stage of logic gives $p_1 = 4p^3$. The second stage gives $P = 4p_1^3 = 256p^9$. This configuration is used in the Heysham 2 power station.

The periodic testing of the reactor protective system requires skilled technicians and is a rather lengthy process. Experience with French reactors shows that this requires two technicians and takes some 10 hours to complete. The same operation prior to restarting the reactor requires 2 days and involves a significant delay in restarting the reactor after a shutdown. In order to improve this activity attention is now being devoted to the automatic testing of all sensing amplifiers and signal conditioning equipment. In most nuclear power stations the control and safety shutdown systems are separate. The safety circuits and their logic are normally hardwired, but increasingly, the control function is handed over to a computer. This basically involves regulating the reactor output so that the heat output and the electrical power generated match the load demanded by the transmission system which the station feeds. This is a tedious and difficult task to perform manually, and a computer can ease the burden on the operator. The major problem is the need to control various sectors of the reactor to ensure a fairly even temperature distribution throughout the core. Early stations used a central computer with standby, but later AGR stations such as Heysham 2 used 11 distributed microprocessors with direct digital control to replace the former analog control loops. The advantage of the digital technique is that more complex control algorithms can be used, and it is easier to modify and optimize the various control loops. The availability of computing power enables some of the safety monitoring to be undertaken by the computer, the rest being performed by hardwired logic. This gives a useful diversity of techniques for the protection circuits.

43.9.3 Types of Failure

The distinction between safe and dangerous failures is of critical concern in nuclear power stations; a safe failure will shut the reactor down without cause. This involves a considerable loss of revenue as most nuclear stations are used to supply the base load and are run as nearly as possible at full power continuously. A further problem arises from the build-up of neutron-absorbing products such as xenon after a reactor has been shut down. This element builds up for about 10 hours and then decays, effectively "poisoning" the reactor so that unless it is restarted soon after shutdown the operators must wait for perhaps 20 hours before there is sufficient reactivity to permit sustained fission.

The accident at Chernobyl which emitted a large volume of radioactive products has shown the world the enormous cost of a nuclear meltdown. Thus there is a great impetus to make all monitoring equipment very reliable, and to ensure that as many of the failures as possible are safe failures. Some steps can be taken in the design of the equipment to this end, but it is impossible to eliminate all dangerous faults.

If we use directly coupled amplifiers, either with discrete components or as IC operational amplifiers in general, faults will result in the output either falling to a voltage near to the lowest power supply or rising towards the most positive supply. One of these will represent a safe failure and the other a dangerous failure, and they will usually occur in equal proportions. One method of reducing the proportion of undetected dangerous failures is to convert the incoming signal to a square wave which is subsequently amplified by an ac amplifier and rectified. Almost all faults will result in the disappearance of the ac signal, and this can be monitored and cause an alarm.

The early Magnox stations used relays for the logic of the safety system; they are fast enough and regarded as very reliable. To reduce the chance of dangerous failures they were connected so that the most reliable operation (opening a normally closed contact) corresponds to the action needed to cope with a reactor fault. The contacts of the various monitoring groups are connected in series with two contactors and the power supply, and all contacts are normally closed. The series circuit is triplicated. Two guard relays are used in each guard line, one next to each pole of the power supply to ensure that an earth fault cannot cause a dangerous failure. The contacts of the guard relays are connected in a 2/3 circuit to power contactors which are also normally operated and connect a power supply to the electromagnets which hold up the reactor control rods. The scheme is shown in Figure 43.23. When a fault occurs in the reactor the current in all relay and contactor coils is cut off and their contacts open. The final result is that the control rods are released from their supports and fall into the reactor core, so reducing the neutron flux and the generation of heat.

Although the telephone type relays originally proposed for these circuits have a long history of reliable operation, some design engineers thought that this could be improved. Their object was to remove a possible source of failure—pick-up of dust and contamination by the contacts—by sealing the relay in an airtight enclosure. This should in principle afford some improvement, but in fact it produced a number of dangerous failures which were detected during routine tests in which a relay occasionally failed to open its contact when de-energized. In order to investigate the fault the relay was removed and the enclosure cut away. Unfortunately, the handling disturbed the relay and cleared the fault, and it was necessary to take an X-ray photograph of the relay *in situ* to discover that the relay contacts seemed to be stuck together. This was traced to the effect of the varnish used to impregnate the relay coil. Although the coil is baked to dry the varnish before the relay is assembled it is very difficult to remove all traces of solvent. As the relay coil is normally energized when it is in use the heat evolved would evaporate any remaining traces of solvent and these were found condensed on the contacts. The problem was solved by sealing the contact stack separately so that no contamination from the coil could reach them.

43.9.4 Common Mode Faults

The need for extremely high reliability in nuclear safety systems requires great care to avoid any common mode faults. The first consideration is the integrity of the power supplies; the arrangement of the relay logic is basically fail-safe in that the removal of the power supply would shut the reactor down. However, this will involve considerable loss of income since the nuclear plant is generally used to supply base load and is thus delivering near to full output for 24 hours a day. It is thus worth providing typically two different rectified ac supplies, a stand-by battery and a separate generator with automatic switching to ensure a very low probability of loss of power. Another possibility is to supply the equipment from float-charged batteries which will ensure operation for some hours if there is a mains failure. Usually

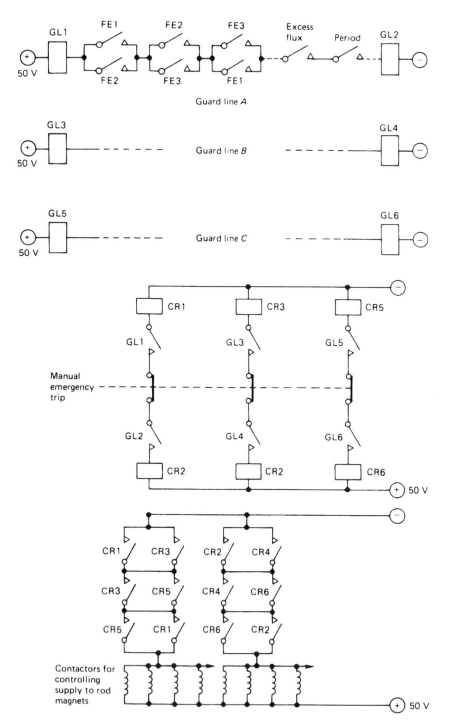

FIGURE 43.23 Redundant logic controlling safety shutdown rods.

a stand-by engine-driven generator is started a few minutes after a mains failure has occurred to continue float-charging. There may also be a second generator or an alternative mains supply.

Another possible common mode failure was revealed by the Three Mile Island accident; the power and control cables were all taken through the same tunnel between the plant and the control room. Consequently, when a fire occurred initially in the power cables this very soon damaged the control cables, so preventing important details of the plant's condition from reaching the operators. Some of the safety circuits were also damaged. It is essential that the redundancy in the instrumentation should not be nullified by bunching together all power and signal cables in the same duct or trunking. Ideally, each of the three or four channels involved in every measurement should be physically separated from the others, with an appropriate fireproof barrier. A worst-case possibility is a major damage to the main control room; to cope with this a separate emergency stand-by facility may be provided some distance away with minimum instrumentation and sufficient controls to monitor the reactor state and shut it down safely. For example, at the Heysham 2 AGR station the emergency center is some 100 m away from the main control room.

43.9.5 Reactor Protection Logic

The early Magnox stations used standard telephone-type relays for switching logic; their characteristics had been established over some decades of use and their reliability was considered to be adequate. However, as the monitoring equipment became more reliable, mainly through the change from valves to transistors, a comparable increase in the reliability of the switching logic was sought. One possibility was to move to some static apparatus rather than electromechanical devices such as relays. Early transistors had poor reliability and magnetic devices offered much higher reliability, consisting only of magnetic cores and copper windings.

The design principles adopted were as follows:

1. Each parameter checked by the safety system should be measured by three separate channels whose outputs are combined on a majority voting basis. If only one channel trips, an alarm should be given.
2. No single fault or credible combination of two equipment faults should prevent the reactor being shut down when a demand occurs.
3. No single fault in the safety circuits should cause a reactor trip.
4. If three guard lines are used they should all be opened by a reactor fault condition, if all safety equipment is working correctly.
5. Once a guard line has tripped, it should remain tripped until restored manually.
6. The guard lines should be segregated from one another, and only one should be accessible at a time.
7. A shorting plug is provided which can be inserted to maintain the continuity of the guard line when any piece of equipment is removed for maintenance. Only one item should be able to be removed at a time.
8. The control rods should be operated by two distinct methods. This is usually ensured by providing two groups of rods, controlled separately but by similar means.
9. To allow for overriding operator action a manual trip should be provided which is as close as possible to the final control rod holding circuits.

The component selected for the logic function was a multi-aperture ferrite device known as a "Laddic." This had the geometry of a ladder with several rungs, with a number of different windings. After experiments a seven-aperture device was adopted. This is energized by two interleaved pulse trains at a frequency of 1 kHz, called "set" and "reset." The pulses required for each guard line are about 1 A peak with a duration of 10 μs. An output is obtained only when the dc "hold" signals are present, so giving a three-input AND logic function. The hold currents are obtained directly from the various monitoring units, and are cut off when any parameter reaches its threshold value. By splitting the hold windings in two, either of them can be energized to provide an output, so giving an OR logic operation. With suitable connections, the AND and OR functions can be combined to give an output $X = (A + B) \cdot (B + C) \cdot (C + A)$. This is logically equivalent to $X = A \cdot B + B \cdot C + C \cdot A$, which is a "two-out-of-three" majority vote, the function required for a triplicated guard line. The arrangement of core and windings is shown in Figure 43.24. The output signal from the Laddic is a pulse of about 100 mA peak and 2 μs long. It requires only a single-transistor amplifier to couple this to the next Laddic when a chain of them is used to form a guard line.

As the output is a pulse train, a pulse-to-dc converter is used to turn this into a continuous signal which can hold up the safety rods. The converter does not reset automatically, so that once its output has disappeared it will not return until a manual reset is operated.

One way of guarding against common mode failures is to use more than one set of safety equipment, with different technologies and physical separation. This concept led to the development of pulse-coded logic for reactor safety circuits, which initially used standard integrated circuits and was designed for 2/3 voting. Three guard lines are used in a 2/3 configuration so that if any two or all three guard lines trip, all six shutdown rods fall into the core as the supply to

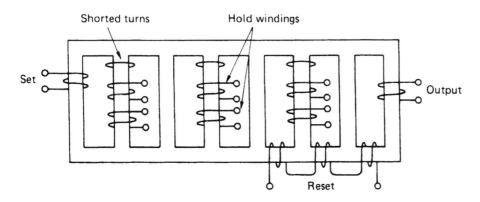

FIGURE 43.24 Core and windings of Laddic.

the magnets which holds them up is cut off. Each guard line has its own pulse generator which generates a train of pulses which examine the state of each trip instrument. The three pulse generators are synchronized, and the pulse train from each line is communicated to the others via optical fibers.

In one version of the scheme the pulse trains consist of 10 bits at 7.8 kbit/s and a "stuck at 1" or a "stuck at 0" fault in any instrument will produce a unique code which can be used to identify the faulty unit. A 14-parameter version of the scheme was attached to the Oldbury power station protection system between 1978 and 1982 and then transferred to the DIDO materials testing reactor, acting in a passive role. It has now been expanded for active operation, and is considered as adequate for use in a commercial power reactor.

43.10 PROCESS AND PLANT CONTROL

43.10.1 Additional Hazards in Chemical Plants

Although the equipment mentioned in previous sections is often required to operate in hostile environments involving salt spray, dust, humidity, etc., the atmosphere which envelopes it is generally inert. In contrast, many chemical plants handle corrosive, toxic, flammable, or explosive substances, and any design for reliability must tackle these extra hazards. This affects two system areas, first, the major plant activity which usually involves transporting liquid or gaseous materials around the plant while it undergoes various processes. The probability of various modes of failure must be examined and appropriate methods devised to counteract them. This often involves providing alternative ways of performing various operations.

At the same time most of the control and measurement information is handled as electrical signals, and output devices may include solenoid- or motor-operated valves and electrically driven pumps. Any switch, relay, or contactor which involves moving contacts and handles an appreciable current can cause a spark when interrupting current. The temperature of the spark can reach thousands of degrees Centigrade, sufficient to ignite any flammable gas or vapor. Thus any such equipment used in a location where flammable gas or vapor may be present must be surrounded by a flameproof enclosure. The enclosure must have suitable access covers which can be removed for inspection, maintenance, or the connection of power supply cables. The covers must have sufficiently wide flange couplings, and small enough airgaps to prevent a flame propagating from the inside of the enclosure to the outside and so causing a fire or an explosion. In addition to this requirement for normal working conditions it is necessary to ensure that no accident can occur when the covers are removed. The main possibilities arise from "hot spots," that is, parts of the equipment which are hot enough to ignite a flammable gas or vapor, and the discharge time of capacitors. There must be enough delay between switching off the power and allowing access to capacitors to ensure that they have no residual charge which could cause a spark. BS 5501 (EN 50014) specifies the maximum charge on the capacitors which is permitted when the enclosure is opened. If the charge has not decayed to a safe value when the case is opened normally a label is required showing the delay needed after disconnecting supplies before the case should be opened. A similar delay may be needed if any component has an excessive surface temperature, to allow it to cool.

BS 5000 and BS 5501 specify various ways in which electrical apparatus can be made safe for use in explosive atmospheres. These include:

1. *Type "d"* Flameproof enclosures. These can withstand internal explosion of an explosive mixture without igniting an explosive atmosphere surrounding the enclosure.
2. *Type "e"* Increased security against possibility of excessive temperature and occurrence of arcs and sparks.
3. *Type "i"* Intrinsic safety. This specifies electrical systems in which the circuits are incapable of causing ignition of the surrounding atmosphere. The maximum nominal system voltage is restricted to 24 V (limit of 34 V) and a normal current of 50 mA, short-circuit value 100 mA.
4. *Type "m"* Encapsulated to prevent contact with atmosphere.
5. *Type "p"* Pressurized enclosure. In this a protective gas (usually air) is maintained at a pressure greater than that of the surrounding atmosphere. The enclosure must withstand a pressure of 1.5 times the internal pressure.
6. *Type "o"* Oil immersed. Here the oil will prevent any arcs or sparks igniting an external flammable gas or vapor.
7. *Type "q"* Powder filling. This specifies apparatus using voltages of 6.6 kV or less with no moving parts in contact with the filling. The preferred filling is quartz granules; no organic material is permitted.

The hazard posed by liquids is mainly dependent upon their flash point; those with flash points below 66°C are classed as flammable and those with flash points below 32°C as highly flammable. Many industrial gases and vapors will ignite with a concentration of only 1 percent by volume in air, and some mixtures are flammable with a wide range of concentrations. For example, any mixture of air and hydrogen with between 4 percent and 74 percent of hydrogen by volume is flammable. For a mixture of acetylene and air the limits are 2.5 percent and 80 percent. The limits for an explosive mixture are generally somewhat narrower than for flammability.

43.10.2 Hazardous Areas

An important factor in designing equipment for use in areas which may contain flammable gases or vapors is the distance from the point of release within which a dangerous concentration may exist.

Some degree of classification is given in BS 5345: Part 2 (IEC 79–10). The grade of release is designated as *continuous* where it is expected to occur for fairly long periods, as *primary* where it is expected to occur periodically or occasionally during normal working, and as *secondary* where it is not expected to occur in normal operation, and if it does so only infrequently and for short periods.

Areas of continuous release are normally graded as Zone 0, areas of primary release as Zone 1, and areas of secondary release as Zone 2.

BS 5345 also classifies the ventilation, an important factor in deciding the extent of the hazardous areas, according to its effectiveness in clearing away dangerous gases or vapors and the proportion of working time during which it is in operation.

An area classification code applied to petroleum has been proposed by the Institute of Petroleum (Jones 1988). This puts liquefied petroleum gas in category 0 and classifies other petroleum products according to their flash points (FP):

- *Class I*: liquids with an FP below 21°C.
- *Class II(1)*: liquids with an FP from 21°C to 55°C, handled below their FP.
- *Class II(2)*: liquids with an FP from 21°C to 55°C, handled at or above their FP.
- *Class III(1)*: liquids with an FP above 55°C to 100°C, handled below their FP.
- *Class III(2)*: liquids with an FP above 55°C to 100°C, handled at or above their FP.
- *Unclassified*: liquids with an FP above 100°C.
- A further area classification specifically for all installations handling natural gas at all pressures is based upon the British Gas Engineering Standard BG/PS SHA1.

A factor which greatly affects the hazard posed by gases or vapors is their density relative to air. If they are heavier than air they will tend to collect near the ground and are much more likely to ignite than lighter gases, such as hydrogen, which will rise and disperse readily, particularly out doors.

43.10.3 Risks to Life

Inadequate reliability of plant and equipment can involve high expenditure in repairing damage caused by failure and loss of income while the plant is inoperative. Increased expenditure on measures to improve reliability can be justified if their cost is less than the reduction they are likely to bring in the cost of failure. The calculation is not difficult although the data required may be known only imprecisely. For example, if the probability of a dangerous occurrence is estimated as once in 400 years, and its estimated cost is £800,000, the average annual cost is £2000. If some measure can reduce the probability of the occurrence to once in 1000 years, this means a reduced average annual cost of £800. Should the cost of this measure, converted to an annual charge, be £1500, it would not be worthwhile since it would save only £1200. If, however, its annual charge were only £1000 the expenditure would be justified.

The same procedure could be adopted for situations involving risks to people if there were any agreed value placed on human life. There is, however, no such agreement, and the usual procedure is to design for an agreed maximum level of risk. This is expressed as a fatal accident frequency rate (FAFR) or fatal accident rate (FAR). It is the number of fatal accidents in a group of 1000 people during their working life, usually taken as 10^8 man-hours. The figure for the U.K. chemical industry is 4 if the Flixborough accident is ignored, and about 5 if it is included in a 10-year average. The figure for all premises covered by the U.K. Factories Act is also 4 (Green 1982). Since about half of the accidents in the chemical industry are unconnected with the material being handled, and involve falling down stairs, or vehicles, the FAFR for a chemical plant should be no more than 2.

This figure represents the total risk; where it is difficult to predict each individual risk it is suggested that the figure for any particular risk should not exceed 0.4. To give an idea of the import of this value, it is estimated that we all accept an FAFR of about 0.1 when engaged in driving, flying, or smoking (Green 1982). This figure is somewhat different from that quoted elsewhere; for example, the FAFR for hunting, skiing, and smoking has been reported as 10–100.

A detailed analysis depends upon a knowledge of the likely failure rates of the various components of the plant which are best obtained from previous experience with similar equipment. Typical figures quoted are as follows:

1. The failure rate for natural gas pipelines in the U.S. is about 47×10^{-5} per mile per year.
2. The rate for sudden failure of a pump (including cable, motor, and gearbox) is about 0.4 per year.
3. The rate for failure of a level controller is about 0.5 per year.
4. The rate for a control valve failing shut is about 0.5 per year.

43.10.4 The Oil Industry

The oil industry is particularly susceptible to fire and explosion hazards since its raw material and nearly all of its products are flammable. Reliable plant designs for land-based equipment have been developed over many years, but the exploitation of the North Sea oilfields revealed several problems arising from the hostile environment which had not previously been encountered. Some of the oilfields were several hundred miles offshore where the sea is over 600 feet (180 m) deep and waves often 30–50 feet (9–15 m) high.

The height of a wave expected once in a hundred years is over 100 ft (30 m). The lack of previous experience in working under these conditions meant that some structures failed; for example, in March 1980 the *Alexander Keilland*, an oil rig in the Norwegian sector, turned turtle in some 20 minutes with the loss of 123 lives. The cause was ascribed to the collapse of one of the five supporting columns. The column was held to the platform by six bracings, and an opening had been cut into one of them to house a hydrophone positioning control which was welded in place. The reduction in strength caused fatigue fractures in the welds and the bracing, which eventually failed, so throwing extra load on to the other bracings which caused them to fail in turn. The column became detached from the platform, giving the rig a list of some 30°. The damage caused to the deck and the lack of compliance with instructions for watertight bulkheads and ventilators allowed much of the deck to be flooded, and the whole structure turned over in some 20 minutes.

In this case the accident had a number of contributory causes: the effect of cutting an opening in a bracing had not been investigated fully, the spread of fatigue cracks in various welds and the structure itself had not been observed, and instructions about watertight doors and ventilators were ignored. In view of the harsh environment it is clear that some degree of redundancy should have been built into the structure so that it would survive if at least one, and preferably two, of the bracings failed. Also, some interlocking mechanism could be provided which would prevent use of the rig in certain circumstances unless the watertight doors were closed.

By the mid-1970s the likelihood of a worker on an offshore installation in the British sector of the North Sea being killed was about 11 times greater than that of a construction worker and nearly six times greater than that of a miner. These figures do not include the 167 killed in the 1988 *Piper Alpha* explosion; 63 were killed in the period 1969–1979, so the inclusion of the 167 would increase the 10-year average from 63 to 230, a factor of nearly four.

For over a century it has been realized that many aspects of reliability which affect the safety of industrial workers or the general public cannot be left to industry without government regulation and monitoring. Generally, legislation is introduced some years after new practices or processes have been developed, often due to public alarm after some fatal accident. This is clearly evident in the history of the railways and coal mines, and is equally true of the offshore oil industry. Here matters are complicated because the platforms are often outside territorial waters, and legislation was needed to extend government control (Continental Shelf Act 1964). This was passed in a great hurry and made little provision for safety measures. In the UK the Health and Safety at Work Act of 1974 made great improvements in the safety legislation in factories and other workplaces but was not extended to the Continental Shelf until 1977. A further complication not envisaged originally was that some investigations such as inquests would take place under Scottish law, which has many differences from English law. An example of the legal difficulties arose in 1976 when the Grampian Police received a report that some fires which could have been started deliberately had occurred on a Panamanian-registered barge which was owned by a Dutch company and was on charter to an American company and operating within the safety zone of a production platform in the North Sea. At the time of the incident the barge was lying outside the safety zone because of bad weather. Although the police visited the barge, it was subsequently concluded that they had no jurisdiction.

43.10.5 Reliability of Oil Supply

A typical oil platform contains a large number of separate units, not all of which contribute directly to the oil supply. In order to enhance the system reliability and to allow for routine maintenance, much of the equipment is replicated. The overall reliability can then be estimated in terms of the reliabilities of the individual units, with allowance for any duplicate units.

Figure 43.25 shows a fault tree representing part of the pumping system of a typical oil platform having two supply

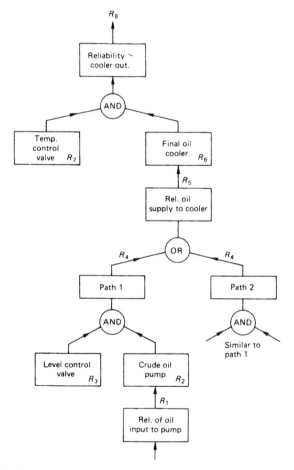

FIGURE 43.25 Fault tree of part of an oil platform.

paths. In the figure the symbols for unit reliability, such as R_2, R_3 and R_6, are shown within the blocks which represent the units, and the reliabilities of the supply path to a particular point, such as R_4 and R_5, are shown outside the blocks.

For path 1 to be operative, we require a supply of oil to the transfer pump and the transfer pump and the level control valve to be operating correctly. Thus the reliability of path 1 is

$$R_4 = R_1 \cdot R_2 \cdot R_3$$

Production can be sustained if either path 1 or path 2 is working. Thus the reliability up to the final cooler is

$$R_5 = 2R_4 + (R_4)^2$$

assuming that both paths gave the same reliability.

Beyond this we need both the temperature control valve and the final oil cooler to be operative, so that the reliability of supply up to the storage facility is

$$R_9 = R_7 \cdot R_8 = R_5 \cdot R_6 \cdot R_8$$
$$= \left[2R_4 - (R_4)^2\right] R_5 \cdot R_6 \cdot R_8$$

where

$$R_4 = R_1 \cdot R_2 \cdot R_3$$

On a large platform there may be four paths obtained by duplicating the equipment shown in Figure 43.25 which will give greater reliability if only one path need be operative.

43.10.6 Electrostatic Hazards

In many chemical plants and in the transport and refinery of oil the hazard of igniting flammable vapors and gases is countered by flame-proof enclosures and the segregation of potential spark-generating equipment from the flammable materials. Where the sparks or arcs are generated by current-carrying conductors this is a comparatively straightforward procedure, but in certain circumstances electrostatic potentials may be generated, and these are sometimes difficult to predict. The most likely cause is friction between insulators; under dry conditions potential of 10 kV or more can easily be generated. In conjunction with a capacitance of 50 pF (typical for a metal bucket) the stored energy is given by

$$E = \frac{1}{2C \cdot V^2} = 2.5 \text{ mJ}$$

This energy is sufficient to ignite hydrocarbons, solvent vapors, and ethylene. Figures for minimum spark ignition energy are:

Vapor-oxygen mixtures	0.002–0.1 mJ
Vapor-air mixtures	0.1–1.0 mJ
Chemical dust clouds	5–5000 mJ

Many plastic materials such as nylon are good insulators and readily generate static, so where sparks could be dangerous it is essential to earth all conducting objects in the vicinity.

Insulating liquids flowing in pipelines can also carry charge, generating currents of up to 10^{-6} A, and powders emerging from grinding machines can generate currents of 10^{-4} to 10^{-8} A. If we have a current of 10^{-7} A flowing into an insulated metallic container with a capacitance of 100 pF (e.g., a bucket), its potential will rise at a rate of 1 kV per second. In practice there will be some leakage and the rise in potential will be somewhat slower; it is nevertheless clear that potentials sufficient to cause ignition of flammable mixtures can be produced quickly.

Apart from earthing all metal objects in the working area, a number of other steps can be taken to reduce the risk of ignition:

1. Where flammable substances are being transported all vehicles and pipes used for the transfer must be bonded to earth before transfer starts.
2. Liquids having a very high resistivity can retain charge for some time even when passed through an earthed pipe. Their resistivity can be markedly reduced by introducing a few parts in a million of an ionic agent, so much reducing the hazard.
3. As the charging current generated by liquid flowing through a pipe is roughly proportional to the square of its velocity, the hazard can be reduced by ensuring that flow velocities are low.
4. Flammable atmospheres can be avoided by using an inert gas to dilute the concentration of flammable gas or vapor.
5. As static can be generated by free-falling liquids, entry and discharge pipes should be taken to the bottom of storage tanks to avoid this.
6. In humid conditions insulators attract a conducting layer on their surface which provides paths for charges to leak away to earth. Thus static hazards can be much reduced by operating the plant in air with a relative humidity greater than about 60 percent.
7. Most float-operated level indicators fitted to storage tanks use metal parts which must be firmly earthed.
8. People walking on synthetic carpets or flooring can easily become charged to a potential 10 kV or more. Before handling flammable liquids they should be earthed, for example, by using conducting footwear and a conducting floor (BS 3187, BS 3389).

It is generally accepted that static electricity is likely to be generated only in a dry environment, and high humidity is a valuable preventative. Surprisingly, three explosions occurred in oil supertankers in 1969 while their tanks were being washed out with a high-pressure jet of sea water. After small-scale tests it was concluded that charges were liberated

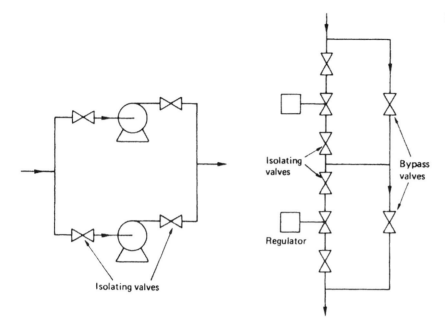

FIGURE 43.26 Parallel and series redundancy.

by friction between the water and the tank walls, and accumulated on falling water masses called "water slugs," causing spark discharges between them.

One hazard which has arisen only recently is the ignition of explosive and flammable atmospheres by radio transmissions. These occur in the vicinity of very high-power transmitters when the voltages induced in conductors are sufficient to create a spark or arc. Since the fields typical of normal broadcasting are only of the order of millivolts per meter it may be thought that cases of ignition should be very rare. However, with some high-power transmitters now delivering powers of a megawatt or more, cases have occurred and it is recognized that oil refineries should not be located too close to high-power transmitters. BS 6656 and BS 6657 deal with this situation and suggest safe distances beyond which ignition should not occur.

43.10.7 The Use of Redundancy

As with electronic systems, redundancy is a widely used method of improving the reliability of many industrial plants and processes. Since many of the units in these plants, such as pumps, valves, and compressors involve moving parts, they are prone to wear and require periodic inspection and maintenance. Where the plant is required to operate continuously, isolating valves are required on either side of the unit so that it can be taken off-stream for inspection, maintenance, or exchange. The arrangement depends upon the way in which redundancy is used.

Where a gas or liquid is pumped into a container and a failure of supply would be dangerous, two or more pumps may be installed in parallel, as shown in Figure 43.26, with isolating valves which allow each pump to be disconnected. Sometimes the reverse situation obtains when in an emergency the flow must be cut off. To allow for failure, two or more valves can be installed in series; if they are motor operated, separate or stand-by power supplies may be desirable. In order to take a valve offstream it is necessary to provide a bypass as well as isolating valves, as shown in Figure 43.26.

Other devices may also be connected in series for greater reliability; for example, in some satellites pressure regulators are used for the propellant supply. Their main failure mode is to open; that is no regulation, so two regulators are used in series for better reliability.

REFERENCES

Andeen, G. B. (ed.), *Robot Design Handbook*, McGraw-Hill, New York (1988).
Asahi, Y., et al., "Conceptual design of the integrated reactor with inherent safety (IRIS)," *Nucl. Technol.*, **91** (1990).
Asami, K., et al., "Super-high reliability fault-tolerant system," *IEEE Trans. Ind. Electron.*, **IE-33**, 148 (1988).
Asher, H., and Feingold, H., *Repairable Systems Reliability*, Marcel Dekker, New York (1984).
Atallar, S., *Chem. Eng.*, 94 (8 September 1980).
Ballard, D. R., "Designing fail-safe microprocessor systems," *Electronics* (4 January 1979).
Baber, R. L., *Error-free Software*, Wiley, Chichester, U.K. (1991).
Brozendale, J., "A framework for achieving safety-integrity in software," *IEE Conference Proceedings No. 314* (1989).
Celinski, K., "Microcomputer controllers introduce modern technology in fail-safe signaling," *IEE Conference Publication No. 279* (1987).
Clark, A. P., *Principles of Digital Data Transmission*, 2nd ed., Pentech Press, London (1983).

Cluley, J. C., *Electronic Systems Reliability*, 2nd ed., Macmillan, London (1981).

Cohen, E. M., "Fault-tolerant processes," *Chem. Eng.*, 73 (16 September 1985).

Dalgleish, D. J., *An Introduction to Satellite Communications*, Peter Peregrinus, London (1989).

Dijkstra, E. W., *Formal Development of Programs and Proofs*, Addison-Wesley, Reading, Mass. (1990).

Diller, A., *An Introduction to Formal Methods*, Wiley, London (1990).

Dittmar, R., "The Viper microprocessor," *Electron Power*, 723 (October 1986).

Green, A. E. (ed.), *High Risk Safety Technology*, Wiley, Chichester, U.K. (1982).

Hamming, R. W., "Error detecting and error correcting codes," *Bell System Technical J.*, **29**, 147 (1950).

Jennings, F., *Practical Data Communications*, Black-well, London (1986).

Jones, J. V., *Engineering Design, Reliability, Maintain-ability and Testability*, Tab Books, Blue Ridge Summit, Pa. (1988).

Klassen, H. B., and J. C. L. van Peppen, *System Reliability*, Edward Arnold, London (1989).

Lambert, B., *How Safe is Safe?* Unwin Hyman, London (1990).

Musa, J. D., A. Iannino, and K. Okiumoto, et al., *Software Reliability*, Macmillan, London (1987).

Ould, M. A., and Unwin, C., *Testing in Software Development*, Cambridge University Press, Cambridge (1986).

Sefton, B., "Safety related systems for the process industries," *IEE Conference Proceedings No. 314*, 41 (1989).

Sennett, C., *High-Integrity Software*, Pitman, London (1989).

Sibley, M. J. N., *Optical Communication*, Macmillan, London (1990).

Smith, D., "Failure to safety in process-control systems," *Electron. Power*, **30** (March 1984).

Swain, A. D., *The Human Element in System Safety*, Incomtech House, Camberley, U.K. (1974).

Ward, M., *Software that Works*, Academic Press, London (1990).

Wiggert, D., *Codes for Error Control and Synchronization*, Artech House, Norwood, Mass. (1988).

BRITISH STANDARDS

BS 787 Specification for mining type flame-proof gate end boxes. Parts 1–4: 1968–72

BS 889: 1965 (1982) Specification for flameproof electric lighting fittings

BS 2915: 1960 Specification for bursting disc and bursting disc devices for protection of pressure systems from excess pressure or vacuum

BS 3187: 1978 Specification for electrically conducting rubber flooring

BS 3395: 1989 Specification for electrically bonded rubber hoses and hose assemblies for dispersing petroleum fuels

BS 4137: 1967 Guide to the selection of electric equipment for use in division 2 areas

BS 4200 Guide on reliability of electronic equipment and parts used therein. Parts 1–8, 1967–87

BS 4683 Specification for electrical apparatus for explosive atmospheres (to be replaced by BS 5501)

BS 4778 Quality vocabulary. Part 1: 1987 International terms. Part 2: 1979 National terms

BS 4891: 1972 A guide to quality assurance

BS 5000 Rotating electrical machines of particular types or for particular applications. Parts 1–17

BS 5345 Code of practice for the selection, installation and maintenance of electrical apparatus for use in potentially explosive atmospheres (other than mining or explosive manufacture and processing). Parts 1–8: 1978–90.

See also EN 50014–20 (IEC 79), EN 50028 and EN 50039

BS 5420: 1977 (1988). Specification for degrees of protection of enclosures of switchgear and control gear for voltages up to 1,000 V a.c. and 1,200 V d.c. Now superseded by BS EN 60947–1: 1992

BS 5501 Electrical apparatus for potentially explosive atmospheres. See also EN 50014 and EN 50020

BS 5750 Quality systems. Parts 0–6: 1981–7

BS 5760 Reliability of constructed or manufactured products, systems, equipments, and components. Parts 0–4: 1981–6

BS 6132: 1983 Code of practice for safe operation of alkaline secondary cells and batteries

BS 6133: 1985 Code of practice for the safe operation of lead and secondary cells and batteries

BS 6387: 1983 Specification for performance requirements for cable required to maintain circuit integrity under fire conditions

BS 6467 Electrical apparatus with protection by enclosure for use in the presence of combustible dusts. Parts 1 and 2: 1985 and 1988

BS 6656: 1986 Guide to the prevention of inadvertent ignition of flammable atmospheres by radio-frequency radiations

BS 6657: 1986 Guide for prevention of inadvertent initiation of electro-explosive devices by radio frequency radiation

BS 6713 Explosion prevention systems. Parts 1–4: 1986

BS 6941: 1988 Specification for electrical apparatus for explosive atmospheres with type of protection "N" (Replaces BS 4683: Part 3)

BS 9400: 1970 (1985) Specification for integrated electronic circuits and micro-assemblies of assessed quality

BS 9401–94 deals with detail specifications for particular forms of integrated circuit

BRITISH STANDARD CODES OF PRACTICE

BS CP 1003 Electrical apparatus and associated equipment for use in explosive atmospheres of gas or vapor other than mining applications (largely replaced by BS 5345)

BS CP 1013: 1965 Earthing

BS CP 1016 Code of practice for use of semiconductor devices. Part 1: 1968 (1980) General considerations. Part 2: 1973 (1980) Particular considerations

EUROPEAN AND HARMONIZED STANDARDS

BS QC 16000–763000 Harmonized system of quality assurance for specific components

BS CECC 00009–96400 Quality assessment of specific classes of component

BS E9007: 1975 Specification for harmonized system of quality assessment for electronic components. Basic specification: sampling plans and procedures for inspection by attributes

BS E9063–377 deals with specific classes of component

British Standards are available from the BSI Sales Department, Linford Wood, Milton Keynes MK 14 6LE, U.K.

Chapter 44

Safety

L. C. Towle

44.1 INTRODUCTION

The interactions between the design and application of instrumentation and safety are many and diverse. The correct utilization of instrumentation for monitoring and control reduces risk. An obvious example is a fire detection and control system, but even a simple cistern control which prevents a water tank from overflowing affects overall safety. Any instrumentation which contributes to maintaining the designed status of an installation can arguably affect safety. However, instrumentation can increase the danger in an installation, usually by being incorrectly designed or used. The principal direct risks from electrical instrumentation are electrocution and the possibility of causing a fire or explosion by interaction between the electricity and flammable materials, which range from various insulating materials used on cables to the more sensitive oxygen-enriched hydrogen atmosphere of a badly ventilated battery charging room. Some aspects of the safety of lasers and the risks from radiation are dealt with elsewhere in this reference book, Part 3, Chapters 21, 22, and 24. Toxic materials should also be considered (see *Substances Hazardous to Health* in the References). These risks pale into insignificance when compared with the full range of possibilities of misapplying instrumentation to a process plant, but nevertheless, in an overall safety analysis all risks must be minimized.

It is important to recognize that nowhere is absolute safety achievable, and that the aim is to achieve a socially acceptable level of safety. Quite what level has to be achieved is not well defined; it is perhaps sufficient to say that people are even more reluctant to be killed at work than elsewhere, and hence the level of safety must be higher than is generally accepted. For example, the risk level accepted by a young man riding a motorcycle for pleasure would not be acceptable to a process operator in a petrochemical plant. There are similar problems in determining how much financial expenditure is justified in achieving safety.

As well as the moral responsibilities implicit in not wishing to harm fellow mortals there are, in the majority of countries, strong legal sanctions, both civil and criminal, which can be used to encourage all designers to be careful. In the United Kingdom, the Health and Safety at Work Act 1974, together with the Electricity Regulations, provides a framework for prosecuting anyone who carelessly puts at risk any human being, including himself. (In the United States, the same functions derive from the Occupational Safety and Health Administration, part of the federal government, with similar agencies in each state and some municipal authorities.) The Act places responsibilities on manufacturers, users, and individuals in some considerable detail, and the requirements are applied in almost all circumstances which can conceivably be regarded as work. For example, manufacturers are required to sell only equipment which is safe for its intended use, test it to check that it is safe, provide adequate installation instructions and be aware of the "state of the art." The Act was derived from the Robens Report, which is a very readable, well argued discussion document which sets a reasonable background to the whole subject of industrial safety. The Act lays great stress on the need to recognize, record, and evaluate levels of danger and the methods of reducing the risk to an acceptable level, and consequently, there is a need for adequate documentation on the safety aspects of any installation. In the majority of installations the enforcing organization is the Factory Inspectorate, who have awesome powers to enter, inspect, and issue various levels of injunction to prevent hazards. Fortunately, the majority of factory inspectors recognize that they do not have quite the infinite wisdom required to do their job, and proceed by a series of negotiated compromises to achieve a reasonable level of safety without having to resort to extreme measures. It is important to realize that the legal requirement in most installations is to take "adequate precautions." However, in the real world the use of certified equipment applied to the relevant British

Standard Code of Practice is readily understood, easy to document, and defensible, and is consequently the solution most frequently adopted. In the United States, the National Electrical Code, promulgated by the National Fire Prevention Association, is the controlling set of specifications for electrical safety.

In addition, the reader is referred to ANSI/ISA standards as follows:

ANSI/ISA84.01-1966 "Application of Safety Instrumented Systems to the Process Industries"
ANSI/ISA91.01-1995 "Identification of Emergency Shutdown Systems & Controls That Are Critical to Maintain Safety in the Process Industries"
ANSI/ISA RP12.6-1995 "Recommended Practice for Hazardous (Classified) Locations..."

44.2 ELECTROCUTION RISK

In designing any electrical equipment it is necessary to reduce the risk of electrocution as far as possible. Many sectors of industry have special standards of construction and inspection combined with certification schemes to take into account their particular risks. For example, electro-medical equipment has to meet stringent standards, particularly in cases where sensors are inserted in the body.

It is useful to try to assess the equivalent circuit of the human body, and there are a large number of references on the subject which show quite wide discrepancies between experimental results. A few facts appear to be common. Figure 44.1 shows the generally accepted figures for the ability to detect the presence of current, and the level of current which causes muscular contraction, although it must again be stressed that individuals vary considerably. Muscular contraction is a fascinating process, involving an electrical impulse signal releasing a chemical which causes the mechanical movement. The currents required are about 15 mA, and to maintain a muscle contracted it requires about 10 pulses/s. When a direct current is applied it causes the muscle to contract once and then relax; consequently direct current tends to be safer. However, at higher levels direct current does cause paralysis, since variation in body resistance due to burns, etc., causes the current to fluctuate and hence contract the muscles. The 50–60 Hz normally used for domestic supplies is ideally chosen to make certain that paralysis occurs.

Body resistance is quite a complex picture, since much of the initial resistance is in the skin. A dry outer layer of skin, particularly in the areas which are calloused, gives quite high resistance at low voltage, typically 10–100 kΩ, but this falls to 1 kΩ at 500 V. Other, more sensitive areas of the body, such as elbows, have a much lower resistance (2 kΩ). Once the outer layer of skin is broken, the layer immediately below it has many capillaries filled with body fluid and has very low resistance. The bulk resistance of humans is mostly concentrated in the limbs and is taken to be 500 Ω. Figure 44.2 shows one curve of body resistance and a possible equivalent circuit of a human being at low voltage when the skin resistance is converted to a threshold voltage.

The process of killing someone directly by electricity is also quite complex. Generally, it is agreed that a current of 20–30 mA applied to the right muscles of the heart would stop it functioning. Just how to persuade this current to flow in the practical problem of hand-to-hand electrocution

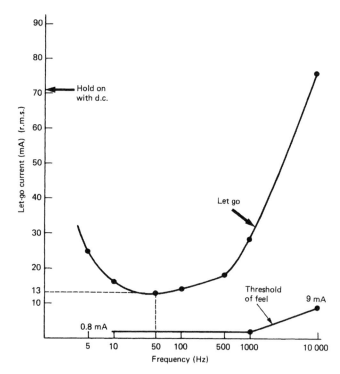

FIGURE 44.1 Variation with frequency of let-go current and threshold of feel.

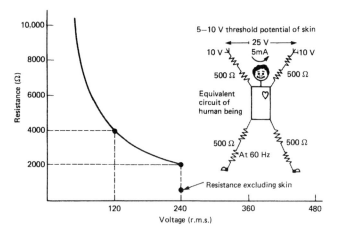

FIGURE 44.2 (a) Apparent increase of body resistance (hand to hand—dry) with reduction of voltage; (b) equivalent circuit of human being.

is widely discussed. Some sources suggest currents of the order of 10 A are necessary and others suggest there is a possibility of 40 mA being enough. The level of current is further complicated because there is a time factor involved in stopping the heart, and some protection techniques rely at least partially on this effect to achieve safety. The change is quite dramatic. For example, one reference suggests that heart fibrillation is possible at 50 mA if applied for 5 s and 1 A if applied for 10 ms. There seems little doubt, however, that the conventional 250 V 50 Hz supply used in the United Kingdom is potentially lethal, and that standing chest deep in a swimming pool with a defective under-water low-voltage lighting system is one very effective way of shortening a human being's life span.

The majority of modern instrumentation systems operate at 30 V or below, which to most people is not even detectable and is generally well below the accepted level of paralysis. There are, however, circumstances where even this voltage might be dangerous. Undersea divers are obviously at risk, but people working in confined hot spaces where sweat and moisture are high also need special care. Once the skin is broken, the danger is increased, and the possibilities of damage caused by electrodes fastened to the skull are so horrendous that only the highest level of expertise in the design of this type of equipment is acceptable. However, for the majority of conventional apparatus a level of 30 V is usable and is generally regarded as adequately safe. The design problem is usually to prevent the mains supply from becoming accessible, either by breaking through to the low-voltage circuitry, making the chassis live, or some other defect developing.

44.2.1 Earthing (Grounding) and Bonding

It follows from the previous discussion that if all objects which can conduct electricity are bonded together so that an individual cannot become connected between two points with a potential difference greater than 30 V, then the installation is probably safe. The pattern of earthing (grounding) and bonding varies slightly with the type of electrical supply available. Figure 44.3 illustrates the situation which arises if U.K. practice is followed. The supply to the instrument system is derived from the conventional 440 V three-phase neutral earthed distribution system, the live side being fused. A chassis connection to the neutral bond provides an adequate fault path to clear the fuse without undue elevation of the instrument chassis. All the adjacent metalwork, including the handrail, is bonded to the instrument chassis and returned separately (usually by several routes) to the neutral star point. Any personnel involved in the loop as illustrated are safe, because they are in parallel with the low-resistance bond XX′ which has no significant resistance. If the bond XX′ were broken then the potential of the handrail would be determined by the ill-defined resistance of the earth (ground) path. The instrument system would be elevated by the effects of the transient fault current in the chassis earth (ground) return, and the resultant potential difference across the human being might be uncomfortably high.

The fundamental earthing (grounding) requirements of a safe system are therefore that there should be an adequate fault return path to operate any protective device which is incorporated, and that all parts of the plant should be bonded together to minimize potential differences.

There are, however, a number of circumstances where earthing (grounding) is not used as a means of ensuring protection. Large quantities of domestic portable equipment are protected by "double insulation," in which the primary insulation is reinforced by secondary insulation and there would need to be a coincident breakdown of two separate layers of insulation for danger to arise. Similarly, some areas for work on open equipment are made safe by being constructed entirely of insulating material, and the supplies derived from isolating transformers so as to reduce the risk of electrocution.

Where the environment is harsh or cables are exposed to rough treatment there is always the need to reduce working voltage, and there are many variants on the method of electrical protection, all of which have their particular advantages. Figure 44.4 shows the type of installation which is widely used in wet situations and, provided that the tools and cables are subject to frequent inspection, offers a reasonable level of protection.

The transformer is well designed to reduce the available voltage to 110 V, which is then center tapped to earth (ground), which further reduces the fault voltage to earth (ground) to 55 V. Both phases of the supply are fused, but a more sensitive detection of fault current is achieved by using an earth (ground) leakage circuit breaker (ELCB) which monitors the balance of the phase currents and if they differ by more than 20 mA triggers the circuit breaker. This sensitive fast detection combined with the lower voltage produces a reasonably safe system for most circumstances.

There are therefore many different techniques for reducing electrical shock risk. They all require consideration to be given to the nature of the supply, the design of the equipment, the environment, use, the method of installation, and the frequency and effectiveness of inspection. These factors all interact so strongly that any safe installation must consider all these aspects.

44.3 FLAMMABLE ATMOSPHERES

A large proportion of process control instrumentation is used in the petrochemical industry, where there is a possible risk of explosion if the equipment comes into contact with a flammable atmosphere. In practice, similar risks occur in all petrochemical and gas distribution sites, printing works, paint-spray booths, and the numerous small stores of

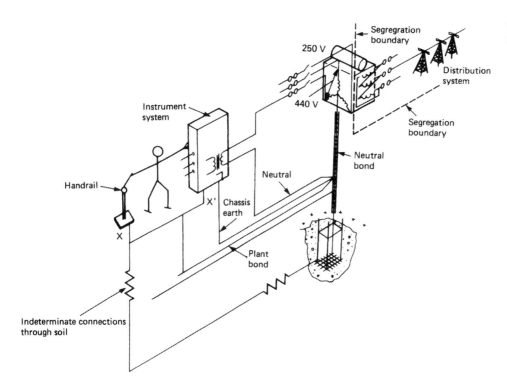

FIGURE 44.3 Normal U.K. installation with bonded neutral.

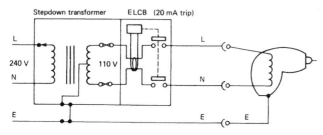

FIGURE 44.4 Isolating transformer supplying 110 Vcenter tapped to earth (ground) with earth (ground) leakage circuit breaker.

varnish, paint, and encapsulating compounds which exist on most manufacturing sites.

The other related risk is that of dust explosions, which tend to attract less interest but are possibly more important. Almost any finely divided material is capable of being burned (most people are familiar with the burning steelwool demonstration) and, in particular, finely divided organic substances such as flour, sugar, and animal feedstuffs all readily ignite. Dust explosions tend to be dramatic, since a small explosion normally raises a further dust cloud and the explosion rolls on to consume the available fuel. However, in general dusts need considerably more energy than gas to ignite them (millijoules rather than microjoules) and are usually ignited by temperatures in the region of 200°C. Frequently the instrumentation problem is solved by using T4 (135°C) temperature-classified intrinsically safe equipment in a dust-tight enclosure.

The basic mechanism of a gas explosion requires three constituents: the flammable gas, oxygen (usually in the form of air), and a source of ignition (in this context an electrical spark or hot surface). A gas–air mixture must be mixed in certain proportions to be flammable. The boundary conditions are known as the lower and upper flammable limits, or in some documents the lower and upper explosive limits. The subject of explosion prevention concentrates on keeping these three constituents from coming together. The usual approach is to attempt to decide on the probability of the gas–air mixture being present and then to choose equipment which is protected adequately for its environment.

The study of the probability of gas–air mixture being present within the flammable limits is called "area classification," and is without doubt the most difficult aspect of this subject. Expertise on all aspects of the plant and the behavior of the gases present is required to carry out area classification well, and hence it is usually done by a committee on which the instrument engineer is only one member. Present practice is to divide the hazardous area according to the IEC Standard 79–10, as follows:

Zone 0: in which an explosive gas–air mixture is continuously present or present for long periods.
(*Note*: The vapor space of a closed process vessel or storage tank is an example of this zone.)
Zone 1: in which an explosive gas–air mixture is likely to occur in normal operation.
Zone 2: in which an explosive gas–air mixture is not likely to occur, and if it occurs it will only exist for a short term.

By inference, any location which is not a hazardous area is a safe area. Many authorities prefer the use of "non-hazardous area," for semantic and legalistic reasons. The use of "safe" is preferred in this document since it is a shorter, more distinctive word than "non-hazardous."

In the USA, the relevant standard is Article 504 of the National Electrical Code, and the ANSI/ISA standards that explain it. There are minor differences between Article 504 at this writing and IEC Standard 79–10.

American common practice is still to divide hazardous areas into two divisions. Division 1 is the more hazardous of the two divisions and embraces both Zone 0 and Zone 1. Zone 2 and Division 2 are roughly synonymous. However, this practice is being overtaken by the changes in the National Electrical Code to conform to IEC standards.

The toxicity of many industrial gases means that an analysis of a plant from this aspect must be carried out. The two problems are frequently considered at the same time.

Having decided the risk of the gas being present, then the nature of the gas from a spark ignition or flame propagation viewpoint is considered.

One of the better things that has happened in recent years is the almost universal use of the IEC system of grouping apparatus in a way which indicates that it can safely be used with certain gases. Pedantically, it is the apparatus that is grouped, but the distinction between grouping gases or equipment is an academic point which does not affect safety. The international gas grouping allocates the Roman numeral I to the underground mining activity where the predominant risk is methane, usually called firedamp, and coal dust. Historically, the mining industry was the initial reason for all the work on equipment for flammable atmospheres, and it retains a position of considerable influence. All surface industry equipment is marked with Roman numeral II and the gas groups are subdivided into IIA (propane), IIB (ethylene), and IIC (hydrogen). The IIC group requires the smallest amount of energy to ignite it, the relative sensitivities being approximately 1:3:8. The representative gas which is shown in parentheses is frequently used to describe the gas group.

This gas classification has the merit of using the same classification for all the methods of protection used. The boundaries of the gas groupings have been slightly modified to make this possible.

Unfortunately, the USA and Canada have opted to maintain their present gas and dust classification. The classifications and subdivisions are:

CLASS I: Gases and vapors
Group A (acetylene)
Group B (hydrogen)
Group C (ethylene)
Group D (methane)
CLASS II: Dusts
Group E (metal dust)
Group F (coal dust)
Group G (grain dust)

CLASS III: Fibers
(No subgroups)

Gas–air mixtures can be ignited by contact with hot surfaces, and consequently, all electrical equipment used in hazardous atmospheres must be classified according to its maximum surface temperature. BS 4683: Part 1 is the relevant standard in the United Kingdom, and this is almost identical to IEC 79–8. The use of temperature classification was introduced in the United Kingdom in the late 1960s, and one of the problems of using equipment which was certified prior to this (e.g., equipment certified to BS 1259) is that somehow a temperature classification has to be derived.

For intrinsically safe circuits the maximum surface temperature is calculated or measured, including the possibility of faults occurring, in just the same way as the electrical spark energy requirements are derived. The possibility that flameproof equipment could become white hot under similar fault conditions is guarded against by generalizations about the adequate protective devices. All temperature classifications, unless otherwise specified, are assessed with reference to a maximum ambient temperature of 40°C. If equipment is used in a temperature higher than this, then its temperature classification should be reassessed. In the majority of circumstances, regarding the temperature classification as a temperature-rise assessment will give adequate results. Particular care should be exercised when the 'ambient' temperature of a piece of apparatus can be raised by the process temperature (e.g., a pilot solenoid valve thermally connected to a hot process pipe). Frequently, equipment has a specified maximum working temperature at which it can safely be used, determined by insulating material, rating of components, etc. This should not be confused with the temperature classification, and both requirements must be met.

When the probability of gas being present and the nature of gas has been established then the next step is to match the risk to the equipment used. Table 44.2 shows the alternative methods of protection which are described in the CENELEC standards and the areas of use permitted in the United Kingdom.

TABLE 44.1 Temperature classification

Class	Maximum surface temperature (°C)
T1	450
T2	300
T3	200
T4	135
T5	100
T6	85

TABLE 44.2 Status of standards for methods of protection kill (as of January 1984)

Technique	IEC symbol Ex	Standard IEC 79–	Standard CENELEC EN 50	Standard BRITISH BS 5501 Part	U.K. code of BS 5501 part of BS5345	Permitted zone of use in U.K.	ATEX
General requirement		Draft	014	1	1		
Oil immersion	o	6	015	2	None	2	
Pressurization	p	2	016	3	5	1 or 2	
Powder filling	q	5	017	4	None	2	
Flameproof enclosure	d	1	018	5	3	1	
Increased safety	e	7	019	6	6	1 or 2	
Intrinsic safety	ia or ib	3 Test apparatus 11 Construction	020 Apparatus 020 System	7 9	4	0 ia 1 ib	
Non-incendive	n(N)	Voting draft	021 (Awaits IEC)	BS 4683 Pt3	7	2	
Encapsulation	m	None	028 (Voting draft)	None	None	1	
Special	s	None	None	SFA 3009	8	1	

In light current engineering the predominant technique is intrinsic safety, but flameproof and increased safety are also used. The flameproof technique permits the explosion to occur within the enclosure but makes the box strong enough and controls any apertures well enough to prevent the explosion propagating to the outside atmosphere. Increased safety uses superior construction techniques and large derating factors to reduce the probability of sparking or hot spots occurring to an acceptable level. The other technique which is used to solve particular problems is pressurization and purging. This achieves safety by interposing a layer of air or inert gas between the source of ignition and the hazardous gas.

Where it can be used, intrinsic safety is normally regarded as the technique which is relevant to instrumentation. Intrinsic safety is a technique for ensuring that the electrical energy available in a circuit is too low to ignite the most easily ignitable mixture of gas and air. The design of the circuit and equipment is intended to ensure safety both in normal use and in all probable fault conditions.

There is no official definition of intrinsic safety. EN 50 020, the relevant CENELEC apparatus standard, defines an intrinsically safe circuit as:

> A circuit in which no spark or any thermal effects produced in the test conditions prescribed in this standard (which include normal operation and specified fault conditions) is capable of causing ignition of a given explosive atmosphere.

There are now two levels of intrinsic safety: "ia" being the higher standard where safety is maintained with up to two-fault and "ib," where safety is maintained with up to one-fault. Equipment certified to "ib" standards is generally acceptable in all zones except Zone 0, and "ia" equipment is suitable for use in all zones.

Intrinsic safety is, for all practical purposes, the only acceptable safety technique in Zone 0 (continuously hazardous) and the preferred technique in Zone 1 (hazardous in normal operation).

This technique is frequently used in Zone 2 (rarely hazardous) locations to ease the problems of live maintenance, documentation, and personnel training. Intrinsic safety is essentially a low-power technique, and hence is particularly suited to industrial instrumentation. Its principal advantages are low cost, more flexible installations, and the possibility of live maintenance and adjustment. Its disadvantages are low available power and its undeserved reputation of being difficult to understand. In general, if the electrical requirement is less than 30 V and 50 mA, then intrinsic safety is the preferred technique. If the power required is in excess of 3 W or the voltage greater than 50 V, or the current greater than 250 mA, the probability is that some other technique would be required. The upper limit is a rash generalization, because, with ingenuity, intrinsically safe systems can safely exceed these limits. Between these two sets of values intrinsically safe systems can frequently be devised.

When there is interconnection between more than one intrinsically safe apparatus, an analysis of the interactions and their combined effect on safety reveals that intrinsic safety is essentially a system concept. It can be argued that the other techniques rely on correct interconnection and the choice of the method of electrical protection. For example, a flameproof motor depends for its safety on having correctly rated switchgear for starting overload and fault protection, adequate provision for earthing (grounding), and a satisfactory means of isolation, all of which constitute a system. However, the danger resulting from the failure of unsatisfactory safe-area equipment in an intrinsically safe system is more immediate and obvious, and hence there is a requirement for a more detailed consideration of all safety aspects which results in a system certificate and documentation. Where a system comprises intrinsically safe apparatus in the hazardous area and a certified source of power and receiving apparatus in the safe area, then the combination can be assessed against the CENELEC system standard EN 50039. The agreed term for equipment intended for mounting in the safe area which is certified as having terminals which may be connected to the hazardous area is "associated electrical apparatus." This inelegant and quite forgettable expression is very rarely used by anyone other than writers of standards, but it does distinguish certified safe-area equipment from equipment which can be mounted in the hazardous area.

Where an instrument loop is relatively simple, self-contained, and comprises the same equipment in the majority of applications, then it is usual for both the hazardous-area and safe-area equipment to be certified, and a system certificate for the specific combination to exist as illustrated in Figure 44.5.

In practice, there are only a few completely self-contained circuits, since the signal to or from the hazardous area is usually fed into or supplied from complex equipment. In these circumstances there is no real possibility of certifying the safe-area apparatus since it is complex, and there is a need to maintain flexibility in its choice and use. The solution in these circumstances is to introduce into the circuit an intrinsically safe interface which cannot transmit a dangerous level of energy to the hazardous area (see Figure 44.6).

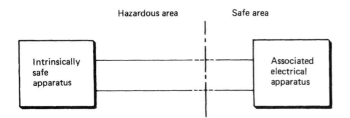

FIGURE 44.5 System with certified safe area equipment (associated apparatus).

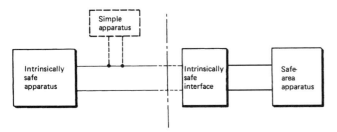

FIGURE 44.6 System with certified intrinsically safe interface.

The majority of interfaces are designed to be safe with 250 V with respect to earth (ground) applied to them (i.e., the 440 three-phase neutral earth (ground) system commonly used in the United Kingdom).

Whatever the cause of the possible danger and the technique used to minimize it, the need to assess the risk, and to document the risk analysis and the precautions taken, is very important. There is a legal requirement to produce the documentation. There is little doubt that if the risks are recognized and documentary proof that they have been minimized is established, then the discipline involved in producing that proof will result in an installation which is unlikely to be dangerous and is infinitely easier to maintain in a safe condition.

44.4 OTHER SAFETY ASPECTS

The level of integrity of any interlock or instrument system depends upon the importance of the measurement and the consequences of a failure. It is not surprising that some of the most careful work in this area has been related to the control of atomic piles and similar sources of potential catastrophic failure. The majority of systems are less dramatic, and in the United Kingdom an excellent Code of Practice, BS 5304: 1975, discusses the techniques generally used for safeguarding machinery in non-hazardous circumstances. The general principles to be applied can be summarized as:

1. The failure of any single component (including power supplies) of the system should not create a dangerous situation.
2. The failure of cabling to open or short circuit or short circuiting to ground of wiring should not create a dangerous situation. Pneumatic or electro-optic systems have different modes of failure but may have particular advantages in some circumstances.
3. The system should be easily checked and readily understood. The virtue of simplicity in enhancing the reliability and serviceability of a system cannot be overstressed.
4. The operational reliability of the system must be as high as possible. Foreseeable modes of failure can usually be arranged to produce a "fail-safe" situation, but if the system fails and produces spurious shutdowns too frequently, the temptation to override interlocks can become overwhelming. An interlock system, to remain credible, must therefore be operationally reliable and, if possible, some indication as to whether the alarm is real or a system fault may also be desirable.

These basic requirements, following up a fundamental analysis of the level of integrity to be achieved, form a framework upon which to build an adequate system.

44.5 CONCLUSION

It is difficult to adequately summarize the design requirements of a safe system. The desire to avoid accidents and in particular to avoid injuring and killing people is instinctive in the majority of engineers and hence does not need to be emphasized. Accident avoidance is a discipline to be cultivated, careful documentation tends to be a valuable aid, and common sense is the aspect which is most frequently missing.

The majority of engineers cannot experience or have detailed knowledge of all aspects of engineering, and safety is not different from any other factor in this respect. The secret of success must therefore be the need to recognize the danger so as to know when to seek advice. This chapter has attempted to provide the background for recognizing the need to seek expert advice; it is not comprehensive enough to ensure a safe design.

REFERENCES

Bass, H. G., *Intrinsic Safety*, Quartermaine House, Gravesend, Kent, U.K. (1984).

Cooper, W. F., *Electrical Safety Engineering*, Butter-worth-Heinemann, Oxford (1993).

Electrical Safety in Hazardous Environments, Conferences, Institution of Electrical Engineers (1971), (1975) and (1982).

Garside, R. H., *Intrinisically Safe Instrumentation: A Guide*, Safety Technology (1982). Predominantly applications, strong on U.K. and U.S. technology and standards.

Hall, J., *Intrinsic Safety*, Institution of Mining Electrical and Mining Mechanical Engineers (1985). A comprehensive treatise on mining applications of the art ICI Engineering Codes and Regulations, ROSPA Publications No. IS 91. Now unfortunately out of print. Slightly dated but the most useful publication in this area. Beg, borrow, or steal the first copy you find. Essential.

Magison, E. C., *Electrical Instruments in Hazardous Locations*, 3rd ed. Instrument Society of America (1978). Comprehensive book portraying American viewpoint.

Olenik, H., et al., *Explosion Protection Manual*, 2nd ed., Brown Boveri & Cie (1984). An excellent book on West German practice.

Redding, R. J., *Intrinsic Safety*, McGraw-Hill, New York (1971). Slightly dated but still relevant.

Robens, Lord (chairman), *Safety and Health at Work*, Report of the Committee HMSO Cmnd. 5034 (1972) *Safety in Universities—Notes for Guidance*, Association of Commonwealth Universities (1978).

Substances Hazardous to Health, Croner Publications, New Malden, Surrey, U.K. (1986 with updates).

Towle, C., *Intrinsically Safe Installations of Ships and Offshore Structures*, Institute of Marine Engineers TP 1074 (1985).

Many British Standards, IEC Standards, and ANSI/ISA Standards refer to safety. With the wide availability of these standards on the World Wide Web, the reader is referred to these agencies for an up-to-date listing of relevant standards.

FURTHER READING

Buschart, R. J., *Electrical and Instrumentation Safety for Chemical Processes*, Van Nostrand Reinhold, New York (1991).

Chapter 45

EMC

T. Williams

45.1 INTRODUCTION

Electromagnetic interference (EMI) is a serious and increasing form of environmental pollution. Its effects range from minor annoyances due to crackles on broadcast reception to potentially fatal accidents due to corruption of safety-critical control systems. Various forms of EMI may cause electrical and electronic malfunctions, can prevent the proper use of the radio frequency (rf) spectrum, can ignite flammable or other hazardous atmospheres, and may even have a direct effect on human tissue. As electronic systems penetrate more deeply into all aspects of society, so both the potential for interference effects and the potential for serious EMI-induced incidents will increase.

Some reported examples of electromagnetic incompatibility are:

1. New electronic push-button telephones installed near the Brookmans Park medium wave transmitter in North London were constantly afflicted with BBC radio programs.
2. Mobile phones have been found to interfere with the readings of certain types of gasoline pump meters.
3. Interference to aeronautical safety communications at a U.S. airport was traced to an electronic cash register a mile away.
4. The instrument panel of a well known airliner was said to carry the warning "ignore all instruments while transmitting h.f."
5. Electronic point-of-sale units used in shoe, clothing, and optician shops (where thick carpets and nylon-coated assistants were common) would experience lock-up, false data, and uncontrolled drawer openings.
6. When a piezoelectric cigarette lighter was lit near the cabinet of a car park barrier control box, the radiated pulse caused the barrier to open, and drivers were able to park free of charge.
7. Lowering the pantographs of electric locomotives at British Rail's Liverpool Street station interfered with newly installed signaling control equipment, causing the signals to "fail safe" to red.
8. Hearing aids are severely affected by the pulse modulated radio frequency injected when their wearers use digital cellular telephones.

45.1.1 Compatibility between Systems

The threat of EMI is controlled by adopting the practices of electromagnetic *compatibility* (EMC). This is defined as: "The ability of a device, unit of equipment, or system to function satisfactorily in its electromagnetic environment without introducing intolerable electromagnetic disturbances to anything in that environment." The term EMC has two complementary aspects:

1. It describes the ability of electrical and electronic systems to operate without interfering with other systems.
2. It describes the ability of such systems to operate as intended within a specified electromagnetic environment.

Thus it is closely related to the environment within which the system operates. Effective EMC requires that the system is designed, manufactured, and tested with regard to its predicted operational electromagnetic environment: that is, the totality of electromagnetic phenomena existing at its location. Although the term "electromagnetic" tends to suggest an emphasis on high-frequency field-related phenomena, in practice the definition of EMC encompasses all frequencies and coupling paths, from d.c. to 400 GHz.

45.1.1.1 Subsystems within an Installation

There are two approaches to EMC. In one case the nature of the installation determines the approach. EMC is especially problematic when several electronic or electrical systems are packed into a very compact installation, such as on board aircraft, ships, satellites, or other vehicles. In these cases susceptible systems may be located very close to powerful emitters, and special precautions are needed to maintain compatibility. To do this cost-effectively calls for a detailed knowledge of both the installation circumstances and the characteristics of the emitters and their potential victims. Military, aerospace, and vehicle EMC specifications have evolved to meet this need and are well established in their particular industry sectors.

45.1.1.2 Equipment in Isolation

The second approach assumes that the system will operate in an environment which is electromagnetically benign within certain limits, and that its proximity to other sensitive equipment will also be controlled within limits. So, for example, most of the time a control system will not be operated in the vicinity of a high-power radar transmitter, nor will it be located next to a mobile radio receiving antenna. This allows a very broad set of limits to be placed on both the permissible emissions from a device and on the levels of disturbance within which the device should reasonably be expected to continue operating. These limits are directly related to the class of environment—domestic, commercial, industrial, etc.—for which the device is marketed. The limits and the methods of demonstrating that they have been met form the basis for a set of standards, some aimed at emissions and some at immunity, for the EMC performance of any given product in isolation.

Compliance with such standards will not guarantee electromagnetic compatibility under all conditions. Rather, it establishes a probability (hopefully very high) that equipment will not cause interference nor be susceptible to it when operated under *typical* conditions. There will inevitably be some special circumstances under which proper EMC will not be attained—such as operating a computer within the near field of a powerful transmitter—and extra protection measures must be accepted.

45.1.2 The Scope of EMC

The principal issues which are addressed by EMC are discussed below. The use of microprocessors in particular has stimulated the upsurge of interest in EMC. These devices are widely responsible for generating radio frequency interference and are themselves susceptible to many interfering phenomena. At the same time, the widespread replacement of metal chassis and cabinets by molded plastic enclosures has drastically reduced the degree of protection offered to circuits by their housings.

45.1.2.1 Malfunction of Systems

Solid-state and especially processor-based control systems have taken over many functions which were earlier the preserve of electromechanical or analog equipment such as relay logic or proportional controllers. Rather than being hardwired to perform a particular task, programmable electronic systems rely on a digital bus-linked architecture in which many signals are multiplexed onto a single hardware bus under software control. Not only is such a structure more susceptible to interference, because of the low level of energy needed to induce a change of state, the effects of the interference are impossible to predict; a random pulse may or may not corrupt the operation, depending on its timing with respect to the internal clock, the data that are being transferred, and the program's execution state. Continous interference may have no effect as long as it remains below the logic threshold, but when it increases further the processor operation will be completely disrupted. With increasing functional complexity comes the likelihood of system failure in complex and unexpected failure modes.

Clearly, the consequences of interference to control systems will depend on the value of the process that is being controlled. In some cases disruption of control may be no more than a nuisance, in others it may be economically damaging or even life-threatening. The level of effort that is put into assuring compatibility will depend on the expected consequences of failure.

Phenomena Electromagnetic phenomena which can be expected to interfere with control systems are:

1. Supply voltage interruptions, dips, surges, and fluctuations.
2. Transient overvoltages on supply, signal, and control lines.
3. Radio frequency fields, both pulsed (radar) and continuous, coupled directly into the equipment or onto its connected cables.
4. Electrostatic discharge (ESD) from a charged object or person.
5. Low-frequency magnetic or electric fields.

Note that we are not directly concerned with the phenomenon of component damage due to ESD, which is mainly a problem of electronic production. Once the components are assembled into a unit they are protected from such damage unless the design is particularly lax. But an ESD transient can corrupt the operation of a microprocessor or clocked circuit just as a transient coupled into the supply or signal ports can, without actually damaging any components (although this may also occur), and this is properly an EMC phenomenon.

Software Malfunctions due to faulty software may often be confused with those due to EMI. Especially with real-time systems, transient coincidences of external conditions with critical software execution states can cause operational failure which is difficult or impossible to replicate, and may survive development testing to remain latent for years in fielded equipment. The symptoms—system crashes, incorrect operation, or faulty data—can be identical to those induced by EMI. In fact, you may only be able to distinguish faulty software from poor EMC by characterizing the environment in which the system is installed.

45.1.2.2 Interference with Radio Reception

Bona fide users of the radio spectrum have a right to expect their use not to be affected by the operation of equipment which has nothing to do with them. Typically, received signal strengths of wanted signals vary, from less than a microvolt to more than a millivolt, at the receiver input. If an interfering signal is present on the same channel as the wanted signal then the wanted signal will be obliterated if the interference is of a similar or greater amplitude. The acceptable level of co-channel interference (the "protection factor") is determined by the wanted program content and by the nature of the interference. Continuous interference on a high-fidelity broadcast signal would be unacceptable at very low levels, whereas a communications channel carrying compressed voice signals can tolerate relatively high levels of impulsive or transient interference.

Field Strength Level Radiated interference, whether intentional or not, decreases in strength with distance from the source. For radiated fields in free space, the decrease is inversely proportional to the distance provided that the measurement is made in the far field (see below for a discussion of near and far fields). As ground irregularity and clutter increase, the fields will be further reduced because of shadowing, absorption, scattering, divergence, and defocusing of the diffracted waves. Annex D of EN 55 011 suggests that for distances greater than 30 m over the frequency range 30–300 MHz, the median field strength varies as $1/d^n$, where n varies from 1.3 for open country to 2.8 for heavily built-up urban areas. An average value of $n = 2.2$ can be taken for approximate estimations; thus increasing the separation by ten times would give a drop in interfering signal strength of 44 dB.

Limits for unintentional emissions are based on the acceptable interfering field strength that is present at the receiver—that is, the minimum wanted signal strength for a particular service modified by the protection ratio—when a nominal distance separates it from the emitter. This will not protect the reception of very weak wanted signals, nor will it protect against the close proximity of an interfering source, but it will cover the majority of interference cases, and this approach is taken in all those standards for emission limits that have been published for commercial equipment by CISPR. *CISPR Publication No. 23* gives an account of how such limits are derived, including the statistical basis for the probability of interference occurring.

Below 30 MHz the dominant method of coupling out of the interfering equipment is via its connected cables, and therefore, the radiated field limits are translated into equivalent voltage or current levels that, when present on the cables, correspond to a similar level of threat to high-and medium-frequency reception.

45.1.2.3 Malfunction Versus Spectrum Protection

It should be clear from the foregoing discussion that radio frequency (rf) emission limits are not determined by the need to guard against malfunction of equipment which is not itself a radio receiver. As discussed in the previous section, malfunction requires fairly high energy levels—for example, rf field strengths in the region of 1–10 V/m. Protection of the spectrum for radio use is needed at much lower levels, of the order of 10–100 µV/m, i.e., 10,000 to 100,000 times lower. Radio frequency incompatibility between two pieces of equipment, neither of which intentionally uses the radio frequency spectrum, is very rate. Normally, equipment immunity is required from the local fields of intentional radio transmitters, and unintentional emissions must be limited to protect the operation of intentional radio receivers. The two principal EMC aspects of emissions and immunity therefore address two different issues.

Free Radiation Frequencies Certain types of equipment generate high levels of rf energy but use it for purposes other than communication. Medical diathermy and rf heating apparatus are examples. To place blanket emission limits on this equipment would be unrealistic. In fact, the International Telecommunications Union (ITU) has designated a number of frequencies specifically for this purpose, and equipment using only these frequencies (colloquially known as the "free radiation" frequencies) is not subject to emission restrictions. Table 45.1 lists these frequencies. In the U.K. certain other frequencies are permitted with a fixed radiation limit.

TABLE 45.1 ITU designated industrial, scientific, and medical free-radiation frequencies (EN 55011:1991)

Center frequency (MHz)	Frequency range (MHz)
6,780	6,765–6,795*
13,560	13,553–13,567
27,120	26,957–27,283
40,680	40,66–40,70
433,920	433,05–434,79*
2,450	2,400–2,500
5,800	5,725–5,875
24,125	24,000–24,250
61,250	61,000–61,500*
122,500	122,000–123,000*
245,000	244,000–246,000*

*Maximum radiation limit under consideration; use subject to special authorization.

45.1.2.4 Disturbances on the Line-Voltage Supply

Line-voltage electricity suffers a variety of disturbing effects during its distribution. These may be caused by sources in the supply network or by other users, or by other loads within the same installation. A pure, uninterrupted supply would not be cost effective; the balance between the cost of the supply and its quality is determined by national regulatory requirements, tempered by the experience of the supply utilities. Typical disturbances are:

1. *Voltage variations.* The distribution network has a finite source impedance, and varying loads will affect the terminal voltage. Including voltage drops within the customer's premises, an allowance of ±10 percent on the nominal voltage will cover normal variations in the U.K.; proposed limits for all CENELEC countries are +12 percent, −15 percent. Under the CENELEC voltage harmonization regime the European supply voltage at the point of connection to the customer's premises will be 230 V +10 percent, −6 percent.
2. *Voltage fluctuations.* Short-term (subsecond) fluctuations with quite small amplitudes are annoyingly perceptible on electric lighting, though they are comfortably ignored by electronic power supply circuits. Generation of flicker by high power load switching is subject to regulatory control.
3. *Voltage interruptions.* Faults on power distribution systems cause almost 100 percent voltage drops but are cleared quickly and automatically by protection devices, and throughout the rest of the distribution system the voltage immediately recovers. Most consumers therefore see a short voltage dip. The frequency of occurrence of such dips depends on location and seasonal factors.
4. *Waveform distortion.* At the source, the a.c. line-voltage is generated as a pure sine wave but the reactive impedance of the distribution network, together with the harmonic currents drawn by non-linear loads, causes voltage distortion. Power converters and electronic power supplies are important contributors to non-linear loading. Harmonic distortion may actually be worse t points remote from the non-linear load because of resonances in the network components. Not only must non-linear harmonic currents be limited, but equipment should be capable of operating with up to 10 percent total harmonic distortion in the supply waveform.
5. *Transients and surges.* Switching operations generate transients of a few hundred volts as a result of current interruption in an inductive circuit. These transients normally occur in bursts and have risetimes of no more than a few nanoseconds, although the finite bandwidth of the distribution network will quickly attenuate all but local sources. Rarer high amplitude spikes in excess of 2 kV may be observed due to fault conditions. Even higher voltage surges due to lightning strikes occur, most frequently on exposed overhead line distribution systems in rural areas.

All these sources of disturbance can cause malfunction in systems and equipment that do not have adequate immunity.

Line-Voltage Signaling A further source of incompatibility arises from the use of the line-voltage distribution network as a telecommunications medium, or line-voltage signaling (MS). MS superimposes signals on the line-voltage in the frequency band 3–150 kHz and is used both by the supply industry itself and by consumers. Unfortunately, this is also the frequency band in which electronic power converters—not just switch-mode power supplies, but variable speed motor drives, induction heaters, fluorescent lamp inverters, and similar products—operate to their best efficiency. There are at present no pan-European standards which regulate conducted emissions on the line-voltage below 150 kHz, although EN 50 065: Part 1 (BS 6839: Part 1) sets the frequency allocations and output and interference limits for MS equipment itself. The German radio frequency emission standard VDE 0871 (now superseded) extends down to 9 kHz for some classes of equipment. Overall, compatibility problems between MS systems and such power conversion equipment can be expected to increase.

45.1.2.5 Other EMC Issues

The issues discussed above are those which directly affect product design to meet commercial EMC requirements, but there are two other aspects which should be mentioned briefly.

EEDs and Flammable Atmospheres The first is the hazard of ignition of flammable atmospheres in petrochemical plants, or the detonation of electro-explosive devices in places such as quarries, due to incident radio frequency energy. A strong electromagnetic field will induce currents in large metal structures which behave as receiving antennas. A spark will occur if two such structures are in intermittent contact or are separated. If flammable vapor is present at the location of the spark, and if the spark has sufficient energy, the vapor will be ignited. Different vapors have different minimum ignition energies, hydrogen/air being the most sensitive. The energy present in the spark depends on the field strength, and hence on the distance from the transmitter, and on the antenna efficiency of the metal structure. BS 656 discusses the nature of the hazard and presents guidelines for its mitigation.

Similarly, electro-explosive devices (EEDs) are typically connected to their source of power for detonation by a long wire, which can behave as an antenna. Currents induced in it by a nearby transmitter could cause the charges to explode prematurely if the field was strong enough. As with ignition of flammable atmospheres, the risk of premature detonation depends on the separation distance from the transmitter and the efficiency of the receiving wire. EEDs can if necessary be filtered to reduce their susceptibility to radio frequency energy. BS 6657 discusses the hazard to EEDs.

Data Security The second aspect of EMC is the security of confidential data. Low-level radio-frequency emissions from data-processing equipment may be modulated with the information that the equipment is carrying—for instance, the video signal that is fed to the screen of a VDU. These signals could be detected by third parties with sensitive equipment located outside a secure area and demodulated for their own purposes, thus compromising the security of the overall system. This threat is already well recognized by government agencies, and specifications for emission control (under the Tempest scheme) have been established for many years. Commercial institutions, particularly in the finance sector, are now beginning to become aware of the problem.

45.2 INTERFERENCE COUPLING MECHANISMS

45.2.1 Source and Victim

Situations in which the question of electromagnetic compatibility arises invariably have two complementary aspects. Any such situation must have a source of interference emissions and a victim which is susceptible to this interference. If either of these is not present, there is no EMC problem. If both source and victim are within the same piece of equipment we have an "intrasystem" EMC situation; if they are two different items, such as a computer monitor and a radio receiver, it is said to be an "intersystem" situation. The same equipment may be a source in one situation and a victim in another.

Knowledge of how the source emissions are coupled to the victim is essential, since a reduction in the coupling factor is often the only way to reduce interference effects, if a product is to continue to meet its performance specification. The two aspects are frequently reciprocal, that is, measures taken to improve emissions will also improve the susceptibility, though this is not invariably so. For analysis, they are more easily considered separately.

Systems EMC Putting source and victim together shows the potential interference routes that exist from one to the other (Figure 45.1). When systems are being built, it is necessary to know the emission's signature and susceptibility of the component equipment, in order to determine whether problems are likely to be experienced with close coupling. Adherence to published emission and susceptibility standards does not guarantee freedom from systems EMC problems. Standards are written from the point of view of protecting a particular service—in the case of emissions standards, this is radio broadcast and telecommunications—and they have to assume a minimum separation between source and victim.

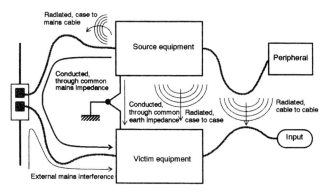

FIGURE 45.1 Coupling paths.

Most electronic hardware contains elements which are capable of antenna-like behavior, such as cables, PCB tracks, internal wiring, and mechanical structures. These elements can unintentionally transfer energy via electric, magnetic, or electromagnetic fields which couple with the circuits. In practical situations, intrasystem and external coupling between equipment is modified by the presence of screening and dielectric materials, and by the layout and proximity of interfering and victim equipment and especially their respective cables. Ground or screening planes will enhance an interfering signal by reflection or attenuate it by absorption. Cable-to-cable coupling can be either capacitive or inductive and depends on orientation, length, and proximity. Dielectric materials may also reduce the field by absorption, though this is negligible compared with the effects of conductors in most practical situations.

45.2.1.1 Common Impedance Coupling

Common impedance coupling routes are those which are due to a circuit impedance which the source shares with the victim. The most obvious common impedances are those in which the impedance is physically present, as with a shared conductor; but the common impedance may also be due to mutual inductive coupling between two current loops, or to mutual capacitive coupling between two voltage nodes. Philosophically speaking, every node and every loop is coupled to all others throughout the universe. Practically, the strength of coupling falls off very rapidly with distance. Figure 45.4 shows the variation of mutual capacitance and inductance of a pair of parallel wires versus their separation.

Conductive Connection When an interference source (output of system A in Figure 45.2) shares a ground connection with a victim (input of system B) then any current due to A's output flowing through the common impedance section X–X develops a voltage in series with B's input. The common impedance need be no more than a length of wire or PCB track. High-frequency or high di/dt components in the output will couple more efficiently because of the inductive nature of the impedance. The output and input may be part of the same system, in which case there is a spurious feedback path through the common impedance which can cause oscillation.

The solution as shown in Figure 45.2 is to separate the connections so that there is no common current path, and hence no common impedance, between the two circuits. The only "penalty" for doing this is the need for extra wiring or track to define the separate circuits. This applies to any circuit which may include a common impedance, such as power rail connections. Grounds are the most usual source of common impedance because the ground connection, often not shown on circuit diagrams, is taken for granted.

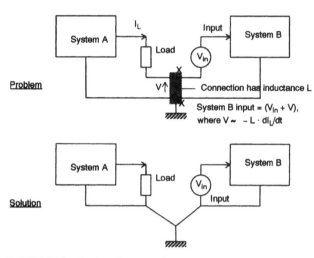

FIGURE 45.2 Conducted common impedance coupling.

Magnetic Induction Alternating current flowing in a conductor creates a magnetic field which will couple with a nearby conductor and induce a voltage in it (Figure 45.3(a)). The voltage induced in the victim conductor is given by:

$$V = \frac{-M \cdot dI_L}{dt} \quad (45.1)$$

where M is the mutual inductance (henry). M depends on the areas of the source and victim current loops, their orientation and separation distance, and the presence of any magnetic screening. Typical values for short lengths of cable loomed together lie in the range 0.1–3 μH. The equivalent circuit for magnetic coupling is a voltage generator in series with the victim circuit. Note that the coupling is unaffected by the presence or absence of a direct connection between the two circuits; the induced voltage would be the same if both circuits were isolated or connected to ground.

Electric Induction Changing voltage on one conductor creates an electric field which may couple with a nearby conductor and induce a voltage on it (Figure 45.3(b)). The voltage induced on the victim conductor in this manner is

$$V = \frac{C_C \cdot dV_L}{dt \cdot Z_{in}} \quad (45.2)$$

where C_C is the coupling capacitance, and Z_{in} is the impedance to ground of the victim circuit. This assumes that the impedance of the coupling capacitance is much higher than that of the circuit impedances. The noise is injected as if from

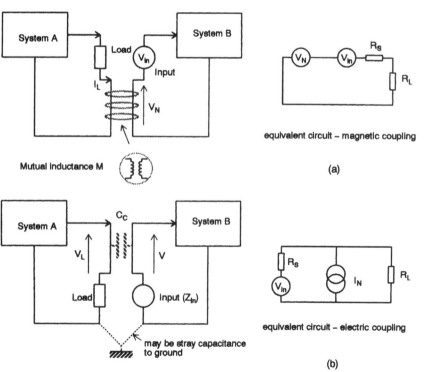

FIGURE 45.3 Magnetic and electric induction.

a current source with a value of $C_C \cdot dV_L/dt$. The value of C_C is a function of the distance between the conductors, their effective areas, and the presence of any electric screening material. Typically, two parallel insulated wires 2.5 mm apart show a coupling capacitance of about 50 pF per meter; the primary-to-secondary capacitance of an unscreened medium power line-voltage transformer is 100–1000 pF.

In the case of *floating circuits*, both circuits need to be referenced to ground for the coupling path to be complete. But if either is floating, this does not mean that there is no coupling path: the floating circuit will exhibit a stray capacitance to ground, and this is in series with the direct coupling capacitance. Alternatively, there will be stray capacitance direct from the low-voltages nodes of A to B even in the absence of any ground node. The noise current will still be injected across R_L, but its value will be determined by the series combination of C_C and the stray capacitance.

Effect of Load Resistance Note that the difference in equivalent circuits for magnetic and electric coupling means that their behavior with a varying circuit load resistance is different. Electric field coupling increases with an increasing R_L, while magnetic field coupling decreases with an increasing R_L. This property can be useful for diagnostic purposes; if you vary R_L while observing the coupled voltage, you can deduce which mode of coupling predominates. For the same reason, magnetic coupling is more of a problem for low-impedance circuits while electric coupling applies to high-impedance circuits.

Spacing Both mutual capacitance and mutual inductance are affected by the physical separation of source and victim conductors. Figure 45.4 shows the effect of spacing on mutual capacitance of two parallel wires in free space, and on mutual inductance of two conductors over a ground plane (the ground plane provides a return path for the current).

45.2.1.2 Line-Voltage Coupling

Interference can propagate from a source to a victim via the line-voltage distribution network to which both are connected. This is not well characterized at high frequencies, although the impedance viewed at any connection is reasonably predictable. The radio frequency impedance presented by the line-voltage can be approximated by a network of 50 Ω in parallel with 50 μH. For short distances such as between adjacent outlets on the same ring, coupling via the line-voltage connection of two items of equipment can be represented by the equivalent circuit in Figure 45.5.

Over longer distances, power cables are fairly low loss transmission lines of around 150–200 Ω characteristic impedance up to about 10 MHz. However, in any local power distribution system the disturbances and discontinuities introduced by load connections, cable junctions, and distribution components will dominate the radio frequency transmission characteristic. These all tend to increase the attenuation.

45.2.1.3 Radiated Coupling

To understand how energy is coupled from a source to a victim at a distance with no intervening connecting path, you need to have a basic understanding of electromagnetic wave propagation. This section will do no more than introduce the necessary concepts. The theory of EM waves has been well covered in many other works (e.g., Hayt, 1988).

Field Generation An electric field (E field) is generated between two conductors at different potentials. The field is measured in volts per meter and is proportional to the applied voltage divided by the distance between the conductors.
A magnetic field (H field) is generated around a conductor carrying a current, is measured in amps per meter, and is proportional to the current divided by the distance from the conductor.

When an alternating voltage generates an alternating current through a network of conductors, an electromagnetic (EM) wave is generated which propagates as a combination of E and H fields at right angles. The speed of propagation is determined by the medium; in free space it is equal to the speed of light (3×10^8 m/s). Near to the radiating source the geometry and strength of the fields depend on the characteristics of the source. Further away only the

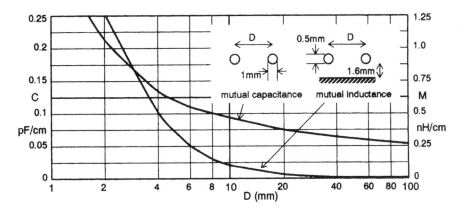

FIGURE 45.4 Mutual capacitance and inductance versus spacing.

orthogonal fields remain. Figure 45.6 demonstrates these concepts graphically.

Wave Impedance The ratio of the electric to magnetic field strengths (E/H) is called the wave impedance (Figure 45.7).

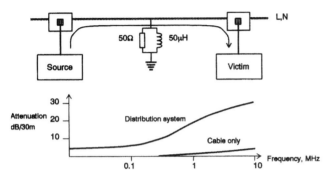

FIGURE 45.5 Coupling via the line-voltage network.

The wave impedance is a key parameter of any given wave as it determines the efficiency of coupling with another conducting structure, and also the effectiveness of any conducting screen which is used to block it. In the far field, d > $\lambda/2\pi$, the wave is known as a plane wave, and its impedance is constant and equal to the impedance of free space given by

$$Z_o = \left(\frac{\mu_o}{\varepsilon_o}\right)^{0.5} = 120\pi = 377 \, \Omega \qquad (45.3)$$

where μ_o is $4\pi \times 10^{-7}$ H/m, and ε_o is 8.85×10^{-12} F/m.

In the near field, d > $\lambda/2\pi$, the wave impedance is determined by the characteristics of the source. A low-current, high-voltage radiator (such as a rod) will generate mainly an electric field of high impedance, while a high-current, low-voltage radiator (such as a loop) will generate mainly a magnetic field of low impedance. The region around $\lambda/2\pi$,

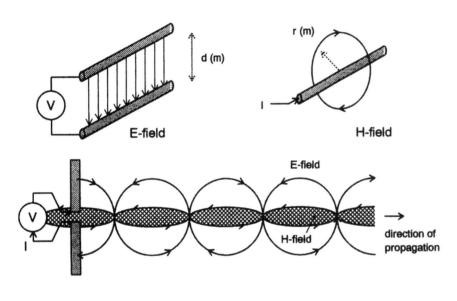

FIGURE 45.6 Electromagnetic fields.

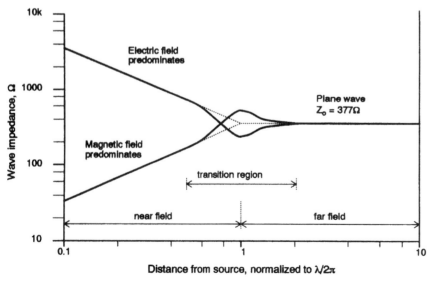

FIGURE 45.7 The wave impedance.

or approximately one-sixth of a wavelength, is the transition region between near and far fields.

Coupling Modes The concepts of differential mode, common mode, and antenna mode radiated field coupling are fundamental to an understanding of EMC and will crop up in a variety of guises throughout this chapter. They apply to coupling of both emissions and incoming interference.

Consider two items of equipment interconnected by a cable (Figure 45.8). The cable carries signal currents in *differential mode* (go and return) down the two wires in close proximity. A radiated field can couple to this system and induce differential mode interference between the two wires; similarly, the differential current will induce a radiated field of its own. The ground reference plane (which may be external to the equipment or may be formed by its supporting structure) plays no part in the coupling.

The cable also carries currents in *common mode*, that is, all flowing in the same direction on each wire. These currents very often have nothing at all to do with the signal currents. They may be induced by an external field coupling to the loop formed by the cable, the ground plane, and the various impedances connecting the equipment to ground, and may then cause internal differential currents to which the equipment is susceptible. Alternatively, they may be generated by internal noise voltages between the ground reference point and the cable connection, and be responsible for radiated emissions. Note that the stray capacitances and inductances associated with the wiring and enclosure of each unit are an integral part of the common mode coupling circuit, and play a large part in determining the amplitude and spectral distribution of the common mode currents. These stray reactances are incidental rather than designed into the equipment and are therefore much harder to control or predict than parameters such as cable spacing and filtering which determine differential mode coupling.

Antenna Mode currents are carried in the same direction by the cable and the ground reference plane. They should not arise as a result of internally generated noise, but they will flow when the whole system, ground plane included, is exposed to an external field. An example would be when an aircraft flies through the beam of a radar transmission; the aircraft structure, which serves as the ground plane for its internal equipment, carries the same currents as the internal wiring. Antenna mode currents only become a problem for the radiated field susceptibility of self-contained systems when they are converted to differential or common mode by varying impedances in the different current paths.

45.2.2 Emissions

When designing a product to a specification without knowledge of the system or environment in which it will be installed, one will normally separate the two aspects of emissions and susceptibility, and design to meet minimum requirements for each. Limits are laid down in various standards, but individual customers or market sectors may have more specific requirements. In those standards which derive from CISPR, emissions are subdivided into radiated emissions from the system as a whole, and conducted emissions present on the interface and power cables. Conventionally, the breakpoint between radiated (high frequency) and conducted (low frequency) is set at 30 MHz. Radiated emissions can themselves be separated into emissions that derive from internal PCBs or other wiring, and emissions from common-mode currents that find their way onto external cables that are connected to the equipment.

45.2.2.1 Radiated Emissions

Radiation from the PCB In most equipment, the primary emission sources are currents flowing in circuits (clocks, video and data drivers, and other oscillators) that are mounted on PCBs.

Radiated emission from a PCB can be modeled as a small loop antenna carrying the interference current (Figure 45.9). A small loop is one whose dimensions are smaller than a quarter wavelength ($\lambda/4$) at the frequency of interest (e.g., 1 m at 75 MHz). Most PCB loops count as "small" at emission frequencies of up to a few hundred megahertz. When the dimensions approach ($\lambda/4$) the currents at different points on the loop appear out of phase at a distance, so that the effect is to reduce the field strength at any given

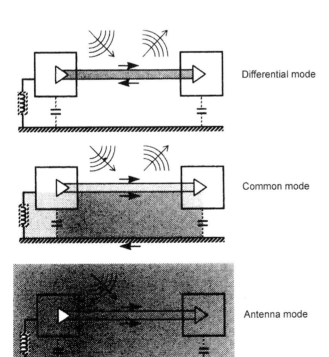

FIGURE 45.8 Radiated coupling modes.

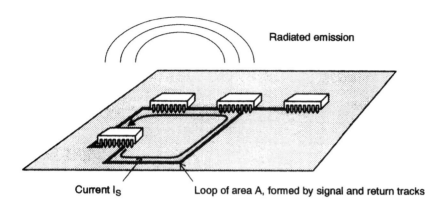

FIGURE 45.9 PCB radiated emissions.

point. The maximum electric field strength from such a loop over a ground plane at 10 m distance is proportional to the square of the frequency (Ott, 1988):

$$E = 263 \times 10^{-12} \left(f^2 A I_S \right) \text{ volts/meter} \quad (45.4)$$

where A is the loop area (cm²), and f (MHz) is the frequency of I_S, the source current (mA).

In free space, the field falls off proportionally to distance from the source. The figure of 10 m is used as this is the standard measurement distance for the European radiated emissions standards. A factor of 2 is allowed for worst-case field reinforcement due to reflection from the ground plane, which is also a required feature of testing to standards.

The loop whose area must be known is the overall path taken by the signal current and its return. Equation (45.4) assumes that I_S is at a single frequency. For square waves with many harmonics, the Fourier spectrum must be used for I_S. These points are taken up again in Section 45.3.2.2.

Assessing PCB Design You can use equation (45.4) to indicate roughly whether a given PCB design will need extra screening. For example, if $A = 10$ cm², $I_S = 20$ mA and $f = 50$ MHz, then the field strength E is 42 dBμV/m, which is 12 dB over the European Class B limit. Thus if the frequency and operating current are fixed, and the loop area cannot be reduced, screening will be necessary.

The converse, however, is not true. Differential mode radiation from small loops on PCBs is by no means the only contributor to radiated emissions; common mode currents flowing on the PCB and, more important, on attached cables can contribute much more. Paul (1989) goes so far as to say:

> ... predictions of radiated emissions based solely on differential-mode currents will generally bear no resemblance to measured levels of radiated emissions. Therefore, basing system EMC design on differential-mode currents and the associated prediction models that use them exclusively while neglecting to consider the (usually much larger) emissions due to common-mode currents can lead to a strong "false sense of security."

Common-mode currents on the PCB itself are not at all easy to predict, in contrast with the differential mode currents which are governed by Kirchhoff's current law. The return path for common-mode currents is via stray capacitance (displacement current) to other nearby objects, and therefore a full prediction would have to take the detailed mechanical structure of the PCB and its case, as well as its proximity to ground and to other equipment, into account. Except for trivial cases this is for all intents and purposes impossible. It is for this reason more than any other that EMC design has earned itself the distinction of being a "black art."

Radiation from Cables Fortunately (from some viewpoints) radiated coupling at VHF tends to be dominated by cable emissions, rather than by direct radiation from the PCB. This is for the simple reason that typical cables resonate in the 30–100 MHz region and their radiating efficiency is higher than PCB structures at these frequencies. The interference current is generated in common mode from ground noise developed across the PCB or elsewhere in the equipment and may flow along the conductors, or along the shield of a shielded cable. The model for cable radiation at lower frequencies (Figure 45.10) is a short ($L < \lambda/4$) monopole antenna over a ground plane. (When the cable length is resonant the model becomes invalid.) The maximum field strength allowing +6 dB for ground plane reflections at 10 m due to this radiation is directly proportional to frequency (Ott, 1985):

$$E = 1.26 \times 10^{-4} \left(f L I_{CM} \right) \text{ volt/meter [25]} \quad (45.5)$$

where L is the cable length (meters), and I_{CM} is the common-mode current (mA) at f(MHz) flowing in the cable.

For a 1 m cable, I_{CM} must be less than 20 μA for a field strength at 10 m of 42 dBμV/m, i.e., a thousand times less than the equivalent differential mode current!

Common-Mode Cable Noise At the risk of repetition, it is vital to appreciate the difference between common-mode and differential-mode cable currents. Differential-mode current, I_{CM} in Figure 45.10, is the current which flows in one direction along one cable conductor and in the reverse direction along another. It is normally equal to the signal or power current, and is not present on the shield. It contributes little to the net radiation as long as the total loop area formed by the two conductors is small; the two currents tend to cancel each other. Common mode current I_{CM} flows equally in the same direction along all conductors in the cable, potentially including the shield, and is only related to the differential signal currents insofar as these are converted to common mode by unbalanced external impedances, and may be quite *unrelated* to them. It returns via the associated ground network, and therefore the radiating loop area is large and uncontrolled. As a result, even a small I_{CM} can result in large emitted signals.

45.2.2.2 Conducted Emissions

Interference sources within the equipment circuit or its power supply are coupled onto the power cable to the equipment.

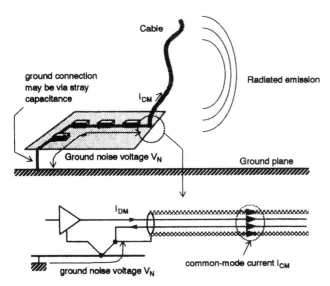

FIGURE 45.10 Cable radiated emissions.

Interference may also be coupled either inductively or capacitively from another cable onto the power cable. Until recently, attention has focused on the power cable as the prime source of conducted emissions since CISPR-based standards have only specified measurements on this cable. However, signal and control cables can and do also act as coupling paths, and amendments to the standards will apply measurements to these cables as well.

The resulting interference may appear as differential mode (between live and neutral, or between signal wires) or as common mode (between live/neutral/signal and ground) or as a mixture of both. For signal and control lines, only common-mode currents are regulated. For the line-voltage port, the voltages between live and ground and between neutral and ground at the far end of the line-voltage cable are measured. Differential mode emissions are normally associated with low-frequency switching noise from the power supply, while common-mode emissions can be due to the higher frequency switching components, internal circuit sources, or intercable coupling.

Coupling Paths The equivalent circuit for a typical product with a switch-mode power supply, shown in Figure 45.11, gives an idea of the various paths these emissions can take. (Section 45.3.2.4 looks at SMPS emissions in more detail.) Differential-mode current I_{DM} generated at the input of the switching supply is converted by imbalances in stray capacitance, and by the mutual inductance of the conductors in the line-voltage cable, into interference voltages with respect to earth at the measurement point. Higher frequency switching noise components $V_{Nsupply}$ are coupled through C_c to appear between L/N and E on the line-voltage cable, and C_s to appear with respect to the ground plane. Circuit ground noise V_{Ncct} (digital noise and clock harmonics) is referenced to ground by C_s and coupled out via signal cables as I_{CMsig} or via the safety earth as I_{CME}.

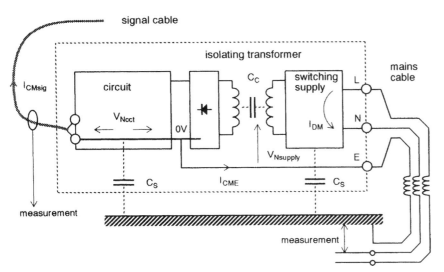

FIGURE 45.11 Coupling paths for conducted emissions.

The problem in a real situation is that all these mechanisms are operating simultaneously, and the stray capacitances C_s are widely distributed and unpredictable, depending heavily on proximity to other objects if the case is unscreened. A partially screened enclosure may actually worsen the coupling because of its higher capacitance to the environment.

45.2.2.3 Line-Voltage Harmonics

One EMC phenomenon, which comes under the umbrella of the EMC Directive and is usually classified as an "emission," is the harmonic content of the line-voltage input current. This is mildly confusing since the equipment is not actually "emitting" anything: it is simply drawing its power at harmonics of the line frequency as well as at the fundamental.

The Supplier's Problem The problem of line-voltage harmonics is principally one for the supply authorities, who are mandated to provide a high-quality electricity supply. If the aggregate load at a particular line-voltage distribution point has a high harmonic content, the non-zero distribution source impedance will cause distortion of the voltage waveform at this point. This in turn may cause problems for other users connected to that point, and the currents themselves may also create problems (such as overheating of transformers and compensating components) for the supplier. The supplier does, of course, have the option of uprating the distribution components or installing special protection measures, but this is expensive, and the supplier has room to argue that the users should bear some of the costs of the pollution they create.

Harmonic pollution is continually increasing, and it is principally due to low-power electronic loads installed in large numbers. Between them, domestic TV sets and office information technology equipment account for about 80 percent of the problem. Other types of load which also take significant harmonic currents are not widely enough distributed to cause a serious problem yet, or are dealt with individually at the point of installation as in the case of industrial plant. The supply authorities are nevertheless sufficiently worried to want to extend harmonic emission limits to all classes of electronic products.

Non-Linear Loads A plain resistive load across the line-voltage draws current only at the fundamental frequency (50 Hz in Europe). Most electronic circuits are anything but resistive. The universal rectifier-capacitor input draws a high current at the peak of the voltage waveform and zero current at other times; the well known triac phase control method for power control (lights, motors, heaters, etc.) begins to draw current only partway through each half-cycle. These current waveforms can be represented as a Fourier series, and it is the harmonic amplitudes of the series that are subject to regulation. The relevant standard is EN 60 555: Part 2, which in its present (1987) version applies only to household products.

There is a proposal to extend the scope of EN 60 555 to cover a wide range of products, and it will affect virtually all line-voltage powered electronic equipment above a certain power level which has a rectifier-reservoir input. The harmonic limits are effectively an additional design constraint on the values of the input components, most notably, the input series impedance (which is not usually considered as a desirable input component at all). With a typical input resistance of a few ohms for a 100 W power supply, the harmonic amplitudes are severely in excess of the proposed revision to the limits of EN 60 555: Part 2.

Increasing input series resistance to meet the harmonic limits is expensive in terms of power dissipation except at very low powers. In practice, deliberately dissipating between 10 percent and 20 percent of the input power rapidly becomes unreasonable above levels of 50–100 W. Alternatives are to include a series input choke which, since it must operate down to 50 Hz, is expensive in size and weight; or to include electronic power factor correction (PFC), which converts the current waveform to a near-sinusoid but is expensive in cost and complexity. PFC is essentially a switch-mode converter on the front-end of the supply and therefore is likely to contribute extra radio frequency switching noise at the same time as it reduces input current harmonics. It is possible to combine PFC with the other features of a direct-off-line switching supply, so that if you are intending to use a SMPS anyway there will be little extra penalty. It also fits well with other contemporary design requirements such as the need for a "universal" (90–260 V) input voltage range. Such power supplies can already be bought off the shelf, but unless you are a power supply specialist, to design a PFC-SMPS yourself will take considerable extra design and development effort.

Phase Control Power control circuits which vary the switch-on point with the phase of the line-voltage waveform are another major source of harmonic distortion on the input current. Lighting controllers are the leading example of these. Figure 45.12 shows the harmonic content of such a waveform switched at 90° (the peak of the cycle, corresponding to half power). The maximum harmonic content occurs at this point, decreasing as the phase is varied either side of 90°. Whether lighting dimmers will comply with the draft limits in EN 60 555-2 without input filtering or PFC depends at present on their power level, since these limits are set at an absolute value.

45.2.3 Susceptibility

Electronic equipment will be susceptible to environmental electromagnetic fields and/or to disturbances coupled into

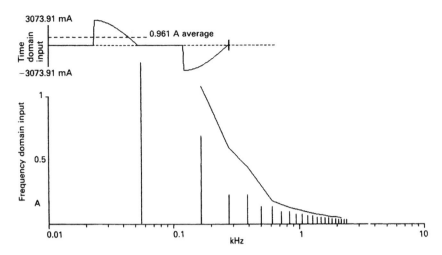

FIGURE 45.12 Mains input current harmonics for 500 W phase control circuit at half power.

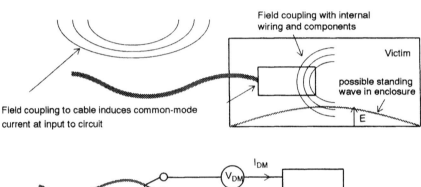

FIGURE 45.13 Radiated field coupling.

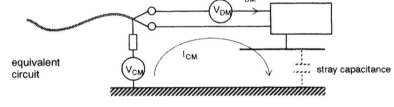

its ports via connected cables. An electrostatic discharge may be coupled in via the cables or the equipment case, or a nearby discharge can create a local field which couples directly with the equipment. The potential threats are:

1. Radiated radio frequency fields
2. Conducted transients
3. Electrostatic discharge (ESD)
4. Magnetic fields
5. Supply voltage disturbances

Quite apart from legal requirements, equipment that is designed to be immune to these effects—especially ESD and transients—will save its manufacturer considerable expense through preventing field returns. Unfortunately, the shielding and circuit suppression measures that are required for protection against ESD or radio frequency interference may be more than you need for emission control.

45.2.3.1 Radiated Field

An external field can couple either directly with the internal circuitry and wiring in differential mode or with the cables to induce a common mode current (Figure 45.13). Coupling with internal wiring and PCB tracks is most efficient at frequencies above a few hundred megahertz, since wiring lengths of a few inches approach resonance at these frequencies.

Radio frequency voltages or currents in analog circuits can induce non-linearity, overload, or d.c. bias, and in digital circuits can corrupt data transfer. Modulated fields can have a greater effect than unmodulated ones. Likely sources of radiated fields are walkie-talkies, cellphones, high-power broadcast transmitters, and radars. Field strengths between 1 and 10 V/m from 20 MHz to 1 GHz are typical, and higher field strengths can occur in environments close to such sources.

Cable Resonance Cables are most efficient at coupling radio frequency energy into equipment at the lower end of the vhf spectrum (30–100 MHz). The external field induces a common mode current on the cable shield or on all the cable conductors together, if it is unshielded. The common mode current effects in typical installations tend to dominate the direct field interactions with the equipment as long as the equipment's dimensions are small compared with half the wavelength of the interfering signal.

A cable connected to a grounded victim equipment can be modeled as a single conductor over a ground plane, which appears as a transmission line (Figure 45.14). The current induced in such a transmission line by an external field increases steadily with frequency until the first resonance is reached, after which it exhibits a series of peaks and nulls at higher resonances (Smith, 1977). The coupling mechanism is enhanced at the resonant frequency of the cable, which depends on its length and on the reactive loading of whatever equipment is attached to its end. A length of 2 m is quarter-wave resonant at 35.5 MHz, half-wave resonant at 75 MHz.

Cable Loading The dominant resonant mode depends on the radio frequency impedance (high or low) at the distant end of the cable. If the cable is connected to an ungrounded object such as a hand controller it will have a high rf impedance, which will cause a high coupled current at quarter-wave resonance and high coupled voltage at half-wave. Extra capacitive loading such as body capacitance will lower its apparent resonant frequency.

Conversely, a cable connected to another grounded object, such as a separately earthed peripheral, will see a low impedance at the far end, which will generate high coupled current at half-wave and high coupled voltage at quarter-wave resonance. Extra inductive loading, such as the inductance of the earth connection, will again tend to lower the resonant frequency.

These effects are summarized in Figure 45.15. The rf common-mode impedance of the cable varies from around 35 Ω at quarter-wave resonance to several hundred ohms maximum. A convenient average figure (and one that is taken in many standards) is 150 Ω. Because cable configuration, layout, and proximity to grounded objects are outside the designer's control, attempts to predict resonances and impedances accurately are unrewarding.

Current Injection A convenient method for testing the radio frequency susceptibility of equipment without reference to its cable configuration is to inject radio frequency as a common mode current or voltage directly onto the cable port. This represents real-life coupling situations at lower frequencies well, until the equipment dimensions approach a half wavelength. It can also reproduce the fields (E_{RF} and H_{RF}) associated with radiated field coupling. The route taken by the interference currents, and hence their effect on the circuitry, depends on the various internal and external radio frequency impedances to earth, as shown in Figure 45.16. Connecting other cables will modify the current flow to a marked extent, especially if the extra cables interface to a physically different location on the PCB or equipment. An applied voltage of 1 V, or an injected current of 10 mA, corresponds in typical cases to a radiated field strength of 1 V/m.

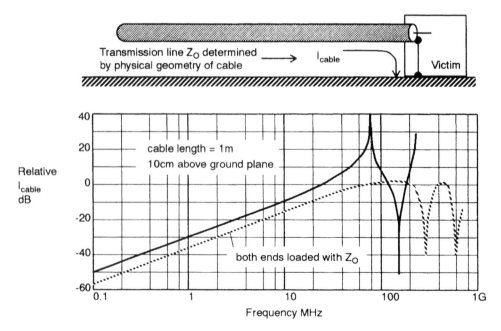

FIGURE 45.14 Cable coupling to radiated field.

Cavity Resonance A screened enclosure can form a resonant cavity; standing waves in the field form between opposite sides when the dimension between the sides is a multiple of a half-wavelength. The electric field is enhanced in the middle of this cavity while the magnetic field is enhanced at the sides. This effect is usually responsible for peaks in the susceptibility versus frequency profile in the UHF region.

Predicting the resonant frequency accurately from the enclosure dimensions is rarely successful because the contents of the enclosure tend to "detune" the resonance. But for an empty cavity, resonances occur at

$$F = 150\sqrt{\left[(k/l)^2 + (m/h)^2 + (n/w)^2\right]} \text{ MHz} \quad (45.6)$$

where l, h, and w are the enclosure dimensions (meters), and k, m, and n are positive integers, but no more than one at a time can be zero.

For approximately equal enclosure dimensions, the lowest possible resonant frequency is given by

$$F(\text{MHz}) \approx 212/l \approx 212/h \approx 212/w \quad (45.7)$$

45.2.3.2 Transients

Transient overvoltages occur on the line-voltage supply leads due to switching operations, fault clearance, or lightning strikes elsewhere on the network. Transients over 1 kV account for about 0.1 percent of the total number of transients observed. A study by the German ZVEI (Goedbloed, 1987) made a statistical survey of 28,000 live-to-earth transients exceeding 100 V, at 40 locations over a total measuring time of about 3400 h. Their results were analyzed for peak amplitude, rate of rise, and energy content. Table 45.2 shows the average rate of occurrence of transients for four classes of location, and Figure 45.17 shows the relative number of transients as a function of maximum transient amplitude. This shows that the number of transients varies roughly in inverse proportion to the cube of peak voltage.

High-energy transients may threaten active devices in the equipment power supply. Fast-rising edges are the most disruptive to circuit operation, since they are attenuated least by the coupling paths, and they can generate large voltages in inductive ground and signal paths. The ZVEI study found that rate of rise increased roughly in proportion to the square root of peak voltage, being typically 3 V/ns for 200 V pulses and 10 Vns for 2 kV pulses. Other field experience has shown that mechanical switching produces multiple transients (bursts) with risetimes as short as a few nanoseconds and peak amplitudes of several hundred volts. Attenuation through the line-voltage network (see Section 45.2.1.2) restricts fast risetime pulses to those generated locally.

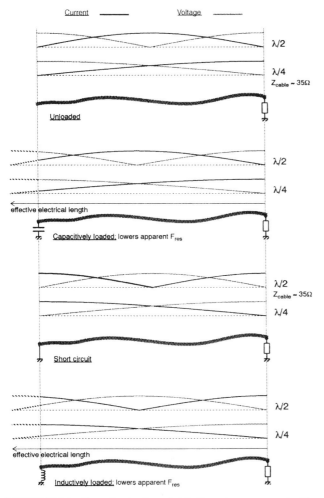

FIGURE 45.15 Current and voltage distributions along a resonant cable.

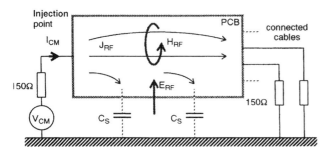

FIGURE 45.16 Common-mode radiofrequency injection. J_{RF} represents common-mode radiofrequency current density through the PCB.

TABLE 45.2 Average rate of occurrence of mains transients (Goedbloed, 1987)

Area class	Average rate of occurrence (transients/hour)
Industrial	17.5
Business	2.8
Domestic	0.6
Laboratory	2.3

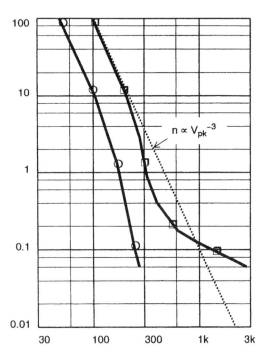

FIGURE 45.17 Relative number of transients (percent) versus maximum transient amplitude (volts). Line-voltage lines ($V_T = 100$ V); O, telecommunication lines ($V_T = 50$ V) (Goedbloed, 1987, 1990).

Analog circuits are almost immune to isolated short transients, whereas digital circuits are easily corrupted by them. As a general guide, microprocessor equipment should be tested to withstand pulses at least up to 2 kV peak amplitude. Thresholds below 1 kV will give unacceptably frequent corruptions in nearly all environments, while between 1–2 kV thresholds occasional corruption will occur. For a belt-and-braces approach for high-reliability equipment, a 4–6 kV threshold is recommended.

Coupling Mode Line-voltage transients may appear in differential mode (symmetrically between live and neutral) or common mode (asymmetrically between live neutral and earth). Coupling between the conductors in a supply network tends to mix the two modes. Differential-mode spikes are usually associated with relatively slow rise times and high energy, and require suppression to prevent input circuit damage but do not, provided this suppression is incorporated, affect circuit operation significantly. Common-mode transients are harder to suppress because they require connection of suppression components between live and earth, or in series with the earth lead, and because stray capacitances to earth are harder to control. Their coupling paths are very similar to those followed by common mode radio frequency signals. Unfortunately, they are also more disruptive because they result in transient current flow in ground traces.

Transients on Signal Lines Fast transients can be coupled, usually capacitively, onto signal cables in common mode, especially if the cable passes close to or is routed alongside an impulsive interference source. Although such transients are generally lower in amplitude than line-voltage-borne ones, they are coupled directly into the I/O ports of the circuit and will therefore flow in the circuit ground traces, unless the cable is properly screened and terminated or the interface is properly filtered.

Other sources of conducted transients are telecommunication lines and the automotive 12 V supply. The automotive environment can regularly experience transients that are many times the nominal supply range. The most serious automotive transients are the load dump, which occurs when the alternator load is suddenly disconnected during heavy charging; switching of inductive loads, such as motors and solenoids; and alternator field decay, which generates a negative voltage spike when the ignition switch is turned off. A recent standard (ISO 7637) has been issued to specify transient testing in the automotive field.

Work on common mode transients on telephone subscriber lines (Goedbloed, 1990) has shown that the amplitude versus rate of occurrence distribution also follows a roughly inverse cubic law as in Figure 45.17. Actual amplitudes were lower than those on the line-voltage (peak amplitudes rarely exceeded 300 V). A transient ringing frequency of 1 MHz and rise times of 10–20 ns were found to be typical.

45.2.3.3 Electrostatic Discharge

When two non-conductive materials are rubbed together or separated, electrons from one material are transferred to the other. This results in the accumulation of triboelectric charge on the surface of the material. The amount of the charge caused by movement of the materials is a function of the separation of the materials in the triboelectric series (Figure 45.18(a)). Additional factors are the closeness of contact, rate of separation, and humidity. The human body can be charged by triboelectric induction to several kilovolts.

When the body (in the worst case, holding a metal object such as a key) approaches a conductive object, the charge is transferred to that object normally via a spark, when the potential gradient across the narrowing air gap is high enough to cause breakdown. The energy involved in the charge transfer may be low enough to be imperceptible to the subject; at the other extreme it can be extremely painful.

The ESD Waveform When an electrostatically charged object is brought close to a grounded target the resultant discharge current consists of a very fast (sub-nanosecond) edge followed by a comparatively slow bulk discharge curve. The characteristic of the hand/metal ESD current waveform is a function of the approach speed, the voltage, the geometry of the electrode, and the relative humidity. The equivalent circuit for such a situation is shown in Figure 45.18(c). The capacitance C_D (the typical human body capacitance is around 150 pF) is charged via a high resistance up to the electrostatic voltage V. The actual value of V will vary as the charging and leakage paths change with the environmental

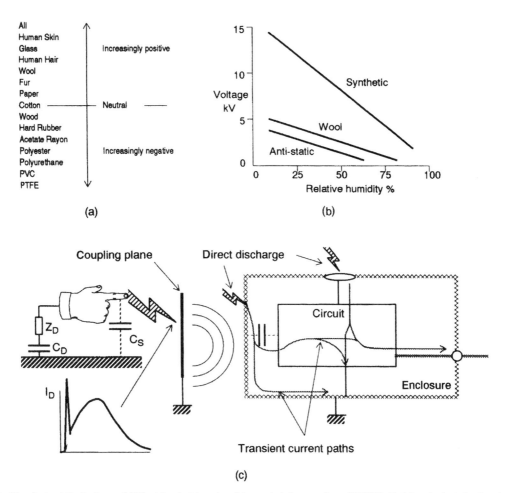

FIGURE 45.18 The electrostatic discharge; (a) The triboelectric series; (b) expected charge voltage (IEC 801–2); (c) equivalent circuit and current waveform.

circumstances and movements of the subject. When a discharge is initiated, the free space capacitance C_s, which is directly across the discharge point, produces an initial current peak, the value of which is determined only by the local circuit stray impedance, while the main discharge current is limited by the body's bulk inductance and resistance Z_D.

The resultant sub-nanosecond transient equalizing current of several tens of amps follows a complex route to ground through the equipment and is very likely to upset digital circuit operation if it passes through the circuit tracks. The paths are defined more by stray capacitance, case bonding, and track or wiring inductance than by the designer's intended circuit. The high magnetic field associated with the current can induce transient voltages in nearby conductors that are not actually in the path of the current. Even if not discharged directly to the equipment, a nearby discharge such as to a metal desk or chair will generate an intense radiated field which will couple into unprotected equipment.

ESD Protection Measures When the equipment is housed in a metallic enclosure this itself can be used to guide the ESD current around the internal circuitry. Apertures or seams in the enclosure will act as high-impedance barriers to the current, and transient fields will occur around them, so they must be minimized. All metallic covers and panels must be bonded together with a low impedance connection (<2.5 mΩ at d.c.) in at least two places; long panel-to-panel "bonding" wires must be avoided since they radiate high fields during an ESD event. I/O cables and internal wiring may provide low-impedance paths for the current, in the same way as they are routes into and out of the equipment for common mode radio frequency interference. The best way to eliminate susceptibility of internal harnesses and cables is not to have any, through economical design of the board interconnections. External cables must have their shields well decoupled to the ground structure, following the rules in Section 45.4.1.5, i.e., 360° bonding of cable screens to connector backshells and no pigtails (Staggs, 1989).

Insulated enclosures make the control of ESD currents harder to achieve, and a well designed and low-inductance circuit ground is essential. But if the enclosure can be designed to have no apertures which provide air gap paths to the interior, then no direct discharge will be able to occur, provided the material's dielectric strength is high enough. However, you will still need to protect against the field of an indirect discharge.

45.2.3.4 Magnetic Fields

Magnetic fields at low frequencies can induce interference voltages in closed wiring loops, their magnitude depending on the area that is intersected by the magnetic field. Non-toroidal line-voltage transformers and switchmode supply transformers are prolific sources of such fields, and they will readily interfere with sensitive circuitry or components within the same equipment. Any other equipment needs to be immune to the proximity of such sources. Particular environments may result in high low-frequency or d.c. magnetic field strengths, such as electrolysis plants where very high currents are used, or certain medical apparatus. The voltage developed in a single-turn loop is

$$V = \frac{A \cdot dB}{dt} \quad (45.8)$$

where A is the loop area (m^2), and B is the flux density normal to the plane of the loop (tesla; 1 mT = 795 A/m = 10 Gauss).

It is rare for such fields to affect digital or large signal analog circuits, but they can be troublesome with low-level circuits where the interference is within the operating bandwidth, such as audio or precision instrumentation. Specialized devices which are affected by magnetic fields, such as photomultiplier or cathode ray tubes, may also be susceptible.

Magnetic Field Screening Conventional screening is ineffective against low-frequency magnetic fields, because it relies on reflection rather than absorption of the field. Due to the low source impedance of magnetic fields, reflection loss is low. Since it is only the component of flux normal to the loop which induces a voltage, changing the relative orientation of source and loop may be effective. Low-frequency magnetic shielding is only possible with materials which exhibit a high absorption loss such as steel, mumetal or permalloy. As the frequency rises these materials lose their permeability and hence shielding efficiency, while non-magnetic materials such as copper or aluminum become more effective. Around 100 kHz shielding efficiencies are about equal. Permeable metals are also saturated by high field strengths, and are prone to lose their permeability through handling.

45.2.3.5 Supply Voltage Fluctuations

Brown-outs (voltage droops) and interruptions are a feature of all line-voltage distribution networks, and are usually due to fault clearing or load switching elsewhere in the system (Figure 45.19). Such events will not be perceived by the equipment if its input reservoir hold-up time is sufficient, but if this is not the case then restarts and output transients can be experienced. Typically, interruptions (as opposed to power cuts) can last for 10–500 ms.

Load and line voltage fluctuations are maintained between +10 percent and –15 percent of the nominal line voltage in most industrialized countries. The CENELEC harmonization document HD472 specifies 230 V, –6 percent +10 percent. Slow changes in the voltage within these limits occur on a diurnal pattern as the load on the power system varies. The declared voltage does not include voltage drops within the customer's premises, and so you would be wise to design stabilized power supplies to meet at least the −15 percent limit. Dips exceeding 10 percent of nominal voltage occur up to four times per month for urban consumers and more frequently in rural areas where the supply is via overhead lines (Hensman, 1989). Much wider voltage (and frequency) fluctuations and more frequent interruptions are common in those countries which do not have a well developed supply network. They are also common on supplies which are derived from small generators.

Harmonic distortion of the supply voltage is a function of loads which draw highly distorted current waveforms such as those discussed above. Most electronic power supplies should be immune to such distortion, although it can have severe effects at high levels on power factor correction capacitors, motors and transformers, and may also interfere with audio systems.

45.3 CIRCUITS, LAYOUT, AND GROUNDING

Designing for good EMC starts from the principle of controlling the flow of interference into and out of the equipment. You must assume that interference will occur to and will be generated by any product which includes active electronic

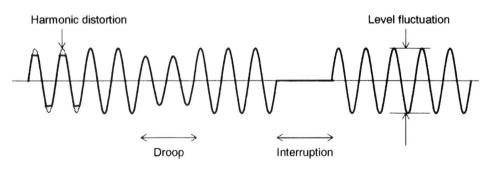

FIGURE 45.19 Line-voltage supply fluctuations.

devices. To improve the electromagnetic compatibility of the product you place barriers and route currents such that incoming interference is diverted or absorbed before it enters the circuit, and outgoing interference is diverted or absorbed before it leaves the circuit.

You can conceive the EMC control measures as applying at three levels, primary, secondary, and tertiary, as shown in Figure 45.20. Control at the primary level involves circuit design measures such as decoupling, balanced configurations, bandwidth, and speed limitation, and also board layout and grounding. For some low-performance circuits, and especially those which have no connecting cables, such measures may be sufficient in themselves. At the secondary level you must always consider the interface between the internal circuitry and external cables. This is invariably a major route for interference in both directions, and for some products (particularly where the circuit design has been frozen) all the control may have to be applied by filtering at these interfaces. Choice and mounting of connectors form an important part of this exercise.

Full shielding (the tertiary level) is an expensive choice to make, and you should only choose this option when all other measures have been applied. But since it is difficult or impossible to predict the effectiveness of primary measures in advance, it is wise to allow for the possibility of being forced to shield the enclosure. This means adapting the mechanical design so that a metal case could be used, or if a molded enclosure is essential, you should ensure that apertures and joints in the moldings can be adequately bonded at radio frequency, that ground connections can be made at the appropriate places, and that the molding design allows for easy application of a conductive coating.

45.3.1 Layout and Grounding

The most cost-effective approach is to consider the equipment's layout and ground regime at the beginning. No unit cost is added by a designed-in ground system. About 90 percent of post-design EMC problems are due to inadequate layout or grounding: a well designed layout and ground system can offer both improved immunity and protection against emissions, while a poorly designed one may well be a conduit for emissions and incoming interference. The most important principles are:

1. Partition the system to allow control of interference currents.
2. Consider ground as a path for current flow, both of interference into the equipment and conducted out from it; this means both careful placement of grounding points and minimizing ground impedance.
3. Minimize radiated emissions from and susceptibility of current loops by careful layout of high di/dt loop areas.

45.3.1.1 System Partitioning

The first design step is to partition the system. A poorly partitioned or non-partitioned system (Figure 45.21) may have its component subsystems separated into different areas of the board or enclosure, but the interfaces between them will be ill-defined and the external ports will be dispersed around the periphery. This makes it difficult to control the common-mode currents that will exist on the various interfaces. Dispersal of the ports means that the distances between ports on opposite sides of the system are large, leading to high induced ground voltages as a result of incoming interference, and efficient coupling to the cables of internally generated emissions.

Usually the only way to control emissions from and immunity of such a system is by placing an overall shield around it and filtering each interface. In many cases it will be difficult or impossible to maintain integrity of the shield and still permit proper operation—the necessary apertures and access points will preclude effective attenuation through the shield.

The Partitioned System Partitioning separates the system into critical and non-critical sections from the point of view of EMC. Critical sections are those which contain radiating sources such as microprocessor logic or video circuitry, or which are particularly susceptible to imported interference: microprocessor circuitry and low-level analog circuits. Non-critical sections are those whose signal levels, bandwidths, and circuit functions are such that they are not susceptible to interference nor capable of causing it: non-clocked logic,

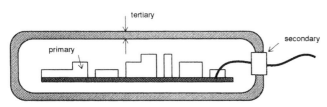

FIGURE 45.20 EMC control measures.

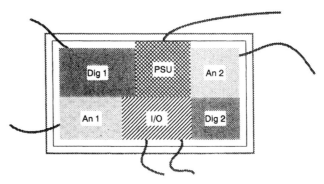

FIGURE 45.21 The haphazard system.

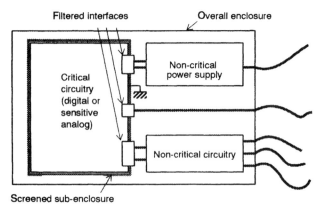

FIGURE 45.22 System partitioning.

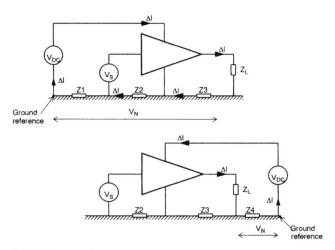

FIGURE 45.23 Ground current paths.

linear power supplies, and power amplifier stages are typical examples. Figure 45.22 shows this method of separation.

Control of Critical Sections Critical sections can then be enclosed in a shielded enclosure into and out of which all external connections are carefully controlled. This enclosure may encase the whole product or only a portion of it, depending on the nature of the circuits: your major design goal should be to minimize the number of controlled interfaces, and to concentrate them physically close together. Each interface that needs to be filtered or requires screened cabling adds unit cost to the product. A system with no electrical interface ports—such as a pocket calculator or infrared remote controller—represents an ideal case from the EMC point of view.

Note that the shield acts both as a barrier to radiated interference and as a reference point for ground return currents. In many cases, particularly where a full ground plane PCB construction is used, the latter is the more important function and it may be possible to do without an enclosing shield.

45.3.1.2 Grounding

Once the system has been properly partitioned, you can then ensure that it is properly grounded. There are two accepted purposes for grounding: one is to provide a route (the "safety earth") for hazardous fault currents, and the other is to give a reference for external connections to the system. The classical definition of a ground is "an equi-potential point or plane which serves as a reference for a circuit or system." Unfortunately this definition is meaningless in the presence of ground current flow. Even where singal currents are negligible, induced ground currents due to environmental magnetic or electric fields will cause shifts in ground potential. A good grounding system will minimize these potential differences by comparison with the circuit operating levels, but it cannot eliminate them. It has been suggested that the term "ground" as conventionally used should be dropped in favor of "reference point" to make the purpose of the node clear.

An alternative definition for a ground is "a low impedance path by which current can return to its source (Ott, 1979)." This emphasizes current flow and the consequent need for low impedance, and is more appropriate when high frequencies are involved. Ground currents always circulate as part of a loop. The task is to design the loop in such a way that induced voltages remain low enough at critical places. You can only do this by designing the ground circuit to be as compact and as local as possible.

Current Through the Ground Impedance When designing a ground layout you must know the actual path of the ground return current. The amplifier example in Figure 45.23 illustrates this. The high-current output ΔI returns to the power supply from the load; if it is returned through the path Z1-Z2-Z3 then an unwanted voltage component is developed across Z2 which is in series with the input V_S, and depending on its magnitude and phase, the circuit will oscillate. This is an instance of common impedance coupling, as was covered in Section 45.2.1.1.

A simple reconnection of the return path to Z4 eliminates the common impedance. For EMC purposes, instability is not usually the problem; rather it is the interference voltages V_N which are developed across the impedances that create emission or susceptibility problems. At high frequencies (above a few kilohertz) or high rates of change of current, the impedance of any connection is primarily inductive and increases with frequency ($V = -L \cdot di/dt$), hence ground noise increases in seriousness as the frequency rises.

45.3.1.3 Ground Systems

Ignoring for now the need for a safety earth, the grounding system as intended for a circuit reference can be configured as single-point, multipoint or as a hybrid of these two.

Single-Point The single-point grounding system (Figure 45.24(a)) is conceptually the simplest, and it eliminates

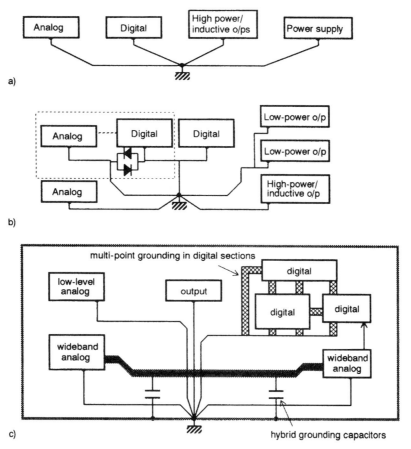

FIGURE 45.24 Grounding systems, (a) Single-point ground; (b) modified single-point ground; (c) multi-point and hybrid ground.

common impedance ground coupling and low-frequency ground loops. Each circuit module has its own connection to a single ground, and each rack or subunit has one bond to the chassis. Any currents flowing in the rest of the ground network do not couple into the circuit. This system works well up to frequencies in the megahertz region, but the distances involved in each ground connection mean that common mode potentials between circuits begin to develop as the frequency is increased. At distances greater than a quarter wavelength, the circuits are effectively isolated from each other.

A modification of the single-point system (Figure 45.24(b)) ties together those circuit modules with similar characteristics and takes each common point to the single ground. This allows a degree of common impedance coupling between those circuits where it will not be a problem, and at the same time allows grounding of high-frequency circuits to remain local. The noisiest circuits are closest to the common point in order to minimize the effect of the common impedance. When a single module has more than one ground, these should be tied together with back-to-back diodes to prevent damage when the circuit is disconnected.

Multi-Point Hybrid and multi-point grounding (Figure 45.24(c)) can overcome the radio frequency problems associated with pure single-point systems. Multipoint grounding is necessary for digital and large signal high-frequency systems. Modules and circuits are bonded together with many short ($<0.1\lambda$) links to minimize ground-impedance-induced common mode voltages. Alternatively, many short links to a chassis, ground plane, or other low-impedance conductive body are made. This is not appropriate for sensitive analog circuits, because the loops that are introduced are susceptible to magnetic field pickup. It is very difficult to keep 50/60 Hz interference out of such circuits. Circuits which operate at higher frequencies or levels are not susceptible to this interference. The multi-point subsystem can be brought down to a single-point ground in the overall system.

Hybrid Hybrid grounding uses reactive components (capacitors or inductors) to make the grounding system act differently at low frequencies and at radio frequency. This may be necessary in sensitive wideband circuits. In the example shown in Figure 45.24(c), the sheath of a relatively long cable is grounded directly to chassis via capacitors to prevent rf standing waves from forming. The capacitors block d.c. and low-frequencies and therefore prevent the formation of an undesired extra ground loop between the two modules.

When using such reactive components as part of the ground system, you need to take care that spurious resonances (which could enhance interference currents) are not introduced into it. For example if 0.1 μF capacitors are used to decouple a cable whose self-inductance is 0.1 μH, the resonant frequency of the combination is 1.6 MHz. Around this frequency the cable screen will appear to be anything but grounded!

When using separate d.c. grounds and a radio frequency ground plane (such as is offered by the chassis or frame structure), each subsystem's d.c. ground should be referenced to the frame by a 10–100 nF capacitor. The two should be tied together by a low-impedance link at a single point where the highest di/dt signals occur, such as the processor motherboard or the card cage backplane.

Grounding of Large System Large systems are difficult to deal with because distances are a significant fraction of a wavelength at lower frequencies. This can be overcome to some extent by running cables within cabinets in shielded conduit or near to the metal chassis. The distributed capacitance that this offers allows the enclosure to act as a high-frequency ground plane and to keep the impedance of ground wires low.

At least two separate grounds not including the safety ground should be incorporated within the system (Figure 45.25), an electronics ground return for the circuits and a chassis ground for hardware (racks and cabinets). These should be connected together only at the primary power ground. The chassis provides a good ground for high-frequency return currents, and the circuit grounds should be referenced to chassis locally through 10–100 nF capacitors. Safety earths for individual units can be connected to the rack metalwork. All metalwork should be solidly bonded together—it is not enough to rely on hinges, slides, or casual contact for ground continuity. Bonding may be achieved by screw or rivet contact, provided that steps are taken to ensure that paint or surface contamination does not interfere with the bond, but where panels or other structural components are removable then bonding should be maintained by a separate short, easily identified connecting strap.

The electronics ground can if necessary be subdivided further into "clean" and "dirty" ground returns for sensitive and noisy circuits, sometimes allocated as "signal" and "power" grounds. This allows low-frequency single-point grounding either at each rack or at the overall cabinet ground.

The Impedance of Ground Wires When a grounding wire runs for some distance alongside a ground plane or chassis before being connected to it, it appears as a transmission line. This can be modeled as an LCR network with the L and C components determining the characteristic impedance Z_o of the line (Figure 45.26). As the operating frequency rises, the inductive reactance exceeds the resistance of the wire and the impedance increases up to the first parallel resonant point. At this point the impedance seen at the end of the wire is high, typically hundreds of ohms (determined by the resistive loss in the circuit). After first resonance, the impedance for a lossless circuit follows the law

$$Z = Z_o \cdot \tan\left(\omega x \sqrt{L/C}\right) \quad (45.9)$$

where x is the distance along the wire to the short, and successive series (low-impedance) and parallel (high-impedance) resonant frequencies are found. As the losses rise due to skin effect, so the resonant peaks and nulls become less pronounced. To stay well below the first resonance and hence remain an effective conductor, the ground wire should be less than 1/20 of the shortest operating wavelength.

The Safety Earth From the foregoing discussion, you can see that the safety ground (the green and yellow wire) is not a radio frequency ground at all. Many designers may argue that everything is connected to earth via the green and yellow wire, without appreciating that this wire has a high and variable impedance at radio frequency. A good low impedance ground to a chassis, frame, or plate is also necessary and in many cases must be provided *in parallel with* the safety earth. It may even be necessary for you to take the safety earth *out* of the circuit deliberately, by inserting a choke of the appropriate current rating in series with it.

Summary For frequencies below 1 MHz, single-point grounding is possible and preferable. Above 10 MHz a single-point ground system is not feasible because wire and track inductance raises the ground impedance unacceptably, and stray capacitance allows unintended ground

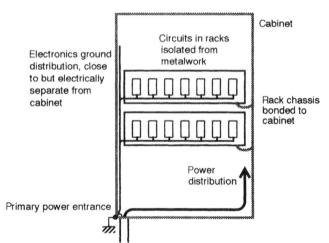

FIGURE 45.25 Grounding of a rack system.

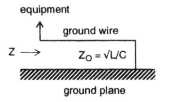 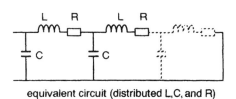

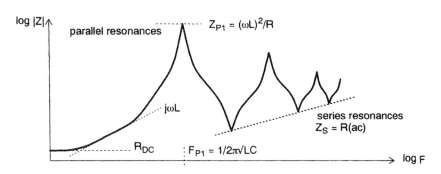

FIGURE 45.26 The impedance of long ground wires.

return paths to exist. For high frequencies multipoint grounding to a low-inductance ground plane or shield is essential. This creates ground loops which may be susceptible to magnetic field induction, so should be avoided or specially treated in a hybrid manner when used with very sensitive circuits.

For EMC purposes, even a circuit which is only intended to operate at low frequencies must exhibit good immunity from radio frequency interference. This means that those aspects of its ground layout which are exposed to the interference—essentially all external interfaces—must be designed for multi-point grounding. At the bare minimum, some low-inductance ground plate or plane must be provided at the interfaces.

45.3.1.4 PCB Layout

The way in which you design a printed circuit board (PCB) makes a big difference to the overall EMC performance of the product which incorporates it. The principles outlined above must be carried through onto the PCB, particularly with regard to partitioning, interface layout, and ground layout. This means that the circuit designer must exert tight control over the layout draftsman, especially when CAD artwork is being produced. Conventional CAD layout software works on a node-by-node basis, which if allowed to will treat the entire ground system as one node, with disastrous results for the high-frequency performance if left uncorrected.

The safest way to lay out a PCB (if a multilayer construction with power and ground planes is ruled out) is to start with the ground traces, manually if necessary, then to incorporate critical signals such as high-frequency clocks or sensitive nodes which must be routed near to their ground returns, and then to track the rest of the circuitry at will. As much information should be provided with the circuit diagram as possible, to give the layout draftsman the necessary guidance at the beginning. These notes should include:

1. Physical partitioning of the functional submodules on the board.
2. Positioning requirements of sensitive components and I/O ports.
3. Marked up on the circuit diagram, the various different ground nodes that are being used, together with which connections to them must be regarded as critical.
4. Where the various ground nodes may be commoned together, and where they must not be.
5. Which signal tracks must be routed close to the ground tracks.

Track Impedance Careful placement of ground connections goes a long way towards reducing the noise voltages that are developed across ground impedances. But on any non-trivial PCB it is impractical to eliminate circulating ground currents entirely. The other aspect of ground design is to minimize the value of the ground impedance itself.

Track impedance is dominated by inductance at frequencies higher than a few kilohertz (Figure 45.27). You can reduce the inductance of a connection in two ways:

1. Minimizing the length of the conductor and, if possible, increasing its width.
2. Running its return path parallel and close to it.

The inductance of a PCB track is primarily a function of its length, and only secondarily a function of its width. For a single track of width w and height h over a ground plane, the inductance is given by

$$L = 0.005 \ln(2\pi \cdot h/w) \text{ microhenry/inch} \quad (45.10)$$

Because of the logarithmic relationship of inductance and width, doubling the width does not go so far as to halve the

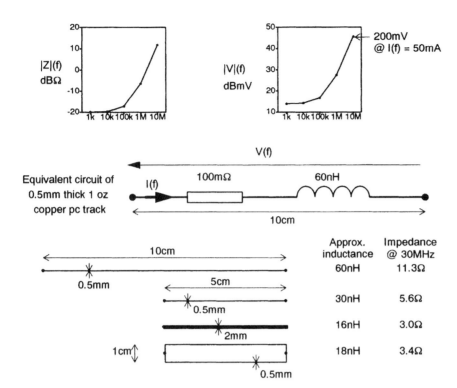

FIGURE 45.27 Impedance of printed circuit tracks.

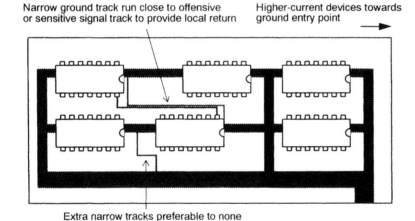

FIGURE 45.28 The gridded ground structure.

inductance. Paralleling tracks will reduce the inductance *pro rata* provided that they are separated by enough distance to neutralize the effect of mutual inductance (see Figure 45.4). For narrow conductors spaced more than 1 cm apart, mutual inductance effects are negligible.

Gridded Ground The logical extension to paralleling ground tracks is to form the ground layout in a grid structure (Figure 45.28). This maximizes the number of different paths that a ground return current can take and therefore minimizes the ground inductance for any given signal route. Such a structure is well suited to digital layout with multiple packages, when individual signal/return paths are too complex to define easily (German, 1985).

A wide ground track is preferred to narrow for minimum inductance, but even a narrow track linking two otherwise widely separated points is better than none. The grid layout is best achieved by putting the grid structure down first, before the signal or power tracks are laid out. You can follow the X–Y routing system for double-sided boards, where the X direction tracks are all laid on one side and the Y direction tracks all on the other, provided the via hole impedance at junctions is minimized. Offensive (high di/dt) signal tracks can then be laid close to the ground tracks to keep the overall loop area small; this may call for extra ground tracking, which should be regarded as an acceptable overhead.

Ground Style Versus Circuit Type A gridded ground is not advisable for low-frequency precision analog circuits, because in these cases it is preferable to define the ground paths accurately to prevent common impedance coupling. Provided that the bandwidth of such circuits is low then high-frequency noise due to ground inductance is less of a problem. For reduced ESD susceptibility, the circuit ground needs to remain stable during the ESD event. A low-inductance ground network is essential, but this must also be coupled (by capacitors or directly) to a master reference ground structure.

45.3.1.5 Ground Plane

The limiting case of a gridded ground is when an infinite number of parallel paths are provided and the ground conductor is continuous, and it is then known as a ground plane. This is easy to realize with a multilayer board and offers the lowest possible ground path inductance. It is essential for radiofrequency circuits and digital circuits with high clock speeds, and offers the extra advantages of greater packing density and a defined characteristic impedance for all signal tracks (Motorola, 1990). A common four-layer configuration includes the power supply rail as a separate plane, which gives a low power ground impedance at high frequencies.

Note that the main EMC purpose of a ground plane is to provide a low-impedance ground and power return path to minimize induced ground noise. Shielding effects on signal tracks are secondary and are in any case nullified by the component lead wires, when these stand proud of the board. There is little to be gained from having power and ground planes outside the signal planes on four-layer boards, especially considering the extra aggravation involved in testing, diagnostics, and rework.

Figure 45.29 compares the impedance between any two points (independent of spacing) on an infinite ground plane with the equivalent impedance of a short length of track. The impedance starts to rise at higher frequencies because the skin effect increases the effective resistance of the plane, but this effect follows the square root of frequency (10 dB/decade) rather than the inductive wire impedance which is directly proportional to frequency (20 dB/decade). For a finite ground plane, points in the middle will see the ideal impedance while points near the outside will see up to four times this value.

Ground Plane on Double-Sided PCBs A partial ground plane is also possible on double-sided PCBs. This is not achieved merely by filling all unused space with copper and connecting it to ground; as the purpose of the ground plane is to provide a low-inductance ground path, it must be positioned under (or over) the tracks which need this low-inductance return. At high frequencies, return current does not take the geometrically shortest return path but will flow preferentially in the neighborhood of its signal trace. This is because such a route encloses the least area and hence has the lowest overall inductance. Thus the use of an overall ground plane ensures that the optimum return path is always available, allowing the circuit to achieve minimum inductive loop area by its own devices (Swainson, 1990).

A Partial Ground Plane Not all of the copper area of a complete ground plane needs to be used, and it is possible to reduce the ground plane area by keeping it only under offending tracks. Figure 45.30 illustrates the development of the ground plane concept from the limiting case of two parallel identical tracks. To appreciate the factors which control inductance, remember that the total loop inductance of two parallel tracks which are carrying current in opposite directions (signal and return) is given by

$$L = L1 + L2 - 2M \quad (45.11)$$

where $L1$ and $L2$ are the inductances of each track, and M is the mutual inductance between them. M is inversely proportional to the spacing of the tracks; if they were co-located it would be equal to L and the loop inductance would be zero. In contrast, the inductance of two identical tracks carrying current in the same direction is given by

$$L = \left(\frac{L1 + M}{2}\right) \quad (45.12)$$

so that a closer spacing of tracks increases the total inductance. Since the ground plane is carrying the return current for signal tracks above it, it should be kept as close as possible to the tracks to keep the loop inductance to a minimum. For a continuous ground plane this is set only by the thickness of the intervening board laminate.

Breaks in the Ground Plane What is essential is that the plane remain unbroken in the direction of current flow. Any deviations from an unbroken plane effectively increase the loop area and hence the inductance. If breaks are necessary it is preferable to include a small bridging track next to a critical signal track to link two adjacent areas of plane (Figure 45.31). A slot in the ground plane will nullify the beneficial effect of the plane if it interrupts the current, however narrow it is. This is why a multilayer construction with an unbroken internal ground plane is the easiest to design, especially for fast logic which requires a closely controlled track characteristic impedance. Where double-sided board with a partial ground plane is used, bridging tracks as shown in Figure 45.31 should accompany all critical tracks, especially clocks.

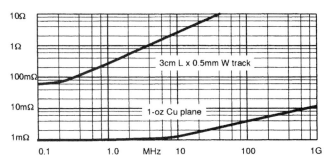

FIGURE 45.29 Impedance of ground plane versus track.

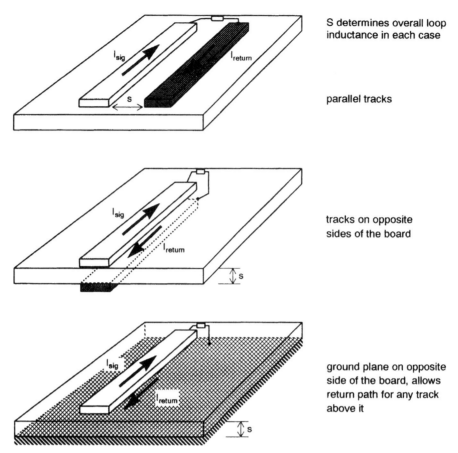

FIGURE 45.30 Return current paths.

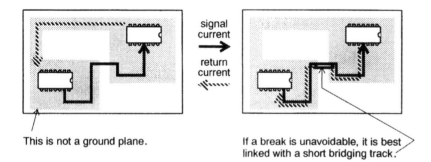

FIGURE 45.31 A broken ground plane.

45.3.1.6 Loop Area

The major advantage of a ground plane is that it allows for the minimum area of radiating loop. This ensures the minimum differential-mode emission from the PCB and also the minimum pick-up of radiated fields. If you do not use a ground plane, it is still possible to ensure minimum loop area by keeping the tracks or leads that carry a high di/dt circuit, or a susceptible circuit, close to one another (Ott, 1981). This is helped by using components with small physical dimensions, and by keeping the separation distance as small as possible consistent with layout rules. Such circuits (e.g., clocks or sensitive inputs) should not have their source and destination far apart. The same also applies to power rails. Transient power supply currents should be decoupled to ground close to their source, by several low-value capacitors well distributed around the board.

The Advantage of Surface Mount Surface mount technology (SMT) offers smaller component sizes and therefore should give a reduction in interference coupling, since the overall circuit loop area can be smaller. This is in fact the case, but to take full advantage of SMT a multilayer board construction with ground plane is necessary. There is a slight improvement when a double-sided board is relaid out

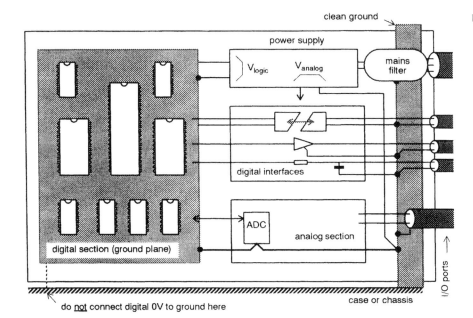

FIGURE 45.32 Grounding at the interfaces.

to take SMT components, which is mainly due to shrinking the overall board size and reducing the length of individual tracks. The predominant radiation is from tracks rather than components.

But when a multilayer board is used, the circuit loop area is reduced to the track length times the track-to-ground plane height. Now, the dominant radiation is from the extra area introduced by the component lead-outs. The reduction in this area afforded by SMT components is therefore worthwhile. For EMC purposes, SMT and multilayer groundplane construction are complementary.

A further advantage of surface mount is that, rather than taking advantage of the component size reduction to pack more functions into a given board area, you can reduce the board area needed for a given function. This allows you more room to define quiet I/O areas and to fit suppression and filtering components when these prove to be needed.

45.3.1.7 Configuring I/O Grounds

Decoupling and shielding techniques to reduce common-mode currents appearing on cables both require a "clean" ground point, not contaminated by internally generated noise. *Filtering at high frequencies is next to useless without such a ground*. Unless you consider this as part of the layout specification early in the design phase, such a ground will not be available. Provide a clean ground by grouping all I/O leads in one area and connecting their shields and decoupling capacitors to a separate ground plane in this area. The clean ground can be a separate area of the PCB (Ott, 1985), or it can be a metal plate on which the connectors are mounted. The external ground (which may be only the line-voltage safety earth) and the metal or metallized case, if one is used, are connected here as well, via a low-inductance link. Figure 45.32 shows a typical arrangement for a product with digital, analog, and interface sections.

This Clean Ground Must Only Connect to the Internal Logic Ground at One Point This prevents logic currents flowing through the clean ground plane and "contaminating" it. No other connections to the clean ground are allowed. As well as preventing common mode emissions, this layout also shunts incoming interference currents (transient or radio frequency) to the clean ground and prevents them flowing through susceptible circuitry. If, for other reasons, it is essential to have leads interfacing with the unit or PCB at different places, you should still arrange to couple them all to a clean ground, that is, one through which no circuit currents are flowing. In this case a chassis plate is mandatory.

For ESD protection the circuit ground *must* be referenced to the chassis ground. This can easily be done by using a plated-through hole on the ground track and a metallic standoff spacer. If there has to be d.c. isolation between the two grounds at this point, use a 10–100 nF radio frequency (ceramic or polyester) capacitor. You can provide a clean I/O ground on plug-in rack mounting cards by using wiping finger-style contacts to connect this ground track directly to the chassis.

Separate Circuit Grounds Never extend a digital ground plane over an analog section of the PCB as this will couple digital noise into the analog circuitry. A single-point connection between digital and analog grounds can be made at the system's analog-to-digital converter. It is very important not to connect the digital circuitry separately to an external ground (Catherwood, 1989). If you do this, extra current paths are set up which allow digital circuit noise to circulate in the clean ground.

Interfaces directly to the digital circuitry (for instance, to a port input or output) should be buffered so that they do not need to be referenced to the digital 0 V. The best interface is an opto-isolator or relay, but this is of course expensive. When you cannot afford isolation, a separate buffer 1C which can be referenced to the I/O ground is preferable; otherwise, buffer the port with a series resistor or choke and decouple the line *at the board interface* (not somewhere in the middle of the board) with a capacitor and/or a transient suppressor to the clean ground. More is said about I/O filtering in Section 45.4.2.4.

Note how the system partitioning, discussed in Section 45.3.1.1, is essential to allow you to group the I/O leads together and away from the noisy or susceptible sections. Notice also that the line-voltage cable, as far as EMC is concerned, is another I/O cable. Assuming that you are using a block line-voltage filter, fit this to the "clean" ground reference plate directly.

45.3.1.8 Rules for Ground Layout

Because it is impractical to optimize the ground layout for all individual signal circuits, you have to concentrate on those which are the greatest threat. These are the ones which carry the highest di/dt most frequently, especially clock lines and data bus lines, and square-wave oscillators at high powers, especially in switching power supplies. From the point of view of susceptibility, sensitive circuits (particularly edge-triggered inputs, clocked systems, and precision analog amplifiers) must be treated similarly. Once these circuits have been identified and partitioned you can concentrate on dealing with their loop inductance and ground coupling. The aim should be to ensure that circulating ground noise currents do not get the opportunity to enter or leave the system.

Ground Map A fundamental tool for use throughout the equipment design is a ground map. This is a diagram which shows all the ground reference points and grounding paths (via structures, cable screens, etc., as well as tracks and wiring) for the whole equipment. It concentrates on grounding only; all other circuit functions are omitted or shown in block form. Its creation, maintenance, and enforcement throughout the project design should be the responsibility of the EMC design authority.

45.3.2 Digital and Analog Circuit Design

Digital circuits are prolific generators of electromagnetic interference. High-frequency square-waves, rich in harmonics, are distributed throughout the system. The harmonic frequency components reach into the part of the spectrum where cable resonance effects are important. Analog circuits are in general much quieter because high-frequency squarewaves are not normally a feature. A major exception is wide-bandwidth video circuits, which transmit broadband signals up to several megahertz, or several tens of megahertz for high-resolution video. Any analog design which includes a high-frequency oscillator or other high di/dt circuits must follow high-frequency design principles, especially with regard to ground layout.

Some low-frequency amplifier circuits can oscillate in the megahertz range, especially when driving a capacitive load, and this can cause unexpected emissions. The switching power supply is a serious cause of interference at low to medium frequencies since it is essentially a high-power squarewave oscillator.

Because the microprocessor is a state machine, processor-based circuits are prone to corruption by fast transients which can cause the execution of false states. Great care is necessary to prevent any clocked circuit (not just microprocessor-based) from being susceptible to incoming interference. Analog signals are more affected by continuous interference, which is rectified by non-linear circuit elements and causes bias or signal level shifts. The immunity of analog circuits is improved by minimizing amplifier bandwidth, maximizing the signal level, using balanced configurations, and electrically isolating I/O that will be connected to "dirty" external circuits.

45.3.2.1 The Fourier Spectrum

The Time Domain and the Frequency Domain Basic to an understanding of why switching circuits cause interference is the concept of the time domain/frequency domain transform. Most circuit designers are used to working with waveforms in the time domain, as viewed on an oscilloscope, but any repeating waveform can also be represented in the frequency domain, for which the basic measuring and display instrument is the spectrum analyzer. Whereas the oscilloscope shows a segment of the waveform displayed against time, the spectrum analyzer will show the same waveform displayed against frequency. Thus the relative amplitudes of different frequency components of the signal are instantly seen.

Figure 45.33 shows the spectral amplitude compositions of various types of waveform (phase relationships are rarely of any interest for EMC purposes). The sine wave has only a single component at its fundamental frequency. A squarewave with infinitely fast rise and fall times has a series of odd harmonics (multiples of the fundamental frequency) extending to infinity. A sawtooth contains both even and odd harmonics.

Switching waveforms can be represented as trapezoidal; a digital clock waveform is normally a squarewave with defined rise and fall times. The harmonic amplitude content of a trapezoid decreases from the fundamental at a rate of 20 dB/decade until a breakpoint is reached at $1/\pi t_r$, after which it decreases at 40 dB/decade (Figure 45.34(a)). Of related interest is the differentiated trapezoid, which is an impulse with finite rise and fall times. This has the same

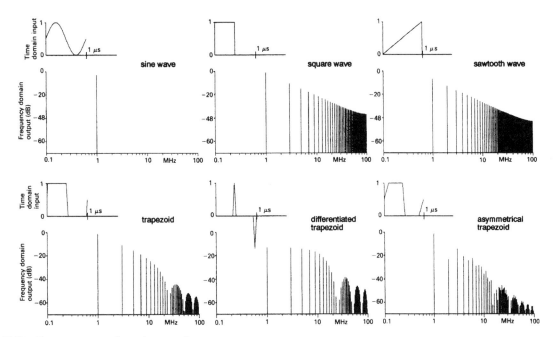

FIGURE 45.33 Frequency spectra for various waveforms.

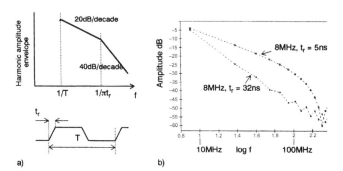

FIGURE 45.34 Harmonic envelope of a trapezoid.

spectrum as the trapezoid at higher frequencies, but the amplitude of the fundamental and lower-order harmonics is reduced and flat with frequency. (This property is intuitively obvious as a differentiator has a rising frequency response of +20 dB/decade.) Reducing the trapezoid's duty cycle from 50 percent has the same effect of decreasing the fundamental and low-frequency harmonic content.

Asymmetrical slew rates generate even as well as odd harmonics. This feature is important, since differences between high- and low-level output drive and load currents mean that most logic circuits exhibit different rise and fall times, and it explains the presence and often preponderance of even harmonics at the higher frequencies.

Choice of Logic Family The damage as far as emissions are concerned is done by switching edges which have a fast rise or fall time (note that this is not the same as propagation delay and is rarely specified in data sheets; where it is, it is usually a maximum figure). Using the slowest rise time compatible with reliable operation will minimize the amplitude of the higher-order harmonics where radiation is more efficient. Figure 45.34(b) shows the calculated harmonic amplitudes for an 8 MHz clock with rise times of 5 and 32 ns. An improvement approaching 20 dB is possible at frequencies around 100 MHz by slowing the rise time.

The advice based on this premise must be: Use the slowest logic family that will do the job; do not use fast logic when it is unnecessary. Treat with caution any proposal to substitute devices from a faster logic family, such as replacing 74 HC parts with 74 AC. Where parts of the circuit must operate at high speed, use fast logic only for those parts and keep the clock signals local. This preference for slow logic is unfortunately in direct opposition to the demands of software engineers for ever-greater processing speeds.

The graph in Figure 45.35 shows the measured harmonics of a 10 MHz square wave for three devices of different logic families in the same circuit. Note the emphasis in the harmonics above 200 MHz for the 74 AC and 74 F types. From the point of view of immunity, a slow logic family will respond less readily to fast transient interference (see below).

Some IC manufacturers are addressing the problem of radio frequency emissions at the chip level. By careful attention to the internal switching regime of VLSI devices, noise currents appearing at the pins can be minimized. The transition times can be optimized rather than minimized for a given application (Gilbert, 1990). Revised package design and smaller packages can allow the decoupling capacitor to be placed as close as possible to the chip, without the internal leadframe's inductance negating its effect; also, the reduction in operating silicon area gained from shrinking silicon design rules can be used to put a

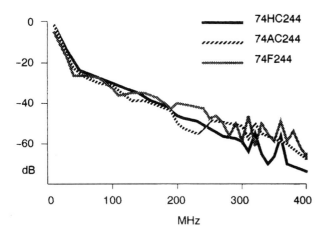

FIGURE 45.35 Comparison of harmonic spectra of different logic families.

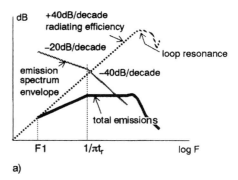

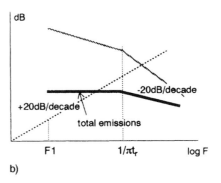

FIGURE 45.36 Emissions from digital trapezoid waves via different paths. (a) Differential-mode radiation. (b) Common-mode radiation.

respectable-sized decoupling capacitor (say 1 nF) actually on the silicon.

45.3.2.2 Radiation from Logic Circuits

Differential-Mode Radiation The radiation efficiency of a small loop is proportional to the square of the frequency (+40 dB/decade). This relationship holds good until the periphery of the loop approaches a quarter wavelength, for instance, about 15 cm in epoxy-glass PCB at 250 MHz, at which point the efficiency peaks. Superimposing this characteristic onto the harmonic envelope of a trapezoidal waveform shows that differential-mode emissions (primarily due to current loops) will be roughly constant with frequency (Figure 45.36(a)) above a breakpoint determined by the rise time (Ott, 1988). The actual radiated emission envelope at 10 m can be derived from equation (45.4) provided the peak-to-peak squarewave current, rise time, and fundamental frequency are known. The Fourier coefficient at the fundamental frequency $F1$ is 0.64, therefore the emission at $F1$ will be

$$E = 20 \log_{10} \left[119 \times 10^{-6} \left(f^2 A I_{pk} \right) \right] \text{dB}\mu\text{V/m} \quad (45.13)$$

from which the +20 dB/decade line to the breakpoint at $1/\pi t_r$ is drawn. In fact, by combining the known rise and fall times and transient output current capability for a given logic family with the trapezoid Fourier spectrum at various fundamental frequencies, the maximum radiated emission can be calculated for different loop areas. If these figures are compared with the EN Class B radiated emissions limit (30 dBµV/m at 10 m up to 230 MHz), a table of maximum allowable loop area for various logic families and clock frequencies can be derived. Table 45.3 shows such a table.

The ΔI figure in Table 45.3 is the dynamic switching current that can be supplied by the device to charge or discharge the node capacitance. It is not the same as the steady-state output high or low load current capability. Some manufacturers' data will give these figures as part of the overall family characteristics. The same applies to t_r and t_f, which are typical figures, and are not the same as maximum propagation delay specifications.

The implication of these figures is that for clock frequencies above 30 MHz, or for fast logic families (AC, AS, or F), a ground plane layout is essential as the loop area restrictions cannot be met in any other way. Even this is insufficient if you are using fast logic at clock frequencies above 30 MHz. The loop area introduced by the device package dimensions exceeds the allowed limit, and extra measures (shielding and filtering) are unavoidable. This information can be useful in the system definition stages of a design, when the extra costs inherent in choosing a higher clock speed (versus, for example, multiple processors running at lower speeds) can be estimated.

Table 45.3 applies to a single radiating loop. Usually, only a few tracks dominate the radiated emissions profile. Radiation from these loops can be added on a root mean square basis, so that for n similar loops the emission is proportional to $\sqrt{n}$. If the loops carry signals at different frequencies then their emissions will not add.

Do not make the mistake of thinking that if your circuit layout satisfies the conditions given in Table 45.3 then your radiated emissions will be below the limit. Total radiation is frequently dominated by common-mode emissions,

TABLE 45.3 Differential mode emission: allowable loop area*

Logic family	t_r/t_f (ns)	ΔI (mA)	Loop area (cm²) at clock frequency (MHz)			
			4	10	30	100
4000B CMOS @ 5 V	40	6	1000	400	–	–
74HC	6	20	45	18	6	–
74LS	6	50	18	7.2	2.4	–
74ALS	3.5	50	10	4	1.4	0.4
74AC	3	80	5.5	2.2	0.75	0.25
74F	3	80	5.5	2.2	0.75	0.25
74AS	1.4	120	2	0.8	0.3	0.15

*Loop area is for 30 dBμV/m, and 37 dBμV/m 230–1000 MHz at 10 m.

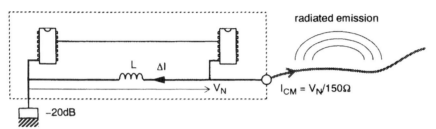

FIGURE 45.37 Common-mode emission model.

as we are about to discover, and Table 45.3 only relates to differential-mode emissions. But if the circuit does *not* satisfy Table 45.3, then extra shielding and filtering will definitely be needed.

Common-Mode Radiation Common-mode radiation which is due mainly to cables and large metallic structures increases at a rate linearly proportional to frequency, as shown in equation (45.5). Thus combining the harmonic spectrum and the radiating efficiency as before gives a radiated field which is constant up to the $1/\pi t_r$ breakpoint and then decreases at a rate of 20 dB/decade (Figure 45.36(b)). It might appear from this that this coupling mode is less important. This comparison is misleading for two reasons:

1. Cable radiation is much more effective than from a small loop, and so a smaller common-mode current (of the order of microamps) is needed for the same field strength.
2. Cable resonance usually falls within the range 30–100 MHz, and radiation is enhanced over that of the short cable model. In fact, common-mode coupling is usually the major source of radiated emissions.

A similar calculation to that done for differential mode can be done for cable radiation on the basis of the model shown in Figure 45.37. This assumes that the cable is driven by a common-mode voltage developed across a ground track which forms part of a logic circuit. The ground track carries the current ΔI which is separated into its frequency components by Fourier analysis, and this current then generates a noise voltage differential V_N of $\Delta I\, j\omega. L$ between the ground reference (assumed to be at one end of the track) and the cable connection (assumed to be at the other). A factor of –20 dB is allowed for lossy coupling to the ground reference. The cable impedance is assumed to be a resistive 150 Ω and constant with frequency—this is a crude approximation generally borne out in practice.

Track Length Implications The inductance L is crucial to the level of noise that is emitted. In the model, it is calculated from the length of a 0.5 mm wide track separated from its signal track by 0.5 mm, so that mutual inductance cancellation reduces the overall inductance. Table 45.4 lists the resulting allowable track lengths versus clock frequency and logic family as before, for a radiated field strength corresponding to the EN Class B limits.

This model should not be taken too seriously for prediction purposes. Too many factors have been simplified: cable resonance and impedance variations with frequency and layout, track and circuit resonance and self-capacitance, and resonance and variability of the coupling path to ground have all been omitted. The purpose of the model is to demonstrate that logic circuit emissions are normally dominated by common-mode factors. Common-mode currents can be combatted by:

1. Ensuring that logic currents do not flow between the ground reference point and the point of connection to external cables.

TABLE 45.4 Common-mode emission: allowable track length*

Logic family	t_r/t_f (ns)	ΔI (mA)	Loop area (cm²) at clock frequency (MHz)			
			4	10	30	100
4000B CMOS @ 5 V	40	6	180	75	–	–
74HC	6	20	8.5	3	1	–
74LS	6	50	3.25	1.3	0.45	–
74ALS	3.5	50	1.9	0.75	0.25	0.08
74AC	3	80	1.0	0.4	0.14	0.05
74F	3	80	1.0	0.4	0.14	0.05
74AS	1.4	120	0.4	0.15	0.05	–

*Allowable track length is for 30 dBµV/m 30–230 MHz and 37 dBµV/m 230–1000 MHz at 10 m; cable length 1 m; layout is parallel 0.5 mm tracks 0.5 mm apart (2.8 nH/cm).

2. Filtering all cable interfaces to a "clean" ground.
3. Screening cables with the screen connection made to a "clean" ground.
4. Minimizing ground noise voltages by using low-inductance ground layout or, preferably, a ground plane.

Table 45.4 shows that the maximum allowable track length for the higher frequencies and faster logic families is impracticable (fractions of a millimeter!). Therefore one or a combination of the above techniques will be essential to bring such circuits into compliance.

Clock and Broadband Radiation The main source of radiation in digital circuits is the processor clock (or clocks) and its harmonics. All the energy in these signals is concentrated at a few specific frequencies, with the result that clock signal levels are 10–20 dB higher than the rest of the digital circuit radiation. Since the commercial radiated emissions standards do not distinguish between narrowband and broadband, these narrowband emissions should be minimized first, by proper layout and grounding of clock lines. Then pay attention to other broadband sources, especially data/address buses and backplanes, and video or high-speed data links.

Backplanes Buses which drive several devices or backplanes which drive several boards carry much higher switching currents (because of the extra load capacitance) than circuits which are compact and/or lightly loaded. Products which incorporate a backplane are more prone to high radiated emissions. A high-speed backplane should always use a multilayer board with a ground plane, and daughterboard connectors should include a ground pin for every high-speed clock, data, or address pin (Figure 45.38). If this is impractical, multiple distributed ground returns can be used to minimize loop areas. The least significant data/address bit usually has the highest frequency component of a bus and should be run closest to its ground return. Clock distribution tracks must *always* have an adjacent ground return.

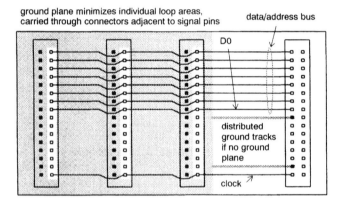

FIGURE 45.38 Backplane layout.

Ringing on Transmission Lines If you transmit data or clocks down long lines, these must be terminated to prevent ringing. Ringing is generated on the transitions of digital signals when a portion of the signal is reflected back down the line due to a mismatch between the line impedance and the terminating impedance. A similar mismatch at the driving end will re-reflect a further portion towards the receiver, and so on. Severe ringing will affect the data transfer if it exceeds the device's input noise margin.

Aside from its effect on noise margins, ringing may also be a source of interference in its own right. The amplitude of the ringing depends on the degree of mismatch at either end of the line while the frequency depends on the electrical length of the line (Figure 45.39). A digital driver/receiver combination should be analyzed in terms of its transmission line behavior if

$$2 \times t_{PD} \times \text{Line length} > \text{Transition time} \quad (45.14)$$

where t_{PD} is the line propagation delay (ns per unit length) (Motorola, 1990). Line propagation delay itself depends on

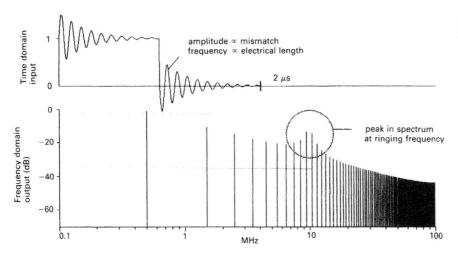

FIGURE 45.39 Ringing due to a mismatched transmission line.

the dielectric constant of the board material and can be calculated from:

$$T_{pd} = 1.017\sqrt{(0.475\varepsilon_r + 0.67)} \text{ ns/ft} \qquad (45.15)$$

where ε_r is the board dielectric constant, typically 4.5 for fiberglass.

This means matching the track's characteristic impedance to the source and load impedances, and may require extra components to terminate the line at the load. Most digital circuit data and application handbooks include advice and formulae for designing transmission line systems in fast logic. Table 45.5 is included as an aid to deciding whether the particular circuit you are concerned with should incorporate transmission line principles.

TABLE 45.5 Critical transmission line length

Logic family	t_r/t_f (ns)	Critical line length
4000B CMOS @ 5 V	40	12 ft
74HC	6	1.75 ft
74LS	6	1.75 ft
74ALS	3.5	1 ft
74AC	3	10 in
74F	3	10 in
74AS	1.4	5 in

Line length calculated for dielectric constant = 4.5 (FR4 epoxy glass), t_{PD} = 1.7 ns/ft.

Digital Circuit Decoupling No matter how good the V_{CC} and ground connections are, track distance will introduce an impedance which will create switching noise from the transient switching currents. The purpose of a decoupling capacitor is to maintain a low dynamic impedance from the individual IC supply voltage to ground. This minimizes the local supply voltage droop when a fast current pulse is taken from it, and more important it minimizes the lengths of track which carry high di/dt currents. Placement is critical; the capacitor must be tracked close to the circuit it is decoupling. "Close" in this context means less than half an inch for fast logic such as AS-TTL, AC, or ECL, especially when high current devices such as bus drivers are involved, extending to several inches for low-current, slow devices such as 4000B-series CMOS.

Components The crucial factor when selecting capacitor type for high-speed logic decoupling is lead inductance rather than absolute value. Minimum lead inductance offers a low impedance to fast pulses. Small disc or multilayer ceramics, or polyester film types (lead pitch 2.5 or 5 mm), are preferred; chip capacitors are even better. The overall inductance of each connection is the sum of both lead and track inductances. Flat ceramic capacitors, matched to the common dual-in-line pinouts and intended for mounting directly beneath the IC package, minimize the pin-to-pin inductance and offer superior performance above about 50 MHz. They are appropriate for extending the usefulness of double-sided boards (with a gridded ground layout but no ground plane) to clock frequencies approaching 50 MHz. A recommended decoupling regime (Williams, 1991) for standard logic (74HC) is

1. One 22 μF bulk capacitor per board at the power supply input.
2. One 1 μF tantalum capacitor per 10 packages of SSI/MSI logic or memory.
3. One 1 μF tantalum capacitor per 2–3 LSI packages.
4. One 22 nF ceramic or polyester capacitor for each octal bus buffer/driver IC or for each MSI/LSI package.
5. One 22 nF ceramic or polyester capacitor per 4 packages of SSI logic.

The value of 22 nF offers a good trade-off between medium frequency decoupling ability and high self-resonant frequency (see below). The minimum required value can be calculated as

$$C = \Delta I \cdot \Delta t / \Delta V \quad (45.16)$$

ΔI and Δt can to a first order be taken from the data in Tables 45.3 and 45.4, while ΔV depends on your judgment of permissible supply voltage drop at the capacitor. Typically a power rail drop of 0.25 V is reasonable; for an octal buffer taking 50 mA per output and switching in 6ns, the required capacitance is 9.6 nF. For smaller devices and faster switching times, less capacitance is required, and often the optimum capacitance value is as low as 1nF. The lower the capacitance, the higher will be its self-resonant frequency, and the more effectively will it decouple the higher-order harmonics of the switching current.

Small tantalum capacitors are to be preferred for bulk decoupling because due to their non-wound construction their self inductance is very much less than for an aluminum electrolytic of the same value.

45.3.2.3 Analog Circuit Emissions

In general, analog circuits do not exhibit the high di/dt and fast rise times that characterize digital circuits, and are therefore less responsible for excessive emissions. Analog circuits which deliberately generate high-frequency signals (remembering that the emissions regulatory regime currently begins at 150 kHz, and may be extended downwards) need to follow the same layout and grounding rules as already outlined. It is also possible for low frequency analog circuits to operate unintentionally outside their design bandwidth.

Instability Analog amplifier circuits may oscillate in the megahertz region and thereby cause interference for a number of reasons:

1. Feedback-loop instability
2. Poor decoupling
3. Output stage instability

Capacitive coupling due to poor layout and common-impedance coupling are also sources of oscillation. Any prototype amplifier circuit should be checked for high-frequency instability, whatever its nominal bandwidth, in its final configuration. Feedback instability is due to too much feedback near the unity-gain frequency, where the amplifier's phase margin is approaching a critical value. It may be linked with incorrect compensation of an uncompensated op-amp.

Decoupling Power supply rejection ratio falls with increasing frequency, and power supply coupling to the input at high frequencies can be significant in wideband circuits. This is cured by decoupling, but typical 0.01–0.1 μF decoupling capacitors may resonate with the parasitic inductance of long power leads in the megahertz region, so decoupling-related instability problems usually show up in the 1–10 MHz range. Paralleling a low-value capacitor with a 1–10 μF tantalum capacitor will drop the resonant frequency and stray circuit Q to a manageable level. Note that the tantalum's series inductance could resonate with the ceramic capacitor and actually worsen the situation. To cure this, a few ohms resistance in series with the tantalum is necessary. The input stages of multi-stage high-gain amplifiers may need additional resistance or a ferrite bead suppressor in series with each stage's supply to improve decoupling from the power rails.

Output Stage Instability Capacitive loads cause a phase lag in the output voltage by acting in combination with the operational amplifier's openloop output resistance (Figure 45.40). This increased phase shift reduces the phase margin of a feedback circuit, possibly by enough to cause oscillation. A typical capacitive load, often invisible to the designer because it is not treated as a component, is a length of coaxial cable. Until the length starts to approach a quarter-wavelength at the frequency of interest, coaxial cable looks like a capacitor: for instance, 10 m of the popular RG58C/U 50 Ω type will be about 1000 pF. To cure output instability, decouple the capacitance from the output with a low-value series resistor, and add high-frequency feedback with a small direct feedback capacitor C_F which compensates for the phase lag caused by C_L. When the critical frequency is high a ferrite bead is an acceptable substitute for R_S.

45.3.2.4 The Switching Power Supply

Switching supplies present extreme difficulties in containing generated interference (Wimmer, 1986). Typical switching frequencies of 50–200 kHz can be emitted by both differential and common-mode mechanisms. Lower frequencies are more prone to differential mode emission while higher frequencies are worse in common mode. Figure 45.41 shows a typical direct-off-line switching supply with the

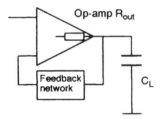
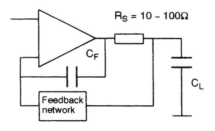

FIGURE 45.40 Instability due to capacitive loads.

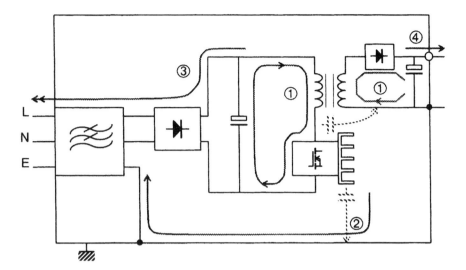

FIGURE 45.41 Switching supply emission paths. (1) H field radiation from di/dt loop. (2) Capacitive coupling of E-field radiation from high dv/dt node to earth. (3) Differential mode current conducted through d.c. (4) Conducted and/or radiated on output.

major emission paths marked; topologies may differ, or the transformer may be replaced by an inductor, but the fundamental interference mechanisms are common to all designs.

Radiation from a High di/dt Loop Magnetic field radiation from a loop which is carrying a high di/dt can be minimized by reducing the loop area or by reducing di/dt. With low output voltages, the output rectifier and smoothing circuit may be a greater culprit in this respect than the input circuit. Loop area is a function of layout and physical component dimensions (see Section 45.3.1.6). di/dt is a trade-off against switching frequency and power losses in the switch. It can, to some extent, be controlled by slowing the rate of rise of the drive waveform to the switch.

The trend towards minimizing power losses and increasing frequencies goes directly against the requirements for low EMI. The lowest di/dt for a given frequency is given by a sinewave: sinusoidal converters have reduced EM emissions.

Magnetic Component Construction Note that, as explained earlier, screening will have little effect on the magnetic field radiation due to this current loop although it will reduce the associated electric field. The transformer (or inductor) core should be in the form of a closed magnetic circuit in order to restrict magnetic radiation from this source. A toroid is the optimum from this point of view, but may not be practical because of winding difficulties or power losses; if you use a gapped core such as the popular E-core type, the gap should be directly underneath the windings since the greatest magnetic leakage flux is to be found around the gap.

Direct radiation from the circuit via this route has usually fallen well below radiated emission limits at the lowest test frequency of 30 MHz, unless the circuit rise times are very fast. On the other hand, such radiation can couple into the output or line-voltage leads and be responsible for conducted

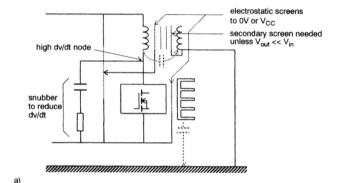

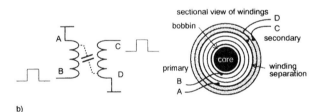

FIGURE 45.42 Common-mode capacitive coupling. (a) Reduction of capacitive coupling. (b) Transformer construction—the interwinding capacitance is dominated by the low dv/dt layers A and D.

interference (over 150 kHz to 30 MHz) if the overall power supply layout is poor. You should always keep any wiring which leaves the enclosure well away from the transformer or inductor.

Capacitive Coupling to Earth High dv/dt at the switching point (the collector or drain of the switching transistor) will couple capacitively to ground and create common mode interference currents. The solution is to minimize dv/dt, and minimize coupling capacitance or provide a preferential route for the capacitive currents (Figure 45.42).

dv/dt is reduced by a snubber and by keeping a low transformer leakage inductance and di/dt. These objectives are also desirable, if not essential, to minimize stress on the switching device, although they increase power losses. The snubber capacitor is calculated to allow a defined dv/dt with the maximum load as reflected through the transformer; the series resistor must be included to limit the discharge current through the switching device when it switches on. You can, if necessary, include a diode in parallel with the resistor to allow a higher resistor value and hence lower switching device ratings.

Capacitive Screening Capacitive coupling is reduced by providing appropriate electrostatic screens, particularly in the transformer and on the device heat sink. Note the proper connection of the screen: to either supply rail, which allows circulating currents to return to their source, not to earth. Even if the transformer is not screened, its construction can aid or hinder capacitive coupling from primary to secondary (Figure 45.42(b)). Separating the windings onto different bobbins reduces their capacitance but increases leakage inductance. Coupling is greatest between nodes of high dv/dt; so the end of the winding which is connected to V_{CC} or ground can screen the rest of the winding in a multi-layer design. Physical separation of parts carrying high dv/dt is desirable (Wimmer, 1986), although hard to arrange in compact products. Extra screening of the offending component(s) is an alternative.

Differential Mode Interference Differential mode interference is caused by the voltage developed across the finite impedance of the reservoir capacitor at high di/dt. It is nearly always the dominant interference source at the lower switching harmonics. Choosing a capacitor with low equivalent series impedance (ESL and ESR) will improve matters, but it is impossible to obtain a low enough impedance in a practical capacitor to make generated noise negligible.

Extra series inductance and parallel capacitance on the input side will attenuate the voltage passed to the input terminals. A capacitor on its own will be ineffective at low frequencies because of the low source impedance. Series inductors of more than a few tens of microhenries are difficult to realize at high d.c. currents (remembering that the inductor must not saturate at the peak ripple current, which is much higher than the d.c. average current), and multiple sections with smaller inductors will be more effective than a single section. When several parallel reservoir capacitors are used, one of these may be separated from the others by the series inductor; this will have little effect on the overall reservoir but will offer a large attenuation to the higher frequency harmonics at little extra cost.

Figure 45.43 demonstrates filtering arrangements. The LC network may also be placed on the input side of the rectifier. This will have the advantage of attenuating broadband noise caused by the rectifier diodes switching at the line frequency. The line-voltage input filter itself (see Section 45.4.2.3) will offer some differential mode rejection. It is also possible to choose switching converter topologies with input inductors which obviate fast di/dt transitions in the input and/or output waveforms. When testing the performance of a differential mode filter, be sure always to check it at the maximum operating input power. Not only do the higher switching currents generate more noise, but the peak line-voltage input current may drive the filter inductor(s) into saturation and make it ineffective.

Output Noise Switching spikes are a feature of the d.c. output of all switching supplies, again mainly because of the finite impedance of the output reservoir. Such spikes are conducted out of the unit on the output lines in both differential and common mode, and may re-radiate onto other

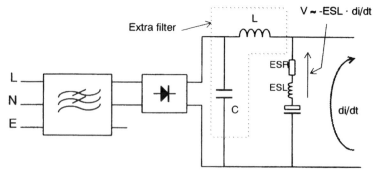

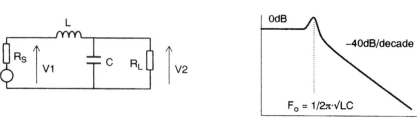

FIGURE 45.43 Differential mode filtering. The insertion loss of a simple LC filter is:

$$|V_1/V_2| = |1 - \omega^2 LC \cdot [R_L/(R_S + R_L)] + j\omega \cdot [(CR_L R_S + L)/(R_S + R_L)]|$$

For low R_S and high R_L this reduces to:

$$|V_1/V_2| = |1 - \omega^2 LC|$$

Standard line-voltage impedance R_L is approximately by 100 Ω/100 μH differentially, which approaches 100 Ω above 300 kHz. Note that resonance of LC at low frequency gives insertion gain.

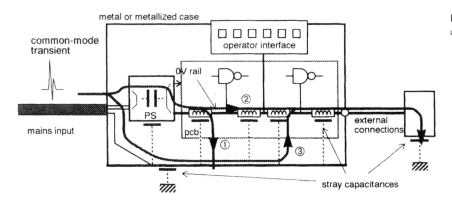

FIGURE 45.44 Representative high-frequency equivalent circuit: transients.

leads or be coupled to the ground connection and generate common mode interference. A low-ESL reservoir capacitor is preferable, but good differential mode suppression can be obtained, as with the input, with a high frequency L-section filter; 20–40 dB is obtainable with a ferrite bead and 0.1 μF capacitor above 1 MHz. Common-mode spikes will be unaffected by adding a filter, and this is a good way to diagnose which mode of interference is being generated.

The abrupt reverse recovery characteristic of the output rectifier diode(s) can create extra high-frequency ringing and transients. These can be attenuated by using soft recovery diodes or by paralleling the diodes with an RC snubber.

45.3.2.5 Design for Immunity: Digital Circuits

The first principle with microprocessor susceptibility is that because the logic threshold is nearer to 0 V than to V_{CC}, most of the critical interference is ground-borne, whether it is common-mode radiofrequencies or transients. Differential-mode interference will not propagate far into the circuit from the external interfaces. Therefore, lay out the circuit to keep ground interference currents away from the logic circuits. If layout is not enough, filter the I/O leads or isolate them, to define a preferential safe current path for interference. Radiated radio frequency fields that generate differential mode voltages internally are dealt with in the same way as differential radio frequency emissions, by minimizing circuit loop area, and by restricting the bandwidth of susceptible circuits where this is feasible.

Use the highest noise threshold logic family that is possible. 74HC is the best all-rounder. 4000B-series is useful for slow circuits, but beware of capacitively coupled interference. Use synchronous design, and try to avoid edge triggered data inputs wherever possible. However good the circuit's immunity, there will always be a transient that will defeat it. Every microprocessor should include a watchdog, and software techniques should be employed that minimize the effects of corruption.

Interference Paths: Transients A typical microprocessor-based product, including power supply, operator interface, processor board, enclosure, and external connections can be represented at high frequency (Barlow, 1990) by the layout shown in Figure 45.44. Note that the 0 V rail will appear as a network of inductances with associated stray capacitances to the enclosure. An incoming common-mode transient on the line-voltage can travel through the circuit's 0 V rail, generating ground differential spikes as it goes, through any or all of several paths as shown (observe the influential effect that stray capacitance has on these paths):

1. Stray capacitance through the power supply to 0 V, through the equipment, and then to case.
2. As above, but then out via an external connection.
3. Direct to case, then via stray capacitance to 0 V and out via an external connection.

If there are no external connections, (1) is the only problem and can be cured by a line-voltage filter and/or by an electrostatic screen in the line-voltage transformer. (2) arises because the external connection can provide a lower impedance route to ground than case capacitance. You cannot control the impedance to ground of external connections, so you have to accept that this route will exist and ensure that the transient current has a preferential path via the case to the interface which does not take in the circuit. This is achieved by ensuring that the case structure is well bonded together, i.e., it presents a low-impedance path to the transient, and by decoupling interfaces to the case at the point of entry/exit (see Sections 45.4.2.4 and 45.3.1.7). If the enclosure is non-conductive then transient currents would have no choice but to flow through the circuit, and local grouping of interfaces is essential.

With external connections, route (3) can actually be *caused* by a line-voltage filter, since at high frequencies parts of the enclosure can float with respect to true ground. This is sometimes the hardest concept to grasp—that even large conducting structures can exhibit high impedances, and hence voltage differentials when subjected to fast transients. The safety earth—the green and yellow wire—is *not* a reference point at high frequency (see above for ground wire impedances) and if this is the only "earth" connection, the case's potential with respect to reference ground is defined by a complex network of inductances (connected cables)

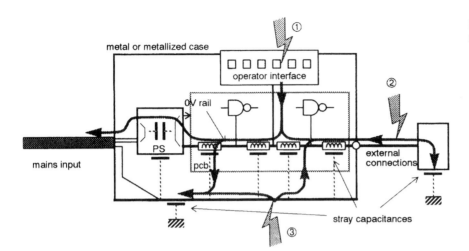

FIGURE 45.45 Representative high-frequency equivalent circuit: ESD.

and stray capacitances which are impossible to predict. In all cases, grouping all I/O leads together with the line-voltage lead will offer low-inductance paths that bypass the circuit and prevent transient currents from flowing through the PCB tracks.

Interference Paths: ESD An electrostatic discharge can occur to any exposed part of the equipment. Common trouble spots as shown in Figure 45.45 are keyboards and controls (1), external cables (2), and accessible metalwork (3). A discharge to a nearby conductive object (which could be an ungrounded metal panel on the equipment itself) causes high local transient currents which will then also induce currents within the equipment by inductive or common impedance coupling.

Because there are many potential points of discharge, the possible routes to ground that the discharge current can take are widespread. Many of them will include part of the PCB ground layout, via stray capacitance, external equipment, or exposed circuitry, and the induced transient ground differentials will cause maloperation. The discharge current will take the route (or routes) of least inductance. If the enclosure is well bonded to ground then this will be the natural sink point. If it is not, or if it is non-conductive, then the routes of least inductance will be via the connecting cables. If the edge of the PCB may be exposed, as in card frames, then a useful trick is to run a "guard trace" around it, unconnected to any circuitry, and separately bond this to ground.

When the enclosure consists of several conductive panels, then these must all be well bonded together, following the rules described in Section 45.4.3 for shielded enclosures. If this is not done, then the edges of the panels will create very high transient fields as the discharge current attempts to cross them. If they are interconnected by lengths of wire, the current through the wire will cause a high magnetic field around it which will couple effectively with nearby PC tracks.

The discharge edge has an extremely fast rise time (sub-nanosecond; see above) and so stray capacitive coupling is essentially transparent to it, whilst even short ground connectors of a few nanohenries will present a high impedance. For this reason the presence or absence of a safety ground wire (which has a high inductance) will often make little difference to the system response to ESD.

Transient and ESD Protection Techniques to guard against corruption by transients and ESD are generally similar to those used to prevent radio frequency emissions, and the same components will serve both purposes. Specific strategies aim to prevent incoming transient and radio frequency currents from flowing through the circuit, and instead to absorb or divert them harmlessly and directly to ground (Figure 45.46). To achieve this:

1. Keep all external interfaces physically near each other.
2. Filter all interfaces to ground at their point of entry.
3. If this is not possible, isolate susceptible interfaces with a common-mode ferrite choke or optocouplers.
4. Use screened cable with the screen connected directly to ground.
5. Screen PCBs from exposed metalwork or external discharge points with extra internally grounded plates.

The Operator Interface Keyboards, for example, present an operator interface which is frequently exposed to ESD. Keyboard cables should be foil-and-braid shielded, which is 360° grounded at both ends to the low-inductance chassis metal-work. Plastic key caps will call for internal metal or foil shielding between the keys and the base PCB which is connected directly to the cable shield, to divert transients away from the circuitry. The shield ground should be coupled to the circuit ground at the cable entry point via a 10–100 nF capacitor to prevent ground potential separation during an ESD event. A membrane keyboard with a polyester surface material has an inherently high dielectric strength and is therefore resistant to ESD, but it should incorporate a ground plane to provide a bleed path for the accumulated

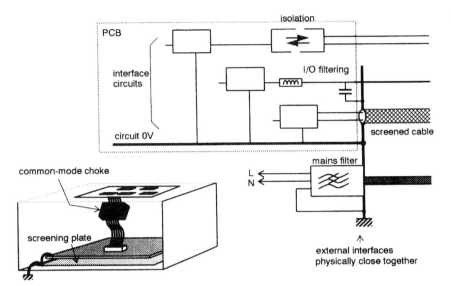

FIGURE 45.46 ESD protection.

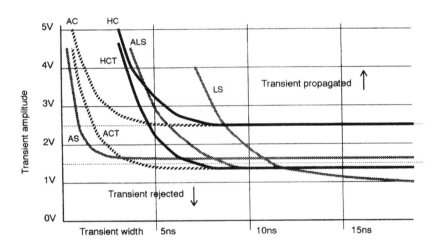

FIGURE 45.47 Dynamic noise margins.

charge and to improve radio-frequency immunity; this ground plane must be "hidden" from possible discharges by sealing it behind the membrane surface.

45.3.2.6 Logic Noise Immunity

The ability of a logic element to operate correctly in a noisy environment involves more than the commonly quoted static noise margins. To create a problem an externally generated transient must cause a change of state in an element which then propagates through the system. Systems with clocked storage elements or those operating fast enough for the transient to appear as a signal are more susceptible than slow systems or those without storage elements (combinational logic only).

Dynamic Noise Margin The effect of a fast transient will depend on the peak voltage coupled into the logic input, and also on the speed of response of the element. Any pulse positive-going from 0 V but below the logic switching threshold (typically 1.8 V for TTL circuits) will not cause the element input to switch from 0 to 1 and will not be propagated into the system. Conversely, a pulse above the threshold will cause the element to switch. But a pulse which is shorter than the element's response time will need a higher voltage to cause switchover, and therefore the graph shown in Figure 45.47 can be constructed, which illustrates the susceptibility of different logic families versus pulse width and amplitude. Bear in mind that switching and ESD transients may lie within the 1–5 ns range. Here is another argument for slow logic!

With synchronous logic, the time of arrival of the transient with respect to the system clock (assuming it corrupts the data line rather than the clock line, due to the former's usually greater area) is important. If the transient does not coincide with the active clock edge then an incorrect value on the data line will not propagate through the system. Thus you can expand the graphs of Figure 45.47 to incorporate another dimension of elapsed time from the clock edge. Tront (1991) has simulated a combinational logic circuit with a flip-flop in 3 μm CMOS technology and generated a series of "upset windows" in this way to describe the susceptibility of that particular circuit to interference. Such a simulation process, using the simulation package SPICE3, can pinpoint those parts of a circuit which have a high degree of susceptibility.

Transient Coupling The amplitude of any pulse coupled differentially into a logic input will depend on the loop area of the differential coupling path which is subjected to the transient field $H_{transient}$ due to transient ground currents $I_{transient}$ and also on the impedance of the driving circuit—less voltage is coupled into a lower impedance. For this reason the lower logic O threshold voltage of LSTTL versus HCMOS is somewhat compensated by the higher logic O output sink capability of LSTTL. If sensitive signal tracks are run close to their ground returns as recommended for emissions, then the resulting loop area is small and little interference is coupled differentially into the sensitive input (Figure 45.48).

Susceptibility Prediction IC susceptibility to radio frequency, as opposed to transient interference, tends to be most marked in the 20–200 MHz region. Susceptibility at the component level is broadband, although there are normally peaks at various frequencies due to resonances in the coupling path. As the frequency increases into the microwave region, component response drops as parasitic capacitances offer alternative shunt paths for the radio frequency energy, and the coupling becomes less efficient. Prediction of the level of radio frequency susceptibility of digital circuits using simulation is possible for small scale integrated circuits (Whalen, 1985), but the modeling of the radio frequency circuit parameters of VLSI devices requires considerable effort, and the resources needed to develop such models for microprocessors and their associated peripherals is overwhelmed by the rate of introduction of new devices.

45.3.2.7 The Microprocessor Watchdog

Circuit techniques to minimize the amplitude and control the path of disruptive interference go a long way towards "hardening" a microprocessor circuit against corruption. But they cannot *eliminate* the risk. The coincidence of a sufficiently high-amplitude transient with a vulnerable point in the data transfer is an entirely statistical affair. The most cost-effective way to ensure the reliability of a microprocessor-based product is to accept that the program *will* occasionally be corrupted, and to provide a means whereby the program flow can be automatically recovered, preferably transparently to the user. This is the function of the microprocessor watchdog (Williams, 1991).

Some of the more up-to-date microprocessors on the market include built-in watchdog devices, which may take the form of an illegal-opcode trap, or a timer which is repetitively reset by accessing a specific register address. If such a watchdog is available, it should be used, because it will be well matched to the processor's operation; otherwise, one must be designed-in to the circuit.

Basic Operation The most serious result of a transient corruption is that the processor program counter or address register is upset, so that it starts interpreting data or empty memory as valid instructions. This causes the processor to enter an endless loop, either doing nothing or performing a few meaningless or, in the worst case, dangerous instructions. A similar effect can happen if the stack register or memory is corrupted. Either way, the processor will appear to be catatonic, in a state of "dynamic halt."

A watchdog guards against this eventuality by requiring the processor to execute a specific simple operation regularly, regardless of what else it is doing, on pain of consequent reset. The watchdog is actually a timer whose output is linked to the $\overline{RESET}$ input, and which itself is being constantly retriggered by the operation the processor performs, normally writing to a spare output port. This operation is shown schematically in Figure 45.49.

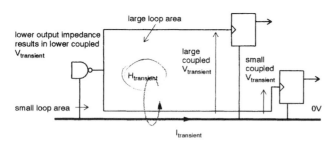

FIGURE 45.48 Transient coupling via signal/return current loops.

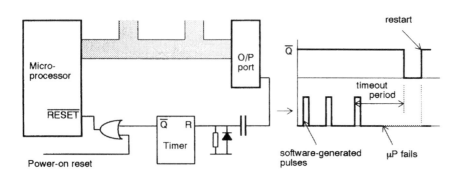

FIGURE 45.49 Watchdog operation.

Time-Out Period If the timer does not receive a "kick" from the output port for more than its time-out period, its output goes low ("barks") and forces the microprocessor into reset. The time-out period must be long enough so that the processor does not have to interrupt time-critical tasks to service the watchdog, and so that there is time for the processor to start the servicing routine when it comes out of reset (otherwise it would be continually barking and the system would never restart properly). On the other hand, it must not be so long that the operation of the equipment could be corrupted for a dangerous period. There is no one time-out period which is right for all applications, but usually it is somewhere between 10 ms and 1 s.

Timer Hardware The watchdog circuit has to exceed the reliability of the rest of the circuit and so the simpler it is, the better. A standard timer IC is quite adequate, but the time-out period may have an unacceptably wide variation in tolerance, besides needing extra discrete components. A digital divider such as the 4060B fed from a high-frequency clock and periodically reset by the report pulses is a more attractive option, since no other components are needed. The divider logic could instead be incorporated into an ASIC if this is present for other purposes. The clock has to have an assured reliability in the presence of transient interference, but such a clock may well already be present or could be derived from the unsmoothed a.c. input at 50/60 Hz.

An extra advantage of the digital divider approach is that its output in the absence of retriggering is a stream of pulses rather than a one-shot. Thus if the microprocessor fails to be reset after the first pulse, or more probably is derailed by another burst of interference before it can retrigger the watchdog, the watchdog will continue to bark until it achieves success (Figure 45.50). This is far more reliable than a monostable watchdog that only barks once and then shuts up.

A programmable timer must not be used to fulfill the watchdog function, however attractive it may be in terms of component count. It is quite possible that the transient corruption could result in the timer being programed off, thereby completely silencing the watchdog. Similarly, it is unsafe to disable the watchdog from the program while performing long operations; corruption during this period will not be recoverable. It is better to insert extra watchdog "kicks" during such long sequences.

Connection to the Microprocessor Figure 45.49 shows the watchdog's $\overline{Q}$ output being fed directly to the $\overline{\text{RESET}}$ input along with the power-on reset (POR) signal. In many cases it will be possible and preferable to trigger the timer's output from the POR signal, in order to assure a defined reset pulse width at the microprocessor on power-up.

It is essential to use the RESET input and not some other signal to the microprocessor such as an interrupt, even a non-maskable one. The processor may be in any conceivable state when the watchdog barks, and it must be returned

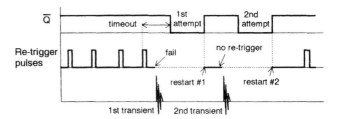

FIGURE 45.50 The advantage of an astable watchdog.

to a fully characterized state. The only state which can guarantee a proper restart is RESET. If the software must know that it was the watchdog that was responsible for the reset, this should be achieved by reading a separate latched input port during initialization.

Source of the Retrigger Pulse Equally important is that the microprocessor should not be able to carry on kicking the watchdog when it is catatonic. This demands a.c. coupling to the timer's retrigger input, as shown by the R–C–D network in Figure 45.49. This ensures that only an edge will retrigger the watchdog, and prevents an output which is stuck high or low from holding the timer off. The same effect is achieved with a timer whose retrigger input is edge- rather than level-sensitive.

Using a programmable port output in conjunction with a.c. coupling is attractive for two reasons. It needs two separate instructions to set and clear it, making it very much less likely to be toggled by the processor executing an endless loop; this is in contrast to designs which use an address decoder to produce a pulse whenever a given address is accessed, a practice which is susceptible to the processor rampaging uncontrolled through the address space. Second, if the programmable port device is itself corrupted but processor operation otherwise continues properly, then the retrigger pulses may cease even though the processor is attempting to write to the port. The ensuing reset will ensure that the port is fully reinitialized. As a matter of software policy, programmable peripheral devices should be periodically reinitialized anyway.

Generation of the Retrigger Pulses in Software If possible, two independent software modules should be used to generate the two edges of the report pulse (Figure 45.51). With a port output as described above, both edges are necessary to keep the watchdog held off. This minimizes the chance of a rogue software loop generating a valid retrigger pulse. At least one edge should only be generated at one place in the code; if a real-time "tick" interrupt is used, this could be conveniently placed at the entry to the interrupt service routine, whilst the other is placed in the background service module. This has the added advantage of guarding against the interrupt being accidentally masked off.

Placing the watchdog retrigger pulse(s) in software is the most critical part of watchdog design and repays careful

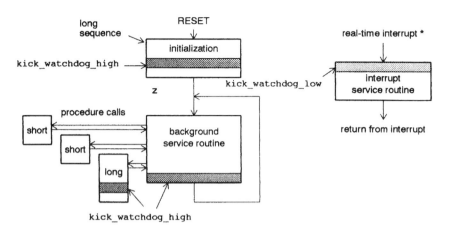

FIGURE 45.51 Software routine for watchdog retrigger. The real-time interrupt must not be masked off for longer than the time-out period less the longest period between successive calls to kick.watchdog.high.

analysis. On the one hand, too many calls in different modules to the pulse-generating routine will degrade the security and efficiency of the watchdog; but, on the other hand, any non-trivial application software will have execution times that vary and will use different modules at different times, so that pulses will have to be generated from several different places. Two frequent critical points are on initialization, and when writing to non-volatile (EEPROM) memory. These processes may take several tens of milliseconds. Analyzing the optimum placement of retrigger pulses, and ensuring that under all correct operating conditions they are generated within the time-out period, is not a small task.

Testing the Watchdog This is not at all simple, since the whole of the rest of the circuit design is bent towards making sure the watchdog never barks. Creating artificial conditions in the software is unsatisfactory because the tested system is then unrepresentative. An adequate procedure for most purposes is to subject the equipment to repeated transient pulses which are of a sufficient level to corrupt the processor's operation predictably, if necessary using specially "weakened" hardware. For safety-critical systems you may have to perform a statistical analysis to determine the pulse repetition rate and duration of test that will establish acceptable performance. A LED on the watchdog output is useful to detect its barks. A particularly vulnerable condition is the application of a burst of spikes, so that the processor is hit again just as it is recovering from the last one. This is unhappily a common occurrence in practice.

As well as testing the reliability of the watchdog, a link to disable it must be included in order to test new versions of software.

45.3.2.8 Defense Programming

Some precautions against interference can be taken in software. Standard techniques of data validation and error correction should be widely used. Hardware performance can also be improved by well-thought-out software. Some means of disabling software error-checking is useful when optimizing the equipment hardware against interference, as otherwise weak points in the hardware will be masked by the software's recovery capabilities. For example, software which does not recognize digital inputs until three polls have given the same result will be impervious to transients which are shorter than this. If your testing uses only short bursts or single transients the equipment will appear to be immune, but longer bursts will cause maloperation which might have been prevented by improving the hardware immunity.

Not all microprocessor faults are due to interference. Other sources are intermittent connections, marginal hardware design, software bugs, metastability of asynchronous circuits, etc. Typical software techniques are to:

1. Type-check and range-check all input data.
2. Sample input data several times and either average it, for analog data, or validate it, for digital data.
3. Incorporate parity checking and data checksums in all data transmission.
4. Protect data blocks in volatile memory with error detecting and correcting algorithms.
5. Wherever possible, rely on level-rather than edge-triggered interrupts.
6. Periodically reinitialize programmable interface chips (PIAs, ACIAs, etc.).

Input Data Validation and Averaging If known limits can be set on the figures that enter as digital input to the software, then data which are outside those limits can be rejected. When, as in most control or monitoring applications, each sensor inputs a continuous stream of data, this is simply a question of taking no action on false data. Since the most likely reason for false data is corruption by a noise burst or transient, subsequent data in the stream will probably be correct, and nothing is lost by ignoring the bad item. Data-logging applications might require a flag on the bad data rather than merely to ignore it.

This technique can be extended if there is a known limit to the maximum rate-of-change of the data. An input which exceeds this limit can be ignored even though it may be still within the range limits. Software averaging on a stream of data to smooth out process noise fluctuations can also help remove or mitigate the effect of invalid data.

When using sophisticated software for error detection, care should be taken not to lock out genuine errors which need flagging or corrective action, such as a sensor failure. The more complex the software algorithm is, the more it needs to be tested to ensure that these abnormal conditions are properly handled.

Digital Inputs A similar checking process should be applied to digital inputs. In this case, there are only two states to check so range testing is inappropriate. Instead, given that the input ports are being polled at a sufficiently high rate, compare successive input values with each other and take no action until two or three consecutive values agree. This way, the processor will be "blind" to occasional transients which may coincide with the polling time slot. This does mean that the polling rate must be two or three times faster than the minimum required for the specified response time, which in turn may require a faster microprocessor than originally envisaged.

Interrupts For similar reasons to those outlined above, it is preferable not to rely on edge-sensitive interrupt inputs. Such an interrupt can be set by a noise spike as readily as by its proper signal. Undoubtedly edge-sensitive interrupts are necessary in some applications, but in these cases you should treat them in the same way as clock inputs to latches or flip-flops and take extra precautions in layout and drive impedance to minimize their noise susceptibility. If there is a choice in the design implementation, then favor a level-sensitive interrupt input.

Data and Memory Protection Volatile memory (RAM as distinct from ROM) is susceptible to various forms of data corruption. Frequently this is due to noise occurring as data are being written to the memory, and hence EEPROM may be as susceptible as RAM. This can be prevented by placing critical data in tables in the memory. Each table is then protected by a checksum, which is stored with the table. Checksum-checking diagnostics can be run by the background routine automatically at whatever interval is deemed necessary to catch corruption, and an error can be flagged or a software reset can be generated as required. The absolute values of stored data do not need to be known provided that the checksum is recalculated every time a table is modified. Beware that the diagnostic routine is not interrupted by a genuine table modification, or vice versa, or errors will start appearing from nowhere! Of course, the actual partitioning of data into tables is a critical system design decision, as it will affect the overall robustness of the system.

Unused Program Memory One of the threats discussed in the section on watchdogs (Section 45.3.2.7) was the possibility of the microprocessor accessing unused memory space due to corruption of its program counter. If it does this, it will interpret whatever data it finds as a program instruction. In such circumstances it would be useful if this action had a predictable outcome.

Normally a bus access to a non-existent address returns the data $\#FF_H$, provided there is a passive pull-up on the bus, as is normal practice. Nothing can be done about this. However, unprogrammed ROM also returns $\#FF_H$, and this can be changed. A good approach is to convert all unused $\#FF_H$ locations to the processor's one-byte NOP (no operation) instruction (Figure 45.52). The last few locations in ROM can be programmed with a JMP RESET instruction, normally three bytes, which will have the effect of resetting the processor. Then, if the processor is corrupted and accesses anywhere in unused memory, it finds a string of NOP instructions and executes these (safely) until it reaches the JMP RESET, at which point it restarts.

The effectiveness of this technique depends on how much of the total possible memory space is filled with NOPs, since the processor can in theory be corrupted to a random address. If the processor accesses an empty bus, its action will depend on the meaning of the $\#FF_H$ instruction. The relative cheapness of large ROMs and EPROMs means that you could consider using these, and filling the entire memory map with ROM, even if your program requirements are small.

Reinitialization As well as RAM data, you must remember to guard against corruption of the setup conditions of programmable devices such as I/O ports or UARTs. Many programers assume erroneously that once an internal device

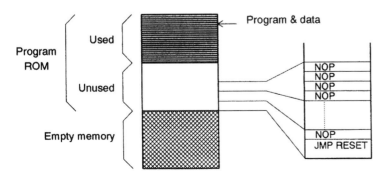

FIGURE 45.52 Protecting unused program memory with NOPs.

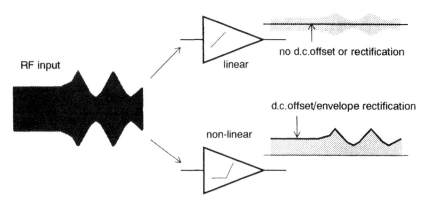

FIGURE 45.53 Radiofrequency demodulation by non-linear circuits.

control register has been set up (usually in the initialization routine) it will stay that way forever. Experience shows that control registers can change their contents, even though they are not directly connected to an external bus, as a result of interference. This may have consequences that are not obvious to the processor: for instance, if an output port is reprogrammed as an input, the processor will happily continue writing data to it oblivious of its ineffectiveness.

The safest course is to periodically reinitialize all critical registers, perhaps in the main idling routine if one exists. Timers, of course, cannot be protected in this way. The period between successive reinitializations depends on how long the software can tolerate a corrupt register, versus the software overhead associated with the reinitialization.

45.3.2.9 Transient and Radio Frequency Immunity: Analog Circuits

Analog circuits in general are not as susceptible to transient upset as digital, but may be more susceptible to demodulation of rf energy. This can show itself as a d.c. bias shift which results in measurement non-linearities or non-operation, or as detection of modulation, which is particularly noticeable in audio and video circuits. Such bias shift does not affect digital circuit operation until the bias is enough to corrupt logic levels, at which point operation ceases completely. Other interference effects occur when the interfering signal is not demodulated but is within or near the signal operating bandwidth; these can include heterodyning (beating), saturation on transients, and false locking of frequency-selective circuits such as phase lock loops.

Improvements in immunity result from attention to the areas as set out below. The greatest radio frequency signal levels are those coupled in via external interface cables, so interface circuits should receive the first attention.

Audio Rectification This is a term used rather loosely to describe the detection of radio frequency signals by low-frequency circuits. It is responsible for most of the ill effects of radio-frequency susceptibility of both analog and digital products.

When a circuit is fed a rf signal that is well outside its normal bandwidth, the circuit can respond either linearly or non-linearly (Figure 45.53). If the signal level is low enough for it to stay linear, it will pass from input to output without affecting the wanted signals or the circuit's operation. If the level drives the circuit into non-linearity, then the envelope of the signal (perhaps severely distorted) will appear on the circuit's output. At this point it will be inseparable from the wanted signal and indeed the wanted signal will itself be affected by the circuit's forced non-linearity. The response of the circuit depends on its linear dynamic range and on the level of the interfering signal. All other factors being equal, a circuit which has a wide dynamic range will be more immune to rf than one which has not.

Bandwidth, Level, and Balance The level of the interfering signal can be reduced by restricting the operating bandwidth to the minimum acceptable. This can be achieved (referring to Figure 45.54) by input RC or LC filtering (1), feedback RC filtering (2), and low value (10–33 pF) capacitors directly at the input terminals (3). RC filters may degrade stability or worsen the circuit's common-mode rejection (CMR) properties, and the value of C must be kept low to avoid this, but an improvement in rf rejection of between 10 and 35 dB over the range 0.15–150 MHz has been reported (Sutu and Whalen, 1985) by including a 27 pF feedback capacitor on an ordinary inverting operational-amplifier circuit. High-frequency CMR is determined by the imbalance between capacitances on balanced inputs. If R would be too high and might affect circuit d.c. conditions, a lossy ferrite-cored choke or bead is an alternative series element.

You should design for signal level to be as high as possible throughout, consistent with other circuit constraints, but at the same time impedances should also be maintained as low as possible to minimize capacitive coupling, and these requirements may conflict. The decision will be influenced by whether inductive coupling is expected to be a major interference contributor. If it is (because circuit loop areas cannot be made acceptably small), then higher impedances will result in lower coupled interference levels. Refer to the discussion on inductive and capacitive coupling (Section 45.2.1.1).

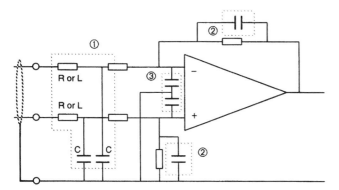

FIGURE 45.54 Bandwidth limitation.

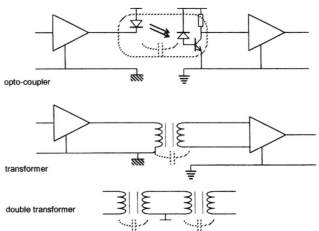

FIGURE 45.55 Signal isolation.

Balanced circuit configurations allow maximum advantage to be taken of the inherent common-mode rejection of operational-amplifier circuits. But note that CMR is poorer at high frequencies and is affected by capacitive and layout imbalances, so it is unwise to trust too much in balanced circuits for good radio frequency and transient immunity. It has also been observed (Whalen, 1985) that in the frequency range 1–20 MHz, mean values of demodulated RFI are 10–20 dB lower for BiFET operational amplifiers than for bipolar ones.

Isolation Signals may be isolated at input or output with either an opto-coupler or a transformer (Figure 45.55). The ultimate expression of the former is fiber optic data transmission, which with the falling costs of fiber optic components is becoming steadily more attractive in a wide range of applications. Given that the major interference coupling route is via the connected cables, using optical fiber instead of wire completely removes this route. This leaves only direct coupling to the enclosure, and coupling via the power cable, each of which is easier to deal with than electrical signal interfaces.

Signal processing techniques will be needed to ensure accurate transmission of precision a.c. or d.c. signals, which increases the overall cost and board area. One problem with opto-couplers that is not suffered by transformers is that large transients may saturate the photo-transistor, effectively "blinding" it for several microseconds and hence stretching the apparent duration of the transient.

Coupling Capacitance Isolation breaks the electrical ground connection and therefore substantially removes common-mode noise injection, as well as allowing a d.c. or low-frequency a.c. potential difference to exist. However, there is still a residual coupling capacitance which will compromise the isolation at high frequencies or high rates of common-mode dv/dt. This capacitance is typically 3 pF per device for an optocoupler; where several channels are isolated the overall coupling capacitance (from one ground to the other) rises to several tens of picofarads. This common-mode impedance is a few tens of ohms at 100 MHz, which is not much of a barrier!

Electrostatically screened transformers and opto-couplers are available where the screen reduces the coupling of common-mode signals into the receiving circuit, and hence improves the common mode transient immunity *of that (local) circuit*. This improvement is gained at the expense of increasing the overall capacitance across the isolation barrier and hence reducing the impedance of the transient or radio frequency coupling path to the rest of the unit. A somewhat expensive solution to this problem is to use two unscreened transformers in series, with the intervening coupled circuit separately grounded.

It is best to minimize the number of channels by using serial rather than parallel data transmission. Do not compromise the isolation further by running tracks from one circuit near to tracks from the other.

45.4 INTERFACES, FILTERING, AND SHIELDING

45.4.1 Cables and Connectors

The most important sources of radiation from a system, or of coupling into a system, are the external cables. Due to their length these are more efficient at interacting with the electromagnetic environment than enclosures, PCBs, or other mechanical structures. Cables, and the connectors which form the interface to the equipment, must be carefully specified. The main purpose of this is to ensure that differential-mode signals are prevented from radiating from the cables, and that common-mode cable currents are neither impressed on the cable by the signal circuit nor are coupled into the signal circuit from external fields via the cable.

In many cases you will have to use screened cables. Exceptions are the line-voltage power cable (provided a line-voltage filter is fitted), and low-frequency interfaces which can be properly filtered to provide transient and radio frequency immunity. An unfiltered, unscreened interface will provide a path for external emissions and for undesired inward coupling. The way that the cable screen is terminated at the connector interface is critical in maintaining the screening properties of the cable.

45.4.1.1 Cable Segregation and Returns

To minimize cross-talk effects within a cable, the signals carried by that cable should all be approximately equal (within, say, ±10 dB) in current and voltage. This leads to the grouping of cable classifications shown in Figure 45.56. Cables carrying high-frequency interfering currents should be kept away from other cables, even within shielded enclosures, as the interference can readily couple to others nearby and generate conducted common-mode emissions. See Figure 45.4 for the effect on mutual capacitance and inductance of the spacing between cables; good installation practice maintains a segregation of at least 125 mm between power and signal cable runs, 300 mm from cables carrying power to fluorescent and are lighting. The breakpoint at which "low frequency" becomes "high frequency" is determined by cable capacitance and circuit impedances and may be as low as a few kilohertz.

All returns should be closely coupled to their signal or power lines, preferably by twisting, as this reduces magnetic field coupling to the circuit. Returns should never be shared between power and signal lines, and preferably not between individual signal lines, as this leads to common impedance coupling. Extra uncommitted grounded wires can help to reduce capacitive crosstalk within cables.

Having advised segregation of different cable classes, it is still true that the best equipment design will be one which puts no restrictions on cable routing and mixing—i.e., one where the major EMC design measures are taken internally. There are many application circumstances when the installation is carried out by unskilled and untrained technicians who ignore your carefully specified guidelines, and the best product is one which works even under these adverse circumstances.

Return Currents It is not intuitively obvious that return currents will necessarily flow in the conductor which is local to the signal wires, when there are several alternative return paths for them to take. At d.c., the return currents are indeed shared only by the ratio of conductor resistances. But as the frequency increases the mutual inductance of the coupled pair (twisted or coaxial) tends to reduce the impedance presented to the return current by its local return compared to other paths, because the enclosed loop area is smallest for this path (Figure 45.57). This is a major reason for the use of twisted-pair cable for data transmission. This effect is also responsible for the magnetic shielding property of coaxial cable, and is the reason why current in a ground plane remains local to its signal track (see Section 45.3.1.5).

45.4.1.2 Cable Screens at Low Frequencies

Optimum screening requires different connection regimes for interference at low frequencies (audio to a few hundred kilohertz) and at radio frequency.

Screen Currents and Magnetic Shielding An overall screen, grounded only at one end, provides good shielding from capacitively coupled interference (Figure 45.58(a)) but none at all from magnetic fields, which induce a noise voltage in the loop that is formed when both source and load are grounded. (Beware: different principles apply when either source or load is not grounded!) To shield against a magnetic field, *both* ends of the screen must be grounded. This allows an induced current (I_S in Figure 45.58(b)) to flow in the screen, which will negate the current induced in the center conductor. The effect of this current begins to become apparent only above the cable cut-off frequency, which is a function of the screen inductance and resistance and is around 1–2 kHz for braided screens or 7–10 kHz for aluminum foil screens. Above about five times the cutoff frequency, the current induced in the center conductor is

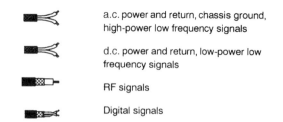

a.c. power and return, chassis ground, high-power low frequency signals

d.c. power and return, low-power low frequency signals

RF signals

Digital signals

FIGURE 45.56 Cable classification.

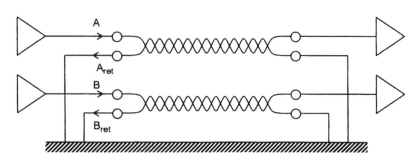

FIGURE 45.57 Signal return current paths. Signal return currents A_{ret} and B_{ret} flow through their local twisted pair return path rather than through ground because this offers the lowest overall path inductance.

constant with frequency (Figure 45.58(c)). The same principle applies when shielding a conductor to prevent magnetic field emission. The return current must flow through the screen, and this will only occur (for a circuit which is grounded at both ends) at frequencies substantially above the shield cut-off frequency.

Where to Ground the Cable Screen The problem with grounding the screen at both ends in the previous circuit is that it becomes a circuit conductor, and any voltage dropped across the screen resistance will be injected in series with the signal. Whenever a *circuit* is grounded at both ends, only a limited amount of magnetic shielding is possible because of the large interference currents induced in the screen-ground loop, which develop an interference voltage along the screen. To minimize low-frequency magnetic field pickup, one end of the circuit should be isolated from ground, the circuit loop area should be small, and the screen should not form part of the circuit. You can best achieve this by using shielded twisted-pair cable with the screen grounded at only one end. The screen then takes care of capacitive coupling while the twisting minimizes magnetic coupling.

For a circuit with an ungrounded source the screen should be grounded at the input common, whereas if the input is floating and the source is grounded then the screen should be grounded to the source common. These arrangements (Figure 45.59) minimize capacitive noise coupling from the screen to the inner conductor(s), since they ensure the minimum voltage differential between the two. Note, though, that as the frequency increases, stray capacitance at the nominally ungrounded end reduces the efficiency of either arrangement by allowing undesired ground and screen currents to flow.

45.4.1.3 Cable Screens at Radio Frequency

Once the cable length approaches a quarter wavelength at the frequency of interest, screen currents due to external fields become a fact of life. An open circuit at one end of the cable becomes transformed into a short circuit a quarter wavelength away, and screen currents flow in a standing wave pattern whether or not there is an external connection (Figure 45.60). The magnitude of the current is related to the characteristic impedance of the transmission line formed by the cable and the ground plane (this behavior was discussed

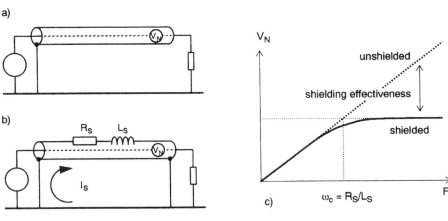

FIGURE 45.58 Magnetic shielding effectiveness versus screen grounding. (a) Good capacitive shielding, no magnetic shielding. (b) Good magnetic shielding.

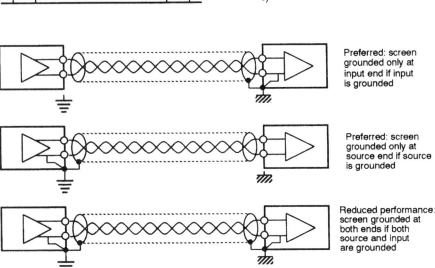

FIGURE 45.59 Screen grounding arrangements versus circuit configuration.

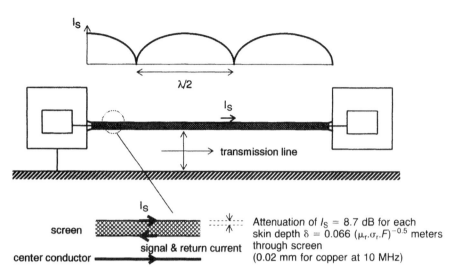

FIGURE 45.60 The cable screen at radio frequency.

above). Even below resonant frequencies, stray capacitance can allow screen currents to flow.

Separation of Inner and Outer Screen Currents At high frequencies the inner and outer of the screen are isolated by skin effect, which forces currents to remain on the surface of the conductor. Signal currents on the inside of the screen do not couple with interference currents on the outside. Thus multiple grounding of the screen, or grounding at both ends, does not introduce interference voltages on the inside to the same extent as at low frequencies.

This effect is compromised by a braided screen due to its incomplete optical coverage and because the strands are continuously woven from inside to out and back again. It is also more seriously compromised by the quality of the screen ground connection at either end, as discussed in Section 45.4.1.5.

45.4.1.4 Types of Cable Screen

The performance of cable screens depends on their construction. Figure 45.61 shows some of the more common types of screen available commercially at reasonable cost; for more demanding applications, specialized screen constructions, such as optimized or multiple braids, are available at a premium. Of course, unscreened cable can also be run in a shielded conduit, in a separate braided screen, or can be wrapped with screening or permeable material. These options are most useful for systems or installation engineers. In any case, proper termination of the screening material to the appropriate grounded structure is vital.

1. *Lapped wire* screens consist of wires helically wound onto the cable. They are very flexible, but have poor screening effectiveness and are noticeably inductive at high frequency, and so are restricted to audio use.
2. *Single braid* screens consist of wire woven into a braid to provide a metallic frame covering the cable, offering

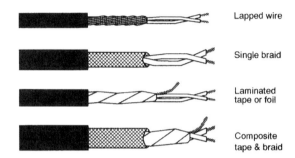

FIGURE 45.61 Common screen types.

80–95 percent coverage and reasonable high-frequency performance. The braid adds significantly to cable weight and stiffness.

3. *Laminated tape* or foil with drain wire provides a full cover, but at a fairly high resistance and hence only moderate screening efficiency. Light weight, flexibility, small diameter, and low cost are retained. Making a proper termination to this type of screen is difficult; screen currents will tend to flow mainly in the drain wire, making it unsuitable for magnetic screening, although its capacitive screening is excellent.
4. *Composite tape and braid* combines the advantages of both laminated tape and single braid to optimize coverage and high-frequency performance. Multiple braid screens improve the performance of single braids by separating the inner and outer current flows, and allowing the screens to be dedicated to different (low- and high-frequency) purposes.

Surface Transfer Impedance The screening performance of shielded cables is best expressed in terms of surface transfer impedance (STI). This is a measure of the voltage induced on the inner conductor(s) of the cable by an interference current flowing down the cable outer shield, which will vary with frequency and is normally expressed in milliohms per meter length. A perfect screen would not allow

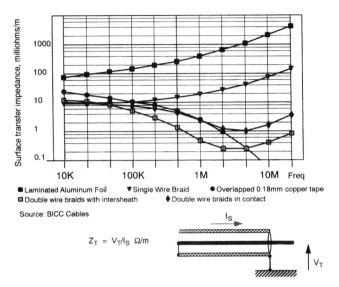

FIGURE 45.62 Surface transfer impedance of various screen types.

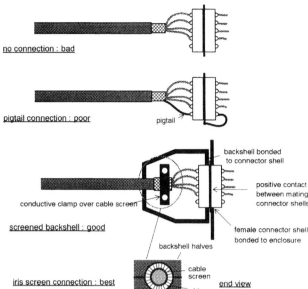

FIGURE 45.63 Cable screen connection methods at radiofrequency.

any voltage to be induced on the inner conductors and would have an STI of zero, but practical screens will couple some energy onto the inner via the screen impedance. At low frequencies it is equal to the d.c. resistance of the screen.

Figure 45.62 compares STI versus frequency for various types of cable screen construction. The decrease in STI with frequency for the better-performance screens is due to the skin effect separating signal currents on the inside of the screen from noise currents on the outside. The subsequent increase is due to field distortion by the holes and weave of the braid. Note that the inexpensive types have a worsening STI with increasing frequency. Once the frequency approaches cable resonance, then STI figures become meaningless; figures are not normally quoted above 30 MHz. Note that the laminated foil screen is approximately 20 dB worse than a single braid, due to its higher resistance and to the field distortion introduced by the drain wire, which carries the major part of the longitudinal screen current.

45.4.1.5 Screened Cable Connections

How to Ground the Cable Shield The overriding requirement for terminating a cable screen is a connection direct to the metal chassis or enclosure ground which exhibits the lowest possible impedance. This ensures that interference currents on the shield are routed to ground without passing through or coupling to other circuits. The best connection in this respect is one in which the shield is extended up to and makes a solid 360° connection with the ground plane or chassis (Figure 45.63). This is best achieved with a hardwired cable termination using a conductive gland and ferrule which clamps over the cable screen. A connector will always compromise the quality of the screen-to-chassis bond, but some connectors are very much better than others.

Connector Types Military-style connectors allow for this construction, as do the standard ranges of radio frequency coaxial connectors such as N type or BNC. Of the readily available commercial multi-way connectors, only those with a connector shell that is designed to make positive 360° contact with its mate are suitable. Examples are the subminiature D range with dimpled tin-plated shells. Connector manufacturers are now introducing properly designed conductive shells for other ranges of mass-termination connector as well.

The Importance of the Backshell The cable screen must make 360° contact with a screened, conductive backshell which must itself be positively connected to the connector shell. The 360° contact is best offered by an iris or ferrule arrangement, although a well-made conductive clamp to the backshell body is an acceptable alternative. A floating cable clamp or a backshell which is not tightly mated to the connector shell are not adequate. The backshell itself can be conductively coated plastic rather than solid metal with little loss of performance, because the effect of the 360° termination is felt at the higher frequencies where the skin depth allows the use of very thin conductive surfaces. On the other hand, the backshell is *not* primarily there to provide electric field screening; simply using a metal or conductively coated shell without ensuring a proper connection to it is pointless.

The Effect of the Pigtail A pigtail connection is one where the screen is brought down to a single wire and extended through a connector pin to the ground point. Because of its ease of assembly it is very commonly used for connecting the screens of data cables. Unfortunately, it may be almost as bad as no connection at high frequencies because of the pigtail inductance (Paul, 1980; Jones, 1985). This can be

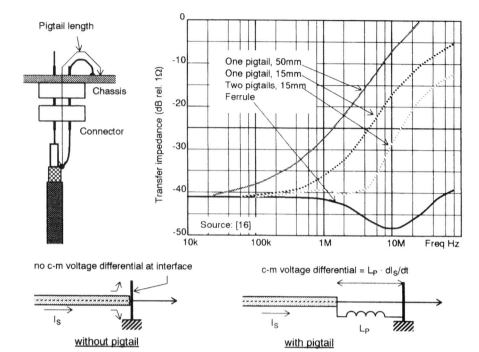

FIGURE 45.64 The pigtail.

visualized as being a few tens of nanohenries in series with the screen connection (Figure 45.64), which develops a commonmode voltage on the cable screen at the interface as a result of the screen currents.

The equivalent surface transfer impedance of such a connection rises rapidly with increasing frequency until it is dominated by the pigtail inductance, and effectively negates the value of a good high-frequency screened cable. At higher frequencies resonances with the stray capacitances around the interface limit the impedance, but they also make the actual performance of the connection unpredictable and very dependent on construction and movement. If a pigtail connection is unavoidable then it must be as short as possible, and preferably doubled and taken through two pins on opposite ends of the connector so that its inductance is halved.

Effective Length Note that the effective length of the pigtail extends from the end of the cable screen through the connector and up to the point of the ground plane or chassis connection. The common practice of mounting screened connectors on a PCB with the screening shell taken to ground via a length of track (which sometimes travels the length of the board) is equivalent to deliberate insertion of a pigtail on the opposite side of the connection. Screened connectors must always be mounted so that their shells are bonded directly to chassis.

45.4.1.6 Unscreened Cables

You are not always bound to use screened cable to combat EMC problems. The various unscreened types offer major advantages in terms of cost and the welcome freedom from the need to terminate the screen properly. In situations where the cable carries signal circuits that are not in themselves susceptible or emissive, and where common-mode cable currents are inoffensive or can be controlled at the interface by other means such as filtering, unscreened cables are quite satisfactory.

Twisted Pair Twisted pair is a particularly effective and simple way of reducing both magnetic and capacitive interference pickup. Twisting the wires tends to ensure a homogeneous distribution of capacitances. Both capacitance to ground and to extraneous sources are balanced. This means that common-mode capacitive coupling is also balanced, allowing high common-mode rejection provided that the rest of the circuit is also balanced.

Twisting is most useful in reducing low-frequency magnetic pick-up because it reduces the magnetic loop area to almost zero. Each twist reverses the direction of induction, so assuming a uniform external field, two successive twists cancel the interaction of the wires with the field. Effective loop pick-up is now reduced to the small areas at each end of the pair, plus some residual interaction due to non-uniformity of the field and irregularity in the twisting. If the termination area is included in the field, the number of twists per unit length is unimportant (Cowdell, 1979; Cathey and Keith, 1981). Clearly, the untwisted termination area or length should be minimized. If the field is localized along the cable, performance improves as the number of twists per unit length increases.

Ribbon Cable Ribbon is widely used for parallel data transmission within enclosures. It allows mass termination

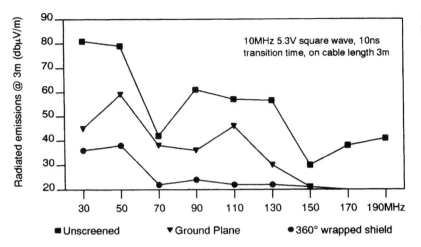

FIGURE 45.65 Coupling from different types of ribbon cable (source: Palmgren, 1981).

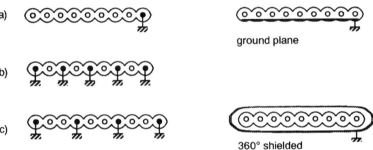

FIGURE 45.66 Ribbon cable configurations.

to the connector and is therefore economical. It should be shielded if it carries high-frequency signals and is extended outside a screened enclosure, but you will find that proper termination of the shield is usually incompatible with the use of a mass-termination connector. Ribbon cable can be obtained with an integral ground plane underneath the conductors, or with full coverage screening. Figure 45.65 shows the relative merits of each in reducing emissions from a typical digital signal. However, the figures for ground plane and shielded cables assume a low-inductance termination, which is difficult to achieve in practice; typical terminations via drain wires will worsen this performance, more so at high frequencies.

The performance of ribbon cables carrying high-frequency data is very susceptible to the configuration of the ground returns. The cheapest configuration is to use one ground conductor for the whole cable (Figure 45.66(a)). This creates a large inductive loop area for the signals on the opposite side of the cable, and cross-talk and ground impedance coupling between signal circuits. The preferred configuration is a separate ground return for each signal (Figure 45.66(b)). This gives almost as good performance as a properly terminated ground plane cable, and is very much easier to work with. Cross-talk and common impedance coupling is virtually eliminated. Its disadvantage is the extra size and cost of the ribbon and connectors. An acceptable alternative is the configuration shown in (Figure 45.66(c)), two signal conductors per return. This improves cable utilization by over 50 percent (Figure 45.66(b)) and maintains the small inductive loop area, at the expense of possible cross-talk and ground coupling problems. The optimum configuration shown in Figure 45.66(b)) can be improved even more by using twisted pair configured into the ribbon construction.

Ferrite-Loaded Cable Common-mode currents in cable screens are responsible for a large proportion of overall radiated emission. A popular technique for reducing these currents is to include a common-mode ferrite choke around the cable, typically just before its exit from the enclosure (see below). Such a choke effectively increases the high-frequency impedance of the cable to common-mode currents without affecting differential mode (signal) currents.

An alternative to discrete chokes is to surround the screen with a continuous coating of flexible ferrite material. This has the advantage of eliminating the need for an extra component or components, and since it is absorptive rather than reflective it reduces discontinuities and hence possible standing waves at high frequencies. This is particularly useful for minimizing the effect of the antenna cable when making radiated field measurements. It can also be applied to unscreened cables such as line-voltage leads. Such "ferrite-loaded" cable is unfortunately expensive, not widely available and like other ferrite applications is only really effective at very high frequencies. It use is more suited to one-off or ad hoc applications than as a production item. It can be especially useful when transients or ESD conducted along the cable are troublesome in particular situations.

45.4.2 Filtering

It is normally impossible to completely eliminate noise being conducted out of or into equipment along connecting leads. The purpose of filtering is to attenuate such noise to a level either at which it meets a given specification, for exported noise, or at which it does not result in malfunction of the system, for imported noise. If a filter contains lossy elements, such as a resistor or ferrite component, then the noise energy may be absorbed and dissipated within the filter. If it does not (i.e., if the elements are purely reactive) then the energy is reflected back to its source and must be dissipated elsewhere in the system. This is one of the features which distinguishes EMI filter design from conventional signal filter design, that in the stop-band the filter should be as lossy as possible.

45.4.2.1 Filter Configuration

In EMC work, "filtering" almost always means low-pass filtering. The purpose is normally to attenuate high-frequency components while passing low-frequency ones. Various simple low-pass configurations are shown in Figure 45.67, and filter circuits are normally made up from a combination of these. The effectiveness of the filter configuration depends on the impedances seen at either end of the filter network.

The simple inductor circuit will give good results (better than 40 dB attenuation) in a low-impedance circuit but will be quite useless at high impedances. The simple capacitor will give good results at high impedances but will be useless at low ones. The multi-component filters will give better results provided that they are configured correctly; the capacitor should face a high impedance and the inductor a low one.

Real-World Impedances Conventionally, filters are specified for terminating impedances of 50 Ω at each end because this is convenient for measurement and is an accepted rf standard. In the real application, Z_S and Z_L are complex and perhaps unknown at the frequencies of interest for suppression. If either or both has a substantial reactive component then resonances are created which may convert an insertion loss into an insertion gain at some frequencies. Differential mode impedances may be predictable if the components which make up the source and load are well characterized at radio frequency, but commonmode impedances such as are presented by cables or the stray reactances of mechanical structures are essentially unpredictable. Practically, cables have been found to have common-mode impedances in the region of 100–400 Ω except at resonance, and a figure of 150 Ω is commonly taken for a rule of thumb.

Parasitic Reactances Filter components, like all others, are imperfect. Inductors have self-capacitance, capacitors have self-inductance. This complicates the equivalent circuit at high frequencies and means that a typical filter using discrete components will start to lose its performance above about 10 MHz. The larger the components are physically, the lower will be the break frequency. For capacitors, as the frequency increases beyond capacitor self-resonance the impedance of the capacitors in the circuit actually rises, so that the insertion loss begins to fall. This can be countered by using special construction for the capacitors (see below). Similarly, inductors have a self-resonant frequency beyond which their impedance starts to fall. Filter circuits using a single choke are normally limited in their performance by the self-resonance of the choke (see Figure 45.68) to 40 or 50 dB. Better performance than this requires multiple filter sections.

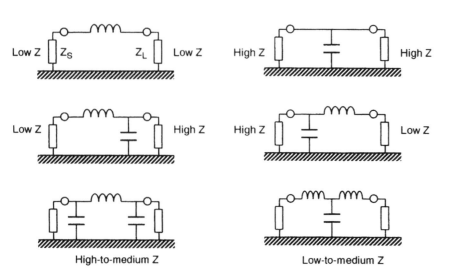

FIGURE 45.67 Filter configuration versus impedance.

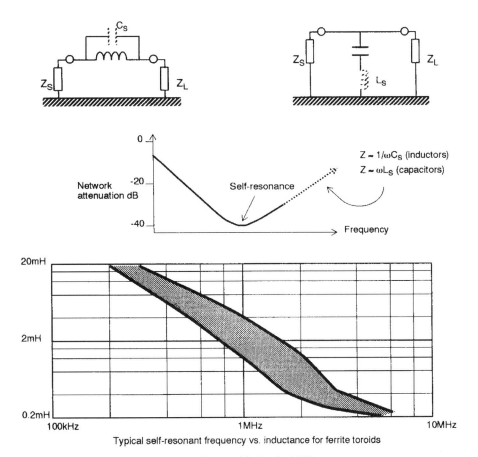

FIGURE 45.68 Self-resonant effects due to parasitic reactances (Crane and Srebranig, 1985).

Capacitors Ceramic capacitors are usually regarded as the best for radio frequency purposes, but in fact, the subminiature polyester or polystyrene film types are often perfectly adequate, since their size approaches that of the ceramic type. Small capacitors with short leads (the ideal is a chip component) will have the lowest self-inductance. For EMI filtering, lossy dielectrics such as X7R, Y5V, and Z5U are an advantage.

Inductors The more turns an inductor has, the higher will be its inductance but also the higher its self-capacitance. The number of turns for a given inductance can be reduced by using a high-permeability core, but these also exhibit a high dielectric constant which tends to increase the capacitance again, for which reason you should always use a bobbin on a high-permeability core rather than winding directly onto the core. For minimum self-capacitance the start and finish of a winding should be widely separated; winding in sections on a multi-section bobbin is one way to achieve this. A single layer winding exhibits the lowest self-capacitance. If you have to use more turns than can be accommodated in a single layer, progressive rather than layer winding (see Figure 45.69) will minimize the capacitance.

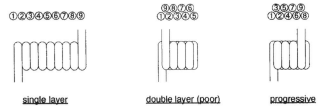

FIGURE 45.69 Inductor winding techniques.

Component Layout Lead inductance and stray capacitance degrade filter performance markedly at high frequency. Two common faults in filter applications are not to provide a low-inductance ground connection, and to wire the input and output leads in the same loom or at least close to or passing each other. This construction will offer low-frequency differential mode attenuation, but high-frequency common-mode attenuation will be minimal.

A poor ground offers a common impedance which rises with frequency and couples high-frequency interference straight through via the filter's local ground path. Common input-output wiring does the same thing through stray capacitance or mutual inductance, and it is also possible for the "clean" wiring to couple with the unfiltered side through inappropriate routing. The cures (Figure 45.70) are to mount the filter so that its ground node is directly coupled to the lowest inductance ground of the equipment, preferably the chassis, and to keep the I/O leads separate, preferably screened from each other. The best solution is to position the filter so that it straddles the equipment shielding, where this exists.

Component layout within the filter itself is also important. Input and output components should be well separated from each other for minimum coupling capacitance, while all tracks and in particular the ground track should be short and substantial. It is best to lay out the filter components exactly as they are drawn on the circuit diagram.

45.4.2.2 Components

There are a number of specialized components which are intended for EMI filtering applications.

Ferrites A very simple, inexpensive, and easily fitted filter is obtained by slipping a ferrite sleeve around a wire or cable (Figure 45.71). The effect of the ferrite is to concentrate the magnetic field around the wire and hence to increase its inductance by several hundred times. The attractiveness of the ferrite choke is that it involves no circuit redesign, and often no mechanical redesign, either. It is therefore very popular for retro fit applications. Several manufacturers offer kits which include halved ferrites, which can be applied to cable looms immediately to check for improvement.

If a ferrite is put over a cable which includes both signal and return lines, it will have no effect on the signal (differential-mode) current, but it will increase the impedance to common-mode currents. The effectiveness can be increased by looping the cable several times through the core, or by using several cores in series. Stray capacitance limits the improvement that can be obtained by extra turns.

Ferrite effectiveness increases with frequency. The impedance of a ferrite choke is typically around 50 Ω at 10 MHz, rising to hundreds of ohms above 100 MHz (the actual value depends on shape and size, more ferrite giving more impedance). The impedance variation with frequency differs to some extent between manufacturers and between different grades of ferrite material. Figure 45.72 shows this for two grades of ferrite with the same geometry.

A useful property of ferrites is that they are lossy at high frequencies, so that interference energy tends to be absorbed rather than reflected. This reduces the Q of inductive suppressor circuits and minimizes resonance problems. In fact, ferrites for suppression purposes are especially developed to have high losses, in contrast to those for low-frequency

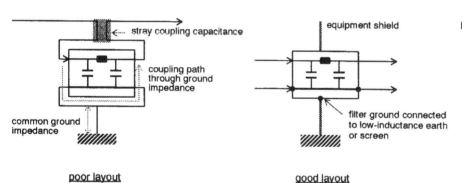

FIGURE 45.70 The effect of filter layout.

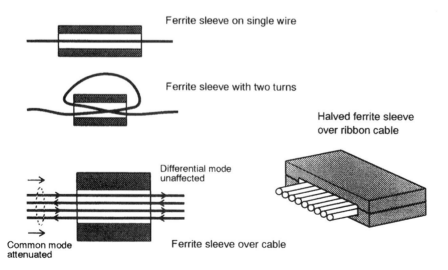

FIGURE 45.71 Uses of ferrites.

or power inductors, which are required to have minimum losses. Figure 45.73 shows how the resistive component dominates the impedance characteristic at the higher frequencies.

Because a ferrite choke is no more than a lossy inductor, it only functions usefully between low impedances. A ferrite included in a high-impedance line will offer little or no attenuation (the attenuation can be derived from the equivalent circuit of Figure 45.67). Most circuits, and especially cables, show impedances that vary with frequency in a complex fashion but normally stay within the bounds of 10–1000 Ω, so a single ferrite will give modest attenuation factors averaging around 10 dB and rarely better than 20 dB.

A ferrite choke is particularly effective at slowing the fast rate of rise of an electrostatic discharge current pulse which may be induced on internal cables. The transient energy is absorbed in the ferrite material rather than being diverted or reflected to another part of the system.

Three-Terminal Capacitors Any low-pass filter configuration except for the simple inductor uses a capacitor in parallel with the signal path. A perfect capacitor would give an attenuation increasing at a constant 20 dB per decade as the frequency increased, but a practical wire-ended capacitor has some inherent lead inductance which in the conventional configuration puts a limit to its high-frequency performance as a filter. The impedance characteristics show a minimum at some frequency and rise with frequency above this minimum.

This lead inductance can be put to some use if the capacitor is given a three-terminal construction (Figure 45.74), separating the input and output connections. The lead inductance now forms a T-filter with the capacitor, greatly improving its high-frequency performance. A ferrite bead on each of the upper leads will further enhance the lead inductance and increase the effectiveness of the filter when it is used with a relatively low impedance source or load. The three-terminal configuration can extend the range of a small ceramic capacitor from below 50 MHz to beyond 200 MHz, which is particularly useful for interference in the VHF band. To fully benefit from this approach, you must terminate the middle (ground) lead directly to a low-inductance ground such as a ground plane; otherwise the inductance remaining in this connection will defeat the capacitor's purpose.

Feedthrough Capacitors Any leaded capacitor is still limited in effectiveness by the inductance of the connection to the ground point. For the ultimate performance, and especially where penetration of a screened enclosure must be protected at uhf and above, then a feedthrough (or lead-through) construction (Figure 45.75) is essential. Here, the ground

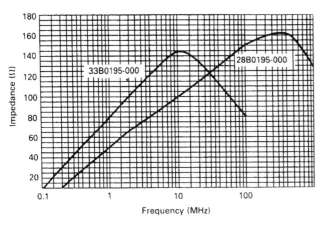

FIGURE 45.72 Impedance versus frequency for two grades of material.

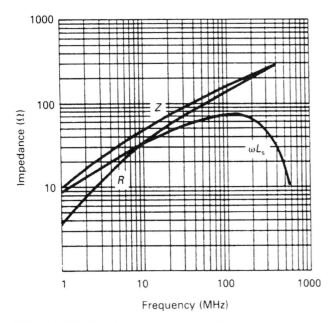

FIGURE 45.73 Impedance components versus frequency.

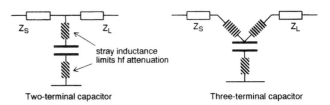

FIGURE 45.74 Two-and three-terminal capacitor.

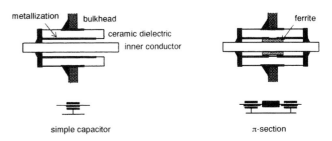

FIGURE 45.75 The feedthrough capacitor.

connection is made by screwing or soldering the outer body of the capacitor directly to the metal screening or bulkhead. Because the current to ground can spread out for 360° around the central conductor, there is effectively no inductance associated with this terminal, and the capacitor performance is maintained well into the gigahertz region. This performance is compromised if a 360° connection is not made or if the bulkhead is limited in extent. The inductance of the through lead can be increased, thereby creating a π-section filter, by separating the ceramic metallization into two parts and incorporating a ferrite bead within the construction. Feedthrough capacitors are available in a wide range of voltage and capacitance ratings, but their cost increases with size.

Chip Capacitors Although they are not seen principally as EMI filter components, surface-mounting chip capacitors offer an extra advantage for this use, which is that their lead inductance is zero. The overall inductance is reduced to that of the component itself, which is typically three to five times less than the lead plus component inductance of a conventional part. Their self-resonant frequency can therefore be double that of a leaded capacitor of the same value. Tracks to capacitors used for filtering and decoupling should be short and direct, in order not to lose this advantage through additional track inductance.

45.4.2.3 Line-Voltage Filters

RFI filters for line-voltage supply inputs have developed as a separate species and are available in many physical and electrical forms from several specialist manufacturers. A typical "block" filter for European line-voltage supplies with average insertion loss might cost around $5–10. Some of the reasons for the development and use of block line-voltage filters are:

1. Mandatory conducted emission standards concentrate on the line-voltage port, hence there is an established market for filter units.
2. Add-on "fit and forget" filters can be retrofitted.
3. Safety approvals for the filter have already been achieved.
4. Many equipment designers are not familiar with radio frequency filter design.

In fact, the market for line-voltage filters really took off with the introduction of VDE and FCC standards regulating conducted line-voltage emissions, compounded by the rising popularity of the switchmode power supply. With a switching supply, a line-voltage filter is essential to meet these regulations. EMC has historically tended to be seen as an afterthought on commercial equipment, and there have been many occasions on which retrofitting a single component line-voltage filter has brought a product into compliance, and this has also encouraged the development of the line-voltage filter market. A real benefit is that safety approvals needed for all components on the line-voltage side of the equipment have been already dealt with by the filter manufacturer if a single-unit filter is used.

Application of Line-Voltage Filters Merely adding a block filter to a line-voltage input will improve low-frequency emissions such as the low harmonics of a switching power supply. But high-frequency emissions (above 1 MHz) require attention to the layout of the circuitry around the filter (see above). Treating it like any other power supply component will not give good high-frequency attenuation and may actually worsen the coupling, through the addition of spurious resonances and coupling paths. Combined filter and CEE22 inlet connector modules are a good method of ensuring correct layout, providing they are used within a grounded conducting enclosure.

A common layout fault is to wire the line-voltage switch in before the filter, and then to bring the switch wiring all the way across the circuit to the front panel and back. This ensures that the filter components are only exposed to the line-voltage supply while the equipment is switched on, but it also provides a ready-made coupling path via stray induction to the unfiltered wiring. Preferably, the filter should be the first thing the line-voltage input sees. If this is impossible, then mount switches, fuses, etc., immediately next to the inlet so that unfiltered wiring lengths are minimal, or use a combined inlet/switch/fuse/filter component. Wiring on either side of the filter should be well separated and extend straight out from the connections. If this also is impossible, try to maintain the two sections of wiring at 90° to each other to minimize coupling.

Typical Line-Voltage Filter A typical filter (Figure 45.76) includes components to block both common-mode and differential-mode components. The common-mode choke L consists of two indentical windings on a single high-permeability, usually toroidal, core, configured so that differential (line-to-neutral) currents cancel each other. This allows high inductance values, typically 1–10 mH, in a small volume without fear of choke saturation caused by the line-voltage frequency supply current. The full inductance of each winding is available to attenuate common-mode currents with respect to earth, but only the leakage inductance L_{1kg} will attenuate differential mode interference. The performance of the filter in differential mode is therefore severely affected by the method of construction of the choke, since this determines the leakage inductance. A high L_{1kg} will offer greater attenuation, but at the expense of a lower saturation current of the core. Low L_{1kg} is achieved by bifilar winding, but safety requirements normally preclude this, dictating a minimum separation gap between the windings.

Common Mode Capacitors Capacitors C_{Y1} and C_{Y2} attenuate common mode interference, and if C_{X2} is large, have

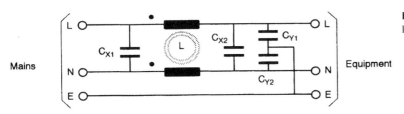

FIGURE 45.76 Typical line-voltage filter and its equivalent circuit.

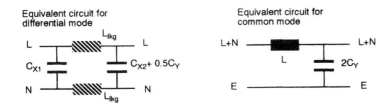

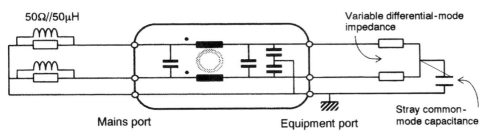

FIGURE 45.77 Impedances seen by the line-voltage filter.

no significant effect on differential mode. The effectiveness of the C_Y capacitors depends very much on the common mode source impedance of the equipment (Figure 45.77). This is usually a function of stray capacitance coupling to earth which depends critically on the mechanical layout of the circuit and the primary-to-secondary capacitance of the line-voltage transformer, and can easily exceed 1000 pF. The attenuation offered by the potential divider effect of C_Y may be no more than 15–20 dB. The common-mode choke is the more effective component, and in cases where C_Y is very severely limited more than one common-mode choke may be needed.

Differential-Mode Capacitors Capacitors C_{X1} and C_{X2} attenuate differential mode only but can have fairly high values, 0.1–0.47 μF being typical. Either may be omitted depending on the detailed performance required, remembering that the source and load impedances may be too low for the capacitor to be useful. For example, a 0.1 μF capacitor has an impedance of about 10 Ω at 150 kHz, and the differential mode source impedance seen by C_{X2} may be considerably less than this for a power supply in the hundreds of watts range, so that a C_{X2} of this value would have no effect at the lower end of the frequency range where it is most needed.

Safety Considerations C_{Y1} and C_{Y2} are limited in value by the permissible continuous current which may flow in the safety earth, due to the line-voltage operating voltage impressed across C_{Y1} (or C_{Y2} under certain fault conditions). Values for this current range from 0.25 to 5 mA, depending on the approvals authority, safety class, and use of the apparatus. Medical equipment has an even lower leakage requirement, typically 0.1 mA. Note that this is the *total* leakage current due to the apparatus; if there are other components (such as transient suppressors) which also form a leakage path to earth, the current due to them must be added to that due to C_Y, putting a further constraint on the value of C_Y.

BS 613, which specifies EMI filters in the U.K., allows a maximum value for C_Y of 5000 pF with a tolerance of ±20 percent for safety class I or II apparatus, so this value is frequently found in general-purpose filter units.

Both C_X and C_Y carry line-voltage voltages continuously and must be specifically rated to do this. Failure of C_X will result in a fire hazard, while failure of C_Y will result in both a fire hazard and a potential shock hazard. X and Y class components to BS 6201: Part 3 (IEC384–14) are marketed specifically for these positions.

Insertion Loss Versus Impedance Ready-made filters are universally specified between 50 Ω source and load impedances. The typical filter configuration outlined above is capable of 40–50 dB attenuation up to 30 MHz in both common and differential modes. Above 30 MHz stray component reactances limit the achievable loss and also make it more difficult to predict behavior. Below 1 MHz the attenuation falls off substantially as the effectiveness of the components reduces.

The 50 Ω termination does not reflect the real situation. The line-voltage port high-frequency impedance can be generalized for both common and differential mode by a 50 Ω M/50 μH network as provided by the common commercial test specifications; when the product is tested for compliance, this network will be used anyway. The equipment port impedance will vary substantially depending on load and on the high-frequency characteristics of the input components such as the line-voltage transformer, diodes, and reservoir. Differential-mode impedance is typically a few ohms for small electronic products, while common-mode impedance as discussed above can normally be approximated by a capacitive reactance of around 1000 pF. The effect of these load impedances differing from the nominal may be to enhance resonances within the filter and thus to achieve insertion gain at some frequencies.

Core Saturation Filters are specified for a maximum working rms current, which is determined by allowable heating in the common-mode choke. This means that the choke core will be designed to saturate above the sinusoidal peak current, about 1.5 times the rms current rating. Capacitor input power supplies have a distinctly non-sinusoidal input current waveform (see the discussion on line-voltage harmonics in Section 45.2.2.3), with a peak current of between 3 and 10 times the rms. If the input filter design does not take this into account, and the filter is not de-rated, then the core will saturate on the current peaks (Figure 45.78(a)) and drastically reduce the filter's effectiveness. Some filter manufacturers now take this into account and over-rate the inductor, but the allowable peak current is rarely specified on data sheets.

The core will also saturate when it is presented with a high-voltage, high-energy common mode surge, such as a switching transient on the line-voltage (Figure 45.78(b)). The surge voltage will be let through delayed and with a slower rise time but only slightly attenuated with attendant ringing on the trailing edge. Standard line-voltage filters designed only for attenuating frequency-domain emissions are inadequate to cope with large incoming common-mode transients, though some are better than others. Differential-mode transients require considerably more energy to saturate the core, and these are more satisfactorily suppressed.

Extended Performance In some cases the insertion loss offered by the typical configuration will not be adequate. This may be the case when, for example, a high-power switching supply must meet the most stringent emission limits, or there is excessive coupling of common-mode interference, or greater incoming transient immunity is needed. The basic filter design can be extended in a number of ways (Figure 45.79):

1. *Extra differential line chokes:* These are separate chokes in L and N lines which are not cross-coupled and therefore present a higher impedance to differential-mode signals, giving better attenuation in conjunction with C_X. Because they must not saturate at the full a.c. line current they are much larger and heavier for a given inductance.
2. *An earth line choke:* This increases the impedance to common-mode currents flowing in the safety earth and may be the only way of dealing with common-mode interference, both incoming and outgoing, when C_Y is already at its maximum limit and nothing can be done about the interference at source. Because it is in series with the safety earth its fault current carrying capability must satisfy safety standards. It is necessary to ensure that it is not short circuited by an extra earth connection to the equipment case.
3. *Transient suppressors:* A device such as a voltage dependent resistor (VDR) across L and N will clip incoming differential mode surges (see also Section 45.4.2.5). If it is placed at the line-voltage port then it must be rated for the full expected transient energy, but it will prevent the choke from saturating and protect the filter's C_X; if it is placed on the equipment side then it can be substantially downrated since it is protected by the impedance of the filter. Note that a VDR in these positions has no effect on common-mode transients.

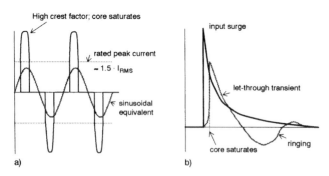

FIGURE 45.78 Core saturation effects. (a) High crest factor effect. (b) Incoming transient effect.

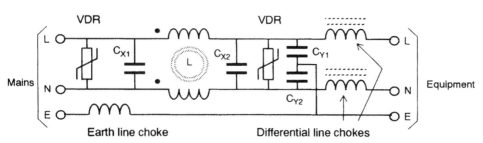

FIGURE 45.79 Higher-performance line-voltage filter.

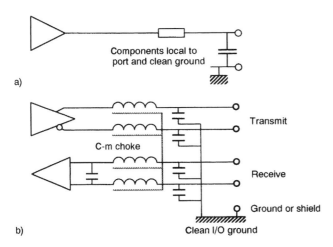

FIGURE 45.80 I/O filtering techniques. (a) Simple RC filter. (b) Common-mode choke/capacitor filter.

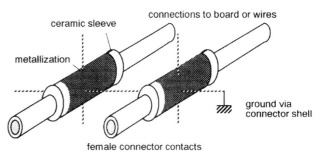

FIGURE 45.81 Filtered connector pins.

In addition to these extra techniques the basic filter π-section can be cascaded with further similar sections, perhaps with intersection screens and feedthroughs to obtain much higher insertion loss. For these levels of performance the filter must be used in conjunction with a well screened enclosure to prevent high-frequency coupling around it. Large values of C_X should be protected with a bleeder resistor in parallel, to prevent a hazardous charge remaining between L and N when the power is removed (detailed requirements can be found in safety specifications such as IEC 335/EN 60 335).

45.4.2.4 I/O Filtering

If I/O connections carry only low bandwidth signals and low current it is possible to filter them using simple RC low-pass networks (Figure 45.80). The decoupling capacitor must be connected to the clean I/O ground (see Section 45.3.1.7) which may not be the same as circuit 0 V.

This is not possible with high-speed data links, but it is possible to attenuate common-mode currents entering or leaving the equipment without affecting the signal frequencies by using a discrete common-mode choke arrangement. The choke has several windings on the same core such that differential currents appear to cancel each other whereas common-mode currents add, in the same fashion as the line-voltage common-mode choke described above. Such units are available commercially (sometimes described as "data line filters") or can be custom designed. Stray capacitance across each winding will degrade high-frequency attenuation.

Filtered Connectors A convenient way to incorporate both the capacitors and to a lesser extent the inductive components of Figure 45.80 is within the external connector itself. Each pin can be configured as a feedthrough capacitor with a ceramic jacket, its outside metallization connected to a matrix which is grounded directly to the connector shell (Figure 45.81). Thus the inductance of the ground connection is minimized, provided that the connector shell itself is correctly bonded to a clean ground, normally the metal backplate of the unit. Any series impedance in the ground path not only degrades the filtering but will also couple signals from one line into another, leading to designed-in cross-talk.

The advantage of this construction is that the insertion loss can extend to over 1 GHz, the low-frequency loss depending entirely on the actual capacitance (typically 50–2000 pF) inserted in parallel with each contact. With some ferrite incorporated as part of the construction, a π-filter can be formed as with the conventional feed-through (see above). No extra space for filtering needs to be provided. The filtered connector has obvious attractions for retrofit purposes, and may frequently solve interface problems at a stroke. You can also now obtain ferrite blocks tailored to the pin-out dimensions of common multiway connectors, which effectively offer individual choking for each line with a single component.

The disadvantage is the significant extra cost over an unfiltered connector; if not all contacts are filtered, or different contacts need different capacitor values, you will need a custom part. Voltage ratings may be barely adequate, and reliability may be worsened. Its insertion loss performance at low to medium frequencies can be approached with a small "piggy-back" board of chip capacitors mounted immediately next to the connector.

Circuit Effects of Filtering When using any form of capacitive filtering, the circuit must be able to handle the extra capacitance to ground, particularly when filtering an isolated circuit at radio frequencies. Apart from reducing the available circuit signal bandwidth, the radio frequency filter capacitance provides a ready-made a.c. path to ground for the signal circuit and will seriously degrade the a.c. isolation, to such an extent that a radio frequency filter may actually increase susceptibility to lower-frequency common-mode interference. This is a result of the capacitance imbalance between the isolated signal and return lines, and it may restrict the allowable radio-frequency filter capacitance to a few tens of picofarads.

Capacitive loading of low-frequency analog amplifier outputs may also push the output stage into instability (see above).

45.4.2.5 Transient Suppression

Incoming transients on either line-voltage or signal lines are reduced by non-linear devices: the most common are varistors (voltage dependent resistors (VDRs), Zeners, and spark gaps (gas discharge tubes). The device is placed in parallel with the line to be protected (Figure 45.82) and to normal signal or power levels it appears as a high impedance—essentially determined by its self-capacitance and leakage specifications. When a transient which exceeds its breakdown voltage appears, the device changes to a low impedance which diverts current from the transient source away from the protected circuit, limiting the transient voltage (Figure 45.83). It must be sized to withstand the continuous operating voltage of the circuit, with a safety margin, and to be able to absorb the energy from any expected transient.

The first requirement is fairly simple to design to, although it means that the transient clamping voltage is usually 1.5–2 times the continuous voltage, and circuits that are protected by the suppressor must be able to withstand this. The second requirement calls for a knowledge of the source impedance Z_{Strans} and probable amplitude of the transients, which is often difficult to predict accurately, especially for external connections. This determines the amount of energy which the suppressor will have to absorb. Cherniak (1982) gives details of how to determine the required suppressor characteristics from a knowledge of the circuit parameters, and also suggests design values for the energy requirement for suppressors on a.c. power supplies. These are summarized in Table 45.6. Table 45.7 compares the characteristics of the most common varieties of transient suppressor.

Combining Types You may sometimes have to parallel different types of suppressor in order to achieve a performance which could not be given by one type alone. The disadvantages of straightforward Zener suppressors, that their energy handling capability is limited because they must dissipate the full transient current at their breakdown voltage, are overcome by a family of related suppressors which integrate a thyristor with a Zener. When the over-voltage breaks down the Zener, the thyristor conducts and limits the applied voltage to a low value, so that the power dissipated is low, and a given package can handle about ten times the current of a Zener on its own. Provided that the operating circuit current is less than the thyristor holding current, the thyristor stops conducting once the transient has passed.

Layout of Transient Suppressors Short and direct connections to the suppressor (including the ground return path) are vital to avoid compromising the high-speed performance by undesired extra inductance. Transient edges have very fast rise times (a few nanoseconds for switching-induced interference down to subnanosecond for ESD), and any inductance in the clamping circuit will generate a high differential voltage during the transient plus ringing after it, which will defeat the purpose of the suppressor.

The component leads must be short (suppressors are available in SM chip form), and they must be connected locally to the circuit that is to be clamped (Figure 45.84). Any common impedance coupling, via ground or otherwise, must be avoided. When the expected transient source impedance is low (less than a few ohms), it is worthwhile raising the radiofrequency impedance of the input circuit with a lossy component such as a ferrite bead. Where suppressors are to be

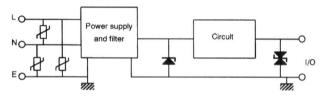

FIGURE 45.82 Typical locations for transient suppressors.

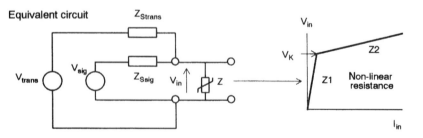

FIGURE 45.83 The operation of a transient suppressor.

TABLE 45.6 Suggested transient suppressor design parameters

Type of location	Waveform	Amplitude	Energy deposited (J) in a suppressor with clamping voltage of:	
			500 V (120 V system)	1000 V (240 V)
Long branch circuits and outlets	0.5 μs/100 kHz oscillatory	6 kV/200 A	0.8	1.6
Major feeders and short branch circuits	0.5 μs/100 kHz oscillatory	6 kV/500 A	2	4
	8/20 μs surge	6 kV/3 kA	40	80

TABLE 45.7 Comparison of transient suppressor types

Device	Leakage	Follow-on current	Clamp voltage	Energy capability	Capacitance	Response time	Cost
ZnO varistor	Moderate	No	Medium	High	High	Medium	Low
Zener	Low	No	Low to medium	Low	Low	Fast	Moderate
Spark gap GDT	Zero	Yes	High ignition, low clamp	High	Very low	Slow	Moderate to high

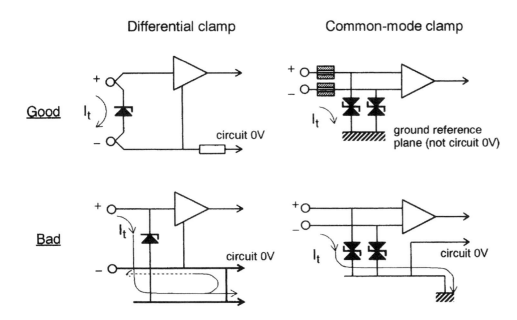

FIGURE 45.84 Layout and configuration of I/O transient suppressors.

combined with I/O filtering you may be able to use the three-terminal varistor/capacitor devices that are now available.

45.4.2.6 Contact Suppression

An opening contact which interrupts a flow of current (typically a switch or relay) will initiate an arc across the contact gap. The arc will continue until the available current is not enough to sustain a voltage across the gap (Figure 45.85). The stray capacitance and inductance associated with the contacts and their circuit will in practice cause a repetitive discharge until their energy is exhausted, and this is responsible for considerable broadband interference. A closure can also cause interference because of contact bounce.

Any spark-capable contact should be suppressed. The criteria for spark capability are a voltage across the contacts of greater than 320 V, and/or a circuit impedance which allows a dV/dt of greater than typically 1 V/μs—this latter criterion being met by many low-voltage circuits. The conventional suppression circuit is an RC network connected directly across the contacts. The capacitor is sized to limit the rate of rise of voltage across the gap to below that which initiates an arc. The resistor limits the capacitor discharge current on contact closure; its value is a compromise between

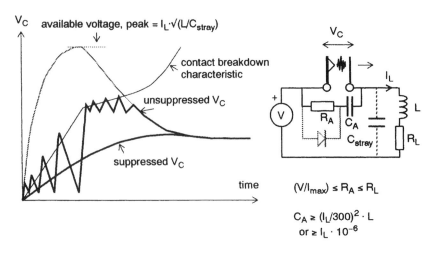

FIGURE 45.85 Contact noise generation and suppression.

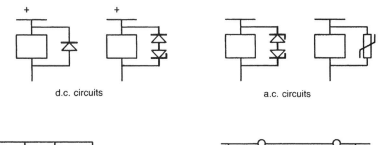

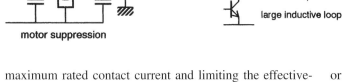

FIGURE 45.86 Inductive load suppression.

maximum rated contact current and limiting the effectiveness of the capacitor. A parallel diode can be added in d.c. circuits if this compromise cannot be met.

Suppression of Inductive Loads When current through an inductance is interrupted a large transient voltage is generated, governed by $V = -L \cdot di/dt$. Theoretically, if di/dt is infinite then the voltage is infinite, too; in practice it is limited by stray capacitance if no other measures are taken, and the voltage waveform is a damped sinusoid (if no breakdown occurs), the frequency of which is determined by the values of inductance and stray capacitance. Typical examples of switched inductive loads are motors, relay coils, and transformers, but even a long cable can have enough distributed inductance to generate a significant transient amplitude. Switching can either be via an electro mechanical contact or a semiconductor, and the latter can easily suffer avalanche breakdown due to the overvoltage if the transient is unsuppressed. Radio frequency interference is generated in both cases at frequencies determined by stray circuit resonances and is usually radiated from the wiring between switch and load.

The *RC* snubber circuit can be used in some cases to damp an inductive transient. Other circuits use diode, Zener, or varistor clamps as shown in Figure 45.86. Motor interference may appear in common mode with respect to the housing as a result of the high stray capacitance between the housing and the windings, hence you will often need to use common-mode decoupling capacitors. In all cases the suppression components must be mounted immediately next to the load terminals, otherwise a radiating current loop is formed by the intervening wiring. Protection of a driver transistor mounted remotely must be considered as a separate function from radio frequency suppression.

45.4.3 Shielding

Shielding and filtering are complementary practices. There is little point in applying good filtering and circuit design practice to guard against conducted coupling if there is no return path for the filtered currents to take. The shield provides such a return, and also guards against direct field coupling with the internal circuits and conductors. Shielding involves placing a conductive surface around the critical parts of the circuit so that the electromagnetic field which couples to it is attenuated by a combination of reflection and absorption. The shield can be an all-metal enclosure if protection down to low frequencies is needed, but if only high-frequency (>30MHz) protection will be enough, then a thin conductive coating deposited on plastic is adequate.

Will a Shield Be Necessary? Shielding is often an expensive and difficult-to-implement design decision, because many other factors (aesthetic, tooling, accessibility) work against it. A decision on whether or not to shield should be made as early as possible in the project. Section 45.4.2 showed that interference coupling is via interface cables and direct induction to/from the PCB. You should be able to calculate to a rough order of magnitude the fields generated by PCB tracks and compare these to the desired emission limit (see Section 45.3.2.2). If the limit is exceeded at this point and the PCB layout cannot be improved, then shielding is essential. Shielding does not of itself affect common-mode cable coupling, and so if this is expected to be the dominant coupling path, a full shield may not be necessary. It does establish a "clean" reference for decoupling common-mode currents to, but it is also possible to do this with a large area ground plate if the layout is planned carefully.

45.4.3.1 Shielding Theory

An a.c. electric field impinging on a conductive wall of infinite extent will induce a current flow in that surface of the wall, which in turn will generate a reflected wave of the opposite sense. This is necessary in order to satisfy the boundary conditions at the wall, where the electric field must approach zero. The reflected wave amplitude determines the reflection loss of the wall. Because shielding walls have finite conductivity, part of this current flow penetrates into the wall, and a fraction of it will appear on the opposite side of the wall, where it will generate its own field (Figure 45.87). The ratio of the impinging to the transmitted fields is one measure of the shielding effectiveness of the wall.

The thicker the wall, the greater the attenuation of the current through it. This absorption loss depends on the number of "skin depths" through the wall. The skin depth (defined by equation (45.17)) is an expression of the electromagnetic property which tends to confine a.c. current flow to the surface of a conductor, becoming less as frequency, conductivity, or permeability increases. Fields are attenuated by 8.6 dB ($1/e$) for each skin depth of penetration.

Typically, skin depth in aluminum at 30 MHz is 0.015 mm. The skin depth d (meters) is given by:

$$d = (\pi \cdot F \cdot \mu \cdot \sigma)^{-0.5} \qquad (45.17)$$

where F is frequency (MHz), μ is material permeability, and σ is material conductivity.

Shielding Effectiveness Shielding effectiveness of a solid conductive barrier describes the ratio between the field strength without the barrier in place to that when it is present. It can be expressed as the sum (in decibels) of reflection (R), absorption (A), and re-reflection (B) losses (Figure 45.88):

$$SE = R + A + B \qquad (45.18)$$

The *reflection loss* (R) depends on the ratio of wave impedance to barrier impedance. The concept of wave impedance has been described in Section 45.2.1.3. The impedance of the barrier is a function of its conductivity and permeability, and of frequency. Materials of high conductivity such as copper and aluminum have a higher E-field reflection loss than do lower-conductivity materials such as steel. Reflection losses decrease with increasing frequency for the E field (electric) and increase for the H field (magnetic). In the near field, closer than $\lambda/2\pi$, the distance between source and barrier also affects the reflection loss. Near to the source, the electric field impedance is high, and the reflection loss is also correspondingly high. Conversely, the magnetic field impedance is low, and the reflection loss is low. When the barrier is far enough away to be in the far field, the impinging wave is a plane wave, the wave impedance is constant,

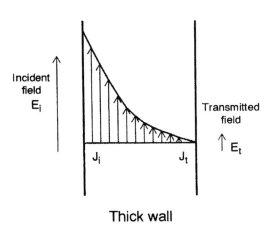
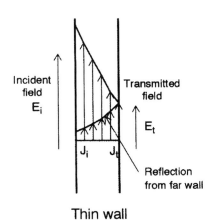

FIGURE 45.87 Variation of current density with wall thickness.

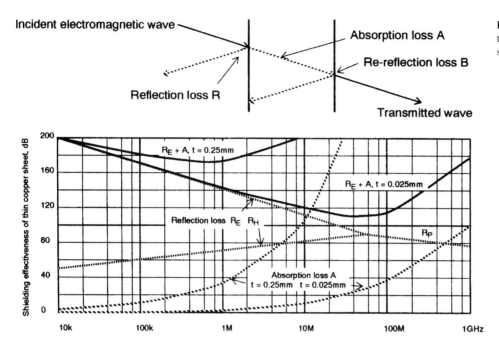

FIGURE 45.88 Shielding effectiveness versus frequency for copper sheet.

and the distance is immaterial. Refer back to Figure 45.7 for an illustration of the distinction between near and far fields.

The re-reflection loss B is insignificant in most cases where absorption loss A is greater than 10 dB, but becomes important for thin barriers at low frequencies.

Absorption Loss (A) depends on the barrier thickness and its skin depth and is the same whether the field is electric, magnetic, or plane wave. The skin depth, in turn, depends on the properties of the barrier material; in contrast to reflection loss, steel offers higher absorption than copper of the same thickness. At high frequencies, as Figure 45.88 shows, it becomes the dominant term, increasing exponentially with the square root of the frequency.

Shielding against Magnetic Fields at Low Frequencies is for all intents and purposes impossible with purely conductive materials. This is because the reflection loss to an impinging magnetic field (R_H) depends on the mismatch of the field impedance to barrier impedance. The low field impedance is well matched to the low barrier impedance, and the field is transmitted through the barrier without significant attenuation or absorption. A high-permeability material such as mumetal or its derivatives can give low-frequency magnetic shielding by concentrating the field within the bulk of the material, but this is a different mechanism from that discussed above, and it is normally only viable for sensitive individual components such as *CRT*s or transformers.

45.4.3.2 The Effect of Apertures

The curves of shielding effectiveness shown in Figure 45.88 suggest that upwards of 200 dB attenuation is easily achievable using reasonable thicknesses of common materials. In fact, the practical shielding effectiveness is not determined by material characteristics but is limited by necessary apertures and discontinuities in the shielding. Apertures are required for ventilation, for control and interface access, and for viewing indicators. For most practical purposes shielding effectiveness (SE) is determined by the apertures.

There are different theories for determining SE due to apertures. The simplest (Figure 45.89) assumes that SE is directly proportional to the ratio of longest aperture dimension and frequency, with zero SE when $\lambda = 2L$: $SE = 20\log(\lambda/2L)$. Thus the SE increases linearly with decreasing frequency up to the maximum determined by the barrier material, with a greater degradation for larger apertures. A correction factor can be applied for the aspect ratio of slot-shaped apertures. Another theory, which assumes that small apertures radiate as a combination of electric and magnetic dipoles, predicts a constant SE degradation with frequency in the near field and a degradation proportional to F^2 in the far field. This theory predicts a SE dependence on the *cube* of the aperture dimension, and also on the distance of the measuring point from the aperture. Neither theory accords really well with observations. However, as an example based on the simpler theory, for frequencies up to 1 GHz (the present upper limit for radiated emissions standards) and a minimum shielding of 20 dB, the maximum hole size allowable is 1.6 cm.

Windows and Ventilation Slots Viewing windows normally involve a large open area in the shield, and you have to cover the window with a transparent conductive material, which must make good continuous contact with the surrounding screen, or accept the penalty of shielding at lower frequencies only. You can obtain shielded window components which are laminated with fine blackened copper mesh, or which are coated with an extremely thin film of gold. In either case, there is a trade-off in viewing quality over a clear window, due to reduced light transmission (between 60 percent and 80 percent) and diffraction effects of the mesh. Screening effectiveness of a transparent conductive coating is significantly less than a solid shield, since the coating will have a resistance of a few ohms per square, and attenuation will be entirely due to reflection loss. This is not the case with a mesh, but shielding effectiveness of better than 40–50 dB may be irrelevant anyway because of the effect of other apertures.

Using a Subenclosure An alternative method which allows a clear window to be retained is to shield behind the display with a subshield (Figure 45.90), which must, of course, make good all-around contact with the main panel.

The electrical connections to the display must be filtered to preserve the shield's integrity, and the display itself is unshielded and must therefore not be susceptible nor contain emitting sources. This alternative is frequently easier and cheaper than shielded windows.

Mesh and Honeycomb Ventilation holes can be covered with a perforated mesh screen, or the conductive panel may itself be perforated. If individual equally sized perforations are spaced close together (hole spacing $<\lambda/2$) then the reduction in shielding over a single hole is approximately proportional to the square root of the number of holes. Thus a mesh of 100 4-mm holes would have a shielding effectiveness 20 dB worse than a single 4-mm hole. Two similar apertures spaced greater than a half-wavelength apart do not suffer any significant extra shielding reduction.

You can if necessary gain improved shielding of vents, at the expense of thickness and weight, by using "honeycomb" panels in which the honeycomb pattern functions as a waveguide below cut-off (Figure 45.91). In this technique the shield thickness is several times that of the width of each individual aperture. A common thickness/width ratio is 4:1, which offers an intrinsic shielding effectiveness of over 100 dB. This method can also be used to conduct insulated control spindles (*not* conductive ones!) through a panel.

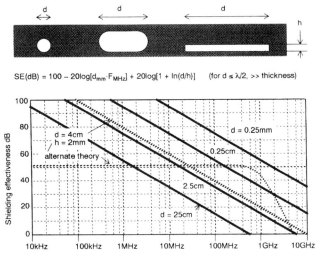

FIGURE 45.89 Shielding effectiveness degradation due to apertures.

The Effect of Seams An electromagnetic shield is normally made from several panels joined together at seams. Unfortunately, when you join two sheets the electrical conductivity across the joint is imperfect. This may be because of distortion, so that surfaces do not mate perfectly, or because of painting, anodizing, or corrosion, so that an insulating layer is present on one or both metal surfaces.

Consequently, the shielding effectiveness is reduced by seams almost as much as it is by apertures (Figure 45.92). The ratio of the fastener spacing d to the seam gap h is high enough to improve the shielding over that of a large aperture

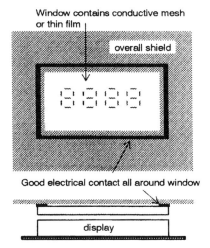
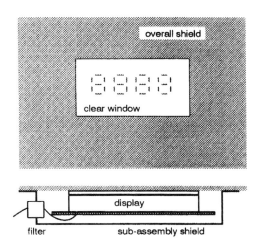

FIGURE 45.90 Alternative ways of shielding a display window.

by 10–20 dB. The problem is especially serious for hinged front panels, doors, and removable hatches that form part of a screened enclosure. It is mitigated to some extent if the conductive sheets overlap, since this forms a capacitor which provides a partial current path at high frequencies. Figure 45.93 shows preferred ways to improve joint conductivity. If these are not available to you then you will need to use extra hardware as outlined in the next section.

Seam and Aperture Orientation The effect of a joint discontinuity is to force shield current to flow around the discontinuity. If the current flowing in the shield were undisturbed then the field within the shielded area would be minimized, but as the current is diverted, so a localized discontinuity occurs, and this creates a field coupling path through the shield. The shielding effectiveness graph shown in Figure 45.89 assumes a worst-case orientation of current flow. A long aperture or narrow seam will have a greater effect on current flowing at right angles to it than on parallel current flow. (Antenna designers will recognize that this describes a slot antenna, the reciprocal of a dipole.) This effect can be exploited if you can control the orientation of susceptible or emissive conductors within the shielded environment (Figure 45.94).

The practical implication of this is that if all critical internal conductors are within the same plane, such as on a PCB, then long apertures and seams in the shield should be aligned parallel to this plane rather than perpendicular to it. Generally, you are likely to obtain an advantage of no more than 10 dB by this trick, since the geometry of internal conductors is never exactly planar. Similarly, cables or wires, if they must be routed near to the shield, should be run along or parallel to apertures rather than across them. But because the leakage field coupling due to joint discontinuities is large near the discontinuity, internal cables should preferably not be routed near apertures or seams. Apertures on different surfaces or seams at different orientations can be treated separately since they radiate in different directions.

45.4.3.3 Shielding Hardware

Many manufacturers offer various materials for improving the conductivity of joints in conductive panels. Such materials can be useful if properly applied, but they must be used with an awareness of the principles discussed above, and their expense will often rule them out for cost-sensitive applications except as a last resort.

Gaskets and Contact Strip Shielding effectiveness can be improved by reducing the spacing of fasteners between different panels. If you need effectiveness to the upper limit of 1 GHz, then the necessary spacing becomes unrealistically small when you consider maintenance and accessibility.

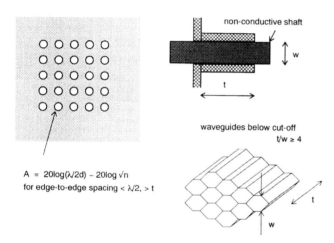

FIGURE 45.91 Mesh panels and the waveguide below cut-off.

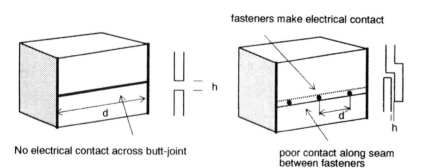

FIGURE 45.92 Seams between enclosure panels. The size of d determines the shielding effectiveness, modified by the seam gap h.

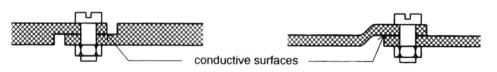

FIGURE 45.93 Cross-sections of joints for good conductivity.

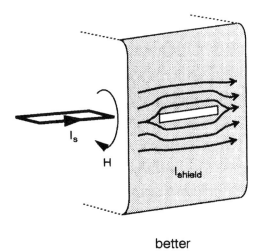
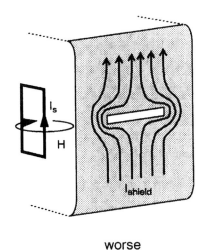

FIGURE 45.94 Current loop versus aperture orientation.

In these cases the conductive path between two panels or flanges can be improved by using any of the several brands of conductive gasket, knitted wire mesh, or finger strip that are available. The purpose of these components is to be sandwiched in between the mating surfaces to ensure continuous contact across the joint, so that shield current is not diverted (Figure 45.95). Their effectiveness depends entirely on how well they can match the impedance of the joint to that of the bulk shield material.

A number of factors should be borne in mind when selecting a conductive gasket or finger material:

1. *Conductivity:* This should be of the same order of magnitude as the panel material.
2. *Ease of mounting:* Gaskets should normally be mounted in channels machined or cast in the housing, and the correct dimensioning of these channels is important to maintain an adequate contact pressure without overtightening. Finger strip can be mounted by adhesive tape, welding, riveting, soldering, fasteners, or by just clipping in place. The right method depends on the direction of contact pressure.
3. *Galvanic compatibility with the host:* To reduce corrosion, the gasket metal and its housing should be close together and preferably of the same group within the electrochemical series (Table 45.8). The housing material should be conductively finished: alochrome or alodine for aluminum, nickel or tin plate for steel.
4. *Environmental performance:* Conductive elastomers will offer combined electrical and environmental protection, but may be affected by moisture, fungus, weathering, or heat. If you choose to use separate environmental and conductive gaskets, the environmental seal should be placed *outside* the conductive gasket and mounting holes.

Conductive Coatings Many electronic products are enclosed in plastic cases for aesthetic or cost reasons. These can be made to provide a degree of electromagnetic shielding by covering one or both sides with a conductive coating (Rankin, 1986). Normally this involves both a molding supplier and a coating supplier. Conductively filled plastic composites can also be used to obtain a marginal degree of shielding (around 20 dB); it is debatable whether the extra material cost justifies such an approach, considering that better shielding performance can be offered by conductive coating at lower overall cost (Bush, 1989). Conductive fillers affect the mechanical and aesthetic properties of the plastic, but their major advantage is that no further treatment of the molded part is needed. Another problem is that the molding process may leave a "resin-rich" surface which is not conductive, so that the conductivity across seams and joints is not ensured. As a further alternative, metallized fabrics are now becoming available which can be incorporated into some designs of compression molding.

Shielding Performance The same dimensional considerations apply to apertures and seams as for metal shields. Thin coatings will be almost as effective against electric fields at high frequencies as solid metal cases, but are ineffective against magnetic fields. The major shielding mechanism is reflection loss (Figure 45.88) since absorption is negligible except at very high frequencies, and re-reflection (B) will tend to reduce the overall reflection losses. The higher the resistivity of the coating, the less its efficiency. For this reason conductive paints, which have a resistivity of around 1 Ω/square, are poorer shields than the various types of metallization (see Table 45.9) which offer resistivities below 0.1 Ω/square.

Enclosure Design Resistivity will depend on the thickness of the coating, which in turn is affected by factors such as the shape and sharpness of the molding—coatings will adhere less to, and abrade more easily from, sharp edges and corners than rounded ones. Ribs, dividing walls, and other mold features that exist inside most enclosures make application of sprayed-on coatings (such as conductive paint or zinc arc spray) harder and favor the electroless plating methods. Where coatings must cover such features, the molding

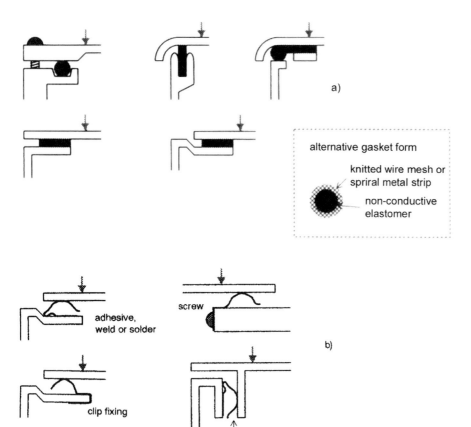

FIGURE 45.95 Usage of gaskets and finger strip. (a) Conductive elastomer gaskets. (b) Beryllium copper finger strip.

TABLE 45.8 The electrochemical series: corrosion occurs when ions move from the more anodic metal to the more cathodic, facilitated by an electrolytic transport medium such as moisture or salts

	Anodic—most easily corroded →				
	Group I	Group II	Group III	Group IV	Group V
Magnesium	Aluminum + alloys				
	Zinc	Carbon steel			
	Chromium	Iron	Nickel		
	Galvanized iron	Cadmium	Tin, solder	Copper + alloys	
			Lead	Silver	
			Brass	Palladium	
			Stainless steel	Platinum	
				Gold	
				→ Cathodic least easily corroded	

design should include generous radii, no sharp corners, adequate space between ribs, and no deep or narrow crevices.

Coating Properties Environmental factors, particularly abrasion resistance and adhesion, are critical in the selection of the correct coating. Major quality considerations are:

1. Will the coating peel or flake off into the electrical circuitry?
2. Will the shielding effectiveness be consistent from part to part?
3. Will the coating maintain its shielding effectiveness over the life of the product?

Adhesion is a function of thermal or mechanical stresses, and is checked by a non-destructive tape or destructive cross-hatch test. Typically, the removal of any coating as an immediate result of tape application constitutes a test failure. During and at completion of thermal cycling or humidity testing, small flakes at the lattice edges after a cross-hatch test should not exceed 15 percent of the total coating removal.

Electrical properties should not change after repeated temperature/humidity cycling within the parameters agreed with the molding and coating supplies. Resistance measurements should be taken from the farthest distances of the test piece and also on surfaces critical to the shielding performance, especially mating and grounding areas.

Table 45.9 compares the features of the more commonly available conductive coatings (others are possible but are more expensive and little used). These will give shielding effectiveness in the range of 30–70 dB if properly applied. It is difficult to compare shielding effectiveness figures given by different manufacturers unless the methods used to perform their shielding effectiveness tests are very clearly specified; different methods do not give comparable results. Also, laboratory test methods do not necessarily correlate with the performance of a practical enclosure for a commercial product.

45.5 THE REGULATORY FRAMEWORK

45.5.1 Customer Requirements

Historically, the EMC aspects of instrumentation have largely been driven by the customer's need for a reliable instrument. An instrument must function correctly in whatever environment it is supplied for, and one aspect of that environment is electromagnetic interference. Therefore, a technically alert customer will specify that correct operation must be maintained under given conditions of interference.

Several large organizations have over the past decades developed their own standard requirements, sometimes based on published International Standards and sometimes not. These requirements are enforced contractually. Internally developed requirements are usually a response to known or anticipated problems that are special to the environment in which the equipment will operate, for example, the railway track-side environment is particularly aggressive with respect to transient interference, and instrumentation for railways must meet specifications laid down by the Railway Industries Association. Wherever possible, use is made of test methods developed and published internationally, since these assure the widest availability and greatest cost-effectiveness of the test program.

Until recently, customer requirements in the instrumentation sector have been almost invariably for immunity only; no control has been placed on radio frequency or supply harmonic emissions. The concern has been mainly that instrument accuracy and operation remains unaffected by interference coupled in from an inevitably noisy environment. Because the environment is so aggressive, the instrument's own emissions are insignificant. Historical exceptions to this have generally been in the military, aerospace, automotive, and naval sectors, where equipment has had to share platforms with radio receivers, and the proximity of the two has called for tight control over radio frequency emissions. With the increasing penetration of mobile communications into all areas, this need has become more widespread. Nevertheless, control of emissions is generally left to legislation.

Until the arrival of the EMC Directive, in most cases the specification for immunity of instruments has been left to the customer. A particular exception to this has been in instruments for legal metrology, in which field the immunity requirements are now covered by the Directive on non-automatic weighing instruments (90/384/EEC). The EMC Directive now sets minimum requirements for immunity (via the generic immunity standard) which apply to all equipment, including instrumentation. However, these requirements are not particularly onerous, and the customer is still at liberty to impose his or her own specifications over and above the legal minimum.

45.5.2 The EMC Directive

Of the various aims of the creation of the Single Market, the free movement of goods between European states is

TABLE 45.9 Comparison of conductive coating techniques

	Cost (£/m^2)	E-field shielding	Thickness	Adhesion	Scratch resistance	Maskable	Comments
Conductive paint (nickel, copper)	5–5	Poor/average	0.05 mm	Poor	Poor	Yes	Suitable for prototyping
Zinc arc spray (zinc)	5–10	Average/good	0.1–0.15 mm	Depends on surface preparation	Good	Yes	Rough surface, inconsistent
Electroless plate (copper, nickel)	10–15	Average/good	1–2 μm	Good	Poor	No	Cheapter if entire part plated
Vacuum metallization (aluminum)	10–15	Average	2–5 μ	Depends on surface preparation	Poor	Yes	Poor environmental qualities

fundamental. All member states impose standards and obligations on the manufacture of goods in the interests of quality, safety, consumer protection, and so forth. Because of detailed differences in procedures and requirements, these act as technical barriers to trade, fragmenting the European market and increasing costs because manufacturers have to modify their products for different national markets.

For many years the EC tried to remove these barriers by proposing directives which gave the detailed requirements that products had to satisfy before they could be freely marketed throughout the Community, but this proved difficult because of the detailed nature of each directive and the need for unanimity before it could be adopted. In 1985 the Council of Ministers adopted a resolution setting out a "New Approach to Technical Harmonization and Standards."

Under the "new approach," directives are limited to setting out the essential requirements which must be satisfied before products may be marketed anywhere within the EC. The technical detail is provided by standards drawn up by the European standards bodies CEN, CENELEC, and ETSI. Compliance with these standards will demonstrate compliance with the essential requirements of each directive. All products covered by each directive must meet its essential requirements, but all products which do comply, and are labeled as such, may be circulated freely within the Community; no member state can refuse them entry on technical grounds. Decisions on new approach directives are taken by qualified majority voting, eliminating the need for unanimity and speeding up the process of adoption.

The EMC Directive is possibly the most significant and wide-ranging of the new approach directives.

45.5.2.1 Background to the Legislation

In the U.K., previous legislation on EMC has been limited in scope to radio communications. Section 10 of the *Wireless Telegraphy Act 1949* enables regulations to be made for the purpose of controlling both radio and non-radio equipment which might interfere with radio communications. These regulations have taken the form of various statutory instruments (SIs) which cover interference emissions from spark ignition systems, electromedical apparatus, radio frequency heating, household appliances, fluorescent lights, and CB radio. The SIs invoke British Standards which are closely aligned with international and European standards.

This previous legislation is not comparable in scope to the EMC Directive, which covers far more than just interference to radio equipment, and extends to include immunity as well as emissions.

45.5.2.2 Scope and Requirements

The EMC Directive, 89/336/EEC, was adopted in 1989 to come into force on 1 January 1992. A subsequent amending directive, 92/31/EEC, extended the original transitional period to 31 December 1995. The EMC Directive applies to apparatus which is liable to cause electromagnetic disturbance or which is itself liable to be affected by such disturbance. "Apparatus" is defined as all electrical and electronic appliances, equipment, and installations. Essentially, anything which is powered by electricity is covered, regardless of whether the power source is the public supply line-voltage, a battery source, or a specialized supply.

An electromagnetic disturbance is any electromagnetic phenomenon which may degrade performance, without regard to frequency or method of coupling. Thus radiated emissions as well as those conducted along cables, and immunity from electromagnetic fields, line-voltage disturbances, conducted transients and radio frequency, electrostatic discharge, and lightning surges are all covered. *No* specific phenomena are *excluded* from the Directive's scope.

Essential Requirements The essential requirements of the Directive (Article 4) are that the apparatus shall be so constructed that:

1. The electromagnetic disturbance it generates does not exceed a level allowing radio and telecommunications equipment and other apparatus to operate as intended.
2. The apparatus has an adequate level of intrinsic immunity to electromagnetic disturbance to enable it to operate as intended.

The intention is to protect the operation not only of radio and telecommunications equipment, but also of any equipment which might be susceptible to electromagnetic disturbances, such as information technology or control equipment. At the same time, all equipment must be able to function correctly in whatever environment it might reasonably be expected to occupy.

Sale and Use of Products The Directive applies to all apparatus that is placed on the market or taken into service. The definitions of these two conditions do not appear within the text of the Directive but have been the subject of an interpretative document issued by the Commission (DTI 1991).

The "market" means the market in any or all of the EC member states; products which are found to comply within one state are automatically deemed to comply within all others. "Placing on the market" means the *first* making available of the product within the EC, so that the Directive covers not only new products manufactured within the EC but also both new and used products imported from a third country. Products sold secondhand within the EC are outside its scope. Where a product passes through a chain of distribution before reaching the final user, it is the passing of the product from the manufacturer into the distribution chain which constitutes placing on the market. If the product is manufactured in or imported into the EC for subsequent export to a third country, it has not been placed on the market.

The Directive applies to each individual item of a product type, regardless of when it was designed, and whether it is a one-off or a high-volume product. Thus items from a product line that was launched at any time before 1996 must comply with the provisions of the Directive after 1 January 1996. Put another way, there is no "grandfather clause" which exempts designs that were current before the Directive took effect. However, products already *in use* before 1 January 1996 do not have to comply retrospectively.

"Taking into service" means the first *use* of a product in the EC by its final user. If the product is used without being placed on the market, if, for example, the manufacturer is also the end user, then the protection requirements of the Directive still apply. This means that sanctions are still available in each member state to prevent the product from being used if it does not comply with the essential requirements or if it causes an actual or potential interference problem. On the other hand, it should not need to go through the conformity assessment procedures to demonstrate compliance (Article 10, which describes these procedures, makes no mention of taking into service). Thus an item of special test gear built up by a laboratory technician for use within the company's design department must still be designed to conform to EMC standards, but should not need to follow the procedure for applying the CE mark.

If the manufacturer resides outside the EC, then the responsibility for certifying compliance with the Directive rests with the person placing the product on the market for the first time within the EC, i.e., the manufacturer's authorized representative or the importer. Any person who produces a new finished product from already existing finished products, such as a system builder, is considered to be the manufacturer of the new finished product.

45.5.2.3 Systems, Installations, and Components

An area of concern to system builders is how the Directive applies to two or more separate pieces of apparatus sold together or installed and operating together. It is clear that the Directive applies in principle to systems and installations. The Commission's interpretative document (DTI 1991) defines a *system* as several items of apparatus combined to fulfill a specific objective and intended to be placed on the market as a single functional unit. An *installation* is several combined items of apparatus or systems put together at a given place to fulfill a specific objective but not intended to be placed on the market as a single functional unit. Therefore a typical system would be a personal computer workstation comprising the PC, monitor, keyboard, printer, and any other peripherals. If the units were to be sold separately they would have to be tested and certified separately; if they were to be sold as a single package, then they would have to be tested and certified as a package.

Any other combination of items of apparatus, not initially intended to be placed on the market together, is considered to be not a system but an installation. Examples of this would appear to be computer suites, telephone exchanges, electricity substations or television studios. Each item of apparatus in the installation is subject to the provisions of the Directive individually, under the specified installation conditions.

As far as it goes, this interpretation is useful, in that it allows testing and certification of installations to proceed on the basis that each component of the installation will meet the requirements on its own. The difficulty of testing large installations *in situ* against standards that were never designed for them is largely avoided. Also, if an installation uses large numbers of similar or identical components, then only one of these needs to be actually tested.

Large Systems The definition unfortunately does not help system builders who will be "placing on the market" (i.e., supplying to their customer on contract) a single installation, made up of separate items of apparatus but actually sold as one functional unit. Many industrial, commercial, and public utility contracts fall into this category. According to the published interpretation, the overall installation would be regarded as a system and therefore should comply as a package. As it stands at present, there are no standards which specifically cover large systems, i.e., ones for which testing on a test site is impractical, although some emissions standards do allow measurements *in situ*. Neither are there any provisions for large systems in the immunity standards. Therefore the only compliance route available to system builders is the technical construction file (see Section 45.5.2.5), but guidance as to how to interpret the Directive's essential requirements in these cases is lacking. The principal dilemma of applying the Directive to complete installations is that to make legally relevant tests is difficult, but the nature of EMC phenomena is such that to test only the constituent parts without reference to their interconnection is largely meaningless.

Components The question of when a "component" (which is not within the scope of the Directive) becomes "apparatus" (which is) remains problematical. The Commission's interpretative document defines a component to be "any item which is used in the composition of an apparatus and which is not itself an apparatus with an intrinsic function intended for the final consumer." Thus individual small parts such as ICs and resistors are definitely outside the Directive. A component may be more complex provided that it does not have an intrinsic function and its only purpose is to be incorporated inside an apparatus, but the manufacturer of such a component must indicate to the equipment manufacturer how to use and incorporate it. The distinction is important for manufacturers of board-level products and other subassemblies which may appear to have an intrinsic function and are marketed separately, yet cannot be used separately from the apparatus in which they will be installed. The Commission has indicated that it regards subassemblies which are to be tested as part of a larger apparatus as outside the Directive's scope. On the other hand, subassemblies such as plug-in cards which are supplied by a third party to be inserted by the user, *should* be tested and certified separately.

45.5.2.4 The CE Mark and the Declaration of Conformity

The manufacturer or his authorized representative is required to attest that the protection requirements of the Directive have been met. This requires two things:

1. That he issues a declaration of conformity which must be kept available to the enforcement authority for 10 years following the placing of the apparatus on the market.
2. That he affixes the CE mark to the apparatus, or to its packaging, instructions, or guarantee certificate.

A separate Directive concerning the affixing and use of the CE mark has been published (EEC 1993). The mark consists of the letters CE as shown in Figure 45.96. The mark should be at least 5 mm in height and be affixed "visibly, legibly and indelibly," but its method of fixture is not otherwise specified. Affixing this mark indicates conformity not only with the EMC Directive but also with the requirements of any other new approach Directives relevant to the product—for instance, electrical machinery with the CE mark indicates compliance both with the Machinery Directive and the EMC Directive.

The EC declaration of conformity is required whether the manufacturer self-certifies to harmonized standards or follows the technical file route (Section 45.5.2.5). It must include the following components:

1. A description of the apparatus to which it refers.
2. A reference to the specifications under which conformity is declared, and where appropriate, to the national measures implemented to ensure conformity.
3. An identification of the signatory empowered to bind the manufacturer or his authorized representative.
4. Where appropriate, reference to the EC type examination certificate (for radiocommunications apparatus only).

45.5.2.5 Compliance with the Directive

Self-Certification The route which is expected to be followed by most manufacturers is self-certification to harmonized standards (Section 45.5.3). Harmonized standards are those CENE-LEC or ETSI standards which have been announced in the *Official Journal of the European Communities* (OJEC). In the U.K. these are published as dual-numbered BS and EN standards.

The potential advantage of certifying against standards from the manufacturer's point of view is that there is no mandatory requirement for testing by an independent test house. The only requirement is that the manufacturer makes a declaration of conformity (see Section 45.5.2.4) which references the standards against which compliance is claimed. Of course, the manufacturer will normally need to test the product to assure himself that it actually does meet the requirements of the standards, but this could be done in-house. Many firms will not have sufficient expertise or facilities in-house to do this testing, and will therefore have no choice but to take the product to an independent test house. But the long-term aim ought to be to integrate the EMC design and test expertise within the rest of the development or quality department, and to decide which standards apply to the product range, so that the prospect of self-certification for EMC is no more daunting than the responsibility of functionally testing a product before shipping it.

The Technical Construction File The second route available to achieve compliance is for the manufacturer or importer to generate a technical construction file (TCF). This is to be held at the disposal of the relevant competent authorities as soon as the apparatus is placed on the market and for ten years after the last item has been supplied. The Directive specifies that the TCF should describe the apparatus, set out the procedures used to ensure conformity with the protection requirements, and should contain a technical report or certificate obtained from a competent body. This last requirement means that, in contrast to the self-certification route, the involvement of a third party (who must have been appointed by the national authorities) is mandatory.

The purpose of the TCF route is to allow compliance with the essential requirements of the Directive to be demonstrated when harmonized or agreed national standards do not exist, or exist only in part, or if the manufacturer chooses not to apply existing standards for his own reasons. Since the generic standards are intended to cover the first two of these cases, the likely usage of this route will be under the following circumstances:

1. When existing standards cannot be applied because of the nature of the apparatus or because it incorporates advanced technology which is beyond their breadth of concept.
2. When testing would be impractical because of the size or extent of the apparatus, or because of the existence of many fundamentally similar installations.
3. When the apparatus is so simple that it is clear that no testing is necessary.

FIGURE 45.96 The CE mark.

4. When the apparatus has already been tested to standards that have not been harmonized or agreed but which are nevertheless believed to meet the essential requirements.

45.5.3 Standards Relating to the EMC Directive

45.5.3.1 Product Specific Standards

Where they are available, the preferred method of complying with the EMC Directive is to comply with the requirements of harmonized product or product-family EMC standards. These are drawn up by CENELEC, and their reference numbers are published in the *Official Journal of the European Communities*. There are no such standards for instrumentation in general at the time of writing. Several standards for specific classes of instrumentation are in preparation.

Meanwhile, it may be possible to apply one or more of the already published harmonized standards. Table 45.10 lists those which may be applicable within the instrumentation sector. The decision to use any of these standards should be based on a careful reading of its scope.

45.5.3.2 The Generic Standards

There are many industry sectors for which no product-specific standards have been developed. This is especially so for immunity. In order to fill this gap wherever possible, CENELEC has given a high priority to developing the generic standards. These are standards with a wide application, not related to any particular product or product family, and are intended to represent the essential requirements of the Directive. They are divided into two standards, one for immunity and one for emissions, each of which has separate parts for different environment classes (Table 45.10).

Where a relevant product-specific standard does exist, this takes precedence over the generic standard. It will be common, though, for a particular product to be covered by one product standard for line-voltage harmonic emissions, another for radio frequency emissions and the generic standard for immunity. All these standards must be satisfied before compliance with the Directive can be claimed. Other mixed combinations will occur until a comprehensive range

TABLE 45.10 Relevant standards for instrumentation

Reference	Scope	Requirements
Emissions		
EN50081	Generic Part 1: residential, commercial, and light industry	Tests as EN55022 Class B, EN55014, EN60555-2,3 when applicable
	Generic Part 2: industrial	Tests as EN55011
EN55011	Industrial, scientific, and medical apparatus	Conducted r.f. 150 kHz–30 MHz, radiated r.f. 30–1000 MHz
EN55014	Motor operated and thermal appliances for household and similar purposes	Conducted r.f. 150 kHz–30 MHz, disturbance power 30–300 MHz, discontinuous disturbances
EN55022	Information technology equipment	Conducted r.f. 150 kHz–30 MHz, radiated r.f. 30–1000 MHz
EN60555	Supply system disturbances caused by household appliances and similar equipment	Part 2: mains harmonic currents Part 3: flicker
Immunity		
EN50082	Generic Part 1: residential, commercial, and light industry	Tests as IEC801 Parts 2, 3, and 4 where applicable
	Generic Part 2: industrial	Tests as IEC801 Parts 2, 3, and 4 where applicable
EN55024	Information technology equipment	Part 2: electrostatic discharge (draft) Part 3: radiated r.f. (draft) Other parts in preparation
IEC801-X (IEC1000–4-X)	Originally intended for industrial process measurement and control equipment.	EN60801-2: Part 2: electrostatic discharge ENV50140: Part 3: radiated r.f.
	Now denoted a "basic" standard	Part 4: electrical fast transients
		Part 5: surge (draft) ENV50141
		Part 6: conducted f.f.
IEC-1000-4-X	EMC testing and measurement techniques (not relevant for EMC Directive)	Part 8: power frequency magnetic field
		Part 9: pulse magnetic field
		Part 10: damped oscillatory magnetic field

of product standards has been developed—a process which will take several years.

Environmental Classes The distinction between environmental classes is based on the electromagnetic conditions that obtain in general throughout the specified environments. The inclusion of the "light industrial" environments (workshops, laboratories, and service centers) in Class 1 has been the subject of some controversy, but studies have shown that there is no significant difference between the electromagnetic conditions at residential, commercial, and light industrial locations. Equipment for the Class 2 "industrial" environment is considered to be connected to a dedicated transformer or special power source, in contrast to the Class 1 environment which is considered to be supplied from the public line-voltage network.

Referenced Standards The tests defined in the generic standards are based only on internationally approved, already existing standards. For each electromagnetic phenomenon a test procedure given by such a standard is referenced, and a single test level or limit is laid down. No new tests are defined in the body of any generic standard. Since the referenced standards are undergoing revision to incorporate new tests, these are noted in an "informative annex" in each generic standard. The purpose of this is to warn users of those requirements that will become mandatory in the future, when a new standard or the revision to the referenced standard is agreed and published.

REFERENCES

Barlow, S. J., *Improving the Line-voltage Transient Immunity of Microprocessor Equipment*, Cambridge Consultants, IEE Colloquium, Interference and Design for EMC in Microprocessor Based Systems, London (1990).

Bond, A. E. J., "Implementation of the EMC Directive in the U.K.," *DTI, IEE 8th International Conference on EMC*, 21–24 September 1992, Edinburgh, *IEE Conference Publication No. 362*.

BS 6656 *Guide to prevention of inadvertent ignition of flammable atmospheres by radio frequency radiation.*

BS 6657 *Guide for prevention of inadvertent initiation of electro-explosive devices by radio frequency radiation.*

BS 6839: Part 1 *Line-voltage signalling equipment: specification for communication and interference limits and measurements* (EN 50065: Part 1).

Bush, D. R., "A simple way of evaluating the shielding effectiveness of small enclosures," *8th Symposium on EMC*, Zurich, 5–7 March 1989.

Catherwood, M., "Designing for EMC with HCMOS microcontrollers," *Motorola Application Note AN1050* (1989).

Cathey, W. T., and R. M. Keith, "Coupling reduction in twisted wire," *International Symposium on EMC*, IEEE, Boulder, CO, August 1981.

Chernisk, S., "A review of transients and their means of suppression," *Motorola Application Note AN-843* (1982).

CISP R23 *Determination of limits for industrial, scientific and medical equipment.*

Cowdell, R. B., "Unscrambling the mysteries about twisted wire," *International Symposium on EMC*, IEEE, San Diego, 9–11 October 1979.

Crane, L. F., and Srebranig, S. F., "Common mode filter inductor analysis," *Coilcraft Data Bulletin* (1985).

DTI, *Electromagnetic Compatibility: European Commission Explanatory Document on Council Directive 89/336 EEC*, Department of Trade & Industry (November 1991).

EEC, "Council Directive of 3rd May 1989 on the approximation of the laws of the Member States relating to Electromagnetic Compatibility (89/336/EEC)," *Off. J. Eur. Commun.*, **L 139** (23 May 1989) (Amended by Directives 92/31/EEC and 93/68/EEC).

EEC, "Council Directive 92/31/EEC (Transitional period)," *Off. J. Eur. Commun.*, **L 126** (12 May 1992).

EEC, "Council Directive 93/68/EEC (the CE marking Directive)," *Off. J. Eur. Commun.*, **L 220** (30 August 1993).

EN 55011 *Limits and methods of measurement of radio disturbance characteristics of industrial, scientific and medical (ISM) radio-frequency equipment.*

EN 55022 *Limits and methods of measurement of radio interference characteristics of information technology equipment.*

EN 60555 *Disturbances in supply systems caused by household appliances and similar electrical equipment.*

EN 50081 *Electromagnetic compatibility—generic emission standard.*

EN 50082 *Electromagnetic compatibility—generic immunity standard.*

pr EN 55024 *Immunity requirements for information technology equipment*

German, R. F., "Use of a ground grid to reduce printed circuit board radiation," *6th Symposium on EMC*, Zurich, 5–7 March 1985.

Gilbert, M. J., "Design innovations address advanced CMOS logic noise considerations," *National Semiconductor Application Note AN-690* (1990).

Goedbloed, J. J., "Transients in low-voltage supply networks," *IEEE Trans. Electromag. Compatibility*, **EMC-29**, 104–115 (1987).

Goedbloed, J. J., and W. A. Pasmooij, "Characterization of transient and CW disturbances induced in telephone subscriber lines," *IEE 7th International Conference on EMC*, York, 211–218, 28–31 August 1990.

Hayt, W. H., *Engineering Electromagnetics*, 5th ed., McGraw-Hill, New York (1988).

Hensman, G. O., "EMC related to the public electricity supply network," *EMC 89—Product Design for Electromagnetic Compatibility (ERA Technology Seminar Proceedings 89–0001)*, Electricity Council (1989).

Howell, E. K., "How switches produce electrical noise," *IEEE Trans. Electromag. Compatibility*, **EMC-21**, 162–170 (1979).

IEC801 (BS 6667) *Electromagnetic compatibility for industrial-process measurement and control equipment.*

IEC 1000 *Electromagnetic compatibility.*

Jones, J. W. E., "Achieving compatibility in interunit wiring," *6th Symposium on EMC*, Zurich, 5–7 March 1985.

Kay, R., "Co-ordination of IEC standards on EMC and the importance of participating in standards work," *IEE 7th International Conference on EMC*, 28–31 August 1990, York, IEC, 1–6 (1990).

Marshall, R. C., "Convenient current-injection immunity testing," *IEE 7th International Conference on EMC*, York, 28–31 August, 1990, 173–176 (1990).

Mazda, F. F. (ed.), *Electronic Engineer's Reference Book*, 5th ed., Butterworth-Heinemann, Oxford (1983).

Morrison, R., *Grounding and Shielding Techniques in Instrumentation*, 3d ed., Wiley, Chichester, U.K. (1986).

Motorola, "Transmission line effects in PCB applications," *Application Note AN1051* (1990).

Ott, H. W., "Ground—a path for current flow," *International Symposium on EMC*, IEEE, San Diego, 9–11 October 1979.

Ott, H. W., "Digital circuit grounding and interconnection," *International Symposium on EMC*, IEEE, Boulder, Colo., August 1981.

Ott, H. W., "Controlling EMI by proper printed wiring board layout," *6th Symposium on EMC*, Zurich, 5–7 March 1985.

Ott, H. W., *Noise Reduction Techniques in Electronic Systems*, 2d ed., Wiley, Chichester, U.K. (1988).

Palmgren, C. M., "Shielded flat cables for EMI and ESD reduction," *International Symposium on EMC*, IEEE, Boulder, Colo., August 1981.

Paul, C. R., "Effect of pigtails on crosstalk to braided-shield cables," *IEEE Trans. Electromag. Compatibility*, **EMC-22**, No. 3 (1980).

Paul, C. R., "A comparison of the contributions of common-mode and differential-mode currents in radiated emissions," *IEEE Transactions on Electromagnetic Compatibility*, **EMC-31**, No. 2 (1989).

Rankin, I., "Screening plastics enclosures," *Suppression Components, Filters and Screening for EMC (ERA Technology Seminar Proceedings 86–0006)* (1986).

Smith, A. A., *Coupling of External Electromagnetic Fields to Transmission Lines*, Wiley, Chichester, U.K. (1977).

Staggs, D. M., "Designing for electrostatic discharge immunity," *8th Symposium on EMC*, Zurich, March 1989.

Sutu, Y.-H., and J. J. Whalen, "Demodulation RFI in inverting and non-inverting operational amplifier circuits," *6th Symposium on EMC*, Zurich, March 1985.

Swainson, A. J. G., "Radiated emission, susceptibility and crosstalk control on ground plane printed circuit boards," *IEEE 7th International Conference on EMC*, York, 28–31 August 1990, 37–41.

Tront, J. G., "RFI susceptibility evaluation of VLSI logic circuits," *9th Symposium on EMC*, Zurich, March 1991.

Whalen, J. J., "Determining EMI in microelectronics—a review of the past decade," *6th Symposium on EMC*, Zurich, 5–7 March 1985.

Williams, T., *The Circuit Designer's Companion*, Butterworth-Heinemann, Oxford (1991).

Wimmer, M., "The development of EMI design guidelines for switched mode power supplies—examples and case studies," *5th International Conference on EMC, York (IERE Conference Publication No. 71)* (1986).

FURTHER READING

Morgan, D., *Handbook for EMC Testing and Measurement*, Institution of Electrical Engineers (1994).

Appendix A

General Instrumentation Books

For the most part, individual technical chapters of the *Instrumentation Reference Book* give references for further reading to books with a particular relevance to the topic of that chapter. Here we list some more general books that each give an overview of a wide range of instrumentation subjects. To help the reader decide whether any particular book will help with a particular problem, we include a table of contents for each of these books.

Abnormal Situation Consortium, *ASM Consortium Guidelines: Effective Operator Display Design*, ASM Consortium, Phoenix, AZ (2008).
Introduction
Preventing and Responding to Abnormal Situations
Guideline Detail and Examples
Guidelines Conformance Examples
Aftermatter and Appendices

Battikha, N., *The Condensed Handbook of Measurement and Control*, ISA Press, Research Triangle Park, NC (3rd ed., 2006).
Symbols
Measurement
Control Loops
Control Valves
Tables for Unit Conversion
Corrosion Guide
Enclosure Ratings
Resources

Berge, Jonas, *Software for Automation: Architecture, Integration, and Security*, ISA Press, Research Triangle Park, NC (2005).
Introduction and Overview
Benefits, Savings and Doubts
Setup
Configuration and Scripting
Enterprise Integration and System Migration
Troubleshooting
Application
Engineering and Design
Management and Administration
Safety, Availability and Security

Bolton, W., *Industrial Control and Instrumentation*, Longman, Harlow, U.K. (1991).
Measurement systems
Control systems
Transducers
Signal conditioning and processing
Controllers
Correction units
Data display
Measurement systems
Control systems

Bolton, W., *Instrumentation and Process Measurements*, Longman, Harlow, U.K. (1991).
Basic instrument systems
Sensing elements
Signal converters
Displays
Pressure measurement
Measurement of level and density
Measurement of flow
Measurement of temperature
Maintenance

Considine, D. M., *Industrial Instruments and Controls Handbook*, McGraw-Hill, New York (1993).
Introductory review
Control system fundamentals
Controllers
Process variables—field instrumentation
Geometric and motion sensors
Physicochemical and analytical systems
Control communications
Operator interface
Valves, servos, motors and robots

Corripio, A. B., *Tuning of Industrial Control Systems*, 2nd ed., ISA Press, Research Triangle Park, NC (2001).
Feedback controllers
Open-loop characterization of process dynamics
How to select feedback controller modes
How to tune feedback controllers
Computer feedback control

Tuning cascade control systems
d-forward, ratio, multivariable, adaptive, and self-tuning control

Dally, J. and W. Riley, *Instrumentation for Engineering Measurements*, 2nd ed., Wiley, New York (1993).
Applications of electronic systems
Analysis of circuits
Analog recording instruments
Digital recording systems
Sensors for transducers
Signal conditioning circuits
Resistance-type strain gauges
Force, torque and pressure measurements
Displacement, velocity and acceleration measurements
Analysis of vibrating systems
Temperature measurements
Fluid flow measurements
Statistical methods
(*Note*: Summary and problems appear in every chapter)

Dieck, R. H., *Measurement Uncertainty,* 4th ed., ISA Press, Research Triangle Park, NC (2007).
Fundamentals of measurement uncertainty analysis
The measurement uncertainty model
How to do it summary
Uncertainty (error) propagation
Weighting method for multiple results
Applied considerations
Presentation of results

Figliola, R. S. and D. E. Beasley, *Theory and Design for Mechanical Measurements*, Wiley, New York (1991).
Basic concepts of measurement methods
Static and dynamic characteristics of signals
Measurement system behavior
Probability and statistics
Uncertainty analysis
Electrical devices, signal processing and data acquisition
Temperature measurements
Pressure and velocity measurements
Flow measurements
Metrology, displacement and motion measurements
Strain measurement

Finkelstein, L. and K. T. V. Grattan, *Concise Encyclopedia of Measurement and Instrumentation*, Pergamon, Oxford (1993).
General theoretical principles of measurement and instrumentation
Instrument and instrument systems in relation to their life cycles
Instrument system elements and general technology
Measurement information systems classified by measurand
Applications
History of measurement and instrumentation

Gifford, Charlie (editor and contributing author), *The Hitchhiker's Guide to Manufacturing Operations Management: ISA95 Best Practices Book 1.0,* ISA Press (2007).
ISA95 best practices and business case evolve through manufacturing application
An overview and comparison of ISA95 and OAGIS
OAGIS, ISA95 and related manufacturing integration standards, a survey
ISA95—as-is/to-be study
Manufacturing information systems—ISA88/95 based functional definitions
ISA95 implementation best practices, workflow descriptions using B2MML
ISA95 based operations and KPI metrics assessment and analysis
ISA95 the SAP Enterprise-Plant link to achieve adaptive manufacturing analysis
ISA95 based change management

Gillum, Donald R., *Industrial Pressure, Level and Density Measurement*, 2nd ed., ISA Press, Research Triangle Park, NC (2009).
Introduction to Measurements
Pressure Measurement and Calibration Principles
Pressure Transducers and Pressure Gages
Transmitters and Transmission Systems
Level Measurement Theory and Visual Measurement Techniques
Hydrostatic Head Level Measurement
Electrical Level Measurement
Liquid Density Measurement
Hydrostatic Tank Gaging
Instrument Selection and Applications
Deadweight Gage Calibration
Pressure Instruments Form ISA20.40a1

Gruhn, P. and H. Cheddie, *Safety Instrumented Systems,* 2nd ed., ISA Press, Research Triangle Park, NC (2005).
Introduction
Design life cycle
Risk
Process control vs.
Safety control
Protection layers
Developing the safety requirement specifications
Developing the safety integrity level
Choosing a technology
Initial system evaluations
Issues relating to field devices
Engineering a system
Installing a system
Functional testing
Managing changes to a system
Justification for a safety system
SIS design checklist
Case study

APPENDIX | A General Instrumentation Books

Horne, D. F., *Measuring Systems and Transducers for Industrial Applications*, Institute of Physics Publishing, London (1988).
Optical and infra-red transmitting systems
Photogrammetry and remote earth sensing
Microwave positioning and communication systems
Seismic field and seabed surveying and measurement of levels

Hughes, T. A., *Programmable Controllers*, 4th ed., ISA Press, Research Triangle Park, NC (2005).
Introduction
Numbering systems and binary codes
Digital logic fundamentals
Electrical and electronic fundamentals
Input/output systems
Memory and storage devices
Ladder logic programming
High-level programming languages
Data communication systems
System design and applications
Installation, maintenance and troubleshooting

ISA, *Dictionary of Measurement and Control*, 3rd ed., ISA Press, Research Triangle Park, NC (1999).

Liptak, B., *Instrument Engineer's Handbook* 4th ed., Chilton, U.S.; ISA Press, Butterworth-Heinemann, U.K. and rest of world (2003).

Volume One: Process Measurement and Analysis
1. Instrument terminology and performance
2. Flow measurement
3. Level measurement
4. Temperature measurement
5. Pressure measurement
6. Density measurement
7. Safety, weight and miscellaneous sensors
8. Analytical instrumentation
9. Appendix

Volume Two: Process Control
1. Control theory
2. Controllers, transmitters converters and relays
3. Control centers, panels and displays
4. Control valves, on-off and throttling
5. Regulators and other throttling devices
6. PLCs and other logic devices
7. DCS and computer based systems
8. Process control systems
9. Appendix

Love, Jonathan, *Process Automation Handbook: A Guide to Theory and Practice*, Springer Verlag, London (2007).
Technology and Practice
Instrumentation
Final Control Elements
Conventional Control Strategies
Process Control Schemes
Digital Control Systems
Control Technology
Management of Automation Projects
Theory and Technique
Maths and Control Theory
Plant and Process Dynamics
Simulation
Advanced Process Automation
Advanced Process Control

Magison, Ernest C., *Electrical Instruments in Hazardous Locations*, 4th ed., ISA Press, Research Triangle Park, NC (1998).
Historical background and perspective
Combustion and explosion fundamentals
Classification of hazardous locations and combustible materials
Practice and principles of hazard reduction practice
Explosion proof enclosures
Reduction of hazard by pressurization
Encapsulation, sealing and immersion
Increased safety, type of protection *e*
Ignition of gases and vapors by electrical means
Intrinsically safe and nonincendive systems
Design and evaluation of intrinsically safe apparatus, intrinsically safe systems and nonincendive systems
Ignition by optical sources
Dust hazards
Human safety
Degree of protection by enclosures

Mandel, J., *Evaluation and Control of Measurements*, Marcel Dekker, New York (1991).
Measurement and statistics
Basic statistical concepts
Precision and accuracy: the central limit theorem, weighting
Sources of variability
Linear functions of a single variable
Linear functions of several variables
Structured two-way tables of measurements
A fit of high precision two-way data
A general treatment of two-way structured data
Interlaboratory studies
Control charts
Comparison of alternative methods
Data analysis: past, present and future

Marshall, Perry S., and Rinaldi, John S., *Industrial Ethernet*, 2nd ed., ISA Press, Research Triangle Park, NC (2005).
What Is Industrial Ethernet?
A Brief Tutorial on Digital Communication
Ethernet Hardware Basics
Ethernet Protocol and Addressing
Basic Ethernet Building Blocks
Network Health, Monitoring and System Maintenance
Installation, Troubleshooting and Maintenance Tips

Basic Precautions for Network Security
Power over Ethernet (PoE)
Wireless Ethernet

Nachtigal, C. L., *Instrumentation and Control: Fundamentals and Applications*, Wiley, New York (1990).
Introduction to the handbook
Systems engineering concepts
Dynamic systems analysis
Instrument statics
Input and output characteristics
Electronic devices and data conversion
Grounding and cabling techniques
Bridge transducers
Position, velocity and acceleration measurement
Force, torque and pressure measurement
Temperature and flow transducers
Signal processing and transmission
Data acquisition and display systems
Closed-loop control system analysis
Control system performance modification
Servoactuators for closed-loop control
Controller design
General purpose control devices
State-space methods for dynamic systems analysis
Control system design using state-space methods

Scholten, Bianca, *The Road to Integration: A guide to applying the ISA95 Standard in manufacturing*, ISA Press, Research Triangle Park, NC (2007).
Getting acquainted with ISA95
Applying ISA95 as an Analysis Tool
Understanding and Applying the ISA95 Object Models
Applying ISA95 to Vertical Integration

Sherman, R. E., *Analytical Instrumentation*, ISA Press, Research Triangle Park, NC (1996).
Introduction to this technology
Typical analyzer application justifications
Interfacing analyzers with systems
Specification and purchasing of analyzers
Calibration considerations
Training aspects
Spc/sqc for analyzers
Personnel and organizational issues
Validation of process analyzers
Sample conditioning systems
Component specific analyzers
Electrochemical analyzers
Compositional analyzers, spectroscopic analyzers
Physical property

Spitzer, D. W., *Flow Measurement*, 3rd ed., ISA Press, Research Triangle Park, NC (2004).
Physical properties of fluids
Fundamentals of flow measurement
Signal handling
Field calibration
Installation and maintenance
Differential pressure flowmeters
Magnetic flowmeters
Mass flowmeters—open channel flow measurement
Oscillatory flowmeters
Positive displacement flowmeters
Target flowmeters
Thermal mass flowmeters and controllers
Tracer dilution measurement; turbine flowmeters
Ultrasonic flowmeters
Variable area flowmeters
Insertion (sampling) flow measurement
Custody transfer measurement
Sanitary flowmeters
Metrology, standards, and specifications

Spitzer, D. W., *Regulatory and Advanced Regulatory Control: Application Techniques*, ISA Press, Research Triangle Park, NC (1994).
Introduction
Manual Control
Field Measurement Devices
Controllers
Final Control Elements
Field Equipment Tuning
Regulatory Controller Features
PID Control
Controller Tuning
Regulatory Control Loop Pairing
The Limitations of Regulatory Control
Advanced Regulatory Control Tools
Advanced Regulatory Control Design
Applying Advanced Regulatory Control

Sydenham, P. H., N. H. Hancock, and R. Thorn, *Introduction to Measurement Science and Engineering*, Wiley, Chichester (1992).
Introduction
Fundamental concepts
Signals and information
The information machine
Modeling and measurement system
Handling and processing information
Creating measurement systems
Selecting and testing of instrumentation

Trevathan, Vernon L., ed., *A Guide to the Automation Body of Knowledge*, 2nd ed., ISA Press, Research Triangle Park, NC (2006).
Process Instrumentation
Analytical Instrumentation
Continuous Control
Control Valves
Analog Communications
Control System Documentation

Control Equipment
Discrete Input/Output Devices and General Manufacturing
    Measurements
Discrete and Sequencing Control
Motor and Drive Control
Motion Control
Process Modeling
Advanced Process Control
Industrial Networks
Manufacturing Execution Systems and Business Integration
System and Network Security

Operator Interface
Data Management
Software
Custom Software
Operator Training
Checkout, System Testing and Startup
Troubleshooting
Maintenance, Long term Support and System Management
Automation Benefits and Project Justifications
Project Management and Execution
Interpersonal Skills for Automation Professionals

Appendix B

Professional Societies and Associations

American Association for Laboratory Accreditation (A2LA), 656 Quince Orchard Rd., Gaithersburg, MD 20878-1409. 301-670-1377. A2LA is a nonprofit, scientific membership organization dedicated to the formal recognition of testing and calibration organizations that have achieved a demonstrated level of competence.

The American Automatic Control Council (AACC) is an association of the control systems divisions of eight member societies in the United States.

American Electronics Association (AEA), 5201 Great America Pkwy, Suite 520, Santa Clara, CA 95054. 408-987-4200. The AEA is a high-tech trade association representing all segments of the electronics industry.

The American Institute of Physics includes access to their publications.

American National Standards Institute (ANSI), 11 W. 42nd St., New York, NY 10036. 212-642-4900.

ASHRAE, the American Society of Heating, Refrigerating, and Air-Conditioning Engineers.

American Society for Nondestructive Testing (ASNT), 1711 Arlingate Lane, P.O. Box 28518, Columbus, OH 43228-0518. 614-274-6003. ASNT promotes the discipline of nondestructive testing (NDT) as a profession, facilities NDT research, and the application of NDT technology, and provides its 10,000 members with a forum for exchange of NDT information. ASNT also provides NDT educational materials and training programs.

American Society for Quality Control (ASQC), 611 E. Wisconsin Ave, Milwaukee, WI 53202. 414-272-8575.

American Society for Testing and Materials (ASTM), 1916 Race St., Philadelphia, PA 19103. 215-299-5400. ASTM is an international society of 35,000 members (representatives of industry, government, academia, and the consumer) who work to develop high-quality, voluntary technical standards for materials, products, systems, and services.

American Society of Test Engineers (ASTE), P.O. Box 389, Nutting Lake, MA 01865-0389. 508-765-0087. The ASTE is dedicated to promoting test engineering as a profession.

ASM International for management of materials.

Association of Independent Scientific, Engineering, and Testing Firms (ACIL), 1659 K St., N.W., Suite 400, Washington, DC 20006. 202-887-5872. ACIL is a national association of third-party scientific and engineering laboratory testing and R and D companies serving industry and the public through programs of education and advocacy.

The Automatic Meter Reading Association Advancing utility technology internationally.

British Institute of Non-Destructive Testing, 1 Spencer Parade, Northampton, NN1 5 AA, U.K. 01604 30124.

British Society for Strain Measurement, Dept of Civil Engineering, University of Surrey, Guild-ford GU2 5XH, U.K. 01483 509214.

British Standards Institution, 2 Park Street, London, W1A 2BS, U.K. 0171 629 9000.

Canadian Standards Association (CSA), 178 Rexdale Blvd., Rexdale, Ontario M9W 1R3, Canada. 416-747-4007.

China Instrument and Control Society (CIS), based in Beijing.

The Computer Society is the leading provider of technical information and services to the world's computing professionals.

Electronic Industries Association (EIA), 2001 Pennsylvania Ave. N.W., Washington, D.C. 20006. 202-457-4900.

The Embedded Software Association (ESOFTA) provides its members with assorted marketing and communications services, a framework for member-initiated standards activities, and a forum for software creator and user communications.

The European Union Control Association (EUCA) In 1990, a number of prominent members of the systems and control community from countries of the European Union decided to set up an organization, EUCA, the main purpose of which is to promote initiatives aiming at enhancing scientific exchanges, disseminating information, coordinating research networks, and technology transfer in the field of systems and control within the Union.

The Fabless Semiconductor Association (FSA) mission is to stimulate technology and foundry capacity by communicating the future needs of the fabless semiconductor segment in terms of quantity and technology; to provide interactive forums for the mutual benefit of all FSA members; and to be a strong, united voice on vital issues affecting the future growth of fabless semiconductor companies.

The Federation of the Electronics Industry is a useful body and a source for information on standards such as the European Directives on EMC (the CE Mark).

GAMBICA Association Ltd., Leicester House, 8 Leicester Street, London, WC2H 7BN, U.K. 0171 4370678. GAMBICA is the trade association of the British Instrumentation Control and Automation Industry.

IMEKO (International Measurement Confederation) Forum for advancements in measurement science and technology.

The Industrial Automation Open Networking Alliance (IAONA) is for industrial automation leaders committed to the advancement of open networking from sensing devices to the boardroom via Internet- and Ethernet-based networks.

The Institute of Electrical and Electronics Engineers (IEEE) is the world's largest technical professional society. A nonprofit organization, which promotes the development and application of electrotechnology and allied sciences for the benefit of humanity, the advancement of the profession, and the wellbeing of its members.

Institute of Environmental Sciences (IES), 940 E. Northwest Hwy., Mount Prospect, IL 60056. 708-255-1561.

Institute of Measurement and Control, 87 Gower Street, London, WC1E 6AA, U.K. 0171 387 4949.

The Institute of Physics (IOP) gives some information on publications and has some interesting links.

Institute of Quality Assurance, 8-10 Grosvenor Gardens, London, SW1W 0DQ, U.K. 0171 730 7154.

Institution of Chemical Engineers, 12 Gayfere Street, London, SW1P 3HP, U.K. 0171 222 2681.

The Institution of Electrical Engineers (IEE) lists all its services and books, with brief reviews, conference proceedings, etc.

The International Electrotechnical Commission (IEC) is the worldwide standards organization charged with development and control of global standards. 3, rue de Varembé, P.O. Box 131, CH-1211, Geneva 20, Switzerland.

The Institute of Instrumentation and Control Australia is the professional body serving those involved in the field of instrumentation and control in Australia.

The Institute of Measurement and Control is a British-based organization, with some international branches in Ireland and Hong Kong, for instance.

Institution of Mechanical Engineers, 1 Birdcage Walk, London, SW1H 9JJ, U.K. 0171 222 7899.

ISA-The International Society of Automation, formerly the Instrument Society of America and the Instrumentation Systems and Automation Society, P.O. Box 12277, Research Triangle Park, NC 27709. 919-549-8411. The ISA is an international society of more than 49,000 professionals involved in instrumentation, measurement, and control. The ISA conducts training programs, publishes a variety of literature, and organizes an annual conference and exhibition of instrumentation and control. The ISA page, in constant development, is now an excellent resource. ISA was formed some 50 years ago and boasts almost 30,000 members.

International Electronics Packaging Society (IEPS), P.O. Box 43, Wheaton, IL 60189–0043. 708-260-1044.

International Federation of Automatic Control, founded in September 1957, is a multinational federation of National Member Organizations (NMOs), each one representing the engineering and scientific societies concerned with automatic control in its own country.

International Frequency Sensor Association (IFSA) The main aim of IFSA is to provide a forum for academicians, researchers, and engineers from industry to present and discuss the latest research results, experiences, and future trends in the area of design and application of different sensors with digital, frequency (period), time interval, or duty-cycle output. Very fast advances in IC technologies have brought new challenges in the physical design of integrated sensors and micro-sensors. This development is essential for developing measurement science and technology in this millennium.

The International Instrument Users' Association has been set up "for cooperative instrument evaluations by member-users and manufacturers."

International Organization for Standardization (ISO), 1 rue de Varembe, CH-1211, Geneva 20, Switzerland, +41-22-749-01-11. The ISO promotes standardization and related activities with a view to facilitating the international exchange of goods and services and to developing cooperation in the spheres of intellectual, scientific, technological, and economic activity.

APPENDIX | B Professional Societies and Associations

The SPIE—The International Society for Optical Engineering, formed in 1955 as the Society for Photo-Optical Instrumentation Engineers, serves more than 11,000 professionals in the field throughout the world, as does the **Optical Society of America**. The former is perhaps more applications and engineering oriented, the latter more in the field of research—although both areas overlap these days. Another site of interest to optical engineers is the **Laser Institute of America.**

International Telecommunications Union (CCITT), Place des Nations, 1121 Geneva 20, Switzerland. +41-22-99-51-11.

Japan Electric Measuring Instruments Manufacturers' Association (JEMIMA), 1-9-10, Torano-mon, Minato-ku, Tokyo 103, Japan. +81-3-3502-0601. JEMIMA was established in 1935 as a nonprofit industrial organization authorized by the Japanese government, JEMIMA is devoted to a variety of activities: cooperation with the government, providing statistics about the electronics industry, and sponsoring local and overseas exhibitions.

The Low Power Radio Association (LPRA) is an association for companies involved in deregulated radio anywhere in the world. It is believed to be the only such association for users and manufacturers of low-power radio devices in the deregulated frequency bands.

National Conference of Standards Laboratories (NCSL), 1800 30th St., Suite 305B, Boulder, CO 80301-1032. 303-440-3339.

National Institute of Standards and Technology (NIST), Publications and program inquiries, Gaithersburg, MD 20899. 301-975-3058. As a non-regulatory agency of the U.S. Department of Commerce Technology Administration, NIST promotes U.S. economic growth by working with industry to develop and apply technology, measurements, and standards.

National ISO 9000 Support Group, 9964 Cherry Valley, Bldg. 2, Caledonia, MI 49316. 616-891-9114. The National ISO 9000 Support Group is a nonprofit network of companies that serves as a clearinghouse of ISO 9000 information and certified assessors. The group provides full ISO 9000 implementation support for 150 per year.

National Technical Information Service (NTIS), 5285 Port Royal Road, Springfield, VA 22161. 703-487-4812. The NTIS is a self-supporting agency of the U.S. Department of Commerce and is the central source for public sale of U.S. government-sponsored scientific, technical, engineering, and business-related information.

Optical Society of America (OSA), 2010 Massachusetts Ave., N.W., Washington, D.C. 20036. 202-223-8130.

Precision Measurements Association (PMA), 3685 Motor Ave., Suite 240, Los Angeles, CA 90034. 310-287-0941.

Process Industry Practices (PIP) is a consortium of process industry owners and engineering construction contractors who serve the industry.

Reliability Analysis Center, IIT Research Institute, 201 Mill St., Rome, NY 13440-6916. 315-337-0900.

Process engineers will be interested in the site of **The Royal Society of Chemistry**.

SAE, 400 Commonwealth Dr., Warrendale, PA 15096-0001. 412-776-4841.

Semiconductor Equipment and Materials International (SEMI), 805 E. Middlefield Road, Mountain View, CA 94043. 415-964-5111.

Semiconductor Industry Association (SIA), 4300 Stevens Creek Blvd., Suite 271, San Jose, CA 95129. 408-246-2711.

Society for Information Display (SID), 1526 Brookhollow Dr., Suite 82, Santa Ana, CA 92705-5421. 714-545-1526. SID is a nonprofit international society devoted to the advancement of display technology, manufacturing, integration, and applications.

The Society of Manufacturing Engineers has developed the Global Manufacturing Network to help users find information on manufacturers, suppliers, etc.

Society of Women Engineers (SWE), 120 Wall St., New York, NY 10005. 212-509-9577.

Software Engineering Institute, Carnegie-Mellon University, Pittsburgh, PA 15213-3890. 412-269-5800.

TAPPI is the technical association for the pulp and paper industry.

Telecommunications Industry Association (TIA), 2001 Pennsylvania Ave., N.W., Suite 800, Washington, D.C. 20006-1813. 202-457-4912.

Underwriters Laboratories, 333 Pfingsten Road, Northbrook, IL 60062. 708-272-8800. Underwriters Laboratories (UL) is an independent, nonprofit certification organization that has evaluated products in the interest of public safety for 100 years.

VDI/VDE-GMA Society for Measurement and Automatic Control in Germany.

World Batch Forum offers a noncommercial venue for the dissemination of information batch engineers need.

Appendix C

The Institute of Measurement and Control

ROLE AND OBJECTIVES

The science of measurement is very old and has advanced steadily in the precision with which measurements may be made and by the variety and sophistication of the methods available. In the last century the rate of advance was very rapid, stimulated in particular by the needs of industry. Control engineering has a much more recent origin: with the advent of the complex requirements, such as those of the process and aerospace industries, there has been a veritable explosion of new theory and application during the last 50 years. In this period there has been a correspondingly rapid increase in the number of people working in these fields.

The theory and application of measurement and control characteristically require a multidisciplinary approach and so do not fit into any of the single disciplinary professional institutes.

The Institute brings together thinkers and practitioners from the many disciplines which have a common interest in measurement and control. It organizes meetings, seminars, exhibitions, and national and international conferences on a large number of topics. It has a very strong local section activity, providing opportunities for interchange of experience and for introducing advances in theory and application.

It provides qualifications in a rapidly growing profession and is one of the few chartered engineering institutions which qualifies incorporated engineers and engineering technicians as well as chartered engineers.

In its members' journal, *Measurement and Control*, the Institute publishes practical technical articles, product and business news, and information on technical advances; in the newsletter, *Interface*, the activities of the Institute, its members, and the engineering profession in general are reported. In addition the Institute provides a whole range of learned and other publications.

The objects of the Institute, expressed in the Royal Charter, are: "To promote for the public benefit, by all available means, the general advancement of the science and practice of measurement and control technology and its application."

To further its objects the Institute acts as a qualifying body, conferring membership only on those whose qualifications comply with the Institute's standards. It acts as a learned society by disseminating and advancing the knowledge of measurement and control and its application at all levels. It is the academic and professional body for the profession, requiring members to observe a code of conduct.

HISTORY

Like many professional bodies, the Institute of Measurement and Control arose through the need for a group of like-minded people to meet and exchange ideas. They first met at the Waldorf Hotel in London during October 1943, and a society of instrument technology was proposed.

The Institute was founded in May 1944 as the Society of Instrument Technology (SIT) to cater to the growing body of instrument technologists whose interests transcended the fields of existing institutions.

During the late 1940s and the 1950s the Society progressed steadily. By 1960 the number of members had grown to over 2500 and local sections had been formed in the main industrial areas in the United Kingdom.

Control engineering, as opposed to measurement, began to be recognized as a distinct discipline only after the establishment of SIT. The evidence of the relationship between the two topics stimulated the formation of a control section of SIT, and the large and enthusiastic participation in that section's first meeting more than vindicated its creation.

In 1957 the importance of the computer was acknowledged through the formation of a data processing section, created to serve the large and growing interest in data handling related to process control, a combination outside the scope of any other learned society.

By 1965 there were four specialized sections concerned with measurement technology, control technology, systems engineering, and automation. At that time it was realized that in a field developing as rapidly as that of measurement and control, a more flexible structure would be required to deal with the steadily advancing and changing interests of the Institute's members. Consequently, a national technical committee was set up overseeing the work of panels which at present include: a physical measurements panel, a systems and control technology panel, a systems and management panel, an industrial analytical panel, an educational activities panel, and a standards policy panel.

Since 1986, the work of the national technical committee has been taken over by a learned society board, to which, in addition to the above technical panels, the publications executive committees report.

Members who have particular interests in specialized fields are encouraged to set up new panels within the framework of the Institute through which their work can be advanced at a professional level.

In 1975 the Institute was confirmed as a representative body of the United Kingdom for those engaged in the science and practice of measurement and control technology through the granting, by the Privy Council, of a Royal Charter of incorporation.

QUALIFICATIONS

The Institute influences educational courses from the broadly based Full Technological Certificate of the City and Guilds of the London Institute, through BTEC and SCOTVC2 certificates and diploma to the first degrees as the Engineering Council-authorized course-accreditation institution. Standards of courses are maintained by the education training and qualification committee and its subsidiary accreditation council.

CHARTERED STATUS FOR INDIVIDUALS

Corporate members of the Institute, those with the grade of fellow or member, all bear the title "chartered measurement and control technologist." In addition, those with appropriate engineering qualifications can be registered by the Institute on the chartered engineers section of the register maintained by the Engineering Council, thus becoming chartered engineers (CEng). Registration as a European engineer (EurIng) with FEANI (European Federation of National Engineering Associations) is also possible for CEng members of the Institute, through the Institute.

INCORPORATED ENGINEERS AND ENGINEERING TECHNICIANS

Licentiates and associates of the Institute may also be entered on the Engineering Council's register by the Institute as incorporated engineers and engineering technicians, IEng and Eng Tech respectively. Both titles are becoming increasingly recognized as significant qualifications in British industry and there is additionally a route for IEng registration with FEANI.

MEMBERSHIP

The Institute has a wide range of grades of membership available in two basic forms, *corporate* or *non-corporate*.

CORPORATE MEMBERS

Corporate members with accredited UK degrees or equivalent can be nominated by the Institute for registration on the Engineering Council's register as chartered engineers. They may use the designatory letters "CEng." All corporate members are entitled to use the exclusive and legally protected title "chartered measurement and control technologist." There are three classes of corporate membership: honorary fellows (HonFInstMC), fellows, (FInstMC) and members (MInstMC). The following briefly summarizes the requirements for these grades of membership:

HONORARY FELLOW

The council of the Institute from time to time invites eminent professional engineers to become honorary fellows. They are fellows or members of the Institute who have achieved exceptionally high distinction in the profession.

FELLOWS

Members over 33 years of age who have carried superior responsibility for at least five years may be elected to fellow. Persons who have achieved eminence through outstanding technical contributions or superior professional responsibility may be directly elected as fellows by the council.

MEMBERS

Engineers over 25 years of age who have an approved degree or equivalent with at least four years' professional experience and responsibility, of which two years should be professional training, may be elected as members of the Institute.

Exceptionally, there are mature routes for those over 35 years of age who have 15 years' experience and insufficient academic qualifications. Written submissions and interviews are required.

Information and advice is available from the Institute about the appropriate educational qualifications and the mature route. There is a specific syllabus for the Engineering

Council Examination, success in which provides the necessary level of qualification.

NONCORPORATE MEMBERS

There are seven classes of non-corporate member: companion, graduate, licentiate, associate, student, affiliate, and subscriber. The following briefly summarizes the requirements.

COMPANIONS

Persons who, in the opinion of Council, have acquired national distinction in an executive capacity in measurement and control and are at least 33 years of age may be elected as companions. There is no particular academic requirement for this class of membership.

GRADUATES

The requirement for graduate membership is an accredited degree or equivalent. Information and advice is available from the Institute about educational qualifications.

LICENTIATES

Persons of at least 23 years of age who have an accredited BTEC or SCOTVEC Higher National Award or equivalent plus five years' experience, of which two must be approved practical training, may be elected as licentiates.

Licentiates can register through the Institute as incorporated engineers. Registration allows the use of the designatory letters "IEng."

Exceptionally, for those who have not achieved the academic qualification, there are mature routes. Candidates must be at least 35 years of age and have 15 years' experience. Written submissions and interviews are required.

ASSOCIATES

Persons who are least 21 years of age and have attained the academic standards of an accredited BTEC or SCOTVEC National Award or equivalent plus five years' experience, including two years' approved practical training, may be elected as associates.

Associates can register through the Institute as engineering technicians. Registration allows the use of the designatory letters "EngTech."

Exceptionally, for those who have not achieved the academic qualifications, there are mature routes. Candidates must be at least 35 years of age and have 15 years' experience. Written submissions and interviews are required.

STUDENTS

Students who are at least 16 years of age and following a relevant course of study may be elected as student members of the Institute.

AFFILIATES

Anyone wishing to be associated with the activities of the Institute who is not qualified for other classes of membership may become an affiliate.

SUBSCRIBERS

Companies and organizations concerned with measurement and control may become subscribers.

APPLICATION FOR MEMBERSHIP

Full details of the requirements for each class of membership, including the rules for mature candidates, examinations, and professional training are available from the Institute.

NATIONAL AND INTERNATIONAL TECHNICAL EVENTS

The Institute organizes a range of technical events—from one-day colloquia to multinational conferences held over several days—either on its own account or on behalf of international federations. The wide nature of the Institute's technical coverage means that many events are in association with other, more narrowly based, institutions and societies.

LOCAL SECTIONS

Members meet on a local basis through the very active local sections. There are more than 20 local sections in the UK, with one also covering Ireland and one in Hong Kong. Each local section is represented on the Institute's council, providing a direct link between the members and the council. Normally, about 200 local section meetings take place annually.

PUBLICATIONS

In addition to the monthly journal *Measurement and Control* and newsletter *Interface*, the Institute publishes *Transactions* which contains primary, refereed material. Special issues of the *Transactions*, covering particular topics, are published within the five issues a year.

In addition, the Institute publishes texts, conference proceedings, and information relevant to the profession. There is also the *Instrument Engineer's Yearbook*, a main information source for measurement and control practitioners.

ADVICE AND INFORMATION

The Institute plays its part in policy formulation through its representation on such bodies as the Parliamentary and Scientific Committee, the Engineering Council, the Business and Technician Education Council, the British Standards Institution, the United Kingdom-Automatic Control Council, the City and Guilds of London Institute, and numerous other national and local groups and committees.

AWARDS AND PRIZES

The institute has a considerable number of awards and prizes ranging from the high-prestige Sir George Thomson Gold Medal awarded every five years to a person whose contribution to measurement and science has resulted in fundamental improvements in the understanding of the nature of the physical world, to prizes for students in measurement and control on national courses and to school students.

GOVERNMENT AND ADMINISTRATION

The Institute is governed by its council, which consists of the president, three most recent past presidents, up to four vice-presidents, honorary treasurer, honorary secretary, and 36 ordinary members. The president, vice presidents, honorary treasurer, and honorary secretary are elected by council. Twenty-four ordinary members of the council are elected by regional committees. Twelve ordinary members of the council are nationally elected by all corporate members. Additional non-voting members are co-opted by the council (some chairmen of local sections and at least two non-corporate members).

In addition to the council there is a management board and four standing committees which report to the council. These are: the learned society board, education, training, and qualifications committee, local sections committee, and membership committee. The Institute has a full-time secretariat of 12 staff.

In 1984 the Institute purchased a building for its headquarters containing committee rooms, a members' room, and administration and office facilities for the secretariat.

—The Institute of Measurement and Control
87 Gower Street
London WC1E 6AA, U.K.
Tel: 0171 387 4949
Fax: 0171 388 8431

International Society of Automation, Formerly Instrument Society of America

ISA was founded in 1945 as the Instrument Society of America to advance the application of instrumentation, computers, and systems of measurement for control of manufacturing and other continuous processes. The Society is a nonprofit educational organization serving more than 49,000 members. In 2008, the Society renamed itself the International Society of Automation, to better reflect its broadened purview of the entire practice of automation, not simply continuous process.

ISA is recognized worldwide as the leading professional organization for instrumentation practitioners. Its members include engineers, scientists, technicians, educators, sales engineers, managers, and students who design, use, or sell instrumentation and control systems.

Members are affiliated with local sections that are charted by the Society. The sections are grouped into 12 geographic districts in the United States and Canada; non-North American members and their sections are affiliated with ISA through ISA International, a nonprofit subsidiary. ISA International was established in 1988 to meet the special needs of instrumentation and control practitioners outside the United States and Canada.

The Society provides a wide range of activities and offers members the opportunity for frequent interaction with other instrumentation specialists in their communities. By joining special interest divisions, ISA members share ideas and expertise with their peers throughout the world. These divisions are classified under the Industries and Sciences Department and the Automation and Technology Department.

The members of each local section elect delegates to the district council and the council of society delegates. These delegates elect the ISA officers and determine major policies of the Society.

ISA's governing body is the executive board. The board is responsible for enacting policies, programs, and financial affairs. Executive board members are the president, past president, president-elect secretary, treasurer, and district and department vice presidents elected by the Councils of District and Department Vice Presidents. A professional staff manages the daily business of ISA and implements the executive board's program and policies. Administrative offices are located in Research Triangle Park and Raleigh, North Carolina.

The Society held its first major conference, Instrumentation and the University, in Philadelphia, Pennsylvania, in 1945.

ISA has become the leading organizer of conferences and exhibitions for measurement and control. The society hosts a large annual instrument and control conference in North America, attracting more than 13,000 people.

ISA has cosponsored events with other organizations and regular exhibitions in China and Europe and regularly embraces other conferences within its overall technical program.

TRAINING

ISA is a leading training organization and producer of training products and services. This year the Society will reach over 4,000 people through 300 training courses and customized training programs offered internationally. In addition to this direct training, ISA produces electronic packages, videos, and online and offline interactive multimedia instruction.

STANDARDS AND PRACTICES

ISA actively leads in the standardization of instrumentation and control devices under the auspices of the American National Standards Institute (ANSI). The Society regularly issues a compendium of all its standards and practices for measurement and control. The multiple-volume set includes copies of more than 90 ISA standards. Nearly 3,500 people on 140 committees are currently involved in developing more than 80 new ISA standards.

PUBLICATIONS

ISA is a major publisher of books and papers and offers over 600 titles, written by the leading experts in the field. The Society's first publications included *Basic Instrumentation Lecture Notes*, in 1960, and the first edition of *Standards and Practices for Instrumentation*, in 1963. Today ISA publishes *INTECH* magazine. Other major publications include *ISA Transactions* and the *ISA Directory of Instrumentation*.

The Instrument Society of America
U.S. Office: P.O. Box 12277
Research Triangle Park, NC 27709 USA
Tel: (919) 549 8411
Fax: (919) 549 8288
U.K. Office: P.O. Box 628, Croydon, U.K. CR9 2ZG

Index

A

Abbé refractometer, 515
Abrasion measurement with nuclear techniques, 559
Absolute pressure, 145, 149, 162
Absorption, 534
　in UV, visible and IR, 346–348
Absorption coefficient, 550
AC potential difference (AC/PD) in underwater testing, 589
Acceleration
　accelerometers, 115
　　calibration, 117
　　closed-loop, 119
　　mass motion in, 119
　　single-and three-axis, 119
　measurement, 121–124, 129
Accessibility of instruments, 655–656
Accuracy/precision of measurements, 551
AC/DC conversion, 461–462
Acids, properties in solution, 363
Acoustic emission
　inspection systems, 580, 581
　underwater, 589–590
Acoustic holography, 580
Acoustic radiation, use of, 87–90
Acoustic wave technology, 218
Adaptive control, 24
ADCs. See Analog-to-digital converters
Adder unit, 545
Adhesives, 646
Adiabatic expansion, 35
Adsorption, 670
　methods, 189
Advanced encryption standard (AES), 258
Advanced gas-cooled reactors (AGR), 777
Advanced process control (APC), 24, 619
　characteristics, 628
　constraint-pushing strategy, 623
　development, 622
　goal of, 626
　hierarchical approach to, 624–625, 627
Advanced regulatory control (ARC) vs. MPC, 623–624
Air supplies to instruments, 656
Aircraft autopilot controls, 767
Aircraft control systems, 772–773

Alexander Keilland, 784
Alkalis, properties in solution, 363–364
Alpha-detector systems, 534–536
Ambient temperature, 279–280
American military aircraft, position of controls in, 763
Ampere (unit), 439
Ampere's law, 569
Amplifiers, 543
　for piezoelectric sensors, 123–124
Analog circuits
　coupling capacitance, 841
　decoupling, 830
　instability, 830
　signal isolation, 841
　transient and radio frequency immunity, 840–841
Analog coding, 749
Analog instruments, 249
Analog oscilloscopes, 706
Analog redundancy, 758–759
Analog signal, transmission of, 678, 689–690
Analog square root extractor, pneumatic, 729, 730
Analog-to-digital converters (ADCs), 544
　dual-ramp, 457–458
　precision pulse-width, 458, 460
　pulse-width, 458–460
　successive-approximation, 456–457
　voltage references in, 460
Analysis equipment parts, sampling, 662
Analytical columns in HPLC system, 330
AND logic function, 781
Andersen cascade impactor, 188
Andreasen's pipette, 183–184
Anemometers, 51
Aneroid barometer, 150, 151
Angled-propeller meters, 46
Anodic stripping voltammetry, 334–335
Antennas, 258
　connection, 260–261
　directional, 258–259
　gains and losses, 259
　mounting, 259–260
　omnidirectional, 258
　reciprocity, 258
　selecting, 261

Anticoincidence circuit, 545–546
Antimony electrode, 387–388
Antithixotropy, 70
APC. See Advanced process control
Apertures effects of shielding, 860–862
Array detectors, 505–506
Ashcroft Model XLdp low pressure transmitter, 152, 153
Assembly screening, 746
Asset management system, 21
Asset optimization system, 21
Atomic absorption spectroscopy, 351–352
Atomic emission spectroscopy, 349–351
　applications, 350–351
Atomic fluorescence spectroscopy, 352–353
Attenuation characteristics of optical fiber, 682–683
Audio rectification, 840
Automatic control valves, 620–621
Automatic microscope size analyzers, 183
Automatic psychrometers, 433
Automation. See Industrial automation

B

Backing-fluorescence gauge, 562, 563
Backplane buses, 828
Backscattering, 534
Balancing units, 545
Bar graph indicator, 700, 702
Barringer remote sensing correlation spectrometer, 348
Base metal thermocouples, 299
Batch controls, 26
Batch processes, 629
Batch-switch system, 728
Bathtub curve for component failure rates, 742
Bayard-Alpert ionization gauge, 172–173
Beam chopping, 506
　and phase-sensitive detection, 507
Bearings, instrument, 646
Beer's law, 347
Bellows elements, 151–152, 160
Bench test, 261
Bench-mounting instruments, 647–649
Bending machines, 644
Bendix oxygen analyzer, 424
Bernoulli's theorem, 33–34

889

Beta particles detection, 536–537
Beta-backscatter gauge, 562
Beta-transmission gauge, differential, 562
Bimetal strip thermometer, 285–286
Bit-error ratio (BER), 255, 256
Blackbody radiation, 306–307, 519
Blondel's theorem, 468
Bluetooth, 263
Boeing aircraft control, 774
Bonded resistance strain gauges, 93–94
Bourdon tubes, 149–150, 158–160, 167, 278, 279
   for resonant wire sensor, 161
   spiral and helical, 150
Boxcar detector, 507–508
Boyle's law, 35
Bragg cell, 194, 206, 207
Bragg gratings, holographic in-fiber, 212–215
   optical configuration for, 213
   in strain-monitoring applications, 213, 214
   structure of, 212, 213
Bragg wavelength, 213, 214
Bragg's law, 355
Bridge
   definition of, 474
   rectifier, 542
Brinkmann probe colorimeter, 348
Brittle lacquer technique, 99
Brookfield viscometer, 73
BS 9000, 755, 756
BS 9001, 755
BS 9002, 755
BS 5345 classification on hazardous areas, 783
BS 5000/5501 safety of electrical apparatus, 782
Buffer solutions, 378
Built-in test equipment, 754
Bulk ultrasonic scanning, 589
Buoyancy measurements, pneumatic instrumentation for, 721
Buoyancy transducer, 136, 137
Buoyancy transmitters, 136
Business system integration, 26
Bypass meters, 46–47

C

Cables
   common-mode noise, 806–807
   and connectors, 261, 841–842
   ferrite-loaded, 847–848
   fiber-optic, 681
   radiation from, 806, 807, 827
   requirements, 657
   resonance, 810
   return currents, 842
   ribbon, 846–847
   screening
      composite tape and braid, 844
      connector types, 845
      currents, separation, 844
      grounding screens, 843, 845
      laminated tape, 844
      lapped wire, 844
      at low frequency, 842–843
      magnetic shielding, 842–843
      pigtail connection, 845–846
      at radio frequency, 843–844
      single braid, 844
      surface transfer impedance, 842–843
      types, 844–845
   segregation, 658
   and returns, 842
   testing, 659
   thermocouple compensating, 304
   twisted pair, 846
   types, 658
   unscreened, 846–848
Calibration
   accelerometer, 117
   amplitude, 117
   of electrical conductivity cells, 367
   field, 609
   field-system, 609
   force, 117–118
   formal, 609
   of gas analyzer, 426–428
   length measurement, 77
   of level-measuring systems, 106
   of microphones, 609
   moisture measurement, 435–436
   of neutron-moisture gauges, 557
   of radiation thermometer, 312
   of radiocarbon time scale, 565
   shock, 117
   of sound-level meters, 609
   strain-gauge load cells, 132–133
   system, 21, 609
Calomel electrode, 379, 382–384
Campbell mutual inductor, 443–444
Capacitance length-sensing structures, 87
Capacitance level sensor, 109
Capacitance manometers, 152–154, 167
Capacitance probes, 108–109
Capacitance sensing, 57
Capacitance type pressure sensors, 161, 162
Capacitive strain gauges, 99
Capacitors
   equivalent circuits of, 477–478, 480
   for filtering, 849
      chip, 852
      feedthrough, 851–852
      three-terminal, 851
Capillary tube, 279–280
Capillary viscometer, 71–72
Carbon dating. See Radiocarbon dating
Carrier gas, 402, 416
Carrier gas effect, 422
Carrier wave modulation, 687–688
Cascade control, 24
Casting in instrument construction, 645–646
CAT. See Combined actuator transducer
Catalytic detector, 411
Catastrophic failure, 743, 765
Cathode ray tubes (CRTs), 574, 575, 704
CCDs. See Charge-coupled devices
CE mark, 868
Cell conductance measurement using Wheatstone bridge, 369
Cellular wireless, 263
Celsius temperature scale, 272

CENELEC Electronic Components Committee (CECC) standards, 756, 793, 794
Central fringe identification method, 201
   centroid method, 201
   two-wavelength beat method, 202
Centrifugal pump, 668
Ceramics in instrument construction, 643
Certified Automation Professional (CAP) classification system, ISA, 5–13
Charge-coupled devices (CCDs), 505
Charge-injection devices (CIDs), 505
Charles's law, 35
Chemical milling, 645
Chemical plants, hazards in, 782
Chemiluminescence, 349
Cherenkov detectors, 524
Chernobyl nuclear reactor, explosion in, 764, 776, 779
Chip capacitors, 852
Cholesteric compounds, 324
Chromatography
   definition, 328
   gas. See Gas chromatography
   high-performance liquid. See High-performance liquid chromatography
   ion. See Ion chromatography
   paper, 328–329
   process. See Process chromatography
   thin-layer, 328–329
CIDs. See Charge-injection devices
Circuit fault analysis, 767–768
Circular chart recorders, 711
Citizen's band (CB) radio, 256
Clamp-on transit-time flow sensors, 66
Climet method, 189
Cloud chambers, 524
CMOS digital circuits, operating voltage effect on failure rates, 745
Coalescers, 666
Coanda effect meters, 57–58
Coating thickness, nuclear measurement for, 561–562
Coating-fluorescence gauge, 562, 563
Coaxial cables, 260, 261, 680
Coincidence circuit, 545–546
Cold junction compensation in thermocouples, 297–298
Collinear antennas, 258, 259
Color displays, 705–706
Color measurement, 512–514
Colorimetric measurements, 513
Color-temperature meters, 510
Combined actuator transducer (CAT), 652–653
Commissioning instruments. See Instrument installation and commissioning
Commissioning system, 21
Common impedance coupling, 801–803
Common ion effect, 378
Common mode capacitors, 852
Common mode faults, 759–760, 779–781
Common mode radiation, 827
   track length implications, 827–828
Common mode signals in bridge measurements, 462

Index

Communication channels, 679–684
 radio frequency (rf) transmission, 681
 sources and detectors, 683
 transmission lines, 679–681
Compact source lamps, 501
Component layout for filtering, 849–850
Components
 failure rates, 738, 742, 746
  effect of operating voltage on, 745
  effect of temperature on, 743–744
  estimating, 747–748
 for filtering, 850–852
 sampling, 662
 screening, 746
 temperature, estimating, 744–745
 tolerances, 751–752
Compressible fluids, critical flow of, 36
Computer automated measurement and control (CAMAC), 695–696
Computer integrated manufacturing (CIM)
 enterprises, modeling, 13–14
 standard, 629
Concentric orifice plate, 37
Condensate analyzer, 372–373
Condenser microphone, 596–600
Conducted emissions, 807–808
Conductive coating techniques, 863, 865
Conductivity. *See* Electrical conductivity
Conductivity probes, 111
Conductivity ratio monitors, 373
Cone-and-plate viscometer, 72–73
Confidence level, 746
Confidence limits, 746
Construction of instruments. *See* Design and construction of instruments
Continental Shelf Act 1964, 784
Continuous advanced control, 26
Continuous analog methods, 470
Control methodology, 20
Control modules, 26
Control system faults, 767
Controller, 20
Conventional light source, characteristics, 499
Converters
 electropneumatic, 734–735
 pneumatic-to-current, 730–732
Coolers, 666
Cooling, detector, 506–507
Cooling profile, liquid cast iron, 336
Core saturation, 854
Coriolis effect, 231
Coriolis mass flowmeters, 229–233
Cosmic rays, 563
Couette viscometer, 72
Coulometric instruments, 431–432
Coulter counter, 188
Coupling compliance in vibration sensor, 116
Couplings, 647
Crest factor, 462
Critical flow of compressible fluids, 36
Cross-axis coupling in transducers, 116
Cross-correlation system, 58
CRTs. *See* Cathode ray tubes
Crystals, analyzing, 354
CTs. *See* Current transformers

Curie temperature, 313, 421
Curie-Weiss law, 421
Current measurement, 462
Current transformers (CTs). *See also* Voltage transformers (VTs)
 AC range extension
  current/ratio error, 453
  equivalent circuit, 452
  phase-angle error/phase displacement, 453
  principle of, 452
  ratio errors, 454
Cyber security, ICS, 28
Cyclosizer analyzer, 188

D

D series sensors, 231, 232
DAC. *See* Digital-to-analog converter
Dall tube, 38–39
Damage potential, definition, 765
Damping adjustments, 224
Damping ratio, 115
Daniell cell, EMF of, 379
Data integrity, 245
Data interfacing standards, 693–697
Data loggers, 713
Data sensitivity, 255
Data transmission, 245, 693–697
DataGator flowmeter, 38, 63
DC common-mode rejection, 461
DC Wagner earthing arrangement, 478, 482
Deadband control, 23
Deadweight testers, 147–149
Decision-tree approach, 456
Deer rheometer, 73
Deflecting-vane flowmeter, 50–51
Degassing system in HPLC system, 329
Demodulation, radiofrequency, 840
Density measurement
 buoyancy methods, 136
 gas, 141–143
 hydrostatic-head method, 137
 liquid, 140–141
 radiation methods, 140
 using resonant elements, 140–143
 weight methods, 135–136
Density transmitters, 236–239
Depletion region, 533
Derating process, 744
Design and construction of instruments
 electronic instruments, 647–650
 elements of construction
  ceramics, 643
  electronic components and printed circuits, 640–642
  epoxy resins, 644
  interconnections, 642–643
  mechanical manufacturing processes, 644–646
  metals, 643
  plastics and polymers, 644
  surface-mounted assemblies, 642
 functional components, 646–647
 mechanical instruments, 650–653
 viewpoint of designer's, 639–640

Design automation, 753–754
Design software, 25–26
Desorption, 670
Detectors, 502
 applications, 533–541
 array, 505–506
 characteristics of, 502
 circuit time constants, 506
 classification of, 524
  Cherenkov detectors, 524
  cloud chambers, 524
  gas detectors, 526–528
  plastic film detectors, 524
  scintillation, 528–532
  solid-state, 532–533, 538
  thermoluminescent detectors, 525
 cooling, 506–507
 in HPLC system, 330–331
 photomultipliers, 502–503
 photovoltaic and photoconductive, 503–504
 pyroelectric, 504–505
 sources and, 683
 techniques, 506–508
 x-ray, 559
Deuterium lamps, 501
Dewpoint, 430
 instruments, 431
Diaflex, 217
Dial thermometers, 320
Diamagnetics, 421
Diaphragm gauge, 167
Diaphragm meters, 48
Diaphragm motor actuators, pneumatic, 732–733
Diaphragm pressure elements, 150–151, 160, 161
Diaphragm pump, 668
Differential beta-transmission gauge, 562
Differential pressure (DP)
 devices, 36–37
  installation requirements for, 39
  pressure loss in, 39
 measurement, 37
 transmitter methods, 137–138
  bubble tubes, 139–140
  flanged/extended diaphragm, 139
  overflow tank, 138
  pressure repeater, 139
  pressure seals, 139
  wet leg, 138–139
Differential pulse polarography, 333–334
Differential refractometers, 331
Differential transmitters, 226–229
Differential-mode capacitors, 853
Differential-mode radiation, 826–827
Diffraction, x-ray, 355
Diffusion-tube calibrators, 427–428
Digital circuits
 components, 829–830
 decoupling, 829
 failure modes, 743
 immunity, design for, 833–835
Digital communication, 223
Digital frequency counter, 491

Digital multimeters (DMMs)
    AC voltage and current measurement, 462
    DC input stage and guarding, 461
    elements of, 462
        control, 463
        specifications, 464–465
        visual display of, 463
Digital multiplexing, 241–242
Digital pressure transducers, 162–163
Digital quartz crystal pressure sensors, 156–158
    design and performance requirements, 157
    mechanisms, 157, 158
Digital signal transmission, 678, 690
    modems, 692–693
Digital transient recorders, 607–608
Digital voltmeters (DVMs)
    AC voltage and current measurement, 461
    DC input stage and guarding, 461
    elements of, 462
        control, 463
        visual display of, 463
Digitally coded systems, 750
Digital-to-analog converter (DAC)
    most significant bit (MSB) of, 456
    N-bit R-2R ladder network, 457
Digitizing oscilloscopes, 708–709
Dilution gauging, 64, 67
Dipole antenna, 258, 259
Dipping refractometers, 515
Direct sequence spread spectrum (DSSS), 257
Direct-indicating analog wattmeters, 465–467
Directional antennas, 258–259
Discharge coefficient, 34
Discharge lamps, 500–501
Discharge-tube gauge, 170
Dispersive infrared analysis, 345–346
Dispersive x-ray fluorescence analysis, 553
Displacement measurement, 120
Display elements, 26
Display techniques, 700
    cathode ray tubes, 704
    color, 705–706
    graphical, 709
    light-emitting diodes, 700–702
    liquid crystal displays, 702–703
    plasma, 703, 704
Disposable glass microfiber element, 665
Dissolved oxygen analyzer, water-sampling system for, 676
Distillation point analyzer, gas oil sampling for, 675–676
DL series sensors, 231, 232
DMMs. See Digital multimeters
Doppler anemometry, 202–203
    fluid flow, 204–206
    particle size, 203–204
    vibration monitoring, 206–210
Doppler blood-flow velocity measurement, dual-fiber method for, 205
Doppler flowmeters, 54
Doppler shift, 203, 206, 207
Double 2/3 logic, 777, 778
Double ratio transformer bridge, 485
Double-beam techniques, 510
Double-ended tuning fork force sensor, 157

Drilling in mechanical manufacturing, 644
Dry gases, 35–36
DT series sensors, 231, 233
Dual gauges, 557
Dual-beam oscilloscopes, 706
Dual-ramp ADCs, 457–458
Dual-slope integrated-circuit chip set, 459
Dust explosions, 792
DVMs. See Digital voltmeters
Dynamic matrix control (DMC), 623
Dynamic viscosity of fluid, 32–33
Dynamometer instruments, 454
    operation of, 454
    steadystate deflection, 454
    torque, 454
Dynamometer wattmeter, 466
    connection of, 467–468
    correction factors, 466
    errors in, 466
    FSD, 467
Dynode resistor chains, 544

E

Earth leakage circuit breaker (ELCB), 791, 792
Earthing. See Grounding
Eddy-current damping, 445
Eddy-current testing, 570–571, 589
Eductor or aspirator type pumps, 666
Effective radiated power (ERP), 258
Elastic elements, 129
Electret microphones, 600
Electric induction, 802–803
Electrical capacitance sensors, 86–87
Electrical conductivity
    of liquids, 364–365
    measurement
        alternating current cells with contact electrodes, 365–367
        applications of, 372–375
        direct measurement method, 369, 370
        electrodeless method, 371
        multiple-electrode cells, 369–370
        temperature compensation, 370–371
        Wheatstone bridge method, 369
    of pure water, 372
    of solutions, 365
Electrical conductivity cells, 365
    calibration of, 367
    cell constant, 366
    cleaning and maintenance of, 368
    construction, 367–368
    retractable, 368
    stainless steel and monel, 369
Electrical energy measurement, 472–473
Electrical impedance, 434
Electrical magnetic inductive processes, 84–86
Electrical sensor instruments, 432
Electrical SI units, 439
Electrochemical analyzers, 393–399
Electrochemical milling, 645
Electrocution risk, 790–791
Electrode potentials
    Daniell cell, 379
    general theory, 378–379
    variation with ion activity, 380

Electrodeless method, 371
Electrodes
    annular graphitic, 369
    antimony, 387–388
    calomel, 379, 382–384
    gas-sensing membrane, 381–382
    glass. See Glass electrodes
    heterogeneous membrane, 381
    hydrogen, 387
    ion-selective. See Ion-selective electrodes
    liquid ion exchange, 381
    pH, 383, 385
    platinized, 368
    redox, 382, 390
    reference, 382–384
    silver/silver chloride, 382
    solid-state, 381
    thalamid, 384
Electro-explosive devices (EEDs), 800–801
Electrolytes, 365
Electromagnetic compatibility (EMC)
    analog circuit design, 824–841
    control measures, 815
    customer requirements of instrumentation, 865
    and data security, 801
    digital circuit design, 824–841
    directive, 865
        CE mark and declaration of conformity, 868
        compliance with, 868–869
        legislation of, 866
        sale and use of products, 866–867
        scope and requirements, 866–867
        standards relating to, 869–870
        systems, installations, and components, 867–868
    and EEDs, 800–801
    emissions, 805–808
    equipment in isolation, 798
    filtering, 848–858
    grounding and layout, 815–824
    interference with radio reception, 799
    line-voltage coupling, 803, 804
    line-voltage signaling, 800
    line-voltage supply disturbances, 800
    malfunctions, 798
    malfunctions vs. spectrum protection, 799
    phenomena, 798
    regulatory framework for, 865–870
    shielding, 858–865
    source emissions and victim, 801–805
    subsystems within installation, 797–798
    susceptibility, 808–814
    system partitioning, 815–816
    between systems, 797–798
Electromagnetic (EM) energy, 258
Electromagnetic (EM) wave, 253, 803
Electromagnetic flowmeters, 51–54, 63, 233–234
    accuracy of, 54
    application, 53
    field-coil excitation, 52
    installation, 54
    nonsinusoidal excitation, 53

Index

principle of operation of, 51
sinusoidal AC excitation, 52–53
Electromagnetic force (EMF), Weston standard cell, 442
Electromagnetic incompatibility, examples of, 797
Electromagnetic interference (EMI), 797, 824
Electromagnetic radiation, use of, 87–90
Electromagnetic spectrum, 581
Electromagnetic velocity probe, 65
Electron capture detector, 409
Electron paramagnetic resonance (EPR) spectroscopy, 356
Electron probe microanalysis, 354
Electronic assemblies, 541–542
Electronic components
 construction of, 640–642
 selection, 755–756
Electronic conduction, 364
Electronic flow controller, operating principle of, 416, 417
Electronic instruments, 299
 construction
  bench-mounting, 647–649
  panel mounting, 647
  portable mounting, 649–650
  rack-mounting, 649
  site mounting, 647
Electronic monitoring system, designing, 765
Electronic multimeters, 448
 specification, 450–451
Electronic sources, 501
Electronic wattmeters, 469–470
 specification, 472
Electropneumatic converters, 734–735
Electrostatic discharge (ESD)
 interference paths, 834
 operator interface, 834
 protection measures, 813, 834, 835
 waveform, 812–813
Electrostatic hazards, 785–786
Electrostatic voltmeter
 principle of, 455
 torque, 455
Electrostatic wattmeter, torque in, 467
ELITE (CMF) series sensors, 231
Elutriation, centrifugal, 187–188
EMC. *See* Electromagnetic compatibility
Emergency core cooling system, 764
EMI. *See* Electromagnetic interference
Emissivity corrections for radiation thermometer, 309–310
Encapsulation, electronic components, 650
Enclosure design, shielding, 863–864
Engine instruments, layout in Boeing 737-400 aircraft, 763
Enrico Fermi pile, 776
Enterprise transformation, 20-point checklist for, 15–18
Environmental data corporation stack-gas monitoring system, 346, 347
Environmental testing, 748–749
Epoxy resins in instrument construction, 644
Equilibrium relative humidity, 434
Equipment modules, 26

Equivalent circuit
 capacitors, 477–478
 current transformers, 452
 four-arm AC bridge, 483
 inductors, 477–478
 Kelvin double bridge, 478
 for resistance, capacitance, and inductance, 479
 resistors, 477–478
 thermistor system, 470
 of transformer ratio bridges, 483
 voltage transformers, 453
Error detection, 688–689
Error rates, human operator, 761
ESD. *See* Electrostatic discharge
Expansibility factor, 36
Exponential dilution technique, 428
Exponential failure law, 738–739
Extensional viscosity measurement, 74
Extent of error, 44
Extruding construction material, 645

F

Fabry–Perot sensor cavity, 197, 198, 210–212
Fade margin, 255
Fahrenheit temperature scale, 273–274
Fail-dangerous errors, 765, 767
Fail-safe error, 765, 767
Fail-safe features of railway signaling, 775
Fail-safe systems, designing, 765–766
Failure
 intensities, 769
 modes, 743
 probability, 741, 756, 758, 773
 types of, 765, 779
Failure density function, 747
Failure effort and mode analysis (FEMA), 17
Failure rates
 for aircraft controls, 774
 by bathtub curve, 742
 chemical plant, 783
 component, 738, 746
  effect of operating voltage, 745
  effect of temperature on, 743–744
  variation with time, 742–743
 dangerous, 777
  probability of, 778
 for transistors and capacitors, 757
Failure-tolerant systems, 771
Faradaic current, 333
Faraday-effect polarimeter, 517–518
Faraday's law, 51
Fast Fourier transform analyzers, 607
Fatal accident frequency rate (FAFR), 783
Fatal accident rate (FAR), 783
Fault tree of oil platform, 784
Faults
 common mode, 759–760, 779–781
 estimating number of, 769–770
 vs. failures, 769
FC052 ultra-low-pressure transmitter, 152, 153
FDM. *See* Frequency-division multiplexing
Federal Communications Commission (FCC), 256

Feedback amplifiers, temperature effects on, 752
Feedback control, 23, 621
Feed-forward control, 24, 622, 623
Feedthrough capacitors, 851–852
Ferguson spin-line rheometer, 74
Ferrites, 850–851
Ferromagnetics, 421
Fiber optic communication, 681–684
Fiber optic Doppler anemometer (FODA), 203–205
Fiber optic sensors. *See* Optical fiber sensors
FID. *See* Flame ionization detector
Field calibration methods, 23
Field control system (FCS), 249
Field controllers, 617, 618
Field effect transistor (FETs), 386
Fieldbus, 241
 analog instruments, 249
 concept of, 241
 diagnostics for, 249
 digital multiplexing, 241–242
 field-mounted control, 248
 FOUNDATION, 247–248
 function and benefits, 247–251
 HART protocol, 243–246
 layer implementation of, 242
 network, 221
 sensor validation and, 249
Field-coil excitation, 52
Field-survey radiation instruments, 563
Filter, manual self-cleaning, 666
Filtered connectors, 855
Filtering, 848
 circuit effects of, 855
 components, 850–852
 contact suppression, 857–858
 filter configuration, 848–850
 I/O, 855
 line-voltage, 852–855
 transient suppression, 856–857
Final control element, 20
Finishes in instrument construction, 644
Fisher torque tube, 109
Fixed frequency, 256
Flame ionization detector (FID), 406–407, 417
Flame photometric detector (FPD), 409–410
Flammable atmospheres, 791–795, 800
Flapper/nozzle system, 716, 717, 724
Flash point (FP), 782, 783
Float-driven instruments, 107–108
Flow calorimeters, 470
Flow cells, 532
Flow equations, 34–35
 to apply to gases, 35–36
 for volume rate, 62, 63
Flow measurement, 669–670
 dilution method, 560
 in open channel
  by head/area method, 60–63
  by velocity/area methods, 63–64
 plug method, 561
 streamlined, 31
 turbulent, 32
Flow nozzle, 38

Flowmeters
    calibration methods
        for gases, 67–68
        for liquids, 66–67
    Coriolis mass, 229–233
    electromagnetic, 233–234
    Foxboro 8000 series, 235
    microprocessor-based transmitter and, 229–236
    valve as, 635–636
    vortex, 234–236
Fluid flow rate, 323
Fluid motion, energy of, 32
Fluids in containers measurements, 551–552
Fluorescence spectroscopy, atomic, 352–353
Fluoroscopic images
    advantages, 585
    characteristics of, 585–586
Fluoroscopy, 585–586
Fluted-spiral-rotor flowmeter. *See* Rotating-impeller flowmeter
Flux-leakage detection, 569–570
Foil gauges, 94
Force calibration, 117–118
Force measurement methods
    acceleration measurement, 129
    compound lever balance, 128
    concepts of, 127
    equal-lever balance, 127–128
    force-balance, 128–129
    hydraulic load cells, 129
    piezoelectric transducers, 130
    proving rings, 129–130
    spring balances, 129
    strain-gauge load cells, 130–133
    unequal-lever balance, 128
Force-balance controllers, 725–729
Force-balance mechanism, 717
    in buoyancy transmitter, 722
    for force measurement, 128–129
    forces for, 719
    in level transmitter, 721
    in pressure transmitter, 719, 720
    in speed transmitter, 722
    in target flow transmitter, 722
    for temperature measurement, 718
Force-measuring pressure transmitters, 160–162
Form factor (FF), current waveform and errors, 447, 448
Forward-error correction (FEC) techniques, 256, 688
Fossil-fuel effect, 565
FOUNDATION fieldbus, 247–248
Four-arm AC bridge
    classification of, 479–480
    equivalent circuit of, 483
    impedance in, 478–482
    for measurement of capacitance and inductance, 481–482
    stray capacitances in, 483
Fourier spectrum
    logic family, 825–826
    time domain/frequency domain transform, 824–825

Four-quadrant analog multiplier wattmeter, 471
Four-quadrant multiplier, 469
FPD. *See* Flame photometric detector
Free air ionization chamber, 538
Free radiation frequencies, 799
French reactors, 778
Frequency analyzers
    fast Fourier transform, 607
    narrow-band, 606–607
    octave band, 604–605
    output options with digital, 606
    third-octave, 605–606
Frequency bands for industrial wireless applications, 254
Frequency counters, 489–491
Frequency hopping spread spectrum (FHSS), 256–257
Frequency measurements
    relative accuracy of, 492
    relative resolution, 490
    using Lissajous figures, 497
Frequency modulation (FM) recording systems, 713
Frequency modulation spectroscope, 577
Frequency ratio measurement, 493
Frequency shift keying (FSK), 228, 229, 692
Frequency telemetry, 678
Frequency transmission, 690
Frequency-division multiplexing (FDM), 684–685
Fritsch Analysette sieve shaker, 181
Frostpoint, 430
Fuel cell oxygen-measuring instruments, 397
Full-scale deflection (FSD), 446–447
    of moving-iron instruments, 451
    of permanent magnet-moving coil instruments, 447
    of thermocouple instruments, 452
Full-wave bridge rectifier, 447, 448
Fullwave rectifier, 542
Fundamental interval, 272
Furnace gas probes, 664
Furnace temperature by radiation thermometer, 312
Fuzzy logic control, rules for, 24

G

Galvanometer, 298, 734
    recorders, 711–712
Gamma ray, 581, 582
    detection, 537–538
    spectroscopy, 357
GAMP4 Guidelines, 26
Gas
    detectors, 401, 524, 526–528
        applications of, 413
        catalytic, 411
        electron capture, 409
        flame ionization detector, 406–407, 417
        flame photometric detector, 409–410
        helium ionization, 408, 417
        photo-ionization detector (PID), 407–408
        properties of, 413

        semiconductor, 411–412
        thermal conductivity, 404–406, 417
        ultrasonic, 410–411, 418
    explosions, 792
    pumps, 666–668
    sampling, 661, 672–674
Gas analyzer, 401
    Bendix oxygen analyzer, 424
    calibration of, 426
        dynamic methods, 427–428
        static methods, 427
    magnetic wind oxygen analyzer, 422–423
    magnetodynamic oxygen analyzer, 423–424
    measurement principles of, 426
    nitrogen oxides analyzer, 425–426
    ozone analyzer, 424–425
    paramagnetic oxygen analyzers, 421–424
    Quincke oxygen analyzer, 423
    Servomex oxygen analyzer, 424
Gas chromatography, 402–404
    apparatus for, 402
    backflush method, 403
    heart-cut method, 403
    process. *See* Process chromatography
Gas density
    measurements, 141–143
    transducer, 142
Gas proportional counters, 535
    radiocarbon dating by, 563–565
Gas-cooled nuclear reactors, 429
Gaseous mixtures
    preparation of, 426, 427
    separation methods of
        chemical reactions, 402
        physical methods, 402
        physico-chemical methods, 402
    thermal conductivity of, 404
Gas-filled instruments, 281–282
Gaskets and contact strip, shielding, 862–863
Gas-sampling valve, 415
Gas-sensing membrane electrodes, 381–382
Gate meter, 39–42
Gauge factor, 172
Gauges
    absolute, 165, 166–168
    backing-fluorescence, 562, 563
    beta-backscatter, 562
    coating-fluorescence, 562, 563
    differential beta-transmission, 562
    dual, 557
    foil, 94
    level, 559, 560
    mechanical, 166–167
    neutron-moisture, 557
    nonabsolute, 165, 166, 169–173
    preferential absorption, 562, 563
    properties of, 166
    self-balancing level, 110
    semiconductor, 94–95
    sight, 107
    strain. *See* Strain gauges
    surface-neutron, 557
    wire, 94
Gauging systems, automatic, 91–92
Gaussian distributions, 178–179

Index

Gaussian fringe envelope, 201
Gear pumps, 668
Geiger counters, 535, 536
Geiger–Mueller detectors, 527–528
Generating a solvent gradient, 329
Geometry of radiation, 533–534
Gilflo primary sensor, 42
Glass electrodes, 381, 384–385
    cleaning, 389–390
    continuous-flow type of assembly, 388–389
    electrical circuits for, 385–387
    immersion type, 389–390
Glass thermometers, mercury filled, 274–278
Global support system (GSS), designing, 17
Golay pneumatic detector, 346
Gradient devices in HPLC system, 329
Graphical displays, 709
Graphical recorders, 709–710
Gravimetric calibration method, 67, 68
Gravitrol density meter, 136
Grinding in mechanical manufacturing, 645
Ground plane
    breaks in, 821, 822
    on double-sided PCBs, 821
    partial, 821
    radiating loop area, 822
    vs. track, impedance of, 821
Grounding, 658, 791, 816
    configuring I/O, 823–824
    current through ground impedance, 816
    gridded structure, 820
    ground style *vs.* circuit type, 820–821
    hybrid, 817–818
    at interfaces, 823
    of large system, 818
    layout rules, 824
    map, 824
    multi-point, 817
    safety earth, 818
    single-point, 816–817
    wire impedance, 818, 819
Ground-up testing and training, 26
Gyroscopic/Coriolis mass flowmeters, 59

H

Hacking in wireless system, 257
Hagen-Poiseuille law, 71
Hall-effect technique, 470
Hall-effect wattmeter, 471
Handheld interfaces, 249–250
Handheld terminal (HHT), 233
Hardware, shielding, 862–865
Hardware reliability *vs.* software reliability, 768–769
Harmonic noise, 255
HART protocol, 243
    connection and length limitations in, 246
    operating conditions, 245
    structure, 244
    technical data, 245–246
Head of a weir, 60
Health and Safety at Work Act 1974, 784, 789
Heat, definition, 269
Helium ionization detector, 408–409, 417
Helix meters, 47

Henry's law, 430
Hersch cell for oxygen measurement, 397, 398
Heterodyne converter counter, 496, 497
Heterodyne interferometry, 194
Heterogeneous membrane electrodes, 381
Hiac automatic particle sizer, 188–189
High-current ammeters, 446
High-frequency impedance measurement, 487–488
High-frequency power measurement, 470–472
High-gain antennas, 259
High-performance liquid chromatography (HPLC)
    advantages, 329
    application, 331
    system components, 329–330
High-pressure sample, 672
High-reliability software, 769
High-resistance measurement, 476–477
High-temperature ceramic sensor oxygen probes, 396–397
High-temperature thermometers, 277–278
Holography, 90
    acoustic, 580
Honing in mechanical manufacturing, 645
Hooke's law, 129
Hook-type level indicator, 108
Hot-cathode ionization gauge, 171–172
Hot-metal thermocouples, 303
Hotwire gauges, 169
HPLC. *See* High-performance liquid chromatography
Human operator
    error rates, 761
    features of, 760–762
    speech communication, 761
Hydraulic flumes, 62–63
Hydraulic load cells, 129
Hydraulic pressure measurement, 129
Hydrocarbons, sulfur contents measurement in liquid, 557–558
Hydrocracking severity, 622
Hydrogen electrode, 387
Hydrogen ion activity. *See* pH
Hydrolysis, 376, 378
Hydrostatic-head methods for density measurement, 137
Hygrometer, 435
Hysteresis control. *See* Deadband control
Hysteresis effects, 171, 450

I

ICI mond system, principle of, 343
Ideal gas law, 35
    departure from, 36
IEEE-488 interface, pin assignments for, 696
IEEE standards, 19
Ignition
    of explosive and flammable atmospheres, 786
    risk reduction, 785
Image indicator (IQI), 583
    step/hole type, 584
    wire type, 584

Image-intensification systems, 586
Impaction, 188
Impedance
    for filtering, 848
    vs. insertion loss, 853
Impulse lines, 656–657, 659
Incandescent lamps, 500
Indicator devices, 699–700
Induction wattmeter, torque in, 467
Inductive load suppression, 858
Inductively coupled bridges. *See* Transformer ratio bridges
Inductors
    equivalent circuits of, 477–478, 480
    filtering, 849
    temperature effects on, 752
Industrial, scientific, and medical (ISM) bands, 254
Industrial automation
    analysis of security needs of, 28
    APA for, 24
    asset management, 21
    asset optimization, 21
    career and career paths in, 4–5
    definition, 3
    field calibration methods for, 23
    life-cycle optimization, 21
    measurement methods for, 23
    plant optimization, 21
    process control for, 23
    Purdue model, 14
    reliability engineering, 21
    scope for, 760
    security
        problem, 27–28
        recommendations for, 28
    simulation system in, 25–26
    standards, 19
Industrial control systems (ICS), cyber security, 28
Industrial environment, 870
Industrial viscometer, 73
Influence errors in vibration sensor, 116
Infrared absorption meter, 343–345
Infrared analyzers
    dispersive, 345–346
    nondispersive, 341–342
Infrared discrete-position level sensor, 111
Infrared instruments, 432, 433
Inlet systems, 359
In-line filters, 665
Inorganic scintillators, 528–531
Insertion loss *vs.* impedance, 853
Insertion turbine, 65–66
Insertion vortex, 66
Insertion-point velocity calibration, 66–67
Installation design, 20–21
Installing system, 21
Institute of Electrical and Electronics Engineers (IEEE), 262
Institute of Measurement and Control, 883–886
Instrument installation and commissioning
    cabling, 657–658
    grounding, 658
    loop testing, 659–660

mounting and accessibility, 655–656
piping and cable testing, 659
piping systems, 656–657
plant commissioning, 660
pre-installation testing, 658–659
requirements, 655
storage and protection, 655
Instrument management systems, 250–251
Instrument mounting, 655–656
Integrated circuits, accelerated life tests for, 746
Integration process, 31
Integrators, pneumatic, 729, 730
Intelligent transmitter, 222
definition, 221
developments of, 250
features, 223–224
integration of, 250–251
microprocessor-based and, 222–224
pressure and differential, 226–229, 238
span and zero adjustment, 223
temperature, 224–226
INTELSAT V, 772
Intensity measurement, 508–510
color-temperature meters, 510
photometers, 509
ultraviolet intensity measurements, 509–510
Intensity of magnetization, 421
Interferometric sensing techniques, 193–202
central fringe identification, 201–202
heterodyne interferometry, 194
pseudoheterodyne interferometry, 194–195
white-light interferometry, 195–201
Interferometry, 89–90
Internal energy of fluid in motion, 32
Internal reflux control (IRC), 624
International Bureau of Weights and Measures (BIPM), 440
International Practical Temperature Scale of 1968 (IPTS-68), 273
fixed points of, 274
International Society of Automation (ISA) standards, 19, 887–888
I/O filtering, 855
Ion chromatography, 373–375
Ion-exchange resins, scintillating, 532
Ionic conductivity, 365, 366
Ionization chambers, 526, 534, 536, 537–538
Ionization current, 406
Ionization gauges, 170–173
Ionization of water, 364
Ion-selective electrodes, 380–382
applications of, 392–393
conditioning and storage of, 392
determination of ions by, 390–391
flow cell for, 391
pH and pIon meters, 391
practical arrangements, 391–392
Ion-selective monitor, 392, 393
IPTS. *See* International Practical Temperature Scale of 1968
ISA S88 batch standard, 629
ISA100.11a standard, 263
Isotopes in gamma radiography, 582
IT security, 28

J
Jamming in wireless system, 257
Josephson effect, 443

K
Karl Fischer titration, 433
Katharometer, 405–406
Kelvin double bridge, 475, 478
Kelvin temperature scale, 272
Kinematic design
advantages of, 651
of mechanical instruments, 650–651
Kinematic viscosity of fluid, 32–33
Kinetic energy correction, 71
Kinetic energy of fluid in motion, 32

L
Ladder logic, 618
Laddic, core and windings of, 781
Laminar flow. *See* Streamlined flow
Lapping in mechanical manufacturing, 645
Laser Doppler anemometer, 64
Laser Doppler velocimetry (LDV), 206
Lasers, 501
interferometer, 80, 81, 89
safety precautions, 502
Latent heat, 270
Lateral effect cell, 88
Laurent polarimeter, 516–517
LCDs. *See* Liquid crystal displays
Leak detection with nuclear techniques, 559
LEDs. *See* Light-emitting diodes
Length measurement
contacting and noncontacting, 78
derived, 79
electrical capacitance sensors for, 86–87
electrical magnetic inductive processes, 84–86
electrical resistance for, 83–84
electromagnetic and acoustic radiation for, 87–90
electronic, 82–83
mechanical, 81–82
nature of length, 78–79
ranges and methods of, 77
standards and calibration of length, 77
Level controllers, 620
Level measurement
calibration of systems, 106
error sources, 104–105
full-range, methods of, 107–110
capacitance probes, 108–109
float-driven instruments, 107–108
force or position balance, 110
microwave and ultrasonic time-transit methods, 109–110
pressure sensing, 109
sight gauges, 107
upthrust buoyancy, 109
instrument installations, 103–104
neutrons for, 560
pneumatic instrumentation for, 721
sensor selection, guidelines, 106
short-range detection, methods of
electrical conductivity, 110–111
infrared, 111
magnetic, 110
radio frequency, 111–112
using x-rays/gamma rays, 559–560
Level recorders, 607
Lever-balance methods
compound, 128
equal, 127–128
unequal, 128
Life-cycle optimization, 21
Light industrial environments, 870
Light sources, 499–502
discharge lamps, 500–501
electronic sources, 501
incandescent lamps, 500
lasers, 501–502
Light-emitting diodes (LEDs), 501, 700–702, 764
and *p-i-n* diode detector, 683
Light-intensity measurement, 508
Linear variable-differential transformer (LVDT), 85, 86
Line-voltage
coupling, 803, 804
filters, 852–855
harmonics, 808
signaling, 800
supply disturbances, 800
supply fluctuations, 814
transients, 812
Liquid
crystals, 324–325
and gases, 270
pumps, 668–669
sampling, 661, 674–676
Liquid crystal displays (LCDs), 702–703
Liquid density measurement, 140–141
Liquid ion exchange electrodes, 381
Liquid level measurement systems, 239
Liquid manometers, 167
Liquid metal thermocouples, 303
Liquid sealed drum flowmeters, 49–50
Liquid-filled dial thermometers, 278–281
comparison, 284
gas-filled, 281–282
liquid-in-metal, 281
vapor pressure, 282–284
Liquid-in-glass thermometers, 274
Liquid-in-metal thermometers, 281
Liquid-scintillation counting, radiocarbon dating by, 563–565
Lissajous figures for phase and frequency measurement, 497
Lithium fluoride (LiF), 525
Load cells, 651–652
hydraulic, 129
spring mechanism, 652
strain-gauge. *See* Strain-gauge load cells
Local area networks (LANs), 253
Lockheed L 1011-500 airliner, 773
Logic noise immunity
dynamic noise margin, 835–836
susceptibility prediction, 836
transients coupling, 836

Index

Logic voting circuit, 757
Log-normal distributions, 179–180
Loop testing, 659–660
Low-resistance measurement, 475–476
Luft-type infrared gas analyzer, 342

M

Machine health monitoring, 116–117
Mach–Zehnder bulk optic interferometer, 206, 207
Mackereth oxygen sensor assemblies, 397–398
Magnetic field screening, 814
Magnetic flux surface-inspection methods, 569–570
Magnetic induction, 802
Magnetic level indicator, 108
Magnetic moment, 421
Magnetic particle inspection (MPI) in underwater, 588
Magnetic recording, 712–713
Magnetic wind oxygen analyzer, 422–423
Magnetic-reluctance proximity sensor, 85
Magnetization methods, 569
Magnetodynamic oxygen analyzer, 423–424
Magnetoelectric sensors, 86
Magnox gas-cooled reactors, 777
    fault sensors in, 777
    protection logic, 781
Manipulated variables (MVs), 623
Manometer
    capacitance, 152–154, 167
    with limbs of different diameters, 147
    liquid, 167
    U-tube, 145
    with wet leg connection, 147
Manufacturing execution infrastructure (MEI) plan, 17
Manufacturing Execution Systems Association (MESA), 14
Manufacturing execution systems (MES), 15, 26
Manufacturing operations management (MOM), 15
Mass flow rate, 43
Mass measurement, 561
Mass spectrometer, 357–358
    inlet systems, 359
    ion sources, 359
    principle, 358–359
    quadrupole, 361–362
    separation of ions, 359–361
    time-of-flight, 361
Mass-spring seismic sensors, open-loop, 118–119
Master meter, 67
McLeod gauge, 167–168
    Ishii effect, 168
    pressure calculation, 168
Mean time between failures (MTBF), 21, 773
    and reliability, 737–738
    for system assemblies, 747
Measured disturbances, 621–622
Measurement systems, 240
    liquid level, 239

Mechanical fail-safe devices, 766
Mechanical gas pump, 667–668
Mechanical manufacturing processes, 644–646
Mechatronics, 4
Melinex, 550
MEMS. See Microelectromechanical systems
Mercury-in-glass thermometers, 274–278
Mercury-in-steel thermometer, 278–280
Mercury/mercurous chloride. See Calomel electrode
Mesh technologies, 262
Message recovery, 256
Metal film resistors, temperature effects on, 752
Metallic conduction, 364
Meter prover. See Pipe prover
Michelson interferometer, 195, 197, 199, 209
Microelectromechanical systems (MEMS), 217–218
    pressure sensor, 218
    viscosity sensor, 218
Microphones
    calibration of, 609
    condenser, 596–600
    electret, 600
    at low frequencies, 600
    positions for, 612
Microprocessor
    defense programming, 838
        data and memory protection, 839
        digital inputs, 839
        input data validation and averaging, 838–839
        interrupts, 839
        reinitialization, 839–840
        unused program memory, 839
Microprocessor techniques, 53, 58
Microprocessor watchdog
    operation, 836
    retrigger pulse, 837
    testing, 838
    timer hardware, 837
Microprocessor-based transmitter
    definition, 220
    features, 222–223
    and flowmeters, 229–236
    and intelligent transmitter, 222–224
        density transmitters, 236–239
        pressure and differential transmitter, 226–229, 238
        temperature transmitter, 224–226
        user experience with, 246–247
    measurement systems, 240
        liquid level, 239
Microprocessors, 753
Microscope counting, basic methods, 181–182
Microwave instruments, 433–434
Microwave spectroscopy, 355–356
Microwave time-transit methods, 109–110
Microwave-frequency measurement, 495–497
Milling, 644
    chemical and electrochemical, 645
Mineral-insulated thermocouples, 302–303
Miniature in-line filter, 666
Mixer unit, 545

Mobile phase in HPLC system, 330
Model 3051C transmitters, 227–229
Model predictive control (MPC), 24, 619
    emergence of, 623
    licensors of software, 625
    problems with, 625–626
    recommendations for using, 627
    vs. ARC, 623–624
Model RTF9739 transmitter, 231, 233
Modems, 692–693
Modified Wheatstone bridges, 476
Moiré fringe position-sensing methods, 88, 89
Moiré-type fringe pattern, 200
Moisture
    in gases, 429–430
    in liquids and solids, 430–431
Moisture measurement, 348
    calibration, 435–436
    definitions, 429–430
    in gases, 399, 431–433
    in liquids, 399, 433–434
    by neutrons, 555–557
    in solids, 434–435
Moisture meter, 556
Monochromator, 346
Moore's law, 617
Motion-balance controllers, 717, 723–725
    configuration of, 724
    for temperature measurement, 718
Moving-iron instruments
    damping of, 448
    errors, 451
    friction in, 450
    inductance effects in, 450, 451
    steady-state deflection, 450
Moving-pointer indicators, 699, 702
Moving-scale indicators, 700, 702
MPC. See Model predictive control
MTBF. See Mean time between failures
MTL 418 transmitters, 224–225
Muller bridge, 477
Multichannel analyzer (MCA), 544
Multichannel spectrometer, 354
Multidecade ratio transformers, 484
Multimeters, 448
    specification, 449
Multiple-electrode cells, 369–370
Mutual-inductance methods, 85

N

Nanotechnology
    definition of, 217
    for pressure transmitters, 217
Narrow-band analyzers, 606–607
National Physical Laboratory (NPL), 440
NDT. See Non-destructive testing
Negative temperature coefficient thermistors, 290–291
Nernst equation, 380
Neutron activation, 357, 552–553
Neutron moderation, 435
Neutron sources, 522
Neutron-moisture gauges, calibration, 557
Neutron-proton reactions, 540

Neutrons
    for level measurement, 560
    for moisture measurement, 555–557
Neutrons detection, 538–541
Newspeak, 771
Newtonian behavior, 69–70
Nickel resistance thermometers, 289
Nitrogen oxides analyzer, 425–426
Noise labeling, 596
Noise measurement
    calibration of instruments, 609
    for diagnostic purposes, 596
    for engineering design/noise-control decisions, 595–596
    for evaluating noise effect on human beings, 595
    frequency analyzers, 604–607
    humidity and rain effect on, 613
    instrumentation for
        acoustic calibrators, 603–604
        frequency weighting networks and filters, 600–601
        microphones, 596–599
        noise-exposure/dose meters, 603
        sound-level meters, 601–603
    other noises effect on, 613
    recorders, 604–608
    standards for, 597
    temperature effect on, 613
    wind effect on, 613
Noise-exposure/dose meters, 603
Noncontacting sensor, 78
Non-destructive testing (NDT)
    certification of personnel, 590
    developments in, 590
    purpose of, 567
    radiography, 580–586
    standards in ultrasonic testing, 590–591
    surface inspection, 567–571
    ultrasonic. *See* Ultrasonic non-destructive testing
    underwater. *See* Underwater non-destructive testing
    visual inspection, 568
Nondispersive infrared analyzers, 341–342
Non-dispersive X-ray fluorescence analysis, 554–555
Nonintrusive simulation interfaces, 25
Non-Newtonian behavior, 69–70
Nonsinusoidal excitation, 53
Normal distributions. *See* Gaussian distributions
North Sea oilfields, 783
Nuclear gauging instrumentation, 521
    electronics for, 541–546
Nuclear gauging system, 105
Nuclear magnetic resonance spectroscopy, 357
Nuclear measurement techniques
    materials analysis, 552–559
    mechanical measurements, 559–563
    optimum time of measurement, 551
Nucleonic belt weigher, 561
Numerical aperture (NA) of optical fiber, 682
Nutating-disc flowmeter, 45

O

Octave band analyzers, 604–605
Ohm, absolute determination, 443–444
Omnidirectional antennas, 258
One-shot time-interval measurement, resolution, 495
On/off control, 23
Open circuit failure, 766
Open loop control. *See* Feed forward control
Open System Interconnection (OSI) reference model, 244
Operational amplifiers, voltage effect on failure rates, 745
Optical fiber, 681–683
    in anemometry applications, 202–203
    multimode, 192, 197
    numerical aperture of, 203
    single mode, 192, 195, 197
Optical fiber sensors. *See also* Doppler anemometry; Interferometric sensing techniques
    classification, 192
    distributed sensor, 192
    extrinsic sensor, 192
    in-fiber sensing structures, 210–215
    intrinsic sensor, 192
    low-coherence high temperature, 197
    modulation parameters, 192–193
    performance criteria, 193
    performance-and market related factors, 191
    point sensor, 192
    principles of, 192–193
    two-wavelength intensity modulated, 193
Optical instruments, 499
Optical interferometer, 79, 81
Optical position-sensitive detectors, 87, 88
Optical properties measurement, 514–518
Optical radiation thermometers, 314
Opto-couplers, 841
OR logic function, 781
Organic scintillators, 530, 531
Orifice plate, 37
    advantages and disadvantages, 37
Oscillating disc sensing, 57
Oscillatory fluidic flowmeters, types, 56–58
Oscilloscope
    analog, 706
    digitizing, 708–709
    dual-beam, 706
    frequency and phase measurement, 497
    sampling, 708
    storage, 707–708
Ostwald viscometer, 71
Oval-gear flowmeter, 46
Oven-stabilized oscillators, 490
Oxygen analyzers, paramagnetic, 421–424. *See also specific oxygen analyzers*
Oxygen probes, high-temperature ceramic sensor, 396–397
Oxygen sensor
    galvanic Mackereth electrode, 397–398
    microfuel cell, 397
    polarographic, 395–396
Ozone analyzer, 424–425

P

Paints in instrument construction, 644
Panel mounting instruments, 647, 648
Paper chromatography, 328–329
Paper dielectric capacitors, voltage effect on failure rates, 745
Parallel systems, reliability, 748
Parallel-plate rheometer, 73
Parallel-plate viscometer, 73
Paramagnetic oxygen analyzers, 421–424
Parasitic reactances, 848
Parity-bit coding, 689
Particle sizing
    Ferêt diameter, 176, 177, 182
    Gaussian distributions, 178–179
    image shear diameter, 176
    log-normal distributions, 179–180
    Martin's diameter, 176
    projected area diameter, 176
    relative frequency plot, 178
    Rosin–Rammler distributions, 180
    sampling, 175
    statistical mean diameters, 176
    Stokes diameters, 176
Particles
    characterization of, 175–176
    measuring size of, 180–183
    measuring terminal velocity of, 183–188
    methods for characterizing groups of, 178–180
    optical effects caused by, 177
    shape, 177–178
Passenger lifts, 766
PCB. *See* Printed circuit board
PCM. *See* Pulse code modulation
Pellistor, 411
Peltier effect, 294–295
Penning ionization gauge, 170–171
Period measurement, relative accuracy of, 492
Peristaltic pumps, 669
Permanent magnet-moving coil instruments, 445–448
    AC voltage and current measurement, 447
    characteristics of, 447
    construction, 445
    FSD, 446–447
    torque in, 445, 447
Permeability of gases, 671
Permeation-tube calibrators, 427–428
pH
    buffer solutions, 378
    of common acids, bases, and salts, 377
    common ion effect, 378
    electrode, 383, 385
    general theory, 375–376
    hydrolysis, 376, 378
    measurements, 384–390
    measuring circuit using field effect transistor, 386
    and Na ion error, relationship of, 387
    neutralization, 376
    scale, 376
    standards, 376
pH meters, 381, 387

Index

Phase measurement using Lissajous figures, 497
Phase shift keying (PSK), 692
Phaselock loop (PLL), 207, 208
Phase-sensitive detection system, 86
Phasor diagram for three-voltmeter method, 465
Phosphors characteristics, 705
Photo-acoustic spectroscopy, 355
Photoconductive detectors, 503–504
Photodiode array, 505
Photoelastic visualization methods, 579
Photoelasticity of strain gauges, 100–101
Photoelectric radiation thermometers, 316
Photography, 587
Photo-ionization detector (PID), 407–408
Photometers, 509
Photomultipliers, 502–503, 532
Photon counting detector, 508
Photopotentiometer, 88
Photosedimentation, 184
Photovoltaic detectors, 503–504
pH-to-current converter, 386
Physical vapor deposition (PVD), 217
Piezoelectric ceramic transducers, 55
Piezoelectric effect, 154, 155
Piezoelectric force sensor
    oscillator mode for, 157
    resonant, 157
Piezoelectric humidity instrument, 433
Piezoelectric (PZT) sensors, 122–123
    amplifiers for, 123–124
Piezoelectric transducer, 130
Piezometer ring, 38
Piezoresistive pressure sensors, 155–156
Pin assignments
    for IEEE-488 interface, 696
    for RS-232 standard, 694
    for RS-449 standard, 695
PIon meters, 391
Pipe flow, hydraulic conditions for, 33
Pipe joining methods, 671
Pipe prover, 67
Piper Alpha explosion, 784
Piping systems, 656–657, 659
Pirani gauge, 169–170
Pistonphone, 604, 605
Pitot tube for pressure measurement, 64–65
Plain-wire thermocouples, 301
Planck's radiation law, 518
Plant commissioning, 660
Plant optimization, 21
Plasma displays, 703, 704
Plastic film detectors, 524
Plastic scintillators, 531–532
Plastics in instrument construction, 644
Plating, 335
Platinum resistance thermometers, 287
Pneumatic force-balance pressure transmitters, 159–160
Pneumatic instrumentation
    analog square root extractor, 729, 730
    automatic manual transfer switch, 727, 728
    batch-switch system, 727–729
    buoyancy measurements, 721
    characteristics of, 715
    controllers, 723–729
    diaphragm motor actuators, 732–733
    electropneumatic positioners, 735
    flapper/nozzle system, 716, 717, 724
    force-balance controllers, 725–729
    integrators, 729, 730
    level measurements, 721
    measurement and control systems, 716–717
    motion-balance controllers, 717, 723–725
    pneumatic-to-current converters, 730–732
    pressure measurement, 718–721
    speed, 722
    summing unit and dynamic compensator, 729–732
    temperature measurement, 717–718
    transmission, 722–723
    valve positioner, 733, 734
Pneumatic motion-balance pressure transmitters, 159
Pneumatic signals, 656
Pneumatic transmission systems, 677
Point velocity measurement
    electromagnetic velocity probe, 65
    hotwire anemometer, 64
    insertion turbine, 65–66
    insertion vortex, 66
    laser Doppler anemometer, 64
    pitot tube, 64–65
    propeller current meter, 66
    ultrasonic Doppler velocity probe, 66
Poisson distribution, 521
Poisson's ratio, 93
Polarimeters, 516–518
Polarizing beam splitter (PBS), 206, 207
Polarographic process oxygen analyzer, 395–396
Polarography
    applications of, 334
    differential pulse, 333–334
    direct current (DC), 331–332
    pulse, 333
    sampled DC, 332
    single-sweep cathode ray, 332
Polyester capacitor, voltage effect on failure rates, 745
Polymers in instrument construction, 644
Portable instruments, 649–650
Portable thermocouple instruments, 304
Position-sensitive photocells, 87–89
Positive displacement meters
    applications, 43
    fluted-spiral-rotor, 45
    liquid sealed drum, 49–50
    nutating-disc, 45
    oval-gear, 46
    principle of measurement, 44
    reciprocating piston, 44–45
    rotary-piston, 44
    rotating-impeller, 50
    rotating-vane, 46
    sliding-vane, 45–46
Positive temperature coefficient (PTC) thermistors, 291
Positive-displacement pumps, 668–669
Post-deflection acceleration (PDA), 704
Potential drop surface-inspection methods, 570
Potential energy of fluid motion, 32
Potentiometric instruments, 298–299
Power dissipation, 446, 465
    in three-phase measurement system, 468–469
Power factor correction (PFC), 808
Power failure, probability of, 768
Power loss, 467
Power management, 265
Power measurement
    high-frequency, 470–472
    three-phase, 468–469
    three-voltmeter method of, 465
    two-wattmeter method of, 469
Power Reactor Inherently Safe Module (PRISM), 776
Power supplies, 542
    high voltage, 543
Power surges, 260
Power-assisted steering, 767, 768
Power-factor measurement, 473–474
Preamplifiers, 116, 543
Precious metal thermocouples, 301
Predictive maintenance, 21
Preferential absorption gauge, 562, 563
Pressure
    absolute, 145, 149, 162, 165
    definition, 145, 165
    differential, 145, 162
    gauge, 145, 162
    units, 165
Pressure energy of fluid motion, 32
Pressure measurement, pneumatic instrumentation for, 718–721
Pressure measurements
    bellows elements, 151–152, 160
    bourdon tubes, 149–150, 158–160
    capacitance manometers, 152–154
    deadweight testers, 147–149
    diaphragm pressure elements, 150–151, 160, 161
    digital quartz crystal pressure sensors, 156–158
    low pressure range elements, 152
    piezoresistive pressure sensors, 155–156
    quartz electrostatic pressure sensors, 154–155
    Schaffer gauge, 150
    strain-gauge pressure sensors, 156
    types, 145
    units for, 145, 146
Pressure reduction, 662, 670
Pressure sensor
    feature of, 720
    MEMS, 218
    white-light interferometry, 198
Pressure tappings, 38, 41
Pressure transducers, 133
Pressure transmitter, 226–229, 238
    range limits, 229

Pressure transmitters, 158–159
    force-measuring, 160–162
    nanotechnology for, 217
    pneumatic force-balance, 159–160
    pneumatic motion-balance, 159
Prevost's theory, 307
Printed circuit board (PCB)
    construction of, 640–642
    design, 806
    ground plane on double-sided, 821
    layout, 819–821
    radiated emissions, 805–806
    track impedance, 819–820
Probes, 664–665
Process chromatography
    carrier gas, 416
    chromatographic data display, 418–419
    columns in, 417
    components of, 412, 414
    controlled temperature enclosures, 417
    data-processing systems, 418–419
    detectors, 417–418
    gas-chromatographic integrators, 419
    operation of, 419–421
    programmers, 418
    sampling system, 414–415
    switching and logic steps, 420, 421
Process control, 23–24
Process disturbance
    measured, 621–622
    unmeasured, 620
Process variable (PV), 621
Profibus-PA, 247–248
Programmable automation controller (PAC), 618
Programmable logic controllers (PLCs), 617
Prompt gamma-ray analysis, 553
Propeller current meter, 66
Proportional-integral-derivative (PID) control, 23–24
    algorithm, 622, 626
Proton-recoil counters, 540
Proving rings, 129–130
Proximity transducer, 651
Pseudoheterodyne interferometry, 194–195
Pseudo-plasticity, 70
Psychrometers, automatic, 433
Pulse amplitude modulation (PAM), 685
Pulse code modulation (PCM), 685–686
    bit error probability of, 688
Pulse code modulation (PCM) techniques, 713
Pulse output, 223
Pulse polarography, 333
Pulse telemetry, 678
Pulsed frequency-modulation spectroscope, 578
Pulsed laser grating method, 213
Pulse-duration modulation (PDM), 686
Pulse-echo spectroscopy, 578
Pulse-height analyzers, 543–544
Pulse-position modulation (PPM), 686
Pulse-width ADCs, 458–460
    effect of time-varying input on, 461
Pumps
    gas, 666–668
    in HPLC system, 329–330
    liquid, 668–669

Purdue model, 14
Pyroelectric detectors, 313, 504–505
Pyrometric cones, 324

Q

Q meter, 488, 489
Quadrupole mass spectrometer, 361–362
Quality factor, 478
Quartz crystal oscillator, 489
    characteristics, 492
    frequency stability of, 490
    instrument, 432–433
Quartz crystals, 157
Quartz electrostatic pressure sensors, 154–155
Quartz spiral gauge, 167
Quincke oxygen analyzer, 423

R

Race hazard phenomenon, 754
Rack-mounting instruments, 649
Radiant emission effect, 316
Radiated coupling, 809
    cable loading, 810
    cable resonance, 810
    cavity resonance, 811
    current injection, 810
    field generation, 803–804
    modes, 805
    wave impedance, 804–805
Radiation, 271
    acoustic, use of, 87–90
    from cables, 806, 807
    clock and broadband, 828
    for density measurement, 140
    electromagnetic, use of, 87–90
    errors, 322
    from logic circuits, 826–830
    from PCB, 805–806
    reflected, 348
Radiation pyrometers. See Radiation thermometers
Radiation shield, 534
Radiation sources, 522
Radiation surveys, undersea, 563
Radiation thermometers, 306
    applications, 318–319
    calibration, 312
    optical, 314
    photoelectric, 316
    pyroelectric technique, 313
    signal conditioning for, 318
    surface, 310–312
    total, 309–312
    types, 307
Radio bands, 254
Radio frequency (RF)
    bridges, 487
    signals, 254
    transmission, 681
Radio interference, 255
Radio noise, 255
Radio spectrum, 254
Radio test, 261
Radio transmitters, AM, 771–772
Radioactive decay, 523–524

Radioactive measurement relations, 550
Radioactive source, health and safety, 525–526
Radiocarbon dating, 563–565
    calculation of, 564
    statistics of, 564
Radiocarbon time scale, calibration of, 565
Radiofrequency demodulation, 840
Radiography
    application, 580
    fluoroscopic method, 585–586
    gamma rays, 581, 582
    image-intensification method, 585–586
    in non-destructive testing, 581
    sensitivity and image indicator, 582–585
    xerography, 585
    X-rays, 582
Radioisotope calcium monitor, 558–559
Radioisotopes, applications of, 549
Radiometrical surveys, land-based, 563
Railway signaling and control, 774–775
Rankine temperature scale, 273–274
Ratio control, 24
Reactor control
    principles of, 776–779
    requirements for, 776
Reactor protection logic, 781–782
Real-time rigorous optimization (RETRO), 627–628
Reciprocal frequency counters, 490, 491
    vs. time graph, 738
Reciprocating piston flowmeter, 44–45
Reciprocating piston pump, 668
Recorders, 701
    circular chart, 711
    data loggers, 713
    digital transient, 607–608
    galvanometer, 711–712
    graphical, 709–710
    level, 607
    magnetic, 712–713
    strip chart, 710–711
    tape, 608
    transient/waveform, 713
    XY plotters, 607
Rectangular notch, 60–61
Redox electrodes, 382, 390
Redundancy
    analog, 758–759
    level of, 758
    with majority voting circuit, 757–758
    parallel and series, 786
    of railway signalling, 775
    three-channel, 775
    use of, 756–757, 786
Redundant control system, 767
Redundant power supply switching, 760
Reference electrodes, 382–384
Reference voltage sources, 461
Reflected radiation, 348
Refractive index, 515
Refractometers, 514–516
    Abbé refractometer, 515
    of solid samples, 515–516
Relay contacts, 766
Relay power supply, failure of, 766

Index

Relay tripping circuits, 766
Relay-driving circuit, 767
Reliability
    budgets, 755
    conditions for, 737
    cost of, 776
    definition of, 737
    hardware and software, 768–769
    and MTBF, 737–738
    of oil supply, 784–785
    optimum, 739–740
    parallel systems, 748
    program, 770
    of signaling system, 774
    of system components, 740–741
    and total life cost, 740
Reliability engineering, 21
Remote reading thermometers, 319
Reservoir in HPLC system, 329
Residual chlorine analyzer, 393–395
Resistance measurement, 462
Resistance thermometers, 286–290
    connections, 289–290
    construction of, 289
    nickel, 289
    platinum, 287
    temperature/resistance relationship of, 287–289
Resistivities, 287
Resistors
    equivalent circuits of, 477–478
    equivalent series/parallel, 480
Resonant elements, density measurement using, 140–143
Resonant wire pressure transmitter, 161
Response time, 224
Reynolds number, 32, 35, 57
RF. *See* Radio frequency
Robotic systems, 775–776
Rod thermostat, 285
Root mean square (RMS) measurement, 462
Rosette of gauges, 95
Rosin–Rammler distributions, 180
Rotameter, 39
Rotary gas meter assemblies, 51
Rotary pumps, 667
Rotary vane pumps, 669
Rotary-piston flowmeter, 44
Rotating mechanical meters
    for gases, 48–51
    for liquids, 43–48
Rotating-impeller flowmeter, 45, 50
Rotating-vane flowmeter, 46, 51
Routing algorithms, 262
Royal Air Force Institute of Aviation Medicine (IAM), 763
RS-232 standard, pin assignments for, 694
RS-449 standard, pin assignments for, 695
RS-232/RS-485 interfaces, 163
Rubidium oscillator, 489

S

Safe failure, 779
Safety
    earthing and bonding, 791
    electrocution risk, 790–791
    flameproof, 794
    flammable atmospheres, 791–795
    in hazardous area, 792, 793, 795
    intrinsic, 794
    in non-hazardous area, 792, 793, 795
    standards for, 790, 792
Safety features, robotic systems, 775–776
Safety procedures, 764–765
Salt-in-crude-oil monitor, 375
Sample disposal, 662
Sample probe, 662, 664
Sample systems
    components, 662
        coalescers, 666
        coolers, 666
        filters, 665–666
        gas pumps, 666–668
        liquid pumps, 668–669
        pressure-reduction stage in, 670
        probes, 664–665
    construction, 663
    flow measurement, 669–670
Sample-conditioning system, 662
Sampled DC polarography, 332
Sample-transport system, 662
Sampling, 661
    analysis equipment parts, 662
    gas oil, 675–676
    mixed-phase, 662
    representative, 661–662
    steam, 673, 674
    time lags, 662–663
Sampling oscilloscopes, 708
Sampling wattmeter, 469–471
Satellite links, 772
Scattering, 534
Scattering coefficient, 177
Schaffer pressure gauge, 150
Schlieren visualization methods, 579
Scintillation counters, 524, 535
Scintillation detectors, 528–532
Sealers, 543
Sedimentation, 183
    cumulative methods
        centrifugal methods, 187
        decanting, 186
        sedimentation balance, 184–185
        sedimentation columns, 185–186
        two-layer methods, 186
    incremental methods
        Andreasen's pipette, 183–184
        density-measuring methods, 184
        photosedimentation, 184
Seebeck effect, 294
Segmental orifice plate, 37
Seismic sensors, mass-spring, 118
    open-loop, 118–119
Self-absorption, 534
Self-balancing level gauge, 110
Self-tuning controllers, 24
Semiautomatic microscope size analyzers, 182–183
Semiconductor, temperature measurement, 291–293
Semiconductor detector. *See* Solid-state detectors
Semiconductor gauges, 94–95
Sensor, 220
    basics of, 20–21
    validation, 249
Sensor instrumentation, fiber optics in. *See* Optical fiber sensors
Sensor/transmitter combination, 20
Sequence control, 26
Series mode rejection (SMR), 458
Servo accelerometers, 119
Servomex oxygen analyzer, 424
Shear strain, 93
Shear thickening, 70
Shear viscosity measurement, 71–73
Shear-thinning behavior, 69
Sheathed thermocouples, 302
Shielding, 858
    apertures effects, 860–862
    coating properties, 864–865
    with conductive coating, 863, 865
    enclosure design, 863–864
    hardware, 862–865
    mesh and honeycomb, 861
    performance, 863
    seam and aperture orientation, 862
    seams effects on, 861
    theory, 859–860
    use of subenclosure, 861
    windows and ventilation slots, 860
Shielding effectiveness (SE), 859–860
Shock
    calibration, 117
    measurement of, 124
Shop-floor viscometers, 73–74
Short circuit failure, 766
Shunt leakage resistance, 476
Shunt meter. *See* Bypass meters
Shuttle ball sensing, 57
SI units
    base units, 439–440
    electrical, 439
Sieving, 180–181
Sight gauges, 107
Sight-glass level indicator, 107
Signal coding, 749–750
Signal multiplexing, 684–685
Signal processing techniques, 841
Signaling system, reliability of, 774
Signal-to-noise ratio (SNR), 255
Silicon junction diode, 291–292, 504
Silver/silver chloride electrode, 382
Simulation system
    in industrial automation, 25–26
    selection, 26
Single side-band modulation (SSB), 687
Single-beam technique, 512
Single-shot time interval measurements, 492, 493
Single-sweep cathode ray polarography, 332
Sinusoidal AC excitation, 52–53
Site mounting instruments, 647
Size analysis methods, 180–182
Sliding-contact sensors, 83

Sliding-vane flowmeter, 45–46
Small-volume sample probe, 664
Smart grid technology, 28
Smart transmitter., 221
Smart valves, 635
Smith bridge, 477
Sneak circuits, 754–755
Snell's law, 350
Soap-film burette, 68
Soft X-ray effect, 172–173
Software reliability *vs.* hardware reliability, 768–769
Solar radiation, 519
Solid, sampling, 661
Solid expansion, 270, 271, 285–286
Solid-state amplifiers, 772
Solid-state detectors, 411–412, 524, 532–533, 538
Solid-state electrodes, 381
Sound
    nature of, 593–594
    power determination, 611–612
        by sound intensity, 612–613
Sound energy level (SEL), 602, 603
Sound pressure level, 593, 594. *See also* Noise measurement
    calibrator, operating principle of, 604
    measurement of, 609–610
        instrumentation for, 596–604
        sound power determination, 611–613
        space averaging, 611
        statistical distribution and percentiles, 611
        time averaging, 610–611
Sound source/field, quantities characterizing, 594
Sound waves, velocity of propagation, 594–595
Sound-intensity analyzers, 608–609
Sound-level meters
    calibration, 609
    integrating, 601–603, 611
    statistical, 603
    time weighting in, 610
    types of, 601
Spatial fringe pattern, 198, 199
SPD. *See* Spectral power distribution
Specific heat capacity, 269
Specific heats of gas, 35
Spectral power distribution (SPD), 512
Spectra-Tek Sentinel Flow Computer, 240
Spectrometer, 563
    multichannel, 354
Spectrophotometer, 346, 510–512
Spectroradiometers, 512
Spectroscopy
    absorption and reflection techniques
        chemiluminescence, 349
        infrared, 341–346
        radiation, reflected, 348–349
        visible and ultraviolet, 346–348
    atomic absorption, 351–352
    atomic emission, 349–351
    electron paramagnetic resonance, 356
    gamma ray, 357
    microwave, 355–356
    nuclear magnetic resonance, 357
    photo-acoustic, 355
    x-ray fluorescence, 353–355
Speech communication, human operator, 761
Sphere fast-neutron detector, 541
Spread spectrum, 256–257
Spring balances method, 129
Spring torque motor, 108
Springs, 647
Square-wave excitation, 53
Stabilizing collector current using emitter resistor, 752
Stable capacitors, temperature effects on, 752
Standard resistors, 443
Standards
    for instrumentation, 869
    and specifications for wireless, 263
Static calorimetric techniques, 470
Static electricity, elimination of, 565
Stationary phases in HPLC system, 330
Statistics of counting, 521–524
    nonrandom errors, 523
Status information, 224
Statutory instruments (SIs), 866
Steam ejection probe, 665
Steam sampling for conductivity, 673, 674
Steam-injection probe, 672
Steel thermometer, mercury filled, 278–280
Stefan–Boltzmann law, 307
Storage oscilloscopes, 707–708
Strain, 93
Strain gauge transducer, 751
Strain gauges
    bonded resistance, 93–95
    capacitive, 99
    characteristics of, 95–96
    circuits for, 98
    cross-sensitivity, 96
    foil, 94
    installation of, 96–97
    for measuring residual stress, 95
    photoelasticity of, 100–101
    rosette, 95
    semiconductor, 94–95
    surveys of surfaces, 99–100
    temperature sensitivity, 96
    vibrating wire, 98–99
    wire, 94
Strain sensing, 57
Strain-gauge load cells
    applications, 132
    calibration, 132–133
    design, 130–131
    selection and installation, 131–132
Strain-gauge pressure sensors, 156
Stray capacitances in four-arm AC bridge, 483
Stray impedances
    in AC bridges, 480–482
    effect on transformer ratio bridges, 486
Streamlined flow, 32
    velocity profile, 32
Strip chart recorders
    methods, 710
    specifications for, 711
Structured programming, 770–771
    ADA, 771
    flowcharts for, 770
    Newspeak, 771
Successive-approximation ADCs, 456–457
Sulfur contents measurement in liquid hydrocarbons, 557–558
Superconducting quantum interferometric detector (SQUID), 443
Surface contact thermocouples, 303
Surface mount technology (SMT), 822–823
Surface radiation thermometer, 310–312
Surface temperature measurement, 323
Surface transfer impedance (STI) of cables screening, 844–845
Surface-inspection methods
    eddy-current, 570–571
    magnetic flux, 569–570
    potential drop, 570
    visual, 568–569
Surface-neutron gauges, 557
Swirlmeter, 57
Switching power supply, 830
    capacitive coupling, 831–832
    capacitive screening, 832
    differential mode interference, 832
    magnetic component construction, 831
    output noise, 832–833
    radiation from di/dt loop, 831
Synchros, 86
System components
    failure rates of, 745
    of reliability, 740–741
System design
    basics of, 20–21
    performance margins in, 750
System interfaces, 262–263
System management, 262
System parameter tolerance, coping with, 751
Systematic error (SE), 492
Systems Analysis and Design Methodology (SSADM), 770

T

Tape recorders, 608
Target flow transmitter, 721
Target flowmeter, 42–43
TCD. *See* Thermal conductivity detector
TDM. *See* Time-division multiplexing
Technical construction file (TCF), 868
Telemetry, 677
    in instrumentation, 678
    radio frequency, 681
Telemetry transmitting system, reliability diagram of, 748
Temperature compensated oscillator, 490
Temperature controllers, 320
Temperature measurement, 269
    computer-compatible, 320
    considerations, 319
    definitions, 269
    instruments, 269
    pneumatic instrumentation for, 717–718
    realization of, 274
    semiconductor, 291–293

Index

sensor location considerations, 320–324
surface, 323
techniques, 276
    electrical, 286–293
    gas-filled instruments, 281–282
    liquid crystals, 324–325
    liquid-filled dial thermometers, 278–281
    liquid-in-glass thermometers, 274–278
    pyrometric cones, 324
    radiation thermometers, 306–319
    solid expansion, 285–286
    temperature-sensitive pigments, 324
    thermal imaging, 325
    thermocouples, 293–306
    turbine blade temperatures, 325–326
    vapor pressure thermometers, 282–284
in vessels, 320
Temperature scales, 272
    Celsius, 272
    comparison, 275
    Fahrenheit and Rankine, 273–274
    IPTS-68, 273
    thermodynamic, 272–273
Temperature sensitivity, strain gauges, 96
Temperature sensors, 60
Temperature transmitter, 224–226
Temperature-sensing integrated circuits, 292–293
Temperature-sensitive pigments, 324
Temporal fringe method, 196
Terminal velocity, 176–177
    measurement in particles, 183–188
Thalamid electrodes, 384
Thermal analysis, 335–339
Thermal conductivity, 270
Thermal conductivity detector (TCD), 404–406, 417
    multistream chromatograph with, 416
Thermal conductivity gauges, 169–170
Thermal EMF noise, 255
Thermal expansion, 270
Thermal imaging, 325
    techniques, 518–519
Thermal mass flowmeters, 60
Thermal neutrons, 521
Thermal sensing, 57
Thermal stability of organic materials, 337
Thermistor, 406
    gauge, 170
Thermistor power meter, 470
    RF power, 470, 471
Thermistor system, equivalent circuit of, 470
Thermistors, 290–291
Thermocouple gauge, 169
Thermocouple instruments, 454–455
Thermocouple wattmeter, 467
Thermocouples, 293–306
    accuracy, 305
    availability, 301
    British Standards, 300
    circuit, 298–299
    cold junction compensation in, 297–298
    construction, 301–306
    materials, 299–301
Thermodynamic temperature scale, 272–273

Thermoelectric diagram, 295–296
Thermoelectric effects, 293–299
Thermoelectric EMFs, addition of, 296–297
Thermoelectric inversion, 296
Thermoluminescence (TL), 525
Thermoluminescent detectors, 525
Thermometer bulb, 279
Thermometer pockets, 322–323
Thermometers
    bimetal strip, 285–286
    gas-filled, 281–282
    high-temperature, 277–278
    liquid-filled, 278–281
    liquid-in-glass, 274
    liquid-in-metal, 281
    mercury-in-glass, 274–278
    radiation, 306–319
    remote reading, 319
    resistance, 286–290
    vapor pressure, 282–284
Thermo-oxidative stability of organic materials, 337
Thermopiles, 303, 309
Thermostats, 320
Thermowells, 322–323
Thin-layer chromatography, 328–329
    apparatus for, 329
Third-octave analyzers, 605–606
Thixotropy, 70, 72
Thompson-Lampard capacitor, 444
Thomson effect, 295
Three Mile Island (TMI-2) reactor, 776
    accident, common mode failure, 780
Three-phase power measurement, 468–469
Three-terminal capacitors, 851
Three-voltmeter method of power measurement, 465
Three-winding voltage transformer, 485
Time constant, 224
Time lags, sampling, 662–663
Time-base error (TBE), 492
Time-division multiplexing (TDM), 684–685
Time-division multiplication wattmeter, 471
Time-interval averaging (TIA), 492–495
Time-lapse pulse holography method, 90
Time-of-flight mass spectrometer, 361
Time-of-flight methods, 89
Time-transit methods, microwave and ultrasonic, 109–110
Total energy of fluid motion, 32
Total radiation thermometer, 309–312
Tracer measurement method, 67
Transducers, 129
    buoyancy, 136, 137
    combined actuator, 652–653
    cross-axis coupling, 116
    gas density, 142
    liquid density, 141
    piezoelectric, 130
    pressure, 133
    proximity, 651
Transfer oscillator counter, 496, 497
Transformer ratio bridges, 482
    advantage, 482
    autobalancing, 487

balance condition of, 486
configurations, 484–486
effect of stray impedances on, 486
equivalent circuit of, 483
loading effect, 484
sensitivity of, 486–487
in unbalanced condition, 486–487
windings, 483
Transient suppression device, 856–857
Transients, 811–812
    coupling mode, 812
    interference paths, 833–834
    protection, 834
    on signal lines, 812
Transient/waveform recorders, 713
Transistor
    accelerated life tests for, 745
    dissipation, 745
    junction temperature, 744–745
    stabilizing current in, 753
Transistor-transistor logic (TTL) circuit, 750
Transit-time methods, 89–90
Transmission
    analog signal, 678, 689–690
    data, 693–697
    digital signal, 678, 690–697
    frequency, 690
    radio frequency, 681
Transmission lines, 679–681
    distributed primary constants of, 679
    ringing on, 828–829
Transmissive flowmeters
    principle of operation, 54–55
    velocity measurement, 56
Transmitters, 219, 220
    temperature, 319
Traveling-wave tube (TWT), 772
Triangular notch, 61–62
Triboelectric charge, 812
Trigger error (TE), 491
Tristimulus colorimeters, 513
Troubleshooting tips, 261
True mass-flow measurement methods, 58–60
    fluid-momentum, 58–59
    pressure differential, 59–60
Tube joining methods, 672
Tungsten lamp, 500
Turbidity-nephelometer, 434
Turbine blade temperatures, 325–326
Turbine current meter, 63
Turbine meters, 47–48, 51
Turbulent flow, 32
    velocity profile, 32
Two-dimensional statistical mean diameters, 176
Two-terminal passive network, 465

U

U.K. primary standards
    DC and low-frequency, 441
    of resistance, 443
    RF and microwave, 441–442
    of voltage, 442
        monitoring absolute value, 443
Ultrasonic detector, 410–411, 418
Ultrasonic Doppler velocity probe, 66

Ultrasonic flaw detectors, 574
Ultrasonic flow measurement method, 64
Ultrasonic flowmeters, types of, 54–56
Ultrasonic non-destructive testing, 588
    acoustic emission, 580, 581
    automated testing, 580
    equipment controls and visual presentation, 574–575
    principles of, 571–574
    probe construction, 576
    underwater, 588
Ultrasonic scanning, bulk, 589
Ultrasonic sensing, 57
Ultrasonic spectroscopy, 577–578
    applications of, 578–579
Ultrasonic time-transit methods, 109–110
Ultraviolet intensity measurements, 509–510
Undersea radiation surveys, 563
Underwater non-destructive testing, 586–587
    AC potential difference (AC/PD), 589
    acoustic emission, 589–590
    bulk ultrasonic scanning, 589
    corrosion protection, 588
    diver operations and communication, 587
    eddy-current, 589
    magnetic particle inspection, 588
    photography, 587
    ultrasonics, 588
    visual examination, 587
Uninterrupted power supply (UPS), 759
Units, electronic, 544–546
Universal asynchronous receiver transmitters (UARTS), 690, 691
Universal timer/counters, 489
    specifications, 494
Universal transformer bridge, 485
Unmeasured disturbance, 620
U.S. Nuclear Regulatory Commission (USNRC), 541
User-friendly design, 762–764
U-tube manometer, 145

V

Vacuum measurements
    absolute gauges, 165, 166–168
    ionization gauges, 170–173
    liquid manometers, 167
    McLeod gauge, 167–168
    mechanical gauges, 166–167
    methods, 165
    nonabsolute gauges, 165, 166, 169–173
    thermal conductivity gauges, 169–170
Vacuum spectrographs, 350
Valves, control
    characteristics, 633–634
    as flowmeters, 635–636
    loop tuning, 634–635
    positioning, 635
    rangeability of, 634
    smart, 635
    types of, 631–633

Vapor pressure methods, 433
Vapor pressure thermometers, 282–284
    liquids used in, 283
Vaporization, 670
Variable-orifice meter, 39–42, 669
Variable-reluctance methods, 85
VDUs. *See* Visual display units
Velocity
    of approach factor, 34
    measurement of, 120–121
    sensor, 121
Velocity profile, 32
Velometers. *See* Deflecting-vane flowmeter
Venturi flume, 62
Venturi nozzle, 38
Venturi tube
    advantages and disadvantages, 38
    application, 38
    components, 37
    pressure tappings in, 38
Verifiable integrated processor for enhanced reliability (VIPER), 753
Vestigial side-band modulation (VSM), 687
Vibration
    parameters, frequency spectrum and magnitude of, 114
    physical considerations, 113–116
    wire strain gauge, 94, 98–99
Vibration measurement, areas of application, 116–117
Vibration monitoring
    frequency modulated laser diode, 209–210
    heterodyne modulation, 206–208
    pseudoheterodyne modulation, 208–209
    using Bragg cell frequency shifter, 206
    using phaselock loop demodulator, 206, 207
Vibration sensor
    practical problems of installation, 116
    reference grade, 208
Vibrometer, 120
Viscometer, 69
    Brookfield, 73
    capillary, 71–72
    cone-and-plate, 72–73
    Couette, 72
    industrial, 73
    Ostwald, 71
    parallel-plate, 73
    shop-floor, 73–74
Viscosity measurement
    accuracy and range, 74–75
    extensional, 74
    kinetic-energy correction, 71
    online, 74
    shear, 71–73
    under temperature and pressure, 74
Viscosity sensor, MEMS, 218
Visual display units (VDUs), 704, 709
Visual surface-inspection methods, 568–569
Voltage dependent resistor (VDR), 854

Voltage derating, 745
Voltage regulator, 542
Voltage surges. *See* Power surges
Voltage transformers (VTs)
    AC range extension
        equivalent circuit, 453
        phase-angle error/phase displacement, 453, 454
        phasor diagram, 453
        ratio errors, 454
        voltage/ratio error, 453, 454
Voltage-controlled oscillator (VCO), 55, 198
Voltage-doubling circuit, 542
Volume flow rate, 43
Volume rate, flow equation for, 62
Volumetric calibration method, 67
Vortex flowmeter, 234–236
    calibration factor for, 57
    installation parameters for, 57
    principle of operation, 56
Vortices, sensing methods for, 57
Voting operation, 773

W

Wagner earthing arrangement, 478, 482, 483
Wallmark, 88
Water purity and conductivity, 372
Water-displacement method, 68
Water-sampling system, 676
Water-wash probe, 664–665
Watson image-shearing eyepiece, 182
Watt-hour meter
    phasor diagram of
        eddy currents in, 473
        fluxes, 473
    torque in, 472–473
Wave impedance, 804–805
Wavelengths
    and color, 510–514
    of thermal radiation, 272
    transmission, 305
    vs. radiant emission effect, 316
Wear measurement with nuclear techniques, 559
Wear-out phase, failure rate, 742, 747
Weight-loss curves, 336
Weirs, 60–62
Westinghouse instrument, 586
Weston mercury cadmium cells, 442–443
Wet gases
    density, 36
    humidity, 36
Wheatstone bridge
    cell conductance measurement by, 369
    modified, 476
    resistance measurement, 474
    self-heating in, 475
    sensitivity, 474
    three-lead measurements using, 476
    with three-terminal high resistances, 478
    unbalanced mode, 475

Index

Whip antennas, 259
White-light interferometry, 195–201
    electronically scanned method, 198–201
    for force measurement, 200
    pressure sensor, 198
    temporally scanned method, 196–198
Wien's laws, 307
WiFi 802.11, 262
WiFi Wired Equivalent Privacy (WEP) encryption, 257
WiMax, 263
Wire strain gauge, 94, 98–99
Wireless
    cellular, 263
    communication, 254
    functionality and applications, 264
    history of, 253–254
    information, 254
    LANs, 253
    mesh technologies, 262
    planning for, 263–265
    reliability, 256
    standards and specifications for, 263
    system security, 257–258
WirelessHART, 263
Wire-wound resistors, temperature effects, 752
Wollaston prism, 200
World Batch Forum (WBF), 630

X

Xerography, 585
X-ray fluorescence analysis
    coating and backing measurement by, 562
    dispersive, 553
    non-dispersive, 554–555
    sources for, 554
X-ray fluorescence spectroscopy, 353–355
X-ray sedimentation, 184
X-rays, 582
    detector, 559
    diffraction, 355
    fluorescence gauge, 562, 563
XY plotters, 607
x–y recorders, 712

Y

"Y" strainers, 665
Yagi antenna, 258–259
Yield stress, 70
Yokogawa vortex flowmeter, 234, 236

Z

Zeiss–Endter analyzer, 183
Zener diode, 460, 542
ZigBee, 263